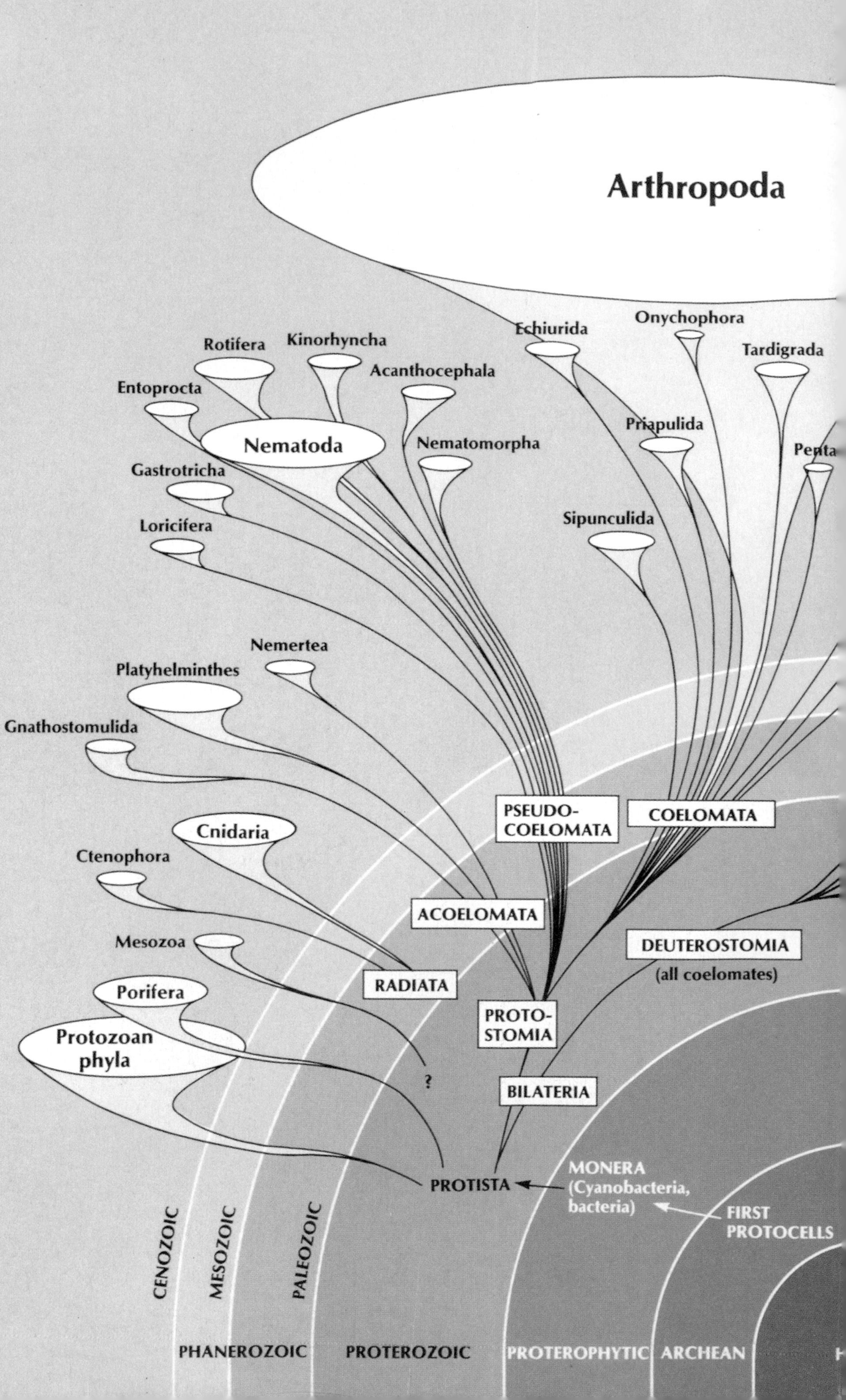
Arthropoda
Onychophora
Echiurida
Tardigrada
Rotifera
Kinorhyncha
Acanthocephala
Entoprocta
Priapulida
Nematoda
Nematomorpha
Gastrotricha
Loricifera
Sipunculida
Nemertea
Platyhelminthes
Gnathostomulida
PSEUDO-
COELOMATA
COELOMATA
Cnidaria
Ctenophora
ACOELOMATA
Mesozoa
DEUTEROSTOMIA
(all coelomates)
RADIATA
Porifera
PROTO-
STOMIA
Protozoan
phyla
?
BILATERIA
PROTISTA
MONERA
(Cyanobacteria,
bacteria)
FIRST
PROTOCELLS
CENOZOIC
MESOZOIC
PALEOZOIC
PHANEROZOIC
PROTEROZOIC
PROTEROPHYTIC
ARCHEAN

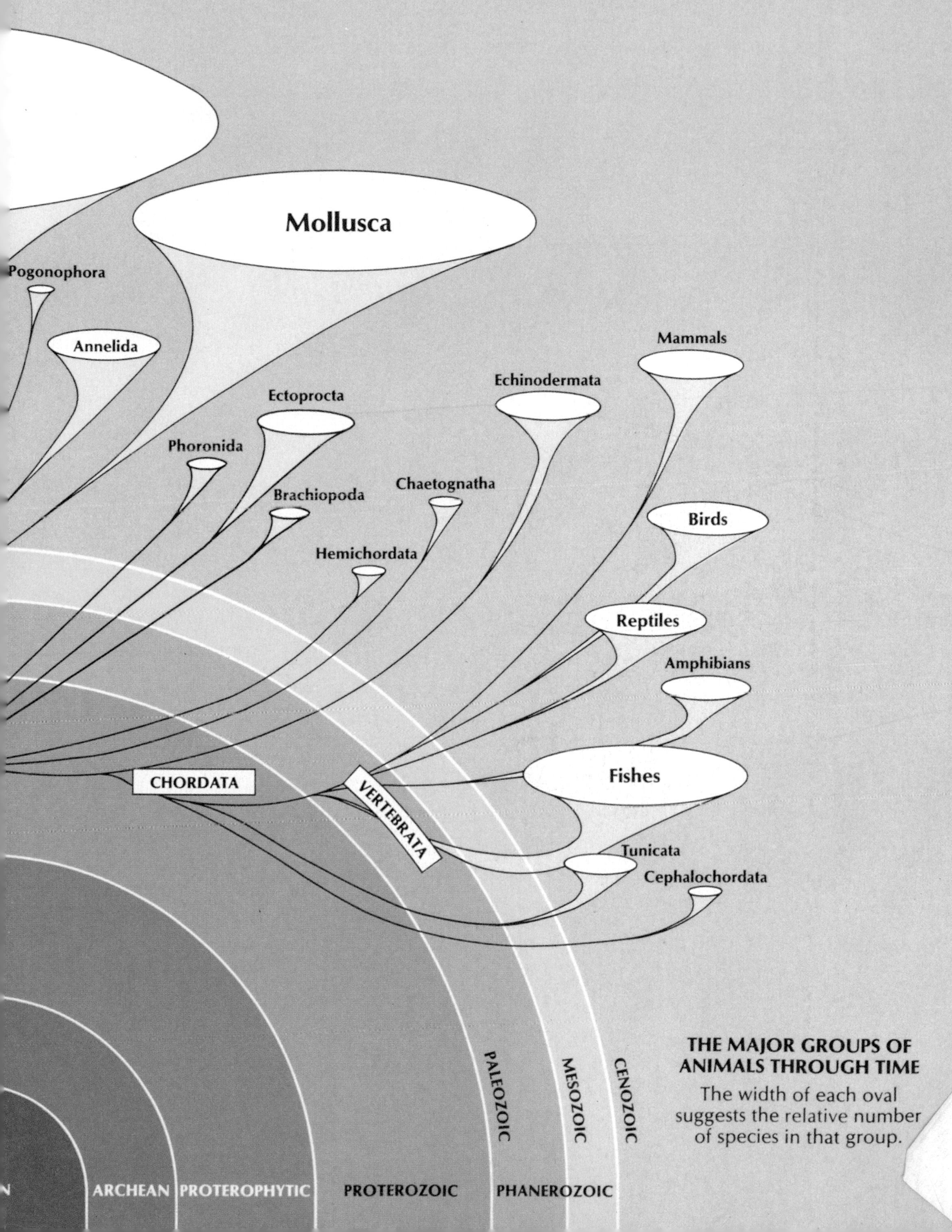

THE MAJOR GROUPS OF ANIMALS THROUGH TIME

The width of each oval suggests the relative number of species in that group.

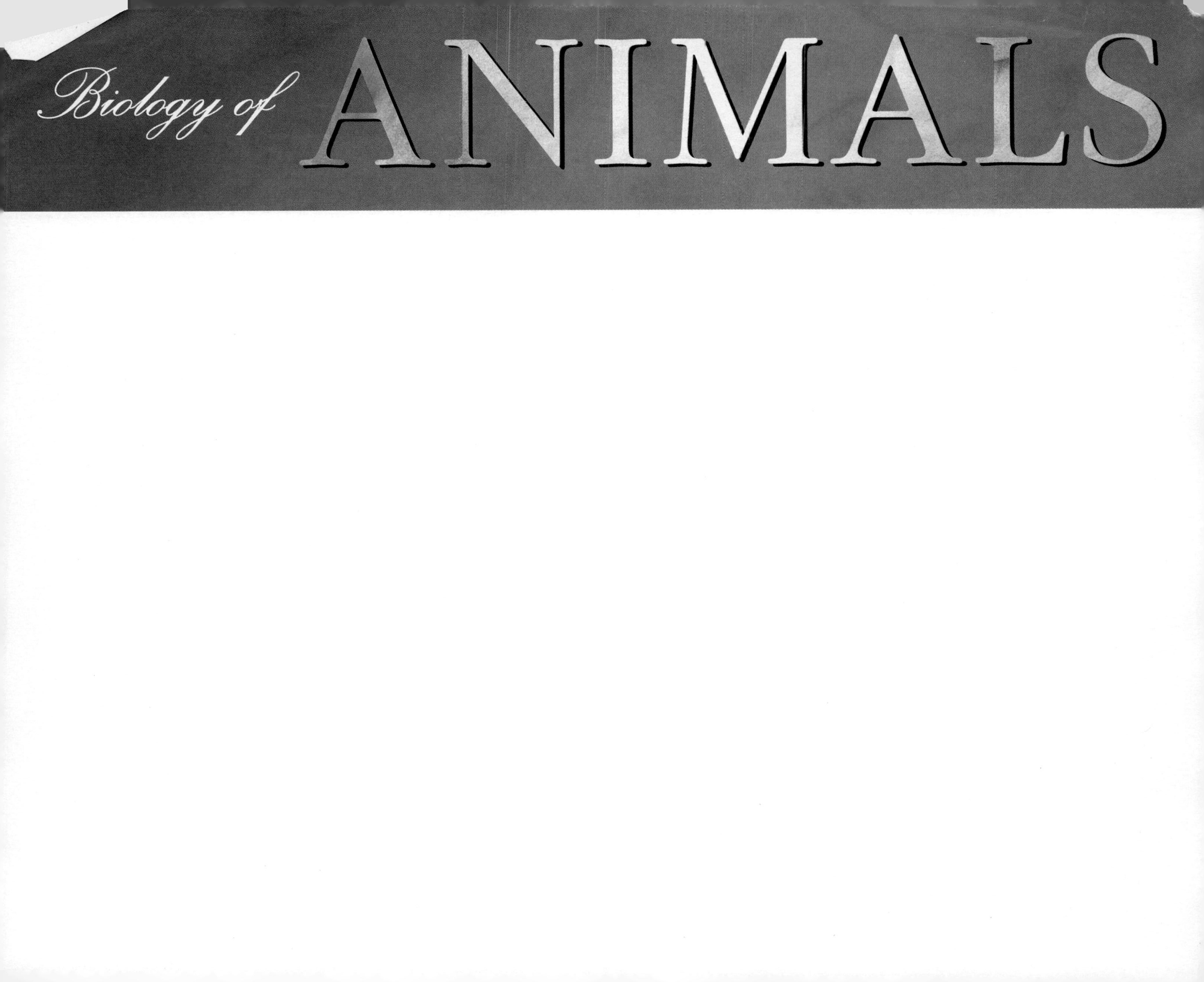
Biology of ANIMALS

Seventh Edition

CLEVELAND P. HICKMAN, JR.
Washington and Lee University

LARRY S. ROBERTS
University of Miami

ALLAN LARSON
Washington University

Original Artwork by
WILLIAM C. OBER *and* CLAIRE GARRISON

Boston, Massachusetts Burr Ridge, Illinois Dubuque, Iowa
Madison, Wisconsin New York, New York San Francisco, California St. Louis, Missouri

WCB/McGraw-Hill

A Division of The **McGraw-Hill** *Companies*

BIOLOGY OF ANIMALS

This book is printed on recycled, acid-free paper containing 10% postconsumer waste.

1 2 3 4 5 6 7 8 9 0 QPD/QPD 9 0 9 8 7

ISBN 0-697-28933-8

Publisher: Michael D. Lange
Sponsoring editor: Margaret J. Kemp
Developmental editor: Adora L. Pozolinski
Marketing manager: Thomas C. Lyon
Project manager: Ann Fuerste
Production supervisor: Sandra Hahn
Designer: Katherine Farmer
Cover image: ©Kim Westerkov/Tony Stone Images
Photo research coordinator: Lori Hancock
Art editor: Joyce Watters
Compositor: Graphic World Inc.
Typeface: 10/12 Garamond Book
Printer: Quebecor Dubuque

Library of Congress Cataloging-in Publication Data

Hickman, Cleveland P.
Biology of animals / Cleveland P. Hickman, Jr., Larry S. Roberts
Allan Larson ; original artwork by William C. Ober and Claire
Garrison. — 7th ed.
p. cm.
Includes bibliographical references and index.
ISBN 0-697-28933-8
1. Zoology I. Roberts, Larry S., 1935- . II. Larson, Allan.
III. Title.
QL47.2.H528 1997
590—dc21 97-21836
CIP

INTERNATIONAL EDITION

1 2 3 4 5 6 7 8 9 0 QPD QPD 90987

When ordering this title, use ISBN 0-07-115327-6

http://www.mhhe.com

brief contents

part one
Evolution of Animal Life

1 **Life:** General Considerations, Basic Molecules, and Origins, **2**
2 **The Cell as the Unit of Life, 23**
3 **Genetic Basis of Evolution, 56**
4 **Evolution of Animal Diversity, 83**
5 **Ecology and Distribution of Animals, 112**

part two
Animal Form and Function

6 **Animal Architecture:** Body Organization, Support, and Movement, **138**
7 **Homeostasis:** Osmotic Regulation, Excretion, and Temperature Regulation, **166**
8 **Internal Fluids and Respiration 187**
9 **Immunity 210**
10 **Digestion and Nutrition, 222**
11 **Nervous Coordination:** Nervous System and Sense Organs, **240**
12 **Chemical Coordination:** Endocrine System, **266**
13 **Animal Behavior, 290**
14 **Reproduction and Development, 308**

part three
The Invertebrate Animals

15 **Classification and Phylogeny of Animals, 342**
16 **The Animal-like Protista, 361**
17 **Sponges:** Phylum Porifera **381**
18 **Radiate Animals:** Cnidarians and Ctenophores, **392**
19 **Acoelomate Animals:** Flatworms, Ribbon Worms, and Jaw Worms, **416**
20 **Pseudocoelomate Animals, 433**
21 **Molluscs, 448**
22 **Segmented Worms:** The Annelids, **474**
23 **Arthropods, 490**
24 **Lesser Protostomes and Lophophorates, 531**
25 **Echinoderms, Hemichordates, and Chaetognaths, 542**
General References to Part III, 562

part four
The Vertebrate Animals

26 **Vertebrate Beginnings:** The Chordates, **563**
27 **Fishes, 582**
28 **The Early Tetrapods and Modern Amphibians, 606**
29 **Reptiles, 621**
30 **Birds, 640**
31 **Mammals, 666**

Appendix: Basic Structure of Matter, 691
Glossary, 699
Credits, 733
Index, 739

preface, xix

part one Evolution of Animal Life, 1

chapter one

Life: General Considerations, Basic Molecules, and Origins, **2**

Principles of Science, 3
- Nature of Science, 3
- Scientific Method, 3
- Physiological and Evolutionary Sciences, 4

What is Life?, 5
Chemistry of Life, 6
- Organic Molecules, 7

Origin of Life, 11
- Historical Perspective, 12
- Modern Experimentation, 13
- Formation of Polymers, 15

Origin of Living Systems, 16
- Origin of Metabolism, 16
- Appearance of Photosynthesis and Oxidative Metabolism, 17

Precambrian Life, 17
- Prokaryotes and the Age of Cyanobacteria, 18
- Appearance of the Eukaryotes, 19

Summary, 20
Review Questions, 21
Selected References, 22

chapter two

The Cell as the Unit of Life, 23

How Cells Are Studied, 24
Organization of Cells, 26
- Prokaryotic and Eukaryotic Cells, 26
- Components of Eukaryotic Cells and Their Functions, 27
- Surfaces of Cells and Their Specializations, 30
- Membrane Function, 31

Cell Division, 35
- Structure of Chromosomes, 35
- Stages in Mitosis, 36
- Cytokinesis: Cytoplasmic Division, 37
- Cell Cycle, 37
- Flux of Cells, 38

Cellular Metabolism, 39
- Energy and Life, 39
- Role of Enzymes, 40
- Cellular Energy Transfer, 42
- Cellular Respiration, 43
- Metabolism of Lipids, 50
- Metabolism of Proteins, 51
- Management of Metabolism, 52

Summary, 52
Review Questions, 54
Selected References, 55

chapter three

Genetic Basis of Evolution, 56

Mendel's Investigations, 57
Chromosomal Basis of Inheritance, 57
- Meiosis: Maturation Division of Gametes, 58
- Sex Determination, 59

Mendelian Laws of Genetics, 61
- Mendel's First Law, 61
- Mendel's Second Law, 63
- Multiple Alleles, 65
- Sex-Linked Inheritance, 66
- Autosomal Linkage and Crossing Over, 66
- Chromosomal Aberrations, 68

Gene Theory, 69
- Gene Concept, 69

Storage and Transfer of Genetic Information, 70
- Nucleic Acids: Molecular Basis of Inheritance, 70
- Transcription and the Role of Messenger RNA, 72
- Translation: Final Stage in Information Transfer, 75
- Regulation of Gene Function, 75
- Genetic Engineering, 77

Sources of Phenotypic Variation, 78
- Gene Mutations, 79

Molecular Genetics of Cancer, 79
- Oncogenes and Tumor Suppressor Genes, 79

Summary, 80
Review Questions, 81
Selected References, 82

chapter four

Evolution of Animal Diversity, 83

History of Evolutionary Theory, 84
Pre-Darwinian Evolutionary Ideas, 84
Darwin's Great Voyage of Discovery, 85
Darwin's Theory of Evolution, 88
Evidence for Darwins' Five Theories of Evolution, 91
Perpetual Change, 91
Common Descent, 94
Multiplication of Species, 98
Gradualism, 100
Natural Selection, 102
Revisions of Darwinian Evolutionary Theory, 103
Neo-Darwinism, 103
Emergence of Modern Darwinism: The Synthetic Theory, 103
Microevolution: Genetic Variation and Change within Species, 104
Genetic Equilibrium, 105
Processes of Evolution: How Genetic Equilibrium Is Upset, 106
Macroevolution: Major Evolutionary Events, 107
Speciation and Extinction through Geological Time, 108
Mass Extinctions, 108
Summary, 109
Review Questions, 110
Selected References, 111

chapter five

Ecology and Distribution of Animals, 112

The Hierarchy of Ecology, 113
Environment and the Niche, 113
Populations, 114
Community Ecology, 118
Ecosystems, 123
Distribution of Life on Earth, 127
The Biosphere and Its Subdivisions, 127
Terrestrial Environments: Biomes, 128
Aquatic Environments, 132
Summary, 133
Review Questions, 134
Selected References, 135

part two Animal Form and Function, 137

chapter six

Animal Architecture: Body Organization, Support and Movement, **138**

The Hierarchical Organization of Animal Complexity, 140
Complexity and Body Size, 140
Extracellular Components of the Metazoan Body, 140
Types of Tissues, 141
Epithelial Tissue, 142
Connective Tissue, 142
Muscular Tissue, 144
Nervous Tissue, 145
Integument Among Various Groups of Animals, 145
Invertebrate Integument, 147
Vertebrate Integument, 149
Skeletal Systems, 150
Hydrostatic Skeletons, 150
Rigid Skeletons, 150
Plan of the Vertebrate Skeleton, 153
Effect of Body Size on Bone Stress, 155
Animal Movement, 156
Ameboid Movement, 156
Ciliary and Flagellar Movement, 156
Muscular Movement, 157
Muscle Performance, 163
Summary, 164
Review Questions, 164
Selected References, 165

chapter seven

Homeostasis: Osmotic Regulation, Excretion, and Temperature Regulation, **166**

Water and Osmotic Regulation, 167
How Marine Invertebrates Meet Problems of Salt and Water Balance, 167
Invasion of Fresh Water, 168
Return of Fishes to the Sea, 168
How Terrestrial Animals Maintain Salt and Water Balance, 170
Invertebrate Excretory Structures, 171
Contractile Vacuole, 171
Nephridium, 171
Arthropod Kidneys, 172
Vertebrate Kidney, 173
Ancestry and Embryology, 173
Vertebrate Kidney Function, 173
Glomerular Filtration, 175
Tubular Reabsorption, 176
Tubular Secretion, 177
Water Excretion, 177

Temperature Regulation, 179
Ectothermy and Endothermy, 180
How Ectotherms Achieve Temperature Independence, 180
Temperature Regulation in Endotherms, 181
Adaptive Hypothermia in Birds and Mammals, 183
Summary, 185
Review Questions, 185
Selected References, 186

chapter eight

Internal Fluids and Respiration, 187

Internal Fluid Environemnt, 188
Composition of the Body Fluids, 188
Composition of Blood, 188
Hemostasis: Prevention of Blood Loss, 190
Circulation, 191
Open and Closed Circulations, 192
Plan of the Vertebrate Circulatory Systems, 193
Mammalian Heart, 194
Arteries, 195
Capillaries, 197
Veins, 198
Lymphatic System, 198
Respiration, 199
Problems of Aquatic and Aerial Breathing, 199
Respiratory Organs, 200
Structure and Function of the Mammalian Respiratory System, 202
Summary, 207
Review Questions, 208
Selected References, 209

chapter nine

Immunity, 210

Innate Defense Mechanisms, 211
Phagocytosis, 211
Immunity in Invertebrates, 212
Innate Immunity in Vertebrates, 212
Acquired Immune Response in Vertebrates, 213
Basis of Self and Nonself Recognition, 213
T-Cell Receptors, 215
Cells of the Immune Response, 215
Cytokines, 215
Generation of a Humoral Response, 216
The Cell-Mediated Response, 218
Acquired Immune Deficiency Syndrome (AIDS), 218
Inflammation, 218
Blood Group Antigens, 219
ABO Blood Types, 219
Rh Factor, 219
Summary, 220
Review Questions, 220
Selected References, 221

chapter ten

Digestion and Nutrition, 222

Feeding Mechanisms, 223
Feeding on Particulate Matter, 223
Feeding on Food Masses, 225
Feeding on Fluids, 226
Digestion, 226
Action of Digestive Enzymes, 227
Motility in the Alimentary Canal, 228
Organization and Regional Function of the Alimentary Canal, 228
Receiving Region, 228
Conduction and Storage Region, 229
Region of Grinding and Early Digestion, 229
Region of Terminal Digestion and Absorption: The Intestine, 231
Region of Water Absorption and Concentration of Solids, 234
Regulation of Food Intake, 234
Nutritional Requirements, 235
Summary, 238
Review Questions, 238
Selected References, 239

chapter eleven

Nervous Coordination: Nervous System and Sense Organs, 240

The Neuron: Functional Unit of the Nervous System, 241
Nature of the Nerve Impulse, 242
Synapses: Junction Points between Nerves, 244
Evolution of the Nervous System, 244
Invertebrates: Development of Centralized Nervous Systems, 244
Vertebrates: Fruition of Encephalization, 246
Sense Organs, 253
Classification of Receptors, 253
Chemoreception, 253
Mechanoreception, 254
Photoreception: Vision, 260
Summary, 263
Review Questions, 264
Selected References, 265

chapter twelve

Chemical Coordination: Endocrine System, 266

Mechanism of Hormone Action, 267
Membrane-Bound Receptors and the Second Messenger Concept, 267
Nuclear Receptors, 268
How Secretion Rates of Hormones are Controlled, 268
Invertebrate Hormones, 269
Vertebrate Endocrine Glands and Hormones, 270

Hormones of the Pituitary Gland, Hypothalamus, and Pineal Gland, 270
Nonendocrine Hormones, 275
Hormones of Metabolism, 276
Hormones of Digestion, 282
Hormones of Vertebrate Reproduction, 282
Hormonal Control of the Timing of Reproductive Cycles, 282
The Gonadal Steroid and Their Control, 283
The Menstrual Cycle, 284
Hormones of Human Pregnancy and Birth, 285
Summary, 287
Review Questions, 288
Selected References, 289

chapter thirteen

Animal Behavior, 290

The Science of Animal Behavior 291
Describing Behavior: Principles of Classical Ethology, 292
Control of Behavior, 293
The Genetics of Behavior, 294
Learning and the Diversity of Behavior, 295
Social Behavior, 297
Advantages of Sociality, 297
Aggression and Dominance, 299
Territoriality, 301
Animal Communication, 302
Chemical Sex Attraction in Moths, 302
Language of the Bees, 303
Communication by Displays, 303
Communication between Humans and Other Animals, 304
Summary, 306
Review Questions, 306
Selected References, 307

chapter fourteen

Reproduction and Development, 308

Nature of the Reproductive Process, 309
Asexual Reproduction: Reproduction without Gametes, 310
Sexual Reproduction: Reproduction with Gametes, 310
What Good Is Sex? 313
Formation of Reproductive Cells, 313
Gametogenesis, 314
Plan of Reproductive Systems, 314
Invertebrate Reproductive Systems, 315
Vertebrate Reproductive Systems, 316
The Developmental Process, 318
Fertilization, 319
Cleavage and Early Development, 320
Blastula Formation, 323
Gastrulation and the Formation of Germ Layers, 324
Development of Systems and Organs, 326
Derivatives of Ectoderm: Nervous System and Nerve Growth, 326
Derivatives of Endoderm: Digestive Tube and Survival of Gill Arches, 328
Derivatives of Mesoderm: Support, Movement, and Beating Heart, 328
Mechanisms of Development, 329
Nuclear Equivalence, 329
Cytoplasmic Localization: Significance of the Cortex, 329
Nuclear Transplantation Experiments, 330
Embryonic Induction, 330
Gene Expression during Development, 331
Strategies for Protection of the Developing Embryo, 333
Vertebrate Development, 333
The Common Vertebrate Heritage, 333
Amniotes and the Amniotic Egg, 334
The Mammalian Placenta and Early Mammalian Development, 335
Summary, 337
Review Questions, 338
Selected References, 339

part three The Invertebrate Animals, 341

chapter fifteen

Classification and Phylogeny of Animals, 342

Linnaeus and the Development of Classification, 343
Taxonomic Characters and Phylogenetic Reconstruction, 344
Using Character Variation to Reconstruct Phylogeny, 345
Sources of Phylogenetic Information, 346
Theories of Taxonomy, 346
Traditional Evolutionary Taxonomy, 346
Phylogenetic Systematics/Cladistics, 348
Current State of Animal Taxonomy, 349
Species, 350
Criteria for Recognition of Species, 350
Concepts of Species, 350
Major Divisions of Life, 351
Major Subdivisions of the Animal Kingdom, 352
Body Plans of Animals, 353
Symmetry of Animals, 353
Body Cavities, 354
Metamerism (Segmentation), 358
Cephalization, 358
Summary, 359
Review Questions, 359
Selected References, 360

chapter sixteen

The Animal-like Protista, 361

Form and Function, 363
Locomotor Organelles, 363
Nutrition and Digestion, 365
Excretion and Osmoregulation, 366
Reproduction, 367
Life Cycles, 368
Phylum Sarcomastigophora, 369
Subphylum Mastigophora, 369
Subphylum Sarcodina, 371
Phylum Apicomplexa, 373
Class Sporozoea, 373
Phylum Ciliophora, 375
Reproduction and Life Cycles, 376
Phylogeny and Adaptive Radiation, 376
Phylogeny, 376
Adaptive Radiation, 377
Summary, 379
Review Questions, 379
Selected References, 380

chapter seventeen

Sponges: Phylum Porifera, 381

Ecological Relationships, 382
Form and Function, 382
Types of Canal Systems, 383
Types of Cells, 384
Types of Skeletons, 385
Sponge Physiology, 386
Reproduction and Development, 387
Brief Survey of Sponges, 387
Class Calcarea (Calcispongiae), 387
Class Hexactinellida (Hyalospongiae), 388
Class Demospongiae, 388
Class Sclerospongiae, 388
Phylogeny and Adaptive Radiation, 389
Phylogeny, 389
Adaptive Radiation, 389
Summary, 390
Review Questions, 390
Selected References, 391

chapter eighteen

Radiate Animals: Cnidarians and Ctenophores, 392

Phylum Cnidaria, 393
Ecological Relationships, 394
Form and Function, 394
Class Hydrozoa, 399
Class Scyphozoa, 402
Class Cubozoa, 404
Class Anthozoa, 404
Phylum Ctenophora, 411
Form and Function, 411
Phylogeny and Adaptive Radiation, 412
Phylogeny, 412
Adaptive Radiation, 413
Summary, 414
Review Questions, 414
Selected References, 415

chapter nineteen

Acoelomate Animals: Flatworms, Ribbon Worms, and Jaw Worms, 416

Phylum Platyhelminthes, 417
Ecological Relationships, 417
Form and Function, 418
Class Turbellaria, 420
Class Trematoda, 421
Class Monogenea, 423
Class Cestoda, 424
Phylum Nemertea (Rhynchocoela), 427
Form and Function, 428
Phylum Gnathostomulida, 429
Phylogeny and Adaptive Radiation, 429
Phylogeny, 429
Adaptive Radiation, 430
Summary, 431
Review Questions, 431
Selected References, 432

chapter twenty

Pseudocoelomate Animals, 433

Phylum Rotifera, 434
Phylum Gastrotricha, 435
Phylum Kinorhyncha, 436
Phylum Loricifera, 436
Phylum Priapulida, 436
Phylum Nematoda: Roundworms, 437
Form and Function, 437
Some Nematode Parasites, 438
Phylum Nematomorpha, 442
Phylum Acanthocephala, 442
Phylum Entoprocta, 444
Phylogeny and Adaptive Radiation, 446
Phylogeny, 446
Adaptive Radiation, 446
Summary, 446
Review Questions, 447
Selected References, 447

chapter twenty-one

Molluscs, 448

Ecological Relationships, 449
Economic Importance, 449
Form and Function, 451
Internal Structure and Function, 452

Classes Caudofoveata and Solenogastres, 453
Class Monoplacophora, 453
Class Polyplacophora: Chitons, 453
Class Scaphopoda, 454
Class Gastropoda, 454
Form and Function, 455
Major Groups of Gastropods, 458
Class Bivalvia (Pelecypoda), 459
Form and Function, 460
Class Cephalopoda, 464
Form and Function, 465
Phylogeny and Adaptive Radiation, 469
Summary, 472
Review Questions, 473
Selected References, 473

chapter twenty-two

Segmented Worms: The Annelids, **474**

Ecological Relationships, 475
Economic Importance, 475
Body Plan, 476
Class Polychaeta, 477
Class Oligochaeta, 480
Earthworms, 480
Form and Function, 481
Freshwater Oligochaetes, 484
Class Hirudinea, 485
Form and Function, 485
Phylogeny and Adaptive Radiation, 486
Phylogeny, 486
Adaptive Radiation, 488
Summary, 488
Review Questions, 489
Selected References, 489

chapter twenty-three

Arthropods, 490

Ecological Relationships, 491
Why Have Arthropods Achieved Such Great Diversity and Abundance?, 491
Subphylum Trilobita, 492
Subphylum Chelicerata, 493
Class Merostomata, 493
Class Pyconogonida: Sea Spiders, 493
Class Arachnida, 493
Subphylum Crustacea, 498
Form and Function, 499
Class Branchiopoda, 504
Class Maxillopoda, 504
Class Malacostraca, 505
Subphylum Uniramia, 508
Class Chilopoda: Centipedes, 508
Class Diplopoda: Millipedes, 508
Class Insecta: Insects, 508
Phylogeny and Adaptive Radiation, 526
Phylogeny, 526
Adaptive Radiation, 526
Summary, 528
Review Questions, 529
Selected References, 530

chapter twenty-four

Lesser Protostomes and Lophophorates, 531

The Lesser Protostomes, 532
Phylum Sipuncula, 532
Phylum Echiura, 532
Phylum Pogonophora, 533
Phylum Pentastomida, 533
Phylum Onychophora, 534
Phylum Tardigrada, 535
The Lophophorates, 535
Phylum Phoronida, 536
Phylum Ectoprocta, 536
Phylum Brachiopoda, 538
Phylogeny, 539
Summary, 540
Review Questions, 541
Selected References, 541

chapter twenty-five

Echinoderms, Hemichordates, and Chaetognaths 542

Phylum Echinodermata, 543
Ecological Relationships, 543
Class Asteroidea: Sea Stars, 543
Class Ophiuroidea: Brittle Stars, 549
Class Echinoidea: Sea Urchins, Sand Dollars, and Heart Urchins, 550
Class Holothuroidea: Sea Cucumbers, 551
Class Crinoidea: Sea Lilies and Feather Stars, 553
Class Concentricycloidea: Sea Daisies, 554
Phylogeny and Adaptive Radiation, 554
Phylum Hemichordata: Acorn Worms, 557
Class Enteropneusta, 557
Class Pterobranchia, 558
Phylogeny, 558
Phylum Chaetognatha: Arrowworms, 560
Summary, 560
Review Questions, 561
Selected References, 561

part four The Vertebrate Animals, 563

chapter twenty-six

Vertebrate Beginnings: The Chordates, **564**

Traditional and Cladistic Classification of the Chordates, 565
Four Chordate Hallmarks, 566
Notochord, 569
Dorsal, Tubular Nerve Cord, 569
Pharyngeal Pouches and Gill Slits, 569
Postanal Tail, 570
Ancestry and Evolution of the Chordates, 570
Subphylum Urochordata (Tunicata), 570
Subphylum Cephalochordata, 571
Subphylum Vertebrata, 573
Adaptations That Have Guided Vertebrate Evolution, 573
The Search for the Vertebrate Ancestral Stock, 575
The Earliest Vertebrates: Jawless Ostracoderms, 576
Early Jawed Vertebrates, 577
Evolution of Modern Fishes and Tetrapods, 578
Summary, 580
Review Questions, 580
Selected References, 581

chapter twenty-seven

Fishes, 582

Ancestry and Relationships of Major Groups of Fishes, 583
Superclass Agnatha: Jawless Fishes, 583
Hagfishes: Class Myxini, 583
Lampreys: Class Cephalaspidomorphi, 587
Cartilaginous Fishes: Class Chondrichthyes, 587
Sharks and Rays: Subclass Elasmobranchii, 588
Chimaeras: Subclass Holocephali, 590
Bony Fishes: Class Osteichthyes, 591
Origin, Evolution, and Diversity, 591
Ray-Finned Fishes: Subclass Actinopterygii, 591
The Fleshy-Finned Fishes: Subclass Sarcopterygii, 593
Structural and Functional Adaptations of Fishes, 595
Locomotion in Water, 595
Neutral Buoyancy and the Swim Bladder, 596
Respiration, 597
Migration, 598
Reproduction and Growth, 600
Summary, 604
Review Questions, 604
Selected References, 605

chapter twenty-eight

The Early Tetrapods and Modern Amphibians, 606

Movement onto Land, 607
Early Evolution of Terrestrial Vertebrates, 607
Devonian Origin of the Tetrapods, 607
Carboniferous Radiation of the Tetrapods, 609
The Modern Amphibians, 609
Caecilians: Order Gymnophiona (Apoda), 611
Salamanders: Order Caudata (Urodela), 612
Frogs and Toads: Order Anura (Salientia), 615
Summary, 619
Review Questions, 620
Selected References, 620

chapter twenty-nine

Reptiles, 621

Origin and Adaptive Radiation of Reptiles, 622
Changes in Traditional Classification of Reptiles, 622
Characteristics of Reptiles That Distinguish Them from Amphibians, 625
Characteristics and Natural History of Reptilian Orders, 627
Anapsid Reptiles: Subclass Anapsida, 627
Diapsid Reptiles: Subclass Diapsida, 628
Summary, 638
Review Questions, 639
Selected References, 639

chapter thirty

Birds, 640

Origin and Relationships, 641
Adaptations of Bird Structure and Function for Flight, 645
Feathers, 645
Skeleton, 646
Muscular System, 646
Digestive System, 648
Circulatory System, 649
Respiratory System, 649
Excretory System, 649
Nervous and Sensory System, 650
Flight, 652
Bird Wing as a Lift Device, 652
Basic Forms of Bird Wings, 652
Flapping Flight, 653
Migration and Navigation, 654
Migration Routes, 654
Stimulus for Migration, 654
Direction Finding in Migration, 655
Social Behavior and Reproduction, 656
Reproductive System, 656
Mating Systems, 657
Nesting and Care of Young, 658
Bird Populations, 659
Summary, 664
Review Questions, 664
Selected References, 665

chapter thirty-one

Mammals, 666

Origin and Evolution of Mammals, 667
Structural and Functional Adaptations of Mammals, 671
Integument and Its Derivatives, 671
Food and Feeding, 674
Migration, 676
Flight and Echolocation, 677
Reproduction, 679
Mammal Populations, 680
Human Evolution, 682
Summary, 688
Review Questions, 689
Selected References, 690

Appendix: Basic Structure of Matter, 691
Glossary, 699
Credits, 733
Index, 739

preface

Biology of Animals is a textbook for use in introductory zoology courses and is helpful for students of varying backgrounds. You have before you the seventh edition of the text, and like some animals, the book has evolved dramatically through its history. Nevertheless, we have retained the overall organization and distinctive features that have found favor with so many students and professors. These features include emphasis on evolution, animal diversity and adaptations, principles of zoological science, and not least, readability. We also continue the learning aids that have helped students: opening prologues for each chapter that relate a theme or topic drawn from the chapter to engage student interest; chapter summaries and review questions to aid student comprehension and study; accurate and visually appealing illustrations; in-text derivations of generic names; boxed notes and essays that enhance and enlarge on text material; pronunciations of taxa in the tables of classification; and an extensive glossary providing pronunciation, derivation, and definition of terms used in the text.

New For This Edition

Throughout the book we updated and sometimes rewrote sections of text, and we updated references. Many illustrations were revised or replaced. We rewrote many review questions, seeking to provoke thought and to reduce emphasis on rote memorization. We replaced the end paper on Origin of Life and Geologic Time Table with a revised version in full color.

Many readers of this text know that it has a larger sibling, *Integrated Principles of Zoology.* Because we alternate revising the two books, changes instituted in a revision of one can be incorporated into and benefit the next revision of the other. For example, full-color cladograms with illustrations of representatives of animal groups were introduced in *Integrated Principles,* tenth edition, and we use them now in this edition of *Biology of Animals.* Both books emphasize the importance of cladistics in modern taxonomy. Another feature we brought from *Integrated Principles* was not new to the last edition of that book, but it has been popular with students: succinct statements of "Position in the Animal Kingdom" and "Biological Contributions" at the beginning of each survey chapter.

Organization and Coverage

Part I: Evolution of Animal Life

The chapters in Part I introduce the evolution of animals. In chapter 1, we explain the principles of science, provide a brief review of the chemistry of life (students lacking introductory chemistry should refer to the appendix on Basic Structure of Matter), and trace the early evolution of life on earth from its primitive beginning some 3 billion years ago to the appearance of eukaryotes toward the end of the Precambrian. We added a boxed essay on the animal rights controversy. Chapter 2 moves to the organization of eukaryotic cells and discusses mitosis and control of cell division. We added new sections on exocytosis, apoptosis, cyclins, and cdk's, and updated and reillustrated the discussion of oxidative phosphorylation and electron transport. Chapter 3 focuses on the basic principles of heredity and molecular genetics. We added a paragraph on translational control of gene expression and revised the discussion of the genetics of cancer to include the Ras protein and p53. Encoded by a tumor-suppressor gene, p53 has reached public attention because of its connection with smoking and lung cancer. Chapter 4 (Evolution of Animal Diversity) begins with an historical account of Charles Darwin's life and discoveries. The five components of Darwin's evolutionary theory are presented, together with important challenges and revisions to this theory and an assessment of its current scientific status. The chapter ends with discussion of microevolution and macroevolution. Chapter 5 (Ecology and Distribution of Animals) was completely rewritten and reorganized for this edition with greater emphasis on principles of populational and community ecology and animal distribution.

Part II: Animal Form and Function

The nine chapters in this part treat animal form and function, animal behavior, and reproduction and development. In chapter 6 on animal architecture we begin with the basic uniformity of organization of animals and follow with discussion of the integumentary, skeletal, and muscular systems. We updated several sections (complexity and body size; injurious effects of sunlight; ameboid movement; myoneural junction; energy of muscle contraction), added invertebrate examples to the discussion of rigid skeletons, and added a new illustration of a fish skeleton. In chapter 7 (Homeostasis) we emphasize the importance of homeostasis, which permeates all physiological thinking. This chapter includes discussions of some of the accessible examples of homeostasis: osmotic regulation among animals in different habitats (here we revised and updated the explanation of contractile vacuole function), the prominent role of the kidney in body fluid regulation, and temperature regulation. In chapter 8 (Internal Fluids and Respiration) we revised and reillustrated the sections on respiration and gas exchange.

New with this edition is an entire chapter devoted to immunity (chapter 9). The chapter provides in-depth treatment of both vertebrate and invertebrate immunity, with emphasis on the importance of cytokines.

Chapter 10 is a comparative treatment of feeding mechanisms, digestion and the organization of the alimentary canal, and the nutritional requirements of animals. Chapters 11 and 12 cover nervous and endocrine coordination in detail. In chapter 11 (Nervous Coordination) we revised and updated sections on the sense of smell, memory, flatworm nervous system, and the chemistry of vision. Several illustrations were replaced with new, full-color art. In chapter 12 (Chemical Coordination) we added new sections on the pineal gland and melatonin function, prostaglandins, and cytokines. Chapter 13 (Animal Behavior) begins with distinction between proximate and ultimate causation and explains the different experimental approaches to animal behavior. It continues with discussions of control of behavior (genetic and learned) and social behavior, including communication among animals. We reorganized chapter 14 (Reproduction and Development) and inserted a new section on invertebrate reproductive systems, broadened the explanation of the proposed benefits of sex, and explained nongenetic sex determination. In the development section of this chapter we moved the section on development of systems and organs to follow the treatment of early development, and we added a new section on the common vertebrate heritage with a new illustration of phylotypic stages in vertebrate embryogenesis.

Part III: The Invertebrate Animals

The 17 chapters of Parts III and IV are a comprehensive, modern, and thoroughly researched coverage of the phyla of animals. We emphasize the unifying architectural and functional theme of each group. The structure and function of representative forms are described, together with their ecological, behavioral, and evolutionary relationships. We aid the student's comprehension of each chapter by drawing out the underlying themes and distinctive features of each group.

The invertebrate chapters were thoroughly updated and many fine new color photographs were chosen to replace existing illustrations. Following are some of the more significant changes in these chapters. A new note in chapter 16 (Animal-like Protista) mentions *Toxoplasma* as an opportunistic infection in AIDS; the importance of *Cryptosporidium* as a global diarrheal infection, important in AIDS patients, and the newly emerging disease caused by *Cyclospora*. The popular media have carried recent accounts on these emerging infections. Several changes in chapter 19 (Acoelomate Animals) include a table showing the important trematode parasites of humans. Changes in chapter 20 (Pseudocoelomate Animals) include revision of *Trichinella* coverage and the addition of a paragraph on dog heartworm. In the mollusc chapter (chapter 21) we have revised the classification of gastropods. In chapter 23 (Arthropods) we revised the section on arthropod phylogeny giving more emphasis to insects and included a new cladogram. Included in this chapter for the first time are many fine photographs of insects by photographer Jim Castner. A number of new illustrations were added or substituted in chapter 24 (Lesser Protostomes and Lophophorates) and chapter 25 (Echinoderms, Hemichordates,and Chaetognaths). We changed the title and content of chapter 25 because of the current uncertainty regarding the position of the chaetognaths.

Part IV: The Vertebrate Animals

The six chordate chapters were updated with new information and rewritten in many places to enhance clarity. Many new illustrations were added or replaced old illustrations. In chapter 26 (Vertebrate Beginnings) we added a new section on conodonts. In chapter 27 (Fishes) we revised the sections on elasmobranch biology and on fish reproduction and growth, and we added a new illustration of a shark. In chapter 28 (The Early Tetrapods and Modern Amphibians) we added a new illustration showing the evolution of the tetrapod limb. We also revised the discussion of amphibian reproduction to include tropical frogs. In chapter 29 (Reptiles) we added a discussion and a brief description of worm lizards (amphisbaenians); added a new essay and illustration on dinosaurs; and added new illustrations of reptile skin and the amniotic egg. In chapter 30 (Birds) we rewrote parts of the section on digestion, added a comment on the numbers of song birds killed by house cats, and had several illustrations redrawn. A significant change in chapter 31 (Mammals) is the transfer of the discussion of human evolution from chapter 4 to this chapter. We feel that the narrative flow is better here, as well as being more appropriate in this location.

Learning Aids

Vocabulary Development

Key words are boldfaced, and the derivations of generic names of animals are given where they first appear in the text. In addition, the derivations of many technical terms are provided in the text; in this way students gradually become familiarized with the more common roots that recur in many technical terms. An extensive glossary of more than 1000 terms provides pronunciation, derivation, and definition of each term.

Chapter Prologues

A distinctive feature of this text is an opening essay at the beginning of each chapter. Each essay draws out some theme or topic relating to the subject of the chapter. Some prologues present biological, particularly evolutionary, principles; others (especially those in the survey sections) illuminate distinguishing characteristics of the group treated in the chapter. Each one is intended to present an important concept drawn from the chapter in an interesting manner that will facilitate learning by students, as well as engage their interest and pique their curiosity.

Boxed Notes

Boxed notes, which appear throughout the book, augment the text material and offer interesting sidelights without interrupting the narrative. We prepared several new notes for this edition and revised many of the existing ones.

For Review

Each chapter ends with a concise summary, a list of review questions, and annotated selected references. The review questions enable students to self-test retention and understanding of the more important chapter material.

Art Program

The appearance and usefulness of this edition have been further enhanced with many new full-color paintings by William C. Ober and Claire W. Garrison. Bill's artistic skills, knowledge of biology, and experience gained from an earlier career as a practicing physician have enriched this text through six of its editions. Claire practiced pediatric and obstetric nursing before turning to scientific illustration as a full-time career. Texts illustrated by Bill and Claire have received national recognition and have won awards from the Association of Medical Illustrators, American Institute of Graphic Arts, Chicago Book Clinic, Printing Industries of America, and Bookbuilders West. They are also recipients of the Art Directors Award.

Supplements

Instructor's Resource Guide and Test Bank

The Resource Guide provides a chapter outline, test bank, commentary and lesson plan, and resource listing for each chapter. We expect this supplement to be particularly useful to first-time users of the text.

Laboratory Manual

The laboratory manual by Cleveland P. Hickman, Jr., Frances M. Hickman, and Lee B. Kats, *Laboratory Studies in Integrated Zoology,* now in its ninth edition, has been extensively rewritten and reillustrated. It was designed to accompany a year-long course in zoology but can be adapted conveniently for semester or term courses by judicious selection of exercises. The popular wall chart, "Chief Taxonomic Subdivisions and Organ Systems of Animals," is available on request.

Computerized Test Bank

The test questions contained in the Instructors' Resource Guide are available as a computerized test generation system for IBM-compatible and Macintosh computers. Using this system, instructors can create tests and quizzes quickly and easily. Instructors can sort questions by type or level of difficulty, and can add their own questions to the bank of questions provided.

Transparency Acetates

A set of full-color transparency acetates of important textual illustrations is available with this edition of *Biology of Animals.* Labeling is clear, dark, and bold for easy reading.

Animal Diversity Slides

A set of animal diversity slides, photographed by the authors (CPH and LSR) and Bill Ober on their various excursions, are offered in this unique textbook supplement. Both invertebrates and vertebrates are represented. Descriptions, including specific names of each animal and a brief overview of the animal's ecology and/or behavior, accompany the slides.

Student Study Guide

This guide, new in 1997, provides students with additional help in testing their comprehension of difficult concepts. The study guide will feature brief introductions, objective questions/answers and essay questions.

NetQuest: Exploring Zoology

This **new** supplement provides a variety of Internet addresses to correlate to the key zoological topics in this text. Now you can easily locate interesting and current information on the web, regarding almost any subject in zoology. The text includes a brief overview of each topic and suggested "activities."

Life Science Animation Video Series

These videotapes feature physiological processes that occur at the cellular/molecular level. For the scope of this book, the following are appropriate: Tape #1 Chemistry, The Cell, and Energetics; Tape #2 Cell Division, Heredity, Genetics, Reproduction and Development; Tape #3 Animal Biology #1; and Tape #4 Animal Biology #2.

Life Science Living Lexicon CD-ROM

This is an interactive program that features a glossary of common biological roots, prefixes, and suffixes; a categorized glossary, and a section describing the classification system.

BioSource Videodisc

The videodisc contains 10,000 images organized by phyla and provides a complete tour of the animal and plant kingdom. It is supported by a reference manual that contains a sequential listing of all images and their correlating bar codes.

Acknowledgments

We wish to thank the following zoologists who suggested numerous improvements and whose comments were of the greatest assistance to us as we approached this revision. We are especially indebted to Lawrence Hurd of Washington and Lee University, Lexington, Virginia, who made a major contribution to the revision of chapter 5.

Barbara J. Abraham, Hampton University
John S. Addis, Carroll College
William B. Ahern, Greenville College
Julia W. Albright, George Washington University Medical Center
Joe Arruda, Pittsburgh State University
Steven Bassett, Southeast Community College
Bryan C. Bates, Coconino Community College
Richard D. Bates, Rancho Santiago College
Gerald Bergman, Northwest State College
Franklyn F. Bolander, Jr., University of South Carolina
Joseph W. Camp, Jr., Purdue University North Central
Wade L. Collier, Manatee Community College
Mario W. Caprio, Volunteer State Community College
Suzzette Chopin, Texas A&M University–Corpus Christi
Jay P. Clymer III, Marywood College
Tom Dale, Kirtland Community College
Charles J. Dick, Pasco-Hernando Community College
Peter Ducey, State University of New York at Cortland
David A. Easterla, N.W. Missouri State University
Phillip Eichman, University of RioGrande
Eugene J. Fenster, Longview Community College
Edward R. Fliss, Missouri Baptist College
Mark R. Flood, Fairmont State College
G. L. Forman, Rockford College
Merrill W. Foster, Bradley University
Douglas B. Gibson, Mount Vernon College
Susan K. Gilmore, University of Pittsburgh at Bradford
Glenn A. Gorelick, Citrus College
Peter M. Grant, Southwestern Oklahoma State University
Dr. Robert Gregson, Lyon College
Leon E. Hallacher, University of Hawaii at Hilo
Clare Hays, Metropolitan St. College
Sherry Hickman, Hillsborough Community College
Jeffery Hill, Southern Utah University–Darton College
Christine Holler-Dinsmore, Fort Park Community College
Ken Hoover, Jacksonville University
Howard L. Wosick, Washington State University
Arthur B. Jantz, Western Oklahoma State College
Susan Keys, Springfield College
Maria G. Kortier Davis, Olivet College
Robert M. Kruger, Mayville State University
Stephen C. Landers, Troy State University
Roger M. Loyd, Florida Community College
Monica Marquez, Grinnell College
Karen E. McCracken, Defiance College
Sharon C. McDonald, Henry Ford Community College
Vicky McMillan, Colgate University
C. Neal McReynolds, Blue Mountain College
Alan D. Maccarone, Friends University
David J. Mense, Southeastern Community College
Margaret Nordlie, University of Mary
Robert Powell, Avila College
Tricia A. Reichert, Colby Community College
Donald C. Rizzo, Marygrove College
Douglas P. Schelhaas, University of Mary
John Shiber, University of Kentucky-Prestonberg Community College
Neil Schanker, College of the Siskiyous
Fred Schindler, Indian Hills Community College
Brain R. Shmaefsky, Kingwood College
Debra L. Stamper, Kings College
Paul Keith Small, Eureka College
Danny Wann, Carl Albert State College
David K. Webb, University of Vermont
Don R. Yeltman, Davis and Elkins College
Kenneth Thomas, Hillsborough Community College

The production of this textbook was a group effort requiring the skills of many. The authors express their gratitude to the able and conscientious staff of WCB/McGraw-Hill who made this book possible. We especially thank Sponsoring Editor Marge Kemp and Developmental Editor Adora Pozolinski who played vital roles in shaping this seventh edition and providing support throughout. Project Manager Ann Fuerste bore the responsibility of keeping people, text, and art moving in the proper directions and at the right time. Others who played key roles and to whom we are indebted are: Cathy Dipasquale

Conroy, who copy edited the manuscript; Lori Hancock, who coordinated the photography; and Joyce Watters, who managed the art program. The book was designed by Kaye Farmer.

Although we make every effort to bring to you an error-free text, errors of many kinds, including those of fact and emphasis, inevitably find their way into a text book of this scope and complexity. We will be grateful for readers who have comments or suggestions concerning content to send their remarks to Adora Pozolinski, Developmental Editor, WCB/McGraw-Hill, 2460 Kerper Boulevard, Dubuque, IA 52001. You can also contact her at http://www.mhhe.com/zoology.

Cleveland P. Hickman, Jr.
Larry S. Roberts
Allan Larson

part one

Evolution of Animal Life

1 Life: General Considerations, Basic Molecules, and Origins

2 The Cell as the Unit of Life

3 Genetic Basis of Evolution

4 Evolution of Animal Diversity

5 Ecology and Distribution of Animals

Life:
General Considerations, Basic Molecules, and Origins

chapter | one

"That Mystery of Mysteries . . ."

Cosmologists endeavor excitedly to understand the origins of the universe as a whole and our own little solar system in particular. Scientists agree that the universe is about 15 billion years old, and the earth itself came into existence about 4.5 billion years ago. Difficult as the question of cosmic origin is, more challenging still is the issue of the origin of life itself. The most intriguing of recent findings is the consistent picture of very early life on earth as revealed by the fossil record. The oldest fossils yet discovered are filamentous and sheath-enclosed colonial microorganisms from Western Australia that have been firmly dated radioactively at 3.3 to 3.5 billion years old. These are the earliest known life forms, among the first so far as we know to have closed themselves off from the rest of the universe as living, self-reproducing units. Responding then to the forces of natural selection, protocells such as these began to evolve, eventually yielding the web of life on earth today. Throughout time, life and environment have evolved together, each deeply marking the other. The primitive earth, with its reducing atmosphere of ammonia, methane, and water, was superbly fit for the prebiotic synthesis that led to life's beginnings. Yet, it was totally unsuited, indeed lethal, for the kinds of organisms that inhabit the earth today, just as early forms of life could not survive in our present environment.

Charles Darwin pondered the origins of life and species, that "mystery of mysteries," as he called it. Since Darwin's day, we have learned more and more about the history of life forms that have evolved or are evolving on this earth.

Zoology (Gr. *zōon,* animal, + *logos,* discourse on, study of) is the scientific study of animals. It is a subdivision of an even broader science, biology (Gr. *bios,* life, + *logos,* discourse on, study of), the study of all life. The panorama of animal life, how animals function, live, reproduce, and interact with their environment, is exciting, fascinating, and awe inspiring. A complete understanding of all phenomena included in zoology is beyond the ability of any single person, perhaps of all humanity, but the satisfaction of knowing as much as possible is worth the effort. In the chapters to follow, we hope to give you an introduction to this science and to share our excitement in the pursuit of it.

Why study zoology? One of the best reasons is curiosity: curiosity about animals, how they function, and how they relate to one another and to their environment. Understanding why things happen or how they have come to be as they are is satisfying to us. Study of zoology helps not only to understand animals in general, it helps us to understand ourselves, the kind of animals we call humans.

Every educated person should have at least a modest grasp of human function to live better and more happily in the world. Certainly, a study of zoology helps us to understand current issues related to such phenomena as environmental destruction and pollution, extinction of species, genetic engineering, in vitro fertilization, the hazards of radiation, and many others. Some people want to study certain aspects in depth and contribute in a tangible way to human knowledge and to the solution of problems in the world. Many students take courses in zoology to help them prepare for a medical, veterinary, or other health-care career, either in clinical practice or as researchers on diseases of humans or other animals.

Principles of Science

Nature of Science

A basic understanding of zoology requires an understanding of what science is, what it is not, and how knowledge is gained by use of the scientific method.

In this section we examine the methodology that zoology shares with science as a whole. These features distinguish the sciences from those activities that we exclude from the realm of science, such as art and religion.

Despite the enormous impact that science has had on our lives, many people have only a minimal understanding of the real nature of science. For example, on March 19, 1981, the governor of Arkansas signed into law the Balanced Treatment for Creation-Science and Evolution-Science Act (Act 590 of 1981). This act falsely presented "creation-science" as a valid scientific endeavor. "Creation-science" is actually a religious position advocated by a minority of the American religious community, and it does not qualify as science. Enactment of this law led to a historic lawsuit tried in December 1981 in the court of Judge William R. Overton, U.S. District Court, Eastern District of Arkansas. The suit was brought by the American Civil Liberties Union on behalf of 23 plaintiffs, including a number of religious leaders and groups representing several denominations, individual parents, and educational associations. The plaintiffs contended that the law was a violation of the First Amendment to the U.S. Constitution, which prohibits "establishment of religion" by the government. This prohibition includes passing a law that would aid one religion or prefer one religion over another. On January 5, 1982, Judge Overton permanently enjoined the state of Arkansas from enforcing Act 590.

Considerable testimony during the trial dealt with the nature of science. On the basis of testimony by scientists, Judge Overton was able to state explicitly these essential characteristics of science:

1. *It is guided by natural law.*
2. *It has to be explanatory by reference to natural law.*
3. *It is testable against the observable world.*
4. *Its conclusions are tentative, that is, are not necessarily the final word.*
5. *It is falsifiable.*

The pursuit of scientific knowledge must be guided by the physical and chemical laws that govern the state of existence. Scientific knowledge must explain what is observed by reference to natural law without requiring the intervention of any supernatural being or force. We must be able to observe events in the real world, directly or indirectly, to test hypotheses about nature. If we draw a conclusion relative to some event, we must be ready to discard or modify our conclusion if further observations contradict it. As Judge Overton stated, "While anybody is free to approach a scientific inquiry in any fashion they choose, they cannot properly describe the methodology used as scientific, if they start with a conclusion and refuse to change it regardless of the evidence developed during the course of the investigation." Science is neutral on the question of religion, and the results of science do not favor one religious position over another.

Scientific Method

These essential criteria of science form the basis for an approach known as the **hypothetico-deductive method.** The first step of this method is the generation of **hypotheses** or potential answers to the question being asked. These hypotheses are usually based on prior observations of nature (Figure 1-1), or they are derived from theories based on such observations. Scientific hypotheses often constitute general statements about nature that may explain a large number of diverse observations. Darwin's hypothesis of natural selection (Chapter 4), for example, explains the observations that many different species have properties that adapt them to their environments. On the basis of the hypothesis, the scientist must say, "If my hypothesis is a valid explanation of past observations, then future observations ought to have certain characteristics." The best hypotheses are those that mak

A

B

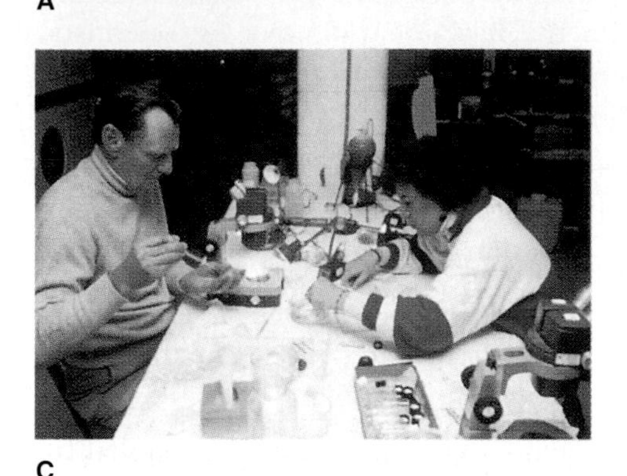

C

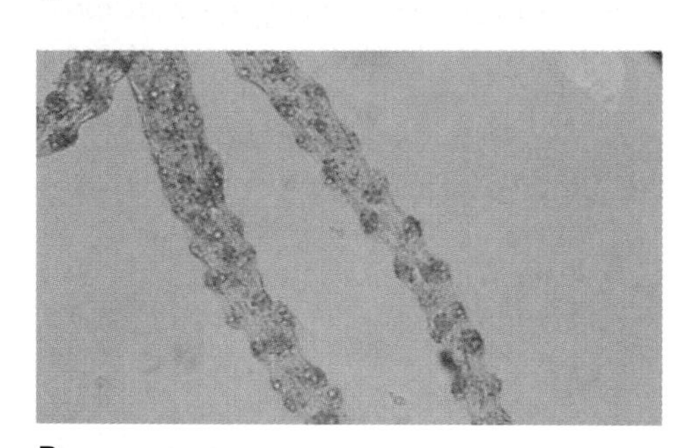

D

E

figure 1-1

A few of the many dimensions of zoological research: **A,** Observing coral growth in the Caribbean; **B,** studying insect larvae collected from an arctic pond on Canada's Baffin Island; **C,** separating growth stages of crab larvae at a marine laboratory; **D,** observing nematocyst discharge from hydrozoan tentacles **(E)**.

many predictions which, if found erroneous, will lead to rejection, or falsification, of the hypothesis.

If an hypothesis is very powerful in explaining a wide variety of related phenomena, it attains the status of a **theory.** Natural selection is a good example. Natural selection provides a potential explanation for the occurrence of many different traits distributed among virtually all animal species. Each of these instances constitutes a specific hypothesis generated from the theory of natural selection. However, falsification of a specific hypothesis does not necessarily lead to rejection of the theory as a whole. Natural selection may fail to explain the origins of human behavior, for example, but it provides an excellent explanation for many structural modifications of the pentadactyl (five-fingered) vertebrate limb for diverse functions. The most useful theories are those that can explain the largest array of different natural phenomena.

We emphasize that the meaning of the word "theory," when used by scientists, is not "speculation" as it is in ordinary English usage. The failure to make this distinction has been prominent in the creationism versus evolution controversy. The creationists have spoken of evolution as "only a theory," as if it were little better than a guess. In fact, the theory of evolution is supported by such massive evidence that most biologists view repudiation of evolution as tantamount to repudiation of reality. Nonetheless, evolution, along with all other theories in science, has not been proven in a mathematical sense, but it is testable, tentative, and falsifiable.

Physiological and Evolutionary Sciences

The many questions that people have asked about the animal world since the time of Aristotle can be grouped into two major categories.* The first category seeks to understand the **proximate** or **immediate causes** that underlie the functioning of biological systems. These include the problems of explaining how animals perform their metabolic, physiological and behavioral functions at the molecular, cellular, organismal, and even population levels. For example, how is genetic information expressed to guide the synthesis of proteins? What causes cells to divide to produce new cells? How does population density affect the physiology and behavior of organisms?

The biological sciences that address proximate causes are the **physiological sciences,** and they proceed using the **experimental method.** This method consists of three steps: (1) predicting how a system being studied will respond to a disturbance, (2) making the disturbance, and then (3) comparing the observed results with the predicted ones. Experimental conditions are repeated to eliminate chance occurrences that might produce errors. **Controls** (repetitions of the experimental procedure that lack the disturbance) are established to protect against any unperceived factors that may bias the outcome of the experiment.

*Mayr, E. 1982. *The Growth of Biological Thought.* Cambridge, Harvard University Press. pp. 67–71.

The processes by which animals maintain a body temperature under different environmental conditions, digest their food, migrate to new habitats, or store energy are some additional examples of physiological phenomena that are studied with the experimental method (Chapters 6 through 14). Subfields of biology that constitute physiological sciences include molecular biology, cell biology, endocrinology, developmental biology, and community ecology.

In contrast to questions concerning the proximate causes of biological systems, the **evolutionary sciences** address questions of **ultimate causes** that have produced these systems and their properties through evolutionary time. For example, what are the evolutionary factors that caused some birds to acquire complex patterns of seasonal migration between North and South America? Why do different species of animals have different numbers of chromosomes in their cells? Why do some animal species maintain complex social systems, whereas the animals of other species are largely solitary?

The evolutionary sciences proceed largely using the **comparative method** rather than experimentation. Characteristics of molecular biology, cell biology, organismal structure, development, and ecology are compared among related species to identify their patterns of variation. The patterns of similarity and dissimilarity can then be used to test hypotheses of relatedness, and thereby to reconstruct the evolutionary tree that relates the species being studied. Clearly, the evolutionary sciences rely on results of the physiological sciences for comparison. Evolutionary sciences include comparative biochemistry, molecular evolution, comparative cell biology, comparative anatomy, comparative physiology, and phylogenetic systematics.

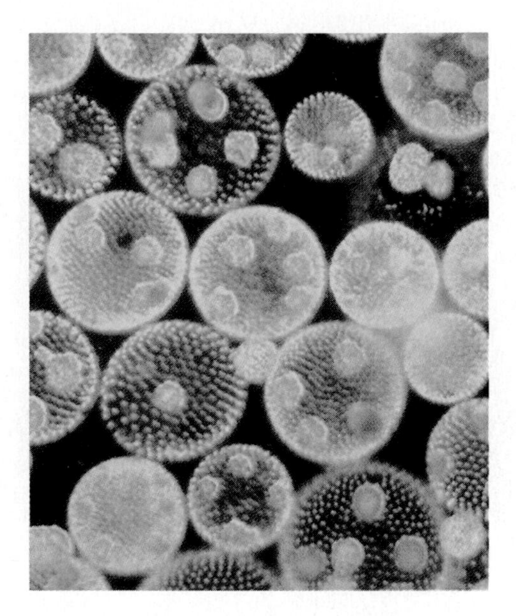

figure 1-2

Three different levels of the biological hierarchy, as illustrated by *Volvox:* cellular, organismal, and populational. Each spheroid is an individual organism containing cells embedded in a gelatinous matrix. The larger cells function in reproduction and the smaller ones perform the general metabolic functions of the organism. These organisms reproduce both sexually and asexually, and collectively they form a population.

What Is Life?

This is a very difficult question. Life can be defined only in terms of the characteristics we attribute to it. However, having introduced the basic molecules of life, and to preface our discussion of life's origins, let us briefly examine these properties.

1. **Chemical uniqueness.** *Living systems demonstrate a unique and complex molecular organization.* Living systems feature macromolecules, much larger than the small molecules that constitute nonliving matter. The four major categories of macromolecules, described below, are carbohydrates, lipids, proteins, and nucleic acids. Although they are composed of common subunits, there is enormous variety in ways in which the subunits are combined, giving living systems both a biochemical unity and a great potential for diversity.

2. **Complexity and hierarchical organization.** *Living systems demonstrate a unique and complex hierarchical organization.* Atoms and molecules are combined into patterns in the living world that do not exist in the nonliving world. In living systems, we find a hierarchy of levels that includes, in ascending order of complexity, macromolecules, cells, organisms, populations, and species (Figure 1-2).

The cell is the smallest unit of biological hierarchy that is semiautonomous in its ability to conduct its basic functions, including reproduction. The cell is therefore viewed as the basic unit of living systems (Chapter 2). Each successively higher level of the biological hierarchy is composed of units of the preceding lower level in the hierarchy. An important characteristic of this hierarchy is that the properties of any given level cannot be obtained from even the most complete knowledge of the properties of its component parts. For example, systems of social interaction, as observed in bees, occur at the level of the population; it would not be possible to infer the properties of this social system by knowing only the properties of individual bees.

The appearance of new characteristics at a given level of organization is called **emergence,** and these characteristics are known as **emergent properties.** They arise from the interactions that occur among the component parts of the system.

3. **Reproduction.** *Living systems can reproduce themselves.* Life does not arise spontaneously but comes only from prior life, through a process of reproduction. Life originated from nonliving matter at least once (p. 16), but this required enormously long periods of time and conditions very different

A

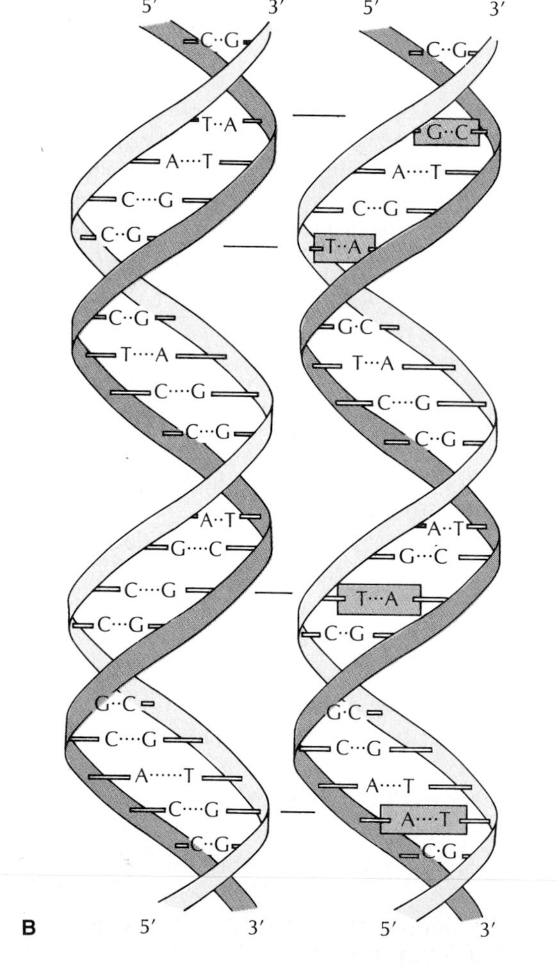

figure 1-3

James Watson and Francis Crick, codiscoverers of the structure of DNA, with the first model of the DNA double helix (**A**). Genetic information is coded in the nucleotide base sequence inside the DNA molecule. Genetic variation is shown (**B**) in DNA molecules that are similar in base sequence but differ from each other at four positions. Such differences can encode alternative traits, such as different eye colors.

from those of the modern earth. Cells, organisms, and populations can reproduce themselves. Although individuals at each level of the hierarchy may fail to reproduce, reproduction is nonetheless a potential in living systems.

Reproduction at each level features the complementary, yet seemingly contradictory, phenomena of **heredity** and **variation.** We will see later in this book that the interaction of heredity and variation in the reproductive process is the basis for organic evolution.

4. **Possession of a genetic program.** *A genetic program provides fidelity of inheritance* (Figure 1-3). The structures of the protein molecules needed for organismal development and functioning are coded in nucleic acids (Chapter 3). For animals and most other organisms, the genetic information is contained in DNA.

The genetic code was established early in the history of life and the same code is present in bacteria and in the nuclear genomes of almost all animals and plants. The near constancy of this code among living forms provides strong evidence for a single origin of life.

5. **Metabolism.** *Living organisms maintain themselves by obtaining nutrients from their environments* (Figure 1-4). The nutrients are broken down to obtain chemical energy and molecular components for use in building and maintaining the living system (Chapter 2). We call these essential chemical processes **metabolism.** They include digestion, production of energy (respiration), and synthesis of molecules and structures.

6. **Development.** *All organisms pass through a characteristic life cycle.* Development describes the characteristic changes that an organism undergoes from its origin (usually the fertilization of the egg) to its final adult form (Chapter 14). Development usually involves changes in size and shape, and the differentiation of structures within the organism. Even the simplest one-celled organisms grow in size and replicate their component parts until they divide into two or more cells.

7. **Environmental interaction.** *All animals interact with their environment* (Chapter 5). All organisms respond to stimuli in their environment (Figure 1-5), and this property is called **irritability.** The stimulus and response may be simple, such as a unicellular organism moving from or toward light or away from a noxious substance, or it may be quite complex, such as a bird responding to a complicated series of signals in a courtship ritual. Life and environment are inseparable. We cannot isolate the evolutionary history of a lineage of organisms from the environments in which it occurred.

Chemistry of Life

A first principle of biology is that living systems and their constituents obey physical and chemical laws. Within the cells of any organism, the living substance is composed of a

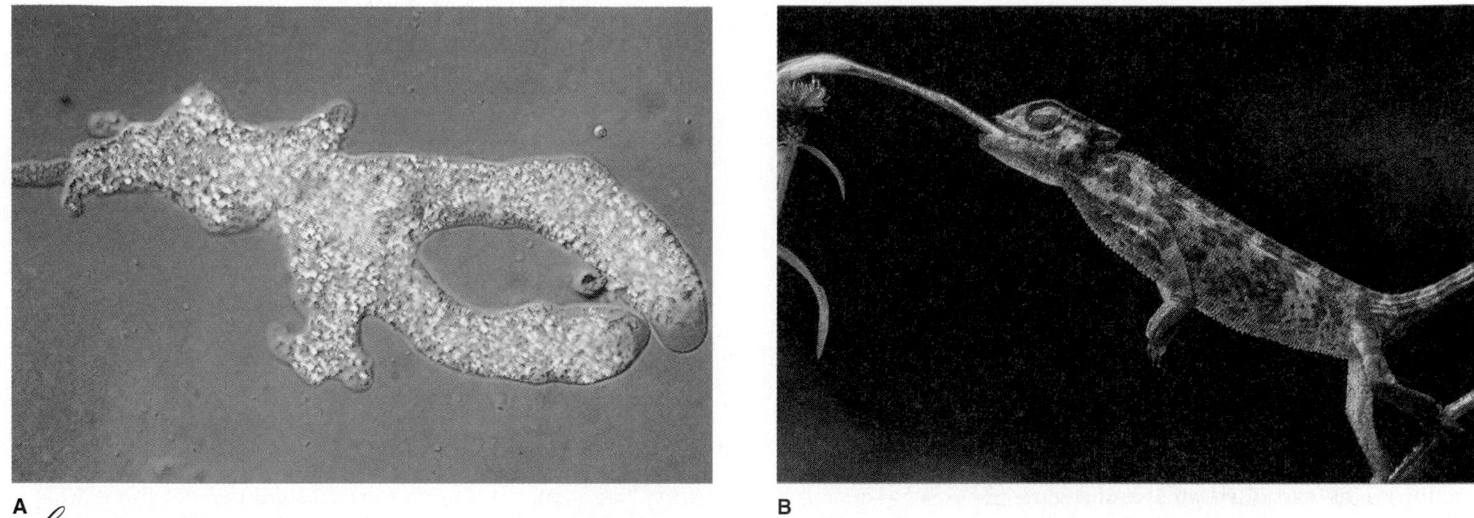

figure 1-4

Feeding processes illustrated by (**A**) an ameba surrounding food and (**B**) a chameleon capturing insect prey with its projectile tongue.

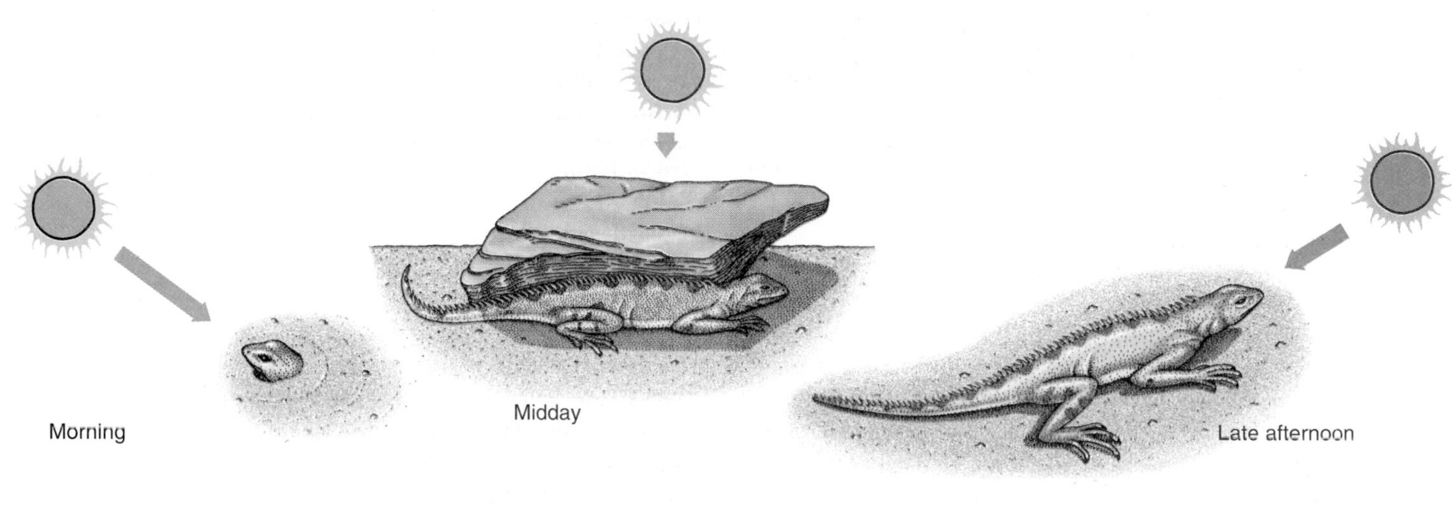

figure 1-5

A lizard regulates its body temperature by choosing different environments at different times of the day.

multitude of chemical constituents: proteins, nucleic acids, fats, carbohydrates, waste metabolites, crystalline aggregates, pigments, and many others. Physical and chemical interactions of such substances account for the many processes essential to life, including digestion and absorption of nutrients, derivation of energy, removal of waste, communication of cells with each other, conduction of nerve impulses, and transmission of genetic information from one generation to the next. Because these phenomena will be discussed in later pages, we present some basic information on biochemistry here. For those unfamiliar with, or wishing review of, basic chemistry, we include a basic explanation of atoms, elements, and molecules; chemical bonds; acids, bases, and salts in Appendix A at the end of the book.

Organic Molecules

The term *organic compounds* has been applied to substances derived from plants and animals. All organic compounds contain carbon, but many also contain hydrogen, oxygen, nitrogen, sulfur, phosphorus, salts, and other elements. Organic compounds are specifically those carbon compounds in which the principal bonds are carbon to carbon and carbon to hydrogen.

Carbon has a great ability to bond with other carbon atoms in chains of varying lengths and configurations. More than a million organic compounds have been identified; more are being added daily. Carbon-to-carbon combinations introduce the possibility of enormous complexity and variety into molecular structure. In living organisms organic compounds perform structural and energy-storage functions, provide a means of heredity from parent to offspring, and facilitate the myriad biochemical reactions needed to sustain life.

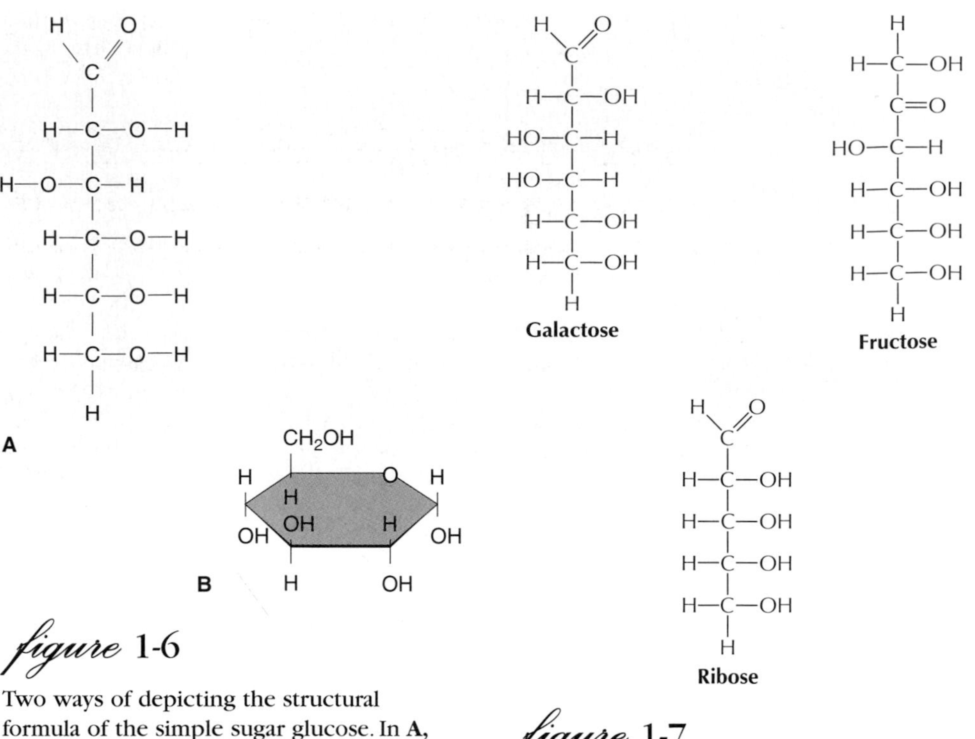

figure 1-6

Two ways of depicting the structural formula of the simple sugar glucose. In **A,** the carbon atoms are shown in open-chain form. When dissolved in water, glucose tends to assume a ring form as in **B.** In this ring model the carbon atoms located at each turn in the ring are usually not shown.

figure 1-7

These two hexoses and a pentose are common monosaccharides. Galactose and ribose are aldehyde sugars; fructose is a ketone sugar.

Carbohydrates: Nature's Most Abundant Organic Substance

Carbohydrates are compounds of carbon, hydrogen, and oxygen. They are usually present in the ratio of 1 C : 2 H : 1 O and are grouped as H–C–OH. Familiar examples of carbohydrates are sugars, starches, and cellulose (the woody structure of plants). There is more cellulose on earth than all other organic materials combined. Glucose, a simple carbohydrate, is synthesized by green plants from water and carbon dioxide, with the aid of the sun's energy. This process, called **photosynthesis,** is a reaction upon which all life depends, for it is the starting point in the formation of food.

Carbohydrates are usually divided into the following three classes: (1) **monosaccharides,** or simple sugars; (2) **disaccharides,** or double sugars; and (3) **polysaccharides,** or complex sugars. Simple sugars are most commonly composed of carbon chains containing four carbons (tetroses), five carbons (pentoses), or six carbons (hexoses). Simple sugars, such as glucose, galactose, and fructose, all contain a free sugar group,

```
 OH O
 |  ||
—C—C—
 |
 H
```

in which the double-bonded O may be attached to the terminal C of a chain or to a nonterminal C. The hexose **glucose** (also called dextrose) is the most important carbohydrate in the living world. Glucose is often shown as a straight chain (Figure 1-6A), but in water tends to form a ring structure (Figure 1-6B). Galactose and fructose are also biologically important hexoses, and ribose is a vital pentose (Figure 1-7).

Disaccharides are double sugars formed by the bonding of two simple sugars. An example is maltose (malt sugar), composed of two glucose molecules. As shown in Figure 1-8, the two glucose molecules are bonded together by the removal of a molecule of water. This **condensation reaction,** with the sharing of an oxygen atom by the two sugars, characterizes the formation of all disaccharides. (A reaction in the reverse direction, adding a molecule of water, is a **hydrolysis,** or **hydrolytic reaction.**) Two other common disaccharides are sucrose (ordinary cane, or table, sugar), formed by the linkage of glucose and fructose, and lactose (milk sugar), composed of glucose and galactose.

Polysaccharides are made up of many molecules of simple sugars (usually glucose) linked together in long chains. Such substances with many units linked together are called **polymers.** Their empirical formula is usually written $(C_6H_{10}O_5)_n$, where n stands for the unknown number of simple sugar molecules of which they are composed. Starch is the common storage form of sugar in most plants and is an important food for animals. **Glycogen** is an important storage form for sugar in animals. It is found mainly in liver and muscle cells in vertebrates. When needed, glycogen is converted into glucose and is delivered by the blood to the tissues. Another polymer of glucose is **cellulose,** which is the principal structural carbohydrate of plants. **Chitin** (p. 147) is a polysaccharide containing nitrogen and is found in many animals.

The main role of carbohydrates in cells is to serve as a source of chemical energy, that is, as **fuels.** Glucose is the most important of these energy carbohydrates. Some carbohydrates

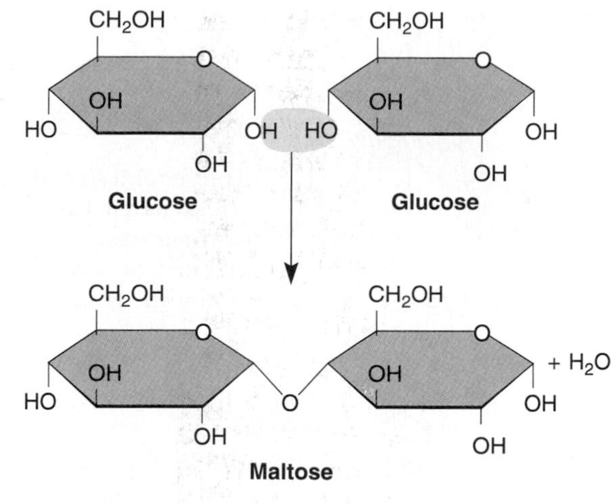

figure 1-8

Formation of a double sugar (disaccharide maltose) from two glucose molecules with the removal of one molecule of water. This type of reaction is a condensation reaction. The reverse reaction, adding a molecule of water to maltose and forming two glucose molecules, is a hydrolytic reaction.

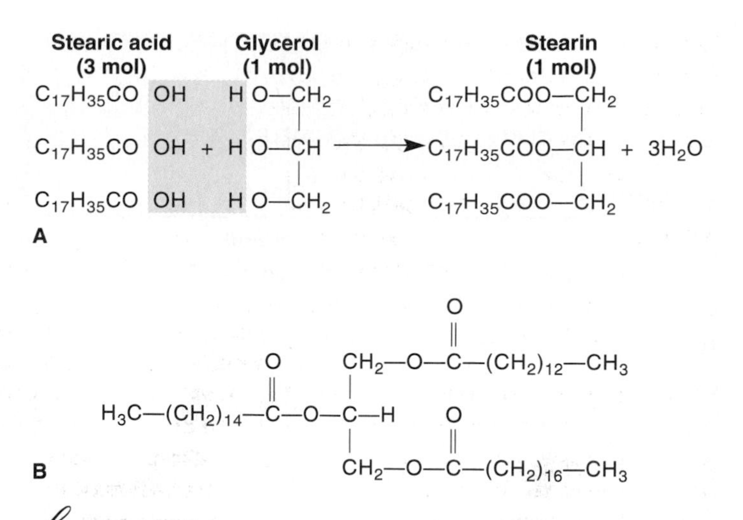

figure 1-9

Neutral fats. **A,** Formation of a neutral fat, stearin, from three molecules of stearic acid (a fatty acid) and one molecule of glycerol. **B,** A neutral fat bearing three different fatty acids.

$$CH_3\text{—}(CH_2)_7\text{—}CH{=}CH\text{—}(CH_2)_7\text{—}COOH$$

Oleic acid

$$CH_3\text{—}(CH_2)_4\text{—}CH{=}CH\text{—}CH_2\text{—}CH{=}CH\text{—}(CH_2)_7\text{—}COOH$$

Linoleic acid

figure 1-10

Unsaturated fatty acids: oleic acid having one double bond and linoleic acid having two double bonds. The remainder of the hydrocarbon chains of both acids is saturated.

become basic components of protoplasmic structure, such as the pentoses that form constituent groups of nucleic acids and of nucleotides.

Lipids: Fuel Storage and Building Material

Lipids are fats and fatlike substances. They are composed of molecules of low polarity; consequently, they are virtually insoluble in water but are soluble in organic solvents such as acetone and ether. Three principle groups of lipids are neutral fats, phospholipids, and steroids.

> *In general, organic substances with a high proportion of hydroxyl (–OH) or other polar groups compared to the number of carbon atoms in the molecule are soluble in water, as are those with ionizable groups, such as acids. As the number of polar groups compared to the carbons decreases, solubility in water decreases, and solubility in nonpolar solvents increases.*

Neutral Fats The neutral, or "true" fats are major fuels of animals. Stored fat may be derived directly from dietary fat or indirectly from dietary carbohydrates that are converted to fat for storage. Fats are oxidized and released into the bloodstream as needed to meet tissue demands, especially the demands of active muscle.

Triglycerides are neutral fats that consist of glycerol and three molecules of fatty acids (Figure 1-9). The fatty acids in triglycerides are simply long-chain monocarboxylic acids; they vary in size but are commonly 14 to 24 carbons long. The production of a typical fat by a condensation reaction of glycerol and stearic acid is shown in Figure 1-9A. In this reaction we can see that the three fatty acid molecules have bonded with the OH group of the glycerol to form stearin (a neutral fat), with the production of three molecules of water. A **monoglyceride** is a glycerol molecule with only one fatty acid, and a **diglyceride** has two fatty acids.

Most triglycerides contain two or three different fatty acids attached to glycerol, bearing ponderous names such as myristoyl stearoyl glycerol (Figure 1-9B). The fatty acids in this triglyceride are **saturated;** that is, every carbon within the chain holds two hydrogen atoms. Saturated fats, more common in animals than in plants, are usually solid at room temperature. **Unsaturated** fatty acids, typical of plant oils, have two or more carbon atoms joined by double bonds; that is, the carbons are not "saturated" with hydrogen atoms and are able to form additional bonds with other atoms. Two common unsaturated fatty acids are oleic acid and linoleic acid (Figure 1-10). Plant fats such as peanut oil and corn oil tend to be liquid at room temperature. (See Table 1-1.)

table 1-1
Some Types of Neutral Fats

Animal	Plant
Lard	Corn oil
Dairy butter	Peanut oil
"Fat"	Cottonseed oil
	Soybean oil
	Olive oil
	Safflower oil
	Palm oil*
	Coconut oil*

**These plant oils are considered saturated.*

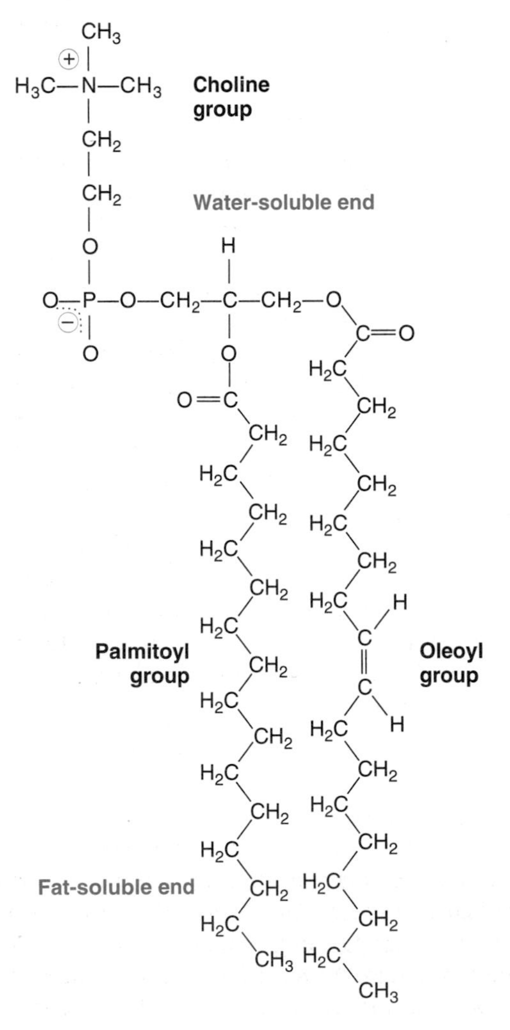

figure 1-11
Lecithin (phosphatidyl choline), an important phospholipid of nerve membranes.

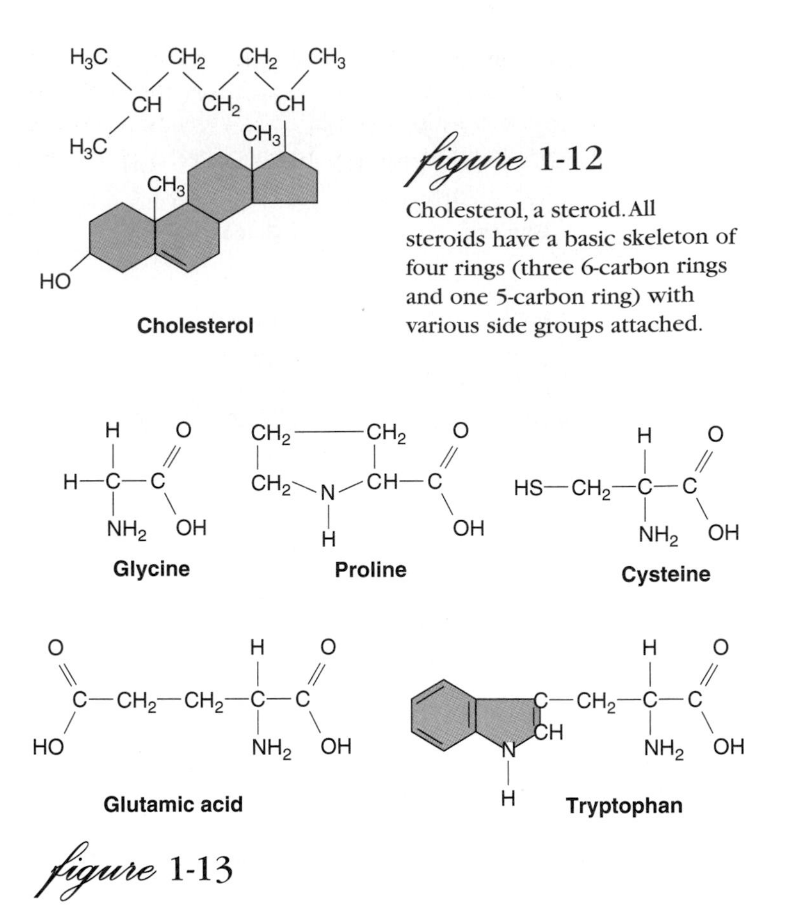

figure 1-12
Cholesterol, a steroid. All steroids have a basic skeleton of four rings (three 6-carbon rings and one 5-carbon ring) with various side groups attached.

figure 1-13
Five of the 20 amino acids that normally occur in proteins.

Phospholipids Unlike the fats that are fuels and serve limited structural roles in the cell, phospholipids are important components of the molecular organization of cells, especially membranes. They resemble triglycerides in structure, except that one of the three fatty acids is replaced by phosphoric acid and an organic base containing nitrogen. An example is lecithin, an important phospholipid of nerve membrane (Figure 1-11). Because the phosphate group on phospholipids is charged and polar and therefore soluble in water, and the remainder of the molecule is nonpolar, phospholipids can bridge two environments and bind water-soluble molecules such as proteins to water-insoluble materials.

Steroids Steroids are complex alcohols; although they are structurally unlike fats, they have fatlike properties. The steroids are a large group of biologically important molecules, including cholesterol (Figure 1-12), vitamin D, many adrenocortical hormones (Figure 12-14, p. 280), and the sex hormones (Figure 12-18, p. 283).

Amino Acids and Proteins

Proteins are large, complex molecules built from 20 commonly occurring kinds of amino acids (Figure 1-13). The amino acids are linked together by **peptide bonds** to form long, chainlike polymers. In the formation of a peptide bond, the carboxyl group of one amino acid is linked by a covalent bond to the amino group of another, with the elimination of water, as follows:

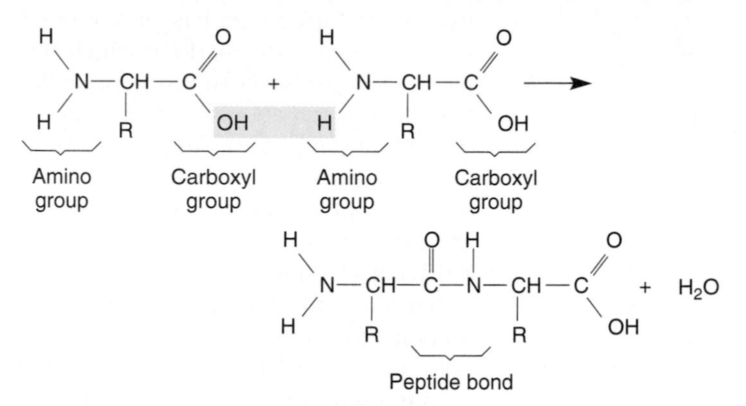

The combination of two amino acids by a peptide bond forms a dipeptide, and, as is evident, there is still a free amino group on one end and a free carboxyl group on the other; therefore additional amino acids can be joined to both ends until a long chain is produced. A molecule consisting of many joined amino acids, but not as complex as a protein, is a **polypeptide.** The 20 different kinds of amino acids can be arranged in an enormous variety of sequences of up to hundreds of amino acid units; therefore, it is not difficult to account for the practically countless varieties of proteins among living organisms.

A **protein** is not just a long string of amino acids; it is a highly organized molecule. For convenience, biochemists have recognized four levels of protein organization called primary, secondary, tertiary, and quaternary.

The **primary structure** of a protein refers to the identity and sequence of amino acids making up the polypeptide chain. Because the bonds between the amino acids in the chain are characterized by a limited number of stable angles, certain recurrent structural patterns are assumed by the chain. This is called the **secondary structure,** and it is often that of an **alpha-helix,** that is, helical turns in a clockwise direction, like a screw (Figure 1-14). The spirals of the chains are stabilized by hydrogen bonds, usually between a hydrogen atom of one amino acid and the peptide-bond oxygen of another amino acid in an adjacent turn of the helix.

Not only does the polypeptide chain (primary structure) spiral into helical configurations (secondary structure), but also the helices themselves bend and fold, giving the protein its complex, yet stable, three-dimensional **tertiary structure** (Figure 1-14). The folded chains are stabilized by the interactions between side groups of amino acids. One of these interactions is the **disulfide bond,** a covalent bond between the sulfur(s) atoms in pairs of cysteine (sis′tee-in) units that are brought together by folds in the polypeptide chain. Other kinds of bonds that help stabilize the tertiary structure of proteins are hydrogen bonds, ionic bonds, and hydrophobic bonds.

The term **quaternary structure** describes those proteins that contain more than one polypeptide chain unit. For example, hemoglobin (the oxygen-carrying substance in blood) of higher vertebrates is composed of four polypeptide subunits nested together into a single protein molecule. The tertiary and quaternary structures of a protein are vitally important to its function, for example, as an enzyme; if the shape is wrong, the enzyme will not fit the substrate (see p. 41).

Proteins as Enzymes Proteins perform many functions in living things. They serve as the structural framework of cells and form many cell components. However, the most important role of proteins by far is as **enzymes,** the biological catalysts required for almost every chemical reaction in the body (p. 40).

Enzymes lower the energy required for specific reactions and enable life processes to proceed at moderate temperatures. They control the reactions by which food is digested, absorbed, and metabolized. They promote the synthesis of structural materials for growth, maintenance, and repair of the body. They determine the release of energy used in respiration, muscle contraction, physical and mental activities, and many other activities. Enzyme action is described in Chapter 2 (p. 40).

Nitrogen Base Derivatives

Many vital compounds in cells contain units of **nitrogenous bases,** which are ring compounds containing nitrogen as well as carbon (p. 71). Some of these, such as ATP, NAD, and FAD, play critical roles in energy metabolism (Chapter 2). Others are organized into **nucleic acids,** which are complex substances of high molecular weight.

The sequence of nitrogenous bases in nucleic acids encodes the genetic information necessary for all aspects of biological inheritance. Nucleic acids not only direct the synthesis of enzymes and other proteins but are also the only molecules that have the power (with the help of the right enzymes) to replicate themselves. The two kinds of nucleic acids in cells are **deoxyribonucleic acid (DNA)** and **ribonucleic acid (RNA).** They are polymers of repeated units called **nucleotides,** each containing a sugar, nitrogenous base, and phosphate group. Because the structure of nucleic acids is crucial to the mechanism of inheritance and protein synthesis, this subject is discussed further in Chapter 3.

Origin of Life

Considering the exquisite organization of living organisms and the complexity of the molecules and the reactions that result in the properties of life, how could life have originated from nonliving substances? To most biologists the question of life's beginnings is one of profound interest. Despite the complexity, the biologist is struck by a remarkable unity at the molecular and cellular levels. All organisms, from humans to the simplest microbes, share two kinds of basic biomolecules: nucleic acids and proteins.

We must admit at the outset that we do not know how life originated on earth. However, in the last 40 years or so a multidisciplinary effort of scientists from several specialties has made it possible to construct a scenario in which simple living organisms evolved from nonliving constituents about 4 billion years BP (before present). These studies are not attempts to prove or disprove any religious or philosophical belief, but rather they are endeavors to provide an intellectually satisfying account of how life could have arisen on earth by natural means. In the sections to follow, the student should note how scientific methodology was used to shed light on these questions.

Historical Perspective

From ancient times it was commonly believed that life could arise by **spontaneous generation** from dead material, in addition to arising from parental organisms by reproduction **(biogenesis).** Frogs appeared to arise from damp earth, mice from putrefied matter, insects from dew, maggots from decaying meat, and so on.

The question of spontaneous generation fell under the scrutiny of experimental science in the sixteenth and seventeenth centuries. However, the doctrine was too firmly entrenched to be disbelieved. It remained for the great French scientist Louis Pasteur in 1860 to silence all but the most stubborn proponents of spontaneous generation (Figure 1-15). The most famous of an elegant series of experiments involved the use of flasks with the necks drawn out into a long "swan neck" (Figure 1-16). It was known that microorganisms would appear "spontaneously" in nutrient broth left open to the air.

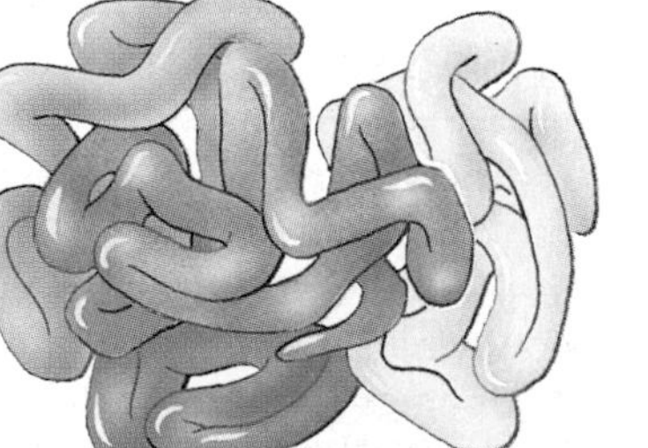

figure 1-14

Structure of proteins. The amino acid sequence of a protein *(primary structure)* encourages the formation of hydrogen bonds between nearby amino acids, producing coils and foldbacks (the *secondary structure*). Bends and helices cause the chain to fold back on itself in a complex manner *(tertiary structure).* Individual polypeptide chains of some proteins aggregate together to form the functional molecule composed of several subunits *(quaternary structure).*

figure 1-15

Louis Pasteur, who refuted the idea of spontaneous generation.

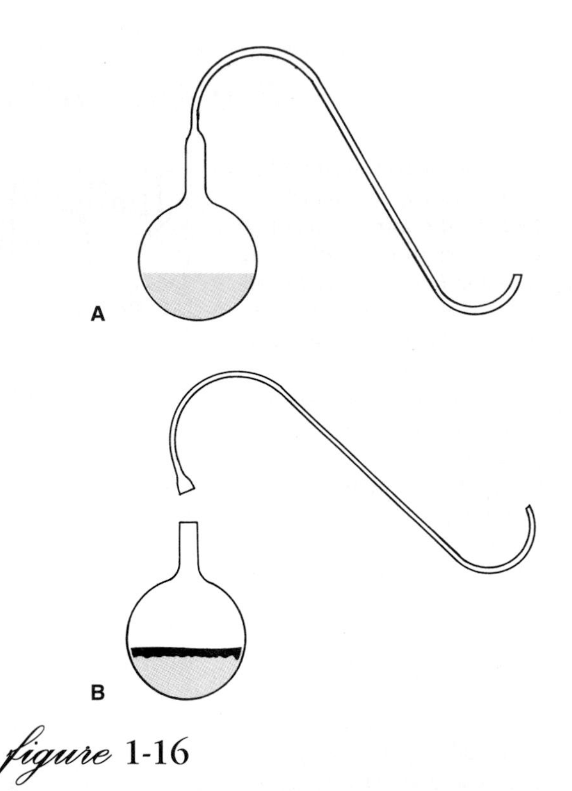

figure 1-16

Louis Pasteur's swan-neck flask experiment. **A,** Sugared yeast water boiled in swan-neck flask remains sterile until the neck is broken. **B,** Within 48 hours, the flask is swarming with living microorganisms.

Pasteur's hypothesis was that these microorganisms were carried to the broth by dust particles in the air. Broth was placed in the swan-neck flask and thoroughly boiled. His control was broth in a straight-neck flask similarly treated. Dust particles entering the swan-neck flask were trapped at the bottom of the neck; thus the broth in this flask remained sterile. The broth in the straight-neck flask was soon teeming with microorganisms, and the hypothesis was supported. As a control on the ability of the broth in the swan-neck flask to support growth, the neck was broken off, and colonies of microorganisms quickly began to grow in the flask.

Pasteur's experiments were so convincing that they ended further inquiry into the spontaneous origins of life for a long period. The rebirth of interest into the origins of life occurred in the 1920s. The Russian biochemist Alexander I. Oparin and the British biologist J.B.S. Haldane independently proposed that life originated on earth after an inconceivably long period of "abiogenic molecular evolution." They suggested that the simplest living units came into being gradually by the progressive assembly of organic molecules into more complex organic molecules. These molecules would react with each other to form living microorganisms. Although their proposals differed somewhat on the composition of the earth's early atmosphere, they agreed that the atmosphere lacked free oxygen.

Modern Experimentation

In 1953 Stanley Miller, while a Ph.D. candidate at the University of Chicago, made the first attempt to simulate with laboratory apparatus the conditions believed to prevail on the primitive earth. His experiment was designed to test the Oparin-Haldane hypothesis by simulating conditions that would have prevailed on the earth, then determining whether biologically important molecules could be produced. Miller built an apparatus that would circulate a mixture of methane, hydrogen, ammonia, and water (representing the atmosphere of the early earth) past an electric spark (Figure 1-17). The spark represented lightning, an energy source to provide necessary energy for the chemical reactions. Water in the flask was boiled to produce steam that helped circulate the gases. The products formed in the electrical discharge were condensed in the condenser and collected in the U-tube and small flask (representing the ocean). The control was an apparatus containing the same materials but with no sparking.

After a week of continuous sparking, the water containing the products was analyzed. The results were surprising. Approximately 15% of the carbon that was originally in the "atmosphere" had been converted into organic compounds that collected in the "ocean." The most striking finding was that many compounds related to life were synthesized. These included four amino acids commonly found in proteins; urea; and several simple fatty acids. No such compounds were found in the control apparatus. Thus the hypothesis was supported.

Miller discovered that amino acids were not formed directly in the spark, but rather were produced by th

The Animal Rights Controversy

In recent years, the debate surrounding the use of animals to serve human needs has intensified. Most controversial of all is the issue of animal use in biomedical and behavioral research and in commercial product testing.

A few years ago, Congress passed a series of amendments to the Federal Animal Welfare Act, a body of laws covering animal care in laboratories and other facilities. These amendments have become known as the three R's: **Reduction** in the number of animals needed for research; **Refinement** of techniques that might cause stress or suffering; **Replacement** of live animals with simulations or cell cultures whenever possible. As a result, the total number of animals used each year in research and in testing of commercial products has declined steadily as scientists and businesses have become more concerned and more accountable. The animal rights movement, largely made up of vocal antivivisectionists, has helped to create an awareness of the needs of laboratory research animals and has stretched the resources and creativity of the researchers to discover cheaper, more efficient, and more humane alternatives to experimentation on animals, particularly mammals.

However, computers and cell cultures—the alternatives—can simulate the effects on organismal systems of, for instance, drugs, only when the principles are well known. When the principles are themselves being scrutinized and tested, computer modeling is insufficient. A recent report by the National Research Council concedes that while the search for alternatives to animal use in research and testing will continue, "the chance that alternatives will completely replace animals in the foreseeable future is nil." Realistic immediate goals, however, are reduction in numbers of animals used, replacement of mammals with lower vertebrates, and refinement of experimental procedures that lead to reduction of the pain and discomfort of the animals being tested.

According to the U.S. Department of Health and Human Services, animal research has helped extend our life expectancy by 20.8 years.

Medical and veterinary progress depend on animal research. Every drug and every vaccine developed to improve the human condition has first been tested on an animal. Animal research has enabled medical science to eliminate smallpox and polio; provided immunization against diseases previously common and often deadly, such as diphtheria, mumps, and rubella; helped create treatments for cancer, diabetes, heart disease, and manic-depressive psychoses; and helped in the development of surgical procedures such as heart surgery, blood transfusions, and cataract removal. AIDS research is wholly dependent on animal studies largely because the similarity of simian AIDS, identified in rhesus monkeys, to human AIDS has permitted the disease in monkeys to serve as a model for the human disease. Recent work indicates that cats, too, may prove to be useful models for the development of an AIDS vaccine. Skin-grafting experiments, first done with cattle and later with other animals, opened a new era in immunological research with vast ramifications for health and treatment of disease in humans and other animals.

Animal research also has benefited *other animals* through discovery of veteri-

---

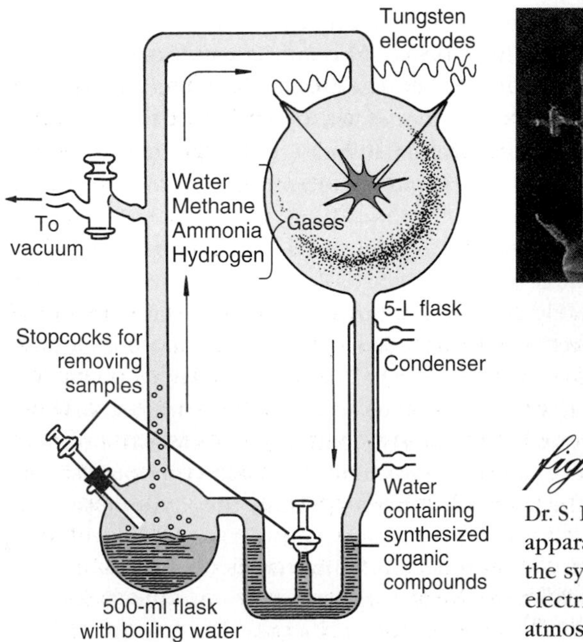

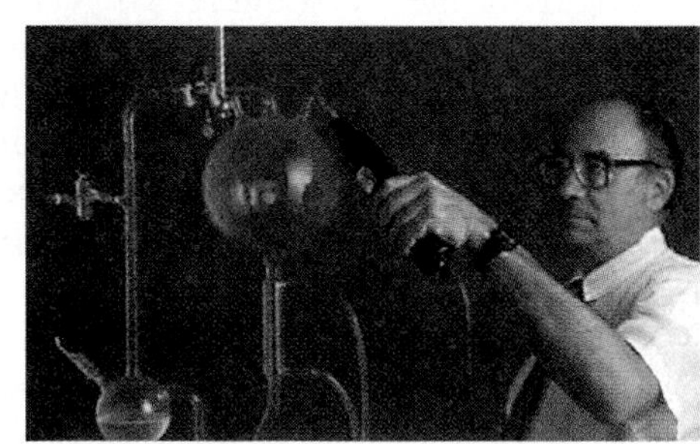

figure 1-17

Dr. S. L. Miller with a replica of the apparatus used in his 1953 experiment on the synthesis of amino acids with an electric spark in a strongly reducing atmosphere.

condensation of certain reactive intermediates, especially hydrogen cyanide and formaldehyde reacting with ammonia. He found that hydrogen cyanide would react with ammonia under prebiotic conditions to form adenine, one of the four bases found in nucleic acids and a component of adenosine triphosphate (ATP), the universal energy intermediate of living systems. Adenine is a purine, a complex molecule chemically (p. 71). The ease with which it was produced under prebiotic conditions suggests that it came to occupy a central position in biochemistry because it was abundant on the primitive earth. Other nucleic acid bases and sugars have been synthesized under conditions likely to have prevailed on the primitive earth.

nary cures. The vaccines for feline leukemia and canine parvovirus were first introduced to other cats and dogs. Many other vaccinations for serious animal diseases were developed through animal research: for example, rabies, distemper, anthrax, hepatitis, and tetanus. No endangered species is used in general research (except to protect that species from total extinction). Thus animal research has provided enormous benefits to humans and other animals. Still, much remains to be learned about treatment of diseases such as cancer, AIDS, diabetes, and heart disease, and research with animals will be required for this investigation.

Despite the remarkable benefits produced by animal research, animal rights advocates continue to present an inaccurate and emotionally distorted picture of animal research. The ultimate goal of most animal rights activists, who have focused specifically on the use of animals in science rather than on the treatment of animals in all contexts, remains the total abolition of all forms of animal research. The scientific community is deeply concerned about the impact of these attacks on the ability of scientists to conduct important experiments that will benefit human and animal well-being. They believe that if we are justified in the use of animals for food and fiber, as well as for personal use, we must be justified in experimentation for the benefit of human survival when these studies are conducted humanely and with full acceptance of our ethical responsibilities in the protectorship of animals.

References on Animal Rights Controversy

Commission on Life Sciences, National Research Council. 1988. Use of laboratory animals in biomedical and behavioral research. Washington, D.C., National Academy Press. *Statement of national policy on guidelines for the use of animals in biomedical research. Includes a chapter on the benefits derived from the use of animals.*

Goldberg, A.M., and J.M. Frazier. 1989. Alternatives to animals in toxicity testing. Sci. Am. **261:**24–30 (Aug.). *Describes alternatives that are being developed for the costly and time-consuming use of animals in the testing of thousands of chemicals that each year must be evaluated for potential toxicity to humans.*

Pringle, L. 1989. The animal rights controversy. San Diego, California, Harcourt Brace Jovanovich, Publishers. *Although no one writing about the animal rights movement can honestly claim to be totally objective and impartial on such an emotionally charged issue, this book comes as close as any to presenting a balanced treatment.*

Roush, W. 1996. Hunting for animal alternatives. Science **274:**168–171. *The number of animals used in research has fallen, but many hurdles remain on the road to animal-free experiments.*

Rowan, A.N. 1984. Of mice, models, and men: a critical evaluation of animal research. Albany, New York, State University of New York Press. *Good review of the issues. Chapter 7 deals with the use of animals in education, points out that our educational system provides little help in resolving the contradiction of teaching kindness to animals on the one hand and using animals in experimentation in biology classes on the other.*

Sperling, S. 1988. Animal liberators: research and morality. Berkeley, University of California Press. *Thoughtful and carefully researched study of the animal rights movement, its ideological roots, and the passionate idealism of animal rights activists.*

Since Miller's original experiments were conducted it has become increasingly clear that the earth's early atmosphere was not as strongly reducing (conditions in which all compounds tend to be reduced, see Appendix A, p. 693) as thought previously, but intermediate or slightly reducing, although free oxygen was absent in either case. (Our present atmosphere is strongly oxidizing.) However, Miller and others showed that amino acids and other organic compounds were produced when carbon dioxide was substituted for methane and molecular nitrogen for ammonia; that is, oxidized compounds were substituted for reduced compounds.

Formation of Polymers

Need for Concentration

The next stage in chemical evolution involved the condensation of amino acids, purines, pyrimidines, and sugars to yield larger molecules that resulted in proteins and nucleic acids. Such condensations do not occur easily in dilute solutions because the presence of excess water tends to drive reactions toward decomposition (hydrolysis). Although the primitive ocean has been called a primordial soup, it was probably a rather dilute one containing organic material that was approximately one-tenth to one-third as concentrated as chicken bouillon.

Prebiotic synthesis must have occurred in restricted regions where concentrations were higher, and modern experimentation has shown that any of a variety of mechanisms could have been effective. Violent weather and impacts of asteroids could have lofted great amounts of dust into the atmosphere. The dust particles could have become foci of water droplets, where the salt concentration would have been high enough to provide a concentrated medium for chemical reactions. Alternatively, prebiotic molecules might have been concentrated by adsorption on the surface of clay and other minerals. The surface of iron pyrite (FeS_2) has also been suggested as a site for the evolution of biochemical pathways.

Thermal Condensations

Most biological polymerizations are condensation (dehydration) reactions, that is, monomers are linked together by the removal of water (p. 8). In living systems, condensation reactions always take place in an aqueous (cellular) environment in the presence of appropriate enzymes.

One of the ways in which dehydration reactions could have occurred in primitive earth conditions without enzymes is by thermal condensation. The simplest dehydration is accomplished by driving off water from solids by direct heating.

The thermal synthesis of polypeptides to form "proteinoids" has been studied by the American scientist Sidney Fox. When a mixture of all 20 amino acids was heated to 180°C in water, a good yield of a protein concentrate was obtained, and boiling the water caused the proteinoids to form enormous numbers of hard, minute spherules called microspheres. The proteinoid microspheres (Figure 1-18) possessed certain characteristics of living systems.

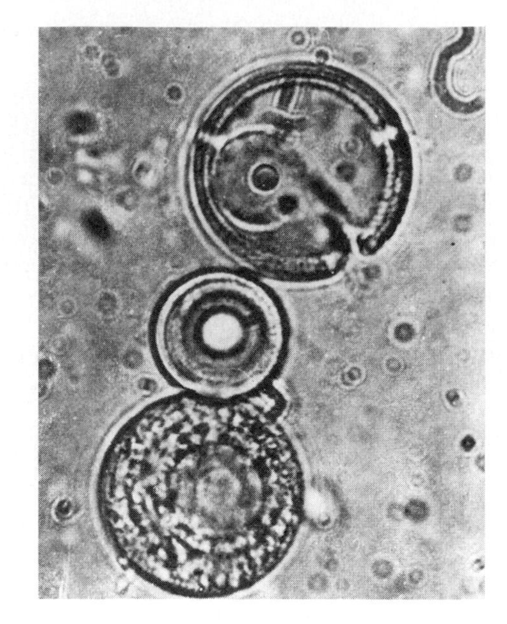

 1-18

Electron micrograph of proteinoid microspheres. These proteinlike bodies can be produced in the laboratory from polyamino acids and may represent precellular forms. They have definite internal ultrastructure. (× 1700)

Origin of Living Systems

The first living organisms were cells: autonomous membrane-bound units with a complex functional organization that permitted the essential activity of self-reproduction. The primitive chemical systems we have described lack this essential property. The principal problem in understanding the origin of life is explaining how primitive chemical systems could have become organized into living, autonomous, self-reproducing cells.

As we have seen, a lengthy chemical evolution on the primitive earth produced several molecular components of living forms. In a later stage of evolution, nucleic acids (DNA and RNA) began to behave as simple genetic systems that directed the synthesis of proteins, especially enzymes. However, this scheme led to a troublesome chicken-egg paradox: (1) How could nucleic acids have appeared without enzymes to synthesize them? (2) How could enzymes have evolved without nucleic acids to direct their synthesis? These questions were based on a long-accepted dogma that only proteins could act as enzymes. We now know that catalytic RNA (ribo-zymes) can mediate processing of messenger RNA, splicing out introns (p. 74), and can catalyze formation of peptide bonds. Evidence is strong that translation of mRNA by ribosomes (p. 75) is catalyzed by their RNA, not protein, content.

Therefore the earliest enzymes could have been RNA. Investigators are now calling this stage the "RNA world." Nevertheless, proteins have several important advantages over RNA as catalysts, and the first protocells with protein enzymes would have had a powerful selective advantage over those with only RNA.

Once this stage of organization was reached, natural selection (p. 102) began acting on these primitive self-replicating systems. This was a critical point. Before this stage, biogenesis was shaped by the favorable physical and chemical conditions on the primitive earth and by the nature of the reacting elements themselves. When self-replicating systems became responsive to changing conditions in the environment, their subsequent evolution became directed. The more rapidly replicating and more successful systems were favored and they replicated even faster. In short, the most efficient forms survived. From this evolved the genetic code and fully-directed protein synthesis (p. 75). The entity could be called a living organism.

Origin of Metabolism

Living cells today are organized systems that possess complex and highly ordered sequences of enzyme-mediated reactions. How did such vastly complex metabolic schemes develop?

Organisms that depend on organic molecules they have not synthesized for their food supplies are known as **heterotrophs** (Gr. *heteros,* another, + *trophos,* feeder),

figure 1-19

Koala, a heterotroph, feeding on a eucalyptus tree, an autotroph. All heterotrophs depend for their nutrients directly or indirectly on autotrophs that capture the sun's energy to synthesize their own nutrients.

whereas those organisms that can synthesize organic molecules from inorganic sources using light or another source of energy are called **autotrophs** (Gr. *autos,* self, + *trophos,* feeder) (Figure 1-19). The earliest microorganisms are sometimes called **primary heterotrophs** because they existed before there were any autotrophs. They were probably anaerobic, bacteria-like organisms similar to modern *Clostridium,* and they obtained all their nutrients directly from the environment. Chemical and physical processes of the primordial earth had already supplied generous stores of nutrients in the prebiotic soup. There would be neither advantage nor need for the earliest organisms to synthesize their own compounds, as long as they were freely available from the environment.

Once the supply of a required compound was exhausted or became precarious (perhaps because of an increase in the numbers of organisms using it), those protocells able to convert a precursor to the required compound would have a tremendous advantage over those that lacked this ability. As this situation was repeated over and again, long reaction sequences could develop.

An enzyme is normally required to catalyze each one of these reactions. So, when we say that early protocells developed a reaction sequence as we have described (A made from B, B from C, and so on), we are really assuming that the appropriate enzymes appeared to catalyze these reactions. The numerous enzymes of cellular metabolism appeared when cells became able to use proteins for catalytic functions and thereby gained a selective advantage. No planning was required; the results were achieved through natural selection.

Appearance of Photosynthesis and Oxidative Metabolism

Eventually, almost all usable energy-rich nutrients of the prebiotic soup were consumed. This ushered in the next stage of biochemical evolution: the use of readily available solar radiation to provide metabolic energy. In an age of increasing scarcity of nutrient molecules, it is easy to appreciate what an advantage the autotrophs had over the primary heterotrophs.

In plant photosynthesis, water is the source of the electrons that are used to reduce carbon dioxide to sugars (that is, to add hydrogen), and molecular oxygen is liberated:

$$6\ CO_2 + 6\ H_2O \xrightarrow[\text{Light}]{} C_6H_{12}O_6 + 6\ O_2$$

This equation is a summary of the many reactions now known to take place in photosynthesis. Undoubtedly, these reactions did not appear all at once, and other reduced compounds, such as hydrogen sulfide (H_2S), were probably the early electron donors, rather than H_2O. As these reducing agents were used up, oxygen-evolving photosynthesis appeared. Ozone (a unique compound of three oxygen atoms) began to accumulate in the atmosphere, and, as a result, strong ultraviolet radiation was screened out.

At this important juncture, accumulating oxygen began to interfere with cellular metabolism, which up to this point had evolved under anaerobic conditions. As the atmosphere slowly changed from a somewhat reducing to a highly oxidizing one, a new and highly efficient kind of energy metabolism appeared: **oxidative (aerobic) metabolism.** By using the available oxygen as a terminal electron acceptor (see p. 46) and oxidizing glucose completely to carbon dioxide and water, much of the bond energy stored by photosynthesis could be recovered. Most living forms became wholly dependent on oxidative metabolism.

Precambrian Life

As depicted on the inside back cover of this book, the Precambrian period spanned the geological time before the beginning of the Cambrian period 600 million years ago. Thus about seven-eighths of the age of the earth is encompassed by the Precambrian period. At the beginning of the Cambrian period, most of the major phyla of invertebrate animals made their appearance within a few million years.

This appearance has been called the "Cambrian explosion," because before this time fossil deposits were apparently rare and almost devoid of anything more complex than single-celled algae. We now recognize that the apparent rarity of Precambrian fossils was because they escaped notice owing to their microscopic size (Figure 1-20). What were the forms of life that existed on earth before the burst of evolutionary

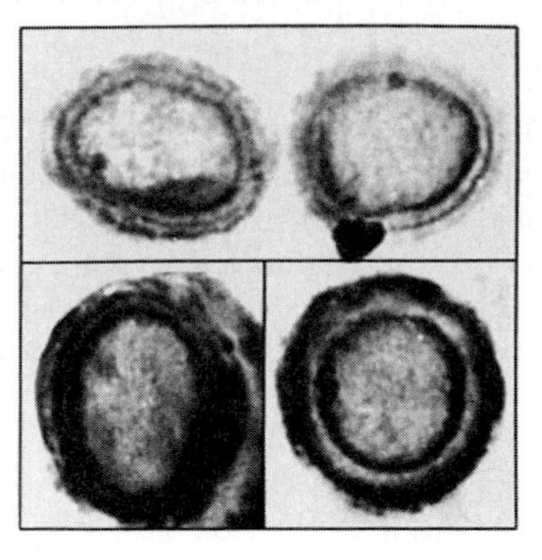

figure 1-20

Fossil bacteria from the Bitter Springs Formation in Australia. These fossils are too small to be visualized except with an electron microscope. They are about 850 million years old.

activity in the early Cambrian world, and what organisms were responsible for the momentous change from a reducing to an oxidizing atmosphere?

Prokaryotes and the Age of Cyanobacteria

The earliest bacteria-like organisms proliferated, giving rise to a great variety of bacterial forms, some of which were capable of photosynthesis. From these arose the oxygen-evolving **cyanobacteria** (blue-green algae) some 3 billion years ago.

> *Cyanobacteria were formerly called "blue-green algae," but this name is misleading because it suggests a relationship to the eukaryotic green algae. These were the organisms responsible for producing oxygen released into the atmosphere. Study of the biochemical reactions in present cyanobacteria suggests that they evolved in a time of fluctuating oxygen concentration. For example, although they can tolerate atmospheric concentrations of oxygen (21%), the optimum concentration for many of their metabolic reactions is only 10%.*

Bacteria and cyanobacteria are called **prokaryotes,** meaning literally "before nucleus." They carry their genetic material on a single, very large loop of DNA not located in a membrane-bound nucleus, but found in a nuclear region, or **nucleoid.** The DNA is associated with proteins different from those associated with eukaryotic DNA, and prokaryotes lack membranous organelles such as mitochondria, plastids, Golgi apparatus, and endoplasmic reticulum (Chapter 2). They reproduce by fission or budding, never by true mitotic cell division.

Bacteria and especially cyanobacteria ruled the earth's oceans unchallenged for some 1.5 to 2 billion years. The cyanobacteria reached the zenith of their success approximately 1 billion years ago when filamentous forms produced great floating mats on the ocean surface. This long period of cyanobacterial dominance, encompassing approximately two-thirds of the history of life, has been called with justification the "age of blue-green algae." Bacteria and cyanobacteria are so completely different from forms of life that evolved later that they are placed in a separate kingdom, Monera (p. 351).

More recently, however, Carl Woese and his colleagues have discovered that the prokaryotes actually comprise two distinct lines of descent: the eubacteria ("true" bacteria) and the archaebacteria. Although members of these two groups look very much alike, they are biochemically distinct. Using the technique of molecular sequencing (see note), Woese found that the sequence of bases in one kind of RNA, ribosomal RNA, is sharply different in archaebacteria from that of all other bacteria as well as from that found in the eukaryotes (see the following discussion). Thus Woese places the archaebacteria in a separate kingdom, and the Monera then comprise only the true bacteria.

> *Molecular sequencing has emerged as a very successful approach to unraveling the genealogies of ancient forms of life. The sequences of nucleotides in the DNA of an organism's genes are a record of evolutionary relationship because every gene that exists today is an evolved copy of a gene that existed millions, even billions, of years ago. Genes become altered by mutations through the course of time, but vestiges of the original gene usually persist. With modern techniques, one can determine the sequence of nucleotides in an entire molecule of DNA or in short segments of the molecule. When genes for the same function in two different organisms are compared, the extent to which they differ can be correlated with the time elapsed since the two organisms diverged from a common ancestor.*

There is evidence that one group of archaebacteria, the **eocytes,** may share a common ancestor with the eukaryotes (see following discussion). If this finding is supported by further work, the eocytes are a third prokaryotic group and cannot be retained taxonomically within the Archaebacteria.

Appearance of the Eukaryotes

Approximately 1.4 billion years ago, after the accumulation of an oxygen-rich atmosphere, organisms with nuclei appeared. These **eukaryotes** (true nucleus) have cells with membrane bound nuclei containing DNA complexed with certain characteristic proteins. Eukaryotes are generally larger than prokaryotes, contain much more DNA, and usually divide by some form of mitosis (p. 36). Within their cells are numerous membranous organelles, including mitochondria in which the enzymes for oxidative metabolism are packaged. Protista (including protozoa and eukaryotic algae), fungi, plants, and animals are composed of eukaryotic cells. There is fossil evidence that single-celled eukaryotes arose at least 1.5 billion years ago (Figure 1-21).

Prokaryotes and eukaryotes are profoundly different from each other (Figure 1-22) and clearly represent a marked dichotomy in the evolution of life. The ascendancy of the eukaryotes resulted in a rapid decline in the dominance of cyanobacteria as the eukaryotes proliferated and fed on them.

Why were the eukaryotes immediately so successful? Probably because they developed an important process facilitating rapid evolution: sex. Sex promotes great genetic variability by mixing the genes of each two individuals that mate. By preserving favorable genetic variants, natural selection encourages rapid evolutionary change. Prokaryotes propagate effectively and efficiently, but their mechanisms for interchange of genes, which do occur in some cases, lack the systematic genetic recombination characteristic of sexual reproduction.

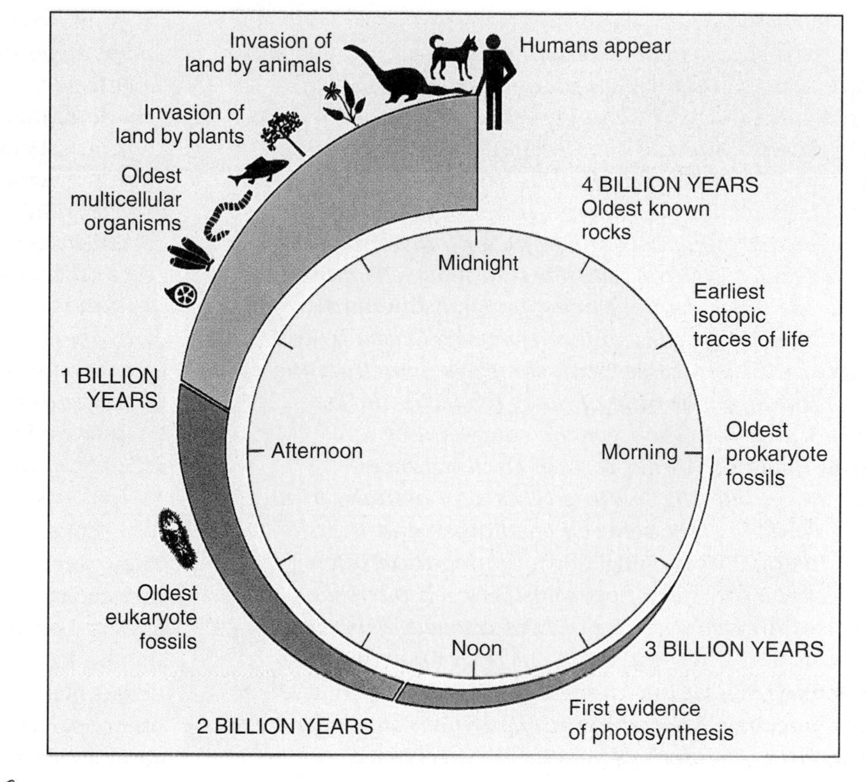

figure 1-21

The clock of biological time. A billion seconds ago it was 1961, and most students using this text had not yet been born. A billion minutes ago the Roman empire was at its zenith. A billion hours ago Neanderthals were alive. A billion days ago the first bipedal hominids walked the earth. A billion months ago the dinosaurs were at the climax of their radiation. A billion years ago no creature had ever walked on the surface of the earth.

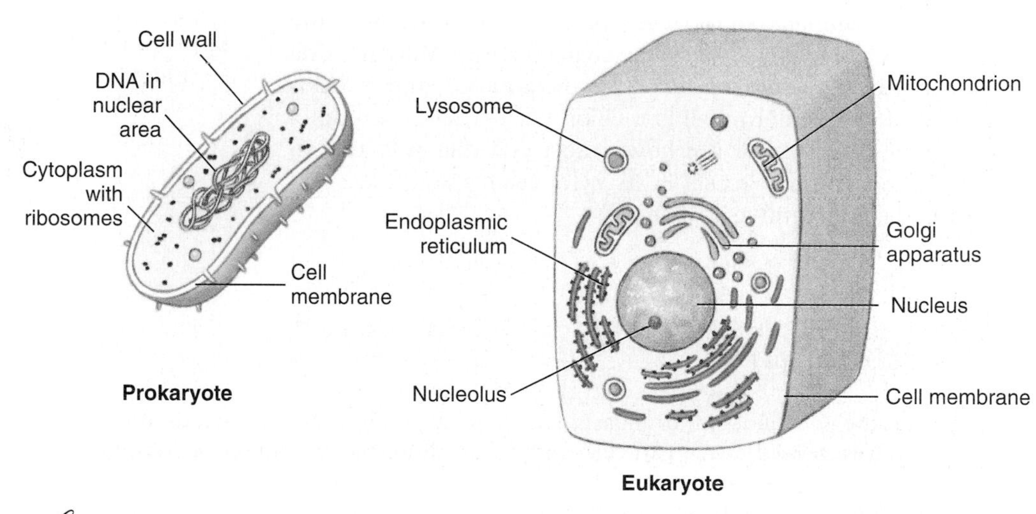

figure 1-22

Comparison of prokaryotic and eukaryotic cells. The prokaryotic cell is about one-tenth the size of the eukaryotic cells.

In the past it was believed that the asexual prokaryotes could only stamp out carbon copies of the parental cells, and that genetic change occurred solely when a mutation intervened. However, several processes of genetic recombination are now known in bacteria. For example, when certain viruses infect bacterial cells, the viruses may pick up a fragment of the bacterial DNA and transfer it to new cells during the course of subsequent infection (transduction). *Sometimes, DNA in the environment (released by cell death or other processes) can somehow penetrate the cell wall and membrane of other bacteria and be incorporated into their genetic complement* (transformation). *Under certain circumstances, bacteria of differing "mating types" can actually form cytoplasmic bridges between each other and transfer segments of DNA* (conjugation). *Conjugation often involves the transfer of* plasmids *between bacteria. Plasmids are small circles of DNA carried by bacteria in addition to their large loop of DNA. They are about 1/1000 the size of the large loop and carry only a few genes, but they are an important element in the new and exciting methods of genetic engineering.*

The organizational complexity of the eukaryotes compared to that of the prokaryotes is so much greater that it is difficult to visualize how a eukaryote could have arisen from any known prokaryote. Margulis and others proposed that eukaryotes did not, in fact, arise from any single prokaryote, but were derived from a **symbiosis** (life together) of two or more types. Mitochondria and plastids, for example, each contain their own complement of DNA (apart from the nucleus of the cell), which has some prokaryote characteristics. Mitochondria contain the enzymes of oxidative metabolism, and plastids (a plastid with chlorophyll is a chloroplast) carry out photosynthesis. It is easy to see how a host cell that was able to accommodate such guests in its cytoplasm would have had enormous competitive advantages.

The hypothesis of symbiotic origin of eukaryotes described here is supported by studies of the amino acid sequence in certain proteins and base sequence in nucleic acids. Furthermore, there is some evidence that microtubules in eukaryotes were acquired symbiotically. Microtubules are tiny tubular structures characteristic of eukaryotes that are important in cell motility (as in cilia and flagella, for example) and mitotic and meiotic cell division. Margulis and coworkers found microtubules in certain spirochaete bacteria (which are prokaryotes) that were constituted of a protein (tubulin) apparently similar to that in microtubules of all eukaryotes.

Eukaryotes may have originated more than once. They were no doubt unicellular and many were photosynthetic autotrophs. Some of these lost their photosynthetic ability and became heterotrophs, feeding on the autotrophs and the prokaryotes. As the cyanobacteria were cropped, their dense filamentous mats began to thin, providing space for other species. Carnivores appeared and fed on the herbivores. Soon a balanced ecosystem of carnivores, herbivores, and primary producers appeared. This was ideal for evolutionary diversity. By freeing space, cropping herbivores encouraged a greater diversity of producers, which in turn promoted the evolution of new and more specialized croppers. An ecological pyramid appeared with carnivores at the top.

The eukaryotic cell made possible the richness and diversity of life on earth today. The burst of evolutionary activity that followed at the end of the Precambrian period and beginning of the Cambrian period was unprecedented; nothing approaching it has occurred since. Nearly all animal phyla appeared and established themselves within a relatively brief period of a few million years.

Summary

Zoology is the scientific study of animals. Science is characterized by a particular approach to the acquisition of human knowledge. It is guided by, and is explanatory with reference to, natural law, and it is testable, tentative, and falsifiable. These criteria form the basis for the scientific method, which may be described as hypothetico-deductive in character. On the basis of prior observations, the scientist formulates an explanatory hypothesis. Predictions about future observations, based on the hypothesis, can support or falsify the hypothesis. A hypothesis for which there is a great deal of supporting data, particularly one that explains a very large number of observations, may be elevated to the status of a theory.

The physiological sciences seek to understand proximate or immediate causes in biological systems and use the experimental method. Support for hypotheses is sought by means of experiments, which must include controls. The evolutionary sciences seek ultimate causes in biological systems by using the comparative method.

Life is characterized by its chemical uniqueness, complexity and hierarchical organization, reproduction, possession of a genetic program, metabolism, development, and interaction with the environment.

The chemical uniqueness of life lies in its use of carbon to build several classes of macromolecules. Carbon is especially versatile in bonding with itself or with other atoms and is the only element capable of forming the variety of molecules found in living things. Carbohydrates are composed primarily of carbon, hydrogen, and oxygen grouped as H—C—OH. Sugars serve as immediate sources of energy in living systems. Monosaccharides, or simple sugars, may bond together to form disaccharides or polysaccharides, which serve as storage forms of sugar or perform structural roles. Lipids exist principally as fats, phospholipids, and steroids.

Proteins are large molecules composed of amino acids linked together by peptide bonds. Proteins have a primary, secondary, tertiary, and often, quaternary structure. Proteins perform many functions, especially as enzymes (biological catalysts).

Nucleic acids are polymers of nucleotide units, each composed of a sugar, a nitrogenous base, and a phosphate group. They contain the coding for inheritance and function in protein synthesis.

A remarkable uniformity in the chemical constituents of all living things and many of their cellular reactions suggests that life on earth may have had a common origin. A.I. Oparin and J.B.S. Haldane proposed a long "abiogenic molecular evolution" on earth in which organic molecules slowly accumulated. The atmosphere of the primitive earth was somewhat reducing, and little or no free oxygen was present. The main source of energy to power chemical reactions was probably lightning. Miller showed the plausibility of the Oparin-Haldane hypothesis by simple, but ingenious, experiments. Although the compounds could have accumulated in rather diluted solutions, condensation reactions could have occurred after concentration and heating by various mechanisms. When self-replicating systems became responsive to the forces in the environment, evolution proceeded more rapidly.

The first organisms were the primary heterotrophs, living on the energy stored in molecules dissolved in the primordial soup. As such molecules were used up, autotrophs had a great selective advantage. Molecular oxygen began to accumulate in the atmosphere as an end product of photosynthesis, and finally the atmosphere became oxidizing. The organisms responsible for this change in atmosphere were apparently cyanobacteria (blue-green algae). The cyanobacteria and true bacteria are prokaryotes, organisms that lack a membrane-bound nucleus and other membranous organelles in their cytoplasm. The eukaryotes originated after the atmophere was oxidizing and apparently arose from symbiotic unions of two or more types of prokaryotes. Eukaryotes have most of their genetic material (DNA) borne in a membrane-bound nucleus and have mitochondria and other organelles. They include the protistans, fungi, plants, and animals. Their evolutionary success is due in great degree to the variability conferred by sexual reproduction.

Review Questions

1. What are the basic chemical differences that distinguish living from nonliving systems?
2. Name some common, everyday event, and tell how you would investigate it by using the hypothetico-deductive method.
3. What are the essential characteristics of science? Describe how evolutionary studies fit these characteristics whereas "scientific creationism" does not.
4. What is the relationship between a hypothesis and a theory?
5. Describe the hierarchical organization of life. How does this organization lead to the emergence of new properties at different levels of biological complexity?
6. Explain how biologists distinguish between physiological and evolutionary sciences.
7. Name two simple carbohydrates, two storage carbohydrates, and a structural carbohydrate.
8. What are characteristic differences in molecular structure between lipids and carbohydrates?
9. Explain the difference between primary, secondary, tertiary, and quaternary structure of a protein.
10. What are the important nucleic acids in a cell, and of what units are they constructed?
11. Name and briefly explain the characteristics of life.
12. In regard to the experiments of Louis Pasteur and Stanley Miller described in this chapter, explain what constituted the following in each case: observations, hypothesis, deduction, prediction, data, control.
13. Explain the significance of the Miller experiments.
14. What are several mechanisms by which organic molecules in the prebiotic ocean could have been concentrated so that further reactions could occur?
15. Distinguish among the following: primary heterotroph, autotroph, secondary heterotroph.
16. What is the origin of the oxygen in the present-day atmosphere, and what is its metabolic significance to most organisms living today?
17. Distinguish between prokaryotes and eukaryotes as completely as you can.
18. Describe Margulis' view on the origin of eukaryotes from prokaryotes.
19. What was the "Cambrian explosion" and how might you account for it?

Selected References

Cairns-Smith, A.G. 1985. The first organisms. Sci. Am. **252:**90–100 (June). *Argues that a genetic system based on nucleic acids was too complex for the earliest organisms to have evolved. Perhaps the earliest organisms depended on clay crystals with genetic properties.*

Conway Morris, S. 1993. The fossil record and the early evolution of the Metazoa. Nature **361:**2199–2225. *An important summary correlating fossil and molecular evidence.*

de Duve, C. 1991. Blueprint for a cell: the nature and origin of life. Burlington, North Carolina, Carolina Biological Supply Company, Neil Patterson Publishers. *Describes this Nobel laureate's hypothesis that cellular metabolism originated before nucleic acids and genetic code.*

Futuyma, D. J. 1995. Science on trial: the case for evolution. Sunderland Massachusetts, Sinauer Associates. *A defense of evolutionary biology as the exclusive scientific approach to the study of life's diversity.*

Gesteland, R. F., and J. F. Atkins, editors. 1993. The RNA world. Cold Spring Harbor, New York, Cold Spring Harbor Laboratory Press. *Evidence that there was a period when RNA served in both catalysis and transmission of genetic information.*

Horgan, J. 1991. In the beginning . . . Sci. Am. **264:**116–125 (Feb.). *Excellent summary of current hypotheses on the origin of life, a field in ferment.*

Kitcher, P. 1982. Abusing science: the case against creationism. Cambridge, Massachusetts, MIT Press. *A treatise on how knowledge is gained in science and why creationism does not qualify as science.*

Lodish, H., D. Baltimore, A. Berk, S. L. Zipursky, P. Matsudira, and J. Darnell. 1995. Molecular cell biology, ed. 2. New York, Scientific American Books, Inc. *Thorough treatment, begins with fundamentals such as energy, chemical reactions, bonds, pH, and biomolecules, then proceeds to advanced molecular biology.*

Margulis, L. 1993. Symbiosis in cell evolution, ed. 2 New York, W. H. Freeman. *An important updating of the author's 1981 book on this topic.*

Medawar, P. B. 1989. Induction and intuition in scientific thought. London, Methuen & Company. *A commentary on the basic philosophy and methodology of science.*

Moore, J. A. 1993. Science as a way of knowing: the foundations of modern biology. Cambridge, Massachusetts, Harvard University Press. *A lively, wide-ranging account of the history of biological thought and the workings of life.*

Overton, W. R. 1982. Judgment, injunction, and memorandum opinion in the case of McLean vs. Arkansas Board of Education. Science **215:**934–943. *Judge Overton's opinion is reprinted verbatim. It is highly recommended reading.*

Perutz, M. F. 1989. Is science necessary? Essays on science and scientists. New York, E. P. Dutton. *A general discussion of the utility of science.*

Rivera, M. C., and J. A. Lake. 1992. Evidence that eukaryotes and eocyte prokaryotes are immediate relatives. Science **257:**774–776. *A controversial paper indicating that eukaryotes may share a common ancestor with a prokaryote classified among the Archaebacteria.*

Sagan, C. 1995. The demon-haunted world. Science as a candle in the dark. New York Random House. *A thoughtful work on the importance of science in our world and how superstition and pseudoscience remain frighteningly prevalent.*

Vidal, G. 1984. The oldest eukaryotic cells. Sci. Am. **250:**48–57 (Feb.). *Careful examination of microscopic fossils shows that eukaryotes evolved in the form of unicellular plankton about 1.4 billion years ago.*

Weinberg, R. A. 1985. The molecules of life. Sci. Am. **253:**48–57 (Oct.). *This entire issue of Scientific American is devoted to findings of molecular biology.*

Woese, C. R. 1984. The origin of life. Carolina Biology Readers, no. 13. Burlington, North Carolina, Carolina Biological Supply Company. *Thought-provoking account of the author's views; critique of traditional concepts, with focus on problems.*

The Cell as the Unit of Life

chapter | two

The Fabric of Life

It is a remarkable fact that living forms, from amebas and unicellular algae to whales and giant redwood trees, are formed from a single type of building unit: the cell. All animals and plants are composed of cells and cell products. Thus the cell theory, or cell doctrine as it is more appropriately called, is another of the great unifying concepts of biology.

New cells come from division of preexisting cells, and the activity of a multicellular organism as a whole is the sum of the activities of its constituent cells and their interactions. The energy to support virtually all life activities flows from sunlight that is captured by green plants and transformed by the process of photosynthesis into chemical bond energy. Chemical bond energy is a form of potential energy that can be released when the bond is broken; the energy is used to perform electrical, mechanical, and osmotic tasks in the cell. Ultimately, all energy is dissipated, little by little, into heat. This is in accord with the second law of thermodynamics, which states that there is a tendency in nature to proceed toward a state of greater molecular disorder, or entropy. Thus the high degree of molecular organization in living cells is attained and maintained only as long as energy fuels the organization.

More than 300 years ago the English scientist and inventor Robert Hooke, using a primitive compound microscope, observed boxlike cavities in slices of cork and leaves. He called these compartments "little boxes or cells." In the years that followed Hooke's first demonstration of the remarkable powers of the microscope before the Royal Society of London in 1663, biologists gradually began to realize that cells were far more than simple containers filled with "juices."

Cells are the fabric of life. Even the most primitive cells are enormously complex structures that form the basic units of all living matter. All tissues and organs are composed of cells. In humans an estimated 40 trillion cells interact, each performing its specialized role in an organized community. In single-celled organisms, all the functions of life are performed within the confines of one microscopic package. There is no life without cells.

With the exception of some eggs, which are the largest cells (in volume) known, cells are small and mostly invisible to the unaided eye. Consequently, our understanding of cells is paralleled by technical advances in the resolving power of microscopes. The Dutch microscopist A. van Leeuwenhoek, using high-quality single lenses that he had made, sent letters to the Royal Society of London containing detailed descriptions of the numerous organisms he had observed (1673 to 1723). In the early nineteenth century, the improved design of the microscope permitted biologists to distinguish objects only 1 μm apart. This advance was quickly followed by new discoveries that laid the groundwork for the **cell theory**—a theory stating that all living organisms are composed of cells and that all cells are derived from preexisting cells.

In 1838 Matthias Schleiden, a German botanist, announced that all plant tissue was composed of cells. A year later one of his countrymen, Theodor Schwann, described animal cells as being similar to plant cells, which is an understanding that had been long delayed because the animal cell is bounded only by a nearly invisible plasma membrane and because it lacks the distinct cell wall characteristic of the plant cell. In 1858 Rudolf Virchow realized that all cells came from preexisting cells; thus Schleiden, Schwann, and Virchow are credited with the unifying cell theory that ushered in a new era of productive exploration in cell biology.

In 1840 J. Purkinje introduced the term **protoplasm** to describe the cell contents. Protoplasm was at first believed to be a granular, gel-like mixture with special and elusive life properties of its own; the cell was thus viewed as a bag of thick soup containing a nucleus. Later, the interior of the cell became increasingly visible as microscopes were improved and better tissue-sectioning and staining techniques were introduced. Rather than being a uniform granular soup, the cell interior is composed of numerous unique structures called **organelles,** each performing a specific function in the life of the cell. Today we realize that the components of a cell are so highly organized, structurally and functionally, that describing its contents as "protoplasm" is like describing the contents of an automobile engine as "autoplasm."

How Cells Are Studied

The light microscope, with all its variations and modifications, has contributed more to biological investigation than any other instrument developed by humans. It has been a powerful exploratory tool for 300 years and continues to be so more than 50 years after the invention of the electron microscope. But, until the electron microscope was perfected, our concept of the cell was limited to that which could be seen with magnifications of 1000 to 2000 diameters. This is the practical limit of the light microscope. In addition to microscopy, modern biochemical, immunological, physical, and molecular techniques have contributed enormously to our understanding of cell structure and function.

The electron microscope employs high voltages to direct a beam of electrons through the object to be examined. The wavelength of the electron beam is approximately 0.00001 that of ordinary white light, thus permitting far greater magnification and resolution (compare A and B of Figure 2-1). In preparation for viewing, specimens are cut into extremely thin sections and treated with "electron stains" (ions of elements such as osmium, lead, or uranium) to increase contrast between different structures. Images are seen on a fluorescent screen and photographed (Figure 2-2). Because the electrons pass through the specimen to the photographic plate, the instrument is called a **transmission electron microscope.**

Units of measurement commonly used in microscopic study are micrometers, nanometers, and angstroms: 1 micrometer (μm) = 1/1,000,000 meter, 1 nanometer (nm) = 1/1,000,000,000 meter, 1 angstrom (Å) = 1/10,000,000,000 meter. Thus 1 m = 10^3 mm = 10^6 μm = 10^9 nm = 10^{10} Å.

In contrast, specimens prepared for the **scanning electron microscope** are not sectioned, and electrons do not pass through them. The whole specimen is bombarded with electrons, causing secondary electrons to be emitted. An apparent three-dimensional image is recorded in the photograph. Although the magnification capability of the scanning instrument is not as great as the transmission microscope, a great deal has been learned about the surface features of organisms and cells. Examples of scanning electron micrographs are shown on pages 175 and 191.

Advances in the techniques of cell study were not limited to improvements in microscopes but have included new methods of tissue preparation, staining for microscopic study, and the great contributions of modern biochemistry. For example, the various organelles of cells have differing, characteristic densities. Cells can be broken with most of the organelles remaining intact, then centrifuged in a density gradient (Figure 2-3), and relatively pure preparations of each organelle may be recovered. Thus the biochemical functions of the various organelles may be studied separately. The DNA and various types of RNA can be

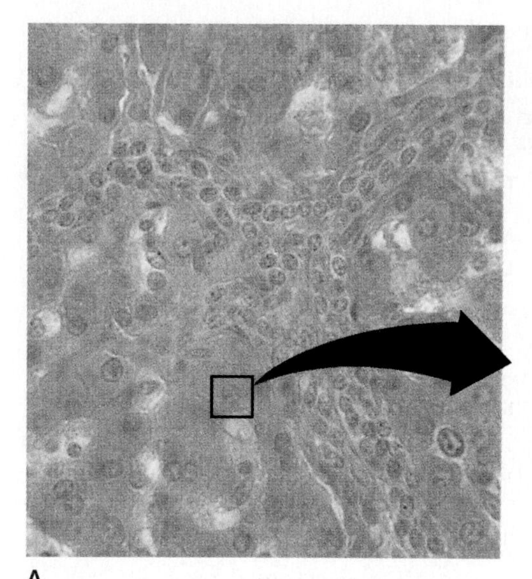

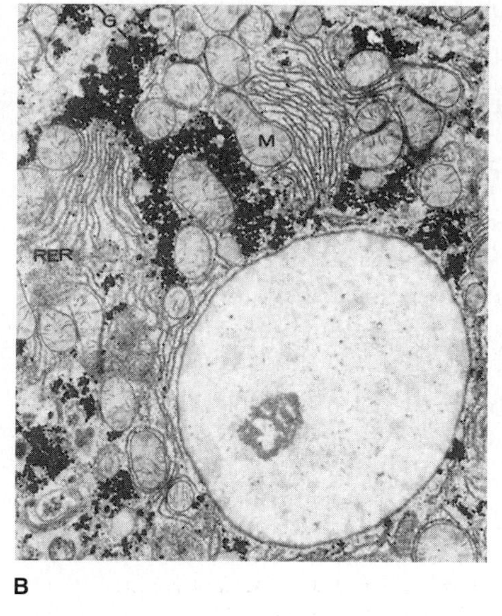

figure 2-1

Liver cells of rat. **A,** Magnified approximately 400 times through light microscope. Note prominently stained nucleus in each polyhedral cell. **B,** Portion of a single liver cell, magnified approximately 5000 times through an electron microscope. Single large nucleus dominates the field; mitochondria *(M),* rough endoplasmic reticulum *(RER),* and glycogen granules *(G)* are also seen.

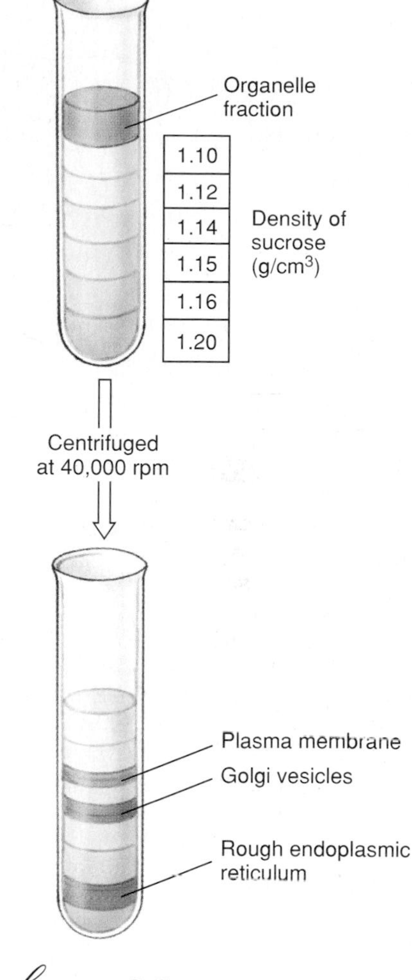

figure 2-3

Separation of organelles in a density gradient by ultracentrifugation. The gradient is formed by layering sucrose solutions in a centrifuge tube, then carefully placing a preparation of mixed organelles on top. The tube is centrifuged at about 40,000 revolutions per minute for several hours, and the organelles become separated down the tube according to their density.

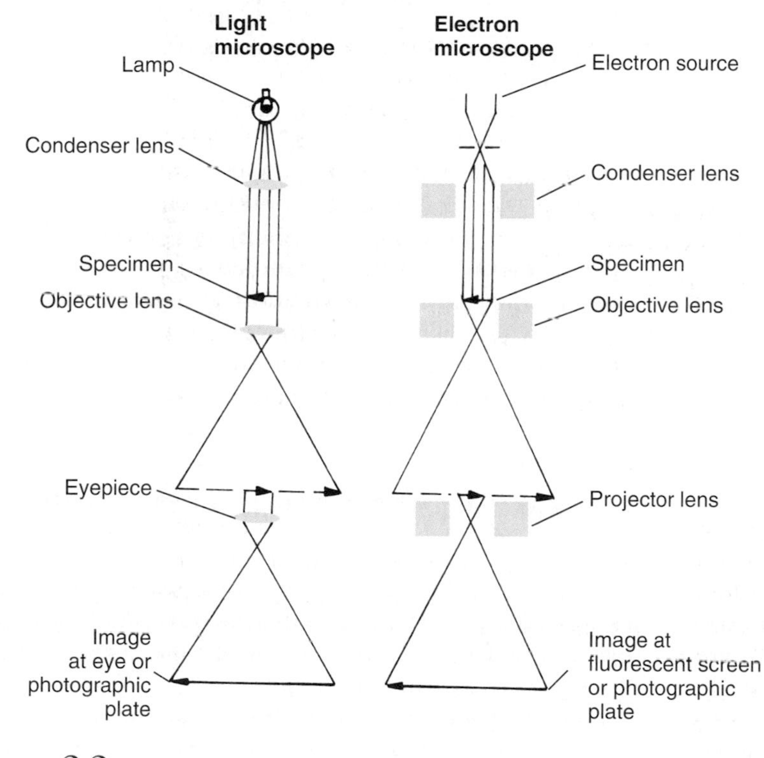

figure 2-2

Comparison of optical paths of light and electron microscopes. Note that to facilitate comparison, the light microscope is shown inverted from its usual orientation with light source below and image above.

table 2-1

Comparison of Prokaryotic and Eukaryotic Cells

Characteristic	Prokaryotic cell	Eukaryotic cell
Cell size	Mostly small (1–10 μm)	Mostly large (10–100 μm)
Genetic system	DNA with some nonhistone protein; simple, circular genome in nucleoid; nucleoid not membrane bound	DNA complexed with histone and nonhistone proteins in complex chromosomes within nucleus with membranous envelope
Cell division	Direct by binary fission or budding; no mitosis	Some form of mitosis; centrioles in many; mitotic spindle present
Sexual system	Absent in most; genetic exchange by conjugation in some	Present in most; male and female partners; gametes that fuse
Nutrition	Absorption by most; photosynthesis by some	Absorption, ingestion; photosynthesis by some
Energy metabolism	No mitochondria; oxidative enzymes bound to cell membrane, not packaged separately; great variation in metabolic pattern	Mitochondria in most; oxidative enzymes packaged therein; more unified pattern of oxidative metabolism
Organelles	None	Many
Intracellular movement	None	Cytoplasmic streaming, phagocytosis, pinocytosis
Flagella/cilia	Not with "9 + 2" microtubular pattern	With "9 + 2" microtubular pattern
Cell wall	Contains disaccharide chains cross-linked by peptides	If present, not with disaccharide polymers linked with peptides

extracted and studied. Many enzymes can be purified and their characteristics determined. The use of radioactive isotopes has allowed elucidation of many metabolic reactions and pathways in the cell. Modern chromatographic techniques can separate chemically similar intermediates and products. Many more examples could be cited, and these have contributed enormously to our present understanding of cell structure and function.

Organization of Cells

If we were to restrict our study of cells to fixed and sectioned tissues, we would be left with the erroneous impression that cells are static, quiescent, rigid structures. In fact, the interior of the cell is in a constant state of upheaval. Most cells are continually changing shape, pulsing and heaving; their organelles twist and regroup in a cytoplasm teeming with granules, fat globules, and vesicles of various sorts. This description is derived from studies with time-lapse photography of living cell cultures. If we could see the swift shuttling of molecular traffic through gates in the cell membrane and the metabolic energy transformations with-in organelles, we would have an even stronger impression of internal turmoil. However, the cell is anything but a bundle of disorganized activity. There is order and harmony in the cell's functioning that represents the elusive phenomenon we call life. Studying this dynamic miracle of evolution through the microscope, we realize that as we gradually comprehend more and more about this unit of life and how it operates, we are gaining a greater understanding of the nature of life itself.

Prokaryotic and Eukaryotic Cells

The radically different cell plan of prokaryotes and eukaryotes was described in Chapter 1. A fundamental distinction, expressed in their names, is that prokaryotes lack the membrane-bound nucleus present in all eukaryotic cells. Table 2-1 summarizes other differences.

Prokaryotic organisms are bacteria separated into the kingdoms Archaea and Bacteria. All other organisms are eukaryotes distributed among four kingdoms: Protista (protozoa and eukaryotic algae), Plantae, Fungi, and Animalia. The kingdom classifications are discussed in Chapter 15. The following discussion is restricted to eukaryotic cells, of which all animals are composed.

Components of Eukaryotic Cells and Their Functions

Typically, the eukaryotic cell is enclosed within a thin, selectively permeable **cell membrane** (Figure 2-4). The most prominent organelle is the spherical or ovoid **nucleus,** enclosed within *two* membranes to form the double-layered **nuclear envelope** (Figure 2-4). The region outside the nucleus is regarded as cytoplasm. Within the cytoplasm are many organelles, such as mitochondria, Golgi complexes, centrioles, and endoplasmic reticulum. Plant cells typically contain **plastids,** some of which are photosynthetic organelles, and plant cells bear a cell wall containing cellulose outside the cell membrane.

The **fluid-mosaic model** is the currently accepted concept of the cell membrane. With the electron microscope, the cell membrane appears as two dark lines, each approximately 3 nm thick, at each side of a light zone. The entire membrane is 8 to 10 nm thick. This image is the result of a phospholipid bilayer, that is, two layers of phospholipid molecules, all oriented with their water-soluble ends toward the outside and their fat-soluble portions toward the inside of the membrane (Figure 2-5). An important characteristic of the phospholipid bilayer is that it is a liquid, giving the membrane flexibility and allowing the phospholipid molecules to move sideways freely within their own monolayer. Molecules of cholesterol are interspersed in the lipid portion of the bilayer (Figure 2-5). They make the membrane even less permeable and decrease its flexibility.

Glycoproteins (proteins with carbohydrates attached) are essential components of cell membranes. Some of these proteins catalyze the transport of substances such as negatively charged ions across the membrane. Others act as specific receptors for various molecules or as highly specific markings. For example, the self/nonself recognition that enables the immune system to react to invaders (Chapter 9) is based on proteins of this type. Some aggregations of protein molecules form pores through

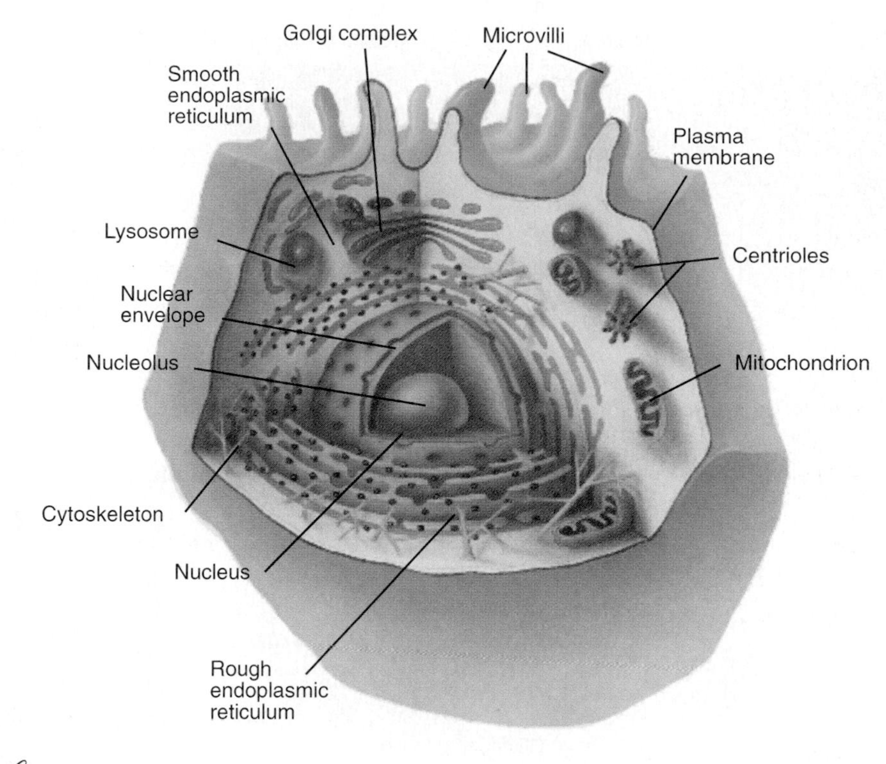

figure 2-4

Generalized cell with principal organelles, as might be seen with the electron microscope. No single cell contains all these organelles, but many cells contain a large number of them.

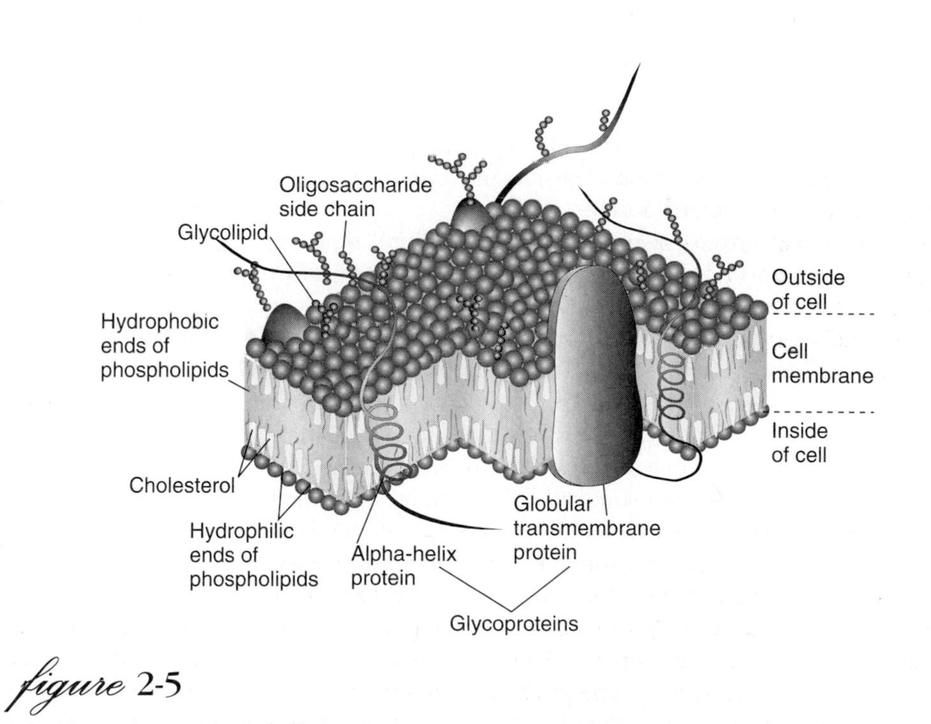

figure 2-5

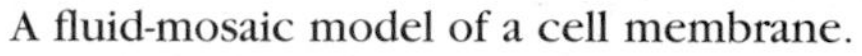

A fluid-mosaic model of a cell membrane.

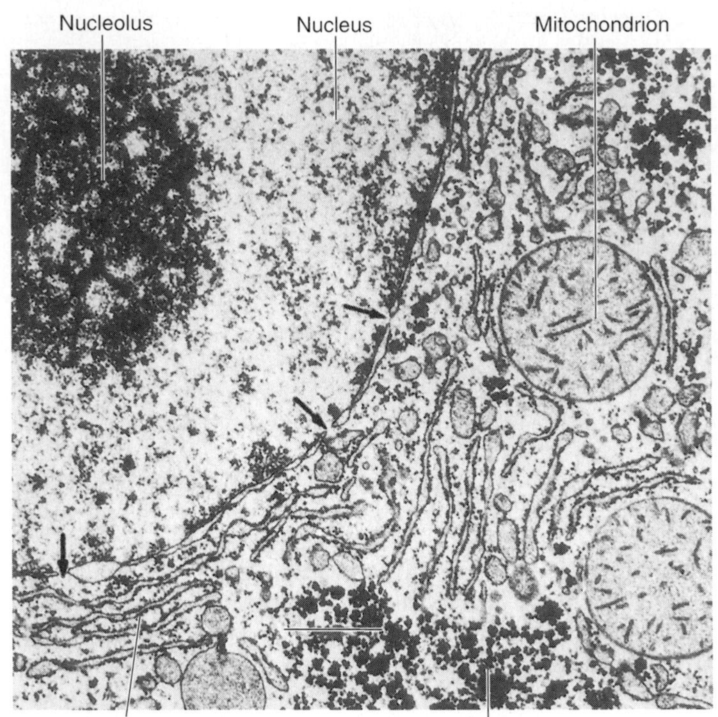

figure 2-6

Electron micrograph of part of a hepatic cell of a rat showing a portion of nucleus *(left)* and surrounding cytoplasm. Endoplasmic reticulum and mitochondria are visible in cytoplasm, and pores *(arrows)* can be seen in the nuclear envelope. (× 14,000)

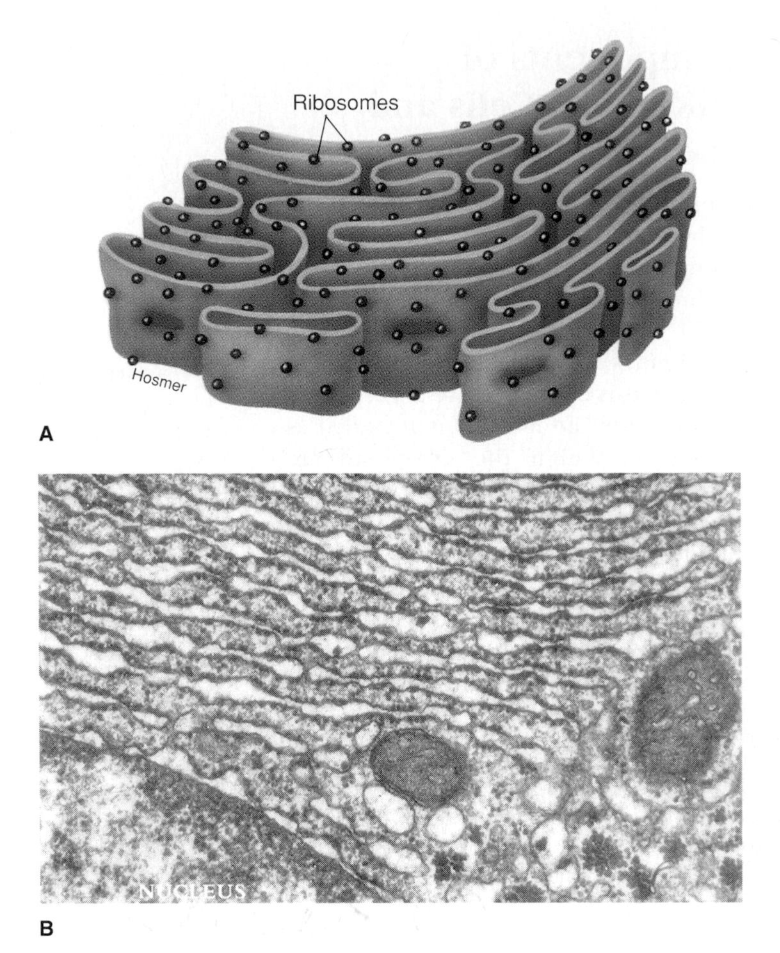

figure 2-7

Endoplasmic reticulum. **A,** Endoplasmic reticulum is continuous with the nuclear envelope. It may have associated ribosomes (rough endoplasmic reticulum) or not (smooth endoplasmic reticulum). **B,** Electron micrograph including rough endoplasmic reticulum. (× 28,000)

which small polar molecules may enter. Like the phospholipid molecules, most of the glycoproteins can move laterally in the membrane, although more slowly.

The nuclear envelope contains less cholesterol than the cell membrane, and pores in the envelope (Figure 2-6) allow molecules to move between the nucleus and cytoplasm. The nucleus contains the **chromatin,** a complex of DNA, basic proteins called **histones,** and nonhistone protein. The chromatin carries the genetic information; that is, the code that results in most of the components characteristic of the cell after transcription and translation (see Chapter 3). **Nucleoli** are specialized parts of certain chromosomes that stain in a characteristically dark manner. They carry multiple copies of the DNA information to synthesize ribosomal RNA. After transcription from the DNA, the ribosomal RNA combines with protein to form a **ribosome,** detaches from the nucleolus, and passes to the cytoplasm through the pores in the nuclear envelope.

The outer membrane of the nuclear envelope is continuous, with extensive membranous elements in the cytoplasm called **endoplasmic reticulum (ER)** (Figures 2-6 and 2-7). The space between the membranes of the nuclear envelope communicates with the channels **(cisternae)** in the ER. The ER is a complex of membranes that separates some of the products of the cell from the synthetic machinery that produces them, apparently functioning as routes for transport of proteins within the cell. The membranes of the ER may be lined on their outer surfaces with ribosomes and are thus designated **rough ER,** or they may lack the ribosomal lining and be called **smooth ER.** Smooth ER functions in the synthesis of lipids and phospholipids. Protein synthesized by the ribosomes on the rough ER enters the cisternae and from there is transported to the Golgi apparatus or complex.

The **Golgi complex** (Figures 2-8 and 2-9) is composed of a stack of membranous vesicles that functions in storage, modification, and packaging of protein products, especially secretory products. The vesicles do not synthesize protein but may add complex carbohydrates to the molecules. Small vesicles of ER containing protein pinch off and then fuse with sacs on the "forming face" of the Golgi. After modification, the proteins bud off vesicles on the "maturing face" of the complex (Figure 2-9). The contents of some of these vesicles may be expelled to the outside of the cell, as secretory products destined to be exported from a glandular cell. Others may contain digestive enzymes that remain in the same cell that produces

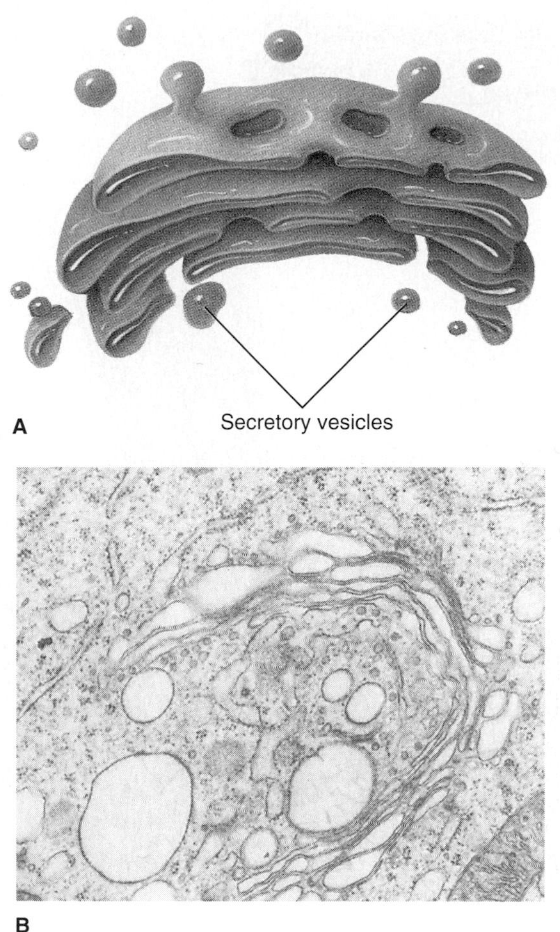

figure 2-8

Golgi complex (= Golgi body, Golgi apparatus). **A,** The smooth cisternae of the Golgi complex have enzymes that modify proteins synthesized by the rough endoplasmic reticulum. **B,** Electron micrograph of a Golgi complex. (× 46,000)

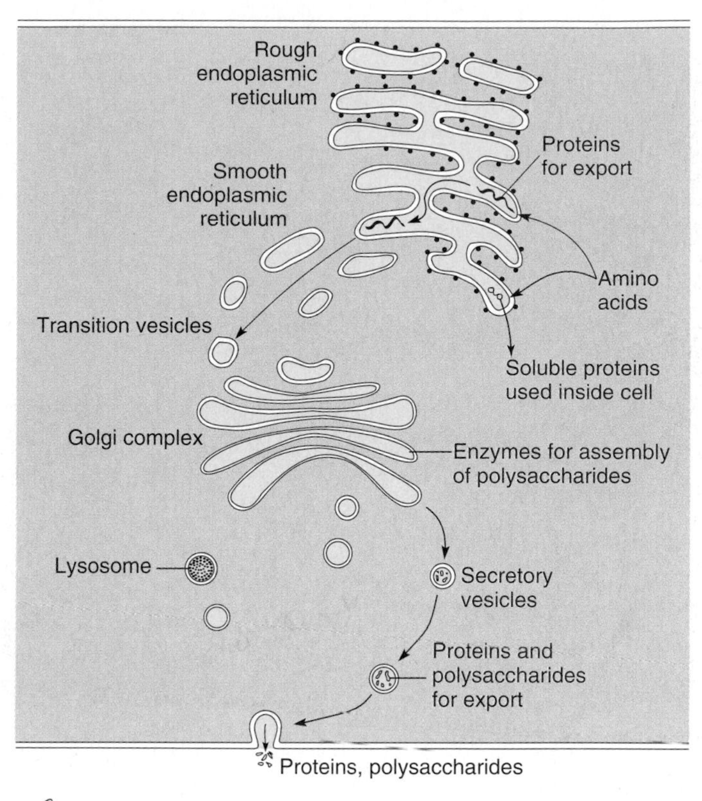

figure 2-9

System for assembling, isolating, and secreting proteins for export from, and use inside, a eukaryotic cell.

them. Such vesicles are called **lysosomes** (literally "loosening body," a body capable of causing lysis, or disintegration). The enzymes that they contain are involved in the breakdown of foreign material, including bacteria engulfed by the cell. Lysosomes also are capable of breaking down injured or diseased cells and worn-out cellular components, since the enzymes they contain are so powerful that they kill the cell that formed them if the lysosome membrane ruptures. In normal cells the enzymes remain safely enclosed within the protective membrane. Lysosomal vesicles may pour their enzymes into a larger membrane-bound body containing an ingested food particle, the **food vacuole** or **phagosome.** Other vacuoles may contain only fluid and be important in water and ion regulation, for example, **contractile vacuoles** in protistans (p. 366).

Mitochondria (Figure 2-10) (sing., **mitochondrion**) are conspicuous organelles present in nearly all eukaryotic cells. They are diverse in shape, size, and number; some are rodlike, and others are more or less spherical in shape. They may be scattered uniformly through the cytoplasm, or they may be localized near cell surfaces and other regions of unusual metabolic activity. The mitochondrion is composed of a double membrane. The outer membrane is smooth, whereas the inner membrane is folded into numerous platelike projections called **cristae** (Figure 2-10), which increase the internal surface area where chemical reactions take place as if in an assembly line. Mitochondria are often called "powerhouses of the cell" because enzymes located on the cristae carry out the energy-yielding steps of aerobic metabolism. ATP (adenosine triphosphate), the most important energy-transfer molecule of all cells, is produced in this organelle. Mitochondria are self-replicating. They have a tiny, circular genome, much like the genomes of prokaryotes except that it is much smaller. It contains DNA that specifies some, but not all, of the proteins of the mitochondrion.

Eukaryotic cells characteristically have a system of tubules and filaments that form the **cytoskeleton** (Figure 2-11). These provide support and maintain the form of the cell, and in many cells they provide a means of locomotion and translocation of organelles within the cell. **Microfilaments** are thin, linear structures, first observed distinctly in muscle cells, where they are responsible for the ability of the cell to contract. They are made of a protein called **actin.** Several dozen other proteins are known that bind with actin and determine its configuration and behavior in particular cells. One of these is **myosin,** whose interaction with actin causes contraction in muscle and other

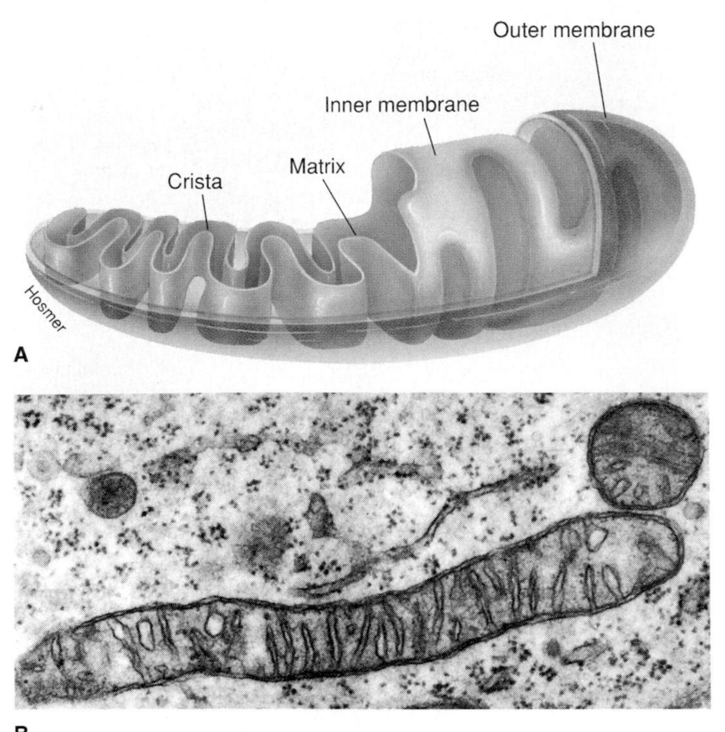

figure 2-10

Mitochondria. **A,** Structure of a typical mitochondrion. **B,** Electron micrograph of mitochondria in cross and longitudinal section. (× 30,000)

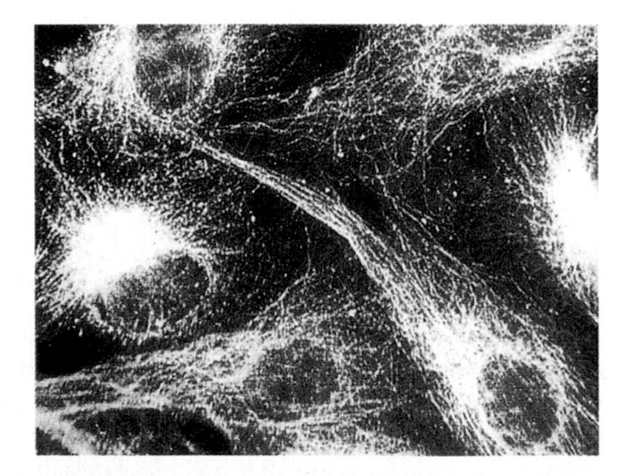

figure 2-11

The microtubules in kidney cells of a baby hamster have been rendered visible by treating them with a preparation of fluorescent proteins that specifically bind to tubulin. The large dark bodies within the cells are the nuclei; the cell boundaries are not visible.

cells. **Microtubules,** somewhat larger than microfilaments, are tubular structures composed of a protein called **tubulin.** They play a vital role in moving the chromosomes toward the daughter cells during cell division as we will see later, and they are important in intracellular architecture, organization, and transport.

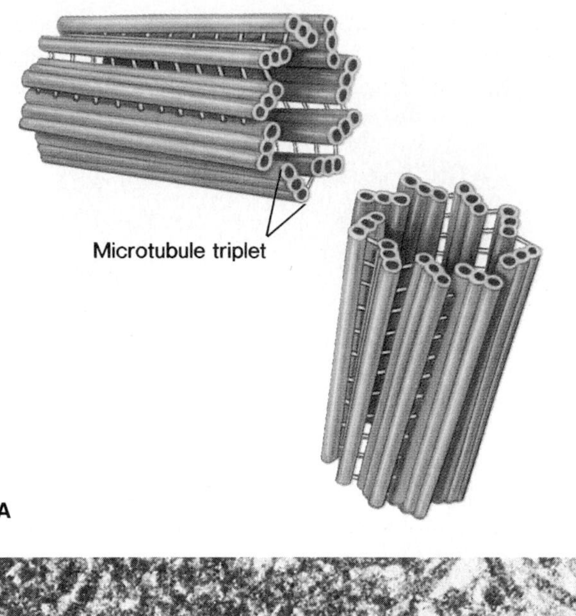

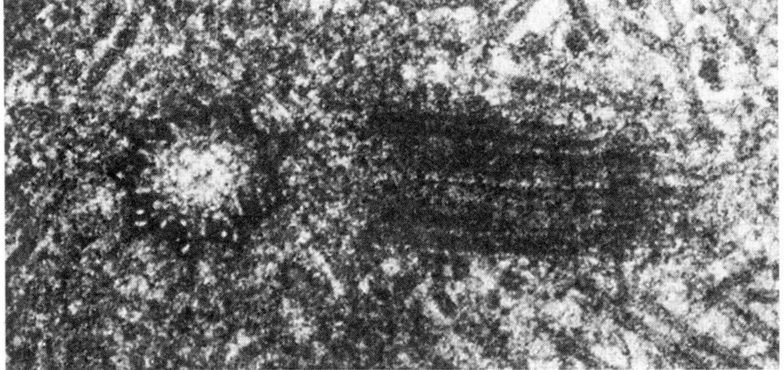

figure 2-12

Centrioles. **A,** Each centriole is composed of nine triplets of microtubules arranged as a cylinder. **B,** Electron micrograph of a pair of centrioles, one in longitudinal *(right)* and one in cross section *(left)*. The normal orientation of centrioles is at right angles to each other.

In addition, microtubules form essential parts of the structures of cilia and flagella. Microtubules radiate from a microtubule organizing center near the nucleus, sometimes called the **centrosome.** Within the centrosome are found a pair of **centrioles** (Figure 2-12), and the microtubules radiating from the centrioles form the **aster.** The centrioles are themselves composed of microtubules. Each centriole of a pair lies at right angles to the other and is a short cylinder of nine triplets of microtubules. They replicate before cell division. **Intermediate filaments** are larger than microfilaments but smaller than microtubules. There are five biochemically distinct types of intermediate filaments, and their composition and arrangement depend on the cell type in which they are found.

Surface of Cells and Their Specializations

The free surface of epithelial cells (cells that cover the surface of a structure or line a tube or cavity) sometimes bear either

cilia or **flagella** (sing., **cilium, flagellum).** These are motile extensions of the cell surface that serve to sweep materials past the cell. In many single-celled organisms and some small multicellular forms, they propel the entire organism through a liquid medium. Flagella provide the means of locomotion for the male reproductive cells of most animals and many plants.

Cilia and flagella have different beating patterns (described in Chapter 6), but their internal structure is the same. With few exceptions, the internal structures of locomotory cilia and flagella are composed of a long cylinder of nine pairs of microtubules enclosing a central pair (see Figure 16-2, p. 364). At the base of each cilium or flagellum is found a **basal body (kinetosome),** which is identical in structure to a centriole. Basal bodies play an important role in forming these locomotory structures.

Cells covering the surface of a structure or cells packed together in a tissue may have specialized junctional complexes between them. Nearest the free surface, the membranes of two cells next to each other appear to fuse, forming a **tight junction** (Figure 2-13). Tight junctions function as seals to prevent the passage of molecules between cells from one side of a layer of cells to another, because there is usually a space of about 20 nm between the cell membranes of adjacent cells. At various points small ellipsoid discs occur, just beneath the cell membrane in each cell. These appear to act as "spot-welds" and are called **desmosomes.** From each desmosome a tuft of intermediate filaments extends into the cytoplasm, and linker proteins extend through the cell membrane into the intercellular space to bind the discs together. Desmosomes are not seals but seem to increase the strength of the tissue. **Gap junctions,** rather than serving as points of attachment, provide a means of intercellular communication. They form tiny canals between cells, so that the cytoplasm becomes continuous, and small molecules can pass from one cell to the other.

Indeed, cilia and flagella are so alike in the details of their structure that it seems highly likely that they had a common evolutionary origin. Whether their origin was the symbiosis of a spirochete-like bacterium and host cell, as suggested by Margulis, is more conjectural. Margulis and others prefer the term undulipodia to include both cilia and flagella, and it is less awkward to use one word for structures that are alike in structure and origin. However, the terms "cilia" and "flagella" are so common and widely used that the student should be familiar with them.

Another specialization of the cell surfaces is the lacing together of adjacent cell surfaces, where the cell membranes of the cells infold and interdigitate very much like a zipper. They are especially common in the epithelial cells of kidney tubules. The distal or apical boundaries of some epithelial cells, as seen with the electron microscope, show regularly arranged **microvilli.** They are small, fingerlike projections consisting of tubelike evaginations of the cell membrane with a core of cytoplasm (Figure 2-13). They are seen clearly in the lining of the intestine, where they greatly increase the absorptive and digestive surface. Such specializations appear as brush borders by the light microscope.

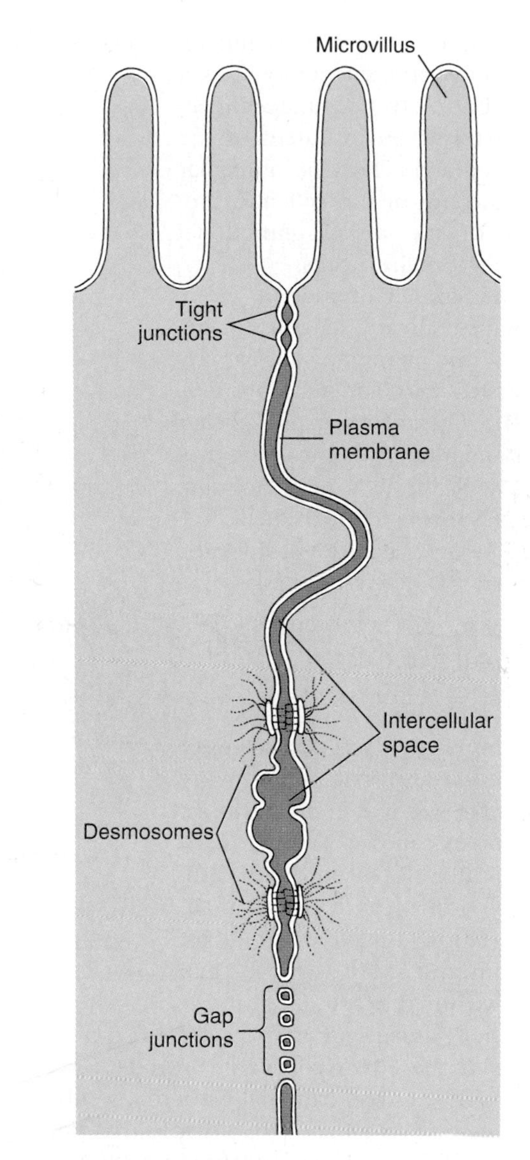

figure 2-13

Cell membranes forming the boundary between two epithelial cells. Various kinds of junctional complexes are found. The tight junction is a firm, adhesive band completely encircling the cell. Desmosomes are isolated "spot-welds" between cells. Gap junctions serve as sites of intercellular communication. Intercellular space may be greatly expanded in epithelial cells of some tissues.

Membrane Function

The incredibly thin, yet sturdy, cell membrane that encloses every cell is vitally important in maintaining the cell's integrity. Once thought a rather static entity that defined cell boundaries and kept cell contents from spilling out, the cell membrane is actually a dynamic structure having remarkable activity and selectivity. It is a permeability barrier that separates the internal and external environment of the cell, regulates the vital flow of molecular traffic into and out of the cell, and provides many of the unique functional properties of specialized cells.

The fluid-mosaic model of membrane structure already has been described. In life the membrane appears remarkably restless and fluid, with proteins constantly moving and reorganizing their molecular configuration. Some of the proteins that extend through the membrane act as channels through which small molecules and ions such as sodium, chloride, and potassium are allowed to pass.

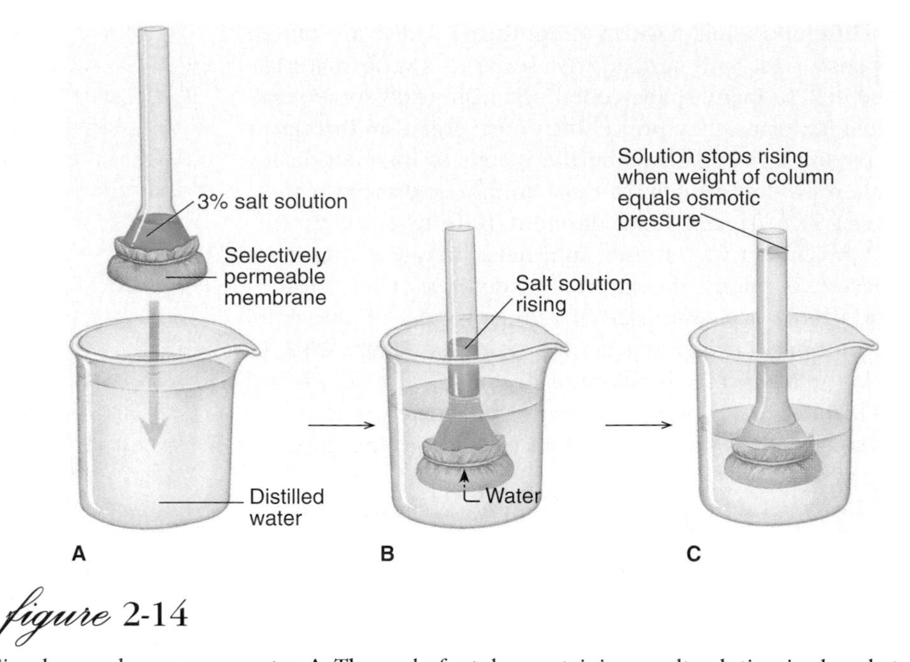

figure 2-14

Simple membrane osmometer. **A,** The end of a tube containing a salt solution is closed at one end by a selectively permeable membrane. The membrane is permeable to water molecules but not to salt. **B,** When the tube is immersed in pure water, water molecules diffuse through the membrane into the tube. Water molecules are in higher concentration in the beaker because they are diluted inside the tube by salt ions. Because the salt cannot diffuse out through the membrane, the volume of fluid inside the tube increases, and the level rises. **C,** When the weight of the column of water inside the tube exerts a downward force (hydrostatic pressure) causing water molecules to leave through the membrane in equal number to those that enter, the volume of fluid inside the tube stops rising. At that point, the osmotic pressure of the fluid in the funnel equals the hydrostatic pressure.

Function of the Cell Membrane

The cell membrane acts as a gatekeeper for the entrance and exit of the many substances involved in cell metabolism. Some substances can pass through with ease, others enter slowly and with difficulty, and still others cannot enter at all. Because conditions outside the cell are different from and more variable than conditions within the cell, it is necessary that the passage of substances across the membrane be rigorously controlled.

We recognize three principal ways that a substance may traverse the cell membrane: (1) by **simple diffusion** along a concentration gradient; (2) by a **mediated-transport system,** in which the substance binds to a specific site that in some way assists it across the membrane; and (3) by **endocytosis,** in which the substance is enclosed within a vesicle that forms on and detaches from the surface, then enters the cell.

Diffusion and Osmosis If a living cell is immersed in a solution having more solute molecules than the fluid inside the cell, a **concentration gradient** instantly exists between the two fluids. Assuming that the membrane is **permeable** to the solute, there is a net movement of solute toward the inside, the side having the lower concentration. The solute diffuses "downhill" across the membrane until its concentrations on each side are equal.

Most cell membranes are **selectively permeable,** that is, permeable to water (the solvent) but variably permeable or impermeable to solutes. In free diffusion, it is this selectivity that regulates molecular traffic. As a rule, gases (such as oxygen and carbon dioxide), urea, and lipid-soluble molecules (such as hydrocarbons and alcohol) are the only solutes that can diffuse through biological membranes with any degree of freedom. Since many water-soluble molecules readily pass through membranes, such movements cannot be explained by simple diffusion. Sugars, as well as many electrolytes and macromolecules, move across membranes by carrier-mediated processes (p. 33).

If a membrane is placed between two unequal concentrations of solutes to which the membrane is impermeable, water flows through the membrane from the more dilute to the more concentrated solution. The water molecules move across the membrane down a concentration gradient from an area where the *water* molecules are more concentrated to an area on the other side of the membrane where they are less concentrated. This is **osmosis.**

Osmosis can be demonstrated by a simple experiment in which a selectively permeable membrane such as cellophane is tied tightly over the end of a funnel, making an **osmometer.** The funnel is filled with a salt solution and placed in a beaker of pure water so that the water levels inside and outside the funnel are equal. In a short time the water level in the glass tube of the funnel rises, indicating that water is passing through the cellophane membrane into the salt solution (Figure 2-14).

Inside the funnel are salt molecules, as well as water molecules. In the beaker outside the funnel are only water mole-

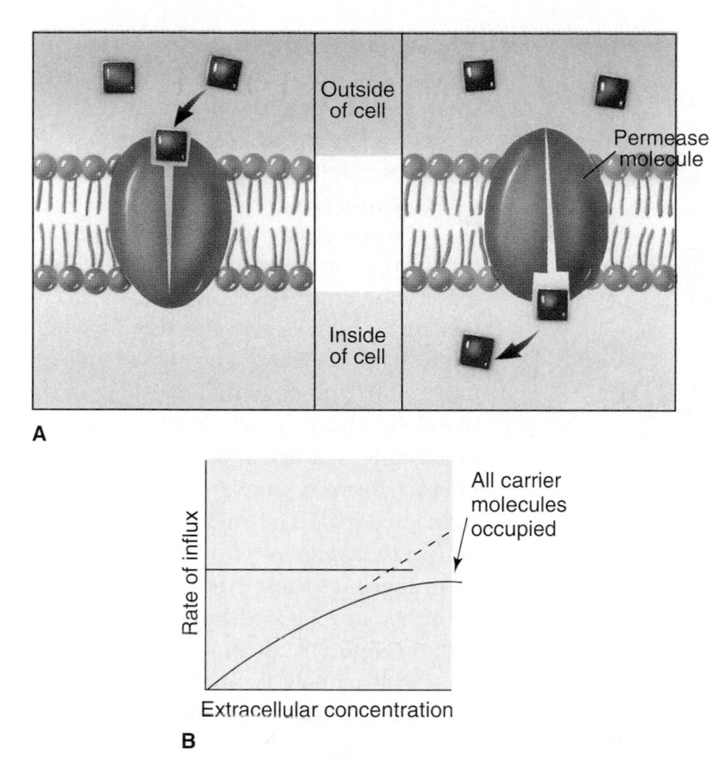

figure 2-15

Facilitated transport. **A,** The permease (transporter) molecule binds with a molecule to be transported (substrate) on one side of the cell membrane, changes shape, and releases the molecule on the other side. Facilitated transport takes place in the direction of a concentration gradient. **B,** Rate of transport increases with increasing substrate concentration until all permease molecules are occupied.

cules. Thus the concentration of water is less on the inside because some of the available space is occupied by the larger, nondiffusible salt molecules. A concentration gradient is said to exist for water molecules in the system. Water diffuses from the region of greater concentration of water (pure water outside) to the region of lesser concentration (salt solution inside).

As water enters the salt solution, the fluid level in the funnel rises, creating a hydrostatic pressure because of gravity inside the osmometer. Eventually the hydrostatic pressure produced by the increasing weight of solution in the funnel pushes water molecules out as fast as they enter. The level in the funnel becomes stationary and the system is in equilibrium. The **osmotic pressure** of the solution is equal to the hydrostatic pressure necessary to prevent further net entry of water.

The concept of osmotic pressure is not without problems. A solution reveals an osmotic "pressure" only when it is separated from solvent (usually water in biology) by a selectively permeable membrane. It can be disconcerting to think of an isolated bottle of salt solution or your blood serum as having "pressure" much as compressed gas in a bottle (*hydrostatic* pressure). However, the osmotic pressure is really the hydrostatic pressure that would have to be applied to a solution to keep it from gaining water *if* it is or were separated from a solution of lower osmotic pressure by a selectively permeable membrane. Consequently, biologists frequently use the term **osmotic potential** rather than osmotic pressure. However, since the term "osmotic pressure" is so firmly fixed in our vocabulary, it is necessary to understand the usage despite its potential confusion.

Mediated Transport We have seen that the cell membrane is an effective barrier to the free diffusion of most molecules of biological significance. Yet it is essential that such materials enter and leave the cell. Nutrients such as sugars and materials for growth such as amino acids must enter the cell, and the wastes of metabolism must leave. Such molecules are moved across the membrane by special proteins called **transporters** or **permeases.** Permeases form a small passageway through the membrane, enabling the solute molecule to cross the phospholipid bilayer (Figure 2-15A). Permeases are usually quite specific, recognizing only a limited group of chemical substances or perhaps even a single substance.

At high concentrations of solute, mediated transport systems show a saturation effect. This means simply that the rate of influx reaches a plateau beyond which increasing the solute concentration has no further effect on influx rate (Figure 2-15B). This is evidence that the number of transporters available in the membrane is limited. When all become occupied by solutes, the rate of transport is at a maximum and it cannot increase. Simple diffusion shows no such limitation; the greater the difference in solute concentrations on the two sides of the membrane, the faster the influx.

We recognize two distinctly different kinds of transporter-mediated mechanisms: (1) **facilitated transport,** in which the transporter assists a molecule to diffuse through the membrane that it cannot otherwise penetrate, and (2) **active transport,** in which energy is supplied to the transporter system to transport molecules in the direction opposite to the gradient (Figure 2-16). Facilitated transport therefore differs from active transport in that it sponsors movement in a downhill direction (in the direction of the concentration gradient) only and requires no metabolic energy to drive the transporter system.

In many animals facilitated transport is important for the transport of glucose (blood sugar) into body cells that burn it as a principal energy source for the synthesis of ATP. The concentration of glucose is greater in the blood than in the cells that consume it, favoring inward diffusion, but glucose is a polar molecule that does not, by itself, penetrate the membrane rapidly enough to support the metabolism of many cells; the carrier system increases the inward flow of glucose.

In active transport, molecules are moved uphill against the forces of passive diffusion. Active transport always involves the expenditure of energy because materials are pumped against a concentration gradient. Among the most important active transport systems in all animals are those which

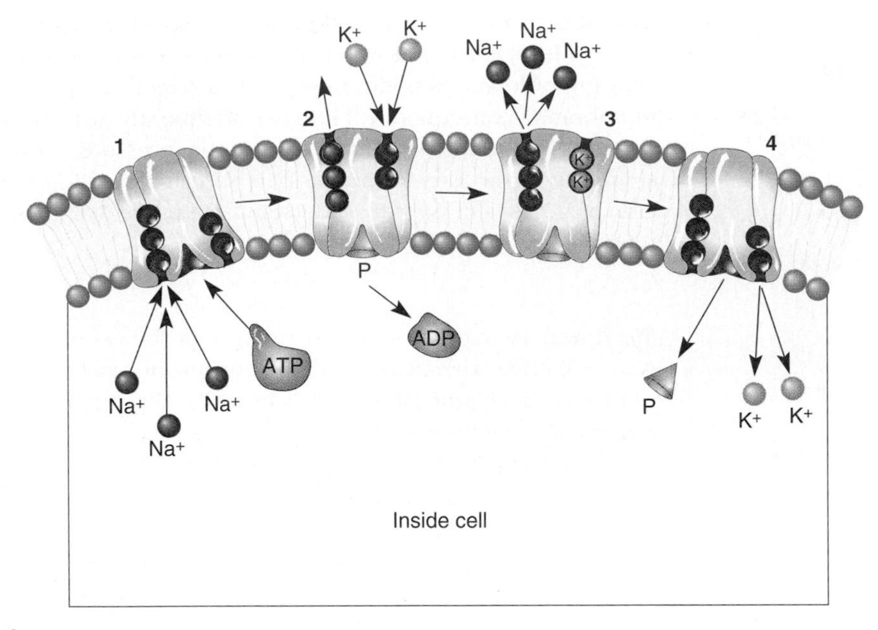

figure 2-16

Sodium-potassium pump, powered by bond energy of ATP, maintains the normal gradients of these ions across the cell membrane. The pump works by a series of conformational changes in the permease: *Step 1.* Three ions of Na^+ bind to the interior end of the permease, producing a conformational change in the protein complex. *Step 2.* The complex binds a molecule of ATP and cleaves it. *Step 3.* The binding of the phosphate group to the complex induces a second conformational change, passing the three Na^+ ions across the membrane, where they are now positioned facing the exterior. This new conformation has a very low affinity for the Na^+ ions, which dissociate and diffuse away, but it has a high affinity for K^+ ions and binds two of them as soon as it is free of the Na^+ ions. *Step 4.* Binding of the K^+ ions leads to another conformational change in the complex, this time leading to dissociation of the bound phosphate. Freed of the phosphate, the complex reverts to its original conformation, with the two K^+ ions exposed on the interior side of the membrane. This conformation has a low affinity for K^+ ions so that they are now released, and the complex has the original conformation, having a high affinity for Na^+ ions.

maintain sodium and potassium gradients between cells and the surrounding extracellular fluid or external environment. Most animal cells require a high internal concentration of potassium for protein synthesis at the ribosome and for certain enzymatic functions. The potassium concentration may be 20 to 50 times greater inside the cell than outside. Sodium, on the other hand, may be 10 times more concentrated outside the cell than inside. Both of these electrolyte gradients are maintained by the active transport of potassium into and sodium out of the cell. In many cells the outward pumping of sodium is linked to the inward pumping of potassium; the same transporter molecule does both. As much as 10% to 40% of all the energy produced by the cell is consumed by the **sodium-potassium exchange pump** (Figure 2-16).

Endocytosis Endocytosis, the ingestion of material by cells, is a collective term that describes the processes of **phagocytosis, potocytosis,** and **receptor-mediated endocytosis** (Figure 2-17). They are pathways for specifically internalizing solid particles, small molecules and ions, and macromolecules, respectively. All require energy and thus may be considered forms of active transport.

Phagocytosis, which literally means "cell eating," is a common method of feeding among animal-like protistans and certain metazoa. It is also the way in which white blood cells (leukocytes) engulf cellular debris and uninvited microbes in the blood. By phagocytosis, an area of the cell membrane, coated internally with actin-myosin, forms a pocket that engulfs the solid material. The membrane-enclosed vesicle then detaches from the cell surface and moves into the cytoplasm, where its contents are digested by intracellular enzymes (Figure 9-1, p. 211).

Potocytosis is similar to phagocytosis except that small areas of the surface membrane are invaginated into cells to form tiny vesicles. The invaginated pits and vesicles are called **caveolae** (ka-vee´o-lee). Specific binding receptors for the molecule or ion to be internalized are concentrated on the cell surface of caveolae. Potocytosis apparently functions for intake of at least some vitamins, and similar mechanisms may be important in translocating substances from one side of a cell to the other (see "exocytosis," following) and in internalizing signal molecules, such as some hormones or growth factors.

Receptor-mediated endocytosis is a specific mechanism for bringing large molecules within the cell. Proteins of the cell membrane specifically bind particular molecules (referred to as **ligands** in this process), which may be present in the extracellular fluid in very low concentrations. The invaginations of the cell surface that bear the receptors are coated within the cell with a protein called **clathrin,** hence they are described as **clathrin-coated pits.** As a clathrin-coated pit with its receptor-bound ligand invaginates and is brought within the cell, it is uncoated, the receptor and the ligand are dissociated, and the receptor and membrane material are recycled back to the surface membrane. Some important proteins and peptide hormones are brought into cells in this manner.

Exocytosis Just as materials can be brought into the cell by invagination and formation of a vesicle, the membrane of a vesicle can fuse with the cell membrane and extrude its contents to the surrounding medium. This is the process of **exocytosis.** This process occurs in various cells to remove undigestible residues of

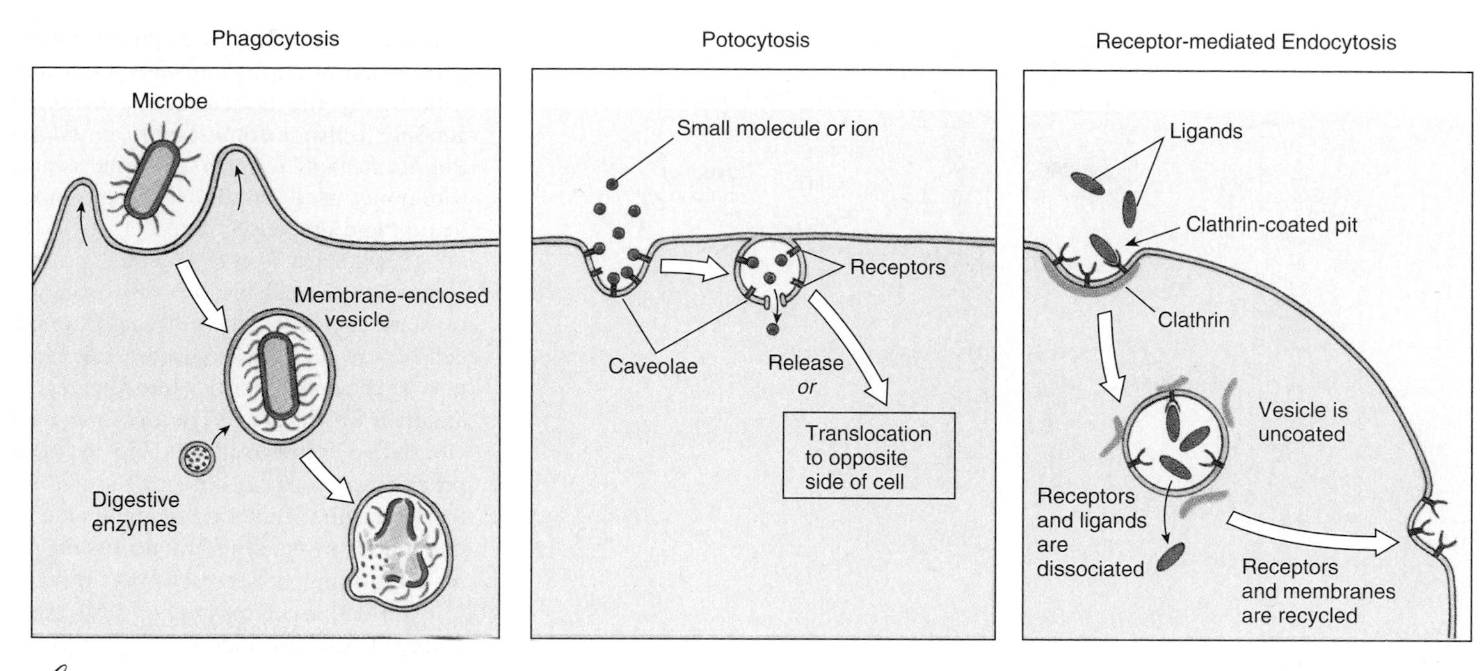

figure 2-17

Three types of endocytosis. In phagocytosis the cell membrane binds to a large particle and extends to engulf it. In potocytosis small areas of cell membrane, bearing specific receptors for a small molecule or ion, invaginate to form caveolae. Receptor-mediated endocytosis is a mechanism for selective uptake of large molecules in clathrin-coated pits. Binding of the ligand to the receptor on the surface membrane stimulates invagination of pits.

substances brought in by endocytosis, to secrete substances such as hormones (Figure 2-9), and to transport a substance completely across a cellular barrier, as we just mentioned. For example, a substance may be picked up on one side of the wall of a blood vessel by potocytosis, moved across the cell, and released by exocytosis.

Cell Division

All cells of the body arise from the division of preexisting cells. All the cells found in most multicellular organisms originate from the division of a single cell, the **zygote,** which is the product of union (fertilization) of an **egg** and a **sperm** (the **gametes).** Cell division provides the basis for one form of growth, both sexual and asexual reproduction, and transmission of hereditary qualities from one cell generation to another cell generation.

In the formation of **body cells (somatic cells)** for growth, maintenance, and repair, the nuclear division is referred to as **mitosis.** The cell divisions that occur in the formation of **germ cells** (ova and sperm, the gametes) are called **meiotic divisions;** we describe meiosis in Chapter 3.

Structure of Chromosomes

As mentioned earlier, DNA in the eukaryotic cell occurs in chromatin, a complex of DNA with histone and nonhistone protein. In fact, the chromatin is organized into a number of discrete bodies called **chromosomes** (color bodies), so named because they stain deeply with certain biological dyes when the chromosomes are condensed. In cells that are not dividing, the chromatin is loosely organized and dispersed, so that the individual chromosomes cannot be distinguished. Prior to division, the chromatin condenses, and the chromosomes can be recognized and their individual morphological characteristics determined. They are of varied lengths and shapes, some bent and some rodlike. Their number is constant for the species, and every body cell (but not the germ cells) has the same number of chromosomes regardless of the cell's function. A human being, for example, has 46 chromosomes in each body (somatic) cell.

During mitosis (nuclear division) the chromosomes shorten and become increasingly condensed and distinct, and each assumes a characteristic shape. At some point on the chromosome is a **centromere** (Figure 2-18). The centromere is the location of the **kinetochore,** a disc of proteins specialized to interact with the spindle fibers during mitosis.

Chromosomes always occur in pairs in somatic cells, or two of each kind. Of each pair, one was inherited from one parent and the other from the other parent. Thus there are 23 pairs in the human species. Each pair usually has certain characteristics of shape and form that aid in identification. A biparental organism begins with the union of two gametes, each of which furnishes a **haploid** set of chromosomes (23 in humans) to produce a somatic or **diploid** number of chromosomes (46 in humans). The chromosomes of a haploid set are also called a

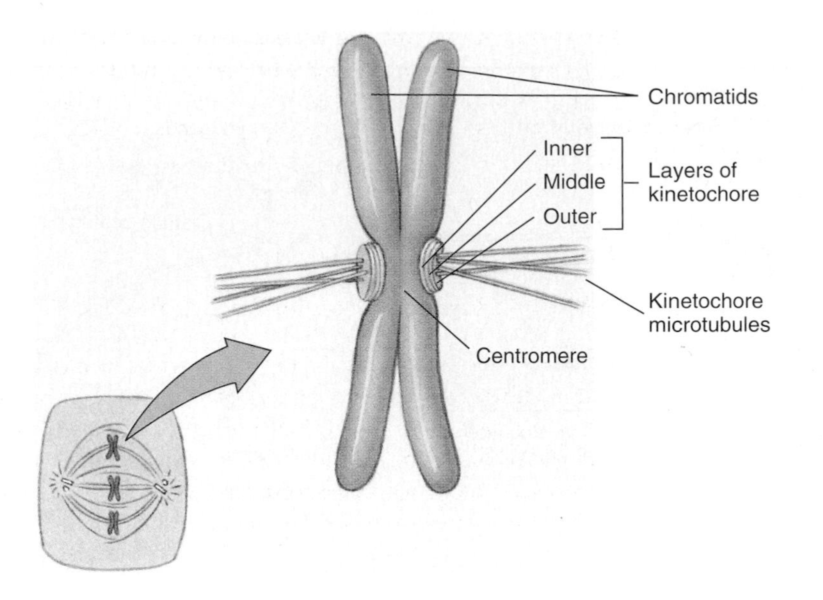

figure 2-18

Structure of a metaphase chromosome. The sister chromatids are still attached at their centromere. Each chromatid has a kinetochore, to which the kinetochore fibers are attached. Kinetochore microtubules from each chromatid run to one of the centrosomes, which are located at opposite poles.

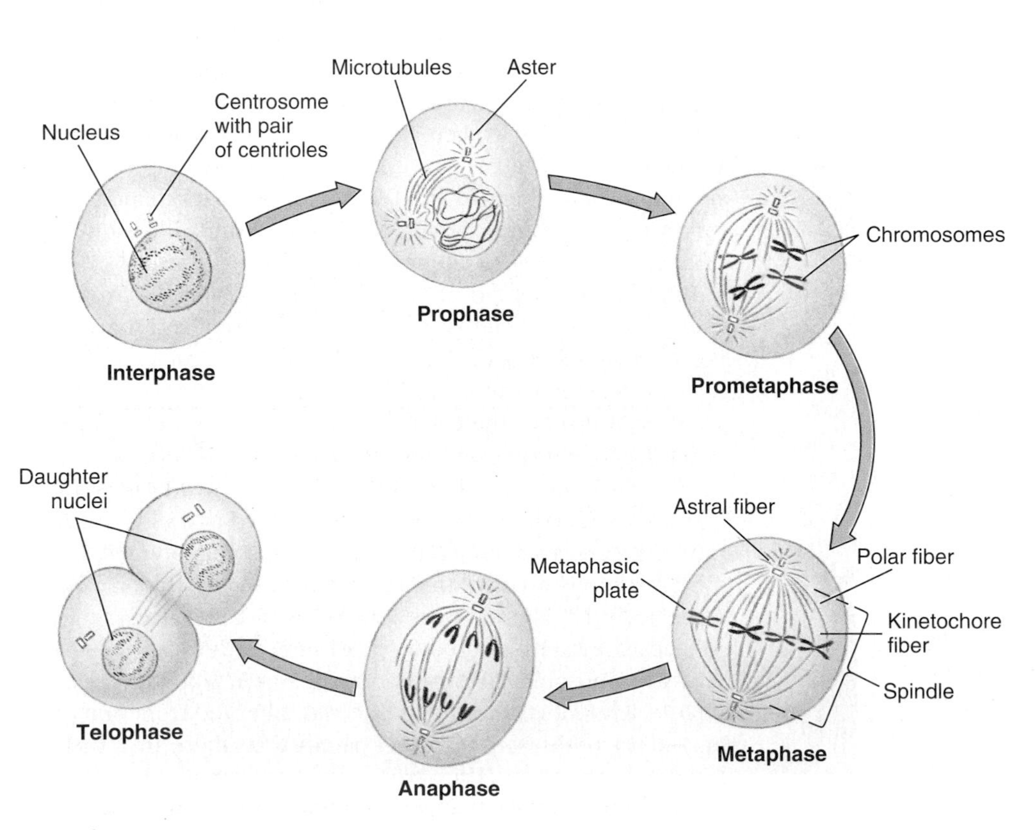

figure 2-19

Stages of mitosis, showing division of a cell with two pairs of chromosomes. One chromosome of each pair is shown in red.

genome. Thus a fertilized egg consists of a paternal genome (chromosomes contributed by the father) and a maternal genome (chromosomes contributed by the mother). (The term "genome" is also sometimes used for all the chromosomes in a diploid nucleus.)

Packaging of the DNA in chromosomes is a formidable problem because, by some estimates, the DNA in a human cell is nearly 4 m in length, yet is packed into 46 chromosomes having an aggregate length of only 200 μm. The DNA must obviously be coiled or folded up in some precise way to fit into a small space. Yet the packaging must be ar-ranged so that the genetic instructions are accessible for reading during the transcription process (the formation of messenger RNA from nuclear DNA, described in Chapter 3).

Stages in Mitosis

There are two distinct phases of cell division: the division of the nuclear chromosomes **(mitosis)** and the division of the cytoplasm **(cytokinesis).** Cytokinesis normally immediately follows mitosis, although there are occasions when the nucleus may divide a number of times without a corresponding division of the cytoplasm. In such a case the resulting mass of protoplasm containing many nuclei is called a **multinucleate cell.** An example is the giant resorptive cell type of bone (osteoclast) that may contain 15 to 20 nuclei. Sometimes a multinucleate mass is formed by cell fusion rather than nuclear proliferation. This arrangement is called a **syncytium.** An example is vertebrate skeletal muscle, which is composed of multinucleate fibers that have arisen by the fusion of numerous embryonic cells.

The process of mitosis is arbitrarily divided for convenience into four successive stages or phases, although one stage merges into the next without sharp lines of transition. These phases are prophase, metaphase, anaphase, and telophase (Figure 2-19). When the chromosomes are not actively undergoing mitosis, the cell is in **interphase,** during which the DNA content of the nucleus is being duplicated. Thus, when the cell begins "active" mitosis, it already has a double set of chromosomes.

Prophase

Before the beginning of prophase, the centrioles replicate, and each pair of centrioles migrates toward opposite sides of the cell. At the same time, fine fibers (microtubules) known as **spindle fibers** appear between the two pairs of centrioles to form a football-shaped **spindle,** so named because of its resemblance to nineteenth-century wooden spindles used to twist thread together in spinning. Other microtubules radiate outward from each centriole to form **asters.**

At the same time, the diffuse nuclear chromatin condenses to form visible chromosomes. These actually consist of two identical sister **chromatids** formed during a period of interphase. The sister chromatids are joined together at their centromere. Dynamic spindle fibers repeatedly extend and retract from the centrosome. When a fiber encounters a kinetochore, it binds to the kinetochore and ceases extending and retracting. It is as if the centrosome were sending out "feelers" to find the chromosomes. During prophase the nuclear envelope fragments into small vesicles and disappears.

Metaphase

Each centromere has two kinetochores, each kinetochore being attached to one of the centrosomes by microtubules. By a kind of tug-of-war during metaphase, the condensed sister chromatids are moved to the middle of the nuclear region to form a **metaphasic plate** (Figure 2-19). The centromeres line up precisely on the plate with the arms of the chromatids trailing off randomly in various directions.

Microtubules are long, hollow, inelastic cylinders composed of the protein tubulin. Each tubulin molecule is actually a doublet composed of two globular proteins. The molecules are attached head-to-tail to form a strand, and thirteen strands aggregate to form a microtubule. Because the tubulin subunits in a microtubule are always attached head-to-tail, the ends of the microtubule differ chemically and functionally. One end (called the plus end) both adds and deletes tubulin subunits more rapidly than the other end (the minus end). In a mitotic spindle, the plus ends of the kinetochore fibers are away from the centrosome, and the minus ends are at the centrosome. The microtubule grows when the rate of adding subunits exceeds that of removing them, and it becomes shorter when the rate of removal exceeds that of addition.

Anaphase

The single centromere that has held the two chromatids together now splits so that two independent chromosomes, each with its own centromere, are formed. The chromosomes part more, apparently pulled by the spindle fibers attached to their kinetochores. The arms of each chromosome trail along behind as though the chromosome were being dragged through a resisting medium. Present evidence indicates that the force moving the chromosomes is disassembly of the tubulin subunits at the kinetochore end of the microtubules (see note).

Telophase

When the daughter chromosomes reach their respective poles, telophase has begun. The daughter chromosomes are crowded together and stain intensely with histological stains. The spindle fibers disappear and the chromosomes lose their identity, reverting to the diffuse chromatin network characteristic of the interphase nucleus. The small vesicles of nuclear membrane begin to rejoin and reassemble the nuclear envelope prior to chromosome decondensation.

Cytokinesis: Cytoplasmic Division

During the final stages of nuclear division a **cleavage furrow** appears on the surface of the dividing cell and encircles it at the midline of the spindle. The cleavage furrow deepens and pinches the cell membrane as though it were being tightened by an invisible rubber band. Microfilaments of actin are present just beneath the surface in the furrow between the cells. Interaction with myosin, similar to that which occurs when muscle cells contract (p. 159), draws the furrow inward. Finally, the infolding edges of the cell membrane meet and fuse, completing cell division.

Cell Cycle

The complete sequence of events resulting in mitosis, the actual cell division, and the events that follow comprise the cell cycle. A cell prepares to divide before the actual division occurs. Actual nuclear division occupies only about 5% to 10% of the cell cycle, the rest of the cell's time is spent in interphase.

The usual constraints that operate to switch off normal cell division when tissue growth is complete are lacking in uncontrolled cancer cell growth. Whereas normal cells "know" when to stop dividing, cancer cells not only divide rapidly but also grow between division. Part of the current research on cancer is focused on how the cell division switch is overridden by internal and external influences during tumor growth (p. 79).

For many years biologists thought that interphase was a period of rest because the nucleus appeared inactive when observed with the ordinary light microscope. With new techniques for revealing DNA replication, along with appreciation of DNA as the genetic material, biologists discovered

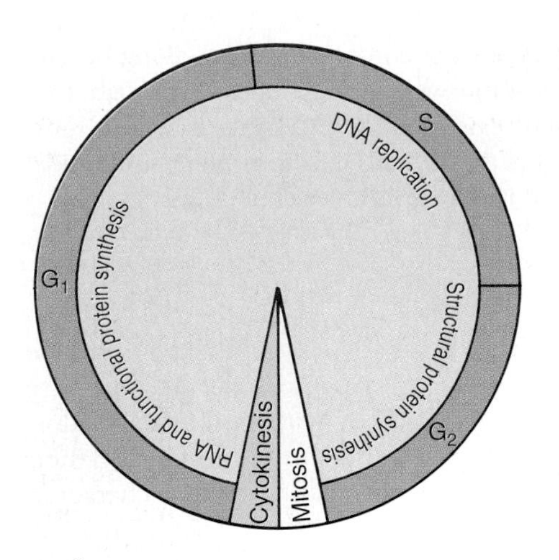

figure 2-20

Cell cycle, showing relative duration of recognized periods. S, G_1, and G_2 are periods within interphase: S, synthesis of DNA; G_1, presynthetic period; G_2, postsynthetic period. Actual duration of the cycle and the different periods varies considerably in different cell types. After mitosis and cytokinesis, the cell may go into an arrested, quiescent stage known as G_0.

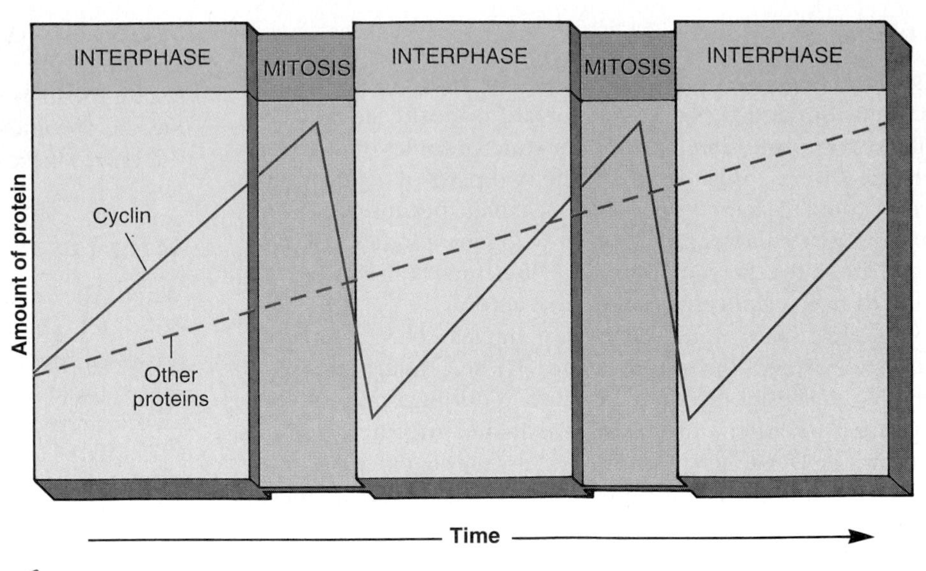

figure 2-21

Variations in the levels of cyclin in dividing cells of early sea urchin embryos. Cyclin binds with its cyclin-dependent kinase to activate the enzyme.

that DNA replication occurred during the interphase stage. Furthermore, they found that cells synthesized many other protein and nucleic acid components essential to normal cell growth and division during the seemingly quiescent interphase period.

The most important preparation—replication of DNA—occurs during the latter part of interphase, termed the S period (period of synthesis) (Figure 2-20). The S period is preceded and succeeded by G_1 and G_2 periods, respectively (*G* stands for gap), when no DNA synthesis is occurring. Enzymes and substrates are being prepared during the G_1 period for the DNA replication that follows. During G_2, spindle and aster proteins are being synthesized. The energy demands of the cell are high during the G_2 period, and cellular metabolism continues throughout the cell cycle.

Events in the cell cycle are exquisitely regulated. Important transitions are mediated by **cyclin-dependent kinases (cdk's)** and activating subunits of the cdk's called **cyclins.** Kinases are enzymes that add phosphate groups to other proteins to activate or inactivate them, and kinases themselves may require activation. The cdk's become active only when they are bound with the appropriate cyclin, and cyclins are synthesized and degraded during the cell cycle (Figure 2-21).

Flux of Cells

Cell division is important for growth, for replacement of cells lost to natural attrition and wear and tear, and for wound healing. Cell division is especially rapid during the early development of the organism, then slows with age. Different cell populations divide at widely different rates. Cells in the central nervous system stop dividing altogether after the early months of fetal development and persist without further division for the life of the individual. Muscle cells also stop dividing during the third month of fetal development, and future growth depends on enlargement of fibers already present.

In other tissues that are subject to wear and tear, lost cells must be constantly replaced, for example cells shed from the skin, the lining of the alimentary canal, and spent blood cells. Such losses of cells are made up by mitosis.

Normal development, however, does entail cell death in which the cells are not replaced. They may become senescent or undergo a programmed cell death, or **apoptosis** (Gr. *apo-*, from, away from, + *ptosis,* a falling), which is in many cases necessary for the continued health and development of the organism. For example, during embryonic development of vertebrates, excess nerve cells and immune cells that would attack the body's own tissues "commit suicide" in this manner.

> *Apoptosis currently is receiving a great deal of attention from researchers. One of the most valuable laboratory models is a tiny free-living nematode,* Caenorhabditis elegans *(see p. 438). The effects of apoptosis are not always beneficial to an organism. For example, an important disease mechanism in AIDS (acquired immune deficiency syndrome) seems to be an inappropriate triggering of programmed cell death among important cells of the immune system.*

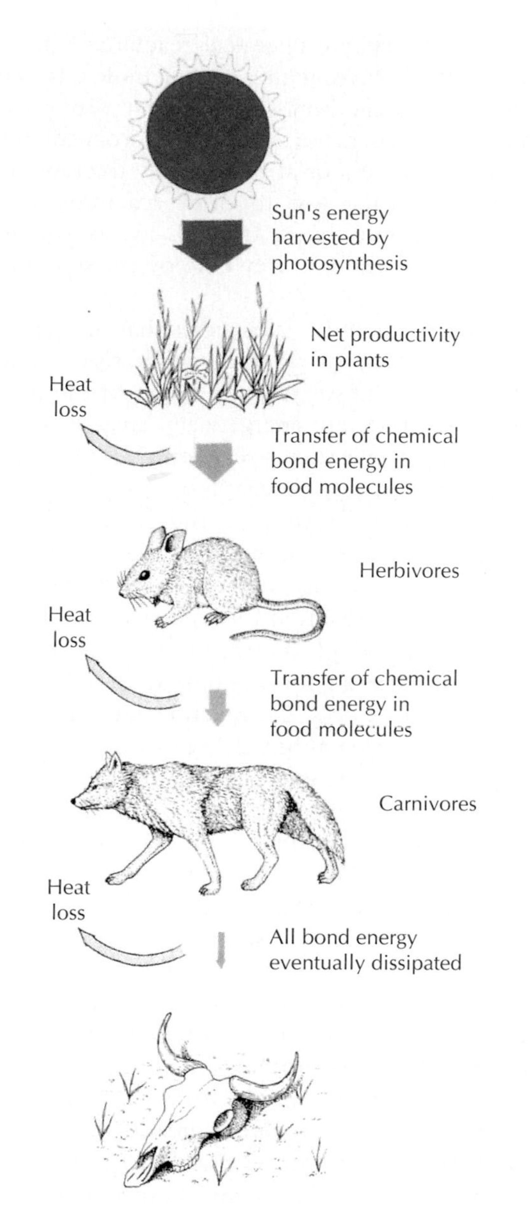

figure 2-22

Solar energy sustains virtually all life on earth. With each energy transfer, however, much energy is lost as heat.

Cellular Metabolism

All cells must obtain energy, synthesize their own internal structure, control much of their own activity, and guard their boundaries. **Cellular metabolism** refers to the collective total of chemical processes that occur within living cells to accomplish these activities. Although the enormous number of reactions in their aggregate are extremely complex, the central metabolic routes through which matter and energy are channeled are not difficult to understand.

Energy and Life

Energy is the capacity for performing work. It cannot be created or destroyed, but it can be changed in form. One form is potential energy, or energy of position, as opposed to kinetic energy, or energy of motion. Energy in the bonds of organic molecules is a kind of potential energy. Energy is required to form the bonds, and energy is released when they are broken, but until they are broken, the bond energy remains potential.

In living organisms the cells gradually liberate the chemical potential energy in organic molecules and couple the release to a variety of energy-consuming processes. As living organisms convert stored energy into other forms, and eventually into heat, less and less remains in reserve. How is it replenished? The ultimate source of energy for life on earth is the sun. The energy of sunlight is captured by green plants and algae (autotrophs), which transform a portion of this energy into chemical-bond energy (food energy). Herbivorous animals (heterotrophs) eat autotrophs and convert a small part of the potential chemical energy into animal tissue and a large part into heat that is invisibly dissipated into space. Carnivorous animals eat the herbivores, and they in turn are eaten by other carnivores. At each step a large part of the energy is lost as heat (Figure 2-22).

At the cellular level, animals and plants use chemical energy stored in food to carry out hundreds of activities. Thus animals are totally dependent on autotrophs, which fortunately accumulate enough energy to sustain both themselves and the animals that feed on them. Plant photosynthesis and respiration of all organisms are inextricably woven together in a vast cycle that is driven by the constant flow of energy from the sun.

Free Energy

To describe the energy changes that take place in chemical reactions, biochemists use the concept of **free energy.** Free energy is simply the energy in a system available for doing work. In a molecule, free energy equals the energy present in chemical bonds minus the energy that cannot be used. The majority of reactions in cells release free energy and are said to be **exergonic** (Gr. *ex,* out, + *ergon,* work). Such reactions are spontaneous and always proceed "downhill" since free energy is lost from the system. Thus,

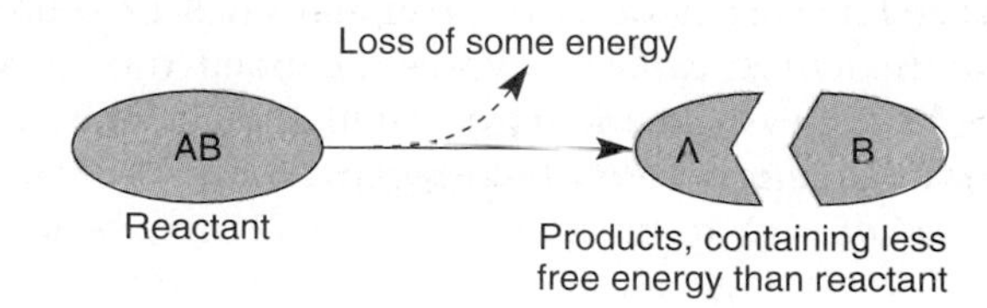

However, many important reactions in cells require the addition of free energy and are said to be **endergonic** (Gr. *endon,* within, + *ergon,* work). Such reactions have to

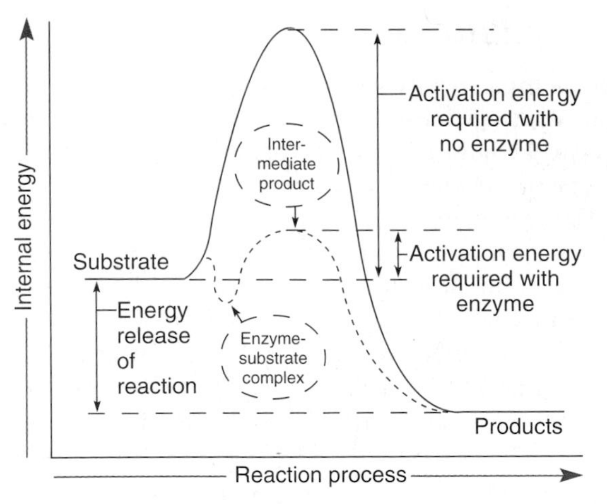

figure 2-23

Energy changes during enzyme catalysis of a substrate. The overall reaction proceeds with a net release of energy (exergonic). In the absence of an enzyme, substrate is stable because of the large amount of activation energy needed to disrupt strong chemical bonds. The enzyme reduces the energy barrier by forming a chemical intermediate with a much lower internal energy state.

be "pushed uphill" because the reactants end up with more energy than they started with:

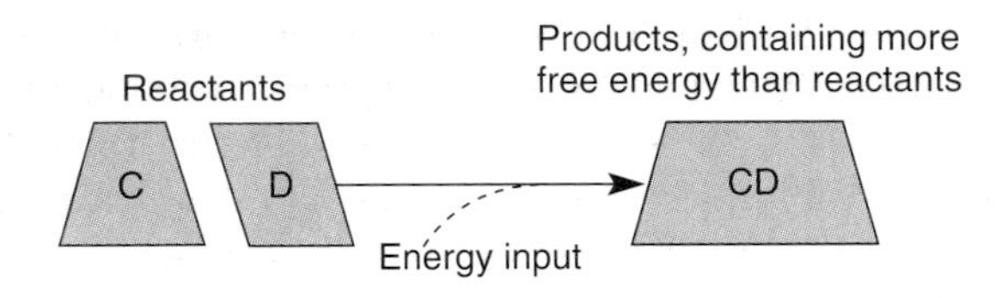

As we will see in a later section, ATP is the ubiquitous, energy-rich intermediate used by organisms to power important uphill reactions such as those required for active transport and cellular synthesis.

Role of Enzymes

Enzymes and Activation Energy

For any reaction to occur, even exergonic ones that tend to proceed spontaneously, the chemical bonds must first be destabilized. For example, if the reaction involves splitting a covalent bond, the atoms forming the bond must first be stretched apart to make them less stable. Some energy, termed the **activation energy,** must be supplied before the bond will be stressed enough to break. Only then will there be an overall loss of free energy and the formation of reaction products. This requirement can be likened to the energy needed to push a cart over the crest of a hill before it will roll spontaneously down the other side, liberating its potential energy as the cart descends.

One way to activate chemical reactants is to raise the temperature. By increasing the rate of molecular collisions and pushing chemical bonds apart, heat can provide the activation energy to make a reaction proceed. But metabolic reactions must occur at biologically tolerable temperatures, temperatures too low to allow reactions to proceed beyond imperceptible rates. Instead, living systems have evolved a different strategy: they employ a system of organic **catalysts.**

Catalysts are chemical substances that increase reaction rates without affecting the products of the reaction and without being altered or destroyed as a result of the reaction. A catalyst cannot make an energetically impossible reaction happen; it simply accelerates a reaction that would have proceeded at a very slow rate otherwise.

Enzymes are the catalysts of the living world. The special catalytic talent of an enzyme is its power to reduce the amount of activation energy required for a reaction. In effect, an enzyme steers the reaction through one or more intermediate steps, each of which requires much less activation energy than that required for a single-step reaction (Figure 2-23). Enzymes affect only the reaction rate; they do not in any way alter the free-energy change of a reaction, nor do they change the proportions of reactants and products in a reaction.

Nature of Enzymes

Enzymes are complex molecules varying in size from small, simple proteins with a molecular weight of 10,000 to highly complex molecules with molecular weights up to 1 million. Many enzymes are pure proteins—delicately folded and interlinked chains of amino acids. Other enzymes require the participation of small nonprotein groups called **cofactors** to perform their enzymatic function. In some cases these cofactors are metallic ions (such as ions of iron, copper, zinc, magnesium, potassium, and calcium) that form a functional part of the enzyme. Examples are carbonic anhydrase, which contains zinc; the cytochromes, which contain iron; and troponin (a muscle-contracting enzyme), which contains calcium.

Another class of cofactors, called **coenzymes,** is organic. All coenzymes contain groups derived from vitamins, compounds that must be supplied in the diet. All of the B complex vitamins are coenzyme compounds. Since animals have lost the ability to synthesize the vitamin components of coenzymes, it is obvious that a vitamin deficiency can be serious. However, unlike dietary fuels and nutrients that must be replaced, once burned or assembled into structural materials, vitamins are recovered in their original form and used repeatedly. Examples of coenzymes that contain vitamins are nicotinamide adenine dinucleotide (NAD), which contains nicotinic acid; coenzyme A, which contains pantothenic acid; and flavin adenine dinucleotide (FAD), which contains riboflavin.

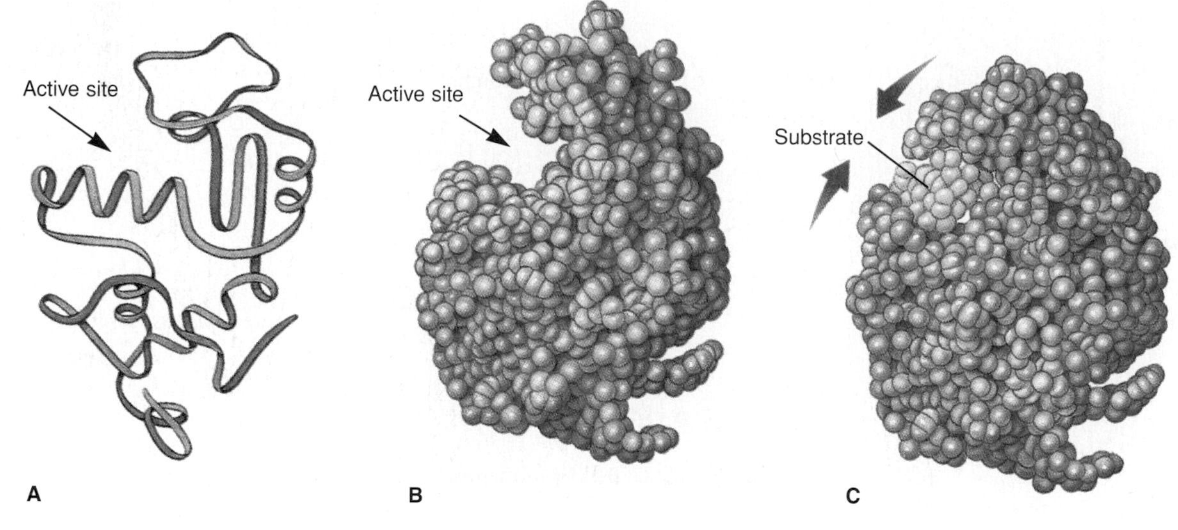

figure 2-24

How an enzyme works. This space-filling model shows that the enzyme lysozyme bears a pocket containing the active site. When a chain of sugars (substrate) enters the pocket, the enzyme changes shape slightly so that the pocket enfolds the substrate and conforms to its shape. This positions the active site (an amino acid in the protein) next to a bond between adjacent sugars in the chain, causing the sugar chain to break.

Action of Enzymes

An enzyme functions by combining in a highly specific way with its **substrate,** the molecule on which it acts. The enzyme bears an active site located within a cleft or pocket and contains a unique molecular configuration. The active site has a flexible surface that enfolds and conforms to the substrate (Figure 2-24). The binding of enzyme to substrate forms an **enzyme-substrate complex (ES complex)** in which the substrate is secured by weak bonds to several points in the active site of the enzyme. The ES complex is not strong and will quickly dissociate, but during this fleeting moment the enzyme provides a unique chemical environment that stresses certain chemical bonds in the substrate so that much less energy is required to complete the reaction.

How can biochemists be certain that an enzyme-substrate (ES) complex exists? The original evidence offered by Leonor Michaelis in 1913 is that, when the substrate concentration is increased while the enzyme concentration is held constant, the reaction rate reaches a maximum velocity. This saturation effect *is interpreted to mean that all catalytic sites become filed at high substrate concentration. It is not seen in uncatalyzed reactions. Other evidence includes the observation that the ES complex displays unique spectroscopic characteristics not displayed by either the enzyme or the substrate alone. Furthermore, some ES complexes can be isolated in pure form, and at least one kind (nucleic acids and their polymerase enzymes) has been directly visualized with the electron microscope.*

Enzymes that engage in important main-line sequences—such as the crucial energy-providing reactions of the cell that go on constantly—seem to operate in enzyme sets rather than in isolation. For example, the conversion of glucose to carbon dioxide and water proceeds through 19 reactions, each requiring a specific enzyme. Main-line enzymes are found in relatively high concentrations in the cell, and they may implement quite complex and highly integrated enzymatic sequences. One enzyme carries out one step, then passes the product to another enzyme, which catalyzes another step, and so on. The reactions may be said to be coupled.

Specificity of Enzymes

One of the most distinctive attributes of enzymes is their high specificity. This is a consequence of the exact molecular fit that is required between enzymes and substrates. Furthermore, an enzyme catalyzes only one reaction; unlike reactions carried out in the organic chemist's laboratory, no side reactions or by-products result. Specificity of both substrate and reaction is obviously essential to prevent a cell from being overwhelmed with useless by-products.

However, there is some variation in degree of specificity. Some enzymes will catalyze the oxidation (dehydrogenation) of only one substrate; for example, succinic dehydrogenase catalyzes the oxidation of succinic acid only. Others, such as proteases (for example, pepsin and trypsin), will act on almost any protein, but each protease has its own particular point of attack in the protein (Figure 2-25). Usually an enzyme will take

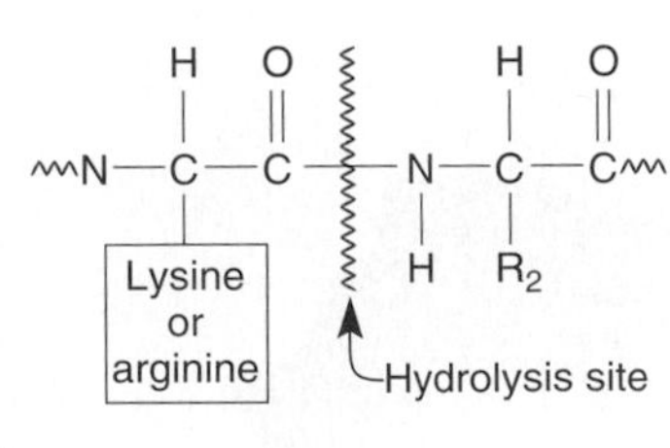

figure 2-25

High specificity of trypsin. It splits only peptide bonds adjacent to lysine or arginine.

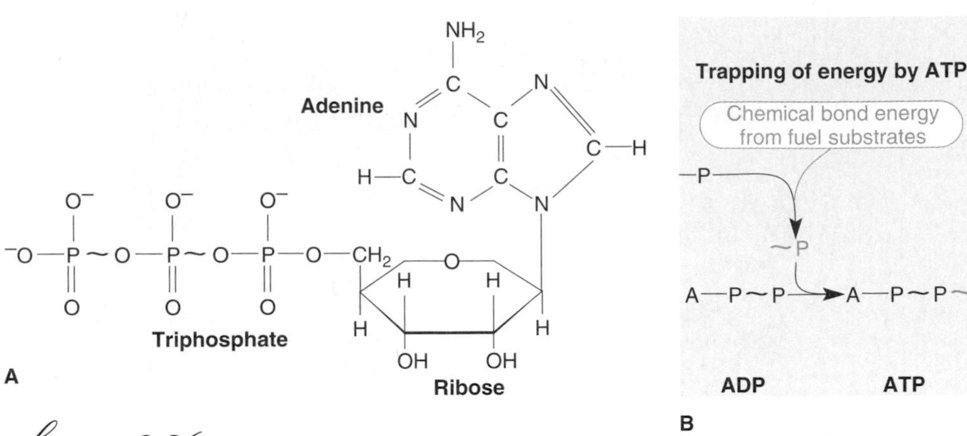

figure 2-26

A, Structure of ATP. **B,** ATP formation from ADP.

on one substrate molecule at a time, catalyze its chemical change, release the product, and then repeat the process with another substrate molecule. The enzyme may repeat this process billions of times until it is finally worn out (after a few hours to several years) and is broken down by scavenger enzymes in the cell. Some enzymes undergo successive catalytic cycles at dizzying speeds of up to a million cycles per minute, but most operate at slower rates.

Enzyme-Catalyzed Reactions

Enzyme-catalyzed reactions are theoretically reversible. This is signified by the double arrows between substrate and products.

$$\text{Fumaric acid} + H_2O \leftrightarrow \text{Malic acid}$$

However, for various reasons, the reactions catalyzed by most enzymes tend to go in one direction. For example, the proteolytic enzyme pepsin can degrade proteins into amino acids, but it cannot accelerate the rebuilding of amino acids into any significant amount of protein. The same is true of most enzymes that catalyze the hydrolysis of large molecules such as nucleic acids, polysaccharides, lipids, and proteins. There is usually one set of reactions and enzymes that break them down, but they must be resynthesized by a different set of reactions that are catalyzed by different enzymes. This apparent irreversibility exists because the chemical equilibrium usually favors the formation of the smaller degradation products.

The net **direction** of any chemical reaction depends on the relative energy contents of the substances involved. If there is little change in the chemical-bond energy of the substrate and products, the reaction is more easily reversible. However, if large quantities of energy are released as the reaction proceeds in one direction, more energy must be provided in some way to drive the reaction in the reverse direction. Thus many enzyme-catalyzed reactions are in practice irreversible, unless the reaction is coupled to another that makes energy available. In the cell, both reversible and irreversible reactions are combined in complex ways to make possible both synthesis and degradation.

Cellular Energy Transfer

Unlike the combustion of fuel in a fire, which is an explosive event with rapid release of energy as heat, metabolic oxidations are flameless and of low temperature. They proceed gradually, and the energy liberated is coupled to a great variety of energy-consuming reactions. Although metabolic energy exchanges proceed with impressive efficiency, part of the energy inevitably is liberated as heat. Heat can be put to some use, of course; the endothermic vertebrates (birds and mammals) use it to elevate and maintain a constant internal body temperature. But for the most part, heat is a useless commodity to a cell, since it is a nonspecific form of energy that cannot be captured and redistributed to power metabolic processes. There is actually only one way in which the oxidative release of energy is made available for use by cells: it is coupled to the production of high-energy phosphate bonds, usually in the form of **ATP (adenosine triphosphate)** by addition of inorganic phosphate to **ADP (adenosine diphosphate)** (Figure 2-26).

The ATP molecule consists of a purine (adenine), a 5-carbon sugar (ribose), and three molecules of phosphoric acid linked together by two pyrophosphate bonds to form a triphosphate group. The pyrophosphate bonds are called **high-energy bonds** because a great deal of free energy in the bonds is liberated when ATP is hydrolyzed to adenosine diphosphate (ADP) and inorganic phosphate. The high-energy pyrophosphate bonds of ADP and ATP are designated by the "tilde" symbol ~. A high-energy phosphate bond is shown as ~P and a low-energy bond (such as the bond linking the triphosphate group to adenosine) as —P. ATP may be symbolized as A—P~P~P and ADP as A—P~P.

The way that ATP can act to drive a coupled reaction is shown in Figure 2-27. A coupled reaction is really a system in-

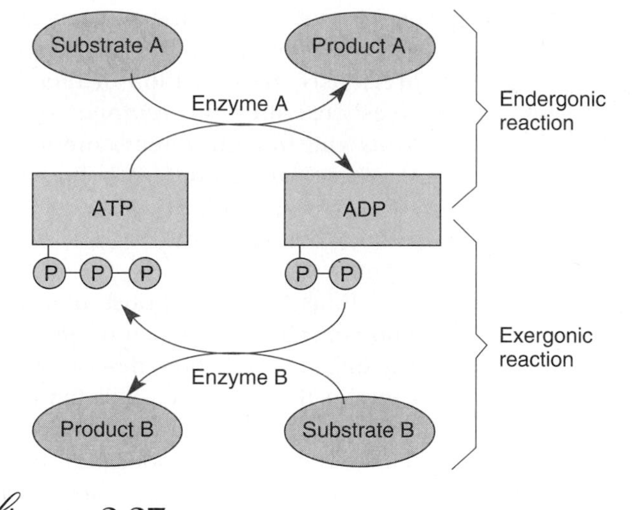

figure 2-27

A coupled reaction. The endergonic conversion of substrate A to product A will not occur spontaneously but requires an input of energy from another reaction involving a large release of energy. ATP is the intermediate through which the energy is shuttled.

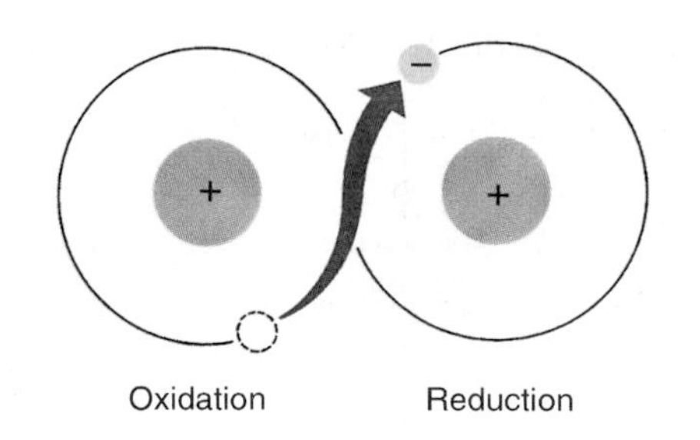

figure 2-28

A redox pair. The molecule at left is oxidized by the loss of an electron. The molecule at right is reduced by gaining an electron.

volving two reactions linked by an energy shuttle (ATP). The conversion of substrate A to product A is endergonic because the product contains more free energy than the substrate. Therefore energy must be supplied by coupling the reaction to one that is exergonic, the conversion of substrate B to product B. Substrate B in this reaction is commonly called a **fuel** (for example, glucose or a lipid). The bond energy that is released in reaction B is transferred to ADP, which in turn is converted to ATP. ATP now contributes its phosphate-bond energy to reaction A, and ADP is produced again.

The high-energy bonds of ATP are actually rather unstable bonds. Because they are unstable, the energy of ATP is readily released when ATP is hydrolyzed in cellular reactions. Note that ATP is an **energy-coupling agent** and *not* a fuel. It is not a storehouse of energy set aside for some future need. Rather it is produced by one set of reactions and is almost immediately consumed by another. ATP is formed as it is needed, primarily by oxidative processes in the mitochondria. Oxygen is not consumed unless ADP and phosphate molecules are available, and these do not become available until ATP is hydrolyzed by some energy-consuming process. *Energy metabolism is therefore mostly self-regulating.*

Cellular Respiration

How Electron Transport Is Used to Trap Chemical-Bond Energy

Having seen that ATP is the one common energy denominator by which all cellular machines are powered, we are in a position to ask how this energy is captured from fuel substrates. This question directs us to an important generalization: *all cells obtain their chemical energy requirements from oxidation-reduction reactions.* This means simply that during the degradation of fuel molecules, hydrogen atoms (electrons and protons) are passed from electron donors to electron acceptors with a release of energy. A portion of this energy is trapped and used to form the high-energy bonds at ATP. The release of energy during electron transfer and its conservation as ATP is the mainspring of cell activity and was a crucial evolutionary achievement.

Because they are so important, let us review what we mean by oxidation-reduction ("redox") reactions. In these reactions there is a transfer of electrons from an electron donor (the reducing agent) to an electron acceptor (the oxidizing agent). As soon as the electron donor loses its electrons, it becomes oxidized. As soon as the electron acceptor accepts electrons, it becomes reduced (Figure 2-28). In other words, a reducing agent becomes oxidized when it reduces another compound, and an oxidizing agent becomes reduced when it oxidizes another compound. Thus for every oxidation there must be a corresponding reduction. When electrons are accepted by the oxidizing agent, energy is liberated because the electrons move to a more stable position. In the cell, the electrons flow through a series of carriers. Each carrier is reduced by accepting electrons and then is reoxidized by passing electrons to the next carrier in the series. By transferring electrons stepwise in this manner, energy is gradually released, and a maximum yield of ATP is realized.

Aerobic versus Anaerobic Metabolism

Ultimately, the electrons are transferred to a **final electron acceptor.** The nature of this final acceptor is the key that determines the overall efficiency of cellular metabolism. The heterotrophs can be divided into two great groups: **aerobes,** those which use molecular oxygen as the final electron acceptor, and **anaerobes,** those which employ some other molecule as the final electron acceptor.

As we have discussed (Chapter 1), life originated in the absence of oxygen, and the abundance of atmospheric oxygen developed after photosynthetic organisms evolved. Some strictly anaerobic organisms still exist and indeed play some

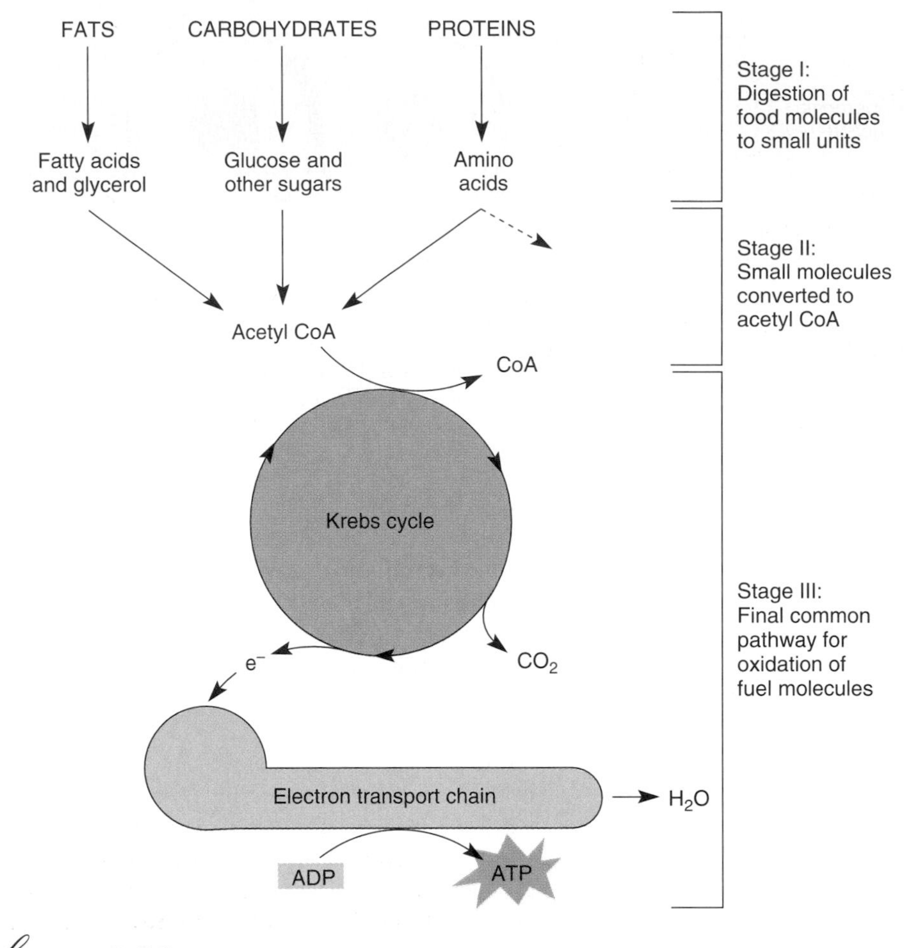

figure 2-29

Overview of cellular respiration, showing the three stages in the complete oxidation of food molecules to carbon dioxide and water.

important roles in specialized habitats. However, evolution has favored aerobic metabolism, not only because oxygen became available, but also because aerobic metabolism is vastly more efficient than anaerobic metabolism. In the absence of oxygen, only a very small fraction of the bond energy present in foodstuffs can be released. For example, when an anaerobic microorganism degrades glucose, the final electron acceptor (such as pyruvic acid) still contains most of the energy of the original glucose molecule. On the other hand, an aerobic organism, using oxygen as the final electron acceptor, can completely oxidize glucose to carbon dioxide and water. Almost 20 times as much energy is released when glucose is completely oxidized as when it is degraded only to the stage of lactic acid. An obvious advantage of aerobic metabolism is that a much smaller quantity of foodstuffs is required to maintain a given rate of metabolism.

General Description of Respiration

Aerobic metabolism is more familiarly known as true **cellular respiration,** defined as the oxidation of fuel molecules with molecular oxygen as the final electron acceptor. As mentioned previously, the oxidation of fuel molecules describes the *removal of electrons* and *not* the direct combination of molecular oxygen with fuel molecules. Let us look at this process in general before considering it in more detail.

Hans Krebs, the British biochemist who contributed so much to our understanding of respiration, described three stages in the complete oxidation of fuel molecules to carbon dioxide and water (Figure 2-29). In stage I, foodstuffs passing through the intestinal tract are digested into small molecules that can be absorbed into the circulation. There is no useful energy yield during digestion, which we will discuss in Chapter 10. In stage II, most of the degraded foodstuffs are converted into two 3-carbon units (pyruvic acid). This occurs in the cytoplasm. The pyruvic acid molecules then enter mitochondria, where in another reaction they join with a coenzyme (coenzyme A) to form acetyl-CoA. Stage II generates some ATP, but the yield is small compared with that obtained in the final stage of respiration. In stage III the final oxidation of fuel molecules occurs, with a large yield of ATP. This stage takes place entirely in the mitochondria. Acetyl coenzyme A is channeled into the Krebs cycle where the acetyl group is completely oxidized to carbon dioxide. Electrons released from the acetyl groups are transferred to special carriers that pass them to electron acceptor compounds in the electron transport chain. At the end of the chain the electrons (and the protons accompanying them) are accepted by molecular oxygen to form water.

Glycolysis

We begin our journey through the stages of respiration with glycolysis, a nearly universal pathway in living organisms that converts glucose into pyruvic acid. In a series of reactions, glucose and other 6-carbon monosaccharides are split into 3-carbon fragments, **pyruvic acid** (Figure 2-30). A single oxidation occurs during glycolysis, and each molecule of glucose yields two molecules of ATP. In this pathway the carbohydrate molecule is phosphorylated twice by ATP, first to glucose-6-phosphate (not shown in Figure 2-30) and then to fructose-1,6-diphosphate. The fuel has now been "primed" with phosphate groups in these preparatory reactions and is sufficiently reactive to enable subsequent reactions to pro-

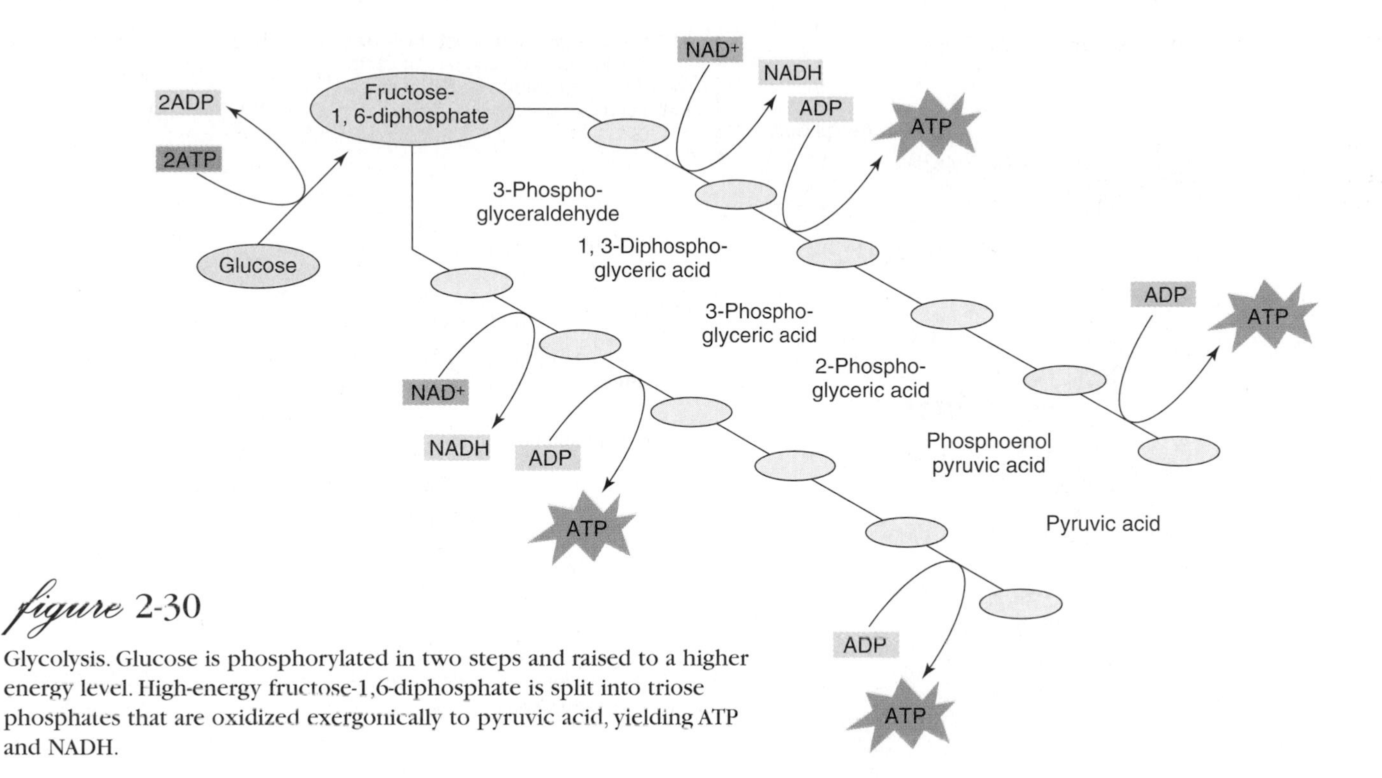

figure 2-30

Glycolysis. Glucose is phosphorylated in two steps and raised to a higher energy level. High-energy fructose-1,6-diphosphate is split into triose phosphates that are oxidized exergonically to pyruvic acid, yielding ATP and NADH.

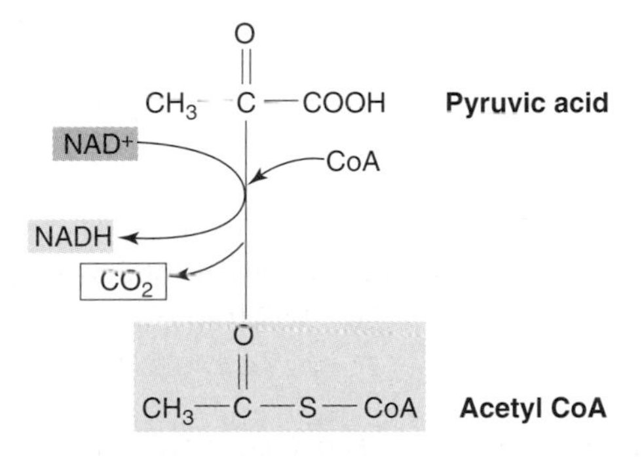

figure 2-31

Formation of acetyl coenzyme A from pyruvic acid.

ceed. This is a kind of deficit financing that is required for an ultimate energy return many times greater than the original energy investment.

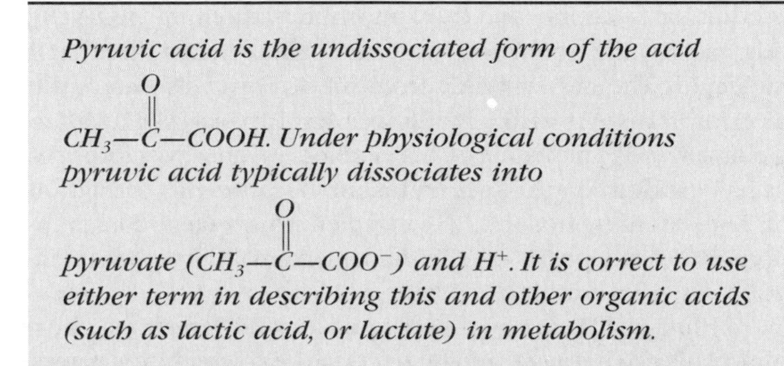

Pyruvic acid is the undissociated form of the acid $CH_3—\overset{\overset{\displaystyle O}{\|}}{C}—COOH$*. Under physiological conditions pyruvic acid typically dissociates into pyruvate* ($CH_3—\overset{\overset{\displaystyle O}{\|}}{C}—COO^-$) *and* H^+*. It is correct to use either term in describing this and other organic acids (such as lactic acid, or lactate) in metabolism.*

Fructose-1, 6-diphosphate next cleaves into two 3-carbon sugars, which undergo an oxidation, with the electrons being accepted by nicotinamide adenine dinucleotide (NAD), a derivative of the vitamin niacin. NAD serves as a carrier molecule to convey high-energy electrons to the final electron transport chain, where ATP will be produced.

The two 3-carbon sugars next undergo several reactions, ending with the formation of two molecules of pyruvic acid (Figure 2-30). In two of these steps, a molecule of ATP is produced. In other words, each 3-carbon sugar yields two ATP molecules, and since there are two 3-carbon sugars, four ATP molecules are generated. Recalling that two ATP molecules were used to prime the glucose initially, the net yield up to this point is two ATP molecules.

Acetyl Coenzyme A: Strategic Intermediate in Respiration

In aerobic metabolism the two molecules of pyruvic acid formed during glycolysis enter the mitochondrion. There, each molecule is oxidized, and one of the carbons is released as carbon dioxide (Figure 2-31). The 2-carbon residue condenses with **coenzyme A** to form **acetyl coenzyme A** (acetyl-CoA).

Acetyl coenzyme A is a critically important compound. Its oxidation in the Krebs cycle (below) provides energized electrons to generate ATP, and it is a crucial intermediate in lipid metabolism (p. 50).

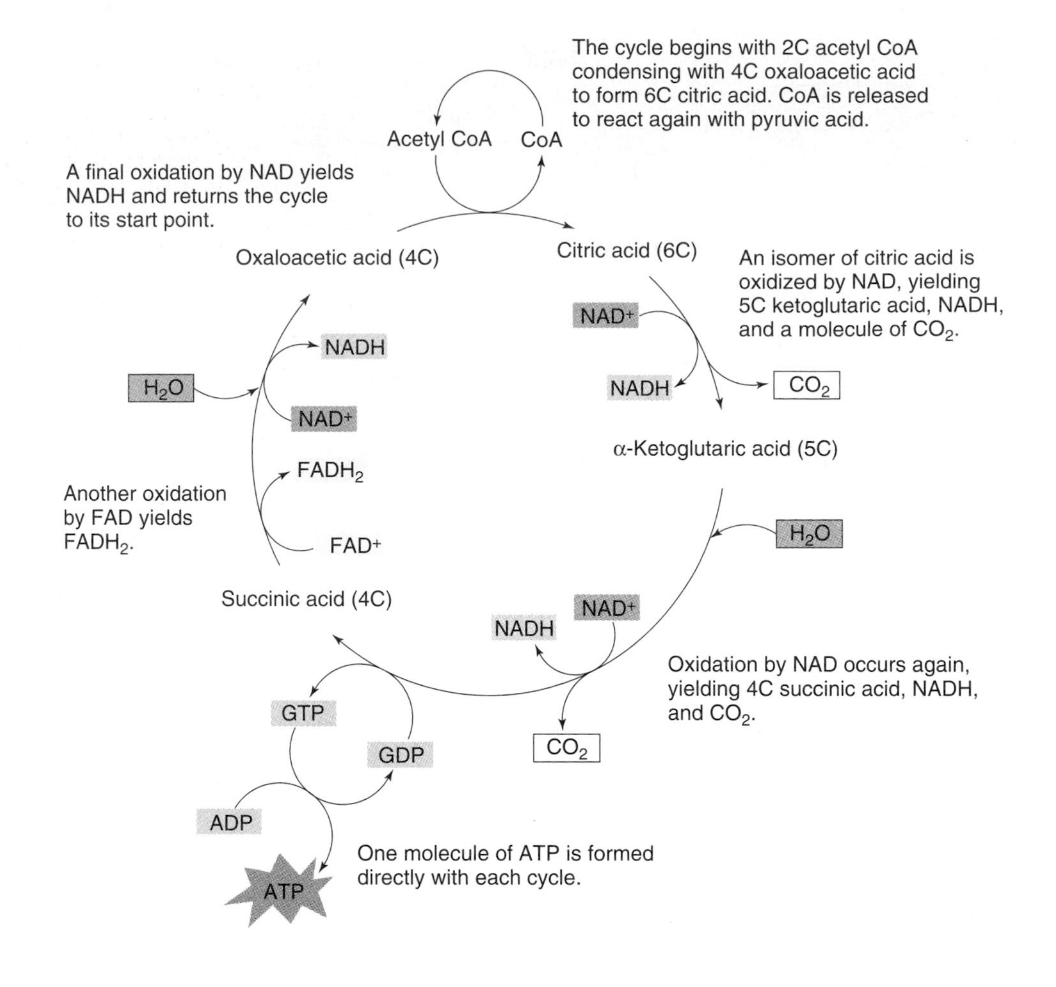

figure 2-32

Krebs cycle in outline form showing the production of three molecules of reduced NAD, one molecule of reduced FAD, and two molecules of carbon dioxide. The molecules of NADH and $FADH_2$ will yield 11 molecules of ATP when oxidized in the electron transport system.

Krebs Cycle: Oxidation of Acetyl Coenzyme A

The degradation (oxidation) of the 2-carbon acetyl group of acetyl coenzyme A occurs in a cyclic sequence called the **Krebs cycle** (also called the tricarboxylic acid cycle [TCA cycle] and citric acid cycle) (Figure 2-32). The acetyl coenzyme A condenses with a 4-carbon acid (oxaloacetic acid), releasing the coenzyme A to react again with pyruvic acid. Through a series of reactions the two carbons from the acetyl group are released as carbon dioxide, and the oxaloacetic acid is regenerated. Electrons in the oxidations transfer to NAD^+ and to FAD (flavine adenine dinucleotide, another electron acceptor), and a pyrophosphate bond is generated in the form of guanosine triphosphate (GTP). This high-energy phosphate is readily transferred to ADP to form ATP.

Electron Transport Chain

The transfer of electrons from reduced NAD and FAD to the final electron acceptor, molecular oxygen, is accomplished in an elaborate electron transport chain embedded in the inner membrane of the mitochondria (Figure 2-33, see also p. 29). Each carrier molecule in the chain (labelled I to IV in Figure 2-33) is a large protein-based complex that accepts and releases electrons at lower energy levels than the carrier preceding it in the chain. As electrons pass from one carrier molecule to the next, free energy is released. Some of this energy is used to drive the synthesis of ATP by setting up a H^+ gradient across the mitochondrial membrane. At three points along the electron transport system, ATP production occurs by the phosphorylation of ADP. By this means, the oxidation of one NADH yields three ATP molecules. The reduced FAD from the Krebs cycle enters the electron transport chain at a lower level than NADH and so yields two ATP molecules. This method of energy capture is called **oxidative phosphorylation** because the formation of high-energy phosphate is coupled to oxygen consumption, and this depends on the demand for ATP by other metabolic activities within the cell.

How is ATP actually generated during oxidative phosphorylation? The most widely accepted explanation is a pro-

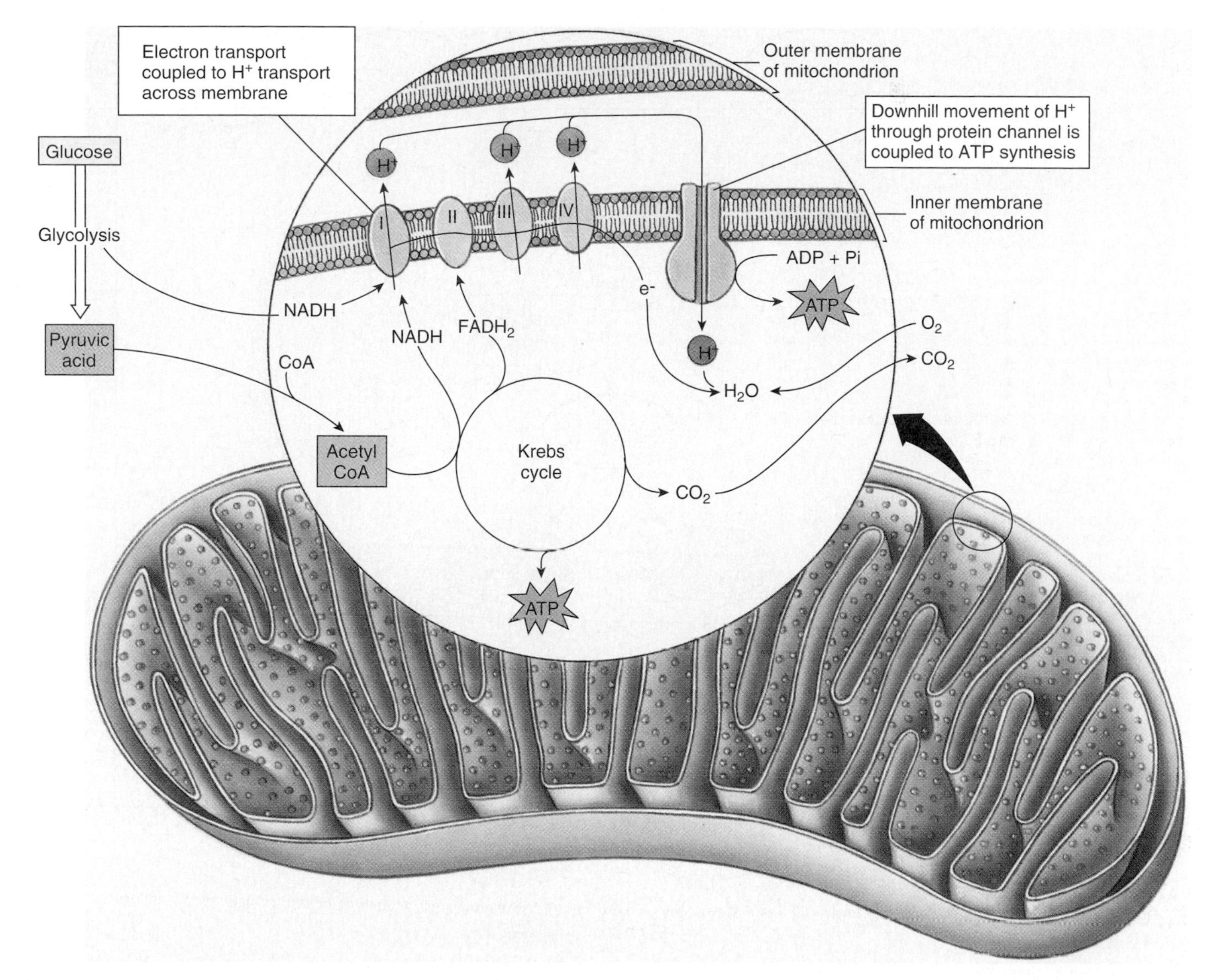

figure 2-33

Oxidative phosphorylation. Most of the ATP in living organisms is produced in the electron transport chain. Electrons removed from fuel molecules in cellular oxidations (glycolysis and the Krebs cycle) flow through the electron transport chain, the major components of which are four protein complexes (I, II, III, and IV). Electron energy is tapped by the major complexes and used to push H^+ outward across the inner mitochondrial membrane. The H^+ gradient created drives H^+ inward through proton channels, which couple H^+ movement to ATP synthesis.

cess called chemiosmotic coupling* (Figure 2-33). According to this model, as the electrons contributed by NADH and $FADH_2$ are carried down the electron transport chain, they activate proton pumping channels which pump protons (hydrogen ions) outward and into the space between the two mitochondrial membranes. This causes the proton concentration outside to rise, producing a diffusion pressure that drives the protons back into the mitochondrion through special proton channels. These channels are ATP-forming protein complexes that use the inward passage of protons to induce the formation of ATP. Exactly how proton movement is coupled to ATP synthesis is not yet understood.

*Hinkle, P., and R. McCarty. 1978. How cells make ATP. Sci. Am. **238:**104–123 (Mar.).

Efficiency of Oxidative Phosphorylation

We are now in a position to calculate the ATP yield from the complete oxidation of glucose (Figure 2-34). The overall reaction is:

$$\text{Glucose} + 2\,\text{ATP} + 36\,\text{ADP} + 36\,\text{P} + 6\,O_2 \rightarrow 6\,CO_2 + 2\,\text{ADP} + 36\,\text{ATP} + 6\,H_2O$$

ATP has been generated at several points along the way (Table 2-2). The cytoplasmic NADH generated in glycolysis requires a

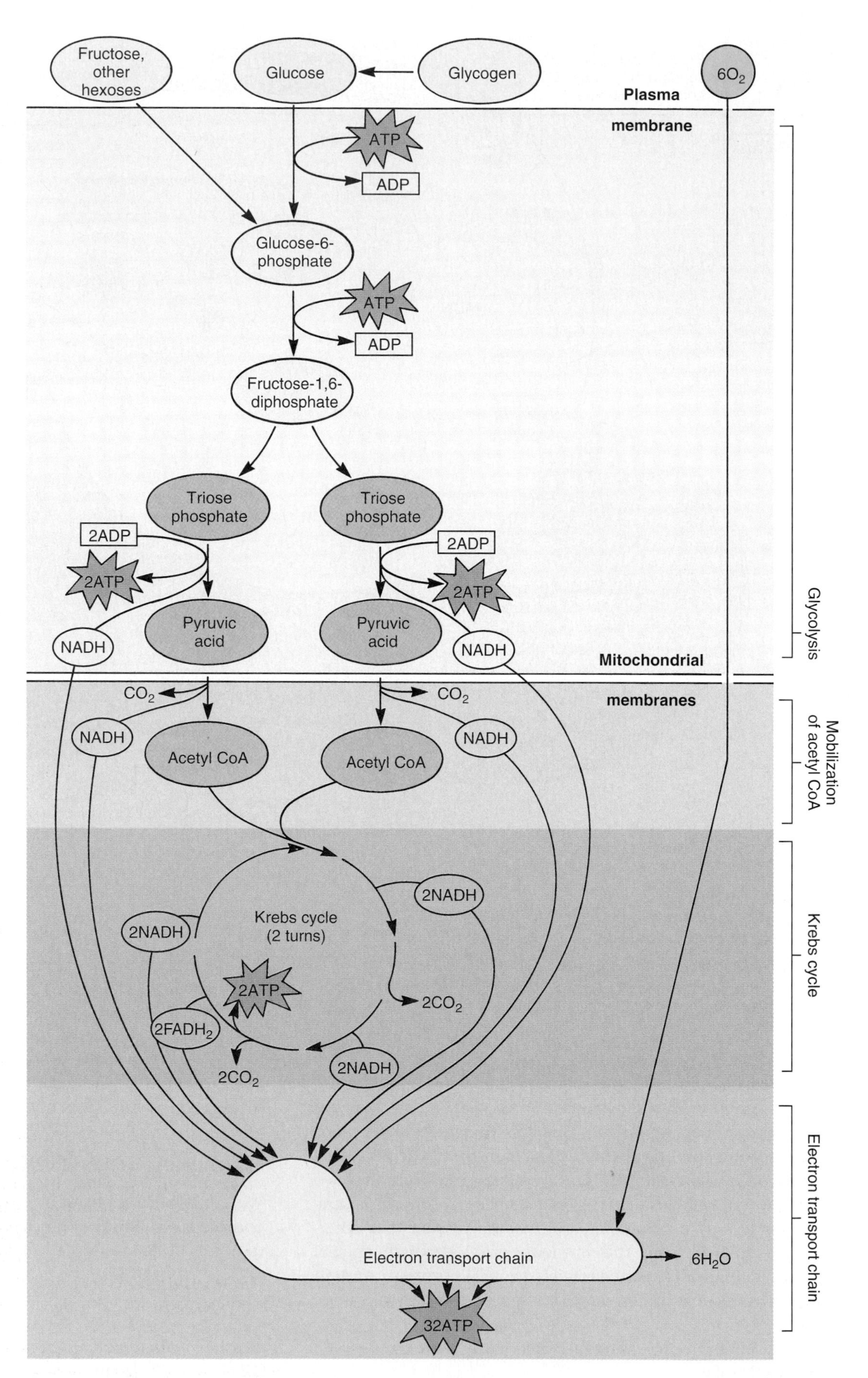

figure 2-34

Pathway for oxidation of glucose and other carbohydrates. Glucose is degraded to pyruvate by cytoplasmic enzymes (glycolytic pathway). Acetyl-CoA is formed from pyruvate and feeds into the Krebs cycle. An acetyl group (two carbons) is oxidized to two molecules of carbon dioxide with each turn of the cycle. Pairs of electrons (2H) are removed from the carbon skeleton of the substrate at several points in the pathway and are carried by oxidizing agents (NADH or $FADH_2$, not shown) to the electron transport chain where 32 molecules of ATP are generated. The glycolytic pathway generates also four molecules of ATP by substrate phosphorylation, yielding a total of 38 molecules of ATP (36 molecules net) per glucose molecule. Molecular oxygen is involved only at the very end of the pathway.

table 2-2

Calculation of Total ATP Molecules Generated in Respiration

ATP generated	Source
4	Directly in glycolysis
2	As GTP in citric acid cycle
4	From NADH in glycolysis
6	From NADH produced in pyruvic acid to acetyl coenzyme A reaction
4	From reduced FAD in citric acid cycle
18	From NADH produced in citric acid cycle
38 Total	
−2	Used in priming reactions in glycolysis
36 Net	

molecule of ATP to fuel transport of each molecule of NADH into the mitochondrion; therefore, each NADH from glycolysis results in only 2 ATP (total of 4), compared to the 3 ATP per NADH (total of 6) formed within the mitochondria. Accounting for the 2 ATP used in the priming reactions in glycolysis, the net yield may be as high as 36 molecules of ATP per molecule of glucose. The overall efficiency of aerobic oxidation of glucose is about 38%, comparing very favorably with human-designed energy conversion systems, which seldom exceed 5% to 10% efficiency.

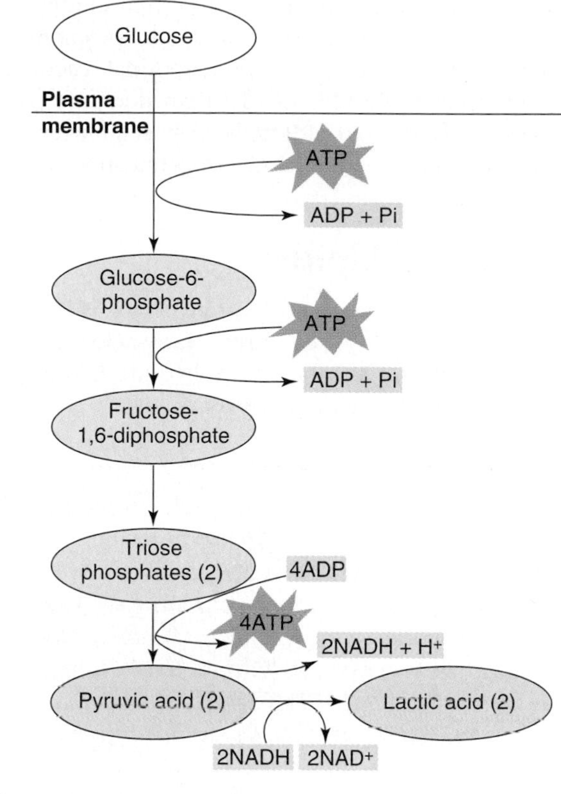

figure 2-35

Anaerobic glycolysis, a process that proceeds in the absence of oxygen. Glucose is broken down to two molecules of pyruvic acid, generating four molecules of ATP and yielding two, since two molecules of ATP are used to produce fructose-1,6-diphosphate. Pyruvic acid, the final electron acceptor for the hydrogen atoms and electrons released during pyruvic acid formation, is converted to lactic acid.

Anaerobic Glycolysis: Generating ATP without Oxygen

Up to this point we have been describing aerobic cellular respiration. We will now consider how animals generate ATP without oxygen, that is, anaerobically.

Under anaerobic conditions, glucose and other 6-carbon sugars are first broken down stepwise to a pair of 3-carbon pyruvic acid molecules, yielding two molecules of ATP and four atoms of hydrogen (four reducing equivalents, represented by 2 NADH + H^+). But in the absence of molecular oxygen, further oxidation of pyruvic acid cannot occur because the Krebs cycle and electron transport system cannot operate and cannot, therefore, provide a mechanism for reoxidation of the NADH produced in glycolysis. The problem is neatly solved in most animal cells by reducing pyruvic acid to lactic acid (Figure 2-35). Pyruvic acid becomes the final electron acceptor and lactic acid the end product of anaerobic glycolysis. In **alcoholic fermentation** (as in yeast, for example) the steps are identical to glycolysis down to pyruvic acid. One of its carbons is then released as carbon dioxide, and the resulting 2-carbon compound is reduced to ethanol, thus regenerating the NAD.

Anaerobic glycolysis is only one-eighteenth as efficient as the complete oxidation of glucose to carbon dioxide and water, but its key virtue is that it provides *some* high-energy phosphate in situations where oxygen is absent or in short supply. Many microorganisms live in places where oxygen is severely depleted, such as waterlogged soil, in mud of lake or sea bottom, or within a decaying carcass. Vertebrate skeletal muscle may rely heavily on glycolysis during short bursts of activity when contraction is too rapid and too powerful to be sustained by oxidative phosphorylation. The lactic acid that accumulates in the muscle diffuses out into the blood and is carried to the liver where it is metabolized.

Some animals rely heavily on anaerobic glycolysis during normal activities. For example, diving birds and mammals fall back on glycolysis almost entirely to give them the energy needed to sustain a long dive. Salmon would never reach their spawning grounds were it not for anaerobic glycolysis providing almost all the ATP used in the powerful muscular bursts needed to carry them up rapids and falls. Many parasitic animals have dispensed with oxidative phosphorylation entirely. They secrete relatively reduced end products of their energy metabolism, such as succinic acid, acetic acid, and propionic acid. These compounds are produced in mitochondrial reactions that derive several more molecules of ATP than does the pathway from glycolysis to lactic acid, although these sequences are still far less efficient than the classical electron transport system.

Metabolism of Lipids

The first step in the breakdown of a triglyceride is the hydrolysis of glycerol from the three fatty-acid molecules (Figure 2-36). Glycerol, a 3-carbon carbohydrate, is phosphorylated and enters the glycolytic pathway.

Glycogen reserves (mainly in the liver) provide about one day's supply of glucose. Lipids and proteins are used when the carbohydrate reserves are diminishing. Stored fats are the greatest reserve fuel in the body. Most of the usable fat resides in adipose tissue that is composed of specialized cells packed with globules of triglycerides. Adipose tissue is widely distributed in the abdominal cavity, in muscles, around deep blood vessels, and especially under the skin. Women average about 30% more fat than men, and this is responsible in no small measure for the curved contours of the female figure. However, its aesthetic contribution is strictly subsidiary to its principal function as an internal fuel depot. Indeed, humans can only too easily deposit large quantities of fat, generating personal unhappiness and hazards to health.

The physiological and psychological aspects of obesity are now being investigated by many researchers. There is increasing evidence that body fat deposition is regulated by a feeding control center located in the lateral and ventral regions of the hypothalamus, an area in the floor of the forebrain. The set point of this regulator determines the normal weight for the individual, which may be rather persistently maintained above or below what is considered normal for the human population. Thus, obesity is not always due to overindulgence and lack of self-control, despite popular notions to the contrary.

The remainder of the triglyceride molecule consists of fatty acids. One of the abundant naturally occurring fatty acids is **stearic acid.**

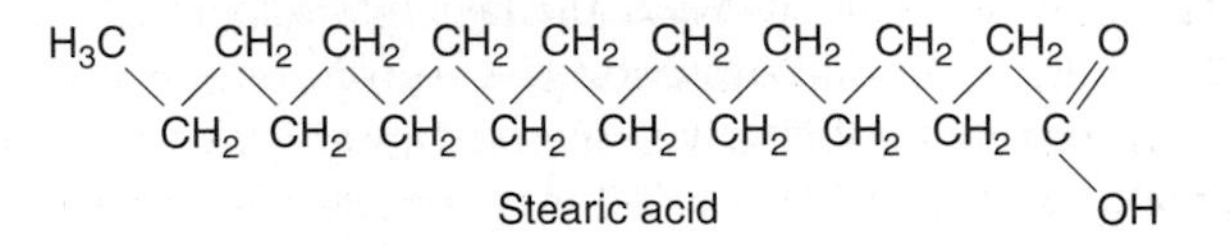

The long hydrocarbon chain of a fatty acid is sliced up by oxidation, two carbons at a time; these are released from the end of the molecule as acetyl coenzyme A. Although two high-energy phosphate bonds are required to prime each 2-carbon fragment, energy is derived both from the reduction of NAD^+ and FAD in the oxidations and from the acetyl group as it is degraded in the Krebs cycle. It can be calculated that the complete oxidation of 18-carbon stearic acid will net 146 ATP molecules. By comparison, three molecules of glucose (also totaling 18 carbons) yield 108 ATP molecules. Since there are three fatty acids in each triglyceride molecule, a total of 440 ATP molecules is formed. An additional 22 molecules of ATP are generated in the breakdown of glycerol, giving a grand total of 462 molecules of ATP. Little wonder that fat is considered the king of animal fuels!

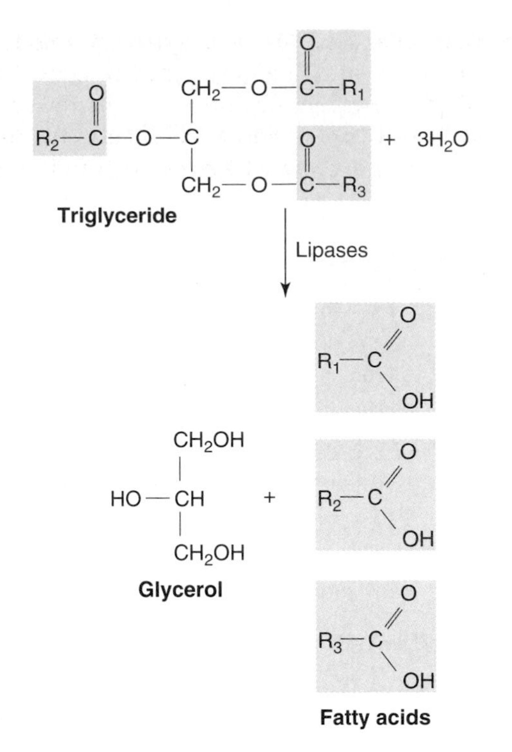

figure 2-36

Hydrolysis of a triglyceride (neutral fat) by intracellular lipase. The R groups of each fatty acid represent a hydrocarbon chain.

Fats are more concentrated fuels than carbohydrates because fats are almost pure hydrocarbons; they contain more hydrogen per carbon atom than sugars do, and it is the energized electrons of hydrogen that generate high-energy bonds, when they are carried through the mitochondrial electron transport system.

Fat stores are derived principally from surplus fats and carbohydrates in the diet. Acetyl coenzyme A is the source of carbon atoms used to build fatty acids. Since all major classes of organic molecules (carbohydrates, fats, and proteins) can be degraded to acetyl coenzyme A, all can be converted into stored fat. The biosynthetic pathway for fatty acids resembles a reversal of the catabolic pathway already described but requires an entirely different set of enzymes. From acetyl coenzyme A, the fatty-acid chain is assembled two carbons at a time. Because fatty acids release energy when they are oxidized, they obviously require an input of energy for their synthesis. This is provided principally by electron energy from glucose degradation. Thus the total ATP derived from oxidation of a molecule of triglyceride is not as great as previously calculated, because varying amounts of energy are required for synthesis and storage.

Metabolism of Proteins

Since proteins are composed of amino acids, 20 kinds in all (p.10), the central topic of our consideration is amino-acid metabolism. Amino-acid metabolism is complex. For one thing, each of the 20 amino acids requires a separate pathway of biosynthesis and degradation. For another, amino acids are precursors to tissue proteins, enzymes, nucleotide bases of nucleic acids, and other nitrogenous constituents that form the very fabric of the cell. The central purpose of carbohydrate and fat oxidation is to provide energy needed to construct and maintain these vital macromolecules.

Let us begin with the **amino-acid** pool in the blood and extracellular fluid from which the tissues draw their requirements. When animals eat proteins, these are digested in the gut, releasing the constituent amino acids, which are then absorbed (Figure 2-37). Tissue proteins also are hydrolyzed during normal growth, repair, and tissue restructuring; their amino acids join those derived from protein foodstuffs to enter the amino-acid pool. A portion of the amino-acid pool is used to rebuild tissue proteins, but most animals ingest a protein surplus. Such amino acids are not excreted as such in any significant amounts; they must be disposed of in some way. In fact, amino acids can be and are metabolized through oxidative pathways to yield high-energy phosphate. In short, excess proteins serve as fuel as do carbohydrates and fats. Their importance as fuel obviously depends on the nature of the diet. In carnivores that ingest a diet of almost pure protein and fat, nearly half of their high-energy phosphate is derived from amino-acid oxidation of amino acids.

Before entering the fuel depot, nitrogen must be removed from the amino-acid molecule. This can be by **deamination** (the amino group splits off to form ammonia and a keto acid) or by **transamination** (the amino group is transferred to a keto acid to yield a new amino acid). Thus degradation of amino acids yields two main products, ammonia and carbon skeletons, which are handled in different ways.

Once the nitrogen atoms are removed, the carbon skeletons of amino acids can be completely oxidized, usually by way of pyruvic acid or acetic acid. These residues then enter regular routes of carbohydrate and fat metabolism.

The other product of amino-acid degradation is ammonia. Ammonia is highly toxic, mainly because it reacts with α-ketoglutaric acid to form glutamic acid (an amino acid). Any accumulation of ammonia effectively removes α-ketoglutarate from the Krebs cycle (see Figure 2-32) and inhibits respiration.

Disposal of ammonia offers little problem to aquatic animals because it is soluble and readily diffuses into the surrounding medium through the respiratory surfaces. Terrestrial forms cannot get rid of ammonia so conveniently and must detoxify it by converting it to a relatively nontoxic compound.

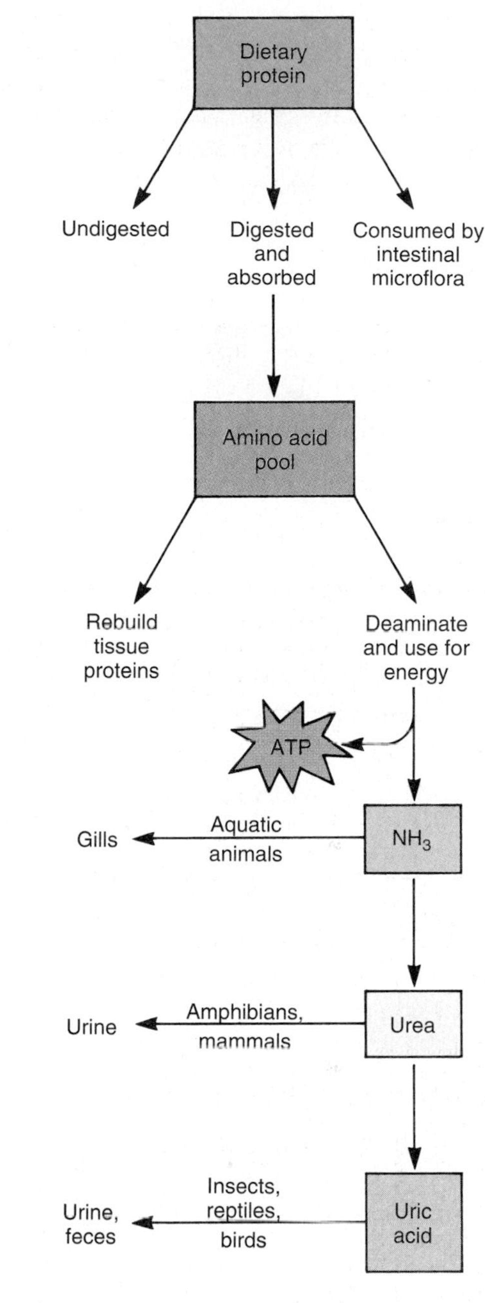

figure 2-37

Fate of dietary protein.

The two principal compounds formed are **urea** and **uric acid,** although a variety of other detoxified forms of ammonia are excreted by different animals. Among vertebrates, amphibians and especially mammals produce urea. Reptiles and birds, as well as many terrestrial invertebrates, produce uric acid.

The key feature that seems to determine the choice of nitrogenous waste is the availability of water in the environment. When water is abundant, the chief nitrogenous waste is ammonia. When water is restricted, it is urea. And for animals living in truly arid habitats, it is uric acid. Uric acid is highly insoluble and easily precipitates from solution, allowing its

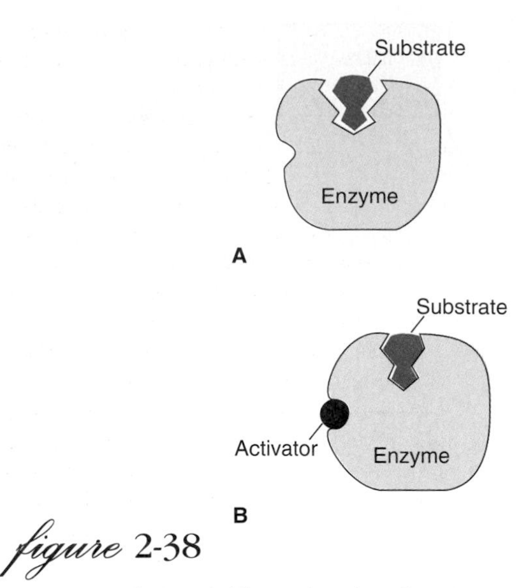

figure 2-38

Enzyme regulation. **A,** The active site of an enzyme may only loosely fit its substrate in the absence of an activator. **B,** With the regulatory site of the enzyme occupied by an activator, the enzyme binds the substrate, and the site becomes catalytically active.

removal in solid form. The embryos of birds and reptiles benefit greatly from excretion of nitrogenous waste as uric acid because the waste cannot be eliminated through their shells. During embryonic development, the harmless, solid uric acid is retained in one of the extraembryonic membranes. When the hatchling emerges into its new world, it discards the accumulated uric acid, along with the shell and membranes that supported development.

Management of Metabolism

The complex pattern of enzyme reactions that constitutes metabolism cannot be explained entirely in terms of physicochemical laws or chance happenings. Although some enzymes do indeed "flow with the tide," the activity of others is rigidly controlled. In the former case, suppose the purpose of an enzyme is to convert A to B. If B is removed by conversion into another compound, the enzyme will tend to restore the original ratio of B to A. Since many enzymes act reversibly, this can result, according to the metabolic situation prevailing, in synthesis or degradation. For example, an excess of an intermediate in the Krebs cycle would result in its contribution to glycogen synthesis; a depletion of such a metabolite would lead to glycogen breakdown. This automatic compensation (equilibration) is not, however, sufficient to explain all that actually takes place in an organism, as for example, what happens at branches in a metabolic pathway.

Mechanisms exist for critically regulating enzymes in both *quantity* and *activity.* Enzyme induction in bacteria is an example of quantity regulation. The genes leading to synthesis of the enzyme are switched on or off, depending on the presence or absence of a substrate molecule. In this way the *amount* of an enzyme can be controlled. It is a relatively slow process.

Mechanisms that alter activity of enzymes can quickly and finely adjust metabolic pathways to changing conditions in a cell. The presence or increase in concentration of some molecules can alter the shape (conformation) of particular enzymes, thus activating or inhibiting the enzyme (Figure 2-38). For example, phosphofructokinase, which catalyzes the phosphorylation of glucose-6-phosphate to fructose-1,6-diphosphate (Figure 2-35), is inhibited by high concentrations of ATP or citric acid. Their presence means that sufficient precursors have reached the Krebs cycle and additional glucose is not needed.

As well as being subject to alteration in physical shape, some enzymes exist in both an active and an inactive form, often by addition or deletion of a phosphate group (p. 34). Enzymes that degrade glycogen (phosphorylase) and synthesize it (synthase) are examples. Conditions that activate the phosphorylase tend to inactivate the synthase and vice versa.

Many cases of enzyme regulation are known, but these selected examples must suffice to illustrate the importance of enzyme regulation in the integration of metabolism.

Summary

All living organisms are composed of one or more cells, which are the basic structural and functional units of life. Cells are studied by light microscopes; transmission and scanning electron microscopes; and biochemical, physical, immunological, and molecular methods. Cells are surrounded by a cell membrane, and eukaryotic cells have a nucleus surrounded by a double membrane. Besides the chromatin, one or more nucleoli are usually found in the nucleus. The fluid-mosaic model of the membrane is a phospholipid bilayer with a mosaic of proteins as functional elements in the membrane. Outside the nuclear membrane is the cytoplasm, subdivided by a membranous network, the endoplasmic reticulum, which is often associated with ribosomes and probably functions in transport of materials within the cell. Among the organelles within the cell are the mitochondria, which contain the enzymes of oxidative energy metabolism. The Golgi complex functions in storage and packaging of proteins. Lysosomes, other membrane-bound vesicles, microfilaments, and microtubules are often found in the cytoplasm. Centrioles are short cylinders of nine triplet microtubules that help organize the mitotic spindle. Various specialized junctions occur between cells.

Substances can enter cells by diffusion, mediated transport, and endocytosis. Diffusion is the movement of molecules from an area of higher concentration to one of lower concentration, and osmosis is diffusion of water through a selectively permeable membrane due to osmotic pressure. Osmotic pressure is not the same as hydrostatic pressure but is defined in terms of equilibrium hydrostatic pressure. Solutes to which the membrane is not permeable require a carrier molecule to traverse it; carrier-mediated systems include facilitated transport (in the direction of a concentration gradient) and active transport (against a concentration gradient, which requires energy). Endocytosis includes bringing small molecules, ions, and macromolecules (potocytosis and receptor-mediated endocytosis) or particles (phagocytosis) into the cell.

The capacity to grow by cell multiplication is a fundamental characteristic of living systems. Ordinary somatic cells contain two of each kind of chromosome (hence are called diploid) and divide by mitosis. In mitosis the chromosomes replicate during interphase, and the replicated chromosomes (sister chromatids) are joined by a centromere. At the beginning of mitosis (prophase) the nuclear envelope disperses, and the chromosomes condense into recognizable bodies. At metaphase the replicated centrioles have moved to opposite poles of the cell, the spindle and asters of microtubules have formed, the chromosomes are on the median plane of the cell, and the kinetochores of each chromosome are attached to a spindle fiber. The centromeres divide, and one of each kind of chromosome is pulled toward the centriole by the attached spindle fiber (anaphase). At telophase the chromosomes are at the position of the new nucleus in each cell, and division of the cytoplasm (cytokinesis) begins. Mitosis itself is only a small part of the total cell cycle. In interphase, G_1, S, and G_2 periods are recognized, and the S period is the time when DNA is synthesized (the chromosomes are replicated). Division of some kinds of cells ceases, but other kinds of cells continue division throughout the life of an animal. Cells may become senescent and die or undergo genetically programmed death (apoptosis).

Enzymes are pure proteins or proteins associated with nonprotein cofactors that vastly accelerate reaction rates in living systems. An enzyme does this by temporarily binding its reactant (substrate) onto an active site in a highly specific fit. In this configuration, internal activation energy barriers are lowered enough to disrupt and split the substrate, and the enzyme is restricted to its original form.

Cells use the energy stored in chemical bonds of organic fuels by breaking the fuels down through a series of enzymatically controlled steps. This bond energy is transferred to ATP and packaged in the form of "high-energy" phosphate bonds. ATP is produced in cells as it is required to power various synthetic, secretory, or mechanical processes.

Glucose is an important source of energy for cells. In aerobic metabolism (respiration), the 6-carbon glucose is split into two 3-carbon molecules of pyruvate. Pyruvate is decarboxylated to form 2-carbon acetyl coenzyme A, a strategic intermediate that leads to the Krebs cycle. Acetyl coenzyme A can also be derived from fat breakdown. In the Krebs cycle, acetyl coenzyme A is oxidized in a series of reactions to carbon dioxide, yielding, in the course of the reactions, energized electrons that are passed to electron acceptor molecules (NAD and FAD). In the final stage, the energized electrons pass along an electron transport chain consisting of a series of electron carriers located in the inner membranes of the mitochondrion. ATP is generated at three points along the chain as the electrons pass from carrier to carrier and finally to oxygen. A net total of 36 molecules of ATP is generated from one molecule of glucose.

In the absence of oxygen (anaerobic glycolysis), glucose is degraded to two 3-carbon molecules of lactate, yielding two molecules of ATP. Although anaerobic glycolysis is vastly less efficient than respiration, it provides essential energy for muscle contraction when heavy expenditure of energy outstrips the oxygen-delivery system of an animal; it also is the only source for microorganisms living in oxygen-free environments.

Triglycerides (neutral fats) are especially rich depots of metabolic energy because the fatty acids of which they are composed are highly reduced and anhydrous. Fatty acids are degraded by sequential removal of 2-carbon units, which enter the Krebs cycle through acetyl-CoA.

Amino acids in excess of requirements for synthesis of proteins and other biomolecules are used as fuel. They are degraded by deamination or transamination to yield ammonia and carbon skeletons. The latter enter the Krebs cycle to be oxidized. Ammonia is a highly toxic waste product that aquatic animals quickly dispose of through respiratory surfaces. Terrestrial animals, however, convert ammonia into much less toxic compounds, urea or uric acid, for disposal.

The integration of metabolic pathways is finely regulated by mechanisms that control both the amount and activity of enzymes. The quantity of some enzymes is regulated by certain molecules that switch on or off enzyme synthesis in the nucleus. Enzyme activity may be altered by the presence or absence of metabolites that cause conformational changes in enzymes and thus improve or diminish their effectiveness as catalysts.

Review Questions

1. Explain the difference (in principle) between a light microscope and an electron microscope.
2. Give a one-sentence definition of each of the following: cell membrane, chromatin, nucleus, nucleolus, rough endoplasmic reticulum (rough ER), Golgi complex, lysosomes, mitochondria, microfilaments, microtubules, centrioles, basal body (kinetosome), tight junction, gap junction, microvilli.
3. You place some red blood cells in a solution and observe that they swell and burst. You place some cells in another solution, and they shrink and become wrinkled. Explain what has happened in each case.
4. Explain the difference between osmotic and hydrostatic pressure.
5. Distinguish between two kinds of mediated transport.
6. Distinguish between three kinds of endocytosis.
7. Define the following: chromosome, haploid, diploid, centromere, genome, mitosis, cytokinesis, syncytium, cyclins, apoptosis.
8. Name the stages of mitosis in order, and describe the behavior of the chromosomes at each stage.
9. When does the "S period" of a cell cycle occur, and what is happening at that period?
10. Briefly explain how enzymes are believed to work.
11. What happens in the formation of an enzyme-substrate complex that favors the disruption of substrate bonds?
12. What is an oxidation-reduction reaction and why are such reactions considered so important in cellular metabolism?
13. Why is aerobic metabolism more efficient than anaerobic metabolism?
14. With respect to glycolysis, answer the following questions. What molecule does the pathway begin with? What are the products of the pathway? How many molecules of ATP are generated (gross and net)? Where in the cell does it occur?
15. Answer the questions in number 14 with respect to the Krebs cycle.
16. Answer the questions in number 14 with respect to the electron transport chain.
17. What is the importance of anaerobic glycolysis?
18. Why is acetyl coenzyme A considered a "strategic intermediate" in respiration?
19. How are amino acids oxidized as energy sources?
20. Explain the relationship of ammonia, urea, and uric acid as nitrogenous wastes to the amount of water in an organism's environment.
21. Give three ways in which enzymes are regulated in cells.

Selected References

Anderson, R. G. W., B. A. Kamen, K. G. Rothberg, and S. W. Lacey. 1992. Potocytosis: sequestration and transport of small molecules by caveolae. Science **255:**410–413. *Describes the mechanism of cell internalization of small molecules.*

Bretscher, M. S. 1985. The molecules of the cell membrane. Sci. Am. **253:**100–108 (Oct.). *Good presentation of molecular structure of cell membranes, junctions, and mechanism of receptor-mediated endocytosis.*

Bretscher, M. S., and S. Munro. 1993. Cholesterol and the Golgi apparatus. Science **261:**1280–1281. *Good description of behavior of cholesterol in the cell and how it is concentrated in cell membranes by the Golgi.*

Glover, D. M., C. Gonzalez, and J. W. Raff. 1993. The centrosome. Sci. Am. **268:**62–68 (June). *The centrosome of animal cells serves as an organizing center for the cytoskeleton.*

Hartwell, L. H., and M. B. Kastan. 1994. Cell cycle control and cancer. Science **266:**1821–1828. *Genetic changes in the coordination of cyclin-dependent kinases, checkpoint controls, and repair pathways can lead to uncontrolled cell division.*

Koshland, D. R., Jr. 1989. The cell cycle. Science **246:**545. *An editorial introducing an issue with five major reviews on various aspects of the cell cycle and mitosis.*

Lodish, H., D. Baltimore, A. Berk, S. L. Zipursky, P. Matsudira, and J. Darnell. 1995. Molecular cell biology, ed. 2. New York, Scientific American Books, Inc. *Up-to-date, thorough, and readable. Includes both cell biology and molecular biology. Advanced, but highly recommended.*

McIntosh, J. R., and K. L. McDonald. 1989. The mitotic spindle. Sci. Am. **261:**48–56 (Oct.). *Current knowledge and hypotheses on the function of the microtubules of mitosis.*

Murray, A., and T. Hunt. 1993. The cell cycle. An introduction. New York, Oxford University Press. *A good review of our present understanding of the cell cycle.*

Murray, A. W., and M. W. Kirschner. 1991. What controls the cell cycle. Sci. Am. **264:**56–63 (Mar.). *Presents the evidence for the fascinating role of cdc2 kinase and cyclin in the cell cycle.*

Sheeler, P., and D. E. Bianchi. 1987. Cell biology: structure, biochemistry, and function, ed. 2. New York, John Wiley & Sons, Inc. *Well-written, well-illustrated introductory cell biology textbook.*

Sundell, C. L., and R. H. Singer. 1991. Requirement of microfilaments in sorting of actin messenger RNA. Science **253:**1275–1277. *Specific messenger RNA is found in particular cellular locations, and it is transported there from the nucleus by microfilaments, not microtubules or intermediate filaments.*

Wolfe, S. L. 1993. Molecular and cellular biology. Belmont, California, Wadsworth Publishing Company. *Chapter 9 provides a well-organized explanation of energy metabolism.*

Genetic Basis of Evolution

chapter | three

A Code for All Life

The principle of hereditary transmission is a central tenet of life on earth: all organisms inherit a structural and functional organization from their progenitors. What is inherited by an offspring is not necessarily an exact copy of the parent but a set of coded instructions that gives rise to a certain expressed organization. These instructions are in the form of genes, the fundamental units of inheritance. One of the great triumphs of modern biology was the discovery in 1953 by James Watson and Francis Crick of the nature of the coded instructions in genes. This was followed by the discovery of the way in which the code is translated into the expression of characteristics. The genetic material (deoxyribonucleic acid, DNA) is composed of nitrogenous bases arranged on a backbone of sugar-phosphate units. The genetic code lies in the linear order or sequence of bases in the DNA strand.

Because the DNA molecules replicate themselves in their passage from generation to generation, genetic variations can persist once they have happened. Such molecular alterations, called mutations, are the ultimate source of biological variation and the raw material of evolution.

A basic principle of modern evolutionary theory is that organisms attain their diversity of form, function, and behavior through hereditary modifications of preexisting lines of ancestors. It means that all known lineages of plants and animals are related by descent from common ancestral groups.

Heredity establishes the continuity of life forms. Although offspring and parents in a particular generation may look different, there is nonetheless a basic sameness that runs from generation to generation for any species of plant or animal. In short, "like begets like." Yet children are not precise replicas of their parents. Some of their characteristics show resemblances to one or both parents, but they also demonstrate many traits not found in either parent. What is actually inherited by an offspring from its parents is a certain type of germinal organization **(genes)** that, under the influence of environmental factors, guides the orderly sequence of differentiation of the fertilized egg into a human being, bearing the unique physical characteristics as we see them. Each generation hands on to the next the instructions required for maintaining continuity of life.

The gene is the unit entity of inheritance, the germinal basis for every characteristic that appears in an organism. The study of what genes are and how they work is the science of genetics. It is the science that deals with the underlying causes of *resemblance,* as seen in the remarkable fidelity of reproduction, and of *variation,* which is the working material for organic evolution. Genetics has shown that all living forms use the same information storage, transfer, and translation system, and thus it has provided an explanation for both the stability of all life and its probable descent from a common ancestral form. This is one of the most important unifying concepts of biology.

Mendel's Investigations

The first person to formulate the cardinal principles of heredity was Gregor Johann Mendel (1822–1884), who was an Augustinian monk living in Brünn (Brno), Moravia. At that time Brünn was a part of Austria, but now it is in the Czech Republic. While conducting breeding experiments in a small monastery garden from 1856 to 1864, he examined with great care the progeny of many thousands of plants. He worked out in elegant simplicity the laws governing the transmission of characters from parent to offspring. His discoveries, published in 1866, were of great potential significance, coming just after Darwin's publication of *The Origin of Species.* Yet these discoveries remained unappreciated and forgotten until 1900—some 35 years after the completion of the work and 16 years after Mendel's death.

Mendel's classic observations were based on the garden pea because it had been produced in pure strains by gardeners over a long period of time by careful selection. For example, some varieties were definitely dwarf and others were tall. A second reason for selecting peas was that they were self-fertilizing, but also capable of cross-fertilization. To simplify his problem he chose single characters and characters that were sharply contrasted. Mere qualitative and intermediate characters were carefully avoided. Mendel selected pairs of contrasting characters, such as tall plants, dwarf plants, smooth seeds, and wrinkled seeds (Figure 3-1).

Mendel crossed plants having one of these characters with others having the contrasting character. He removed the stamens (male part, containing the pollen) from a flower to prevent self-fertilization and then placed on the stigma (female part) of this flower pollen from the flower of the plant that had the contrasting character. He also prevented the experimental flowers from being pollinated from other sources such as wind and insects. When the cross-fertilized flower bore seeds, he noted the kind of plants (hybrids) that were produced from the planted seeds. Subsequently he crossed these hybrids among themselves to see what would happen.

A giant stride in chromosomal genetics was made when the great American geneticist Thomas Hunt Morgan and his colleagues selected the fruit fly Drosophila melanogaster *for their studies (1910–1920). It was cheaply and easily reared in bottles in the laboratory, fed on a simple medium of bananas and yeast. Most important, it produced a new generation every 10 days, enabling Morgan to proceed at least 25 times more rapidly than with organisms that take a year to mature, such as garden peas. Morgan's work led to the mapping of genes on chromosomes and founded the discipline of cytogenetics.*

Mendel knew nothing of the cytological basis of heredity, since chromosomes and genes were unknown to him. Although we can admire Mendel's power of intellect in his discovery of the principles of inheritance without knowledge of chromosomes, the principles are certainly easier to understand if we first examine chromosomal behavior, especially in meiosis.

Chromosomal Basis of Inheritance

In sexually reproducing organisms, special **sex cells,** or **gametes** (ova and sperm), are responsible for providing the genetic information to the offspring. Scientific explanation of genetic principles required a study of germ cells and their behavior, which meant working backward from certain visible results of inheritance to the mechanism responsible for such results. The nuclei of sex cells, especially the chromosomes, were early suspected of furnishing the real answer to the mechanism. Chromosomes are apparently the only entities inherited in equal quantities from both parents to offspring.

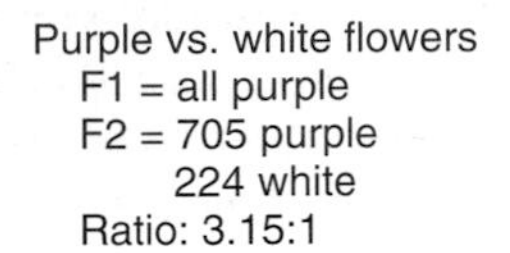

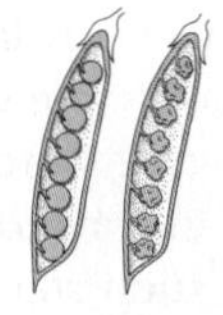

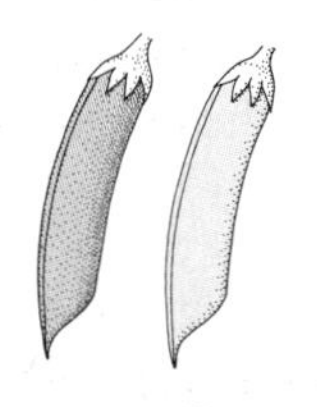

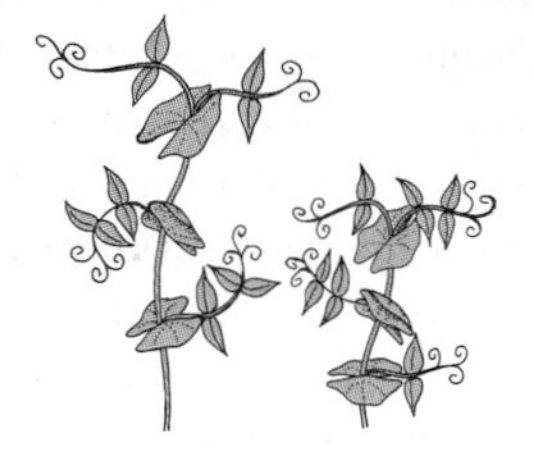

figure 3-1

Seven experiments of Gregor Mendel based on his postulates. These are the results of monohybrid crosses for first and second generations.

When Mendel's laws were rediscovered in 1900, their parallelism with the cytological behavior of the chromosomes was obvious. Later experiments showed that the chromosomes carried the hereditary material.

Meiosis: Maturation Division of Gametes

Every body cell contains *two* chromosomes bearing genes for the same set of characteristics, and the two members of each pair usually, but not always, have the same size and shape. The members of such a pair are called **homologous** chromosomes; one of each pair comes from the mother and the other from the father.

Thus each cell normally has two genes coding for a given trait, one on each of the homologs. These may be alternative forms of the same gene, and, if so, they are **allelic genes,** or **alleles.** Sometimes only one of the alleles has an effect on the organism, although both are present in each cell, and either may be passed to the progeny as a result of meiosis and subsequent fertilization.

> *Alleles are alternative forms of the same gene that have arisen by mutation of the DNA sequence. Like a baseball team with several pitchers, only one of whom can occupy the pitcher's mound at a time, only one allele can occupy a chromosomal locus. Alternative alleles for the locus may be on homologous chromosomes of a single individual, making that individual heterozygous for the gene in question. Numerous allelic forms of a gene may be found among different individuals in the population of the species.*

During an individual's growth, all the chromosomes of the mitotically dividing cells contain the double set of chromosomes. In the reproductive organs, the gametes (germ cells) are formed by a kind of maturation division, called meiosis, which *separates* the homologous pairs of chromosomes. If it were not for this reductional division, the union of ovum (egg) and sperm would produce an individual with twice as many chromosomes as the parents. Continuation of this process in just a few generations could yield astronomical numbers of chromosomes per cell.

Meiosis consists of *two* nuclear divisions in which the chromosomes divide only once (Figure 3-2). The result is that mature gametes have only *one* member of each homologous chromosome pair, or a **haploid** (n) number of chromosomes. In humans the zygotes and all body cells normally have the **diploid** number (2n), or 46 chromosomes; the gametes have the haploid number (n), or 23.

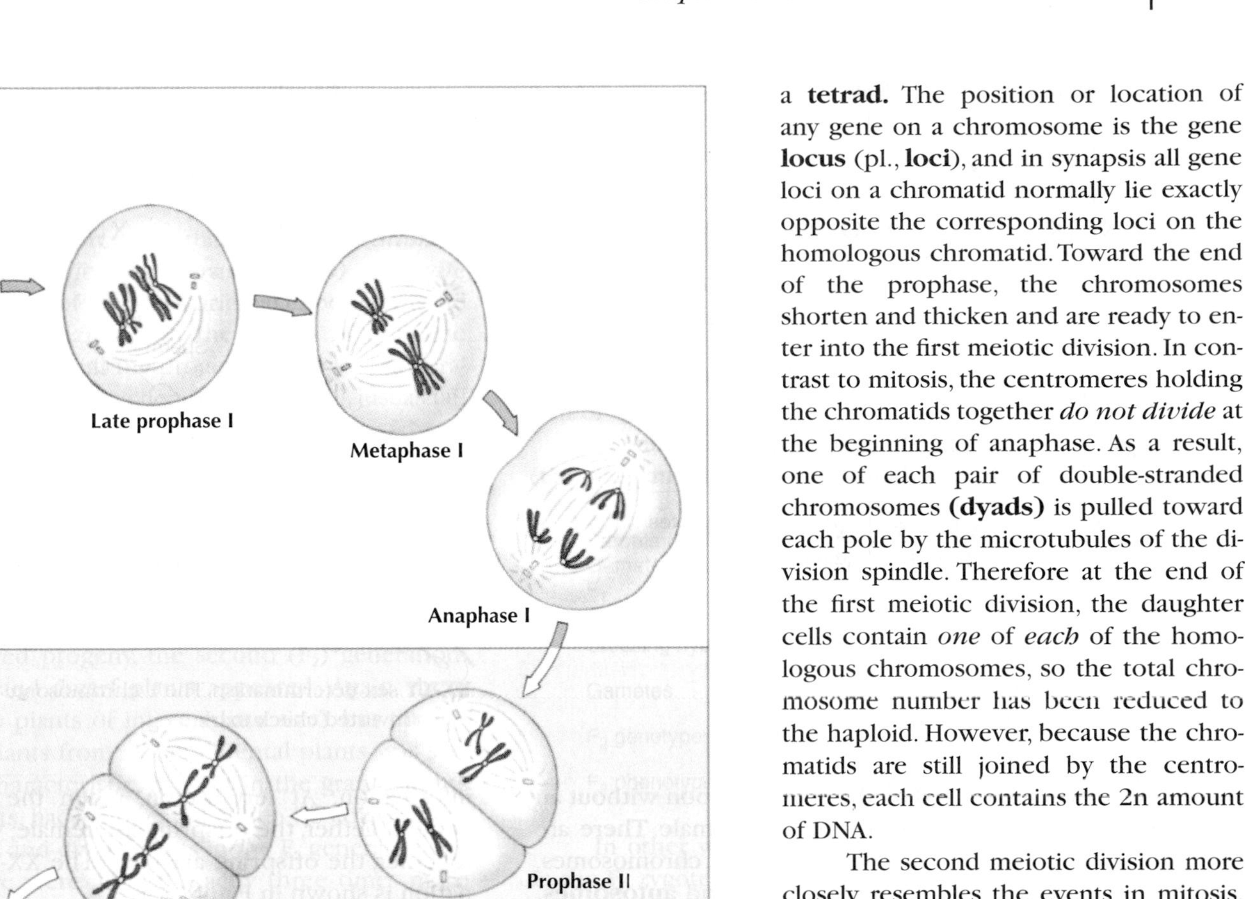

figure 3-2

Meiosis in a sex cell with two pairs of chromosomes.

Most of the unique features of meiosis occur during the prophase of the first meiotic division (Figure 3-2). The two members of each pair of homologous chromosomes come into side-by-side contact **(synapsis)** to form a **bivalent.** Each chromosome of the bivalent has already replicated to form two chromatids, each of which will become a new chromosome. The two chromatids are joined at one point, the centromere, so that each bivalent is made up of two pairs of chromatids, or *four* future chromosomes, and is thus called a **tetrad.** The position or location of any gene on a chromosome is the gene **locus** (pl., **loci**), and in synapsis all gene loci on a chromatid normally lie exactly opposite the corresponding loci on the homologous chromatid. Toward the end of the prophase, the chromosomes shorten and thicken and are ready to enter into the first meiotic division. In contrast to mitosis, the centromeres holding the chromatids together *do not divide* at the beginning of anaphase. As a result, one of each pair of double-stranded chromosomes **(dyads)** is pulled toward each pole by the microtubules of the division spindle. Therefore at the end of the first meiotic division, the daughter cells contain *one* of *each* of the homologous chromosomes, so the total chromosome number has been reduced to the haploid. However, because the chromatids are still joined by the centromeres, each cell contains the 2n amount of DNA.

The second meiotic division more closely resembles the events in mitosis. The dyads are split at the beginning of anaphase by division of the centromeres, and single-stranded chromosomes move toward each pole. Thus by the end of the second meiotic division, the cells have the haploid number of chromosomes and the n amount of DNA. Each chromatid of the original tetrad exists in a separate nucleus. Four products are formed, each containing one complete haploid set of chromosomes and only one allele of each gene.

Sex Determination

Before the importance of chromosomes in heredity was realized in the early 1900s, how gender was determined was totally unknown. The first really scientific clue to the determination of sex came in 1902 when C. McClung observed that bugs (Hemiptera) produced two kinds of sperm in approximately equal numbers. One kind contained among its regular set of chromosomes a so-called accessory chromosome that was lacking in the other kind of sperm. Since all the eggs of these species had the same number of haploid chromosomes, half the sperm would have the same number of chromosomes as the eggs, and half of them would have one chromosome less. When an egg was fertilized by a spermatozoon carrying the accessory (sex) chromosome, the resulting offspring

Multiple alleles arise through mutations at the gene locus over long periods of time. Any gene may mutate (p. 79) if given time and thus can give rise to slightly different alleles at the same locus.

Sex-Linked Inheritance

It is known that the inheritance of some characters depends on the sex of the parent carrying the gene and the sex of the offspring. An example is red-green color blindness in humans, in which red and green colors are indistinguishable to varying degrees. Color-blind men greatly outnumber color-blind women. When color blindness does appear in women, their fathers are color blind. Furthermore, if a woman with normal vision who is a carrier of color blindness (a **carrier** has the gene but is phenotypically normal) bears sons, half of them are likely to be color blind, regardless of whether the father had normal or affected vision. How are these observations explained?

The color-blindness defect is a recessive trait, the gene for which is carried on the X chromosome. It is visibly expressed either when both genes are defective in the female or when only one defective gene is present in the male. The inheritance pattern is shown in Figure 3-7. When the mother is a carrier and the father is normal, half of the sons but none of the daughters are color blind. However, if the father is color blind and the mother is a carrier, half of the sons *and* half of the daughters are color blind (on the average and in a large sample). It is easy to understand then why the defect is much more prevalent in males: a single sex-linked recessive gene in the male has a visible effect. What would be the outcome of a mating between a homozygous normal woman and a color-blind man?

Another example of a sex-linked character was discovered by Morgan in *Drosophila.* The normal eye color of this fly is red, but mutations for white eyes do occur. The genes for eye color are known to be carried in the X chromosome. If a white-eyed male and a red-eyed female are crossed, all the F_1 offspring have red eyes because this trait is dominant (Figure 3-8). If these F_1 offspring are interbred, all the females of F_2 have red eyes, half the males have red eyes, and the other half have white eyes. No white-eyed females are found in this generation; only the males have the recessive character (white eyes). The gene for being white eyed is recessive and should appear in a homozygous condition. However, since the male has only one X chromosome (the Y does not carry a gene for eye color), white eyes appear whenever the X chromosome carries the gene for this trait.

If the reciprocal cross is made in which the females are white eyed and the males red eyed, all the F_1 females are red eyed and all the males are white eyed (Figure 3-9). If these F_1 offspring are interbred, the F_2 generation shows equal numbers of red-eyed and white-eyed males and females.

Autosomal Linkage and Crossing Over

Linkage

Since Mendel's laws were rediscovered in 1900, it became apparent that, contrary to Mendel's second law, not all factors segregate independently. Indeed, many traits are inherited together. Since the number of chromosomes in any organism is relatively small compared to the number of traits, each chromosome must contain many genes. All genes present on a chromosome are said to be **linked.** Linkage simply means that the genes are on the same chromosome, and all genes present on homologous chromosomes belong to the same linkage groups. Therefore there should be as many linkage groups as there are chromosome pairs.

In *Drosophila,* in which this principle has been worked out most extensively, there are four linkage groups that correspond to the four pairs of chromosomes found in these fruit flies. Usually, small chromosomes have small linkage groups, and large chromosomes have large groups.

Crossing Over

Linkage, however, is usually not complete. If we perform an experiment in which animals such as *Drosophila* are crossed, we find that linked traits separate in some percentage of the offspring. Separation of genes located on the same chromosome occurs because of **crossing over.**

As described earlier, during the protracted prophase of the first meiotic division, chromatids of homologous chromosomes come to lie alongside each other, so that the locus of each gene is exactly opposite the same locus in the chromatid of the homologous chromosome. Sometimes the chromatids break and exchange equivalent portions; genes "cross over" from one chromosome to its homolog, and vice versa (Figure 3-10). Breaks and exchanges occur at corresponding points on nonsister chromatids. (Breaks and exchanges also occur between sister chromatids but usually have no genetic significance because sister chromatids are identical.) Crossing over then is a means for exchanging genes between homologous chromosomes and as such greatly increases the amount of genetic recombination. The frequency of crossing over varies with the species, but usually at least one and often several crossovers occur each time chromosomes pair in humans.

Because the frequency of recombination is proportional to the distance between loci, the comparative linear position of each locus can be determined. Genes located far apart on very large chromosomes may assort independently because the probability of a crossover occurring between them in each meisosis is close to 100%. Such genes are found to be carried on the same chromosome only because each one is genetically linked to additional loci located physically between them on the chromosome. Laborious genetic experiments over many years have produced gene maps that indicate the positions of more than 500 genes distributed on the four chromosomes of *Drosophila melanogaster.*

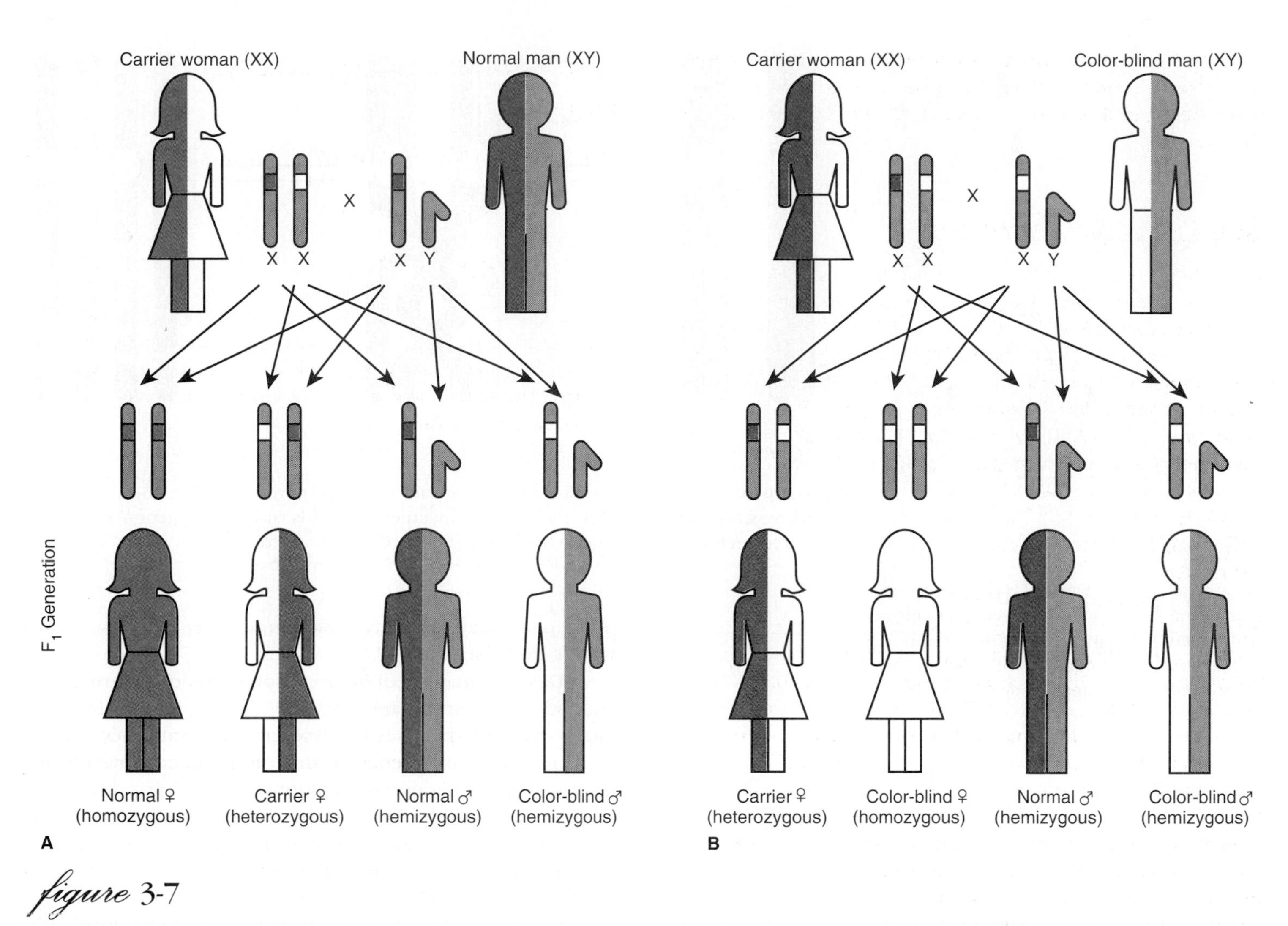

figure 3-7

Sex-linked inheritance of red-green color blindness in humans. (*Hemizygous* refers to a gene present in a single dose, such as genes on the X chromosomes in males of humans.) **A,** Carrier mother and normal father produce color blindness in one-half of their sons but in none of their daughters. **B,** Half of both sons and daughters of carrier mother and color-blind father are color blind.

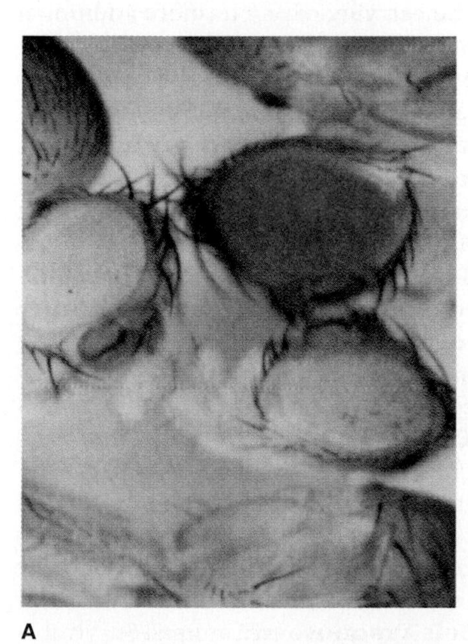

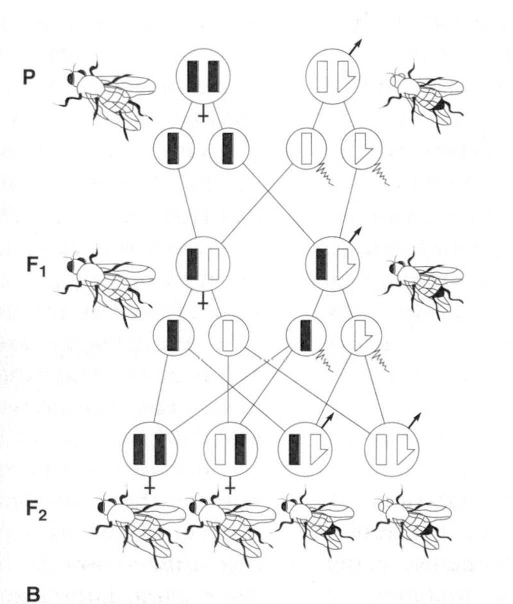

figure 3-8

Sex-linked inheritance of eye color in fruit fly *(Drosophila melanogaster).* **A,** White and red eyes of *Drosophila melanogaster.* **B,** Genes for eye color are carried on X chromosome; Y carries no genes for eye color. Normal red is dominant to white. Homozygous red-eyed female mated with white-eyed male gives all red-eyed in F_1 F_2 ratios from F_1 cross are one homozygous red-eyed female and one heterozygous red-eyed female to one red-eyed male and one white-eyed male.

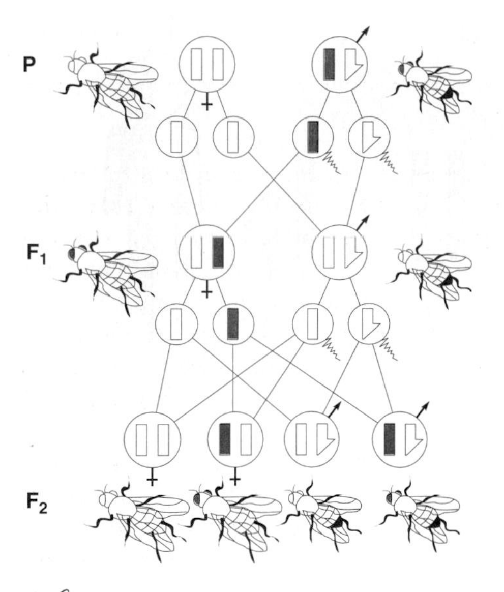

figure 3-9

Reciprocal cross of Figure 3-8 (homozygous white-eyed female with red-eyed male) gives white-eyed males and red-eyed females in F_1. F_2 shows equal numbers of red-eyed and white-eyed females and red-eyed and white-eyed males.

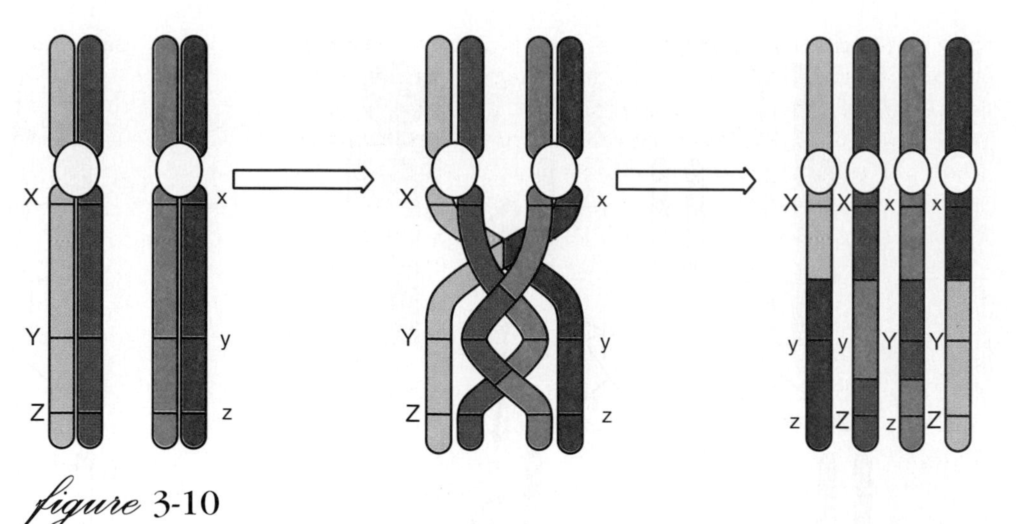

figure 3-10

Crossing over during meiosis. Nonsister chromatids exchange portions, so that none of the resulting gametes is genetically the same as any other. Gene X is farther from gene Y than Y is from Z; therefore gene X is more frequently separated from Y in crossing over than Y is from Z.

Geneticists commonly use the word "linkage" in two somewhat different senses. Sex-linkage refers to inheritance of a trait on the sex chromosomes, and thus its phenotypic expression depends on the sex of the organism and the factors already discussed. Autosomal linkage, or simply, linkage, refers to inheritance of the genes on a given autosomal chromosome. Letters used to represent such genes are normally written without a slash mark between them, indicating that they are on the same chromosome. For example, AB/ab *shows that genes* A *and* B *are on the homologous chromosome. Interestingly, Mendel studied seven characteristics of garden peas, which assorted independently because they were on seven different chromosomes. If he had studied eight characteristics, he would not have found independent assortment in two of the traits because garden peas have only seven pairs of homologous chromosomes.*

Chromosomal Aberrations

Structural and numerical deviations from the norm that affect many genes at once are called chromosomal aberrations. They are sometimes called chromosomal mutations, but most cytogeneticists prefer to use the term "mutation" to refer to qualitative changes within a gene; gene mutations will be discussed on p. 79.

Despite the incredible precision of meiosis, chromomal aberrations do occur, and they are more common than one might think. They are responsible for great economic benefit in agriculture. Unfortunately, they are also responsible for many human genetic malformations. It is estimated that five out of every 1000 humans are born with *serious* genetic defects, attributable to chromosomal anomalies. An even greater number of embryos with chromosomal defects are aborted spontaneously, far more than ever reach term.

Changes in chromosome numbers are called **euploidy** when there is the addition or deletion of complete haploid sets and **aneuploidy** when a single chromosome is added or subtracted from a diploid set. The most common kind of euploidy is **polyploidy,** the carrying of one or more additional sets of chromosomes. Such aberrations are much more common in plants than animals. Animals are much less tolerant of chromosomal aberrations because sex determination requires a delicate balance between the numbers of sex chromosomes and autosomes. Many domestic plant species are polyploid (cotton, wheat, apples, oats, tobacco, and others), and perhaps over 40% of flowering plants probably originated in this manner. Horticulturists favor polyploids and often try to develop them because they have more intensely colored flowers and more vigorous vegetative growth.

Aneuploidy is usually caused by nondisjunctional separation of chromosomes during meiosis. If a pair of homologous chromosomes fails to separate during the first meiotic division or sister chromatids fail to separate during the second meiotic division, both members go to one pole and none to the other. This results in one gamete having $n-1$ number of chromosomes and another having $n+1$ number of chromosomes. If the $n-1$ gamete is fertilized by a normal n gamete, the result is a **monosomic** animal. Survival is rare because the lack of one chromosome gives an uneven balance of genetic instructions. **Trisomy,** the result of the fusion of a normal n gamete and an n+1 gamete, is much more common, and several kinds of trisomic conditions are known in humans. Perhaps the most familiar is **trisomy 21,** or **Down syndrome.** As the name indicates, it involves an extra chromosome 21 combined with the chromosome pair 21, and it is caused by nondisjunction of that pair during meiosis. It occurs spontaneously, and there seldom is any family history of the abnormality. However, the risk of its appearance rises dramatically with increasing age of the mother; it occurs 40 times as often in pregnancies of women over 40 years old than among women between the ages of 20 and 30.

Structural aberrations involve whole sets of genes within a chromosome. A portion of a chromosome may be reversed, placing the linear arrangement of genes in reverse order **(inversion);** nonhomologous chromosomes may exchange sections **(translocation);** entire blocks of genes may be lost **(deletion);** or an extra section of chromosome may attach to a normal chromosome **(duplication).** These are all structural changes that usually do not produce phenotypic changes. Duplications, although rare, are important for evolution because they supply additional genetic information that may enable new functions.

A syndrome *is a group of symptoms associated with a particular disease or abnormality, although every symptom is not necessarily shown by every patient with the condition. An English physician, John Langdon Down, described the syndrome in 1866 that we now know is caused by trisomy 21. Because of Down's belief that the facial features of affected individuals were mongoloid in appearance, the condition has been known as mongolism. The resemblances are superficial, however, and the currently accepted names are trisomy 21 and Down syndrome. Among the numerous characteristics of the condition, the most disabling is severe mental retardation. Down syndrome, as well as other conditions caused by chromosomal aberrations and several other birth defects, can be diagnosed* prenatally *by a procedure involving* amniocentesis. *The physician inserts a hypodermic needle through the abdominal wall of the mother and into the fluids surrounding the fetus (not into the fetus) and withdraws some of the fluid, which contains some fetal cells. The cells are grown in culture, their chromosomes are examined, and other tests done. If a severe birth defect is found, the mother has the option of having an abortion performed. As an extra "bonus," the sex of the fetus is learned after amniocentesis. How?*

Gene Theory

Gene Concept

The term "gene" (Gr. *genos,* descent) was coined by W. Johannsen in 1909 to refer to the hereditary factors of Mendel. Initially, they were regarded as indivisible units of the chromosomes on which they were located. Later studies with multiple mutant alleles demonstrated that alleles are in fact divisible by recombination; that is, *portions* of a gene are separable. Furthermore, parts of many genes in eukaryotes are separated by sections of DNA that do not specify a part of the finished product **(introns).**

As the chief unit of genetic information, genes encode products essential for specifying the basic architecture of every cell, the nature and life of the cell, the specific protein syntheses, the enzyme formation, the self-reproduction of the cell, and, directly or indirectly, the entire metabolic function of the cell. Because of their ability to mutate, to be assorted and shuffled in different combinations, genes have become the basis for our modern interpretation of evolution. Genes are units of molecular information that can maintain their identities for many generations, can be self-duplicated in each generation, and can control processes by allowing their specificities to be copied.

One Gene–One Enzyme Hypothesis

Since genes act to produce different phenotypes, we may infer that their action follows the scheme: gene→gene product→phenotypic expression. Furthermore, we may suspect that the gene product is usually a protein, because proteins act as enzymes, antibodies, hormones, and structural elements throughout the body.

The first clear, well-documented study to link genes and enzymes was carried out on the common bread mold *Neurospora* by Beadle and Tatum in the early 1940s. This organism was ideally suited to a study of gene function for several reasons: these molds are much simpler to handle than fruit flies, they grow readily in well-defined chemical media, and they are haploid organisms that are consequently unencumbered with dominance relationships. Furthermore, mutations were readily induced by irradiation with ultraviolet light. Ultraviolet light-induced mutants, grown and tested in specific nutrient media, had single-gene mutations that were inherited in accord with Mendelian principles of segregation. Each mutant strain was defective in one enzyme, which prevented that strain from synthesizing one or more complex molecules. Putting it another way, the ability to synthesize a particular enzyme was controlled by a single gene.

From these experiments Beadle and Tatum set forth an important and exciting formulation: **one gene produces one enzyme.** For this work they were awarded the Nobel Prize in 1958. The new hypothesis was soon validated by the research of others who studied other biosynthetic pathways. Hundreds of inherited disorders, including dozens of human

table 3-2

Chemical Components of DNA and RNA

	DNA	RNA
Purines	Adenine	Adenine
	Guanine	Guanine
Pyrimidines	Cytosine	Cytosine
	Thymine	Uracil
Sugar	2-Deoxyribose	Ribose
Phosphate	Phosphoric acid	Phosphoric acid

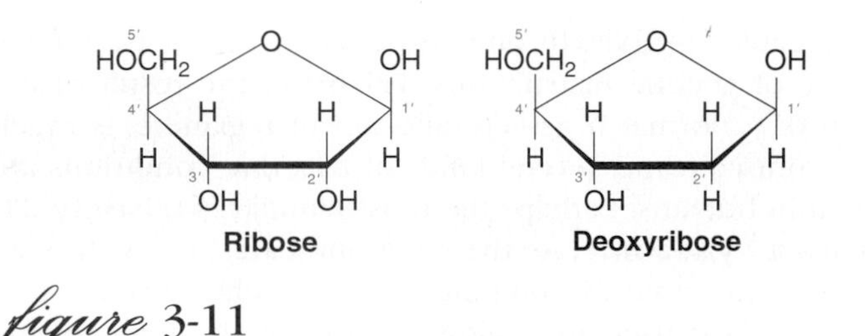

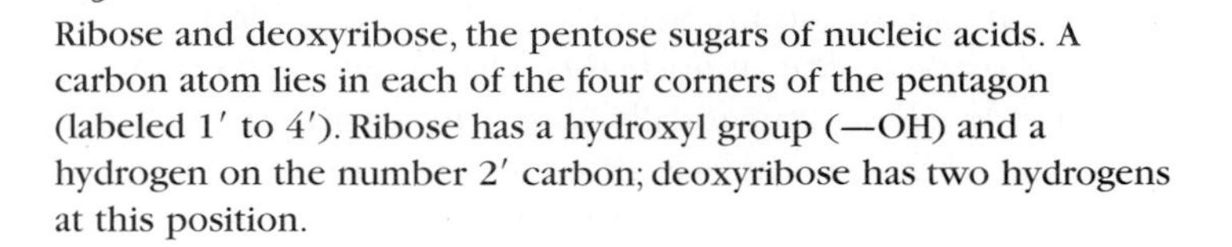

figure 3-11

Ribose and deoxyribose, the pentose sugars of nucleic acids. A carbon atom lies in each of the four corners of the pentagon (labeled 1′ to 4′). Ribose has a hydroxyl group (—OH) and a hydrogen on the number 2′ carbon; deoxyribose has two hydrogens at this position.

hereditary diseases, are caused by single mutant genes that result in the loss of a specific essential enzyme. We now know that a particular protein may be made of several chains of amino acids (polypeptides), each of which may be specified by a different gene, and not all proteins specified by genes are enzymes (for example, antibodies and hormones). Furthermore, genes direct the synthesis of various kinds of RNA. Therefore a gene now may be defined as a **nucleic acid sequence (usually DNA) that encodes a functional polypeptide or RNA sequence.**

Storage and Transfer of Genetic Information

Nucleic Acids: Molecular Basis of Inheritance

Cells contain two kinds of nucleic acids: deoxyribonucleic acid (DNA), which is the genetic material of the chromosomes of cells, and ribonucleic acid (RNA), which functions in protein synthesis. Both DNA and RNA are polymers built of repeated units called **nucleotides.** Each nucleotide contains three parts: a **sugar,** a **nitrogenous base,** and a **phosphate group.** The sugar is a pentose (5-carbon) sugar; in DNA it is **deoxyribose** and in RNA it is **ribose** (Figure 3-11).

The nitrogenous bases of nucleotides are of two types: pyrimidines, which consist of a single, 6-membered ring, and purines, which are composed of two fused rings. Both of these types of compounds contain nitrogen as well as carbon in their rings, which is why they are called "nitrogenous" bases. The purines in both RNA and DNA are adenine and guanine (Table 3-2). The pyrimidines in DNA are thymine and cytosine, and in RNA they are uracil and cytosine. The carbon atoms in the bases are numbered (for identification) according to standard biochemical notation (Figure 3-12). The carbons in the ribose and deoxyribose are also numbered, but to distinguish them from the carbons in the bases, the numbers for the carbons in the sugars are given prime signs (see Figure 3-11).

The sugar, phosphate group, and nitrogenous base are linked as shown in the generalized scheme for a nucleotide:

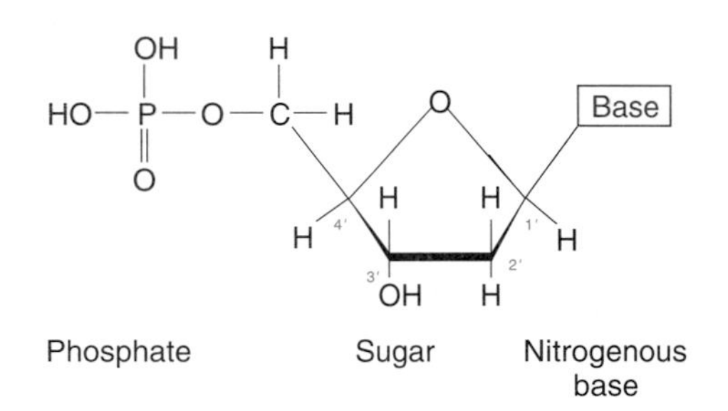

In DNA the backbone of the molecule is built of phosphoric acid and deoxyribose; to this backbone are attached the nitrogenous bases (Figure 3-13). The **5′ end** of the backbone has a free phosphate group on the **5′** carbon of the ribose, and the **3′ end** has a free hydroxyl group on the **3′** carbon. However, one of the most interesting and important discoveries about nucleic acids is that DNA is not a single polynucleotide chain; rather it consists of *two* complementary chains that are precisely cross-linked by specific hydrogen bonding between purine and pyrimidine bases. The number of adenines is equal to the number of thymines, and the number of guanines equals the number of cytosines. This fact suggested a pairing of bases: adenine with thymine (AT) and guanine with cytosine (GC) (Figure 3-14). Because the strands are complementary, the 3′ end of one strand lies at the 5′ end of the other and vice versa; the strands are thus antiparallel.

The result is a ladder structure (Figure 3-15). The upright portions are the sugar-phosphate backbones, and the connecting rungs are the paired nitrogenous bases, AT or GC. However, the ladder is twisted into a **double helix** with approximately 10 base pairs for each complete turn of the helix (Figure 3-16).

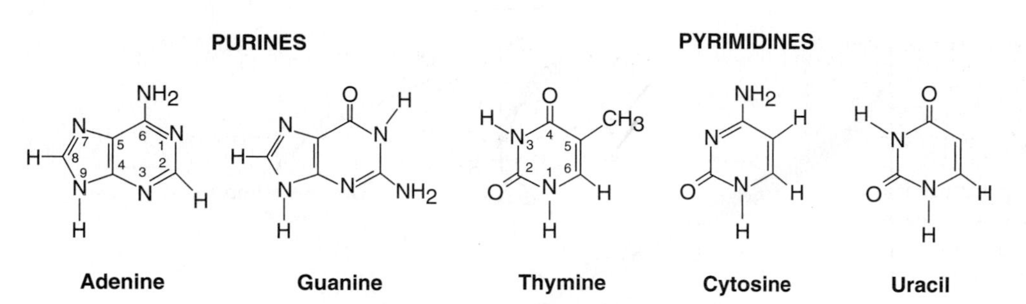

figure 3-12

Purines and pyrimidines of DNA and RNA.

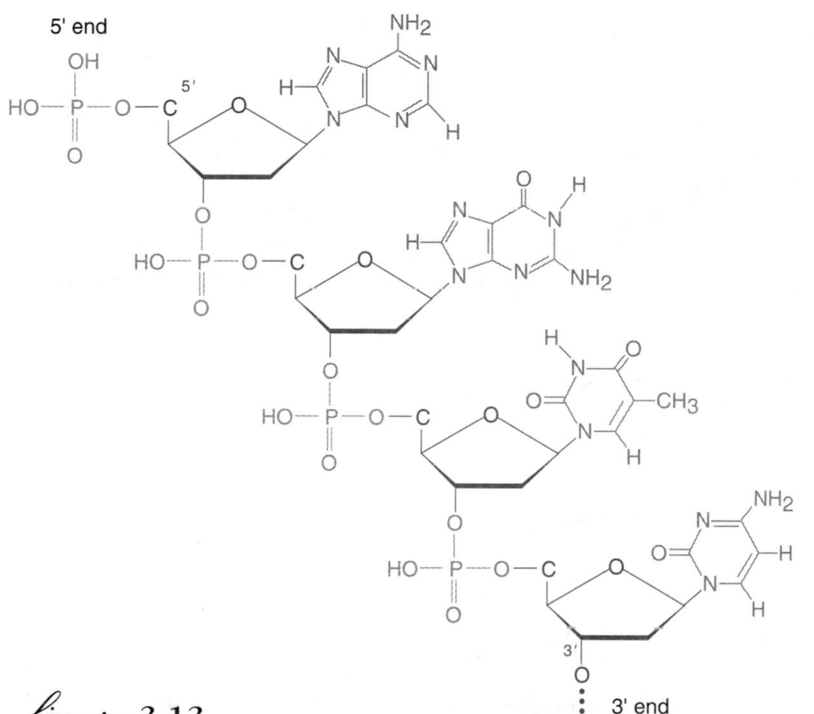

figure 3-13

Section of a strand of DNA. Polynucleotide chain is built of a backbone of phosphoric acid and deoxyribose sugar molecules. Each sugar holds a nitrogenous base side arm. Shown from top to bottom are adenine, guanine, thymine, and cytosine.

The determination of the structure of DNA has been widely acclaimed as the single most important biological discovery of this century. It was based on the x-ray diffraction studies of Maurice H. F. Wilkins and Rosalind Franklin and on the ingenious proposals of Francis H. C. Crick and James D. Watson published in 1953. Watson, Crick, and Wilkins were later awarded the Nobel Prize for Medicine and Physiology for their momentous work.

RNA is similar to DNA in structure except that it consists of a *single* polynucleotide chain, has ribose instead of deoxyribose, and has uracil instead of thymine. The three kinds of RNA (ribosomal, transfer, and messenger) are described below.

Every time a cell divides, the structure of DNA must be precisely copied in the daughter cells. This is called **replication** (Figure 3-17). During replication, the two strands of the double helix unwind, and each separated strand serves as a **template** against which a complementary strand is synthesized. That is, an enzyme (DNA polymerase) assembles a new strand of polynucleotides with a thymine group going next to the adenine group in the template strand, a guanine group next to the cytosine group, and so on.

DNA Coding by Base Sequence

Since DNA is the genetic material and is composed of a linear sequence of base pairs, an obvious extension of the

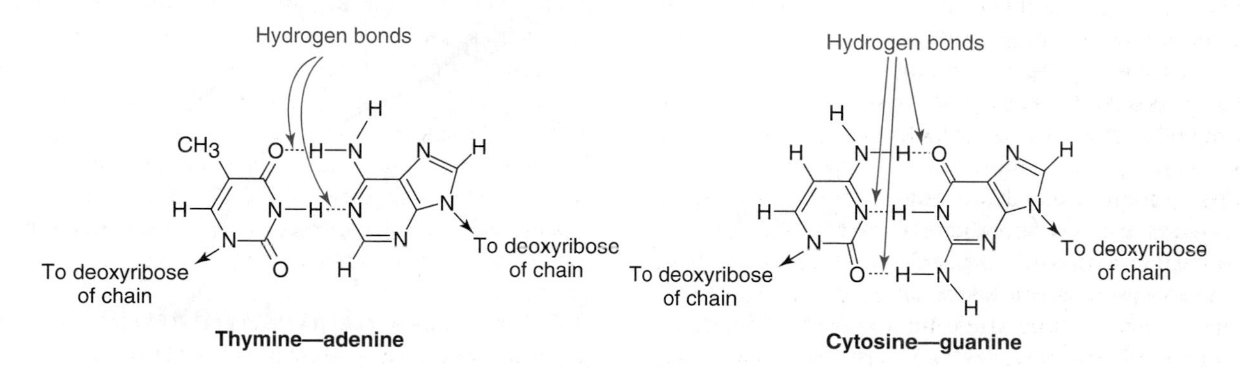

figure 3-14

Positions of hydrogen bonds between thymine and adenine and between cytosine and guanine in DNA.

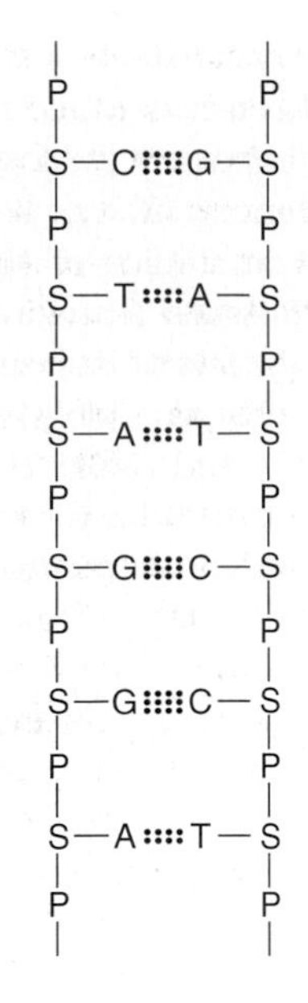

figure 3-15

DNA, showing how the complementary pairing of bases between the sugar-phosphate "backbones" keeps the double helix at a constant diameter for the entire length of the molecule. Dotted lines represent the three hydrogen bonds between each cytosine and guanine and the two hydrogen bonds between each adenine and thymine.

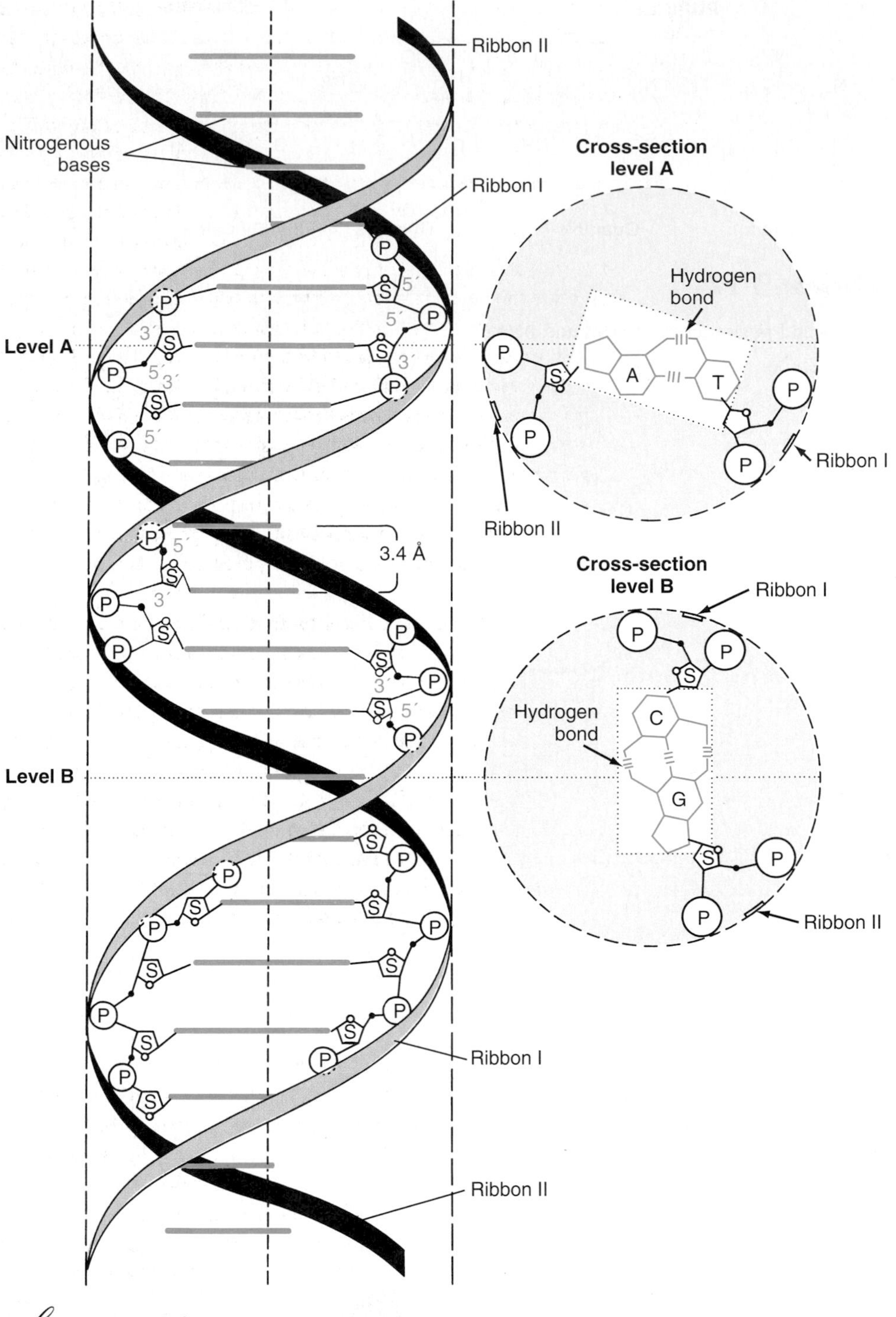

figure 3-16

DNA molecule.

Watson-Crick model is that the sequence of base pairs in DNA codes for, and is colinear with, the sequence of amino acids in a protein. The coding hypothesis had to account for the way a string of four different bases—a four-letter alphabet—could dictate the sequence of 20 different amino acids.

In the coding procedure, obviously there cannot be a 1:1 correlation between four bases and 20 amino acids. If the coding unit (often called a word, or **codon**) consisted of two bases, only 16 words (4^2) could be formed, which could not account for 20 amino acids. Therefore the protein code had to consist of at least three bases or three letters because 64 possible words (4^3) could be formed by four bases when taken as triplets. This meant that there could be a considerable redundancy of codons, since DNA codes for just 20 amino acids. Later work by Crick confirmed that nearly all of the amino acids are specified by more than one triplet code. The triplet codes for each amino acid are shown for messenger RNA in Table 3-3.

Transcription and the Role of Messenger RNA

Information is coded in DNA, but DNA does not participate directly in protein synthesis. It is obvious that an intermediary is required. This intermediary is another nucleic acid called

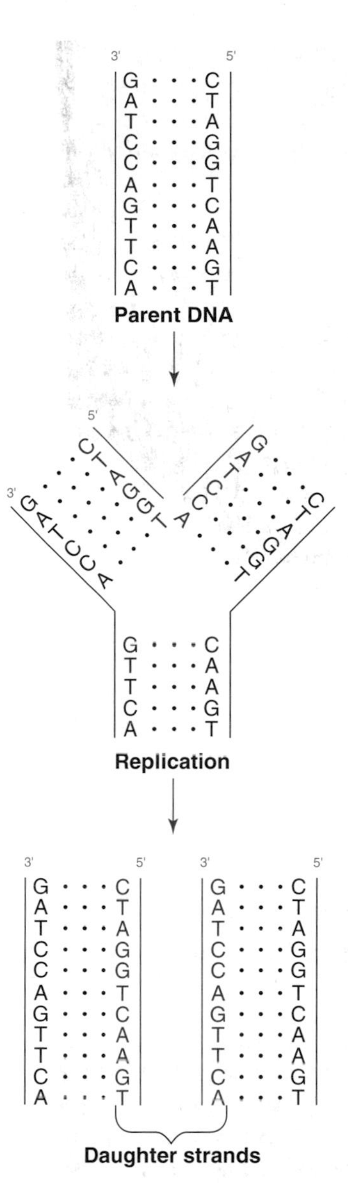

figure 3-17

Replication of DNA. The parent strands of DNA part, and DNA polymerase synthesizes daughter strands using the base sequence of parent strands as a template. The diagram shows unidirectional replication, but most DNA replication is bidirectional; that is, it proceeds in both directions at once.

messenger RNA (mRNA). The triplet codes in DNA are **transcribed** into mRNA, with uracil substituting for thymine (Table 3-3).

Ribosomal, transfer, and messenger RNAs are transcribed directly from DNA, each encoded by different sets of genes. In this process of making a complementary copy of one strand or gene of DNA in the formation of RNA, an enzyme, **RNA polymerase,** is needed. (In eukaryotes each type of RNA [ribosomal, transfer, and messenger] is transcribed by a specific type of RNA polymerase.) The mRNA contains a sequence of bases that complements the bases in one of the two DNA strands, just as the DNA strands complement each other. Thus, A in the coding DNA strand is replaced by U in RNA; C is replaced by G, G is replaced by C, and T is replaced by A. Only one of the two chains is used as the template for RNA synthesis. The reason why only one strand of the double-stranded DNA is a "coding strand" is that mRNA otherwise would always be formed in complementary pairs, and enzymes also would be synthesized in complementary pairs. In other words, two different enzymes would be produced for every DNA coding sequence instead of one. This would certainly lead to metabolic chaos.

Only one strand of DNA serves as the coding strand in all DNA except that found in plasmids (see p. 77). Messenger RNA can be transcribed from both DNA strands in one region of plasmid DNA, and this is the only known example of proteins being encoded in both DNA strands.

Genes on the DNA of prokaryotes are coded on a continuous stretch of DNA, which is transcribed into mRNA and then translated (see the following section). It was assumed that this was also the case for eukaryotic genes until the surprising discovery that stretches of DNA are transcribed in the nucleus but are not found in the corresponding mRNA in the cytoplasm. In other words, pieces of the nuclear mRNA were spliced out in the nucleus before the finished mRNA was transported to the cytoplasm (Figure 3-18). It was thus discovered that many genes are split, interrupted by sequences of bases that do not code for the final product, and the mRNA transcribed from them must be edited or "matured" before translation in the cytoplasm. The intervening segments of DNA are now known as **introns,** while those that code for part of the mature RNA and are translated into gene products are called **exons.** Before the mRNA leaves the nucleus, the introns are spliced out and a methylated guanine **cap** is added at the 5′ end, while a tail of adenine nucleotides **(poly-A)** is added at the 3′ end (Figure 3-18). The cap and the poly-A tail are characteristic of mRNA molecules.

In mammals the genes coding for the histones and for interferon are on continuous stretches of DNA. However, we now know that genes coding for many proteins are split. In lymphocyte differentiation the parts of the split genes coding for immunoglobulins are actually *rearranged* during development, so that different proteins result from subsequent transcription and translation. This partly accounts for the enormous diversity of antibodies manufactured by the descendants of the lymphocytes (p. 214).

Base sequences in some introns are complementary to other base sequences in the intron, suggesting that the intron could fold so that the complementary sequences would pair. This may be necessary to control proper alignment of intron boundaries before splicing. Most surprising of all has been the discovery that, at least in some cases, the RNA can "self-catalyze" the excision of introns. The ends of the intron join; the intron thus becomes a small circle of RNA, and the

table 3-3

The Genetic Code: Amino Acids Specified by Codons of Messenger RNA

First letter	Second letter: U	Second letter: C	Second letter: A	Second letter: G	Third letter
U	UUU Phenylalanine	UCU Serine	UAU Tyrosine	UGU Cysteine	U
U	UUC Phenylalanine	UCC Serine	UAC Tyrosine	UGC Cysteine	C
U	UUA Leucine	UCA Serine	UAA End chain	UGA End chain	A
U	UUG Leucine	UCG Serine	UAG End chain	UGG Tryptophane	G
C	CUU Leucine	CCU Proline	CAU Histidine	CGU Arginine	U
C	CUC Leucine	CCC Proline	CAC Histidine	CGC Arginine	C
C	CUA Leucine	CCA Proline	CAA Glutamine	CGA Arginine	A
C	CUG Leucine	CCG Proline	CAG Glutamine	CGG Arginine	G
A	AUU Isoleucine	ACU Threonine	AAU Asparagine	AGU Serine	U
A	AUC Isoleucine	ACC Threonine	AAC Asparagine	AGC Serine	C
A	AUA Isoleucine	ACA Threonine	AAA Lysine	AGA Arginine	A
A	AUG Methionine*	ACG Threonine	AAG Lysine	AGG Arginine	G
G	GUU Valine	GCU Alanine	GAU Aspartic acid	GGU Glycine	U
G	GUC Valine	GCC Alanine	GAC Aspartic acid	GGC Glycine	C
G	GUA Valine	GCA Alanine	GAA Glutamic acid	GGA Glycine	A
G	GUG Valine	GCG Alanine	GAG Glutamic acid	GGG Glycine	G

**Also, begin chain.*

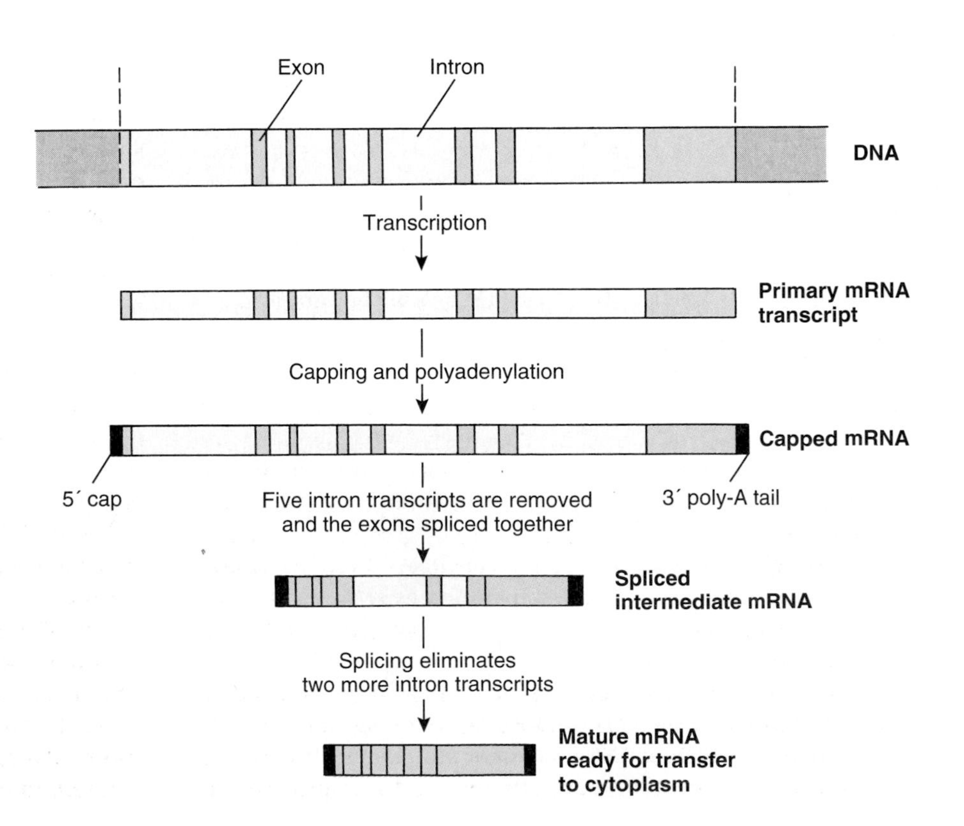

figure 3-18

Transcription and maturation of ovalbumin gene of chicken. The entire gene of 7700 base pairs is transcribed to form the primary mRNA, then the 5′ cap of methyl guanine and the 3′ polyadenylate tail are added. After the introns are spliced out, the mature mRNA is transferred to the cytoplasm.

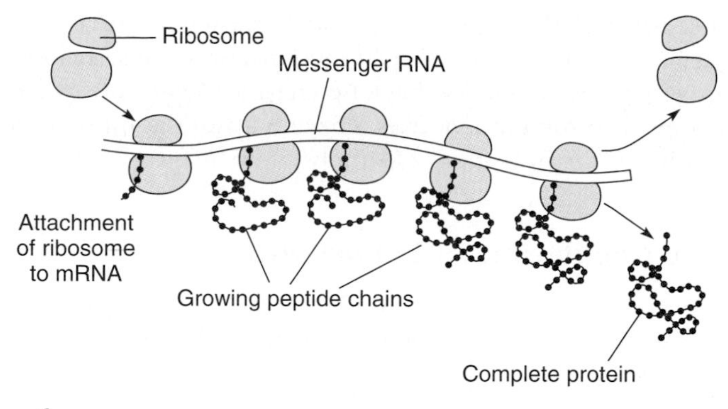

figure 3-19

How the protein chain is formed. As ribosomes move along messenger RNA, the amino acids are added stepwise to form the polypeptide chain.

exons are spliced together. This process does not fit the classical definition of an enzyme or other catalyst since the molecule itself is changed in the reaction.

Translation: Final Stage in Information Transfer

The **translation** process takes place on **ribosomes,** granular structures composed of protein and **ribosomal RNA (rRNA).** Ribosomal RNA is composed of a large and a small subunit, and the small subunit comes to lie in a depression of the large subunit to form the functional ribosome (Figure 3-19). The mRNA molecules fix themselves to the ribosomes to form a messenger RNA-ribosome complex. Since only a short section of an mRNA molecule is in contact with a single ribosome, the mRNA usually fixes itself to several ribosomes at once. The entire complex, called a **polyribosome** or **polysome,** allows several molecules of the same kind of protein to be synthesized at once, one on each ribosome of the polysome (Figure 3-19).

The assembly of proteins on the mRNA-ribosome complex requires the action of another kind of RNA called **transfer RNA (tRNA).** The tRNAs are surprisingly large molecules that are folded in a complicated way in the form of a cloverleaf (Figure 3-20). The tRNA molecules collect the free amino acids from the cytoplasm and deliver them to the polysome, where they are assembled into a protein. There are special tRNA molecules for every amino acid. Furthermore, each tRNA is accompanied by a specific tRNA synthetase. The tRNA synthetases are enzymes that are necessary to sort and attach the correct amino acid to a site on the end of each tRNA by a process called **charging.**

On the cloverleaf-shaped molecule of tRNA, a special sequence of three bases (the **anticodon**) is exposed in just the right way to form base pairs with complementary bases (the codon) in the mRNA. The codons are read and proteins as-

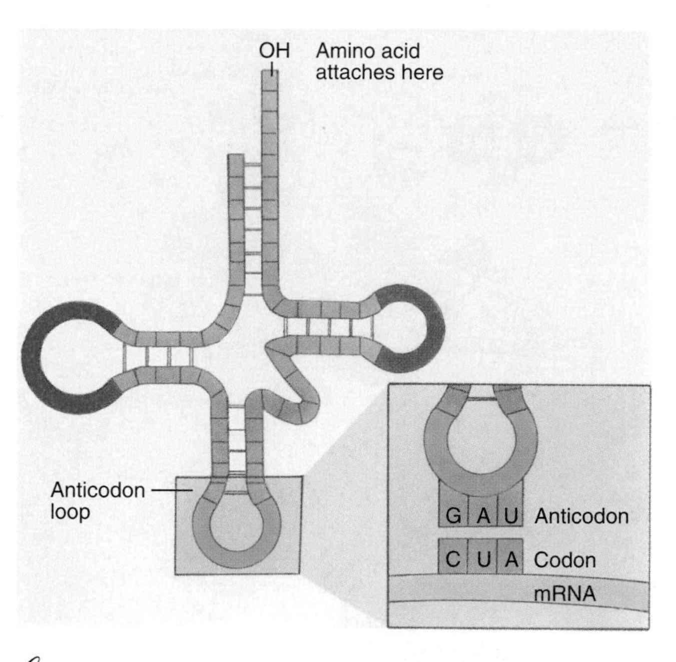

figure 3-20

Structure of a tRNA molecule. The anticodon loop bears bases complementary to those in the mRNA codon. The other two loops function in binding to the ribosomes in protein synthesis. The amino acid is added to the free single-stranded—OH end by tRNA synthetase.

sembled along the mRNA in a 5′ to 3′ direction. The anticodon of the tRNA is the key to the correct sequencing of amino acids in the protein being assembled.

For example, alanine is assembled into a protein when it is signaled by the codon GCG in an mRNA. The translation is accomplished by alanine tRNA in which the anticodon is CGC. The alanine tRNA is first charged with alanine by its tRNA synthetase. The alanine tRNA complex enters the ribosome where it fits precisely into the right place on the mRNA stand. Then the next charged tRNA specified by the mRNA code (glycine tRNA, for example) enters the ribosome and attaches itself beside the alanine tRNA. The two amino acids are united with a peptide bond (with the energy from a molecule of guanosine triphosphate), and the alanine tRNA falls off. The process continues stepwise as the protein chain is built (Figure 3-21). A protein of 500 amino acids can be assembled in less than 30 seconds.

Regulation of Gene Function

In Chapter 14 we will see how the orderly differentiation of an organism from fertilized ovum to adult requires the involvement of genetic material at every stage of development. Developmental biologists have provided convincing evidence that every cell in a developing embryo is genetically equivalent. Thus it is clear that as tissues differentiate (change developmentally), they use only a part of the genetic instruction present in every cell. Certain genes express themselves only at

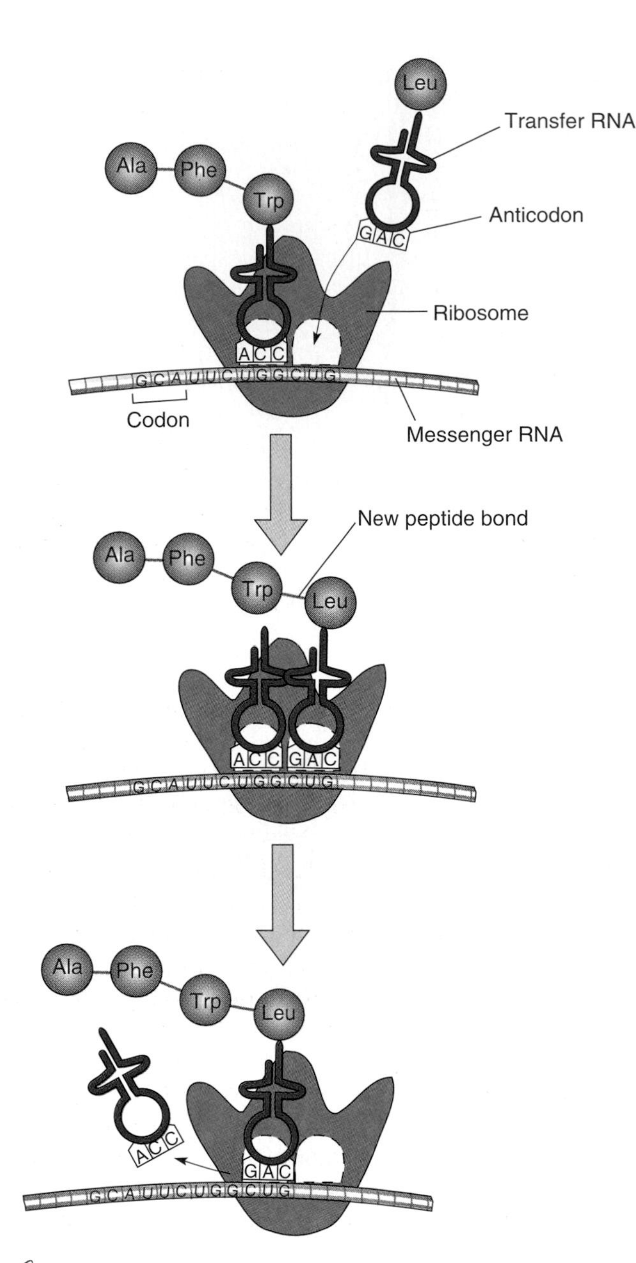

figure 3-21

Formation of polypeptide chain on messenger RNA. As ribosome moves down the messenger RNA molecule, transfer RNA molecules with attached amino acids enter the ribosome *(top)*. Amino acids are joined together into a polypeptide chain, and transfer RNA molecules leave the ribosome *(bottom)*.

certain times and not at others. Indeed, there is reason to believe that in a particular cell or tissue, most of the genes are inactive at any given moment. The problem in development is to explain how, if every cell has a full gene complement, certain genes are "turned on" and produce proteins that are required for a particular developmental stage while the other genes remain silent.

Actually, although the developmental process brings the question of gene activation clearly into focus, gene regulation is necessary throughout an organism's existence. The cellular enzyme systems that control all functional processes obviously require genetic regulation because enzymes have powerful effects even in minute amounts. Enzyme synthesis must be responsive to the influences of supply and demand.

Gene Regulation in Eukaryotes

There are a number of different phenomena in eukaryotic cells that can serve as control points, and the following are a few examples.

Transcriptional Control This may be the most important mechanism. **Transcriptional factors** are molecules that may have a positive or a negative effect on transcription of RNA from the DNA of the target genes. The factors may act within the cells that produce them or they may be transported to different parts of the body prior to action. An example of a positive transcriptional factor is a steroid receptor. Steroid hormones produced by endocrine glands elsewhere in the body enter the cell and bind with a receptor protein in the nucleus. The steroid-receptor complex then binds with DNA near the target gene (p. 268). Progesterone, for example, binds with a nuclear receptor in cells of the chicken oviduct; the hormone-receptor complex then activates the transcription of genes encoding egg albumin and other substances.

Translational Control Genes can be transcribed and the mRNA sequestered in some way so that translation is delayed. This commonly happens in the development of eggs of many animals. The oocyte accumulates large quantities of messenger RNA during its development, then fertilization activates metabolism and initiates translation of maternal mRNA.

Gene Rearrangement Vertebrates contain cells called lymphocytes that bear genes coding for proteins called antibodies (p. 214). Each type of antibody has the capacity to bind specifically with a particular foreign substance (antigen). Because the number of different antigens is enormous, the genetic diversity of antibody genes must be equally great. One source of this diversity is rearrangement of DNA sequences coding for the antibodies during the development of lymphocytes.

DNA Modification An important mechanism for turning genes off appears to be methylation of cytosine residues, adding a methyl group (CH_3-) to the carbon in the 5 position in the cytosine ring (Figure 3-22A). This usually happens when the cytosine is next to a guanine residue; thus, the bases in the complementary DNA strand would also be a cytosine and a guanine (Figure 3-22B). When the DNA is replicated, an enzyme recognizes the CG sequence and quickly methylates the daughter strand, maintaining the gene in an inactive state.

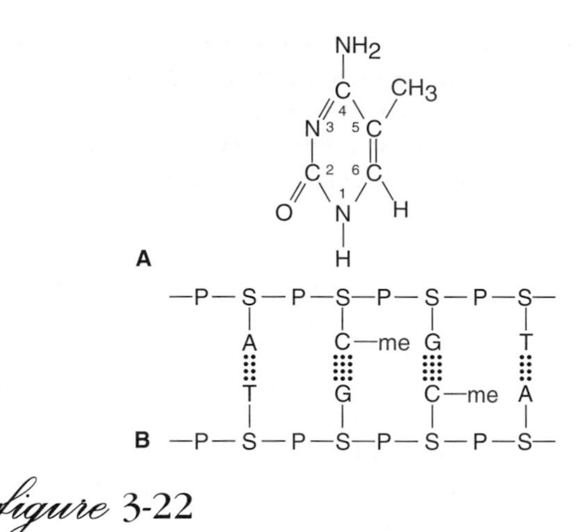

figure 3-22

Some genes in eukaryotes are apparently turned off by the methylation of some cytosine residues in the chain. **A,** Structure of 5-methyl cytosine. **B,** Cytosine residues next to guanine are those that are methylated in a strand, thus allowing both strands to be symmetrically methylated.

Genetic Engineering

Progress in our understanding of genetic mechanisms on the molecular level, as discussed in the last few pages, has been almost breathtaking in the last few years. We can expect many more discoveries in the near future. This progress has been due largely to the effectiveness of many biochemical techniques now used in molecular biology. We have space to describe only a few briefly.

One of the most important tools in this technology is a series of enzymes called **restriction endonucleases.** Each of these enzymes, derived from bacteria, cleaves double-stranded DNA at particular sites determined by the particular base sequences at that point. Many of these endonucleases cut the DNA strands so that one has several bases projecting farther than the other strand (Figure 3-23), leaving what are called "sticky ends." When these DNA fragments are mixed with others that have been cleaved by the same endonuclease, they tend to anneal (join) by the rules of complementary base pairing. They are sealed into their new position by the enzyme **DNA ligase.**

Besides their chromosomes, most prokaryotic and at least some eukaryotic cells have small circles of double-stranded DNA called plasmids. *Though comprising only 1% to 3% of the bacterial genome, they may carry important genetic information, for example, resistance to an antibiotic. Plastids in plant cells (for example, chloroplasts) and mitochondria, found in most eukaryotic cells, are self-replicating and have their own complement of DNA in the form of small circles, reminiscent of plasmids. The DNA of mitochondria and plastids codes for some of their proteins, and some of their proteins are specified by nuclear genes.*

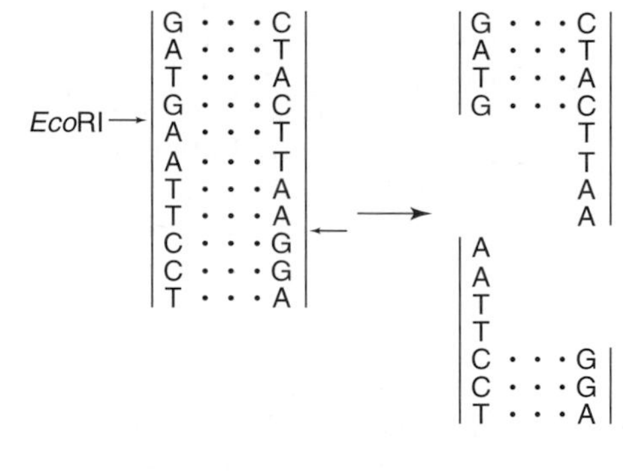

figure 3-23

Action of restriction endonuclease, *Eco*RI. Such enzymes recognize specific base sequences that are palindromic (a palindrome is a word spelled the same backward and forward). *Eco*RI leaves "sticky ends," which are easily matched to other DNA fragments cleaved by the same enzyme and linked by DNA ligase.

If the DNA annealed after cleavage by the endonuclease is from two different sources, for example, a plasmid (see marginal note) and a mammal, the product is **recombinant DNA.** To make use of the recombinant DNA, the modified plasmid must be cloned in bacteria. The bacteria are treated with dilute calcium chloride to make them more susceptible to taking up the recombinant DNA, but the plasmids do not enter most of the cells present. The bacterial cells that have taken up the recombinant DNA can be identified if the plasmid has a marker, for example, resistance to an antibiotic. Then, only the bacteria that can grow in the presence of the antibiotic are those that have absorbed the recombinant DNA. Some bacteriophages (bacterial viruses) also have been used as carriers for recombinant DNA. Plasmids and bacteriophages that carry recombinant DNA are called **vectors.** The vectors retain the ability to replicate in the bacterial cells; therefore the recombinant insert is amplified.

The techniques of molecular biology have allowed scientists to accomplish feats of which few could dream only a decade or so ago. These accomplishments will bring enormous benefits for humanity in the form of enhanced food production and treatment of disease. Progress with crop plants has been so rapid that we can expect genetically engineered soybean, cotton, rice, corn, sugarbeet, tomato, and alfalfa to reach the market before the year 2000. Development of transgenic animals of potential use has not progressed as far as development of such plants. Gene therapy for inherited diseases presents many difficulties, but research in this area is vigorous, and clinical trials for certain conditions are just beginning.

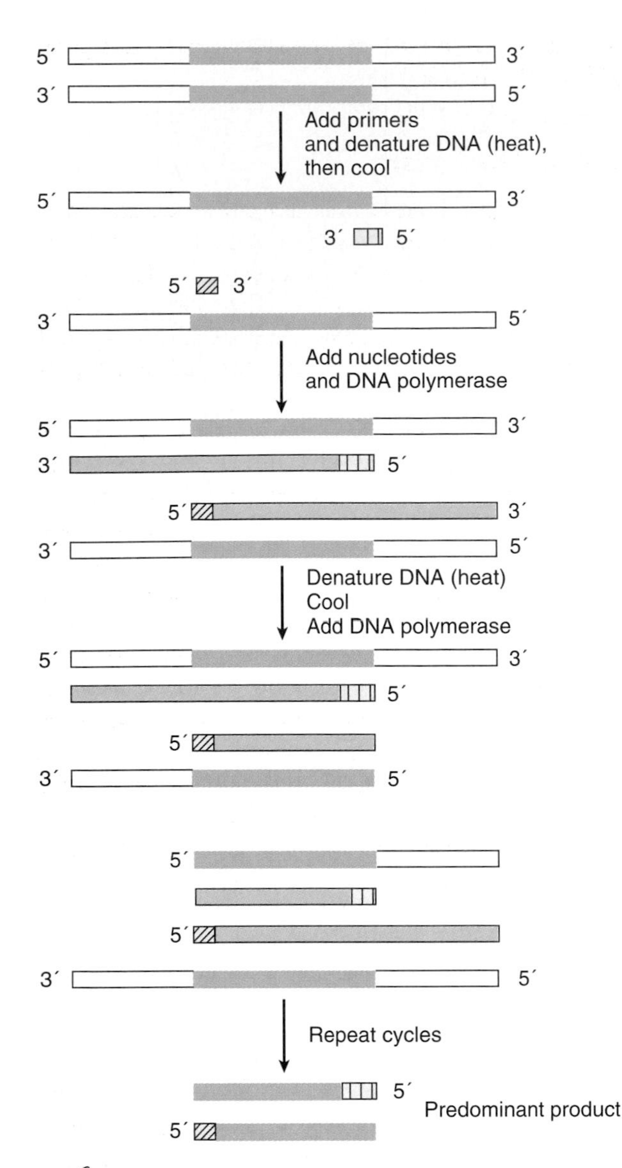

figure 3-24

Steps in the polymerase chain reaction (PCR).

Recent advances have made it a simple task to clone a specific gene enzymatically from any organism as long as part of the sequence of that gene is known. The technique is called the **polymerase chain reaction (PCR).** Two short chains of nucleotides called primers are synthesized; primers are complementary to different DNA strands in the known sequence. A large excess of each primer is added to a sample of DNA from the organism, and the mixture is heated to separate the double helix into single strands. When the mixture is cooled, there is a much greater probability that each strand of the gene of interest will anneal to a primer rather than to the other strand of the gene—because there is so much more primer present. DNA polymerase is added along with the four deoxyribonucleotide triphosphates, and DNA synthesis proceeds from the 3′ end of each primer. If the primers are chosen so that each anneals toward the 3′ end of each of the complementary strands, entire new complementary strands will be synthesized, and the number of copies of the gene has doubled (Figure 3-24). The reaction mixture is then reheated and cooled again to allow more primers to bind original and new copies of each strand. With each cycle of DNA synthesis, the number of copies of the gene doubles. Since each cycle can take less than 5 minutes, the number of copies of a gene can increase from one to over one million in less than 2 hours! The PCR allows cloning a known gene from an individual patient, identification of a drop of dried blood at a crime scene, or cloning the DNA of a 40,000-year-old woolly mammoth.

Recombinant DNA technology and the PCR are currently in use in many areas with great positive potential and many practical applications.

A clone is a collection of individuals or cells all derived by asexual reproduction from a single individual. When we speak of cloning a gene or plasmid in bacteria, we mean that we isolate a colony or group of bacteria derived from a single ancestor into which the gene or plasmid was inserted.

Sources of Phenotypic Variation

The creative force of evolution is natural selection acting on biological variation. Without variability among individuals, there could be no continued adaptation to a changing environment and no evolution.

There are actually several sources of variability, some of which we have already described. The independent assortment of chromosomes during meiosis is a random process that creates new chromosomal recombinations in the gametes. In addition, chromosomal crossing over during meiosis allows recombination of linked genes between homologous chromosomes, further increasing variability. The random fusion of gametes from both parents produces still another source of variation.

There is a story that George Bernard Shaw once received a letter from a famous actress who suggested that they conceive a perfect child who would combine her beauty and his brains. He declined the offer, pointing out that the child could just as well inherit her brains and his beauty. Shaw was correct; the fusion of parental gametes is random and thus unpredictable.

Thus sexual reproduction multiplies variation and provides the diversity and plasticity necessary for a species to survive environmental change. Sexual reproduction with its sequence of gene segregation and recombination, generation after generation, is, as the geneticist T. Dobzhansky has said, the "master adaptation which makes all other evolutionary adaptations more readily accessible."

Although sexual reproduction re-shuffles and amplifies whatever genetic diversity exists in the population, there must be ways to generate *new* genetic material. This happens through gene mutations and, sometimes, through chromosomal aberrations.

Gene Mutations

Gene mutations are chemicophysical changes in genes resulting in an alteration of the sequence of bases in the DNA. These mutations can be studied directly by determining the DNA sequence and indirectly through their effects on organismal phenotype, if such effects are present. A mutation may result in a codon substitution as, for example, in the condition in humans known as **sickle cell anemia.** Homozygotes with sickle cell trait usually die before the age of 30 because the ability of their red blood cells to carry oxygen is greatly impaired, a result of the substitution of only a single amino acid in the amino-acid sequence of their hemoglobin. Other mutations may involve the deletion of one or more bases from a codon or the insertion of additional bases into the DNA chain. The transcription of mRNA will thus be shifted, leading to codons that specify incorrect amino acids.

Once a gene is mutated, it faithfully reproduces its new self just as it did before it was mutated. Many mutations are harmful, many are neither helpful nor harmful, and sometimes mutations are advantageous. Helpful mutations are of great significance to evolution because they furnish new possibilities on which natural selection works. Natural selection determines which new genes merit survival; the environment imposes a screening process that passes the beneficial and eliminates the harmful.

When an allele of a gene is mutated to the new allele, it tends to be recessive and its effects are normally masked by its partner allele. Only in the homozygous condition can such mutant genes be expressed. Thus a population carries a reservoir of mutant recessive genes, some of which are lethal when homozygous but which are rarely present in the homozygous condition. Inbreeding encourages the formation of homozygotes and increases the probability of recessive mutants being expressed in the phenotype.

Most mutations are destined for a brief existence. There are cases, however, in which mutations may be harmful or neutral under one set of environmental conditions and helpful under a different set. The earth's changing environment has provided numerous opportunities for new gene combinations and mutations, as evidenced by the great diversity of animal life today.

Frequency of Mutations

Although mutation occurs randomly, different mutation rates prevail at different loci. Some *kinds* of mutations are more likely to occur than others, and individual genes differ considerably in length. A long gene (more base pairs) is more likely to have a mutation than a short gene. Nevertheless, it is possible to estimate average spontaneous rates for different organisms and traits.

Relatively speaking, genes are extremely stable. In the well-studied fruit fly *Drosophila melanogaster* there is approximately one detectable mutation per 10,000 loci (rate of 0.01% per locus per generation). The rate for humans is one per 10,000 to one per 100,000 loci per generation. If we accept the latter, more conservative figure, then a single normal allele is expected to go through 100,000 generations before it is mutated. However, since human chromosomes contain 100,000 loci, every person carries approximately one new mutation. Similarly, each ovum or spermatozoon produced contains, on the average, one mutant allele.

Since most mutations are deleterious, these statistics are anything but cheerful. Fortunately, most mutant genes are recessive and are not expressed in heterozygotes. Only a few will by chance increase enough in frequency for homozygotes to be produced.

Molecular Genetics of Cancer

The crucial defect in cancer cells is that they proliferate in an unrestrained manner **(neoplastic growth).** The mechanism that controls the rate of division of normal cells has somehow broken down, and the cancer cells multiply much more rapidly, invading other tissues in the body. Cancer cells originate from normal cells that lose their constraint on division and become dedifferentiated (less specialized) to some degree. Thus there are many kinds of cancer, depending on the original founder cells of the tumor. In recent years mounting evidence has indicated that the change in many cancerous cells, perhaps all, has a genetic basis, and investigation of the genetic damage that causes cancer is now a major thrust of cancer research.

Oncogenes and Tumor Suppressor Genes

We now recognize that cancer is a result of a series of specific genetic changes that take place in a particular clone of cells. These include alterations in two types of genes: **oncogenes** and **tumor suppressor genes,** and there are numerous specific genes of each type now known.

Of the many ways that cellular DNA can sustain damage, the three most important are ionizing radiation, ultraviolet radiation, and chemical mutagens. The high energy of ionizing radiation (x rays and gamma rays) causes electrons to be ejected from the atoms it encounters, resulting in ionized atoms with unpaired electrons (free radicals). The free radicals (principally from water) are highly reactive chemically, and they react with molecules in the cell, including DNA. Some damaged DNA is repaired, but if the repair is inaccurate, a mutation results. Ultraviolet radiation is of much lower energy than ionizing radiation and does not produce free radicals. It is absorbed by pyrimidines in DNA and causes formation of a double covalent bond between the adjacent pyrimidines. UV repair mechanisms can also be inaccurate. Chemical mutagens react with the DNA bases and cause mispairing during replication.

Oncogenes (Gr. *onkos,* bulk, mass, + *genos,* descent) are genes whose activity has been associated for some time with the production of cancer. They are genes that are normally found in cells, and in their normal form they are called **proto-oncogenes.** One of these codes for a protein known as **Ras.** Ras protein is a guanosine triphosphatase (GTPase) that is located just beneath the cell membrane. When a receptor on the cell surface binds a growth factor, Ras is activated and initiates a cascade of reactions, ultimately leading to cell division. The oncogene form codes for a protein that initiates the cell-division cascade even when the growth factor has not bound to the surface receptor; that is, the growth factor is absent.

Gene products of tumor suppressor genes act as a constraint on cell proliferation. One such product is called **p53** (for "53-kilodalton protein," a reference to its molecular weight). Mutations in the gene coding for p53 are present in about half of the 6.5 million cases of human cancer diagnosed each year. Normal p53 has a number of crucial functions, depending on the circumstances of the cell. It can trigger apoptosis (p. 38), act as a transcription activator or repressor (turning genes on or off), control progression from G_1 to S phase in the cell cycle, and promote repair of damaged DNA. Many of the mutations known in p53 interfere with its binding to DNA and thus its function.

Summary

In sexual animals the genetic material is distributed to the offspring in the gametes (ova and sperm), produced in the process of meiosis. Each somatic cell in an organism has two chromosomes of each kind (homologous chromosomes) and is thus diploid. Meiosis separates the homologous chromosomes, so that each gamete has half the somatic chromosome number (haploid). In the first meiotic division, the centromeres do not divide, and each daughter cell receives one of each of the replicated homologous chromosomes with the chromatids still attached to the centromere. At the beginning of the first meiotic division, the replicated homologous chromosomes come to lie alongside each other (synapsis), forming a bivalent. The gene loci on one set of chromatids lie opposite the corresponding loci on the homologous chromatids. Portions of the adjacent chromatids can exchange with the nonsister chromatids (crossing over) to produce new genetic combinations. At the second meiotic division, the centromeres divide, completing the reduction in chromosome number and amount of DNA. The diploid number is restored when the male and female gametes fuse to form the zygote. Gender is determined in most animals by the sex chromosomes; in humans, fruit flies, and many other animals, females have two X chromosomes, and males have an X and a Y.

Genes may be dominant, recessive, or intermediate; the recessive genes in the heterozygous genotype will not be expressed in the phenotype, but require the homozygous condition for overt expression. In a monohybrid cross involving a dominant gene and its recessive allele (both parents homozygous), the F_1 generation will all be heterozygous, whereas the F_2 genotypes will occur in a 1:2:1 ratio, and the phenotypes in a 3:1 ratio. This demonstrates Mendel's law of segregation. Heterozygotes in intermediate inheritance show phenotypes intermediate between the homozygous phenotypes, or sometimes they show a different phenotype altogether, with corresponding alterations in the phenotypic ratios. Dihybrid crosses (in which the alleles for the two traits are carried on separate pairs of homologous chromosomes) demonstrate Mendel's law of independent assortment, and the phenotypic ratios will be 9:3:3:1 with dominant and recessive characters. The ratios for monohybrid and dihybrid crosses can be determined by construction of a Punnett square, but the laws of probability allow calculation of the ratios in crosses of two or more characters more easily.

A gene on the X (sex) chromosome shows sex-linked inheritance and will produce an effect in the male, even if it is recessive, because the Y chromosome does not carry a corresponding allele. All genes on a given autosomal chromosome are linked and do not assort independently unless crossing over occurs. Crossing over increases the amount of genetic recombination in a population.

Occasionally, a pair of homologous chromosomes may fail to disjoin in meiosis, and one of the gametes ends up with one chromosome too many and the other with $n - 1$ chromosomes. Resulting zygotes usually do not survive; humans with $2n + 1$ chromosomes may live, but they are born with serious abnormalities, such as Down syndrome.

One gene most commonly controls the production of one protein or polypeptide via an mRNA molecule (one gene–one polypeptide hypothesis), but the ribosomal and transfer RNA also are encoded on the genes.

The nucleic acids in the cell are DNA and RNA, which are large polymers of nucleotides composed of a nitrogenous base, pentose sugar, and phosphate group. The nitrogenous bases in DNA are adenine (A), guanine (G), thymine (T), and cytosine (C), and those in RNA are the same except that uracil (U) is substituted for thymine. DNA is a double-stranded, helical molecule in which the bases extend toward each other from the sugar-phosphate backbone: A always pairs with T and G with C. Thus the strands are antiparallel and complementary, being held in place by hydrogen bonds between the paired bases. In DNA replication the strands part, and the enzyme DNA polymerase synthesizes a new strand along each parent strand, using the parent strand as a template.

The sequence of the bases in DNA is a code for the amino-acid sequence in the ultimate product protein. Each triplet of three bases specifies a particular amino acid.

Proteins are synthesized by transcription of the information coded into DNA into the base sequence of a molecule of messenger RNA (mRNA), which functions in concert with ribosomes (containing ribosomal RNA [rRNA] and protein) and transfer RNA (tRNA). Ribosomes attach to the strand of mRNA and move along it, assembling the amino-acid sequence of the protein. Each amino acid is brought into position for assembly by a molecule of tRNA, which itself bears a base sequence (anticodon) complementary to the respective codons of the mRNA. In eukaryotes the sequences of bases in DNA coding for amino acids in a protein (exons) are interrupted by intervening sequences (introns). The introns are spliced out of the primary mRNA before it leaves the nucleus, and the protein is synthesized in the cytoplasm.

Genes, and the synthesis of the products for which they are responsible, must be regulated: turned on or off in response to varying environmental conditions or cell differentiation. Gene regulation in eukaryotes is complex, and a number of possible mechanisms are known. Transcriptional control is probably the most important.

Modern methods in molecular genetics have made spectacular advances possible. Restriction endonucleases cleave DNA at specific base sequences, and such cleaved DNA from different sources can be rejoined to form recombinant DNA. Combining mammalian with plasmid or viral DNA, a mammalian gene can be introduced into bacterial cells, which then multiply and express the mammalian gene. The polymerase chain reaction (PCR) makes it relatively simple to clone specific genes if only a small sequence of the gene is known.

A mutation is a physicochemical alteration in the bases of the DNA that changes the effect of the gene. Although rare and usually detrimental to the survival and reproduction of the organism, mutations are occasionally beneficial and provide new genetic material on which natural selection can work.

Cancer (neoplastic growth) is associated with a series of genetic changes in a clone of cells that allow unrestrained proliferation of those cells. Oncogenes (such as the gene coding for Ras protein) and inactivation of tumor suppressor genes (such as that coding for p53 protein) have been implicated in many cancers.

Review Questions

1. What is the relationship between homologous chromosomes and alleles?
2. Describe or diagram the sequence of events in meiosis (both divisions).
3. What are the designations of the sex chromosomes in males of bugs, humans, and butterflies? How do the chromosomal mechanisms of determining sex differ in these three taxa?
4. Define the following: dominant, recessive, zygote, heterozygote, homozygote, phenotype, genotype, monohybrid cross, dihybrid cross.
5. Diagram by Punnett square a cross between individuals with the following genotypes: $A/a \times A/a$; $A/a\ B/b \times A/a\ B/b$.
6. Concisely state Mendel's law of segregation and his law of independent assortment.
7. Assuming brown eyes *(B)* are dominant over blue eyes *(b)*, determine the genotypes of all the following individuals. The blue-eyed son of two brown-eyed parents marries a brown-eyed daughter whose mother was brown eyed and whose father was blue eyed. Their child is blue eyed.
8. Recall that red color *(R)* in four-o'clock flowers is incompletely dominant over white *(R')*. In the following crosses, give the genotypes of the gametes produced by each parent and the flower color of the offspring: $R/R' \times R/R'$; $R'/R' \times R/R'$; $R/R \times R/R'$; $R/R \times R'/R'$.
9. A brown mouse is mated with two female black mice. When each female has produced several litters of young, the first female has had 48 black and the second female has had 14 black and 11 brown young. Can you deduce the pattern of inheritance of coat color and the genotypes of the parents?
10. Rough coat *(R)* is dominant over smooth coat *(r)* in guinea pigs, and black coat *(B)* is dominant over white *(b)*. If a homozygous rough black is mated with a homozygous smooth white, give the appearance of each of the following: F_1; F_2; offspring of F_1 mated with smooth, white parent; offspring of F_1 mated with rough, black parent.
11. Assume right-handedness *(R)* dominates over left-handedness *(r)* in humans, and that brown eyes *(B)* are dominant over blue *(b)*. A right-handed, blue-eyed man marries a right-handed, brown eyed woman. Their two children are right handed, blue eyed and left handed, brown eyed. The man marries again, and this time the woman is right handed and brown eyed. They have 10 children, all right handed and brown eyed. What are the genotypes of the man and his two wives?
12. In *Drosophila melanogaster,* red eyes are dominant to white and the recessive characteristic is on the X chromosome. Vestigal wings *(v)* are recessive to normal *(V)*. Give the phenotypes of and phenotypic ratios expected for the following crosses: $X^W/X^w\ V/v \times X^w/Y\ v/v$, $X^w/X^w V/v \times X^W/Y\ V/v$.
13. Assume that color blindness is a recessive character on the X

chromosome. A man and woman with normal vision have the following offspring: daughter with normal vision who has one color-blind and one normal son; daughter with normal vision who has six normal sons; and a color-blind son who has a daughter with normal vision. What are the probable genotypes of all the individuals?

14. Distinguish the following: euploidy, aneuploidy, and polyploidy; monosomy and trisomy.
15. Name the purines and pyrimidines in DNA and tell which pair with each other in the double helix. What are the purines and pyrimidines in RNA and to what are they complementary in DNA?
16. Explain how DNA is replicated.
17. Why is it not possible for a codon to consist of only two bases?
18. Explain the transcription and processing of mRNA in the nucleus.
19. Explain the role of mRNA, tRNA, and rRNA in protein synthesis.
20. What are some ways that genes can be regulated in eukaryotes?
21. In modern molecular genetics, what is recombinant DNA, and how is it prepared?
22. Name three sources of phenotypic variation.
23. Distinguish between proto-oncogene and oncogene. What are two mechanisms whereby cancer can be caused by genetic changes?
24. What are Ras protein and p53? How can mutations in the genes for these proteins contribute to cancer?
25. Outline the essential steps in the procedure for the polymerase chain reaction.

Selected References

Cavenee, W. K., and R. L. White. 1995. The genetic basis of cancer. Sci. Am. **272:**72–79 (Mar.). *Describes mutations in cells of colorectal cancer and brain tumors.*

Culotta, E., and D. E. Koshland, Jr. 1993. p53 sweeps through cancer research. Science **262:**1958–1961. *p53 was discovered in 1979, but it was 10 years before scientists began to uncover its importance.*

Erlich, H. A., D. Gelfand, and J. J. Sninsky. 1991. Recent advances in the polymerase chain reaction. Science **252:**1643–1651. *A review of recent developments in methods and applications of the PCR.*

Friend, S. 1994. p53: a glimpse at the puppet behind the shadow play. Science **265:**334–335. *A short summary of the crucial roles of p53 protein and how mutations in the gene coding for it lead to inactivation.*

Hall, A. 1994. A biochemical function for Ras—at last. Science **264:**1413–1414. *Ras protein is an enzyme in a signal transduction cascade stimulating a cell to divide.*

Klug, W. S. 1991. Concepts of genetics, ed. 3. New York, Macmillan Publishing Company. *A shorter text.*

Koshland, D. E., Jr. 1989. The engineering of species. Science **244:**1233. *This is the lead editorial in an issue of the journal containing several reviews on genetic engineering.*

Mange, A. P., and E. J. Mange. 1994. Basic human genetics. Sunderland, Massachusetts, Sinauer Associates. *A readable, introductory text concentrating on the genetics of the animal species of greatest concern to most of us.*

Marx, J. 1994. Oncogenes reach a milestone. Science **266:**1942–1944. *Research on oncogenes has helped us understand many normal cell processes.*

Mullis, K. B. 1990. The unusual origin of the polymerase chain reaction. Sci. Am. **262:**56–65 (Apr.). *How the author had the idea for the simple production of unlimited copies of DNA while driving through the mountains of California.*

Russell, P. J. 1992. Genetics, ed. 3. New York, Harper Collins Publishers. *Popular general genetics text.*

Verma, I. M. 1990. Gene therapy. Sci. Am. **263:**68–84 (Nov.). *A review of prospects for treating and preventing genetic diseases by putting healthy genes into the body.*

Weinberg, R. A. 1991. Tumor suppressor genes. Science **254:**1138–1146. *How inactivaton of tumor suppressor genes is a step in production of cancer.*

Evolution of Animal Diversity

chapter | four

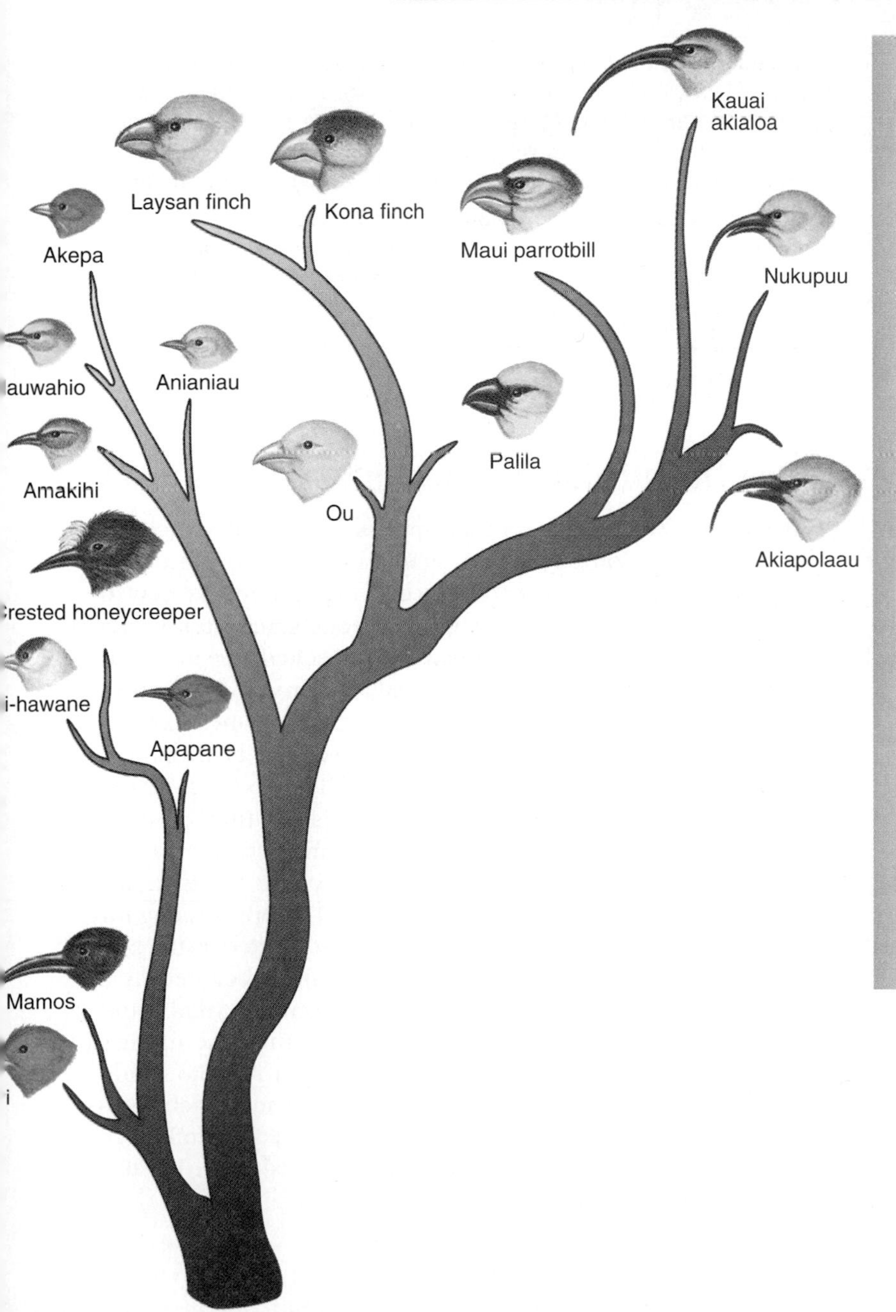

A Legacy of Change

The predominant feature of life's history is the legacy of perpetual change. Despite the apparent permanence of the natural world, change characterizes all things on Earth and in the universe. Countless kinds of animals and plants have flourished and disappeared, leaving behind an imperfect fossil record of their existence. Many, but not all, have left living descendants that bear a partial resemblance to them.

Life's changes are perceived and measured in many ways. On a short evolutionary timescale, we see changes in the frequencies of different genetic variants within populations. Evolutionary changes in the relative frequencies of light- and dark-colored moths were observed within a single human lifetime in the polluted countryside of industrial England. The fossil record reveals the formation of new species and dramatic changes in the appearances of organisms on longer timescales covering 100,000 to 1 million years. Major evolutionary trends and periodic mass extinctions occur on even larger timescales covering tens of millions of years. The fossil record of horses through the past 50 million years shows a series of different species replacing older ones through time and ending with the familiar horses that we know today. The fossil record of marine invertebrates shows us a series of mass extinctions separated by intervals of approximately 26 million years.

The Earth bears its own record of the irreversible, historical change that we call organic evolution. Because every feature of life as we know it today is a product of the evolutionary process, biologists consider organic evolution the keystone of all biological knowledge.

The wonder of the earth is not only that it harbors life, but that life on earth is so enormously diverse. Even from space, the diversity of life is evident in the kaleidoscopic patterns of grassland, forests, lakes and croplands stretching across the earth's surface, their colors changing with the seasons. Each ecosystem is biologically and physically distinct and is occupied by a great variety of organisms, living independently and manifesting diverse adaptations to their surroundings. Animals are an integral part of nearly every ecosystem on the earth, and their adaptive diversity is astounding.

To understand the diversity of animal life, we must study its long history, which began nearly 600 million years ago. From the earliest animals to the millions of animal species living today, this history demonstrates perpetual change, which we call **evolution.** We can depict the history of animal life as a branching genealogical tree, called a **phylogeny.** We place the earliest species ancestral to all animals at the trunk; then all living animal species fall at the growing tips of the branches. Each successive branching event on the tree represents the formation of new species from an ancestral one. The newly formed species inherit many characteristics from their immediate ancestor, but they also evolve new features that appear for the first time in the history of life. Each branch therefore has its own unique combination of characteristics and contributes a new dimension to the broad spectrum of animal diversity.

The study of animal evolution has two major goals. The first is to reconstruct the phylogeny of animal life and to find where in evolutionary history we see the origins of multicellularity, coelomic organization, spiral cleavage, vertebrae, homeothermy, and the many other features that comprise animal diversity as we know it today. The second major goal is to understand the historical processes that have generated diverse species and adaptations throughout evolutionary history. Darwin's theory of evolution provides the foundation for attaining both of these goals. This chapter is devoted to tracing the history of Darwin's theory, evaluating its major claims, and presenting some of the major discoveries regarding animal evolution to which it led.

figure 4-1

Founders of the theory of natural selection. **A,** Charles Robert Darwin (1809 to 1882), as he appeared in 1881, the year before his death. **B,** Alfred Russel Wallace (1823 to 1913) in 1895. Darwin and Wallace independently developed the same theory. A letter and essay from Wallace written to Darwin in 1858 spurred Darwin into writing *The Origin of Species,* published in 1859.

History of Evolutionary Theory

Charles Robert Darwin and Alfred Russel Wallace (Figure 4-1) were the first to establish evolution as a powerful scientific theory. Today the reality of organic evolution can be denied only by abandoning reason. As the noted English biologist Sir Julian Huxley wrote, "Charles Darwin effected the greatest of all revolutions in human thought, greater than Einstein's or Freud's or even Newton's, by simultaneously establishing the fact and discovering the mechanism of organic evolution." Darwinian theory allows us to understand both the genetics of populations and long-term trends in the fossil record. Darwin and Wallace were not the first, however, to consider the basic idea of organic evolution, which has an ancient history. We review the history of evolutionary thinking as it led to Darwin's theory and then discuss evidence supporting it.

Pre-Darwinian Evolutionary Ideas

Before the eighteenth century, speculation on the origin of species rested on myth and superstition, not on anything resembling a testable scientific theory. Creation myths viewed the world as a constant entity that did not change after its creation. Nevertheless, some thinkers approached the idea that nature has a long history of perpetual and irreversible change.

Early Greek philosophers, notably Xenophanes, Empedocles, and Aristotle, developed a primitive idea of evolutionary change. They recognized fossils as evidence for a former life that they believed had been destroyed by natural catastrophe. Despite their spirit of intellectual inquiry, the Greeks failed to establish an evolutionary concept, and the issue declined well before the rise of Christianity. The opportunity for evolutionary thinking became even more restricted as the biblical account of the earth's creation became accepted as a tenet of faith. The year 4004 B.C. was fixed by Archbishop James Ussher (mid-seventeenth century) as the time of life's creation. Evolutionary views were considered rebellious and heretical. Still, some speculation continued. The French naturalist Georges Louis Buffon (1707–1788) stressed the influence of environment on the modifications of animal type. He also extended the age of the earth to 70,000 years.

Lamarckism: The First Scientific Explanation of Evolution

The first complete explanation of evolution was authored by the French biologist, Jean Baptiste de Lamarck (1744–1829) (Figure 4-2) in 1809, the year of Darwin's birth. He made the first convincing case for the idea that fossils were the remains of extinct animals. Lamarck's evolutionary mechanism, **inheritance of acquired characteristics,** was engagingly simple: organisms, by striving to meet the demands of their environments, acquire adaptations and pass them by heredity to their offspring. According to Lamarck, the giraffe evolved its long neck because its ancestors lengthened their necks by stretching to obtain food and then passed the lengthened neck to their offspring. Over many generations, these changes accumulated to produce the long neck of the modern giraffe.

figure 4-2

Jean Baptiste de Lamarck (1744–1829), French naturalist who offered the first scientific explanation of evolution. Lamarck's hypothesis that evolution proceeds by the inheritance of acquired characteristics has been disproven.

figure 4-3

Sir Charles Lyell (1797–1875), English geologist and friend of Darwin. His book *Principles of Geology* greatly influenced Darwin during Darwin's formative period. This photograph was made about 1856.

We call Lamarck's concept of evolution **transformational,** because it claims that individual organisms transform their appearance to produce evolution. We now reject transformational theories because genetic studies show that traits acquired by an organism during its lifetime, such as strengthened muscles, are not inherited by offspring (Chapter 3). Darwin's evolutionary theory differs from Lamarck's in being a **variational** theory. Evolutionary change is caused by the differential survival and reproduction of genetically variable organisms, not by inheritance of acquired characteristics.

Charles Lyell and Uniformitarianism

The geologist Sir Charles Lyell (1797–1875) (Figure 4-3) established in his *Principles of Geology* (1830–1833) the principle of **uniformitarianism.** Uniformitarianism encompasses two important principles that guide the scientific study of the history of nature. These principles are (1) that the laws of physics and chemistry remain the same throughout the history of the earth, and (2) that past geological events occurred by natural processes similar to those that we observe in action today. Lyell showed that natural forces, acting over long periods of time, could explain the formation of fossil-bearing rocks. Lyell's geological studies led him to conclude that the earth's age must be reckoned in millions of years. These principles were important for discrediting miraculous and supernatural explanations of the history of nature and replacing them with scientific explanations. Lyell also stressed the gradual nature of geological changes that occur through time, and he argued further that such changes have no inherent directionality. We will see that both of these claims left important marks on Darwin's evolutionary theory.

Darwin's Great Voyage of Discovery

"After having been twice driven back by heavy southwestern gales, Her Majesty's ship *Beagle,* a ten-gun brig, under the command of Captain Robert FitzRoy, R.N., sailed from Devonport on the 27th of December, 1831." Thus began Charles Darwin's account of the historic five-year voyage of the *Beagle* around the world (Figure 4-4). Darwin, not quite 23 years old, had been asked to serve as naturalist without pay on the *Beagle,* a small vessel only 90 feet in length, which was about to depart on an extensive surveying voyage to South America and the Pacific. It was the beginning of one of the most important voyages of the nineteenth century.

During the voyage (1831–1836) (Figure 4-5), Darwin endured seasickness and the erratic companionship of the authoritarian Captain FitzRoy. But Darwin's youthful physical strength and early training as a naturalist equipped him for his work. The *Beagle* made many stops along the harbors and coasts of South America and adjacent regions. Darwin made extensive collections and observations on the fauna and flora of these regions. He unearthed numerous fossils of animals long extinct and noted the resemblance between fossils of the South American pampas and the known fossils of North America. In the Andes he encountered seashells embedded in rocks at 13,000 feet. He experienced a severe earthquake and watched the mountain torrents that relentlessly wore away the earth. These observations strengthened his conviction that natural forces were responsible for the geological features of the earth.

In mid-September of 1835, the *Beagle* arrived at the Galápagos Islands, a volcanic archipelago straddling the equator 600 miles west of Ecuador (Figure 4-6). The fame of the islands

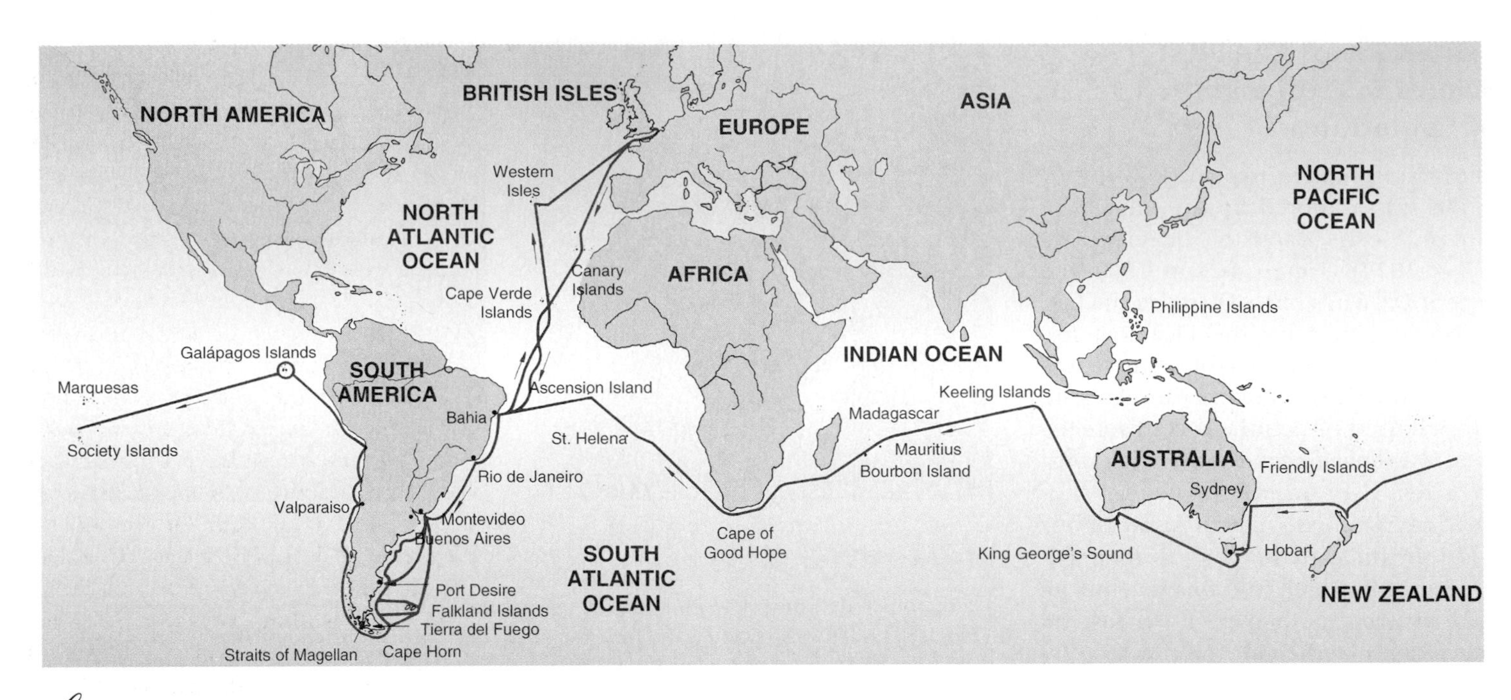

figure 4-4

Five-year voyage of H.M.S. *Beagle.*

figure 4-5

Charles Darwin and H.M.S. *Beagle.* **A,** Darwin in 1840, four years after the *Beagle* returned to England, and a year after his marriage to his cousin, Emma Wedgwood. **B,** The H.M.S. *Beagle* sails in Beagle Channel, Tierra del Fuego, on the southern tip of South America in 1833. This watercolor was painted by Conrad Martens, one of two official artists during the voyage of the *Beagle.*

stems from their infinite strangeness. They are unlike any other islands on earth. Some visitors today are struck with awe and wonder; others with a sense of depression and dejection. Circled by capricious currents, surrounded by shores of twisted lava, bearing skeletal brushwood baked by the equatorial sun, almost devoid of vegetation, inhabited by strange reptiles and by convicts stranded by the Ecuadorian government, the islands indeed had few admirers among mariners. By the middle of the seventeenth century, the islands were already known to the Spaniards as "Las Islas Galápagos"—the tortoise islands. The giant tortoises, used for food first by buccaneers and later by American and British whalers, sealers, and ships of war, were the islands' principal attraction. At the time of Darwin's visit, the tortoises already were heavily exploited.

figure 4-6

The Galápagos Islands viewed from the rim of a volcano.

figure 4-7

Darwin's study at Down House in Kent, England, is preserved today much as it was when Darwin wrote *The Origin of Species.*

During the *Beagle's* five-week visit to the Galápagos, Darwin began to develop his views of the evolution of life on earth. His original observations of the giant tortoises, marine iguanas, mockingbirds, and ground finches, all contributed to the turning point in Darwin's thinking.

Darwin was struck by the fact that, although the Galápagos Islands and the Cape Verde Islands (visited earlier in this voyage of the *Beagle*) were similar in climate and topography, their fauna and flora were altogether different. He recognized that Galápagos plants and animals were related to those of the South American mainland, yet differed from them in curious ways. Each island often contained a unique species of a particular group of animals that was related to forms on other islands. In short, Galápagos life must have originated in continental South America and then undergone modification in the various environmental conditions of the different islands. He concluded that living forms were neither divinely created nor immutable; they were, in fact, the products of evolution. Although Darwin devoted only a few pages to Galápagos animals and plants in his monumental *On the Origin of Species,* published more than two decades later, his observations on the unique character of the animals and plants were, in his own words, the "origin of all my views."

On October 2, 1836, the *Beagle* returned to England, where Darwin conducted the remainder of his scientific work (Figure 4-7). Most of Darwin's extensive collections had preceded him there, as had most of his notebooks and diaries kept during the cruise. Darwin's journal was published three years after the *Beagle's* return to England. It was an instant success and required two additional printings within the first year. In later versions, Darwin made extensive changes and titled his book *The Voyage of the Beagle.* The fascinating account of his observations written in a simple, appealing style has made the book one of the most lasting and popular travel books of all time.

Curiously, the main product of Darwin's voyage, his theory of evolution, did not appear in print for more than 20 years after the *Beagle's* return. In 1838 he "happened to read for amusement" an essay on populations by T.R. Malthus (1766–1834), who stated that animal and plant populations, including human populations, tend to increase beyond the capacity of the environment to support them. Darwin had already been gathering information on the artificial selection of animals under domestication by humans. After reading

Malthus's article, Darwin realized that a process of selection in nature, a "struggle for existence" because of overpopulation, could be a powerful force for evolution of wild species.

He allowed the idea to develop in his own mind until it was presented in 1844 in a still-unpublished essay. Finally in 1856 he began to assemble his voluminous data into a work on the origin of species. He expected to write four volumes, a very big book, "as perfect as I can make it." However, his plans were to take an unexpected turn.

In 1858, he received a manuscript from Alfred Russel Wallace (1823–1913), an English naturalist in Malaya with whom he was corresponding. Darwin was stunned to find that in a few pages, Wallace summarized the main points of the natural selection theory on which Darwin had been working for two decades. Rather than withhold his own work in favor of Wallace as he was inclined to do, Darwin was persuaded by two close friends, the geologist Lyell and the botanist Hooker, to publish his views in a brief statement that would appear together with Wallace's paper in the *Journal of the Linnean Society.* Portions of both papers were read before an unimpressed audience on July 1, 1858.

For the next year, Darwin worked urgently to prepare an "abstract" of the planned four-volume work. This was published in November 1859, with the title *On the Origin of Species by Means of Natural Selection, or the Preservation of Favoured Races in the Struggle for Life.* The 1250 copies of the first printing were sold the first day! The book instantly generated a storm that has never completely abated. Darwin's views were to have extraordinary consequences on scientific and religious beliefs and remain among the greatest intellectual achievements of all time.

Once Darwin's caution had been swept away by the publication of *On the Origin of Species,* he entered an incredibly productive period of evolutionary thinking for the next 23 years, producing book after book. He died April 19, 1882, and was buried in Westminster Abbey. The little *Beagle* had already disappeared, having been retired in 1870 and presumably broken up for scrap.

> *"Whenever I have found that I have blundered, or that my work has been imperfect, and when I have been contemptuously criticized, and even when I have been overpraised, so that I have felt mortified, it has been my greatest comfort to say hundreds of times to myself that 'I have worked as hard and as well as I could, and no man can do more than this.'"*
> Charles Darwin, in his autobiography, 1876.

Darwin's Theory of Evolution

Darwin's theory of evolution is now over 130 years old. Biologists today are frequently asked, "What is Darwinism?" and "Do biologists still accept Darwin's theory of evolution?" These questions cannot be given simple answers because

figure 4-8

Professor Ernst Mayr, a major contributor to our knowledge of evolutionary biology, particularly theories of speciation.

Darwinism encompasses several different, although mutually compatible, theories. Professor Ernst Mayr of Harvard University (Figure 4-8) has argued that Darwinism should be viewed as five major theories. These five theories have somewhat different origins and different fates and cannot be discussed accurately as if they were only a single statement. The theories are (1) **perpetual change,** (2) **common descent,** (3) **multiplication of species,** (4) **gradualism,** and (5) **natural selection.** The first three theories are generally accepted as having universal application throughout the living world. The theories of gradualism and natural selection are controversial among evolutionists. Gradualism and natural selection are clearly part of the evolutionary process, but they might not be as pervasive as Darwin thought. Legitimate controversies regarding gradualism and natural selection often are misrepresented by creationists as challenges to the first three theories whose validity is strongly supported by all relevant facts.

1. **Perpetual change.** This is the basic theory of evolution on which the others are based. It states that the living world is neither constant nor perpetually cycling, but is always changing. The properties of organisms undergo modification across generations throughout time. This theory originated in antiquity but did not gain widespread acceptance until Darwin advocated it in the context of his other four theories. "Perpetual change" is documented by the fossil record, which clearly refutes creationists' claims for a recent origin of all living forms. Because it has withstood repeated testing and is supported by an overwhelming number of observations, we now regard "perpetual change" as a scientific fact.

2. **Common descent.** The second Darwinian theory, "common descent," states that all forms of life descended from a common ancestor through a branching of lineages (Figure 4-9). The opposing argument, that the different forms of life arose independently and descended to the present in linear, unbranched genealogies, has been refuted by comparative

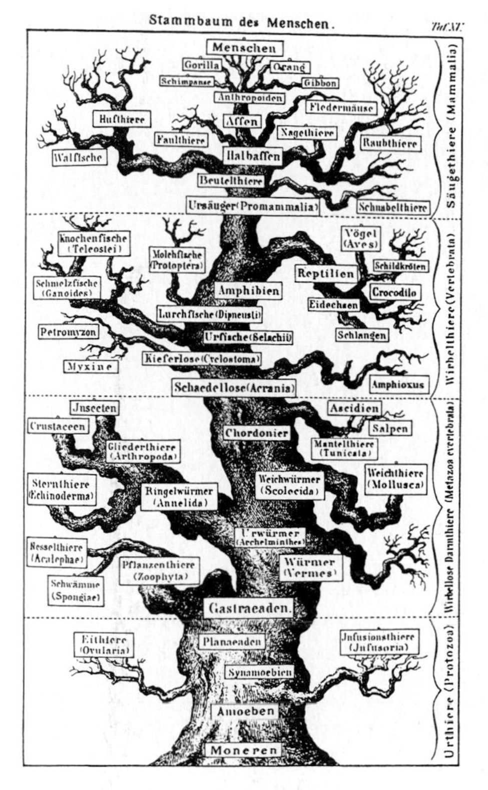

figure 4-9

An early tree of life drawn in 1874 by the German biologist, Ernst Haeckel, who was strongly influenced by Darwin's theory of common descent. Many of the phylogenetic hypotheses shown in this tree, including the unilateral progression of evolution toward humans (= Menschen, *top*), subsequently have been refuted.

studies of organismal form, cell structure, and macromolecular structures (including those of the genetic material, DNA). All of these studies confirm the theory that life's history has the structure of a branching evolutionary tree, known as a **phylogeny.** Species that share relatively recent common ancestry have more similar features at all levels than do species that have only an ancient common ancestry. Much current research is guided by Darwin's theory of common descent toward reconstructing life's phylogeny using the patterns of similarity and dissimilarity observed among species. The resulting phylogeny serves as the basis for our taxonomic classification of animals (Chapter 15).

3. **Multiplication of species.** Darwin's third theory states that the evolutionary process produces new species by the splitting and transformation of older ones. Species are now generally viewed as reproductively distinct populations of organisms that usually but not always differ from each other in organismal form. Once species are fully formed, interbreeding among members of different species does not occur. Evolutionists generally agree that the splitting and transformation of lineages produces new species, although there is still much controversy concerning the details of this process and the precise meaning of the term "species" (Chapter 15). Biologists are actively studying the historical processes that generate new species.

4. **Gradualism.** Darwin's theory of gradualism states that the large differences in anatomical traits that characterize different species originate through the accumulation of many small incremental changes over very long periods of time. This theory opposes the notion that large anatomical differences can evolve by sudden genetic changes. This theory is important because genetic changes having very large effects on organismal form are usually harmful to the organism. It is possible, however, that some genetic variants that have large effects on the organism are nonetheless sufficiently beneficial to be favored by natural selection. Therefore, although gradual evolution is known to occur, it may not explain the origin of all structural differences that we observe among species. Scientists are studying this question actively.

Thomas Henry Huxley (1825–1895), one of England's greatest zoologists, on first reading the convincing evidence of natural selection in Darwin's The Origin of Species *is said to have exclaimed, "How extremely stupid not to have thought of that!" He became Darwin's foremost advocate and engaged in often bitter debates with Darwin's critics. Darwin, who disliked publicly defending his own work, was glad to leave such encounters to his "bulldog," as Huxley called himself.*

5. **Natural selection.** Natural selection, Darwin's most famous theory, explains why organisms are constructed to meet the demands of their environments, a phenomenon called **adaptation.** This theory describes a natural process by which populations accumulate favorable characteristics throughout long periods of evolutionary time. Adaptation was viewed

Darwin's Explanatory Model of Evolution by Natural Selection

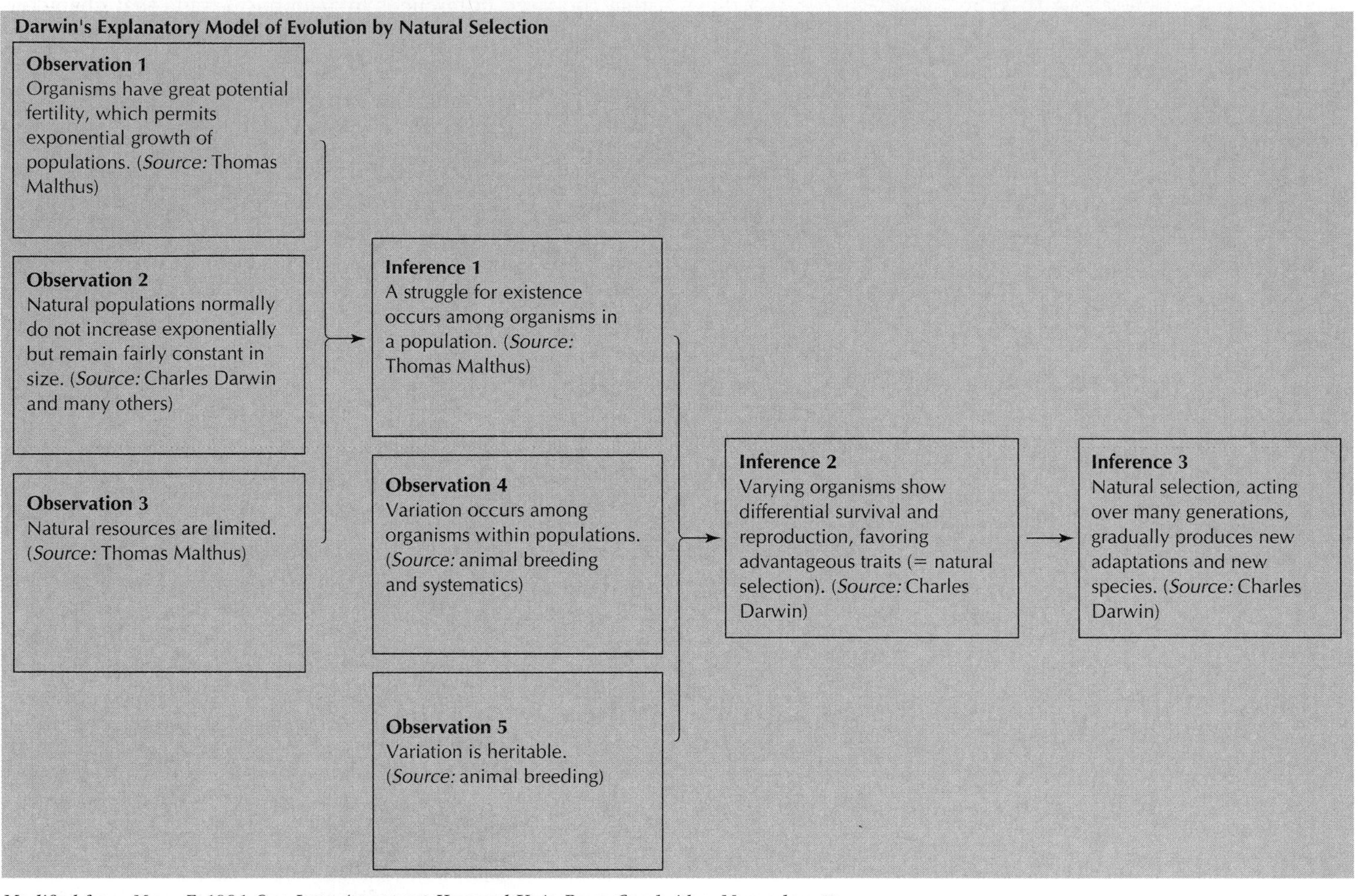

Modified from Mayr, E. 1991. One Long Argument. Harvard Univ. Press. Cambridge, Massachusetts.

previously as strong evidence against evolution. Darwin's theory of natural selection was therefore important for convincing people that a natural process, capable of being studied scientifically, could produce new adaptations and new species. Demonstration that natural processes could produce adaptation was important to the eventual acceptance of all five Darwinian theories. Darwin developed his theory of natural selection as a series of five observations and three inferences from them:

Observation 1—Organisms have great potential fertility. All populations produce large numbers of gametes and potentially large numbers of offspring each generation. Population size would increase exponentially at an enormous rate if all individuals that were produced each generation survived and reproduced. Darwin calculated that, even in a slow-breeding species such as the elephant, a single pair breeding from age 30 to 90 and having only six young could produce 19 million descendants in 750 years.

Observation 2—Natural populations normally remain constant in size, except for minor fluctuations. Natural populations fluctuate in size across generations and sometimes go extinct, but no natural populations show the continued exponential growth that their reproductive capacity theoretically could sustain.

Observation 3—Natural resources are limited. Exponential growth of a natural population would require unlimited natural resources to provide food and habitat for the expanding population, but natural resources are finite.

Inference 1—There exists a continuing struggle for existence among members of a population. Survivors represent only a part, often a very small part, of the individuals produced each generation. Darwin wrote in *The Origin of Species* that "it is the doctrine of Malthus applied with manifold force to the whole animal and vegetable kingdoms." The struggle for food, shelter, and space becomes increasingly severe as overpopulation develops.

Observation 4—All organisms show *variation*. No two individuals are exactly alike. They differ in size, color, physiology, behavior, and many other ways.

Observation 5—Variation is heritable. Darwin noted that offspring tend to resemble their parents, although he did not understand how. The hereditary mechanism discovered by Gregor Mendel would be applied to Darwin's theory many years later.

Inference 2—There is *differential survival and reproduction* among varying organisms in a population. Survival in the struggle for existence is not random with respect to hereditary variation present in the population. Some traits give their possessors an advantage in using the environment for effective survival and reproduction.

Inference 3—Over many generations, differential survival and reproduction generate new adaptations and new species. The differential reproduction of varying organisms gradually transforms species and results in the long-term "improvement" of types. Darwin knew that people often use hereditary variation to produce useful new breeds of livestock and plants. *Natural* selection acting over millions of years should be even more effective in producing new types than the *artificial* selection imposed during a human lifetime. Natural selection acting independently on geographically separated populations would cause them to diverge from each other, thereby generating the reproductive barriers that lead to speciation.

Natural selection can be viewed as a two-step process with a random component and a nonrandom component. The production of variation among organisms is the random component. The mutational process has no inherent tendency to generate traits that are favorable to the organism; if anything, the reverse is probably true. The nonrandom component is the survival of different traits. This is determined by the effectiveness of different traits in permitting their possessors to use environmental resources to survive and to reproduce. The phenomenon of differential survival and reproduction among varying organisms is now called **sorting** and should not be equated with natural selection. We now know that even random processes (genetic drift, p.106) can produce sorting. Darwin's theory of natural selection states that sorting occurs *because certain traits give their possessors advantages in survival and reproduction* relative to others that lack those traits. Selection is therefore a specific cause of sorting.

The popular phrase "survival of the fittest" was not originated by Darwin but was coined a few years earlier by the British philosopher Herbert Spencer, who anticipated some of Darwin's principles of evolution. Unfortunately the phrase later came to be coupled with unbridled aggression and violence in a bloody, competitive world. In fact, natural selection operates through many other characteristics of living things. The fittest animal may be the most helpful or the most caring. Fighting prowess is only one of several means toward successful reproductive advantage.

Evidence for Darwin's Five Theories of Evolution

Perpetual Change

Perpetual change in the form and diversity of animal life throughout its 600- to 700-million-year history is seen most directly in the fossil record. A **fossil** is a remnant of past life uncovered from the crust of the earth (Figure 4-10). Some fossils constitute complete remains (mammoths and amber insects), actual hard parts (teeth and bones), or petrified skeletal parts that are infiltrated with silica or other minerals (ostracoderms and molluscs). Other fossils include molds, casts, impressions, and fossil excrement (coprolites). In addition to documenting organismal evolution, fossils reveal profound changes in the earth's environment, including major changes in the distributions of lands and seas. Because many organisms left no fossils, a complete record of the past is always beyond our reach; nonetheless, discovery of new fossils and reinterpretation of existing ones expand our knowledge of how the form and diversity of animals changed through geological time.

Interpreting the Fossil Record

The fossil record is biased because preservation is selective. Vertebrate skeletal parts and invertebrates with shells and other hard structures left the best record (Figure 4-10). Soft-bodied animals, including the jellyfishes and most worms, are fossilized only under very unusual circumstances such as those that formed the Burgess shale of British Columbia (Figure 4-11). Exceptionally favorable conditions for fossilization produced the Precambrian fossil bed of south Australia, the tar pits of Rancho La Brea (Hancock Park, Los Angeles), the great dinosaur beds (Alberta, Canada, and Jensen, Utah; Figure 4-12) and the Olduvai Gorge of Tanzania.

Fossil remains may on rare occasions include soft tissues preserved so well that recognizable cellular organelles can be viewed with the electron microscope! Insects are frequently found entombed in amber, the fossilized resin of trees. One study of a fly entombed in 40-million-year-old amber revealed structures corresponding to muscle fibers, nuclei, ribosomes, lipid droplets, endoplasmic reticulum, and mitochondria (Figure 4-10D). This extreme case of mummification probably occurred because chemicals in the plant sap diffused into the embalmed insect's tissues.

Fossils are deposited in stratified layers with new deposits forming on top of older ones. If left undisturbed, which is rare, a sequence is preserved with the ages of fossils being directly proportional to their depth in the stratified layers. Characteristic fossils often serve to identify particular layers.

A

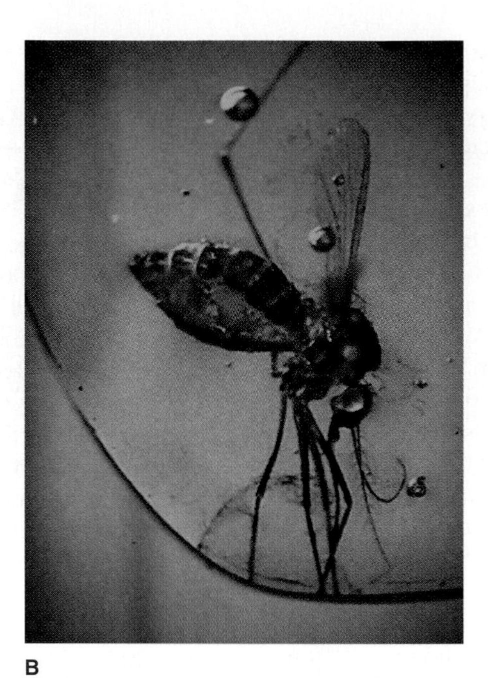

B

C

D

figure 4-10

Four examples of fossil material. **A,** Stalked crinoids (class Crinoidea, phylum Echinodermata p.533) from 85-million-year-old Cretaceous rocks. The fossil record of these echinoderms shows that they reached their peak millions of years earlier and began a slow decline to the present. **B,** An insect fossil that was encased in the resin of a tree 40 million years ago and that has since hardened into amber. **C,** Fish fossil from rocks of the Green River Formation, Wyoming. Such fish swam here during the Eocene epoch of the Tertiary period, approximately 55 million years ago. **D,** Electron micrograph of tissue from a fly fossilized as shown in **B;** the nucleus of a cell is screened in red.

Certain widespread marine invertebrate fossils, including various foraminiferans (p. 372) and echinoderms (Chapter 25), are such good indicators of specific geological periods that they are called "index," or "guide," fossils. Unfortunately, the layers are usually tilted or folded or show faults (cracks). Old deposits exposed by erosion may be covered with new deposits in a different plane. When exposed to tremendous pressures or heat, stratified sedimentary rock metamorphoses into crystalline quartzite, slate, or marble, which destroys fossils.

Geological Time

Long before the earth's age was known, geologists divided its history into a table of succeeding events based on the ordered layers of sedimentary rock. The "law of stratigraphy" produced a relative dating with the oldest layers at the bottom and the youngest at the top of the sequence. Time was divided into eons, eras, periods, and epochs as shown on the endpaper inside the back cover of this book. Time during the last eon (Phanerozoic) is expressed in eras (for example, Cenozoic), periods (for example, Tertiary), epochs (for example, Paleocene), and sometimes smaller divisions of an epoch.

In the late 1940s, radiometric dating methods were developed for determining the absolute age in years of rock formations. Several independent methods are now used, all based on the radioactive decay of naturally occurring elements into other elements. These "radioactive clocks" are independent of pressure and temperature changes and therefore are not affected by often violent earth-building activities.

One method, potassium-argon dating, depends on the decay of potassium-40 (^{40}K) to argon-40 (^{40}Ar) (12%) and calcium-40 (^{40}Ca) (88%).

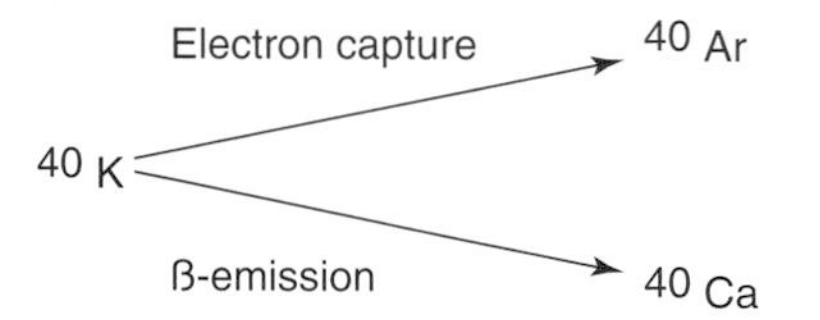

The half-life of potassium-40 is 1.3 billion years. This means that half of the original atoms will decay in 1.3 billion years, and half of the remaining atoms will be gone at the end of the next 1.3 billion years. This decay continues until all radioactive potassium-40 atoms are gone. To measure the age of the rock, one calculates the ratio of remaining potassium-40 atoms to the amount of potassium-40 originally there (the remaining potassium-40 atoms plus the argon-40 and calcium-40 into which they have decayed). Several such isotopes exist for dating purposes, some for dating the age of the earth itself. One of the most useful radioactive clocks depends on the decay of uranium into lead. With this method,

A

figure 4-11

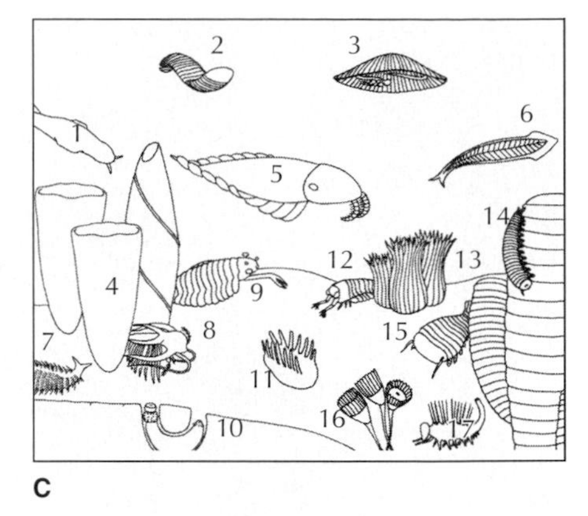

C

A, Fossil trilobites (p. 492) visible at the Burgess Shale Quarry, British Columbia. **B,** Animals of the Cambrian period, approximately 580 million years ago, as reconstructed from fossils preserved in the Burgess Shale of British Columbia, Canada. The main new body plans that appeared rather abruptly at this time established the body plans of animals familiar to us today. **C,** Key to Burgess Shale drawing. *Amiskwia* (1), from an extinct phylum; *Odontogriphus* (2), from an extinct phylum; *Eldonia* (3), a possible echinoderm; *Halichondrites* (4), a sponge; *Anomalocaris canadensis* (5), from an extinct phylum; *Pikaia* (6), an early chordate; *Canadia* (7), a polychaete; *Marrella splendens* (8), a unique arthropod; *Opabinia* (9), from an extinct phylum; *Ottoia* (10), a priapulid; *Wiwaxia* (11), from an extinct phylum, *Yohoia* (12), a unique arthropod; *Xianguangia* (13), an anemone-like animal; *Aysheaia* (14), an onychophoran or extinct phylum; *Sidneyia* (15), a unique arthropod; *Dinomischus* (16), from an extinct phylum; *Hallucigenia* (17), from an extinct phylum.

B

figure 4-12
A hadrosaur skeleton from Dinosaur Provincial Park, Alberta.

rocks over 2 billion years old can be dated with a probable error of less than 1%.

The fossil record of macroscopic organisms begins near the base of the Cambrian period of the Paleozoic era, approximately 600 million years BP. The period before the Cambrian is called the Precambrian era. Although the Precambrian era occupies 85% of all geological time, it has received much less attention than later eras, partly because oil, which provides the commercial incentive for much geological work, seldom exists in Precambrian formations. There is, however, evidence for life in the Precambrian era: well-preserved fossils of bacteria and algae, and casts of jellyfishes, sponge spicules, soft corals, segmented flatworms, and worm trails. Most, but not all, are microscopic fossils.

Evolutionary Trends

The fossil record allows us to view evolutionary change across the broadest scale of time. Species arise and then become extinct repeatedly throughout the fossil record. Animal species typically survive approximately 1 million to 10 million years, although their duration is highly variable. When we study patterns of species or taxon replacement through time, we observe trends. Trends are directional changes in the characteristic features or patterns of diversity in a group of organisms. Fossil trends clearly demonstrate Darwin's principle of perpetual change. We must emphasize that trends are observed only "after the fact." We cannot predict from the early fossils of a group what the appearance or diversity of the later fossils will be. The evolutionary process has no predetermined directions built into it.

A well-studied fossil trend is the evolution of horses from the Eocene epoch to the present (Figure 4-13). Looking back at the Eocene epoch, we see many different genera and species of horses that replaced each other through time (Figure 4-13). George Gaylord Simpson (p. 347) showed that this trend is compatible with Darwinian evolutionary theory. The three characteristics that show the clearest trends in horse evolution are body size, foot structure, and tooth structure. Compared to modern horses, the horses in extinct genera were small, their teeth had a relatively small grinding surface, and their feet had a relatively large number of toes (four). Throughout the subsequent Oligocene, Miocene, Pliocene, and Pleistocene epochs, there were continuing patterns of new genera arising and old ones becoming extinct. In each case, there was a net increase in body size, expansion of the grinding surface of the teeth, and reduction in the number of toes. As the number of toes was reduced, the central digit became increasingly more prominent in the foot and eventually only this central digit remained.

The fossil record shows a net change not only in the characteristics of horses but also variation in the numbers of different horse genera (and numbers of species) that exist through time. The many horse genera of past epochs have been lost to extinction, leaving only a single survivor. Evolutionary trends in diversity are observed in fossils of many different groups of animals (Figure 4-14).

Trends in fossil diversity through time are produced by different rates of species formation versus extinction through time. Why do some lineages generate large numbers of new species whereas others generate relatively few? Why do different lineages undergo higher or lower rates of extinction (of species, genera, or families) throughout evolutionary time? To answer these questions, we must turn to Darwin's other four theories of evolution. Regardless of how we answer these questions, however, the observed trends in animal diversity clearly illustrate Darwin's principle of perpetual change. Because the remaining four theories of Darwinism rely on the theory of perpetual change, evidence supporting these theories strengthens Darwin's theory of perpetual change.

Common Descent

Darwin proposed that all plants and animals have descended from "some one form into which life was first breathed."

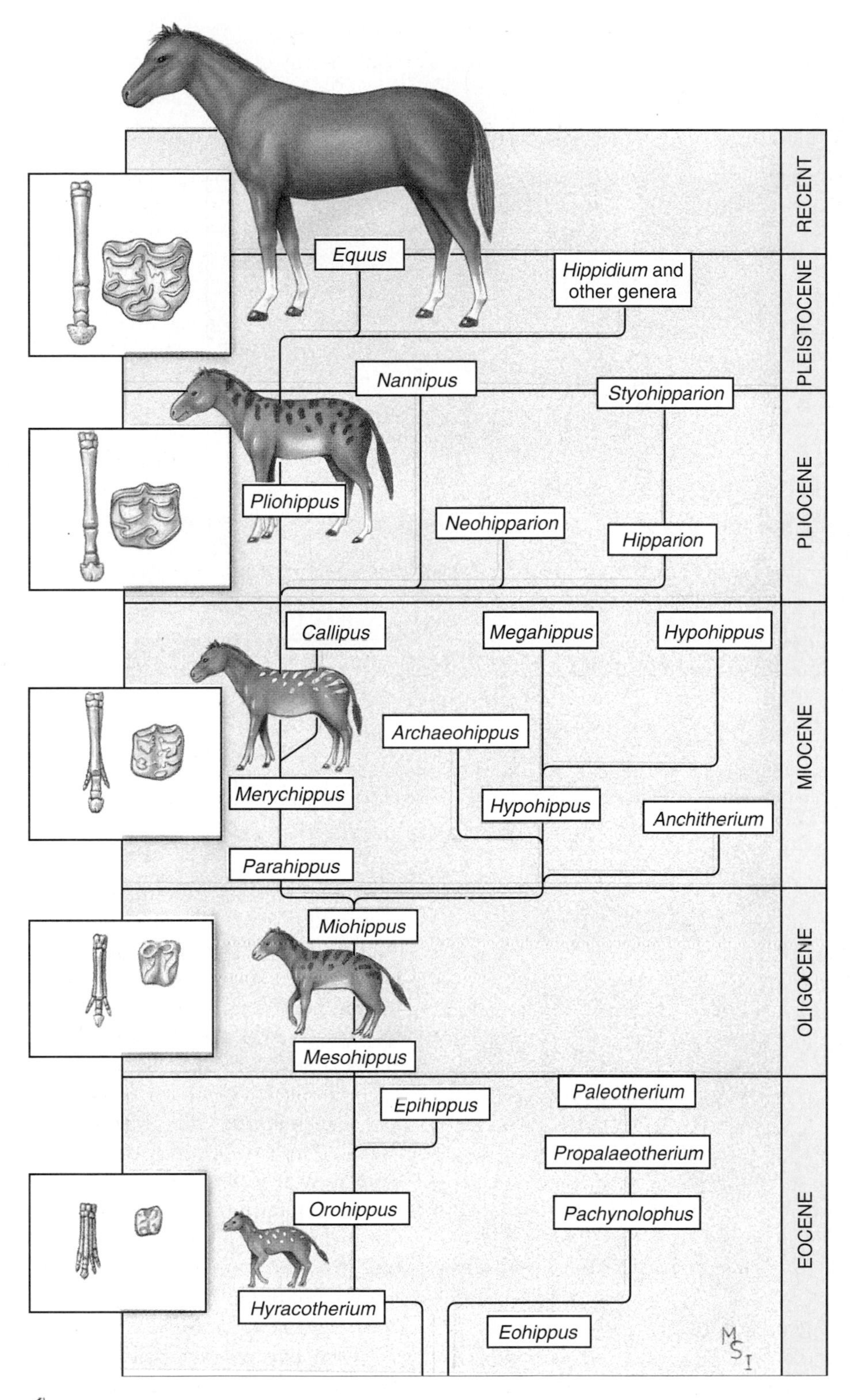

figure 4-13

A reconstruction of the genera of horses from the Eocene to the present. Evolutionary trends toward increased size, elaboration of molars, and loss of toes are shown together with a hypothetical genealogy of extant and fossil genera.

Life's history is depicted as a branching tree, called a **phylogeny,** that gives all of life a unified evolutionary history. Pre-Darwinian evolutionists, including Lamarck, advocated multiple independent origins of life, each of which gave rise to lineages that changed through time without extensive branching. Like all good scientific theories, common descent makes several important predictions that can be tested and potentially used to reject it. According to this theory, we should be able to trace the genealogies of all modern species backward until they converge on ancestral lineages shared with other species, both living and extinct. We should be able to continue this process, moving farther backward through evolutionary time, until we reach the primordial ancestor of all life on earth. All forms of life will connect to this tree somewhere, including many extinct forms that represent dead branches. Although reconstructing the history of life in this manner may seem almost impossible, it has in fact been extraordinarily successful. How has this difficult task been accomplished?

Homology and Phylogenetic Reconstruction

Darwin recognized the major source of evidence for common descent in the concept of **homology.** Darwin's contemporary, Richard Owen (1804–1892), used this term to denote "the same organ in different organisms under every variety of form and function." The classic example of homology is the limb skeleton of vertebrates. The bones of the vertebrate limb maintain characteristic structures and patterns of connection despite diverse modifications for different functions (Figure 4-15). According to Darwin's theory of common descent, the structures that we call homologies represent characteristics inherited with some modification from a corresponding feature in a common ancestor.

Darwin devoted an entire book, *The Descent of Man and Selection in Relation to Sex,* largely to the idea that humans share common descent with apes and other animals. This idea was

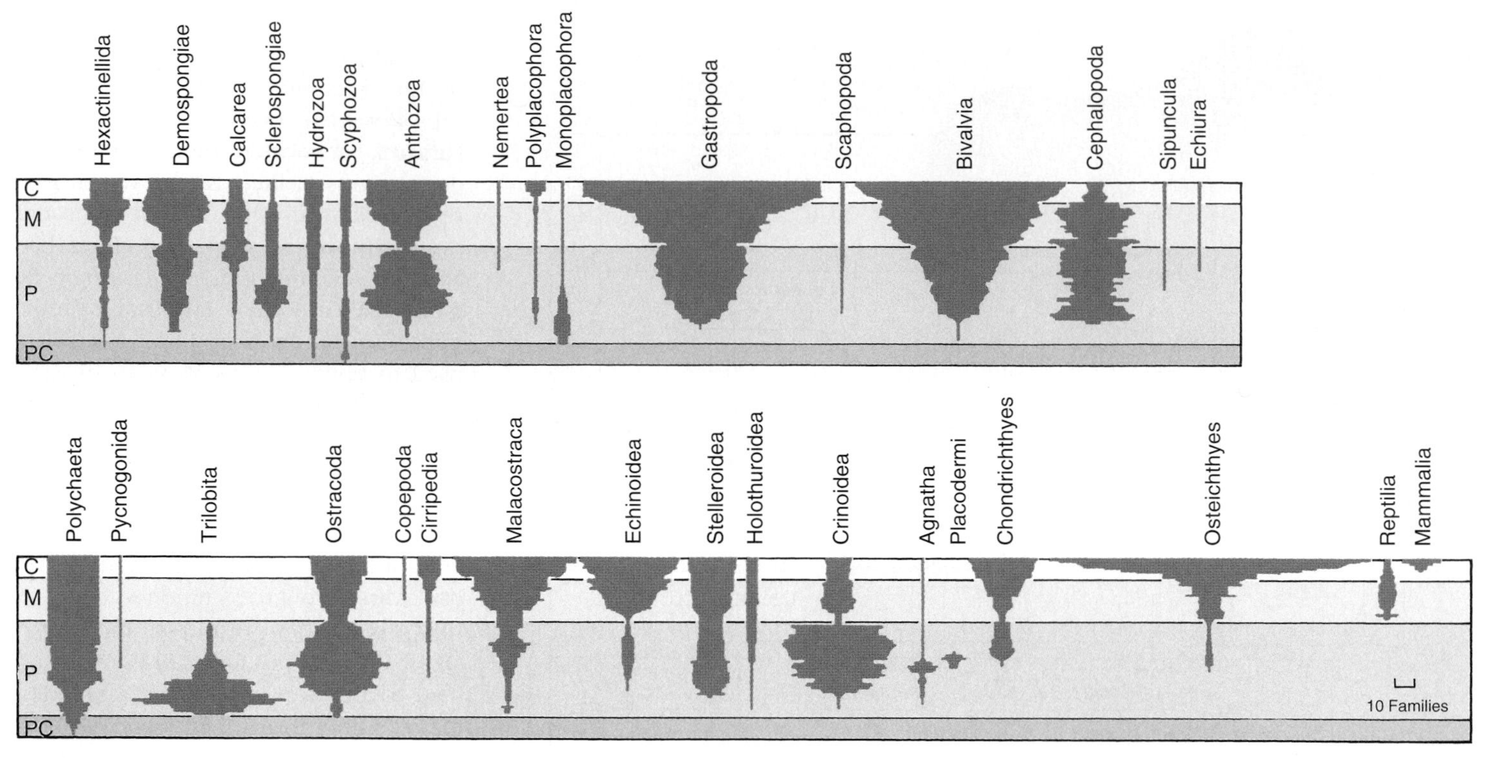

figure 4-14

Diversity profiles of taxonomic families from different animal groups in the fossil record. The scale marks off the Precambrian (PC) and the Paleozoic (P), Mesozoic (M), and Cenozoic (C) eras. The number of families is indicated from the width of the profile.

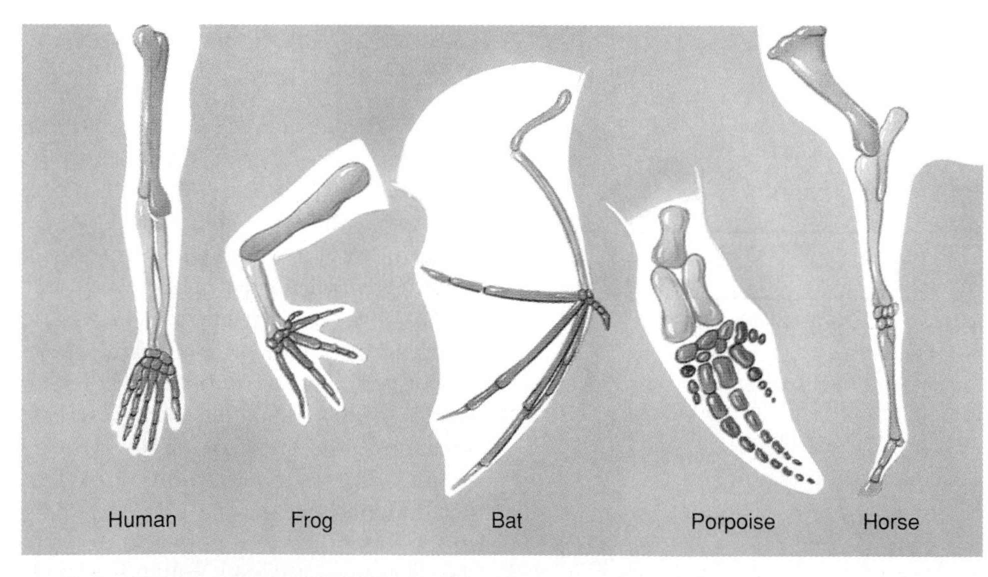

figure 4-15

Forelimbs of five vertebrates show skeletal homologies: green, humerus; yellow, radius and ulna; purple, "hand" (carpals, metacarpals, and phalanges). Homologies of bones and patterns of connection are evident despite evolutionary modification for various particular functions.

repugnant to the Victorian world, which responded with predictable outrage (Figure 4-16). Darwin built his case mostly on anatomical comparisons revealing homology between humans and apes. To Darwin, the close resemblances between apes and humans could be explained only by common descent.

Throughout the history of all forms of life, evolutionary processes generate new features that are transmitted across generations. Every time a new feature arises on a lineage destined to be ancestral to others, we see the origin of a new homology. The pattern formed by these homologies provides evidence for common descent and allows us to reconstruct the branching evolutionary history of life. We can illustrate such evidence using a phylogenetic tree of the ground-dwelling ratite birds (Figure 4-17). A new skeletal homology (see Figure 4-17) arises on each of the lineages shown (descriptions of the homologies are not included because they are highly technical). The different groups

figure 4-16

This 1873 advertisement for Merchant's Gargling Oil ridicules Darwin's theory of the common descent of humans and apes, which hardly received universal acceptance during Darwin's lifetime.

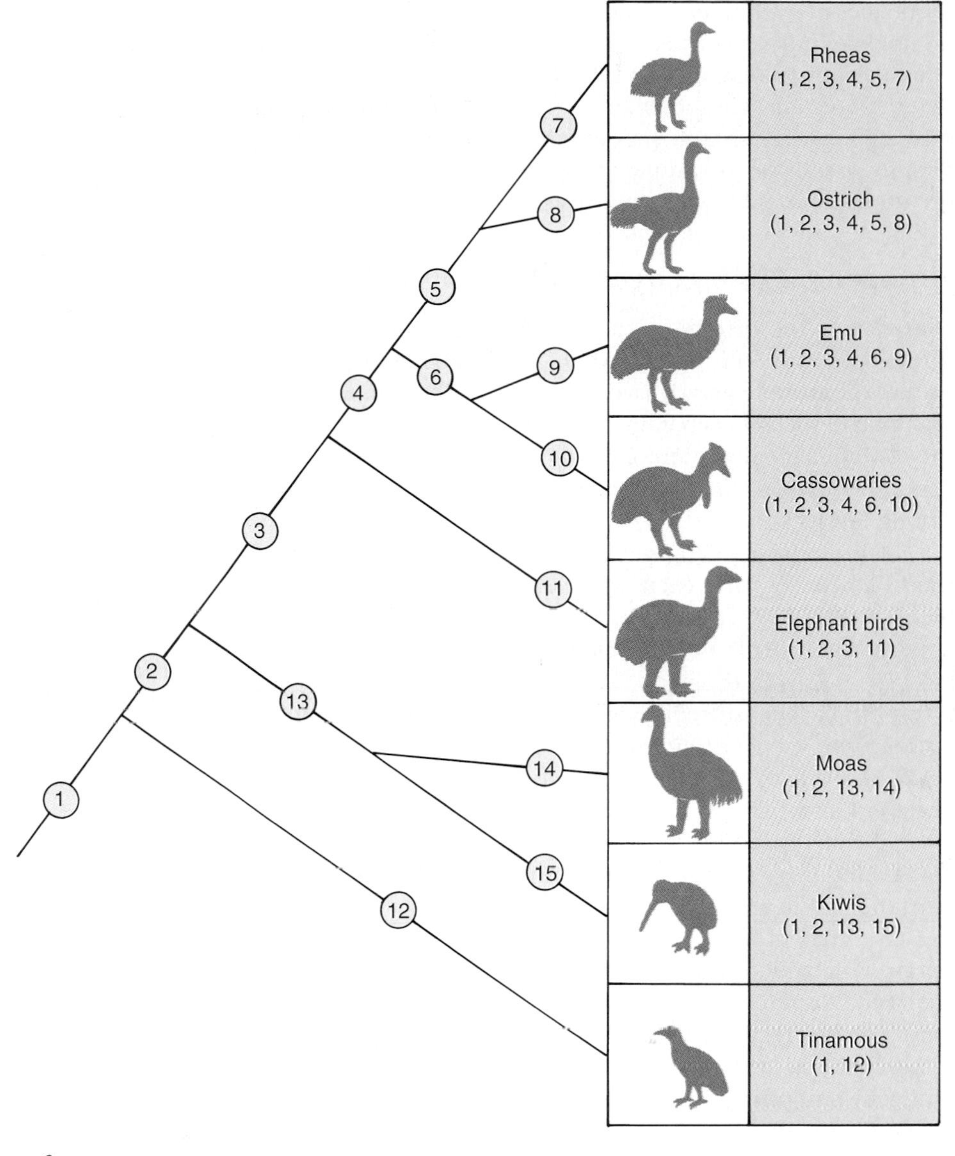

figure 4-17

The phylogenetic pattern specified by fifteen homologous structures in the skeletons of ratite birds. Homologous features are numbered 1 through fifteen and are marked both on the branches of the tree on which they arose and on the birds that have them. If you were to erase the tree structure, you would be able to reconstruct it without error from the distributions of homologous features shown for the birds at the terminal branches.

of species located at the tips of the branches contain different combinations of these homologies that reflect ancestry. For example, the ostriches show homologies 1 through 5 and 8, whereas the kiwis show homologies 1, 2, 13, and 15. The branches of the tree combine these species into a **nested hierarchy** of groups within groups (see Chapter 15). Smaller groups (species grouped near terminal branches) are contained within larger ones (species grouped by basal branches, including the trunk of the tree). If we erase the tree structure but retain the patterns of homology observed in the terminal groups of species, we will still be able to reconstruct the branching structure of the entire tree. Evolutionists test the theory of common descent by observing the patterns of homology present within all groups of organisms. The pattern formed by all homologies taken together should specify a single branching tree structure that represents the evolutionary genealogy of the group.

The nested hierarchical structure of homology is so pervasive in the living world that it forms the basis for our systematic classification of all forms of life (genera grouped into families, families grouped into orders, and so on). Hierarchical classification even preceded Darwin's theory because this pattern is so evident, but it was not explained adequately before Darwin. Once the idea of common descent was accepted, biologists began investigating the structural, molecular, and/or chromosomal homologies of all animal groups. Taken together, the nested hierarchical patterns uncovered by these studies have permitted us to reconstruct the evolutionary trees of many groups and to continue investigating others. The use of Darwin's theory of common descent to reconstruct the evolutionary history of life and to classify animals is the subject of Chapter 15.

Note that the earlier evolutionary hypothesis that life arose many times, forming unbranched lineages, predicts linear sequences of evolutionary change with no nested hierarchy of homologies among species. Because we do observe nested hierarchies of homologies, that hypothesis is rejected. Note also that the creationist argument, which is not a scientific hypothesis, can make no testable predictions about any pattern of homology; it only devises supernatural excuses for whatever observations are made by evolutionists.

Ontogeny, Phylogeny, and Recapitulation

Ontogeny is the history of the development of an organism through its entire life. Early developmental and embryological features contribute greatly to our knowledge of homology and common descent. Comparative studies of ontogeny show how the evolutionary alteration of developmental timing generates new phenotypes, thereby causing evolutionary divergence among lineages.

The German zoologist Ernst Haeckel, a contemporary of Darwin, believed that each successive stage in the development of an individual represented one of the adult forms that appeared in its evolutionary history. The human embryo with gill depressions in the neck was believed, for example, to signify a fishlike ancestor. On this basis Haeckel gave his generalization: *ontogeny (individual development) recapitulates (repeats) phylogeny (evolutionary descent).* This notion later became known simply as **recapitulation** or the **biogenetic law.** Haeckel based his biogenetic law on the flawed premise that evolutionary change occurs by successively adding stages onto the end of an unaltered ancestral ontogeny, compressing the ancestral ontogeny into earlier developmental stages. This notion was based on Lamarck's concept of the inheritance of acquired characteristics (p. 85).

The nineteenth-century embryologist, K.E. von Baer, gave a more satisfactory explanation of the relationship between ontogeny and phylogeny. He argued that early developmental features were simply more widely shared among different animal groups than later ones. For example, Figures 4-18 (p. 333) and 14-29 (p. 334) show early embryological similarities of organisms whose adult forms are very different. The adults of animals with relatively short and simple ontogenies often resemble pre-adult stages of other animals whose ontogeny is more elaborate, but the embryos of descendants do not necessarily resemble the adults of their ancestors. Even early development undergoes evolutionary divergence among groups, however, and it is not quite as stable as von Baer believed.

We now know that there are many parallels between ontogeny and phylogeny, but the features of an ancestral ontogeny can be shifted either to earlier or later stages in descendant ontogenies. Evolutionary change in the timing of development is called **heterochrony,** a term initially used by Haeckel to denote exceptions to recapitulation. Because the lengthening or shortening of ontogeny can change different parts of the body independently, we often see a mosaic of different kinds of developmental evolutionary change in a single lineage. Therefore, cases in which an entire ontogeny recapitulates phylogeny are rare.

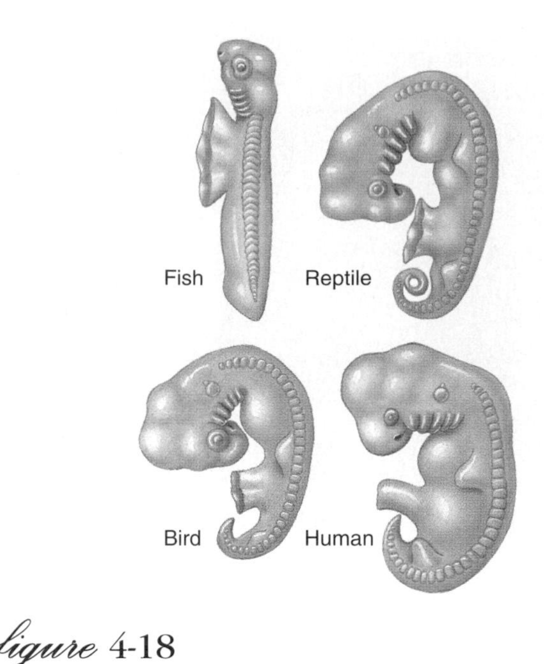

figure 4-18

Comparison of gill arches of different embryos. All are shown separated from the yolk sac. Note the remarkable similarity of the four embryos at this early stage in development.

Despite the many changes that have occurred throughout the years in scientific thinking about the relationship between ontogeny and phylogeny, one important fact remains clear. The theory of common descent is strengthened enormously by the many homologies found among the various developmental stages of organisms belonging to different species.

Multiplication of Species

The multiplication of species through time is a logical corollary to Darwin's theory of common descent. A branch point on the evolutionary tree means that an ancestral species has split into two different ones. Darwin's theory postulates that the variation present within a species, especially variation that occurs between geographically separated populations, provides the material from which new species are produced. Because evolution is a branching process, the total number of species produced by evolution increases through time, although most of these species eventually become extinct. A major challenge for evolutionists is to discover the process by which an ancestral species "branches" to form two or more descendant species.

Before we explore the multiplication of species, we must decide what we mean by "species." As we will see in Chapter 15, there is no consensus regarding the definition of species. Most biologists would agree, however, that important criteria for recognizing species include (1) descent from a common ancestral population, (2) reproductive compatibility (ability to

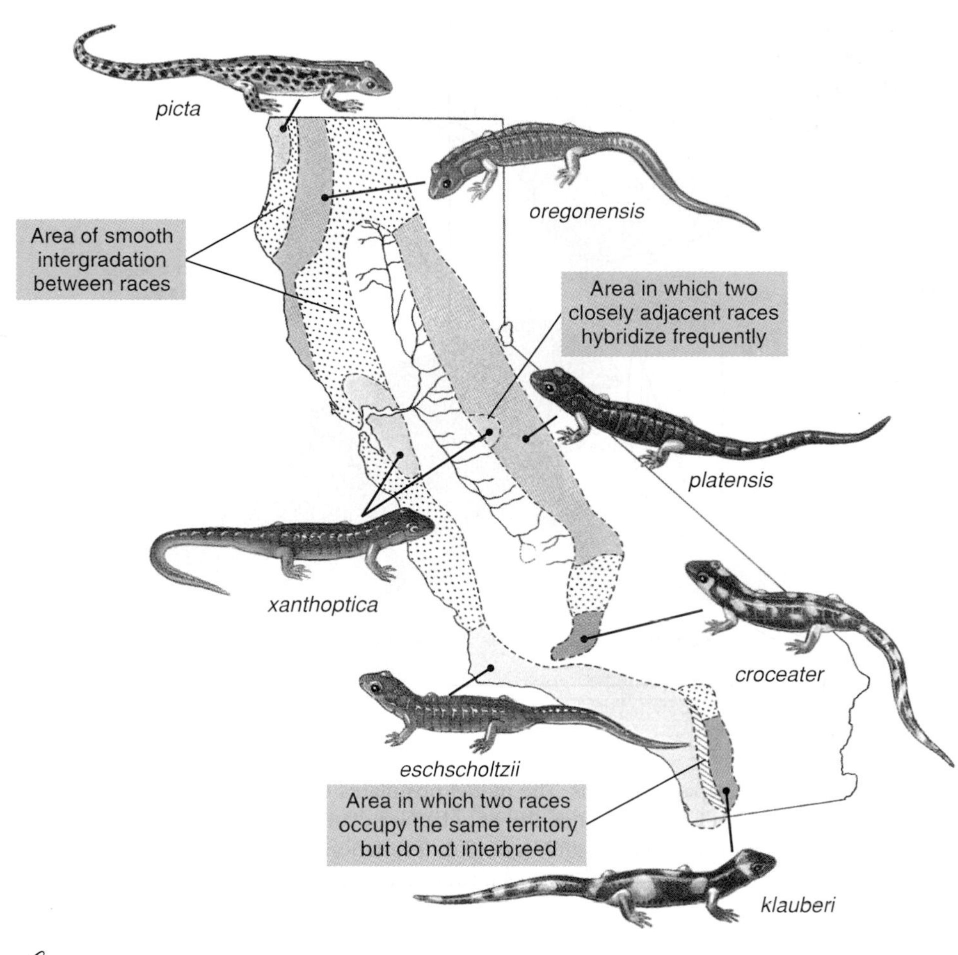

figure 4-19

Speciation in progress: geographic variation of color patterns in the salamander genus *Ensatina.* Populations of *Ensatina eschscholtzii* form a geographic ring around the Central Valley of California. Adjacent, differentiated populations throughout the ring can exchange genes except at the bottom of the ring where the subspecies *E. e. eschscholtzii* and *E. e. klauberi* overlap without interbreeding. These two subspecies would be recognized as distinct species if the intermediate populations linking them across the ring were extinct. This example demonstrates that reproductive barriers between populations can evolve gradually.

interbreed) within and reproductive incompatibility between species, and (3) maintenance within species of genotypic and phenotypic cohesion (lack of abrupt differences among populations in allelic frequencies [see the following text] and organismal appearance). The criterion of reproductive compatibility has received the greatest attention in studies of species formation, also called **speciation.**

The biological factors that prevent different species from interbreeding are called **reproductive barriers.** The primary problem of speciation is to discover how two initially compatible populations evolve reproductive barriers that cause them to become distinct, separately evolving lineages. How do populations diverge from each other in their reproductive properties while maintaining complete reproductive compatibility within each population?

Reproductive barriers between populations usually evolve gradually. Evolution of reproductive barriers requires that diverging populations must be kept physically separate for long periods of time. If the diverging populations were reunited before reproductive barriers were completely formed, interbreeding would occur between the populations and they would merge. Speciation by gradual divergence in animals may require extraordinarily long periods of time, perhaps 10,000 to 100,000 years or more. Geographical isolation followed by gradual divergence is the most effective way for reproductive barriers to evolve, and many evolutionists consider geographical separation a prerequisite for branching speciation. Speciation that results from the evolution of reproductive barriers between geographically separated populations is known as **allopatric speciation,** or geographic speciation.

Evidence for allopatric ("in another land") speciation occurs in many forms, but perhaps the most convincing is the occurrence of geographically separated but adjoining, closely-related populations that illustrate the gradual origin of reproductive barriers. Populations of the salamander, *Ensatina eschscholtzii,* in California are a particularly clear example (Figure 4-19). These populations show evolutionary divergence in color pattern and collectively form a geographic ring around California's central valley. Genetic exchange between differentiated, geographically adjoining populations is evident through the formation of hybrids and occasionally regions of extensive genetic exchange (called zones of introgression). Two populations at the southern tip of the geographical range (called *E. e. eschscholtzii* and *E. e. klauberi)* make contact but do not interbreed. A gradual accumulation of reproductive differences among contiguous populations around the ring is visible, with the two southernmost populations having achieved sufficient divergence to behave as different species.

Additional evidence for allopatric speciation comes from the observation of animal diversification on islands. Oceanic islands that were formed by volcanoes are initially devoid of life. They are gradually colonized by plants and animals from a continent or from other islands in separate invasions. The invaders often encounter situations ideal for evolutionary diversification, because environmental resources that were exploited heavily by other species on the mainland are free for

figure 4-22

The ancon breed of sheep arose from a "sporting mutation" that caused dwarfing of the legs. Many of his contemporaries criticized Darwin for his claim that such mutations are not important in the process of evolution by natural selection.

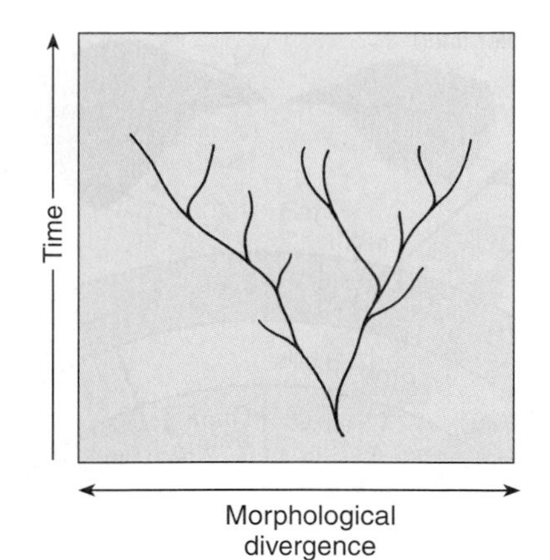

figure 4-23

The gradualist model of evolutionary change in morphology, viewed as proceeding more or less steadily through geological time (millions of years). Bifurcations followed by gradual divergence led to speciation.

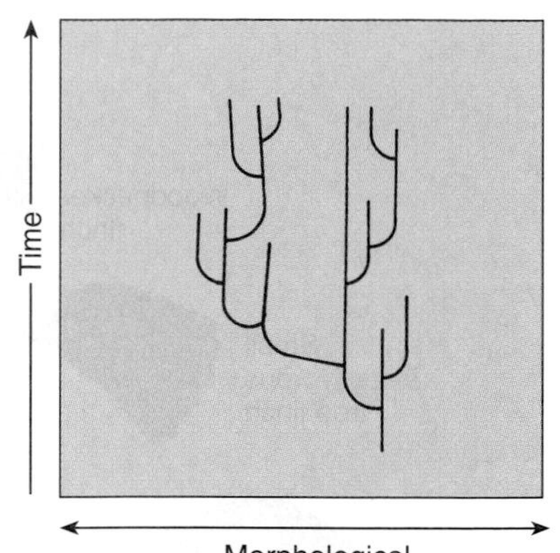

figure 4-24

The punctuated equilibrium model sees evolutionary change being concentrated in relatively rapid bursts of branching speciation (*lateral lines*) followed by prolonged periods of little change throughout geological time (millions of years).

page of fossil history in Africa. Peter Williamson, a British paleontologist working in fossil beds 400 m deep near Lake Turkana, documented a remarkably clear record of speciation in freshwater snails. The geology of the Lake Turkana basin reveals a history of instability. Earthquakes, volcanic eruptions, and climatic changes caused the waters to rise and fall periodically, sometimes by hundreds of feet. Thirteen lineages of snails show long periods of stability interrupted by relatively brief periods of rapid change in shell shape when snail populations were fragmented by receding waters. These populations diverged to produce new species that then remained unchanged through thick deposits before becoming extinct and being replaced by descendant species. The transitions occurred within 5000 to 50,000 years. In the few meters of sediment where speciation occurred, transitional forms were visible. Williamson's study conforms well to the punctuated equilibrium model, which remains an important challenge to gradualism on a geological timescale.

Natural Selection

Many examples show how natural selection alters populations in nature. Sometimes selection can proceed very rapidly as, for example, in the evolution of high resistance to insecticides by insects, especially flies and mosquitoes. Doses that at first killed almost all pests later were ineffective in controlling them. As more insects were exposed to insecticides, those most sensitive were killed, leaving more space and less competition for resistant strains to multiply. Thus, as a result of selection, mutations bestowing high resistance, but previously rare in the population, increased in frequency.

Perhaps the most famous instance of rapid selection is that of industrial melanism (dark pigmentation) in the peppered moth of England (Figure 4-25). Before 1850, the peppered moth was white with black speckling in the wings and body. In 1849, a mutant black form of the species appeared. It became increasingly common, reaching frequencies of 98% in Manchester and other heavily industrialized areas by 1900. The peppered moth, like most moths, is active at night. It rests during the day in exposed places, depending upon its cryptic coloration for protection. Experimental studies have shown that, consistent with the hypothesis of natural selection, birds are able to locate and to eat moths that do not match their surroundings, but that birds in the same area frequently fail to find moths that match their surroundings. The mottled pattern of the normal white form blends well with lichen-covered tree trunks. With increasing industrialization, the soot from thousands of chimneys darkened the bark of trees for miles around centers such as Manchester. Against this dark background, the white moth was conspicuous to predatory birds, whereas the mutant black form was camouflaged. When polluted areas were cleaned, the frequency of lightly pigmented individuals in moth populations increased (Figure 4-25C), consistent with the hypothesis of natural selection.

A recurring criticism of natural selection is that it cannot generate new structures or species but can only modify old ones. Most structures in their early evolutionary stages could not have performed the functions that the fully-formed structures perform, and therefore it is unclear how natural selection could have favored them. What use is half a wing or the rudiment of a feather for a flying bird? To answer this criticism, evolutionists propose that many structures evolved initially for purposes different from the ones that they have

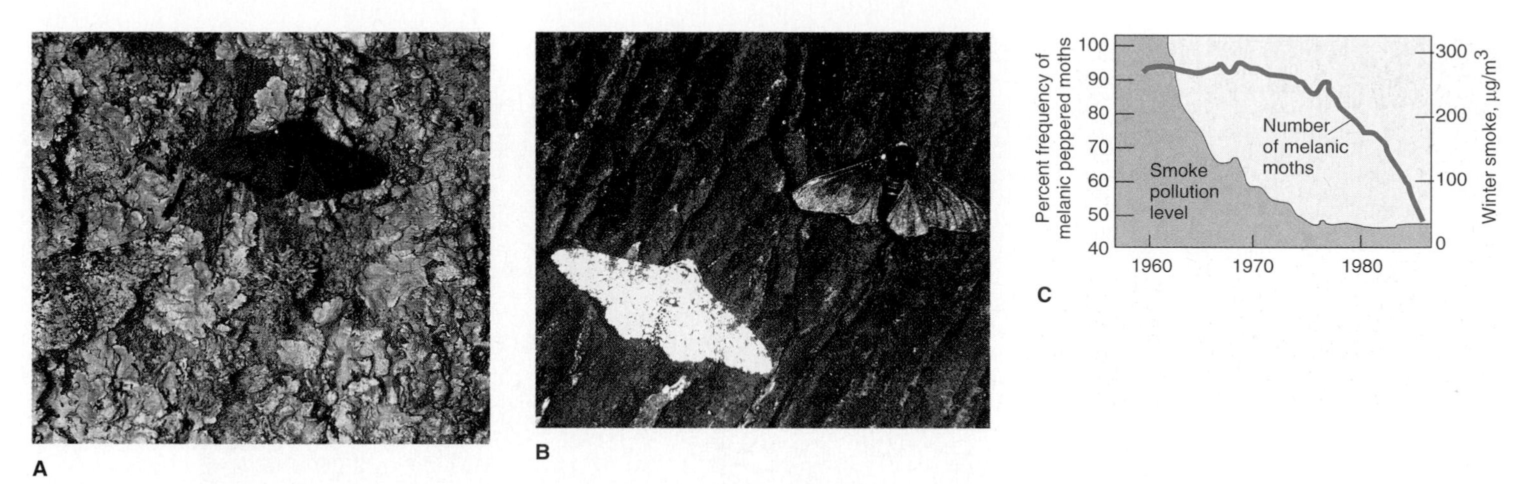

figure 4-25

Light and melanic forms of the peppered moth, *Biston betularia* on, **A,** a lichen-covered tree in unpolluted countryside and, **B,** a soot-covered tree near industrial Birmingham, England. These color variants have a simple genetic basis. **C,** Recent decline in the frequency of the melanic form of the peppered moth with falling air pollution in industrial areas of England. The frequency of the melanic form still exceeded 90% in 1960, when smoke and sulfur dioxide emissions were still high. Later, as emissions fell and light-colored lichens began to grow again on the tree trunks, the melanic form became more conspicuous to predators. By 1986, only 50% of the moths were still of the melanic form, the rest having been replaced by the light form.

today. Rudimentary feathers could have been useful in thermoregulation, for example. Their role in flying would have evolved later after they incidentally had acquired some aerodynamic properties that permitted them to be selected for improvement of flying. Because the anatomical differences observed among organisms from different, closely related species resemble variation observed within species, it is unreasonable to propose that selection will never lead beyond the species boundary.

Revisions of Darwinian Evolutionary Theory

Neo-Darwinism

The most serious weakness in Darwin's argument was his failure to identify correctly the mechanism of inheritance. Darwin saw heredity as a blending phenomenon in which the characteristics of the parents melded together in the offspring. Darwin also invoked the Lamarckian hypothesis that an organism could alter its heredity through the use and disuse of body parts and through the direct influence of the environment. Darwin did not realize that hereditary factors could be discrete and non-blending, and that a new genetic variant therefore could persist unaltered from one generation to the next. The German biologist August Weissmann (1834–1914) rejected Lamarckian inheritance by showing experimentally that modifications of an organism during its lifetime do not change its heredity, and he revised Darwinian evolutionary theory accordingly. We now use the term **neo-Darwinism** to denote Darwinian evolutionary theory as revised by Weissmann. The genetic basis of neo-Darwinism eventually became what is now called the **chromosomal theory of inheritance,** a synthesis of Mendelian genetics and cytological studies of the segregation of chromosomes into gametes (Chapter 3).

Emergence of Modern Darwinism: The Synthetic Theory

In the 1930s a new breed of geneticists began to reevaluate Darwinian evolutionary theory from a different perspective. These were population geneticists, scientists who studied variation in natural populations of animals and plants and who had a sound knowledge of statistics and mathematics. Gradually, a new comprehensive theory emerged that brought together population genetics, paleontology, biogeography, embryology, systematics, and animal behavior in a Darwinian framework.

Population geneticists study evolution as a change in the genetic composition of populations. With the establishment of population genetics, evolutionary biology became divided into two different subfields. **Microevolution** is the field that studies evolutionary changes in the frequencies of different allelic forms of genes within populations. **Macroevolution** studies evolution on a grand scale, encompassing the origin of new organismal structures and designs, evolutionary trends, adaptive radiation, phylogenetic relationships of species, and mass extinction. Macroevolutionary research is based in systematics and

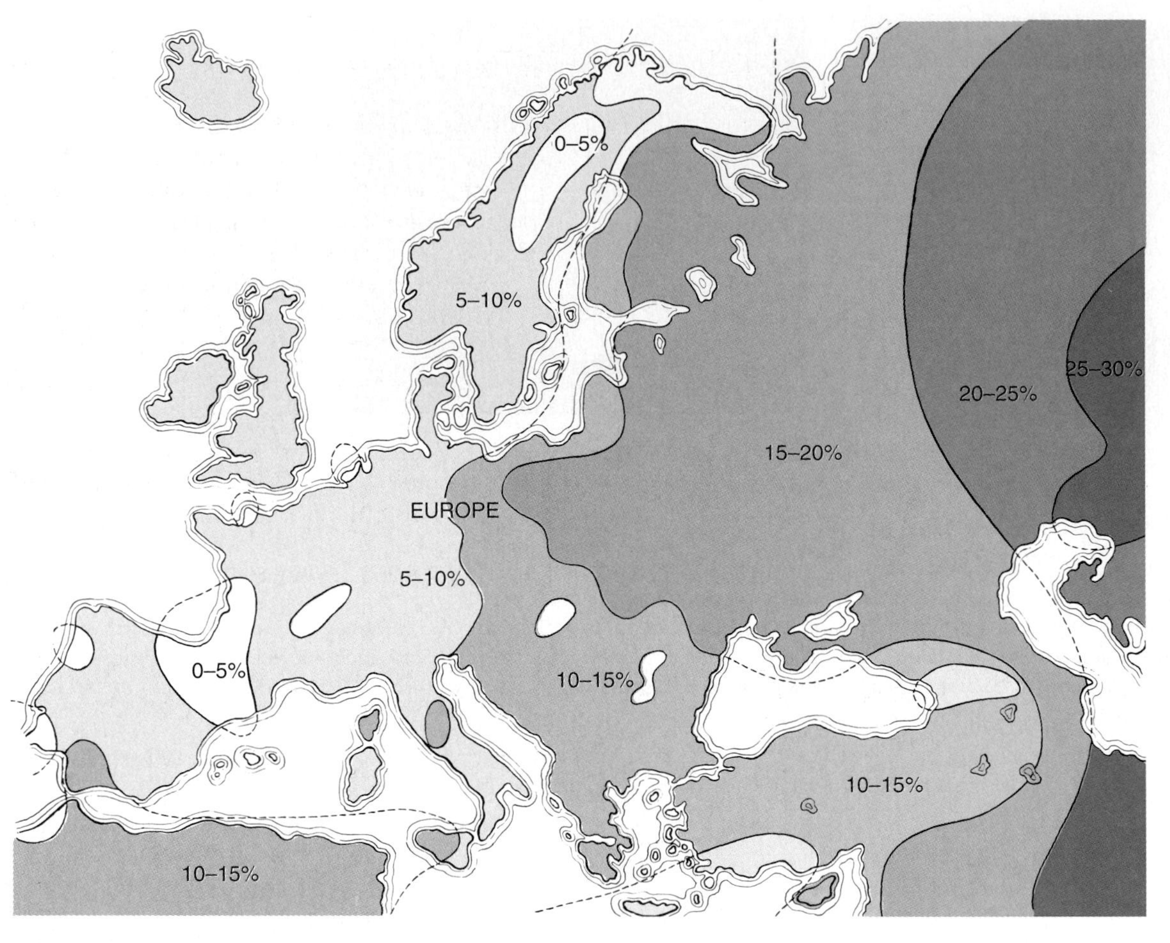

figure 4-26

Frequencies of the blood-type B allele among humans in Europe. The allele is more common in the east and rarer in the west. The allele may have arisen in the east and gradually diffused westward through the genetic continuity of human populations. There is no known selective advantage of this allele, and its changing frequency probably represents the effects of random genetic drift.

the comparative method (p. 346). Following the evolutionary synthesis, both macroevolution and microevolution have operated firmly within the tradition of neo-Darwinism, and both have expanded Darwinian theory in important ways.

Microevolution: Genetic Variation and Change within Species

Microevolution is the study of genetic change occurring within natural populations. The occurrence of different allelic forms of a gene in a population is called **polymorphism.** All of the alleles of all genes possessed by members of a population collectively form the **gene pool.** The amount of polymorphism present in large populations is potentially enormous, because at observed mutation rates, many different alleles are expected for all genes.

Population geneticists study polymorphism by identifying the different allelic forms of a gene that are present in a population and then measuring their relative frequencies in the population. The relative frequency of a particular allelic form of a gene in a population is known as its **allelic frequency.** For example, in the human population, there are three different allelic forms of the gene encoding the ABO blood types: I^A, I^B, and i (p. 219). Because each individual contains two copies of this gene, the total number of copies present in the population is twice the number of individuals. What fraction of this total is represented by each of the three different allelic forms? In the French population, we find the following allelic frequencies: I^A = .46, I^B = .14, and i = .40. In the Russian population, the corresponding allelic frequencies differ (I^A = .38, I^B = .28, and i = .34), demonstrating microevolutionary divergence between these populations (see Figure 4-26). Genetically, alleles I^A and I^B are dominant to i, but i is nearly as frequent as I^A and exceeds the frequency of I^B in both populations. Dominance describes the *phenotypic effect* of an allele in heterozygous individuals, not its relative abundance in a population of individuals. We will demonstrate that Mendelian inheritance and dominance do not alter allelic frequencies directly or produce evolutionary change in a population.

Hardy-Weinberg Equilibrium: Why the Hereditary Process Does Not Change Allelic Frequencies

The Hardy-Weinberg law is a logical consequence of Mendel's first law of segregation and expresses the tendency toward equilibrium inherent in Mendelian heredity.

Let us select for our example a population having a single locus bearing just two alleles *T* and *t*. The phenotypic expression of this gene might be, for example, the ability to taste a chemical compound called phenylthiocarbamide. Individuals in the population will be of three genotypes for this locus, *T/T*, *T/t* (both tasters), and *t/t* (nontasters). In a sample of 100 individuals, let us suppose we have determined that there are 20 of *T/T* genotype, 40 of *T/t* genotype, and 40 of *t/t* genotype. We could then set up a table showing the allelic frequencies as follows (remember that every individual has two copies of the gene):

Genotype	Number of Individuals	Copies of the *T* Allele	Copies of the *t* Allele
T/T	20	40	
T/t	40	40	40
t/t	40		80
TOTAL	100	80	120

Of the 200 copies, the proportion of the *T* allele is 80/200 = 0.4 (40%); and the proportion of the *t* allele is 120/200 = 0.6 (60%). It is customary in presenting this equilibrium to use "p" and "q" to represent the two allelic frequencies. The genetically dominant allele is represented by p, and the genetically recessive by q. Thus:

$$p = \text{frequency of } T = 0.4$$
$$q = \text{frequency of } t = 0.6$$
$$\text{Therefore } p + q = 1$$

Having calculated allelic frequencies in the sample, let us determine whether these frequencies will change spontaneously in a new generation of the population. Assuming the mating is random (and this is important; all mating combinations of genotypes must be equally probable), each individual will contribute an equal number of gametes to the "common pool" from which the next generation is formed. This being the case, the frequencies of gametes in the "pool" will be proportional to the allelic frequencies in the sample. That is, 40% of the gametes will be *T*, and 60% will be *t* (ratio of 0.4:0.6). Both ova and sperm will, of course, show the same frequencies. The next generation is formed as follows:

	Ova	
Sperm	$T = 0.4$	$t = 0.6$
$T = 0.4$	$T/T = 0.16$	$T/t = 0.24$
$t = 0.6$	$T/t = 0.24$	$t/t = 0.36$

Collecting the genotypes, we have:

$$\text{frequency of } T/T = 0.16$$
$$\text{frequency of } T/t = 0.48$$
$$\text{frequency of } t/t = 0.36$$

Next, we determine the values of p and q from the randomly mated populations. From the table, we see that the frequency of *T* will be the sum of genotypes *T/T*, which is 0.16, and one-half of the genotype *T/t*, which is 0.24:

$$T(p) = 0.16 + .5(0.48) = 0.4$$

Similarly, the frequency of *t* will be the sum of genotypes *t/t*, which is 0.36, and one-half the genotype *T/t*, which is 0.24:

$$t(p) = 0.36 + .5(0.48) = 0.6$$

The new generation bears exactly the same allelic frequencies as the parent population! Note that there has been no increase in the frequency of the genetically dominant allele *T*. Thus *in a freely interbreeding, sexually reproducing population, the frequency of each allele would remain constant generation after generation in the absence of natural selection, migration, recurring mutation, and genetic drift* (see text). The more mathematically minded reader will recognize that the genotype frequencies *T/T*, *T/t*, and *t/t* are actually a binomial expansion of $(p + q)^2$:

$$(p + q)^2 = p^2 + 2pq + q^2 = 1$$

Genetic Equilibrium

In many human populations, genetically recessive traits, including the O blood type, blond hair, and blue eyes, are very common. Why have not the genetically dominant alternatives gradually supplanted these recessive traits? It is a common misconception that a characteristic associated with a dominant allele increases in proportion because of its genetic dominance. This is not the case, because there is a tendency in *large* populations for allelic frequencies to remain in equilibrium generation after generation. This principle is based on the **Hardy-Weinberg equilibrium** (see box), which forms the foundation for population genetics. According to this theorem, the hereditary process alone does not produce evolutionary change. In large biparental populations, allelic frequencies and genotypic ratios attain an equilibrium in one generation and remain constant thereafter unless disturbed by recurring mutations, natural selection, migration, nonrandom mating, or genetic drift (random sorting). Such disturbances are the sources of microevolutionary change.

A rare allele, according to this principle, does not disappear from a large population merely because it is rare. That is why certain rare traits, such as albinism and cystic fibrosis, persist for endless generations. For example, albinism in humans is caused by a rare recessive allele *a*. Only one person in 20,000 is an albino, and this individual must be homozygous (*a/a*) for the recessive allele. Obviously there are many carriers in the population with normal pigmentation, that is, people who are heterozygous (*A/a*) for albinism. What is their frequency? A convenient way to calculate the frequencies of genotypes in a population is with the binomial expansion of $(p + q)^2$ (see box). We will let p represent the allelic frequency of *A* and q the allelic frequency of *a*.

figure 4-27
The cheetah, a species whose genetic variability has been depleted to very low levels because of small population size in the past.

Assuming that mating is random (a questionable assumption, but one that we will accept for our example), the distribution of genotypic frequencies is $p^2 = A/A$, $2pq = A/a$, and $q^2 = a/a$. Only the frequency of genotype a/a is known with certainty, 1/20,000; therefore:

$$q^2 = 1/20{,}000$$
$$q = (1/20{,}000)^{1/2} = 1/141$$
$$p = 1 - q = 140/141$$

The frequency of carriers is as follows:

$$A/a = 2pq = 2 \times 140/141 \times 1/141 = 1/70$$

One person in every 70 is a carrier! Although a recessive trait may be rare, it is amazing how common a recessive allele may be in a population. There is a message here for anyone proposing to eliminate a "bad" recessive allele from a population by controlling reproduction. It is practically impossible. Because only the homozygous recessive individuals reveal the phenotype against which artificial selection could act (by sterilization, for example), the allele would continue to surface from heterozygous carriers. For a recessive allele present in 2 of every 100 persons (but homozygous in only 1 in 10,000 persons), it would require 50 generations of complete selection against the homozygotes just to reduce its frequency to one in 100 persons.

Processes of Evolution: How Genetic Equilibrium Is Upset

Genetic equilibrium is disturbed in natural populations by (1) random genetic drift, (2) nonrandom mating, (3) recurring mutation, (4) migration, (5) natural selection, and interactions among these factors. Recurring mutation is the ultimate source of variability in all populations, but it usually requires interaction with one or more of the other factors to upset genetic equilibrium. We will look at these other factors individually.

Genetic Drift

Some species, like the cheetah (Figure 4-27), contain very little genetic variation, probably because their ancestral lineages contained some very small populations. A small population clearly cannot contain large amounts of genetic variation. Each individual organism has at most two different allelic forms of each gene, and a single breeding pair contains at most four different allelic forms of a gene. Suppose that we have such a breeding pair. We know from Mendelian genetics (Chapter 3) that chance decides which of the different allelic forms of a gene gets passed to offspring. It is therefore possible by chance alone that one or two of the parental alleles in this example will not be passed to any offspring. It is highly unlikely that the different alleles present in a small ancestral population are all passed to descendants without any change of allelic frequency. This chance fluctuation in allelic frequency from one generation to the next, including loss of alleles from the population, is called **genetic drift.**

Genetic drift occurs to some degree in all populations of finite size. Perfect constancy of allelic frequencies, as predicted by Hardy-Weinberg equilibrium, occurs only in infinitely large populations, and such populations occur only in mathematical models. All populations of animals are finite and therefore experience some effect of genetic drift, which becomes greater, on average, as population size declines. Genetic drift erodes the genetic variability of a population. If population size remains small for many generations in a row, genetic variation can be greatly depleted. This loss is harmful to a species' evolutionary success because it restricts potential genetic responses to environmental change. Indeed, biologists are concerned that cheetah populations may have insufficient variation for continued survival.

Nonrandom Mating

If mating is nonrandom, genotypic frequencies will deviate from the Hardy-Weinberg expectations. For example, if two different alleles of a gene are equally frequent ($p = q = .5$), we expect half of the genotypes to be heterozygous ($2pq = 2$ [.5] [.5] $= .5$) and one-quarter to be homozygous for each of the respective alleles ($p^2 = q^2 = [.5]^2 = .25$). If we have **positive assortative mating,** individuals mate preferentially with others of the same genotype, such as albinos mating with other albinos. Matings among homozygous parents generate offspring that are homozygous like themselves. Matings among heterozygous parents produce on average 50% heterozygous offspring and 50% homozygous offspring (25% of each alternative type) each generation. This increases the frequency of homozygous genotypes and decreases the frequency of heterozygous genotypes in the population but does not change allelic frequencies.

Preferential mating among close relatives also increases homozygosity and is called **inbreeding.** Whereas positive assortative mating usually affects one or a few traits, inbreeding simultaneously affects all variable traits. Strong inbreeding greatly increases the chances that rare recessive alleles will become homozygous and be expressed.

Because inbreeding and genetic drift are both promoted by small population size, they are often confused with each other. Their effects are very different, however. Inbreeding alone cannot change allelic frequencies in the population, only the ways that alleles are combined into genotypes. Genetic drift changes allelic frequencies and consequently also changes genotypic frequencies. Even very large populations have the potential for being highly inbred if there is a behavioral preference for mating with close relatives, although this rarely occurs in nature. Genetic drift, however, will be relatively weak in very large populations.

figure 4-28

A pair of wood ducks. Brightly colored feathers of male birds probably confer no survival advantage and might even be harmful by alerting predators. Such colors nonetheless confer advantage in attracting mates, which overcomes, on average, the negative consequences of these colors for survival. Darwin used the term "sexual selection" to denote traits that give an individual an advantage in attracting mates, even if the traits are neutral or harmful for survival.

Migration

Migration prevents different populations of a species from diverging. If a species is divided into many small populations, genetic drift and selection acting separately in the different populations can produce evolutionary divergence among them. A small amount of migration between populations each generation keeps the different populations from becoming too different. For example, the French and Russian populations whose ABO allelic frequencies were discussed previously show some genetic divergence, but continuing migration between them prevents them from becoming completely distinct.

Natural Selection

Natural selection can change both allelic frequencies and genotypic frequencies in a population. Although the effects of selection are often reported for particular polymorphic genes, we must stress that natural selection acts on the whole animal, not on isolated traits. The organism that possesses the superior combination of traits will be favored. An animal may have traits that confer no advantage or even a disadvantage, but it is successful overall if its combination of traits is favorable. When we claim that a genotype at a particular gene has a higher **relative fitness** than others, we state that on average that genotype confers an advantage in survival and reproduction in the population. If alternative genotypes have unequal probabilities of survival and reproduction, the Hardy-Weinberg equilibrium will be upset.

Some traits and combinations of traits are advantageous for certain aspects of the organism's survival or reproduction and disadvantageous for others. Darwin used the term, **sexual selection,** to denote the selection of traits that are advantageous for obtaining mates but may be harmful for survival. Bright colors and elaborate feathers may enhance a male bird's competitive ability in obtaining mates while simultaneously increasing his vulnerability to predators (Figure 4-28). Changes in the environment can alter the selective value of different traits. The action of selection on character variation is therefore very complex.

Interactions of Selection, Drift, and Migration

Subdivision of a species into small populations that exchange migrants is an optimal one for promoting rapid adaptive evolution of the species as a whole. The interaction of genetic drift and selection in the different populations permits many different genetic combinations of many polymorphic genes to be tested against natural selection. The migration among the populations permits particularly favorable new genetic combinations to spread throughout the species as a whole. The interaction of selection, genetic drift, and migration in this example produces evolutionary change that is qualitatively different from what would result if any of these three factors acted alone. Natural selection, genetic drift, mutation, nonrandom mating, and migration interact in natural populations to create an enormous opportunity for evolutionary change; the perpetual stability predicted by Hardy-Weinberg equilibrium almost never lasts across any significant amount of evolutionary time.

Macroevolution: Major Evolutionary Events

Macroevolution describes large-scale events in organic evolution. The process of speciation links macroevolution and

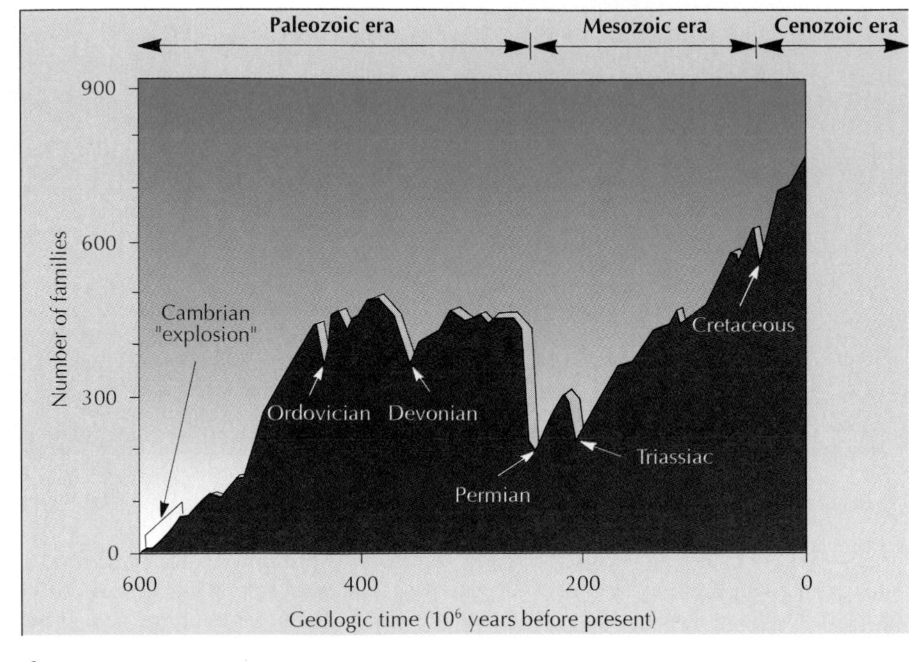

figure 4-29

Changes in numbers of taxonomic families of marine animals through time from the Cambrian period to the present. Sharp drops represent five major extinctions of skeletonized marine animals. Note that despite the extinctions, the overall number of marine families has increased to the present.

microevolution. The major trends in the fossil record described earlier (see Figures 4-13 and 4-14) fall clearly within the realm of macroevolution. The patterns and processes of macroevolutionary change emerge from those of microevolution, but they acquire some degree of autonomy in doing so. The emergence of new adaptations and species, and the varying rates of speciation and extinction observed in the fossil record go beyond the fluctuations of allelic frequencies within populations.

Speciation and Extinction through Geological Time

Evolutionary change at the macroevolutionary level provides a new perspective on Darwin's theory of natural selection. A species has two possible evolutionary fates: it may give rise to new species or become extinct without leaving descendants. Rates of speciation and extinction vary among lineages, and the lineages that have the highest speciation rates and lowest extinction rates produce the greatest diversity of living forms. The characteristics of a species may make it more or less likely than others to undergo speciation or extinction events. Because many characteristics are passed from ancestral to descendant species (analogous to heredity at the organismal level), lineages whose properties enhance the probability of speciation and confer resistance to extinction should come to dominate the living world. This species-level process that produces differential rates of speciation and extinction among lineages is analogous in many ways to natural selection. It represents an expansion of Darwin's theory of natural selection.

Species selection is the differential survival and multiplication of species through geological time based on variation among lineages in emergent, species-level properties. These species-level properties include mating rituals, social structuring, migration patterns, and geographic distribution. Descendant species usually resemble their ancestors for these properties. For example, a "harem" system of mating in which a single male and several females compose a breeding unit characterizes some mammalian lineages but not others. We expect speciation rates to be enhanced by social systems that promote the founding of new populations by small numbers of individuals. Certain social systems may increase the likelihood that a species will survive environmental challenges through cooperative action. Such properties would be favored over geological time by species selection.

Mass Extinctions

When we study evolutionary change on an even larger timescale, we observe periodic events in which large numbers of taxa go extinct simultaneously. These events are called **mass extinctions** (see Figure 4-29). The most cataclysmic of these extinction episodes happened about 225 million years ago, when at least half of the families of shallow-water marine invertebrates, and fully 90% of marine invertebrate species disappeared within a few million years. This was the **Permian extinction.** The **Cretaceous extinction,** which occurred about 65 million years ago, marked the end of the dinosaurs, as well as numerous marine invertebrates and many small reptilian taxa.

The causes of mass extinctions and their occurrence at intervals of approximately 26 million years are difficult to explain. Some have proposed biological explanations for these periodic mass extinctions and others consider them artifacts of our statistical and taxonomic analyses. Walter Alvarez proposed that the earth was periodically bombarded by asteroids, causing these mass extinctions (Figure 4-30). The drastic effects of such a bombardment of a planet were observed in July 1994 when fragments of comet Shoemaker-Levy 9 bombarded Jupiter. The first fragment to hit Jupiter was estimated to have the force of 10 million hydrogen bombs. Twenty additional fragments hit Jupiter within the following week, one of which was 25 times more powerful than the first fragment. This bombardment was the most violent event in the recorded history of the solar system. Bombardments by asteroids or comets could change the earth's climate drastically, sending debris into the atmosphere and blocking sunlight. Temperature changes would have challenged the ecological tolerances of many species. This hypothesis is being tested in several ways, including a search for impact craters left by the asteroids and for altered mineral content of the rock strata where mass extinctions occurred. Atypical concentrations of the rare earth element iridium in some strata imply that this element entered the earth's atmosphere through asteroid bombardment.

figure 4-30

Twin craters of Clearwater Lakes in Canada show that multiple impacts on the earth are not as unlikely as they might seem. Evidence suggests that at least two impacts within a short time were responsible for the Cretaceous mass extinction.

Sometimes, lineages favored by species selection are unusually susceptible to mass extinction. The climatic changes produced by the hypothesized asteroid bombardments could produce selective challenges very different from those encountered at other times in the earth's history. Selective discrimination of particular biological traits by events of mass extinction is termed **catastrophic species selection.** For example, mammals survived the mass extinction at the end of the Cretaceous that destroyed the dinosaurs and other prominent vertebrate and invertebrate groups. Following this event, the mammals were able to use environmental resources that previously had been denied them, leading to their adaptive radiation.

Natural selection, species selection, and catastrophic species selection interact to produce the macroevolutionary trends that we see in the fossil record. The study of these interacting causal processes has made modern paleontology an active and exciting field.

Summary

Organic evolution explains the diversity of living organisms as the historical outcome of gradual change from previously existing forms. Evolutionary theory is strongly identified with Charles Robert Darwin who presented the first credible explanation for evolutionary change. Darwin derived much of the material used to construct his theory from his experiences on a five-year voyage around the world aboard the H.M.S. *Beagle.*

Darwin's evolutionary theory has five major components. Its most basic proposition is *perpetual change,* the theory that the world is neither constant nor perpetually cycling but is steadily undergoing irreversible change. The fossil record amply demonstrates perpetual change in the continuing fluctuation of animal form and diversity following the Cambrian explosion 600 million years ago. Darwin's theory of *common descent* states that all organisms descend from a common ancestor through a branching of genealogical lineages. This theory explains morphological homologies among organisms as characteristics inherited with modification from a corresponding feature in their common evolutionary ancestor. Patterns of homology formed by common descent with modification permit us to classify organisms according to their evolutionary relationships.

A corollary of common descent is the *multiplication of species* through evolutionary time. Allopatric speciation describes the evolution of reproductive barriers between geographically separated populations to generate new species. Adaptive radiation is the proliferation of many adaptively diverse species from a single ancestral lineage. Oceanic archipelagoes, such as the Galápagos Islands, are particularly conducive to the adaptive radiation of terrestrial organisms.

Darwin's theory of *gradualism* states that large phenotypic differences between species are produced by the accumulation through evolutionary time of many individually small changes. Gradualism is still controversial. Mutations that have large effects on the phenotype have been useful in animal breeding, leading some to dispute Darwin's claim that such mutations are not important in evolution. On a macroevolutionary perspective, punctuated equilibrium states that most evolutionary change occurs in relatively brief events of branching speciation, separated by long intervals in which little phenotypic change accumulates.

Darwin's fifth major statement is that *natural selection* is the guiding force of evolution. This principle is founded on observations that all species overproduce their kind, causing a struggle for the limited resources that support existence. Because no two organisms are exactly alike, and because variable traits are at least partially heritable, those whose hereditary endowment enhances their use of resources for survival and reproduction contribute disproportionately to the next generation. Over many generations, the sorting of variation by selection produces new species and new adaptations.

Mutations are the ultimate source of all new variation on which selection acts. Darwin's theory emphasizes that variation is produced at random with respect to the organism's needs and that differential survival and reproduction provide the direction for evolutionary change. Darwin's theory of natural selection was modified in this century by correction of his genetic errors. This modified theory became known as neo-Darwinism.

Population geneticists discovered the principles by which genetic properties of populations change through time. A particularly important discovery, known as Hardy-Weinberg equilibrium, showed that the hereditary process itself does not change the genetic composition of populations. The important sources of evolutionary change include mutation, genetic drift, nonrandom mating, migration, natural selection, and their interactions.

Neo-Darwinism, as elaborated by population genetics, formed the basis for the Evolutionary Synthesis of the 1930s and 1940s. Genetics, natural history, paleobiology, and systematics were brought together under the common goal of expanding our knowledge of Darwinian evolution. Microevolution comprises the study of genetic change within contemporary populations. These studies show that most natural populations contain enormous amounts of variation. Macroevolution comprises the study of evolutionary change on a geological timescale. Macroevolutionary studies measure rates of speciation, extinction, and changes of diversity through time. These studies have expanded Darwinian evolutionary theory to include higher-level processes that regulate rates of speciation and extinction among lineages, including species selection and catastrophic species selection.

Review Questions

1. Briefly summarize Lamarck's concept of the evolutionary process. What is wrong with this concept?
2. What is "uniformitarianism"? How did it influence Darwin's evolutionary theory?
3. Why was the *Beagle's* journey so important to Darwin's thinking?
4. What was the key idea contained in Malthus's essay on populations that was to help Darwin formulate his theory of natural selection?
5. Explain how each of the following contributes to Darwin's evolutionary theory: fossils; geographic distributions of closely related animals; homology; animal classification.
6. How do modern evolutionists view the relationship between ontogeny and phylogeny?
7. What is the main evolutionary lesson provided by Darwin's finches on the Galápagos Islands?
8. How is the observation of "sporting mutations" in animal breeding used to challenge Darwin's theory of gradualism? Why did Darwin reject such mutations as having little evolutionary importance?
9. What does the theory of punctuated equilibrium state about the occurrence of speciation throughout geological time?
10. Describe the observations and inferences that compose Darwin's theory of natural selection.
11. Identify the random and nonrandom components of Darwin's theory of natural selection.
12. Describe some recurring criticisms of Darwin's theory of natural selection. How can these be refuted?
13. It is a common but mistaken belief that because some alleles are dominant and others are recessive, the dominants will eventually replace (drive out) all the recessives. How does the Hardy-Weinberg equilibrium answer this notion?
14. Assume you are sampling a trait in animal populations; the trait is controlled by a single allelic pair *A* and *a,* and you can distinguish all three genotypes *AA, Aa,* and *aa* (intermediate inheritance). Your sample includes:

Population	***AA***	***Aa***	***aa***	**TOTAL**
I	300	500	200	1000
II	400	400	200	1000

    Calculate the distribution of genotypes in each population as expected under Hardy-Weinberg equilibrium. Is population I in equilibrium? Is population II in equilibrium?
15. If, after studying a population for a trait determined by a single pair of alleles, you find that the population is not in equilibrium, what possible reasons might explain the lack of equilibrium?
16. Explain why genetic drift is more powerful in small populations.
17. Is it easier for selection to remove a deleterious recessive allele from a randomly mating population or a highly inbred population? Why?
18. Distinguish between microevolution and macroevolution.

Selected References

Avise, J. C. 1994. Molecular markers, natural history and evolution. New York, Chapman & Hall. *An exciting and readable account of the evolutionary discoveries made using molecular studies, with particular attention to conservation.*

Bowlby, J. 1990. Charles Darwin: a new life. New York, W. W. Norton & Company. *An interpretive biography of Charles Darwin.*

Buss, L.W. 1987. The evolution of individuality. Princeton, New Jersey, Princeton University Press. *An original and provocative thesis on the relationship between development and evolution, with examples drawn from many different animal phyla.*

Darwin, C. 1859. On the origin of species by means of natural selection, or the preservation of favoured races in the struggle for life. London, John Murray. *There were five subsequent editions by the author.*

Endler, J. A. 1986. Natural selection in the wild. Princeton, New Jersey, Princeton University Press. *A review of what we have learned about selection from studying natural populations.*

Futuyma, D. J. 1986. Evolutionary biology. Sunderland, Massachusetts, Sinauer Associates. *A very thorough introductory textbook on evolution.*

Glen, W. 1994. The mass extinction debates: how science works in a crisis. Stanford, Stanford University Press. *A discussion of mass extinction presented in the form of a debate and panel discussion among concerned scientists.*

Gould, S. J. 1977. Ontogeny and phylogeny. Cambridge, Massachusetts, Harvard University Press. *An examination of relationship between ontogeny and phylogeny with a detailed history of this subject.*

Gould, S. J. 1989. Wonderful life: the Burgess Shale and the nature of history. New York, W. W. Norton & Company. *An insightful discussion of what fossils tell us about the nature of life's evolutionary history.*

Hall, B. K. 1992. Evolutionary developmental biology. New York, Chapman & Hall. *A review of the interaction of genetics and development in evolving lineages, with particular attention to issues of heterochrony, homology, and developmental constraints on evolution.*

Hartl, D. L., and A.G. Clark. 1989. Principles of population genetics. Sunderland, Massachusetts, Sinauer Associates. *A current textbook on population genetics.*

Keller, E. F., and E. A. Lloyd (eds.). 1992. Keywords in evolutionary biology. Cambridge, Massachusetts, Harvard University Press. *Many terms used in evolutionary biology are plagued by varying or multiple meanings. Forty keywords are defined in essays by 47 contributors.*

Kohn, D. 1985. The Darwinian heritage. Princeton, New Jersey, Princeton University Press. *An edited volume on Darwin and Darwinism with contributions from many leading historians of evolutionary theory.*

Levinton, J. 1988. Genetics, paleontology and macroevolution. Cambridge, England, Cambridge University Press. *A study of the relationship between microevolution and macroevolution.*

Mayr, E. 1988. Toward a new philosophy of biology. Cambridge, Massachusetts, Harvard University Press. *A collection of essays on many aspects of evolution by a leading evolutionary biologist.*

Otte, D., and J. A. Endler. 1989. Speciation and its consequences. Sunderland, Massachusetts, Sinauer Associates. *An edited volume covering recent issues in the study of speciation, with contributions from many active evolutionary biologists.*

Raff, R. A. 1996. The shape of life: genes, development, and the evolution of animal form. Chicago, University of Chicago Press. *A review of the developmental basis of animal evolution.*

Ross, R. M., and W. D. Allmon. 1990. Causes of evolution: a paleontological perspective. Chicago, University of Chicago Press. *An edited volume of studies on the causes of evolution as viewed from a primarily paleontological perspective.*

Ecology and Distribution of Animals

chapter | five

Spaceship Earth

All life is confined to a thin veneer of the earth called the biosphere. From the first remarkable photographs of Earth taken from the Apollo spacecraft, revealing a beautiful blue and white globe lying against the limitless backdrop of space, the phrase "spaceship earth" became part of our vocabulary. All resources for sustaining life occur on a thin layer of land and sea and a narrow veil of atmosphere above it. If we imagine the earth and its dimensions as a 1-meter sphere, we no longer perceive vertical dimensions on the earth's surface. The highest mountains fail to penetrate a thin coat of paint applied to our shrunken earth; a fingernails' scratch on the surface exceeds the depth of the oceans' deepest trenches.

What resources make Earth uniquely suited for life? Foremost is the presence of liquid water on the earth's surface. Water provided the medium for life's origin and conferred a moderate climate suitable for life's continued evolution. Also critical for life are the steady supply of light and heat from an unfailing sun, the supply of chemical elements required by living matter, and a gravitational force strong enough to hold an extensive gaseous atmosphere.

Earth's biosphere and the organisms in it have evolved together. Life shapes and is shaped by its environments. The barren, stormy, and volcanic Earth of 3.5 billion years ago, with a reducing atmosphere of ammonia, methane, and water, provided for the prebiotic synthesis that led to life's beginnings, but would be lethal for most organisms alive today. The earliest living forms generated the atmospheric oxygen and other environmental conditions necessary for our contemporary fauna and flora. Today the biosphere is changing very rapidly from human impact, one of the greatest agents of biotic disturbance the earth has known. Only the historical bombardment of Earth by asteroids produced more catastrophic disturbances of Earth's biota.

In the mid-nineteenth century, the German zoologist Ernst Haeckel introduced the term ecology, defined as the "relation of the animal to its organic as well as inorganic environment." Environment here includes everything external to the animal but most importantly its immediate surroundings. Although we no longer restrict ecology to animals alone, Haeckel's definition is still basically sound. Animal ecology is now a highly synthetic science that incorporates everything we know about the behavior, physiology, genetics, and evolution of animals to study the interactions between populations of animals and their environments. The major goal of ecological studies is to understand how these diverse interactions determine the geographical distributions and abundance of animal populations. Such knowledge is crucial for ensuring the continued survival of many populations when their natural environments are altered by human activity.

Not infrequently the word "ecology" is misused as a synonym for environment, which often makes biologists wince. As people concerned about the environment, we can be environmentalists; a person engaged in the scientific study of the relationship of organisms and their environment is an ecologist. He or she is usually an environmentalist too, but environment is not the same as ecology.

Ecology is studied as a hierarchy of biological systems in interaction with their environments. We begin by describing the hierarchy of ecology and the kinds of ecological processes that regulate the distributions and abundances of animals. We conclude by surveying the major subdivisions of the biosphere that result from the interactions among ecological processes at all levels of complexity.

The Hierarchy of Ecology

At the base of the ecological hierarchy is the **organism.** To understand why animals are distributed as they are, ecologists must examine the varied physiological and behavioral mechanisms that animals use to survive, grow, and reproduce. A near-perfect physiological balance between production and loss of heat is required for the success of certain endothermic species (such as birds and mammals) under extreme temperatures as found in the Arctic or a desert. Other species succeed in these situations by escaping the most extreme conditions by migration, hibernation, or torpidity. Insects, fishes, and other ectotherms (animals whose body temperature is dependent on heat in the environment) compensate for fluctuating temperatures by altering biochemical and cellular processes involving enzymes, lipid organization, and the neuroendocrine system (p. 180). Thus the animal's physiological capacities permit it to live under changing and often adverse environmental conditions. Behavioral responses are important also for obtaining food, finding shelter, escaping enemies and unfavorable environments, finding a mate, courting, and caring for the young. Physiological mechanisms and behaviors that improve adaptability to the environment assist survival of the species. Ecologists who focus their studies at the organismal level are called physiological ecologists or behavioral ecologists.

Animals in nature coexist with others of the same species; these groups are known as **populations.** Populations have properties that cannot be discovered from studying individual animals alone, including genetic variability among individuals (polymorphism), growth in numbers over time, and factors that limit the density of individuals in a given area. Ecological studies at the populational level help us to predict the future success of endangered species and to discover controls for pest species.

Just as individuals do not exist alone in nature, populations of different species co-occur in more complex associations known as **communities.** The variety of a community is measured as **species diversity.** The populations of species in a community interact with each other in many ways, the most prevalent of which are **predation, parasitism,** and **competition. Predators** obtain energy and nutrients by killing and eating prey. **Parasites** derive similar benefits from their hosts, but usually do not kill the hosts. Competition occurs when food or space are in limited supply. Communities are complex because all of these interactions occur simultaneously, and their individual effects on the whole structure often cannot be isolated.

Most people are aware that lions, tiger, and wolves are predators, but the world of invertebrates also includes numerous predaceous animals. These range from protozoa, jellyfish and their relatives, and various worms to predaceous insects, sea stars, and many others.

Ecological communities are biological components of the even larger, more complex entities called **ecosystems.** An ecosystem consists of all of the populations in a community together with their physical environment. The study of ecosystems helps us to understand two key processes in nature, the flow of energy and the cycling of materials through biological channels. The largest ecosystem is the **biosphere,** the thin veneer of land, water, and atmosphere that envelopes the great mass of the planet, and that supports all life on earth.

Environment and the Niche

An animal's environment is composed of all the conditions that directly affect its chances of survival and reproduction. These factors include space, forms of energy such as sunlight,

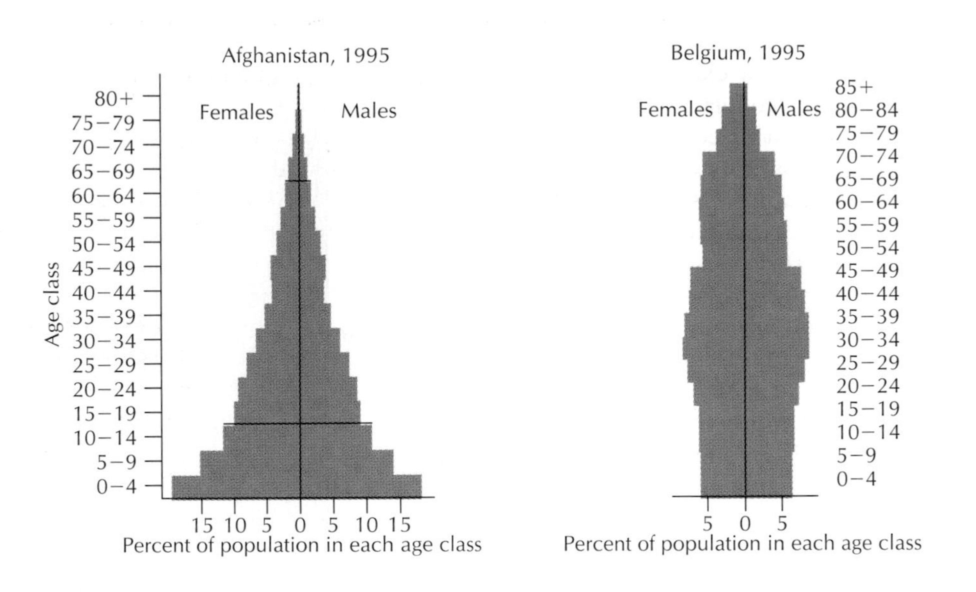

figure 5-3

Age structure profiles of the human populations of Afghanistan and Belgium in 1995 contrast the rapidly growing, youthful population of Afghanistan with the stable population of Belgium, where the fertility rate is below replacement. Countries such as Afghanistan with a large fraction of the population as children are strained to provide adequate child services. With so many children soon to enter their reproductive years, the population will continue to grow rapidly for many years to come.

in Figure 5-4 shows this kind of growth. If species actually grew in this fashion unchecked, earth's resources soon would be exhausted and mass extinction would follow. A bacterium dividing three times per hour could produce a colony a foot deep over the entire earth after a day and a half, and this mass would be over our heads only one hour later. Animals have much lower potential growth rates than bacteria, but could achieve the same kind of result over a longer period of time, given unlimited resources. Many insects lay thousands of eggs each year. A single codfish may spawn 6 million eggs in a season, and a field mouse can produce 17 litters of five to seven young each year. Obviously, unrestricted growth is not the rule in nature.

Even in the most benign environment, a growing population eventually exhausts food or space. Exponential increases such as locust outbreaks or planktonic blooms in lakes must end when food or space is expended. Actually, among all resources that could limit a population, the one in shortest supply relative to the needs of the population will be depleted before others. This one is termed the **limiting resource.** The largest population that can be supported by the limiting resource in a habitat is called the **carrying capacity** of that environment, symbolized **K.** Ideally, a population will slow its growth rate in response to diminishing resources until it just reaches K, as represented by the sigmoid curve in Figure 5-4. The mathematical expression of exponential and sigmoid (or logistic) growth curves are compared in the box on page 117. Sigmoid growth occurs when there is negative feedback between growth rate and population density. This phenomenon is called density dependence, and is the mechanism for intrinsic regulation of populations. We can compare density dependence by negative feedback to the way endothermic animals regulate their body temperatures when the environmental temperature exceeds an optimum. If the resource is expendable, as with food, carrying capacity is reached when the rate of resource replenishment equals the rate of depletion by the population; the population is then at K for that limiting resource. According to the logistic model, when population density reaches K, rates of birth and death are equal and growth of the population ceases. Thus, a population of grasshoppers in a green meadow may be at carrying capacity even though we see plenty of unconsumed food.

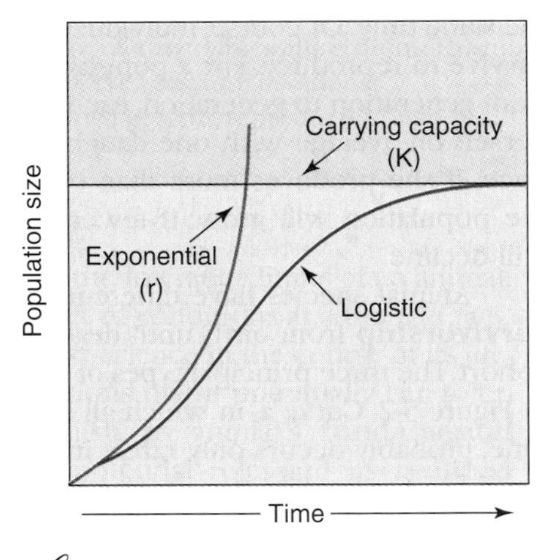

figure 5-4

Population growth, showing exponential growth of a species in an unlimited environment, and logistic growth in a limited environment.

Although experimental populations of protozoa may fit the logistic growth curve closely, most populations in nature tend to fluctuate above and below carrying capacity. For example, after sheep were introduced to Tasmania around 1800, their numbers changed logistically with small oscillations around an average population size of about 1.7

Exponential and Logistic Growth

We can describe the sigmoid growth curve (see Figure 5-4) by a simple model called the logistic equation. The slope at any point on the growth curve is the growth rate, that is, how rapidly the population size is changing with time. If **N** represents the number of organisms and **t** the time, we can, in the language of calculus, express growth as an instantaneous rate:

dN/dt = the rate of change in the number of organisms per time at a particular instant in time.

When populations are growing in an environment of unlimited resources (unlimited food and space, and no competition from other organisms), growth is limited only by the inherent capacity of the population to reproduce itself. Under these ideal conditions growth is expressed by the symbol **r,** which is defined as the intrinsic rate of population growth per capita. The index **r** is actually the difference between the birth rate and death rate per individual in the population at any instant. The growth rate of the population as a whole is then:

$$dN/dt = rN$$

This expression describes the rapid, **exponential growth** illustrated by the early upward-curving portion of the sigmoid growth curve (see Figure 5-4).

Growth rate for populations in the real world slows as the upper limit is approached, and eventually stops altogether. At this point **N** has reached its maximum density; that is, the space being studied has become "saturated" with animals. This limit is called the carrying capacity of the environment and is expressed by the symbol **K.** The sigmoid population growth curve can now be described by the logistic equation, which is written as follows:

$$dN/dt = rN([K - N]/K)$$

This equation states that the rate of increase per unit of time (dN/dt) = rate of growth per capita **(r)** × population size **(N)** × unutilized freedom for growth (**[K − N]/K**). One can see from the equation that when the population approaches the carrying capacity, that is, when **K − N** approaches 0, dN/dt also approaches 0, and the curve will flatten.

Populations occasionally overshoot the carrying capacity of the environment so that **N** exceeds **K.** The population then exhausts some resource (usually food or shelter). The rate of growth, dN/dt, then becomes negative and the population must decline.

million; we thereby infer the carrying capacity of the environment to be 1.7 million sheep (Figure 5-5A). Ring-necked pheasants introduced on an island in Ontario, Canada exhibited wider oscillations (Figure 5-5B).

Why do intrinsically regulated populations oscillate this way? First, the carrying capacity of an environment can change over time, requiring that a population change its density to track the limiting resource. Second, animals always experience a lag between the time that a resource becomes limited and the time that the population responds by reducing its rate of growth. Third, **extrinsic** factors occasionally may limit a population's growth below carrying capacity. We consider extrinsic factors below.

On the global scale, humans have the longest record of exponential population growth (Figure 5-5C). Although famine and war have restrained growth of populations locally, the only dip in global human growth resulted from bubonic plague ("black death"), which decimated much of Europe during the fourteenth century. What then is the carrying capacity for the human population? The answer is far from simple, and several important factors must be considered when estimating the human K.

With the development of agriculture, the carrying capacity of the environment increased, and the human population grew steadily from 5 million around 8000 B.C., when agriculture was introduced, to 16 million around 4000 B.C. Despite the toll taken by terrible famines, disease, and war, the population reached 500 million by 1650. With the coming of the Industrial Revolution in Europe and England in the eighteenth century, followed by a medical revolution, discovery of new lands for colonization, and better agriculture practices, the human carrying capacity increased dramatically. The population doubled to 1 billion around 1850. It doubled again to 2 billion by 1930, to 4 billion in 1976, passed 5.8 billion in 1996, and is expected to reach 10 billion by the year 2025. Thus, the growth has been exponential and remains high (Figure 5-5C).

Recent surveys provide hope that the growth of the human population is slackening. Between 1970 and 1995 the annual growth rate decreased from 1.9% to 1.6%. At 1.6%, it will take 43.3 years for the world population to double rather than 36.5 years at the higher annual growth rate figure. The decrease is credited to better family planning. Despite the drop in growth rate, the greatest surge in population lies ahead, with a projected 3 billion people added within the next three decades, the most rapid increase ever in human numbers.

Although rapid advancements in agricultural, industrial, and medical technology have undoubtedly increased the earth's carrying capacity for humans, it also has widened the difference between birth and death rates to increase our rate of exponential growth. Each day we add 250,000 people (net) to the approximately 5.8 billion people currently alive. Assuming that growth remains constant (certainly not a safe assumption, based on the history of human population growth), by the year 2030 more than half a million people will be added

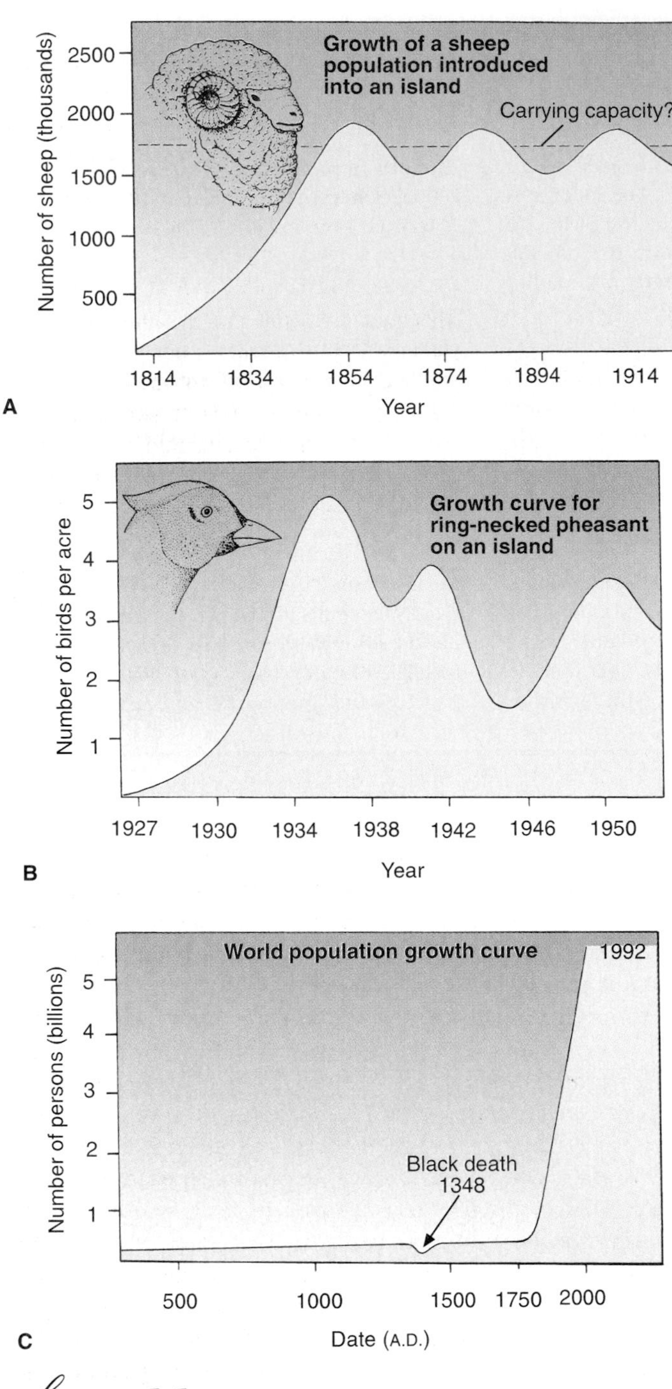

figure 5-5

Growth curves for sheep **(A)**, ring-necked pheasant **(B)**, and world human populations **(C)** throughout history. Note that the sheep population on an island is stable because of human control of the population, but the ring-necked pheasant population oscillates greatly, probably because of large changes in carrying capacity. Where would you place the carrying capacity for the human population?

each day. In other words, less than ten days will be required to replace all people who inhabited the world in 8000 B.C.

In trying to arrive at an estimate of carrying capacity for the human species, we must consider not only quantity of resources, but quality of life. Approximately two billion of the 5.8 billion people alive today are malnourished. At present 99% of our food comes from the land, and the tiny fraction that we derive from the sea is decreasing due to over-exploitation of fish stocks (p.237). Although there is some disagreement on what would constitute the maximum sustainable agricultural output, scientists do not expect food production to keep pace with population growth.

Extrinsic Limits to Growth

We have seen that the intrinsic carrying capacity of a population for an environment prevents unlimited exponential growth of the population. Population growth also can be limited by extrinsic biotic factors, including predation, parasitism (including disease-causing pathogens), and interspecific competition, or by abiotic influences such as floods, fires, and storms. Although abiotic factors certainly can reduce populations in nature, they cannot truly regulate population growth because their effect is wholly independent of population size; abiotic limiting factors are **density-independent.** A single hailstorm can kill most of the young of wading bird populations, and a forest fire can eliminate entire populations of many animals, regardless of how many individuals there may be.

In contrast, biotic factors can and do act in a **density-dependent** manner. Predators and parasites respond to changes in density of their prey and host populations, respectively, to maintain populations at fairly constant sizes. These sizes are below carrying capacity, because populations regulated by extrinsic factors are not limited by their resources. Competition between species for a common limiting resource lowers the effective carrying capacity for each species below that of either one alone.

Community Ecology

Interactions among Populations in Communities

Populations of animals are part of a larger system, known as the **community,** within which populations of different species interact. The number of species that share a habitat is known as **species diversity.** These species interact in a variety of ways that can be detrimental (−), beneficial (+), or neutral (0) to each species, depending on the nature of the interaction. For instance, we can consider a predator's effect on its prey as (−), because the survival of the prey animal is reduced. However, the same interaction benefits the predator (+) because the food obtained from the prey increases the predator's ability to survive and reproduce. Thus, the predator-prey interaction is + −. Ecologists use this shorthand notation to characterize interspecific interactions because it helps us to view the direction in which the interaction affects each species.

We see other kinds of + − interactions. One of these is **parasitism,** in which the parasite benefits by using the host

figure 5-6

Among many examples of mutualism that abound in nature is the whistling thorn acacia of the African savanna and the ants that make their homes in the acacia's swollen galls. The acacia provides both protection for the ants larvae (*right photograph of opened gall*) and honeylike secretions used by the ants as food. In turn, ants protect the tree from herbivores by swarming out as soon as the tree is touched. Giraffes, however, which love the tender acacia leaves, seem immune to the ants' fiery bites.

as a home and source of nutrition, and the host is harmed. **Herbivory,** in which an animal eats a plant, is another + − relationship. **Commensalism** is an interaction that benefits one species and neither harms nor benefits the other (0+). Most bacteria that normally inhabit our intestinal tracts do not affect us (0), but the bacteria benefit (+) by having food and a place to live. A classic example of commensalism is the association of pilot fishes and remoras with sharks. These fishes get the "crumbs" remaining when the host shark makes its kill, but we now know that some remoras also feed on ectoparasites of the sharks. Commensalism therefore grades into **mutualism.**

Organisms engaged in mutualism have a friendlier arrangement than commensalistic species, because the fitness of both is enhanced (++). Biologists are finding mutualistic relationships far more common in nature than previously believed (Figure 5-6). Some mutualistic relationships are not only beneficial, but necessary for survival of one or both species. An example is the relationship between a termite and the protozoa inhabiting its gut. The protozoa can digest the wood eaten by the termite because the protozoa produce an enzyme, lacking in the termite, that digests cellulose; the termite lives on waste products of protozoan metabolism. In return, the protozoa gain a habitat and food supply. Such absolute interdependence among species can be a liability if one of the participants is lost. *Calvaria* trees native to the island of Mauritius have not reproduced successfully for over 300 years, because their seeds germinate only after being eaten and passed through the gut of a dodo bird, now extinct.

Competition between species reduces the fitness of both (−−). Many biologists, including Darwin, considered competition the most common and important interaction in nature. Ecologists have constructed most of their theories of community structure from the premise that competition is the chief organizing factor in species assemblages. Sometimes the effect on one of the species in a competitive relationship is negligible. This condition is called **amensalism,** or **asymmetric competition** (0−). For example, two species of barnacles that commonly occur in rocky intertidal habitats, *Chthamalus stellatus* and *Balanus balanoides,* compete for space. A famous experiment by Joseph Connell[2] demonstrated that *B. balanoides* excluded *C. stellatus* from a portion of the habitat, while *C. stellatus* had no effect on *B. balanoides.*

We have treated interactions as occurring between pairs of species. However, in natural communities containing populations of many species, a predator may have more than one prey and several animals may compete for the same resource. Thus, ecological communities are quite complex and dynamic, a challenge to ecologists who wish to study this level of natural organization.

Competition and Character Displacement

Competition occurs when two or more species share a **limiting resource.** Simply sharing food or space with another species does not produce competition unless the resource is in short supply relative to the needs of the species that share it. Thus, we cannot prove that competition occurs in nature based soley on the sharing of resources. However, we find evidence of competition by investigating the different ways that species exploit a resource.

Competing species may reduce conflict by reducing the overlap of their niches. **Niche overlap** is the portion of the niche's resources shared by two or more species. For example, if two species of birds eat seeds of exactly the same size, competition eventually will exclude one species from the habitat. This example illustrates the principle of **competitive exclusion:** strongly competing species cannot coexist indefinitely. To coexist in the same habitat, species must specialize by partitioning a shared resource and using different portions of it. Specialization of this kind is called **character displacement.**

Character displacement usually appears as differences in organismal morphology or behavior related to exploitation of a resource. For example, in his classic study of the Galápagos finches (p. 100), English ornithologist David Lack noticed that bill sizes of these birds depended on whether they occurred together on the same island (Figure 5-7). On the islands Daphne and Los Hermanos, where *Geospiza fuliginosa* and *G. fortis* occur separately and therefore do not compete with each other, bill sizes are nearly identical; on the island Santa Cruz, where both *G. fuliginosa* and *G. fortis* coexist, their bill

[2]Connell, J.H. 1961. The influence of interspecific competition and other factors on the distribution of the barnacle *Chthamalus stellatus.* Ecology **42:**710–723.

sizes do not overlap. These results suggest resource partitioning, because bill size determines the size of seeds selected for food. Recent work by the American ornithologist Peter Grant has confirmed what Lack suspected: *G. fuliginosa* with its smaller bill selects smaller seeds than does *G. fortis* with its larger bill. Where the two species coexisted, competition between them led to evolutionary displacement of the bill sizes to diminish competition. The absence of competition today has been called appropriately "the ghost of competition past."

Character displacement promotes coexistence by reducing niche overlap. When several species share the same general resources by such partioning, they form a **guild.** Just as a guild in medieval times constituted a brotherhood of men sharing a common trade, species in an ecological guild share a common livelihood. The term guild was introduced to ecology by Richard Root in his 1967 paper on niche patterns of the blue-gray gnatcatcher.[3] A classic example of a bird guild is Robert MacArthur's study of a feeding guild consisting of five species of warblers in spruce woods of the northeastern United States.[4] At first glance, we might ask how five birds, very similar in size and appearance, could coexist by feeding on insects in the same tree. However, on close inspection MacArthur found subtle differences among these birds in sites of foraging (Figure 5-8). One species searched only on the outer branches of spruce

[3] Root, R.B. 1967. The niche exploitation pattern of the blue-gray gnatcatcher. Ecological Monographs **37:**317–350.
[4] MacArthur, R.H. 1958. Population ecology of some warblers of northeastern coniferous forests. Ecology **39:**599–619.

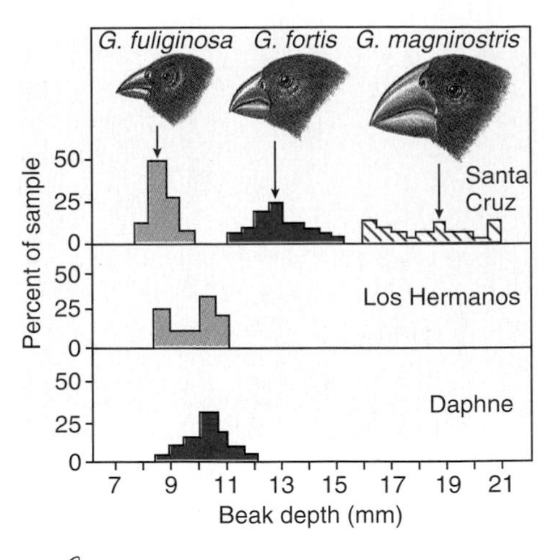

figure 5-7

Displacement of beak sizes in Darwin's finches from the Galápagos Islands. Beak depths are given for the ground finches *Geospiza fuliginosa* and *G. fortis* where they occur together (sympatric) on Santa Cruz Island and where they occur alone on the islands Daphne and Los Hermanos. *G. magnirostris* is another large ground finch that lives on Santa Cruz.

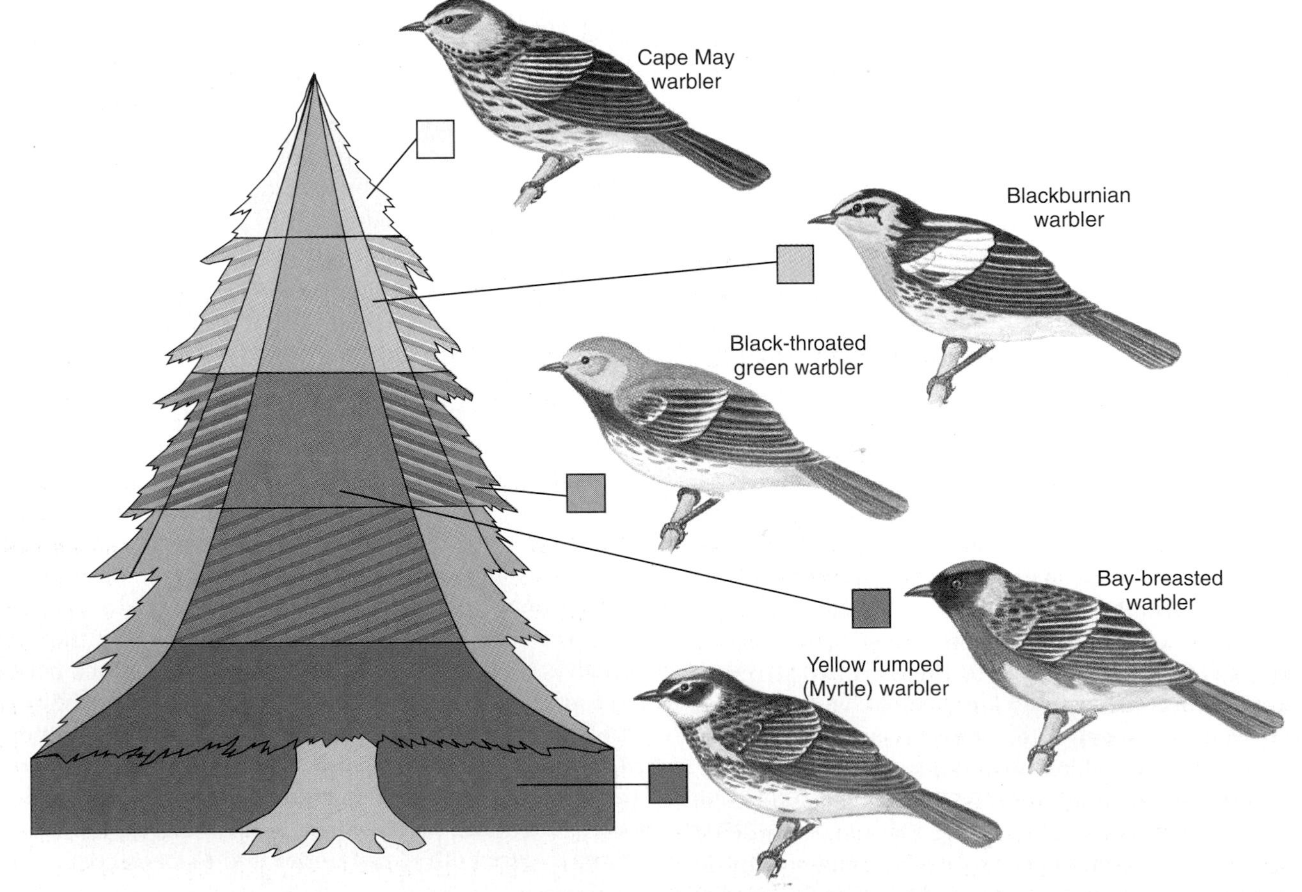

figure 5-8

Distribution of foraging effort among five species of wood warblers in a northeastern spruce forest. The warblers form a feeding guild.

crowns; another species used the top 60% of the tree's outer and inner branches, although not next to the trunk; another species concentrated on the inner branches closer to the trunk; another species used the midsection from the periphery to the trunk; and still another species foraged in the bottom 20% of the tree. These observations suggest that each warbler's niche within this guild is defined by structural differences in the habitat.

Guilds are not limited to birds. For example, a study done in England on insects associated with Scotch broom plants revealed nine different guilds of insects, including three species of stem miners, two gall-forming species, two that fed on seeds and five that fed on leaves. Another insect guild consists of three species of praying mantids that avoid both competition and predation by differing in sizes of their prey, timing of hatching, and height of vegetation in which they forage.

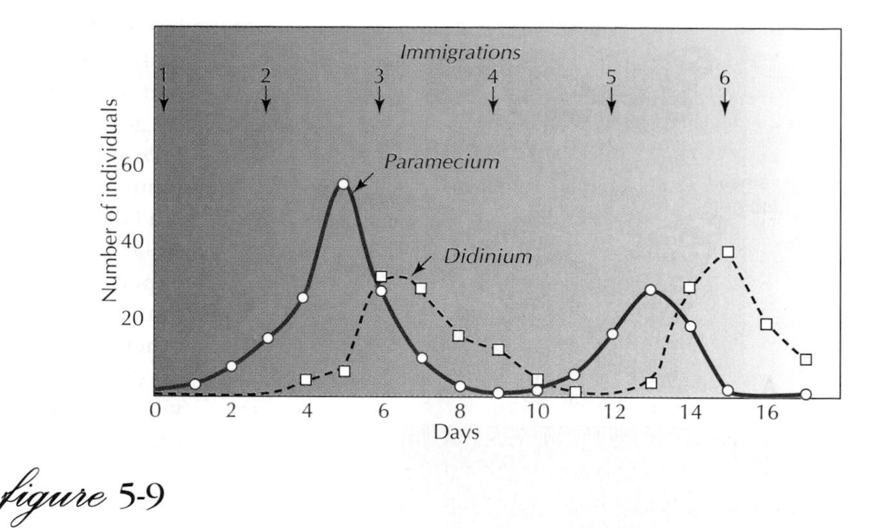

figure 5-9

Classic predator-prey experiment by Russian biologist G.F. Gause in 1934 shows the cyclic interaction between predator (*Didinium*) and prey (*Paramecium*) in laboraory culture. When the *Didinium* find and eat all the *Paramecium,* the *Didinium* themselves starve. Gause could keep the two species coexisting only by occasionally introducing one *Didinium* and one *Paramecium* to the culture (*arrows*); these introductions simulated migration from an outside source.

Predators and Parasites

The ecological warfare by predators against their prey causes coevolution: predators get better at catching prey, and prey get better at escaping predators. This is an evolutionary race that the predator cannot afford to win. If a predator became so efficient that it exterminated its prey, the predator species would become extinct. Because most predators feed on more than a single species, specialization on a single prey to the point of extermination is uncommon.

However, when a predator does rely primarily on a single prey species, both populations tend to fluctuate cyclically. First the prey density increases, then that of the predator until prey become scarce. At that point, predators must adjust their population size downward by leaving the area, lowering reproduction, or dying. When density of the predator population falls enough to allow reproduction by prey to outpace mortality from predation, the cycle begins again. Thus, populations of both predators and their prey show cycles of abundance, but increases and decreases in predator abundance are slightly delayed relative to those of the prey because of the time lag in a predator's response to changing prey density. We can illustrate this process in the laboratory with protozoa (Figure 5-9). Perhaps the longest documented natural example of a predator-prey cycle is between Canadian populations of the snowshoe hare and lynx (see Figure 31-21, p. 682).

The war between predators and prey reaches high art in the evolution of defenses by potential prey. Many animals that are palatable escape detection by matching their background, or by resembling some inedible feature of the environment (such as a bird dropping). Such defenses are called cryptic. In contrast to cryptic defenses, animals that are toxic or distasteful to predators actually advertise their strategy with bright colors and conspicuous behavior. These species are protected because predators learn to recognize and to avoid them after one or more distasteful encounters.

When distasteful prey adopt warning coloration, advantages of deceit arise for palatable prey. Palatable prey can deceive potential predators by mimicking distasteful prey. Coral snakes and monarch butterflies are both brightly colored, noxious prey. The coral snake has a venomous bite, and the butterfly is poisonous because as a caterpillar it stores poison (cardiac glycoside) from the milkweed it eats. Both species serve as **models** for other species, called **mimics,** that do not possess toxins of their own but look like the model species that do (Figure 5-10A and B).

In another form of mimicry, two or more toxic species resemble each other (Figure 5-10C). We can ask why an animal that has its own poison should gain by evolving resemblance to another poisonous animal. The answer is that the predator needs only to experience the toxicity of one species to avoid all similar prey. A predator can learn one warning signal more easily than many!

Sometimes the influence of one population on others is so pervasive that its absence drastically changes the character of the entire community. We call such a population a **keystone species.**[5] For example, in 1983, a mysterious epidemic swept through the Caribbean populations of the sea urchin *Diadema antillarum,* destroying more than 95% of the animals. The

[5]Paine, R.T. 1969. A note on trophic complexity and community stability. American Naturalist **103:**91–93.

provides a vivid impression of the great difference in numbers of organisms involved in each step of the chain, and supports the observation that large predatory animals are rarer than the small animals on which they feed. However, a pyramid of numbers does not indicate the actual mass of organisms at each level.

The concepts of food chains and ecological pyramids were invented and first explained in 1923 by Charles Elton, a young ecologist at Oxford University. Working for a summer on a treeless arctic island, Elton watched the arctic foxes as they roamed, noting what they ate and, in turn, what their prey had eaten, until he was able to trace the complex cycling of nitrogen in food throughout the animal community. Elton realized that life in a food chain comes in discrete sizes, because each form had evolved to be much bigger than the thing it eats. He thus explained the common observation that large animals are rare while small animals are common.

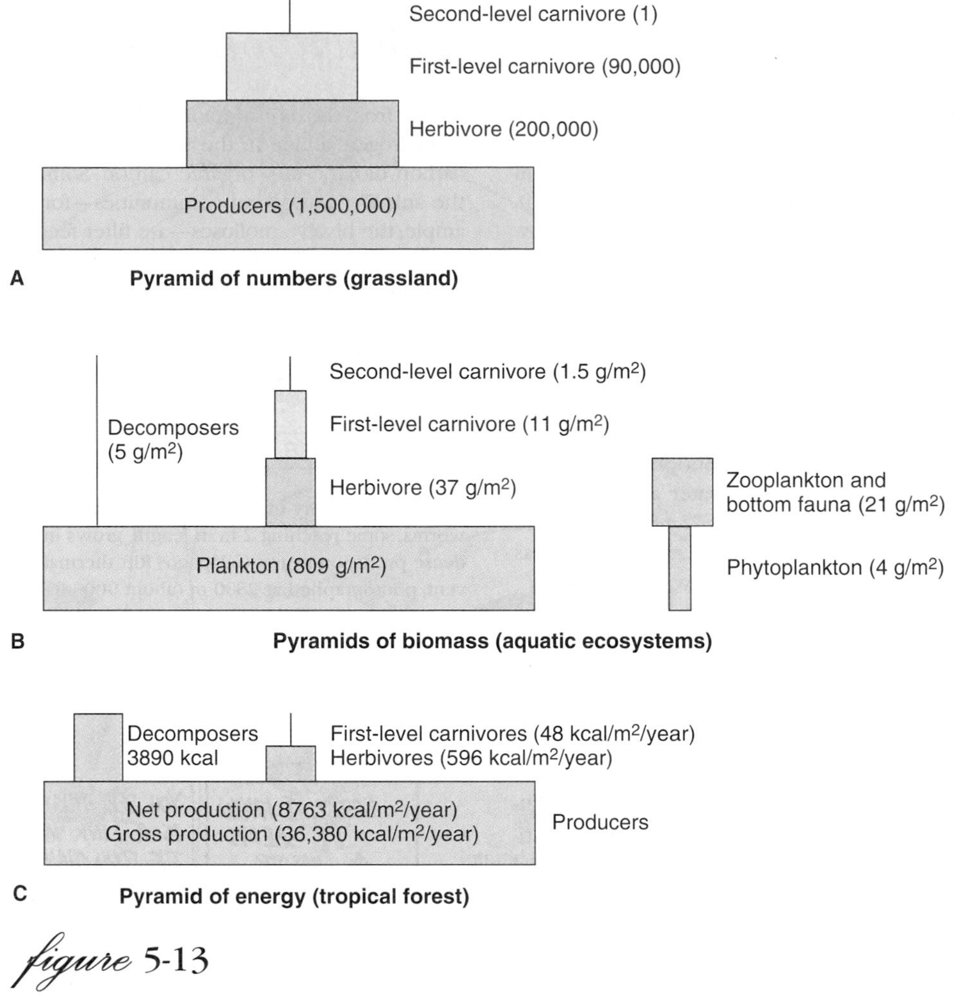

figure 5-13

Ecological pyramids of numbers, biomass, and energy. Pyramids are generalized, because area within each trophic level is not scaled proportionately to quantitative differences in units given.

More instructive are pyramids of biomass (Figure 5-13B), which depict the total bulk, or "standing crop," of organisms at each trophic level. Such pyramids usually slope upward because mass and energy are lost at each transfer. However, in some aquatic ecosystems in which the producers are algae, which have short life spans and rapid turnover, the pyramid is inverted. Algae can tolerate heavy exploitation by the zooplankton consumers. Therefore, the base of the pyramid (biomass of phytoplankton) is smaller than the biomass of zooplankton it supports. We could liken this inverted pyramid to a person who weighs far more than the food in a refrigerator, but who can be sustained from the refrigerator because the food is constantly replenished.

A third type of pyramid is the pyramid of energy, which shows rate of energy flow between levels (Figure 5-13C). An energy pyramid is never inverted because energy transferred from each level is less than what was put into it. A pyramid of energy gives the best overall picture of community structure because it is based on production. In the example above, productivity of phytoplankton exceeds that of zooplankton, even though the biomass of phytoplankton is less than the biomass of zooplankton (because of heavy grazing by the zooplankton consumers).

Nutrient Cycles

All elements essential for life are derived from the environment, where they are present in the air, soil, rocks, and water. When plants and animals die and their bodies decay, or when organic substances are burned or oxidized, the elements and inorganic compounds essential for life processes (nutrients) are released and returned to the environment. Decomposers fulfill an essential role in this process by feeding on the remains of plants and animals and on fecal material. The result is that nutrients flow in a perpetual cycle between the biotic and abiotic components of the ecosystem. Nutrient cycles are often called **biogeochemical cycles** because they involve exchanges between living organisms (bio-) and the rocks, air, and water of the earths' crust (geo-). The continuous input of energy from the sun keeps nutrients flowing and the ecosystem functioning (Figure 5-14).

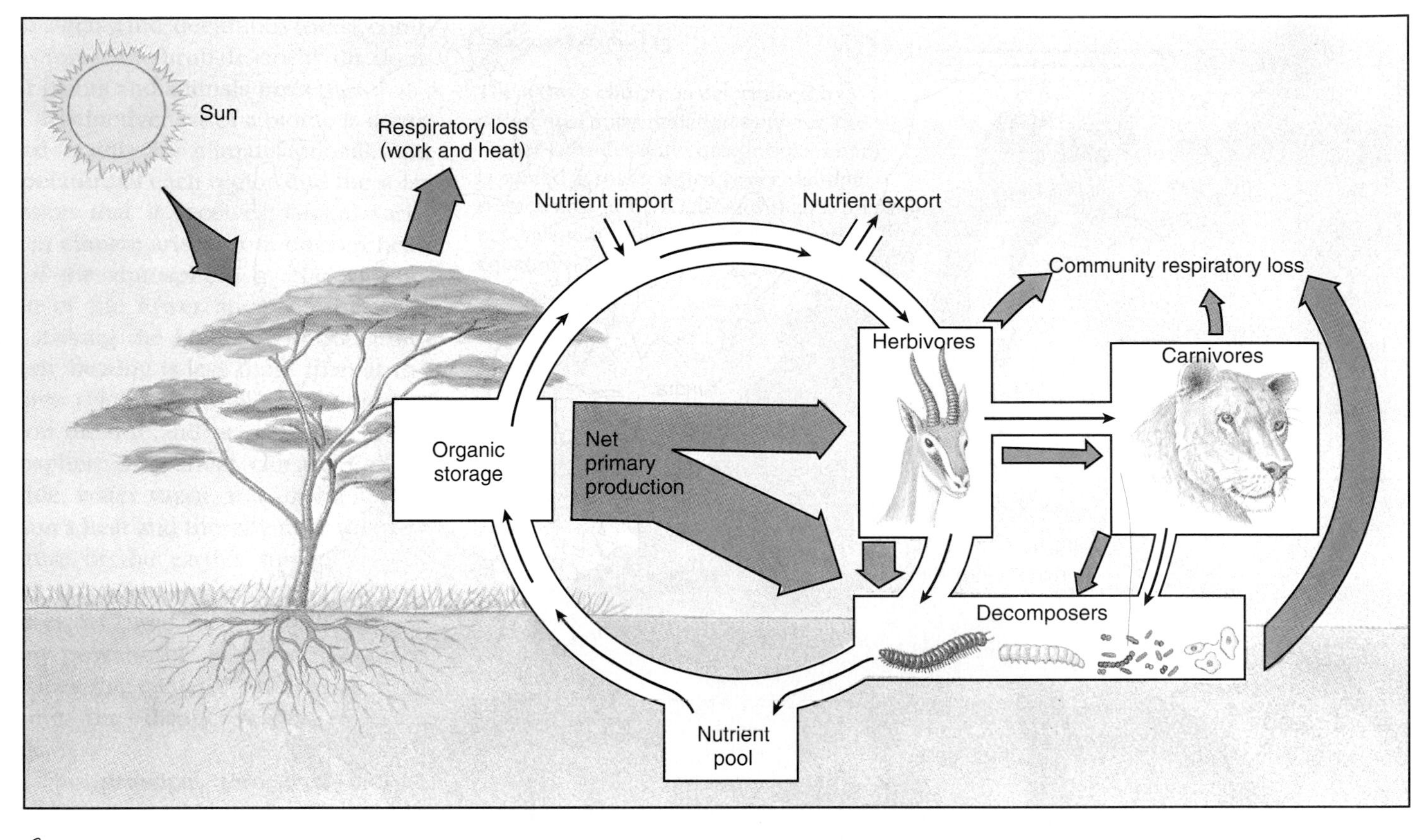

figure 5-14

Nutrient cycles and energy flow in a terrestrial ecosystem. Note that nutrients are recycled, whereas energy flow (*red*) is one way.

We tend to think of biogeochemical cycles in terms of naturally occurring elements, such as water, carbon, and nitrogen. In recent times, however, humans have added synthetic materials to the biosphere that have entered food webs, sometimes with disastrous consequences. Probably the most harmful of these materials, in terms of ecosystemic processes, are pesticides. We currently produce about 2.5 million tons of pesticides worldwide, mainly for protection of crops from insects.

Despite all of this poison being used, more than half of our crops are lost either before or after harvest to pests. The role of pesticides in natural food webs can be insidious for three reasons. First, many pesticides become concentrated as they travel up succeeding trophic levels. The highest concentrations will occur in the biomass of top carnivores such as hawks and owls, diminishing their ability to reproduce. Second, many species that are killed by pesticides are not pests, but merely innocent bystanders, called nontarget species. Nontarget effects happen when pesticides move out of the agricultural field to which they were applied, through rainwater runoff, leaching through the soil, or dispersal by wind. The third problem is persistence; some chemicals used as pesticides have a long life span in the environment, so that nontarget effects persist long after the pesticides have been applied. Scientists are working to create new pesticides that are more specific in their effects and decompose faster in the environment, but we have a long way to go to limit our warfare against those animals that compete with us for our supply of food.

Distribution of Life on Earth

The Biosphere and Its Subdivisions

The biosphere is the thin outer layer of the earth capable of supporting life. It is probably best viewed as a global system that includes all life on earth and the physical environments in which living organisms exist and interact. The nonliving subdivisions of the biosphere include the **lithosphere, hydrosphere,** and **atmosphere.** The lithosphere is the rocky material of the earth's outer shell and is the ultimate source of all mineral elements required by living organisms. The hydrosphere is the water on or near the earths' surface, and it extends into the lithosphere and the atmosphere. The atmosphere extends to 3500 km above the surface of the earth, but all life is confined to the lowest 8 to 15 km (troposphere). The screening layer in the atmosphere of oxygen-ozone is concentrated mostly between 20 and 25 km. The main gases present in the troposphere are (by volume) nitrogen, 78%; oxygen, 21%; argon, 0.93%; carbon dioxide, 0.03%; and variable amounts of water vapor.

All organisms have an energy budget consisting of gross and net productivity, and respiration. For animals, respiration usually is at least 90% of this budget. Thus, the transfer of energy from one trophic level to another is limited to about 10%, which in turn limits the number of trophic levels in an ecosystem. Ecological pyramids of energy depict how productivity decreases in successively higher trophic levels of food webs.

Ecosystem productivity is a result of energy flow and material cycles within ecosystems. All energy is lost as heat, but nutrients and other materials including pesticides are recycled. No ecosystem, including the global biosphere, is closed because they all depend upon imports and exports of energy and materials from outside.

The biosphere is the thin life-containing blanket surrounding the earth. The presence of life on earth is possible because of a steady supply of energy from the sun, presence of water, a suitable range of temperatures, the correct proportion of major and minor elements, and the screening of lethal ultraviolet radiation by atmospheric ozone. The earth's terrestrial environment is composed of biomes that bear a distinctive array of plant life and associated animal life and are mainly determined by annual temperature and rainfall. Aquatic habitats include freshwater rivers, streams, and lakes, although oceans are much more prevalent, occupying 71% of the earth's surface.

Review Questions

1. The term ecology is derived from the Greek meaning "house" or "place to live." However, as used by scientists, the term "ecology" is not the same as "environment." What is the distinction between the terms?
2. How would you distinguish between ecosystem, community, and population?
3. What is the distinction between habitat and environment?
4. Define the niche concept. How does the "realized niche" of a population differ from its "fundamental niche"? How does the concept of niche differ from the concept of guild?
5. Populations of independently living (unitary) animals have a characteristic age structure, sex ratio, and growth rate. However, these properties are difficult to determine for modular animals. Why?
6. Explain which of the three survivorship curves in Figure 5-2 best fits the following: (a) a population in which mortality as a proportion of survivors is constant; (b) a population in which there is little early death and most individuals live to old age; (c) a population that experiences heavy mortality of the very young but with the survivors living to old age. Offer an example from the real world of each survivorship pattern.
7. Contrast exponential and logistic growth of a population. Under what conditions might you expect a population to exhibit exponential growth? Why cannot exponential growth be perpetuated indefinitely?
8. Growth of a population may be hindered by either density-dependent or density-independent mechanisms. Define and contrast these two mechanisms. Offer examples of how growth of the human population might be curbed by either agent.
9. Herbivory is an example of an interspecific interaction that is beneficial for the animal (+) but harmful to the plant it eats (−). What are some +− interactions among animal populations? What is the difference between commensalism and mutualism?
10. Explain how character displacement can ease competition between coexisting species.
11. Define predation. How does the predator-prey relationship differ from the parasite-host relationship? Why is the evolutionary race between predator and prey one that the predator cannot afford to win?
12. Mimicry of the monarch butterfly by the viceroy is an example of a palatable species resembling a toxic one. What is the advantage to the viceroy of this form of mimicry? What is the advantage to a toxic species of mimicing another toxic species?
13. A keystone species has been defined as one whose removal from a community causes the extinction of other species. How does this extinction happen?
14. What is a trophic level, and how does it relate to a food chain?
15. Define *productivity* as the word is used in ecology. What is a primary producer? What is the distinction between gross productivity, net productivity, and respiration? What is the relation of net productivity to biomass (or standing crop)?
16. What is a food chain? How does a food chain differ from a food web?
17. How is it possible to have an inverted pyramid of biomass in which the consumers have a greater biomass than the producers? Can you think of an example of an inverted pyramid of *numbers* in which there are, for example, more herbivores than plants on which they feed?
18. The pyramid of energy has been offered as an example of the second law of thermodynamics (p. 126). Why?
19. Recently discovered animal communities surrounding deep-sea thermal vents apparently exist in total independence of solar energy. How can this existence be possible?
20. What is the biosphere? How would you distinguish between the following subdivisions of the biosphere: lithosphere, hydrosphere, atmosphere?
21. What is a biome? Briefly describe six examples of biomes.
22. What are some very productive marine environments, and why are they so productive?
23. What is the source of nutrients for animals living in the deep-sea habitat?

Selected References

Brooks, D.R., and D.A. McLennan. 1991. Phylogeny, ecology, and behavior. Chicago, University of Chicago Press. *A discussion of how systematic methods can be used to enhance the study of ecology and animal behavior.*

Colinvaux, P. 1993. Ecology 2. New York, John Wiley & Sons. *Comprehensive college textbook.*

Kates, R.W. 1994. Sustaining life on the earth. Sci.Am. **271:** 114–122 (Oct.). *Major technological revolutions—toolmaking, agriculture, and manufacturing—have triggered geometric growth in the human population. The author asks whether we can learn enough about biological, physical, and social reality to fashion a future that our planet can sustain.*

Krebs, C.J. 1993. Ecology: the experimental analysis of distribution and abundance, ed. 4. New York, Harper & Row, Publishers. *Important treatment of population ecology.*

Moore, P.D. (ed.). 1987. The encyclopedia of animal ecology. New York, Facts on File Publications. *Although ecological principles are sparsely treated in this book, the world's major ecosystems are surveyed with extensive text and beautifully illustrated with photographs, paintings, and helpful diagrams.*

Pianka, E.R. 1993. Evolutionary ecology, ed. 5. New York, Harper & Row, Publishers. *An introduction to ecology written from an evolutionary perspective.*

Smith, R.L. 1995. Ecology and field biology, ed 5. HarperCollins, Publishers. *Clearly written, well-illustrated general ecology text.*

Tunnicliffe, V. 1992. Hydrothermal-vent communities of the deep sea. Am. Sci. **80:**336–349 (July-Aug.). *At hot vents along mid-ocean ridges, nuclear and chemical energy make possible exotic ecosystems that have evolved in near-total isolation.*

part two

Animal Form and Function

6 Animal Architecture: Body Organization, Support, and Movement

7 Homeostasis: Osmotic Regulation, Excretion, and Temperature Regulation

8 Internal Fluids and Respiration

9 Immunity

10 Digestion and Nutrition

11 Nervous Coordination

12 Chemical Coordination

13 Animal Behavior

14 Reproduction and Development

Animal Architecture: Body Organization, Support, and Movement

chapter | six

Of Grasshoppers and Superman

"A dog," remarked Galileo in the seventeenth century, "could probably carry two or three such dogs upon his back; but I believe that a horse could not carry even one of its own size." Galileo was referring to the principle of scaling, a procedure that allows us to understand the consequences of changing body size. A grasshopper can jump to a height of 50 times the length of its body, yet a man in a standing jump cannot clear an obstacle that is no higher than he is tall. Without an understanding of scaling, this comparison could easily lead us to the erroneous conclusion that there is something very special about the musculature of insects. To the authors of a nineteenth-century entomology text it seemed that "this wonderful strength of insects is doubtless the result of something peculiar in the structure and arrangement of their muscles, and principally their extraordinary power of contraction." But grasshopper muscles are in fact no more powerful than human muscles because *muscles of small and large animals exert the same force per cross-sectional area.* Grasshoppers leap high in proportion to their size because they are small, not because they possess extraordinary muscles.

The authors of this nineteenth-century text even suggested that it was indeed fortunate that higher animals lacked the powers of insects, for they would surely have "caused the early desolation of the world." More probably, such powers would have led to their own desolation. For earthly mortals would need more than superhuman muscles were they to leap in the proportions of a grasshopper. They would require superhuman tendons, superhuman ligaments, and superhuman bones to withstand the stresses of mighty contractions, not to mention the crushing strains of landing again on earth at terminal velocity. The feats of Superman would be quite impossible were he built of the structural materials available to earthbound animals, rather than the wondrous materials available to inhabitants of the mythical planet Krypton.

The English satirist Samuel Butler proclaimed that the human body was merely "a pair of pincers set over a bellows and a stewpan and the whole thing fixed upon stilts." While human attitudes toward the human body are distinctly ambivalent, most people less cynical than Butler would agree that the body is a triumph of intricate, living architecture. Less obvious, perhaps, is that the architecture of humans and most other animals conforms to the same well-defined plan. The basic uniformity of biological organization derives from the common ancestry of animals and from their basic cellular construction. Despite the vast differences of structural complexity of organisms ranging from the simplest protozoa to humans, all share an intrinsic material design and fundamental functional plan.

We begin our study of animal form and function with the levels of organization in animal complexity. Then we will consider the body's basic structural framework: tissues, which are the "bricks and mortar" that compose organs and systems; the

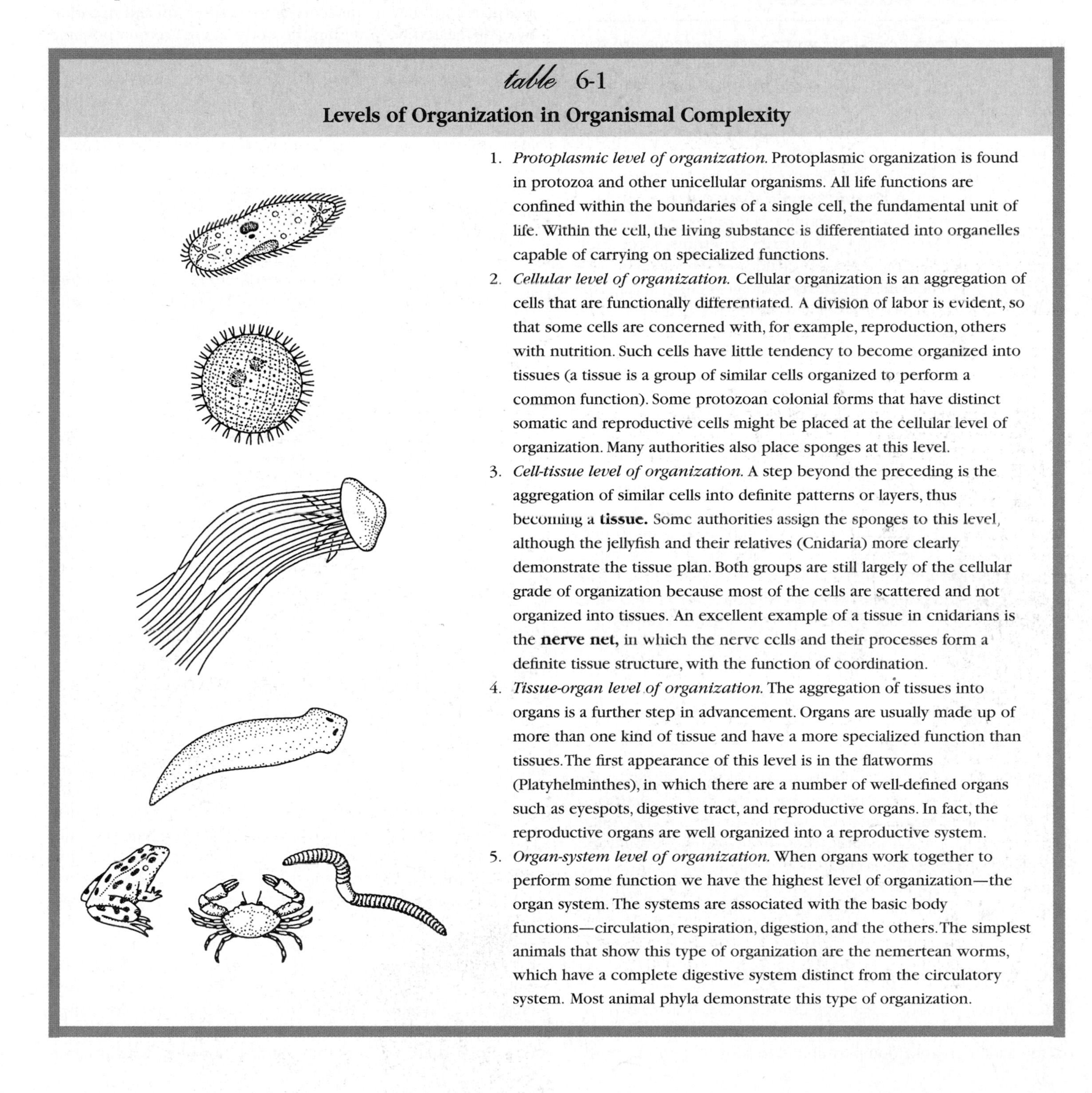

table 6-1

Levels of Organization in Organismal Complexity

1. *Protoplasmic level of organization.* Protoplasmic organization is found in protozoa and other unicellular organisms. All life functions are confined within the boundaries of a single cell, the fundamental unit of life. Within the cell, the living substance is differentiated into organelles capable of carrying on specialized functions.
2. *Cellular level of organization.* Cellular organization is an aggregation of cells that are functionally differentiated. A division of labor is evident, so that some cells are concerned with, for example, reproduction, others with nutrition. Such cells have little tendency to become organized into tissues (a tissue is a group of similar cells organized to perform a common function). Some protozoan colonial forms that have distinct somatic and reproductive cells might be placed at the cellular level of organization. Many authorities also place sponges at this level.
3. *Cell-tissue level of organization.* A step beyond the preceding is the aggregation of similar cells into definite patterns or layers, thus becoming a **tissue.** Some authorities assign the sponges to this level, although the jellyfish and their relatives (Cnidaria) more clearly demonstrate the tissue plan. Both groups are still largely of the cellular grade of organization because most of the cells are scattered and not organized into tissues. An excellent example of a tissue in cnidarians is the **nerve net,** in which the nerve cells and their processes form a definite tissue structure, with the function of coordination.
4. *Tissue-organ level of organization.* The aggregation of tissues into organs is a further step in advancement. Organs are usually made up of more than one kind of tissue and have a more specialized function than tissues. The first appearance of this level is in the flatworms (Platyhelminthes), in which there are a number of well-defined organs such as eyespots, digestive tract, and reproductive organs. In fact, the reproductive organs are well organized into a reproductive system.
5. *Organ-system level of organization.* When organs work together to perform some function we have the highest level of organization—the organ system. The systems are associated with the basic body functions—circulation, respiration, digestion, and the others. The simplest animals that show this type of organization are the nemertean worms, which have a complete digestive system distinct from the circulatory system. Most animal phyla demonstrate this type of organization.

integument, the body's outer enclosure; and the musculoskeletal system, a triumvirate of bone, muscle, and connective tissue organized to support the body, protect its internal organs, and give it mobility.

The Hierarchical Organization of Animal Complexity

Among the different metazoan groups, we can recognize five major grades of organization (Table 6-1). Each grade is more complex than the one before, and builds upon it in a hierarchical manner.

The unicellular protozoa (Protista) are the simplest animal-like organisms. These unicellular forms are nonetheless complete organisms that perform all of the basic functions of life as seen in the more complex animals. Within the confines of their cell, they show remarkable organization and division of labor, possessing distinct supportive structures, locomotor devices, fibrils, and simple sensory structures. The diversity observed among unicellular organisms is achieved by varying the architectural patterns of subcellular structures, organelles and the cell as a whole (Chapter 16).

The **metazoa,** or multicellular animals, evolved greater structural complexity by combining cells into larger units. The metazoan cell is a specialized part of the whole organism and, unlike the protozoan cell, it is not capable of independent existence. The cells of a multicellular organism are specialized for performing the various tasks accomplished by subcellular elements in the protozoa. The simplest metazoans show the **cellular** grade of organization in which cells demonstrate division of labor but are not strongly associated to perform a specific collective function (Table 6-1). In the more complex **tissue** grade, similar cells are grouped together and perform their common functions as a highly coordinated unit. In animals of the tissue-organ grade of organization, tissues are assembled into still larger functional units called **organs.** Usually one type of tissue carries the burden of an organ's chief function, as muscle tissue does in the heart; other tissues—epithelial, connective, and nervous—perform supportive roles.

Most metazoa (nemerteans and all more structurally complex phyla) have an additional level of complexity in which different organs operate together as **organ systems.** Eleven different kinds of organ systems are observed in metazoans: skeletal, muscular, integumentary, digestive, respiratory, circulatory, excretory, nervous, endocrine, immune, and reproductive. The great evolutionary diversity of these organ systems is covered in Chapters 19 through 31.

Complexity and Body Size

The chapter opening essay suggests that size is a major consideration in the design of animals. The most complex grades of metazoan organization permit and to some extent even promote the evolution of large body size (Figure 6-1). Large size confers several important physical and ecological consequences for the organism. As animals become larger, the body surface increases much more slowly than body volume. This happens because surface area increases as the square of body length (length^2), whereas volume (and therefore mass) increases as the cube of body length (length^3). In other words, a large animal will have less surface area relative to its volume than will a small animal of the same shape. The surface area of a large animal may be inadequate for respiration and nutrition by cells located deep within the body. There are two possible solutions to this problem. One solution is to fold or invaginate the body surface to increase the surface area or, as exploited by the flatworms, flatten the body into a ribbon or disc so that no internal space is far from the surface. This solution allows the body to become large without internal complexity. However, most large animals adopted a second solution; they developed internal transport systems to shuttle nutrients, gases, and waste products between the cells and the external environment.

Larger size buffers the animal against environmental fluctuations; it provides greater protection against predation and enhances offensive tactics; and it permits a more efficient use of metabolic energy. A large mammal uses more oxygen than a small mammal but the cost of maintaining its body temperature is less per gram of weight for the large mammal than for a small one. Large animals can also move themselves about at less energy cost than can small animals. A large mammal uses more oxygen in running than a small mammal, but the energy cost of moving 1 g of its body over a given distance is much less for a large mammal than for a small one (Figure 6-2). For all of these reasons, ecological opportunities of larger animals are very different from those of small ones. In subsequent chapters we will describe the extensive adaptive radiations observed in the taxa of large animals.

Extracellular Components of the Metazoan Body

In addition to the hierarchically arranged cellular structures discussed above, the metazoan animal contains two important noncellular components: the body fluids and the extracellular structural elements. In all eumetazoans, the body fluids are subdivided into two fluid "compartments": those that occupy the **intracellular space,** within the body's cells, and those that occupy the **extracellular space,** outside the cells. In animals with closed vascular systems (such as segmented worms and vertebrates), the extracellular fluids are subdivided further into the **blood plasma** (the fluid portion of the blood outside the cells; blood cells are really part of the intracellular compartment) and **interstitial fluid.** The interstitial fluid, also called tissue fluid, occupies the space surrounding the cells. Many invertebrates have open blood systems, however, with no true

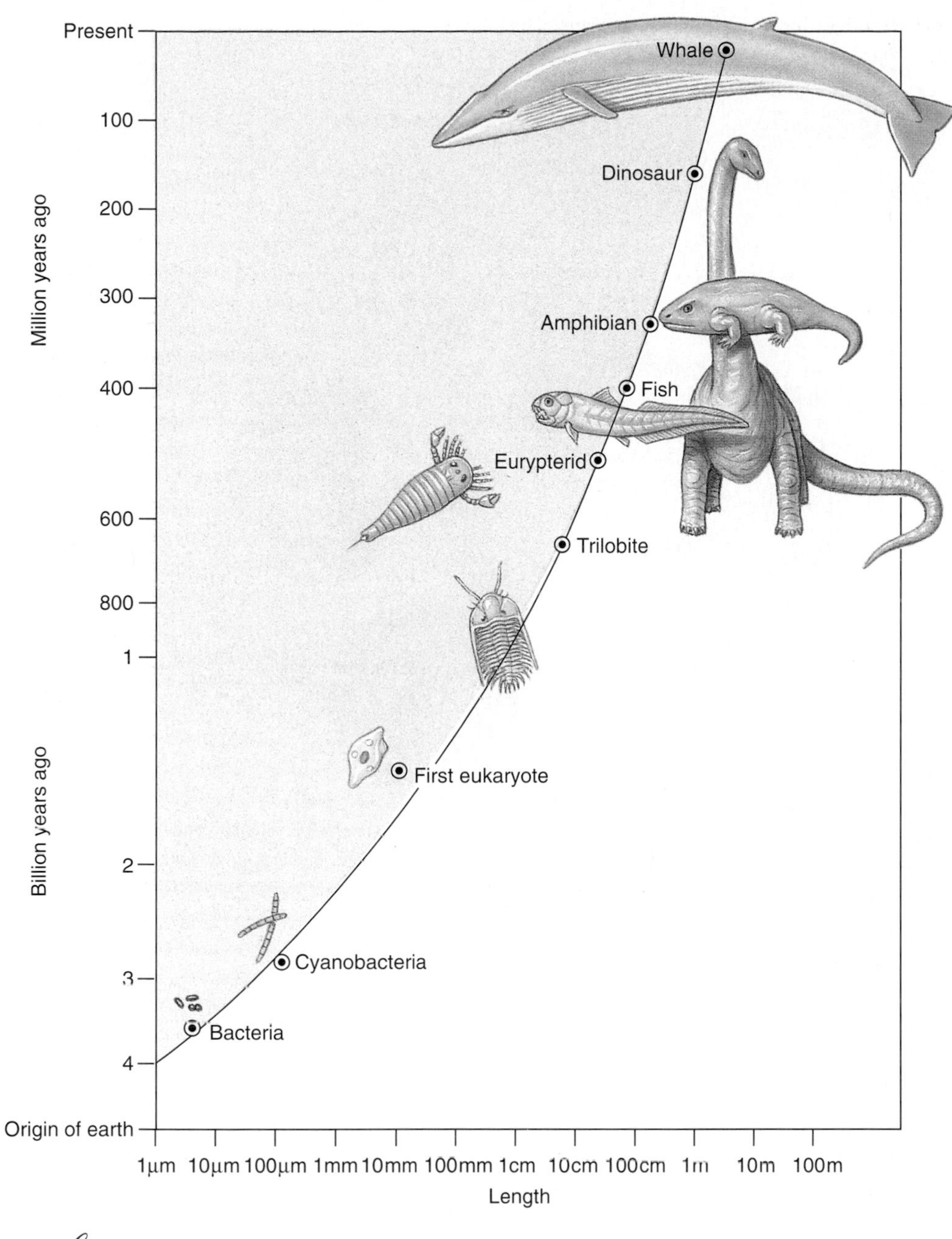

figure 6-1

Graph showing the evolution of length increase in the largest organisms present at different periods of life on earth. Note that both scales are logarithmic.

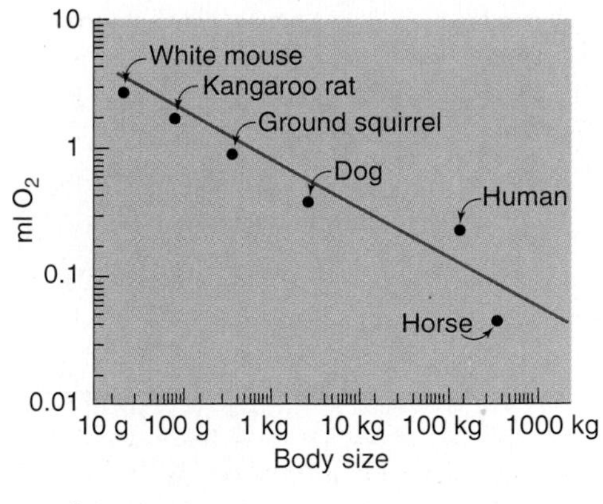

figure 6-2

Net cost of running for mammals of various sizes. Each point represents the cost (measured in rate of oxygen consumption) of moving 1 g of body over 1 km. The cost decreases with increasing body size.

separation of blood plasma from interstitial fluid. We will explore these relationships further in Chapter 8.

> *The term* intercellular, *meaning "between cells," should not be confused with the term* intracellular, *meaning "within cells."*

If we were to remove all the specialized cells and body fluids from the interior of the body, we would be left with the third element of the animal body: extracellular structural elements. This is the supportive material of the organism, including loose connective tissue (especially well developed in vertebrates but present in all metazoa), cartilage (molluscs and chordates), bone (vertebrates), and cuticle (arthropods, nematodes, annelids, and others). These elements provide mechanical stability and protection. In some instances, they act also as a depot of materials for exchange and serve as a medium for extracellular reactions. We will describe the diversity of extracellular skeletal elements characteristic of the different groups of animals in Chapters 17 through 31.

Types of Tissues

A **tissue** is a group of similar cells (together with associated cell products) specialized for the performance of a common function. The study of tissues is called **histology** (Gr. *histos,* tissue, + *logos,* discourse). All cells in metazoan animals take part in the formation of tissues. Sometimes the cells of a tissue may be of several kinds, and some tissues have a great many intercellular materials.

During embryonic development, the germ layers become differentiated into four kinds of tissues. These are epithelial, connective (including vascular), muscular, and nervous tissues (Figure 6-3). This is a surprisingly short list of basic tissue types that are able to meet the diverse requirements of animal life.

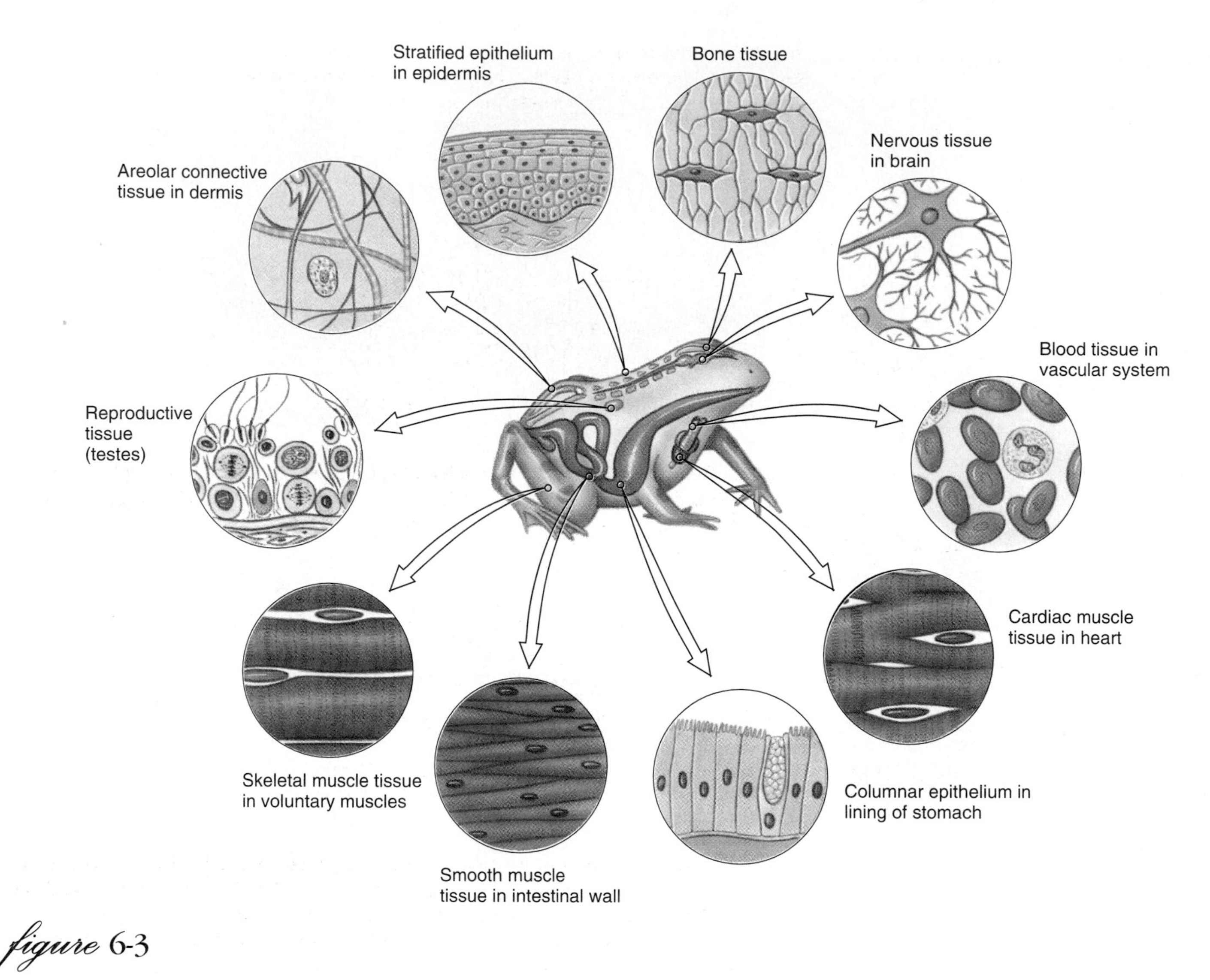

figure 6-3

Types of tissues in a vertebrate, showing examples of where different tissues are located in a frog.

Epithelial Tissue

An **epithelium** (pl., epithelia) is a sheet of cells that covers an external or internal surface. On the outside of the body, the epithelium forms a protective covering. Inside, the epithelium lines all the organs of the body cavity, as well as ducts and passageways through which various materials and secretions move. On many surfaces the epithelial cells are often modified into glands that produce lubricating mucus or specialized products such as hormones or enzymes.

Epithelia are classified on the basis of cell form and number of cell layers. Simple epithelia (Figure 6-4) are found in all metazoan animals, while stratified epithelia (Figure 6-5) are mostly restricted to the vertebrates. All types of epithelia are supported by an underlying basement membrane which is a condensation of the ground substance of connective tissue. Blood vessels never penetrate into epithelial tissues, so they depend on the diffusion of oxygen and nutrients from the underlying tissues.

Connective Tissue

Connective tissues are a diverse group of tissues that serve various binding and supportive functions. They are so widespread in the body that the removal of other tissues would still leave the complete form of the body clearly apparent. Connective tissue is made up of relatively few **cells,** a great many extracellular **fibers,** and a fluid, known as **ground substance** (also called **matrix**), in which the fibers are embedded. We recognize several different types of connective tissue. **Connective tissue proper** in vertebrates includes both **loose connective tissue,** composed of fibers and fixed and wandering cells suspended in a syrupy ground substance, and **dense connective tissue,** such as tendons and ligaments, that is composed largely of densely packed fibers (Figure 6-6). Much of the fibrous tissue of connective tissue is composed of **collagen** (Gr. *kolla,* glue, + *genos,* descent), a protein material of great tensile strength. Collagen is the most abundant protein in the

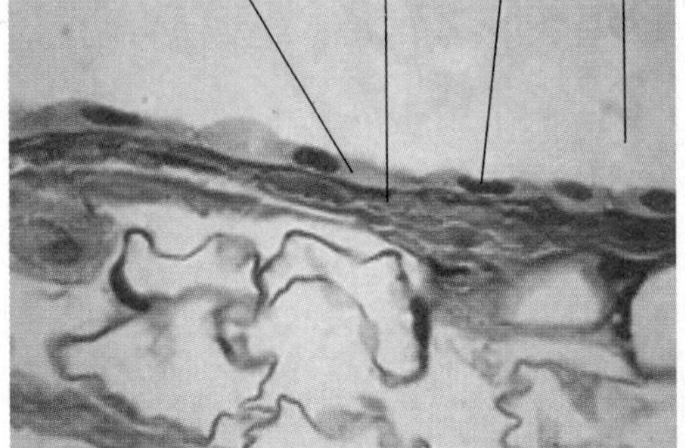

Simple squamous epithelium, composed of flattened cells that form a continuous delicate lining of blood capillaries, lungs, and other surfaces where it permits the passive diffusion of gases and tissue fluids into and out of cavities.

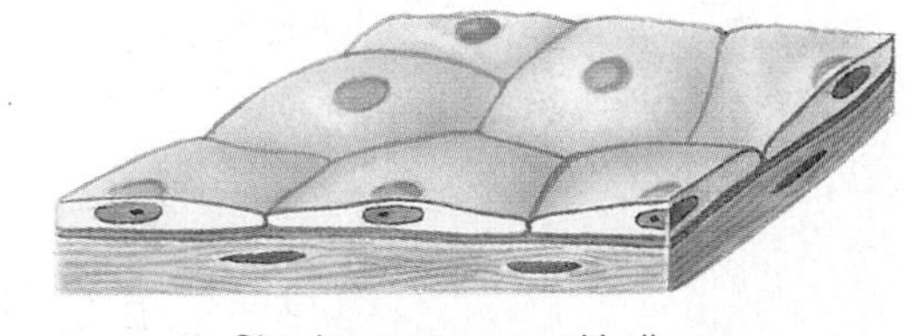

Simple squamous epithelium

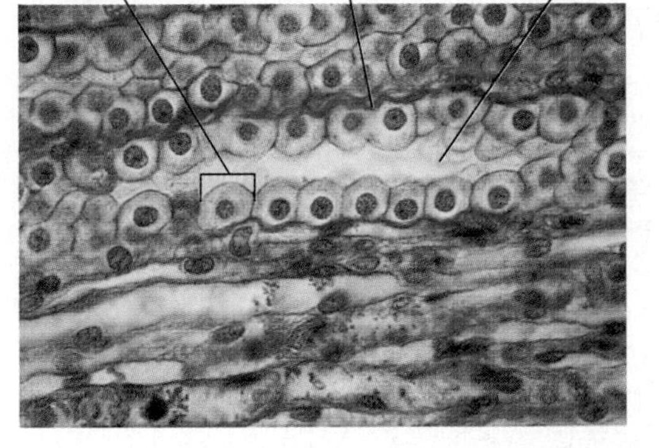

Simple cuboidal epithelium is composed of short, boxlike cells. Cuboidal epithelium usually lines small ducts and tubules, such as those of the kidney and salivary glands, and may have active secretory or absorptive functions.

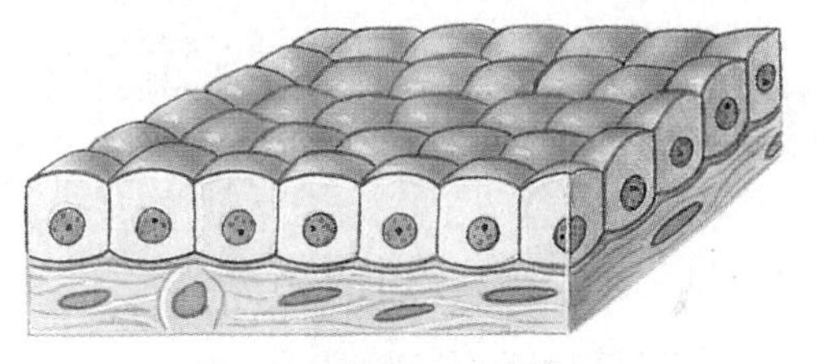

Simple cuboidal epithelium

Simple columnar epithelium resembles cuboidal epithelium but the cells are taller and usually have elongate nuclei. This type of epithelium is found on highly absorptive surfaces such as the intestinal tract of most animals. The cells often bear minute, fingerlike projections called microvilli that greatly increase the absorptive surface. In some organs, such as the female reproductive tract, the cells are ciliated.

Basement membrane
Epithelial cells
Microvilli on cell surface
Nuclei

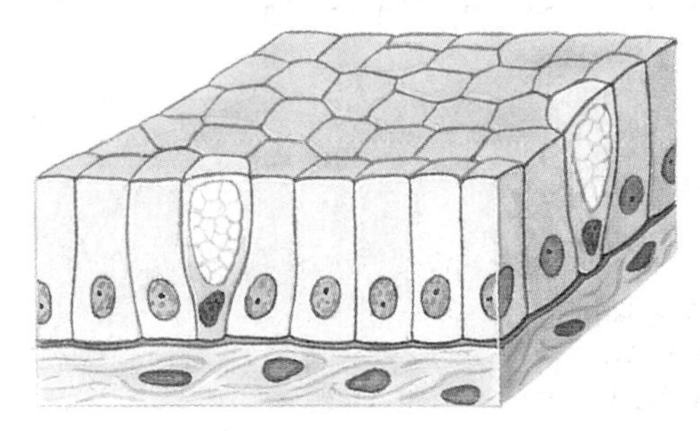

Simple columnar epithelium

figure 6-4

Types of simple epithelium

animal kingdom, found in animal bodies wherever both flexibility and resistance to stretching are required. The connective tissue of invertebrates, as in vertebrates, consists of cells, fibers, and ground substance, but it is not as elaborately developed.

Other types of connective tissue include **blood, lymph** and **tissue fluid** (collectively considered vascular tissue), composed of distinctive cells in a fluid ground substance, the plasma. Vascular tissue lacks fibers under normal conditions. **Cartilage** is a semirigid form of connective tissue with closely

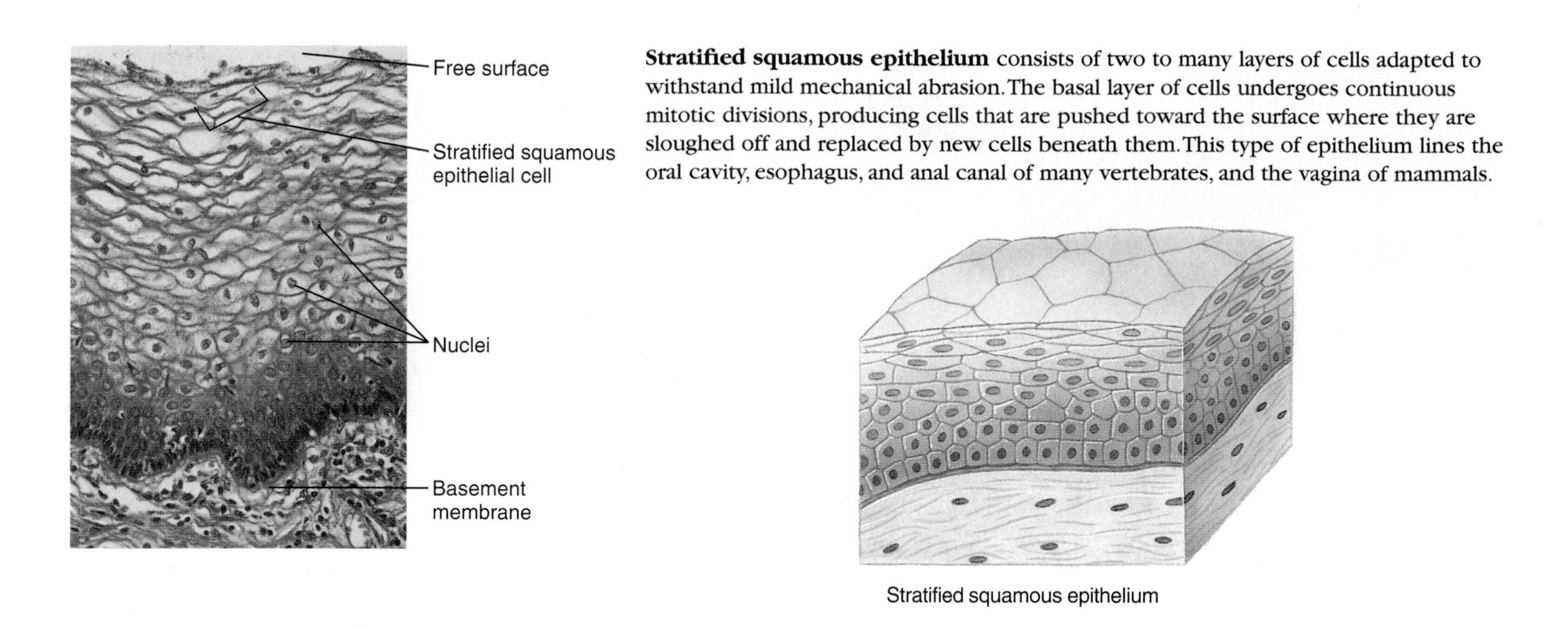

Stratified squamous epithelium consists of two to many layers of cells adapted to withstand mild mechanical abrasion. The basal layer of cells undergoes continuous mitotic divisions, producing cells that are pushed toward the surface where they are sloughed off and replaced by new cells beneath them. This type of epithelium lines the oral cavity, esophagus, and anal canal of many vertebrates, and the vagina of mammals.

Stratified squamous epithelium

Transitional epithelium is a type of stratified epithelium specialized to accommodate great stretching. This type of epithelium is found in the urinary tract and bladder of vertebrates. In the relaxed state it appears to be four or five cell layers thick, but when stretched out it appears to have only two or three layers of extremely flattened cells.

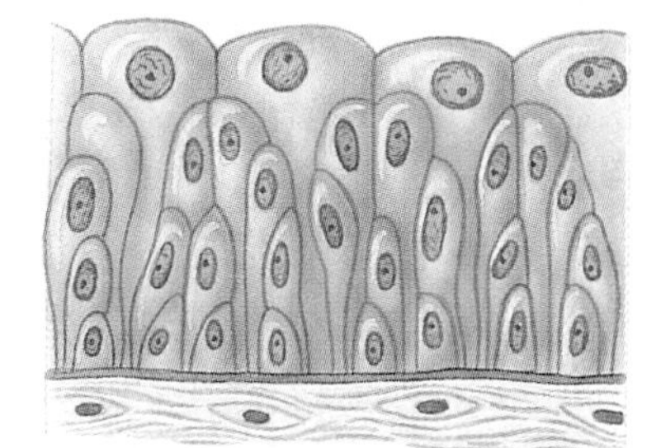

Transitional epithelium—unstretched

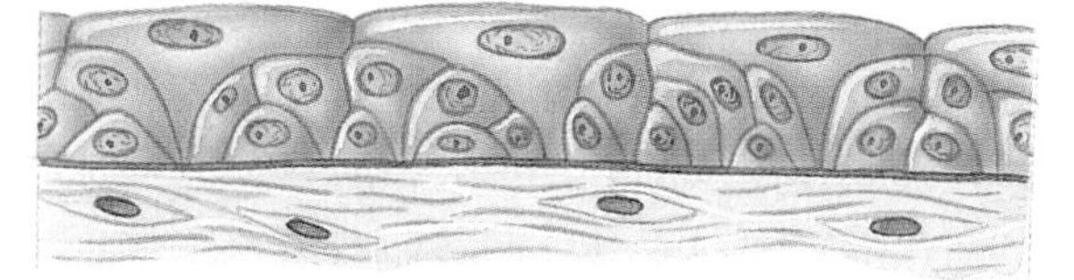

Transitional epithelium—stretched

figure 6-5

Types of stratified epithelium

packed fibers embedded in a gel-like ground substance (matrix). **Bone** is a calcified connective tissue containing calcium salts organized around collagen fibers (see Figure 6-6).

Muscular Tissue

Muscle is the most common tissue in the body of most animals. It is made up of elongated cells often called fibers, specialized for contraction. It originates (with few exceptions) from the mesoderm, and its unit is the cell or **muscle fiber.** When viewed with a light microscope **striated muscle** appears transversely striped (striated), with alternating dark and light bands (Figure 6-7). In vertebrates we can recognize two types of striated muscle: **skeletal** and **cardiac muscle.** A third kind of muscle is **smooth** (or visceral) **muscle,** which lacks the characteristic alternating bands of the striated type (Figure 6-7). The unspecialized cytoplasm of muscles is called sarcoplasm, and the contractile elements within the fiber are the **myofibrils.**

Loose connective tissue, also called areolar connective tissue, is the "packing material" of the body that anchors blood vessels, nerves, and body organs. It contains fibroblasts that synthesize the fibers and ground substance of connective tissue and wandering macrophages that phagocytize pathogens or damaged cells. The different fiber types include strong collagen fibers (thick and red in micrograph) and elastic fibers (black and branching in micrograph) formed of the protein elastin. Adipose (fat) tissue is considered a type of loose connective tissue.

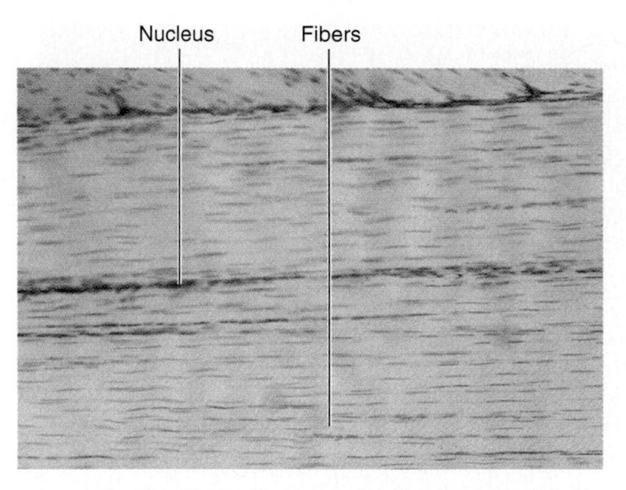

Dense connective tissue forms tendon, ligaments, and fasciae (fa′sha), the latter arranged as sheets or bands of tissue surrounding skeletal muscle. In tendon (shown here) the collagenous fibers are extremely long and tightly packed together.

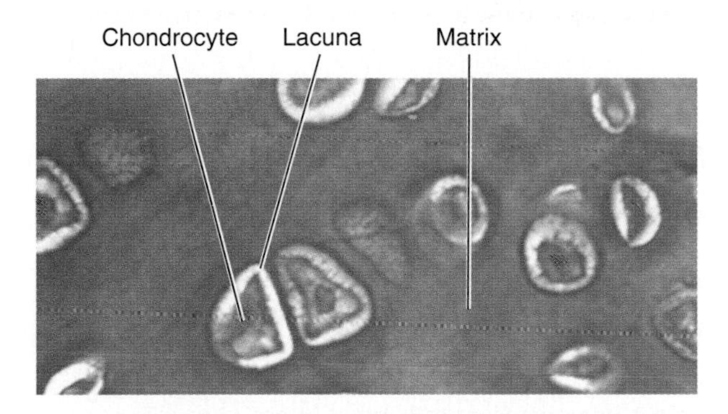

Cartilage is a vertebrate connective tissue composed of a firm gel ground substance (matrix) containing cells (chondrocytes) living in small pockets called lacunae, and collagen or elastic fibers (depending on the type of cartilage). In hyaline cartilage shown here, both collagen fibers and ground substance are stained uniformly purple and cannot be distinguished one from the other. Because cartilage lacks a blood supply, all nutrients and waste materials must diffuse through the ground substance from surrounding tissues.

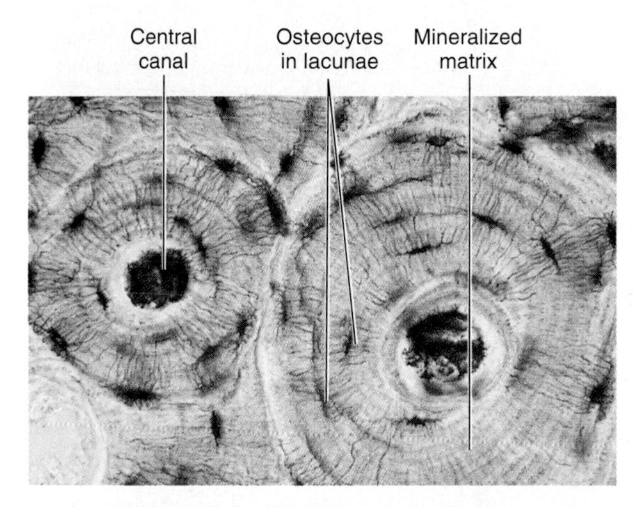

Bone, the strongest of vertebrate connective tissues, contains mineralized collagen fibers. Small pockets (lacunae) within the matrix contain bone cells, called osteocytes. The osteocytes communicate with blood vessels that penetrate into bone by means of a tiny network of channels called canaliculi. Unlike cartilage, bone undergoes extensive remodeling during an animal's life, and can repair itself following even extensive damage.

figure 6-6

Types of connective tissue

Nervous Tissue

Nervous tissue is specialized for the reception of stimuli and the conduction of impulses from one region to another. The two basic types of cells in nervous tissue are **neurons** (Gr. nerve), the basic functional unit of the nervous system, and **neuroglia** (nu-rog′le-a; Gr. nerve, + *glia,* glue), a variety of nonnervous cells that insulate neuron membranes and serve various supportive functions. Figure 6-8 shows the functional anatomy of a typical nerve cell.

Integument Among Various Groups of Animals

The integument is the outer covering of the body, a protective wrapping that includes the skin and all structures derived from or associated with the skin, such as hair, setae, scales, feathers, and horns. In most animals it is tough and pliable, providing mechanical protection against abrasion and puncture

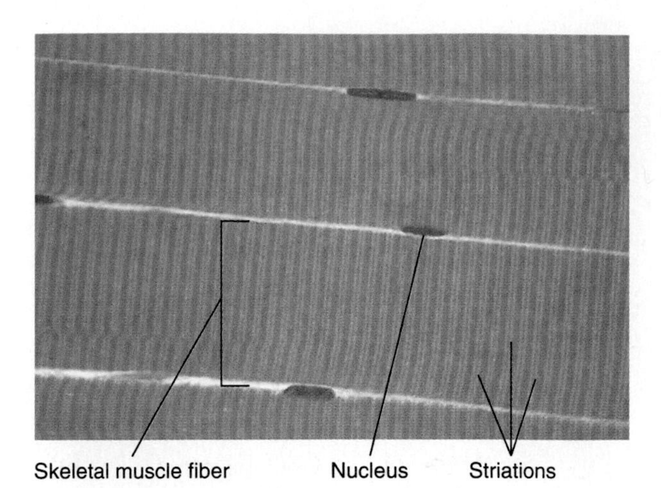

Skeletal muscle is a type of striated muscle found in both invertebrates and vertebrates. It is composed of extremely long, cylindrical fibers, which are multinucleate cells that may reach from one end of the muscle to the other. Viewed through the light microscope, the cells appear to have a series of stripes, called striations, running across them. Skeletal muscle is called voluntary muscle (in vertebrates) because it contracts when stimulated by nerves under conscious cerebral control.

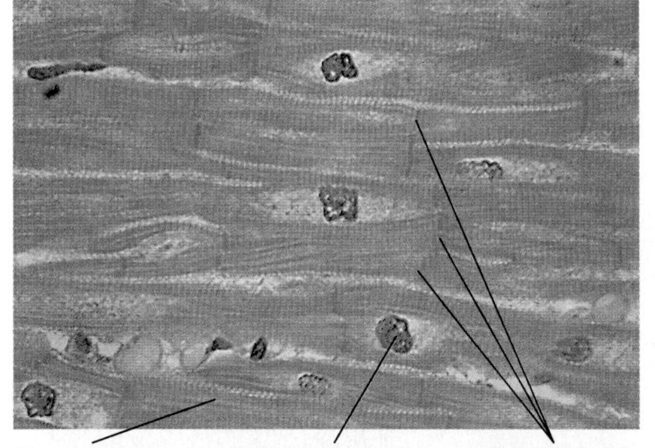

Cardiac muscle is another type of striated muscle found only in the vertebrate heart. The cells are much shorter than those of skeletal muscle and have only one nucleus per cell (uninucleate). Cardiac muscle tissue is a branching network of fibers with individual cells interconnected by junctional complexes called intercalated discs. Cardiac muscle is called involuntary muscle because it does not require nerve activity to stimulate contraction. Instead, heart rate is controlled by specialized pacemaker cells located in the heart itself. However, autonomic nerves from the brain may alter pacemaker activity.

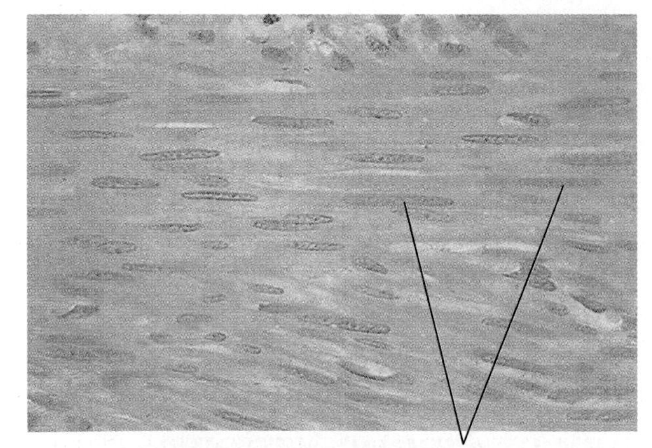

Smooth muscle is nonstriated muscle found in both invertebrates and vertebrates. Smooth muscle cells are long, tapering strands, each containing a single nucleus. Smooth muscle is the most common type of muscle in invertebrates in which it serves as body wall musculature and lines ducts and sphincters. In vertebrates, smooth muscle cells are organized into sheets of muscle circling the walls of the alimentary canal, blood vessels, respiratory passages, and urinary and genital ducts. Smooth muscle is typically slow acting and can maintain prolonged contractions with very little energy expenditure. Its contractions are involuntary and unconscious. The principal functions of smooth muscles are to push the material in a tube, such as the intestine, along its way by active contractions or to regulate the diameter of a tube, such as a blood vessel, by sustained contraction.

figure 6-7
Types of muscle tissue

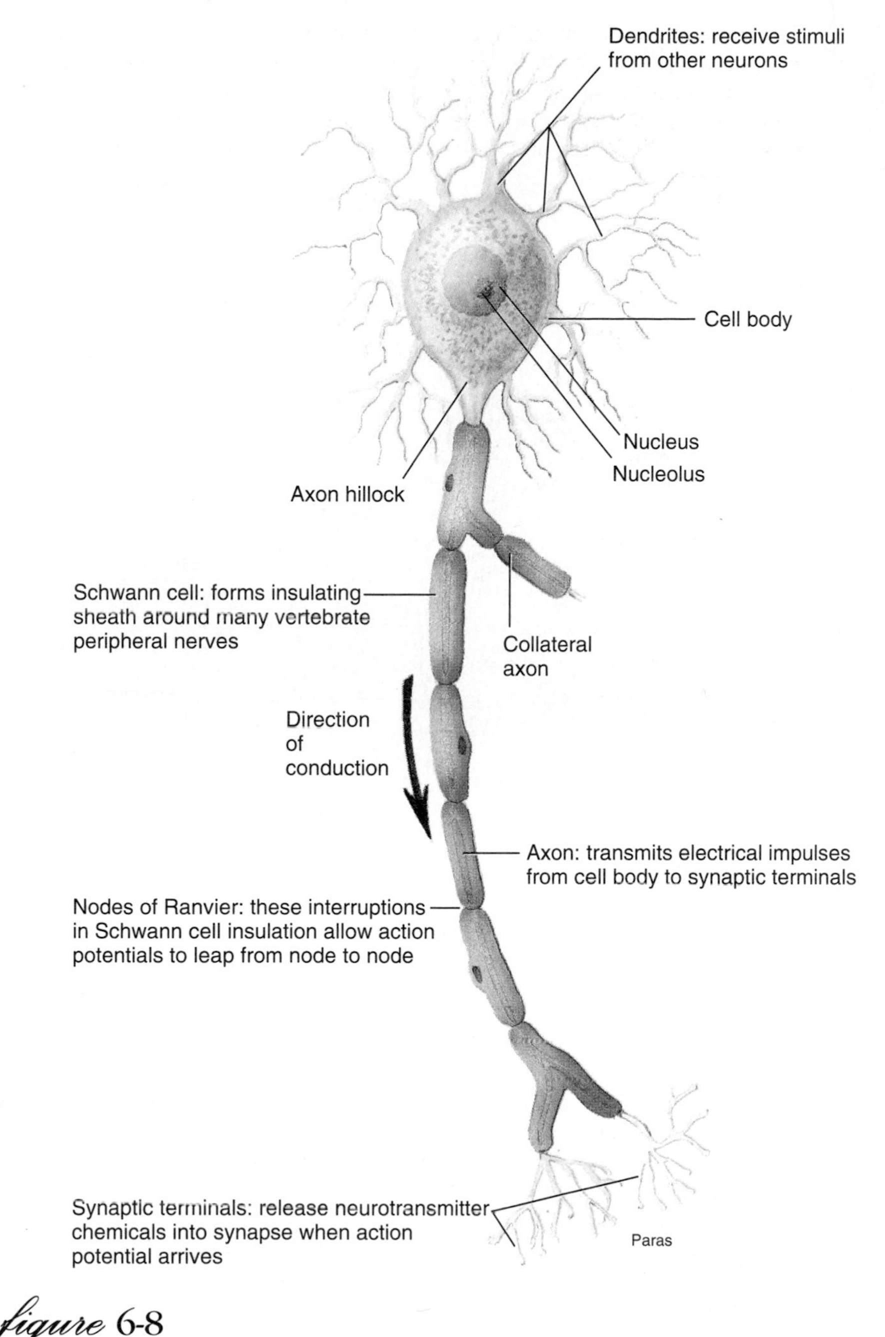

figure 6-8

Functional anatomy of a neuron. From the nucleated body, or **soma,** extend one or more **dendrites** (Gr. *dendron,* tree), which receive electrical impulses from receptors or other nerve cells, and a single **axon** that carries impulses away from the cell body to other nerve cells or to an effector organ. The axon is often called a **nerve fiber.** Nerves are separated from other nerves or from effector organs by specialized junctions called synapses.

and forming an effective barrier against the invasion of bacteria. It may provide moisture proofing against fluid loss or gain. The skin helps protect the underlying cells against the damaging action of the ultraviolet rays of the sun. But, in addition to being a protective cover, the skin serves a variety of important regulatory functions. For example, in endothermic ("warm-blooded") vertebrates, it is vitally concerned with temperature regulation, since most of the body's heat is lost through the skin; it contains mechanisms that cool the body when it is too hot and slow heat loss when the body is too cold. The skin contains sensory receptors that provide essential information about the immediate environment. It has excretory functions and, in some animals, respiratory functions as well. Through skin pigmentation the organism can make itself more or less conspicuous. Skin secretions can make the animal sexually attractive or repugnant or provide olfactory cues that influence behavioral interactions between individuals.

Invertebrate Integument

Many protozoa have only the delicate plasma membrane for external coverings; others, such as *Paramecium,* have developed a more rigid pellicle. Most multicellular invertebrates, however, have more complex tissue coverings. The principal covering is a single-layered **epidermis.** Some invertebrates have added a secreted noncellular **cuticle** over the epidermis for additional protection.

The molluscan epidermis is delicate and soft and contains mucous and other glands, some of which secrete the calcium carbonate of the shell. Cephalopod molluscs (squids and octopuses) have developed a more complex integument, consisting of cuticle, simple epidermis, layer of connective tissue, layer of reflecting cells (iridocytes), and thicker layer of connective tissue.

Arthropods have the most complex of invertebrate integuments, providing not only protection but also skeletal support. The development of a firm **exoskeleton** and jointed appendages suitable for the attachment of muscles has been a key feature in the extraordinary diversity of this phylum, the largest of animal groups. The arthropod integument consists of a single-layered **epidermis** (also called more precisely **hypodermis**), which secretes a complex cuticle of two zones (Figure 6-9A). The inner zone, the **procuticle,** is composed of protein and chitin (a polysaccharide) laid down in layers (lamellae) much like the veneers of plywood. The outer zone of cuticle, lying on the external surface above the procuticle, is the thin **epicuticle.** The epicuticle is a

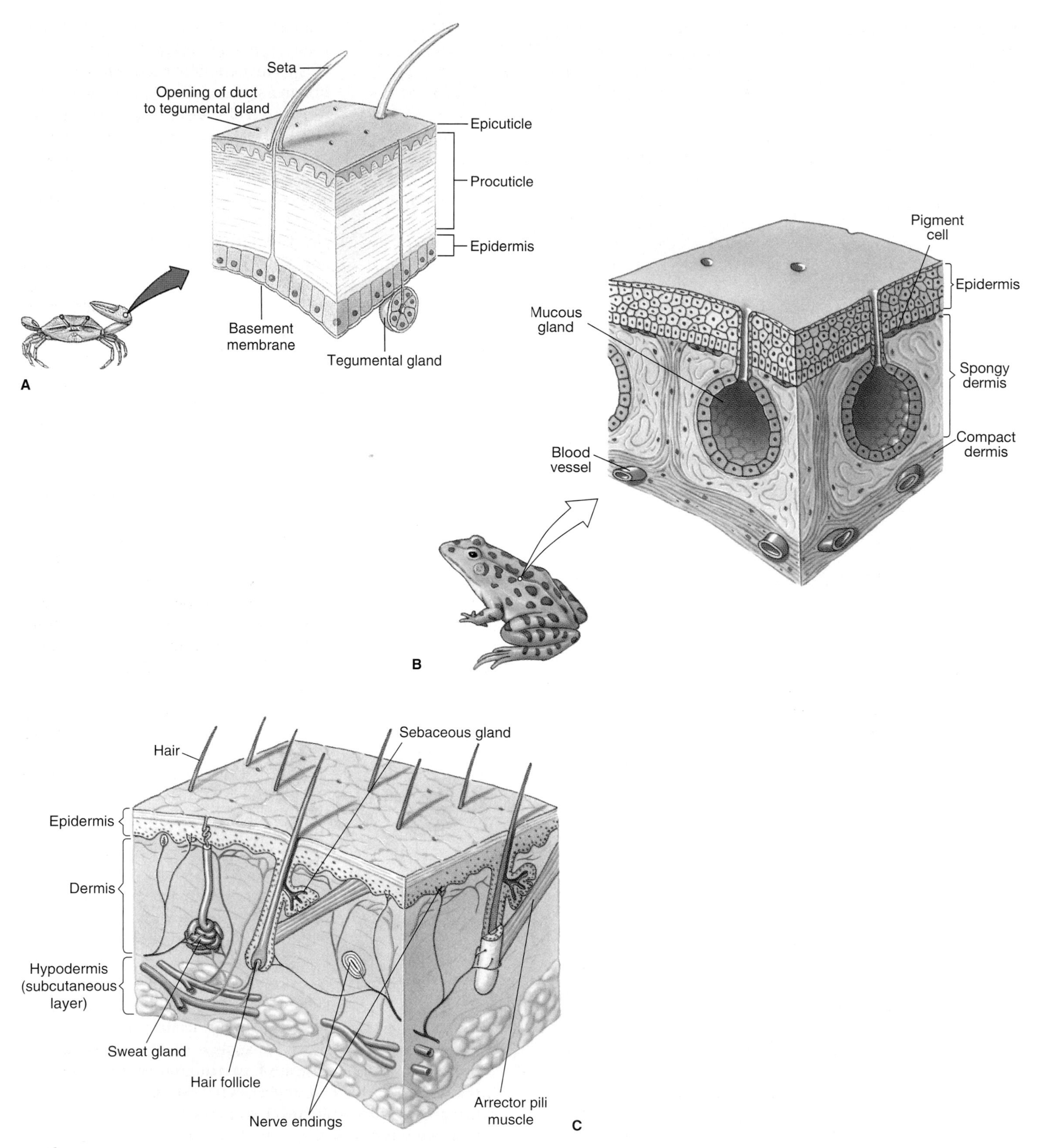

figure 6-9

Integumentary systems of animals, showing the major layers. **A,** Structure of arthropod (crustacean) body wall showing cuticle and epidermis. **B,** Structure of amphibian (frog) integument. **C,** Structure of human integument.

nonchitinous complex of proteins and lipids that provides a protective moisture-proofing barrier to the integument.

The arthropod cuticle may remain as a tough but soft and flexible layer, as it is in many microcrustaceans and insect larvae. However, it may be hardened by either of two ways. In the decapod crustaceans, for example, crabs and lobsters, the cuticle is stiffened by **calcification,** the deposition of calcium carbonate. In insects hardening occurs when the protein molecules bond together with stabilizing cross-linkages, within and between the adjacent lamellae of the procuticle. The result of the process, called **sclerotization,** is the formation of a highly resistant and insoluble protein, **sclerotin.** Arthropod cuticle is one of the toughest materials synthesized by animals; it is strongly resistant to pressure and tearing and can withstand boiling in concentrated alkali, yet it is light, having a specific mass of only 1.3 (1.3 times the weight of water).

When arthropods molt, the epidermal cells first divide by mitosis. Enzymes secreted by the epidermis digest most of the procuticle. The digested materials are then absorbed and consequently are not lost to the body. Then in the space beneath the old cuticle a new epicuticle and procuticle are formed. After the old cuticle is shed, the new cuticle is thickened and calcified or sclerotized.

Vertebrate Integument

The basic plan of the vertebrate integument, as exemplified by frog and human skin (Figure 6-9B and C), includes a thin, outer stratified epithelial layer, the **epidermis,** derived from ectoderm and an inner, thicker layer, the **dermis,** or true skin, which is of mesodermal origin.

Although the epidermis is thin and appears simple in structure, it gives rise to most derivatives of the integument, such as hair, feathers, claws, and hooves. The dermis contains blood vessels, collagenous fibers, nerves, pigment cells, fat cells, and connective tissue cells called fibroblasts. These elements support, cushion, and nourish its overlying partner, which is devoid of blood vessels.

The epidermis is a stratified squamous epithelium (p. 144) consisting usually of several layers of cells. The basal part is made up of cells that undergo frequent mitosis to renew the layers that lie above. As the outer layers of cells are displaced upward by new generations of cells beneath, an exceedingly tough, fibrous protein called **keratin** accumulates in the interior of the cells. Gradually, keratin replaces all metabolically active cytoplasm. The cell dies and is eventually shed, lifeless and scalelike. Such is the origin of dandruff as well as a significant fraction of household dust. This process is called **keratinization,** and the cell, thus transformed, is said to be **cornified.** Cornified cells, highly resistant to abrasion and water diffusion, comprise the outermost **stratum corneum.** This epidermal layer becomes especially thick in areas exposed to persistent pressure or wear, such as calluses and the foot pads of mammals. Such thickenings may involve the dermis as well as the epidermis.

Lizards, snakes, and turtles were among the first to exploit the adaptive possibilities of the remarkably tough protein keratin. The reptilian epidermal scale that develops from keratin is a much lighter and more flexible structure than the bony, dermal scale of fishes, yet it provides excellent protection from abrasion and desiccation. Scales may be overlapping structures, as in snakes and some lizards, or develop into plates as in turtles and crocodilians. Birds dedicated keratin to new uses. Feathers, beaks, and claws, as well as scales, are all epidermal structures composed of dense keratin. Mammals continued to capitalize on keratin's virtues by turning it into hair, hooves, claws, and nails. As a result of its keratin content, hair is by far the strongest material in the body. It has a tensile strength comparable to that of rolled aluminum and is nearly twice as strong, weight for weight, as the strongest bone.

The **dermis,** as already mentioned, mainly serves a supportive role for the epidermis. Nevertheless, true bony structures, where they occur in the integument, are always dermal derivatives. Heavy bony plates were common in primitive ostracoderm and placoderm fishes of the Paleozoic era and persist in some living fishes such as sturgeons. Scales of contemporary fishes are bony dermal structures that have evolved from the bony armor of the Paleozoic fishes but are much smaller and more flexible. Although of dermal origin, fish scales are intimately associated with the thin, overlying epidermis; in some species the scales protrude through the epidermis, but typically the epidermis forms a continuous sheath that is reflected under the overlapping scales (Figure 6-10). Dermal bone also forms some of the flat bones of the skull and gives rise to antlers (p. 672), which are outgrowths of dermal frontal bone.

Injurious Effects of Sunlight

The familiar vulnerability of the human skin to sunburn reminds us of the potentially damaging effects of ultraviolet radiation on cells. Many animals, such as protozoa and flatworms, are damaged or killed by ultraviolet radiation if exposed to the sun in shallow water. Most land animals are protected from such damage by the screening action of special body coverings, for example, the cuticle of arthropods, the scales of reptiles, and the feathers and fur of birds and mammals. Humans, however, are "naked apes" that lack the furry protection of other mammals. We must depend on thickening of the epidermis (stratum corneum) and on epidermal pigmentation for protection. Most ultraviolet radiation is absorbed in the epidermis but about 10% penetrates into the dermis. Damaged cells in both the epidermis and dermis release histamine and other vasodilator substances that cause blood vessel enlargement in the dermis and the

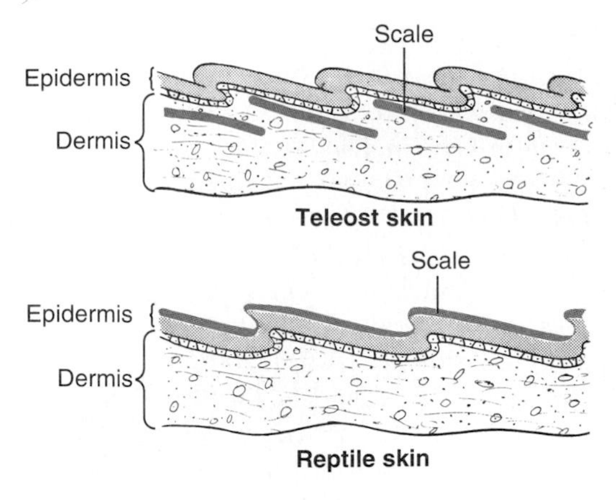

figure 6-10

Integument of bony fishes and lizards. Bony (teleost) fishes have bony scales from dermis, and reptiles have horny scales from epidermis. Dermal scales of fishes are retained throughout life. Since a new growth ring is added to each scale each year, fishery biologists use scales to tell the age of fishes. Epidermal scales of reptiles are shed periodically.

characteristic red coloration of sunburn. Light skins suntan through the formation of the pigment **melanin** in the deeper epidermis and by "pigment darkening," that is, the photooxidative blackening of bleached pigment already present in the epidermis. Unfortunately, tanning does not bestow perfect protection. Sunlight still ages the skin prematurely, and tanning itself causes the skin to become dry and leathery. Sunlight also is responsible for approximately one million new cases of skin cancer annually in the United States alone, making skin cancer the most common of malignancies among Caucasians. There is now strong evidence that genetic mutations caused by high doses of sunlight received during the pre-adult years are responsible for skin cancers that appear after middle age.

Skeletal Systems

Skeletons are supportive systems that provide rigidity to the body, surfaces for muscle attachment, and protection for vulnerable body organs. The familiar bone of the vertebrate skeleton is only one of several kinds of supportive and connective tissues serving various binding and weight-bearing functions, which are described in this discussion.

Hydrostatic Skeletons

Not all skeletons are rigid; many invertebrate groups use their body fluids as an internal hydrostatic skeleton. The muscles in the body wall of the earthworm, for example, have no firm base for attachment but develop muscular force by contracting against the coelomic fluids, which are enclosed within a limited space and are incompressible, much like the hydraulic brake system of an automobile.

Alternate contractions of the circular and longitudinal muscles of the body wall enable the worm to thin and thicken, setting up backward-moving waves of motion that propel the animal forward (Figure 6-11). Earthworms and other annelids are helped by the septa that separate the body into more or less independent compartments. An obvious advantage is that if a worm is punctured or even cut into pieces, each part can still develop pressure and move. Worms that lack internal compartments, for example, the lugworm *Arenicola* (Figure 22-10, p. 481), are rendered helpless if the body fluid is lost through a wound.

There are many examples in the animal kingdom of muscles that not only produce movement but also provide a unique form of skeletal support. The elephant's trunk is an excellent example of a structure that lacks any obvious form of skeletal support, yet is capable of bending, twisting, elongating, and lifting heavy weights (Figure 6-12). The elephant's trunk, as well as the tongues of mammals and reptiles and the tentacles of cephalopod molluscs are examples of **muscular hydrostats.** Like the hydrostatic skeletons of worms, muscular hydrostats work because they are composed of incompressible tissues that remain at constant volume. The remarkably diverse movements of muscular hydrostats depend on many muscles arranged in complex patterns.

Rigid Skeletons

Rigid skeletons differ from hydrostatic skeletons in one fundamental way: rigid skeletons consist of rigid elements, usually jointed, to which muscles can attach. Muscles can only contract; to be lengthened they must be extended by the pull of an antagonistic set of muscles or other forces. Rigid skeletons provide the anchor points required by opposing sets of muscles, such as flexors and extensors.

There are two principal types of rigid skeletons: **exoskeleton,** typical of molluscs, arthropods, and many other invertebrates, and **endoskeleton,** characteristic of echinoderms and vertebrates. The exoskeleton of many invertebrates provides protection from mechanical or chemical injury and from the assaults of predators and competitors. An exoskeleton may take the form of a shell, as in molluscs, brachiopods (p. 538), foraminiferans and other sarcodine protozoa (p. 372); or a calcareous, proteinaceous, or chitinous plate, as in arthropods. It may be rigid, as in molluscs, or jointed and movable, as in arthropods in which it performs a vital role in locomotion. Unlike an endoskeleton, which grows with the animal, an exoskeleton is often a limiting coat of armor that must be periodically molted to make way for an enlarged replacement (molting in crustaceans is described on p. 500). Some invertebrate exoskeletons, such as the shells of snails and bivalves, grow with the animal.

The arthropod-type exoskeleton is perhaps a better arrangement for small animals than a vertebrate-type endoskeleton because a hollow cylindrical tube can support much more weight without collapsing than can a solid cylindrical rod of the same material and weight. Arthropods can thus enjoy both protection and structural support from their exoskeleton. But for larger animals the hollow cylinder would be completely impractical. If made thick enough to support the body weight, it would be too heavy to lift; but if kept thin and light, it would be extremely sensitive to buckling or shattering on impact. Finally, can you imagine the sad plight of an animal the size of an elephant when it shed its exoskeleton to molt?

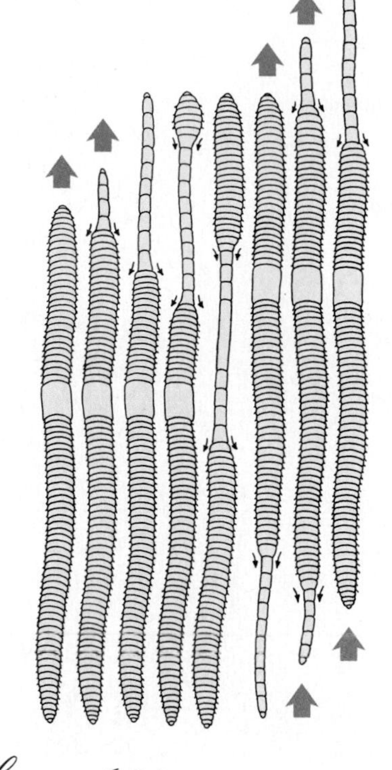

figure 6-11

How an earthworm moves forward. When circular muscles contract, longitudinal muscles are stretched by internal fluid pressure and the worm elongates. Then, by alternate contraction of longitudinal and circular muscles, a wave of contraction passes from anterior to posterior. Bristlelike setae are extended to anchor the animal and prevent slippage.

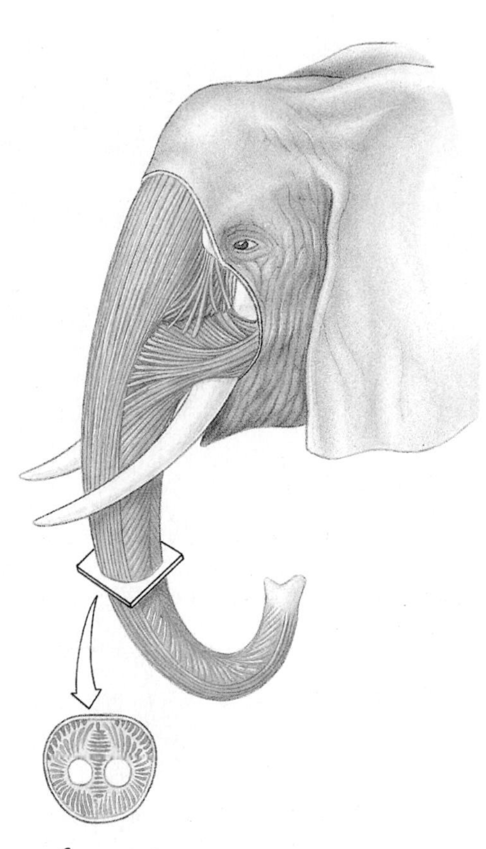

figure 6-12

Muscular trunk of an elephant, an example of a muscular hydrostat.

The vertebrate endoskeleton is formed inside the body and is composed of bone and cartilage, which are forms of dense connective tissue. Bone not only supports and protects but is also the major body reservoir for calcium and phosphorus. In amniote vertebrates the red blood cells and certain white blood cells are formed in the bone marrow.

Notochord and Cartilage

The **notochord** (see Figure 26-1 p. 566) is a semirigid supportive axial rod of the protochordates and all vertebrate larvae and embryos. It is composed of large, vacuolated cells and is surrounded by layers of elastic and fibrous sheaths. It is a stiffening device, preserving body shape during locomotion. Except in the jawless fishes (lampreys and hagfishes), the notochord is surrounded or replaced by the backbone during embryonic development.

Cartilage is a major skeletal element of some vertebrates. The jawless fishes and the elasmobranchs (sharks, skates, and rays) have purely cartilaginous skeletons, which oddly enough is a derived feature, since their Paleozoic ancestors had bony skeletons. Other vertebrates as adults have principally bony skeletons with some cartilage interspersed. Cartilage is a soft, pliable, characteristically deep-lying tissue. Unlike most connective tissues, which are quite variable in form, cartilage is basically the same wherever it is found. The usual form, **hyaline cartilage** (Figure 6-6) has a clear, glossy appearance. It is composed of cartilage cells **(chondrocytes)** surrounded by firm complex protein gel interlaced with a meshwork of collagenous fibers. Blood vessels are virtually absent—the reason that sports injuries involving cartilage heal poorly. In addition to forming the cartilaginous skeleton of some vertebrates and that of all vertebrate embryos, hyaline cartilage makes up the articulating surfaces of many bone joints of most adult vertebrates and the supporting tracheal, laryngeal, and bronchial rings of the respiratory system.

Cartilage similar to hyaline cartilage occurs in some invertebrates, for example the radula of gastropod molluscs and in the lophophore of brachiopods. The cartilage of cephalopod molluscs is of a special type with long, branching processes that resemble the cells of vertebrate bone.

Bone

Bone is a living tissue that differs from other connective and supportive tissues by having significant deposits of inorganic calcium salts laid down in an extracellular matrix. Its structural

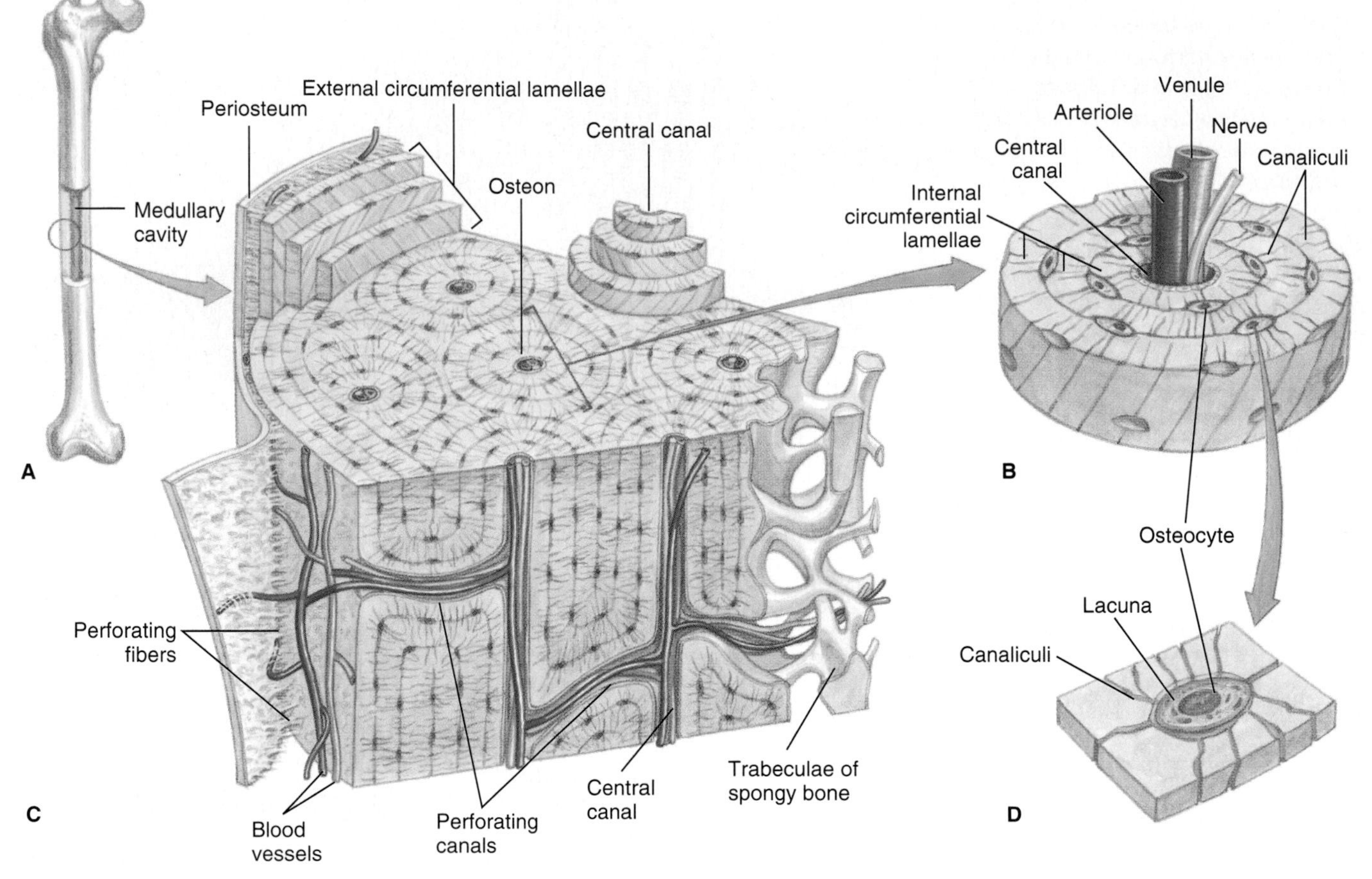

figure 6-13

Structure of compact bone. **A,** Adult long bone with a cut into the medullary cavity. **B,** Enlarged section showing osteons, the basic histological unit of bone. **C,** Enlarged view of an osteon showing the concentric lamellae and the osteocytes (bone cells) arranged within lacunae. **D,** An osteocyte within a lacuna. Bone cells receive nutrients from the circulatory system via tiny canaliculi that interlace the calcified matrix. Bone cells are known as osteoblasts when they are building bone, but, in mature bone shown here, they become resting osteocytes. Bone is covered with compact connective tissue called periosteum.

organization is such that bone has nearly the tensile strength of cast iron, yet is only one third as heavy.

Bone is never formed in vacant space but is always laid down by replacement in areas occupied by some form of connective tissue. Most bone develops from cartilage and is called **endochondral** ("within-cartilage") or **replacement bone.** Embryonic cartilage is gradually eroded leaving it extensively honeycombed; bone-forming cells then invade these areas and begin depositing calcium salts around the strandlike remnants of the cartilage. A second type of bone is **membranous bone,** which develops directly from sheets of embryonic cells. In tetrapod vertebrates, membranous bone is restricted to bones of the face and cranium; the remainder of the skeleton is endochondral bone. Whatever its embryonic origin, once bone is fully formed there is no difference in the histological structure; endochondral and membranous bone look the same.

Fully formed bone, however, may vary in density. **Spongy bone** (cancellous bone) consists of an open, interlacing framework of bony tissue, oriented to give maximum strength under the normal stresses and strains that the bone receives. All bone develops first as spongy bone, but some bones, through further deposition of bone salts, become **compact bone.** Compact bone is dense, appearing solid to the unaided eye. Both spongy and compact bone are found in the typical long bones of tetrapods (Figure 6-13).

Microscopic Structure of Bone

Compact bone is composed of a calcified bone matrix arranged in concentric rings. The rings contain cavities **(lacunae)** filled with bone cells **(osteocytes)** that are interconnected by many minute passages **(canaliculi).** These serve to distribute nutrients throughout the bone. This entire organization of lacunae and canaliculi is arranged into an elongated cylinder called an **osteon** (also called **haversian system)** (Figure 6-13). Bone consists of bundles of osteons cemented together and interconnected with blood vessels and nerves. Because of blood vessels and nerves throughout bone,

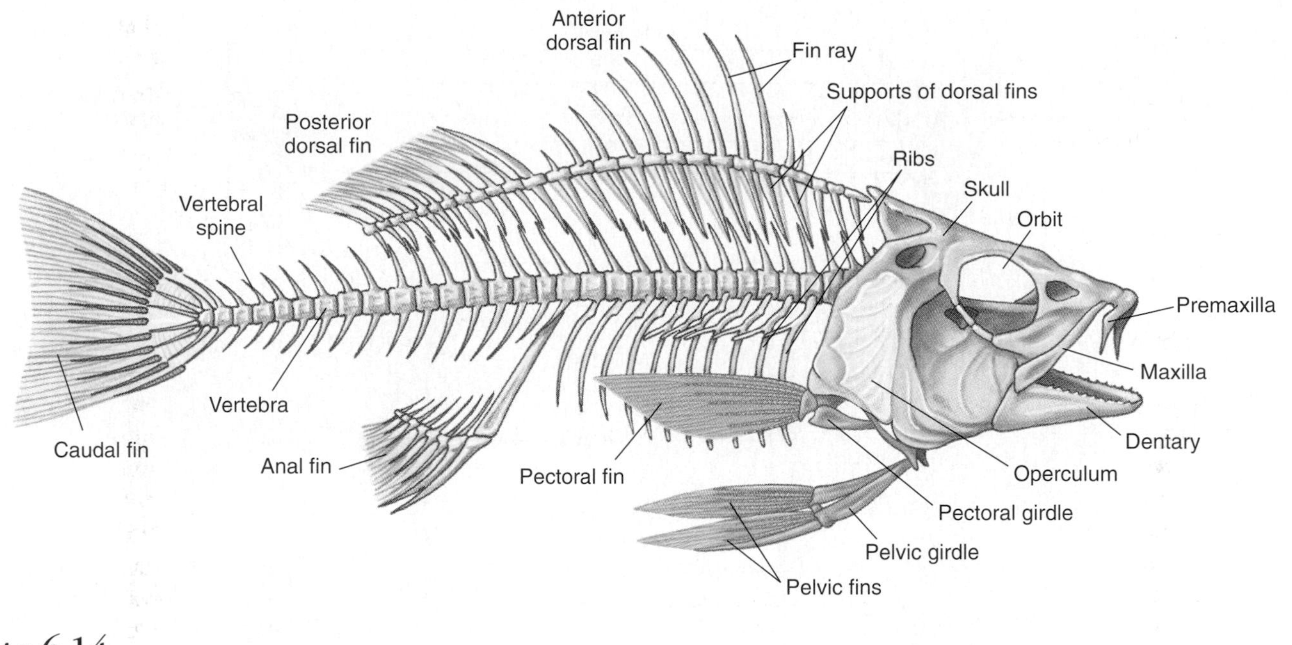

figure 6-14

Skeleton of a perch.

it is living tissue, although nonliving "ground substance" predominates. As a result of its living state, bone breaks can heal and bone diseases can be as painful as any other tissue disease.

Bone growth is a complex restructuring process, involving both its destruction internally by bone-resorbing cells **(osteoclasts)** and its deposition externally by bone-building cells **(osteoblasts).** Both processes occur simultaneously so that the marrow cavity inside grows larger by bone resorption while new bone is laid down outside by bone deposition. Bone growth responds to several hormones, in particular **parathyroid hormone** from the parathyroid gland, which stimulates bone resorption, and **calcitonin** from the thyroid gland, which inhibits bone resorption. These two hormones, together with a derivative of vitamin D, are responsible for maintaining a constant level of calcium in the blood (p. 277).

Following menopause, a woman loses 5% to 6% of her bone mass annually, often leading to the disease osteoporosis and increasing the risk of bone fractures. Dietary supplementation with calcium has been advocated to prevent such losses, but even large doses of calcium alone have little effect in slowing demineralization unless accompanied by therapy with the female sex hormone estrogen (because ovarian production of estrogen drops significantly after menopause). Among animals, only humans, especially females, are troubled with osteoporosis, perhaps a consequence of the long post-reproductive life of the human species.

Plan of the Vertebrate Skeleton

The vertebrate skeleton is composed of two main divisions: the **axial skeleton,** which includes the skull, vertebral column, sternum, and ribs, and the **appendicular skeleton,** which includes the limbs (or fins or wings) and the pectoral and pelvic girdles (Figures 6-14 and 6-15). Not surprisingly, the skeleton has undergone extensive remodeling in the course of vertebrate evolution. The move from water to land forced dramatic changes in body form. With increased cephalization, that is, the further concentration of brain, sense organs, and food-gathering and respiratory apparatus in the head, the skull became the most intricate portion of the skeleton.

Some primitive fishes had as many as 180 skull bones (a source of frustration to paleontologists), but through loss of some bones and fusion of others, skull bones became greatly reduced in number during the evolution of the tetrapods (four-legged vertebrates). Amphibians and lizards have 50 to 95, and mammals, 35 or fewer. Humans have 29.

The vertebral column is the main stiffening axis of the postcranial skeleton. In fishes (Figure 6-14) it serves much the same function as the notochord from which it is partially derived; that is, it provides points for muscle attachment and prevents telescoping of the body during muscle contraction. With the evolution of the amphibious and terrestrial tetrapods, the vertebral column was no longer buoyed by the aquatic environment. The vertebral column became structurally adapted to withstand new regional stresses transmitted to the column by the two pairs of appendages. In the amniote tetrapods (reptiles, birds, and mammals), the vertebrae are differentiated into **cervical** (neck), **thoracic** (chest), **lumbar** (back), **sacral**

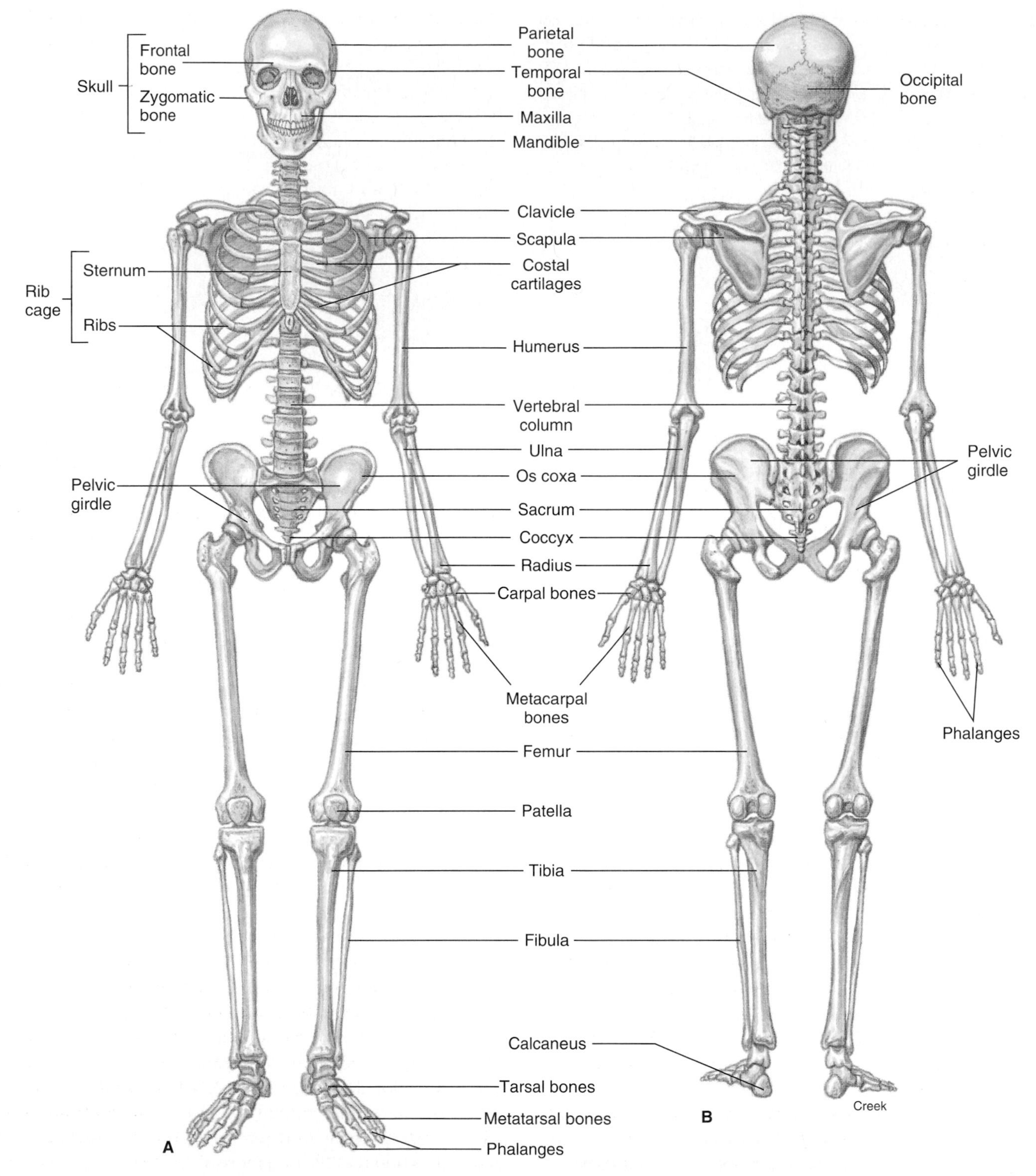

figure 6-15

Human skeleton. **A,** Ventral (front) view. **B,** Dorsal (back) view. In comparison with other mammals, the human skeleton is a patchwork of primitive and specialized parts. Erect posture, brought about by specialized changes in legs and pelvis, enabled the primitive arrangement of arms and hands (arboreal adaptation of human ancestors) to be used for manipulation of tools. Development of the skull and brain followed as a consequence of the premium natural selection put on dexterity and ability to appraise the environment.

(pelvic), and **caudal** (tail) vertebrae. In birds and also in humans the caudal vertebrae are reduced in number and size, and the sacral vertebrae are fused. The number of vertebrae varies among the different vertebrates. The pythons seem to lead the list with over 400. In humans (Figure 6-15) there are 33 in the child, but in the adult 5 are fused to form the **sacrum** and 4 to form the **coccyx.** Besides the sacrum and coccyx, humans have 7 cervical, 12 thoracic, and 5 lumbar vertebrae. The number of cervical vertebrae (7) is constant in nearly all mammals, whether the neck is short as in dolphins, or long as in giraffes.

The first two cervical vertebrae, the **atlas** and the **axis,** are modified to support the skull and permit pivotal movements. The atlas bears the globe of the head much as the mythological Atlas bore the earth on his shoulders. The axis, the second vertebra, permits the head to turn from side to side.

Ribs are long or short skeletal structures that articulate medially with vertebrae and extend into the body wall. The primitive condition is a pair of ribs for every vertebra; ribs serve as stiffening elements in the connective tissue septa that separate the muscle segments and thus improve the effectiveness of the muscle contractions. Many fishes have both dorsal and ventral ribs (Figure 6-14), and some have numerous riblike intermuscular bones as well—all of which increase the difficulty and reduce the pleasure of eating certain kinds of fish. Other vertebrates have a reduced number of ribs, and some, such as frogs and toads, have no ribs at all. Primates except humans have 13 pairs of ribs; humans have 12 pairs, although approximately 1 person in 20 has a thirteenth pair. In mammals the ribs together form the thoracic basket, which supports the chest wall and prevents collapse of the lungs.

Most vertebrates, fishes included, have paired appendages. All fishes except the agnathans have thin pectoral and pelvic fins that are supported by the pectoral and pelvic girdles, respectively (Figure 6-14). Tetrapod vertebrates (except caecilians, snakes, and limbless lizards) have two pairs of **pentadactyl** (five-toed) limbs, also supported by girdles. The pentadactyl limb is similar in all modern tetrapods, having evolved from the fins of Paleozoic fishes (see Figure 28-1, p. 608). Even when highly modified for various modes of life, the elements are rather easily homologized.

Modifications of the basic pentadactyl limb for life in different environments involve the distal elements much more frequently than the proximal, and it is far more common for bones to be lost or fused than for new ones to be added. Horses and their relatives developed a foot structure for fleetness by elongation of the third toe. In effect, a horse stands on its third fingernail (hoof), much like a ballet dancer standing on the tips of the toes. The bird wing is a good example of distal modification. The bird embryo bears 13 distinct wrist and hand bones (carpals and metacarpals), which are reduced to three in the adult. Most of the finger bones (phalanges) are lost, leaving four bones in three digits (see Figure 30-5 p. 647). The proximal bones (humerus, radius, and ulna), however, are little modified in the bird wing.

In nearly all tetrapods the pelvic girdle is firmly attached to the axial skeleton, since the greatest locomotory forces transmitted to the body come from the hind limbs. The pectoral girdle, however, is much more loosely attached to the axial skeleton, providing the forelimbs with greater freedom for manipulative movements.

Effect of Body Size on Bone Stress

As Galileo realized in 1638, the ability of animals' limbs to support a load decreases as animals increase in size (chapter opening essay, p. 138). Imagine two animals, one twice as long as the other, that are proportionally identical. That is, the larger animal is twice as long, twice as wide, and twice as tall as the smaller. The volume (and the weight) of the larger animal will be eight times the volume of the smaller ($2 \times 2 \times 2 = 8$). However, the strength of the larger animal's legs will be only four times the strength of the smaller, because bone, tendon, and muscle strength are proportional to the cross-sectional area. So, as Galileo noted, eight times the weight would have to be carried by only four times the strength. Because the maximum strength of mammalian bone is rather uniform per unit of cross-sectional area, how can animals become larger without placing unbearable stresses on long limb bones? One obvious solution is to make bones stouter and therefore stronger. However, throughout much of their size range, bone shape in different sized mammals does not change much. Instead, mammals have adapted their limb posture so that stresses are shifted to align with the long axis of the bones, rather than transversely. Small animals the size of a chipmunk run in a crouched limb posture, whereas a large mammal, such as a horse, has adopted an upright, rather stiff-legged posture (Figure 6-16). Bones and muscles are capable of carrying far more weight when aligned more closely with the ground reaction force, as they are in a horse's leg. In this way, peak bone stresses during strenuous activity are no greater for a galloping horse than for a running chipmunk or dog.

For animals larger than horses, further mechanical advantage by changing limb posture is not possible because the limbs are fully upright. Instead, the long bones of an elephant weighing 2.5 metric tons, and those of the enormous dinosaur *Apatosaurus,* weighing an estimated 34 metric tons, are (were) extremely thick and robust (Figure 6-16), providing the safety factor these massive animals require(d). However, top running speeds of the largest terrestrial mammals decline with increasing size. Nevertheless, recent calculations of bone stresses in dinosaurs suggest that even the largest were capable of considerable agility.[1]

[1]Alexander, R.M., 1991, "How dinosaurs ran," Sci. Am. **264:**130–136 (Apr.).

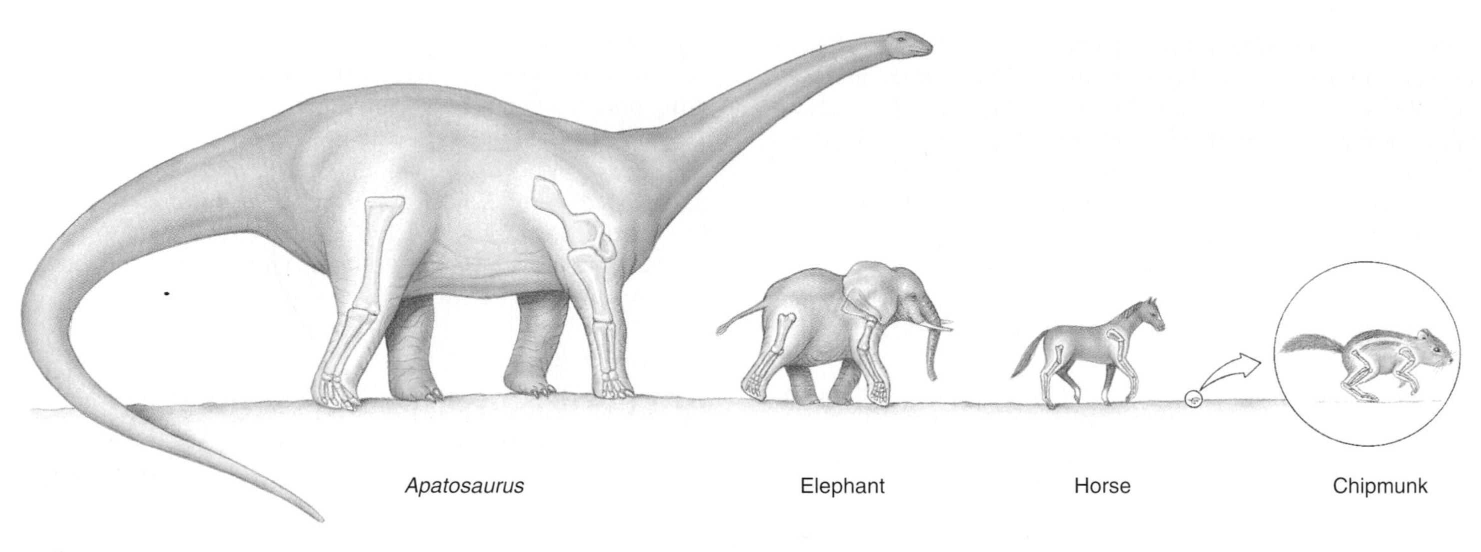

figure 6-16

Comparison of postures in small and large mammals, showing the effect of scale. Because of its more upright posture, bone stresses in the horse are similar to those in the chipmunk. In mammals larger than horses (above about 300 kg), greatly increased stresses require that bones become exceedingly robust and that the animal lose agility.

Animal Movement

Movement is an important characteristic of animals. Animal movement occurs in many forms in animal tissues, ranging from barely discernible streaming of cytoplasm, to extensive movements of powerful striated muscles. Most animal movement depends on a single fundamental mechanism: **contractile proteins,** which can change their form to elongate or contract. This contractile machinery is always composed of ultrafine fibrils—fine filaments, striated fibrils, or tubular fibrils (microtubules)—arranged to contract when powered by **ATP.** By far the most important protein contractile system is the **actomyosin system,** composed of two proteins, **actin** and **myosin.** This is an almost universal biomechanical system found from protozoa to vertebrates; it performs a long list of diverse functional roles. Cilia and flagella, however, are composed of different proteins, and thus are exceptions to the rule. In this discussion we examine the three principal kinds of animal movement: ameboid, ciliary, and muscular.

Ameboid Movement

Ameboid movement is a form of movement especially characteristic of amebas and other unicellular forms; it is also found in many wandering cells of metazoan animals, such as white blood cells, embryonic mesenchyme, and numerous other mobile cells that move through the tissue spaces. Ameboid cells change their shape by sending out and withdrawing **pseudopodia** (false feet) from any point on the cell surface. Beneath the plasmalemma lies a nongranular layer, the gel-like **ectoplasm,** which encloses the more liquid endoplasm (see Figure 16-3, p. 365).

Research with a variety of ameboid cells, including the pathogen-fighting phagocytes present in blood, has produced a consensus model to explain pseudopodial extension and ameboid crawling. Optical studies of an ameba in movement suggest the outer layer of ectoplasm surrounds a rather fluid core of endoplasm. Movement depends on actin and other regulatory proteins. According to one hypothesis[2], as the pseudopod extends, hydrostatic pressure forces actin subunits into the pseudopod where they assemble into a network to form a gel state. At the trailing edge of the gel, where the network disassembles, the freed actin interacts with myosin to create a contractile force that pulls the cell along behind the extending pseudopod. Locomotion is assisted by membrane-adhesion proteins that attach temporarily to the substrate to provide traction, enabling the cell to crawl steadily forward.

Ciliary and Flagellar Movement

Cilia are minute hairlike motile processes that extend from the surfaces of the cells of many animals. They are a particularly distinctive feature of ciliate protozoa, but, except for the nematodes in which motile cilia are absent and the arthropods in which they are rare, cilia are found in all major groups of animals. Cilia perform many roles either in moving small animals such as unicellular ciliates, flagellates, and the ctenophores through their aquatic environment or in propelling fluids and materials across the epithelial surfaces of larger animals.

[2]Stossel, T.P., 1994, "The machinery of cell crawling," Sci. Am. **271:**54–63 (Sept.).

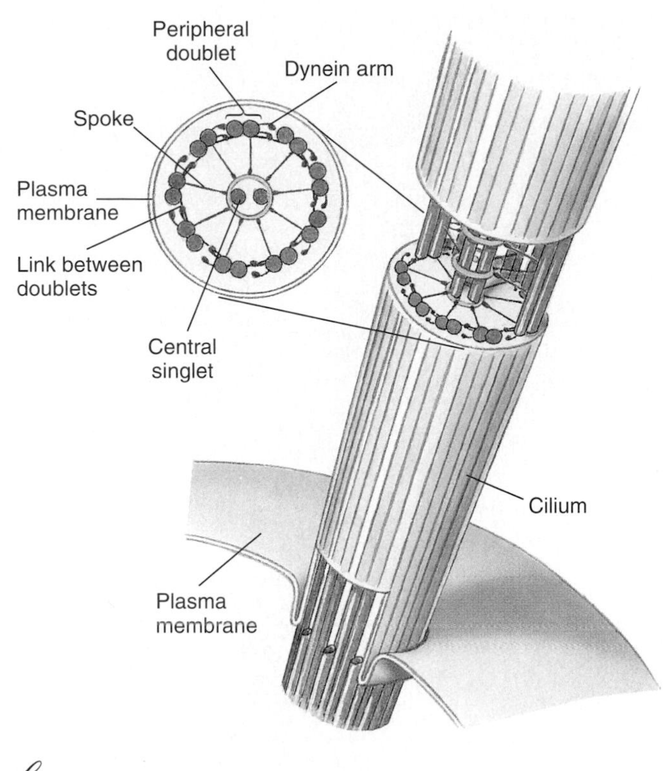

figure 6-17

Cross section of a cilium showing the microtubules and connecting elements of the 9 + 2 arrangement typical of both cilia and flagella.

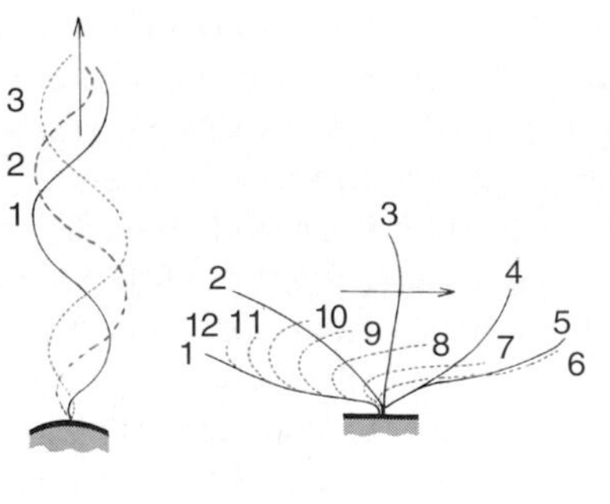

figure 6-18

Flagellum and cilium compared. A flagellum beats in wavelike undulations, propelling water parallel to the main axis of the flagellum. A cilium propels water in direction parallel to the cell surface.

Cilia are of remarkably uniform diameter (0.2 to 0.5 μm) wherever they are found. The electron microscope has shown that each cilium contains a peripheral circle of nine double microtubules and an additional two microtubules in the center (Figure 6-17) (several exceptions to the 9 + 2 arrangement among animals are known). The nine microtubule doublets around the periphery are connected to each other and to the central pair of microtubules by a complex system of connective elements (Figure 6-17). Also extending from each doublet is a pair of arms, composed of the protein dynein, that serve as crossbridges between doublets.

A **flagellum** is a whiplike structure longer than a cilium and usually present singly or in small numbers at one end of a cell. They are found in flagellate protozoa, in animal spermatozoa, and in sponges. The main difference between a cilium and a flagellum is in their beating pattern rather than in their structure, since both look alike internally. A flagellum beats symmetrically with snakelike undulations so that the water is propelled parallel to the long axis of the flagellum. A cilium, in contrast, beats asymmetrically with a fast power stroke in one direction followed by a slow recovery during which the cilium bends as it returns to its original position. The water is propelled parallel to the ciliated surface (Figure 6-18).

Although the mechanism of ciliary movement is not completely understood, it is known that the microtubules behave as "sliding filaments" that move past one another much like the sliding filaments of vertebrate skeletal muscle that is described in the next discussion. During ciliary flexion, the dynein arms link to adjacent microtubules, then swivel and release in repeated cycles, causing microtubules on the concave side to slide outward past microtubules on the convex side. This increases the curvature of the cilium. During the recovery stroke microtubules on the opposite side slide outward to bring the cilium back to its starting position.

Muscular Movement

Contractile tissue is most highly developed in muscle cells called **fibers.** Although muscle fibers themselves can do work only by contraction and cannot actively lengthen, they can be arranged in so many different configurations and combinations that almost any movement is possible.

Types of Invertebrate Muscle

Smooth and striated muscles of invertebrates are similar in structure to vertebrate smooth and striated muscle (Figure 6-7). However, there are many variations of both types and even instances in which the structural and functional features of vertebrate smooth and striated muscle are combined. Striated muscle appears in invertebrate groups as diverse as cnidarians and the arthropods. The thickest muscle fibers known, approximately 3 mm in diameter and 6 cm long and easily seen with the unaided eye, are those of giant barnacles and of Alaska king crabs living along the Pacific coast of North America. These cells are so large that they can easily be penetrated with electrodes or micropipettes for physiological studies and are understandably popular with muscle physiologists.

In the limited space available to treat the great diversity of muscle structure and function in invertebrates, we have selected for discussion two functional extremes: the specialized adductor muscles of molluscs and the fast flight muscles of insects.

Bivalve molluscan muscles contain fibers of two types. One kind is striated muscle that can contract rapidly, enabling the bivalve to snap shut its valves when disturbed. Scallops use these "fast" muscle fibers to swim in their awkward manner. The second muscle type is smooth muscle, capable of slow, long-lasting contractions. Using these fibers, a bivalve can keep its valves tightly shut for hours or even days. Such adductor muscles use very little metabolic energy and receive remarkably few nerve impulses to maintain the activated state. The contracted state has been likened to a "catch mechanism" involving some kind of stable cross-linkage between the contractile proteins within the fiber. However, despite considerable research, there is still much uncertainty about how this adductor mechanism works.

Insect flight muscles are virtually the functional antithesis of the slow, holding muscles of bivalves. The wings of some of the small flies operate at frequencies greater than 1000 beats per second. The so-called **fibrillar muscle,** which contracts at these incredible frequencies—far greater than even the most active of vertebrate muscles—shows unique characteristics. It has very limited extensibility; that is, the wing leverage system is arranged so that the muscles shorten only slightly during each downbeat of the wings. Furthermore, the muscles and wings operate as a rapidly oscillating system in an elastic thorax (see Figure 23-38, p. 511). Since the muscles rebound elastically and are activated by stretch during flight, they receive impulses only periodically rather than one impulse per contraction; one reinforcement impulse for every 20 or 30 contractions is enough to keep the system active. We describe insect flight muscles in more detail in Chapter 23 (p. 510).

Structure and Function of Vertebrate Striated Muscle

Skeletal muscle is typically organized into sturdy, compact bundles or bands. It is called skeletal muscle because it is attached to skeletal elements and is responsible for movements of the trunk, appendages, respiratory organs, eyes, mouthparts, and so on. Skeletal muscle **fibers** are extremely long, cylindrical, multinucleate cells that may reach from one end of the muscle to the other. They are packed into bundles called **fascicles** (L. *fasciculus,* small bundle), which are enclosed by tough connective tissue. The fascicles are in turn grouped into a discrete **muscle** surrounded by a thin connective tissue layer (Figure 6-19). Most skeletal muscles taper at their ends, where they connect by connective tissue bindings, called **tendons,** to bones. Other muscles, such as the ventral abdominal muscles, are flattened sheets.

In most fishes, amphibians, and to some extent lizards and snakes, there is a segmented organization of muscles alternating with the vertebrae. The skeletal muscles of other vertebrates, by splitting, fusion, and shifting, have developed into specialized muscles best suited for manipulating the jointed appendages that have evolved for locomotion on land. Skeletal muscle contracts powerfully and quickly but fatigues more rapidly than does smooth muscle. Skeletal

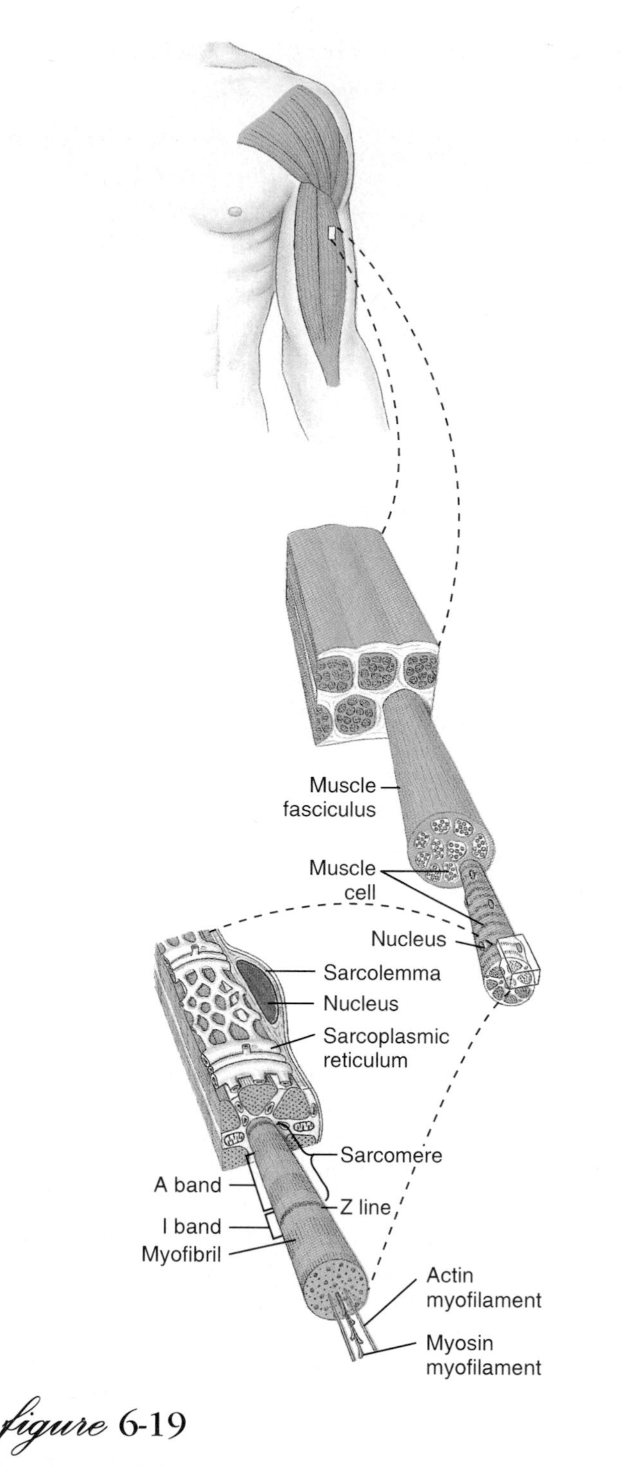

figure 6-19

Organization of skeletal muscle from gross to molecular level. A skeletal muscle (*top*) is composed of thousands of multinucleated muscle fibers (*center*), each containing thousands of myofibrils (*bottom*). Each myofibril contains numerous thick (myosin) and thin (actin) filaments that interact to slide past each other during contraction to shorten the muscle. The sarcoplasmic reticulum is a network of tubules surrounding the myofibrils and serves as a communication system for carrying a depolarization to the filaments within the muscle fiber.

muscle is sometimes called **voluntary muscle** because it is under conscious cerebral control.

Human muscle tissue develops before birth, and a newborn child's complement of skeletal muscle fibers is all that he or she will ever have. But while a weight lifter and a young boy have a similar number of muscle fibers, the weight lifter may be several times the boy's strength because repeated high-intensity, short-duration exercise has induced the synthesis of additional actin and myosin filaments. Each fiber has hypertrophied, becoming larger and stronger. Endurance exercise such as long-distance running produces a very different response. Fibers do not become greatly stronger but develop more mitochondria, more myoglobin, and become adapted for a high rate of oxidative phosphorylation. These changes, together with the development of more capillaries serving the fibers, lead to increased capacity for long-duration activity.

As mentioned earlier (Figure 6-7), striated muscle is so named because of the periodic bands, plainly visible under the light microscope, that pass across the widths of the muscle cells. Each cell, or **fiber,** is a multinucleated tube containing numerous **myofibrils,** packed together and invested by the cell membrane, the **sarcolemma** (Figure 6-19). The myofibril contains two types of **myofilaments:** thick filaments composed of the protein **myosin,** and thin filaments, composed of the protein **actin.** These are the actual contractile proteins of the muscle. The thin filaments are held together by a dense structure called the Z line. The functional unit of the myofibril, the **sarcomere,** extends between successive Z lines. Figure 6-17 shows these anatomical relationships.

Each thick filament is composed of myosin molecules packed together in an elongate bundle (Figure 6-20). Each myosin molecule is composed of two polypeptide chains, each having a club-shaped head. Lined up as they are in a bundle to form a thick filament, the double heads of each myosin molecule face outward from the center of the filament. These heads act as the molecular cross bridges that interact with the thin filaments during contraction.

The thin filaments are more complex because they are composed of three different proteins. The backbone of the thin filament is a double strand of the protein actin, twisted into a double helix. Surrounding the actin filament are two thin strands of another protein, **tropomyosin,** that lie near the grooves between the actin strands. Each tropomyosin strand is itself a double helix as shown in Figure 6-20.

The third protein of the thin filament is **troponin,** a complex of three globular proteins located at intervals along the filament. Troponin is a calcium-dependent switch that acts as the control point in the contraction process.

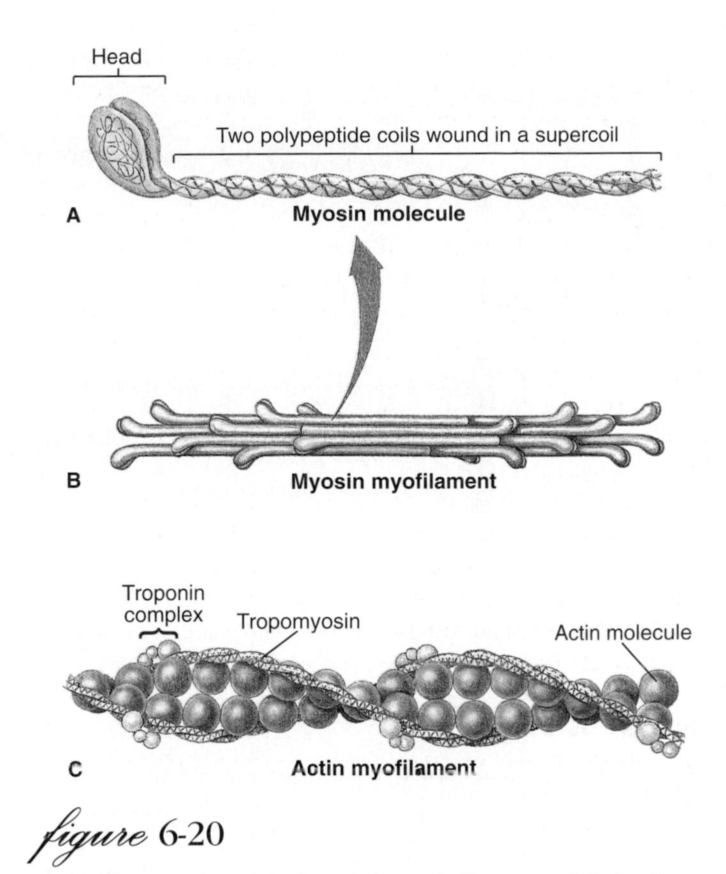

figure 6-20

Molecular structure of thick and thin myofilaments of skeletal muscle. **A,** The myosin molecule is composed of two polypeptides coiled together and expanded at their ends into a globular head. **B,** The thick myofilament is composed of a bundle of myosin molecules with the globular heads extended outward. **C,** The thin myofilament consists of a double strand of actin surrounded by two tropomyosin strands. A globular protein complex, troponin, occurs in pairs at every seventh actin unit. Troponin is a calcium-dependent switch that controls the interaction between actin and myosin.

Sliding Filament Model of Muscle Contraction

In the 1950s the English physiologists A.F. Huxley and H.E. Huxley independently proposed the **sliding filament model** to explain striated muscle contraction. According to this model, the thick and thin filaments become linked together by molecular cross bridges, which act as levers to pull the filaments past each other. During contraction, the cross bridges on the thick filaments swing rapidly back and forth, alternately attaching and releasing to special receptor sites on the thin filaments, and drawing the thin filaments past the thick in a kind of ratchet action. As contraction continues, the Z lines are pulled closer together (Figure 6-21). Thus the sarcomere shortens. Because all of the sarcomere units shorten together, the muscle contracts. Relaxation is a passive process. When the cross bridges between the thick and thin filaments release, the sarcomeres are free to lengthen. This requires some force, which is usually supplied by antagonistic muscles or the force of gravity.

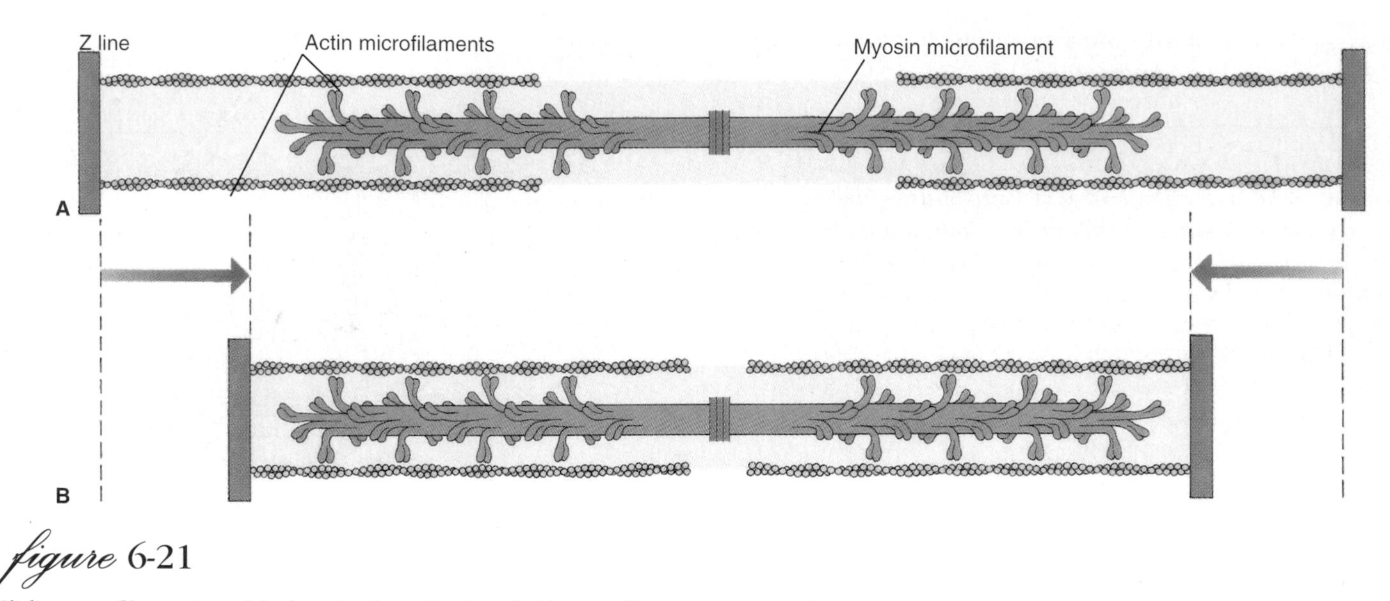

figure 6-21

Sliding myofilament model, showing how thick and thin myofilaments interact during contraction. **A,** Muscle relaxed. **B,** Muscle contracted.

Control of Contraction

Muscle contracts in response to stimulation by nerves. If the nerve supply to a muscle is severed, the muscle **atrophies,** or wastes away. Skeletal muscle fibers are innervated by motor neurons whose cell bodies are located in the spinal cord (Figure 11-9, p. 247). Each cell body gives rise to a motor axon that leaves the spinal cord to travel by way of a peripheral nerve trunk to a muscle where it branches repeatedly into many terminal branches. Each terminal branch innervates a single muscle fiber. Depending on the type of muscle, a single motor axon may innervate as few as three or four muscle fibers (where very precise control is needed, such as the muscles that control eye movement) or as many as 2000 muscle fibers (where precise control is not required, such as the postural muscles of the back). The motor neuron and all the muscle fibers it innervates is called a **motor unit.** The motor unit is the functional unit of skeletal muscle. When a motor neuron fires, the action potential passes to all fibers of the motor unit and each is stimulated to contract simultaneously. The total force exerted by a muscle depends on the number of motor units activated. Precise control of movement is achieved by varying the number of motor units activated at any one time. A smooth and steady increase in muscle tension is produced by increasing the number of motor units brought into play; this is called **motor unit recruitment.**

The Myoneural Junction

The place where a motor axon terminates on a muscle fiber is called the **myoneural junction** (Figure 6-22). At the junction is a tiny gap, or **synaptic cleft,** that thinly separates a nerve fiber and muscle fiber. In the vicinity of the junction, the neuron stores a chemical, **acetylcholine,** in minute vesicles known as **synaptic vesicles.** Acetylcholine is released when a nerve impulse reaches a synapse. This substance is a chemical mediator that diffuses across the narrow junction and acts on the muscle fiber membrane to generate an electrical depolarization. The depolarization spreads rapidly through the muscle fiber, causing it to contract. Thus the synapse is a special chemical bridge that couples together the electrical activities of nerve and muscle fibers.

Built into vertebrate skeletal muscle is an elaborate conduction system that serves to carry the depolarization from the myoneural junction to the densely packed filaments within the fiber. Along the surface of the sarcolemma are numerous invaginations that project as a system of tubules into the muscle fiber. This is called the **T-system** (Figure 6-22). The T-system is continuous with the **sarcoplasmic reticulum,** a system of fluid-filled channels that runs parallel to the myofilaments. The system is ideally arranged for speeding the electrical depolarization from the myoneural junction to the myofilaments within the fiber.

Excitation-Contraction Coupling

How does the electrical depolarization activate the contraction process? In the resting, unstimulated muscle, shortening does not occur because the thin tropomyosin strands lie in a position on the actin filament that prevents the myosin heads from attaching to actin. When the muscle is stimulated and the electrical depolarization arrives at the sarcoplasmic reticulum surrounding the fibrils, calcium ions are released (Figure 6-23). Some of the calcium binds to the control protein troponin. Troponin immediately undergoes changes in shape that allow tropomyosin to move out of its blocking position, exposing active sites on the actin myofilaments. The myosin heads then bind to these sites, forming cross bridges between adjacent thick and thin myofilaments. This binding sets in motion an **attach-pull-release cycle** that occurs in a series of steps as shown in Figure 6-23. The release of bond energy from ATP activates the myosin head, which swings 45 degrees, at the same time releasing a molecule of ADP. This is the power stroke that pulls the actin filament a distance of about 10 nm, and it comes to an end when another ATP molecule binds to the myosin head, inactivating the site. Thus, each cycle requires the expenditure of energy in the form of ATP (Figure 6-23).

Shortening will continue as long as nerve impulses arrive at the myoneural junction and free calcium remains available around the myofilaments. The attach-pull-release cycle can repeat again and again, 50 to 100 times per second, pulling the thick and thin filaments past each other. While the distance each sarcomere can shorten is very small, this distance is multiplied by the thousands of sarcomeres lying end to end in a muscle fiber. Consequently, a strongly contracting muscle may shorten as much as one-third its resting length.

When stimulation stops, the calcium is quickly pumped back into the sarcoplasmic reticulum. Troponin resumes its original configuration, tropomyosin moves back into its blocking position on actin, and the muscle relaxes.

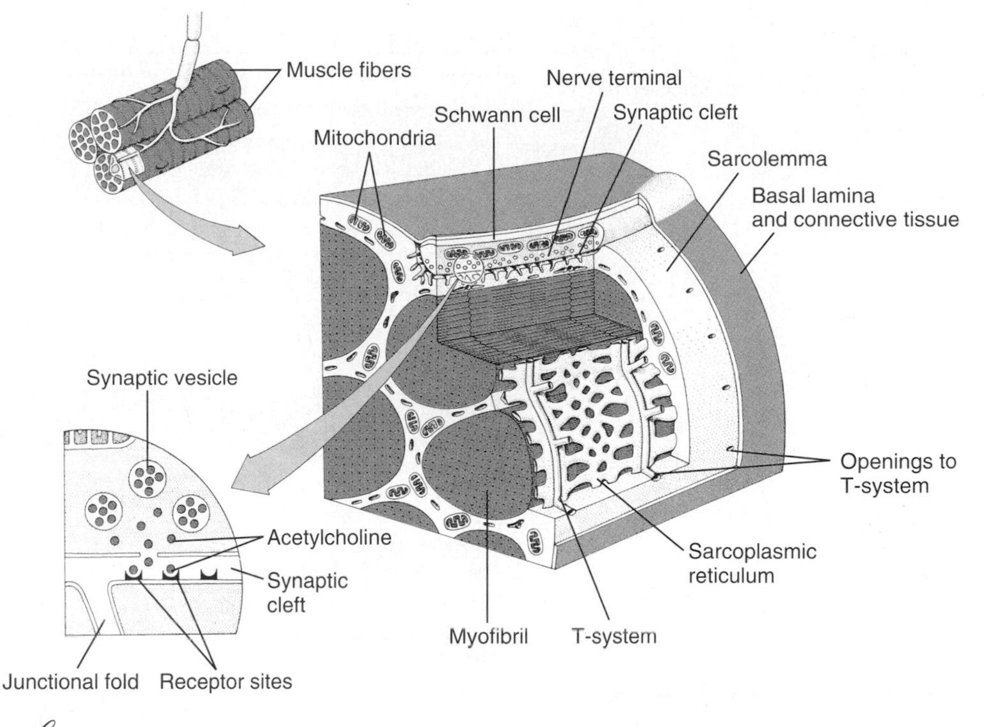

figure 6-22

Section of vertebrate skeletal muscle showing nerve-muscle synapse (myoneural junction), sarcoplasmic reticulum, and connecting transverse tubules (T-tubule system). Arrival of a nerve impulse at the synapse triggers the release of acetylcholine into synaptic cleft (*inset at left*). The binding of transmitter molecules to receptors generates membrane depolarization. This depolarization spreads across the sarcolemma, into the T-tubule system, and to the sarcoplasmic reticulum where the sudden release of calcium sets in motion the contractile machinery of the myofibril.

Energy for Contraction

Muscle contraction requires large amounts of energy. ATP is the immediate source of energy, but the amount present will sustain contraction for only a second or two. Muscle cells immediately call on the second level of energy reserve, **creatine phosphate.** Creatine phosphate is a high-energy phosphate compound that stores bond energy during periods of rest. As ADP is produced during contraction, creatine phosphate releases its stored bond energy to convert ADP to ATP. This reaction can be summarized as:

$$\text{Creatine phosphate} + \text{ADP} \rightarrow \text{ATP} + \text{Creatine}$$

> *The role of creatine phosphate, the high-energy compound of vertebrates that regenerates ATP during muscle contraction, is replaced in most invertebrates by arginine phosphate. Neither creatinine phosphate nor arginine phosphate are stored in quantities sufficient to support contractions for more than a few seconds; sustained contractions depend ultimately on the oxidation of fuels: carbohydrates and fats.*

Within a few seconds—perhaps as long as 30 seconds depending on the rapidity of muscle contraction—the reserves of creatine phosphate are depleted. The contracting muscle now must be fueled from its third and largest store of energy, **glycogen.** Glycogen is a polysaccharide chain of glucose molecules (p. 8) stored in both liver and muscle. Muscle has by far the larger store—some three-fourths of all the glycogen in the body is stored in muscle. As a supply of energy for contraction, glycogen has three important advantages: it is relatively abundant, it can be mobilized quickly, and it can provide energy under anoxic conditions. As soon as the muscle's store of creatine phosphate declines, enzymes break down glycogen, converting it into glucose-6-phosphate, the first stage of glycolysis that leads into mitochondrial respiration and the generation of ATP (p. 42).

If muscular contraction is not too vigorous or too prolonged, the glucose released from glycogen can be completely oxdized to carbon dioxide and water by **aerobic metabolism.** During prolonged or heavy exercise, however, blood flow to the muscles, although greatly increased above the resting level, cannot supply oxygen to the mitochondria rapidly enough to complete oxidation of glucose. The contractile machinery then receives its energy largely by **anaerobic glycolysis,** a process that does not require oxygen (p. 49). The ability to take advantage of this anaerobic pathway, although not nearly as efficient as the aerobic one, is of great importance; without it, all forms of heavy muscular exertion would be impossible.

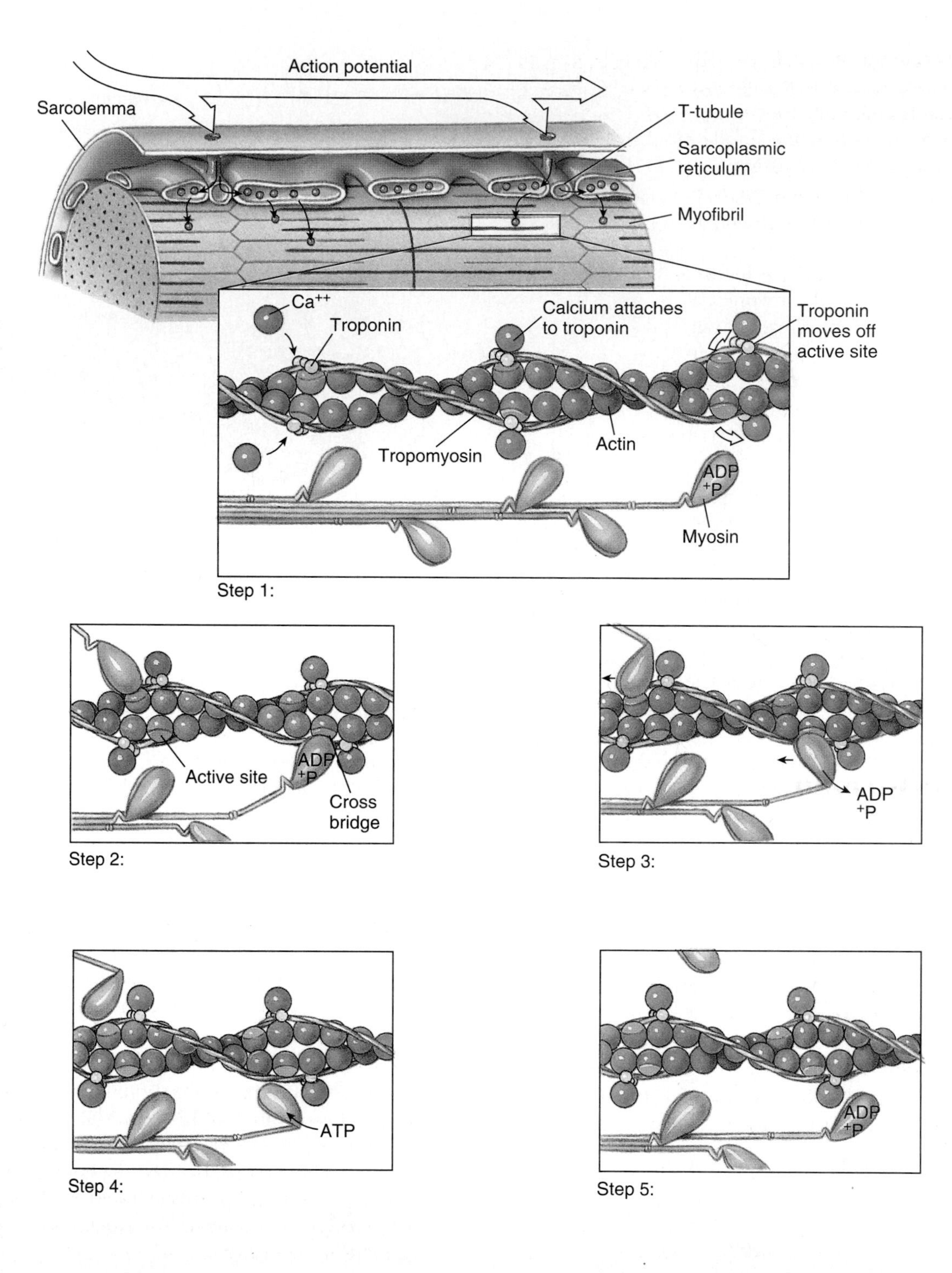

figure 6-23

Coupling of excitation and contraction in vertebrate skeletal muscle. **Step 1:** An action potential spreads along the sarcolemma and is conducted inward to the sarcoplasmic reticulum by way of T-tubules (T-tubule system). Calcium ions released from the sarcoplasmic reticulum diffuse rapidly into the myofibrils and bind to troponin molecules on the actin molecule. Troponin molecules are moved away from the active sites. **Step 2:** Myosin cross bridges bind to the exposed active sites. **Step 3:** Using the energy stored in ATP, the myosin head swings toward the center of the sarcomere. ADP and a phosphate group are released. **Step 4:** The myosin head binds another ATP molecule; this frees the myosin head from the active site on actin. **Step 5:** The myosin head splits ATP, retaining the energy released as well as the ADP and the phosphate group. The cycle can now be repeated as long as calcium is present to open active sites on the actin molecules.

During anaerobic glycolysis, glucose is degraded to lactic acid with release of energy. This is used to resynthesize creatine phosphate, which in turn passes the energy to ADP for the resynthesis of ATP. Lactic acid accumulates in the muscle and diffuses rapidly into the general circulation. If muscular exertion continues, the buildup of lactic acid causes enzyme inhibition and fatigue. Thus, the anaerobic pathway is a self-limiting one, since continued heavy exertion leads to exhaustion. The muscles incur an **oxygen debt** because the accumulated lactic acid must be oxidized by extra oxygen. After a period of exertion, oxygen consumption remains elevated until all of the lactic acid has been oxidized or resynthesized to glycogen.

Not all animals incur an oxygen debt, even during vigorous activity. Insects in flight show the greatest increase in oxygen consumption, as much as 200 times above resting levels of consumption. Bees and butterflies in flight have been shown to consume fifteen times their body volume in oxygen every minute *without incurring an oxygen debt—testimony to the efficiency of the insect tracheal system in delivering oxygen to body tissues.*

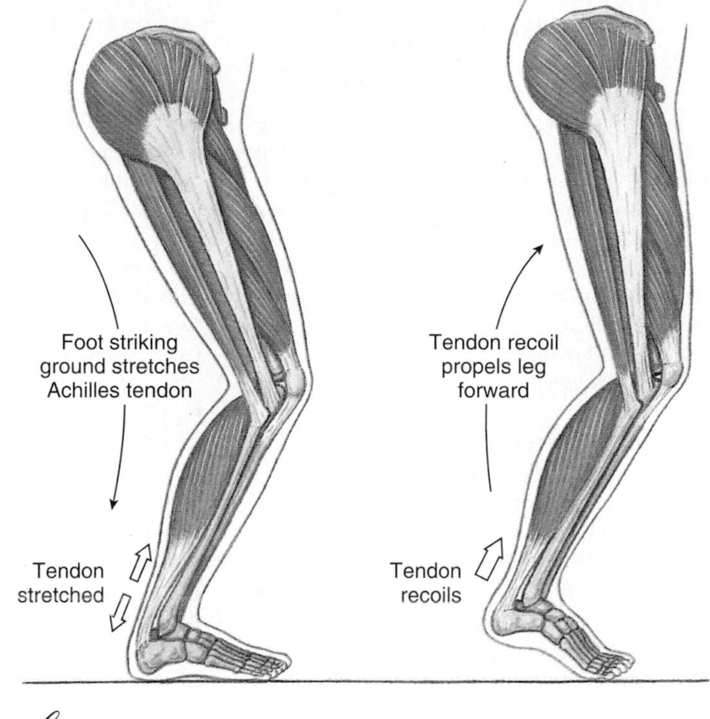

figure 6-24

Energy storage in the Achilles tendon of the human leg. During running, stretching of the Achilles tendon when the foot strikes the ground stores kinetic energy that is released to propel the leg forward.

Muscle Performance

Fast and Slow Fibers

The skeletal muscles of vertebrates consist of more than one type of fiber. Some muscles contain a high proportion of **slow fibers,** which are specialized for slow, sustained contractions without fatigue. Slow fibers are important in maintaining posture in terrestrial vertebrates. Such muscles are often called **red muscles** because they contain an extensive blood supply, a high density of mitochondria for supplying ATP, and abundant stored myoglobin which supplies oxygen reserves, all of which give the muscle a red color. The "dark meat" of a chicken leg is a familiar example.

Other muscles have a high proportion of **fast fibers,** also called twitch fibers, which are capable of fast, powerful contractions. Typically, such muscles cannot maintain their activity for long periods without becoming fatigued, but are essential for fast movements. Lacking an efficient blood supply and high density of mitochondria and myoglobin, fast muscles may be pale in color and are often called **white muscles.** However, not all fast fibers are alike. Some animals, such as dogs and ungulates (hoofed mammals), are capable of running for long periods of time because they have muscles with a high percentage of fast fibers with high oxidative capacity and efficient blood delivery. Such muscles work under aerobic conditions. Members of the cat family, however, have running muscles made up almost entirely of fast fibers that operate anaerobically. During a chase, such muscles build up a substantial oxygen debt that is replenished after the chase. For example, a cheetah after a high-speed chase lasting less than a minute, will pant heavily for 30 to 40 minutes before its oxygen debt is paid off.

Importance of Tendons in Energy Storage

When mammals walk or run, much kinetic energy is stored from step to step as elastic strain energy in the tendons. For example, during running the Achilles tendon is stretched by a combination of the downward force of the body on the foot and the contraction of the calf muscles. The tendon then recoils, extending the foot while the muscle is still contracted, propelling the leg forward (Figure 6-24). An extreme example of this bouncing ball principle is the bounding of a kangaroo, which essentially bounces along on its tendons, utilizing the effect of gravity. This type of movement uses far less energy than would be required if every step relied solely on alternate muscle contraction and relaxation.

There are many examples of elastic storage in the animal kingdom. It is used in the ballistic jumps of grasshoppers and fleas, in the wing hinges of flying insects, in the hinge ligaments of bivalve molluscs, and in the large dorsal tendon (ligamentum nuchae) that pulls the head upright in many large mammals.

Summary

From the relatively simple organisms that make up the beginnings of life on earth, animal evolution has generated many intricately organized forms. Cells are integrated into tissues, tissues into organs, and organs into systems. Whereas a unicellular animal carries out all life functions within the confines of a single cell, a multicellular animal is an organization of subordinate units that are united at successive levels. One correlate of increased anatomical complexity is an increase in body size, which offers certain advantages such as more effective predation, reduced energy cost of locomotion, and improved homeostasis.

The metazoan animal consists of cells, most of which are functionally specialized; body fluids, divided into the intracellular and extracellular fluid compartments; and the extracellular structural elements, which are fibrous or formless elements that serve various structural functions in the extracellular space. The cells of metazoans develop into various tissues made up of similar cells performing common functions. The basic tissue types are nervous, connective, epithelial, and muscular. Tissues are organized into larger functional units called organs, and organs are associated to form systems.

An animal is wrapped in a protective covering, the integument, which may be as simple as the delicate plasma membrane of a protozoan or as complex as the skin of a mammal. The arthropod exoskeleton is the most complex of invertebrate integuments, consisting of multilayered cuticle secreted by a single-layered epidermis. It may be hardened by calcification or sclerotization and must be molted at intervals to permit body growth. Vertebrate integument consists of two layers: the epidermis, which gives rise to various derivatives such as hair, feathers, and claws; and the dermis, which supports and nourishes the epidermis. The dermis also is the origin of bony derivatives such as fish scales and the antlers of deer.

Skeletons are supportive systems that may be hydrostatic or rigid. The hydrostatic skeletons of several soft-walled invertebrate groups depend on body wall muscles that contract against a noncompressible internal fluid of constant volume. Rigid skeletons evolved with attached muscles that act with the supportive skeleton to produce movement. In higher animals, two forms of skeleton appeared. Arthropods have an external skeleton, which must be shed periodically to make way for an enlarged replacement. Vertebrates developed an internal skeleton, a framework formed of cartilage or bone, that can grow with the animal, while, in the case of bone, additionally serving as a reservoir of calcium and phosphate.

Animal movement, whether in the form of cytoplasmic streaming, ameboid movement, or the contraction of an organized muscle mass, depends on specialized contractile proteins. The most important of these is the actomyosin system, which in more complex animals is usually organized into elongated thick and thin filaments that slide past one another during contraction. When a muscle is stimulated, an electrical depolarization passes into the muscle fibers through the sarcoplasmic reticulum, causing the release of calcium. Calcium binds to a protein complex (troponin) associated with the thin actin filament. This causes tropomyosin to shift out of its blocking position and allows the myosin heads to crossbridge with the actin filament. Powered by ATP, the myosin heads swivel back and forth to pull the thick and thin filaments past each other. Phosphate bond energy for contraction is supplied by carbohydrate fuels through a storage intermediate, creatine phosphate.

Vertebrate skeletal muscle consists of variable percentages of both slow fibers, used principally for sustained postural contractions, and fast fibers, used in locomotion. Tendons are important in locomotion because the kinetic energy stored in stretched tendons at one stage of a locomotory cycle is released at a subsequent stage.

Review Questions

1. Name the five levels of organization in animal complexity and explain how each successive level builds upon the one preceding it.
2. Can you suggest why, during the evolutionary history of animals, there has been a tendency for the maximum body size to increase? Do you think it is inevitable that complexity should increase along with body size? Why or why not?
3. Body fluids of eumetazoan animals are separated into fluid "compartments." Name these compartments and explain how compartmentalization may differ in animals with open and closed circulatory systems.
4. What characteristics of epithelial tissues—morphological and functional—distinguish them from other tissue types? What kind of epithelium would you expect to find in the following locations: a) lining of the ducts and tubules of the kidney, b) lining of the oral cavity, c) lining of the urinary bladder, d) inner surface of the intestinal tract, e) lining of blood capillaries.
5. What are the three elements present in all connective tissue? Give some examples of the different types of connective tissue.
6. Why is cardiac muscle considered involuntary muscle even though heart rate is under nervous control?
7. Skeletal and cardiac muscle are both striated muscle. How do they differ morphologically and functionally?
8. The arthropod exoskeleton is the most complex of invertebrate integuments. Describe its structure, and explain the difference in the way the cuticle is hardened in crustaceans and in insects.
9. Distinguish between epidermis and dermis in vertebrate integument, and describe the structural derivatives of these two layers.
10. As "naked apes" humans lack the protective investment of fur that shields other mammals from the damaging effects of sunlight. How does human skin respond to ultraviolet radiation in the short term and with continued exposure?
11. Hydrostatic skeletons have been defined as a mass of fluid enclosed within a muscular wall. How would you modify this definition to make it apply to a muscular hydrostat? Offer examples of both hydrostatic skeleton and muscular hydrostat.

12. One of the special qualities of vertebrate bone is that it is a living tissue that permits continuous remodeling. Explain how the structure of bone allows this to happen.
13. What is the difference between endochondral and membranous bone? Between spongy and compact bone?
14. The laws of scaling tell us that doubling the length of an animal will increase its weight eightfold while the force its bones can bear increase only fourfold. What solutions to this problem have evolved that allow animals to become large, while maintaining bone stresses within margins of safety?
15. Name the major skeletal components included in the axial and in the appendicular skeleton.
16. An unexpected discovery from studies of ameboid movement is that the same proteins found in the contractile system of metazoan muscle—actin and myosin—are present in ameboid cells. Explain how these and other proteins are believed to interact during ameboid movement.
17. A "9 + 2" arrangement of microtubules is typical of both cilia and flagella. Explain how this system is thought to function to produce a bending motion. What is the difference between a cilium and a flagellum?
18. What functional features of molluscan smooth muscle and insect fibrillar muscle set them apart from any known vertebrate muscle?
19. The sliding filament model of skeletal muscle contraction assumes a sliding or slipping of interdigitating filaments of actin and myosin. Electron micrographs show that during contraction the thick and thin filaments remain of constant length while the distance between Z lines shortens. Explain how this happens in terms of the molecular structure of the muscle filaments. What is the role of regulatory proteins in contraction?
20. While the sarcoplasmic reticulum of muscle was first described by nineteenth-century microscopists, its true significance was not appreciated until its intricate structure was revealed much later by the electron microscope. What could you tell a nineteenth-century microscopist about the structure of the sarcoplasmic reticulum and its role in the coupling of excitation and contraction?
21. The filaments of skeletal muscle are moved by free energy derived from the hydrolysis of ATP. Yet the immediately available supply of ATP in muscle is exhausted within the first moments of muscle contraction. Explain where the energy for a sustained contraction originates. Under what circumstances is an oxygen debt incurred during muscle contraction?
22. During evolution, skeletal muscle became adapted to functional demands ranging from sudden, withdrawal movements of a startled worm, to the sustained contractions required to maintain mammalian posture, to supporting a long, fast chase across the African savanna. What are some of the fiber types in vertebrate muscle that evolved to support these kinds of activities?

Selected References

Alexander, R.M. 1982. Locomotion in animals. New York, Chapman & Hall. *Concise, fully comparative treatment. Introduced with a discussion of "sources of power" followed by treatment of mechanism and energetics of locomotion on land, in water, and in the air. Undergraduate level.*

Alexander, R.M. 1992. The human machine. New York, Columbia University Press. *Describes human movement with the human body viewed as an engineered machine. Well chosen illustrations.*

Bonner, J.T. 1988. The evolution of complexity by means of natural selection. Princeton, N.J., Princeton University Press. *Levels of complexity in organisms and how size affects complexity.*

Caplan, A.J. 1984. Cartilage. Sci. Am. **251:** 84–94 (Oct.). *Structure, aging, and development of vertebrate cartilage.*

Hadley, N.F. 1986. The arthropod cuticle. Sci. Am. **255:**104–112 (July). *Properties of this complex covering that account for much of the adaptive success of arthropods.*

Kessel, R.G., and R. H. Kardon. 1979. Tissues and organs: a text-atlas of scanning electron microscopy. San Francisco, W.H. Freeman & Company, Publishers. *Collection of excellent scanning electron micrographs with text.*

Leffell, D. J., and D.E. Brash. 1996. Sunlight and skin cancer. Sci. Am. **275:**52–59 (July). *Skin cancer that appears in older people begins with damage received decades earlier. Many cases are caused by a mutation in a single gene.*

McMahon, T. A. 1984. Muscles, reflexes, and locomotion. Princeton, N.J., Princeton University Press. *Comprehensive, ranging from basic muscle mechanics to coordinated motion. Though sprinkled with mathematical models, the text is lucid throughout.*

McMahon, T. A., and J.T. Bonner. 1983. On size and life. New York, Scientific American Books, Inc. *Wide-ranging, artfully written, and colorfully produced introductory book on the biology and physics of size.*

Nadel, E.R. 1988. Physiological adaptations to aerobic training. Am. Sci. **73(4):** 334–343. *The studies reported here on energy conversion in muscle were crucial to the training of a pilot for the Daedalus project, the successful world record 119-km flight of a human-powered aircraft in April, 1988.*

Shipman, P., A. Walker, and D. Bichell. 1985. The human skeleton. Cambridge, Massachusetts, Harvard University Press. *Comprehensive view of the human skeleton.*

Spearman, R.I.C. 1973. The integument: a textbook of skin biology. Cambridge, Massachusetts, Cambridge University Press. *Comparative treatment, embracing both invertebrates and vertebrates.*

Welsch, U., and V. Storch. 1976. Comparative and animal cytology and histology. London, Sidgwick & Jackson. *Comparative histology with good treatment of invertebrates.*

Homeostasis:
Osmotic Regulation, Excretion, and Temperature Regulation

chapter | seven

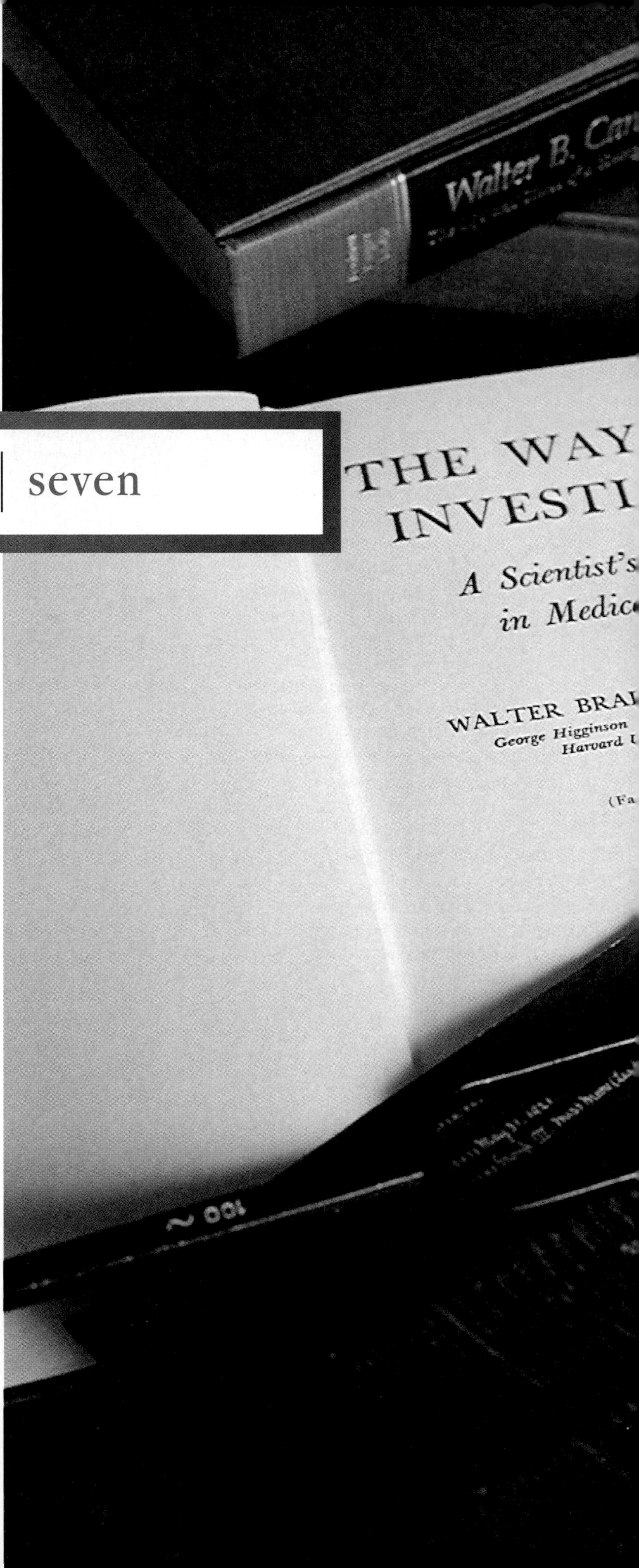

Homeostasis: Birth of a Concept

The tendency toward internal stabilization of the animal body was first recognized by Claude Bernard, the great French physiologist of the nineteenth century who, through his studies of blood glucose and liver glycogen, discovered the first internal secretions. Out of a lifetime of study and experimentation gradually grew the principle for which this retiring and lonely man is best remembered, that of the constancy of the internal environment, a principle that in time would pervade physiology and medicine. Years later at Harvard University, the American physiologist Walter B. Cannon (Figure 7-1) reshaped and restated Bernard's idea. Developed out of his studies of the nervous system and reactions to stress, he described the ceaseless balancing and rebalancing of physiological processes that maintain stability and restore the normal state when it has been disturbed. He also gave it a name: homeostasis. The term soon flooded the medical literature of the 1930s. Physicians spoke of getting their patients back into homeostasis. Even politicians and sociologists saw what they considered deep nonphysiological implications. Cannon enjoyed this broadened application of the concept and later suggested that democracy was the form of government that took the homeostatic middle course. Despite the enduring importance of the homeostasis concept, Cannon never received the Nobel Prize—one of several acknowledged oversights of the Nobel Committee. Near the end of his life, Cannon expressed his ideas about scientific research in his autobiography, *The Way of an Investigator.* This engaging book describes the resourceful career of a homespun man whose life embodied the traits that favor successful research.

The concept of homeostasis, described in the chapter opening essay, permeates all physiological thinking and is the theme of this and the following chapter. While the homeostasis concept was first developed from studies with mammals, it applies to single-celled organisms as well as to the most complex vertebrates. Potential changes in the internal environment arise from two sources. First, metabolic activities require a constant supply of materials such as oxygen, nutrients, and salts, that cells withdraw from their surroundings and that must be replaced. Cellular activity also produces waste products that must be expelled. Second, the internal environment responds to changes in the organism's external environment. Changes from either source must be stabilized by the physiological mechanisms of homeostasis.

In more complex metazoans, homeostasis is maintained by the coordinated activities of the circulatory, nervous, and endocrine systems, and especially by the organs that serve as sites of exchange with the external environment. These last include the kidneys, lungs or gills, digestive tract, and integument. Through these organs oxygen, foodstuffs, minerals, and other constituents of the body fluids enter, water is exchanged, heat is lost, and metabolic wastes are eliminated.

figure 7-1

Walter Bradford Cannon (1871–1945), Harvard professor of physiology who coined the term homeostasis and developed the concept originated by French physiologist Claude Bernard (Figure 8-2, p. 189).

From J. F. Fulton & L. G. Wilson Selected Readings in the History of Physiology, *1966. Courtesy of Charles C. Thomas, Publisher, Springfield.*

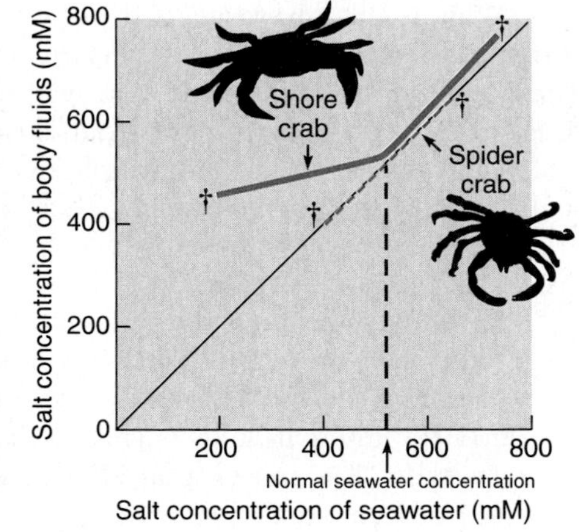

figure 7-2

Salt concentration of body fluids of two crabs as affected by variations in the seawater concentration. The 45-degree line represents equal concentration between body fluids and seawater. Since the spider crab cannot regulate its body-fluid salt concentration, it conforms to whatever changes happen in the external seawater. The shore crab, however, can regulate osmotic concentration of its body fluids to some degree because in dilute seawater the shore crab can hold its body-fluid concentration above seawater concentration. For example, when seawater is 200 mM (millimolar), the shore crab's body-fluid concentration is approximately 430 mM. Crosses at ends of lines indicate tolerance limits for each species.

We will look first at the problems of controlling the internal fluid environment of animals living in aquatic habitats. Next we will briefly examine how these problems are solved by terrestrial animals and consider the function of the organs that regulate the internal state. Finally we will look at the strategies that have evolved for living in a world of changing temperatures.

Water and Osmotic Regulation

How Marine Invertebrates Meet Problems of Salt and Water Balance

Most marine invertebrates are in osmotic equilibrium with their seawater environment. They have body surfaces that are permeable to salts and water so that their body fluid concentration rises or falls in conformity with changes in concentrations of seawater. Because such animals are incapable of regulating the osmotic pressure of their body fluid, they are called **osmotic conformers.** Invertebrates living in the open sea are seldom exposed to osmotic fluctuations because the ocean is a highly stable environment. Oceanic invertebrates have, in fact, very limited abilities to withstand osmotic change. If they should be exposed to dilute seawater, they die quickly because their body cells cannot tolerate dilution and are helpless to prevent it. These animals are restricted to living in a narrow salinity range and are said to be **stenohaline** (Gr. *stenos,* narrow, + *hals,* salt). An example is the marine spider crab, represented in Figure 7-2.

Conditions along the coasts and in estuaries and river mouths are much less constant than those of the open ocean. Here animals must be able to withstand large and often abrupt changes in salinity as the tides ebb and flow and mix with fresh water draining from rivers. These animals are termed **euryhaline** (Gr. *eurys,* broad, + *hals,* salt), meaning that they can survive a wide range of salinity changes, mainly because they demonstrate varying powers of **osmotic regulation.** For example, the brackish-water shore crab can resist body fluid dilution by dilute (brackish) seawater (Figure 7-2). Although the ionic concentration of the body fluid falls, it does so less rapidly than the fall in seawater concentration. This crab is a **hyperosmotic regulator** because in a dilute environment it can maintain the concentration of its blood above that of the surrounding water.

What is the advantage of hyperosmotic regulation over osmotic conformity, and how is this regulation accomplished? The advantage is that by regulating against excessive dilution, thus protecting the body cells from extreme changes, these crabs can successfully live in the physically unstable but biologically rich coastal environment. Their powers of regulation are limited, however, since if the water is highly diluted, their regulation fails and they die.

To understand how the brackish-water shore crab and other coastal invertebrates achieve hyperosmotic regulation, let us examine the problems they face. First, the salt concentration of the internal fluids is greater than in the dilute seawater outside. This causes a steady osmotic influx of water. As with the membrane osmometer placed in a sugar solution (p. 32), water diffuses inward because it is more concentrated outside than inside. The shore crab is not nearly as permeable as a membrane osmometer—most of its shelled body surface is, in fact, almost impermeable to water—but the thin respiratory surfaces of the gills are highly permeable. Obviously the crab cannot insulate its gills with an impermeable hide and still breathe. The problem is solved by removing the excess water through the action of the kidney (the antennal gland, located in the crab's head).

The second problem is salt loss. Again, because the animal is saltier than its environment, it cannot avoid loss of ions by outward diffusion across the gills. Salt is also lost in the urine. The problem is solved by special salt-secreting cells in the gills that can actively remove ions from the dilute seawater and move them into the blood, thus maintaining the internal osmotic concentration. This is an **active transport** process that requires energy because ions must be transported against a concentration gradient, that is, from a lower salt concentration (in the dilute seawater) to an already higher one (in the blood).

Invasion of Fresh Water

Some 400 million years ago, during the Silurian and Lower Devonian periods, the major groups of jawed fishes began to penetrate into brackish-water estuaries and then gradually into freshwater rivers. Before them lay a new, unexploited habitat already stocked with food in the form of insects and other invertebrates, which had preceded them into fresh water. However, the advantages of this new habitat were balanced by a tough physiological challenge: the necessity of developing effective osmotic regulation.

Freshwater animals must keep the salt concentration of their body fluids higher than that of the water. Water enters their bodies osmotically, and salt is lost by diffusion outward. Their problems are similar to those of the brackish-water crab, but more severe and unremitting. Fresh water is much more dilute than are coastal estuaries, and there is no retreat, no salty sanctuary into which the freshwater animal can retire for osmotic relief. It must and has become a permanent and highly efficient hyperosmotic regulator.

The scaled and mucous-covered body surface of a fish is about as waterproof as any flexible surface can be. In addition, freshwater fishes have several defenses against the problems of water gain and salt loss. First, water that inevitably enters by osmosis across the gills is pumped out by the kidney, which is capable of forming a very dilute urine (Figure 7-3). Second, special salt-absorbing cells located in the gills move salt ions, principally sodium and chloride (present in small quantities even in fresh water), from the water to the blood. This, together with salt present in the fish's food, replaces diffusive salt loss. These mechanisms are so efficient that a freshwater fish devotes only a small part of its total energy expenditure to keeping itself in osmotic balance.

Crayfishes, aquatic insect larvae, mussels, and other freshwater animals are also hyperosmotic regulators and face the same hazards as freshwater fishes; they tend to gain too much water and lose too much salt. Not surprisingly, all of these forms solved these problems in the same direct way that fishes did. They excrete the excess water as urine, and they actively absorb salt from the water by some salt-transporting mechanism on the body surface.

Amphibians living in water also must compensate for salt loss by actively absorbing salt from the water (Figure 7-4). They use their skin for this purpose. Physiologists learned some years ago that pieces of frog skin continue to transport sodium and chloride actively for hours when removed and placed in a specially balanced salt solution. Fortunately for biologists, but unfortunately for frogs, these animals are so easily collected and maintained in the laboratory that frog skin became a favorite membrane system for studies of ion transport.

Return of Fishes to the Sea

The marine bony fishes maintain the salt concentration of their body fluids at approximately one-third that of seawater (body fluids = 0.3 to 0.4 gram mole per liter [M]; seawater =

By expressing concentration of salt in seawater or body fluids in molarity, we are saying that the osmotic strength is equivalent to the molar concentration of an ideal solute having the same osmotic strength. In fact, seawater and animal body fluids are not ideal solutions because they contain electrolytes that dissociate in solution. A 1 M solution of sodium chloride (which dissociates in solution) has a much greater osmotic strength than a 1 M solution of glucose, an ideal solute that does not dissociate in solution (nonelectrolyte). Consequently, biologists usually express the osmotic strength of a biological solution in osmolarity rather than in molarity. A 1 osmolar solution exerts the same osmotic pressure as a 1 M solution of a nonelectrolyte.

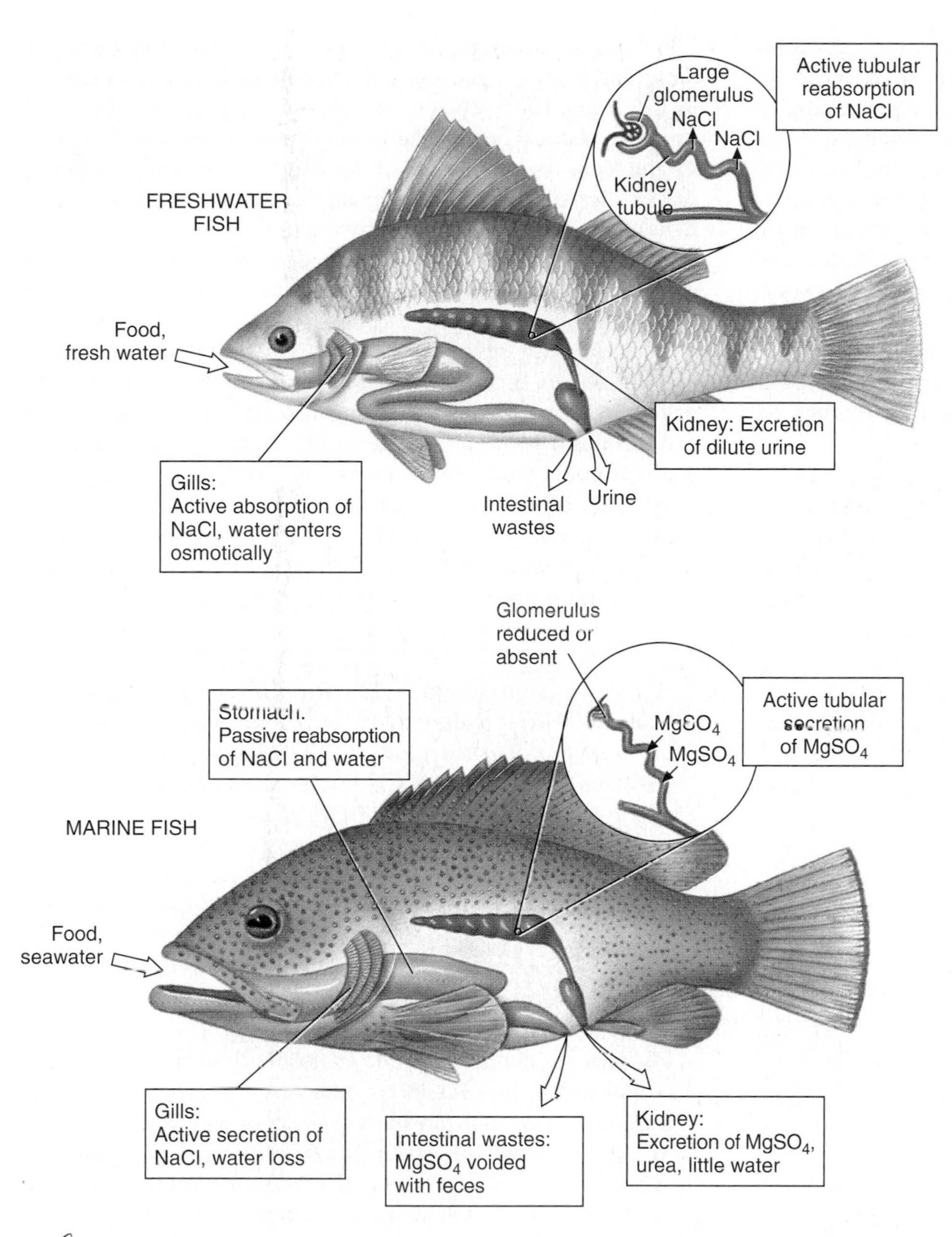

figure 7-3

Osmotic regulation in freshwater and marine bony fishes. A freshwater fish maintains osmotic and ionic balance in its dilute environment by actively absorbing sodium chloride across the gills (some salt enters with food). To expel excess water that constantly enters the body, the glomerular kidney produces a dilute urine by reabsorbing sodium chloride. A marine fish must drink seawater to replace water lost osmotically to its salty environment. Sodium chloride and water are absorbed from the stomach. Excess sodium chloride is secreted outward by the gills. Divalent sea salts, mostly magnesium sulfate, are eliminated with feces and secreted by the tubular kidney.

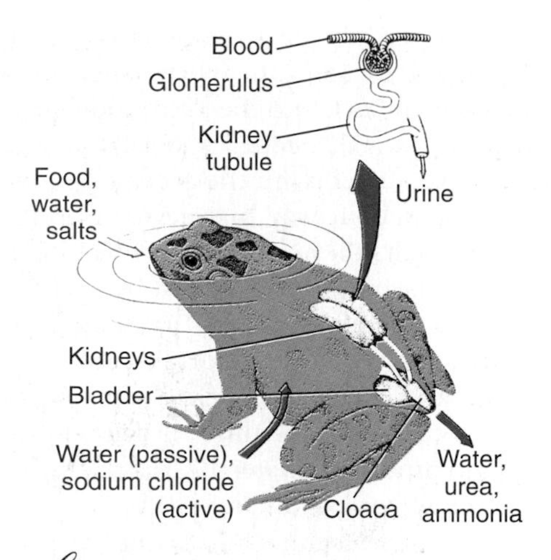

figure 7-4

Exchange of water and solute in a frog. Water enters the highly permeable skin and is excreted by the kidney. The skin also actively transports ions (sodium chloride) from the environment. The kidney forms a dilute urine by reabsorbing sodium chloride. Urine flows into the urinary bladder, where, during temporary storage, most of the remaining sodium chloride is removed and returned to the blood.

1 M). They are **hypoosmotic regulators** because their body fluids are substantially more dilute than their seawater environment. Bony fishes living in the oceans today are descendants of earlier freshwater bony fishes that moved back into the sea during the Triassic period approximately 200 million years ago. The return to their ancestral sea was probably prompted by unfavorable climatic conditions on land and the deterioration of freshwater habitats, but we can only guess at the reasons. During the many millions of years that the freshwater fishes were adapting themselves so well to their environment, they established an ionic concentration in the body fluid equivalent to approximately one-third that of seawater, thus setting the pattern for all the vertebrates that were to evolve later, whether aquatic, terrestrial, or aerial. The ionic composition of vertebrate body fluid is remarkably similar to that of dilute seawater too, a fact that is undoubtedly related to their distant marine heritage.

When some of the freshwater bony fishes of the Triassic period ventured back to the sea, they encountered a new set of problems. Having a much lower internal osmotic concentration than the seawater around them, they lost water and gained salt. Indeed the marine bony fish literally risks drying out, much like a desert mammal deprived of water.

To compensate for water loss, the marine fish drinks seawater (Figure 7-3). This seawater is absorbed from the intestine, and the major sea salt, sodium chloride, is carried by the blood to the gills, where specialized salt-secreting cells transport it back into the surrounding sea. The ions remaining in the intestinal residue, especially magnesium, sulfate, and calcium, are voided with the feces or excreted by the kidney. In this roundabout way, marine fishes rid themselves of the excess sea salts they have drunk, resulting in a net gain of water. This replaces the water lost by osmosis. Samuel Taylor Coleridge's ancient mariner, surrounded by "water, water, everywhere, nor any a drop to drink" undoubtedly would have been tormented even more had he known of the marine fishes' ingenious solution for thirst. A marine fish carefully regulates the amount of seawater it drinks, consuming only enough to replace water loss and no more.

The cartilaginous sharks and rays (elasmobranchs, p. 588) achieve water balance differently. This group is almost totally marine. The salt composition of shark's blood is similar to that of the bony fishes, but the blood also carries a large content of organic compounds, especially **urea** and **trimethylamine oxide.** Urea is a metabolic waste that most animals quickly excrete. The shark kidney, however, conserves urea, allowing it to accumulate in the blood. Added to the usual blood electrolytes, urea and trimethylamine oxide raise the blood's osmotic pressure to exceed slightly that of seawater. In this way the sharks and their kin turn an otherwise useless waste material into an asset, eliminating the osmotic problem encountered by the marine bony fishes.

The high concentration of urea in the blood of sharks and their kin—more than 100 times as high as in mammals—could not be tolerated by most other vertebrates. In the latter, such high concentrations of urea disrupt the peptide bonds of proteins, altering protein configuration. Sharks have adapted biochemically to the presence of the urea that permeates all their body fluids, even penetrating freely into the cells. So accommodated are the elasmobranchs to urea that their tissues cannot function without it, and the heart will stop beating in its absence.

How Terrestrial Animals Maintain Salt and Water Balance

The problems of living in an aquatic environment seem small indeed compared with the problems of life on land. Since animal bodies are mostly water, all metabolic activities proceed in water, and life itself was conceived in water, it might seem that animals were meant to stay in water. Yet many animals, like the plants preceding them, moved onto land, carrying their watery composition with them. Once on land, the terrestrial animals continued their adaptive radiation, solving the threat of desiccation, until they became abundant even in some of the most arid parts of the earth.

Terrestrial animals lose water by evaporation from the lungs and body surface, excretion in the urine, and elimination in the feces. They replace such losses by water in the food, drinking water when available, and forming **metabolic water** in the cells by oxidation of foodstuffs, especially carbohydrates. Certain insects—for example, desert roaches, certain ticks and mites, and the mealworm—are able to absorb water vapor directly from atmospheric air. In some desert rodents, metabolic water may constitute most of the animal's water gain.

Particularly revealing is a comparison (Table 7-1) of water balance in the human being, a nondesert mammal that drinks water, with that of the kangaroo rat, a desert rodent that may drink no water at all. The kangaroo rat acquires all of its water from its food: 90% as metabolic water derived from the oxidation of foodstuffs, and 10% as free moisture in the food. Even though we eat foods with a much higher water content than the dry seeds that make up much of the kangaroo rat's diet, we still must drink half our total water requirement.

Given ample water to drink, humans can tolerate extremely high temperatures while preventing a rise in body temperature. Our ability to keep cool by evaporation was impressively demonstrated more than 200 years ago by a British scientist who remained for 45 minutes in a room heated to 260° F (126° C). A steak he carried in with him was thoroughly cooked, but he remained uninjured and his body temperature did not rise. Sweating rates may exceed 3 liters of water per hour under such conditions and cannot be long tolerated unless the lost water is replaced by drinking. Without water, a human continues to sweat unabatedly until the water deficit exceeds 10% of the body weight, when collapse occurs. With a water deficit of 12% a human is unable to swallow even if offered water, and death occurs when the water deficit reaches about 15% to 20%. Few people can survive more than a day or two in a desert without water. Thus people are not physiologically well adapted for desert climates but prosper there nonetheless by virtue of their technological culture.

The excretion of wastes presents a special problem in water conservation. The primary end product of protein breakdown is ammonia, a highly toxic material. Fishes can easily excrete ammonia across the gills, since there is an abundance of water to wash it away. Terrestrial insects, reptiles, and birds have no convenient way to rid themselves of toxic ammonia; instead, they convert it into uric acid, a nontoxic, almost insoluble compound. This conversion enables them to excrete a semisolid urine with little water loss. The use of uric acid has another important benefit. Reptiles and birds lay amniotic eggs enclosing the embryos (Figure 29-4, p. 626) together with

table 7-1

Water Balance in a Human and a Kangaroo Rat, a Desert Rodent

	Human (%)	Kangaroo rat (%)
Gains		
Drinking	48	0
Free water in food	40	10
Metabolic water	12	90
Losses		
Urine	60	25
Evaporation (lungs and skin)	34	70
Feces	6	5

their stores of food and water and wastes that accumulate during development. By converting ammonia to uric acid, the developing embryo's waste can be precipitated into solid crystals, which are stored harmlessly within the egg until hatching.

Marine birds and turtles have evolved a unique solution for excreting the sizable loads of salt eaten with their food. Located above each eye is a **salt gland** capable of excreting a highly concentrated solution of sodium chloride—up to twice the concentration of seawater. In birds the salt solution runs out the nares (see Figure 30-11, p. 651). Marine lizards and turtles, like Alice in Wonderland's Mock Turtle, shed their salt gland secretion as salty tears. Salt glands are important accessory organs of salt excretion to these animals because their kidney can not produce a concentrated urine, as can the mammalian kidney.

Invertebrate Excretory Structures

Many protozoa and some freshwater sponges have special excretory organelles called contractile vacuoles. The more complex invertebrates have excretory organs that are basically tubular structures that form urine by first producing an ultrafiltrate or fluid secretion of the blood. This enters the proximal end of the tubule and is modified continuously as it flows down the tubule. The final product is urine.

Contractile Vacuole

The tiny, spherical, intracellular vacuole of protozoa and freshwater sponges is not a true excretory organ, since ammonia and other nitrogenous wastes of metabolism readily enter the surrounding water by direct diffusion across the cell membrane. The contractile vacuole is an organ of water balance. Because the cytoplasm of freshwater protozoa is considerably saltier than their freshwater environment, they tend to draw water into themselves by osmosis. This excess water is removed by the contractile vacuole. How the contractile vacuole functions in *Amoeba proteus* and other protists has long remained a mystery. The classical view, illustrated in Figure 16-7, p. 367, is that excess water and ions from the cytoplasm (such as Na^+ and K^+) collect in numerous tiny vesicles surrounding the single thin membrane of the contractile vacuole. The ions are then actively reabsorbed, thus recovering the intracellular ions, and a dilute solution is produced which is discharged into the contractile vacuole. As water accumulates within it the vacuole grows and finally collapses, emptying its contents through a pore on the surface, and the cycle is repeated rhythmically. Recent advances in immunofluorescence and electron microscopy[1] have altered this classical view. Rather than a vacuole surrounded by many independent vesicles, the contractile-vacuole system consists of a network of continuous membranous channels that are populated with numerous proton pumps (proton pumps were described in connection with the electron transport chain in Chapter 2, p. 47). Although the mechanism for filling is still not fully understood, the proton pumps apparently create H^+ and HCO^- gradients that draw water into the vacuole. These ions are excreted along with water when the vacuole empties. Since an isosmotic solution is expelled, this model resolves a problem with the classical view, which required that the contractile vacuole retained water against an osmotic gradient, an almost impossible task for a simple bilayer membrane.

Contractile vacuoles are common in freshwater protozoa, sponges, and radiate animals (such as hydra), but rare or absent in marine forms of these groups, which are isosmotic with seawater and consequently neither lose nor gain too much water.

Nephridium

In planaria and other flatworms the protonephridial system takes the form of two highly branched duct systems distributed throughout the body (Figure 7-5). Fluid enters the system through specialized "flame cells," moves slowly into and down the tubules, and is excreted through pores that open at intervals on the body surface. The rhythmical beat of the flagellar tuft, suggestive of a tiny flickering flame, creates a negative pressure that draws fluid through delicate interdigitations between the flame cell and the tubule cell and drives it into the tubular portion of the system. In the tubule, water and metabolites valuable to the body are recovered by reabsorption, leaving wastes behind to be expelled. Nitrogenous wastes (mainly ammonia) diffuse across the surface of the body.

---

[1]Heuser, J., Q. Zhu, and M. Clarke. 1993. Jour. Cell Biol. **121:**1311–1327.

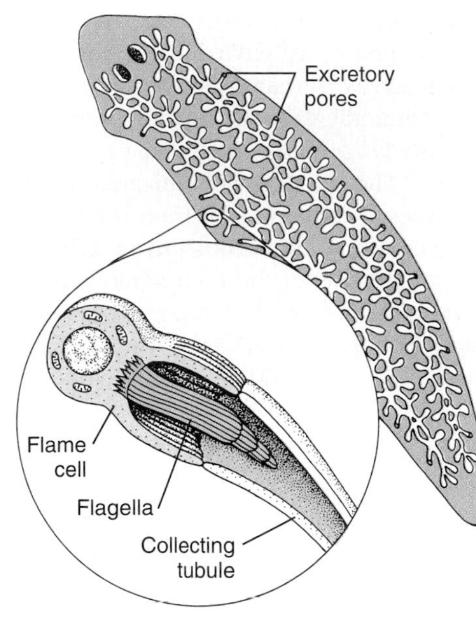

figure 7-5

Flame cell system of a flatworm. Body fluids collected by flame cells (protonephridia) are passed down a system of ducts to excretory pores on the body surface.

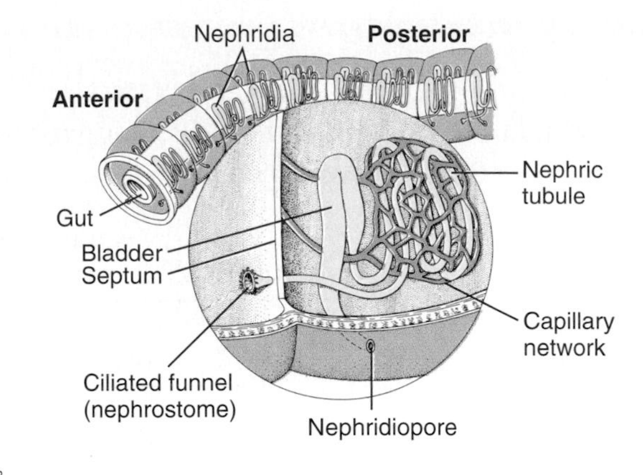

figure 7-6

Excretory system of an earthworm. Each segment has a pair of large nephridia suspended in a fluid-filled coelom. Each nephridium occupies two segments because the ciliated funnel (nephrostome) drains the segment anterior to the segment containing the rest of the nephridium.

The extensive branching of the flame cell system is a consequence of the absence of a circulatory system in these acoelomate animals. This condition is very unlike the condensed kidneys of vertebrates and many invertebrates, which depend on the circulatory system to deliver wastes for excretion.

The **protonephridium** just described is a **closed** system. The tubules are closed on the inner end, and urine is formed from a fluid that must first enter the tubules by being transported across flame cells. A more advanced type of nephridium is the **open,** or "true," nephridium **(metanephridium)** that is found in several of the eucoelomate phyla such as annelids (Figure 7-6), molluscs, and several smaller phyla. The metanephridium is more advanced than the protonephridium in two important ways. First, the tubule is open at *both* ends, allowing fluid to be swept into the tubule through a ciliated funnel-like opening, the **nephrostome.** Second, the metanephridium is surrounded by a network of blood vessels that assists in urine formation by reabsorbing water, salts, sugars, amino acids, and other valuable materials from the tubular fluid.

Despite these differences, the basic process of urine formation is the same in protonephridia and metanephridia: fluid enters and flows continuously through a tubule, where the fluid is selectively modified by (1) withdrawing valuable solutes from it and returning these to the body (reabsorption) and (2) adding waste solutes to it (secretion). The sequence ensures the removal of wastes from the body without the loss of materials valuable to the body. We will see that the kidneys of vertebrates operate in basically the same way.

Arthropod Kidneys

The paired **antennal glands** of crustaceans, located in the ventral part of the head (Figure 7-7), are an advanced design of the basic nephridial organ. However, they lack open nephrostomes. Instead, hydrostatic pressure of the blood forms a protein-free filtrate of the blood (ultrafiltrate) in the end sac. In the tubular portion of the gland, selective reabsorption of certain salts and the active secretion of others modifies the filtrate. Thus crustaceans have excretory organs that are basically vertebrate-like in the functional sequence of urine formation.

Insects and spiders have a unique excretory system consisting of **Malpighian tubules** that operate in conjunction with specialized glands in the wall of the rectum (Figure 7-8). These thin, elastic, blind Malpighian tubules are closed and lack an arterial supply. Urine formation is initiated by the active secretion of salts, largely potassium, into the tubules from the hemolymph. This primary secretion of ions creates an osmotic drag that pulls water, solutes, and nitrogenous wastes, especially uric acid, into the tubule. Uric acid enters the upper end of the tubule as soluble potassium urate, which precipitates as insoluble uric acid in the proximal end of the tubule. Once the formative urine drains into the rectum, most of the water and potassium are reabsorbed by specialized rectal glands, leaving behind uric acid and other wastes that are expelled in the feces. The Malpighian tubule excretory system is ideally suited for life in dry environments and has contributed to the adaptive radiation of insects on land.

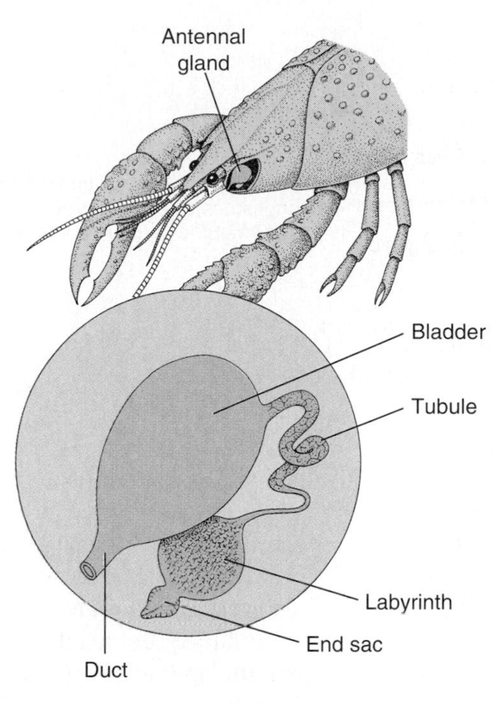

figure 7-7

Antennal glands of a crayfish. These are filtration kidneys, that is, a filtrate of the blood is formed in the end sac. The filtrate is converted into urine as it passes down the tubule toward the bladder.

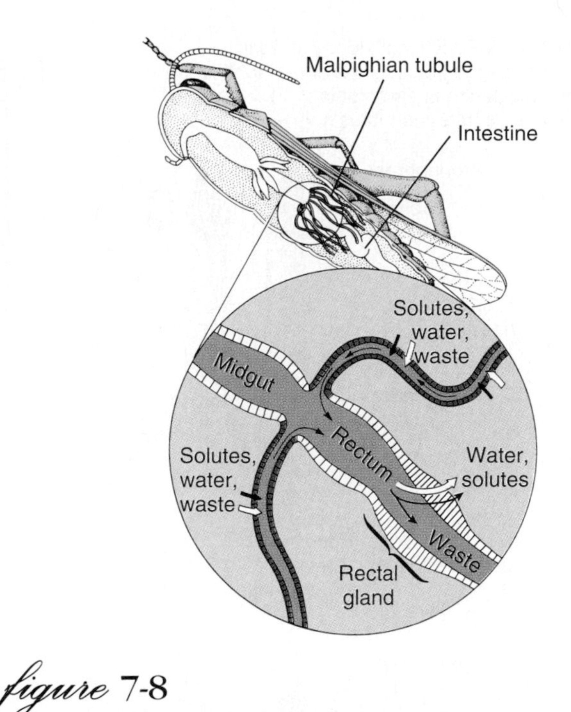

figure 7-8

Malpighian tubules of insects. Malpighian tubules are located at the juncture of the midgut and hindgut (rectum). Solutes, especially potassium, are actively secreted into the tubules from the surrounding arthropod hemolymph. Water and wastes follow. This fluid drains into the rectum, where solutes and water are actively reabsorbed, leaving wastes to be excreted.

Vertebrate Kidney

Ancestry and Embryology

From comparative studies of development, biologists believe that the kidney of the earliest vertebrates extended the length of the coelomic cavity and was made up of segmentally arranged tubules, each resembling an invertebrate nephridium. Each tubule opened at one end into the coelom by a nephrostome and at the other end into a common **archinephric duct.** This ancestral kidney has been called an **archinephros** ("ancient kidney"), and a segmental kidney very similar to the archinephros is found in the embryos of hagfishes and caecilians (Figure 7-9). Almost from the beginning, the reproductive system, which develops beside the excretory system from the same segmental blocks of trunk mesoderm, used the nephric ducts as a convenient conducting system for reproductive products. Thus, even though the two systems have nothing functionally in common, they are closely associated in their use of common ducts.

Kidneys of living vertebrates developed from this primitive plan. During embryonic development of the amniote vertebrates, there is a succession of three developmental stages of kidneys: **pronephros, mesonephros,** and **metanephros** (Figure 7-9). Some, but not all, of these stages are observed also in other vertebrate groups. In all vertebrate embryos, the pronephros is the first kidney to appear. It is located anteriorly in the body and becomes part of the persistent kidney only in adult hagfishes. In all other vertebrates it degenerates during development and is replaced by a more centrally located mesonephros. The mesonephros is the functional kidney of embryonic amniotes (reptiles, birds, and mammals) and contributes to the adult kidney (called an opisthonephros) of fishes and amphibians.

The metanephros, characteristic of adult amniotes, is distinguished in several ways from the pronephros and mesonephros. It is more caudally located and it is a much larger, more compact structure containing a very large number of nephric tubules. It is drained by a new duct, the **ureter,** which developed when the old archinephric duct was relinquished to the reproductive system of the male for sperm transport. Thus the three successive kidney types—pronephros, mesonephros, metanephros—succeed each other embryologically, and to some extent phylogenetically, in amniotes.

Vertebrate Kidney Function

The vertebrate kidney is part of many interlocking mechanisms that maintain homeostasis. The kidney plays a prominent role in this regulatory council because it is the principal

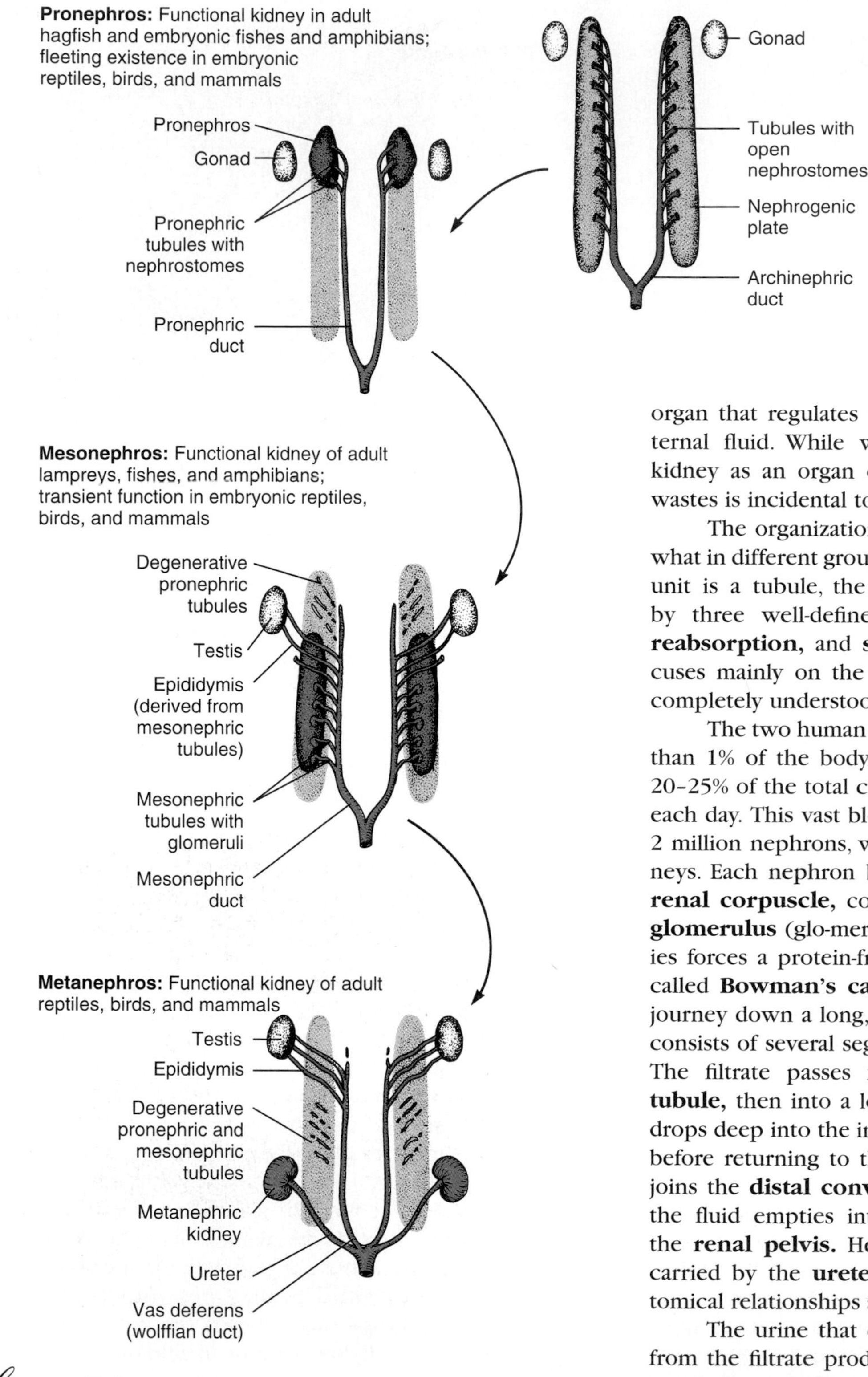

figure 7-9

Comparative development of male vertebrate kidney. *Red,* functional structures. *Light red,* degenerative or undeveloped parts.

organ that regulates the volume and composition of the internal fluid. While we commonly describe the vertebrate kidney as an organ of excretion, the removal of metabolic wastes is incidental to its regulatory function.

The organization of the vertebrate kidney differs somewhat in different groups of vertebrates, but in all the functional unit is a tubule, the **nephron,** and in all urine is formed by three well-defined physiological processes: **filtration, reabsorption,** and **secretion.** The following discussion focuses mainly on the mammalian kidney, which is the most completely understood regulatory organ.

The two human kidneys are small organs comprising less than 1% of the body weight. Yet they receive a remarkable 20–25% of the total cardiac output, some 2000 liters of blood each day. This vast blood flow is channeled to approximately 2 million nephrons, which make up the bulk of the two kidneys. Each nephron begins with an expanded chamber, the **renal corpuscle,** containing a tuft of capillaries called the **glomerulus** (glo-mer′yoo-lus). Blood pressure in the capillaries forces a protein-free **filtrate** into an expanded chamber, called **Bowman's capsule.** From here the filtrate begins a journey down a long, twisted **renal tubule.** The renal tubule consists of several segments that perform different functions. The filtrate passes first into the **proximal convoluted tubule,** then into a long, thin-walled **loop of Henle,** which drops deep into the inner portion of the kidney (the medulla) before returning to the outer portion (the cortex) where it joins the **distal convoluted tubule.** From the distal tubule the fluid empties into a **collecting duct** that drains into the **renal pelvis.** Here the urine is collected before being carried by the **ureter** to the **urinary bladder.** These anatomical relationships are shown in Figure 7-10.

The urine that enters the renal pelvis is very different from the filtrate produced in the renal corpuscle. During its travels through the renal tubule and collecting duct, both the composition and concentration of the original filtrate change. Some solutes such as glucose and sodium have been reabsorbed, while other materials, such as hydrogen ions and urea have been concentrated in the urine.

The nephron, with its pressure filter and tubule, is associated intimately with the blood circulation (Figure 7-11). Blood from the aorta enters each kidney through a large **renal artery,**

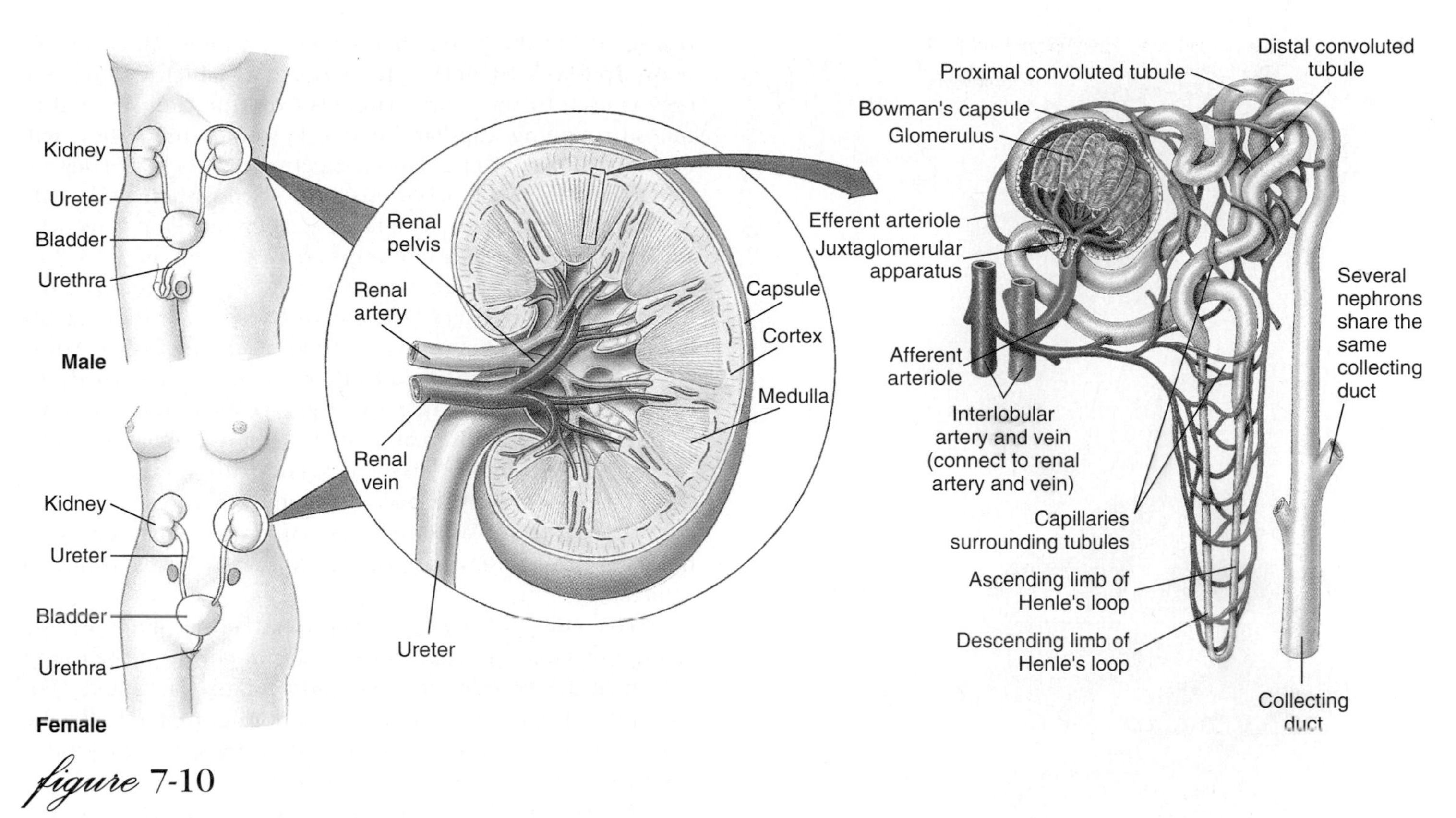

figure 7-10

Urinary system of humans, with enlargements showing detail of the kidney and a single nephron.

which divides into a branching system of smaller arteries. The arterial blood reaches the renal corpuscle through an **afferent arteriole** and leaves by way of an **efferent arteriole.** From the efferent arteriole the blood travels to an extensive capillary network that surrounds and supplies the proximal and distal convoluted tubules and the loop of Henle (Figure 7-10). This capillary network provides a means for the pickup and delivery of materials that are reabsorbed or secreted by the kidney tubules. From these capillaries blood is collected by veins that unite to form the **renal vein.** This vein returns the blood to the vena cava.

Glomerular Filtration

Let us now return to the glomerulus, where the process of urine formation begins. The glomerulus acts as a specialized mechanical filter in which a protein-free filtrate of the plasma is driven by the blood pressure across the capillary walls and into the fluid-filled space of Bowman's capsule. Solute molecules small enough to pass the slit pores of the capillary wall are carried through with the water in which they are dissolved. Red blood cells and the plasma proteins, however, are withheld because they are too large to pass the filter (Figure 7-12). The outward pressure forcing water and solutes through the capillary membrane must be sufficient to exceed the colloid osmotic pressure of the blood (created by the plasma proteins that cannot pass the filter; see p. 198) and the resistance to flow down the tubule.

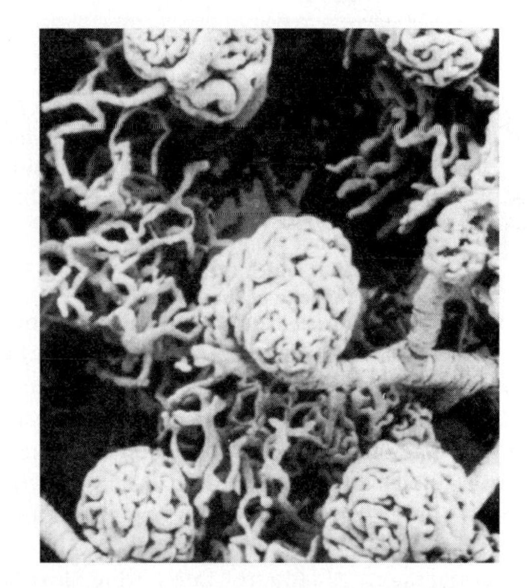

figure 7-11

Scanning electron micrograph of a cast of the microcirculation of the mammalian kidney, showing several glomeruli and associated blood vessels. The Bowman's capsule, which normally surrounds each glomerulus, has been digested away in preparing the cast.

From: Tissues and Organs: A Text Atlas of Scanning Electron Microscopy, *© Richard G. Kessel and Randy H. Kardon, published by W. H. Freeman and Co., 1979.*

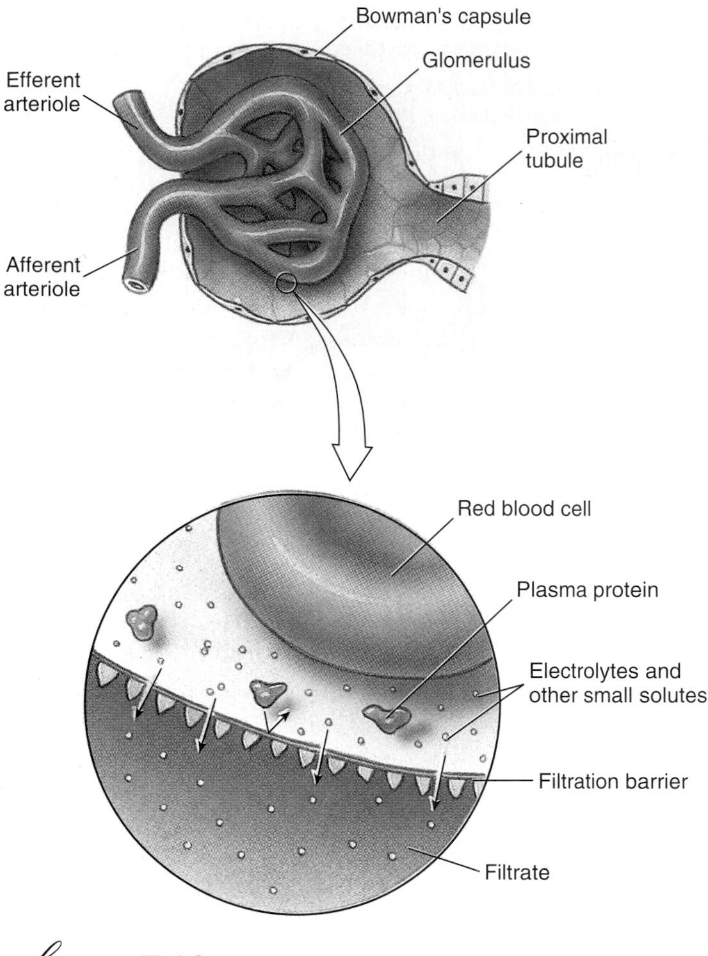

figure 7-12

Bowman's capsule and glomerulus, showing (*enlargement*) the filtration of fluid through the glomerular capillary membrane. Water, electrolytes, and other small molecules pass the porous filtration barrier, but the plasma proteins are too large to pass the barrier. The filtrate is thus protein free.

The filtrate now enters the renal tubular system where it will undergo extensive modification before becoming urine. Human kidneys form approximately 180 liters (nearly 50 gallons) of filtrate each day, a volume many times exceeding the total blood volume. If this volume of water and the valuable nutrients and salts it contains were lost, the body would soon be depleted of these compounds. This depletion does not happen because nearly all of the filtrate is reabsorbed.

The conversion of filtrate into urine involves two processes: (1) modification of the composition of the filtrate through tubular reabsorption and secretion, and (2) changes in the total osmotic concentration of the urine through the regulation of water excretion.

Tubular Reabsorption

Some 60% of the filtrate volume and virtually all of the glucose, amino acids, vitamins, and other valuable nutrients are reabsorbed in the proximal convoluted tubule. Much of this reabsorption is by **active transport,** in which cellular energy is used to transport materials from the tubular fluid to the surrounding capillary network from which they will reenter the blood circulation. Electrolytes such as sodium, potassium, calcium, bicarbonate, and phosphate are reabsorbed by ion pumps, which are carrier proteins driven by the hydrolysis of ATP (ion pumps are described on p. 33–34). Because an essential function of the kidney is to regulate the plasma concentrations of the electrolytes, all are individually reabsorbed by ion pumps specific for each electrolyte. Some are strongly reabsorbed and others weakly reabsorbed, depending on the body's need to conserve each mineral. Some materials are passively reabsorbed. Negatively charged chloride ions, for example, passively accompany the active reabsorption of positively charged sodium ions in the proximal convoluted tubule. Water, too, is withdrawn passively from the tubule, as it follows osmotically the active reabsorption of solutes.

For most substances there is an upper limit to the amount of substance that can be reabsorbed. This upper limit is termed the **transport maximum** for that substance. For example, glucose normally is reabsorbed completely by the kidney because the transport maximum for the reabsorptive mechanism for glucose is poised well above the amount of glucose usually present in the plasma filtrate. Should the plasma glucose concentration exceed this threshold level, as in the disease diabetes mellitus, glucose appears in the urine (Figure 7-13).

In the disease diabetes mellitus ("sweet running through"), glucose rises to abnormally high concentrations in the blood plasma (hyperglycemia) because the hormone insulin, which enables the body cells to take up glucose, is deficient. As the blood glucose rises above a normal level of about 100 mg/100 ml of plasma, the concentration of glucose in the filtrate also rises, and more glucose must be reabsorbed by the proximal tubule. Eventually a point is reached (about 300 mg/100 ml of plasma) at which the reabsorptive capacity of the tubular cells is saturated. This is the transport maximum for glucose. Should the plasma glucose continue to rise, glucose spills over into the urine. In untreated diabetes the victim's urine tastes sweet, thirst is unrelenting, and the body wastes away despite a large food intake. In England the disease for centuries was appropriately called the "pissing evil."

Unlike glucose, most electrolytes are excreted in the urine in variable amounts. The reabsorption of sodium, the dominant cation in the plasma, illustrates the flexibility of the reabsorption process. The human kidney filters approximately 600 g of sodium every 24 hours. Nearly all of this sodium is reabsorbed, but the exact amount is matched pre-

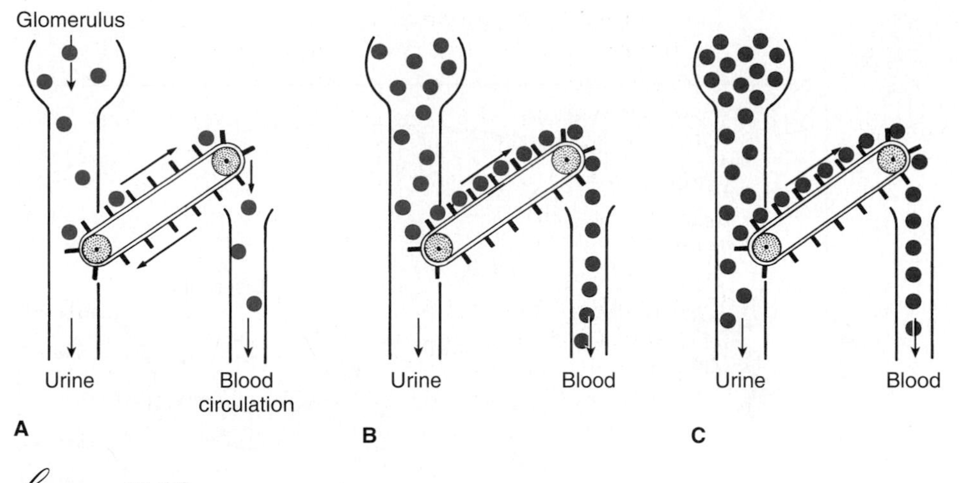

figure 7-13

The mechanism for the tubular reabsorption of glucose can be likened to a conveyor belt running at constant speed. **A,** When the concentration of glucose in the filtrate is low, all is reabsorbed. **B,** When the glucose concentration in the filtrate has reached the transport maximum, all carrier sites for glucose are occupied. If the glucose rises further, **C,** as in the disease diabetes mellitus, some glucose escapes the carriers and appears in the urine.

cisely to sodium intake. With a normal sodium intake of 4 g per day, the kidney excretes 4 g and reabsorbs 596 g each day. A person on a low-salt diet of 0.3 g of sodium per day still maintains salt balance because only 0.3 g escapes reabsorption. But with a very high salt intake, much above 20 g a day, the kidney cannot excrete sodium as fast as it enters. The unexcreted sodium chloride holds additional water in the body fluids, and the person begins to gain weight. (The salt intake of the average North American is about 6 to 18 g a day, approximately 20 times more than the body needs, and three times more than is considered acceptable for those predisposed to high blood pressure.)

The human kidney can excrete large quantities of salt (sodium chloride) under conditions of high salt intake. In societies accustomed to widespread use of foods heavily salted for preservation (for example, salted pork and salt herring) daily intakes may approach or even exceed 100 g. Body weight remains normal under such conditions. However, the acute ingestion of 20–40 g/day by volunteers unadapted to such large intakes of salt caused swelling of tissues, increase in body weight, and some increase in blood pressure.

The distal convoluted tubule carries out the final adjustment of filtrate composition. The sodium reabsorbed by the proximal convoluted tubule—some 85% of the total filtered—is obligatory reabsorption; that is, this amount will be reabsorbed independent of sodium intake. In the distal convoluted tubule, however, sodium reabsorption is controlled by **aldosterone,** a steroid hormone from the adrenal gland. Aldosterone increases the retention of sodium by the distal tubules and thus decreases the loss of sodium in the urine.

The flexibility of distal reabsorption of sodium varies considerably in different animals: it is restricted in humans but very broad in many rodents. These differences have appeared because selective pressures during evolution have resulted in rodents adapted for dry environments. They must conserve water and at the same time excrete considerable sodium. Humans, however, were not designed to accommodate the large salt appetites many have. Our closest relatives, the great apes, are vegetarians with an average salt intake of less than 0.5 g a day.

Tubular Secretion

In addition to reabsorbing materials from the plasma filtrate, the kidney tubules are able to secrete certain substances into the tubular fluid. This process, which is the reverse of tubular reabsorption, enables the kidney to build up the urine concentrations of materials to be excreted, such as hydrogen and potassium ions, drugs, and various foreign organic materials. The distal convoluted tubule is the site of most tubular secretion.

In the kidneys of bony marine fishes, reptiles, and birds, tubular secretion is a much more highly developed process than it is in mammalian kidneys. Marine bony fishes actively secrete large amounts of magnesium and sulfate, seawater salts that are by-products of their mode of osmotic regulation. Reptiles and birds excrete uric acid instead of urea as their major nitrogenous waste. The material is actively secreted by the tubular epithelium. Since uric acid is nearly insoluble, it forms crystals in the urine and requires little water for excretion. Thus the excretion of uric acid is an important adaptation for water conservation.

Water Excretion

The kidney closely regulates the osmotic pressure of the blood. When fluid intake is high, the kidney excretes a dilute urine, saving salts and excreting water. When fluid intake is low, the kidney conserves water by forming a concentrated urine. A dehydrated person can concentrate urine to approximately four times blood osmotic concentration. This important ability to concentrate urine enables us to excrete wastes with minimal loss of water.

The capacity of the kidney of mammals and some birds to produce a concentrated urine involves an interaction between the loop of Henle and the collecting ducts. This interplay results in the formation of an osmotic gradient in the kidney, as shown in Figure 7-14. In the cortex, the interstitial fluid is isosmotic with the blood, but deep in the medulla the osmotic concentration is four times greater than that of the blood (in rodents and desert mammals that can produce highly concentrated urine the osmotic gradient is much greater than in humans). The high osmotic concentrations in the medulla are produced by an exchange of ions in the loop of Henle by **countercurrent multiplication.** "Countercurrent" refers to the opposite directions of fluid movement in the two limbs of the loop of Henle: down in the descending limb and up the ascending limb. "Multiplication" refers to the increasing osmotic concentration in the medulla that results from ion exchange between the two limbs of the loop.

The functional characteristics of this system are as follows. The descending limb of the loop of Henle is permeable to water but impermeable to solutes. The ascending limb is relatively impermeable to both water and solutes. Sodium chloride is actively transported out of the thick portion of the ascending limb and into the surrounding tissue fluid (Figure 7-14). As the interstitium surrounding the loop becomes more concentrated with solute, water moves passively out of the descending limb by osmosis. The tubular fluid in the hairpin, now more concentrated, moves up the ascending limb, where still more sodium chloride is pumped out. In this way the effect of active ion transport in the ascending limb is multiplied as more water is withdrawn from the descending limb and more concentrated fluid is presented to the ascending limb ion pump. Urea also contributes significantly to tissue fluid concentration at the bottom of the hairpin loop; it is reabsorbed from the collecting duct that lies parallel to the loop (Figures 7-14 and 7-15).

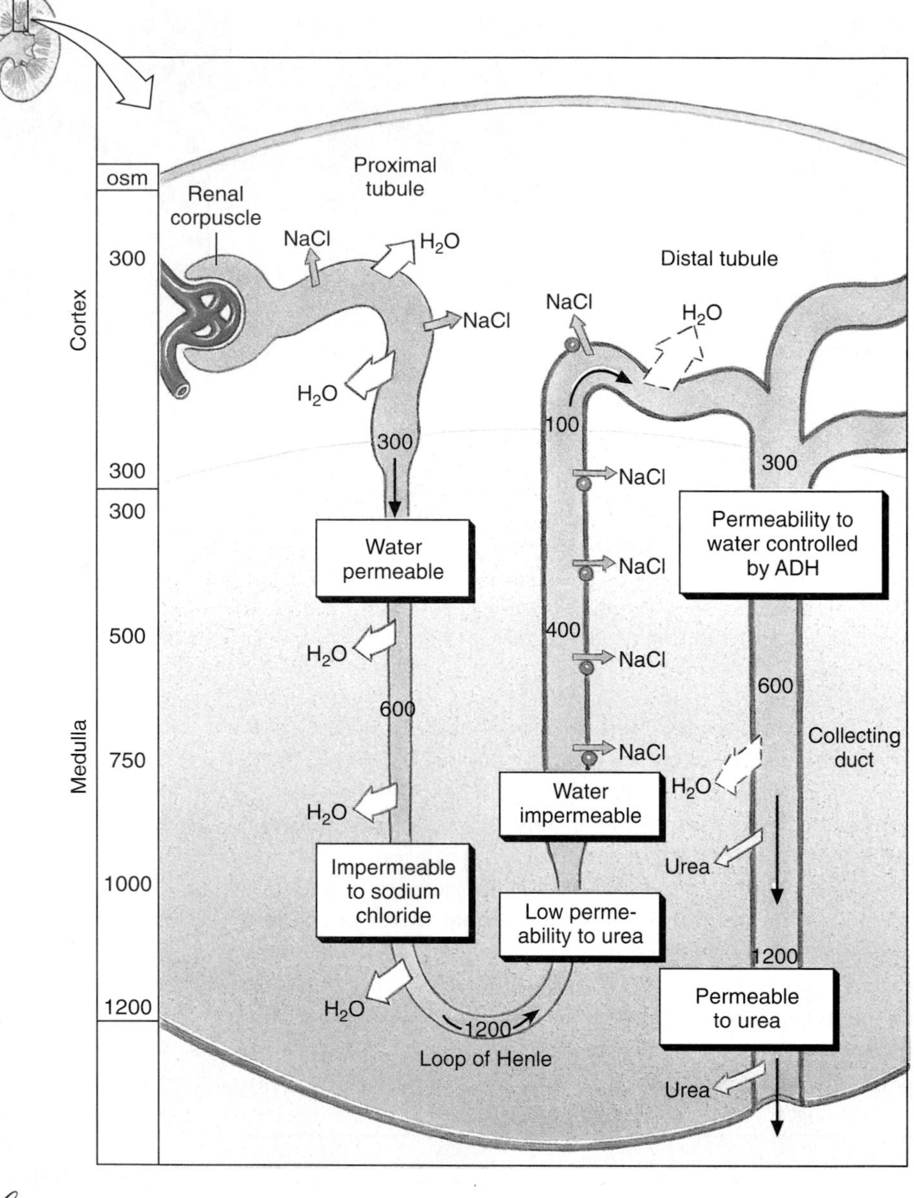

figure 7-14

Mechanism of urine concentration in mammals. Sodium and chloride are pumped from the ascending limb of the loop of Henle, and water is withdrawn passively from the descending limb, which is impermeable to sodium chloride. Sodium chloride and urea reabsorbed from the collecting duct raise the osmotic concentration in the kidney medulla, creating an osmotic gradient for the controlled reabsorption of water from the collecting duct.

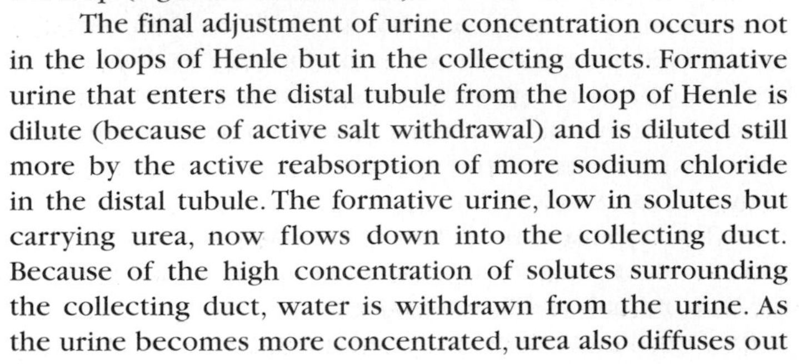

The final adjustment of urine concentration occurs not in the loops of Henle but in the collecting ducts. Formative urine that enters the distal tubule from the loop of Henle is dilute (because of active salt withdrawal) and is diluted still more by the active reabsorption of more sodium chloride in the distal tubule. The formative urine, low in solutes but carrying urea, now flows down into the collecting duct. Because of the high concentration of solutes surrounding the collecting duct, water is withdrawn from the urine. As the urine becomes more concentrated, urea also diffuses out and adds to the high osmotic pressure in the kidney medulla (Figure 7-15).

The amount of water saved and the final concentration of the urine depend on the permeability of the walls of the collecting duct. This is controlled by the **antidiuretic hormone** (ADH, or vasopressin; p. 274), which is released by the posterior pituitary gland (neurohypophysis, p. 272). In turn, special receptors in the brain that constantly sense the osmotic pressure of the blood govern the release of this hormone. When the blood osmotic pressure increases, as during dehydration, the pituitary

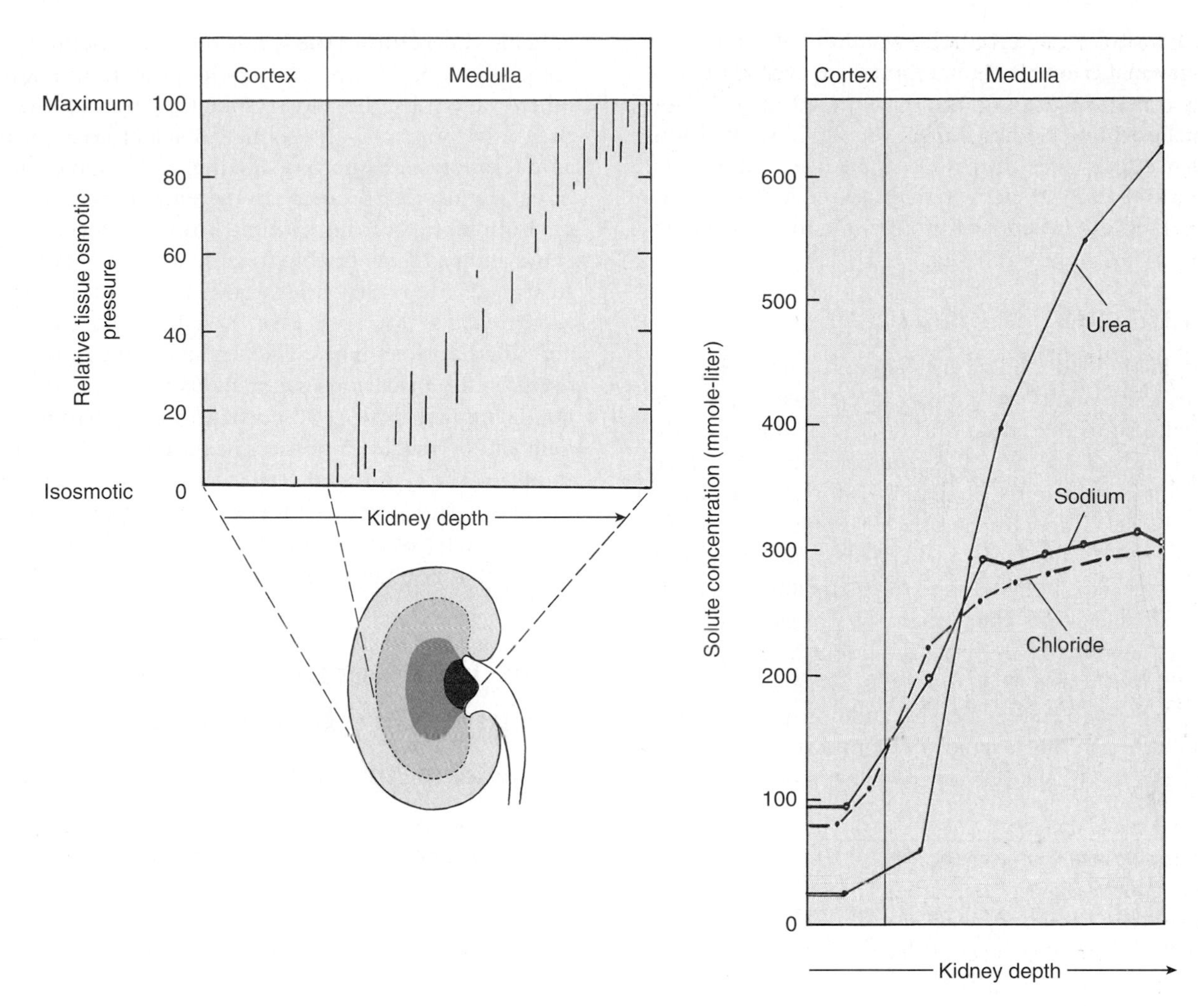

figure 7-15

Osmotic concentration of tissue fluid in the mammalian kidney. Tissue fluid is isosmotic in the kidney cortex (*to left in diagram*) but increases continuously through the medulla, reaching a maximum at the papilla where the urine drains into the ureter.

gland releases more ADH. ADH increases the permeability of the collecting duct, probably by expanding the size of pores in the walls of the duct. Then, as the fluid in the collecting duct passes through the hyperosmotic region of the kidney medulla, water diffuses through the pores into the surrounding interstitial fluid and is carried away by the blood circulation (Figure 7-10). The urine loses water and becomes more concentrated. Given this sequence of events for dehydration, it is not difficult to anticipate how the system responds to overhydration: the pituitary stops releasing ADH, the pores in the collecting duct walls close, and a large volume of dilute urine is excreted.

The varying ability of different mammals to form a concentrated urine correlates closely with the length of the loops of Henle. The beaver, which has no need to conserve water in its aquatic environment, has short loops and can concentrate its urine only to about twice the osmolarity of the blood plasma. Humans, with relatively longer loops, can concentrate urine to 4.2 times the osmolarity of the blood. As we would anticipate, desert mammals have much greater urine concentrating powers. The camel can produce a urine 8 times the plasma concentration, the gerbil 14 times, and the Australian hopping mouse 22 times. In this creature, the greatest urine concentrator of all, the loops of Henle extend to the tip of a long renal papilla that pushes out into the mouth of the ureter.

Temperature Regulation

We have seen that a fundamental problem facing an animal is keeping its internal environment in a state that permits normal cell function. Biochemical activities are sensitive to the chemical environment, and our discussion thus far has examined how the chemical environment is stabilized. Biochemical reactions are also extremely sensitive to temperature. All enzymes have an optimum temperature; at temperatures above or below this optimum, enzyme function is impaired. Temperature therefore is a severe constraint for animals, all of which seek biochemical stability. When body temperatures drop too low, metabolic processes are slowed, reducing the amount of energy the animal can muster for activity and reproduction. If the body temperature rises too high, metabolic reactions become imbalanced, and enzymatic activity is hampered or even destroyed. Thus animals can succeed over only a restricted range of temperature, usually within the limits of 0° to 40° C. Animals must either find a habitat where they do not have to contend with temperature extremes, or they must develop the means of stabilizing their metabolism independent of temperature extremes.

Ectothermy and Endothermy

The terms "cold-blooded" and "warm-blooded" have long been used to divide animals into two groups: invertebrates and vertebrates that feel cold to the touch, and those, like humans, other mammals, and birds, that do not. It is true that the body temperature of mammals and birds is usually (though not always) warmer than the air temperature, but a "cold-blooded" animal is not necessarily cold. Tropical fishes, and insects and reptiles basking in the sun, may have body temperatures equaling or surpassing those of mammals. Conversely, many "warm-blooded" mammals hibernate, allowing their body temperature to approach the freezing point of water. Thus the terms "warm-blooded" and "cold-blooded" are hopelessly subjective and nonspecific but are so firmly entrenched in our vocabulary that most biologists find it easier to accept the usage than try to change people.

A temperature difference of 10° C has become a standard that is used to measure the temperature sensitivity of a biological function. This value, called the Q_{10}, is determined (for temperature intervals of exactly 10° C) simply by dividing the value of a rate function (such as metabolic rate or the rate of an enzymatic reaction) at the higher temperature by the value of the rate function at the lower temperature. In general, metabolic reactions have Q_{10} values of about 2.0 to 3.0. Purely physical processes, such as diffusion, have much lower Q_{10} values, usually close to 1.0.

The terms **poikilothermic** (variable body temperature) and **homeothermic** (constant body temperature) are frequently used by zoologists as alternatives to "cold-blooded" and "warm-blooded," respectively. These terms are more precise and more informative, but they still offer difficulties. For example, deep-sea fishes live in an environment having no perceptible temperature change. Even though their body temperature is absolutely stable, day in and day out, to call such fishes homeotherms would distort the intended application of the term. Furthermore, among the homeothermic birds and mammals there are many that allow their body temperature to change between day and night, or, as with hibernators, between seasons.

Physiologists prefer yet another way to describe body temperatures, one that reflects the fact that an animal's body temperature is a balance between heat gain and heat loss. All animals produce heat from cellular metabolism, but in most the heat is conducted away as fast as it is produced. In these animals, the **ectotherms**—and the overwhelming majority of animals belong to this group—the body temperature is determined solely by the environment. Many ectotherms exploit their environment behaviorally to select areas of more favorable temperature (such as basking in the sun), but the source of energy used to increase body temperature comes from the environment, not from within the body. Alternatively, there are some animals that are able to generate and retain enough heat to elevate their own body temperature to a high but stable level. Because the source of their body heat is internal, they are called **endotherms.** These favored few in the animal kingdom are the birds and mammals, as well as a few reptiles and fast-swimming fishes, and certain insects that are at least partially endothermic. Endothermy allows birds and mammals to stabilize their internal temperature so that biochemical processes and nervous function can proceed at steady high levels of activity. Endotherms can thus remain active in winter and exploit habitats denied to ectotherms.

How Ectotherms Achieve Temperature Independence

Behavioral Adjustments

Although ectotherms cannot control their body temperature physiologically, many are able to regulate their body temperature behaviorally with considerable precision. Ectotherms often have the option of seeking areas in the environment where the temperature is favorable to their activities. Some ectotherms, such as desert lizards, exploit hour-to-hour changes in solar radiation to keep their body temperature relatively constant (Figure 7-16). In the early morning they emerge from their burrows and bask in the sun with their bodies flattened to absorb heat. As the day warms they turn to face the sun to reduce exposure, and raise their bodies from the hot substrate. In the hottest part of the day they may retreat to their burrows. Later they emerge to bask as the sun sinks lower and the air temperature drops.

These behavioral patterns help to maintain a relatively steady body temperature of 36° to 39° C, while the air temperature varies between 29° and 44° C. Some lizards can tolerate intense midday heat without shelter. The desert iguana of the southwestern United States prefers a body temperature of 42° C when active and can tolerate a rise to 47° C, a temperature that is lethal to all birds and mammals and most other lizards. The term "cold-blooded" clearly does not apply to these animals!

Metabolic Adjustments

Even without the help of the behavioral adjustments just described, most ectotherms can adjust their metabolic rates to the prevailing temperature such that the intensity of metabolism remains mostly unchanged. This is called **temperature compensation** and involves complex biochemical and cellular adjustments. These adjustments enable a fish or a salamander, for example, to benefit from the same level of activity in both warm and cold environments. Thus whereas endotherms

figure 7-16

How a lizard regulates its body temperature by its behavior. In the morning the lizard absorbs the sun's heat through its head while keeping the rest of its body protected from the cool morning air. At noon, with its body temperature high, it seeks shade from the hot sun. Later, it emerges and lies parallel to the sun's rays.

achieve metabolic homeostasis by regulating their body temperature independent of environment temperature, ectotherms accomplish much the same by directly regulating their metabolism independent of body temperature. This metabolic regulation also is a form of homeostasis.

Temperature Regulation in Endotherms

Most mammals have body temperatures between 36° and 38° C, somewhat lower than those of birds, which range between 40° and 42° C. This constant temperature is maintained by a delicate balance between heat production and heat loss—not a simple matter when these animals are alternating between periods of rest and bursts of activity.

Heat is produced by the animal's metabolism. This includes the oxidation of foodstuffs, basal cellular metabolism, and muscular contraction. Because much of an endotherm's daily caloric intake is required to generate body heat, especially in cold weather, the endotherm must eat more food than an ectotherm of the same size. Heat is lost by radiation, conduction, and convection (air movement) to a cooler environment and by evaporation of water (Figure 7-17). A bird or mammal

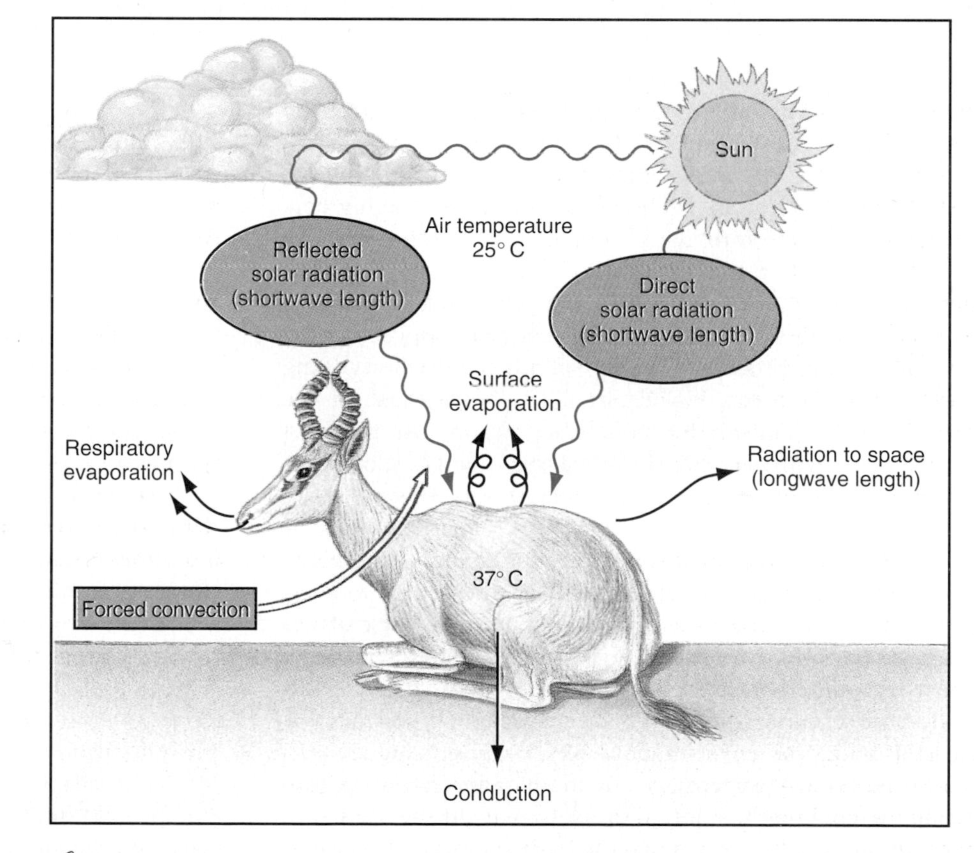

figure 7-17

Exchange of heat between the animal and its environment on a warm day. Blue arrows indicate sources of net heat gain by the animal (all radiation); black arrows are avenues of net heat loss (evaporative cooling, conduction to the ground, longwave radiation into space, and forced convection by the wind). If the air and ground temperatures were warmer than the animal, the arrows for forced convection, conduction, and radiation from space would be reversed. Then the animal could lose heat only by evaporative cooling.

can control both processes of heat production and heat loss within rather wide limits. If the animal becomes too cool, it can increase heat production by increasing muscular activity (exercise or shivering) and by decreasing heat loss by increasing its insulation. If it becomes too warm, it decreases heat production and increases heat loss. We will examine these processes in the following examples.

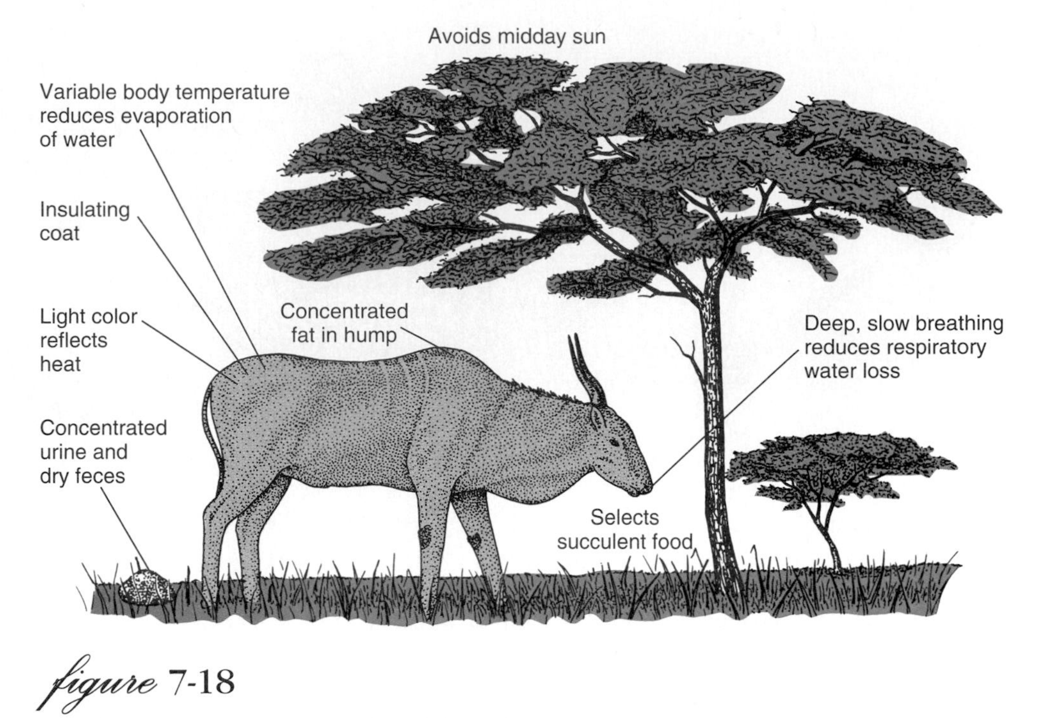

figure 7-18
Physiological and behavioral adaptations of the common eland for maintaining heat balance in the hot, arid savanna of central Africa.

Adaptations for Hot Environments

Despite the harsh conditions of deserts—intense heat during the day, cold at night, and scarcity of water, vegetation, and cover—many kinds of animals live there successfully. The smaller desert mammals are mostly **fossorial** (living mainly in the ground) or **nocturnal,** that is, active at night. The lower temperature and high humidity of burrows help reduce water loss by evaporation. As explained earlier in this chapter (p. 170 and Table 7-1), some desert animals such as the kangaroo rat and the American desert ground squirrels can, if necessary, derive the water they need from their dry food, drinking no water at all. Such animals produce a highly concentrated urine and form almost completely dry feces.

The large desert ungulates (hooved mammals that chew their cud) obviously cannot escape the desert heat by living in burrows. Animals such as camels and desert antelopes (gazelle, oryx, and eland) possess a number of adaptations for coping with heat and dehydration. Figure 7-18 shows those of the eland. The mechanisms for controlling water loss and preventing overheating are closely linked together. The glossy, pallid color of the fur reflects direct sunlight, and the fur itself is an excellent insulation that resists heat transfer. Heat is lost by convection and conduction from the underside of the eland where the pelage is very thin. Fat tissue, an essential food reserve, is concentrated in a single hump on the back of the eland, instead of being uniformly distributed under the skin where it would impair heat loss by radiation. The eland avoids evaporative water loss—the only means an animal has for cooling itself when the environmental temperature is higher than that of the body—by permitting its body temperature to drop during the cool night and then to rise slowly during the day as the body stores heat. Only when the body temperature reaches 41° C must elands prevent further rise through **evaporative cooling** by sweating and panting. They conserve water also by means of concentrated urine and dry feces. Camels also have all these adaptations developed to a similar or even greater degree; they are perhaps the most perfectly adapted of all large desert mammals.

Adaptations for Cold Environments

In cold environments mammals and birds use two major mechanisms to maintain homeothermy: (1) **decreased conductance,** that is, reduction of heat loss by increasing the effectiveness of the insulation and (2) **increased heat production.**

In all mammals living in cold regions of the earth, fur thickness increases in winter, sometimes by as much as 50%. The thick underhair is the major insulating layer, whereas the longer and more visible guard hair serves as protection against wear and for protective coloration.

However, unlike the well-insulated trunk of the body, the body extremities (legs, tail, ears, nose) of arctic mammals are thinly insulated and exposed to rapid cooling. To prevent these parts from becoming major avenues of heat loss, they are allowed to cool to low temperatures, often approaching the freezing point. As warm arterial blood passes into a leg, for example, heat shunts directly from artery to vein and flows back to the core of the body (Figure 7-19). This shunt prevents the loss of valuable body heat through the poorly insulated distal regions of the leg. A consequence of this **peripheral heat exchange system** is that the legs and feet must operate at low temperatures. The temperatures of the feet of the arctic fox and barren-ground caribou are just above the freezing point; in fact, the temperature may be below 0° C in the footpads and hooves. To keep feet supple and flexible at such low temperatures, fats in the extremities have very low melting points, perhaps 30° C lower than ordinary body fats.

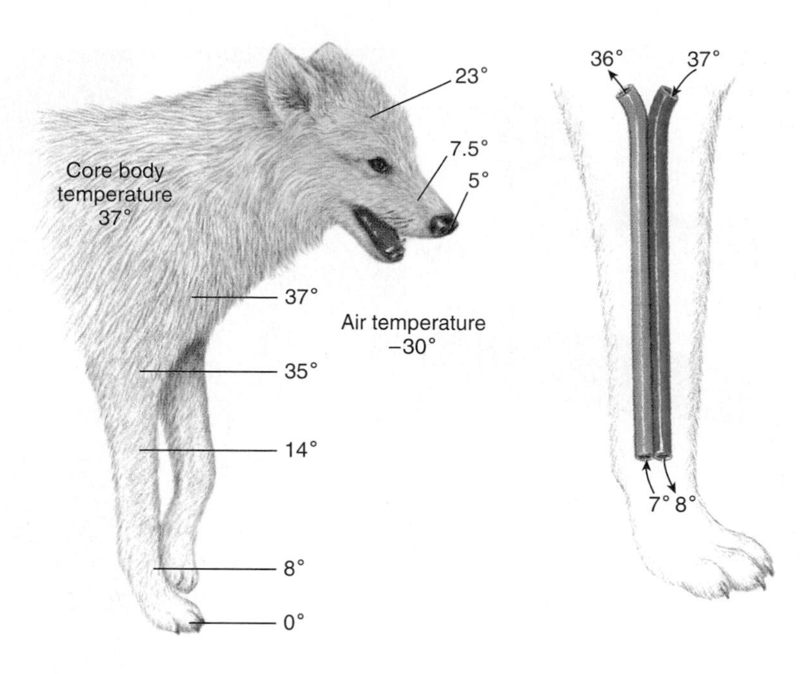

figure 7-19

Countercurrent heat exchange in the leg of an arctic wolf. The diagram on the left shows how the extremities cool when the animal is exposed to low air temperatures. The diagram on the right depicts a portion of the front leg artery and vein, showing how heat is exchanged between outflowing arterial and inflowing venous blood. Heat is thus shunted back into the body core and conserved.

In severely cold conditions all mammals can produce more heat by **augmented muscular activity** through exercise or shivering. We are all familiar with the effectiveness of both activities. A person can increase heat production as much as 18-fold by violent shivering when maximally stressed by cold. Another source of heat is the increased oxidation of foodstuffs, especially brown fat stores (brown fat is described on p. 234). This mechanism is called **nonshivering thermogenesis.**

Small mammals the size of lemmings, moles, and mice meet the challenge of cold environments in a different way. Small mammals are not as well insulated as large mammals because thickness of fur is limited by the need to maintain mobility. Consequently these forms exploit the excellent insulating qualities of snow by living under it in runways on the forest floor, where incidentally their food also is located. In this **subnivean environment** the temperature seldom drops below −5° C even though the air temperature above may fall to −50° C. The snow insulation decreases thermal conductance from small mammals just as thick pelage does for large mammals. Living beneath the snow is really a type of avoidance response to cold.

Adaptive Hypothermia in Birds and Mammals

Endothermy is energetically expensive. Whereas an ectotherm can survive for weeks in a cold environment without eating, an endotherm must always have energy resources to supply its high metabolic rate. The problem is especially acute for small birds and mammals which, because of their intense metabolism, may have to consume food each day equal to their own body weight to maintain homeothermy. It is not surprising then that a few small birds and mammals have evolved ways to abandon homeothermy for periods ranging from a few hours to several months, allowing their body temperature to fall until it equals or remains just above the temperature of the surrounding air.

Some very small mammals, such as bats, maintain normal high body temperatures when active but allow their body temperature to drop profoundly when inactive and asleep. This is called **daily torpor,** an adaptive hypothermia that provides enormous energy savings to small endotherms that are never more than a few hours away from starvation at normal body temperatures. Hummingbirds also may drop their body temperature at night when food supplies are low (Figure 7-20).

Many small and medium-sized mammals in northern temperate regions solve the problem of winter scarcity of food and low temperature by entering a prolonged and controlled state of dormancy: **hibernation.** True hibernators, such as ground squirrels, jumping mice, marmots, and woodchucks (Figure 7-21), prepare for hibernation by building up large amounts of body fat. Entry into hibernation is gradual. After a series of "test drops" during which body temperature decreases a few degrees and then returns to normal, the animal cools to within a degree or less of the temperature of its

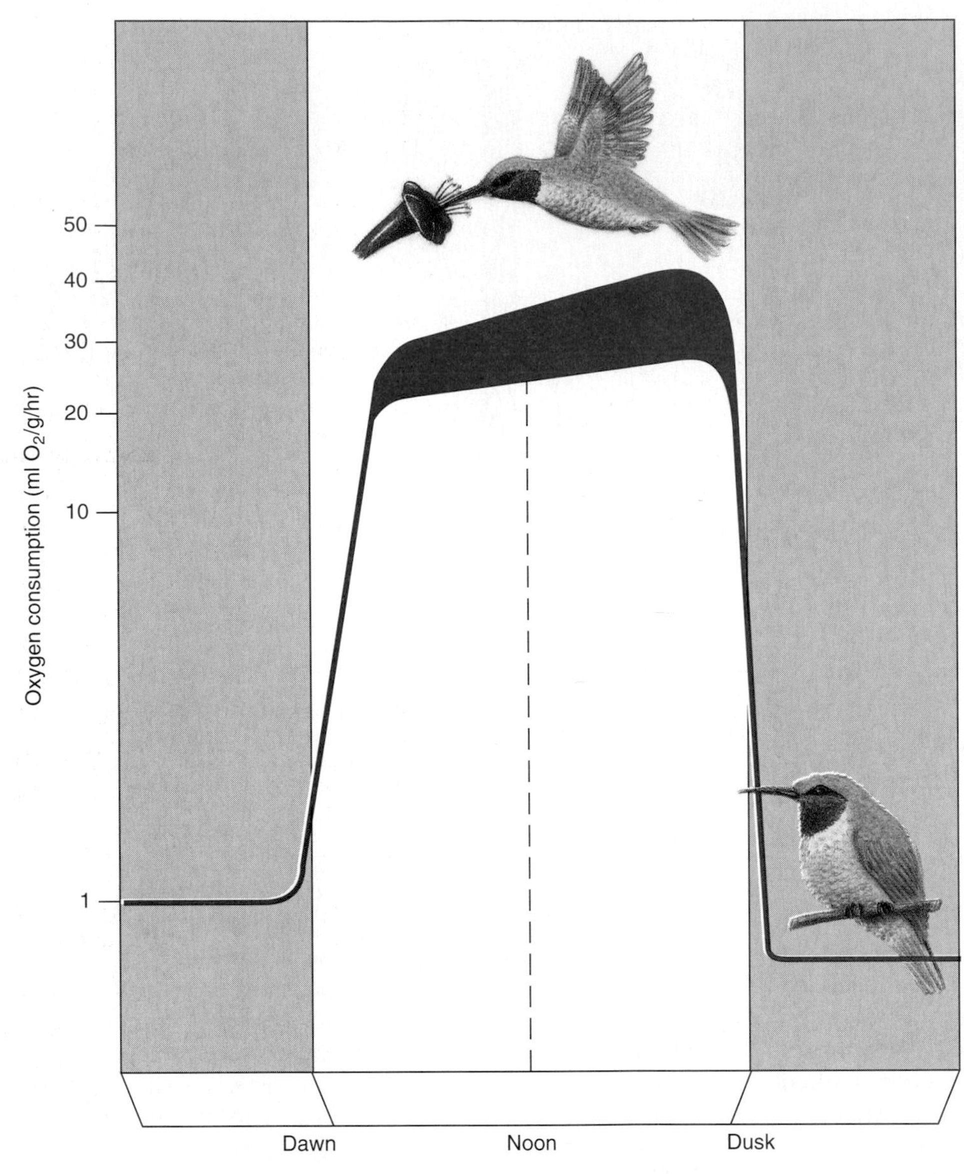

figure 7-20

Daily torpor in hummingbirds. Body temperature and oxygen consumption are high when hummingbirds are active during the day but drop to one-twentieth these levels during periods of food shortage. Torpor vastly lowers demands on the bird's limited energy reserves.

figure 7-21

Hibernating woodchuck *Marmota monax* (order Rodentia) in den exposed by road-building work sleeps on, unaware of the intrusion. Woodchucks begin hibernation in late September while the weather is still warm and may sleep six months. The animal is rigid and decidedly cold to the touch. Breathing is imperceptible, as slow as one breath every five minutes. Although it appears to be dead, it will awaken if the den temperature drops dangerously low.

surroundings. Metabolism decreases to a fraction of normal. In the ground squirrel, for example, the respiratory rate decreases from a normal rate of 200 per minute to 4 or 5 per minute, and the heart rate from 150 to 5 beats per minute. During arousal, the hibernator both shivers violently and employs nonshivering thermogenesis to produce heat.

Some mammals, such as bears, badgers, raccoons, skunks, and opossums, enter a state of prolonged sleep in winter with little or no decrease in body temperature. This is not true hibernation. Bears of the northern forest den-up for several months. Heart rate may decrease from 40 to 10 beats per minute, but body temperature remains normal and the bear is awakened if sufficiently disturbed. One intrepid but reckless biologist narrowly escaped injury when he crawled into a den and attempted to measure the bear's rectal temperature with a thermometer!

Summary

Throughout life, matter and energy pass through the body, potentially disturbing the internal physiological state. Homeostasis, the ability of an organism to maintain internal stability despite such challenges, is a characteristic of all living systems. Homeostasis involves the coordinated activity of several physiological and biochemical mechanisms, and it is possible to relate some major events in animal evolution to increasing internal independence from the consequences of environmental change. In this chapter we have examined two aspects of homeostasis: (1) the varying ability of animals to stabilize the osmotic and chemical composition of the blood, and (2) the capacity of animals to regulate their temperatures in thermally challenging environments.

Most marine invertebrates must either depend on the osmotic stability of the ocean to which they conform, or they must be able to tolerate wide fluctuations in environmental salinity. Some of the latter show limited powers of osmotic regulation, that is, the capacity to resist internal osmotic change, through the evolution of specialized regulatory organs. All animals living in fresh water are hyperosmotic to their environment and have developed mechanisms for recovering salt from the environment and eliminating excess water that enters the body osmotically.

All vertebrate animals, except the hagfishes, show excellent osmotic homeostasis. Marine bony fishes maintain their body fluids distinctly hypoosmotic to their environment by drinking seawater and physiologically distilling it. Elasmobranchs (sharks and their kin) have adopted a strategy of near-osmotic conformity by retaining urea in the blood.

The kidney is the most important organ for regulating the chemical and osmotic composition of the blood. In all metazoa the kidney is some variation on a basic theme: a tubular structure that forms urine by introducing a fluid secretion or filtrate of the blood or interstitial fluid into a tubule in which it is selectively modified to form urine. Terrestrial vertebrates have especially sophisticated kidneys, since they must be able to regulate closely the water content of the blood by balancing gains and expenditures. The basic excretory unit is the nephron, composed of a glomerulus in which an ultrafiltrate of the blood is formed, and a long nephric tubule in which the formative urine is selectively modified by the tubular epithelium. Water, salts, and other valuable materials pass by reabsorption to the peritubular circulation, and certain wastes pass by secretion from the circulation to the tubular urine. All mammals and some birds can produce urine more concentrated than blood by means of a countercurrent multiplier system localized in the loops of Henle, a specialization not found in other vertebrates.

Temperature has a profound effect on the rate of biochemical reactions and, consequently, on the metabolism and activity of all animals. Animals may be classified according to whether body temperature is variable (poikilothermic) or stable (homeothermic), or by the source of body heat, whether external (ectothermic) or internal (endothermic).

Ectotherms partially free themselves from thermal constraints by seeking out habitats with favorable temperatures, by behavioral thermoregulation, or by adjusting their metabolism to the prevailing temperature through biochemical alterations.

The endothermic birds and mammals differ from ectotherms in having a much higher production of metabolic heat and a much lower conductance of heat from the body. They maintain constant body temperature by balancing heat production with loss.

Small mammals in hot environments for the most part escape intense heat and reduce evaporative water loss by burrowing. Large mammals employ several strategies for dealing with direct exposure to heat, including reflective insulation, heat storage by the body, and evaporative cooling.

Endotherms in cold environments maintain body temperature by decreasing heat loss with thickened pelage or plumage, by peripheral cooling, and by increasing heat production through shivering or nonshivering thermogenesis. Small endotherms may avoid exposure to low temperatures by living under the snow.

Adaptive hypothermia is a strategy used by small mammals and birds to blunt energy demands during periods of inactivity (daily torpor) or periods of prolonged cold and minimal food availability (hibernation).

Review Questions

1. Define homeostasis. What evolutionary advantages for a species might result from the successful maintenance of internal homeostasis?
2. The problems of water balance may first have appeared when the early metazoan animals began invading estuaries and rivers. Describe the physiological challenges confronting marine invertebrates entering fresh water and, using crustaceans as an example, suggest solutions to these challenges.
3. Distinguish between the following pairs of terms: osmotic conformity and osmotic regulation; stenohaline and euryhaline; hyperosmotic and hypoosmotic.
4. Young downstream salmon migrants moving from their freshwater natal streams into the sea leave an environment nearly free of salt to enter one containing three times as much salt as their body fluids. Describe the osmotic challenges of each environment and the physiological adjustments salmon must make in moving from fresh water to the sea.
5. Most marine invertebrates are osmotic conformers. How does their body fluid differ from that of the cartilaginous sharks and rays, which are also in near-osmotic equilibrium with their environment?
6. What strategy does the kangaroo rat use that allows it to exist in the desert without drinking any water?
7. In what animals would you expect to find a salt gland? What is its function?
8. Relate the function of the contractile vacuole to the following experimental observations on the time required for protozoa to expel an amount of fluid equal in volume to the volume of the animal: 4 to 53 minutes for freshwater

protozoa and 2 to 5 hours for marine protozoa.

9. How does a protonephridium differ structurally and functionally from a true nephridium (metanephridium)? In what ways are they similar?
10. Describe the developmental stages of the kidney in amniotes. How does the developmental sequence for amniotes differ from that of amphibians and fishes?
11. In what ways does the nephridium of an earthworm parallel the human nephron in structure and function?
12. Describe how each of the following activities in the mammalian kidney contributes to the final composition of the urine: filtration, tubular reabsorption, tubular secretion.
13. Explain how the cycling of sodium chloride between the descending and ascending limbs of the loop of Henle in the mammalian kidney, and the special permeability of these tubules, produces high osmotic concentrations in interstitial fluids in the kidney medulla.
14. Explain how the antidiuretic hormone (vasopressin) controls the excretion of water in the mammalian kidney.
15. Define the following terms and comment on the limitations (if any) of each in describing the thermal relationships of animals: poikilothermy, homeothermy, ectothermy, endothermy.
16. Defend the statement: "Both ectotherms and endotherms achieve metabolic homeostasis in unstable thermal environments, but they do so by employing different physiological strategies."
17. Large mammals live successfully in deserts and in the arctic. Describe the different adaptations mammals use to maintain homeothermy in each environment.
18. Explain why it is advantageous for certain small birds and mammals to abandon homeothermy during brief or extended periods of their lives.

Selected References

Beauchamp, G. K. 1987. The human preference for excess salt. Am. Sci. **75:**27-33. *Humans consume much more salt than nutritionally required; such preference for elevated salt level in food is learned from early dietary experience.*

Cossins, A. R., and K. Bowler. 1987. Temperature biology of animals. London, Chapman & Hall. *Comprehensive treatment of both ectotherms and endotherms.*

Dantzler, W. H. 1989. Comparative physiology of the vertebrate kidney. *Comprehensive review of vertebrate renal function.*

Hardy, R. N. 1983. Homeostasis, ed. 2. The Institute of Biology's Studies in Biology no. 63, London, Edward Arnold. *Introduces the history of the homeostasis concept; temperature and osmotic regulation are treated in the final chapter.*

Rankin, J. C., and J. Davenport. 1981. Animal osmoregulation. New York, John Wiley & Sons, Inc. *Concise and selective treatment.*

Riegel, J. A. 1972. Comparative physiology of renal excretion. New York, Hafner Publishing Company. *Excellent survey of excretory systems both vertebrate and invertebrate.*

Schmidt-Nielsen, K. 1981. Countercurrent systems in animals. Sci. Am. **244:**118-128 (May). *Explains how countercurrent systems transfer heat, gases, or ions between fluids moving in opposite directions.*

Smith, H. W. 1953. From fish to philosopher. Boston, Little, Brown and Company. *Classic account of vertebrate kidney evolution.*

Storey, K. B., and J. M. Storey. 1990. Frozen and alive. Sci. Am. **263:**92-97 (Dec.). *Explains how many animals have evolved strategies for surviving complete or almost complete freezing during the winter months.*

Internal Fluids and Respiration

chapter | eight

William Harvey's Discovery

Ceaselessly, during a human life, the heart pumps blood through the arteries, capillaries, and veins: about 5 liters per minute, until by the end of a normal life the heart has contracted some 2.5 billion times and pumped 300,000 tons of blood. When the heart stops its contractions, life also ends.

The crucial importance of the heart and its contractions for human life has been known since antiquity, probably almost as long as humans have existed. However, the circuit flow of blood, the notion that the heart pumps blood into arteries through the circulation and receives it back in veins, became known only a few hundred years ago. The first correct description of blood flow by the English physician William Harvey initially received vigorous opposition when published in 1628. Centuries before, the Greek physician Galen had taught that air enters the heart from the windpipe and that blood was able to pass from one ventricle to the other through "pores" in the interventricular septum. He also believed that blood first flowed out of the heart into all vessels, then returned—a kind of ebb and flow of blood. Even though there was almost nothing correct about this concept, it was still doggedly trusted at the time of Harvey's publication. Harvey's conclusions were based on sound experimental evidence. He used a variety of animals for his experiments and chided human anatomists, saying that if only they had acquainted themselves with the anatomy of the amphibians and reptiles, they would have understood the blood's circuit. By tying ligatures on arteries, he noticed that the region between the heart and ligature swelled up. When veins were tied off, the swelling occurred beyond the ligature. When blood vessels were cut, blood flowed in arteries from the cut end nearest the heart; the reverse happened in veins. By means of such experiments, Harvey worked out a correct scheme of blood circulation, even though he could not see the capillaries that connected the arterial and venous flows.

Single-celled organisms live in direct contact with their environment. They obtain nutrients and oxygen and release wastes directly across the cell surface. These animals are so small that no special internal system of transport, beyond the normal streaming of the cytoplasm, is required. Even some multicellular forms, such as sponges, cnidarians, and flatworms, lack the internal complexity and metabolic demands that would require a circulatory system. Most of the other multicellular organisms, because of their size, activity, and complexity, require a specialized circulatory system to transport nutrients and respiratory gases to and from all tissues of the body. In addition to serving these primary transport needs, circulatory systems have acquired additional functions; hormones are moved about, finding their way to target organs where they assist the nervous system to integrate body function. Water, electrolytes, and the many other constituents of the body fluids are distributed and exchanged between different organs and tissues. An effective response to disease and injury is vastly accelerated by an efficient circulatory system. Homeothermic birds and mammals depend heavily on the blood circulation to conserve or dissipate heat as required for the maintenance of constant body temperature.

Internal Fluid Environment

The body fluid of a single-celled organism is the cellular cytoplasm, a liquid-gel substance in which the various membrane systems and organelles of the cell are suspended. In multicellular animals the body fluids are divided into two main phases, the **intracellular** and the **extracellular.** The intracellular phase (also called intracellular fluid) is the collective fluid inside all the body's cells. The extracellular phase (or fluid) is the fluid outside and surrounding the cells (Figure 8-1A). Thus the cells, the sites of the body's crucial metabolic activities, are bathed by their own aqueous environment, the extracellular fluid that buffers them from the often harsh physical and chemical changes occurring outside the body. The importance of the extracellular fluid was first emphasized by the great French physiologist Claude Bernard (Figure 8-2). In animals having closed circulatory systems (vertebrates, annelids, and a few other invertebrate groups; see p. 192) the extracellular fluid is further subdivided into **blood plasma** and **interstitial (intercellular) fluid** (Figure 8-1A). The blood plasma is contained within the blood vessels, whereas the interstitial fluid, or tissue fluid as it is sometimes called, occupies the space immediately around the cells. Nutrients and gases passing between the vascular plasma and the cells must traverse this narrow fluid separation. The interstitial fluid is constantly formed from the plasma by filtration through the capillary walls.

Composition of the Body Fluids

All these fluid spaces—plasma, interstitial, and intracellular—differ from each other in solute composition, but all have one feature in common—they are mostly water. Despite their firm appearance, animals are 70% to 90% water. Humans, for example, are approximately 70% water by weight. Of this, 50% is cell water, 15% is interstitial fluid water, and the remaining 5% is in the blood plasma. As Figure 8-1A shows, it is the plasma space that serves as the pathway of exchange between the cells of the body and the outside world. This exchange of respiratory gases, nutrients, and wastes is accomplished by specialized organs (kidney, lung, gill, alimentary canal), as well as by the skin (Figure 8-1A).

The body fluids contain many inorganic and organic substances in solution. Principal among these are the inorganic electrolytes and proteins. **Sodium, chloride,** and **bicarbonate ions** are the chief extracellular electrolytes, whereas **potassium, magnesium,** and **phosphate ions** and **proteins** are the major intracellular electrolytes (Figure 8-1B). These differences are dramatic; they are always maintained despite the continuous flow of materials into and out of the cells of the body. The two subdivisions of the extracellular fluid—plasma and interstitial fluid—have similar compositions except that the plasma has more proteins, which are mostly too large to filter through the capillary wall into the interstitial fluid.

Composition of Blood

Among the invertebrates that lack a circulatory system (such as flatworms and cnidarians) it is not possible to distinguish a true "blood." These forms possess a clear, watery tissue fluid containing some phagocytic cells, a little protein, and a mixture of salts similar to seawater. The "blood" of invertebrates with open circulatory systems, such as arthropods and most molluscs, is more complex and is often called **hemolymph** (Gr. *haimo,* blood, + L. *lympho,* water). Invertebrates with closed circulatory systems, on the other hand, maintain a clear separation between blood contained within blood vessels and tissue (interstitial) fluid surrounding the vessels.

In vertebrates, blood is a complex liquid tissue composed of plasma and formed elements, mostly corpuscles, suspended in the plasma. If we separate the red blood corpuscles and other formed elements from the fluid components by centrifugation, we find that the blood is approximately 55% plasma and 45% formed elements. The composition of mammalian blood is as follows:

Plasma

1. Water 90%
2. Dissolved solids, consisting of the plasma proteins (albumin, globulins, fibrinogen), glucose, amino acids, electrolytes, various enzymes, antibodies, hormones, metabolic wastes, and traces of many other organic and inorganic materials
3. Dissolved gases, especially oxygen, carbon dioxide, and nitrogen

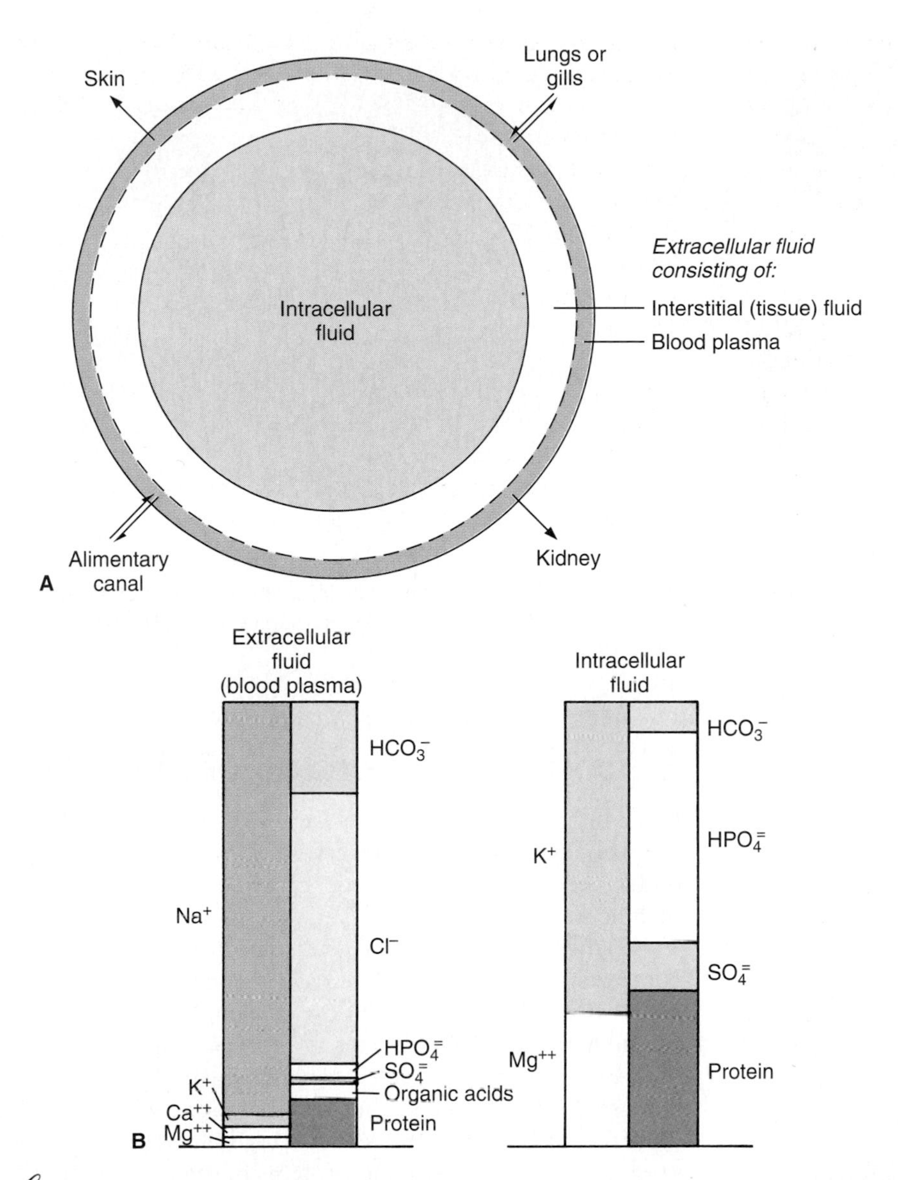

figure 8-1

Fluid compartments of the body. **A,** All body cells can be represented as belonging to a single large fluid compartment that is completely surrounded and protected by extracellular fluid *(milieu intérieur)*. This fluid is further subdivided into plasma and interstitial fluid. All exchanges with the environment occur across the plasma compartment. **B,** Electrolyte composition of extracellular and intracellular fluids. Total equivalent concentration of each major constituent is shown. Equal amounts of anions (negatively charged ions) and cations (positively charged ions) are in each fluid compartment. Note that sodium and chloride, major plasma electrolytes, are virtually absent from intracellular fluid (actually they are present in low concentration). Note the much higher concentration of protein inside the cells.

figure 8-2

French physiologist Claude Bernard (1813–1878), one of the most influential of nineteenth-century physiologists. Bernard believed in the constancy of the *milieu intérieur* ("internal environment"), which is the extracellular fluid bathing the cells. He pointed out that it is through the *milieu intérieur* that foods and wastes and gases are exchanged and through which chemical messengers are distributed. He wrote, "The living organism does not really exist in the external environment (the outside air or water) but in the liquid *milieu intérieur* . . . that bathes the tissue elements."

From J. F. Fulton & L. G. Wilson Selected Readings in the History of Physiology, 1966. *Courtesy of Charles C. Thomas, Publisher, Springfield.*

Formed elements (Figure 8-3)

1. Red blood cells (erythrocytes), containing hemoglobin for the transport of oxygen and carbon dioxide
2. White blood cells (leukocytes), serving as scavengers and as defensive cells
3. Cell fragments (platelets in mammals) or cells (thrombocytes in other vertebrates) that function in blood coagulation

The plasma proteins are a diverse group of large and small proteins that perform numerous functions. The major protein groups are (1) **albumins,** the most abundant group, constituting 60% of the total, which help to keep the plasma in osmotic equilibrium with the cells of the body; (2) **globulins,** a diverse group of high-molecular-weight proteins (35% of total) that includes immunoglobulins (p. 214) and various metal-binding proteins; and (3) **fibrinogen,** a very large protein that functions in blood coagulation. Blood **serum** is plasma minus the proteins involved in clot formation (see below).

Red blood cells, or **erythrocytes,** are present in enormous numbers in the blood, approximately 5.4 billion per milliliter of blood in an adult man and 4.8 billion in adult women. In mammals and birds, they are formed continuously from large nucleated **erythroblasts** in the red bone marrow

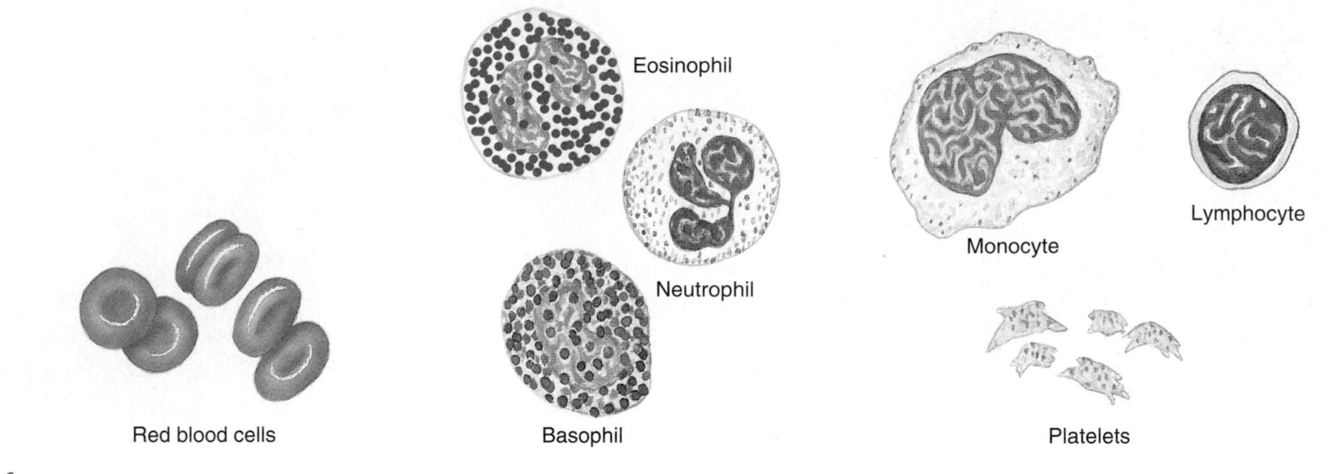

figure 8-3

Formed elements of human blood. Hemoglobin-containing red blood cells of humans and other mammals lack nuclei, but those of all other vertebrates have nuclei. Various leukocytes provide a wandering system of protection for the body. Platelets participate in the body's clotting mechanism.

(in other vertebrates the kidneys and spleen are the principal sites of production of red blood cells). During erythrocyte formation hemoglobin is synthesized and the cells divide several times. In mammals the nucleus shrinks during development to a small remnant and eventually disappears altogether. Many other characteristics of a typical cell also are lost: ribosomes, mitochondria, and most enzyme systems. What is left is a biconcave disc consisting of a baglike membrane packed with about 280 million molecules of the blood-transporting pigment **hemoglobin.** Approximately 33% of the erythrocyte by weight is hemoglobin. The biconcave shape (Figure 8-3) is a mammalian innovation that provides a larger surface for gas diffusion than would a flat or spherical shape. All other vertebrates have nucleated erythrocytes that are usually ellipsoidal in shape.

The erythrocyte enters the circulation for an average life span of approximately four months. During this time it may journey 11,000 km, squeezing repeatedly through the capillaries, which are sometimes so narrow that the erythrocyte must bend to get through. At last it fragments and is quickly engulfed by large scavenger cells called **macrophages** located in the liver, bone marrow, and spleen. The iron from the hemoglobin is salvaged to be used again; the rest of the heme is converted to **bilirubin,** a bile pigment. It is estimated that the human body produces 10 million erythrocytes and destroys another 10 million every second.

The white blood cells, or **leukocytes,** form a wandering system of protection for the body. In adults they number only approximately 7.5 million per milliliter of blood, a ratio of 1 white cell to 700 red cells. There are several kinds of white blood cells: granulocytes (subdivided into **neutrophils, basophils,** and **eosinophils**), and nongranulocytes, the **lymphocytes** and **monocytes** (Figure 8-3). We will discuss the role of the leukocytes in the body's defense mechanisms in Chapter 9.

Hemostasis: Prevention of Blood Loss

It is essential that animals have ways of preventing the rapid loss of body fluids after an injury. Since blood is flowing and is under considerable hydrostatic pressure, it is especially vulnerable to hemorrhagic loss.

When a vessel is damaged, smooth muscle in the wall contracts, which causes the vessel lumen to narrow, sometimes so strongly that blood flow is completely stopped. This simple but highly effective means of preventing hemorrhage is used by invertebrates and vertebrates alike. Beyond this first defense against blood loss, all vertebrates, as well as some of the larger, active invertebrates with high blood pressures, have special cellular elements and proteins in the blood that are capable of forming plugs, or clots, at the injury site.

In vertebrates **blood coagulation** is the dominant hemostatic defense. Blood clots form as a tangled network of fibers from one of the plasma proteins, **fibrinogen.** The transformation of fibrinogen into a **fibrin** meshwork (Figure 8-4) that entangles blood cells to form a gel-like clot is catalyzed by the enzyme thrombin. Thrombin is normally present in the blood in an inactive form called **prothrombin,** which must be activated for coagulation to occur.

In this process, the blood platelets (Figure 8-3) play a vital role. Platelets are formed in the red bone marrow from certain large cells that regularly pinch off bits of their cytoplasm; thus they are fragments of cells. There are 150,000 to 300,000 platelets per cubic millimeter of blood. When the normally smooth inner surface of a blood vessel is disrupted, either by a break or by deposits of a cholesterol-lipid material, the platelets rapidly adhere to the surface and release **thromboplastin** and other clotting factors. These factors, along with factors released from damaged tissue and with calcium ions, initiate the conversion of prothrombin to the active thrombin (Figure 8-5).

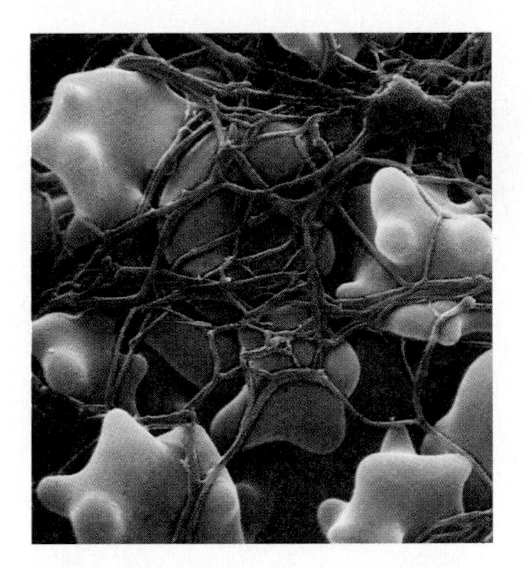

figure 8-4

Human red blood cells entrapped in a fibrin clot. Clotting is initiated after tissue damage by the disintegration of platelets in the blood, resulting in a complex series of intravascular reactions that end with the conversion of a plasma protein, fibrinogen, into long, tough, insoluble polymers of fibrin. Fibrin and entangled erythrocytes form the blood clot, which arrests bleeding. An aggregation of platelets probably underlies the raised mass of fibrin in the center.

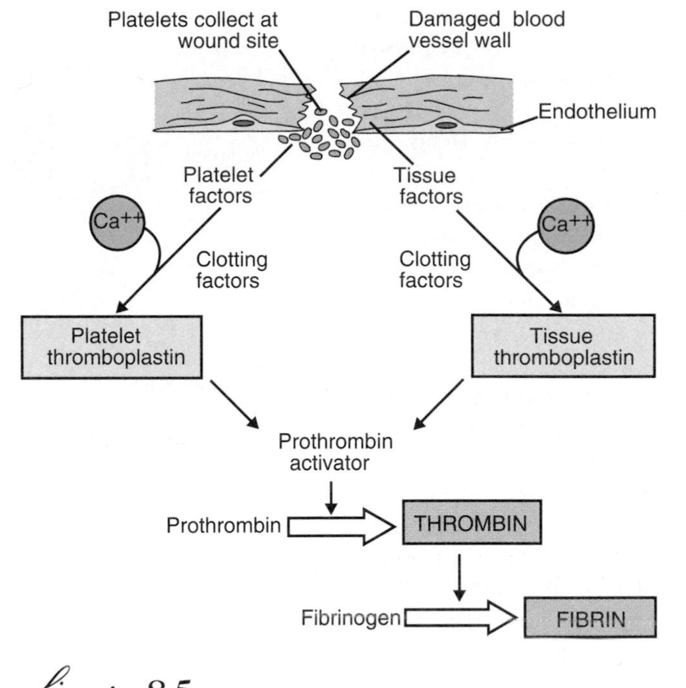

figure 8-5

Stages in the formation of fibrin.

The catalytic sequence in this scheme is unexpectedly complex, involving a series of plasma protein factors, each normally inactive until activated by a previous factor in the sequence. The sequence behaves like a "cascade" with each reactant in the sequence leading to a large increase in the amount of the next reactant. At least 13 different plasma coagulation factors have been recognized. A deficiency of a single factor can delay or prevent the clotting process. Why has such a complex clotting mechanism evolved? Probably it is necessary to provide a fail-safe system capable of responding to any kind of internal or external hemorrhage that might occur and yet a system that cannot be activated into forming dangerous intravascular clots in the absence of injury.

Several kinds of clotting abnormalities in humans are known. One of these, hemophilia, is a condition characterized by the failure of the blood to clot, so that even insignificant wounds can cause continuous severe bleeding. It is due to a rare mutation (the condition occurs in about 1 in 10,000 males) on the X sex chromosome, resulting in an inherited lack of one of the platelet factors in males and in females homozygous for the trait. Called the "disease of kings," it once ran through several interrelated royal families of Europe, apparently having originated from a mutation in one of Queen Victoria's parents.

Hemophilia is one of the best known cases of sex-linked inheritance in humans. Actually two different loci on the X chromosome are involved. Classical hemophilia (hemophilia A) accounts for about 80% of persons with the condition, and the remainder are due to Christmas disease (hemophilia B). An allele at each locus results in the deficiency of a different platelet factor.

Circulation

We pointed out in the opening to this chapter that most animals have evolved mechanisms, in addition to simple diffusion, for transporting materials among various regions of the body. For sponges and radiates the water in which they live provides the medium for transport. Water, propelled by ciliary, flagellar, or body movements, passes through channels or compartments to facilitate the movement of food, respiratory gases, and wastes. True circulatory systems—that is, a system of vessels through which blood moves—become essential with animals so large or so active that diffusion alone cannot supply their oxygen needs. An animal's shape obviously is important. The flattened and leaflike acoelomate flatworms, even though many are relatively large animals, have no need for a circulatory system because the distance of any body part from the surface is short; respiratory gases and metabolic wastes transfer by simple diffusion.

A circulatory system having a full complement of components—propulsive organ, arterial distribution system, capillaries, and venous reservoir and return system—is fully recognizable in the annelid worms. In the earthworm (Figure 8-6) there are two main vessels, a dorsal vessel carrying the blood toward the head, and a ventral vessel that flows posteriorly, delivering blood to all the body by way of segmental vessels and a dense capillary network. The dorsal vessel drives the blood forward by peristalsis (see p. 229) and thus serves as a heart. Five aortic arches that on each side connect the dorsal and ventral vessels are also contractile and serve as accessory hearts to maintain a steady flow of blood into the ventral vessel. Many of the smaller segmental vessels that deliver blood to tissue capillaries are actively contractile as well. We see then that there is no localized pump pushing the blood through a system of passive tubes; instead the power of contraction is widely distributed throughout the vascular system.

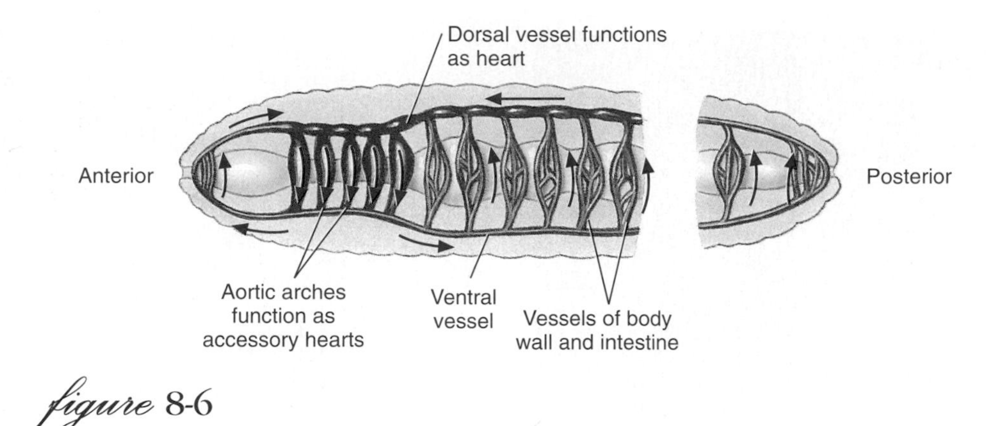

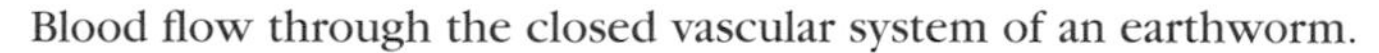

figure 8-6

Blood flow through the closed vascular system of an earthworm.

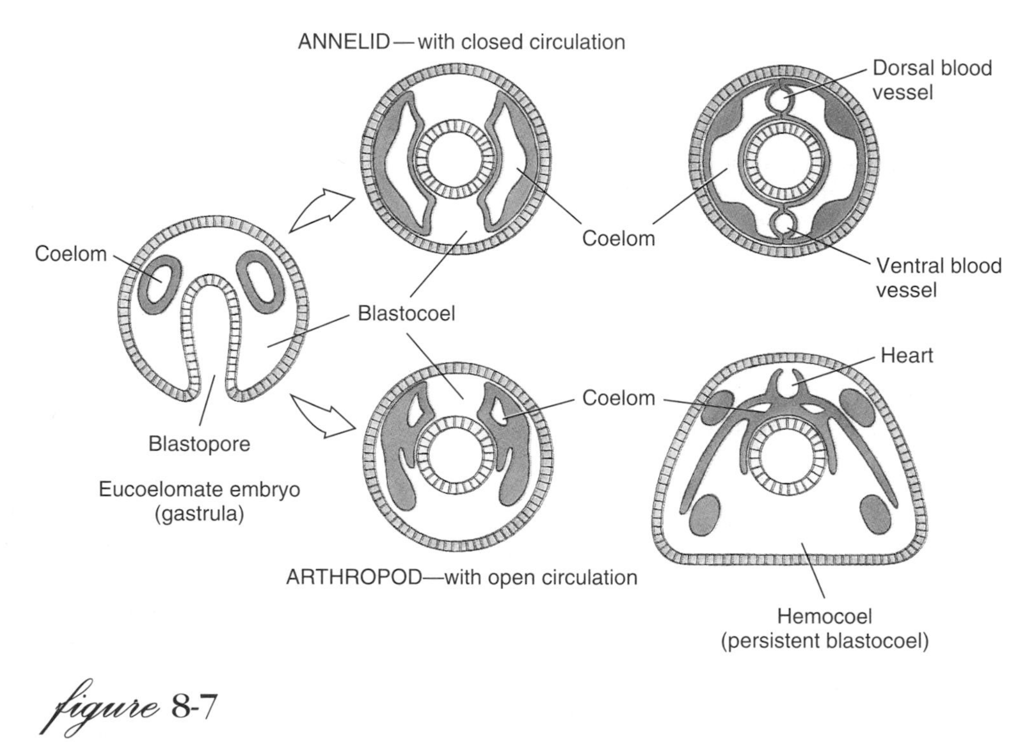

figure 8-7

Diagrams showing how open and closed circulatory systems develop. The principal body cavity of arthropods is the persistent blastocoel which becomes the hemocoel; the true coelom remains mostly undeveloped.

Open and Closed Circulations

The system just described is a **closed circulation** because the circulating medium, the **blood,** is confined to vessels throughout its journey through the vascular system. Many invertebrates have an **open circulation** in which there are no small blood vessels or capillaries connecting arteries with the veins. In insects and other arthropods, in most molluscs, and in many smaller invertebrate groups blood sinuses, collectively called the **hemocoel,** replace the capillary beds found in animals with closed systems. During development of the body cavity in these groups, the blastocoel is not completely obliterated by the expanding mesoderm. This space becomes the hemocoel, which is nothing more than the primary body cavity (persistent blastocoel) through which the blood (also called **hemolymph**) freely circulates (bottom diagrams in Figure 8-7). Since there is no separation of the extracellular fluid into blood plasma and lymph (as there is in a closed circulation, p. 198) the blood volume is large and may constitute 20% to 40% of body volume. By contrast, the blood volume in animals with closed circulations (vertebrates, for example) is only about 5% to 10% of body volume.

In arthropods, the heart and all the viscera lie in the hemocoel, bathed by the blood (Figure 8-8). Blood enters the heart through valved openings, the ostia, and the heart's contractions, which resemble a forward-moving peristaltic wave, propel the blood into a limited arterial system. The blood is distributed to the head and other organs, then escapes into the hemocoel. It is routed through the body and appendages by a system of baffles and longitudinal membranes (septa) before returning to the heart. Because the blood pressure is very low in open systems, seldom exceeding 4 to 10 mm Hg, many arthropods have auxiliary hearts or contractile vessels to boost blood flow (Figure 8-8).

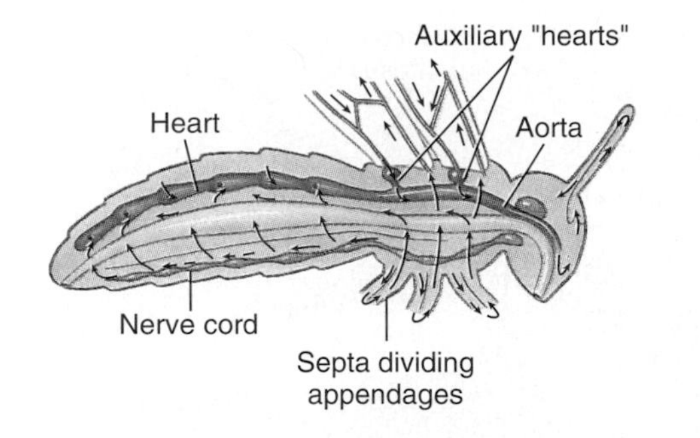

figure 8-8

Circulatory system of an insect. Although the circulatory system is open, the blood is directed through the appendages in channels formed by longitudinal septa. Arrows indicate the course of the circulation.

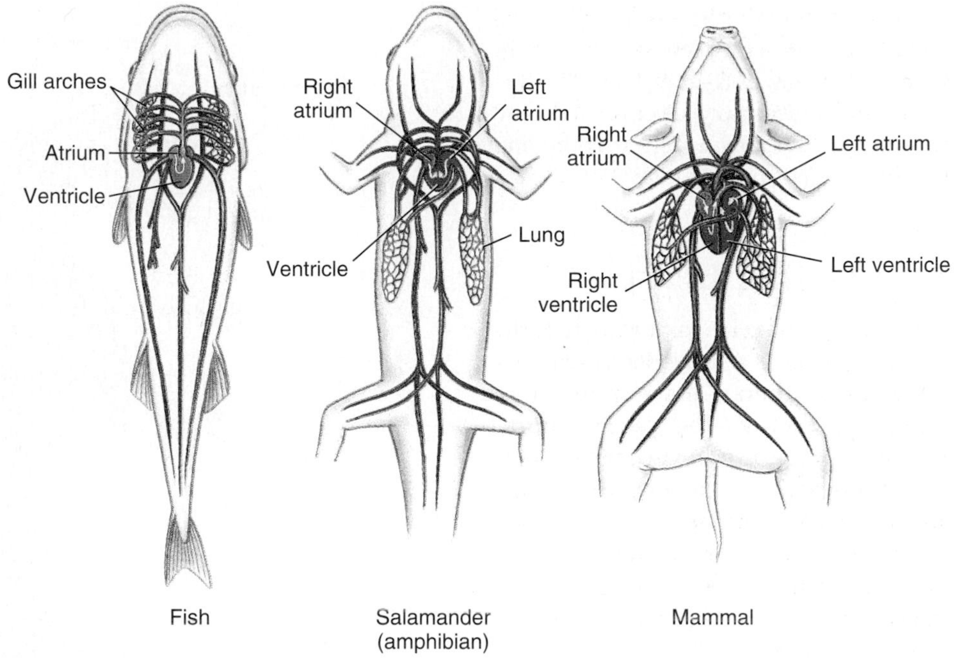

Gill arches
Atrium
Ventricle
Right atrium
Left atrium
Ventricle
Lung
Right atrium
Left atrium
Right ventricle
Left ventricle
Fish
Salamander (amphibian)
Mammal

figure 8-9

Circulatory systems of fish, amphibian, and mammal, showing the evolution of separate systemic and pulmonary circuits in lung-breathing vertebrates.

During embryonic development of animals with closed circulatory systems (most annelids, cephalopod molluscs, and all vertebrates), the coelom increases in size to obliterate the blastocoel and forms a secondary body cavity (top diagrams in Figure 8-7). A system of continuously connected blood vessels develops within the mesoderm. All closed systems have certain features in common. A **heart** pumps the blood into **arteries** that branch and narrow into **arterioles** and then into a vast system of **capillaries.** Blood leaving the capillaries enters **venules** and then **veins** that return the blood to the heart. The capillary walls are thin, permitting rapid rates of transfer of materials between blood and tissues. Closed systems are more suitable for large and active animals because the blood can be moved rapidly to the tissues needing it. In addition, flow to various organs can be readjusted to meet changing needs by varying the diameters of the blood vessels.

Because blood pressures are much higher in closed than in open systems, fluid is constantly filtered across capillary walls into the surrounding tissue spaces. Most of this fluid is drawn back into the capillaries by osmosis (see p. 198). The remainder is recovered by the **lymphatic system,** which has evolved in parallel with the high-pressure system of vertebrates.

Plan of the Vertebrate Circulatory Systems

In vertebrates the principal differences in the blood vascular system involve the gradual separation of the heart into two separate pumps as the vertebrates progressed from aquatic life with gill breathing to fully terrestrial life with lung breathing. These changes are shown in Figure 8-9, which compares the circulation of fish, amphibians, and mammals.

The fish heart contains two main chambers in series, an **atrium** and a **ventricle.** The atrium is preceded by an enlarged chamber, the **sinus venosus,** which collects blood from the venous system to assure a smooth delivery of blood to the heart. Blood makes a single circuit through the fish's vascular system; it is pumped from the heart to the gills, where it is oxygenated, then flows into the dorsal aorta to be distributed to the body organs, and finally returns by veins to the heart. In this circuit the heart must provide sufficient pressure to push the blood through two sequential capillary systems, first the capillaries of the gills, and then the capillaries of the remainder of the body. The principal disadvantage of the single-circuit system is that the gill capillaries offer so much resistance to blood flow that blood pressures to the body tissues are greatly reduced.

With the evolution of lung breathing and the elimination of the gills between the heart and the aorta, vertebrates developed a high pressure **double circulation:** a **systemic circuit** that provides oxygenated blood to the capillary beds of the body organs; and a **pulmonary circuit** that serves the lungs. We see this major evolutionary change in its simplest form in lungfishes and amphibians. In modern amphibians (frogs, toads, salamanders) the atrium is completely separated by a partition into two atria (Figure 8-10). The right atrium receives venous blood from the body while the left atrium receives oxygenated blood from the lungs. The ventricle is undivided, but venous and arterial blood remain mostly separate by the arrangement of the vessels leaving the heart. Separation of the ventricles is nearly complete in some reptiles (crocodilians) and is completely separate in birds and mammals (Figure 8-11). The systemic and pulmonary circuits are now separate circulations, each served by one half of a dual heart (Figure 8-9).

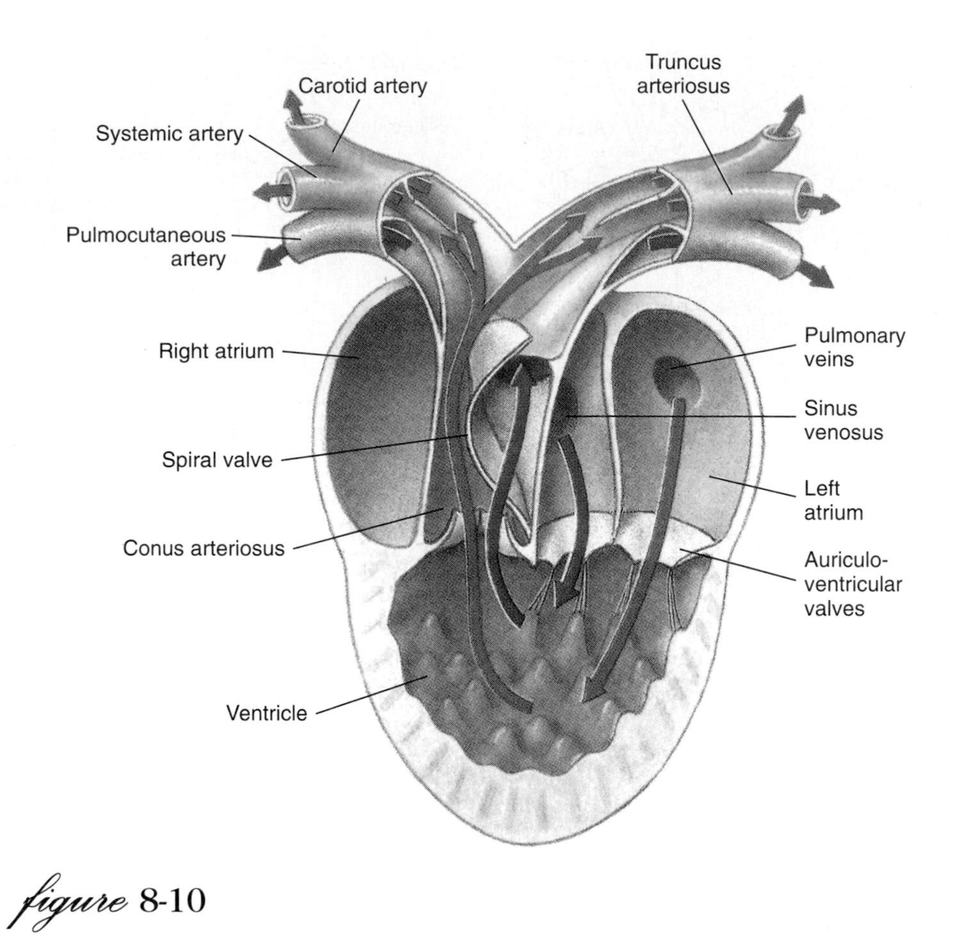

figure 8-10

Route of blood through a frog heart. Atria are completely separated, and the spiral valve helps to route blood to the lungs and systemic circulation.

Mammalian Heart

The four-chambered mammalian heart (Figure 8-11) is a muscular organ located in the thorax and covered by a tough, fibrous sac, the **pericardium.** Blood returning from the lungs collects in the **left atrium,** passes into the **left ventricle,** and is pumped into the body (systemic) circulation. Blood returning from the body flows into the **right atrium,** passes into the **right ventricle,** which pumps it into the lungs. Backflow of blood is prevented by two sets of valves that open and close passively in response to pressure differences between the heart chambers. The **bicuspid** and **tricuspid** valves separate the cavities of the atrium and ventricle in each half of the heart. Where the great arteries, the **pulmonary** from the right ventricle and the **aorta** from the left ventricle, leave the heart, **semilunar valves** prevent backflow into the ventricles.

The contraction of the heart is called **systole** (sis′to-lee), and the relaxation, **diastole** (dy-as′to-lee) (Figure 8-12). The rate of the heartbeat depends on age, sex, and especially exercise. Exercise may increase the **cardiac output** (volume of blood forced from either ventricle each minute) more than fivefold. Both the heart **rate** and the **stroke volume** increase. Heart rates among vertebrates vary with the general level of metabolism and the body size. The ectothermic codfish has a heart rate of approximately 30 beats per minute; an endothermic rabbit of about the same weight has a rate of 200 beats per minute. Small animals have higher heart rates than do large animals. The heart rate in an elephant is 25 beats per minute, in a human 70 per minute, in a cat 125 per minute, in a mouse 400 per minute, and in the tiny 4 g shrew, the smallest mammal, the heart rate approaches a prodigious 800 beats per minute. We must marvel that the shrew's heart can sustain such a frantic pace throughout this animal's life, brief as it is.

Excitation and Control of the Heart

The vertebrate heart is a muscular pump composed of **cardiac muscle.** Cardiac muscle resembles skeletal muscle—both are types of striated muscle—but the cells are branched and joined end-to-end by junctional complexes to form a complex branching network (see Figure 6-7, p. 146). Unlike skeletal muscle, vertebrate cardiac muscle does not depend on nerve activity to initiate a contraction. Instead, regular contractions are established by specialized cardiac muscle cells, called **pacemaker cells.** In the tetrapod heart the pacemaker is in the **sinus node,** a remnant of the sinus venosus in the fishlike ancestor. Electrical activity initiated in the pacemaker spreads over the muscle of the two atria and then, after a slight delay, to the muscle of the ventricles. At this point the electrical activity is conducted rapidly through the **atrioventricular bundle** to the apex of the ventricle and then continues through specialized fibers (**Purkinje fibers**) up the walls of the ventricles (Figure 8-13). This arrangement allows the contraction to begin at the apex or "tip" of the ventricles and spread upward to squeeze out the blood in the most efficient way; it also ensures that both ventricles contract simultaneously. Structural specializations in the Purkinje fibers, such as well-developed intercalated discs (see Figure 6-7, p. 146) and numerous gap junctions facilitate rapid conduction through these fibers.

The **control (cardiac) center** in the brain is in the medulla and sends out two sets of nerves. Impulses sent along

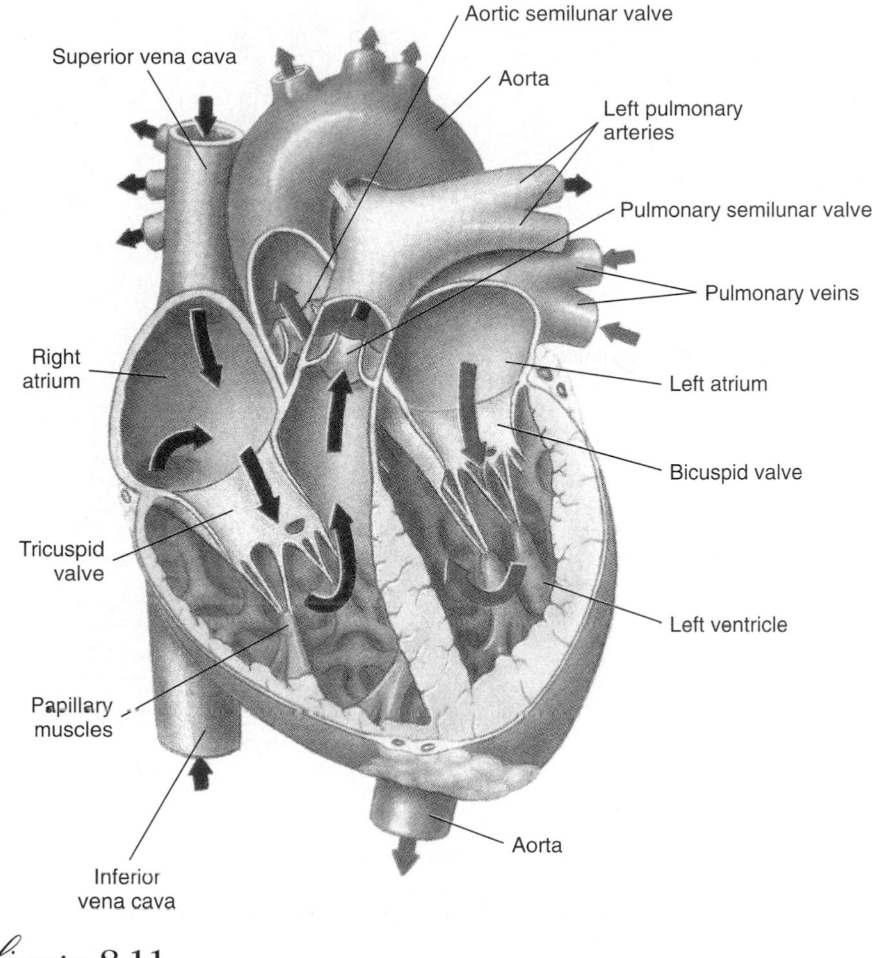

figure 8-11

Human heart. Deoxygenated blood enters the right side of the heart and is pumped to the lungs. Oxygenated blood returning from the lungs enters the left side of the heart and is pumped to the body. The left ventricular wall is thicker than that of the right ventricle, which needs less muscular force to pump blood into the nearby lungs.

one set, the **vagus** nerves, apply a brake action to the heart rate, and impulses sent along the other set, the **accelerator** nerves, speed it up. Both sets of nerves terminate in the sinus node, thus guiding the activity of the pacemaker.

The cardiac center in turn receives sensory information about a variety of stimuli. Pressure receptors (sensitive to blood pressure) and chemical receptors (sensitive to carbon dioxide and pH) are located at strategic points in the vascular system. The cardiac center uses this information to increase or reduce the heart rate and cardiac output in response to activity or changes in body position. Feedback mechanisms thus control the heart and keep its activity constantly attuned to needs of the body.

Because the heartbeat is initiated in specialized muscle cells, vertebrate hearts, together with the hearts of molluscs and several other invertebrates, are called **myogenic** ("muscle origin") hearts. Although the nervous system does alter pacemaker activity to slow down or speed up the heart rate, the myogenic heart will beat spontaneously and involuntarily even if completely removed from the body. An isolated turtle or frog heart will beat for hours if placed in a balanced salt solution. Some invertebrates, for example decapod crustaceans, have **neurogenic** ("nerve origin") hearts. In these, a cardiac ganglion located on the heart serves as pacemaker. If this ganglion is separated from the heart, the heart stops beating, even though the ganglion itself remains rhythmically active.

Coronary Circulation

It is no surprise that an organ as active as the heart needs a generous blood supply of its own. The heart muscle of the frog and other amphibians is so thoroughly channeled with spaces between the muscle fibers that the heart's own pumping action squeezes through sufficient oxygenated blood. In birds and mammals, however, the thickness of the heart muscle and its high rate of metabolism require that the heart have its own vascular supply, the **coronary circulation.** The coronary arteries break up into an extensive capillary network surrounding the muscle fibers and provide them with oxygen and nutrients. Heart muscle has an extremely high oxygen demand. Even at rest it removes 70% of the oxygen from the blood, in contrast to most other body tissues, which remove only approximately 25%. Therefore, an increase in the work of the heart must be met by a massive increase in coronary blood flow—up to nine times the resting level during strenuous exercise. Any reduction in coronary circulation due to partial or complete blockage (coronary artery disease) may cause a heart attack (myocardial infarction) in which heart cells die from lack of oxygen.

Arteries

All vessels leaving the heart are called arteries whether they carry oxygenated blood (aorta) or deoxygenated blood (pulmonary artery). To withstand high, pounding pressures, arteries are invested with layers of both elastic and tough inelastic connective tissue fibers (Figure 8-14). The elasticity of the arteries allows them to yield to the surge of blood leaving the heart during systole and then to squeeze down on the fluid column during diastole. This action keeps the blood pressure

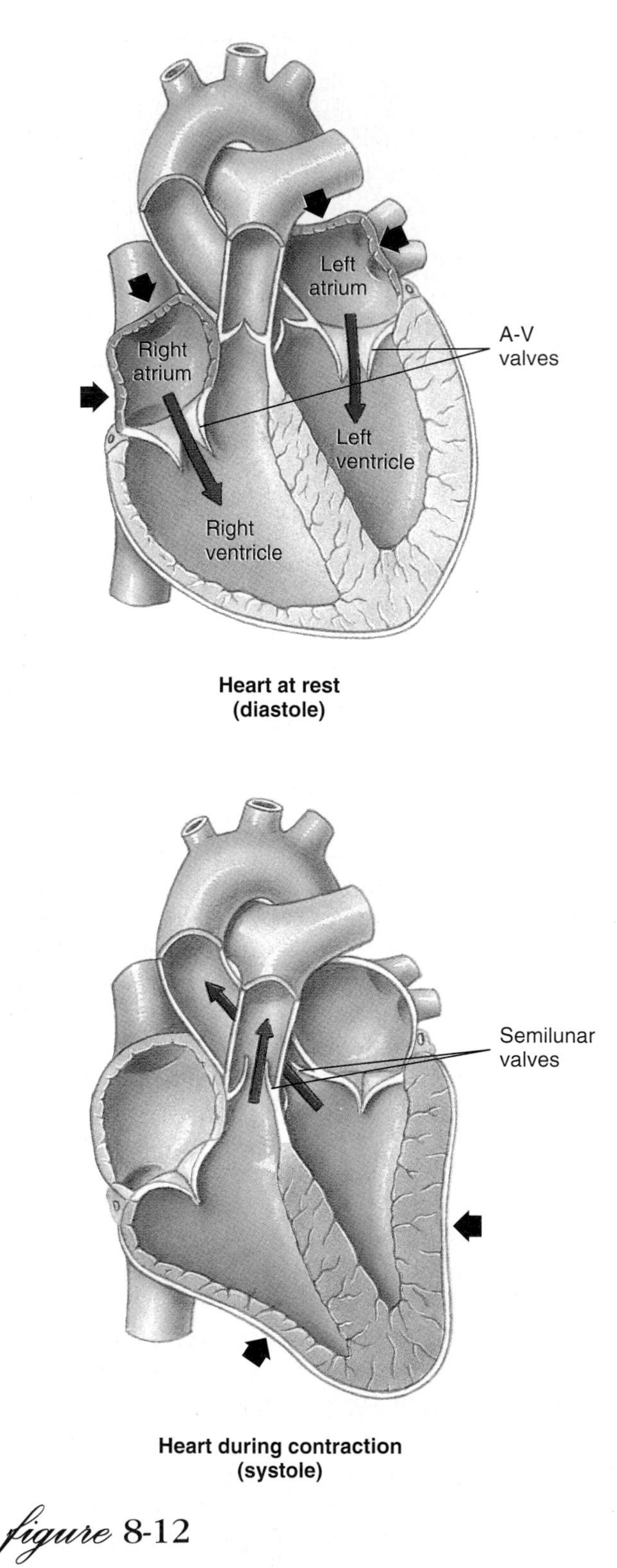

figure 8-12

Human heart in systole and diastole.

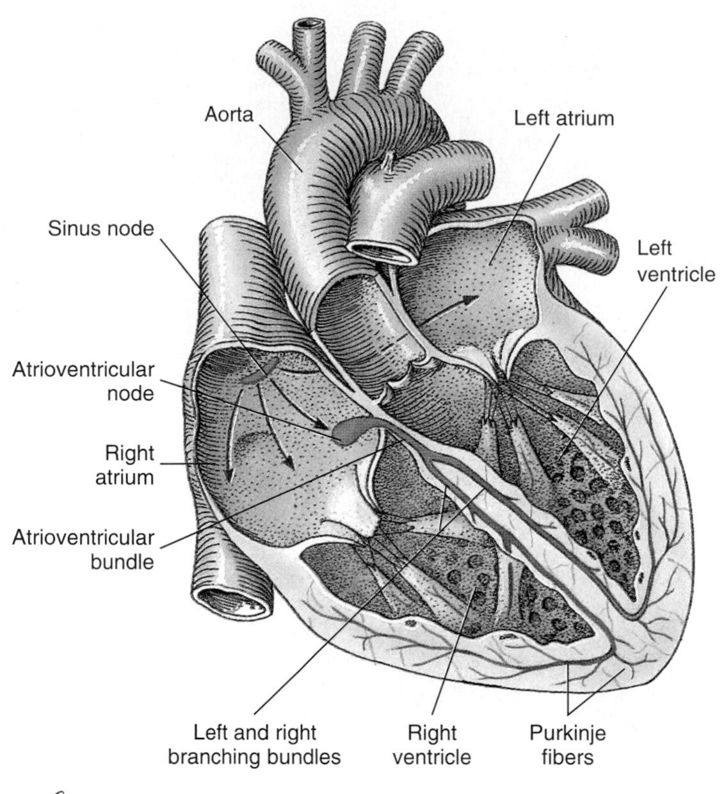

figure 8-13

Neuromuscular mechanisms controlling the heartbeat. Arrows indicate the spread of excitation from the sinus node, across the atria, to the atrioventricular node. The wave of excitation is then conducted very rapidly to the ventricular muscle over the specialized conducting bundles and Purkinje fiber system.

Thickening and loss of elasticity in the arteries is known as arteriosclerosis. *When the arteriosclerosis is caused by fatty deposits of cholesterol in the artery walls, the condition is* atherosclerosis. *Such irregularities in the walls of blood vessels often cause blood to clot around them, forming a* thrombus. *When a bit of the thrombus breaks off and is carried by the blood to lodge elsewhere, it is an* embolus. *If the embolus blocks one of the coronary arteries, the person has a heart attack (a "coronary"). The portion of the heart muscle served by the branch of the coronary artery that is blocked is starved for oxygen. It may be replaced by scar tissue if the person survives.*

relatively even. Thus the normal arterial pressure in humans varies only between a high of 120 mm Hg (systole) and a low of 80 mm Hg (diastole) (usually expressed as 120/80 or 120 over 80), rather than dropping to zero during diastole as we might expect in a fluid system with an intermittent pump.

As the arteries branch and narrow into **arterioles,** the walls become mostly smooth muscle. Contraction of this muscle narrows the arterioles and reduces the flow of blood. The arterioles thus control the blood flow to body organs, diverting it to where it is most needed. The blood must be pumped with a hydrostatic pressure sufficient to overcome the resis-

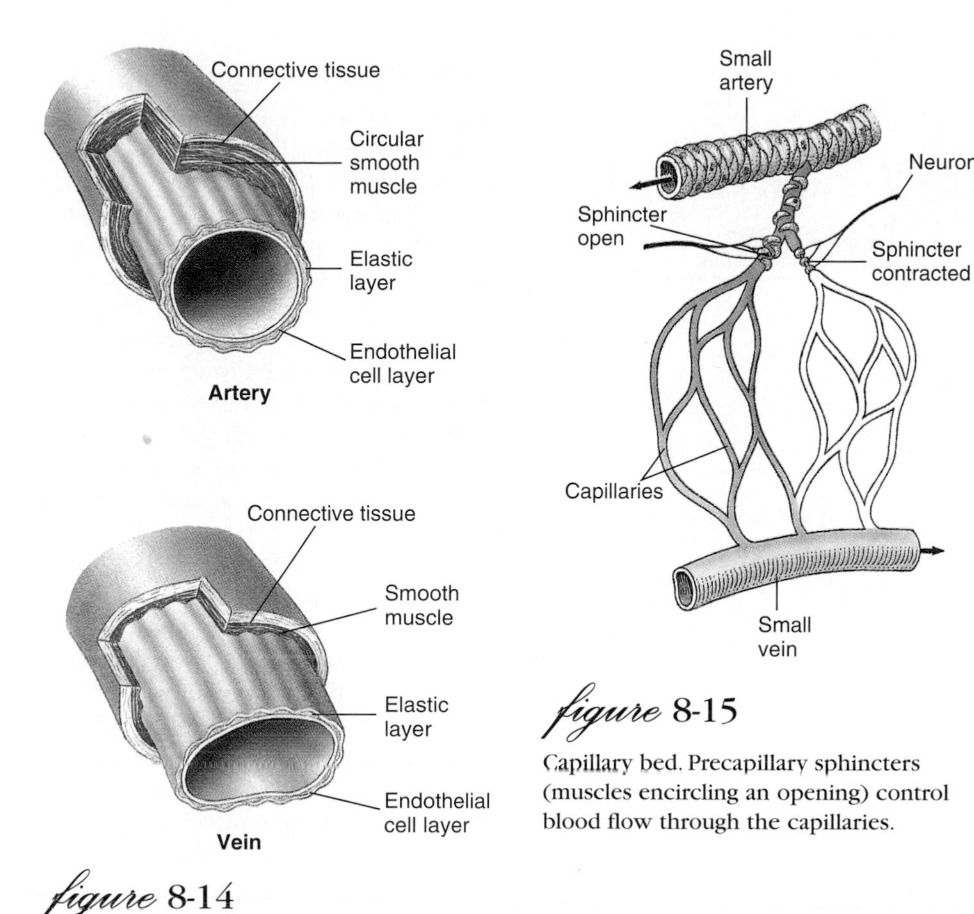

figure 8-14

Artery and vein, showing layers. Note greater thickness of the muscularis layer (tunica media) in the artery.

figure 8-15

Capillary bed. Precapillary sphincters (muscles encircling an opening) control blood flow through the capillaries.

tance of the narrow passages through which the blood must flow. Consequently, large animals tend to have higher blood pressure than do small animals.

Blood pressure was first measured in 1733 by Stephen Hales, an English clergyman with unusual inventiveness and curiosity. He tied his mare, which was "to have been killed as unfit for service," on her back and exposed the femoral artery. This he cannulated with a brass tube, connecting it to a tall glass tube with the windpipe of a goose. The use of the windpipe was both imaginative and practical; it gave the apparatus flexibility "to avoid inconveniences that might arise if the mare struggled." The blood rose 8 feet in the glass tube and bobbed up and down with the systolic and diastolic beats of the heart. The weight of the 8-foot column of blood was equal to the blood pressure. We now express blood pressure as the height of a column of mercury, which is 13.6 times heavier than water. Hales' figures, expressed in millimeters of mercury, indicate that he measured a blood pressure of 180 to 200 mm Hg, about normal for a horse.

Today, we measure blood pressure in humans most commonly and easily with an instrument called a **sphygmomanometer.** We inflate a cuff on the upper arm with air to a pressure sufficient to close the arteries in the arm. Holding a stethoscope over the brachial artery (in the crook of the elbow) and slowly releasing the air from the cuff, we can hear the first spurts of blood through the artery as it opens slightly. This is equivalent to the systolic pressure. As the pressure in the cuff decreases, the sound finally disappears as the blood runs smoothly through the artery. The pressure at which the sound disappears is the diastolic pressure.

Capillaries

The Italian Marcello Malpighi was the first to describe the capillaries in 1661, thus confirming the existence of the minute links between the arterial and venous systems that Harvey knew must exist but could not see. Malpighi studied the capillaries of the living frog's lung, which is still one of the simplest and most vivid preparations for demonstrating capillary blood flow.

The capillaries are present in enormous numbers, forming extensive networks in nearly all tissues (Figure 8-15). In muscle there are more than 2000 per square millimeter (1,250,000 per square inch), but not all are open at once. Indeed, perhaps less than 1% are open in resting skeletal muscle. But when the muscle is active, all capillaries may open to bring oxygen and nutrients to the working muscle fibers and to carry away metabolic wastes.

Capillaries are extremely narrow, averaging about 8 μm in diameter in mammals, which is only slightly wider than the red blood cells that must pass through them. Their walls are formed by a single layer of thin **endothelial** cells, held together by a delicate basement membrane and connective tissue fibers.

Capillary Exchange

Capillaries are quite permeable to small ions, nutrients, and water. Blood pressure within the capillary tends to force fluids out through the capillary walls and into the surrounding interstitial space (p. 140). Because larger molecules such as the plasma proteins cannot pass the capillary wall, an almost protein-free filtrate is forced out. This fluid movement is important in irrigating the interstitial space, in providing tissue cells with oxygen, glucose, amino acids, and other nutrients, and in carrying away metabolic wastes. For capillary exchange to be effective, fluids that leave the capillaries must at some point reenter the circulation. If they did not, fluid would quickly accumulate in tissue spaces, causing edema (p. 218). The delicate balance of fluid exchange across the capillary wall results from the two opposing forces of hydrostatic (blood) pressure and osmotic pressure (Figure 8-16).

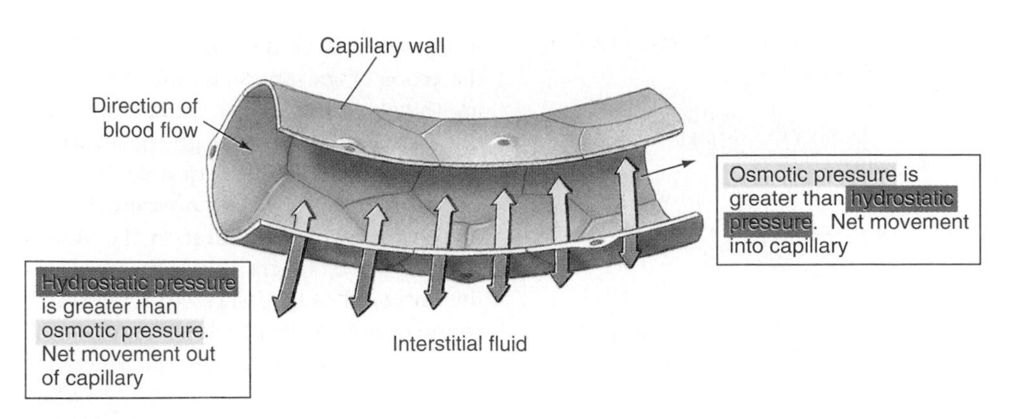

figure 8-16

Fluid movement across the wall of a capillary. At the arterial end of the capillary, hydrostatic (blood) pressure exceeds colloid osmotic pressure contributed by plasma proteins, and a plasma filtrate is forced out. At the venous end, colloid osmotic pressure exceeds the hydrostatic pressure, and fluid is drawn back in. In this way, plasma nutrients are carried out into the interstitial space where they can enter cells, and metabolic end products from the cells are drawn back into the plasma and carried away.

The capillary fluid shift is a bit more complicated than we have presented it because there is a small hydrostatic pressure in the interstitial fluid, and a small amount of protein does leak through the capillary wall. The protein tends to accumulate at the venule end of the capillary, building up a small osmotic pressure there. Although actual calculation of the pressure differences must take into account the interstitial fluid hydrostatic and osmotic pressures, the principle of the shift is as we have described it.

In the capillary, the blood pressure that pushes water molecules and solutes across the capillary wall is greatest at the start of the capillary and declines along its length as the blood pressure falls (Figure 8-16). Opposing the blood hydrostatic pressure is an osmotic pressure created by the proteins that cannot pass the capillary wall. This **colloid osmotic pressure,** which amounts to about 25 mm Hg in mammalian plasma, tends to draw water back into the capillary from the tissue fluid. The result of these two opposing forces is that water and solutes tend to be filtered out of the arteriolar end of the capillary where the hydrostatic pressure exceeds the osmotic pressure, and to be drawn in again at the venous end where the osmotic pressure exceeds the hydrostatic pressure.

The amount of fluid filtered across the capillary wall fluctuates greatly among different capillaries. Usually outflow exceeds inflow, and the excess fluid, called **lymph,** remains in the interstitial spaces between the tissue cells. This excess is picked up and removed by the **lymph capillaries** of the lymphatic system and eventually returned to the circulatory system via larger lymph vessels (see discussion of the lymphatic system, following).

Veins

The venules and veins into which capillary blood drains for its return journey to the heart are thinner walled, less elastic, and of considerably larger diameter than their corresponding arteries and arterioles (Figure 8-14). Blood pressure in the venous system is low, from approximately 10 mm Hg, where capillaries drain into venules, to approximately zero in the right atrium. Because pressure is so low, the venous return gets assistance from valves in the veins, body muscles surrounding the veins, and the rhythmical action of the lungs. If it were not for these mechanisms, the blood might pool in the lower extremities of a standing animal—a very real problem for people who must stand for long periods. Veins that lift blood from the extremities to the heart contain valves that divide the long column of blood into segments. When skeletal muscles contract, as in even slight activity, the veins are squeezed, and blood within them moves toward the heart because the valves within the veins keep the blood from slipping back. The well-known risk of fainting while standing at stiff attention in hot weather usually can be prevented by deliberately pumping the leg muscles. Negative pressure in the thorax created by the inspiratory movement of the lungs also speeds venous return by sucking blood up the large vena cava into the heart.

Lymphatic System

The lymphatic system of vertebrates is an extensive network of thin-walled vessels that arise as blind-ended lymph capillaries in most tissues of the body. These unite to form a treelike structure of increasingly larger lymph vessels, which finally drain into veins in the lower neck (Figure 8-17). A principal function of the lymphatic system is to return to the blood the excess fluid filtered across capillary walls into the interstitial space. This excess fluid is called tissue fluid (lymph), which is similar to plasma but has a much lower concentration of protein. Large molecules, especially fats absorbed from the gut, also reach the circulatory system by way of the lymphatic system. The rate of lymph flow is very low, a minute fraction of the blood flow.

The lymphatic system also plays a central role in the body's defenses. Located at intervals along the lymph vessels are **lymph nodes** (Figure 8-17) that have several defense-related functions. They are effective filters that remove foreign particles, especially bacteria, that otherwise might enter the general circulation. They are also centers (together with the bone marrow and thymus gland) for the production,

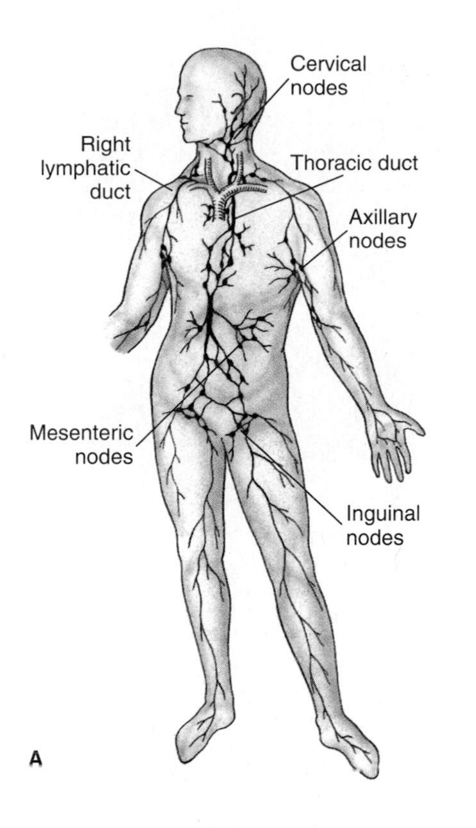

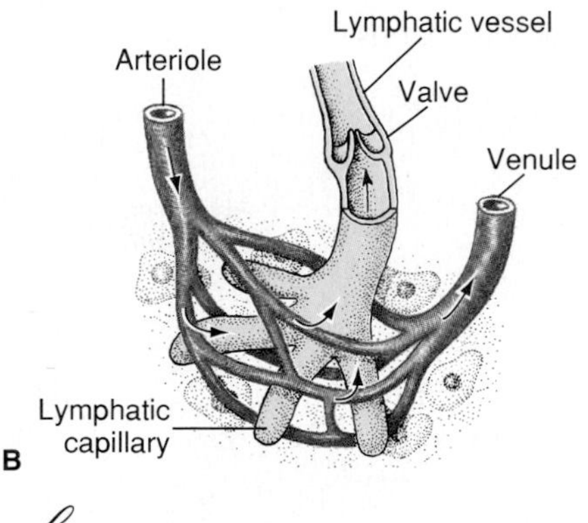

figure 8-17

Human lymphatic system, showing major vessels, **A,** and a detail of the blood and lymphatic capillaries, **B.**

maintenance, and distribution of lymphocytes that produce antibodies—essential components of the body's defense mechanisms (p. 214).

Respiration

Energy bound up in food is released by oxidative processes, usually with molecular oxygen as the terminal electron acceptor. Oxygen for this purpose is taken into the body across some respiratory surface. Physiologists distinguish two separate but interrelated respiratory processes: **cellular respiration,** the oxidative processes that occur within cells (p. 43), and **external respiration,** the exchange of oxygen and carbon dioxide between the organism and the environment. In this section we describe external respiration and the transport of gases from respiratory surfaces to body tissues.

In small organisms such as single-celled protistans, oxygen is acquired and carbon dioxide liberated by direct diffusion across surface membranes. Gas exchange by diffusion alone is possible only for very small organisms less than 1 mm in diameter, where diffusion paths are short and the surface area of the organism is large relative to volume. As animals became larger and evolved a waterproof covering, specialized devices such as lungs and gills evolved that greatly increased the effective surface for gas exchange. But, because gases diffuse so slowly through living tissue, a circulatory system was necessary to distribute the gases to and from the deep tissues of the body. Even these adaptations were inadequate for complex animals with high rates of cellular respiration. The solubility of oxygen in the blood plasma is so low that plasma alone cannot carry enough to support metabolic demands. With the evolution of special oxygen-transporting blood proteins such as hemoglobin, the oxygen-carrying capacity of the blood increased greatly. Thus what began as a simple and easily satisfied requirement resulted in the evolution of several complex and essential respiratory and circulatory adaptations.

Problems of Aquatic and Aerial Breathing

How an animal respires is largely determined by the nature of its environment. The two great arenas of animal evolution—water and land—are vastly different in their physical characteristics. The most obvious difference is that air contains far more oxygen—at least 20 times more—than does water. For example, water at 5° C fully saturated with air contains approximately 9 ml of oxygen per liter (0.9%); by comparison air contains 209 ml of oxygen per 1000 ml (21%). The density and viscosity of water are approximately 800 and 50 times greater, respectively, than that of air. Furthermore, gas molecules diffuse 10,000 times more rapidly in air than in water. These differences mean that aquatic animals had to evolve very efficient ways of removing oxygen from water. Yet even the most advanced fishes with highly efficient gills and pumping mechanisms may use as much as 20% of their energy just extracting oxygen from water. By comparison, the cost for mammals to breathe is only 1% to 2% of their resting metabolism.

> *Note that the terms "gills" and "lungs" are functional descriptions of these structures. Such descriptions do not imply homology of the gills or lungs in the various groups of animals possessing them.*

Respiratory surfaces must be thin and always kept wet with a fine film of fluid to allow diffusion of gases between the environment and the underlying circulation. This is hardly a problem for aquatic animals, immersed as they are in water, but it is a very real challenge for air breathers. To keep the respiratory membranes moist and protected from injury, air breathers have in general developed invaginations of the body surface and then added pumping mechanisms to move air in and out of the body. The lung is the best example of a successful solution to breathing on land. In general **evaginations** of the body surface, such as gills, are most suitable for aquatic respiration; **invaginations,** such as lungs and tracheae, are best for air breathing. We can now consider the specific kinds of respiratory organs employed by animals.

Respiratory Organs

Gas Exchange by Direct Diffusion

Protozoa, sponges, cnidarians, and many worms respire by direct diffusion of gases between the organism and the environment. We have noted that this kind of **cutaneous respiration** is not adequate when the cellular mass exceeds approximately 1 mm in diameter. But, by greatly increasing the surface of the body relative to its mass, many multicellular animals can supply part or all of their oxygen requirements by direct diffusion. Flatworms are an example of this strategy. Cutaneous respiration frequently supplements gill or lung breathing in larger animals such as amphibians and fishes. For example, an eel can exchange 60% of its oxygen and carbon dioxide through its highly vascular skin. During their winter hibernation, frogs and even turtles exchange all their respiratory gases through the skin while submerged in ponds or springs. The lungless salamanders comprise the largest family of salamanders. Some lungless salamanders have larvae with gills, and the gills persist in the adults of some, but adults of most species have neither lungs nor gills.

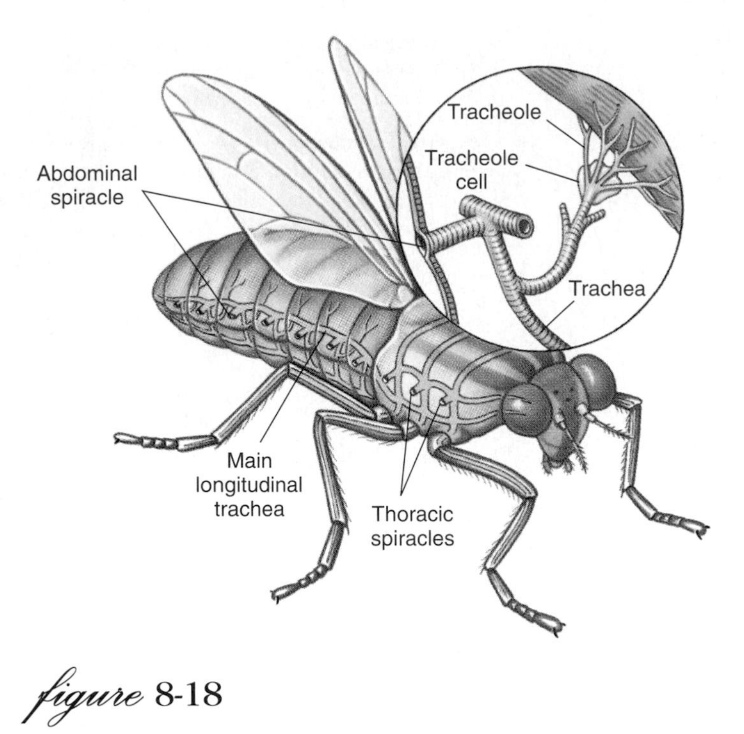

figure 8-18

Tracheal system of insects. Air enters through the spiracles, then travels through tracheae to reach tissues at tracheoles.

Gas Exchange through Tubes: Tracheal Systems

Insects and certain other terrestrial arthropods (centipedes, millipedes, and some spiders) have a highly specialized type of respiratory system, in many respects the simplest, most direct, and most efficient respiratory system found in active animals. It consists of a branching system of tubes (**tracheae**) that extends to all parts of the body (Figure 8-18). The smallest end channels are fluid-filled **tracheoles,** less than 1 μm in diameter, that sink into the cell membranes of the body cells. Air enters the tracheal system through valvelike openings (**spiracles**). Carbon dioxide diffuses out through the same openings. Some insects can ventilate the tracheal system with body movements; the familiar telescoping movement of the bee abdomen is an example. Because the cells have a direct pipeline to the outside, bringing oxygen in and carrying carbon dioxide out, the insect's respiration is independent of the circulatory system. Consequently, insect blood plays no direct role in oxygen transport.

Efficient Gas Exchange in Water: Gills

Gills of various types are effective respiratory devices for life in water. Gills may be simple **external** extensions of the body surface, such as the **dermal papulae** of sea stars (p. 545) or the **branchial tufts** of marine worms (p. 478) and aquatic amphibians (p. 612). Most efficient are the **internal gills** of fishes (p. 597) and arthropods. Fish gills are thin filamentous structures, richly supplied with blood vessels arranged so that blood flow is opposite to the flow of water across the gills. This arrangement, called **countercurrent flow** (p. 183), provides for the greatest possible extraction of oxygen from water. Water flows over the gills in a steady stream, pulled and pushed by an efficient, two-valved, branchial pump (Figure 8-19). Gill ventilation is often assisted by the fish's forward movement through the water.

Lungs

Gills are unsuitable for life in air because, when removed from the buoying water medium, the gill filaments collapse, dry, and stick together; a fish out of water rapidly asphyxiates despite the abundance of oxygen around it. Consequently most air-breathing vertebrates possess lungs, highly vascularized internal cavities. Lungs of a sort are found in certain invertebrates (pulmonate snails, scorpions, some spiders, some small crustaceans), but these structures cannot be very efficiently ventilated.

Lungs that can be ventilated by muscle movements to produce a rhythmic exchange of air are characteristic of most terrestrial vertebrates. The most rudimentary vertebrate lungs are those of lungfishes (Dipneusti), which use them to supplement, or even replace, gill respiration during periods of drought. Although of simple construction, the lungfish lung is supplied with a capillary network in its largely unfurrowed

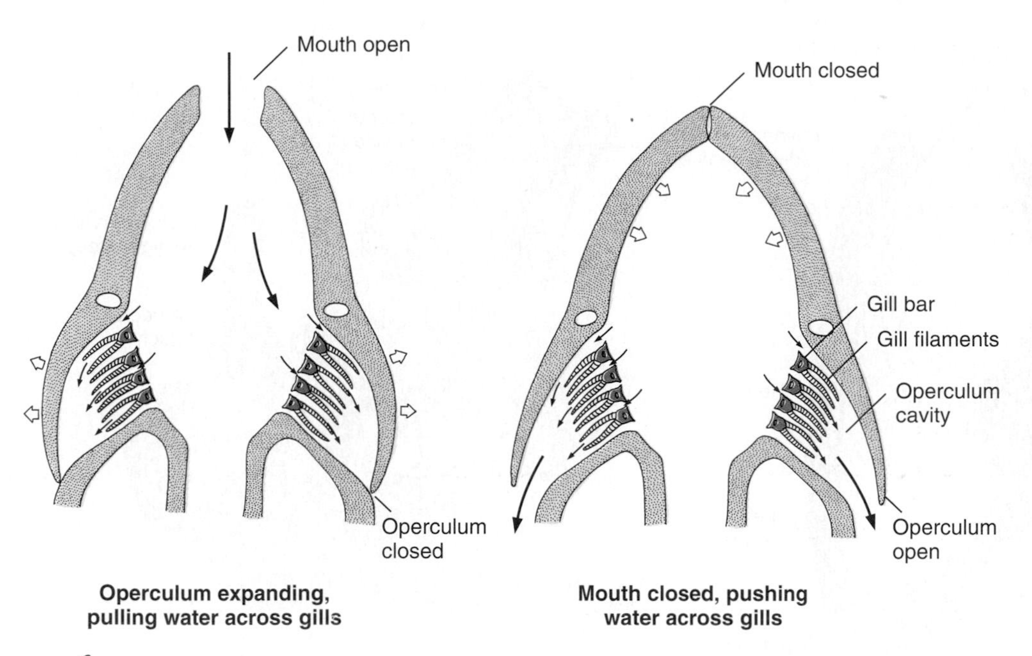

figure 8-19

How a fish ventilates its gills. Through the action of two skeletal muscle pumps, one in the mouth cavity, the other in the opercular cavity, water is drawn into the mouth, passes over the gills, and exits through the gill covers (opercular clefts).

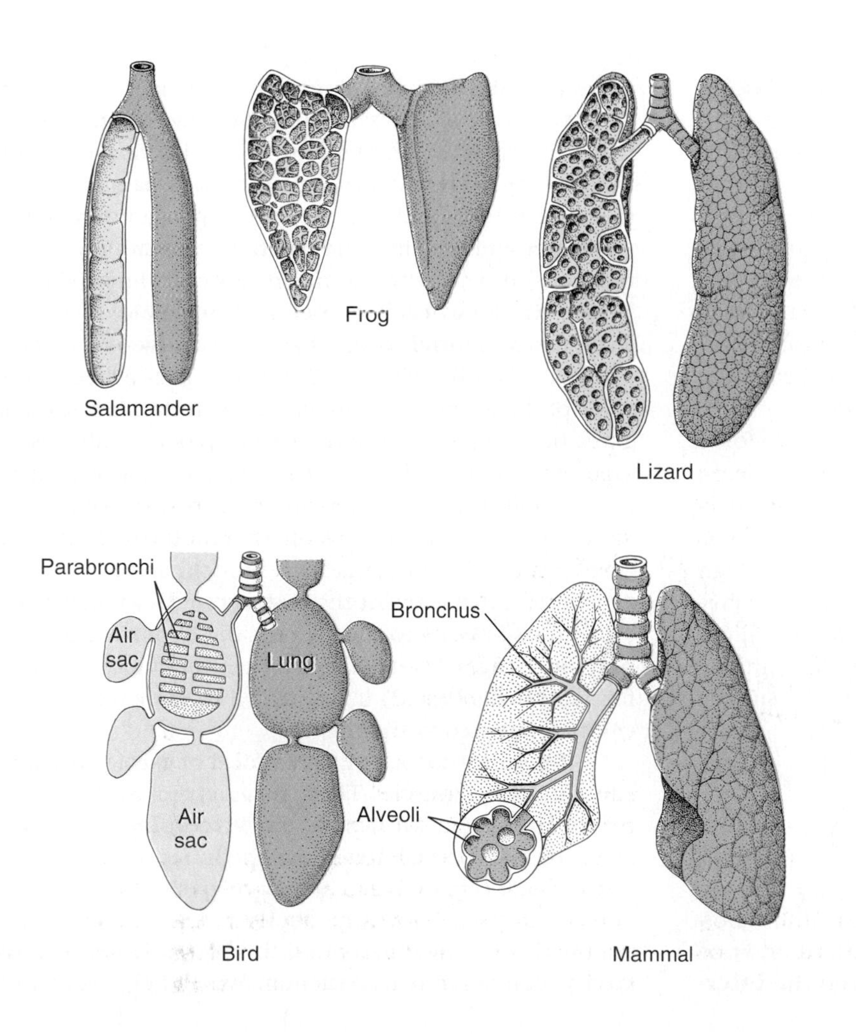

walls, a tubelike connection to the pharynx, and a primitive ventilating system for moving air in and out of the lung.

Amphibian lungs vary from simple, smooth-walled, baglike lungs of some salamanders to the subdivided lungs of frogs and toads (Figure 8-20). The total surface available for gas exchange is much increased in the lungs of reptiles, which are subdivided into numerous interconnecting air sacs. Most elaborate of all is the mammalian lung, a complex of millions of small sacs, called **alveoli** (Figure 8-21), each veiled by a rich vascular network. It has been estimated that human lungs have a total surface area of from 50 to 90 m^2—50 times the area of the skin surface—and contain 1000 km of capillaries. A large surface area is essential for the high oxygen uptake required to support the elevated metabolic rate of endothermic mammals.

A disadvantage of the lung is that gas exchange with the blood can take place only in the alveoli, located at the ends of a branching tree of air tubes (trachea, bronchi, and bronchioles [Figure 8-21]). Unlike the efficient one-way flow of water across fish gills, air must enter and exit a lung through the same channel. After exhalation, the air tubes are filled with "used" air from the alveoli which, during the following inhalation, is pulled back into the lungs. The volume of the air in the lung's passageways is called the **dead space**. This is air that shuttles back and forth with each breath, adding to the difficulty of properly ventilating the lungs. In fact, lung ventilation in humans is so inefficient that in normal breathing only approximately one-sixth of the air in the lungs is replenished with each inspiration. Even after forced expiration, 20% to 35% of the air remains in the lungs.

figure 8-20

Internal structures of lungs among vertebrate groups. Generally, the evolutionary trend has been from simple sacs with little exchange surface between blood and air spaces to complex, lobulated structures, each with complex divisions and extensive exchange surfaces.

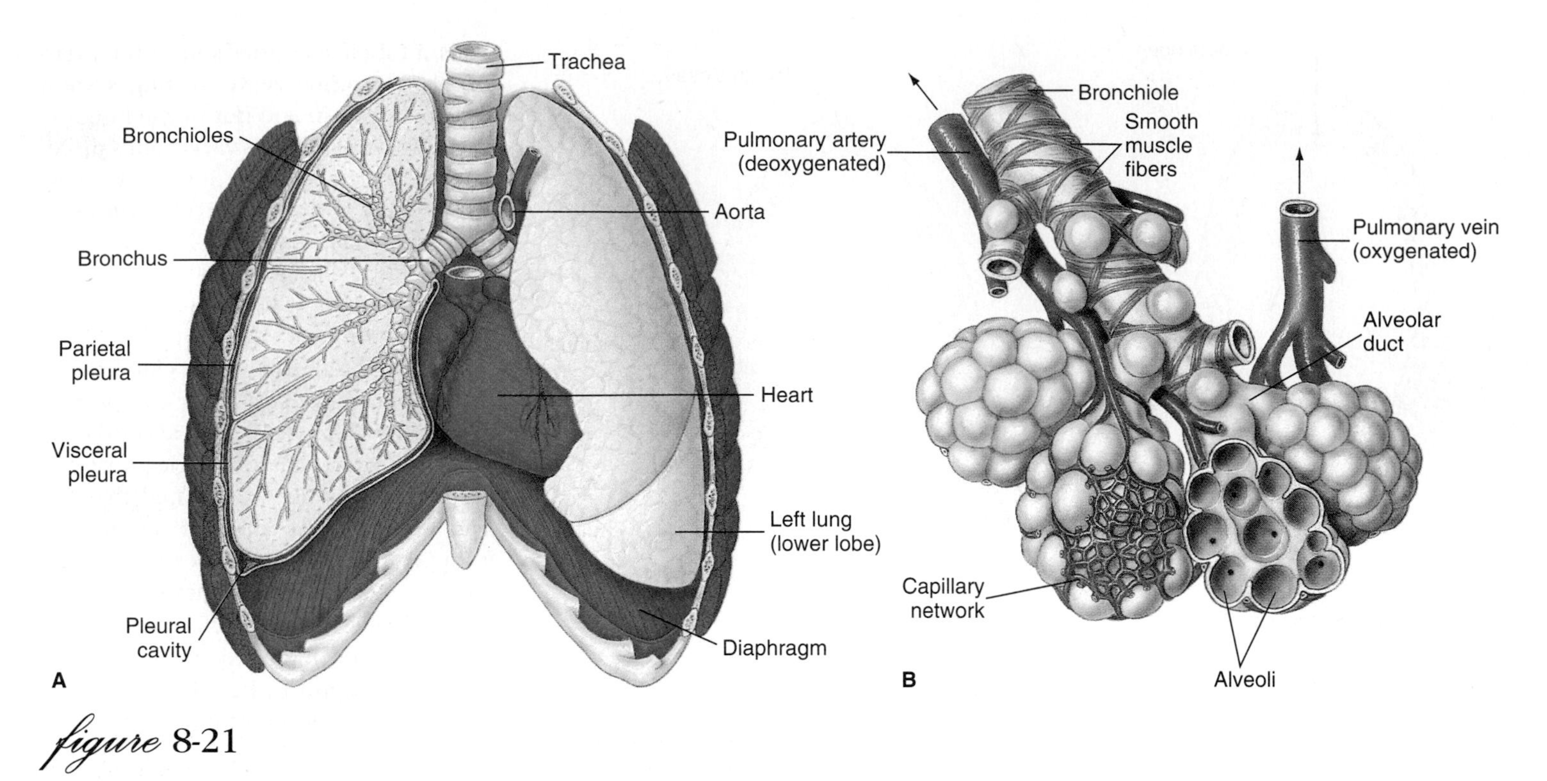

figure 8-21

A, Lungs of human with the right lung shown in section. **B,** Terminal portion of a bronchiole showing air sacs with their blood supply. Arrows show direction of blood flow.

In birds, lung efficiency is improved vastly by adding an extensive system of air sacs (Figure 8-20 and p. 649) that serve as air reservoirs during ventilation. On inspiration, some 75% of the incoming air bypasses the lungs to enter the air sacs (gas exchange does not occur here). At expiration some of this fresh air passes directly through the lung passages and eventually into the one-cell thick **air capillaries** where gas exchange takes place. Thus the air capillaries receive nearly fresh air during both inspiration and expiration. The beautifully designed bird lung is the result of selective pressures during the evolution of flight and its high metabolic demands.

Amphibians employ a **positive pressure** action to force air into their lungs, unlike most reptiles, birds, and mammals, which ventilate the lungs by **negative pressure,** that is, sucking air into the lungs by expansion of the thoracic cavity. Frogs ventilate the lungs by first drawing air into the mouth through the **external nares** (nostrils). Then, closing the nares and raising the floor of the mouth, they drive air into the lungs (Figure 8-22). Much of the time, however, frogs rhythmically ventilate only the mouth cavity, a well-vascularized respiratory surface that supplements pulmonary respiration.

Structure and Function of the Mammalian Respiratory System

Air enters the mammalian respiratory system through the nostrils (external nares), passes through the **nasal chamber,** lined with mucus-secreting epithelium, and then through the **internal nares,** the nasal openings which connect to the **pharynx.** Here, where the pathways of digestion and respiration cross, inhaled air leaves the pharynx by passing into a narrow opening, the **glottis;** food enters the esophagus to pass to the stomach (see Figure 10-10, p. 229). The glottis opens into the **larynx,** or voice box, and then into the **trachea,** or windpipe. The trachea branches into two **bronchi,** one to each lung (Figure 8-21). Within the lungs each bronchus divides and subdivides into small tubes (**bronchioles**) that lead via **alveolar ducts** to the air sacs (**alveoli**) (Figure 8-21). The single-layered endothelial walls of the alveoli are thin and moist to facilitate exchange of gases between air sacs and adjacent blood capillaries. Air passageways are lined with both mucus-secreting and ciliated epithelial cells, which play an important role in conditioning the air before it reaches the alveoli. There are partial, cartilaginous rings in the walls of the tracheae, bronchi, and even some of the bronchioles that prevent those structures from collapsing.

In its passage to the air sacs, the air undergoes three important changes: (1) it is filtered free from most dust and other foreign substances, (2) it is warmed to body temperature, and (3) it is saturated with moisture.

The lungs consist of a great deal of elastic connective tissue and some muscle. They are covered by a thin layer of tough epithelium known as the **visceral pleura.** A similar layer, the **parietal pleura,** lines the inner surface of the walls of the chest (Figure 8-21). The two layers of the pleura are in contact and slide over one another as the lungs expand and contract. The "space" between the pleura, called the **pleural cavity,** contains a partial vacuum, which helps keep the lungs

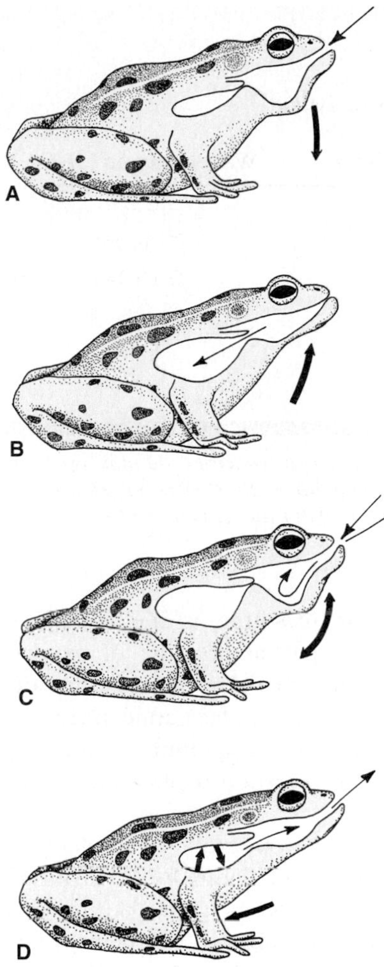

figure 8-22

Breathing in a frog. The frog, a positive-pressure breather, fills its lungs by forcing air into them. **A,** The floor of the mouth is lowered, drawing air in through the nostrils. **B,** With the nostrils closed and glottis open, the frog forces air into its lungs by elevating the floor of its mouth. **C,** The mouth cavity is ventilated rhythmically for a period. **D,** The lungs are emptied by contraction of body wall musculature and by elastic recoil of the lungs.

expanded to fill the pleural cavity. Therefore no real pleural space exists; the two pleura rub together, lubricated by tissue fluid (lymph). The chest cavity is bounded by the spine, ribs, and breastbone, and floored by the **diaphragm,** a dome-shaped, muscular partition between the chest cavity and abdomen. A muscular diaphragm is found only in mammals.

Ventilating the Lungs

The chest cavity is an air-tight chamber. During **inspiration,** muscles between the ribs (**external intercostals**) and some

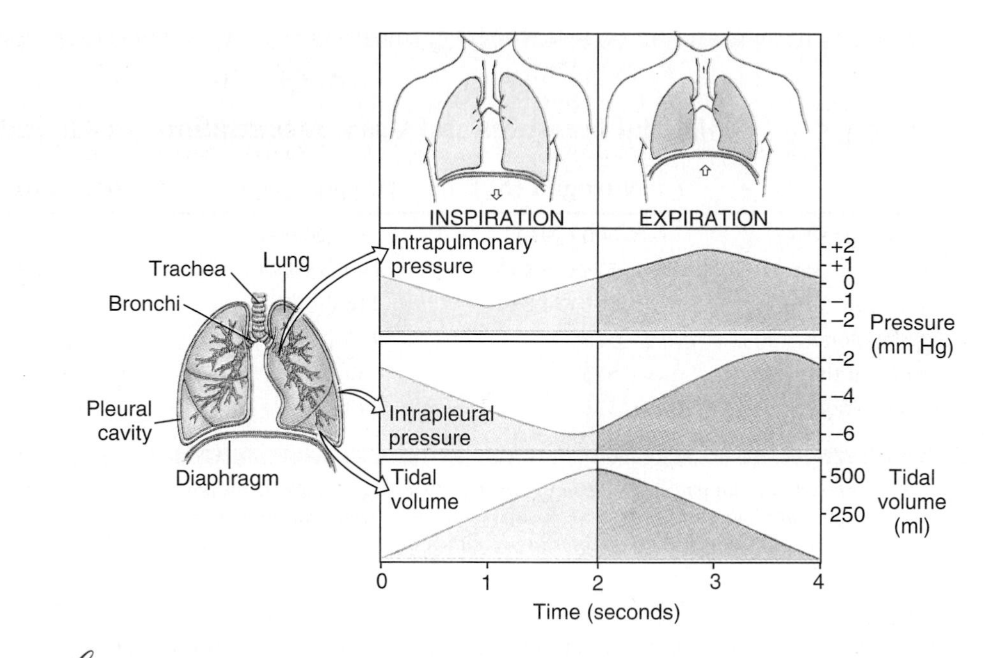

figure 8-23

Mechanism of breathing in humans. On inspiration, muscles between the ribs (external intercostals) and small muscles in the neck contract, raising the rib cage, while the diaphragm contracts. These actions increase the volume of the pleural cavity (which is filled by the lungs), causing air to flow into the lungs. When the intercostals and diaphragm relax, the abdominal muscles pull the rib cage down and push the abdominal contents against the diaphragm. The volume of the pleural cavity becomes smaller, thus expelling air.

small muscles in the neck contract, pulling the rib cage upward. At the same time, contraction of the diaphragm lowers the floor of the pleural cavity (Figure 8-23). The resulting increase in volume of the chest cavity causes the air pressure in the lungs to fall below atmospheric pressure: air rushes in through the air passageways to equalize the pressure. Normal **expiration** is a less active process than inspiration. The external intercostals and diaphragm relax. The abdominal muscles help pull the ribs downward, and the organs in the abdomen push upward against the diaphragm. These actions decrease the size of the chest cavity, the elastic lungs deflate, and air exits (Figure 8-23).

How Breathing Is Coordinated

Breathing is normally involuntary and automatic but can come under voluntary control. Neurons in the medulla of the brain regulate normal, quiet breathing. They spontaneously produce rhythmical bursts that stimulate contraction of the diaphragm and external intercostal muscles. However, respiration must adjust itself to changing requirements of the body for oxygen. Oddly, carbon dioxide rather than oxygen has the greatest effect on respiratory rate because under normal conditions arterial oxygen does not decline enough to stimulate oxygen receptors. Even a small rise in carbon dioxide level in the blood, however,

table 8-1

Partial Pressures and Gas Concentrations in Air and Body Fluids

	Nitrogen (N_2)	Oxygen (O_2)	Carbon dioxide (CO_2)	Water vapor (H_2O)
Inspired air (dry)	600 (79%)	159 (20.9%)	0.2 (0.03%)	—
Alveolar air (saturated)	573 (75.4%)	100 (13.2%)	40 (5.2%)	47 (6.2%)
Expired air (saturated)	569 (74.8%)	116 (15.3%)	28 (3.7%)	47 (6.2%)
Arterial blood	573	100	40	
Peripheral tissues	573	30	50	
Venous blood	573	40	46	

Note: Values expressed in millimeters of mercury (mm Hg). Percentages indicate proportion of total atmospheric pressure at sea level (760 mm Hg). Inspired air is shown as dry, although atmospheric air always contains variable amounts of water. If, for example, atmospheric air at 20° C were half saturated (relative humidity 50%), the partial pressures and percentages would be N_2 593.5 (78.1%); O_2 157 (20.6%); CO_2 0.2 (0.03%); and H_2O 8.75 (1.1%).

has a powerful effect on respiratory activity. Actually, the stimulatory effects of carbon dioxide are due in part to the increase in hydrogen ion concentration in the cerebrospinal fluid.

$$CO_2 + H_2O \leftrightarrow H_2CO_3 \leftrightarrow H^+ + HCO_3^-$$

This reaction shows that carbon dioxide combines with water to form carbonic acid. Carbonic acid then dissociates to release hydrogen ions, making the cerebrospinal fluid more acidic, and stimulating the respiratory receptors in the medulla of the brain. Both rate and depth of respiration increase.

It is well known that swimmers can remain submerged much longer if they vigorously hyperventilate first to blow off carbon dioxide from the lungs. This delays the overpowering urge to surface and breathe. The practice is dangerous because blood oxygen is depleted just as rapidly as without prior hyperventilation, and the swimmer may lose consciousness when the oxygen supply to the brain drops below a critical point. Several documented drownings among swimmers attempting long underwater swimming records have been caused by this practice.

Gaseous Exchange in Lungs and Body Tissues: Diffusion and Partial Pressure

Air (the atmosphere) is a mixture of gases: about 71% nitrogen, 20.9% oxygen, in addition to fractional percentages of other gases, such as carbon dioxide (0.03%). Gravity attracts the mass of the atmosphere to the earth. At sea level the atmosphere exerts a hydrostatic pressure due to gravity equal to the weight of a column of mercury (Hg) 760 mm high. Thus we can speak of atmospheric pressure as being equal to 760 mm Hg. But because air is not a single gas but a mixture, *part* of the 760 mm Hg pressure (**partial pressure**) is due to each component gas. For example, the partial pressure of oxygen is $0.209 \times 760 = 159$ mm, and that for carbon dioxide is $0.0003 \times 760 = 0.23$ mm in dry air. (In fact, atmospheric air is never completely dry, and the varying amount of water vapor present exerts a pressure in proportion to its concentration, like other gases.)

As soon as air enters the respiratory tract, changes in its composition take place (Table 8-1, Figure 8-24). Inspired air becomes saturated with water vapor as it travels through the air-filled passageways toward the alveoli. When inspired air reaches the alveoli, it mixes with residual air remaining from the previous respiratory cycle. Partial pressure of oxygen drops and that of carbon dioxide rises. Upon expiration, air from the alveoli mixes with air in the dead space to produce still a different mixture (Table 8-1). Although no significant gas exchange takes place in the dead space, the air it contains is the first air to leave the body when expiration begins.

Because the partial pressure of oxygen in the lung alveoli is greater (100 mm Hg) than it is in venous blood of lung capillaries (40 mm Hg), oxygen diffuses into the lung capillaries. In a similar manner the carbon dioxide in the blood of the lung capillaries has a higher concentration (46 mm Hg) than has this same gas in the lung alveoli (40 mm Hg), so that carbon dioxide diffuses from the blood in the alveoli.

In the tissues respiratory gases also move along their concentration gradients (Figure 8-24). The partial pressure of oxygen in the blood (100 mm Hg) is greater than in the tissues (0 to 30 mm Hg), and the partial pressure of carbon dioxide in the tissues (45 to 68 mm Hg) is greater than that in blood (40 mm Hg). The gases in each case diffuse from a location of higher concentration to one of lower concentration.

How Respiratory Gases Are Transported in the Blood

In some invertebrates the respiratory gases are simply carried dissolved in the body fluids. However, the solubility of oxygen is

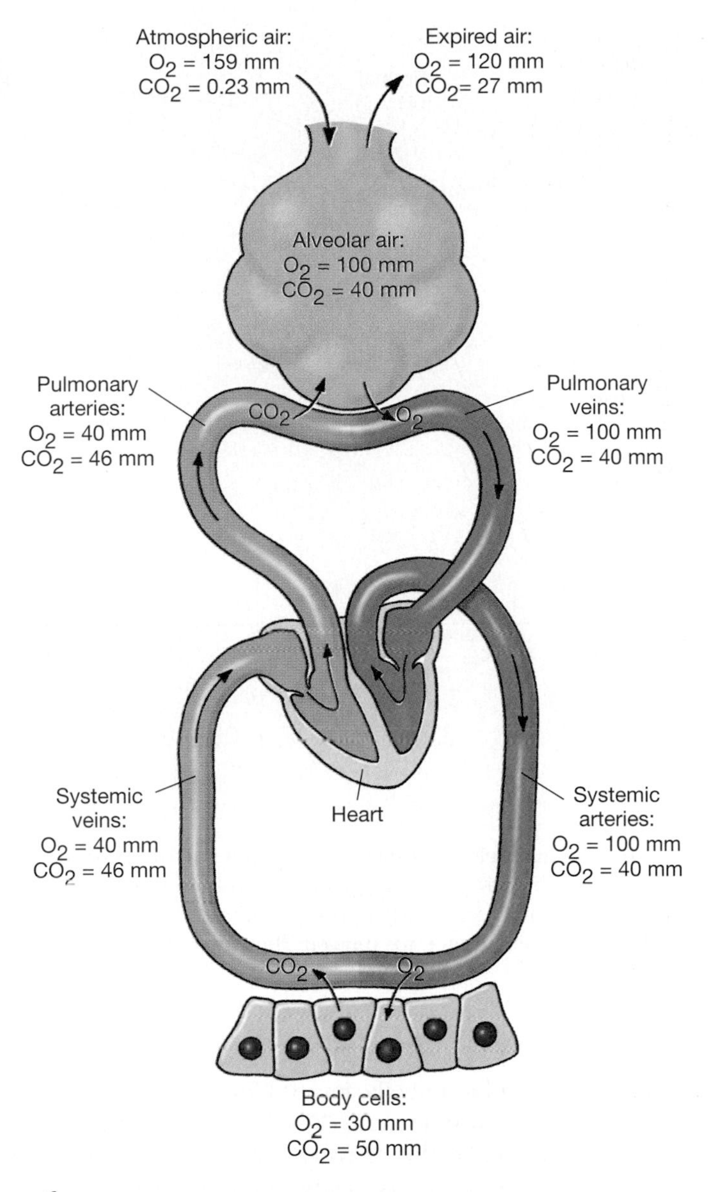

figure 8-24

Exchange of respiratory gases in lungs and tissue cells. Numbers present partial pressures in millimeters of mercury (mm Hg).

so low in water that it is adequate only for animals with low rates of metabolism. For example, only approximately 3% of a human's oxygen requirement can be transported in this way. Consequently in many invertebrates and in virtually all vertebrates, nearly all of the oxygen and a significant amount of carbon dioxide are transported by special colored proteins, or **respiratory pigments,** in the blood. In most animals (all vertebrates) these respiratory pigments are packaged into blood cells.

The most widespread respiratory pigment in the animal kingdom is **hemoglobin,** a red, iron-containing protein present in all vertebrates and many invertebrates. Each molecule of hemoglobin is made up of 5% **heme,** an iron-containing compound giving the red color to blood, and 95% **globin,** a colorless protein. The heme portion of the hemoglobin has a great affinity for oxygen; each gram of hemoglobin can carry a maximum of approximately 1.3 ml of oxygen. Because there are approximately 15 g of hemoglobin in each 100 ml of blood, fully oxygenated blood contains approximately 20 ml of oxygen per 100 ml. Of course, for hemoglobin to be of value to the body it must hold oxygen in a loose, reversible chemical combination so that it can be released to the tissues. The actual amount of oxygen that combines with hemoglobin depends on the shape or conformation of the hemoglobin molecule, which is affected by several factors, including the concentration of oxygen itself. When the oxygen concentration is high, as it is in the capillaries of the lung alveoli, hemoglobin loads up with oxygen; in the tissues where the prevailing oxygen partial pressure is low, hemoglobin releases its stored oxygen reserves (Figure 8-25).

Although hemoglobin is the only vertebrate respiratory pigment, and is widely distributed among invertebrates as well, several other respiratory pigments are known among the invertebrates. Hemocyanin, *a blue, copper-containing protein, is present in the crustaceans and most molluscs. Among other pigments is* chlorocruorin *(klora-croo'o-rin), a green-colored, iron-containing pigment found in four families of polychaete tube worms. Its structure and oxygen-carrying capacity are very similar to those of hemoglobin, but it is carried free in the plasma rather than being enclosed in blood corpuscles.* Hemerythrin *is a red pigment found in some polychaete worms. Although it contains iron, this metal is not present in a heme group (despite the name of the pigment), and its oxygen-carrying capacity is poor.*

We can express the relationship of carrying capacity to surrounding oxygen concentration as **hemoglobin saturation curves** (also called oxygen dissociation curves [Figure 8-25]). As these curves show, the lower the surrounding oxygen tension, the greater the quantity of oxygen released. This important characteristic of hemoglobin allows more oxygen to be released to those tissues that need it most (those having the lowest partial pressure of oxygen).

Another factor that affects the conformation of hemoglobin, and therefore its release of oxygen to the tissues, is the sensitivity of oxyhemoglobin (hemoglobin with bound oxygen) to carbon dioxide. Carbon dioxide shifts the hemoglobin saturation curve to the right (Figure 8-25B), a phenomenon called the **Bohr effect** after the Danish scientist who first described it. As carbon dioxide enters the blood from the respiring tissues, it causes hemoglobin to unload more oxygen. The opposite event occurs in the lungs; as carbon dioxide diffuses from the venous blood into the alveolar space, the hemoglobin saturation curve shifts back to the left, allowing more oxygen to be loaded onto hemoglobin.

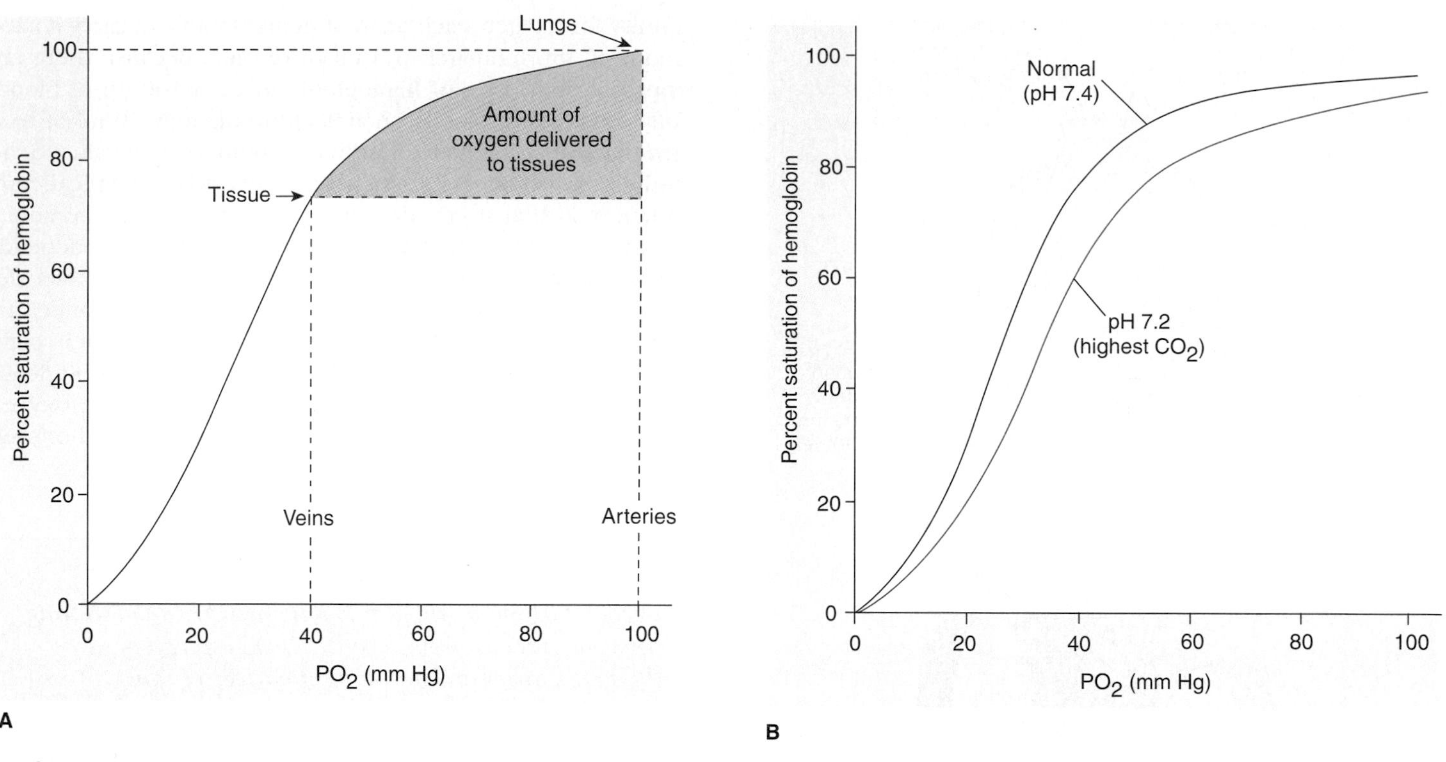

figure 8-25

Hemoglobin saturation curves. Curves show how the amount of oxygen that can bind to hemoglobin is related to oxygen partial pressure. **A,** At the higher partial pressure in the lungs, hemoglobin can load with more oxygen. In the tissues the oxygen concentration is less, so hemoglobin can carry less, that is, it unloads more. **B,** Hemoglobin is also sensitive to carbon dioxide partial pressure (Bohr effect). As carbon dioxide enters blood from the tissues, it shifts the curve to the right, decreasing affinity of hemoglobin for oxygen. Thus the hemoglobin unloads more oxygen in the tissues where carbon dioxide concentration is higher.

Sickle cell anemia is an incurable, inherited condition (p. 79) in which a single amino acid in normal hemoglobin (HbA) is substituted. The ability of sickle cell hemoglobin (HbS) to carry oxygen is severely impaired, and the erythrocytes tend to crumple during periods of oxygen stress (for example, during exercise). Capillaries become clogged with the misshapen red cells; the affected area is very painful, and the tissue may die. About 1 in 10 black Americans carry the trait (heterozygous). Heterozygotes do not have sickle cell anemia and live normal lives, but if both parents are heterozygous, each child has a 25% chance of inheriting the disease.

The same blood that transports oxygen to the tissues from the lungs must carry carbon dioxide back to the lungs on its return trip. However, unlike oxygen that is transported almost exclusively in combination with hemoglobin, the blood transports carbon dioxide in three different forms. A small fraction of the carbon dioxide, only about 7%, is carried as physically dissolved gas in the plasma. The remainder diffuses into the red blood cells. In red blood cells, most of the carbon dioxide, approximately 70%, becomes carbonic acid through the action of the enzyme carbonic anhydrase. Carbonic acid immediately dissociates into hydrogen ion and bicarbonate ion. We can summarize the entire reaction as follows:

$$CO_2 + H_2O \xrightleftharpoons{\text{carbonic anhydrase}} H_2CO_3 \rightleftharpoons H^+ + HCO_3^-$$

Several systems buffer the hydrogen ion in the blood, thus preventing a severe decrease in blood pH. Bicarbonate ions remain in solution in the plasma and red blood cell water since, unlike carbon dioxide, bicarbonate is extremely soluble (Figure 8-26).

Another fraction of the carbon dioxide, approximately 23%, combines reversibly with hemoglobin. Carbon dioxide does not combine with the heme group but with amino groups of several amino acids to form a compound called carbaminohemoglobin.

All of these reactions are reversible. When the venous blood reaches the lungs, carbon dioxide diffuses out of the red blood cells and into the alveolar air.

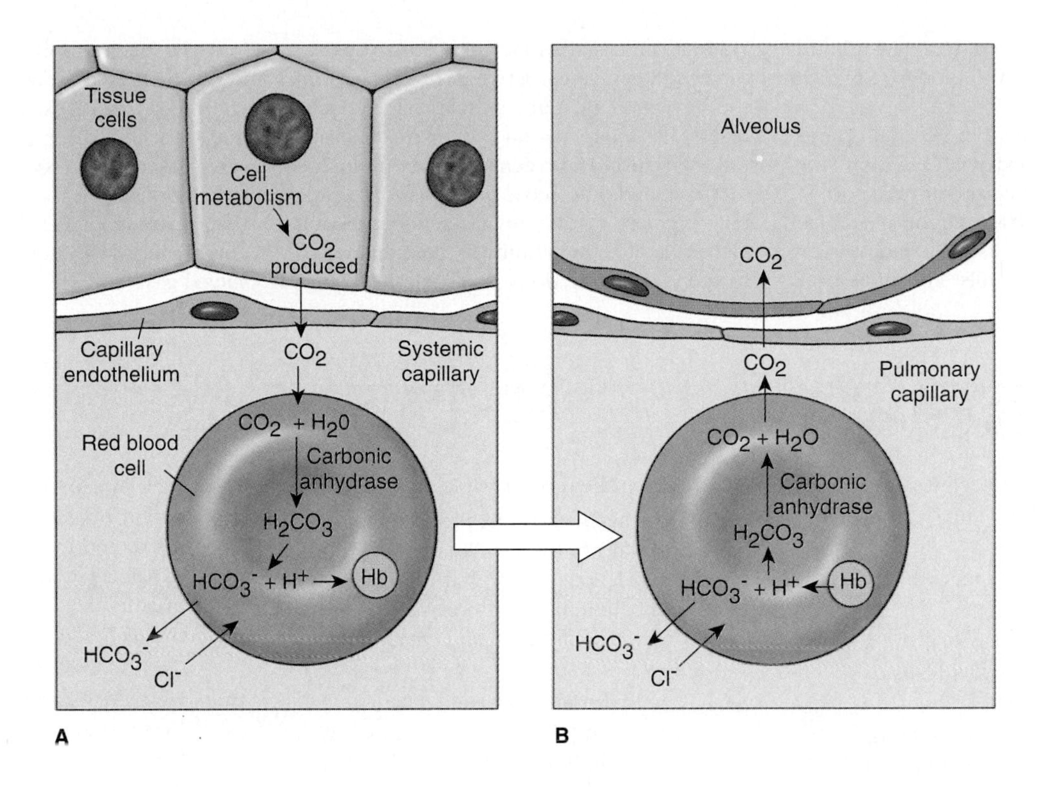

figure 8-26

Transport of carbon dioxide in the blood. **A,** Carbon dioxide produced by metabolic oxidation of glucose diffuses from the tissues into plasma and red blood cells. Carbonic anhydrase in red blood cells catalyzes conversion of carbon dioxide into carbonic acid, then bicarbonate and hydrogen ions. Part of the bicarbonate diffuses out of the cells, and diffusion inward of chloride ions maintains electrical balance. Hydrogen ions mostly associate with hemoglobin. **B,** The lower partial pressure of carbon dioxide in the alveoli of the lungs favors reversal of these reactions.

Summary

The fluid in the body, whether intracellular, plasma, or interstitial, is mostly water, but it has many substances dissolved in it, including electrolytes and proteins. Mammalian blood consists of fluid plasma and formed elements, including red and white blood cells and platelets. The plasma has many dissolved solids, as well as dissolved gases. Mammalian red blood cells lose their nucleus during development and contain the oxygen-carrying pigment, hemoglobin. White blood cells are important defensive elements. Platelets are vital in the process of clotting, necessary to prevent excess blood loss when the blood vessel is damaged. They release a series of factors that activate prothrombin to thrombin, an enzyme that causes fibrinogen to change to the gel form, fibrin.

In open circulatory systems, such as those of arthropods and most molluscs, the blood escapes from arteries into a hemocoel, which is a primary body cavity derived from the blastocoel. In closed circulatory systems, such as those of annelids, vertebrates, and cephalopod molluscs, the heart pumps blood into arteries, then into arterioles of smaller diameter, through the bed of fine capillaries, through venules, and finally through the veins, which lead back to the heart. In fishes, which have a two-chambered heart with a single atrium and a single ventricle, the blood is pumped to the gills and then directly to the systemic capillaries throughout the body without first returning to the heart. With the evolution of lungs, vertebrates developed a double circulation consisting of a systemic circuit serving the body, and a pulmonary circuit serving the lungs. To be fully efficient, this change required partitioning of both the atrium and ventricle to form a double pump; partitioning was begun in the lungfishes and amphibians, which have two atria but an undivided ventricle, and completed in birds and mammals, which have four-chambered hearts.

One-way flow of blood during the heart's contraction (systole) and relaxation (diastole) is assured by valves between the atria and the ventricles and between the ventricles and the pulmonary arteries and the aorta. Although the heart can beat spontaneously, its rate is controlled by nerves from the central nervous system. The heart muscle uses a great deal of oxygen and has a well-developed coronary blood circulation. The walls of arteries are thicker than those of veins, and connective tissue in the walls of arteries allows them to expand during systole and contract during diastole. Normal arterial blood pressure (hydrostatic) of humans in systole is 120 mm Hg and in diastole, 80 mm Hg. Because the capillary walls are permeable to water, a protein-free filtrate crosses capillary walls, its movement determined by the balance between opposing forces of hydrostatic and protein osmotic pressure. Tissue fluid (lymph) that does not reenter the capillary system is collected by the lymphatic system and returned to the blood by lymph ducts.

Very small animals can depend on diffusion between the external environment and their tissues or cytoplasm for transport of respiratory gases, but larger animals require specialized organs, such as gills, tracheae, or lungs, for this function. Gills and lungs provide an increased surface area for exchange of respiratory gases between the blood and the environment. Many animals have special respiratory pigments and other mechanisms to help transport oxygen and carbon dioxide in the blood. The most widespread respiratory pigment in the animal kingdom, hemoglobin, has a higher affinity for oxygen at high oxygen concentrations but releases it at lower concentrations. Vertebrate hemoglobin, which is packaged in red blood cells, combines readily with oxygen in gills or lungs, then releases it in respiring body tissues where the oxygen partial pressure is low. Blood carries carbon dioxide from the tissues to the lungs as the bicarbonate ion, in combination with hemoglobin, and as the dissolved gas.

Review Questions

1. Name the chief intracellular electrolytes and the chief extracellular electrolytes.
2. What is the fate of spent erythrocytes in the body?
3. Outline or briefly describe the sequence of events that leads to blood coagulation.
4. Two distinctly different styles of circulatory systems have evolved among animals: open and closed. What is "open" about an open circulatory system? Closed systems sometimes are cited as adaptive for actively moving animals with (at least at times) high metabolic demand. Can you suggest possible reasons for this assertion?
5. Place the following in correct order to describe the circuit of blood through the vascular system of a fish: ventricle, gill capillaries, sinus venosus, body tissue capillaries, atrium, dorsal aorta.
6. Trace the flow of blood through the heart of a mammal, naming the four chambers, their valves, and explaining where the blood entering each atrium comes from, and where the blood leaving each ventricle goes. When the ventricles contract, what prevents blood from reentering the atria?
7. Explain the origin and conduction of the excitation that leads to a heart contraction. Why is the vertebrate heart said to be a myogenic heart? If the heart is myogenic, how do you account for alterations in rate of heart beat?
8. Define the terms systole and diastole.
9. Explain the movement of fluid across the walls of the capillaries. How does the balance of hydrostatic pressure and colloid osmotic pressure determine the direction of net fluid flow?
10. The hydrostatic pressure at the arterial end of capillaries is about 40 mm Hg in humans. If the hydrostatic pressure at the venous end is about 15 mm Hg, and the colloid osmotic pressure is 25 mm Hg throughout, what is the net effect on fluid movement between the capillaries and tissue spaces?
11. Provide a brief description of the lymphatic system. What are its principal functions? Why is movement of lymph through the lymphatic system very slow?
12. What is an advantage of a fish's gills for breathing in water and a disadvantage for breathing on land?
13. Describe the tracheal system of insects. What is the advantage of such a system for a small animal?
14. Trace the route of inspired air in humans from the nostrils to the smallest chamber of the lungs. What is the "dead air space" of a mammalian lung and how does it affect the partial pressure of oxygen reaching the alveoli?
15. The amount of time that scuba divers can spend underwater is limited by several factors, including the time required to deplete the air supply in their tanks. To make their air last longer, novice divers may be instructed to breathe slowly and exhale as much as possible on each breath. Can you suggest a reason why this behavior would lengthen the diver's air supply?
16. How does a frog ventilate its lungs? Contrast an amphibian's positive pressure breathing with a mammal's negative pressure breathing.
17. What is the role of carbon dioxide in the control of the rate and depth of breathing of a mammal?
18. The air pressure supplied to a scuba diver must equal that exerted by the surrounding seawater, and for each 10 m increase in depth, the pressure of the surrounding seawater increases one full atmosphere. Assuming the partial pressure of oxygen in air at sea level (one atmosphere) is 0.209×760 mm Hg (= 159 mm Hg), what partial pressure of oxygen would the diver be breathing at a depth of 30 m?
19. Explain how oxygen is carried in the blood, including specifically the role of hemoglobin. Answer the same question with regard to carbon dioxide transport.
20. The ability of hemoglobin to bind with oxygen decreases with decreasing oxygen concentration and also decreases with increasing carbon dioxide concentration. What effect do these phenomena have on the delivery of oxygen to the tissues?

Selected References

Feder, M. E., and W. W. Burggren. 1985. Skin breathing in vertebrates. Sci. Am. **253:**126–142 (Nov.) *In many amphibians and reptiles the skin supplements, and may even replace, the work of gills and lungs.*

Golde, D. W. 1991. The stem cell. Sci. Am. **265:**86–93. *Undifferentiated cells in the bone marrow give rise to white and red blood cells, macrophages, and platelets.*

Lawn, R. M., and G. A. Vehar. 1986. The molecular genetics of hemophilia. Sci. Am. **254:**48–54 (Mar.). *The gene coding for the factor in the clotting cascade in the most common form of hemophilia has been isolated and cloned with recombinant DNA techniques. Prospects are good for a safe (virus-free), abundant source of the factor for treating hemophiliacs.*

Lillywhite, H. B. 1988. Snakes, blood circulation and gravity. Sci. Am. **259:**92–98 (Dec.). *How a snake's vascular system is designed to counter the effects of gravity.*

Perutz, M. F. 1978. Hemoglobin structure and respiratory transport. Sci. Am. **240:**92–125 (Dec.). *Hemoglobin transports oxygen and carbon dioxide between the lungs and tissues by clicking back and forth between two structures. Perutz and J.C. Kendrew won the Nobel Prize in 1962 for discovering the structure of hemoglobin.*

Randall, D. J., W. W. Burggren, A. P. Farrell, and M. S. Haswell. 1981. The evolution of air breathing in vertebrates. Cambridge, England, Cambridge University Press. *Traces the physiology of air breathing from aquatic ancestors.*

Robinson, T. F., S. M. Factor, and E. H. Sonnenblick. 1986. The heart as a suction pump. Sci. Am. **254:**84–91 (June). *Suggests that filling of heart in diastole is aided by elastic recoil of energy from systole.*

Zucker, M. B. 1980. The functioning of the blood platelets. Sci. Am. **242:**86–103 (June). *The small blood elements that act to stop blood flow from a wound also perform complex roles in health and disease.*

Immunity

chapter | nine

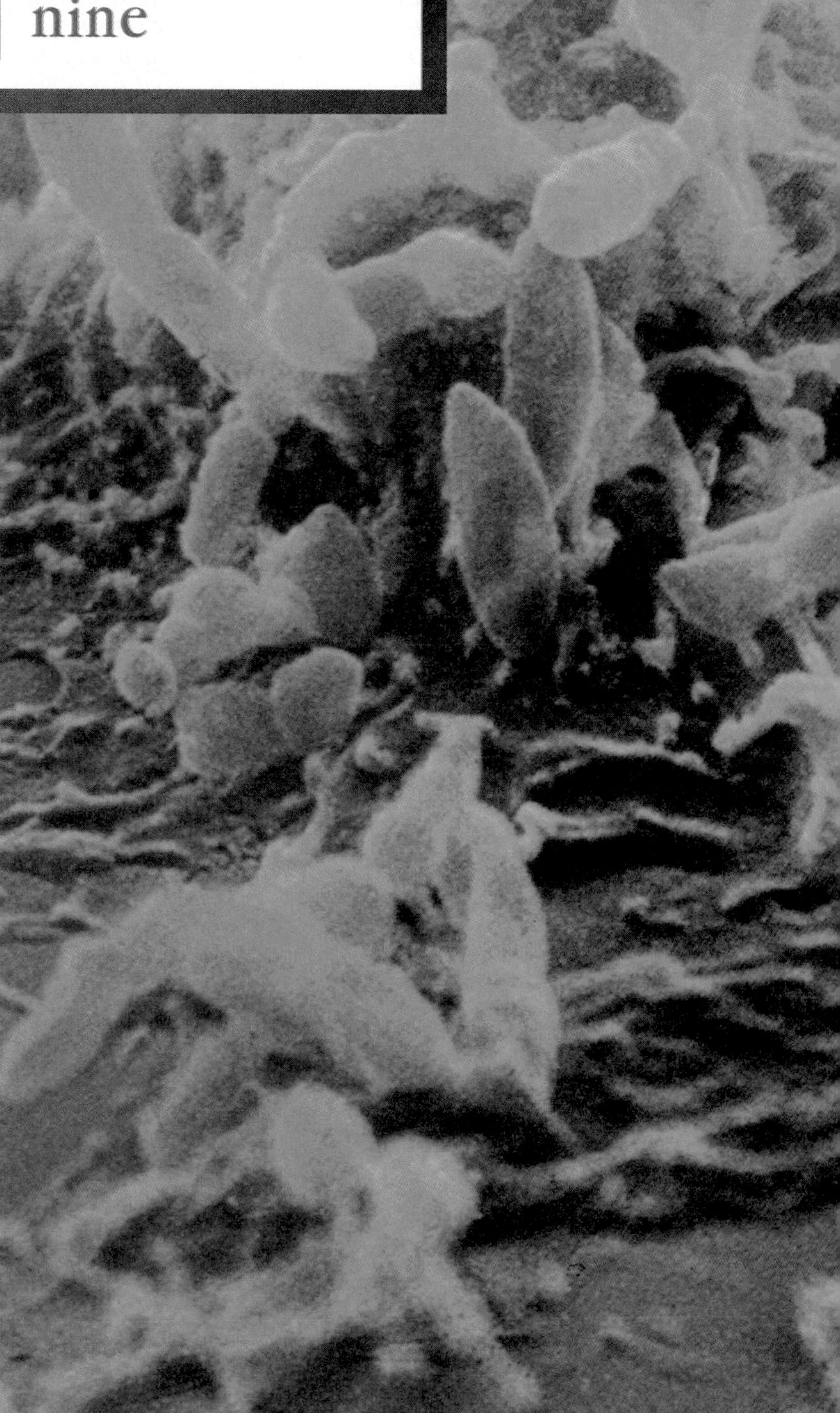

The Language of Cells in Immunity

For some years we have known that certain cells in an animal could secrete substances that affected various processes in other cells, for example, metabolism, physiology, or differentiation, but the means of this communication between cells remained a mystery. Much of the shroud has been lifted by more recent discoveries. Specific signal molecules, often proteins or peptides, are secreted by certain cells. Target cells (which might be the same cells that secreted the signals) have receptors protruding through their outer membranes that specifically bind the signal molecules and only those molecules. Binding of a signal causes changes in the part of the receptor molecule (or in an associated membrane protein) that extends into the cytoplasm, and this sets off a cascade of activations involving protein kinases and phosphorylases. Transcription factors are mobilized. In the nucleus the transcription factors initiate transcription of formerly inactive genes, leading to synthesis of the products they encode.

We now know that hormones affect target cells by this mechanism (see Chapter 12), and it is also the scheme by which the cells of the immune system communicate with each other and with other cells. The signal molecules of the immune system are called *cytokines.* Cytokines and their receptors are the language of communication in the immune system. They perform an intricate and elaborate ballet of activation and regulation, causing some cells to proliferate, suppressing proliferation of others, and stimulating secretion of additional cytokines or defense molecules. Precise signalling among the cells and the exact performance of their duties are essential to maintenance of human health and defense against invading viruses, bacteria, and parasites and for prevention of unrestrained cell division, as in cancer. Successful establishment of invaders in our bodies depends on evasion or subversion of our immune system, and inappropriate response of immune cells may itself produce disease. We have learned to manipulate the immune response so that we can transplant organs between individuals, but complete failure in its cell communication results in profound disease, such as AIDS.

The immune system is spread throughout the body of an animal, and it is as crucial to survival as the respiratory, circulatory, nervous, skeletal, or any other system. Benjamini, Sunshine, and Leskowitz[1] say in their concise text: "The essence of immunology can be . . . stated while standing on one foot. Immunology deals with understanding how the body distinguishes between what is 'self' and what is 'nonself'; all the rest is technical detail." But the devil, as the saying goes, is in the details. Furthermore, while the self/nonself dogma has guided much successful research, some immunologists point to observations it cannot explain. In many cases simple foreignness is insufficient to provoke an immune response; it may be that "danger" alerts the immune system—something that causes tissue injury or cell death by means other than apoptosis (p. 38).

The term **immunity** has been used on the one hand as synonymous with resistance and on the other hand has been associated with the sensitive and specific immune response exhibited by vertebrates. A more general yet concise statement is that an animal demonstrates immunity when it possesses tissues capable of recognizing and protecting the animal against invaders. Most animals show **innate** immunity; that is, it is genetically determined, is potentially fully developed at birth, and is not altered by subsequent exposure to antigens. Vertebrates also show innate immunity, but they additionally develop **acquired immunity,** which is specific to the particular nonself material, requires time for its development, and occurs more quickly and vigorously on secondary response. Acquired immunity has been demonstrated in some invertebrates, but the complex, induced responses based on antibodies and T-cell receptors (p. 215) have been shown only in vertebrates.

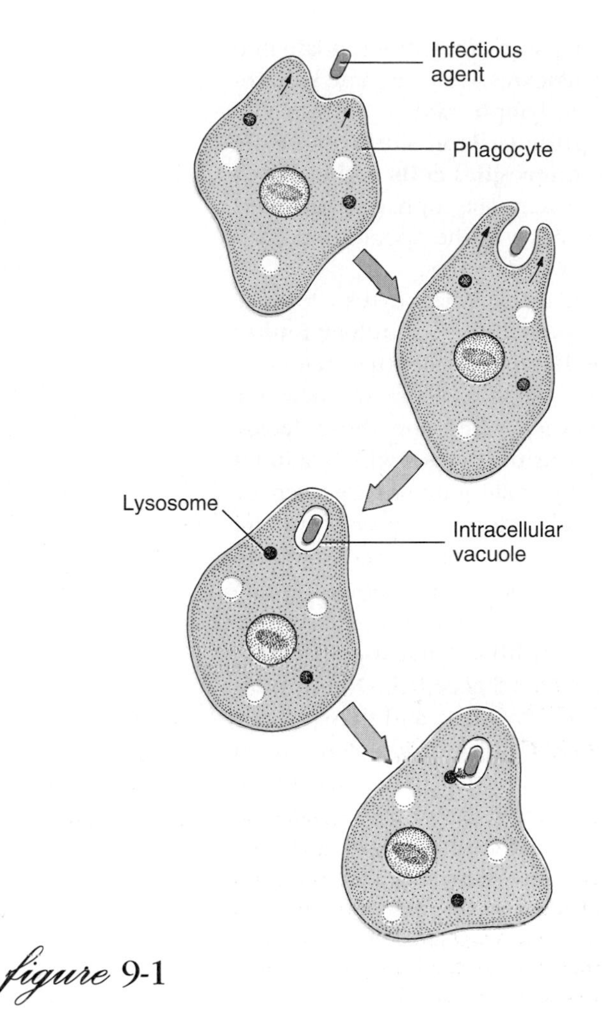

figure 9-1

Phagocytosis. By pseudopodial movement, the phagocyte engulfs the particle. Lysosomes join with the vacuole containing the agent, pouring their contents (digestive enzymes or lysozymes) into the vacuole to destroy the particle.

Innate Defense Mechanisms

Phagocytosis

Most animals have one or more innate mechanisms to protect themselves against invasion of a foreign body or infectious agent. These may be coincidental attributes of certain structures, for example, a tough skin or high stomach acidity, or they may be characteristics evolved as adaptations for defense. For defense against an invader, the cells in an animal must recognize when a substance does not belong in that animal; they must recognize "nonself." **Phagocytosis** illustrates nonself recognition, and it occurs in almost all metazoa and is a feeding mechanism in many single-celled organisms. A cell that has this ability is a **phagocyte.** Phagocytosis is a process of engulfment of the invading particle within an invagination of the phagocyte's cell membrane. The invagination becomes pinched off, and the particle becomes enclosed within an intracellular vacuole (Figure 9-1). **Lysosomes** pour digestive enzymes into the vacuole to destroy the particle. In addition to the digestive enzymes, lysosomes of many phagocytes also contain enzymes that catalyze production of cytotoxic **reactive oxygen intermediates (ROIs)** and **reactive nitrogen intermediates (RNIs).** Examples of ROIs are superoxide radical (O_2^-), hydrogen peroxide (H_2O_2), singlet oxygen ($^1O_2$), and hydroxyl radical (OH). RNIs include nitric oxide (NO) and its oxidized forms, nitrite (NO_2^-) and nitrate (NO_3^-). All such radicals are potentially toxic to invasive microorganisms or parasites.

Phagocytes of Vertebrates

In vertebrates several categories of cells are capable of phagocytosis. Monocytes (see Figure 8-3) arise from stem cells in the bone marrow (Figure 9-2) and give rise to the **mononuclear phagocyte system** or **reticuloendothelial (RE) system.** As the monocytes leave the blood and spread through

[1]Benjamini, E., G. Sunshine, and S. Leskowitz. 1996. Immunology. A short course, ed. 3. New York, Wiley-Liss, Inc.

a variety of tissues, they differentiate into active phagocytes. They become **macrophages** in lymph nodes, spleen, and lung; **Kupffer cells** in sinusoids of the liver; and **microglial cells** in the central nervous system. Macrophages also have important roles in the specific immune response of vertebrates.

Circulating phagocytes in the blood are **polymorphonuclear leukocytes (PMNs),** a name that refers to the highly variable shape of their nucleus. Another name for these leukocytes is **granulocytes,** which alludes to the many small granules that can be seen in their cytoplasm after treatment with appropriate stains. According to the staining properties of the granules, granulocytes are further subdivided into **neutrophils, eosinophils,** and **basophils** (see Figure 8-3). Neutrophils are the most abundant, and they provide the first line of phagocytic defense in an infection. Eosinophils in normal blood account for about 2% to 5% of the total leukocytes, and basophils are the least numerous at about 0.5%. A high **eosinophilia** (eosinophil count in the blood) is often associated with allergic diseases and parasitic infections.

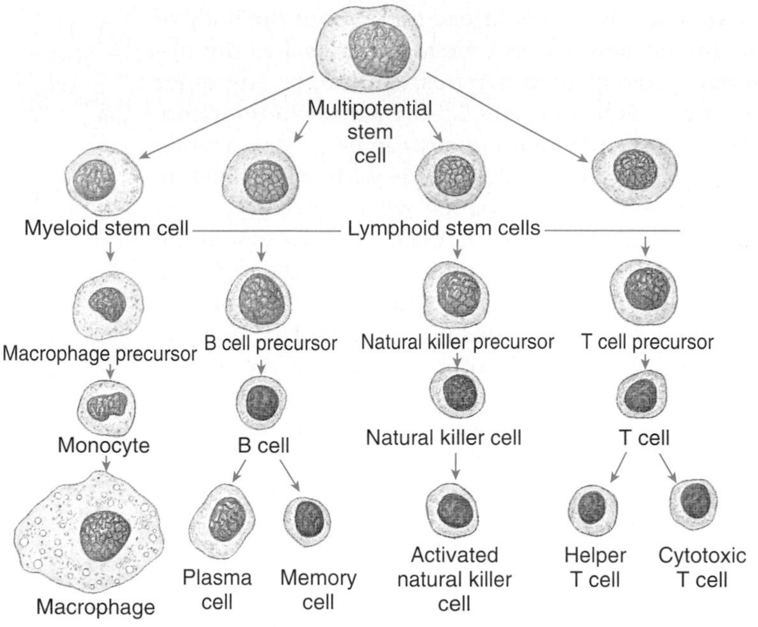

figure 9-2

Lineages of some cells active in immune response. These cells, as well as red blood cells and other white blood cells are derived from multipotential stem cells in the bone marrow. B cells mature in bone marrow and are released into blood or lymph. Precursors of T cells go through a period in the thymus gland. Precursors of macrophages circulate in blood as monocytes.

Immunity in Invertebrates

Many invertebrates have specialized cells that function as itinerant troubleshooters within the body, acting to engulf or wall off foreign material (Table 9-1) and to repair wounds. The cells are variously known as amebocytes, hemocytes, coelomocytes, and others, depending on the animals in which they occur (Table 9-1). If the foreign particle is small, it is engulfed by phagocytosis; but if it is larger than about 10 μm, it is usually encapsulated. Arthropods can wall off the foreign object also by deposition of melanin around it, either from the cells of the capsule or by precipitation from the **hemolymph** (blood).

One of the principal tests of the ability of invertebrate tissues to recognize nonself is by grafting a piece of tissue from another individual of the same species **(allograft)** or a different species **(xenograft)** onto the host. If the graft grows in place with no host response, the host tissue is treating it as self, but if cell response and rejection of the graft occur, the host exhibits immune recognition. Most invertebrates reject xenografts; and sponges, cnidarians, annelids, sipunculids, echinoderms, and tunicates can reject allografts. Interestingly, nemertines, arthropods, and molluscs seem not to reject allografts. Some animals with quite simple body organization, such as sponges (phylum Porifera) and radiates (phylum Cnidaria), can reject allografts. This may be an adaptation to avoid loss of integrity of the individual sponge or cnidarian colony under conditions of crowding, with attendant danger of overgrowth or fusion with other individuals.

Hemocytes of molluscs release degradative enzymes during phagocytosis and encapsulation, and bactericidal substances occur in the body fluids of a variety of invertebrates. Substances that can cause vertebrate erythrocytes to clump together (agglutinate) have been found in many invertebrates, although the function of these molecules in the living animal is unclear. They may act as **opsonins,** enhancing or facilitating phagocytosis, in molluscs and arthropods.

Bacterial infection in some insects stimulates production of antibacterial proteins, but these proteins show broad-spectrum activity and are not specific for a single infective agent. Contact with infectious organisms can bring the defense systems of snails into enhanced levels of readiness that last for up to two months or more. Specific, induced responses that demonstrate memory upon challenge (p. 217), resembling vertebrate immune responses, have been found in cockroaches.

Innate Immunity in Vertebrates

Vertebrates have numerous structural and physiological characteristics that reduce susceptibility to certain invaders. Examples include physical barriers, such as a thick, cornified epidermis or other protective external covering; the ability to repair damaged tissues rapidly; and high acidity in the stomach. A variety of bac-

table 9-1

Some Invertebrate Leukocytes and Their Functions

Group	Cell types and functions	Phagocytosis	Encapsulation	Allograft rejection	Xenograft rejection
Sponges	Archaeocytes (wandering cells that differentiate into other cell types and can act as phagocytes)	+	+	+*	+*
Cnidarians	Amebocytes: "lymphocytes"	+		+	+
Nemertines	Agranular leukocytes; granular macrophage-like cells	+		−	±
Annelids	Basophilic amebocytes (accumulate as "brown bodies"), acidophilic granulocytes	+	+	+	+
Sipunculids	Several types	+	+	±	+
Insects	Several types, depending on family; e.g., plasmatocytes, granulocytes, spherule cells, coagulocytes (blood clotting)	+	+	−	±
Crustaceans	Granular phagocytes; refractile cells that lyse and release contents	+	+	−	+
Molluscs	Amebocytes	+	+	−	+
Echinoderms	Amebocytes, spherule cells, pigment cells, vibratile cells (blood clotting)	+	+	+	+
Tunicates	Many types, including phagocytes; "lymphocytes"	+	+	+	+

**Transplantation reactions occur, but the extent to which the leukocytes are involved is unknown.*

Source: Data from Lackie, A. M. Parasitology 80: 392–412. (See Lackie's article for references.)

tericidal and parasiticidal substances are present in such animal body secretions as tears, mucus, saliva, and urine. In fact at least one of what we previously thought was nonspecific antimicrobial substance we now know is a class of antibody, IgA. IgA can cross cellular barriers easily. It seems to be an important protective agent in the mucus of the intestinal epithelium, and it is present in mucus in the respiratory tract, in tears, in saliva, and in sweat. Other, nonantibody substances in normal human milk can kill intestinal protozoa such as *Giardia lamblia* (p. 371) and *Entamoeba histolytica* (p. 373), and these substances may be important in protection of infants against such infections.

Acquired Immune Response in Vertebrates

Vertebrates have a specialized system of nonself recognition that results in increased resistance to a *specific* foreign substance or invader on repeated exposures. Investigations of the mechanisms involved are currently intense, and our knowledge of them is increasing rapidly.

An **antigen** is any substance that will stimulate an immune response. Antigens may be any of a variety of substances with a molecular weight of over 3000, most commonly proteins, and are usually (but not always) foreign to the host. There are two arms of the immune response, known as **humoral** and **cellular.** Humoral immunity is based on antibodies, most of which are dissolved in and circulate in the blood, whereas cellular immunity is associated entirely with cell surfaces. There is extensive communication and interaction among the cells of the humoral and cellular responses.

Basis of Self and Nonself Recognition

Major Histocompatibility Complex

We have known for many years that nonself recognition is very specific. If tissue from one individual is transplanted into another individual in the same species, the graft grows for a time and then dies as immunity against it arises. Tissue grafts grow successfully only if they are between identical twins or between individuals of highly inbred strains of animals. The molecular basis for this nonself recognition depends on proteins imbedded in the cell surface. These proteins are coded by certain genes, now known as the **major histocompatibility complex (MHC).** The MHC proteins are among the most variable known, and unrelated individuals almost always have different allelic variants of these proteins. There are two types of MHC proteins: class I and class II. Class I proteins are found on the surface of virtually all cells, whereas class II MHC proteins are found only on certain cells participating in the immune responses, such as certain lymphocytes and macrophages.

The capability of an immune response develops over a period of time in the early development of the organism. All substances present at the time the capacity develops are recognized as self in later life. Unfortunately, the system of self and nonself recognition sometimes breaks down, and an animal may begin to produce antibodies against some part of its own body. This leads to one of several known autoimmune diseases, such as rheumatoid arthritis, multiple sclerosis, lupus, and insulin-dependent diabetes mellitus.

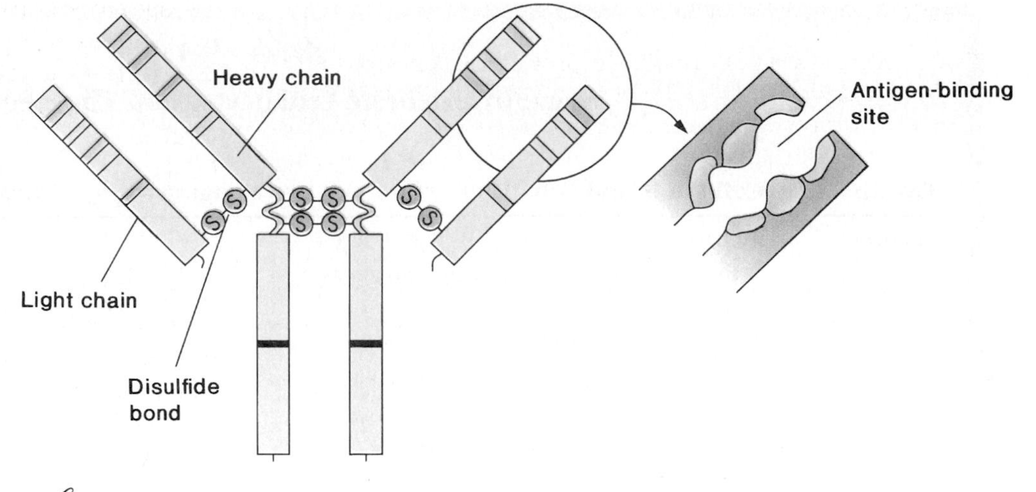

figure 9-3

Antibody molecule is composed of two shorter polypeptide chains (light chains) and two longer chains (heavy chains) held together by covalent disulfide bonds. The light chains may be either of two types. The class of antibody is determined by the type of heavy chain: mu (IgM), gamma (IgG), alpha (IgA), delta (IgD), or epsilon (IgE). The constant portion of each chain does not vary for a given type or class, and the variable portion varies with the specificity of the antibody. Antigen-binding sites are in clefts formed in the variable portions of the heavy and light chains. IgM normally occurs as a pentamer, five of the structures illustrated being bound together by another chain. IgA may occur as a monomer, dimer, or trimer.

We will discuss the role of MHC proteins in nonself recognition in the following text, but they are not themselves the molecules that recognize the foreign substance. This task falls to two basic types of molecules, the genes for which probably evolved from a common ancestor. The types of recognition molecules are **antibodies** and **T-cell receptors.**

Antibodies

Antibodies are proteins called **immunoglobulins.** The basic antibody molecule consists of four polypeptide strands: two identical light chains and two identical heavy chains, held together in a Y-shape by disulfide bonds and hydrogen bonds (Figure 9-3). The amino acid sequence toward the ends of the Y varies in both the heavy and light chains, according to the specific antibody molecule (the **variable region**), and this sequence determines with which antigen the antibody can bind. Each of the ends of the Y forms a cleft that acts as the antigen-binding site (Figure 9-3), and the specificity of the molecule depends on the shape of the cleft and the properties of the chemical groups that line its walls. The diversity in specificity among antibodies is enormous: each individual has more than 1×10^7, perhaps as many as 10^9 structurally different antibody molecules. The remainder of the antibody is known as the **constant region,** but the "constant" region consists of several subtypes. The variable end of the antibody molecule is often called **Fab** (for antigen-binding fragment) region, and the constant end is the **Fc** (for crystallizable fragment) region. The constant region of the light chains can be either of two types; the heavy chains may be any of five types. The type of heavy chain determines the **class** of the antibodies, referred to as **IgM, IgG** (now familiar to many people as "gamma globulin"), **IgA, IgD,** and **IgE,** respectively. The class of the antibody determines the role of the antibody in the immune response (for example, whether the antibody is secreted or held on a cell surface), but not the antigen it recognizes.

Many aspects of immunology have been greatly assisted by the discovery of a method for producing stable clones of cells that will produce only one kind of antibody. Such monoclonal antibodies will bind to only one kind *of antigenic determinant (most proteins bear many different antigenic determinants and thus stimulate the body to produce complex mixtures of antibodies).* Monoclonal *antibodies are made by fusing normal antibody-producing plasma cells with a continuously growing plasma cell line, producing a hybrid of the normal cell with one that can divide indefinitely in culture. This is called a* hybridoma. *Clones are selected from among the hybrids and are grown to become "factories" that produce almost unlimited quantities of one specific antibody. Hybridoma techniques discovered in 1975 have become one of the most important research tools for the immunologist.*

T-cell Receptors

T-cell receptors are transmembrane proteins on the surfaces of T cells (described in the following text). Like antibodies, T-cell receptors have a constant region and a variable region. The constant region extends slightly into the cytoplasm and the variable region, which binds with specific antigens, extends outward. Most T cells also bear other transmembrane proteins closely linked to the T-cell receptors, which serve as **accessory** or **coreceptor** molecules. These are of one of two types: **CD4** or **CD8.**

A major problem of immunology is understanding how the mammalian genome could contain the information needed to produce at least a million different antibodies. The answer seems to be that antibody genes occur in pieces, rather than as continuous stretches of DNA, and that the antigen-recognizing sites (variable regions) of the heavy and light chains of the antibody molecules are pieced together from information supplied by separate DNA sequences, which can be shuffled to increase the diversity of the gene products. The immense repertoire of antibodies is achieved in part by complex gene rearrangements and in part by frequent somatic mutations that produce additional variation in protein structure of the variable regions of the heavy and light antibody chains. Analogous processes occur in the production of genes for T-cell receptors.

Cells of the Immune Response

T Cells and B Cells

We have already mentioned some of the cells that are important in the immune response, such as PMNs and macrophages. A number of others are mostly encompassed by a category of leukocytes known as **lymphocytes. B lymphocytes (B cells)** have antibody molecules in their surface and give rise to cells that actively secrete antibodies into the blood. **T lymphocytes (T cells)** bear T-cell receptors. There is a vast number of different kinds of B cells, each bearing on its surface molecules of antibody that will bind with one particular antigen, even though that antigen may not previously have been present in the body. There is an equally great number of different T cells with receptors for specific antigens.

Lymphocytes are **activated** when they are stimulated to move from their recognition phase, in which they simply bind with a particular antigen, to a phase in which they proliferate and differentiate into cells that function to eliminate that antigen. We also speak of activation of effector cells, such as macrophages, when they are stimulated to carry out their protective function.

Subsets of T Cells Communication between cells in the immune response, regulation of the response, and certain effector functions are carried out by different kinds of T cells. Subsets of T cells can be distinguished by the coreceptor proteins in their surface membranes. Cells with CD4 (for **c**luster of **d**ifferentiation) coreceptors are $CD4^+$ and those with CD8 are described as $CD8^+$. Until recently, immunologists believed that certain $CD4^+$ cells (T helper or T_H) activated immune responses, and certain $CD8^+$ cells suppressed such responses. Evidence now suggests a more complicated web of interactions (Figure 9-4). Some T_H cells (designated T_H1) activate cell-mediated immunity while suppressing the humoral response, and others (called T_H2) activate humoral and suppress cell-mediated immunity.

Cytotoxic T lymphocytes (CTLs) are $CD8^+$ cells that kill target cells expressing a certain antigen. The CTL binds tightly to the target cell and secretes a protein that causes pores to form in the cell membrane. The target cell then lyses.

Other Cells **Natural killer cells (NK)** are lymphocyte-like cells that can kill virus-infected and tumor cells in the absence of antibody. **Mast cells** are basophil-like cells found in the dermis and other tissues. Their surfaces bear receptors for the Fc portions of IgE and IgG.

Cytokines

The 1980s saw rapid advances in our knowledge of how cells of immunity communicate with each other. They do this by means of protein hormones called **cytokines.** Cytokines can produce their effects on the same cells that produce them, on cells nearby, or on cells distant in the body from those that produced the cytokine. Some cytokines important in immune responses are the following:

1. **Interleukin-1 (IL-1).** The interleukins were originally so-called because they are synthesized by leukocytes and have their effect on leukocytes. We now know that some other kinds of cells can produce interleukins, and interleukins produced by leukocytes can affect other kinds of cells. IL-1 is produced by activated macrophages and mediates the host inflammatory response. It also activates T cells and B cells.
2. **Interleukin-2 (IL-2).** IL-2 is produced by $CD4^+$ cells and to a lesser extent by $CD8^+$ cells. It is a major growth factor for T and B cells, and it enhances the cytolytic activity of natural killer cells, causing them to become **lymphocyte-activated killer (LAK)** cells.
3. **Interleukin-3 (IL-3).** IL-3 is produced by $CD4^+$ cells and is a multilineage colony-stimulating factor. It promotes growth and differentiation of all cell types in the bone marrow.
4. **Interleukin-4 (IL-4).** IL-4 is produced mostly by T_H2 $CD4^+$ cells. It is a growth factor for B cells, some $CD4^+$ T cells, and mast cells, but it suppresses T_H1 differentiation.

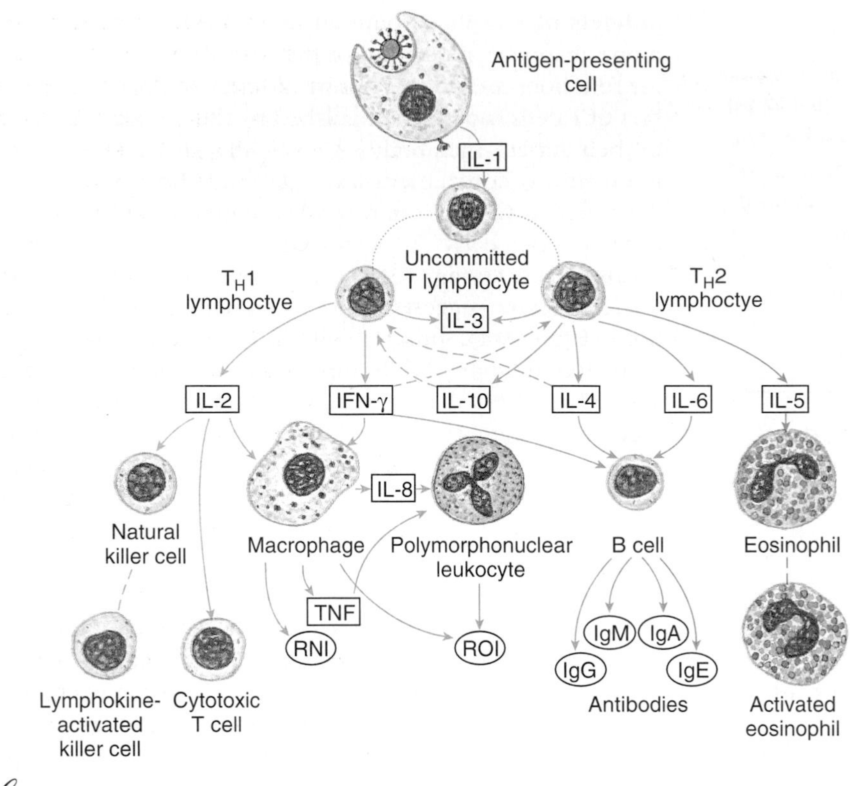

figure 9-4

Major pathways involved in the immune response as mediated by cytokines. Solid arrows indicate positive signals and broken arrows indicate inhibitory signals. Broken lines without arrows indicate path of cellular activation. *IFN-γ*, interferon-γ; *Ig*, immunoglobulin; *IL*, interleukin; *TNF*, tumor necrosis factor; T_H1, helper $CD4^+$ and $CD8^+$ cells that stimulate cell mediated response; T_H2, helper $CD4^+$ and $CD8^+$ cells that stimulate humoral response; *RNI*, reactive nitrogen intermediates; *ROI*, reactive oxygen intermediates.

5. **Interleukin-5 (IL-5).** IL-5 is produced by certain $CD4^+$ cells and stimulates activation of eosinophils so that they can kill some parasitic worms. It also acts with IL-2 and IL-4 to stimulate growth and differentiation of B cells.
6. **Interleukin-6 (IL-6).** IL-6 is produced by macrophages, endothelial cells, fibroblasts, and T_H2 cells. It is an important growth factor for B cells late in the sequence of B-cell differentiation.
7. **Interleukin-8 (IL-8).** IL-8 is one of a family of low molecular weight inflammatory cytokines derived from antigen-activated T cells, activated macrophages, endothelial cells, fibroblasts, and platelets. IL-8 is an activating and chemotactic factor for neutrophils and to a lesser extent for other PMNs.
8. **Interleukin-10 (IL-10).** IL-10 is derived from T_H2 $CD4^+$ cells, and it inhibits T_H1, $CD8^+$, NK, and macrophage cytokine synthesis.
9. **Transforming growth factor-β (TGF-β).** TGF-β is produced by macrophages, lymphocytes, and other cells. It inhibits lymphocyte proliferation, CTL and LAK cell generation, as well as macrophage cytokine production.
10. **Interferon-γ (IFN-γ).** IFN-γ is produced by some $CD4^+$ and almost all $CD8^+$ cells. It is a strong macrophage-activating factor, causes a variety of cells to express class II MHC molecules, promotes differentiation of T and B cells, activates neutrophils and NK cells, and activates endothelial cells to allow lymphocytes to pass through walls of vessels. It activates the T_H1 arm and suppresses the T_H2 arm of the immune response.
11. **Tumor necrosis factor (TNF).** Activated macrophages secrete most TNF. It is a major mediator of inflammation. In low concentrations TNF activates endothelial cells, activates PMNs, and stimulates macrophages and cytokine production (including IL-1, IL-6, and TNF itself). In higher concentrations TNF causes increased synthesis of prostaglandins (p. 275) in the hypothalamus, resulting in fever.

Generation of a Humoral Response

When an antigen is introduced into the body, it binds to a specific antibody on the surface of the appropriate B cell, but this is usually not sufficient to activate the B cell to multiply. Some of the antigen is taken up by **antigen-presenting cells (APCs),** such as macrophages, that partially digest the antigen. The APCs then incorporate portions of the antigen into their own cell surface, bound in the cleft of MHC II protein (Figures 9-4 and 9-5). That portion of the antigen presented on the surface of the macrophage or other APC is called the **epitope** (or **determinant**). The macrophages also secrete IL-1, which stimulates T_H2 cells. The specific T-cell receptor for that particular epitope recognizes the epitope bound to the MHC II protein. Binding of the T-cell receptor to the epitope-MHC II complex is enhanced by the coreceptor CD4, which itself binds to the constant portion of the MHC II protein (Figure 9-6). The bound CD4 molecule also transmits a stimulation signal to the interior of the T cell. Transmission of the stimulation signal also requires binding of certain polypeptides, designated CD3, to the T-cell receptor. Activation of the T cell further requires interaction of additional costimulatory and adhesion signals from other proteins on the surface of the macrophage and T cell. The CD8 coreceptor enhances binding of the T-cell receptor on $CD8^+$ cells and transmits a stimulatory signal into the T cell.

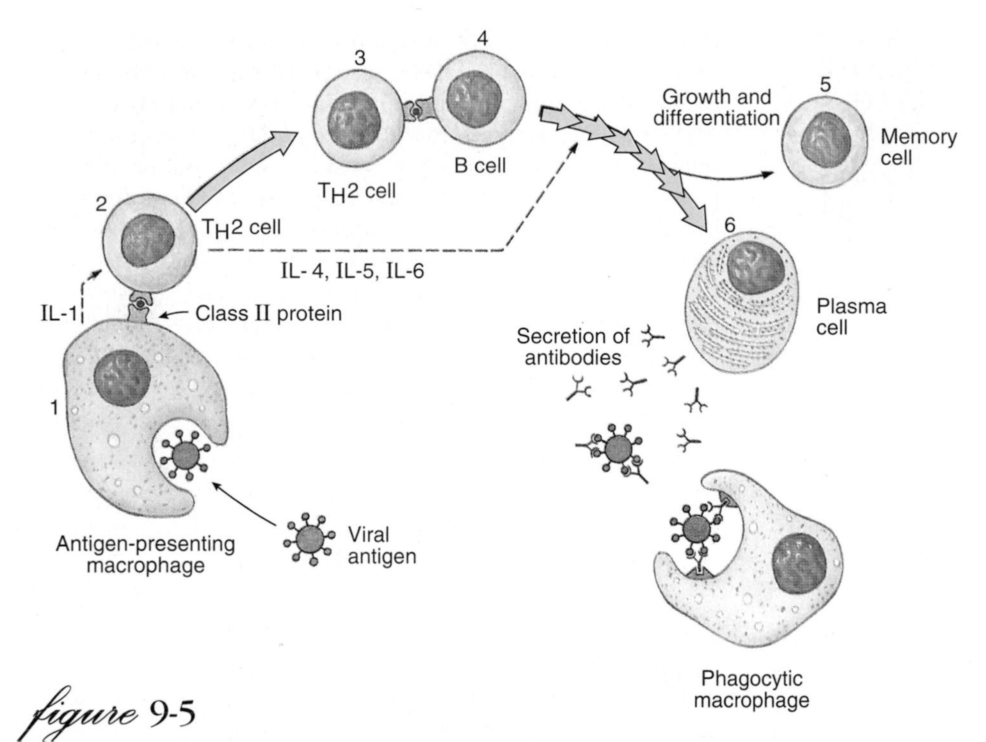

figure 9-5

Humoral immune response. *1,* Macrophage consumes antigen, partially digests it, and displays epitopes on its surface with class II MHC protein, and secretes interleukin-1 (IL-1). *2,* T helper cell, stimulated by IL-1, recognizes antigen and class II protein on macrophage, is activated, and secretes interleukin-2 (IL-2). *3,* T helper then activates B cell, which carries antigen and class II protein on its surface. IL-2 stimulates proliferation of B-cell line and stimulates the T helper and other T cells. *4,* Activated B cells finally produce many plasma cells that secrete antibody. *5,* Some of B-cell progeny become memory cells. *6,* Antibody produced by plasma cells binds to antigen and stimulates macrophages to consume antigen (opsonization).

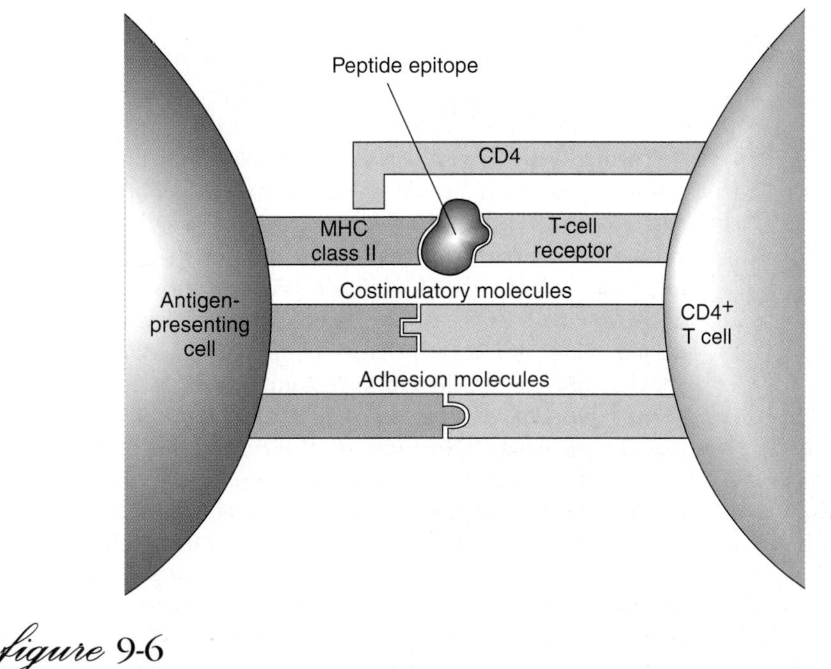

figure 9-6

Interacting molecules during activation of a T helper cell.

The activated T_H2 cell secretes IL-4, IL-5, and IL-6, which activate the B cell that has the same epitope and class II MHC protein on its surface. The B cell multiplies rapidly and produces many **plasma cells,** which secrete large quantities of antibody for a period of time, then die. Thus if we measure the concentration of the antibody **(titer)** soon after the antigen is injected, we can detect little or none. The titer then rises rapidly as the plasma cells secrete antibody, and it may decrease somewhat as they die and the antibody is degraded (Figure 9-7). However, if we give another dose of antigen (the **challenge**), there is no lag, and the antibody titer rises quickly to a higher level than after the first dose. This is the **secondary** or **anamnestic response,** and it occurs because some of the activated B cells gave rise to long-lived **memory cells.** There are many more memory cells present in the body than the original B lymphocyte with the appropriate antibody on its surface, and they rapidly multiply to produce additional plasma cells.

Functions of Antibody in Host Defense

Antibodies can mediate destruction of an invader (antigen) in a number of ways. A foreign particle, for example, becomes coated with antibody molecules as their Fab regions bind to it. Macrophages recognize the projecting Fc regions and are stimulated to engulf the particle. This is the process of **opsonization.**

Another important process, particularly in the destruction of bacterial cells, is interaction with **complement.** Complement is a series of 12 enzymes that are activated by bound antibody, and they actually punch holes in the bacterial cell surface. Complement also plays a role in opsonization.

Antibody bound to the surface of an invader may trigger contact killing of the invader by host cells **(antibody-dependent, cell-mediated cytotoxicity [ADCC]).** Receptors for Fc of bound antibody on a microorganism or tumor

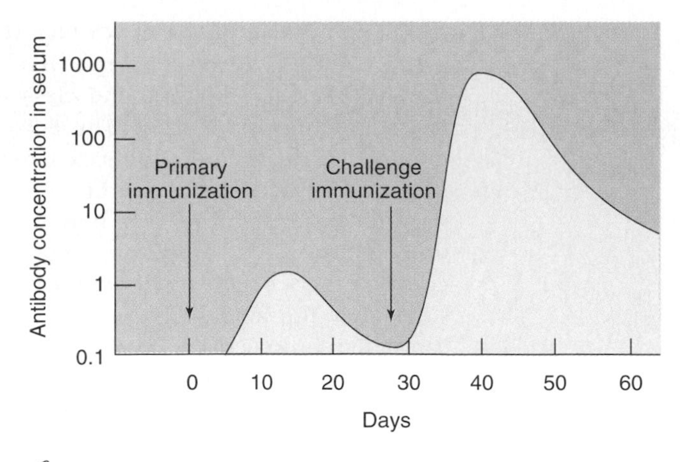

figure 9-7

Typical immunoglobulin response after primary and challenge immunizations. The secondary response is a result of the large numbers of memory cells produced after the primary B-cell activation.

cell cause natural killer cells to adhere to them and pour forth the cytotoxic contents of their vacuoles.

The Cell-Mediated Response

Some immune responses involve little, if any, antibody and depend on the action of cells only. In cell-mediated immunity (CMI) the epitope of the antigen is also presented by macrophages, but the T_H1 arm of the immune response is activated and the T_H2 arm suppressed. Effector cells are macrophages, PMNs, cytotoxic T cells, and activated natural killer cells. The specific interaction of lymphocyte and antigen that generates a CMI greatly influences subsequent events in the nonspecific response we call **inflammation.**

Like humoral immunity, CMI shows a secondary response due to large numbers of memory T cells that were produced from the original activation. For example, a second tissue graft (challenge) between the same donor and host will be rejected much more quickly than the first.

Acquired Immune Deficiency Syndrome (AIDS)

AIDS is an extremely serious disease in which the ability to mount an immune response is disabled completely. It is caused by the **human immunodeficiency virus (HIV).** The first case of what we now know as AIDS was recognized in 1981, and by the end of 1994, over 420,000 individuals had contracted the disease in the United States alone. In 1996 it was estimated that 18.5 million people in the world are infected with HIV, of which 11 million are in sub-Saharan Africa. HIV infection virtually always progresses to AIDS, after a latent period of some years. AIDS is a seriously chronic disease and, to the best of our current knowledge, it is ultimately terminal. AIDS patients are continuously plagued by infections with microbes and parasites that cause insignificant problems in persons with normal immune responses. HIV preferentially invades and destroys $CD4^+$ lymphocytes. CD4 protein is the major surface receptor for the virus. Normally, $CD4^+$ cells make up 60% to 80% of the T-cell population; in AIDS they can become too rare to be detected. T_H1 cells are relatively more depleted than T_H2 cells, which upsets the balance of immunoregulation and results in persistent, nonspecific B-cell activation.

Only a few years ago transplantation of organs from one person to another seemed impossible. Then physicians began to transplant kidneys and depress the immune response to the recipient. It was very difficult to immunosuppress the recipient enough that the new organ would not be rejected and at the same time not leave the patient defenseless against infection. Since cyclosporine, a drug derived from a fungus, was discovered not only kidneys, but also hearts, lungs, and livers can be transplanted. Cyclosporine inhibits IL-2 and affects CTLs more than T_Hs. It has no effect on other white cells or on healing mechanisms, so that a patient can still mount an immune response but not reject the transplant. However, the patient must continue to take cyclosporine, because if the drug is stopped, the body will recognize the transplanted organ as foreign and reject it.

Inflammation

Inflammation is a vital process in mobilizing the body's defenses against an invading organism and in repairing the damage thereafter. Although inflammation is a nonspecific process, it is greatly influenced by the body's prior immunizing experience with the invader. Also, a more noxious foreign substance will produce more intense inflammation. Tissue damage initiates release of pharmacologically active substances (such as histamine) from certain cells in the area. These substances increase the diameter of nearby small blood vessels and also increase their permeability. Thus redness and warmth result from the increased amount of blood in the area (hence "inflammation"), and more proteins and fluid escape into the tissue, causing swelling **(edema).** This swelling and change in permeability of the capillaries allows antibodies and leukocytes to move from the capillaries and easily reach the invader. The first phagocytic line of defense is the neutrophils, which may last only a few days, then macrophages (either fixed or differentiated from monocytes)

table 9-2
Major Blood Groups

Blood type	Genotype	Antigens on red blood cells	Antibodies in serum	Can give blood to	Can receive blood from	Frequency in United States (%)		
						Whites	Blacks	Asians
O	*O/O*	None	Anti-A and anti-B	All	O	45	48	31
A	*A/A, A/O*	A	Anti-B	A, AB	O, A	41	27	25
B	*B/B, B/O*	B	Anti-A	B, AB	O, B	10	21	34
AB	*A/B*	AB	None	AB	All	4	4	10

become predominant. "Pus" of an infection is formed principally from exuded tissue fluid and spent phagocytes.

We are unaware of the majority of minute invasions that are always occurring. Most are efficiently disposed of and heal leaving little or no trace. However, if tissue damage has been more severe, fibrous connective tissue (scar tissue) will be deposited in the area.

Blood Group Antigens

ABO Blood Types

Blood cells differ chemically from person to person, and when two different (incompatible) blood types are mixed, **agglutination** (clumping together) of erythrocytes results. The basis of these chemical differences is naturally occurring antigens on the membranes of red blood cells. The best known of these inherited immune systems is the ABO blood group. The antigens A and B are inherited as codominant alleles of a single gene. Homozygotes for a recessive allele at the same gene have type O blood, which lacks the A and B antigens. Thus, as shown in Table 9-2, an individual with, for example, genes *A/A* or *A/O* develops A antigen (blood type A). The presence of a *B* gene produces B antigens (blood type B), and for the genotype *A/B* both A and B antigens develop on the erythrocytes (blood type AB). Epitopes of A and B also are present on the surfaces of many epithelial and most endothelial cells.

There is an odd feature about the ABO system. Normally we would expect that a type A individual would develop antibodies against type B cells only if cells bearing B epitopes were first introduced into the body. In fact, type A persons acquire anti-B antibodies soon after birth, even without exposure to type B cells. Similarly, type B individuals come to carry anti-A antibodies at a very early age. Type AB blood has neither anti-A nor anti-B antibodies (since if it did, it would destroy its own blood cells), and type O blood has both anti-A and anti-B antibodies. There is evidence that the antibodies develop as a response to A and B epitopes on intestinal microorganisms when the intestine becomes colonized with bacteria after birth. Presumably, small and unnoticed infections with the bacteria occur. The antibodies thus produced cross-react with the A and B epitopes on erythrocytes.

We see then that the blood group names identify their *antigen* content. Persons with type O blood are called universal donors because, lacking antigens, their blood can be infused into a person with any blood type. Even though it contains anti-A and anti-B antibodies, these are so diluted during transfusion that they do not react with A or B antigens in a recipient's blood. In practice, however, clinicians insist on matching blood types to prevent any possibility of incompatibility.

Rh Factor

Karl Landsteiner, an Austrian—later American—physician discovered the ABO blood groups in 1900. In 1940, 10 years after receiving the Nobel Prize, he made still another famous discovery. This was a blood group called the Rh factor, named after the rhesus monkey, in which it was first found. Approximately 85% of white individuals in the United States have the factor (positive) and the other 15% do not (negative). The Rh factor is encoded by a dominant allele at a single gene. Rh-positive and Rh-negative bloods are incompatible; shock and even death may follow their mixing when Rh-positive blood is introduced into an Rh-negative person who has been sensitized by an earlier transfusion of Rh-positive blood. Rh incompatibility accounts for a peculiar and often fatal **hemolytic disease of the newborn (erythroblastosis fetalis).** If an Rh-negative mother has an Rh-positive baby (father is Rh-positive) she can become immunized by the fetal blood during the birth process. Anti-Rh antibodies are predominately IgG and can cross the placenta during a subsequent pregnancy and agglutinate the fetal blood. Erythroblastosis fetalis normally is not a problem in cases of ABO incompatibility because antibodies to ABO antigens are primarily IgM and cannot cross the placenta.

> *The genetics of the Rh factor are very much more complicated than it was believed when the factor was first discovered. Some authorities think that three genes located close together on the same chromosome are involved, whereas others adhere to a system of one gene with many alleles. In 1968 a revision of the single gene concept listed 37 alleles necessary to account for the phenotypes then known. Furthermore, the frequency of the various alleles varies greatly between whites, Asians, and blacks.*

Erythroblastosis fetalis can now be prevented by giving an Rh-negative mother anti-Rh antibodies just after the birth of her first child. These antibodies remain long enough to neutralize any Rh-positive fetal blood cells that may have entered her circulation, thus preventing her own antibody machinery from being stimulated to produce the Rh-positive antibodies. Active, permanent immunity is blocked. The mother must be treated after every subsequent pregnancy (assuming the father is Rh^+). If the mother has already developed an immunity, however, the baby may be saved by an immediate, massive transfusion of blood free of antibodies.

Summary

Phagocytosis, a process of engulfment of a food or foreign particle, is an innate mechanism that demonstrates nonself recognition. Engulfed particles are digested and may be killed by cytotoxic substances produced by the phagocyte. Phagocytes in vertebrates are both fixed in certain tissues and circulate in the blood, such as polymorphonuclear leukocytes.

Many invertebrates show nonself recognition by rejection of xenografts or allografts or both. In some cases they may show enhanced response on repeated exposure.

Vertebrates demonstrate increased resistance to *specific* foreign substances (antigens) on repeated exposure, and the resistance is based on a vast number of specific recognition molecules: antibodies and T-cell receptors. Nonself recognition depends on markers in cell surfaces known as major histocompatibility (MHC) proteins. Antibodies are borne on the surfaces of B lymphocytes (B cells) and in solution in the blood after secretion by the progeny of B cells, the plasma cells. T-cell receptors occur only on the surfaces of T lymphocytes (T cells).

The cells of immunity communicate with each other and with other cells in the body by means of protein hormones called cytokines such as interleukins, tumor necrosis factor, and interferon-γ. The two arms of the vertebrate immune response are the humoral response (T_H2), involving antibodies, and the cell-mediated response (T_H1), involving cell surfaces only. When one arm is activated or stimulated, its cells produce cytokines that tend to suppress activity in the other arm. Activation of either arm requires that the antigen be consumed by an APC (antigen-presenting cell, usually a macrophage), which partially digests the antigen and presents its determinant (epitope) on the surface of the APC along with an MHC class II protein. Extensive communication by cytokines and activation (and suppression) of various cells in the response lead to production of specific antibody or proliferation of T cells with the specific receptors that recognize the antigenic epitope. After the initial response, memory cells remain in the body and are responsible for enhanced response on next exposure to the antigen.

Damage to the immune response done by HIV (human immunodeficiency virus) in production of AIDS (acquired immune deficiency syndrome) is due primarily to destruction of a crucial set of T cells: those bearing the CD4 protein on their surface.

Inflammation is an important part of the body's defense; it is greatly influenced by prior immunizing experience with an antigen.

People have genetically determined antigens in the surfaces of their red blood cells (ABO blood groups and others); blood types must be compatible in transfusions, or the transfused blood will be agglutinated by antibodies in the recipient.

Review Questions

1. Phagocytosis is an important defense mechanism in most animals. How are phagocytes classified? Name several kinds of cells that are phagocytic.
2. Give some evidence that cells of many invertebrates bear molecules on their surface that are specific to the species and even to a particular individual animal.
3. What is the molecular basis of self and nonself recognition in vertebrates?
4. What is the difference between T cells and B cells?
5. What is a cytokine? What are some functions of cytokines?
6. Outline the sequence of events in a humoral immune response from the introduction of antigen to the production of antibody.

7. Define the following: plasma cell, secondary response, memory cell, complement, opsonization, titer, challenge, cytokine, natural killer cell, interleukin-2.
8. What are the functions of CD4 and CD8 proteins on the surface of T cells?
9. In general, what are consequences of activation of the T_H1 arm of the immune response? Activation of the T_H2 arm?
10. Distinguish between class I and class II MHC proteins.
11. Describe a typical inflammatory response.
12. What is a major mechanism by which HIV damages the immune system in AIDS?
13. Give the genotypes of each of the following blood types: A, B, O, AB. What happens when a person with type A gives blood to a person with type B? With type AB? With type O?
14. What causes hemolytic disease of the newborn (erythroblastosis fetalis)? Why does the condition not arise in cases of ABO incompatibility?

Selected References

Abbas, A. K., A. R. Lichtman, and J. S. Pober. 1994. Cellular and molecular immunology, ed. 2. Philadelphia, Saunders. *Good account of current immunology.*

Benjamini, E., G. Sunshine, and S. Leskowitz. 1996. Immunology. A short course, ed. 3. New York, Wiley-Liss, Inc. *An excellent presentation of the essentials without excessive details.*

Cox, F. E. G., and E. Y. Liew. 1992. T-cell subsets and cytokines in parasitic infections. Parasit. Today. **8:**371–374. *Describes the interactions of T cells and cytokines in the T_H1 and T_H2 arms of the immune response.*

Dunn, P. E. 1990. Humoral immunity in insects. BioScience **40:**738–744. *There is evidence that cockroaches are capable of a vertebratelike, specific, adaptive humoral response.*

Engelhard, V. H. 1994. How cells process antigens. Sci. Am. **271:**54–61. *This article focuses on the roles of the MHC proteins.*

Gallo, R. C., and L. Montagnier. 1988. AIDS in 1988. Sci. Am. **259:**40–48 (Oct.). *Authors are the American and French scientists who independently discovered the AIDS virus. Lead article in an issue of* Scientific American *entirely devoted to the status and knowledge of AIDS.*

Garrett, L. 1995. The coming plague: newly emerging diseases in a world out of balance. New York, Penguin Books (orig. publ. 1994, Farrar, Straus, and Giroux, New York). *Like Rachel Carson did on the environment, Garrett sounds a clarion call on infectious diseases. New pathogens are emerging, and familiar ones are developing multidrug resistance at a time when transmission favors the microbes: poor sanitation and increasing crowding in the world's cities and impairment of the immune response in millions of people by malnutrition and AIDS.*

Golde, D. W. 1991. The stem cell. Sci. Am. **265:**86–93. *Undifferentiated cells in the bone marrow give rise to white and red blood cells, macrophages, and platelets.*

Greene, W. C. 1993. AIDS and the immune system. Sci. Am. **269:**98–105 (Sept.). *There is some evidence that HIV triggers widespread apoptosis (programmed cell death) in $CD4^+$ cells.*

Karp, R. D. 1990. Cell-mediated immunity in invertebrates. BioScience **40:**732–737. *Experiments show that allograft rejection in insects has at least a short-term memory component; challenge allografts were rejected more quickly than third-party allograft controls.*

Lichtenstein, L. M. 1993. Allergy and the immune system. Sci. Am. **269:**116–124 (Sept.). *Describes what happens in an allergic response, including the role of mast cells, basophils, cytokines, and chemical mediators.*

Marrack, P., and J. W. Kappler. 1993. How the immune system recognizes the body. Sci. Am. **269:**80–89 (Sept.). *Part of the mechanism by which the immune system tolerates "self" antigens is by a mechanism known as clonal deletion.*

Paul, W. E. 1993. Infectious diseases and the immune system. Sci. Am. **269:**90–97 (Sept.). *Describes the immune response in certain viral, microbial, and parasitic infections.*

Steinman, L. 1993. Autoimmune disease. Sci. Am. **269:**106–114 (Sept.). *In 5% of adults in Europe and North America, the immune system discrimination between "self" and "nonself" breaks down, usually with very serious results.*

Strange, C. 1995. Rethinking immunity. BioScience **45:**663–668. *Some immunologists believe that the self/nonself dogma of immunology is inadequate.*

Weiss, R. 1994. Of myths and mischief. Discover **15**(12):36–42. *Debunks the astonishing myths that have grown up around AIDS, including the most pernicious of all, that HIV is not the cause.*

Digestion and Nutrition

chapter | ten

A Consuming Cornucopia

Sir Walter Raleigh observed that the difference between a rich man and a poor man is that the former eats when he pleases while the latter eats when he can get it. In today's crowded world, with 90 million people added each year to the world's population of 5.8 billion, the separation between the well-fed affluent and the hungry and malnourished poor reminds us that time has not diminished the shrewdness of Sir Walter's remark. Unlike the affluent for whom food acquisition requires only the selection of prepackaged foods at a well-stocked supermarket, the world's poor can appreciate that for them, as for the rest of the animal kingdom, procuring food is fundamental to survival. For most animals, eating is the main business of living.

Potential food is everywhere and little remains unexploited. Animals bite, chew, nibble, crush, graze, browse, shred, rasp, filter, engulf, enmesh, suck, and soak up foods of incredible variety. What an animal eats and how it eats profoundly affect an animal's feeding specialization, its behavior, its physiology, and its internal and external anatomy—in short, both its body form and its role in the web of life. The endless evolutionary jostling between predator and prey has provided compromise adaptations for eating and adaptations for avoiding being eaten. By whatever means food may be secured, there is far less variation among animals in the subsequent digestive simplification of foods. Vertebrates and invertebrates alike use similar digestive enzymes. Even more uniform are the final biochemical pathways for nutrient use and energy transformation. The nourishment of animals can be likened to cornucopia in which the food flows in rather than out. A great diversity of foods procured by countless feeding adaptations streams into the mouth of the horn, is simplified, and finally applied to the common purpose of survival and reproduction.

All organisms require energy to maintain their highly ordered and complex structure. This energy is chemical bond energy that is released by transforming complex compounds acquired from the organism's environment into simpler ones.

The ultimate source of energy for life on earth is the sun. Sunlight is captured by chlorophyll molecules in green plants, which transforms a portion of this energy into chemical bond energy (food energy). Green plants are **autotrophic** organisms; they require only inorganic compounds absorbed from their surroundings to provide the raw material for synthesis and growth. Most autotrophic organisms are the chlorophyll-bearing **phototrophs,** although some, the chemosynthetic bacteria, are **chemotrophs;** they gain energy from inorganic chemical reactions.

Almost all animals are **heterotrophic organisms** that depend on already-synthesized organic compounds of plants and other animals to obtain the materials they will use for growth, maintenance, and reproduction of their kind. Since the food of animals, normally the complex tissues of other organisms, is usually too bulky to be absorbed directly by cells, it must be broken down, or digested, into soluble molecules that are small enough to be used.

Animals may be divided into a number of categories on the basis of dietary habits. **Herbivorous** animals feed mainly on plant life. **Carnivorous** animals feed mainly on herbivores and other carnivores. **Omnivorous** forms eat both plants and animals. **Saprophagous** animals feed on decaying organic matter.

The ingestion of foods and their simplification by digestion are only initial steps in nutrition. Foods reproduced by digestion to soluble, molecular form are **absorbed** into the circulatory system and **transported** to the tissues of the body. There they are **assimilated** into the structure of cells. Oxygen is also transported by blood to the tissues, where food products are **oxidized,** or burned to yield energy and heat. Food not immediately used is **stored** for future use. Wastes produced by oxidation must be **excreted.** Food products unsuitable for digestion are **egested** in the form of feces.

In this chapter we will first examine the feeding adaptations of animals. Next we will discuss digestion and absorption of food. We will close with a consideration of nutritional requirements of animals.

Feeding Mechanisms

Few animals can absorb nutrients directly from their external environments. Some intestinal protozoan parasites, and tapeworms and acanthocephalans that lack a digestive tract, nourish themselves on primary organic molecules absorbed directly across their body surfaces. Most animals, however, must work for their meals. They are active feeders that have evolved numerous specializations for obtaining food. With food procurement as one of the most potent driving forces in animal evolution, natural selection has placed a high priority on adaptations for exploiting new sources of food and the means of food capture and intake. In this brief discussion we consider some of the major food-gathering devices.

Feeding on Particulate Matter

Drifting microscopic particles are found in the upper hundred meters of the ocean. Most of this multitude is **plankton,** organisms too small to do anything but drift with the ocean's currents. The rest is organic debris, the disintegrating remains of dead plants and animals. Although this oceanic swarm of plankton forms a rich life domain, it is unevenly distributed. The heaviest plankton growth occurs in estuaries and areas of updwelling, where there is an abundant nutrient supply. It is consumed by numerous larger animals, invertebrates and vertebrates, using a variety of feeding mechanisms.

One of the most important and widely employed methods for feeding to have evolved is **suspension feeding** (Figure 10-1). The majority of suspension feeders use ciliated surfaces to produce currents that draw drifting food particles into their mouths. Most suspension-feeding invertebrates, such as tube-dwelling polychaete worms, bivalve molluscs, hemichordates, and most protochordates, entrap the particulate food on mucous sheets that convey the food into the digestive tract. Others, such as fairy shrimps, water fleas, and barnacles, use sweeping movements of their setae-fringed legs to create water currents and entrap food, which is transferred to the mouth. In the freshwater developmental stages of certain insect orders, the organisms use fanlike arrangements of setae or spin silk nets to entrap food.

Suspension feeding has evolved frequently as a secondary modification among representatives of groups that are primarily selective feeders. Examples are many of the microcrustaceans, fishes such as herring, menhaden, and basking sharks, certain birds such as the flamingo, and the largest of all animals, baleen (whalebone) whales. The vital importance of one component of plankton, the diatoms, in supporting a great pyramid of suspension-feeding animals is stressed by N.J. Berrill[1]:

> A humpback whale . . . needs a ton of herring in its stomach to feel comfortably full—as many as five thousand individual fish. Each herring, in turn, may well have 6000 or 7000 small crustaceans in its own stomach, each of which contains as many as 130,000 diatoms. In other words, some 400 billion yellow-green diatoms sustain a single medium-sized whale for a few hours at most.

Another type of particulate feeding exploits deposits of disintegrated organic material (detritus) that accumulates on and in the substratum; this is called **deposit feeding.** Some deposit feeders, such as many annelids and some hemichordates,

[1]Berrill, N.J. 1958. You and the universe. New York, Dodd, Mead & Co.

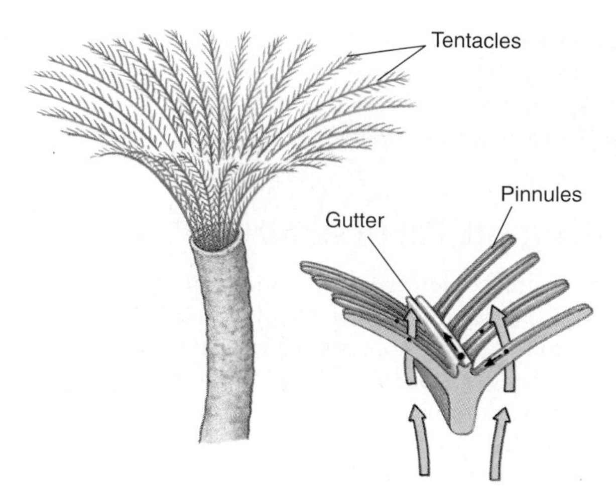

A, Marine fan worms (class Polychaeta, phylum Annelida) have a crown of tentacles. Numerous cilia on the edges of the tentacles draw water (*solid arrows*) between pinnules where food particles are entrapped in mucus; particles are then carried down a "gutter" in the center of the tentacle to the mouth (*broken arrows*).

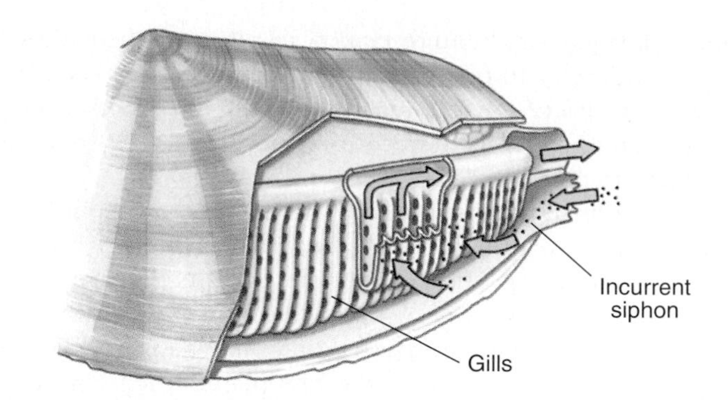

B, Bivalve molluscs (class Bivalvia, phylum Mollusca) use their gills as feeding devices, as well as for respiration. Water currents created by cilia on the gills carry good particles into the current siphon and between slits in the gills where they are entangled in a mucous sheet covering the gill surface. Ciliated food grooves then transport the particles to the mouth (not shown). Arrows indicate direction of water movement.

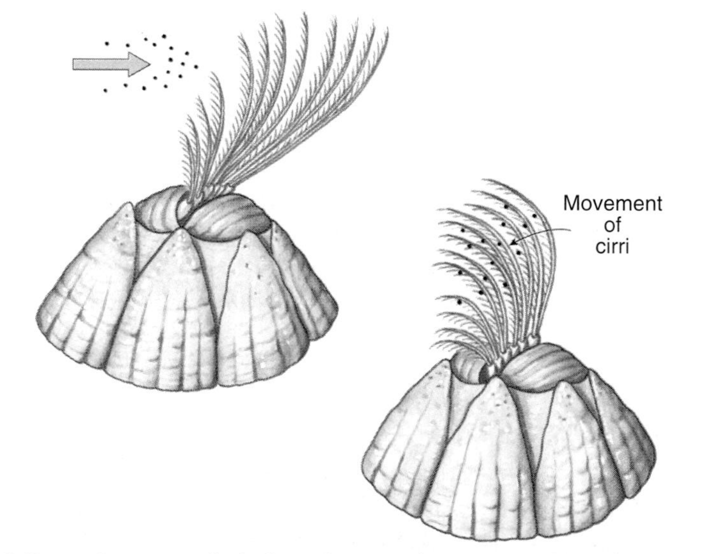

C, Barnacles sweep their thoracic appendages (cirri) through the water to trap plankton and other organic particles on fine bristles that fringe the cirri. Food is transferred to the barnacle's mouth by the first, short cirri. Class Malacostraca, subphylum Crustacea, phylum Arthropoda.

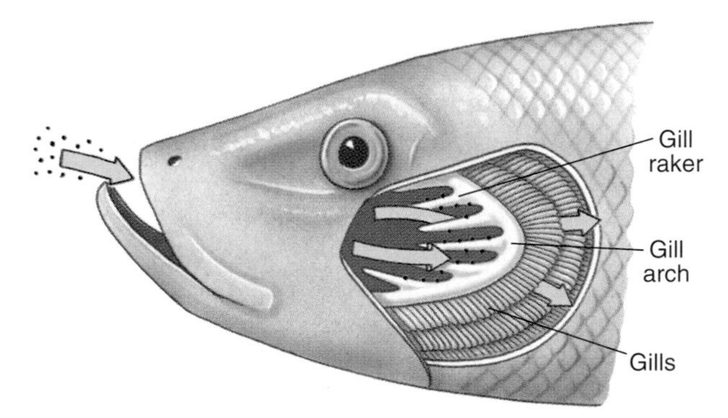

D, Herring and other suspension-feeding fishes (class Osteichthyes, phylum Chordata) use gill rakers that project forward from the gill arches into the pharyngeal cavity to strain plankton. Herring swim almost constantly, forcing water and suspended food into the mouth; food is strained out by the gill rakers, and the water passes through the gill openings.

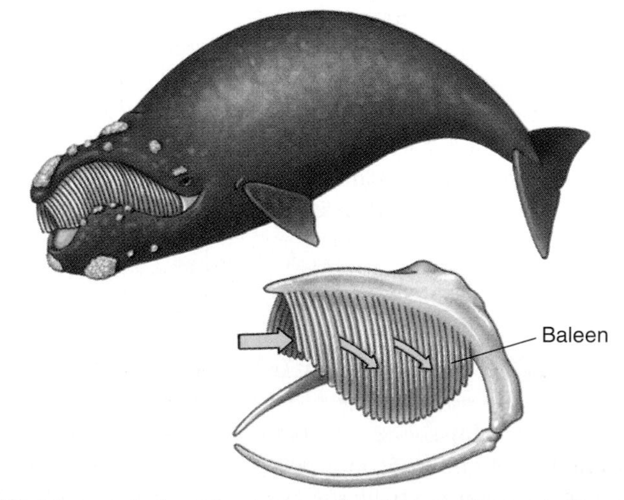

E, Whalebone whales (class Mammalia, phylum Chordata) filter out plankton, principally large crustaceans called krill, with whalebone, or baleen. Water enters the swimming whale's open mouth by the force of the animal's forward motion and is strained out through the more than 300 horny baleen plates that hang like a curtain from the roof of the mouth. Krill and other plankton caught in the baleen are periodically collected by the huge tongue and swallowed.

figure 10-1

Some suspension feeders and their feeding mechanisms.

simply pass the substrate through their bodies, removing from it whatever provides nourishment. Others, such as scaphopod molluscs, certain bivalve molluscs, and some sedentary and tube-dwelling polychaete worms, use appendages to gather organic deposits some distance from the body and move them toward the mouth (Figure 10-2).

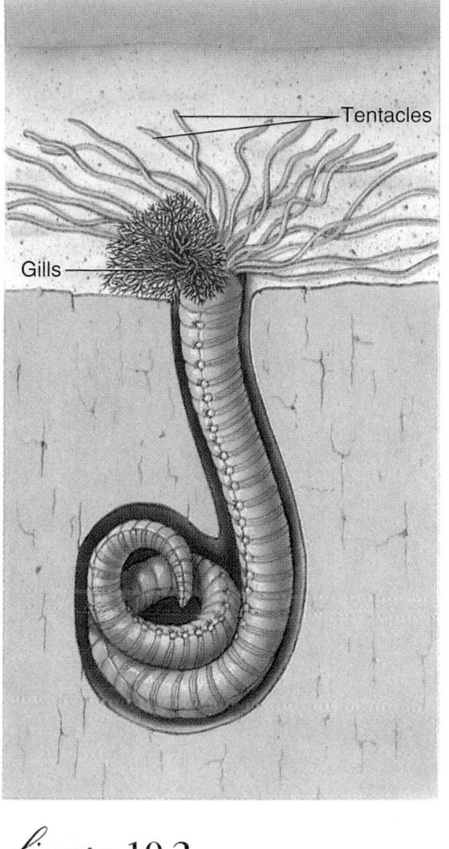

figure 10-2

The annelid *Amphitrite* is a deposit feeder that lives in a mucus-lined burrow and extends long feeding tentacles in all directions across the surface. Food trapped on mucus is conveyed along the tentacles to the mouth.

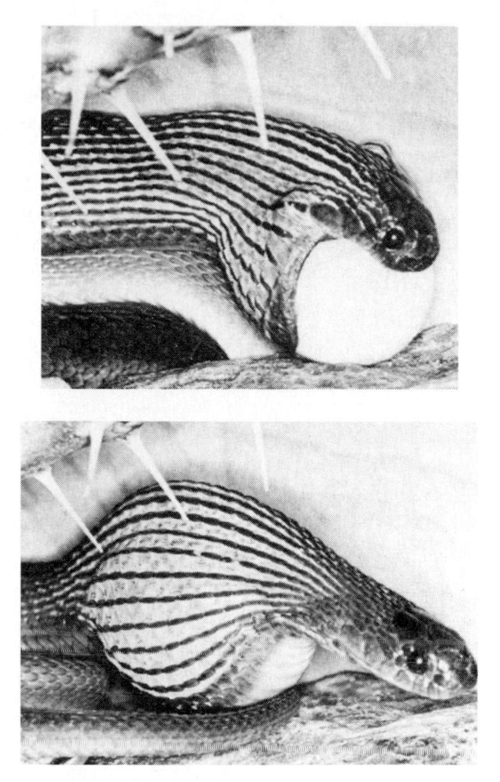

figure 10-3

This African egg-eating snake, *Dasypeltis scaber,* subsists entirely on hard-shelled birds' eggs, which it swallows whole. Its special adaptations are reduced size and number of teeth, enormously expansible jaw provided with elastic ligaments, and teethlike vertebral spurs that puncture the shell. Shortly after the second photograph was taken, the snake punctured and collapsed the egg, swallowed its contents, and regurgitated the crushed shell.

Feeding on Food Masses

Some of the most interesting animal adaptations are those that have evolved for procuring and manipulating solid food. Such adaptations and the animals bearing them are largely shaped by what the animal eats.

Predators must be able to locate, capture, hold, and swallow prey. Most carnivorous animals simply seize food and swallow it intact, although some employ toxins that paralyze or kill prey at the time of capture. Although no true teeth appear among invertebrates, many have beaks or toothlike structures for biting and holding. A familiar example is the carnivorous polychaete *Nereis,* which possesses a muscular pharynx armed with chitinous jaws that can be everted with great speed to seize prey (Figure 22-2A, p. 477). Once a capture is made, the pharynx is retracted and the prey swallowed. Fish, amphibians, and reptiles use their teeth principally to grip the prey and prevent its escape until they can swallow it whole. Snakes and some fishes can swallow enormous meals. This, together with the absence of limbs, is associated with some striking feeding adaptations in these groups: recurved teeth for seizing and holding prey and distensible jaws and stomachs to accommodate their large and infrequent meals (Figure 10-3). Birds lack teeth, but the bills are often provided with serrated edges or the upper bill is hooked for seizing and tearing apart prey.

Many invertebrates are able to reduce food size by shredding devices (such as the shredding mouthparts of many crustaceans) or by tearing devices (such as the beaklike jaws of cephalopod molluscs). Insects have three pairs of appendages on their heads that serve variously as jaws, chitinous teeth, chisels, tongues, or sucking tubes. Usually the first pair serves as crushing teeth; the second as grasping jaws; and the third, as a probing and tasting tongue.

True mastication, that is, the chewing of food as opposed to tearing or crushing, is found only among mammals. Mammals usually have four different types of teeth, each adapted for specific functions. **Incisors** are designed for biting, cutting, and stripping; **canines** are for seizing, piercing, and tearing; **premolars** and **molars,** at the back of the jaw, are for grinding and crushing (Figure 10-4). This basic pattern is often greatly modified in animals having specialized food habits (Figure 10-5; see also Figure 31-9, p. 675). Herbivores have suppressed canines but well-developed molars with enamel ridges for grinding. The well-developed, self-sharpening incisors of rodents grow throughout life and must be worn away by gnawing to keep pace with growth. Some teeth have become so highly modified that they are no longer useful for biting or chewing food. An elephant's tusk (Figure 10-6) is a modified upper incisor used for defense, attack, and rooting, and the male wild boar has modified canines that are used as weapons. Many feeding specializations of mammals are described on pp. 674–676.

Herbivorous, or plant-eating, animals have evolved special devices for crushing and cutting plant material. Some invertebrates have scraping mouthparts, such as the radula of

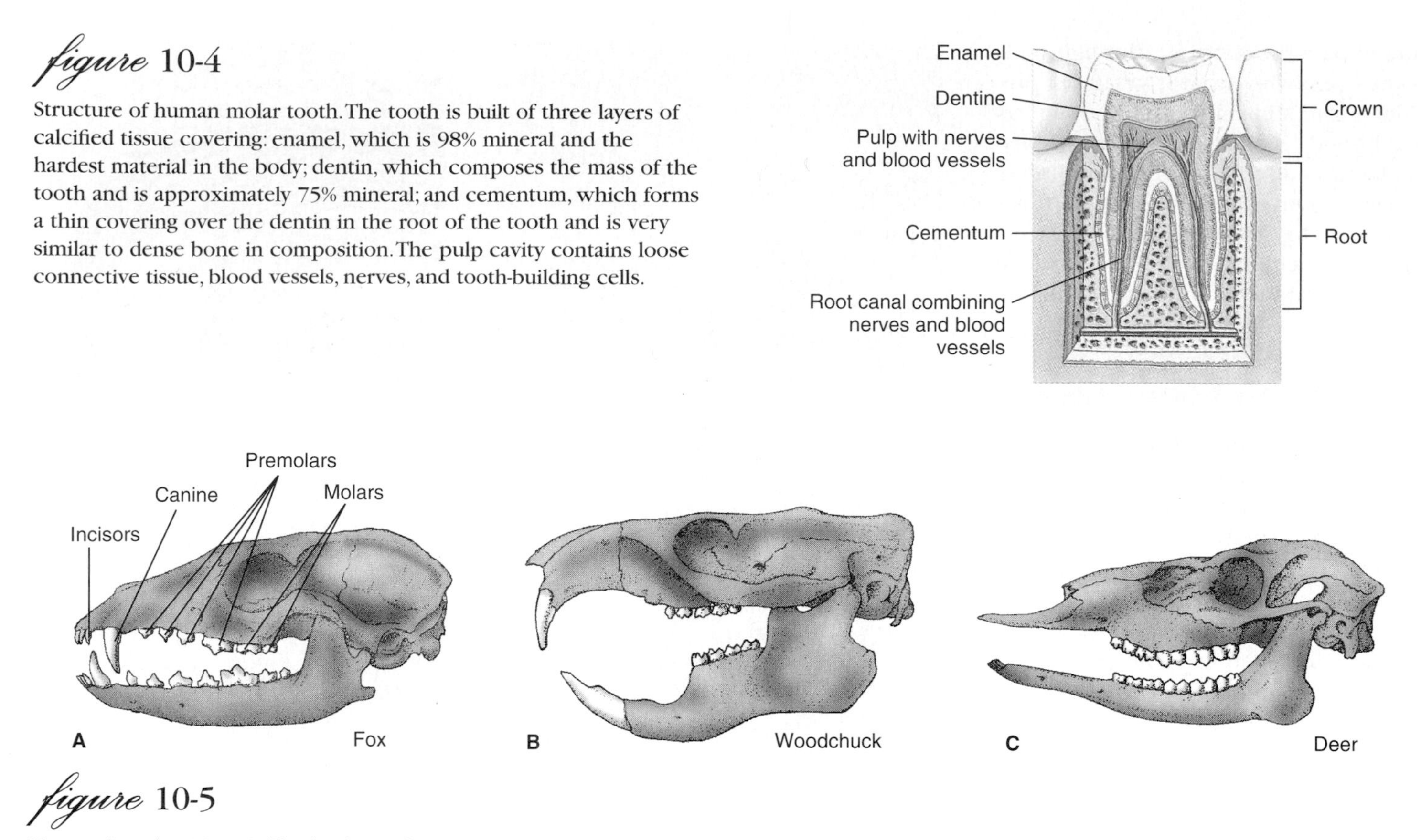

figure 10-4

Structure of human molar tooth. The tooth is built of three layers of calcified tissue covering: enamel, which is 98% mineral and the hardest material in the body; dentin, which composes the mass of the tooth and is approximately 75% mineral; and cementum, which forms a thin covering over the dentin in the root of the tooth and is very similar to dense bone in composition. The pulp cavity contains loose connective tissue, blood vessels, nerves, and tooth-building cells.

figure 10-5

Mammalian dentition. **A,** Teeth of gray fox, a carnivore, showing the four types of teeth; **B,** Woodchuck, a rodent, has chisel-like incisors that continue to grow throughout life to replace wear; **C,** White-tailed deer, a browsing ungulate, with flat molars bearing complex ridges suited for grinding.

snails (Figure 21-3, p. 451). Insects such as locusts have grinding and cutting mandibles; herbivorous mammals such as horses and cattle use wide, corrugated molars for grinding. All these mechanisms disrupt the tough cellulose cell wall to accelerate its digestion by intestinal microorganisms, as well as to release the cell contents for direct enzymatic breakdown. Thus herbivores are able to digest food that carnivores cannot, and in doing so, convert plant material into protein for consumption by carnivores and omnivores.

Feeding on Fluids

Fluid feeding is especially characteristic of parasites, but it is practiced among many free-living forms as well. Some internal parasites (endoparasites) simply absorb the nutrient surrounding them, unwittingly provided by the host. Others bite and rasp host tissue, suck blood, and feed on the contents of the host's intestine. External parasites (ectoparasites) such as leeches, lampreys, parasitic crustaceans, and insects use a variety of efficient piercing and sucking mouthparts to feed on blood or other body fluid. Unfortunately for humans and other warm-blooded animals, the ubiquitous mosquito excels in its bloodsucking habit. Alighting gently, the mosquito sets about puncturing its prey with an array of six needlelike mouthparts (Figure 23-40B, p. 513). One of these is used to inject an anticoagulant saliva (responsible for the irritating itch that follows the "bite" and serving as a vector for microorganisms causing malaria, yellow fever, encephalitis, and other diseases); another mouthpart is a channel through which the blood is sucked. It is of little comfort that only the female dines on blood to obtain the necessary nutrients for the formation of eggs.

Digestion

In the process of digestion, which means literally "carrying asunder," organic foods are mechanically and chemically broken down into small units for absorption. Although food solids consist principally of carbohydrates, proteins, and fats, the very components that make up the body of the consumer, these components must first be reduced to their simplest molecular units and dissolved before they can be assimilated. Each animal reassembles some of these digested and absorbed units into organic compounds of the animal's own unique pattern. Cannibalism confers no special metabolic benefit; victims of an animal's own kind are digested just as thoroughly as food composed of another species.

In protozoa and sponges, digestion is entirely **intracellular** (Figure 10-7). The food particle is enclosed within a food vacuole by phagocytosis (see pp. 34). Digestive enzymes are

figure 10-6

An African elephant loosening soil from a salt lick with its tusk. Elephants use their powerful modified incisors in many ways in the search for food and water: plowing the ground for roots, prying apart branches to reach the edible cambium, and drilling into dry riverbeds for water.

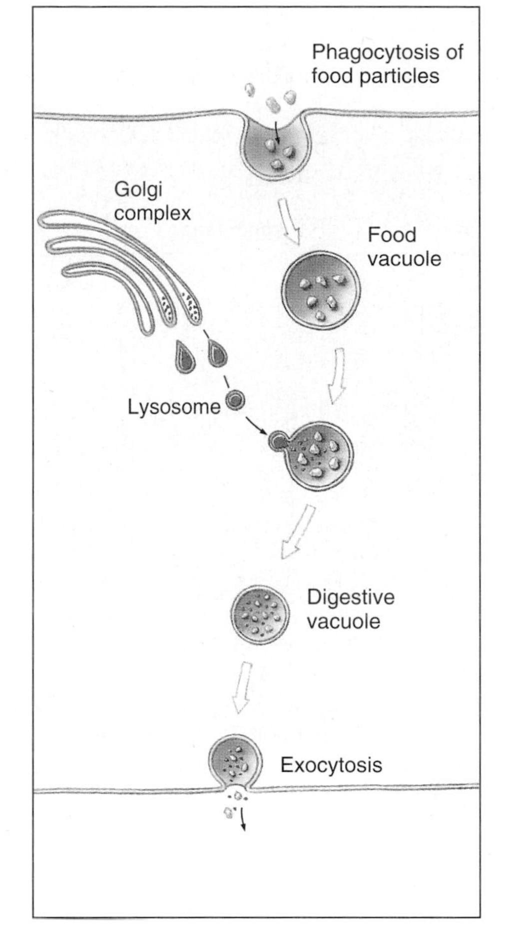

figure 10-7

Intracellular digestion. Lysosomes containing digestive enzymes (lysozymes) are produced within the cell, possibly by the Golgi complex. Lysosomes fuse with food vacuoles and release enzymes that digest the enclosed food. Usable products of digestion are absorbed into the cytoplasm, and indigestible wastes are expelled.

added and the products of digestion, the simple sugars, amino acids, and other molecules, are absorbed into the cell cytoplasm where they may be used directly or, in the case of multicellular animals, may be transferred to other cells. Food wastes are simply extruded from the cell.

There are important limitations to intracellular digestion. Only particles small enough to be phagocytized can be accepted, and every cell must be capable of secreting all of the necessary enzymes, and of absorbing the products into the cytoplasm. These limitations were resolved with the evolution of an **alimentary system** in which **extracellular** digestion of large food masses could take place. In extracellular digestion, certain cells lining the **lumen** (cavity) of the alimentary canal specialize in forming various digestive secretions, whereas others function largely, or entirely, in absorption. Many of the simpler metazoans, such as radiates, turbellarian flatworms, and ribbon worms (nemerteans), practice both intracellular and extracellular digestion. With the evolution of greater complexity and the appearance of complete mouth-to-anus alimentary systems, extracellular digestion became emphasized, together with increasing regional specialization of the digestive tract. For arthropods and vertebrates, digestion is almost entirely extracellular. The ingested food is exposed to various mechanical, chemical, and bacterial treatments, to different acidic and alkaline phases, and to digestive juices that are added at appropriate stages as the food passes through the alimentary canal.

Action of Digestive Enzymes

Mechanical processes of cutting and grinding by teeth and muscular mixing by the intestinal tract are important in digestion. However, the reduction of foods to small, absorbable units relies principally on chemical breakdown by **enzymes,** discussed in Chapter 2 (p. 40). The digestive enzymes are **hydrolytic** enzymes, or **hydrolases,** so called because food molecules are split by the process of **hydrolysis,** that is, the breaking of a chemical bond by adding the components of water across it:

$$R{-}R + H_2O \xrightarrow[\text{enzyme}]{\text{digestive}} R{-}OH + H{-}R$$

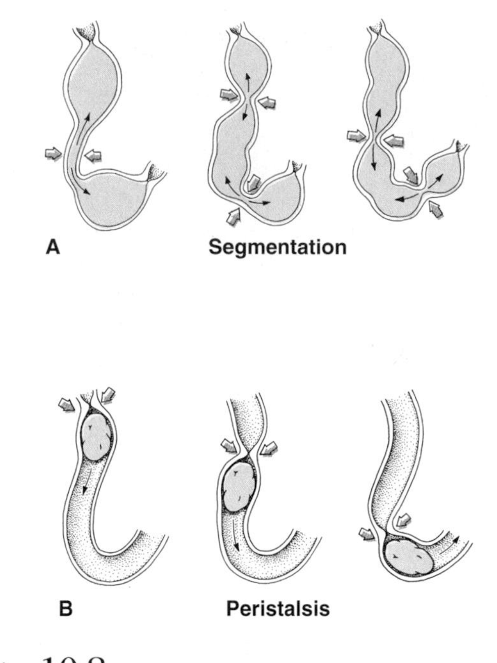

figure 10-8

Movement of intestinal contents by segmentation and peristalsis. **A,** Segmentational movements of food showing how constrictions squeeze the food back and forth, mixing it with enzymes. The sequential mixing movements occur at about 1-second intervals. **B,** Peristaltic movement, showing how food is propelled forward by a traveling wave of contraction.

In this general enzymatic reaction, R—R represents a food molecule that is split into two products, R—OH and R—H. Usually these reaction products must in turn be split repeatedly before the original molecule is reduced to its numerous subunits. Proteins, for example, are composed of hundreds, or even thousands, of interlinked amino acids, which must be completely separated before the individual amino acids can be absorbed. Similarly, carbohydrates must be reduced to simple sugars. Fats (lipids) are reduced to molecules of glycerol, fatty acids, and monoglycerides, although some fats, unlike proteins and carbohydrates, may be absorbed without first being completely hydrolyzed. There are specific enzymes for each class of organic compounds. These enzymes are located in specific regions of the alimentary canal in an "enzyme chain," in which one enzyme may complete what another has started. The product then moves posteriorly for still further hydrolysis.

Motility in the Alimentary Canal

Food is moved through the digestive tract by **cilia** or by specialized **musculature,** and often by both. Movement is usually by cilia in the acoelomate and pseudocoelomate metazoa that lack the mesodermally derived gut musculature of true coelomates. Cilia move intestinal fluids and materials also in some eucoelomates, such as most molluscs, in which the coelom is weakly developed. In animals with well-developed coeloms, the gut is usually lined with two opposing layers of muscle: a longitudinal layer, in which the smooth muscle fibers run parallel with the length of the gut, and a circular layer, in which the muscle fibers embrace the circumference of the gut. The most characteristic gut movement is **segmentation,** the alternate constriction of rings of smooth muscle of the intestine that constantly divide and squeeze the contents back and forth (Figure 10-8A). Walter B. Cannon of homeostasis fame (p. 167), while still a medical student at Harvard in 1900, was the first to use X rays to watch segmentation in experimental animals that had been fed suspensions of barium sulfate. Segmentation serves to mix food but does not move it through the gut. Another kind of muscular action, called **peristalsis,** sweeps the food down the gut with waves of contraction of circular muscle (Figure 10-8B).

Organization and Regional Function of the Alimentary Canal

The metazoan alimentary canal can be divided into five major regions: (1) reception, (2) conduction and storage, (3) grinding and early digestion, (4) terminal digestion and absorption, and (5) water absorption and concentration of solids. Food progresses from one region to the next, allowing digestion to proceed in sequential stages (Figure 10-9).

Receiving Region

The first region of the alimentary canal consists of devices for feeding and swallowing. These include the mouthparts (for example mandibles, jaws, teeth, radula, bills), the **buccal cavity** and muscular **pharynx.** Most metazoans other than suspension feeders have **salivary glands** (buccal glands) that produce lubricating secretions containing mucus to assist swallowing (Figure 10-10). Salivary glands often have other specialized functions such as secretion of toxic enzymes for quieting struggling prey and secretion of salivary enzymes to begin digestion. The salivary secretion of the leech, for example, is a complex mixture containing an anesthetic substance (making its bite nearly painless) and several enzymes that prevent blood coagulation and increase blood flow by dilating veins and dissolving the tissue cement that binds cells together.

Salivary **amylase** is a carbohydrate-splitting enzyme that begins the hydrolysis of plant and animal starches. It is found only in certain herbivorous molluscs, some insects, and in primate mammals, including humans. Starches are long polymers of glucose. Salivary amylase does not completely hydrolyze starch, but breaks it down mostly into two-glucose fragments called **maltose.** Some free glucose and longer fragments of starch are also produced. When the food mass is swallowed,

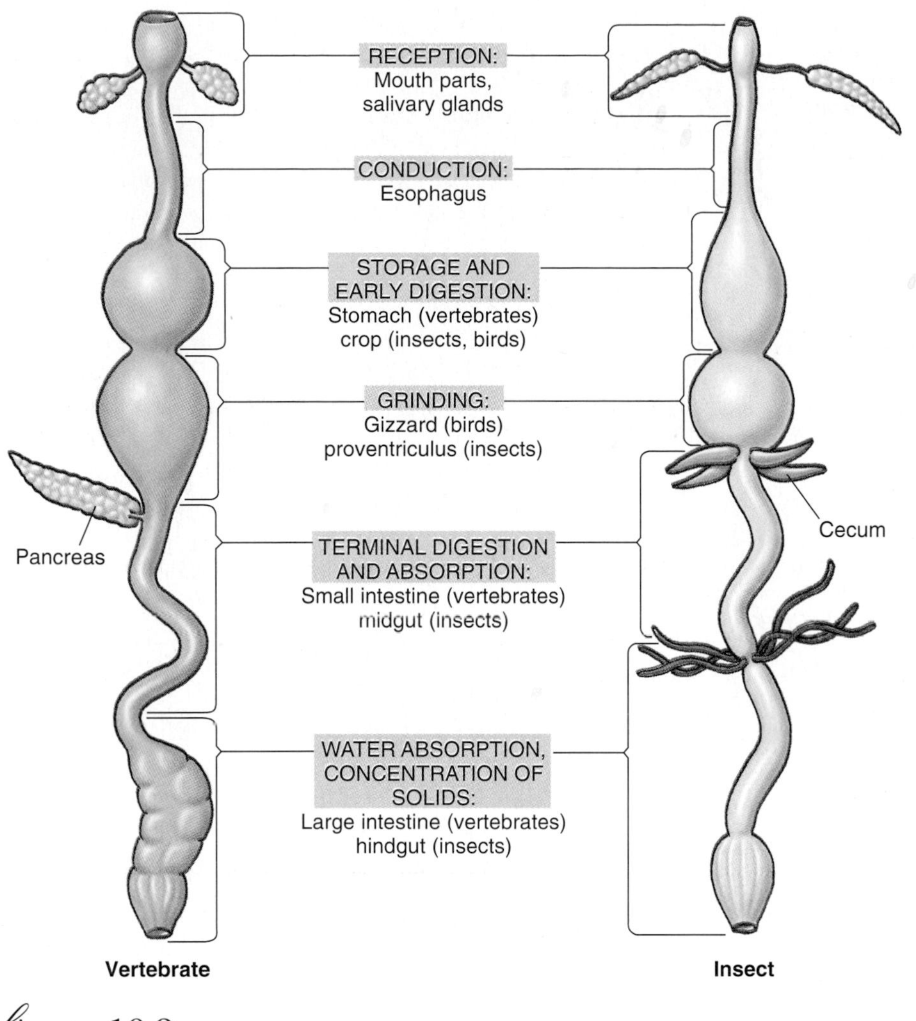

figure 10-9

Generalized digestive tracts of a vertebrate and an insect, showing the major functional regions of the metazoan digestive system.

Conduction and Storage Region

The **esophagus** of vertebrates and many invertebrates serves to transfer food to the digestive region. In many invertebrates (annelids, insects, octopods) the esophagus is expanded into a **crop** (Figure 10-9), used for food storage before digestion. Among vertebrates, only birds have a crop. This crop serves to store and soften food (grains, for example) before it passes to the stomach, or to allow mild fermentation of food before it is regurgitated to feed nestlings.

Region of Grinding and Early Digestion

In most vertebrates, and in some invertebrates, the **stomach** provides for initial digestion as well as for storage and mixing of food with digestive juices. Mechanical breakdown of food, especially plant food with its tough cellulose cell walls, often continues in herbivorous animals by means of grinding and crushing devices in the stomach. The muscular **gizzard** of terrestrial oligochaete worms, many arthropods, and birds, is assisted by stones and grit swallowed along with food (annelids and birds) or by hardened linings (for example, the chitinous teeth of the insect proventriculus, and the calcareous teeth of the gastric mill of crustaceans).

salivary amylase continues to act for some time, digesting perhaps half of the starch before the enzyme is inactivated by the acidic environment of the stomach. Further starch digestion resumes beyond the stomach in the intestine.

The tongue is a vertebrate innovation, usually attached to the floor of the mouth, that assists in food manipulation and swallowing. It may be used for other purposes, however, such as food capture (for example, chameleons, woodpeckers, anteaters) or as an olfactory sensor (many lizards and snakes).

In humans, swallowing begins with the tongue pushing the moistened food toward the pharynx. The nasal cavity closes reflexively by raising the soft palate. As the food slides into the pharynx, the epiglottis tips down over the windpipe, nearly closing it (Figure 10-10). Some particles of food may enter the opening of the windpipe, but contraction of laryngeal muscles prevents it from going farther. Once the food is in the esophagus, peristaltic contraction of esophageal muscles forces it smoothly toward the stomach.

Digestive diverticula—blind tubules or pouches arising from the main passage—often supplement the stomach of many invertebrates. They are usually lined with a multipurpose epithelium having cells specialized for secreting mucus or digestive enzymes, or for absorption or storage. Examples include the ceca of polychaete annelids, digestive glands of bivalve molluscs, the hepatopancreas of crustaceans, and the pyloric ceca of sea stars.

Herbivorous vertebrates have evolved several strategies for exploiting cellulose-splitting microorganisms to get maximal nutrition from plant food. Despite its abundance on earth, the woody cellulose that encloses plant cells can be broken down only by an enzyme, cellulase, that has limited distribution in the living world. No metazoan animals can produce intestinal cellulase for the direct digestion of cellulose. However many herbivorous metazoans harbor microorganisms (bacteria and protozoa) in their gut that do produce cellulase. These ferment cellulose under the anaerobic conditions of the gut,

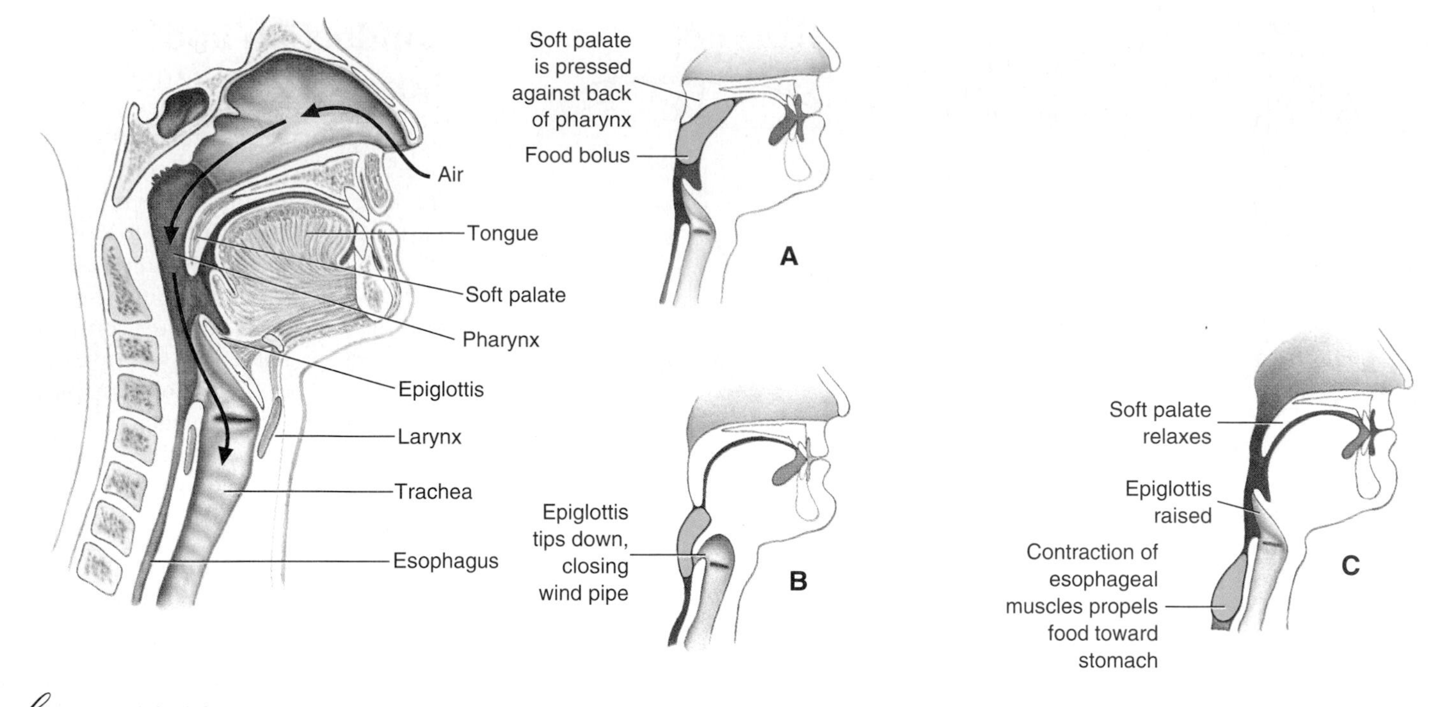

figure 10-10

Sequential events of swallowing in humans.

producing fatty acids and sugars that the herbivore can use. While the ultimate fermentation machine is the multichambered stomach of the cud-chewing ruminants (p. 675), many other animals harbor microorganisms in other parts of the gut, such as the intestine proper or the cecum.

The stomach of carnivorous and omnivorous vertebrates is typically a U-shaped muscular tube provided with glands that produce a proteolytic enzyme and a strong acid, the latter an adaptation that probably arose for killing prey and halting bacterial activity. When food reaches the stomach, the **cardiac sphincter** opens reflexively to allow the food to enter, then closes to prevent regurgitation back into the esophagus. In humans, gentle peristaltic waves pass over the filled stomach at the rate of approximately three each minute. Churning is most vigorous at the intestinal end where food is steadily released into the **duodenum,** first region of the intestine. Deep tubular glands in the stomach wall secrete approximately 2 liters of **gastric juice** in humans each day. Two types of cells line these glands: **chief cells,** which secrete **pepsin,** and **parietal cells,** which secrete **hydrochloric acid.** Pepsin is a **protease** (protein-splitting enzyme) that acts only in an acid medium (pH 1.6 to 2.4). It is a highly specific enzyme that splits large proteins by preferentially breaking down certain peptide bonds scattered along the peptide chain of the protein molecule. Although pepsin, because of its specificity, cannot completely degrade proteins, it effectively hydrolyzes them into smaller polypeptides. Other proteases that together can split all peptide bonds complete digestion of protein in the intestine. Pepsin is present in the stomachs of nearly all vertebrates.

*That the stomach mucosa is not digested by its own powerful acid secretions is a result of another gastric secretion, mucin, a highly viscous organic compound that coats and protects the mucosa from both chemical and mechanical injury. We should note that despite the popular misconception of an "acid stomach" being unhealthy, a notion carefully nourished in advertising, stomach acidity is normal and essential. Sometimes, however, the protective mucous coating fails. This failure is often associated with an infection with a bacterium (*Helicobacter pylori*) that secretes toxins causing inflammation of the stomach's lining. This inflammation may lead to a peptic ulcer.*

Rennin is a milk-curdling enzyme found in the stomach of ruminant mammals. It probably occurs in many other mammals. By clotting and precipitating milk proteins, it slows the movement of milk through the stomach. Rennin extracted from the stomachs of calves is used in making cheese. Human infants, lacking rennin, digest milk proteins with acidic pepsin, just as adults do.

The secretion of the gastric juices is intermittent. Although a small volume of gastric juice is secreted continuously, even during prolonged periods of starvation, secretion normally increases when stimulated by the sight and smell of food, by presence of food in the stomach, and by emotional states such as anxiety and anger.

The most unique and classic investigation in the field of digestion was made by U.S. Army surgeon William Beaumont

figure 10-11

Dr. William Beaumont at Fort Mackinac, Michigan Territory, collecting gastric juice from Alexis St. Martin.

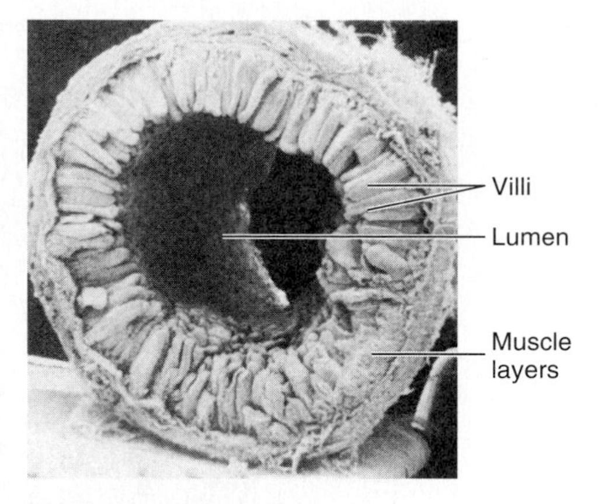

figure 10-12

Scanning electron micrograph of a rat intestine showing the numerous fingerlike villi that project into the lumen and vastly increase the effective absorptive and secretory surface of the intestine. (×21)

From: Tissues and Organs: A Text-Atlas of Scanning Electron Microscopy, © *Richard G. Kessel and Randy H. Kardon, published by W. H. Freeman and Co., 1979.*

during the years 1825 to 1833. His subject was a young, hard-living French Canadian voyageur named Alexis St. Martin, who in 1822 accidentally shot himself in the abdomen with a musket, the blast "blowing off integuments and muscles of the size of a man's hand, fracturing and carrying away the anterior half of the sixth rib, fracturing the fifth, lacerating the lower portion of the left lobe of the lungs, the diaphragm, and perforating the stomach." Miraculously the wound healed, but a permanent opening, or fistula, formed that permitted Beaumont to see directly into the stomach (Figure 10-11). St. Martin became a permanent, although temperamental, patient in Beaumont's care, which included food and housing. Over a period of eight years, Beaumont was able to observe and record how the lining of the stomach changed under different psychological and physiological conditions, how foods changed during digestion, the effect of emotional states on stomach motility, and many other facts about the digestive process of his famous patient.

Region of Terminal Digestion and Absorption: The Intestine

The importance of the intestine varies widely among animal groups. In invertebrates that have extensive digestive diverticula in which food is broken down and phagocytized, the intestine may serve only as a pathway for conducting wastes out of the body. In other invertebrates with simple stomachs, and in all vertebrates, the intestine is equipped for both digestion and absorption.

Devices for increasing the internal surface area of the intestine are highly developed in vertebrates, but are generally absent among invertebrates. Perhaps the most direct way to increase the absorptive surface of the gut is to increase its length. Coiling of the intestine is common among all vertebrate groups and reaches its highest development in mammals, in which the length of the intestine may exceed eight times the length of the body. Although a coiled intestine is rare among invertebrates, other strategies for increasing surface sometimes occur. For example, the **typhlosole** of terrestrial oligochaete worms (see Figure 22-11C, p. 482), an inward folding of the dorsal intestinal wall that runs the full length of the intestine, effectively increases internal surface area of the gut in a narrow body lacking space for a coiled intestine.

Lampreys and sharks have longitudinal or spiral folds in their intestines. Other vertebrates have developed elaborate folds (amphibians and reptiles) or minute fingerlike projections called **villi** (birds and mammals), that give the inner surface of fresh intestinal tissue the appearance of velvet (Figure 10-12). The electron microscope reveals that each cell lining the intestinal cavity additionally is bordered by hundreds of short, delicate processes called **microvilli** (Figure 10-13C and D). These processes, together with larger villi and intestinal folds, may increase the internal surface area of the intestine more than a million times as compared to a smooth cylinder of the same diameter. This enormously facilitates the absorption of food molecules.

Digestion in the Vertebrate Small Intestine

Food is released into the small intestine through the **pyloric sphincter,** which relaxes at intervals to allow entry of acidic

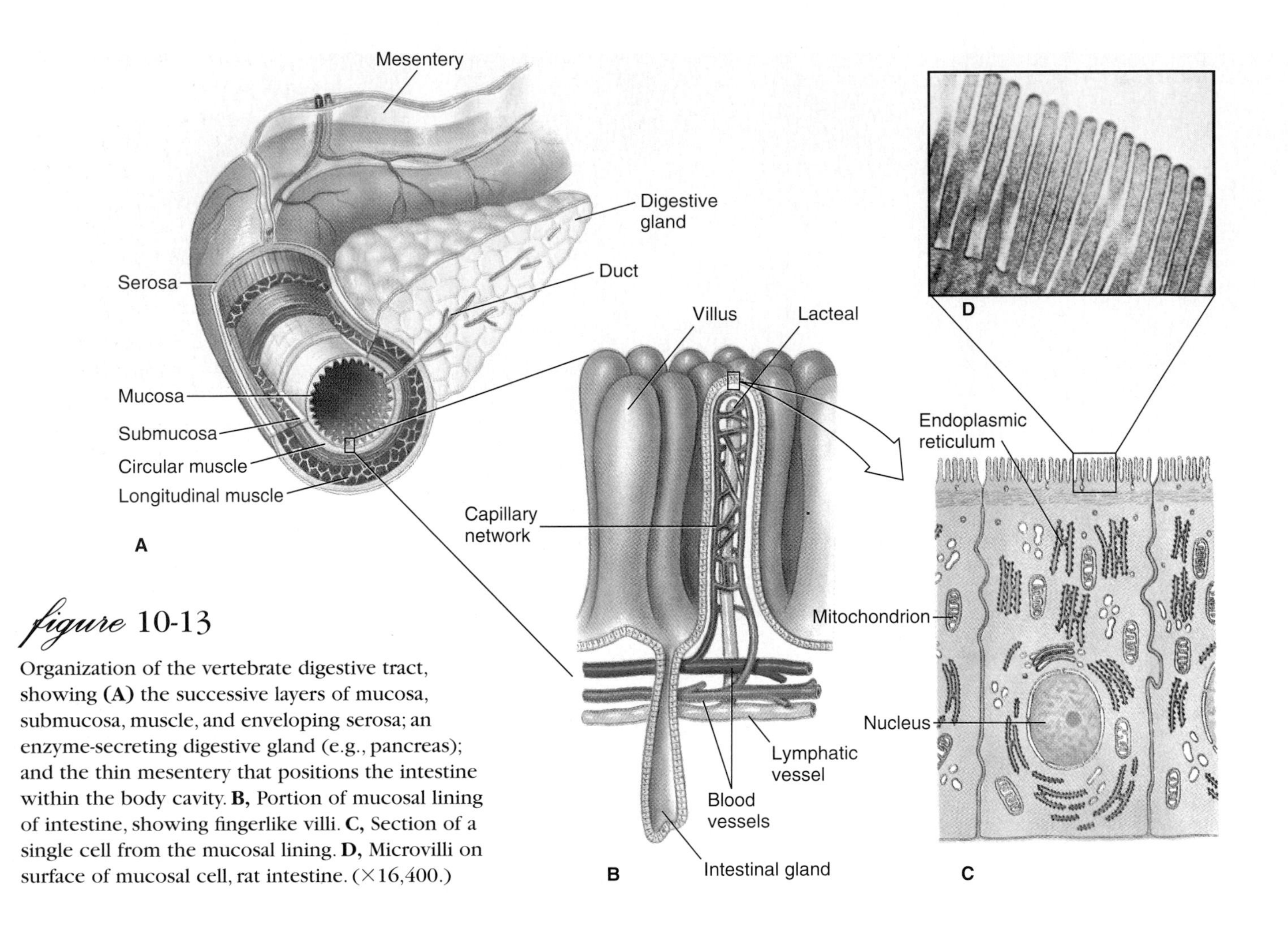

figure 10-13

Organization of the vertebrate digestive tract, showing **(A)** the successive layers of mucosa, submucosa, muscle, and enveloping serosa; an enzyme-secreting digestive gland (e.g., pancreas); and the thin mesentery that positions the intestine within the body cavity. **B,** Portion of mucosal lining of intestine, showing fingerlike villi. **C,** Section of a single cell from the mucosal lining. **D,** Microvilli on surface of mucosal cell, rat intestine. (×16,400.)

stomach contents into the initial segment of the small intestine, the **duodenum.** Two secretions are poured into this region: **pancreatic juice** and **bile** (Figure 10-14). Both of these secretions have a high bicarbonate content, especially pancreatic juice, which effectively neutralizes gastric acid, raising the pH of the liquefied food mass, now called **chyme,** from 1.5 to 7 as it enters the duodenum. This change in pH is essential because all the intestinal enzymes are effective only in a neutral or slightly alkaline medium.

Cells of the intestinal mucosa, like those of the stomach mucosa, are subjected to considerable wear and are constantly undergoing replacement. Cells deep in the crypt between adjacent villi divide rapidly and migrate up the villus. In mammals, the cells reach the tip of the villus in about two days. There they are shed, along with their membrane enzymes, into the lumen at the rate of some 17 billion a day along the length of the human intestine. Before they are shed, however, these cells differentiate into absorptive cells that transport nutrients into the network of blood and lymph vessels, once digestion is complete.

Pancreatic Enzymes The pancreatic secretion of vertebrates contains several enzymes of major importance in digestion (Figure 10-14). Two powerful proteases, **trypsin** and **chymotrypsin,** continue enzymatic digestion of proteins begun by pepsin, which is now inactivated by the alkalinity of the intestine. Trypsin and chymotrypsin, like pepsin, are highly specific proteases that split apart peptide bonds deep inside the protein molecule. The hydrolysis of the peptide linkage may be shown as:

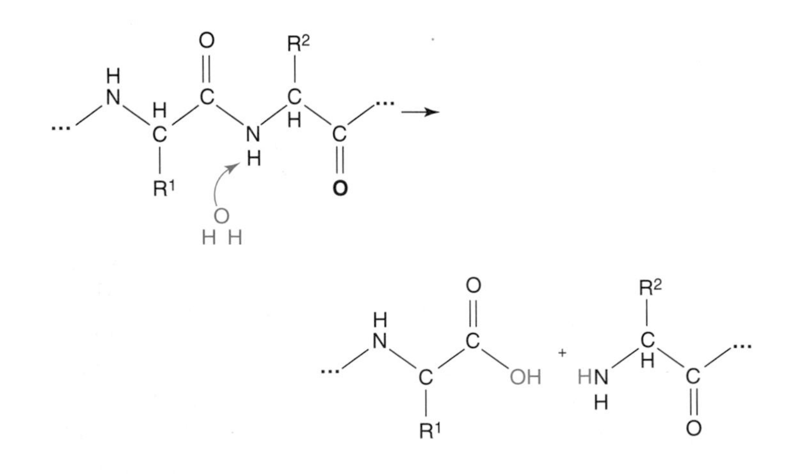

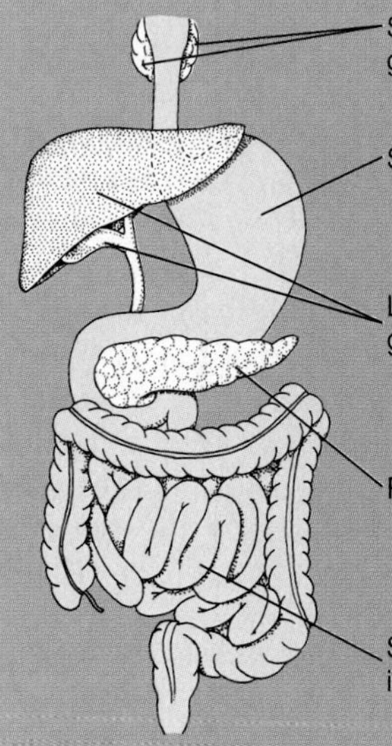

REGION	SECRETION	pH	COMPOSITION
Salivary glands	Saliva	6.5	Amylase Bicarbonate
Stomach	Gastic juice	1.5	Pepsin HCl Rennin in ruminant mammals
Liver and gallbladder	Bile	7–8	Bile salts and pigments Cholesterol
Pancreas	Pancreatic juice	7–8	Trypsin, Chymotrypsin, Carboxypeptidase, Lipase, Amylase, Nucleases Bicarbonate
Small intestine	Membrane enzymes	7–8	Aminopeptidase Maltase Lactase Sucrase Alkaline Phosphatase

figure 10-14

Secretions of the mammalian alimentary canal with the principal components and the pH of each secretion.

Pancreatic juice also contains **carboxypeptidase,** which removes amino acids from the carboxyl ends of polypeptides; **pancreatic lipase,** which hydrolyzes fat into fatty acids and glycerol; **pancreatic amylase,** a starch-splitting enzyme identical to salivary amylase in its action; and **nucleases,** which degrade RNA and DNA to nucleotides.

Membrane Enzymes The cells lining the intestine have digestive enzymes embedded in their surface membrane that continue digestion of carbohydrates, proteins, and phosphate compounds (Figure 10-14). These enzymes of the microvillus membrane (Figure 10-13D) include **aminopeptidase** that removes terminal amino acids from the amino end of short peptides, and several **disaccharidases,** enzymes that split 12-carbon sugar molecules into 6-carbon units. The disaccharidases include **maltase,** which splits maltose into two molecules of glucose; **sucrase,** which splits sucrose to fructose and glucose; and **lactase,** which breaks lactose (milk sugar) into glucose and galactose. Also present is **alkaline phosphatase,** an enzyme that attacks a variety of phosphate compounds.

Bile The liver secretes bile into the **bile duct,** which drains into the upper intestine (duodenum). Between meals the bile collects in the **gallbladder,** an expansible storage sac that releases bile when stimulated by the presence of fatty food in the duodenum. Bile contains water, bile salts, and pigments, but no enzymes. **Bile salts** (mainly sodium taurocholate and sodium glycocholate) are essential for digestion of fats. Fats, because of their tendency to remain in large, water-soluble globules, are especially resistant to enzymatic digestion. Bile salts reduce the surface tension of fat globules, allowing the churning action of the intestine to break up fats into tiny droplets (emulsification). With the total surface exposure of fat particles greatly increased, fat-splitting lipases are able to reach and hydrolyze the triglyceride molecules. The yellow-green color of bile is produced by **bile pigments,** breakdown products of hemoglobin from worn-out red blood cells. The bile pigments also give the feces its characteristic color.

Bile production is only one of the liver's many functions. This highly versatile organ is a storehouse for glycogen, production center for the plasma proteins, site of protein synthesis and detoxification of protein wastes, site for destruction of worn-out red blood cells, and center for metabolism of fat, amino acids, and carbohydrates.

Although milk is the universal food of newborn mammals and one of the most complete human foods, many adult humans cannot digest milk because they are deficient in lactase, the enzyme that hydrolyzes lactose (milk sugar). Lactose intolerance is genetically determined. It is characterized by abdominal bloating, cramps, flatulence, and watery diarrhea, all appearing within 30 to 90 minutes after ingesting milk or its unfermented by-products. (Fermented dairy products, such as yogurt and cheese, create no intolerance problems.)

Northern Europeans and their descendants, which include the majority of North American whites, are most tolerant of milk. Many other ethnic groups are generally intolerant to lactose, including the Japanese, Chinese, Jews in Israel, Eskimos, South American Indians, and most African blacks. Only about 30% of North American blacks are tolerant; those who are tolerant are mostly descendants of slaves brought from east and central Africa where dairying is traditional and tolerance to lactose is high.

Absorption

Little food is absorbed in the stomach because digestion is still incomplete and because of limited absorptive surface area. Some materials, however, such as drugs and alcohol, are absorbed mostly there, which contributes to their rapid action. Most digested food is absorbed from the small intestine where the numerous finger-shaped villi provide an enormous surface area through which materials can pass from the intestinal lumen into the circulation.

Carbohydrates are absorbed almost exclusively as simple sugars (monosaccharides, for example, glucose, fructose, and galactose) because the intestine is virtually impermeable to polysaccharides. Proteins are absorbed principally as their amino acid subunits, although small amounts of small proteins or peptide fragments sometimes may be absorbed. Both passive and active processes transfer simple sugars and amino acids across the intestinal epithelium.

Immediately after a meal these materials are in such high concentration in the gut that they readily diffuse into the blood, where their concentration is initially lower. However, if absorption were passive only, we would expect it to cease as soon as the concentrations of a substance became equal on both sides of the intestinal epithelium. Passive transport alone would permit valuable nutrients to be lost in the feces. In fact, very little is lost because passive transfer is supplemented by an **active transport** mechanism (p. 33), located in the epithelial cells, that transfers food molecules into the blood. Materials thus are moved *against* their concentration gradient, a process requiring the expenditure of energy. Although not all food products are actively transported, those that are, such as glucose, galactose, and most of the amino acids, are handled by transport mechanisms that are specific for each kind of molecule.

As mentioned previously, fat droplets are emulsified by bile salts and then digested by pancreatic lipase. Triglycerides are broken into fatty acids and monoglycerides which, with bile salts, form minute droplets called **micelles.** When the micelles contact the microvilli of the intestinal epithelium, the fatty acids and monoglycerides are absorbed by simple diffusion. They then enter the endoplasmic reticulum of the absorptive cells, where they are resynthesized into triglyercides before passing into the **lacteals** (Figure 10-13B). From the lacteals, fat droplets enter the lymph system (Figure 8-17, p. 199) and eventually pass into the blood through the thoracic duct. After a fatty meal, even a peanut butter sandwich, the presence of numerous fat droplets in the blood imparts a milky appearance to the blood plasma.

Region of Water Absorption and Concentration of Solids

The large intestine consolidates the indigestible remnants of digestion by reabsorption of water to form solid or semisolid feces for removal from the body by **defecation.** Reabsorption of water is of special significance in insects, especially those living in dry environments, which must (and do) conserve nearly all water entering the rectum. Specialized **rectal glands** absorb water and ions as needed, leaving behind fecal pellets that are almost completely dry. In reptiles and birds, which also produce nearly dry feces, most of the water is reabsorbed in the cloaca. A white pastelike feces is formed containing both indigestible food wastes and uric acid.

The colon of humans contains enormous numbers of bacteria, which first enter the sterile colon of the newborn infant with its food. In adults, approximately one-third of the dry weight of feces is bacteria; these include harmless bacilli as well as cocci that can cause serious illness should they escape into the abdomen or bloodstream. Normally the body's defenses prevent invasion of such bacteria. The bacteria break down organic wastes in the feces and provide some nutritional benefit by synthesizing certain vitamins (vitamin K and small quantities of some of the B vitamins), which are absorbed by the body.

Regulation of Food Intake

Most animals unconsciously adjust intake of food to balance energy expenditure. If energy expenditure is increased by greater physical activity, more food is consumed. Most vertebrates, from fish to mammals, eat for calories rather than bulk because, if the diet is diluted with fiber, they respond by eating more. Similarly, intake is adjusted downward following a period of several days when caloric intake is too high.

A "hunger" center located in the hypothalamus of the brain regulates the intake of food in large part. The craving for food normally is stimulated by a drop in the level of glucose in the blood. While most animals seem able to stabilize their weight at normal levels with ease, many humans cannot. Obesity is rising throughout the industrial world and is a major health problem today. According to a 1995 report by the Institute of Medicine, 59% of the adult population in the United States meets the current definition of clinical obesity. It is becoming clear that of these obese people, nearly half do not eat significantly more food than thin people, but rather they have inherited a genetic predisposition to gain weight on a high-fat diet. Many obese people have a reduced capacity to burn off excess calories by "nonshivering thermogenesis" (p. 183). Placental mammals are unique in having a dark adipose tissue called **brown fat,** specialized for the generation of heat. Newborn mammals, including human infants, have much more brown fat than adults. In human infants brown fat is located in the chest, upper back, and near the kidneys. The abundant mitochondria in brown fat contain a membrane protein called **thermogenin** that acts to uncouple oxidative phosphorylation (p. 47). In people of average weight, an increased caloric intake induces brown fat to dissipate excess energy as heat through the uncoupling action of thermogenin. We call this process "diet-induced thermogenesis." In many people tending toward obesity, this capacity is diminished.

The body contains two kinds of adipose tissue that perform completely different functions. White adipose tissue, which comprises the bulk of body fat, is adapted for the storage of fat derived mainly from surplus fats and carbohydrates in the diet. It is distributed throughout the body, particularly in the deep layers of the skin. Brown adipose tissue is highly specialized for mediating nonshivering thermogenesis rather than for the storage of fat. It is brown because it is packed with mitochondria containing large quantities of iron-bearing cytochrome molecules. In ordinary body cells, ATP is generated by the flow of electrons down the respiratory chain (p. 46). This ATP then powers various cellular processes. In brown fat cells, heat is generated instead of ATP. The sympathetic nervous system, which responds to signals from the hypothalamus, activates thermogenesis.

There are other reasons for obesity in addition to the fact that many people simply eat too much and get too little exercise. Fat stores are supervised by the hypothalamus, which may be set to a point higher or lower than the norm. A high set-point can be lowered somewhat by exercise, but as dieters are painfully aware, the body defends its fat stores with remarkable tenacity. In 1995, a hormone produced by fat cells was discovered that cures obesity in mutant mice lacking the gene that produces the hormone. The hormone, called **leptin,** appears to operate through a feedback system that tells the hypothalamus how much fat the body carries. If levels are high, release of leptin by fat cells leads to diminished appetite and increased thermogenesis. The discovery of leptin has initiated a flurry of research on obesity and a resurgence of commercial interest in producing a weight-loss drug based on leptin.

table 10-1

Human Nutrient Requirements

Amino acids	
Phenylalanine	Methionine
Lysine	Cystine
Isoleucine	Tryptophan
Leucine	Threonine
Valine	
Polyunsaturated fatty acids	
Arachidonic	
Linoleic	
Linolenic	
Water-soluble vitamins	
Thiamine (B_1)	Folacin
Riboflavin (B_2)	Vitamin B_{12}
Niacin	Biotin
Pyridoxine (B_6)	Choline
Pantothenic acid	Ascorbic acid (C)
Fat-soluble vitamins	
A, D, E, and K	
Minerals	
Calcium	Silicon
Phosphorus	Vanadium
Sulfur	Tin
Potassium	Nickel
Chlorine	Selenium
Sodium	Manganese
Magnesium	Iodine
Iron	Molybdenum
Fluorine	Chromium
Zinc	Cobalt
Copper	

Adapted from "The Requirements of Human Nutrition" by Nevin S. Scrimshaw and Vernon R. Young. Copyright © September 1976 by Scientific American, Inc. All rights reserved. Adapted by permission.

Nutritional Requirements

The food of animals must include **carbohydrates, proteins, fats, water, mineral salts,** and **vitamins.** Carbohydrates and fats are required as fuels for energy and for the synthesis of various substances and structures. Synthesis of specific proteins and other nitrogen-containing compounds requires proteins (actually the amino acids of which they are composed). Water is the solvent for body chemistry and a major component of all fluids of the body. Inorganic salts are required as the anions and cations of body fluids and tissues and form important structural and physiological components throughout the body. Vitamins are necessary factors from food that are often built into the structure of many enzymes.

A vitamin is a relatively simple organic compound that is not a carbohydrate, fat, protein, or mineral and that is required in very small amounts in the diet for some specific cellular function. Vitamins are not sources of energy but are often associated with the activity of important enzymes that serve vital metabolic roles. Plants and many microorganisms synthesize all of the organic compounds that they need; animals, however, have lost certain synthetic abilities during their long evolution and depend ultimately on plants to supply these compounds. Vitamins therefore represent synthetic gaps in the metabolic machinery of animals.

Vitamins are usually classified as fat soluble (soluble in fat solvents such as ether) or water soluble. The water-soluble vitamins include the B complex and vitamin C (Table 10-1). Vitamins of the B complex, so grouped because the original B vitamin was subsequently found to consist of several distinct molecules, tend to be found together in nature. Almost all animals, vertebrate and invertebrate, require the B vitamins; they are "universal" vitamins. The dietary need for vitamin C and the fat-soluble vitamins A, D, E, and K is mostly restricted to the vertebrates, although some are required by certain invertebrates. Even within groups of close relationship, requirements for vitamins are relative, not absolute. A rabbit does not require vitamin C, but guinea pigs and humans do. Some songbirds require vitamin A, but others do not.

The recognition years ago that many human diseases and those of domesticated animals were caused by or associated with dietary deficiencies led biologists to search for specific nutrients that would prevent such diseases. These studies eventually yielded a list of **essential nutrients** for human beings and other animal species studied. Essential nutrients are those needed for normal growth and maintenance and that *must* be supplied in the diet. In other words, it is "essential" that these nutrients be in the diet because the animal cannot synthesize them from other dietary constituents. Nearly 30 organic compounds (amino acids and vitamins) and 21 elements are essential for humans (Table 10-1). Considering that the body contains thousands of different organic compounds, the list in Table 10-1 is remarkably short. Animal cells have marvelous powers of synthesis, enabling them to build compounds of enormous variety and complexity from a small, select group of raw materials.

In the average diet of North Americans approximately 50% of the total calories (energy content) comes from carbohydrates and 40% comes from lipids. Proteins, essential as they are for structural needs, supply only a little more than 10% of the total calories of the average diet in North Americans. Carbohydrates are widely consumed because they are more abundant and cheaper than proteins or lipids. Actually humans and many other animals can subsist on diets devoid of carbohydrates, provided sufficient total calories and essential nutrients are present. Eskimos, before the decline of their native culture, lived on a diet that was high in fat and protein and very low in carbohydrate.

Lipids are needed principally to provide energy. However, at least three fatty acids are essential for humans because we cannot synthesize them. Much interest and research have been devoted to lipids in our diets because of the association between fatty diets and the disease **atherosclerosis.** The matter is complex, but evidence suggests that atherosclerosis may occur when the diet is high in saturated lipids (lipids with no double bonds in the carbon chains of the fatty acids) but low in polyunsaturated lipids (two or more double bonds in the carbon chains).

Atherosclerosis (Gr. atheroma, *tumor containing gruel-like matter,* + sclerosis, *to harden) is a degenerative disease in which fatty substances are deposited in the lining of arteries, resulting in narrowing of the passage and eventual hardening and loss of elasticity.*

Proteins are expensive foods and limited in the diet. Proteins, of course, are not themselves the essential nutrients but rather contain essential amino acids. Of the 20 amino acids commonly found in proteins, 9 and possibly 11 are essential to humans (Table 10-1). We can synthesize the rest from other amino acids. Generally, animal proteins have more of the essential amino acids than do proteins of plant origin. All nine of the essential amino acids must be present simultaneously in the diet for protein synthesis. If one or more is missing, the use of the other amino acids will be reduced proportionately; they cannot be stored and are broken down for energy. Thus heavy reliance on a single plant source as a diet will inevitably lead to protein deficiency. This problem can be corrected if two kinds of plant proteins having complementary strengths in essential amino acids are ingested together. For example, a balanced protein diet can be prepared by mixing wheat flour, which is deficient only in lysine, with a legume (peas or beans), which is a good source of lysine but deficient in methionine and cysteine. Each plant complements the other by having adequate amounts of those amino acids that are deficient in the other.

Because animal proteins are rich in essential amino acids, they are in great demand in all countries. North Americans eat far more animal protein than do Asians and Africans. In 1989 the annual per capita consumption of red meat was 76 kg in the United States, 27 kg in Japan, 12 kg in Egypt, and 1 kg in India.[2] The high consumption of meat in North America and Europe carries the price of a high death rate from the so-called diseases of affluence: heart disease, stroke, and certain kinds of cancer.

Two different types of severe food deficiency are recognized: marasmus, general undernourishment from a diet low in both calories and protein, and kwashiorkor, protein malnourishment from a diet adequate in calories but deficient in protein. Marasmus (Gr. marasmos, *to waste away) is common in infants weaned too early and placed on low-calorie–low-protein diets; these children are listless, and their bodies waste away. Kwashiorkor is a West African word describing a disease a child gets when displaced from the breast by a newborn sibling. This disease is characterized by retarded growth, anemia, weak muscles, a bloated body with typical pot belly, acute diarrhea, susceptibility to infection, and high mortality.*

[2]Brown, L.R. 1991. State of the world 1991. New York, Worldwatch Institute/W.W. Norton & Company, p. 159.

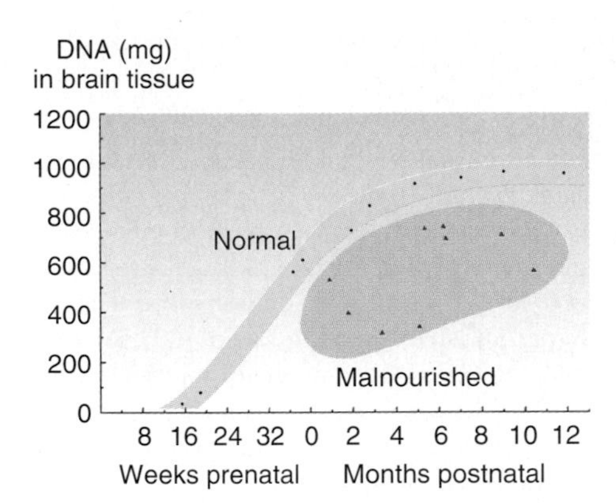

figure 10-15

Effect of early malnutrition on cell number (measured as total DNA content) in the human brain. This graph shows that malnourished infants (*purple oval*) have far fewer brain cells than do normal infants (*green growth curve*).

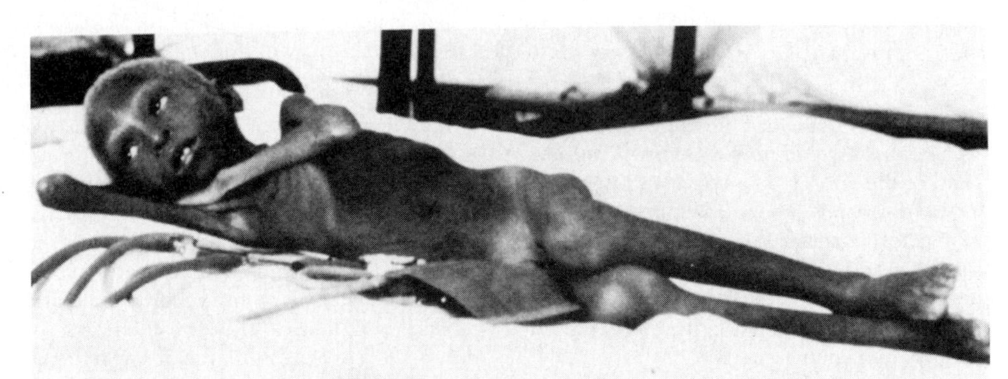

figure 10-16

Biafran refugee child suffering severe malnutrition.

Undernourishment and malnourishment rank as two of the world's oldest problems and remain major health problems today, afflicting an eighth of the human population. Nearly 195 million children under the age of five are undernourished. Growing children and pregnant and lactating women are especially vulnerable to the devastating effects of malnutrition. Cell proliferation and growth in the human brain are most rapid in the terminal months of gestation and the first year after birth. Adequate protein, vitamins, and essential minerals for neuron development are a requirement during this critical time to prevent neurological dysfunction. The brains of children who die of protein malnutrition during the first year of life have 15% to 20% fewer brain cells than those of normal children (Figure 10-15). Malnourished children who survive this period suffer permanent brain damage and often cannot be helped by later corrective treatment (Figure 10-16). Recent studies suggest that poverty, with attendant lack of educational and medical resources, and lowered expectations, exacerbates the effects of malnutrition by delaying intellectual development.[3]

The world's precarious food supply is threatened by rapid population growth. The world population was 2 billion in 1930, reached 3 billion in 1960, passed 5.7 billion in 1995 (Figure 10-17), and is expected to reach 8.9 billion by the year 2030,[4] several years ahead of earlier estimates. Approximately 90 million people are added each year. The equivalent of the total United States population of 250 million people is added to the world every 33 months. Yet, as the demand for food increases, the world per capita production of grain and the world fish catch are declining.[5] Furthermore, the world each year loses billions of tons of topsoil and trillions of gallons of groundwater needed to grow food crops. In the view of many, the exploding human population is a major force driving the global environment crisis.

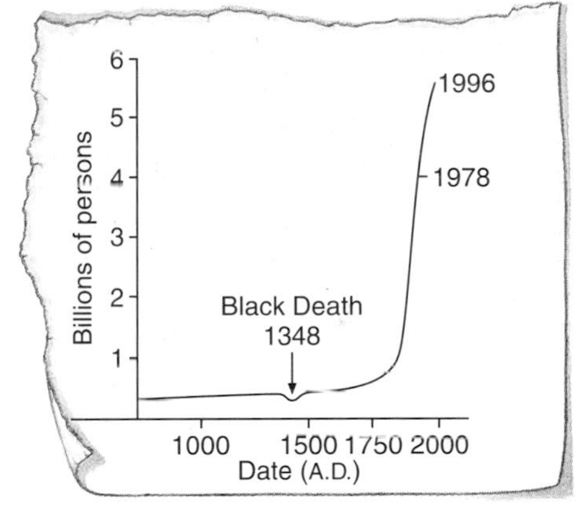

figure 10-17

Portion of a graph for human population growth since A.D. 800, as it appeared in the 1978 edition of this book when the earth's population had passed 4 billion two years earlier, and updated to show the 1996 figure of approximately 5.8 billion.

[3]Brown, J.L., and E. Pollitt, 1996, "Malnutrition, poverty and intellectual development," Sci. Am. **274:**38–43 (Feb.).
[4]Population Division, United Nations, New York, 1993.

[5]According to State of the World 1996, Worldwatch Institute, the world grain harvest has not grown at all since 1990, and 13 of the 15 leading oceanic fisheries are in decline. See also Safina, C., 1995, "The world's imperiled fish," Sci. Am. **273:**46–53 (Nov.).

Summary

Autotrophic organisms (mostly green plants), using inorganic compounds as raw materials, capture the energy of sunlight through photosynthesis and produce complex organic molecules. Heterotrophic organisms (bacteria, fungi, and animals) use the organic compounds synthesized by plants, and chemical bond energy stored therein for their own nutritional and energy needs.

A large group of animals with very different levels of complexity feed by filtering out minute organisms and other particulate matter suspended in water. Others feed on organic detritus deposited in the substrate. Selective feeders, on the other hand, have evolved mechanisms for manipulating larger food masses, including various devices for seizing, scraping, boring, tearing, biting, and chewing. Fluid feeding is characteristic of endoparasites, which may absorb food across the general body surface, and of ectoparasites, herbivores, and predators that have developed specialized mouthparts for piercing and sucking.

Digestion is the process of breaking down food mechanically and chemically into molecular subunits for absorption. Digestion is intracellular in the protozoa and sponges. In more complex metazoans it is supplemented, and finally replaced entirely, by extracellular digestion, which takes place in sequential stages in a tubular cavity, the alimentary canal. The mouth receives food and mixes it with lubricating saliva, then passes it down the esophagus to regions where the food may be stored (crop), or ground (gizzard), or acidified and subjected to early digestion (vertebrate stomach). Among vertebrates, most digestion occurs in the small intestine. Enzymes from the pancreas and intestinal mucosa hydrolyze proteins, carbohydrates, fats, nucleic acids, and various phosphate compounds. The liver secretes bile, containing salts that emulsify fats. Once foods are digested, their products are absorbed as molecular subunits (monosaccharides, amino acids, and fatty acids) into the blood or lymph vessels of the villi of the small intestine. The large intestine (colon) serves mainly to absorb water and minerals from the food wastes as they pass through it. It also contains symbiotic bacteria that produce certain vitamins.

Most animals balance food intake with energy expenditure. Food intake is regulated primarily by a hunger center located in the hypothalamus. In mammals, should caloric intake exceed requirements for energy, specialized brown fat normally dissipates the excess calories as heat. A deficiency in this response is one cause of human obesity.

All animals require a balanced diet containing both fuels (mainly carbohydrates and lipids) and structural and functional components (proteins, minerals, and vitamins). For every multicellular animal, certain amino acids, lipids, vitamins, and minerals are "essential" dietary factors that cannot be produced by the animal's own synthetic machinery. Animal proteins are better-balanced sources of amino acids than are plant proteins, which tend to lack one or more essential amino acids. Undernourishment and protein malnourishment are among the world's major health problems, afflicting millions of people.

Review Questions

1. Distinguish between the following pairs of terms: autotrophic and heterotrophic; phototrophic and chemotrophic; herbivores and carnivores; omnivores and insectivores.
2. Suspension feeding is one of the most important methods of feeding among animals. Explain the characteristics, advantages, and limitations of suspension feeding, and name three different groups of animals that are suspension feeders.
3. An animal's feeding adaptations are an integral part of an animal's behavior and usually shape the appearance of the animal itself. Discuss the contrasting feeding adaptations of carnivores and herbivores.
4. Explain how food is propelled through the digestive tract.
5. Compare intracellular with extracellular digestion and suggest why there has been a phylogenetic trend from intracellular to extracellular digestion.
6. Which structural modifications vastly increase the internal surface area of the intestine (both invertebrate and vertebrate), and why is this large surface area important?
7. Trace the digestion and final absorption of a carbohydrate (starch) in the vertebrate gut, naming the carbohydrate-splitting enzymes, where they are found, the breakdown products of starch digestion, and in what form they are finally absorbed.
8. As in question 7, trace the digestion and final absorption of a protein.
9. Explain how fats are emulsified and digested in the vertebrate gut. Explain how bile aids the digestive process even though it contains no enzymes. Provide an explanation for the following observation: fats are broken down to fatty acids and monoglycerides in the intestinal lumen, but appear later in the blood as fat droplets.
10. Explain the phrase "diet-induced thermogenesis" and relate it to the problem of obesity in some people.
11. Name the basic classes of foods that serve mainly as (1) fuels and as (2) structural and functional components.
12. If vitamins are neither biochemically similar compounds nor sources of energy, what characteristics distinguish vitamins as a distinct group of nutrients? What are the water-soluble and the fat-soluble vitamins?
13. Why are some nutrients considered "essential" and others "nonessential" even though both types of nutrients are used in growth and tissue repair?
14. Explain the difference between saturated and unsaturated lipids, and comment on the current interest in these compounds as they relate to human health.
15. What is meant by "protein complementarity" among plant foods?

Selected References

Blaser, M. J. 1996. The bacteria behind ulcers. Sci. Am. **274:**104–107 (Jan.). *We now know that most cases of stomach ulcers are caused by acid-loving microbes. At least one-third of the human population is infected, although most do not become ill.*

Carr, D. E. 1971. The deadly feast of life. Garden City, New York, Doubleday and Company. *What and how animals eat told with insight and wit.*

Doyle, J. 1985. Altered harvest: agriculture, genetics, and the fate of the world's food supply. New York, Viking Penguin, Inc. *Examines the politics of the agricultural revolution and the environmental and biological costs of the American food-production system.*

Gibbs, W. W. 1996. Gaining on fat. Sci. Am. **275:**88–94 (Aug.). *Summary of research aimed at learning the causes and control of obesity, a disease that has reached epidemic proportions in the world's wealthiest nations.*

Griggs, B. 1986. The food factor. New York, Viking Penguin, Inc. *Packed with facts on nutrition and eating habits with an international perspective and emphasis on food's relation to disease.*

Jennings, J. B. 1973. Feeding, digestion and assimilation to animals, ed. 2. New York, St. Martin's Press, Inc. *A general, comparative approach. Excellent account of feeding mechanisms in animals.*

Magee, D. F., and A. F. Dalley, II. 1986. Digestion and the structure and function of the gut. Basel, Switzerland, S. Karger AG. *Comprehensive treatment of mammalian (mostly human) digestion.*

Milton, K. 1993. Diet and primate evolution. Sci. Am. **269:**86–93 (Aug.). *Studies with primates suggest that modern human diets often diverge greatly from those to which the human body may be adapted.*

Moog, F. 1981. The lining of the small intestine. Sci. Am. **245:**154–176 (Nov.). *Describes how the mucosal cells actively process foods.*

Owen, J. 1980. Feeding strategy. Chicago, University of Chicago Press. *Well-written and generously illustrated book from the series* "Survival in the Wild."

Sanderson, S. L., and R. Wassersug. 1990. Suspension-feeding vertebrates. Sci. Am. **262:**96–101 (Mar.). *A variety of vertebrates, some enormous in size, eat by filtering out small organisms from massive amounts of water passed through a feeding apparatus.*

Stevens, C. E. 1988. Comparative physiology of the vertebrate digestive system. New York, Cambridge University Press. *Lucid and balanced treatment of anatomical characteristics of vertebrate digestive systems and the physiology and biochemistry of food digestion.*

Weindrach, R. 1996. Caloric restriction and aging. Sci. Am. **274:**46–52 (Jan.) *Organisms from single-celled protists to mammals live longer on well-balanced but low-calorie diets. The potential benefits for humans are examined.*

Nervous Coordination: Nervous System and Sense Organs

chapter | eleven

The Private World of the Senses

By any measure, people enjoy a rich sensory world. We are continually assailed by information from the senses of vision, hearing, taste, olfaction, and touch. These classic five senses are supplemented by sensory inputs of cold, warmth, vibration, and pain, as well as by information from numerous internal sensory receptors that operate silently and automatically to help keep our interior domain working smoothly. It is our senses that provide us with our impression of the environment. Yet the world our senses perceive is uniquely human. We share this exclusive world with no other animal, nor can we venture into the sensory world of any other animal except as an abstraction through our imagination.

The idea that each animal enjoys an unshared sensory world was first conceived by Jakob von Uexküll, a seldom-cited German biologist of the early part of this century. Von Uexküll asks us to try to enter the world of a tick through our imagination, supplemented by what we know of tick biology. It is a world of temperature, of light and dark, and the odors of butyric acid and carbon dioxide, chemicals common to all mammals. Insensible to all other stimuli, the tick clambers up a blade of grass to wait, for years if necessary, for the cues that will betray the presence of a potential host. Later, swollen with blood, she drops to the earth, lays her eggs, and dies. The tick's impoverished sensory world, devoid of sensory luxuries and fine-tuned by natural selection for the world she will encounter, has ensured her single goal, reproduction.

A bird and a bat may share for a moment precisely the same environment. The worlds of their perceptions, however, are vastly different, structured by the limitations of the sensory windows each employs and by the brain that garners and processes what it needs for survival. For one, it is a world dominated by vision; for the other, echolocation. The world of each is alien to the other, just as their worlds are to us.

The nervous system originated in a fundamental property of life: **irritability,** the ability to respond to environmental stimuli (Chapter 1, p. 6). The response may be simple, such as a protozoan moving to avoid a noxious substance, or quite complex, such as a vertebrate animal responding to elaborate signals of courtship. A protistan receives and responds to a stimulus, all within the confines of a single cell. The evolution of multicellularity and more complex levels of animal organization required increasingly complex mechanisms for communication between cells and organs. This is accomplished by two principal means: **neural** and **hormonal.** Relatively rapid communication is by neural mechanisms and involves propagated electrochemical changes in cell membranes. The basic plan of the nervous system is to code information and to transmit and process it for appropriate action. These functions will be examined in this chapter. Relatively less rapid or long-term adjustments in animals are governed by hormonal mechanisms, the subject of the next chapter.

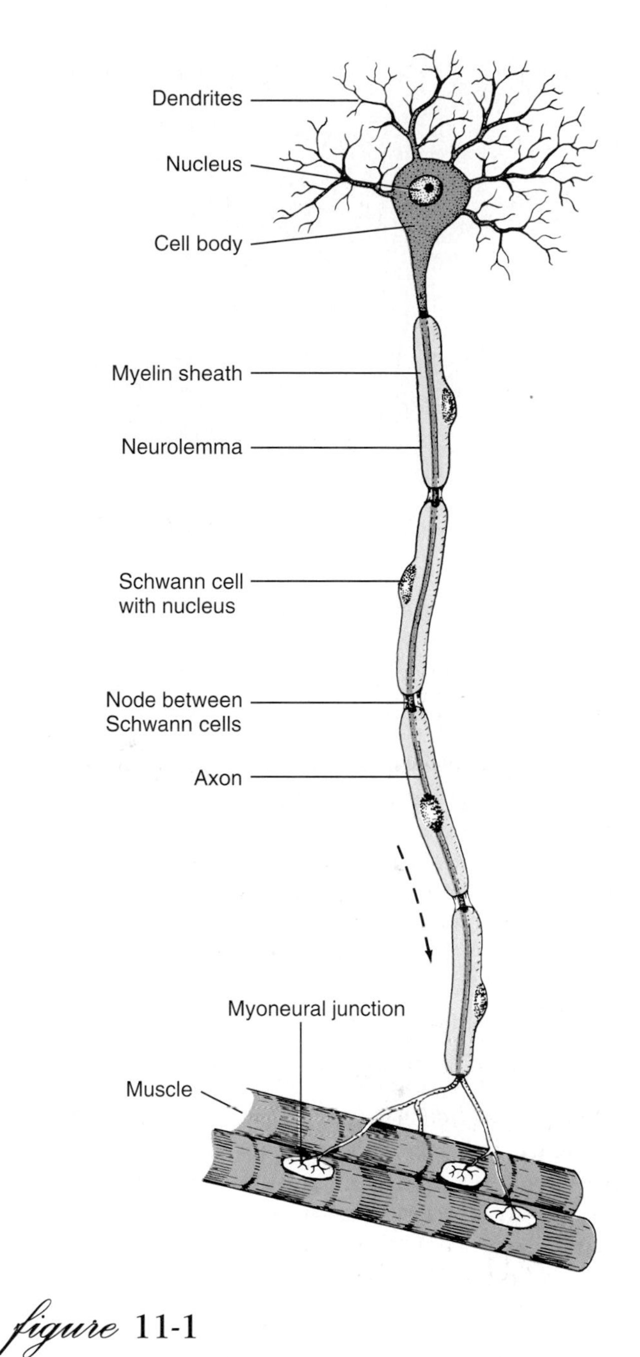

figure 11-1

Structure of a motor (efferent) neuron.

The Neuron: Functional Unit of the Nervous System

The **neuron** is a cell body with all its processes. Although neurons assume many shapes, depending on their function and location, a typical kind is shown diagrammatically in Figure 11-1. From the nucleated body extend **processes** of two types. All but the simplest nerve cells have one or more cytoplasmic **dendrites.** As the name dendrite suggests (Gr. *dendron,* tree), these are often profusely branched. They are the nerve cell's receptive apparatus, often receiving information from several different sources at once. Some of these inputs are excitatory, others inhibitory.

From the nucleated cell body extends a single **axon** (Gr. *axon,* axle), often a long fiber (meters in length in the largest mammals), relatively uniform in diameter, that typically carries impulses away from the cell body. In vertebrates and some complex invertebrates, the axon is usually covered with an insulating sheath.

Neurons are commonly classified as **afferent,** or sensory; **efferent,** or motor; and **interneurons,** which are neither sensory nor motor but connect neurons with other neurons. Afferent and efferent neurons lie mostly outside the central nervous system (brain and nerve cord) while interneurons, which in humans make up 99% of all nerve cells in the body, lie entirely within the central nervous system. Afferent neurons are connected to **receptors,** which function to convert some environmental stimuli into nerve impulses, which are carried by the afferent neurons into the central nervous system. Here, impulses may be perceived as conscious sensation. Impulses also move to efferent neurons, which carry them out by the peripheral system to **effectors,** such as muscles or glands.

In vertebrates, nerve processes (usually axons) usually are bundled together in a well-formed wrapping of connective tissue to form a **nerve** (Figure 11-2). The cell bodies of these nerve processes are located either in the central nervous system or in **ganglia,** which are discrete bundles of nerve cell bodies located outside the central nervous system.

Surrounding the neurons are nonnervous **neuroglial cells** (often simply called "glial" cells) that have a special relationship to the nerve cells. Neuroglial cells are extremely numerous in the vertebrate brain, where they outnumber nerve cells 10 to 1 and may make up almost half the volume of the brain. Some glial cells form intimate insulating sheaths of lipid-containing myelin around nerve fibers. Vertebrate peripheral

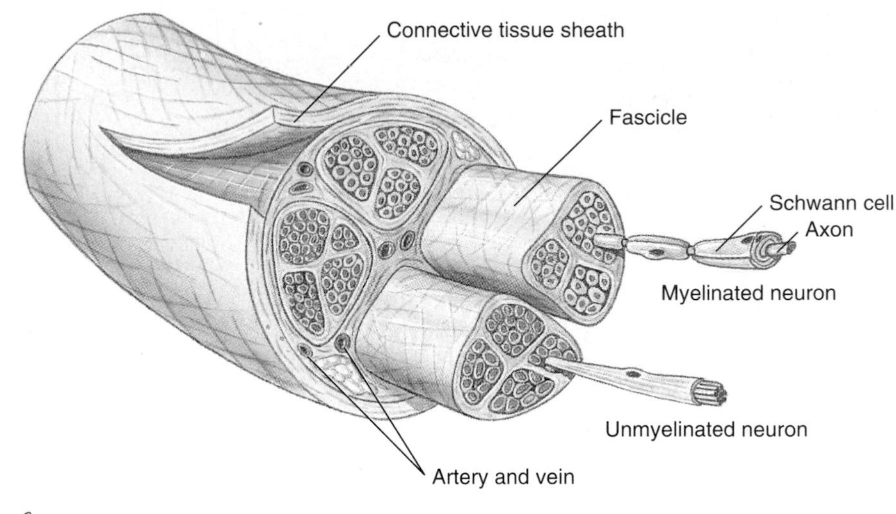

figure 11-2

Structure of a nerve showing nerve fibers surrounded by various layers of connective tissue. Such a nerve may contain thousands of both efferent and afferent fibers.

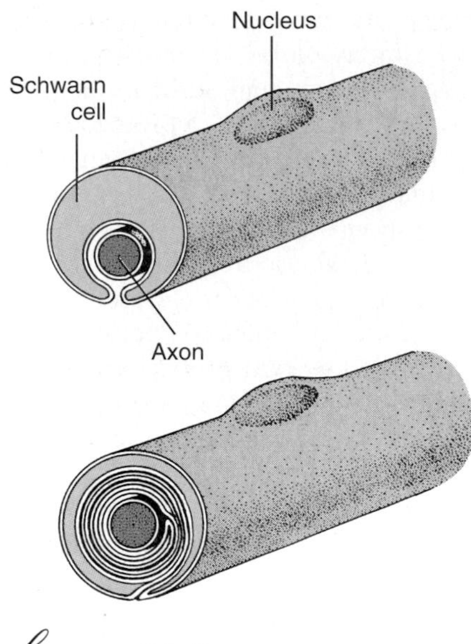

figure 11-3

Development of the myelin sheath. The Schwann cell grows around the axon, then rotates around it, enclosing the axon in a tight, multilayered sheath. The myelin sheath insulates the nerve axon and facilitates the transmission of nerve impulses.

nerves are often enclosed by **myelin,** an insulating sheath laid down in concentric rings by special glial cells called **Schwann cells** (Figure 11-3). Other functional roles of the glial cells are only now becoming known. Certain glial cells, called astrocytes because of their radiating, starlike shape, serve as a scaffold during brain development, enabling migrating neurons to find their way to their destination from points of origin. Astrocytes also remove excess neurotransmitters around synapses (see the following text) to prevent overstimulation, and help to maintain the correct electrolyte balance around nerve cells. Another class of glial cells, the microglia, are extremely mobile cells that migrate to dead neurons where they proliferate and remove dead cells. They are considered the brain's immune system.[1] Unfortunately, both astrocytes and microglia are known to cause or exacerbate several disabling conditions, such as parkinsonism, multiple sclerosis, and Alzheimer's disease.

Nature of the Nerve Impulse

The **nerve impulse** is the chemical-electrical message of nerves, the common functional denominator of all nervous system activity. Despite the incredible complexity of the nervous system of many animals, nerve impulses are basically alike in all nerves and in all animals. An impulse is an "all-or-none" phenomenon; either the fiber is conducting an impulse, or it is not. Because all impulses are alike, the only way a nerve fiber can vary its effect on the tissue it innervates is by changing the frequency of impulse conduction. Frequency change is the language of a nerve fiber. A fiber may conduct no impulses at all or very few per second up to a maximum approaching 1000 per second. The higher the frequency (or rate) of conduction, the greater is the level of excitation.

The Resting Potential

Membranes of nerve cells, like all cellular membranes, have a special permeability that creates ionic imbalances. The interstitial fluid surrounding nerve cells contains relatively high concentrations of sodium (Na^+) and chloride (Cl^-) ions, but a low concentration of potassium ions (K^+). Inside the neuron, the ratio is reversed: the K^+ concentration is high, but the Na^+ and Cl^- concentrations are low (Figure 11-4; see also Figure 8-1B, p. 189). These differences are pronounced; there is approximately 10 times more Na^+ outside than in and 25 to 30 times more K^+ inside than out.

When at rest, the membrane of a nerve cell is selectively permeable to K^+, which can pass the membrane through special potassium channels in the membrane. The permeability to Na^+ and Cl^- is nearly zero because these channels are closed in the resting membrane. Potassium ions tend to diffuse outward through the membrane, following the gradient of potassium concentration. Because chloride ions cannot follow, however, an excess of positively charged potassium ions accumulates outside the cell. This creates a charged membrane that is positive outside and negative inside. Very quickly the positive charge outside reaches a level that prevents any more K^+ from diffusing out of the axon (because like charges repel each other). Now the resting membrane is at equi-

[1]Streit, W.J., and C.A. Kincaid-Colton, 1995, "The brain's immune system," Sci. Am. **273:**54–61 (Nov.).

librium, with a **resting membrane potential** that exactly balances the concentration gradient that forces the K^+ out. The resting potential is usually −70 mV (millivolts), with the inside of the membrane negative to the outside.

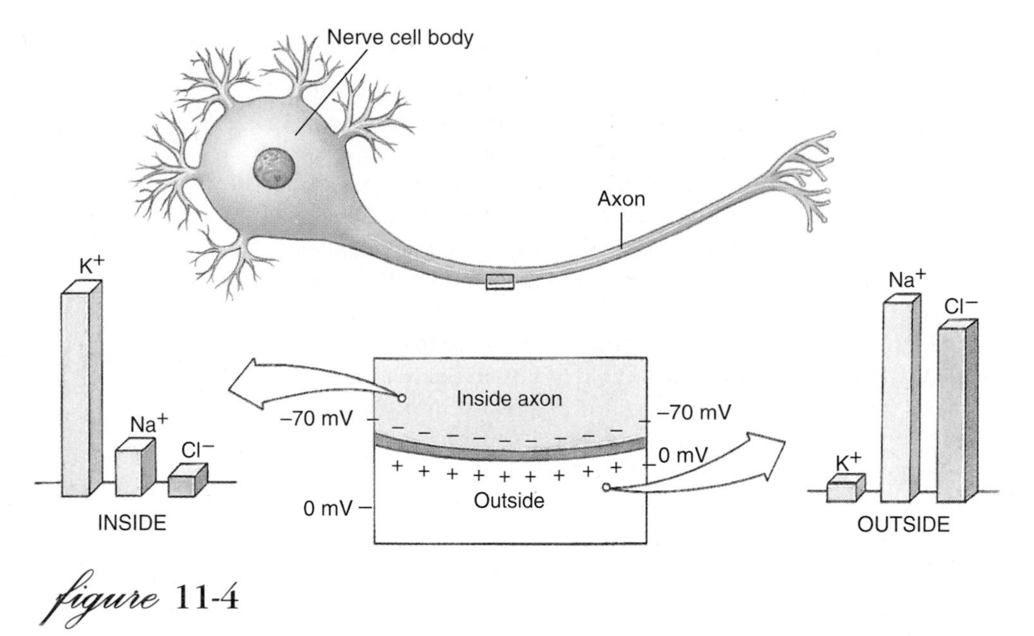

figure 11-4

Ionic composition inside and outside a resting nerve cell. An active sodium-potassium exchange pump located in the cell membrane drives sodium to the outside, keeping its concentration low inside. Potassium concentration is high inside. Although the membrane is "leaky" to potassium, this ion is held inside by the repelling positive charge outside the membrane.

The Action Potential

The nerve impulse is a rapidly moving change in electrical potential called the **action potential** (Figure 11-5). It is a very rapid and brief depolarization of the membrane of the nerve fiber. In most nerve fibers, the action potential does not simply return the membrane potential to zero but instead overshoots zero. In other words, the membrane potential reverses for an instant so that the outside becomes negative as compared with the inside. Then, as the action potential moves ahead, the membrane returns to its normal resting potential ready to conduct another impulse. The entire event occupies approximately a millisecond. Perhaps the most significant property of the nerve impulse is that it is self-propagating; that is, once started the impulse moves ahead automatically, much like the burning of a fuse.

What causes the reversal of polarity in the cell membrane during passage of an action potential? We have seen that the resting potential depends on the high membrane permeability (leakiness) to K^+, some 50 to 70 times greater than the permeability to Na^+. When the action potential arrives at a given point, Na^+ channels suddenly open, permitting a flood of Na^+ to diffuse into the axon from the outside. Actually only a very minute amount of Na^+ moves across the membrane—less than one millionth of the Na^+ outside—but this sudden rush of positive ions wipes out the local membrane resting potential. The membrane is **depolarized,** creating an electrical "hole." Potassium ions, finding their electrical barrier gone, begin to move out. Then, as the action potential passes, the membrane quickly regains its resting properties. It becomes once again practically impermeable to Na^+ and the outward movement of K^+ is checked.

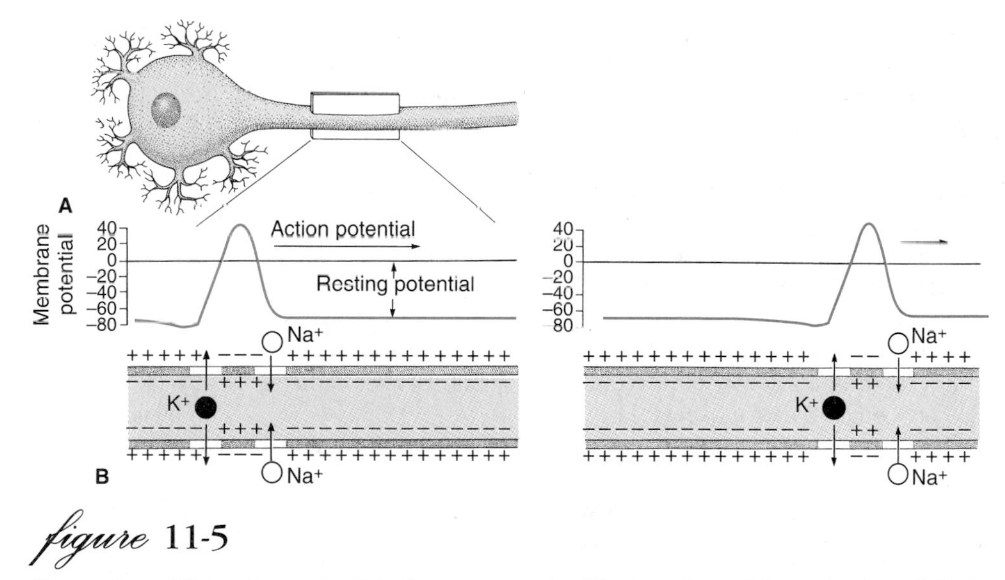

figure 11-5

Conduction of the action potential of a nerve impulse. The impulse originates in the cell body of the neuron **(A)** and moves toward the right. **B** and **C** show the electrical event and associated changes in localized membrane permeability to sodium and potassium. The position of the action potential in **C** is shown about 4 milliseconds after **B.** When the impulse arrives at a point, sodium gates are opened, allowing sodium ions to rush in. Sodium inflow reverses the membrane polarity, making the inner surface of the axon positive and the outside negative. Sodium gates then close and potassium gates open. Potassium ions can now penetrate the membrane and restore the normal resting potential.

The rising phase of the action potential is associated with the rapid influx (inward movement) of Na^+ (Figure 11-5). When the action potential reaches its peak, the Na^+ permeability is restored to normal, and K^+ permeability briefly increases above the resting level. This causes the action potential to drop rapidly toward the resting membrane level.

The Sodium Pump

The resting cell membrane has a very low permeability to Na^+. Nevertheless some Na^+ leaks across, even in the resting condition. When the axon is active, Na^+ flows inward with each passing impulse. If not removed, the accumulation of Na^+ inside the axon would cause the resting potential of the fiber to decay. This is prevented by **sodium pumps,** each a complex of protein subunits embedded in the plasma membrane of the axon (see Figure 2-16, p. 34). Each sodium pump uses the energy stored in ATP to transport sodium from the inside to the outside of the membrane. The sodium pump in nerve axons, as in many other cell membranes, also moves K^+ into the axon while it is moving Na^+ out. Thus it is a **sodium-potassium exchange pump** that helps to restore the ion gradients of both Na^+ and K^+. Recently it was discovered that the astrocytes (mentioned earlier) help to maintain the correct balance of ions surrounding neurons by sweeping away excess potassium produced during neuronal activity.

Synapses: Junction Points between Nerves

When an action potential passes down an axon to its terminal, it must cross a small gap, the synapse (Gr. *synapsis,* contact, union), separating it from another neuron or an effector organ. Two distinct kinds of synapses are known: electrical and chemical.

Electrical synapses, although much less common than chemical synapses, have been demonstrated in several invertebrate groups and are probably rather common in the nervous systems of many vertebrates. Electrical synapses are points at which ionic currents flow directly across a narrow **gap junction** (see Figure 2-13, p. 31) from one neuron to another. Electrical synapses show no time lag and consequently are important for escape reactions.

Much more complex than electrical synapses are **chemical synapses.** These contain packets of specialized chemicals called **neurotransmitters.** Neurons bringing impulses toward chemical synapses are called **presynaptic neurons;** those carrying impulses away are **postsynaptic** neurons. At the synapse, the membranes are separated by a narrow gap, the **synaptic cleft,** having a width of approximately 20 nm.

The axon of most neurons divides at its end into many branches, each of which bears a synaptic knob that sits on the dendrites or cell body of the next neuron (Figure 11-6). The axon terminations of several neurons may almost cover a nerve cell body and its dendrites with thousands of synapses. Because a single impulse coming down a nerve axon splays out into many branches and synaptic endings on the next nerve cell, many impulses converge on the cell body at one instant.

The 20 nm fluid-filled gap between presynaptic and postsynaptic membranes prevents impulses from spreading directly to the postsynaptic neuron. Instead, the synaptic knobs secrete a specific transmitter, usually **acetylcholine,** that communicates chemically with the postsynaptic cell. Inside the synaptic knobs are numerous tiny vesicles, each containing several thousand molecules of acetylcholine. Additional acetylcholine is also present in the cytoplasm of the synaptic knobs. Evidence suggests that when an impulse arrives at the terminal knob, a sequence of events as portrayed in Figure 11-6 occurs. The action potential opens protein channels in the presynaptic membrane, allowing a pulse of acetylcholine to diffuse across the gap in a fraction of a millisecond and bind briefly to receptor molecules on the postsynaptic membrane. This binding creates a voltage change in the postsynaptic membrane. Whether the voltage change is large enough to trigger a postsynaptic potential depends on how many acetylcholine molecules are released and how many channels are opened. The acetylcholine is rapidly destroyed by the enzyme **acetylcholinesterase.** If not inactivated in this way, the transmitter would continue to stimulate indefinitely. The organophosphate insecticides (such as malathion) and certain military nerve gases are poisonous for precisely this reason; they block acetylcholinesterase. The final step in the sequence is the resynthesis of acetylcholine and its storage in vesicles, ready to respond to another impulse.

Several different chemical neurotransmitters have been identified in both vertebrate and invertebrate nervous systems. Some, such as acetylcholine and norepinephrine, depolarize the postsynaptic membrane; these are **excitatory synapses.** Other neurotransmitters, such as gamma aminobutyric acid (GABA), hyperpolarize the postsynaptic membrane; this hyperpolarization tends to stabilize the membrane against depolarization. These are **inhibitory synapses.** Most nerve cells in the central nervous system have both excitatory and inhibitory synapses among the hundreds or thousands of synaptic knobs on the dendrites and cell body of each nerve cell.

It is the net balance of all excitatory and inhibitory inputs received by a postsynaptic cell that determines whether it will generate an action potential. If many excitatory impulses are received at one time, they may reduce the membrane potential enough in the postsynaptic membrane to elicit an action potential. Inhibitory impulses, however, stabilize the postsynaptic membrane, making it less likely that an action potential will be generated. The synapse is of great functional importance because it is a crucial part of the decision-making equipment of the central nervous system. In the synapse information is modulated from one nerve to the next.

Evolution of the Nervous System

Invertebrates: Development of Centralized Nervous Systems

The metazoan phyla reveal variation in complexity of nervous systems that probably reflects in a general way the stages in

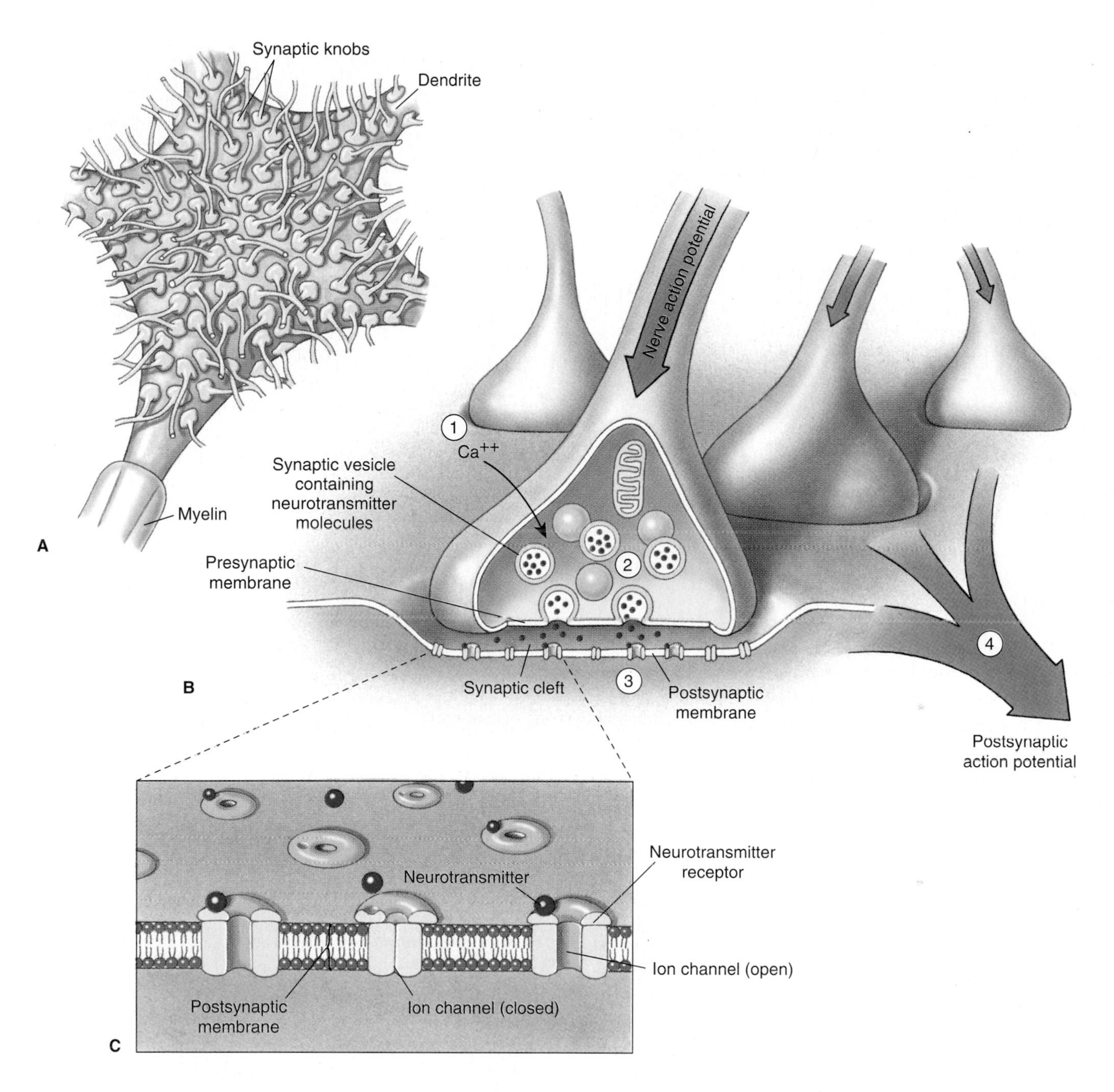

figure 11-6

Transmission of impulses across nerve synapses. **A,** The cell body of a motor nerve is shown with the terminations of interneurons. Each termination ends in a synaptic knob; thousands of synaptic knobs may rest on a single nerve cell body and its dendrites. **B,** A synaptic knob enlarged 60 times more than in **A.** An impulse traveling down the axon causes protein channels in the presynaptic membrane to open, releasing neurotransmitter molecules into the cleft. **C,** Diagram of a synaptic cleft at the ultrastructural level. Neurotransmitter molecules move rapidly across the gap to bind briefly with receptor molecules in the postsynaptic membrane. This binding produces a change in the potential of the postsynaptic membrane.

the evolution of complex nervous systems. The simplest pattern of invertebrate nervous systems is the **nerve net** of radiate animals, such as sea anemones, jellyfish, hydra, and comb jellies (Figure 11-7A). The nerve net is a quantum leap in complexity beyond sensory systems of protozoa, which lack nerves. The nerve net forms an extensive network that is found in and under the epidermis over all the body. An impulse starting in one part of this net is conducted in all directions, since synapses in most radiates do not restrict transmission to one-way movement, as they do in more complex

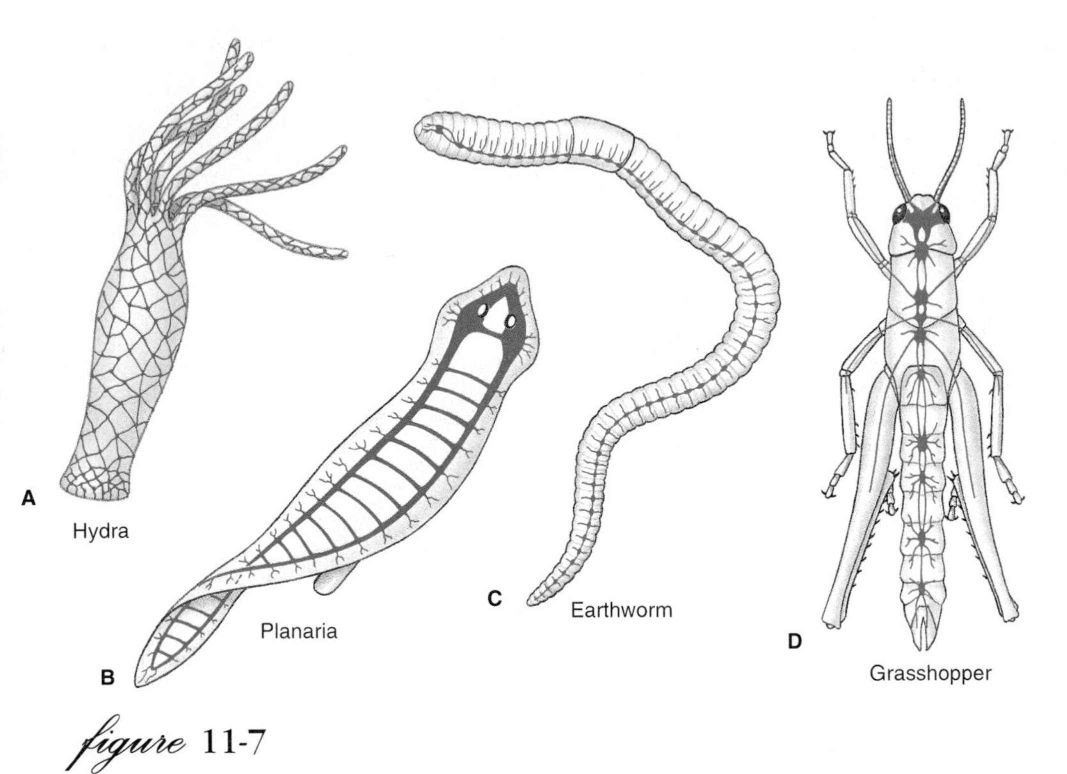

figure 11-7

Invertebrate nervous systems. **A,** The nerve net of radiates, the simplest neural organization. **B,** The flatworm system, the simplest linear-type nervous system of two nerve trunks connected to a complex neuronal network. **C,** The annelid nervous system, organized into a bilobed brain and ventral cord with segmental ganglia. **D,** Arthropod nervous system with large ganglia and with more elaborate sense organs.

animals. There are no differentiated sensory, motor, or connector components in the strict meaning of those terms. Branches of the nerve net connect to receptors in the epidermis and to epithelial cells that have contractile properties. Most responses tend to be generalized, yet many are astonishingly complex for so simple a nervous system. It is interesting that this type of nervous system is found among vertebrates in the form of nerve plexuses located, for example, in the intestinal wall, where they govern generalized intestinal movements such as peristalsis and segmentation.

Bilateral nervous systems, first seen in the flatworms, represent a distinct increase in complexity over the nerve net of radiate animals. Flatworms have two anterior **ganglia,** a clustering of nerve cells with numerous interconnections. The evolution of ganglia, representing the beginnings of a brain, permitted far more complex behavior than possible with a nerve net. From the ganglia two main nerve trunks run posteriorly, with lateral branches extending to the various parts of the body (Figure 11-7B). The flatworm nervous system is the simplest to show differentiation into a **peripheral nervous system** (a communication network extending to all parts of the body) and a **central nervous system,** which coordinates everything. More complex invertebrates have a more centralized nervous system, two longitudinal nerve cords fused (although still recognizable) and many ganglia present. The elaborate nervous system of annelids consists of a bilobed brain, a double nerve cord with ganglia, and distinctive **afferent** (sensory) and **efferent** (motor) neurons (Figure 11-7C). The segmental ganglia serve as relay stations for coordinating regional activity.

The basic plan of the molluscan nervous system is a series of three pairs of well-defined ganglia, but in cephalopods (such as octopus and squid), the ganglia have burgeoned into textured nervous centers of great complexity, such as those of the octopus, which contain more than 160 million cells. Sense organs, too, are highly developed. Consequently, the complexity of cephalopod behavior far outstrips that of any other invertebrate.

The basic plan of the arthropod nervous system (Figure 11-7D) resembles that of annelids, but the ganglia are larger and sense organs are much better developed. Social behavior is often elaborate, particularly in the hymenopteran insects (bees, wasps, and ants), and most arthropods are capable of considerable manipulation of their environment.

Vertebrates: Fruition of Encephalization

The basic plan of the vertebrate nervous system is a hollow, *dorsal* nerve cord terminating anteriorly in a large ganglionic mass, the brain. This pattern contrasts with the nerve cord of bilateral invertebrates, which is solid and ventral to the alimentary canal. By far the most important trend in the evolution of vertebrate nervous systems is the great elaboration of the size, configuration, and functional capacity of the brain, a process called **encephalization.** Vertebrate encephalization has brought to full fruition several functional capabilities including fast responses, great capacity for storage of information, and enhanced complexity and flexibility of behavior. Another consequence of encephalization is the ability to form associations between past, present, and (at least in humans) future events.

The Spinal Cord

The **brain** and **spinal cord** compose the central nervous system. During early embryonic development, the spinal cord and brain begin as an ectodermal neural groove, which by folding and enlarging becomes a long, hollow neural tube. The cephalic end enlarges into the brain vesicles, and the rest becomes the spinal cord. Unlike any invertebrate nerve cord, the segmental nerves of the spinal cord (31 pairs in humans) are separated into dorsal sensory roots and ventral motor roots. The sensory nerve cell bodies are gathered together into dorsal root (spinal) ganglia. Both dorsal (sensory) and ventral (motor) roots meet beyond the spinal cord to form a mixed spinal nerve (Figure 11-8).

The spinal cord is enclosed by the vertebral canal and is additionally wrapped in three layers of membranes called **meninges** (men-in′jeez; Gr. *meningos,* membrane). A cross section of the cord reveals two zones (Figure 11-8). The inner zone of gray matter, resembling in shape the wings of a butterfly, contains the cell bodies of motor neurons and interconnecting interneurons (described in the following text). The outer zone of white matter contains bundles of axons and dendrites linking different levels of the cord with each other and with the brain.

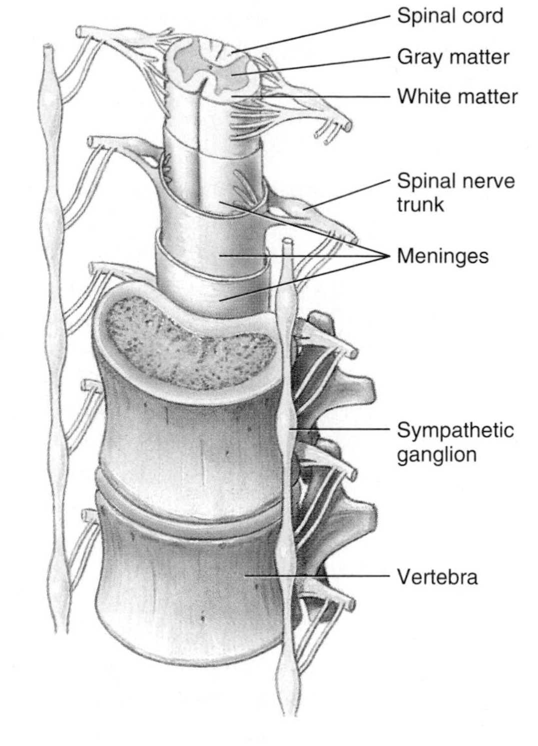

figure 11-8

Human spinal cord and its protection. Two vertebrae show the position of the spinal cord, emerging spinal nerve trunks, and the sympathetic trunk. The cord is wrapped by three layers of membrane (meninges) between two of which lies a protective bath of cerebrospinal fluid.

The Reflex Arc

Many neurons work in groups called **reflex arcs.** There must be at least two neurons in a reflex arc, but usually there are more. The parts of a typical reflex arc (as, for example, in the well-known "knee-jerk" reflex, Figure 11-9) are (1) a **receptor,** a sense organ in the skin, muscle, or other organ; (2) an **afferent,** or sensory, neuron, which carries impulses toward the central nervous system; (3) the **central nervous system,** where synaptic connections are made between the sensory neurons and the interneurons; (4) the **efferent,** or motor, neuron, which makes a synaptic connection with the interneuron and carries impulses from the central nervous system; and (5) the **effector,** by which the animal responds to environmental changes. Examples of effectors are muscles, glands, ciliated cells, nematocysts of the radiate animals, electric organs of fish, and certain pigmented cells called chromatophores.

A reflex arc in its simplest form consists of only two neurons—a sensory (afferent) neuron and a motor (efferent) neuron. Usually, however, interneurons are interposed between sensory and motor neurons (Figure 11-9). Interneurons may connect afferent and efferent neurons on the same side of the spinal cord or on opposite sides, or they may connect them on different levels of the spinal cord, either on the same or opposite sides. In almost any reflex act a number of reflex arcs are involved. For instance, a single afferent neuron may make synaptic connections with many efferent neurons. In a similar way an efferent neuron may receive impulses from many afferent neurons.

A **reflex act** is the response to a stimulus acting over a reflex arc. It is involuntary, meaning that it is not under the control of the will. Many vital processes of the body, such as control of breathing, heartbeat, diameter of blood

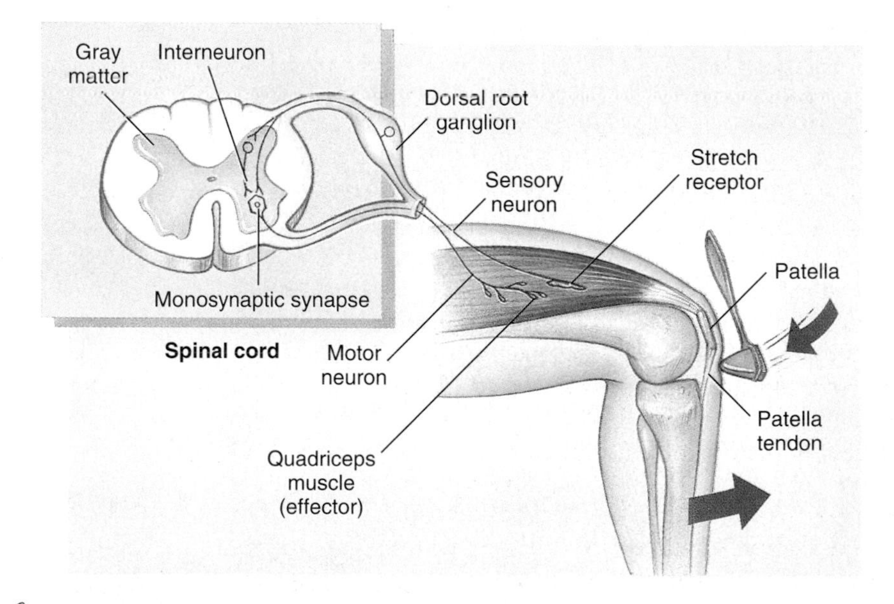

figure 11-9

The knee-jerk reflex, a simple reflex arc. Sudden pressure on the patellar ligament stretches muscles in the upper leg. Impulses generated in stretch receptors are conducted over afferent (sensory) fibers to the spinal cord and relayed by an interneuron to an efferent (motor) nerve cell body. Impulses pass over efferent axons to the leg muscles (effectors), stimulating them to contract.

vessels, and sweat gland secretion are reflex acts. Some reflex acts are innate; others are acquired through learning.

The Brain

Unlike the spinal cord, which has changed little in structure during vertebrate evolution, the brain has changed dramatically. The linear brain of fishes and amphibians contrasts with the expanded, deeply fissured, and enormously intricate brain of mammals (Figures 11-10, 11-12). It reaches its greatest complexity in the human brain, which contains some 35 billion nerve cells, each of which may receive information from tens of thousands of synapses at one time. The ratio between the weight of the brain and that of the spinal cord affords a fair criterion of an animal's intelligence. In fish and amphibians this ratio is approximately 1:1; in humans the ratio is 55:1—in other words, the brain is 55 times heavier than the spinal cord. Although the human brain is not the largest (the sperm whale's brain is seven times heavier) nor the most convoluted (that of the porpoise is even more wrinkled), it is by all odds the best in overall performance. This "great ravelled knot," as the British physiologist Sir Charles Sherrington called the human brain, in fact may be so complex that it will never be able to understand its own function.

Although the large size of their brain undoubtedly makes humans the wisest of animals, it is apparent that they can do without much of it and still remain wise. Brain scans of persons with hydrocephalus show that enlargement of the head is the result of pressure disturbances that cause the brain ventricles to enlarge many times their normal size. While many hydrocephalic persons are functionally disabled, others are nearly normal. The cranium of one person with hydrocephalus was nearly filled with cerebrospinal fluid and the only remaining cerebral cortex was a thin layer of tissue, 1 mm thick, pressed against the cranium. Yet this young man, with only 5% of his brain, had achieved first-class honors in mathematics at a British university and was socially normal. This and other similarly dramatic observations suggest that there is enormous redundancy and spare capacity in corticocerebral function. It also suggests that the deep structures of the brain, which are relatively spared in hydrocephalus, may perform functions once believed to be performed solely by the cortex.

The brain of the early vertebrate fishes had three principal divisions: a forebrain, the **prosencephalon;** a midbrain, the **mesencephalon;** and a hindbrain, the **rhombencephalon** (Figure 11-11). Each of the three parts was concerned with one or more of the special senses: the forebrain

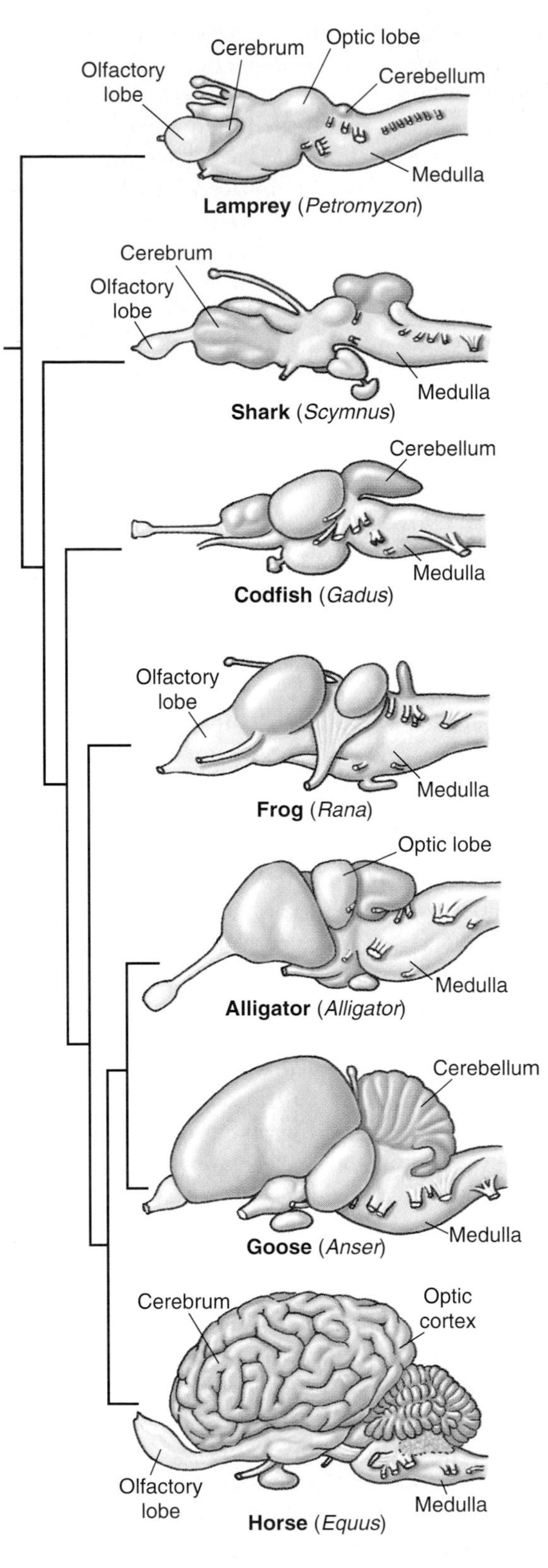

figure 11-10

Variation in the vertebrate brain. Note the differences in size of the cerebrum. The cerebellum, concerned with equilibrium and motor coordination, is largest in animals whose balance and precise motor movements are well developed (fishes, birds, and mammals). Relationships among the vertebrates are indicated by the evolutionary tree.

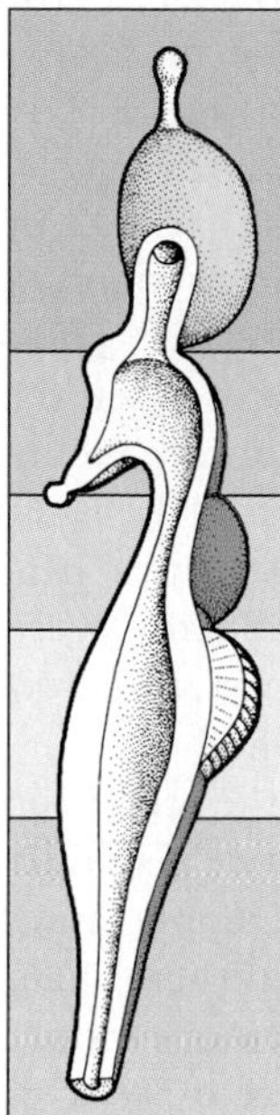

Embryonic vesicle: Early embryo	Embryonic vesicle: Late embryo	Main components in adults	Function
Forebrain (Prosencephalon)	Telencephalon	Cerebrum	Motor area controls voluntary muscle movements; sensory cortex is center of conscious perception of touch, pressure, vibration, pain, temperature, and taste; association areas integrate and process sensory data
	Diencephalon	Thalamus	Part of limbic system; integrates sensory information arriving at thalamus, projects to cerebral frontal lobes
		Hypothalamus	Controls autonomic functions; sets appetitive drives (thirst, hunger, sexual desire) and behavior; sets emotional states; secretes ADH, oxytocin; secretes releasing factors for anterior pituitary
Midbrain (Mesencephalon)	Mesencephalon	Optic lobes (tectum)	Integrates visual information with other sensory inputs; relays auditory information
		Midbrain nuclei	Involuntary control of muscle tone; processing of incoming sensations and outgoing motor commands
Hindbrain (Rhombencephalon)	Metencephalon	Cerebellum	Involuntary coordination and control of outgoing movements for equilibrium, muscle tone, posture
		Pons	Links cerebellum with other brain centers and with medulla and spinal cord; modifies output of respiratory centers in medulla
	Myelencephalon	Medulla oblongata	Regulates heart rate and force of contraction; vasomotor control; sets rate of respiration; relays information to the cerebellum

figure 11-11

Divisions of the vertebrate brain and their functions.

with the sense of smell, the midbrain with vision, and the hindbrain with hearing and balance. These very fundamental concerns of the brain have been in some instances amplified and in others reduced or overshadowed during continued evolution as sensory priorities were shaped by the animal's habitat and way of life.

Hindbrain The **medulla,** the most posterior division of the brain, is really a conical continuation of the spinal cord. The medulla, together with the more anterior midbrain, constitutes the "brain stem," an area that controls numerous vital and largely subconscious activities such as heartbeat, respiration, vascular tone, and swallowing. The **pons,** also a part of the hindbrain, is a thick bundle of fibers that carry impulses from one side of the cerebellum to the other, and to higher brain centers.

The **cerebellum,** literally "little brain," lying dorsal to the medulla, controls equilibrium, posture, and movement (Figure 11-12). Its development is directly correlated with the animal's mode of locomotion, agility of limb movement, and balance. It is usually weakly developed in amphibians and reptiles, forms that live close to the ground, and well developed in the more agile bony fishes. It reaches its apogee in birds and mammals in which it is greatly expanded and folded. The cerebellum does not initiate movement but operates as a precision error-control center, or servomechanism, that programs a movement initiated somewhere else, such as the motor cortex. Primates and especially humans, who possess a manual dexterity far surpassing that of other animals, have the most complex cerebellum. Movements of the hands and fingers may involve the cerebellar coordination of the simultaneous contraction and relaxation of hundreds of individual muscles.

Midbrain The midbrain consists mainly of the **tectum** (including the optic lobes), which contains nuclei that serve as centers for visual and auditory reflexes. (In neurophysiological usage a *nucleus* is a small aggregation of nerve cell bodies within the central nervous system.) The midbrain has undergone little evolutionary change in structure among vertebrates but has changed markedly in function. It mediates the most

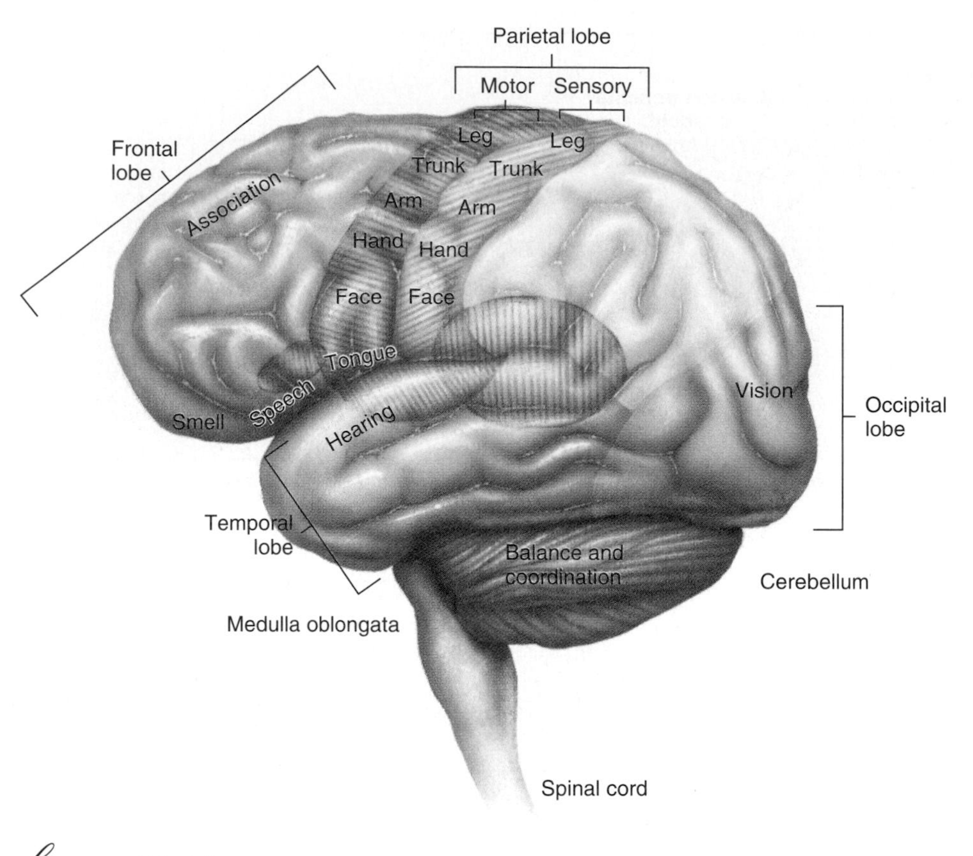

figure 11-12

External view of the human brain, showing lobes of the cerebrum and localization of major function of the cerebrum and cerebellum.

complex behavior of fishes and amphibians. Such integrative functions were gradually assumed by the forebrain in amniotes. In mammals the midbrain is mainly a reflex center for eye muscles and a relay and analysis center for auditory information.

Forebrain Just anterior to the midbrain lie the **thalamus** and **hypothalamus,** the most posterior elements of the forebrain. The egg-shaped thalamus is a major relay station that analyzes and passes sensory information to higher brain centers. In the hypothalamus are several "housekeeping" centers that regulate body temperature, water balance, appetite, and thirst—all functions concerned with the maintenance of internal constancy (homeostasis). Neurosecretory cells located in the hypothalamus produce several pituitary-regulating neurohormones (described in Chapter 12). The hypothalamus also contains centers for diverse emotions such as pleasure, aggression, and sexual arousal.

The anterior portion of the forebrain, the cerebrum (Figures 11-11, 11-12), can be divided into two anatomically distinct areas, the **paleocortex** and **neocortex.** Originally concerned with smell, it became well developed in fishes and early terrestrial vertebrates, which depended on this special sense. In mammals, and especially in primates, the paleocortex is a deep-lying area called the rhinencephalon ("nose brain"), because many of its functions depend on olfaction. Better known as the **limbic system,** it mediates several species-specific behaviors that relate to fulfilling needs such as feeding and sex.

Although a late arrival in vertebrate evolution, the neocortex completely overshadows the paleocortex and has become so expanded that it envelops much of the forebrain and all of the midbrain (Figure 11-12). Almost all of the integrative activities primitively assigned to the midbrain were transferred to the neocortex, or cerebral cortex as it is usually called.

Functions in the cerebrum have been localized by direct stimulation of exposed brains of people and experimental animals, postmortem examination of persons suffering from various lesions, and surgical removal of specific brain areas in experimental animals. The cortex contains discrete motor and sensory areas (Figures 11-12 and 11-13) as well as large "silent" regions, called **association areas,** concerned with memory, judgment, reasoning, and other integrative functions. These regions are not directly connected to sense organs or muscles.

Thus in mammals, and especially in humans, separate parts of the brain mediate conscious and unconscious functions. The unconscious mind, all of the brain except the cerebral cortex, governs numerous vital functions that are removed from conscious control: respiration, blood pressure, heart rate, hunger, thirst, temperature balance, salt balance, sexual drive, and basic (sometimes irrational) emotions. It is also a complex endocrine gland that regulates the body's subservient endocrine system. The other, conscious mind, the cerebral cortex, is the site of higher mental activities (for example, planning and reasoning), memory, and integration of sensory information.

Memory appears to transcend all parts of the brain, although an area of the limbic system called the hippocampus is considered to be essential for memory consolidation. On a cellular level, research with animals with relatively simple nervous systems, such as the marine snail *Aplysia* (see pp. 294–296), has established that long-term memory involves not only the release of neurotransmitters but the expression of genes in neurons of repeatedly activated neural circuits. New proteins are synthesized that induce the growth of new synaptic connections and increase the excitability of existing synaptic connections. There is evidence that these findings apply to the mammalian limbic system as well.

The Peripheral Nervous System

The peripheral nervous system includes all nervous tissue outside of the central nervous system. It consists of two functional divisions: the **afferent division** that brings sensory information to the central nervous system, and the **efferent division** that conveys motor commands to muscles and glands. The efferent division consists of two components, (1) the **somatic nervous system,** which supplies the skeletal muscle, and (2) the **autonomic nervous system,** which supplies smooth muscle, cardiac muscle, and glands.

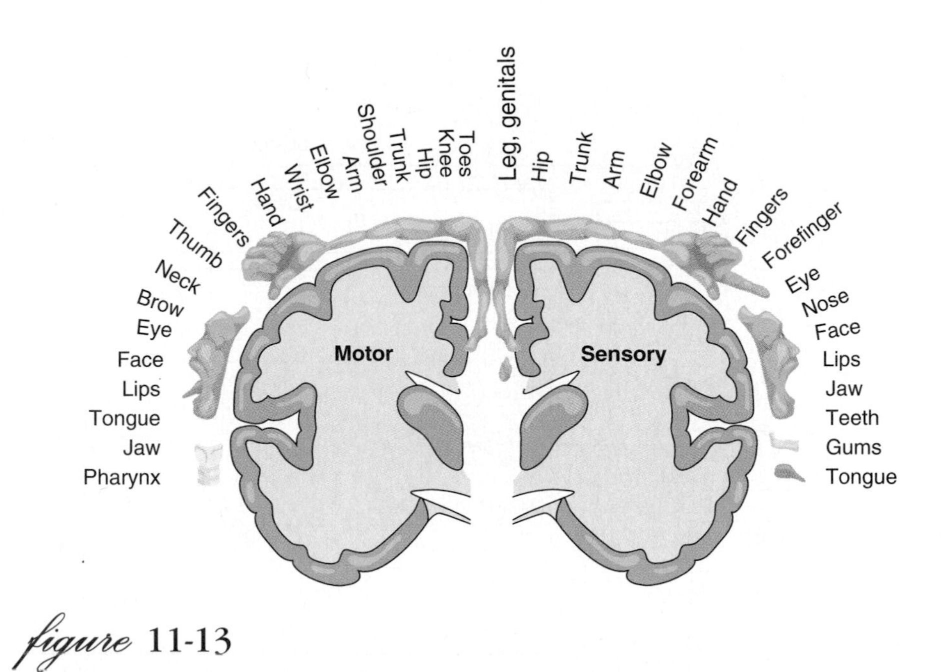

figure 11-13

Arrangement of sensory and motor cortices. Localization of sensory terminations from different parts of the body are shown at left; origin of descending motor pathways are shown at right. The motor cortex lies in front of the sensory cortex, so the two are not superimposed. These maps grew out of the 1930s work of Canadian neurosurgeon Wilder Penfield. Recent research shows that the motor cortex is not as orderly as the map suggests; rather there is a more diffuse correspondence between cortical areas and the areas of the body they control.

Autonomic Nervous System The autonomic system governs the involuntary, internal functions of the body that do not ordinarily affect consciousness, such as the movements of the alimentary canal and heart, the contraction of the smooth muscle of the blood vessels, urinary bladder, iris of the eye, and others, plus the secretions of various glands.

Autonomic nerves originate in the brain or spinal cord as do the nerves of the somatic nervous system, but unlike the latter, the autonomic fibers consist of not one but two motor neurons. They synapse once after leaving the cord and before arriving at the effector organ. These synapses are located outside the spinal cord in ganglia. Fibers passing from the cord to the ganglia are called preganglionic autonomic fibers; those passing from the ganglia to the effector organs are called postganglionic fibers. These relationships are illustrated in Figure 11-14.

Subdivisions of the autonomic system are the **parasympathetic** and the **sympathetic** systems. Most organs in the body are innervated by both sympathetic and parasympathetic fibers, and their actions are antagonistic (Figure 11-15). If one fiber speeds up an activity, the other slows it down. However, neither kind of nerve is exclusively excitatory or inhibitory. For example, parasympathetic fibers inhibit heartbeat but excite peristaltic movements of the intestine; sympathetic fibers increase heartbeat but slow down peristaltic movement.

The parasympathetic system consists of motor neurons, some of which emerge from the brain stem by certain cranial nerves and others of which emerge from the sacral (pelvic) region of the spinal cord (Figures 11-14 and 11-15). In the sympathetic division the nerve cell bodies of all the preganglionic fibers are located in the thoracic and upper lumbar areas of the spinal cord. Their fibers exit through the ventral roots of the spinal nerves, separate from these, and go to the

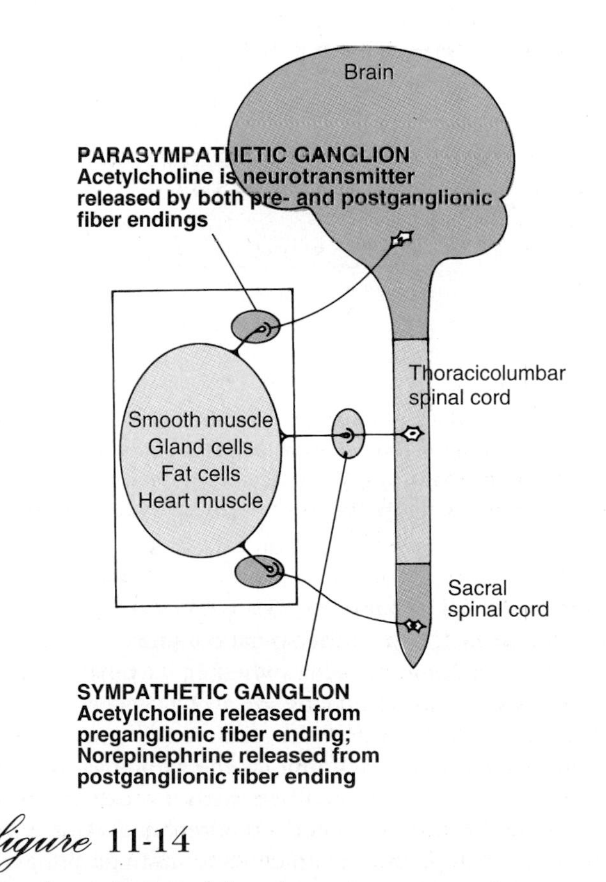

figure 11-14

General organization of the autonomic nervous system.

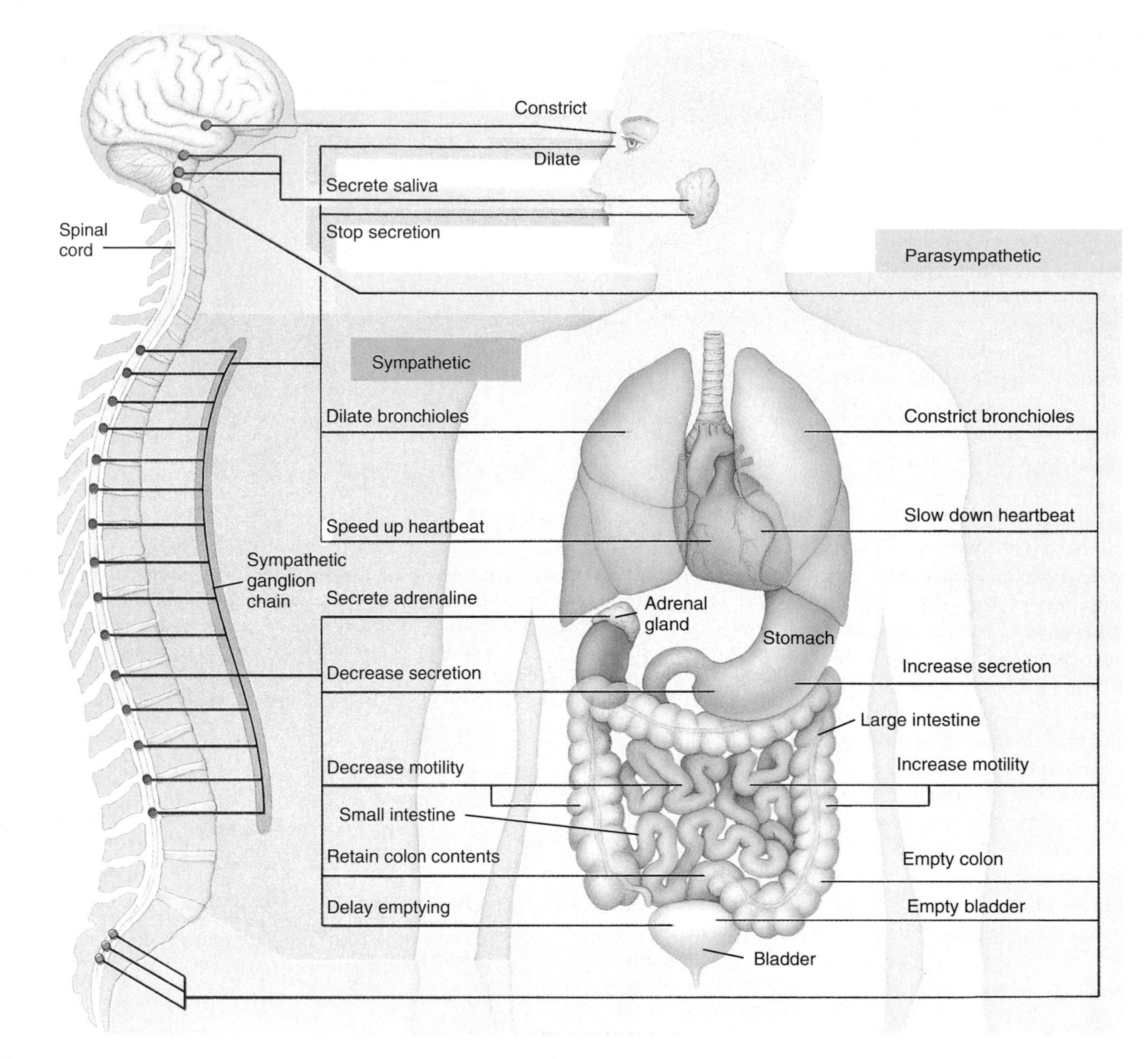

figure 11-15

Autonomic nervous system in humans. The outflow of autonomic nerves from the central nervous system is shown at left. The sympathetic *(red)* outflow is from the thoracic and lumbar areas of the spinal cord by way of a chain of sympathetic ganglia. The parasympathetic *(blue)* outflow is from the cranial and sacral regions of the central nervous system; the parasympathetic ganglia (not shown) are located in or adjacent to the organs innervated. Most organs are innervated by fibers from both sympathetic and parasympathetic divisions.

sympathetic ganglia (Figure 11-15), which are paired and form a chain on each side of the spinal column.

All preganglionic fibers, whether sympathetic or parasympathetic, release acetylcholine at the synapse with the postganglionic cells. However, the parasympathetic postganglionic fibers release acetylcholine at their endings, whereas the sympathetic postganglionic fibers with few exceptions release norepinephrine (also called noradrenaline). This difference is another important characteristic distinguishing the two parts of the autonomic nervous system.

As a general rule the parasympathetic division is active under resting conditions when such functions as eating, digestion, urination, and other vegetative activities are emphasized. The sympathetic division is active under conditions of physical activity and stress. Under such conditions the heart rate increases, blood vessels to the skeletal muscles dilate, blood vessels in the viscera constrict, activity of the intestinal tract decreases, and metabolic rate increases. The importance of these responses in emergency reactions (sometimes called the fight-flight response) are described in the next chapter (p. 280). It should be noted, however, that the sympathetic division is also active during resting conditions in maintaining normal blood pressure and body temperature.

Sense Organs

Animals require a constant inflow of information from the environment to regulate their lives. Sense organs are specialized receptors designed for detecting environmental status and change. An animal's sense organs are its first level of environmental perception; they are channels for bringing information to the brain.

A **stimulus** is some form of energy—electrical, mechanical, chemical, or radiant. The task of the sense organ is to transform the energy form of the stimulus it receives into nerve impulses, the common language of the nervous system. In a very real sense, then, sense organs are biological transducers. A microphone, for example, is a transducer that converts mechanical (sound) energy into electrical energy. Like the microphone that is sensitive only to sound, sense organs are, as a rule, specific for one kind of stimulus. Thus eyes respond only to light, ears to sound, pressure receptors to pressure, and chemoreceptors to chemical molecules. But again, all of these different forms of energy are converted into nerve impulses.

Since all nerve impulses are qualitatively alike, how do animals perceive and distinguish the different sensations of varying stimuli? The answer is that the real perception of sensation is done in localized regions of the brain, where each sensory organ has its own hookup. Impulses arriving at a particular sensory area of the brain can be interpreted in only one way. For example, pressure on the eye causes us to see "stars" or other visual patterns; the mechanical distortion of the eye initiates impulses in the optic nerve fibers that are perceived as light sensations. Although such an operation probably could never be done, a deliberate surgical switching of optic and auditory nerves would cause the recipient literally to see thunder and hear lightning!

Classification of Receptors

Receptors are traditionally classified on the basis of their location. Those near the external surface, called **exteroceptors,** keep the animal informed about the external environment. Internal parts of the body are provided with **interoceptors,** which pick up stimuli from the internal organs. Muscles, tendons, and joints have **proprioceptors,** which are sensitive to changes in the tension of muscles and provide the organism with a sense of body position. Sometimes receptors are classified by the form of energy to which the receptors respond, such as **chemical, mechanical, light,** or **thermal.**

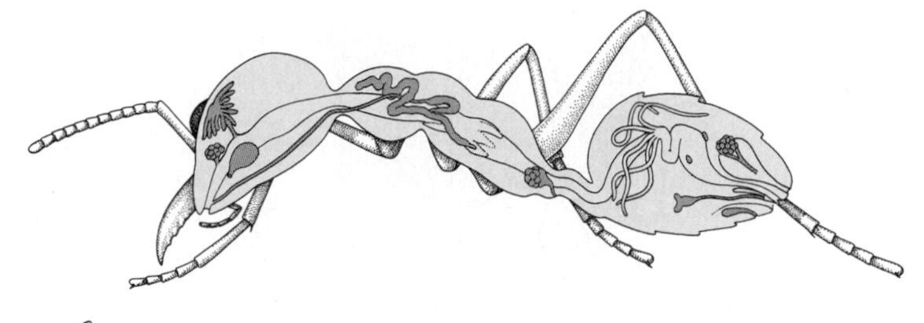

figure 11-16

Pheromone-producing glands of an ant.

Chemoreception

Chemoreception is the oldest and most universal sense in the animal kingdom. It probably guides the behavior of animals more than any other sense. Protozoa use **contact chemical receptors** to locate food and adequately oxygenated water and to avoid harmful substances. These receptors elicit an orientation behavior, called **chemotaxis,** toward or away from the chemical source. Most metazoans have specialized **distance chemical receptors.** These are often developed to a remarkable degree of sensitivity. Distance chemoreception, usually referred to as a sense of smell or olfaction, guides feeding behavior, location and selection of sexual mates, territorial and trail marking, and alarm reactions of numerous animals.

The social insects and some other animals, including mammals, produce species-specific compounds, called **pheromones,** which constitute a highly developed chemical language. Pheromones are a diverse group of organic compounds that an animal releases to affect the physiology or behavior of another individual of the same species. Ants, for example, are walking batteries of glands (Figure 11-16) that produce numerous chemical signals. These include releaser pheromones, such as alarm and trail pheromones, and primer pheromones, which alter the endocrine and reproductive systems of different castes in the colony. Insects bear a variety of chemoreceptors on the surface of the body for sensing specific pheromones, as well as other, nonspecific odors.

In all vertebrates and in insects as well, the senses of **taste** and **smell** are clearly distinguishable. Although there are similarities between taste and smell receptors, in general the sense of taste is more restricted in response and is less sensitive than the sense of smell. Central nervous centers for taste and smell are also located in different parts of the brain.

In vertebrates, taste receptors are found in the mouth cavity and especially on the tongue (Figure 11-17), where they provide a means for judging foods before they are swallowed. A **taste bud** consists of a cluster of several receptor cells surrounded by supporting cells; it is provided with a small external pore through which the slender tips of the sensory cells project. Chemicals being tasted apparently combine with specific receptor sites on the microvilli of the receptor cells. Because they are subject to the wear and tear of

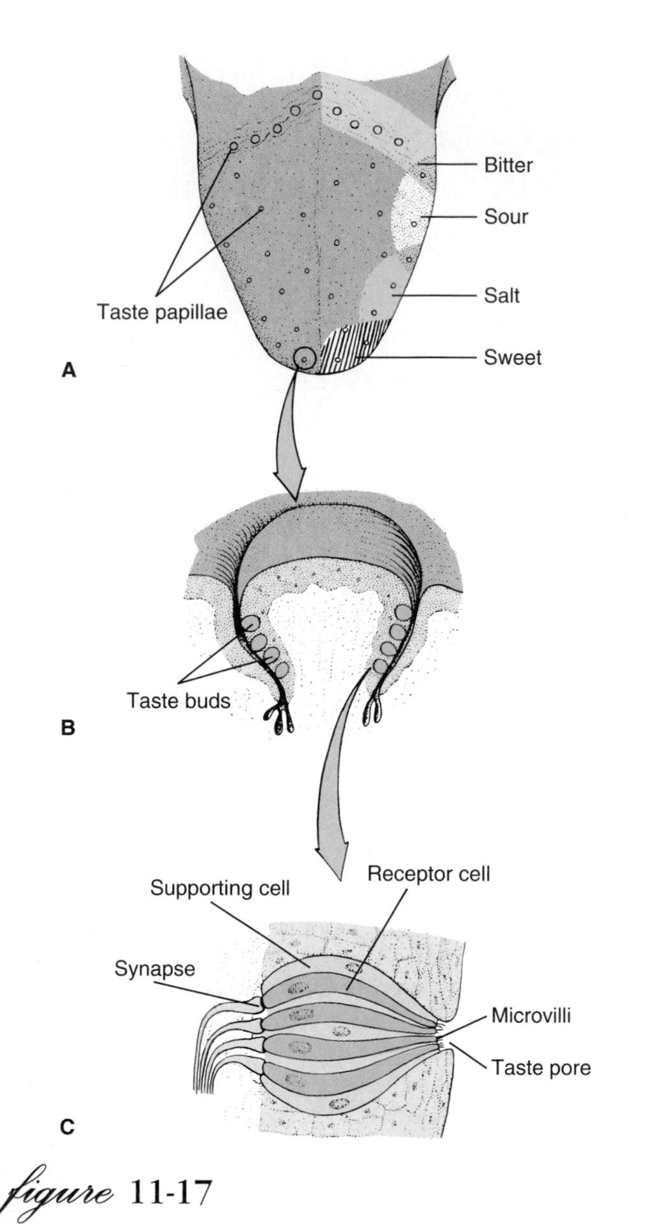

figure 11-17

Taste receptors. **A,** Surface of human tongue showing regions of maximum sensitivity to the four primary taste sensations. **B,** Position of taste buds on a taste papilla. **C,** Structure of a taste bud.

abrasive foods, taste buds have a short life (5 to 10 days in mammals) and are continually being replaced.

The four basic taste sensations possessed by humans—sour, salty, bitter, and sweet—are each attributable to a different kind of taste bud. The tastes for salty and sweet are found mainly on the tip of the tongue, bitter at the base of the tongue, and sour along the sides of the tongue. Of these, the bitter taste is by far the most sensitive, because it provides early warning against potentially dangerous substances, many of which are bitter.

The sense of smell is more complex than taste, and until very recently odor research has lagged behind other areas of sensory physiology. Although the olfactory sense is a primal sense for many animals, used for the identification of food, sexual mates, and predators, olfaction is most highly developed in mammals. Even humans, although a species not celebrated for detecting smells, can discriminate perhaps 10,000 different odors. The human nose can detect 1/25 of one-millionth of 1 mg of mercaptan, the odoriferous substance of the skunk. Even so, our olfactory abilities compare poorly with those of other mammals that rely on olfaction for survival. A dog explores new surroundings with its nose much as we do with our eyes. A dog's nose is justifiably renowned. With some odorous sources a dog's nose is at least a million times more sensitive than ours. Dogs are assisted in their proficiency by having a nose located close to the ground where odors from passing creatures tend to linger.

The olfactory endings are located in a special epithelium covered with a thin film of mucus, positioned deep in the nasal cavity (Figure 11-18). Within the epithelium lie millions of olfactory neurons, each with several hairlike cilia protruding from the free end. Odor molecules entering the nose bind to receptor proteins located in the cilia; this binding generates an electrical signal that travels along axons to the olfactory bulb of the brain. From here odor information is sent to the olfactory cortex where odors are analyzed. Odor information is then projected to higher brain centers to affect emotions, thoughts, and behavior.

Recently, using the techniques of gene cloning and molecular hybridization (p. 78), researchers discovered a large family of genes in mammals (including humans) that appears to code for odor reception. Each of the approximately 1000 genes discovered encodes a separate type of odor receptor. Since mammals can detect at least 10,000 different odors, each receptor must respond to several odor molecules, and each odor molecule must bind with several types of receptors, each of which responds to a part of the molecule's structure. Brain mapping techniques have shown that each olfactory neuron projects to a characteristic location on the olfactory bulb, providing a two-dimensional map that identifies which receptors have been activited in the nose. Projected to the brain, this information is recognized as a unique scent.

Because the flavor of food depends on odors reaching the olfactory epithelium through the throat passage, taste and smell are easily confused. All the various "tastes" other than the four basic ones (sweet, sour, bitter, salty) are really the results of flavor molecules reaching the olfactory epithelium in this manner. Food loses its appeal during a common cold because a stuffy nose blocks odors rising from the mouth.

Mechanoreception

Mechanoreceptors are sensitive to quantitative forces such as touch, pressure, stretching, sound, vibration, and gravity—in short, they respond to motion. Animals require a steady flow of information from mechanoreceptors in order to interact with their environment, feed themselves, maintain normal posture, and to walk, swim, or fly.

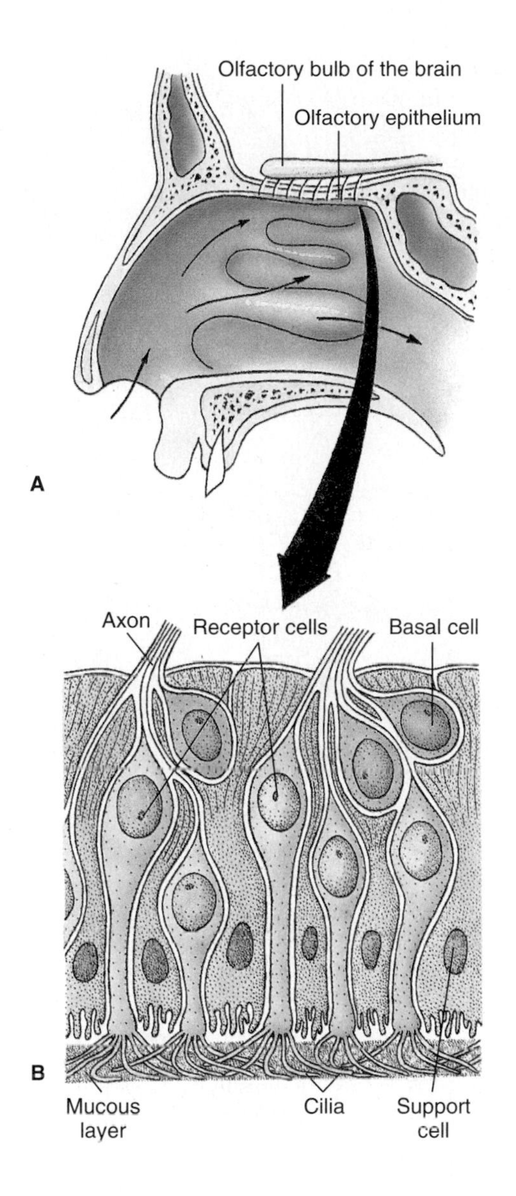

figure 11-18

Human olfactory epithelium. **A,** The epithelium is a patch of tissue positioned in the roof of the nasal cavity. **B,** It is composed of supporting cells, basal cells, and olfactory receptor cells with cilia protruding from their free ends.

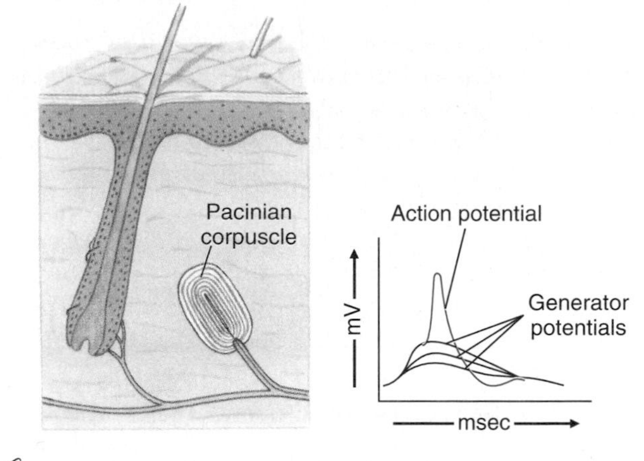

figure 11-19

Response of pacinian corpuscle to applied pressure. Progressively stronger pressure produces stronger receptor potentials. When the threshold stimulus is reached, an all-or-none action potential is generated in the afferent nerve fiber.

Touch and Pain

The **pacinian corpuscle,** a relatively large mechanoreceptor that registers deep touch and pressure in mammalian skin, illustrates the general properties of mechanoreceptors. They are common in the deep layers of the skin, connective tissue surrounding muscles and tendons, and the abdominal mesenteries. Each corpuscle consists of a nerve terminus surrounded by a capsule of numerous, concentric, onionlike layers of connective tissue (Figure 11-19). Pressure at any point on the capsule distorts the nerve ending, producing a graded **receptor potential.** This is a local flow of electric current. Progressively stronger stimuli lead to correspondingly stronger receptor potentials until a **threshold current** is produced; this initiates an action potential in the sensory nerve fiber. Stronger stimuli will produce a burst of action potentials. However, if the pressure applied is sustained, the corpuscle quickly adjusts to the new shape and no longer responds. This adjustment is called **adaptation** (not to be confused with the evolutionary meaning of this term [Chapter 4]) and is characteristic of many kinds of touch receptors, which are admirably suited to detecting a sudden mechanical change but readily adapt to new conditions. We are aware of new pressures when we put on our shoes in the morning, but we are glad not to be reminded of these pressures all day long.

Invertebrates, especially insects, have many kinds of receptors sensitive to touch. Such receptors are well endowed with tactile hairs sensitive to both touch and vibrations. Superficial touch receptors of vertebrates are distributed all over the body but tend to be concentrated in areas especially important for exploring and interpreting the environment. In most vertebrates these areas are on the face and extremities of the limb. Of the more than half million separate sensitive spots on the surface of the human body, most are found on the lips, tongue, and fingertips (Figure 11-13). Many touch receptors are bare nerve-fiber terminals, but there is an assortment of other kinds of receptors of varying shapes and sizes. Each hair follicle is crowded with receptors that are sensitive to touch.

Pain receptors are relatively unspecialized nerve fiber endings that respond to a variety of stimuli signaling possible or real damage to tissues. It is still uncertain whether pain fibers respond directly to injury or indirectly to some substance such as histamine, which is released by damaged cells.

Just as pain is a sign of danger, sensory pleasure is a sign of a stimulus useful to the subject. Pleasure depends on the internal state of the animal and is judged with reference to homeostasis and some physiological set point.

Pain is a distress call from the body signaling some noxious stimulus or internal disorder. Although there is no cortical pain center, discrete areas have been located in the brain stem where pain messages from the periphery terminate. These areas contain two kinds of small peptides, endorphins and enkephalins, that have morphinelike or opiumlike activity. When released, they bind with specific opiate receptors in the midbrain. They are the body's own analgesics.

Lateral Line Systems of Fishes

The lateral line is a distant touch reception system for detecting wave vibrations and currents in water. The receptor cells, called **neuromasts,** are located on the body surface in aquatic amphibians and some fishes, but in many fishes they are located within canals running beneath the epidermis; these canals open at intervals to the surface (Figure 11-20). Each neuromast is a collection of hair cells with the sensory hairs embedded in a gelatinous, wedge-shaped mass known as a **cupula.** The cupula projects into the center of the lateral line canal so that it bends in response to any disturbance of water on the body surface. The lateral line system is one of the principal sensory systems that guide fishes in their movements and in the location of predators, prey, and social partners (p. 589).

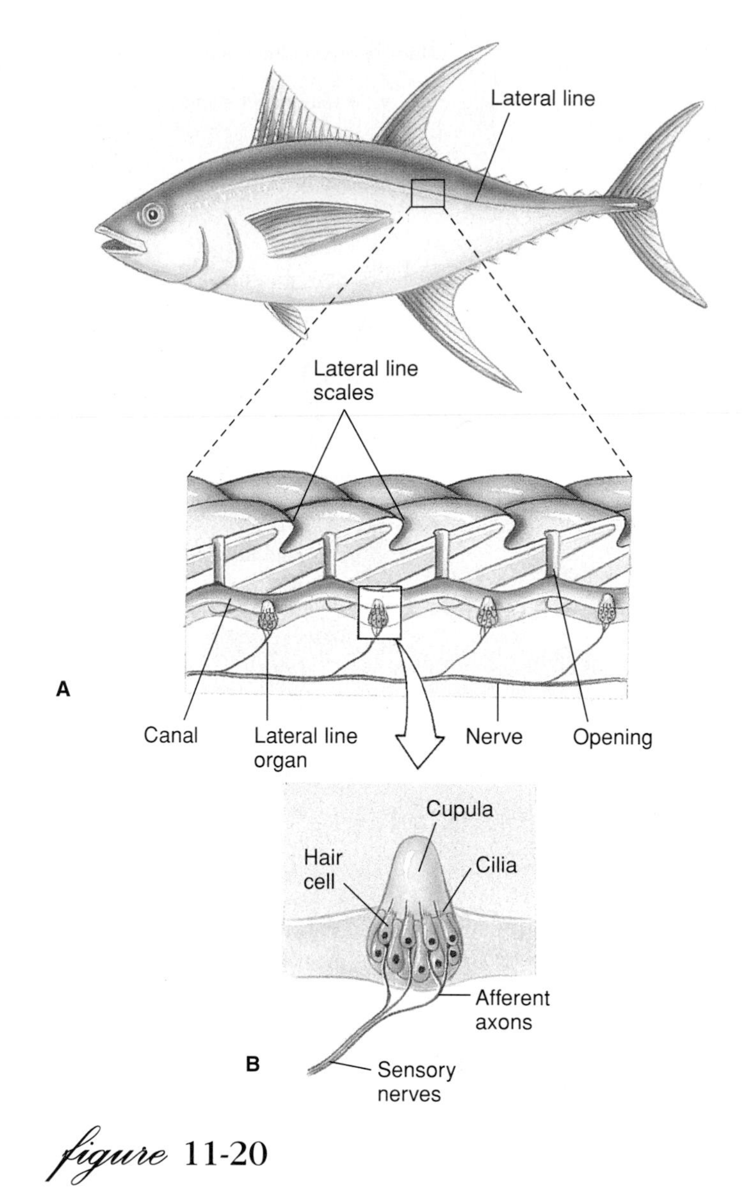

figure 11-20

Lateral line system. **A,** Lateral line of a bony fish with both exposed and hidden neuromasts. **B,** Structure of a neuromast (lateral line organ).

Hearing

The ear is a specialized receptor for detecting sound waves in the surrounding environment. Because sound communication and reception are an integral part of the lives of terrestrial vertebrates, we may be surprised to discover that most invertebrates inhabit a silent world. Only certain arthropod groups—crustaceans, spiders, and insects—have developed true sound-receptor organs. Even among the insects, only the locusts, cicadas, crickets, grasshoppers, and most moths possess ears, and these are of simple design: a pair of air pockets, each enclosed by a tympanic membrane that passes sound vibrations to sensory cells. Despite their spartan construction, insect ears are beautifully designed to detect the sound of a potential mate or a rival male.

Especially interesting are the ultrasonic detectors of certain nocturnal moths. These have evolved specifically to detect approaching bats and thus lessen the moth's chance of becoming a bat's evening meal (echolocation in bats is described on p. 677). Each moth ear possesses just two receptors (Figure 11-21). One of these, known as the A_1 receptor, will respond to the ultrasonic cries of a bat that is still too far away to detect the moth. As the bat approaches and its cries increase in intensity, the receptor fires more rapidly, informing the moth that the bat is coming nearer. Since the moth has two ears, its nervous system can determine the bat's position by comparing firing rates from the two ears. The moth's strategy is to fly away before the bat detects it. But if the bat continues its approach, the second (A_2) receptor in each ear, which responds only to high-intensity sounds, will fire. The moth responds immediately with an evasive maneuver, usually making a power dive to a bush or the ground where it is safe because the bat cannot distinguish the moth's echo from those of the surroundings.

In its evolution, the vertebrate ear originated as a balance organ, the **labyrinth.** In all jawed vertebrates, from fishes to mammals, the labyrinth has a similar structure, consisting of two small chambers called the **saccule** and the **utricle,** and three **semicircular canals.** In fish the base of the saccule is extended into a tiny pocket (the **lagena**) that, during the evolution of the vertebrates, developed into the hearing receptor of tetrapods. With continued elaboration and elongation in the birds and mammals, the fingerlike lagena evolved into the **cochlea.**

The human ear (Figure 11-22) is representative of mammalian ears. The outer, or external, ear collects the sound

figure 11-21

Ear of a moth used to detect approaching bats. See text for explanation.

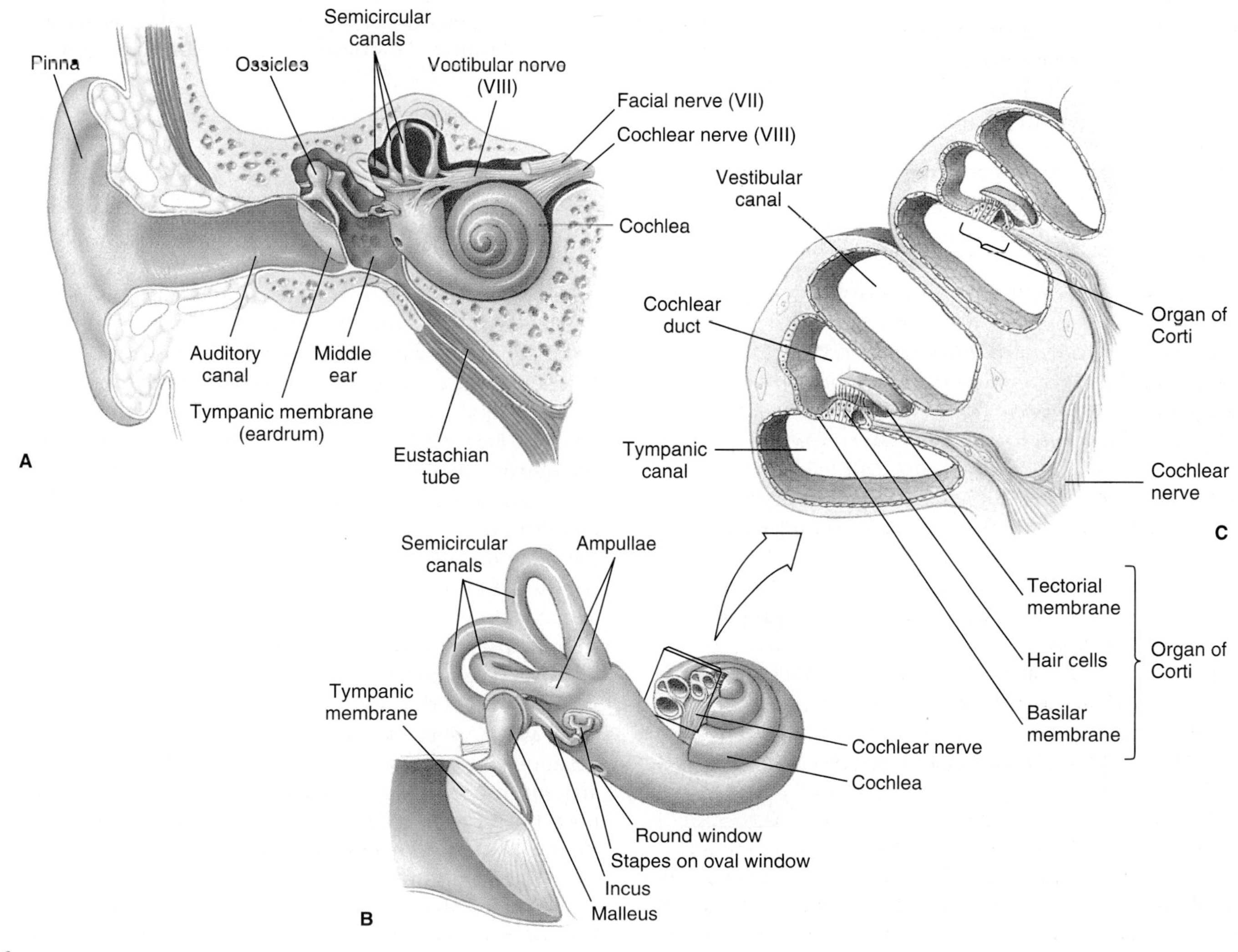

figure 11-22

Human ear. **A,** Longitudinal section showing external, middle, and inner ear. **B,** Enlargement of middle ear and inner ear. The cochlea of the inner ear has been opened to show the arrangement of canals. **C,** Enlarged cross section of cochlea showing the organ of Corti.

waves and funnels them through the **auditory canal** to the eardrum or **tympanic membrane** lying next to the middle ear. The middle ear is an air-filled chamber containing a remarkable chain of three tiny bones, or ossicles, known as the **malleus** (hammer), **incus** (anvil), and **stapes** (stirrup), named because of their fancied resemblance to these objects. These bones conduct the sound waves across the middle ear (Figure 11-22B). The bridge of bones is so arranged that the force of sound waves pushing against the tympanic membrane is amplified as much as 90 times where the stapes contacts the **oval window** of the inner ear. Muscles attached to the middle ear bones contract when the ear receives very loud noises, providing the inner ear some protection from damage. The middle ear connects with the pharynx by means of the **eustachian tube,** which permits pressure equalization on both sides of the tympanic membrane.

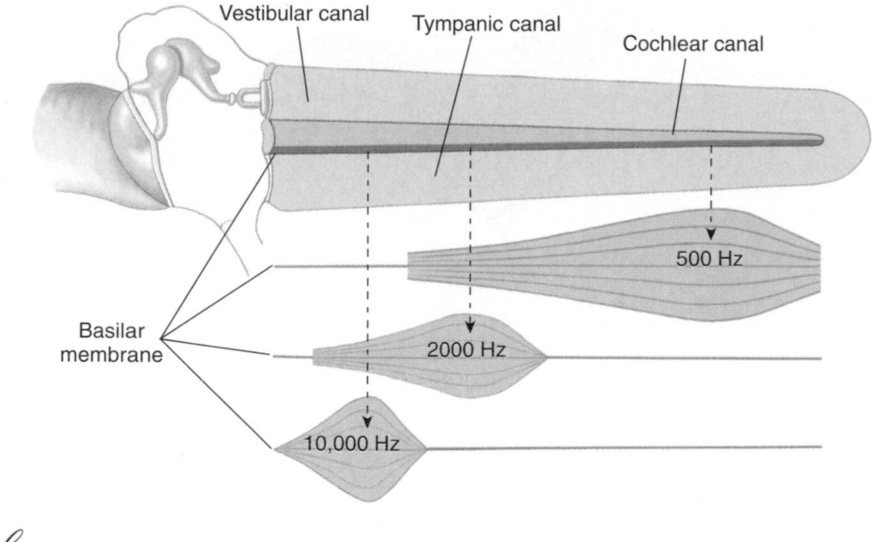

figure 11-23

Frequency localization in the cochlea of the mammalian ear as it would appear with the cochlea stretched out. Sound waves transmitted to the oval window produce vibration waves that travel down the basilar membrane. High-frequency vibrations cause the membrane to resonate perferentially near the oval window. Low-frequency tones travel farther down the basilar membrane.

The origin of the three tiny bones of the mammalian middle ear—the malleus, incus, and stapes—is one of the most extraordinary and well-documented transitions in vertebrate evolution. Amphibians, reptiles, and birds have a single rodlike ear ossicle, the stapes (also called the columella), which originated as a jaw support (the hyomandibular) as seen in fishes (see Figure 26-13, p. 578). With the evolution of the earliest tetrapods, the braincase became firmly sutured to the skull, and the hyomandibular, no longer needed to brace the jaw, became converted into the stapes. In a similar way, the two additional ear ossicles of the mammalian middle ear—the malleus and incus—originated from parts of the jaw of the early vertebrates. The quadrate bone of the reptilian upper jaw became the incus, and the articular bone of the lower jaw became the malleus. The homology of reptilian jaw bones to mammalian ear bones is clearly documented in the fossil record and in the embryological development of mammals.

Within the inner ear is the organ of hearing, the **cochlea** (Gr. *cochlea,* snail's shell), which is coiled in mammals, making two and one-half turns in humans (Figure 11-22B). The cochlea is divided longitudinally into three tubular canals running parallel with one another. This relationship is indicated in Figure 11-23, in which the cochlea is shown stretched out. These canals become progressively smaller from the base of the cochlea to the apex. One of these canals is called the **vestibular canal;** its base is closed by the oval window. The **tympanic canal,** which is in communication with the vestibular canal at the tip of the cochlea, has its base closed by the **round window.** Between these two canals is the **cochlear duct,** which contains the **organ of Corti,** the actual sensory apparatus. Within the organ of Corti are rows of hair cells that run lengthwise from the base to the tip of the cochlea (Figure 11-22C). There are at least 24,000 hair cells in the human ear. The 80 to 100 "hairs" on each cell are actually microvilli and a single large cilium which project into the endolymph of the cochlear canal. Each cell is connected with neurons of the auditory nerve. The hair cells rest on the **basilar membrane,** which separates the tympanic canal and cochlear duct, and they are covered by the **tectorial membrane,** lying directly above them.

When a sound wave strikes the ear, its energy is transmitted through the ossicles of the middle ear to the oval window, which oscillates back and forth, driving the fluid of the vestibular and tympanic canals before it. Because these fluids are noncompressible, an inward movement of the oval window produces a corresponding outward movement of the round window. The fluid oscillations also cause the basilar membrane with its hair cells to vibrate simultaneously.

According to the **place hypothesis of pitch discrimination** formulated by Georg von Békésy, different areas of the basilar membrane respond to different frequencies; that is, for every sound frequency, there is a specific "place" on the basilar membrane where the hair cells respond preferentially to that

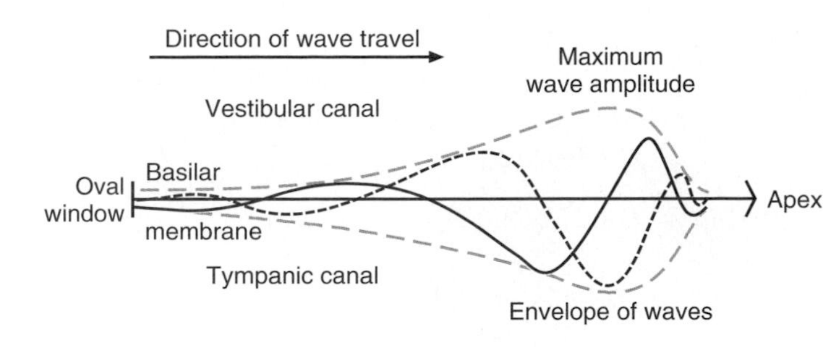

figure 11-24

Profiles of a sound wave traveling along the basilar membrane. The oval window is at left, and the cochlear apex at right. A sound wave beginning at the oval window is shown frozen at two different instances of time (*black solid* and *dashed lines*). The curves in color represent the extreme displacement of the basilar membrane from the traveling wave. Amplitudes are greatly exaggerated; actual displacement of the membrane from a loud sound is only about 1 μm.

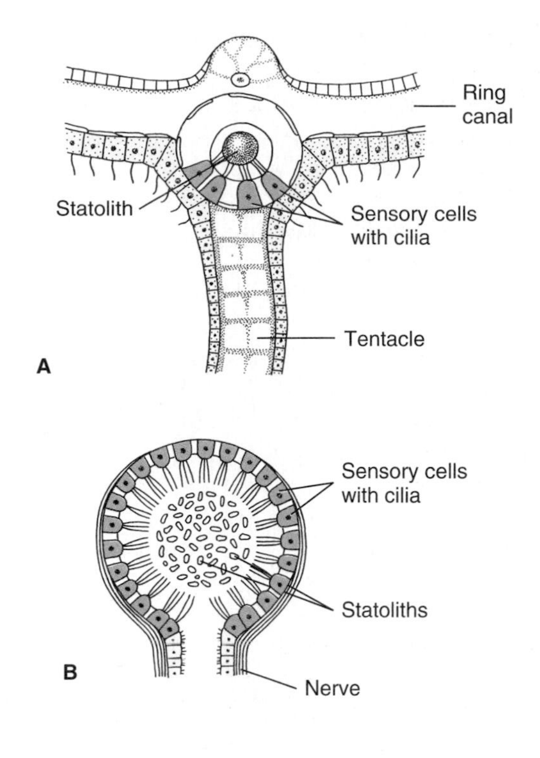

figure 11-25

Types of statocysts, static balance organs of invertebrates. **A,** Statocyst of the medusa of the hydrozoan *Obelia*. **B,** Statocyst of the bivalve mollusc *Pecten*.

frequency. The initial displacement of the basilar membrane starts a wave traveling down the membrane, much as flipping a rope at one end starts a wave moving down the rope (Figure 11-24). The displacement wave increases in amplitude as it moves from the oval window toward the apex of the cochlea, reaching a maximum at the region of the basilar membrane where the natural frequency of the membrane corresponds to the sound frequency. Here, the membrane vibrates with such ease that the energy of the traveling wave is completely dissipated. Hair cells in that region are stimulated and the impulses conveyed to the fibers of the auditory nerve. Those impulses that are carried by certain fibers of the auditory nerve are interpreted by the hearing center as particular tones. The **loudness** of a tone depends on the number of hair cells stimulated, whereas the **timbre,** or quality, of a tone is produced by the pattern of the hair cells stimulated by sympathetic vibration. This latter characteristic of tone enables us to distinguish between different human voices and different musical instruments, although the notes in each case may be of the same pitch and loudness.

Sense of Equilibrium

In invertebrates, specialized sense organs for monitoring gravity and low-frequency vibrations often appear as **statocysts.** Each is a simple sac lined with hair cells and containing a heavy calcareous structure, the **statolith** (Figure 11-25). The delicate, hairlike filaments of the sensory cells are activated by the shifting position of the statolith when the animal changes position. Statocysts are found in many invertebrate phyla from radiates to arthropods. They are all built on similar principles.

In vertebrates, the organ of equilibrium is the **labyrinth.** It consists of two small chambers (**saccule** and **utricle**) and three **semicircular canals** (Figure 11-22B, 11-26). The utricle and saccule are static balance organs that, like invertebrate statocysts, give information about the position of the head or body with respect to the force of gravity. As the head is tilted in one direction or another, stony accretions press on different groups of hair cells; these send nerve impulses to the brain, which interprets this information with reference to head position.

The semicircular canals of vertebrates are designed to respond to **rotational acceleration** and are relatively insensitive to linear acceleration. The three semicircular canals are at right angles to each other, one for each axis of rotation. They are filled with fluid (endolymph), and within each canal is a bulblike enlargement, the **ampulla,** which contains hair cells. The hair cells are embedded in a gelatinous membrane, the **cupula,** which projects into the fluid. When the head rotates, fluid in the canal at first tends not to move because of inertia. Since the cupula is attached, its free end is pulled in the direction opposite to the direction of rotation (Figure 11-26). Bending of the cupula distorts and excites the hair cells embedded in it. The hair cells are excited by acceleration in one direction, causing an increase in the discharge rate over the afferent nerve fibers leading from the ampulla to the brain, and inhibited by acceleration in the opposite direction, causing a drop in nerve discharge rate. This produces the sensation of rotation in one direction or the other. Since the three canals of each ear are in different planes, acceleration in any direction stimulates at least one ampulla.

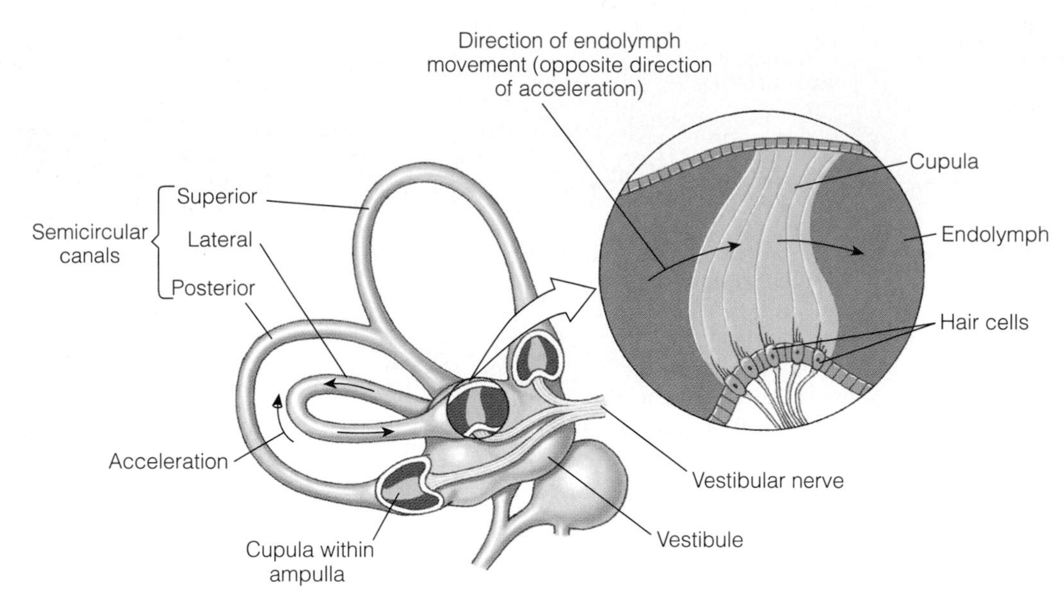

figure 11-26

How the semicircular canals respond to angular acceleration. Because of inertia, the endolymph in the semicircular canal corresponding to the plane of motion moves past the cupula in a direction opposite to the direction of angular acceleration. Movement of the cupula stimulates the hair cells.

Photoreception: Vision

Light-sensitive receptors are called **photoreceptors.** These receptors range all the way from simple light-sensitive cells scattered randomly on the body surface of many invertebrates (dermal light sense) to the exquisitely developed camera-type eye of vertebrates. Eyespots of astonishingly advanced organization appear even in some protozoa. That of the dinoflagellate *Nematodinium* bears a lens, a light-gathering chamber, and a photoreceptive pigment cup—all developed within a single-celled organism (Figure 11-27). The dermal light receptors of many invertebrates are of much simpler design. They are far less sensitive than optic receptors, but they are important in locomotory orientation, pigment distribution in chromatophores, photoperiodic adjustment of reproductive cycles, and other behavior changes.

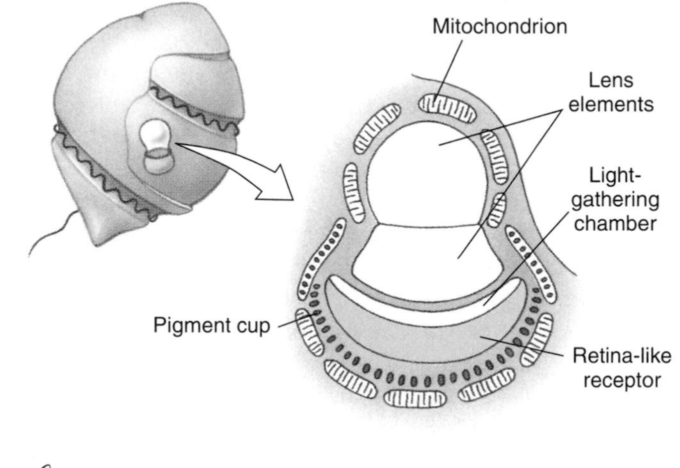

figure 11-27

Eyespot of the dinoflagellate *Nematodinium.*

More highly organized eyes, many capable of excellent image formation, are based on one or the other of two different principles: a single-lens, camera-type eye such as those of cephalopod molluscs and vertebrates; or a multifaceted (compound) eye as in arthropods. Arthropod **compound eyes** are composed of many independent visual units called **ommatidia** (Figure 11-28). The eye of a bee contains about 15,000 of these units, each of which views a separate narrow sector of the visual field. Such eyes form a mosaic of images of varying brightness from the separate units. Resolution (that is, the ability to see objects sharply) is poor as compared with that of a vertebrate eye. A fruit fly, for example, must be closer than 3 cm to see another fruit fly as anything but a single spot. However, the compound eye is especially well suited to detecting motion, as anyone knows who has tried to swat a fly.

The eyes of certain annelids, molluscs, and all vertebrates are built like a camera—or rather we should say that a camera is modeled somewhat after the vertebrate eye. The camera-type eye contains in the front a light-tight chamber and lens system, which focuses an image of the visual field on a light-sensitive surface (the retina) in the back (Figure 11-29). Be-

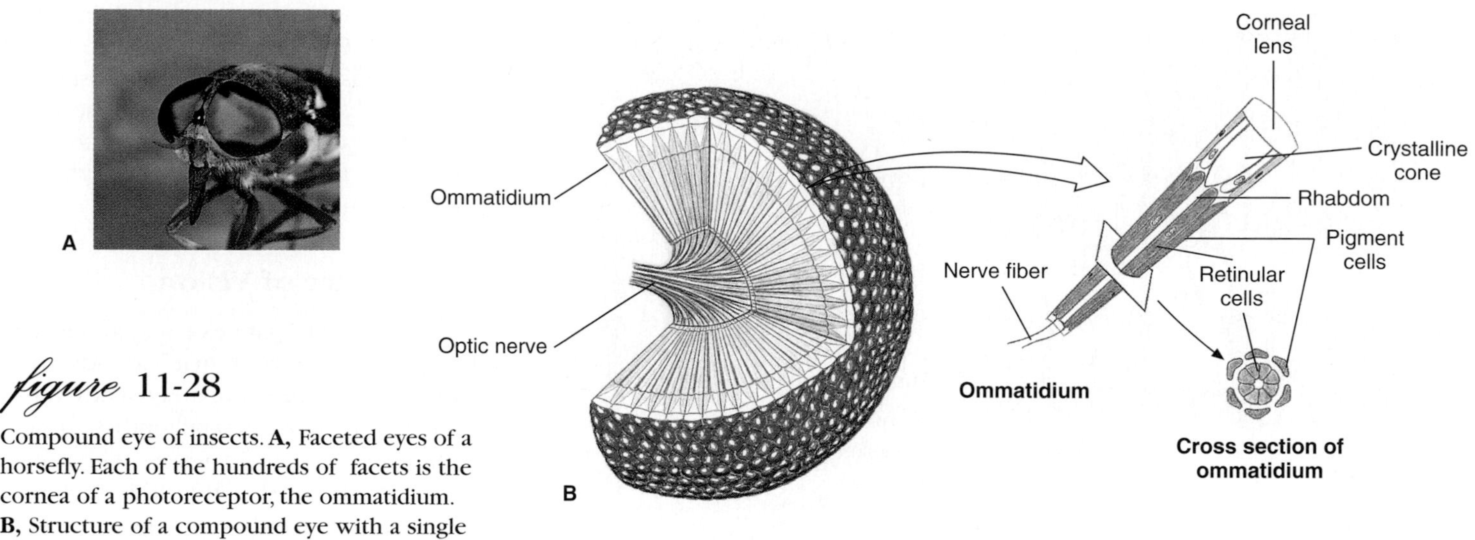

figure 11-28

Compound eye of insects. **A,** Faceted eyes of a horsefly. Each of the hundreds of facets is the cornea of a photoreceptor, the ommatidium. **B,** Structure of a compound eye with a single ommatidium enlarged at right.

cause eyes and cameras are based on the same laws of optics, we can wear eyeglasses to correct optical defects.

The spherical eyeball is built of three layers: (1) a tough outer white **sclera** that provides support and protection, (2) middle **choroid coat,** containing blood vessels for nourishment, and (3) light-sensitive **retina** (Figure 11-29). The **cornea** is a transparent anterior modification of the sclera. A circular, pigmented curtain, the **iris,** regulates the size of the light opening, the **pupil.** Just behind the iris is the **lens,** a transparent, elastic oval disc. **Ciliary muscles** can alter the curvature of the lens to change its focal length and bring objects in focus on the retina. In terrestrial vertebrates the cornea actually does most bending of light, whereas the lens adjusts focus for near and far objects. Between the cornea and the lens is the **outer chamber** filled with watery **aqueous humor;** between the lens and the retina is the much larger **inner chamber** filled with viscous **vitreous humor.**

The retina is composed of several cell layers (Figure 11-30). The outermost layer, closest to the sclera, consists of pigment cells. Adjacent to this layer are the photoreceptors, the **rods** and **cones.** Approximately 125 million rods and 1 million cones are present in each human eye. Cones are primarily concerned with color vision in ample light; rods, with colorless vision in dim light. Next is a network of **intermediate neurons** (bipolar, horizontal, and amacrine cells) that process and relay visual information from the photoreceptors to the **ganglion cells** whose axons form the optic nerve. The network permits much convergence, especially for rods. Information from several hundred rods may converge on a single ganglion cell, an adaptation that greatly increases the effectiveness of rods in dim light. Cones show very little convergence. By coordinating activities between different ganglion cells, and adjusting the sensitivities of bipolar cells, the horizontal and amacrine cells improve overall contrast and quality of the visual image.

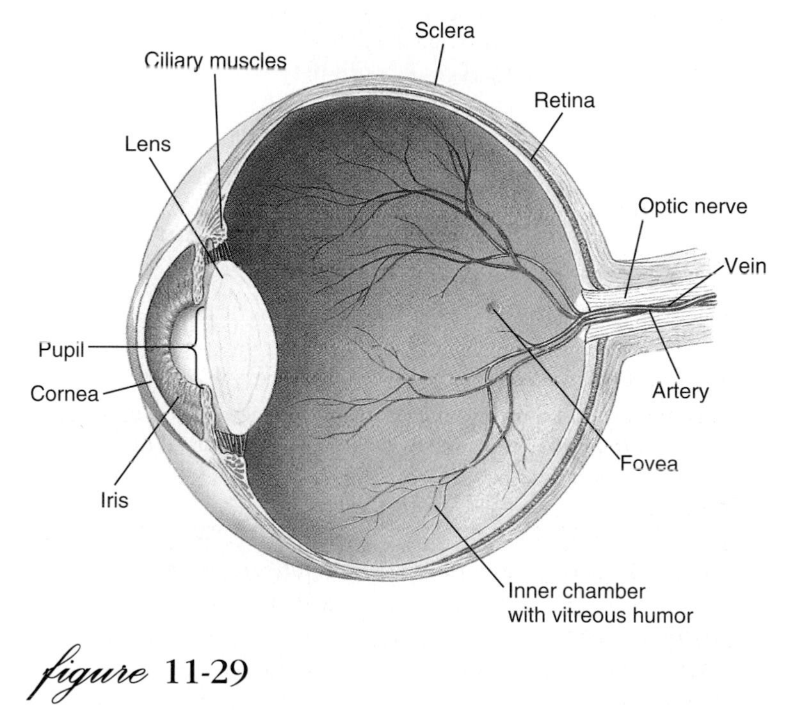

figure 11-29

Structure of the human eye.

The **fovea centralis,** the region of keenest vision, is located in the center of the retina, in direct line with the center of the lens and cornea. It contains only cones, a vertebrate specialization for diurnal (daytime) vision. The acuity of an animal's eyes depends on the density of cones in the fovea. The human fovea and that of a lion contain approximately 150,000 cones per square millimeter. But many water and field birds have up to 1 million cones per square millimeter. Their eyes are as good as our eyes would be if aided by eight-power binoculars.

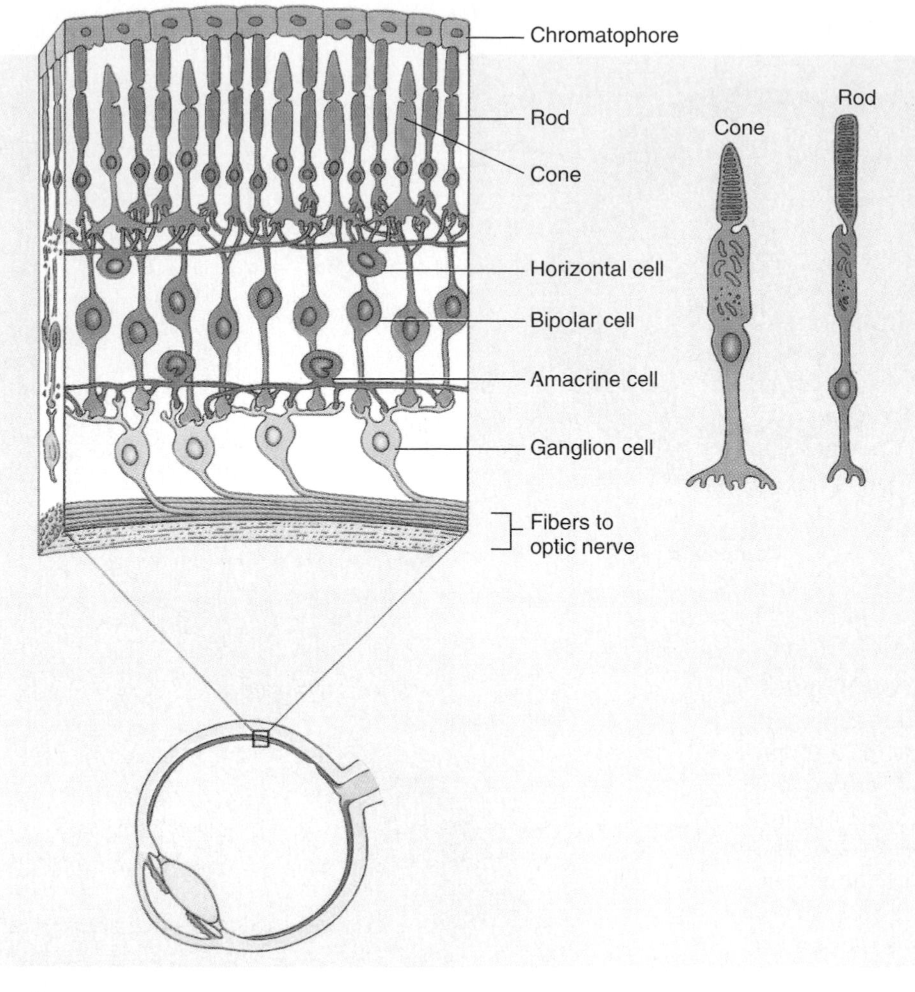

figure 11-30

Structure of the primate retina, showing the organization of intermediate neurons that connect the photoreceptor cells to the ganglion cells of the optic nerve.

> *One of several marvels of the vertebrate eye is its capacity to compress the enormous range of light intensities presented to it into a narrow range that can be handled by the optic nerve fibers. Light intensity between a sunny noon and starlight differs more than 10 billion to 1. Rods quickly saturate with high light intensity, but the cones do not; they shift their operating range with changing ambient light intensity so that a high-contrast image is perceived over a broad range of light conditions. This adjustment is made possible by complex interactions among the network of nerve cells that lie between the cones and the ganglion cells that generate the retinal output to the brain.*

At the peripheral parts of the retina only rods are found. Rods are high-sensitivity receptors for dim light. At night, the cone-filled fovea is unresponsive to low levels of light and we become functionally color blind ("at night all cats are gray"). Under nocturnal conditions, the position of greatest visual acuity is not at the center of the fovea but at its edge. Thus it is easier to see a dim star at night by looking slightly to one side of it.

Chemistry of Vision

What is "light"? Life exists in an electromagnetic spectrum that extends from the cosmic and gamma rays with wavelengths of only a ten-billionth of a centimeter to the radio waves which may be miles in length. Within this spectrum is a narrow band of energy extending from about 380 to 760 nanometers that we call light because of the sensation it creates when it falls on the human eye. This sensation depends entirely upon special pigments—visual pigments—that are chemically altered by the absorption of light to generate nerve action potentials.

The visual pigment of rods is called **rhodopsin.** Each rhodopsin molecule consists of a large protein, **opsin,** which behaves as an enzyme, and a small carotenoid molecule, **retinal,** a derivative of vitamin A. When a quantum of light strikes a rod and is absorbed by the rhodopsin molecule, retinal is isomerized; that is, the shape of the molecule is changed. This triggers the enzymatic activity of opsin, which sets in motion a biochemical sequence of several steps. This complex sequence behaves as an excitatory cascade that vastly amplifies the energy of a single photon to generate a nerve impulse in the rod.

The amount of intact rhodopsin in the retina depends on the intensity of light reaching the eye. A dark-adapted eye contains much rhodopsin and is very sensitive to weak light. Conversely, in a light-adapted eye, most of the rhodopsin is broken down into retinal and opsin. It takes approximately half an hour for the light-adapted eye to accommodate to darkness, while the rhodopsin level gradually rises.

The pigment of cones is similar to rhodopsin, containing retinal combined with a special protein, **cone opsin.** Cones function to perceive color and require 50 to 100 times more light for stimulation than do rods. Consequently, night vision is almost totally rod vision. Unlike humans, who have both day and night vision, some vertebrates specialize for one or the other. Strictly nocturnal animals, such as bats and owls, have pure rod retinas. Purely diurnal forms, such as the common gray squirrel and some birds, have only cones. They are, of course, virtually blind at night.

Color Vision

In 1802 the English physician and physicist Thomas Young speculated that we see color by the relative excitation of three kinds of photoreceptors: one each for red, yellow, and blue. Young's prescient hypothesis was eventually confirmed in 1964 when it became possible to make color absorption measurements on single cones of goldfish. There are three classes of cones, each containing a visual pigment that is maximally sensitive to either blue light at 430 nm, green light at 540 nm, or red light (actually closer to yellow light) at 575 nm (Figure 11-31). As Young suggested, colors are perceived by comparing the levels of excitation of the three different kinds of cones. For example, a light having a wavelength of 530 nm would excite the green cones 95%, the red cones about 70%, and the blue cones not at all. This comparison is made both in nerve circuits in the retina and in the visual cortex of the brain, and the brain interprets this combination as green.

Color vision is present in some members of all vertebrate groups with the possible exception of the amphibians. Bony fishes and birds have particularly good color vision. Surprisingly, most mammals are color blind; exceptions are primates and a few other species such as squirrels.

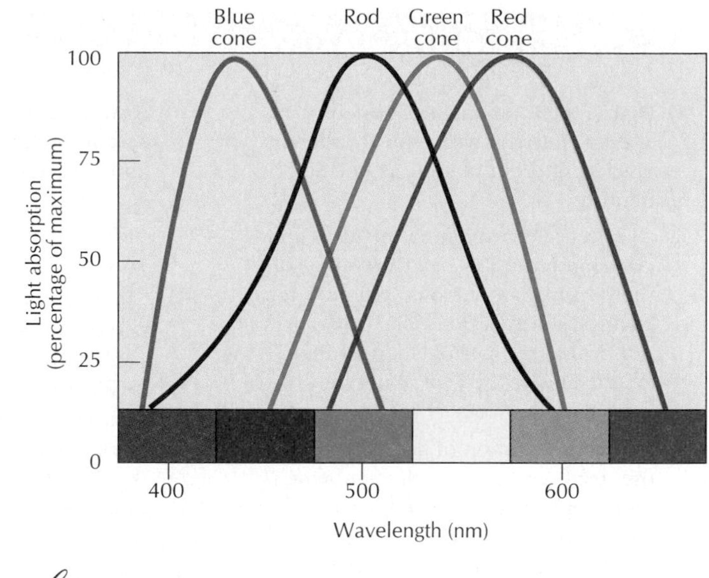

figure 11-31

The absorption spectrum of human vision. Three types of cones absorb maximally at 430 nm (blue cones), 540 nm (green cones), and 575 nm (red cones).

Summary

The nervous system is a rapid communication system that interacts continuously with the endocrine system in the control and coordination of body function. The basic unit of nervous integration in all animals is the neuron, a highly specialized cell designed to conduct self-propagating impulses, called action potentials, to other cells. Action potentials are transmitted from one nerve to another across synapses which may be either electrical or chemical. A chemical transmitter, usually acetylcholine, is released from synaptic knobs to bridge the gap between nerves at chemical synapses.

The simplest organization of neurons into a system is the nerve net of cnidarians. The nerve net is basically a plexus of nerve cells that, with additions, is the basis of the nervous systems of several invertebrate phyla. With the appearance of ganglia (nerve centers) in the bilateral flatworms, nervous systems differentiated into central and peripheral divisions. In vertebrates, the central nervous system consists of the brain and spinal cord. Fishes and amphibians have a three-part linear brain, whereas in mammals, the cerebral cortex has become a vastly enlarged multicomponent structure that has assumed the most important integrative activities of the nervous system. The cerebral cortex completely overshadows the ancient brain, which is consigned to the role of relay center and to serving numerous unconscious but nonetheless vital functions such as breathing, blood pressure, and heart rate.

The peripheral nervous system connects the central nervous system to receptors and effector organs. It is divided broadly into an afferent system, which conducts sensory signals to the central nervous system, and an efferent system, which conveys motor impulses to effector organs. The autonomic nervous system is a motor system with its own separate set of fibers. It is subdivided into anatomically distinct sympathetic and parasympathetic systems, each of which sends fibers to most body organs. Generally the sympathetic system governs excitatory activities, and the parasympathetic system governs maintenance and restoration of body resources.

Sensory organs are receptors designed especially to respond to internal or environmental change. The most primitive and ubiquitous sense is chemoreception. Chemoreceptors may be contact receptors such as the vertebrate sense of taste, or distance receptors such as smell, which detects airborne molecules. Most hypotheses of olfaction postulate some kind of molecular integration between odorant and receptor; the resulting signals are expressed as a spatial pattern that is interpreted by the brain as a particular odor.

The receptors for touch, pain, equilibrium, and hearing are all mechanical force receptors. Touch and pain receptors are characteristically simple structures, but hearing and equilibrium are highly specialized senses based on special hair cells that respond to mechanical deformation. Sound waves received by the ear are mechanically amplified and transmitted to the inner ear where different areas of the cochlea respond to different sound frequencies. Equilibrium receptors, also located in the inner ear, consist of two saclike static balance organs and three semicircular canals that detect angular acceleration.

Vision receptors (photoreceptors) are associated with special pigment molecules that photochemically decompose in the presence of light and, in doing so, trigger nerve impulses in optic fibers. The advanced compound eye of arthropods is especially well suited to detecting motion in the visual field. Vertebrates have a camera eye with focusing optics. The photoreceptor cells of the retina are of two kinds: rods, designed for high sensitivity with dim light, and cones, designed for color vision in daylight. Cones predominate in the fovea centralis of the human eye, the area of keenest vision. Rods are more abundant in the peripheral areas of the retina.

Review Questions

1. Define the following terms: neuron, axon, dendrite, myelin sheath, afferent neuron, efferent neuron, association neuron.
2. Glial cells far outnumber neurons and contribute roughly half the weight of the mammalian nervous system. Offer examples of functions glial cells perform in the peripheral nervous system and in the central nervous system.
3. The concentration of potassium on the inside of a nerve cell membrane is higher than the concentration of sodium on the outside of the membrane. Yet the inside of the membrane (where the cation concentration is higher) is negative to the outside. Explain this observation in terms of the permeability properties of the membrane.
4. What ionic and electrical changes occur during the passage of an action potential along a nerve fiber?
5. Why is the sodium pump *indirectly* important to the action potential and to maintaining the resting membrane potential?
6. Describe the microstructure of a chemical synapse. Summarize what happens when an action potential arrives at a synapse.
7. Describe the cnidarian (radiate) nervous system. How is the tendency toward centralization of the nervous system manifested in flatworms, annelids, molluscs, and arthropods?
8. How does the vertebrate spinal cord differ morphologically from the nerve cord of invertebrates?
9. The knee-jerk reflex is often called a stretch reflex because a sharp tap on patellar ligament stretches the quadriceps femoris, the extensor muscle of the leg. Describe the components and sequence of events that lead to "knee jerk." What is the difference between a reflex arc and a reflex act?
10. Name the major functions associated with the following brain structures: medulla, cerebellum, tectum, thalamus, hypothalamus, cerebrum, limbic system.
11. What is the autonomic nervous system and what activities does it perform that distinguish it from the central nervous system? Why can the autonomic nervous system be described as a "two-neuron" system?
12. Give the meaning of the statement, "The idea that all sense organs behave as biological transducers is a uniting concept in sensory physiology."
13. Knowing that all action potentials are basically alike, how do animals perceive and distinguish the different sensations of varying stimuli?
14. Chemoreception in vertebrates and insects is mediated through the clearly distinguishable senses of taste and smell. Contrast these two senses in humans in terms of anatomical location and nature of the receptors and sensitivity to chemical molecules.
15. Explain how the ultrasonic detectors of certain nocturnal moths are adapted to help them escape an approaching bat.
16. Outline the place theory of pitch discrimination as an explanation of the human ear's ability to distinguish between sounds of different frequencies.
17. Explain how the semicircular canals of the ear are designed to detect rotation of the head in any directional plane.
18. Explain what happens when light strikes a dark-adapted rod that leads to the generation of a nerve impulse. What is the difference between rods and cones in their sensitivity to light?
19. In 1802 Thomas Young hypothesized that we see color because the retina contains three kinds of receptors. What evidence is there to substantiate Young's hypothesis and how is it possible to perceive any color in the visible spectrum when there are only three different classes of color cones in the retina?

Selected References

Agosta, W. C. 1992. Chemical communication: the language of pheromones. New York, Scientific American Library. *Excellent exploration of the chemistry and biology of pheromones, compounds exchanged between members of the same species to govern social interactions.*

Axel, R. 1995. The molecular logic of smell. Sci. Am. **273:**154–159 (Oct.). *Recent research has revealed a surprisingly large family of genes that code for odor molecules. This and other findings help to illuminate how the nose and brain may perceive scents.*

Bullock, T. H., R. Orkland, and A. Grinnell. 1977. Introduction to nervous systems. San Francisco, W. H. Freeman and Company. *Excellent comparative treatment.*

Dunant, Y., and M. Israel. 1985. The release of acetylcholine. Sci. Am. **252:**58–66 (Apr.). *Recent studies have altered prevailing views of the events at a synapse during impulse transmission.*

Freeman, W. J. 1991. The physiology of perception. Sci. Am. **264:**78–85 (Feb.) *How the brain transforms sensory messages almost instantly into conscious perceptions.*

Hudspeth, A. J. 1983. The hair cells of the inner ear. Sci. Am. **248:**54–64 (Jan.). *How these biological transducers work.*

Jacobson, M. 1993. Foundations of neuroscience. New York, Plenum Press. *The historical development of neuroscience and its outstanding personages—and the dangers of hero worship of individual neuroscientists.*

Mind and brain. 1992. Readings from *Scientific American.* New York, W. H. Freeman.

Nathans, J. 1989. The genes for color vision. Sci. Am. **260:**42–49. (Feb.). *The recent isolation of genes that encode color-detecting proteins of the human eye provides clues about the evolution of color vision.*

Schnapf, J. L., and D. A. Baylor. 1987. How photoreceptor cells respond to light. Sci. Am. **256:**40–47 (Apr.). *How rod cells transduce a single photon into a neural signal.*

Snyder, S. H. 1985. The molecular basis of communication between cells. Sci. Am. **253:**132-141 (Oct.). *Describes the different actions of neurotransmitters.*

Stebbins, W. C. 1983. The acoustic sense of animals. Cambridge, Massachusetts, Harvard University Press. *Broadly comparative introduction to the physics, physiology, natural history, and evolution of hearing.*

Stryer, L. 1987. The molecules of visual excitation. Sci. Am. **257:**42–50 (July). *Describes the cascade of molecular events following light absorption by a rod cell that leads to a nerve signal.*

Thompson, R. F. 1993. The brain: a neuroscience primer. New York, W.H. Freeman and Company. *Introduction to the basics of neuroscience.*

Chemical Coordination: Endocrine System

chapter | twelve

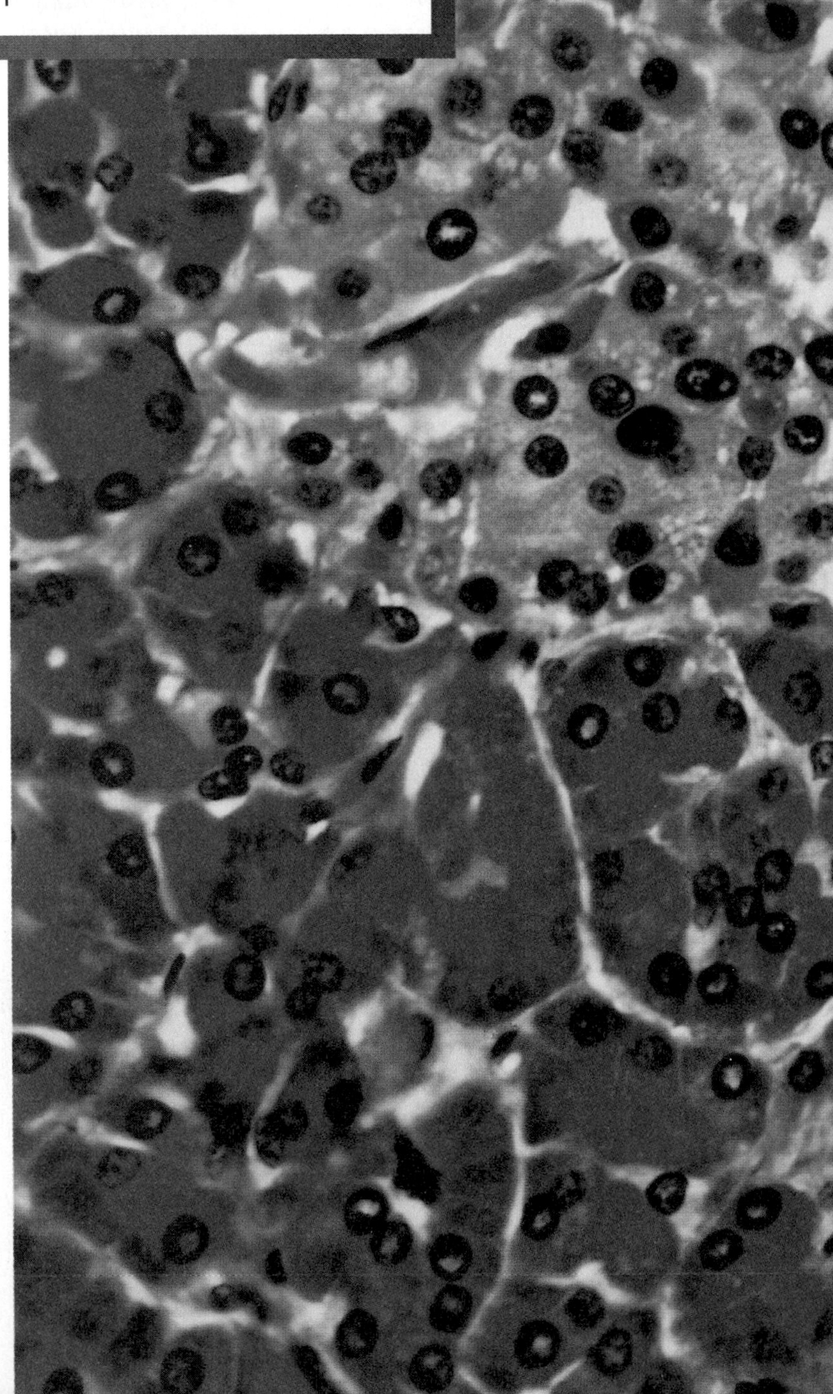

The Crucial Experiment

The birth date of endocrinology as a science is usually given as 1902, the year two English physiologists, W.H. Bayliss and E.H. Starling (Figure 12-1), demonstrated the action of a hormone in a classic experiment that is still considered a model in the use of the scientific method. Bayliss and Starling were interested in determining how the pancreas secreted its digestive juice into the small intestine at the proper time of the digestive process. Only one year earlier, the Russian physiologist Ivan Pavlov had demonstrated that the smell, taste, and thought of food provokes the release of gastric and pancreatic juice. Pavlov suggested that acidic food entering the intestine triggered a nervous reflex that released pancreatic juice. To test this hypothesis, Bayliss and Starling cut away all of the nerves serving a tied-off loop of the small intestine of an anesthetized dog, leaving the isolated loop connected to the body only by its circulation. Injecting acid into the nerveless loop, they saw a pronounced flow of pancreatic juice. Clearly, Pavlov was wrong. Rather than a nervous reflex, some chemical messenger had circulated from the intestine to the pancreas, causing the pancreas to secrete. Yet acid itself could not be the factor because it had no effect when injected directly into the circulation.

Bayliss and Starling then designed the crucial experiment that was to usher in the new science of endocrinology. Suspecting that the chemical messenger originated in the mucosal lining of the intestine, they next prepared an extract of scrapings from the mucosa, injected it into the dog's circulation, and were rewarded with an abundant flow of pancreatic juice. They named the messenger present in the intestinal mucosa *secretin.* Later Starling coined the term *hormone* to describe all such chemical messengers, since he correctly surmised that secretin was only the first of many hormones awaiting discovery.

The endocrine system, the second great integrative system controlling the body's activities, communicates by chemical messengers called **hormones** (Gr. *hormon,* to excite). Hormones are chemical compounds that are released into the blood in small amounts and transported by the circulatory system throughout the body to distant **target cells** where they initiate physiological responses.

Many hormones are secreted by **endocrine glands,** small, well-vascularized ductless glands composed of groups of cells arranged in cords or plates. Since the endocrine glands have no ducts, their only connection with the rest of the body is by the bloodstream; they must capture their raw materials from the extensive blood supply they receive and secrete their finished hormonal products into it. **Exocrine glands,** in contrast, are provided with ducts for discharging their secretions onto a free surface. Examples of exocrine glands are sweat glands and sebaceous glands of skin, salivary glands, and the various enzyme-secreting glands lining the walls of the stomach and intestine.

The classical definitions of hormones and endocrine glands given above, like so many other generalizations in biology, may have to be altered as new information appears. Some hormones, such as certain neurosecretions, may never enter the general circulation at all. Furthermore, there is good evidence that many hormones, such as insulin, are synthesized in minute amounts in a variety of nonendocrine tissues (nerve cells, for example), and some, such as cytokines, are secreted by cells of the immune system (p. 216). Such hormones may function as local tissue factors, called parahormones, substances that stimulate cell growth or some biochemical process. Most hormones, however, are blood borne and therefore diffuse into every tissue space in the body—quite unlike the discrete action of the nervous system with its network of cablelike nerve fibers that selectively send messages to specific points.

Compared with the nervous system, the endocrine system is slow acting because of the time required for a hormone to reach the appropriate tissue, cross the capillary endothelium, and diffuse through tissue fluid to, and sometimes into, cells. The minimum response time is seconds and may be much longer. Hormonal responses in general are long lasting (minutes to days), whereas those under nervous control are short-term (milliseconds to minutes). We expect to find endocrine control where a sustained effect is required, as in many metabolic and growth processes, or where some concentration or rate of secretion must be maintained at a particular level. Despite such differences, the nervous and endocrine systems function as a single, interdependent system. There is no sharp separation between the two. Endocrine glands often receive directions from the brain. Conversely, several hormones act on the nervous system and may significantly affect many kinds of animal behavior.

All hormones are low-level signals. Even when an endocrine gland is secreting maximally, the hormone is so greatly diluted by the large volume of blood it enters that its plasma concentration seldom exceeds 10^{-9} M (one billionth of a 1 M concentration). Some target cells respond to plasma concentrations of hormone as low as 10^{-12} M. Since hormones have far-reaching and often powerful influences on cells, it is evident that their effects are vastly amplified at the cellular level.

A

B

figure 12-1

Founders of endocrinology. **A,** Sir William H. Bayliss (1860–1924). **B,** Ernest H. Starling (1866–1927).

From J. F. Fulton & L. G. Wilson Selected Readings in the History of Physiology, 1966. *Courtesy of Charles C. Thomas, Publisher, Springfield.*

Mechanisms of Hormone Action

The widespread distribution of hormones in the body makes it possible for certain hormones, such as the growth hormone of the pituitary gland, to affect most, if not all, cells during specific stages of cellular differentiation. Other hormones produce highly specific responses only in certain target cells and at certain times. Such specificity is made possible by **receptor molecules** on or in the target cells. A hormone will engage only those cells that display the receptor that, by virtue of its specific molecular shape, will bind with the hormone molecule. Other cells are insensitive to the hormone's presence because they lack the specific receptors. Hormones act through two kinds of receptors: **membrane-bound receptors** and **nuclear receptors.**

Membrane-Bound Receptors and the Second Messenger Concept

Many hormones, such as most amino acid derivatives, and the peptide hormones that are too large to pass cell membranes, bind to receptor sites present on the surface of target cell membranes. The combination of hormone and receptor forms a complex that triggers the release of a molecule from the inner surface of the membrane. The hormone thus behaves as a **first messenger** that causes the release of **second messenger** in the cytoplasm. At least six different molecules have been identified as second messengers, but the best understood is **cyclic AMP.** Cyclic AMP is formed when the binding of hormone to receptor activates the enzyme **adenylate**

cyclase in the target cell membrane. This enzyme converts ATP to cyclic AMP (the second messenger), which then acts to modify the direction and rate of cytoplasmic reactions (Figure 12-2). Since many molecules of cyclic AMP may be manufactured after a single hormone molecule has been bound, the message is amplified, perhaps many thousands of times.

Cyclic AMP mediates the actions of many peptide hormones, including parathyroid hormone, glucagon, adrenocorticotropic hormone (ACTH), thyrotropic hormone (TSH), melanophore-stimulating hormone (MSH), and vasopressin. It also mediates the action of epinephrine (also called adrenaline), an amino-acid derivative.

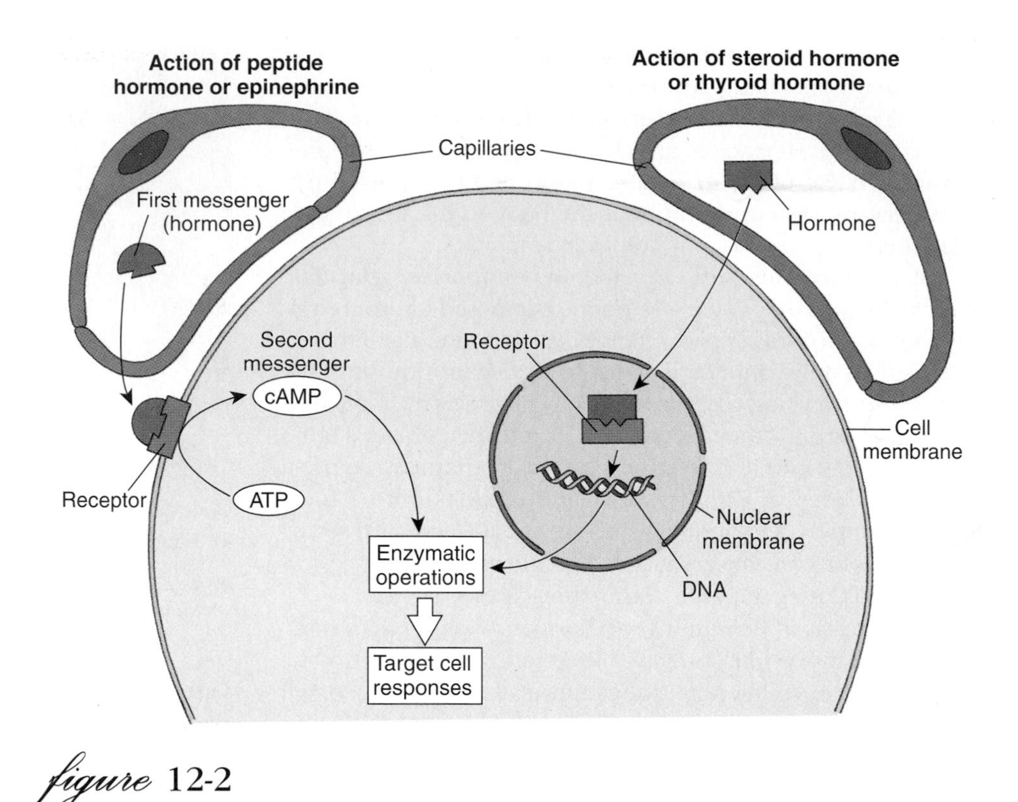

figure 12-2

Mechanisms of hormonal action. Peptide hormones and epinephrine act through cyclic AMP. The combination of the hormone with a membrane receptor stimulates the enzyme adenylate cyclase to catalyze the formation of cyclic AMP (second messenger). Steroid hormones and thyroid hormones penetrate the cell membrane to combine with a nuclear receptor that activates gene transcription.

Nuclear Receptors

Unlike the peptide hormones and epinephrine, which cannot pass through cell membranes, the **steroid hormones** (for example, estrogen, testosterone, and aldosterone), are lipid-soluble molecules that readily diffuse through cell membranes. Once inside the cytoplasm, steroid hormones bind selectively to receptor molecules found in the nucleus of the target cells. The hormone-receptor complex then activates specific genes. As a result, gene transcription is increased, and messenger RNA molecules are synthesized on specific sequences of DNA. Moving from the nucleus into the cytoplasm, the newly formed mRNA acts as a template for the synthesis of key enzymes, thus setting in motion the hormone's observed effect (Figure 12-2). Thyroid hormones and the insect-molting hormone, ecdysone, also act through nuclear receptors.

As compared with the peptide hormones that act *indirectly* through second messenger systems, steroids have a *direct* effect on protein synthesis because they combine with a nuclear receptor that activates specific genes.

How Secretion Rates of Hormones Are Controlled

Hormones influence cellular functions by altering the rates of many different biochemical processes. Many affect enzyme activity and thus alter cellular metabolism, some change membrane permeability, some regulate the synthesis of cellular proteins, and some stimulate the release of hormones from other endocrine glands. Because these are all dynamic processes that must adapt to changing metabolic demands, they must be regulated, not merely activated, by the appropriate hormones. This regulation is achieved by precisely controlled release of a hormone into the blood. However, the concentration of a hormone in the plasma depends on two factors: its rate of secretion and the rate at which it is inactivated and removed from the circulation. Consequently, if secretion is to be controlled correctly, an endocrine gland requires information about the level of its own hormone(s) in the plasma.

Many hormones, especially those of the pituitary gland, are controlled by negative feedback systems that operate between the glands secreting the hormones and the target cells (Figure 12-3). A feedback pattern is one in which the output is constantly compared with a set point, like a thermostat. For example, ACTH, secreted by the pituitary, stimulates the adrenal gland (the target cells) to secrete cortisol. As the cortisol level in the plasma rises, it acts on, or "feeds back" on, the pituitary gland to inhibit the release of ACTH. Thus, any deviation from the set point (a specific plasma level of cortisol) leads to corrective action in the opposite direction (Figure 12-3). Such a negative feedback system is highly effective in preventing extreme oscillations in hormonal output. However, hormonal feedback systems are more complex than a rigid "closed-loop" system such as the thermostat that controls the central heating system in a house, because hormonal feedback may be altered by input from the nervous system or by metabolites or other hormones.

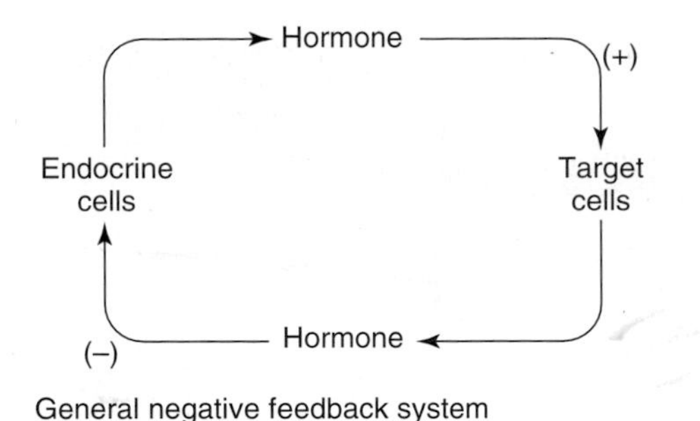

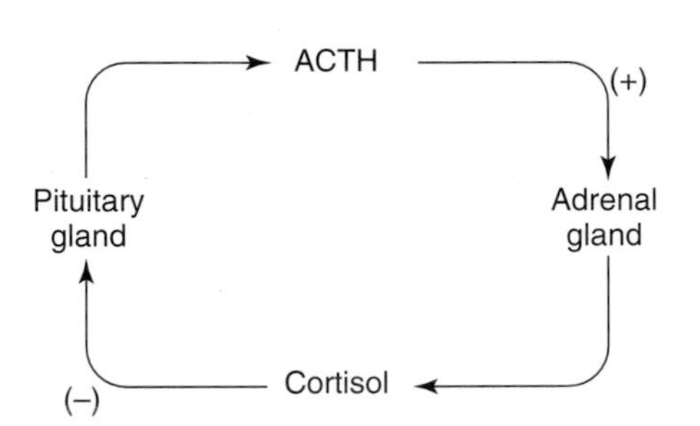

figure 12-3

Negative feedback systems.

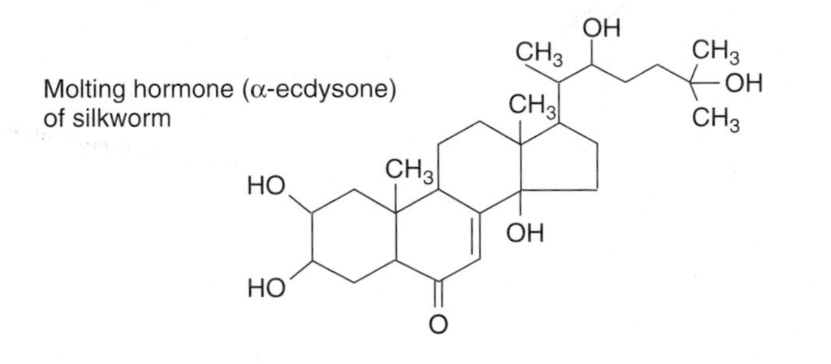

Invertebrate Hormones

In many metazoan phyla, the principal source of hormones is **neurosecretory cells,** specialized nerve cells capable of synthesizing and secreting hormones. Their products, called neurosecretions or neurosecretory hormones, are discharged directly into the circulation. Neurosecretion is an ancient physiological activity. Because it serves as a crucial link between the nervous and endocrine systems, it is believed that hormones first evolved as nerve cell secretions. Later, nonnervous endocrine glands appeared, especially among the vertebrates, but remained chemically linked to the nervous system by the neurosecretory hormones.

Neurosecretory hormones occur in all metazoan groups. The most extensively studied neurosecretory process is the control of development and metamorphosis of insects. In insects, as in other arthropods, growth is a series of steps in which the rigid, nonexpansible exoskeleton is periodically discarded and replaced with a new, larger one. Most insects undergo a process of metamorphosis (p. 515), in which a series of juvenile stages, each requiring the formation of a new exoskeleton, end with a molt. In some orders the change to the adult form is gradual. In others the adult is separated from the larval stages by a quiescent form, the pupa, and the change to the adult is abrupt. Hormonal control of both types is the same.

Insect physiologists have discovered that molting and metamorphosis are controlled by the interaction of two hormones, one favoring growth and the differentiation of adult structures and the other favoring the retention of juvenile structures. These two hormones are the **molting hormone** (also called **ecdysone** [ek′duh-sone or ek-die′sone]), produced by the prothoracic gland, and the **juvenile hormone,** produced by the corpora allata (Figure 12-4). The structure of both hormones has been determined. To show that molting hormone is a steroid, researchers extracted nearly 1000 kg (about 1 ton) of silkworm pupae. Juvenile hormone has an entirely different structure.

The molting hormone is under the control of **brain hormone** (also called ecdysiotropin or prothoracicotropic hormone). This hormone is a polypeptide (molecular weight about 5000) that is produced by neurosecretory cells of the brain. Axons transport brain hormone to the corpora allata, where it is stored. At intervals during juvenile growth, release of brain hormone into the blood stimulates the prothoracic gland to secrete molting hormone. Molting hormone appears to act directly on the chromosomes to set in motion the changes resulting in a molt. The molting hormone favors development of adult structures. This tendency is held in check, however, by juvenile hormone, which favors development of juvenile characteristics. During juvenile life, juvenile hormone predominates and each molt yields another larger juvenile (Figure 12-4). Finally the output of juvenile hormone decreases, allowing final metamorphosis to the adult stage.

The precise location of brain hormone in the brain of pupal tobacco hornworms was revealed by N. Agui by delicate microdissection. Using a human eyebrow hair, he was able to isolate the single cell in each brain hemisphere containing brain hormonal activity. Thus, only two cells, each about 20 μm in diameter, produce this insect's total supply of brain hormone. In an age when sophisticated instrumentation has removed much of the tedium (and some of the creativity) from research, it is refreshing to learn that certain biological mysteries succumb only to skillful use of the human hand.

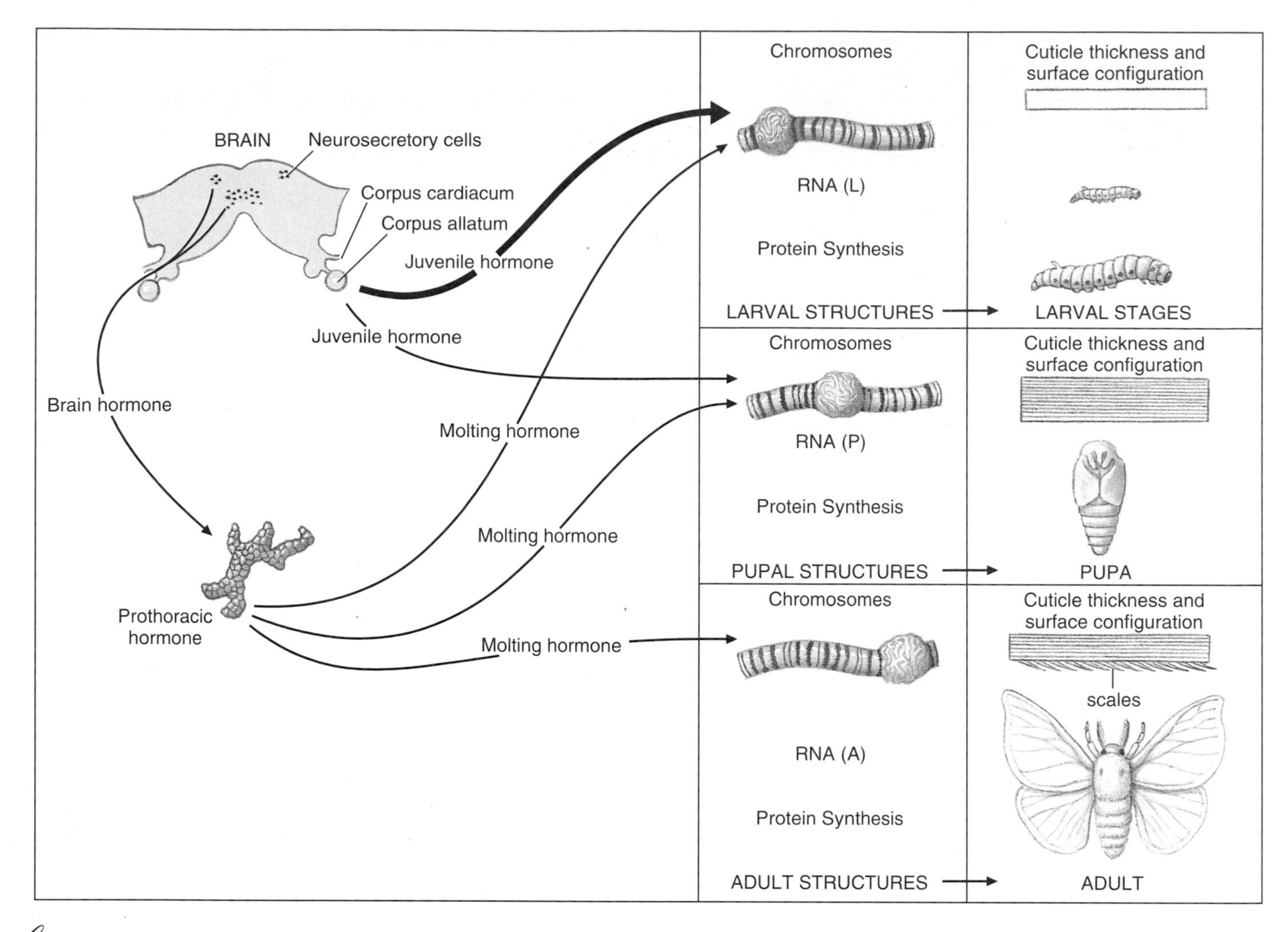

figure 12-4

Endocrine control of molting in a moth, typical of insects having complete metamorphosis. Many moths mate in the spring or summer, and eggs soon hatch into the first of several larval stages, called instars. After the final larval molt, the last and largest larva (caterpillar) spins a cocoon in which it pupates. The pupa overwinters, and an adult emerges in the spring to start a new generation. Juvenile hormone and molting hormone interact to control molting and pupation. Many genes are activated during metamorphosis, as seen by the puffing of chromosomes (center column). Puffs form in sequence during successive molts. Changes in cuticle thickness and surface characteristics are shown at right.

Chemists have synthesized several potent analogs of juvenile hormone, which hold great promise as insecticides. Minute quantities of these synthetic analogs induce abnormal final molts or prolong or block development. Unlike the usual chemical insecticides, they are highly specific and ecologically benign.

Vertebrate Endocrine Glands and Hormones

In the remainder of this chapter we describe some of the best understood and most important of vertebrate hormones. While the following discussion is limited principally to a brief overview of mammalian hormonal mechanisms (since laboratory mammals and humans have always been the objects of the most intensive research), we will note some important differences in the functional roles of hormones among different vertebrate groups.

Hormones of the Pituitary Gland, Hypothalamus, and Pineal Gland

The pituitary gland, or **hypophysis,** is a small gland (0.5 g in humans) lying in a well-protected position between the roof of the mouth and the floor of the brain (Figure 12-5). It is a two-part gland having a double embryonic origin. The **anterior pituitary** (adenohypophysis) is derived embryonically from the

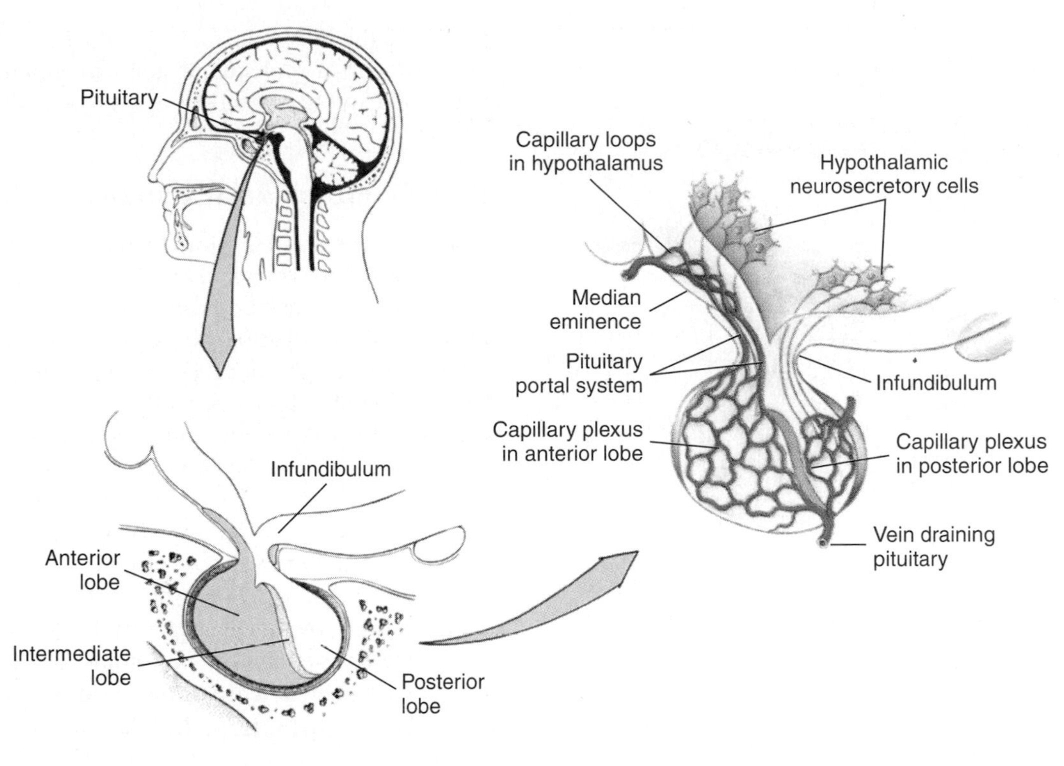

figure 12-5

Human pituitary gland. The posterior lobe is connected directly to the hypothalamus by neurosecretory fibers. The anterior lobe is indirectly connected to the hypothalamus by a portal circulation (shown in red) beginning in the base of the hypothalamus and ending in the anterior pituitary.

roof of the mouth. The **posterior pituitary** (neurohypophysis) arises from a ventral portion of the brain, the **hypothalamus,** and is connected to it by a stalk, the **infundibulum.** Although the anterior pituitary lacks any anatomical connection to the brain, it is functionally connected to it by a special portal circulatory system. A portal circulation is one that delivers blood from one capillary bed to another (Figures 12-5 and 12-6).

Anterior Pituitary

The anterior pituitary consists of an **anterior lobe** (pars distalis) and an **intermediate lobe** (pars intermedia) as shown in Figure 12-5. The anterior pituitary produces seven hormones whose functions have been clearly established. All but one are released by the anterior lobe.

Four hormones of the anterior pituitary are **tropic hormones** (pronounced tropic, from the Greek *tropē,* to turn toward) that regulate other endocrine glands (Table 12-1). **Thyroid-stimulating hormone (TSH)** or **thyrotropin,** stimulates the production of thyroid hormones by the thyroid gland. Two of the tropic hormones are commonly called **gonadotropins** because they act on the gonads (ovary of the female, testis of the male). These are **follicle-stimulating hormone (FSH)** and **luteinizing hormone (LH).** FSH promotes egg production and secretion of estrogen in females. In males FSH supports sperm production. LH induces ovulation and secretion of the female sex hormones progesterone and estrogen. In males LH promotes production of male sex hormones. LH is often called the interstitial cell stimulating hormone (ICSH) in males, but it is the same hormone chemically in males and females. The fourth tropic hormone, **adrenocorticotropic hormone (ACTH),** increases production and secretion of steroid hormones from the adrenal cortex.

Prolactin and the structurally related **growth hormone (GH)** are proteins. Prolactin is essential for preparing the mammary glands for lactation and is required for production of milk after birth. Prolactin has also been implicated in parental behavior in a wide variety of vertebrates. Growth hormone (also called somatotropin) performs a vital role in governing body growth through its stimulatory effect on cellular mitosis and on synthesis of messenger RNA and protein, especially in new tissue of young vertebrates. If produced in excess, growth hormone causes giantism. A deficiency of this hormone in the human child leads to dwarfism. Unlike the tropic hormones, prolactin acts directly on its target tissues rather than through other hormones. Growth hormone acts primarily through a polypeptide hormone, somatomedin, produced by the liver.

The only anterior pituitary hormone produced by the intermediate lobe (Figure 12-5) is **melanophore-stimulating hormone (MSH).** In cartilaginous and bony fishes, amphibians, and reptiles, MSH is a direct-acting hormone that promotes

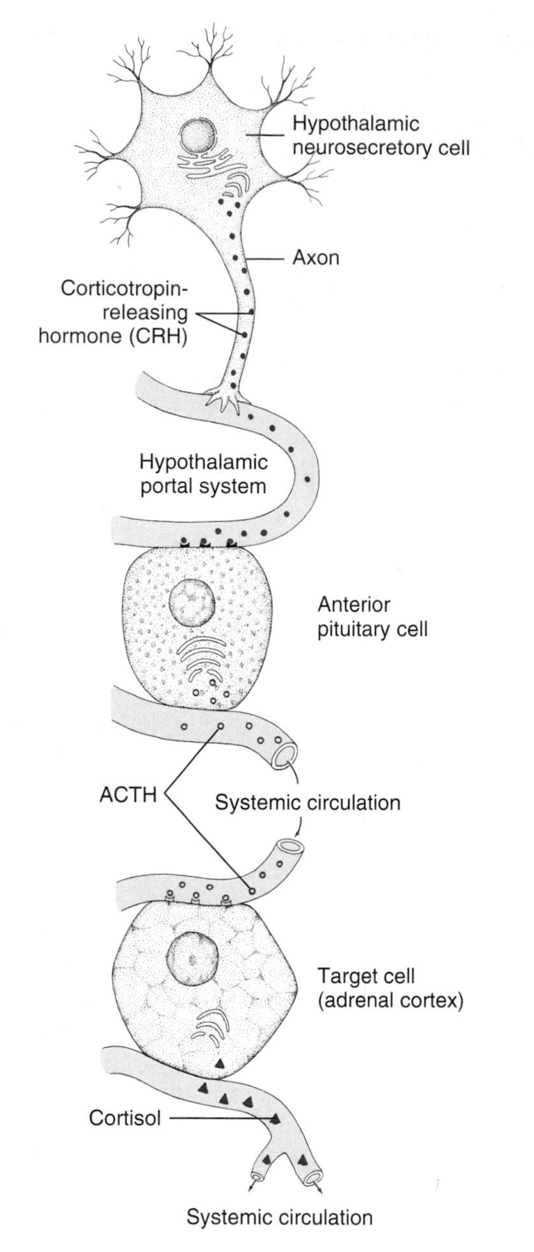

figure 12-6

Relationship of hypothalamic, pituitary, and target-gland hormones. The hormone sequence controlling the release of cortisol from the adrenal cortex is used as an example.

dispersion of the pigment melanin within melanophores, causing darkening of the skin. In birds and mammals, MSH is produced by cells in the anterior pituitary rather than the intermediate lobe (birds and some mammals lack an intermediate lobe altogether), but its physiological function remains unclear. MSH appears unrelated to pigmentation in the endotherms, although it will cause darkening of the skin in humans if injected into the circulation. Until recently, many endocrinologists thought MSH was a vestigial hormone, but interest has been rekindled by studies showing that it enhances memory and growth of the mammalian fetus. MSH and ACTH are derived from a precursor molecule called pro-opiomelanocortin that is transcribed and translated from a single gene.

Hypothalamus and Neurosecretion

Because of the strategic importance of the pituitary in influencing most hormonal activities in the body, the pituitary was once called the "master gland." This description is not appropriate, however, because the tropic hormones are regulated by a higher council, the neurosecretory centers of the hypothalamus. The hypothalamus is itself under ultimate control of the brain. The hypothalamus contains groups of neurosecretory cells, which are specialized giant nerve cells (Figure 12-6). These cells manufacture polypeptide hormones, called **releasing hormones** or **release-inhibiting hormones** (or "factors"), which then travel down nerve fibers to their endings in the median eminence. Here they enter a capillary network to complete their journey to the anterior pituitary by way of the pituitary portal system. The hypothalamic hormones then stimulate or inhibit the release of the various anterior pituitary hormones. Several hypothalamic releasing hormones have been discovered since the demonstration in 1955 of a corticotropin-releasing hormone (Table 12-1). There are one or more releasing hormones regulating each of the seven pituitary hormones. Several of the releasing hormones have now been isolated in pure state and characterized chemically (the identification and action of some of the hypothalamic hormones listed in Table 12-1 is still tentative). All releasing hormones are peptides.

Posterior Pituitary

The hypothalamus is also the source of two hormones of the posterior lobe of the pituitary (Table 12-1). They are formed in neurosecretory cells in the hypothalamus, then transported down the infundibular stalk and into the posterior lobe, ending in proximity to blood capillaries, which the hormones enter when released (see Figure 12-5). In a sense the posterior lobe is not a true endocrine gland, but a storage and release center for hormones manufactured entirely in the hypothalamus. The two posterior lobe hormones of mammals, oxytocin and vasopressin, are chemically very much alike. Both are polypeptides consisting of eight amino acids and are called octapeptides (Figure 12-7). These hormones are among the fastest acting hormones, since they are capable of producing a response within seconds of their release from the posterior lobe.

Oxytocin has two important specialized reproductive functions in adult female mammals. It stimulates contraction of uterine smooth muscles during parturition (birth of the young). In clinical practice, oxytocin is used to induce delivery during a difficult labor and to prevent uterine hemorrhage after birth. The second action of oxytocin is that of milk ejection by the mammary glands in response to suckling. Although present, oxytocin has no known function in males.

table 12-1
Hormones of the Vertebrate Pituitary

	Hormone	Chemical nature	Principal action	Hypothalamic controls
Adenohypophysis				
Anterior lobe	Thyroid-stimulating hormone (TSH)	Glycoprotein	Stimulates thyroid hormone synthesis and secretion	TSH-releasing hormone (TRH)
	Follicle-stimulating hormone (FSH)	Glycoprotein	Female: follicle maturation and estrogen synthesis Male: stimulates sperm production	Gonadotropin-releasing hormone (GnRH)[1]
	Luteinizing hormone (LH, ICSH)	Glycoprotein	Female: stimulates ovulation, corpus luteum formation, estrogen and progesterone synthesis Male: stimulates testosterone secretion	Gonadotropin-releasing hormone (GnRH)[1]
	Prolactin (PRL)	Protein	Mammary gland growth, milk synthesis in mammals; various reproductive and nonreproductive functions in other vertebrates	Prolactin release-inhibiting factor (PIF), and (in birds) Prolactin releasing factor (PRF)
	Growth hormone (GH) (Somatotropin)	Protein	Stimulates growth	Growth hormone release-inhibiting hormone (Somatostatin) and Growth hormone-releasing hormone (Somatocrinin)
	Adrenocorticotropic hormone (ACTH)	Polypeptide	Stimulates steroid hormone synthesis by adrenal cortex	Corticotropin-releasing hormone (CRH)
Intermediate lobe[2]	Melanophore-stimulating hormone (MSH)	Polypeptide	Pigment dispersion in melanophores of ectotherms; function unclear in endotherms	MSH release-inhibiting factor (MIF)
Neurohypophysis				
Posterior lobe	Oxytocin	Octapeptide	In mammals stimulates milk ejection and uterine contractions	
	Vasopressin (Antidiuretic hormone)	Octapeptide	In mammals increases water reabsorption by kidney	
	Vasotocin	Octapeptide	Present in all vertebrate classes; in tetrapods increases water reabsorption	
	Mesotocin	Octapeptide	In lungfish, amphibians, and reptiles decreases water reabsorption by kidney	

[1]*One GnRH hormone regulates both FSH and LH according to recent experimental evidence.*
[2]*Birds and some mammals lack an intermediate lobe. In these forms, MSH is produced by the anterior lobe.*

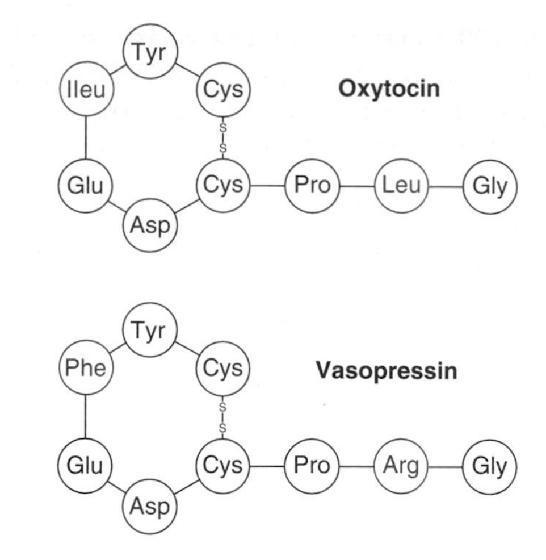

 12-7

Posterior lobe hormones of mammals. Both oxytocin and vasopressin consist of eight amino acids (the two sulfur-linked cysteine molecules are considered a single amino acid, cystine). Oxytocin and vasopressin are identical except for amino acid substitutions in the blue positions. The abbreviations represent amino acids.

Vasopressin, the second posterior lobe hormone, acts on collecting ducts of the kidney to increase water reabsorption and thus restrict urine flow, as already described on p. 178. It is therefore often called the **antidiuretic hormone.** Vasopressin has a second, weaker effect of increasing the blood pressure through its generalized constrictor effect on the smooth muscles of the arterioles. Although the name "vasopressin" unfortunately suggests that vasoconstriction is the hormone's major effect, it is probably of little physiological importance, except perhaps to help sustain blood pressure during a severe hemorrhage.

All jawed vertebrates secrete two posterior lobe hormones that are quite similar to those of mammals. All are octapeptides, but there is some variation in structure because of amino acid substitutions in three of eight amino acid positions in the molecule.

Of all posterior lobe hormones, **vasotocin** (Table 12-1) has the widest phylogenetic distribution and is believed to be the parent hormone from which other octapeptides evolved. It is found in all vertebrate classes except mammals. It is a water-balance hormone in amphibians, especially toads, in which it acts to conserve water by (1) increasing permeability of the skin (to promote water absorption from the environment), (2) stimulating water reabsorption from the urinary bladder, and (3) decreasing urine flow. The action of vasotocin is best understood in amphibians, but it appears to play some water-conserving role in birds and reptiles as well.

Pineal Gland

In all vertebrates the dorsal part of the brain, the diencephalon (Fig. 11-11, p. 249), gives rise to a saclike evagination called the pineal complex, which lies just beneath the skull in a midline position. In ectothermic vertebrates the pineal complex contains glandular tissue and a photoreceptive sensory organ involved in varying pigmentation and in light-dark biological rhythms. In lampreys, many amphibians, lizards, and the tuatara (*Sphenodon,* Figure 29-23 p. 636), the median photoreceptive organ is so well developed, containing structures analogous to the lens and cornea of lateral eyes, that it is often called a third eye. In birds and mammals, the pineal complex has evolved an entirely glandular structure called the pineal gland. The pineal gland produces the hormone **melatonin.** Exposure to light strongly affects secretion of melatonin. Its production is lowest during daylight hours and highest at night. In nonmammalian vertebrates, the pineal gland is responsible for maintaining **circadian rhythms**—self-generated (endogenous) rhythms that are about 24 hours in length. A circadian rhythm serves as a biological clock for many physiological processes that follow a regular pattern.

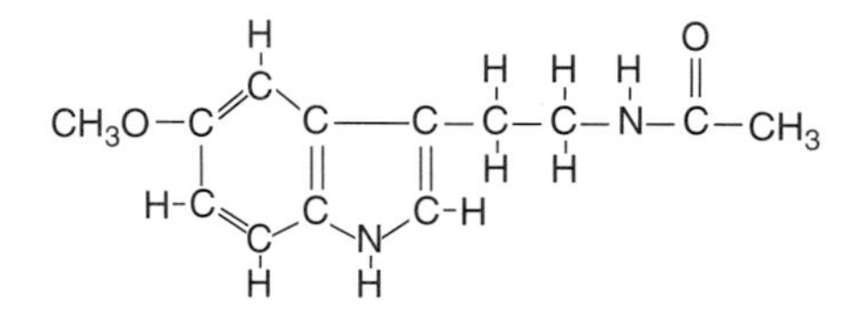

Structure of melatonin

In mammals, the pineal has lost most capacity to pace circadian rhythms. Instead, an area of the hypothalamus called the suprachiasmatic nucleus has become the primary circadian pacemaker in mammals, although the pineal gland still produces melatonin nightly and serves to reinforce the circadian rhythm of the suprachiasmatic nucleus. In mammals in which seasonal rhythms in reproduction are regulated by **photoperiod,** the pineal hormone melatonin plays a critical role in regulating gonadal activity. In long-day breeders, such as ferrets, hamsters, and deer mice, the reduced light stimulation with shortening day length in autumn increases melatonin secretion. Melatonin, by inhibiting synthesis and secretion of the gonadotropin-releasing hormone (GnRH), suppresses reproductive activity during the winter months. The lengthening days in the spring have the opposite effect and reproductive activities are resumed. Short-day breeders, such as white-tailed deer, silver fox, spotted skunk, and sheep, are stimulated by reduced day length in the fall; in these, increasing melatonin levels in the fall stimulates, rather than inhibits, reproductive activity.

Only recently the pineal gland has been shown to produce subtle and incompletely understood effects on circadian and annual rhythms in nonphotoperiodic mammals (such as humans). For example, melatonin secretion has been linked to a sleeping and eating disorder in humans known as seasonal affective disorder (SAD). Some people living in northern latitudes, where day lengths are short in winter and when melatonin production is elevated, become depressed in winter, sleep long periods, and may go on eating binges. Often this wintertime disorder can be treated by exposure to sun lamps with full-spectrum light; such exposure depresses melatonin secretion by the pineal gland.

Nonendocrine Hormones

Brain Neuropeptides

The blurred distinction between the endocrine and nervous systems is nowhere more evident than in the nervous system, where a growing list of hormonelike neuropeptides have been discovered in central and peripheral nervous systems of vertebrates and invertebrates. In mammals, approximately 40 neuropeptides (short chains of amino acids) have been located using immunological labeling with fluorescent antibodies that can be visualized in histological sections under the microscope. Many neuropeptides lead double lives—capable of behaving both as hormones, carrying signals from gland cells to their targets, and as neurotransmitters, relaying signals between nerve cells. For example, both oxytocin and vasopressin have been discovered at widespread sites in the brain by immunochemical methods. Related to this discovery is the fascinating observation that people and experimental animals injected with minute quantities of vasopressin experience enhanced learning and improved memory. This effect of vasopressin in brain tissue is unrelated to its well-known antidiuretic function in the kidney (p. 178). Several hormones, such as gastrin and cholecystokinin (p. 282) (which long had been supposed to function only in the gastrointestinal tract), have been discovered in the cerebral cortex and hippocampus. In addition to its gastrointestinal actions, we know now that cholecystokinin functions in the control of feeding and satiety and may serve other roles as a brain neuroregulator.

The radioimmunoassay technique developed by Solomon Berson and Rosalyn Yalow about 1960 has revolutionized endocrinology and neurochemistry. First, antibodies to the hormone of interest (insulin, for example) are prepared by injecting guinea pigs or rabbits with the hormone. Then, a fixed amount of radioactively labeled insulin and unlabeled insulin antibodies is mixed with the sample of blood plasma to be measured. Native insulin in the blood plasma and radioactive insulin compete for antibodies. The more insulin present in the sample, the less radioactive insulin will bind to the antibodies. Bound and unbound insulin are then separated, and their radioactivities are measured together with those of appropriate standards to determine the amount of insulin present in the blood sample. The method is so incredibly sensitive that it can measure the equivalent of a cube of sugar dissolved in one of the Great Lakes.

Among dramatic developments in this field was the discovery in 1975 of the endorphins and enkephalins, neuropeptides that bind with opiate receptors and influence perception of pleasure and pain (see note on p. 256). The endorphins and enkephalins also are found in brain circuits that modulate several other functions unrelated to pleasure and pain, such as control of blood pressure, body temperature, and body movement. Even more intriguing, the endorphins are derived from the same prohormone that gives rise to the anterior pituitary hormones ACTH and MSH. The discovery in the brain of a complex family of compounds whose functions and interrelationships are not yet clear has spawned an active area of biomedical research.

Prostaglandins

Prostaglandins are a family of long-chain unsaturated fatty acid derivatives that were discovered in seminal fluid in the 1930s. At first they were thought to be produced by the prostate gland (hence the name) but have now been found in virtually all mammalian tissues. Prostaglandins act as local hormones that have diverse actions on many different tissues, making generalizations about their effects difficult. Many of their effects, however, involve smooth muscle. In some tissues prostaglandins regulate vasodilation or vasoconstriction by their often antagonistic action on smooth muscle in the walls of blood vessels. They are known to stimulate contraction of uterine smooth muscle during childbirth. There is also evidence that overproduction of uterine prostaglandins is responsible for the symptoms of painful menstruation (dysmenorrhea) experienced by many women. Several inhibitors of prostaglandins are now available that provide relief from these symptoms. Among other actions of prostaglandins is their intensification of pain in damaged tissues, mediation of the inflammatory response, and involvement in fever. In many instances the physiological significance of the numerous reported effects of prostaglandins is unknown.

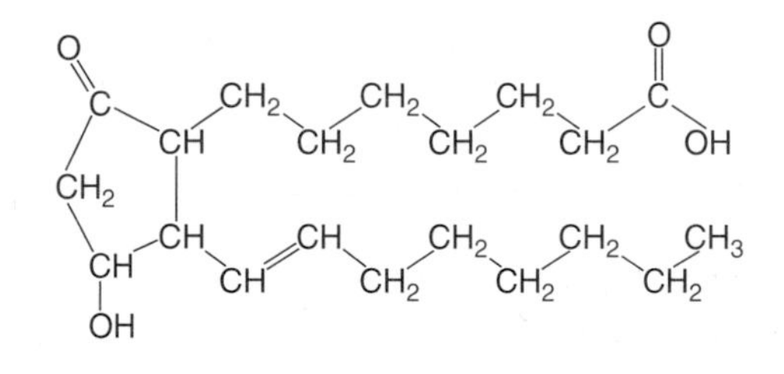

Structure of a typical prostaglandin

Cytokines

For some years we have known that cells of the immune system somehow communicated with each other and that this communication was crucial to the immune response. Now we understand that polypeptide hormones called cytokines (p. 215) mediate communication between the cells of immunity. Cytokines can affect the cells that secrete them, affect nearby cells, and like other hormones, they can affect cells in distant locations. Their target cells bear specific receptors for the cytokine bound to the surface membrane. Cytokines coordinate a complex network, with some target cells being activated, stimulated to divide and often to secrete their own cytokines. The same cytokine that activates some cells may suppress division of other target cells.

Hormones of Metabolism

Another important group of hormones adjusts the delicate balance of metabolic activities. The rates of chemical reactions within cells are often regulated by long sequences of enzymes. Although such sequences are complex, each step in a pathway is mostly self-regulating as long as the equilibrium between substrate, enzyme, and product remains stable. However, hormones may alter the activity of crucial enzymes in a metabolic process, thus accelerating or inhibiting the entire process. It should be emphasized that hormones seldom initiate enzymatic processes; rather they alter their rate, speeding them up or slowing them down. The most important hormones of metabolism are those of the thyroid, parathyroid, adrenal glands, and pancreas.

Thyroid Hormones

Two hormones, **thyroxine** and **triiodothyronine,** are secreted by the thyroid gland. This large endocrine gland is located in the neck of all vertebrates. The thyroid is composed of thousands of tiny spherelike units, called follicles, where thyroid hormone is synthesized, stored, and released into the bloodstream as needed. The size of the follicles, and the amount of stored thyroxine and triiodothyronine they contain, depends on the activity of the gland (Figure 12-8).

Thyroxine

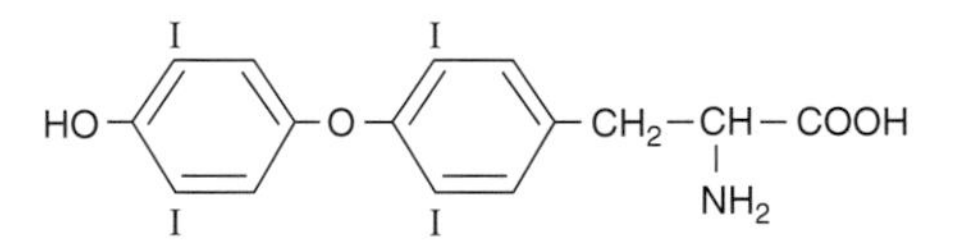

One of the unique characteristics of the thyroid is its high concentration of **iodine;** in most animals this single gland contains well over half the body's store of iodine. The epithelial cells of the thyroid follicles actively trap iodine from the blood and combine it with the amino acid tyrosine, creating the two thyroid hormones. Each molecule of thyroxine contains four atoms of iodine. Triiodothyronine has three instead of four iodine atoms. Thyroxine is formed in much greater amounts than triiodothyronine, but in many vertebrates triiodothyronine is the more physiologically active hormone. Both hormones have two important actions. One is to promote normal growth and development of the nervous system of growing animals. The other is to stimulate the metabolic rate.

Undersecretion of thyroid hormones in fish, birds, and mammals dramatically impairs growth, especially of the nervous system. The human **cretin,** a mentally retarded dwarf, is the result of thyroid malfunction from a very early age. Conversely, the oversecretion of thyroid hormones causes precocious development in all vertebrates, but its effect is particularly prominent in fish and amphibians. In frogs and toads, transformation from aquatic herbivorous tadpole without lungs or legs to semiterrestrial or terrestrial carnivorous adult with lungs and four legs occurs when the thyroid gland becomes active at the end of larval development. Stimulated by a rise in the thyroxine level of the blood, metamorphosis and climax occur (Figure 12-9). Growth hormone directs growth of the frog after metamorphosis.

In birds and mammals, the best known action of the thyroid hormones is control of oxygen consumption and heat production. The thyroid maintains metabolic activity of

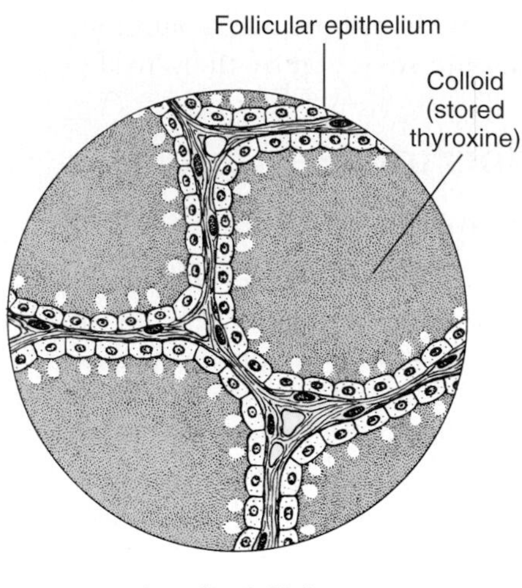

Inactive follicles

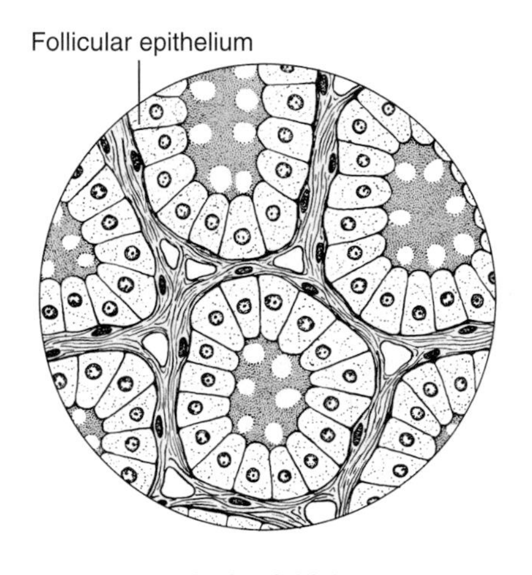

Active follicles

figure 12-8

Appearance of thyroid gland follicles viewed through a microscope (approximately ×350). When the thyroid is inactive, follicles are distended with colloid, the storage form of thyroid hormone, and the epithelial cells are flattened. When the thyroid is active, colloid disappears as thyroid hormone is secreted into the circulation, and the epithelial cells become greatly enlarged.

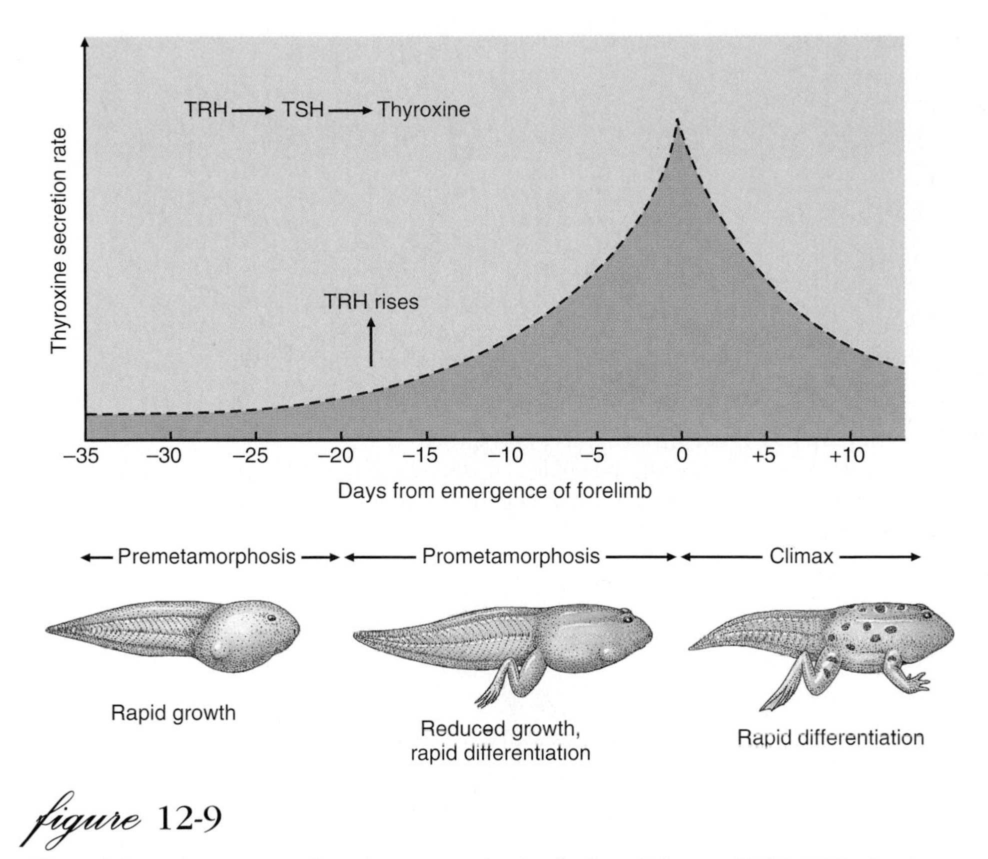

figure 12-9

Effect of thyroxine on growth and metamorphosis of a frog. Release of TRH (TSH-releasing hormone) from the hypothalamus at the end of premetamorphosis sets in motion hormonal changes (increased TSH and thyroid hormone) leading to metamorphosis. Thyroxine levels are maximal at the time the forelimbs emerge.

homeotherms (birds and mammals) at a normal level. Oversecretion of thyroid hormones will speed up body processes as much as 50%, resulting in irritability, nervousness, fast heart rate, intolerance of warm environments, and loss of body weight despite increased appetite. Undersecretion of thyroid hormones slows metabolic activities, which can cause loss of mental alertness, slowing of the heart rate, muscular weakness, increased sensitivity to cold, and weight gain. An important function of the thyroid gland is to promote adaptation to cold environments by increasing heat production. The thyroid hormones stimulate cells to produce more heat and store less chemical energy (ATP); in other words, thyroxine *reduces* the efficiency of cellular oxidative phosphorylation (p. 47). Consequently many cold-adapted mammals have heartier appetites and eat more food in winter than in summer, although their activity is about the same in both seasons. In winter, a larger portion of the food is being converted directly into body-warming heat.

The synthesis and release of the thyroid hormones are governed by **thyrotropic hormone** (TSH) from the anterior pituitary gland (Table 12-1). TSH is in turn regulated by the thyrotropin-releasing hormone (TRH) of the hypothalamus. As noted earlier, TRH is part of a higher regulatory council that controls the tropic hormones of the anterior pituitary. TSH controls thyroid activity through an excellent example of negative feedback. If thyroid hormone level in the blood decreases, more TSH is released, which returns the level to normal. Should thyroid hormone level rise too high, it acts on the anterior pituitary to inhibit TSH release. With declining TSH output, the thyroid is less stimulated and the level of thyroid hormone in the blood returns to normal. Such a system is obviously very effective in damping oscillations in hormonal output by the target gland. It can be overridden, however, by neural stimuli, such as exposure to cold, which can directly stimulate the release of TRH and thus TSH.

Some years ago, a condition called goiter was common among people living in the Great Lakes region of the United States and Canada, as well as some other parts of the world, such as the Swiss Alps. This type of goiter is an enlargement of the thyroid gland caused by a deficiency of iodine in the food and water. By striving to produce thyroid hormone with not enough iodine available, the gland hypertrophies, sometimes so much that the entire neck region becomes swollen (Figure 12-10). Goiter caused by iodine deficiency is seldom seen in North America because of the widespread use of iodized salt. However, it is estimated that even today 200 million people worldwide experience varying degrees of goiter, mostly in the high mountains of South America, Europe, and Asia.

Hormonal Regulation of Calcium Metabolism

Closely associated with the thyroid gland, and in some animals buried within it, are the **parathyroid glands.** These tiny glands occur as two pairs in humans but vary in number and position in other vertebrates. They were discovered at the end of the nineteenth century when the fatal effects of "thyroidectomy" were traced to the unknowing removal of the parathyroid glands together with the thyroid gland. In birds and mammals, including humans, removal of the parathyroid glands causes the level of calcium in the blood to decrease rapidly. This decrease leads to a serious increase in nervous excitability, severe muscular spasms and tetany, and finally death.

Before considering how hormones maintain calcium homeostasis, it is helpful to summarize mineral metabolism in bone, a densely packed storehouse of both calcium and phosphorus. Bone contains approximately 98% of the calcium and 80% of the phosphorus in the body. Although bone is second

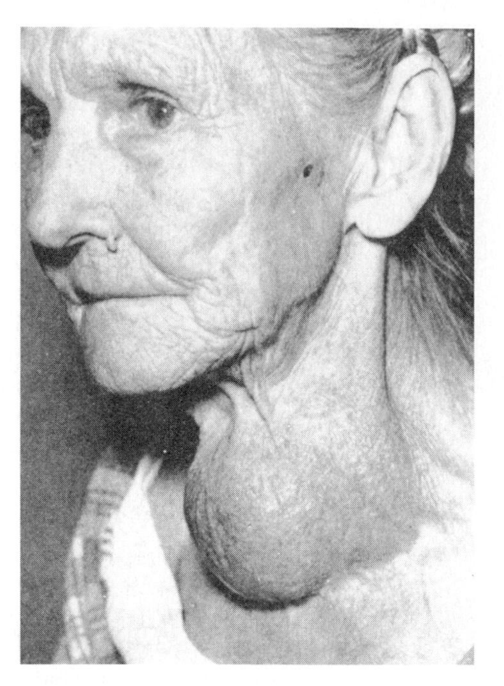

figure 12-10

A large goiter caused by iodine deficiency. By enlarging enormously, the thyroid gland can extract enough iodine from the blood to synthesize the body's requirement for thyroid hormone.

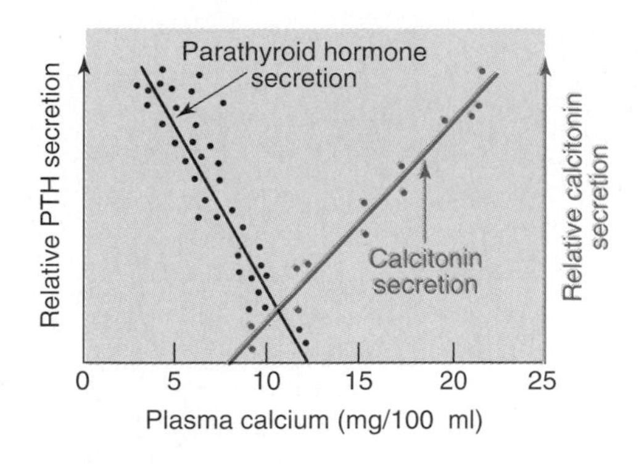

figure 12-11

How the secretions of parathyroid hormone (PTH) and calcitonin respond to changes in the level of calcium in the blood of a mammal.

only to teeth as the most durable material in the body (as evidenced by the survival of fossil bones for millions of years), it is in a state of constant turnover in the living body. Bone-building cells **(osteoblasts)** synthesize the organic fibers of the bone matrix, which later become mineralized with a form of calcium phosphate called hydroxyapatite. Bone-resorbing cells **(osteoclasts)** are giant cells that dissolve the bony matrix, releasing calcium and phosphate into the blood. These opposing activities allow bone constantly to remodel itself, especially in the growing animal, for structural improvements to counter new mechanical stresses on the body. They additionally provide a vast and accessible reservoir of minerals that can be withdrawn as needed for general cellular requirements.

The level of calcium in the blood is maintained by the action of three hormones which coordinate the absorption, storage, and excretion of calcium ions. If blood calcium should decrease slightly, the parathyroid glands increase their secretion of **parathyroid hormone (PTH).** This stimulates the osteoclasts to dissolve bone adjacent to these cells, thus releasing calcium and phosphate into the bloodstream and returning the level of blood calcium to normal. Parathyroid hormone also decreases the rate of calcium excretion by the kidney and increases production of the hormone 1,25-dihydroxyvitamin D (see the following text). The level of parathyroid hormone in the blood varies inversely with blood calcium, as shown in Figure 12-11.

The second hormone involved in calcium metabolism in all tetrapods is derived from vitamin D. Vitamin D, like all vitamins, is a dietary requirement. But unlike other vitamins, vitamin D may also be synthesized in the skin by irradiation with ultraviolet light from the sun. Vitamin D is then converted in a two-step oxidation to a hormonal form, **1,25-dihydroxyvitamin D.** This steroid hormone is essential for active calcium absorption by the gut (Figure 12-12). It also promotes the synthesis of a protein required for calcium uptake from the intestine. The

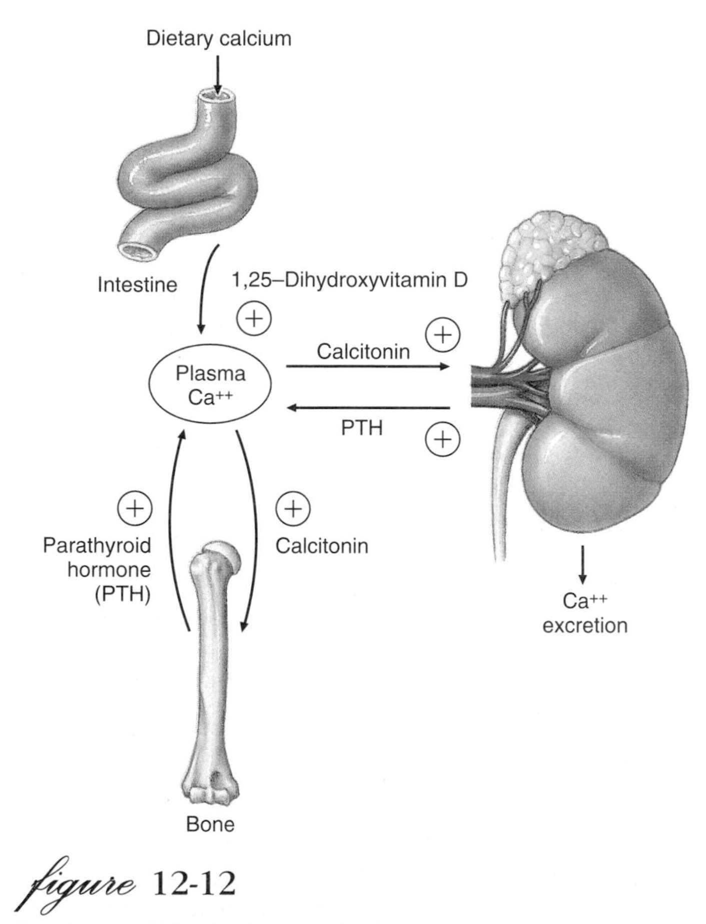

figure 12-12

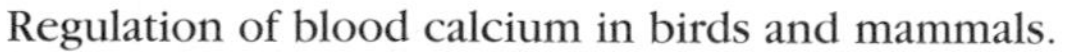

Regulation of blood calcium in birds and mammals.

production of 1,25-dihydroxyvitamin D is stimulated by low plasma phosphate as well as by an increase in PTH secretion.

In humans, a deficiency of vitamin D causes rickets, a disease characterized by low blood calcium and weak, poorly calcified bones that tend to bend under postural and gravitational stresses. Rickets has been called a disease of northern winters, when sunlight is minimal. It was once common in the smoke-darkened cities of England and Europe.

The third calcium-regulating hormone, **calcitonin,** is secreted by specialized cells (C cells) in the thyroid gland of mammals and in the ultimobranchial gland of other vertebrates. Calcitonin is released in response to elevated levels of calcium in the blood. It rapidly suppresses calcium withdrawal from bone, decreases intestinal absorption of calcium, and increases excretion of calcium by the kidneys. Calcitonin thus protects the body against a dangerous increase in the level of calcium in the blood, just as parathyroid hormone protects it from a dangerous decrease in blood calcium (Figure 12-12). Calcitonin has been identified in all vertebrate groups but, except in mammals, its functional role is uncertain.

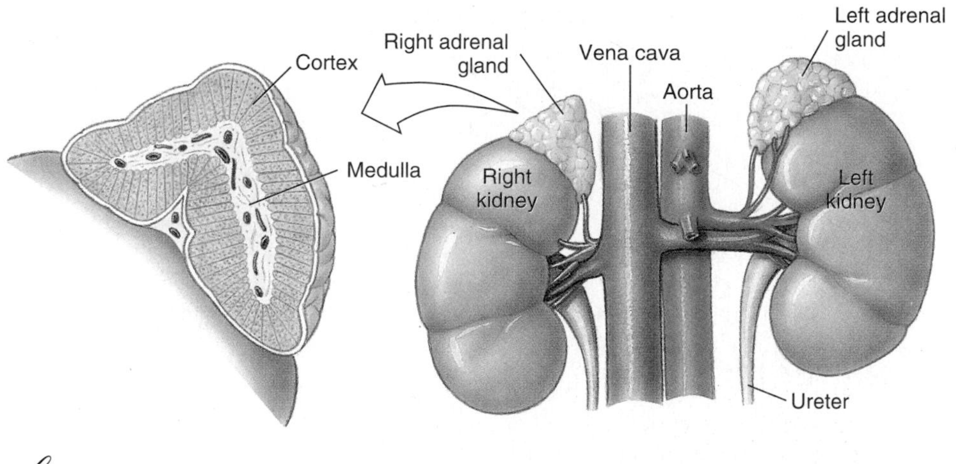

 12-13

Paired adrenal glands of humans, showing gross structure and position on the upper poles of the kidneys. The cortex produces steroid hormones. The medulla produces the sympathetic hormones epinephrine and norepinephrine.

Hormones of the Adrenal Cortex

The mammalian adrenal gland is a double gland composed of two unrelated types of glandular tissue: an outer region of adrenocortical cells, or **cortex,** and an inner region of specialized cells, the **medulla** (Figure 12-13). In nonmammalian vertebrates the homologs of the adrenocortical and medullary cells are organized quite differently; they may be intermixed or distinct, but never arranged in a cortex-medulla relationship as in mammals.

At least 30 different compounds have been isolated from adrenocortical tissue, all of them closely related lipoidal compounds known as steroids. Only a few of these compounds are true steroid hormones; most are various intermediates in the synthesis of steroid hormones from **cholesterol** (Figure 12-14). The corticosteroid hormones are commonly classified into two groups according to their function: glucocorticoids and mineralocorticoids.

Glucocorticoids, such as **cortisol** (Figure 12-14) and **corticosterone,** are concerned with food metabolism, inflammation, and stress. They promote the synthesis of glucose from compounds other than carbohydrates, particularly amino acids and fats. This process is called **gluconeogenesis.** The overall effect is to increase the level of glucose in the blood to provide a quick energy source for muscle and nervous tissue. The glucocorticoids are also important in diminishing the immune response to various inflammatory conditions. Because several diseases of humans are inflammatory diseases (for example, allergies, hypersensitivity, and rheumatoid arthritis), these corticosteroids have important medical applications.

The adrenal steroid hormones, especially the glucocorticoids, are remarkably effective in relieving symptoms of rheumatoid arthritis, allergies, and various disorders of connective tissue, skin, and blood. Following the announcement in 1948 by P.S. Hench and his colleagues at the Mayo Clinic that cortisone dramatically relieved the pain and crippling effects of advanced arthritis, steroid hormones were hailed by the media as "wonder drugs." Optimism was soon dimmed, however, when it became apparent that severe side effects always attended long-term administration of the antiinflammatory steroids. Steroid therapy lulls the adrenal cortex into inactivity and may permanently impair the body's capacity to produce its own steroids. Today steroid therapy is applied with caution, because we realize that the inflammatory response is a necessary part of the body's defenses.

The synthesis and secretion of the glucocorticoids are controlled principally by ACTH of the anterior pituitary (see Figure 12-6). ACTH is also controlled by the corticotropin-releasing hormone (CRH) of the hypothalamus (Table 12-1). As with pituitary control of the thyroid, a negative feedback relationship exists between CRH, ACTH and the adrenal cortex.

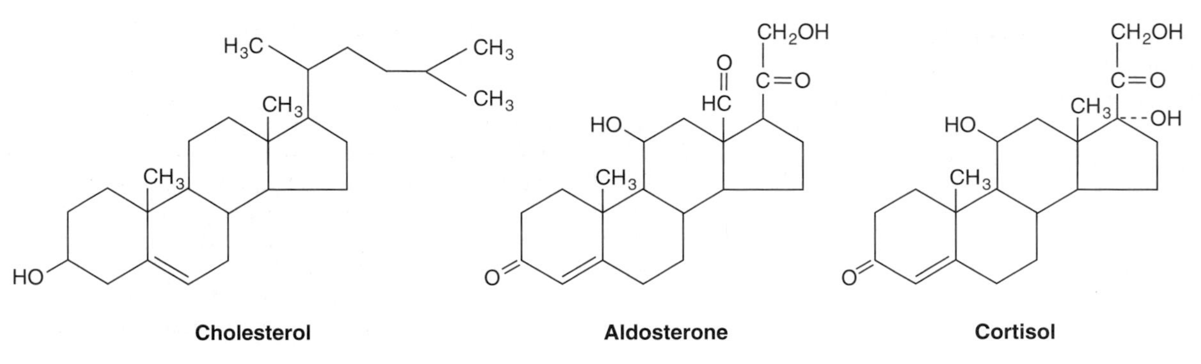

figure 12-14

Hormones of the adrenal cortex. Cortisol (a glucocorticoid) and aldosterone (a mineralocorticoid) are two of several steroid hormones synthesized from cholesterol in the adrenal cortex.

An increase in the release of glucocorticoids suppresses the output of CRH and ACTH; the resulting decline in the blood level of CRH and ACTH then feeds back to the adrenal cortex to inhibit further release of glucocorticoids. An opposite cycle of events happens if the blood level of glucocorticoids drops: ACTH output increases, which in turn stimulates the secretion of glucocorticoids. CRH is known to mediate stressful stimuli through the adrenal axis.

Mineralocorticoids, the second group of corticosteroids, are those that regulate salt balance. **Aldosterone** (Figure 12-14) is by far the most important steroid of this group. Aldosterone promotes the tubular reabsorption of sodium and chloride and the tubular excretion of potassium by the kidney (p. 177). Since sodium usually is in short supply in the diet of many animals and potassium is in excess, the mineralocorticoids play vital roles in preserving the correct balance of blood electrolytes.

The mineralocorticoids *oppose* the antiinflammatory effect of the glucocorticoids. In other words, they promote the *inflammatory* defense of the body to various noxious stimuli. Although these opposing actions of the corticosteroids seem self-defeating, they are not. They are necessary to maintain readiness of the body's defenses for any stress or threat of disease, yet prevent these defenses from becoming so powerful that they turn against the body's own tissues.

The adrenocortical tissue also produces androgens (Gr. *andros,* man, + *genesis,* origin), which, as the name implies, are similar in effect to the male sex hormone, testosterone. The adrenal androgens promote some of the changes that occur just before puberty in the human male and female. The recent development of so-called *anabolic steroids,* synthetic hormones related to testosterone, has led to widespread abuse of steroids among athletes. These synthetics (and testosterone) cause hypertrophy of skeletal muscle and may improve performance that depends on strength. Unfortunately they also have serious side effects, including testicular atrophy, periods of irritability, abnormal liver function, and cardiovascular disease.

The use of anabolic (tissue-building) steroids by athletes became major news following Ben Johnson's drug-fueled win of the 100-meter race at the 1988 Olympics. Despite almost universal condemnation by Olympic, medical, and college sports authorities, an unscientific and clandestine program of experimentation with anabolic steroids continues to be popular with many amateur and professional athletes in many countries. There may be 3 million regular anabolic steroid users in the United States alone. Most anabolic steroids are purchased illegally through a black market, with annual sales of some $400 million. The extensive use of anabolic steroids by Olympic athletes was documented by Robert Voy, who served as Chief Medical Officer for the United States Olympic Committee from 1985 to 1989, when he quit in frustration over a crisis he believed was destroying the Olympic ideal (Voy, R. 1991. Drugs, sport, and politics. *Leisure Press).*

Hormones of the Adrenal Medulla

The adrenal medullary cells secrete two structurally similar hormones: **epinephrine** (adrenaline) and **norepinephrine** (noradrenaline). The adrenal medulla is derived embryologically from the same tissue that gives rise to the postganglionic sympathetic neurons of the autonomic nervous system (p. 251). Norepinephrine serves as a neurotransmitter at the endings of the sympathetic nerve fibers. Thus functionally, as well as embryologically, the adrenal medulla can be considered a very large sympathetic nerve ending.

It is not surprising then that adrenal medullary hormones and the sympathetic nervous system have the same general effects on the body. These effects center on responses to emergencies, such as fear and strong emotional states, flight from danger, fighting, pain, lack of oxygen, blood loss, and ex-

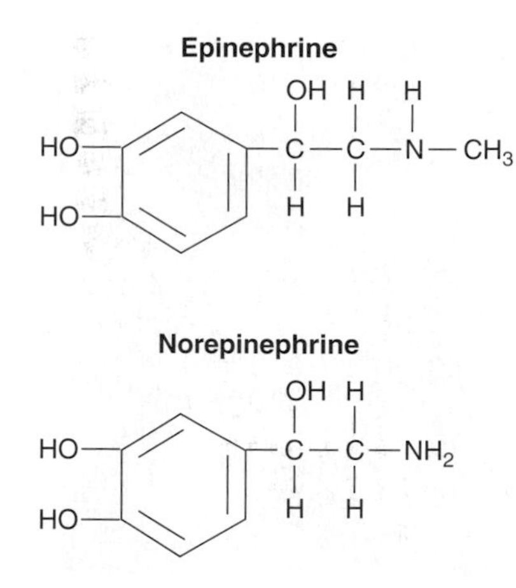

posure to pain. Walter B. Cannon, of homeostasis fame (p. 167), termed these "fight or flight" responses that are appropriate for survival. We are all familiar with the increased heart rate, tightening of the stomach, dry mouth, trembling muscles, general feeling of anxiety, and increased awareness that attends sudden fright or other strong emotional states. These effects are attributable to increased activity of the sympathetic nervous system and to rapid release into the blood of epinephrine from the adrenal medulla.

Epinephrine and norepinephrine have many other effects of which we are not as aware, including constriction of the arterioles (which, together with the increased heart rate, increases the blood pressure), mobilization of liver glycogen and fat stores to release glucose and fatty acids for energy, increased oxygen consumption and heat production, hastening of blood coagulation, and inhibition of the gastrointestinal tract. All these changes prepare the body for emergencies or stressful conditions.

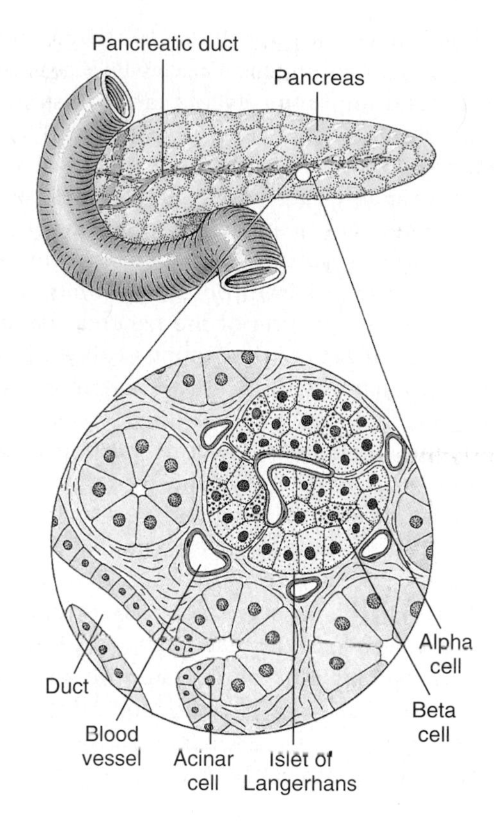

figure 12-15

The pancreas is composed of two kinds of glandular tissue: exocrine acinar cells that secrete digestive juices which enter the intestine through the pancreatic duct, and the endocrine islets of Langerhans. The islets of Langerhans secrete the hormones insulin and glucagon directly into the blood circulation.

Insulin from the Islet Cells of the Pancreas

The pancreas is both an exocrine and an endocrine organ (Figure 12-15). The *exocrine* portion produces pancreatic juice, a mixture of digestive enzymes and bicarbonate ions that is conveyed by a duct to the digestive tract. Scattered within the extensive exocrine portion of the pancreas are numerous small islets of tissue, called **islets of Langerhans** (Figure 12-15). This *endocrine* portion of the gland makes up only 1% to 2% of the total weight of the pancreas. The islets are without ducts and secrete their hormones directly into blood vessels that extend throughout the pancreas.

Two polypeptide hormones are secreted by different cell types within the islets: **insulin,** produced by the **beta cells,** and **glucagon,** produced by the **alpha cells.** Insulin and glucagon have antagonistic actions of great importance in the metabolism of carbohydrates and fats. Insulin is essential for the use of blood glucose by cells, especially skeletal muscle cells. Insulin promotes the entry of glucose into body cells through its action on a glucose transporter molecule found in cell membranes. Without insulin, body cells cannot use glucose. The level of glucose in the blood rises to abnormally high levels (hyperglycemia) to exceed the transport maximum of the kidney (p. 176), and sugar (glucose) appears in the urine. Lack of insulin also inhibits uptake of amino acids by skeletal muscle, and fats and muscle are broken down to provide energy. The body cells starve while the urine abounds in the very substance the body craves. The disease, called diabetes mellitus, afflicts nearly 5% of the human population in varying degrees of severity. If left untreated, it can lead to severe damage to kidneys, eyes, and blood vessels, and greatly shorten life expectancy.

In 1982, insulin became the first hormone produced by genetic engineering (recombinant DNA technology, p. 76) to be marketed for human use. Recombinant insulin has the exact structure of human insulin and therefore will not stimulate an immune response, which has often been a problem for diabetics receiving insulin purified from pig or cow pancreas.

The first extraction of insulin in 1921 by two Canadian scientists, Frederick Banting and Charles Best, was one of the most dramatic and important events in the history of medicine. Many years earlier two German scientists, J. Von Mering andO. Minkowski, discovered that surgical removal of the pancreas of dogs invariably caused severe symptoms of diabetes, resulting in the animal's death within a few weeks. Many attempts were made to isolate the diabetes preventive factor, but all failed because powerful protein-splitting digestive enzymes in the exocrine portion of the pancreas destroyed the hormone during extraction procedures. Following a hunch, Banting, in collaboration with Best and his physiology professor J.J.R. Macleod, tied off the pancreatic ducts of several dogs. This constriction caused the exocrine portion of the gland with its hormone-destroying enzyme to degenerate but left the islets' tissues healthy. Banting and Best then successfully extracted insulin from these glands. Injected into another dog, the insulin immediately lowered the level of sugar in the blood (Figure12-16). Their experiment paved the way for the commercial extraction of insulin from slaughterhouse animals. It meant that millions of people with diabetes, previously doomed to invalidism or death, could look forward to more normal lives.

Glucagon, the second hormone of the pancreas, has several effects on carbohydrate and fat metabolism that are opposite to the effects of insulin. For example, glucagon raises the blood glucose level (by converting liver glycogen to glucose), whereas insulin lowers blood glucose. Glucagon and insulin do not have the same effects in all vertebrates, and in some, glucagon is lacking altogether. Glucagon is an example of a hormone that operates through the cyclic AMP second-messenger system.

figure 12-16

Charles H. Best and Sir Frederick Banting in 1921 with the first dog to be kept alive by insulin.

From J. F. Fulton & L. G. Wilson Selected Readings in the History of Physiology, 1966. *Courtesy of Charles C. Thomas, Publisher, Springfield.*

Hormones of Digestion

The digestive process is coordinated by a family of hormones produced by the body's most diffuse endocrine system, the gastrointestinal tract. These hormones are examples of the many substances produced by the vertebrate body that have hormonal function, yet are not necessarily produced by discrete endocrine glands. Because of their diffuse origins, the gastrointestinal (GI) hormones have been difficult to isolate and study; only recently have they been researched in depth.

Among the principal GI hormones are gastrin, cholecystokinin (CCK), and secretin (Figure 12-17). **Gastrin** is a small polypeptide hormone produced by endocrine cells in the pyloric portion of the stomach. Gastrin is secreted when protein food enters the stomach. Its main actions are to stimulate hydrochloric acid secretion and to increase gastric motility. Gastrin is an unusual hormone in that it exerts its action on the same organ from which it is secreted. **CCK (cholecystokinin)** is also a polypeptide hormone. It bears a striking structural resemblance to gastrin, suggesting that the two arose by duplication of ancestral genes. CCK has at least three distinct functions. It stimulates gallbladder contraction and thus increases the flow of bile salts into the intestine; it stimulates an enzyme-rich secretion from the pancreas; and it acts on the brain to contribute a feeling of well-being after a meal. The principal action of **secretin,** the first hormone to be discovered (see the opening essay on p. 266), is to stimulate the release of an alkaline pancreatic fluid that neutralizes stomach acid as it enters the intestine. It also aids fat digestion by inhibiting gastric motility and increasing bile production.

Several other GI hormones have been isolated recently and their structures determined. All are peptides. It is now well established that several peptide hormones are present in both the GI tract and in the central nervous system. One of these is CCK, which has been found in high concentrations in the cerebral cortex of mammals. By providing a feeling of satiety after eating (mentioned previously), it may play some role in regulating appetite. Several other GI peptides appear to play neurotransmitter roles in the brain. This unexpected versatility has broadened our concept of hormones as molecules capable of functioning in several different ways.

Hormones of Vertebrate Reproduction

Hormonal Control of the Timing of Reproductive Cycles

From fish to mammals, reproduction in vertebrates is usually a seasonal or cyclic activity. Timing is crucial, because the young should appear when food is available and other environmental conditions are optimal for survival. Once set in motion by some environmental cue, such as seasonal change in temperature or

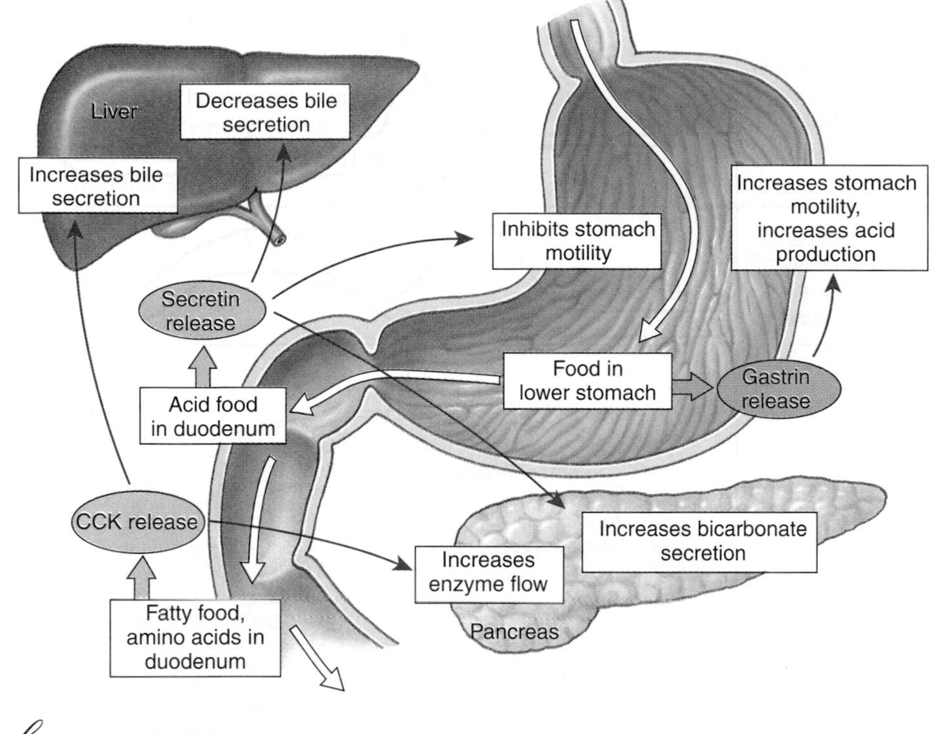

figure 12-17

Three hormones of digestion. Shown are the principal actions of the hormones gastrin, CCK (cholecystokinin), and secretin.

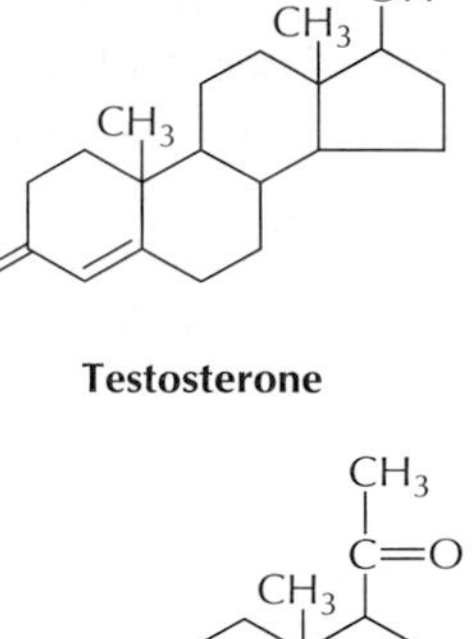

Testosterone

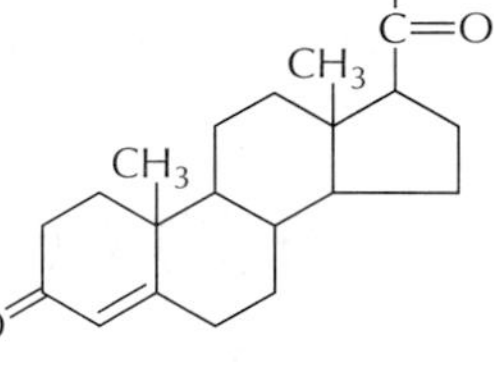

Progesterone

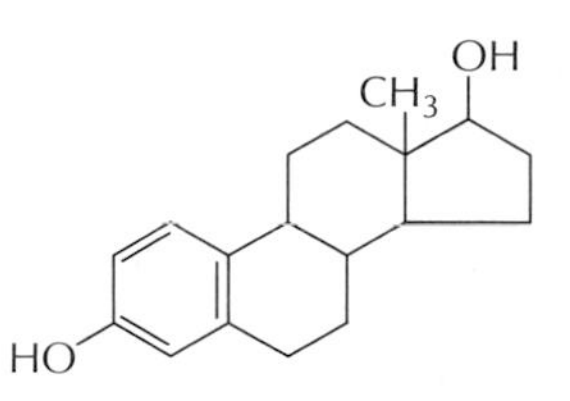

Estradiol

figure 12-18

Sex hormones. These three sex hormones show the basic four-ring steroid structure. The female sex hormone estradiol (an estrogen) is a C_{18} (18-carbon) steroid with an aromatic A ring (first ring to left). The male sex hormone testosterone is a C_{19} steroid with a carbonyl group (C=O) on the A ring. The female pregnancy hormone progesterone is a C_{21} steroid, also bearing a carbonyl group on the A ring.

photoperiod, or some social force, the sexual reproductive process is controlled by hormones. Hormones of the anterior pituitary gland link the neurosecretory centers of the brain to the endocrine tissues of the gonads (Table 12-1). This delicately balanced hormonal system controls the development of the gonads, accessory sex structures, and secondary sexual characteristics (see the following text) as well as the timing of reproduction.

The cyclic reproductive patterns of mammals are of two types: the **estrous cycle,** characteristic of most mammals, and the **menstrual cycle,** characteristic only of the anthropoid primates (monkeys, apes, and humans). The two cycles differ in two important ways. First, in the estrous cycle the female is receptive to the male only during brief periods of **estrus,** or "heat," whereas in the menstrual cycle receptivity may occur throughout the cycle. Second, the menstrual cycle, but not the estrous cycle, ends with collapse and discharge of the inner portion of the endometrium (uterine lining). In the estrous animal, each cycle ends with the uterine lining simply reverting to its original state, without the discharge characteristic of the menstrual cycle.

The Gonadal Steroids and Their Control

The ovaries produce two kinds of steroid sex hormones—**estrogens** and **progesterone** (Figure 12-18). Estrogens are responsible for development of the female accessory sex structures (oviducts, uterus, and vagina) and for stimulating female reproductive activity. The secondary sex characters, that is, characteristics that are not primarily involved in the formation and delivery of ova or sperm but that are essential for the behavioral and functional success of reproduction, are also controlled or maintained by estrogens. The secondary sex characters include characteristics such as distinctive skin or feather coloration, bone development, body size, and, in mammals, the initial development of the mammary glands. In mammals, progesterone is responsible for preparing the uterus to receive the developing embryo. These hormones are controlled by the **pituitary gonadotropins,** follicle-stimulating hormone (FSH), and luteinizing hormone (LH) (Figure 12-19 and Table 12-1). The gonadotropins are in turn governed by the **gonadotropin-releasing hormone (GnRH)** produced by neurosecretory centers in the hypothalamus (Table 12-1). Through this control system environmental factors such as light, nutrition, and stress may influence reproductive cycles.

The male sex hormone **testosterone** (Figure 12-18) is produced by the **interstitial cells** of the testes. Testosterone is necessary for growth and development of the male accessory

Animal Behavior

chapter | thirteen

The Lengthening Shadow of One Man

For as long as people have walked the earth, their lives have been touched by, indeed interwoven with, the lives of other animals. They hunted and fished them, domesticated them, ate them and were eaten by them, made pets of them, revered them, hated and feared them, immortalized them in art, song, and verse, fought them, and loved them. The very survival of ancient people depended on knowledge of wild animals. To stalk animals, people had to know their habits and behaviors. As the hunting societies of primitive people gave way to agricultural civilizations, an awareness was retained of the interrelationship with other animals, and the need to understand their behaviors increased.

This fascination with animals is still evident today. Zoos and public aquaria attract more visitors than ever; wildlife television shows are increasingly popular; game-watching safaris to Africa constitute a thriving enterprise; and millions of pet animals share the cities with us—more than a half million pet dogs live in New York City alone. Although people have always been interested in the behavior of animals, the science of animal behavior is a newcomer to biology. Charles Darwin, with the uncanny insight of genius, prepared for the reception of animal behavior by showing how natural selection would favor specialized behavioral patterns for survival. Darwin's pioneering book, *The Expression of the Emotions of Man and Animals,* published in 1872, mapped a strategy for behavioral research still in use today. However science in 1872 was unprepared for Darwin's central insight that behavioral patterns, no less than bodily structures, are selected and have evolutionary histories. Another 60 years would pass before such concepts would begin to flourish within behavioral science.

It was Ralph Waldo Emerson who said that an institution is the lengthening shadow of one man. For Charles Darwin the shadow is long indeed, for he brought into being entire fields of knowledge, such as evolution, ecology, and finally, after a long gestation, animal behavior. Above all, he altered the way we think about ourselves, the earth we inhabit, and the animals that share it with us.

In 1973, the Nobel Prize in Physiology or Medicine was awarded to three pioneering zoologists, Karl von Frisch, Konrad Lorenz, and Niko Tinbergen (Figure 13-1). The citation stated that these three were the principal architects of the new science of **ethology,** the scientific study of animal behavior, particularly under natural conditions. It was the first time any contributor to the behavioral sciences was so honored, and it meant that the discipline of animal behavior, which has its roots in the work of Charles Darwin, had arrived.

A

B

C

figure 13-1

Pioneers of the science of ethology. **A,** Konrad Lorenz (1903–1989). **B,** Karl von Frisch (1886–1982). **C,** Niko Tinbergen (1907–1988).

The Science of Animal Behavior

Behavioral biologists have traditionally asked two kinds of questions about behavior: *how* animals behave and *why* they behave as they do. "How" questions are concerned with immediate or **proximate causation** (p. 4). For example, a biologist might wish to explain the singing of a male white-throated sparrow in the spring in terms of hormonal or neural mechanisms. Such physiological causes of behavior, that is, the mechanisms the animal uses in its behavior, are proximate factors. Alternatively, a biologist might ask what function singing serves the sparrow, and then seek to understand those events in the ancestry of birds that led to springtime singing. These are "why" questions that focus on **ultimate causation** (p. 5), that is, the evolutionary origin and purpose of a behavior. These are really independent approaches to behavior, because understanding *how* the sparrow sings does not depend on what function singing serves, and vice versa. Students of animal behavior consider this distinction significant. Studies of proximate and ultimate causation are both important, but each may be of limited value in understanding the other.

The study of animal behavior has arisen from several different historical roots, and there is no universally accepted term for the whole subject. Today we can recognize three different experimental approaches to the study of animal behavior: comparative psychology, ethology, and sociobiology. **Comparative psychology** emerged from efforts to find general laws of behavior that would apply to many species, and preferably to humans as well. Early research that depended heavily on inference was later replaced by replicable experimental approaches that concentrated on a few species, particularly white rats, pigeons, dogs, and occasionally primates. Following criticisms that the discipline lacked an evolutionary perspective and focused too narrowly on the white rat as a model for other organisms, many comparative psychologists developed more truly comparative investigations, some of these conducted in the field.

The aim of the second approach, **ethology,** has been to describe the behavior of an animal in its *natural habitat.* Most ethologists have been zoologists. Their laboratory has been the out-of-doors, and early ethologists gathered their data by field observation. They also conducted experiments, often with nature providing the variables, but increasingly ethologists have manipulated the variables for their own purposes by using animal models, playing recordings of animal vocalizations and altering the habitat. Modern ethologists also conduct many experiments in the laboratory where they can test their predictions under closely controlled conditions. However, ethologists usually take pains to compare laboratory observations with observations of free-ranging animals in undisturbed natural environments.

Ethology emphasizes the importance of ultimate factors affecting behavior. One of the great contributions of von mantle cavity and open to the outside by a siphon (Figure 13-8). If one prods the siphon, *Aplysia* withdraws its siphon and gills and folds them up in the mantle cavity. This simple protective response, called the gill-withdrawal reflex, will be repeated when *Aplysia* extends its siphon again. But if the siphon is touched repeatedly, *Aplysia* decreases the gill-withdrawal response and finally comes to ignore the stimulus altogether. This behavioral modification illustrates a widespread form of learning called **habituation.** If now *Aplysia* is given a noxious stimulus (for example, an electric shock) to the head at the same time the siphon is touched, it becomes **sensitized** to the stimulus and withdraws its gills as completely as it did before habituation occurred. Sensitization, then, can reverse any previous habituation.

The mechanisms of habituation and sensitization in *Aplysia* are known because these behaviors constitute a rare instance in which the nervous pathways involved have been completely revealed. Receptors in the siphon are connected through sensory neurons (black pathways in Figure 13-8) to motor neurons (blue pathway in Figure 13-8) that control the gill-withdrawal muscles and muscles of the mantle cavity. Kandel found that repeated stimulation of the siphon caused a decline in the release of synaptic transmitter from the sensory neurons. The sensory neurons continue to fire when the siphon is probed but, with less neurotransmitter being released into the synapse, the system becomes less responsive.

Sensitization requires the action of a different kind of neuron called a facilitating interneuron. These interneurons make connections between sensory neurons in the animal's head and motor neurons that control muscles of the gill and mantle (see Figure 13-8). When sensory neurons in the head are stimulated by an electric shock, they fire the facilitating interneurons, which end on the synaptic terminals of the sensory neurons (red pathways in Figure 13-8). These endings in turn cause an *increase* in the amount of transmitter released by the siphon sensory neurons. This transmitter increases the state of excitation in the excitatory interneurons and motor neurons leading to the muscles of the gill and mantle. The motor neurons now fire more readily than before. The system is now sensitized because any stimulus to the siphon will produce a strong gill-withdrawal response.

The studies of habituation and sensitization on *Aplysia* by Kandel and his colleagues represent the most complete account yet made of the neural and molecular mechanisms involved with learning. Kandel's studies indicate that the strengthening or weakening of the gill-withdrawal reflex involves changes in levels of transmitter in existing synapses. However, we know that some cases of more complex kinds of learning involve the formation of new neural pathways and connections, as well as changes in existing circuits.

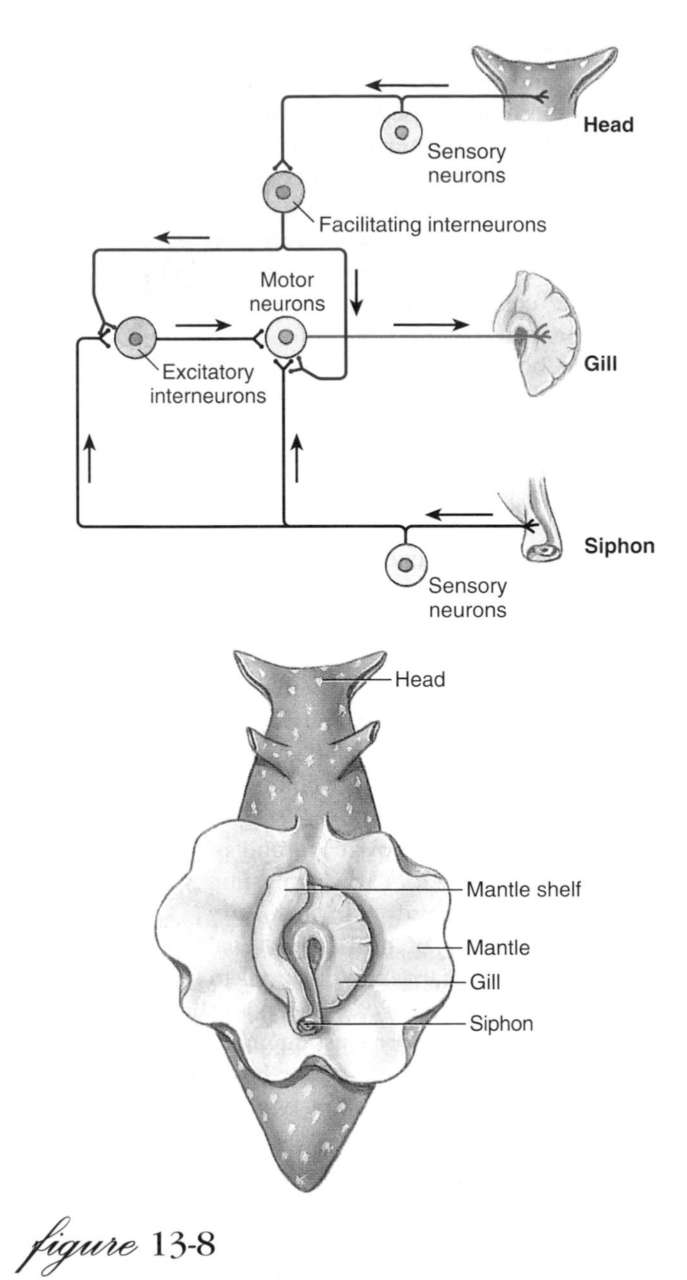

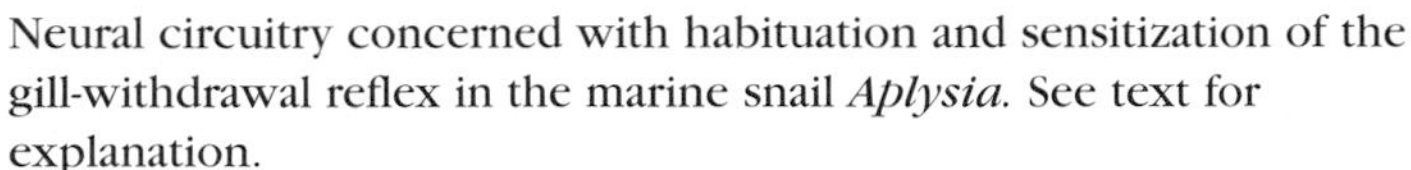

figure 13-8

Neural circuitry concerned with habituation and sensitization of the gill-withdrawal reflex in the marine snail *Aplysia.* See text for explanation.

Imprinting

Another kind of learned behavior is **imprinting,** the imposition of a stable behavior in a young animal by exposure to particular stimuli during a critical period in the animal's development. As soon as a newly hatched gosling or duckling is strong enough to walk, it follows its mother away from the nest. After it has followed the mother for some time it follows no other animal (Figure 13-9). But, if the eggs are hatched in an incubator or if the mother is separated from the eggs as they hatch, the goslings follow the first large object they see. As they grow, the young geese prefer the artificial "mother" to anything else, including their true mother. The goslings are said to be imprinted on the artificial mother.

Imprinting was observed at least as early as the first century A.D. when the Roman naturalist Pliny the Elder wrote of "a goose which followed Lacydes as faithfully as a dog." Konrad Lorenz was the first to study imprinting objectively and sys-

figure 13-9
Canada goose, *Branta canadensis,* with her imprinted young.

tematically. When Lorenz hand-reared goslings, they formed an immediate and permanent attachment to him and waddled after him wherever he went (see Figure 13-1A). They could no longer be induced to follow their own mother or another human being. Lorenz found that the imprinting period is confined to a brief sensitive period in the individual's early life and that once established the imprinted bond usually is retained for life.

What imprinting shows is that the brain of the goose (or the brain of numerous other birds and mammals that show imprinting-like behavior) accommodates the imprinting experience. Natural selection favors evolution of a brain that imprints in this way, in which following the mother and obeying her commands are important for survival. The fact that a gosling can be made to imprint to a mechanical toy duck or a person under artificial conditions is a cost to the system that can be tolerated because goslings seldom encounter these stimuli in their natural environment. The disadvantages of the system's simplicity are outweighed by the advantages of its reliability.

We will cite one final example to complete our consideration of learning. The males of many species of birds have characteristic territorial songs that identify the singers to the other birds and announce territorial rights to other males of that species. Like many other songbirds, the male white-crowned sparrow must learn the song of its species by hearing the song of its father. If the sparrow is hand-reared in acoustic isolation in the laboratory, it develops an abnormal song (Figure 13-10). But if the isolated bird is allowed to hear recordings of normal white-crowned sparrow songs during a critical period of 10 to 50 days after hatching, it learns to sing normally. It even imitates the local dialect it hears.

It might appear from this result that characteristics of the song are determined by learning alone. However, if during the critical learning period, the isolated male white-crowned sparrow is played a recording of another species of sparrow, even a closely related one, it does not learn the song. It learns only the song appropriate to its own species. Thus, although the song must be learned, the brain is constrained to recognize and to learn vocalizations produced by males of its species alone. Learning the wrong song would result in behavioral chaos, and natural selection favors a system that eliminates such errors.

Social Behavior

When we think of "social" animals we are likely to think of highly structured honeybee colonies, herds of antelope grazing on the African plains (Figure 13-11), schools of herring, or flocks of starlings. But social behavior of animals *of the same species* living together is by no means limited to such obvious examples in which individuals influence one another.

In the broad sense, any interaction resulting from the response of one animal to another of the same species represents social behavior. Even a pair of rival males preparing for a fight over possession of a female display a social interaction, despite our perceptual bias as people that might encourage us to label it antisocial. Social aggregations are only one kind of social behavior, and indeed not all aggregations of animals are social.

Clouds of moths attracted to a light at night, barnacles attracted to a common float, or trout gathering in the coolest pool of a stream are groupings of animals responding to environmental signals. Social aggregations, on the other hand, depend on signals from the animals themselves. They remain together and do things together by influencing one another.

Not all animals showing sociality are social to the same degree. While all sexually reproducing species must at least cooperate enough to achieve fertilization, among some animals breeding is about the only adult sociality to occur. Alternatively, swans, geese, albatrosses, and beavers, to name just a few, form strong monogamous bonds that last a lifetime. The most persistent social bonds usually form between mothers and their young, and these bonds for birds and mammals usually terminate at fledging or weaning.

Advantages of Sociality

Living together may be beneficial in many ways. One obvious benefit for social aggregations is defense, both passive and active, from predators. Musk-oxen that form a passive defensive

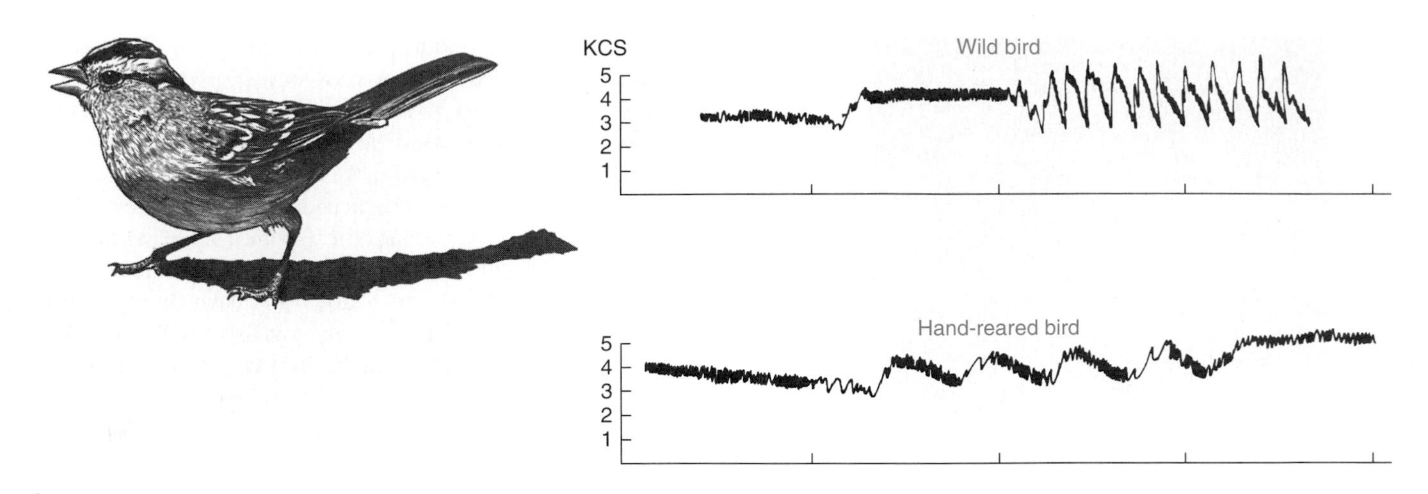

figure 13-10

Sound spectrograms of songs of white-crowned sparrows, *Zonotrichia leucophrys. Top,* natural songs of wild bird; *bottom,* abnormal song of isolated bird.

figure 13-11

Mixed herd of topi and common zebra grazing on the savanna of tropical Africa.

circle when threatened by a wolf pack are much less vulnerable than an individual facing the wolves alone.

As an example of active defense, a breeding colony of gulls, alerted by the alarm calls of a few, attack predators *en masse;* this collective attack is certain to discourage a predator more effectively than individual attacks. Members of a town of prairie dogs, although divided into social units called coteries, cooperate by warning each other with a special bark when danger threatens. Thus every individual in a social organization benefits from the eyes, ears, and noses of all other members of the group.

Sociality offers several benefits to animal reproduction. It facilitates encounters between males and females, which, for solitary animals, may consume much time and energy. Sociality also helps synchronize reproductive behavior through the mutual stimulation that individuals have on one another. Among colonial birds the sounds and displays of courting individuals set in motion prereproductive endocrine changes in other individuals. Because there is more social stimulation, large colonies of gulls produce more young per nest than do small colonies. Furthermore, the parental care that social animals provide their offspring increases survival of the brood (Figure 13-12). Social living provides opportunities for individuals to give aid and to share food with young other than their own. Such interactions within a social network have produced some intricate cooperative behavior among parents, their young, and their kin.

Of the many other advantages of social organization noted by ethologists, we will mention only a few in this brief treatment: cooperation in hunting for food; huddling for mutual protection from severe weather; opportunities for division of labor, which is especially well developed in the social insects with their caste systems; and the potential for learning and transmitting useful information through the society. We discuss this last advantage of social organization in the following example.

Observers of a seminatural colony of macaque monkeys in Japan recount an interesting example of acquiring and passing tradition in a society. The macaques were provisioned with sweet potatoes and wheat at a feeding station on the beach of an island colony. One day a young female named Imo was ob-

figure 13-12

A baby orangutan clings tightly as its mother swings from branches. Parental care, strong at first, gradually declines as the infant ages. Eventually, the mother-offspring contact is broken, usually by the infant, which wanders off to explore.

figure 13-13

Japanese macaque washing sweet potatoes. The tradition began when a young female named Imo began washing sand from the potatoes before eating them. Younger members of the troop quickly imitated the behavior.

served washing the sand off a sweet potato in seawater. The behavior was quickly imitated by Imo's playmates and later by Imo's mother. Still later, when the young members of the troop became mothers, they waded into the sea to wash their potatoes; their offspring imitated them without hesitation. The tradition was firmly established in the troop (Figure 13-13).

Some years later, Imo, an adult, discovered that she could separate wheat from sand by tossing a handful of sandy wheat in the water; allowing the sand to sink, she could scoop up the floating wheat to eat. Again, within a few years, wheat-sifting became a tradition in the troop.

Imo's peers and social inferiors copied her innovations most readily. The adult males, her superiors in the social hierarchy, would not adopt the practice but continued laboriously to pick wet sand grains off their sweet potatoes and scout the beach for single grains of wheat.

Social living also has some disadvantages as compared with a solitary existence for some animals. Species that survive by camouflage from potential predators profit by being dispersed. Large predators benefit from a solitary existence for a different reason, their requirement for a large supply of prey. Thus there is no overriding adaptive advantage to sociality that inevitably selects against the solitary way of life. It depends on the ecological situation.

On the Galápagos Islands, hunting by humans in the last century had so greatly thinned the giant tortoise population on one island that the few surviving males and females seldom, if ever, met. Lichens grew on the females' backs because there were no males to scrub them off during mating! Research personnel saved the tortoise from inevitable extinction by collecting them together in a pen, where they began to reproduce.

Aggression and Dominance

Many animal species are social because of the numerous benefits that sociality offers. Sociality requires cooperation. At the same time animals, like governments, tend to look out for their own interests. In short, they are in competition with one another because of limitations in the common resources that all require for life. Animals may compete for food, water, sexual mates, or shelter when such requirements are limited in quantity and are therefore worth a fight.

Much of what animals do to resolve competition is called **aggression,** which we define as an offensive physical action, or threat, to force others to abandon something they own or might attain. Many ethologists consider aggression part of a somewhat more inclusive interaction called **agonistic** (Gr. contest) **behavior.** Agonistic behavior refers to any activity related to fighting, whether it be aggression, defense, submission, or retreat.

Reproduction and Development

chapter | fourteen

"Omne vivum ex ovo"

In 1651, late in a long life, William Harvey, the English physiologist who earlier had ushered in experimental physiology by explaining the circuit of the blood, published a treatise on reproduction. He asserted that all life developed from the egg—*omne vivum ex ovo.* This insight was particularly impressive since Harvey had no means for visualizing the eggs of many animals, in particular the microscopic mammalian egg, which is no larger than a speck of dust to the unaided eye. Further, argued Harvey, the egg is launched into its developmental course by some influence from the semen, a conclusion that was either remarkably perceptive or a lucky guess, since sperm also were invisible to Harvey. Such ideas differed sharply from existing notions of biogenesis, which saw life springing from many sources of which eggs were but one. Harvey was describing characteristics of sexual reproduction in which two parents, male and female, must come together physically to ensure fusion of gametes from each.

Despite the importance of Harvey's aphorism that all life arises from eggs, it was too sweeping to be wholly correct. Life springs from the reproduction of preexisting life, and reproduction may not be bound up in eggs and sperm. Nonsexual reproduction, the creation of new, genetically identical individuals by budding or fragmentation or fission from a single parent, is common, indeed characteristic, among some phyla. Most animals have found sex to be the winning strategy, probably because sexual reproduction promotes diversity, enhancing long-term survival of the lineage in a world of perpetual change.

Reproduction is one of the ubiquitous miracles of life. Evolution is inextricably linked to reproduction, because the ceaseless replacement of aging predecessors with new life gives animals the means to respond and evolve in a changing environment as the earth itself has changed over the ages. In the first part of this chapter we will distinguish asexual and sexual reproduction and explore the reasons why, for multicellular animals at least, sexual reproduction appears to offer important advantages over asexual. We then examine the components of reproductive systems in sexual animals.

In the second part of the chapter we explore the developmental process and follow the fertilized egg as it is guided by the genetic material from its parents to develop into a functioning and reproducing adult.

Nature of the Reproductive Process

The two modes of reproduction are asexual and sexual. In **asexual** reproduction (Figure 14-1) an animal has only one parent and no special reproductive organs or cells. Each organism is capable of producing genetically identical copies of itself as soon as it becomes an adult. The production of copies is marvelously simple, direct, and typically rapid. **Sexual** reproduction (Figure 14-2) as a rule involves two parents, each of which contributes special **sex cells** (also called germ cells, or **gametes**) that in union (fertilization) develop into a new individual. The **zygote** formed from this union receives genetic material from *both* parents and accordingly is different from both. The combination of genes produces a genetically unique individual, still bearing the characteristics of the species but also bearing traits that make it different from its parents. Sexual reproduction, by recombining the parental characters, tends to multiply variations and makes possible a richer and more diversified evolution.

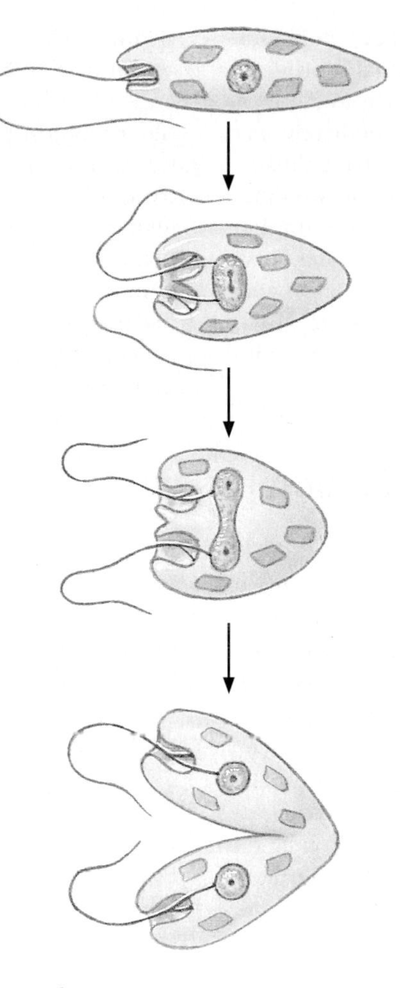

figure 14-1

Examples of asexual reproduction in animals. **A,** Binary fission in *Euglena,* a flagellate protozoan, results in two individuals. **B,** Budding, a simple form of asexual reproduction as shown in a hydra, a radiate animal. The buds eventually detach themselves and grow into fully formed individuals.

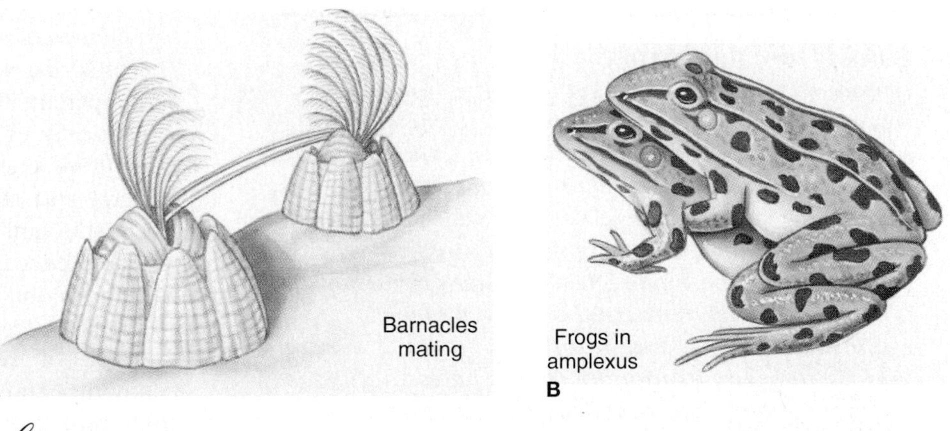

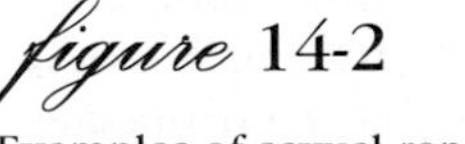

figure 14-2

Examples of sexual reproduction in animals. **A,** Barnacles reproduce sexually, but are hermaphroditic, with each individual bearing both male and female organs. Each barnacle possesses a pair of enormously elongated penises—an obvious advantage to a sessile animal—that can be extended many times the length of the body to inseminate another barnacle some distance away. The partner may reciprocate with its own penises. **B,** Frogs, here in mating position (amplexus), represent biparental reproduction, the most common form of sexual reproduction involving separate male and female individuals.

Mechanisms for interchange of genes between individuals are more limited in organisms with only asexual reproduction. Of course, in asexual organisms that are haploid (bear only one set of genes, p. 58), mutations are immediately expressed and evolution can proceed quickly. In sexual animals, on the other hand, a gene mutation is often not expressed immediately, since it may be masked by its normal partner on the homologous chromosome. (Homologous chromosomes, discussed on p. 58, are those that pair during meiosis and have genes controlling the same characteristics.) There is only a remote chance that both members of a gene pair will mutate in the same way at the same moment.

Asexual Reproduction: Reproduction without Gametes

Asexual reproduction (Figure 14-1) is the production of individuals without gametes, that is, eggs or sperm. It includes a number of distinct processes to be described below, all without involving sex or a second parent. The offspring produced by asexual reproduction of an individual all have the same genotype and are called **clones.**

Asexual reproduction appears in bacteria and protists and in many invertebrate phyla, such as cnidarians, bryozoans, annelids, echinoderms, and hemichordates. In animal phyla in which asexual reproduction occurs, most members employ sexual reproduction as well. In these groups, asexual reproduction ensures rapid increase in numbers when differentiation of the organism has not advanced to the point of forming gametes. Asexual reproduction is absent among the vertebrates (although some forms of parthenogenesis have been interpreted as asexual by some authors; see p. 312).

It would be a mistake to conclude that asexual reproduction is in any way a "defective" form of reproduction relegated to the minute forms of life that have yet to discover the joys of sex. Given the facts of their abundance, that they have persisted on earth for 3.5 billion years, and that they form the roots of the food chain on which all more complex forms depend, the single-celled asexual organisms are both resoundingly successful and supremely important. For these forms, the advantages of asexual reproduction are its rapidity (many bacteria divide every half-hour) and simplicity (no sex cells to produce and no time and energy expended in finding a mate).

The basic forms of asexual reproduction are fission (binary and multiple), budding, gemmulation, and fragmentation. **Binary fission** is common among bacteria and protozoa (Figure 14-1A). In binary fission the body of the parent divides by mitotic cell division into two approximately equal parts, each of which grows into an individual similar to the parent. Binary fission may be lengthwise, as in flagellate protozoa, or transverse, as in ciliate protozoa. In **multiple fission** the nucleus divides repeatedly before division of the cytoplasm, giving rise to many daughter cells simultaneously. Spore formation, called sporogony, is a form of multiple fission that is common among some parasitic protozoa, for example, the malarial parasite (p. 374).

Budding is an unequal division of the organism. The new individual arises as an outgrowth (bud) from the parent, develops organs like those of the parent, and then detaches itself. Failure of buds to detach leads to the formation of a colony. Budding occurs in several animal phyla and is especially prominent in cnidarians (Figure 14-1B).

Gemmulation is the formation of a new individual from an aggregation of cells surrounded by a resistant capsule, called a gemmule. In many freshwater sponges gemmules develop in the fall and survive the winter in the dried or frozen body of the parent. In the spring, the enclosed cells become active, emerge from the capsule, and grow into a new sponge.

In **fragmentation** a multicellular animal breaks into two or more parts, with each fragment capable of becoming a complete individual. Many invertebrates can reproduce asexually by simply breaking in two and then regenerating the missing parts of the fragments.

Sexual Reproduction: Reproduction with Gametes

The essential feature of sexual reproduction is the *production of offspring formed by the union of gametes from two genetically different parents* (Figures 14-2 and 14-3). The offspring will thus have a new genotype different from either parent. The individuals sharing parenthood are characteristically of different **sexes,** male and female (there are exceptions among sexually reproducing organisms, such as bacteria and some protozoa in which sexes are lacking). The distinction between male and female is based, not on any differences in parental size or appearance, but on the size and mobility of the gametes (sex cells) they produce. The **ovum** (egg) is produced by the female. Ova are large (because of stored yolk to sustain early development), nonmotile, and produced in relatively small numbers. The **spermatozoon** (sperm) is produced by the male. Sperm are small, motile, and produced in enormous numbers. Each is a stripped-down package of highly condensed genetic material designed for the single mission of finding and fertilizing the egg.

There is another crucial event that distinguishes sexual from asexual reproduction: **meiosis,** a distinctive type of gamete-producing nuclear division. As described earlier

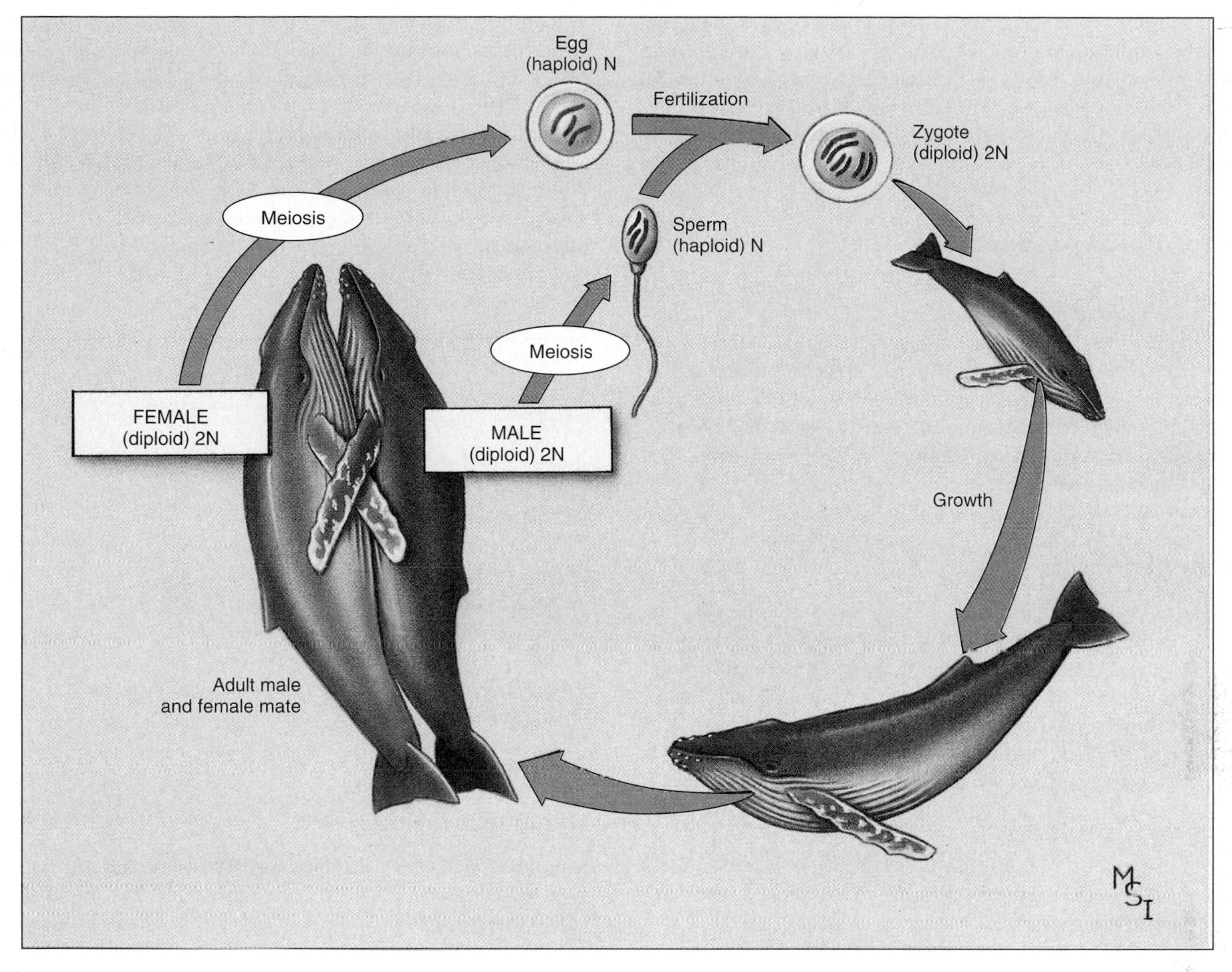

figure 14-3

A sexual life cycle. The life cycle begins with haploid germ cells, formed by meiosis, combining to form a diploid zygote, which grows by mitosis to an adult. Most of the life cycle is spent as a diploid organism.

(p. 58), meiosis differs from ordinary cell division (mitosis) in being a double division. The chromosomes split once, but the cell divides *twice,* producing four cells, each with half the original number of chromosomes (the haploid number). Meiosis is followed by **fertilization** in which two haploid gametes are combined to restore the normal (diploid) chromosomal number of the species.

The new cell (zygote), which now begins to divide by mitosis, has equal numbers of chromosomes from each parent and accordingly is different from each. It is a unique individual bearing a recombination of parental characteristics. Genetic recombination is the great strength of sexual reproduction that keeps feeding new genotypes into the population.

Many protozoa reproduce both sexually and asexually. When sexual reproduction does occur, it may or may not involve male and female gametes. Sometimes two mature sexual parents merely join together to exchange nuclear material or merge cytoplasm (conjugation, p. 376). Sexes cannot be distinguished in these cases.

The male-female distinction is more clearly evident in the metazoa. Organs that produce germ cells are known as **gonads.** The gonad that produces sperm is called the **testis** (see Figure 14-6) and that which forms eggs, the **ovary** (see Figure 14-7). The gonads represent the **primary sex organs,** the only sex organs found in certain groups of animals. Most metazoa, however, have various **accessory sex organs** (such as penis, vagina, oviducts, and uterus) that transfer and receive sex cells. In the primary sex organs the sex cells undergo many complicated changes during their development, the details of which are described later. In our present discussion we will distinguish biparental reproduction from two alternatives: parthenogenesis and hermaphroditism.

Biparental Reproduction

Biparental (or bisexual) reproduction is the common method of sexual reproduction with which we are familiar, involving separate and distinct male and female individuals (Figure 14-3). Each has its own reproductive system and produces only one kind of sex cell, spermatozoon or ovum, but never both. Nearly all vertebrates and many invertebrates have separate sexes, and such a condition is called **dioecious** (Gr. *di-*, two; + *oikos*, house).

Parthenogenesis

Parthenogenesis ("virgin origin") is the development of an embryo from an unfertilized egg or, if a spermatozoon does penetrate the egg, there is no union of male and female pronuclei. There are many patterns of parthenogenesis. In one type, called **ameiotic parthenogenesis,** no meiosis occurs, and the egg is formed by mitotic cell division. This "asexual" form of parthenogenesis is known to occur in some species of flatworms, rotifers, crustaceans, insects, and probably others. In these cases, the offspring are clones of the parent because, without meiosis, there is no reshuffling of genes between chromosomes (this important event during meiosis, called crossing over, is described on p. 66).

In **meiotic parthenogenesis** a haploid ovum is formed by meiosis, and it may or may not be activated by the influence of a male. For example, in some species of fishes, the female may be inseminated by the male of the same or related species, but the sperm serves only to activate the egg; the male's genome is rejected before it can penetrate the egg. In several species of flatworms, rotifers, annelids, mites, and insects, the haploid egg begins development spontaneously; no males are required to stimulate activation of the ovum. The diploid condition is restored by chromosomal duplication. A variant of this type of parthenogenesis occurs in many bees, wasps, and ants. In honeybees, for example, the queen bee can either fertilize the eggs as she lays them or allow them to pass unfertilized. The fertilized eggs become diploid females (queens or workers), and the unfertilized eggs develop parthenogenetically to become haploid males (drones); this type of sex determination is known as **haplodiploidy.** In some animals meiosis may be so severely modified that the offspring are clones of the parent. Certain populations of whiptail lizards of the American Southwest are clones consisting solely of females.[1]

Parthenogenesis is surprisingly widespread in animals. It is an abbreviation of the usual steps required of biparental reproduction. It may have evolved to avoid the problem—which may be great in some animals—of bringing together males and females at the right moment for successful fertilization. The disadvantage of parthenogenesis is that if the environment should suddenly change, as it often does, parthenogenetic species have limited capacity to shift gene combinations to adapt to the new conditions. Biparental species, by recombining parental characteristics, have a better chance of producing variant offspring that can utilize new environments.

From time to time claims arise that spontaneous parthenogenetic development to term has occurred in humans. A British investigation of about 100 cases in which the mother denied having had intercourse revealed that in nearly every case the child possessed characters not present in the mother, and consequently must have had a father. Nevertheless, mammalian eggs very rarely will spontaneously start developing into embryos without fertilization. A remarkable instance of parthenogenetic development has been found in turkeys in which certain strains, selected for their ability to develop without sperm, grow to reproducing adults.

Hermaphroditism

Animals that have both male and female organs in the same individual are called hermaphrodites, and the condition is called hermaphroditism (from a combination of the names of the Greek god Hermes and goddess Aphrodite). In contrast to the dioecious state of separate sexes, hermaphrodites are **monoecious** (Gr. *monos*, single, + *oikos*, house), meaning that both male and female organs are in the same organism. Many sessile, burrowing, or endoparasitic invertebrate animals (for example, most flatworms, some hydroids and annelids, and all barnacles and pulmonate snails) and a few vertebrates (some fishes), are hermaphroditic. Some hermaphrodites fertilize themselves, but most avoid self-fertilization by exchanging germ cells with another member of the same species. An advantage is that with every individual producing eggs, a hermaphroditic species could potentially produce twice as many offspring as could a biparental species in which half the individuals are nonproductive males. In some fishes, called sequential hermaphrodites, the animal experiences a genetically programmed sex change during its life. Many species of reef fishes begin life as either a female or a male (depending on the species) but later become the opposite sex.

[1]Cole, C. J., 1984. Unisexual lizards. Sci. Am. **250:**94–100 (Jan.)

What Good Is Sex?

The question "What good is sex?" appears to have an easy answer: it serves the purpose of reproduction. But if we rephrase the question to ask, "Why do so many animals reproduce sexually rather than asexually?" the answer is not so apparent. Because sexual reproduction is nearly universal among animals, it might be inferred that it must be highly advantageous. Yet it is easier to list disadvantages to sex than advantages. Sexual reproduction is complicated, requires more time, and uses much more energy than asexual reproduction. Mating partners must come together and coordinate their activities to produce young. Many biologists believe that an even more troublesome problem is the "cost of meiosis." A female that reproduces asexually passes all of her genes to each offspring. But when she reproduces sexually the genome is divided during meiosis and only half her genes flow to each offspring. Another cost is wastage in production of males, many of whom fail to reproduce and thus consume resources that could be applied to the production of females. Whiptail lizards of the American Southwest offer a fascinating example of the potential advantage of parthenogenesis. When unisexual and bisexual (biparental) species of the same genus are reared under similar conditions in the laboratory, the population of the unisexual species grows more quickly because all of the unisexual lizards (all females) deposit eggs, whereas only 50% of the bisexual lizards do so (Figure 14-4).

Clearly, the costs of sexual reproduction are substantial. How are they offset? Biologists have disputed this question for years without producing an answer that satisfies everyone. Many biologists believe that sexual reproduction, with its breakup and recombination of genomes, keeps producing novel genotypes that *in times of environmental change* may survive and reproduce, whereas most others die. Variability, advocates of this viewpoint argue, is sexual reproduction's trump card.

But is variability worth the biological costs of sexual reproduction? The underlying problem keeps coming back: asexual organisms, because they can have more offspring in a given time, appear to be more fit in Darwinian terms. And yet most metazoan animals are determinedly committed to sexuality. Considerable evidence suggests that asexual reproduction is most successful in colonizing new environments. When habitats are empty what matters most is rapid reproduction; variability matters little. But as habitats become more crowded, competition between species for resources increases. Selection becomes more intense, and genetic variability—new genotypes produced by recombination in sexual reproduction—furnishes the diversity that permits the population to resist extinction. Therefore, on a geological timescale, asexual lineages, because of the lack of genetic flexibility, may be more prone to extinction than sexual lineages. Sexual reproduction is therefore favored by species selection (species selection is described on p. 108).

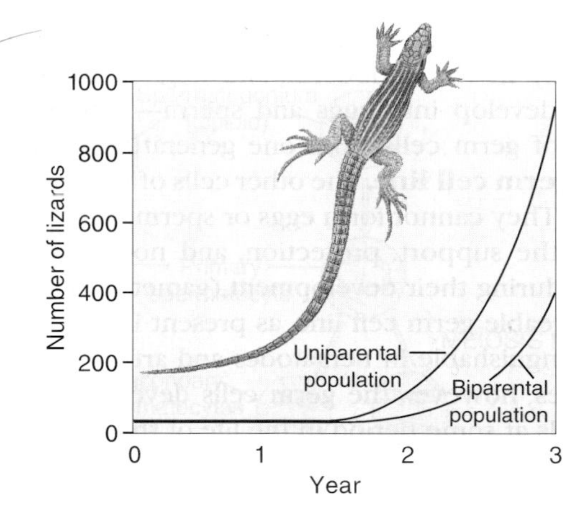

figure 14-4

Comparison of the growth of a population of uniparental whiptail lizards with a population of biparental lizards. Because all individuals of the uniparental population are females, all produce eggs, whereas only half the biparental population are egg-producing females. By the end of the third year the uniparental lizards are more than twice as numerous as the biparental ones.

Variety may make sexual reproduction a winning strategy for the unstable environment, but some biologists believe that for many vertebrates sexual reproduction is unnecessary and may even be maladaptive. In animals in which most of the young survive to reproductive age (humans, for example), there is no demand for novel recombinations to cope with changing habitats. One offspring appears as successful as the next in each habitat. Significantly, parthenogenesis has evolved in several species of fish and in a few amphibians and reptiles. Such species are exclusively parthenogenetic, suggesting that where it has been possible to overcome the numerous constraints to making the transition, biparental reproduction loses out.

There are many invertebrates that use both sexual and asexual reproduction, thus enjoying the advantages each has to offer.

Formation of Reproductive Cells

The vertebrate body has two basically different types of cells: the **somatic cells,** which are differentiated for specialized functions and die with the individual, and the **germ cells,** which form the gametes: eggs and sperm. Germ cells provide continuity of life between generations and ensure the species' survival. Germ cells, or their precursors, the **primordial germ**

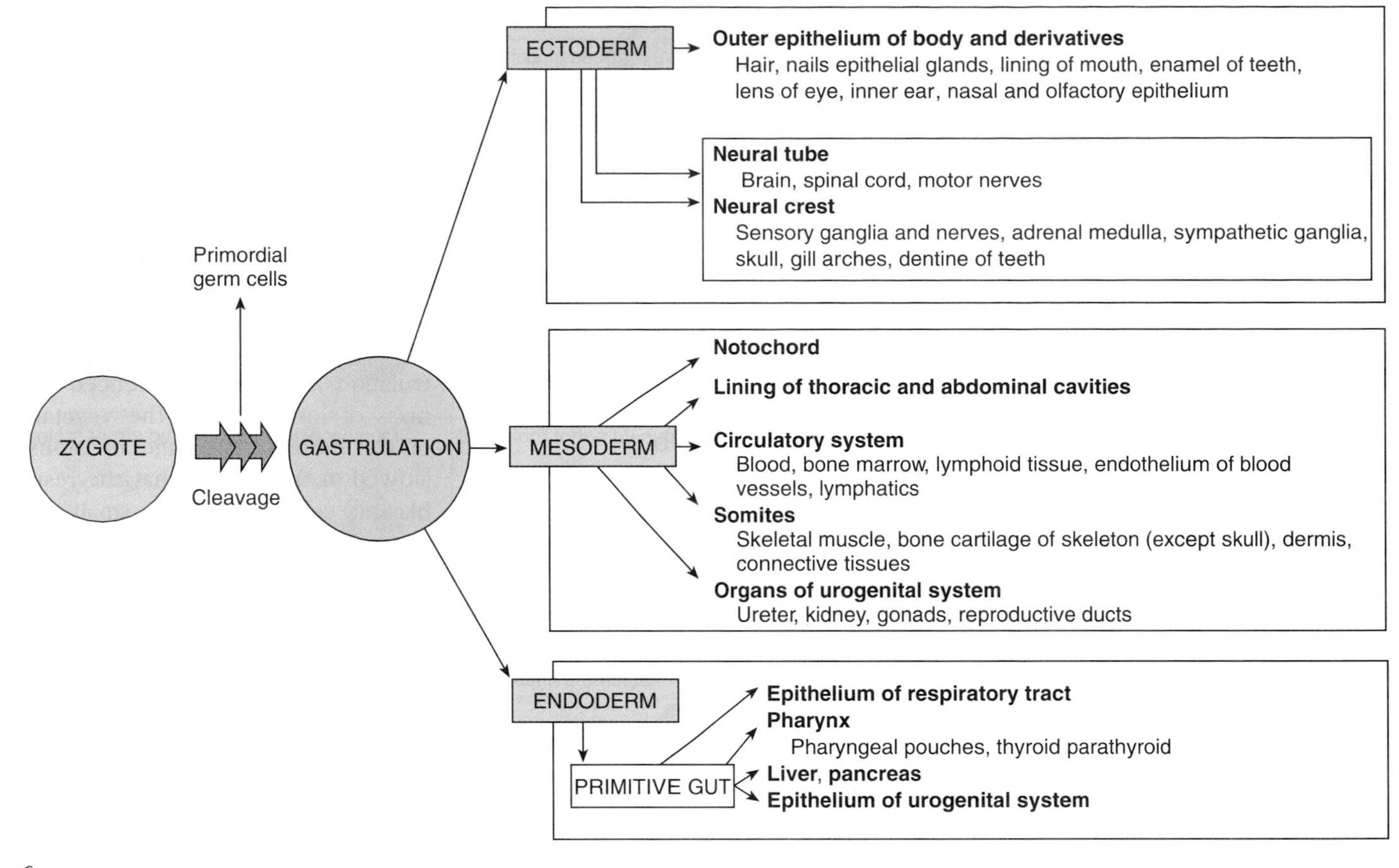

figure 14-19

Derivatives of the primary germ layers in mammals.

In the least derived of the true metazoa, the Cnidaria and the Ctenophora, only two germ layers are formed, the endoderm and ectoderm. These animals are **diploblastic.** In all more derived metazoa the mesoderm also appears, either from pouches of the archenteron or from other cells associated with endoderm formation. This three-layered condition is called **triploblastic.**

Formation of the Coelom

The coelom, or true body cavity that contains the viscera, may be formed by one of two methods—**schizocoelous** or **enterocoelous** (see Figure 14-17)—or by modification of these methods. (The two terms are descriptive, for *schizo* comes from the Greek *schizein,* to split; *entero* is from the Greek *enteron,* gut; and *coelous* comes from the Greek *koilos,* hollow or cavity.) In schizocoelous formation the coelom arises, as the word implies, from the splitting of mesodermal bands that originate from the blastopore region and grow between the ectoderm and endoderm; in enterocoelous formation the coelom comes from pouches of the archenteron, or primitive gut.

These two quite different origins for the coelom are another expression of the deuterostome-protostome dichotomy of bilateral animals. The coelom of protostome animals develops by the schizocoelous method. The deuterostomes primitively follow the enterocoelous plan. Vertebrates, however, are exceptions to this distinction because their coelom is formed by mesodermal splitting (schizocoelous). This is a derived condition that evolved in fishes to accommodate large stores of yolk during early development. In other respects vertebrates develop as deuterostomes, the division to which they are assigned.

Development of Systems and Organs

Derivatives of Ectoderm: Nervous System and Nerve Growth

During vertebrate gastrulation the three germ layers are formed. These layers differentiate, as we have seen, first into primordial cell masses and then into specific organs and tissues. During this process, cells become increasingly committed to specific directions of differentiation. The derivatives of the three germ layers are diagrammed in Figure 14-19.

The brain, spinal cord, and nearly all outer epithelial structures of the body develop from the primitive ectoderm. They are among the earliest organs to appear. Just above the notochord, the ectoderm thickens to form a **neural plate.** The edges of this plate rise up, fold, and join together at the top to create an elongated, hollow **neural tube.** The neural tube gives rise to most of the nervous system. Anteriorly it enlarges and differentiates into the brain and cranial nerves; posteriorly it forms the spinal cord and spinal motor nerves. Much of the rest of the peripheral nervous system is derived from the **neural crest cells,** which pinch off from the neural tube before it closes (Figure 14-20). Among the multitude of different cell types and structures that originate with the neural crest are portions of the cranial nerves, pigment cells, cartilage, and bone of most of the skull, including the jaws, ganglia of the autonomic nervous system, medulla of the adrenal gland, and contributions to several other endocrine glands. Neural crest tissue is unique to vertebrates and was probably of prime importance in the evolution of the vertebrate head and jaws.

How are the billions of nerve axons in the body formed? What directs their growth? Biologists were intrigued with these questions that seemed to have no easy solutions. Since a single nerve axon may be more than a meter in length (for example, motor nerves running from the spinal cord to the toes), it seemed impossible that a single cell could spin out so far. It was suggested that a nerve fiber grew from a series of preformed protoplasmic bridges along its route. The answer had to await the development of one of the most powerful tools available to biologists, the cell culture technique.

In 1907 embryologist Ross G. Harrison discovered that he could culture living neuroblasts (embryonic nerve cells) for weeks outside the body by placing them in a drop of frog lymph hung from the underside of a cover slip. Watching nerves grow for periods of days, he saw that each nerve fiber was the outgrowth of a single cell. As the fibers extended outward, materials for growth flowed down the axon center to the growing tip, where they were incorporated into new protoplasm.

The tissue culture technique developed by Ross G. Harrison is now used extensively by scientists in all fields of active biomedical research, not just by embryologists. The great impact of the technique has been felt only in recent years. Harrison was twice considered for the Nobel Prize (1917 and 1933), but he failed ever to receive the award because, ironically, the tissue culture method was then believed to be "of rather limited value."

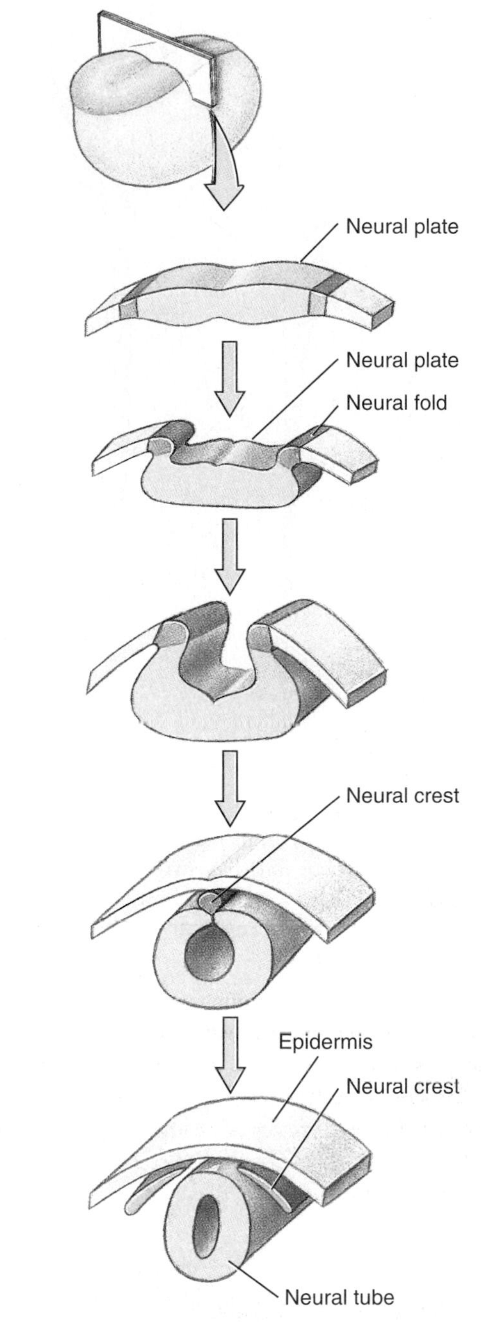

figure 14-20

Development of the neural tube and neural crest cells from the neural plate ectoderm.

The second question—what directs nerve growth—has taken longer to unravel. An idea held well into the 1940s was that nerve growth is a random, diffuse process. A major hypothesis proposed that the nervous system developed as an equipotential network, or blank slate, that later would be shaped by usage into a functional system. The nervous system just seemed too incredibly complex for us to imagine that nerve fibers could find their way selectively to so many predetermined destinations. Yet it appears that this is exactly what they do! Research with invertebrate nervous systems indicated that each of the billions of nerve cell axons acquires a distinct identity that in some way directs it along a specific pathway to its destination. Many years ago Harrison observed that a growing nerve axon terminated in a "growth cone," from which extend numerous tiny threadlike pseudopodial processes

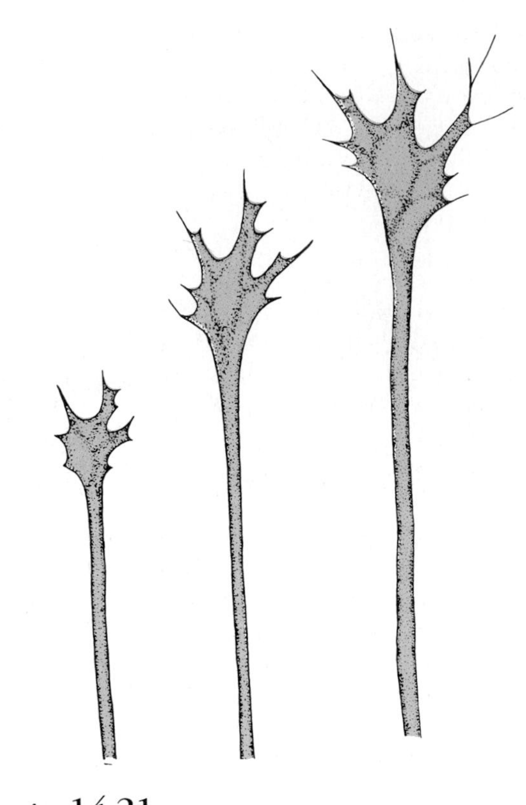

figure 14-21

Growth cone at the growing tip of a nerve axon. Materials for growth flow down the axon to the growth cone from which numerous threadlike filopodia extend. These serve as a pioneering guidance system for the developing axon.

(filopodia) (Figure 14-21). Recent research has shown that the growth cone is steered by an array of guidance molecules secreted along the pathway and by the axon's target. This chemical guidance system, which must, of course, be genetically directed, is just one example of the amazing precision that characterizes the entire process of differentiation.

Derivatives of Endoderm: Digestive Tube and Survival of Gill Arches

In frog embryos the primitive gut makes its appearance during gastrulation with the formation of the archenteron. From this simple endodermal cavity develop the lining of the digestive tract, the lining of the pharynx and lungs, most of the liver and pancreas, the thyroid and parathyroid glands, and the thymus (Figure 14-19).

The **alimentary canal** develops from the primitive gut and is folded off from the yolk sac by the growth and folding of the body wall (Figure 14-22). The ends of the tube open to the exterior and are lined with ectoderm, whereas the rest of the tube is lined with endoderm. The **lungs, liver,** and **pancreas** arise from the foregut.

Among the most intriguing derivatives of the digestive tract are the pharyngeal (gill) arches and pouches, which make

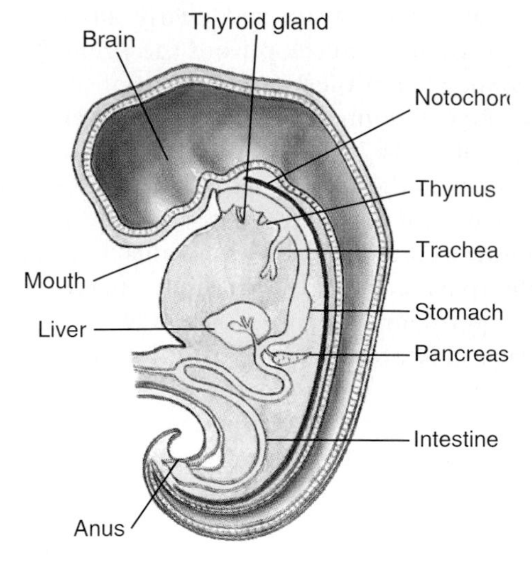

figure 14-22

Derivatives of the alimentary canal of a human embryo.

their appearance in the early embryonic stages of all vertebrates (see Figure 14-29, p. 334). In fishes, the gill arches develop into gills and supportive structures and serve as respiratory organs. When early vertebrates moved onto land, gills were unsuitable for aerial respiration and were replaced by lungs.

Why then do gill arches persist in the embryos of terrestrial vertebrates? Certainly not for the convenience of biologists who use these and other embryonic structures to reconstruct lines of vertebrate descent. Even though the gill arches serve no respiratory function in either embryos or adults of terrestrial vertebrates, they remain as necessary primordia for a variety of other structures. For example, the first arch and its endoderm-lined pouch (the space between adjacent arches) form the upper and lower jaws and inner ear of vertebrates. The second, third, and fourth gill pouches contribute to the tonsils, parathyroid gland, and thymus. We can understand then why gill arches and other fishlike structures appear in early mammalian embryos. Their original function has been abandoned, but the structures are retained for new purposes. The great conservatism of early embryonic development has conveniently provided us with a telescoped evolutionary history.

Derivatives of Mesoderm: Support, Movement, and Beating Heart

The intermediate germ layer, the mesoderm, forms the vertebrate skeletal, muscular, and circulatory structures and the kidney (Figure 14-19). As vertebrates have increased in size and complexity, the mesodermally derived supportive, movement, and transport structures make up an even greater proportion of the body.

Most **muscles** arise from the mesoderm along each side of the spinal cord (Figure 14-23). This mesoderm divides into a linear series of blocklike somites (38 in humans), which by splitting, fusion, and migration become the muscles

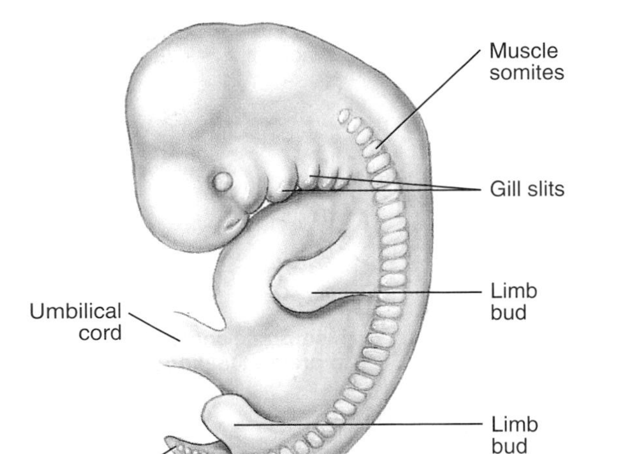

figure 14-23

Human embryo showing muscle somites, which differentiate into skeletal muscles and axial skeleton.

of the body and axial parts of the skeleton. The **limbs** begin as buds from the side of the body. Projections of the limb buds develop into fingers and toes.

Although the primitive mesoderm appears after the ectoderm and endoderm, it gives rise to the first functional organ, the embryonic heart. Guided by the underlying endoderm, clusters of precardiac mesodermal cells move ameba-like into a central position between the underlying primitive gut and the overlying neural tube. Here the heart is established, first as a single, thin tube.

Even while the cells group together, the first twitchings are evident. In the chick embryo, a favorite and nearly ideal animal for experimental embryological studies, the primitive heart begins to beat on the second day of the 21-day incubation period; it begins beating before any true blood vessels have formed and before there is any blood to pump. As the ventricle primordium develops, the spontaneous cellular twitchings become coordinated into a feeble but rhythmical beat. Then, as the atrium develops behind the ventricle, followed by development of the sinus venosus behind the atrium, the heart rate quickens. Each new heart chamber has an intrinsic beat that is faster than its predecessor.

Finally a specialized area of heart muscle called the **sinoatrial** node develops in the sinus venosus and takes command of the entire heartbeat (the role of the sinoatrial node in the excitation of the heart is described on p. 194). This sinoatrial node becomes the heart's **pacemaker.** As the heart builds up a strong and efficient beat, vascular channels open within the embryo and across the yolk. Within the vessels are the first primitive blood cells suspended in plasma.

The early development of the heart and circulation is crucial to continued embryonic development because without a circulation the embryo could not obtain materials for growth. Food is absorbed from the yolk and carried to the embryonic body; oxygen is delivered to all the tissues, and carbon dioxide and other wastes are carried away.The embryo is totally dependent on these extraembryonic support systems, and the circulation is the vital link be tween them.

Mechanisms of Development

Nuclear Equivalence

How does the developing embryo generate the multitude of many cell types of a complete multicellular organism from the single diploid nucleus of the zygote? To many nineteenth-century embryologists there seemed only one acceptable answer: as cell division ensued, the hereditary material had to be broken up and parceled out into smaller and smaller units until only the information required to impart the characteristics of a single cell type remained. This view was refuted early in this century by a series of ingenious experiments which showed conclusively that *all* cells contained the same nuclear information despite their different developmental fates.

Cytoplasmic Localization: Significance of the Cortex

Experimental embryologists now realized that if all the nuclei of a developing embryo are equivalent, each capable of supporting full development, there must be differences in the determinative properties of the cytoplasm that surrounds the nucleus. How else could the dividing cells of the early embryo be started down different developmental paths? However, in the 1930s embryologists discovered that if they centrifuged a sea urchin zygote until the cytoplasm was thoroughly displaced into layers—yolk granules at the bottom and oil droplets at the top—the embryo would still develop normally, or almost so. Even when up to 50% of the zygote's cytoplasm was removed with a micropipette, the embryo would develop perfectly! Only if great centrifugal force was applied, enough to disrupt the pattern of the zygote's outer layer, the cortex, were abnormalities observed.

These surprising results showed conclusively that the cortex, and not the central cytoplasm, contained an invisible but dynamic organization that determined the pattern of cleavage. Cortical organization is at first labile (especially so in regulative eggs) but soon becomes fixed and irreversible. As cleavage progresses, the cortex and the enclosed cytoplasm become segregated into territories with specific determinative properties.

Selected References

Ereshefsky, M. (ed.). 1992. The units of evolution. Cambridge, Massachusetts, MIT Press. *A thorough coverage of concepts of species, including reprints of important papers on the subject.*

Hall, B. K. 1994. Homology: the hierarchical basis of comparative biology. San Diego, Academic Press. *A collection of papers discussing the many dimensions of homology, the central concept of comparative biology and systematics.*

Hillis, D. M., C. Moritz, and B. K. Mable (eds.). 1996. Molecular systematics, ed. 2. Sunderland, Massachusetts, Sinauer Associates, Inc. *A detailed coverage of the biochemical and analytical procedures of comparative biochemistry.*

Hull, D. L. 1988. Science as a process. Chicago, University of Chicago Press. *A study of the working methods and interactions of systematists, containing a thorough review of the principles of evolutionary, phenetic, and cladistic taxonomy.*

Jeffrey, C. 1973. Biological nomenclature. London, Edward Arnold, Ltd. *A concise, practical guide to the principles and practice of biological nomenclature and a useful interpretation of the Codes of Nomenclature.*

Maddison, W. P., and D. R. Maddison. 1992. MacClade version 3.01. Sunderland, Massachusetts, Sinauer Associates, Inc. *A computer program for the MacIntosh that conducts phylogenetic analyses of systematic characters. The instruction manual stands alone as an excellent introduction to phylogenetic procedures. The computer program is user-friendly and excellent for instruction in addition to serving as a tool for analyzing real data.*

Margulis, L., and K. V. Schwartz. 1987. Five kingdoms: an illustrated guide to the phyla of life on earth, ed. 2. San Francisco, W. H. Freeman & Company. *Illustrated catalog and descriptions of all major groups with bibliography and glossary.*

Mayr, E., and P. D. Ashlock. 1991. Principles of systematic zoology. New York, McGraw-Hill. *A detailed survey of systematic principles as applied to animals.*

Panchen, A. L. 1992. Classification, evolution, and the nature of biology. New York, Cambridge University Press. *Excellent explanations of the methods and philosophical foundations of biological classification.*

Wiley, E. O. 1981. Phylogenetics: the theory and practice of phylogenetic systematics. New York, John Wiley & Sons, Inc. *Excellent, thorough presentation of cladistic theory.*

Wiley, E. O., D. Siegel-Causey, D. R. Brooks, and V. A. Funk. 1991. The compleat cladist: a primer of phylogenetic procedures. Lawrence, University of Kansas Printing Service. *A workbook presenting detailed instruction in cladistic concepts and methods.*

Woese, C. R., O. Kandler, and M. L. Wheelis. 1990. Towards a natural system of organisms: proposal for the domains Archaea, Bacteria, and Eucarya. Proceedings of the National Academy of Sciences, USA, **87**:4576–4579. *Proposed cladistic classification for the major taxonomic divisions of life.*

The Animal-Like Protista

chapter | sixteen

Emergence of Eukaryotes and a New Life Pattern

The first reasonable evidence for life on earth dates from approximately 3.5 billion years ago. These first cells were prokaryotic, bacteria-like organisms. After an enormous time span of evolutionary diversificaton at the prokaryotic level, unicellular eukaryotic organisms appeared. Although the origin of single-celled eukaryotes can never be known with certainty, we have good evidence that it came about through a process of symbiosis. Certain aerobic bacteria may have been engulfed by other bacteria unable to cope with the increasing concentrations of oxygen in the atmosphere. The aerobic bacteria had the enzymes necessary for deriving energy in the presence of oxygen, and they would have become the ancestors of mitochondria. Most, but not all, of the genes of the mitochondria would be lost or come to reside in the host cell nucleus. Almost all present-day eukaryotes have mitochondria and are aerobic.

Some ancestral eukaryotic cells engulfed photosynthetic bacteria, which evolved to become chloroplasts, and the eukaryotes thereby were able to manufacture their own food molecules using energy from sunlight. The descendants of one line, the green algae, eventually gave rise to the kingdom Plantae.

Some of the eukaryotes that did not become residences for chloroplasts, and even some that did, evolved animal-like characteristics and gave rise to the protistan subkingdom Protozoa. Protozoa are a diverse assemblage of unicellular organisms with puzzling affinities. They have several characteristics that traditionally have been considered distinctly animal-like: they lack a cell wall, have at least one motile stage in the life cycle, and most ingest their food. Throughout their long history, protozoan groups have radiated to generate a bewildering array of morphological forms within the constraints of a single cell.

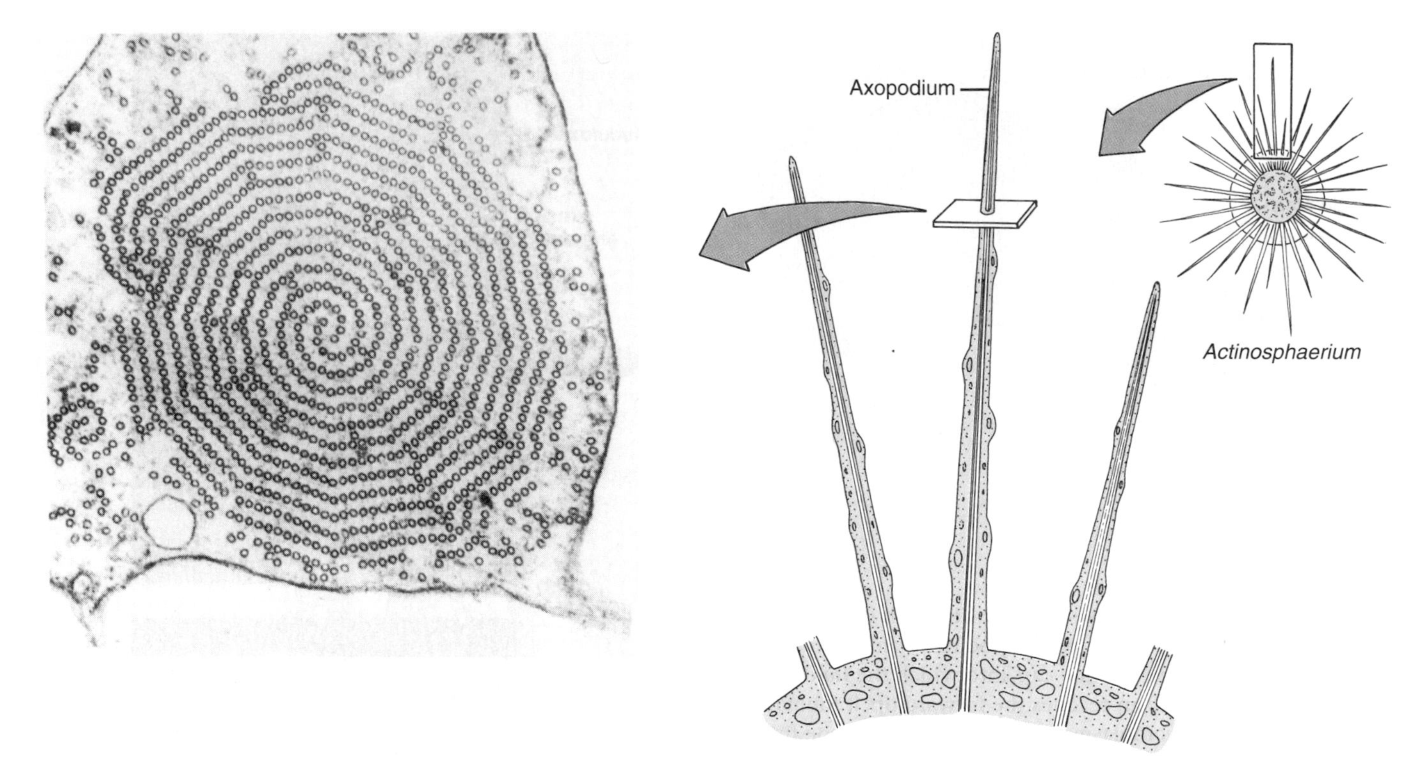

figure 16-5

Diagram of axopodium *(right)* to show orientation of electron micrograph of axopodium (from *Actinosphaerium nucleofilum*) in cross section *(left)*. Protozoa with axopodia are shown in Figure 16-14 (*Actinophrys* and *Clathrulina*). The axoneme of the axopodium is composed of an array of microtubules, which may vary from three to many in number depending on the species. Some species can extend or retract their axopodia quite rapidly. (×99,000)

substrates, and heterotrophs, which must have organic molecules synthesized by other organisms. Another kind of classification, usually applied to heterotrophs, involves those that ingest visible particles of food (**phagotrophs,** or **holozoic** feeders) as contrasted with those ingesting food in a soluble form (**osmotrophs,** or **saprozoic** feeders).

Holozoic nutrition implies phagocytosis (Figures 16-4 and 16-6), in which there is an infolding or invagination of the cell membrane around the food particle. As the invagination extends farther into the cell, it is pinched off at the surface (p. 35). The food particle is thus contained in an intracellular, membrane-bound vesicle, the **food vacuole** or **phagosome.** Lysosomes, small vesicles containing digestive enzymes, fuse with the phagosome and pour their contents into it, where digestion begins. As the digested products are absorbed across the vacuole membrane, the phagosome becomes smaller. Any undigestible material may be released to the outside by exocytosis, the vacuole again fusing with the cell surface membrane. In most ciliates, many flagellates, and many apicomplexans, the site of phagocytosis is a definite mouth structure, the **cytostome** (Figures 16-6, 16-17, and 16-21). In amebas, phagocytosis can occur at almost any point by envelopment of the particle with pseudopodia. Many ciliates have a chracteristic structure for expulsion of waste matter, the **cytopyge** or **cytoproct** (Figure 16-21), found in a characteristic location.

Excretion and Osmoregulation

Water balance, or osmoregulation, is a function of the one or more **contractile vacuoles** (see Figures 16-3, 16-13, and 16-21) possessed by most protozoa, particularly freshwater forms, which live in a hypoosmotic environment. These vacuoles are often absent in marine or parasitic protozoa, which live in a nearly isosmotic medium.

The contractile vacuoles usually are located in the ectoplasm and act as pumps to remove excess water from the cytoplasm. They are filled by droplets fed by a system of collecting canals in some species and from smaller vesicles in others (Figure 16-7). When full, they empty through a canal to the outside. The rate of pulsation varies; in *Paramecium* the posterior vacuole may pulsate faster than the anterior one because of water delivered there along with ingested food. Vacuoles in marine forms pulsate more slowly than in similar freshwater forms.

Nitrogenous wastes from metabolism apparently diffuse through the cell membrane, but some may also be emptied by way of the contractile vacuoles.

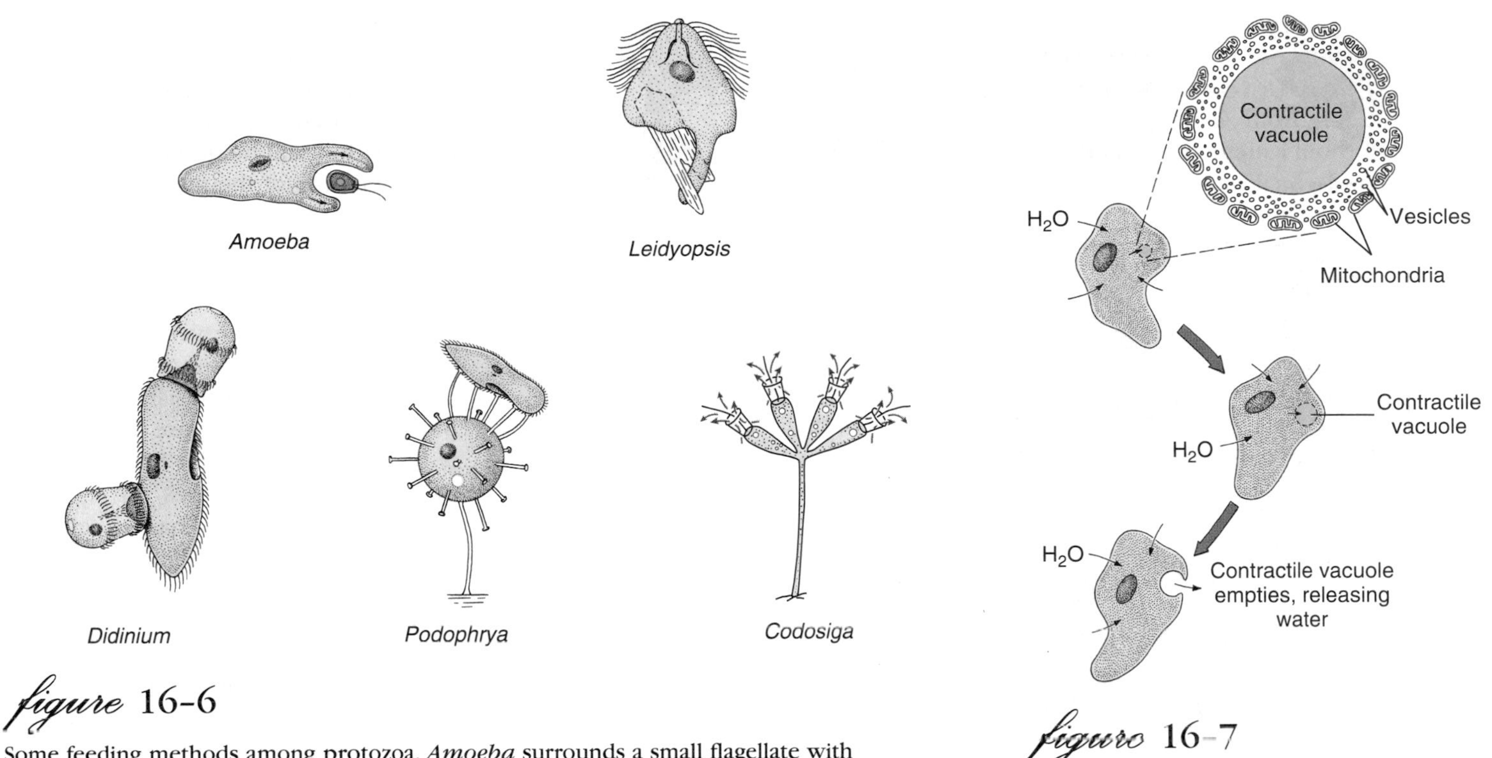

figure 16-6

Some feeding methods among protozoa. *Amoeba* surrounds a small flagellate with pseudopodia. *Leidyopsis,* a flagellate living in the intestine of termites, forms pseudopodia and ingests wood chips. *Didinium,* a ciliate, feeds only on *Paramecium,* which it swallows through a temporary cytostome in its anterior end. Sometimes more than one *Didinium* feed on the same *Paramecium. Podophrya* is a suctorian ciliophoran. Its tentacles attach to its prey and suck prey cytoplasm into its body, where the cytoplasm is pinched off to form food vacuoles. *Codosiga,* a sessile flagellate with a collar of microvilli, feeds on particles suspended in the water drawn through its collar by the beat of its flagellum. The particles are moved to the cell body and ingested (surrounded by small pseudopods). Technically all of these methods are types of phagocytosis.

figure 16-7

How an ameba pumps out water. Water enters the ameba's body by osmosis and is removed by the rhythmic filling and emptying of the contractile vacuole. The contractile vacuole of *Amoeba proteus* is surrounded by tiny vesicles that fill with fluid, which then empties into the vacuole. Note the numerous mitochondria that apparently provide energy needed to adjust the salt content of the tiny vesicles.

Reproduction

Sexual phenomena occur widely among the protozoa, and sexual processes may precede certain phases of asexual reproduction, but embryonic development does not occur; protozoa do not have embryos. The essential features of sexual processes include a reduction division of the chromosome number to half (diploid number to haploid number), the development of sex cells (gametes) or at least gamete nuclei, and usually a fusion of the gamete nuclei (p. 376).

Asexual Reproduction: Fission

The cell multiplication process that results in more individuals in the protozoa is called **fission.** The most common type of fission is **binary,** in which two essentially identical individuals result (Figures 16-8 and 16-9). When the progeny cell is considerably smaller than the parent and then grows to adult size, the process is called **budding.** This process occurs in some ciliates. In **multiple fission,** division of the cytoplasm (cytokinesis) is preceded by several nuclear divisions, so that a number of individuals is produced almost simultaneously (Figure 16-18). Multiple fission, or **schizogony,** is common among the Apicomplexa and some amebas. When multiple fission leads to spore or sporozoite formation, it is called **sporogony.**

Protozoan Colonies Protozoan colonies are formed when the daughter zooids (individuals in the colony), derived asexually, remain associated instead of moving apart and living a separate existence (Figures 16-10 and 16-12). Protozoan colonies vary from individuals embedded together in a gelatinous substance to those having protoplasmic connections among them. The arrangement of the individuals results in a variety of colony types, each characteristic of the protozoan that forms it. Usually the individuals of a colony are structurally and physiologically the same, although there may be some division of labor, such as differentiation of reproductive and somatic zooids.

Sexual Processes

Although all protozoa reproduce asexually, and some are apparently exclusively asexual, the widespread occurrence of sex

Selected References

See also general references for Part Three, p. 562.

Anderson, O. R. 1988. Comparative protozoology: ecology, physiology, life history. New York, Springer-Verlag. *Good treatment of the aspects mentioned in the subtitle.*

Bonner, J.T. 1983. Chemical signals of social amoebae. Sci. Am. **248:**114–120 (April). *Social amebas of two different species secrete different chemical compounds that act as aggregation signals. The evolution of the aggregation signals in these protozoa may provide clues to the origin of the diverse chemical signals (neurotransmitters and hormones) in more complex organisms.*

Fenchel, T. 1987. Ecology of protozoa: the biology of free-living phagotrophic protists. Madison, Wisconsin, Science Tech Publishers.

Harrison, G. 1978. Mosquitoes, malaria and man: a history of the hostilities since 1880. New York, E. P. Dutton. *A fascinating story, well told.*

Kreier, J. P., and J. R. Baker. 1987. Parasitic protozoa. Boston, Allen and Unwin. *Concise summary of parasitic forms.*

Lee, J. 1993. On a piece of chalk—Updated. J. Eukaryotic Microbiol. **40:**395–410. *Excellent summary of current knowledge of foraminiferans.*

Lee, J. J., S. H. Hutner, and E. C. Bovee (eds.). 1985. An illustrated guide to the protozoa. Lawrence, Kansas, Society of Protozoologists. *A comprehensive guide and essential reference for students of the protozoa.*

Lee, J. J., and A. T. Soldo. 1992. Protocols in protozoology. Lawrence, Kansas, Society of Protozoologists. *A treasure trove of experimental and laboratory techniques for protozoa.*

Sleigh, M. A. 1989. Protozoa and other protists. London, Edward Arnold. *Extensively uptdated version of the author's* The biology of protozoa.

Stossel, T. P. 1994. The machinery of cell crawling. Sci. Am. **271:**54–63 (Sept.). *Ameboid movement—how cells crawl—is important throughout the animal kingdom, as well as in Protista. We now understand quite a bit about its mechanism.*

Sponges: Phylum Porifera

chapter | seventeen

The Advent of Multicellularity

Sponges are the simplest multicellular animals. Because the cell is the elementary unit of life, the evolution of organisms larger than unicellular protozoa arose as an aggregate of such building units. Nature has experimented with producing larger organisms without cellular differentiation—certain large, single-celled marine algae, for example—but such examples are rarities. Typically nature has held stubbornly to multicellular construction in the progress toward higher organization. There are many advantages to multicellularity as opposed to simply increasing the mass of a single cell. Since it is at cell surfaces that exchange takes place, dividing a mass into smaller units greatly increases the surface area available for metabolic activities. It is impossible to maintain a workable surface-to-mass ratio by simply increasing the size of a single-celled organism. Thus multicellularity is a highly adaptive path toward increasing body size.

Strangely, although sponges are multicellular, they are phylogenetically distinct from other metazoans, forming a sister group to the Eumetazoa. They apparently split off from the metazoan line before the origin of the radiates (Chapter 18). The sponge body is an assemblage of cells embedded in a gelatinous matrix and supported by a skeleton of minute needlelike spicules and protein. Because sponges neither look nor behave like other animals, it is understandable that they were not completely accepted as animals by zoologists until well into the nineteenth century.

Classification of Phylum Porifera

Class Calcarea (cal-ca′re-a) (L. *calcis,* lime, + Gr. *spongos,* sponge) **(Calcispongiae).** Have spicules of calcium carbonate that often form a fringe around the osculum; spicules needle-shaped or three- or four-rayed; all three types of canal systems (asconoid, syconoid, leuconoid) represented; all marine. Examples: *Sycon, Leucosolenia, Clathrina.*

Class Hexactinellida (hex-ak-tin-el′i-da) (Gr. *hex,* six, + *aktis,* ray) **(Hyalospongiae).** Have six-rayed, siliceous spicules extending at right angles from a central point; spicules often united to form network; body often cylindrical or funnel shaped. Flagellated chambers in simple syconoid or leuconoid arrangement. Habitat mostly deep water; all marine. Examples: Venus' flower basket *(Euplectella), Hyalonema.*

Class Demospongiae (de-mo-spun′je-e) (tolerated misspelling of Gr. *desmos,* chain, tie, bond, + *spongos,* sponge). Have skeleton of siliceous spicules that are not six rayed, or spongin, or both. Leuconoid-type canal systems. One family found in fresh water; all others marine. Examples: *Thenea, Cliona, Spongilla, Myenia,* and all bath sponges.

Class Sclerospongiae (skler′o-spun′je-e) (Gr. *sklēros,* hard, + *spongos,* sponge). Secrete massive basal skeleton of calcium carbonate, with living tissue extending into skeleton from 1 mm to 3 cm or more, extending above skeleton less than 1 mm; have siliceous spicules similar to Demospongiae (sometimes absent), and spongin fibers; leuconoid organization; inhabit caves, crevices, tunnels, and deep water on coral reefs. Examples: *Astrosclera, Calcifibrospongia.*

Summary

The sponges (phylum Porifera) are an abundant marine group with some freshwater representatives. They have various specialized cells, but these cells are not organized into tissues or organs. They depend on the flagellar beat of their choanocytes to circulate water through their bodies for gathering food and exchange of respiratory gases. They are supported by secreted skeletons of fibrillar collagen, collagen in the form of large fibers or filaments (spongin), calcareous or siliceous spicules, or a combination of spicules and spongin in most species.

Sponges reproduce asexually by budding, fragmentation, and gemmules (internal buds). Most sponges are monoecious but produce sperm and oocytes at different times. Embryogenesis is unusual, with a migration of flagellated cells at the surface to the interior (parenchymella). Sponges have great regenerative abilities.

Sponges are an ancient group, remote phylogenetically from other metazoa, but some evidence suggests that they are a sister group to the Eumetazoa. Their adaptive radiation is centered on elaboration of the water circulation and filter feeding system.

Review Questions

1. Give six characteristics of sponges.
2. Briefly describe asconoid, syconoid, and leuconoid body types in sponges.
3. What sponge body type is most efficient and makes possible the largest body size?
4. Define the following: ostia, osculum, spongocel, mesohyl.
5. Define the following: pinacocytes, choanocytes, archaeocytes, sclerocytes, collencytes.
6. What material is found in the skeleton of all sponges?
7. Describe the skeletons of each of the classes of sponges.
8. Describe how sponges feed, respire, and excrete.
9. What is a gemmule?
10. Describe how gametes are produced and the process of fertilization in most sponges.
11. What is the largest class of sponges, and what is its body type?
12. What are the closest relatives of sponges? Justify your answer.
13. It has been suggested that despite being large, multicellular animals, sponges function more like protozoa. What aspects of sponge biology support this statement and how? Consider, for example, nutrition, reproduction, gas exchange, and cellular organization.

Selected References

See also general references for Part Three, p. 562.

Bergquist, P. R. 1978. Sponges. Berkeley, California, University of California Press. *Excellent monograph on sponge structure, classification, evolution, and general biology.*

Gould, S. J. 1995. Reversing established orders. Nat. Hist. **104**(9):12–16. *Describes several anomalous animal relationships, including the sponge that preys on shrimp.*

Hartman, W. D. 1982. Porifera. In S. P. Parker (ed.). Synopsis and classification of living organisms, vol. 1. New York, McGraw-Hill Book Company. *Review of sponge classification.*

Simpson, T. L. 1984. The cell biology of sponges. New York, Springer-Verlag. *A review and synthesis, points out many problems yet to be solved.*

Wainright, P. O., G. Hinkle, M. L. Sogin, S. K. Stickel. 1993. Monophyletic origins of the Metazoa: an evolutionary link with Fungi. Science **260:**340–342. *Reports molecular evidence that the sister group of metazoans is Fungi and that multicellular animals are monophyletic.*

Wood, R. 1990. Reef-building sponges. Am. Sci. **78:**224–235. *The author presents evidence that the known sclerosponges belong to either the Calcarea or the Demospongiae and that a separate class Sclerospongiae is not needed.*

The Radiate Animals: Cnidarians and Ctenophores

chapter | eighteen

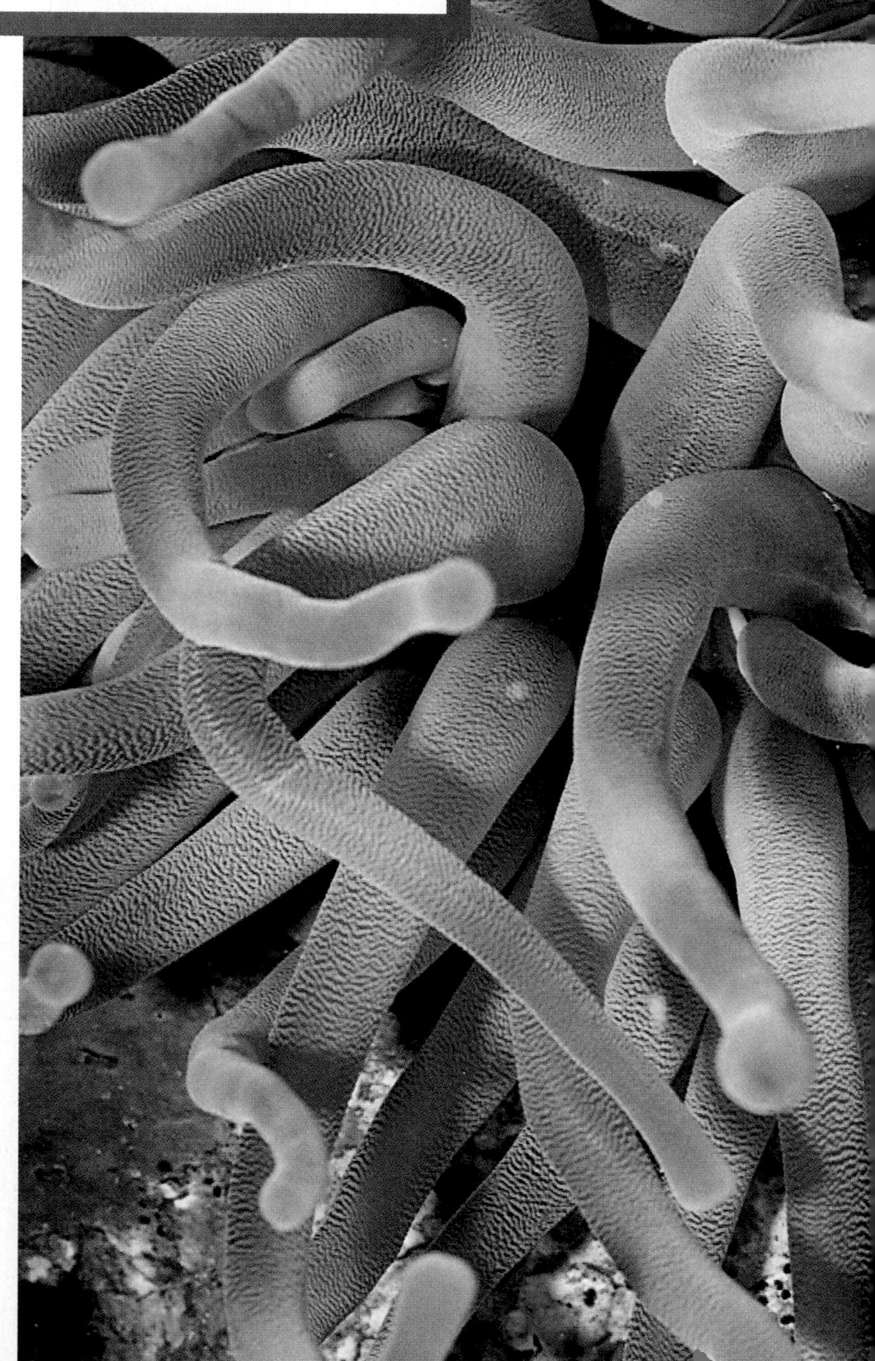

A Fearsome Tiny Weapon

Although members of the phylum Cnidaria are more highly organized than sponges, they are still relatively simple animals. Most are sessile; those that are unattached, such as jellyfish, can swim only feebly. None can chase their prey. Indeed, we might easily get the false impression that the cnidarians were placed on earth to provide easy meals for other animals. The truth is, however, many cnidarians are very effective predators that are able to kill and eat prey that are much more highly organized, swift, and intelligent. They manage these feats because they possess tentacles that bristle with tiny, remarkably sophisticated weapons called nematocysts.

As it is secreted within the cell that contains it, the nematocyst is endowed with potential energy to power its discharge. It is as though a factory manufactured a gun, cocked and ready with a bullet in its chamber, as it rolls off the assembly line. Like the cocked gun, the completed nematocyst requires only a small stimulus to make it fire. Rather than a bullet, a tiny thread bursts from the nematocyst. Achieving a velocity of 2 m/sec and an acceleration of 40,000 $\times$ gravity, it instantly penetrates its prey and injects a paralyzing toxin. A small animal unlucky enough to brush against one of the tentacles is suddenly speared with hundreds or even thousands of nematocysts and quickly immobilized. Some nematocyst threads can penetrate human skin, resulting in sensations ranging from minor irritation to great pain, even death, depending on the species. A fearsome, but wondrous, tiny weapon.

Phylum Cnidaria

The phylum Cnidaria (ny-dar′e-a) (Gr. *knidē,* nettle, + L. *aria* [pl. suffix]; like or connected with) is an interesting group of more than 9000 species. It takes its name from cells called **cnidocytes,** which contain the stinging organelles **(nematocysts)** characteristic of the phylum. Nematocysts are *formed and used* only by cnidarians. Another name for the phylum, Coelenterata (se-len′te-ra′ta) (Gr. *koilos,* hollow, + *enteron,* gut, + L. *ata* [pl. suffix], characterized by), is used less commonly than formerly, and it sometimes now refers to both radiate phyla, since its meaning is equally applicable to both.

The cnidarians are generally regarded as originating close to the basal stock of the metazoan line. They are an ancient group with the longest fossil history of any metazoan, reaching back more than 700 million years. Although their organization has a structural and functional simplicity not found in other metazoans, they form a significant proportion of the biomass in some locations. They are widespread in marine habitats, and there are a few in fresh water. Although they are mostly sessile or, at best, fairly slow moving or slow swimming, they are quite efficient predators of organisms that are much swifter and more complex. The phylum includes some of

Position in Animal Kingdom

The two phyla Cnidaria and Ctenophora make up the radiate animals, which are characterized by **primary radial** or **biradial symmetry,** which we believe is ancestral for the eumetazoans. Radial symmetry, in which the body parts are arranged concentrically around the oral-aboral axis, is particularly suitable for sessile or sedentary animals and for free-floating animals because they approach their environment (or it approaches them) from all sides equally. Biradial symmetry is basically a type of radial symmetry in which only two planes through the oral-aboral axis divide the animal into mirror images because of the presence of some part that is paired. All other eumetazoans have a primary bilateral symmetry; that is, they are bilateral or were derived from an ancestor that was bilateral.

Neither phylum has advanced generally beyond the **tissue level of organization,** although a few organs occur. In general, the ctenophores' structure is more complex than that of the cnidarians.

Biological Contributions

1. Both phyla have developed two well-defined **germ layers,** ectoderm and endoderm; a third, or mesodermal, layer, which is derived embryologically from the ectoderm, is present in some. The body plan is saclike, and the body wall is composed of two distinct layers, epidermis and gastrodermis, derived from the ectoderm and endoderm, respectively. The gelatinous matrix, mesoglea, between these layers may be structureless, may contain a few cells and fibers, or may be composed largely of mesodermal connective tissue and muscle fibers.
2. An internal body cavity, the **gastrovascular cavity,** is lined by the gastrodermis and has a single opening, the mouth, which also serves as the anus.
3. **Extracellular digestion** occurs in the gastrovascular cavity, and intracellular digestion takes place in the gastrodermal cells. Extracellular digestion allows ingestion of larger food particles.
4. Most radiates have **tentacles,** or extensible projections around the oral end, that aid in capturing food.
5. Radiates are the simplest animals to possess true **nerve cells** (protoneurons), but the nerves are arranged as a nerve net, with no central nervous system.
6. Radiates are the simplest animals to possess sense organs, which include well-developed statocysts (organs of equilibrium) and ocelli (photosensitive organs).
7. Locomotion in the free-moving forms is achieved by either **muscular contractions** (cnidarians) or **ciliary comb plates** (ctenophores). However, both groups are still better adapted to floating or being carried by currents than to strong swimming.
8. **Polymorphism**[1] in the cnidarians has widened their ecological possibilities. In many species the presence of both a polyp (sessile and attached) stage and a medusa (free-swimming) stage permits occupation of a benthic (bottom) and a pelagic (open-water) habitat by the same species. Polymorphism also widens the possibilities of structural complexity.
9. Some unique features are found in these phyla, such as **nematocysts** (stinging organelles) in cnidarians and **colloblasts** (adhesive organelles) and **ciliary comb plates** in ctenophores.

[1]Note that polymorphism here refers to more than one structural form of individual within a species, as contrasted with the use of the word in genetics (p. 103), in which it refers to different allelic forms of a gene in a population.

Characteristics of Phylum Cnidaria

1. Entirely aquatic, some in fresh water but mostly marine
2. **Radial symmetry** or biradial symmetry around a longitudinal axis with **oral** and **aboral** ends; no definite head
3. Two basic types of individuals: **polyps** and **medusae**
4. Exoskeleton or endoskeleton of chitinous, calcareous, or protein components in some
5. Body with two layers, epidermis and gastrodermis, with mesoglea **(diploblastic);** mesoglea with cells and connective tissue (ectomesoderm) in some **(triploblastic)**
6. **Gastrovascular cavity** (often branched or divided with septa) with a single opening that serves as both mouth and anus; extensible tentacles usually encircling the mouth or oral region
7. Special stinging cell organelles called **nematocysts** in either epidermis or gastrodermis or in both; nematocysts abundant on tentacles, where they may form batteries or rings
8. **Nerve net** with symmetrical and asymmetrical synapses; with some sensory organs; diffuse conduction
9. Muscular system (epitheliomuscular type) of an outer layer of longitudinal fibers at base of epidermis and an inner one of circular fibers at base of gastrodermis; modifications of this plan in some cnidarians, such as separate bundles of independent fibers in the mesoglea
10. Asexual reproduction by budding (in polyps) or sexual reproduction by gametes (in all medusae and some polyps); sexual forms monoecious or dioecious; **planula larva;** holoblastic indeterminate cleavage
11. No excretory or respiratory system
12. No coelomic cavity

nature's strangest and loveliest creatures: the branching, plantlike hydroids; the flowerlike sea anemones; the jellyfishes; and those architects of the ocean floor, the horny corals (sea whips, sea fans, and others), and the stony corals whose thousands of years of calcareous house-building have produced great reefs and coral islands (p. 408).

We recognize four classes of Cnidaria: Hydrozoa (the most variable class, including hydroids, fire corals, Portuguese man-of-war, and others), Scyphozoa (the "true" jellyfishes), Cubozoa (cube jellyfishes), and Anthozoa (the largest class, including sea anemones, stony corals, soft corals, and others).

Ecological Relationships

Cnidarians are found most abundantly in shallow marine habitats, especially in warm temperatures and tropical regions. There are no terrestrial species. Colonial hydroids are usually found attached to mollusc shells, rocks, wharves, and other animals in shallow coastal water, but some species are found at great depths. Floating and free-swimming medusae are found in open seas and lakes, often far from the shore. Floating colonies such as the Portuguese man-of-war and *Velella* (L. *velum,* veil, + *ellus,* dim. suffix) have floats or sails by which the wind carries them.

Some ctenophores, molluscs, and flatworms eat hydroids bearing nematocysts and use these stinging structures for their own defense. Some other animals, such as some molluscs and fishes, feed on cnidarians, but cnidarians rarely serve as food for humans.

Cnidarians sometimes live symbiotically with other animals, often as commensals on the shell or other surface of their host. Certain hydroids (Figure 18-1) and sea anemones commonly live on snail shells inhabited by hermit crabs, providing the crabs some protection from predators. Algae frequently live as mutuals in the tissues of cnidarians, notably in some freshwater hydras and in reef-building corals. The presence of the algae in reef-building corals limits the occurrence of coral reefs to relatively shallow, clear water where sunlight is sufficient for the photosynthetic requirements of the algae. These corals are an essential component of coral reefs, and reefs are extremely important habitats in tropical waters. Coral reefs are discussed further later in the chapter.

Although many cnidarians have little economic importance, reef-building corals are an important exception. Fish and other animals associated with reefs provide substantial amounts of food for humans, and reefs are of economic value as tourist attractions. Precious coral is used for jewelry and ornaments, and coral rock serves for building purposes.

Planktonic medusae may be of some importance as food for fish that are of commercial value; the reverse is also true—the young fish fall prey to cnidarians.

Form and Function

Dimorphism and Polymorphism in Cnidarians

One of the most interesting—and sometimes puzzling—aspects of this phylum is the dimorphism and often polymorphism displayed by many of its members. All cnidarian forms fit into one of two morphological types (dimorphism): the

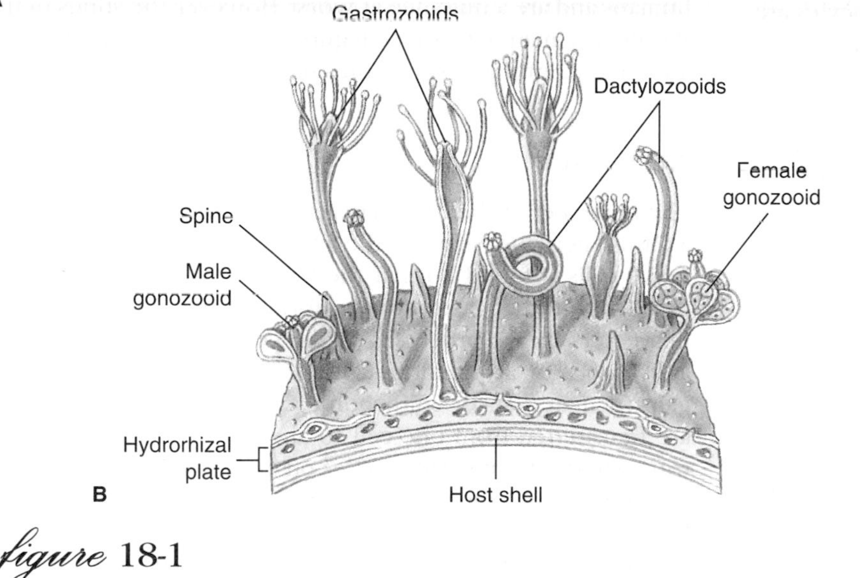

figure 18-1

A, A hermit crab with its cnidarian mutuals. The shell is blanketed with polyps of the hydrozoan *Hydractinia milleri.* The crab gets some protection from predation by the cnidarians, and the cnidarians get a free ride and bits of food from their host's meals. **B,** Portion of a colony of *Hydractinia,* showing the types of zooids and the stolon (hydrorhiza) from which they grow.

polyp, or hydroid form, which is adapted to a sedentary or sessile life, and the **medusa,** or jellyfish form, which is adapted for a floating or free-swimming existence (Figure 18-2).

Most polyps have tubular bodies with a mouth at one end surrounded by tentacles. The aboral end is usually attached to a substratum by a pedal disc or other device. Polyps may live singly or in colonies. Colonies of some species include morphologically differing individuals (polymorphism), each specialized for a certain function, such as feeding, reproduction, or defense (Figure 18-1).

> *The name "medusa" was suggested by a fancied resemblance to the Gorgon Medusa, a mythological lass with snaky tresses that turned to stone any who gazed upon them.*

Medusae are usually free swimming and have bell-shaped or umbrella-shaped bodies and tetramerous symmetry (body parts arranged in fours). The mouth is usually centered on the concave side, and tentacles extend from the rim of the umbrella.

The sea anemones and corals (class Anthozoa) are all polyps: hence, they are not dimorphic. The true jellyfishes (class Scyphozoa) have a conspicuous medusoid form, but many have a polypoid larval stage. The colonial hydroids of class Hydrozoa, however, sometimes have life histories that feature both the polyp, or hydroid, stage and the free-swimming medusa stage—rather like a Jekyll-and-Hyde existence. A species that has both the attached polyp and the floating medusa within its life history can take advantage of the feeding and distribution possibilities of both pelagic (open-water) and benthic (bottom) environments. Many hydrozoans are also polymorphic, with several distinct types of polyps in a colony.

Superficially the polyp and medusa seem very different. But actually each has retained the saclike body plan that is basic to the phylum (Figure 18-2). The medusa is essentially an unattached polyp with the tubular portion widened and flattened into the bell shape.

Both the polyp and the medusa possess the three body wall layers typical of the cnidarians, but the jellylike layer of mesoglea is much thicker in the medusa, constituting the bulk of the animal and making it more buoyant. Because of this mass of mesoglea ("jelly"), the medusae are commonly called jellyfishes.

Nematocysts: The Stinging Organelles

One of the most characteristic structures in the entire cnidarian group is the stinging organelle called the **nematocyst**

figure 18-7

Hydra with developing bud and ovary.

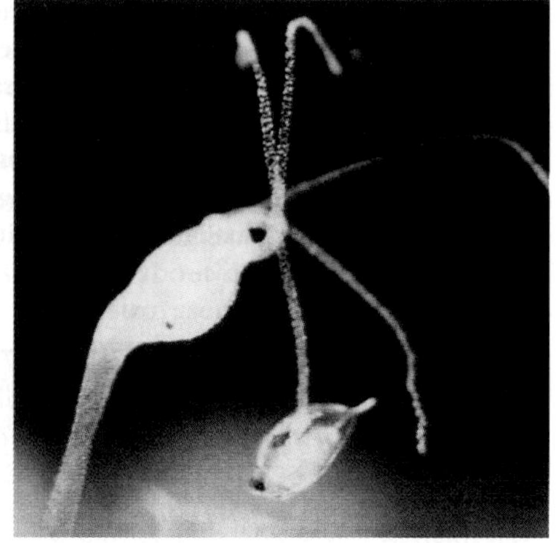

figure 18-8

Hydra catches an unwary water flea with the nematocysts of its tentacles. This hydra already contains one water flea eaten previously.

is encircled by six to ten hollow tentacles that, like the body, are greatly extended when the animal is hungry.

The mouth opens into the **gastrovascular cavity,** which communicates with the cavities in tentacles. In some individuals **buds** may project from the sides, each with a mouth and tentacles like the parent. Testes or ovaries, when present, appear as rounded projections on the surface of the body (Figure 18-7).

Hydras feed on a variety of small crustaceans, insect larvae, and annelid worms. The hydra awaits its prey with tentacles extended (Figure 18-8). The food organism that brushes against its tentacles may find itself harpooned by scores of nematocysts that render it helpless, even though it may be larger than the hydra. The tentacles move the prey toward the mouth, which slowly widens. Well moistened with mucous secretions, the mouth glides over and around the prey, totally engulfing it.

The activator that actually causes the mouth to open is the reduced form of **glutathione,** which is found to some extent in all living cells. Glutathione is released from the prey through the wounds made by the nematocysts, but only those animals releasing enough of the chemical to activate the feeding response are eaten by the hydra. This explains how a hydra distinguishes between *Daphnia,* which it relishes, and some other forms that it refuses. When glutathione is added to water containing hydras, each hydra will go through the motions of feeding, even though no prey is present.

In asexual reproduction, buds appear as outpocketings of the body wall and develop into young hydras that eventually detach from the parent. In sexual reproduction, temporary gonads (Figure 18-7) usually appear in the autumn, stimulated by the lower temperatures and perhaps also by the reduced aeration of stagnant waters. Eggs in the ovary usually mature one at a time and are fertilized by sperm shed into the water. A cyst forms around the embryo before it breaks loose from the parent, enabling it to survive the winter. Young hydras hatch out in spring when the weather is favorable.

Hydroid Colonies

Far more representative of class Hydrozoa than the hydras are those hydroids that have a medusa stage in their life cycle. *Obelia* is often used in laboratory exercises for beginning students to illustrate the hydroid type (Figure 18-9).

A typical hydroid has a base, a stalk, and one or more terminal polyps (zooids). The base by which the colonial hydroids are attached to the substratum is a rootlike stolon, which gives rise to one or more stalks. The living cellular part of the stalks secrete a nonliving chitinous sheath. Attached to the ends of the branches of the stalks are the individual zooids. Most of the zooids are feeding polyps called **hydranths,** or **gastrozooids.** They may be tubular, bottle shaped, or vaselike, but all have a terminal mouth and a circlet of tentacles. In some forms, such as *Obelia,* the chitinous sheath continues as a protective cup around the polyp into which it can withdraw for protection (Figure 18-9). In others the polyp is naked. **Dactylozooids** are polyps specialized for defense.

The hydranths, much like hydras, capture and ingest prey, such as tiny crustaceans, worms, and larvae, thus providing nutrition for the entire colony. After partial digestion in the hydranth, the digestive broth passes into the common gastrovascular cavity where intracellular digestion occurs.

Circulation within the gastrovascular cavity is a function of the ciliated gastrodermis, but rhythmical contractions and pulsations of the body, which occur in many hydroids, also aid circulation.

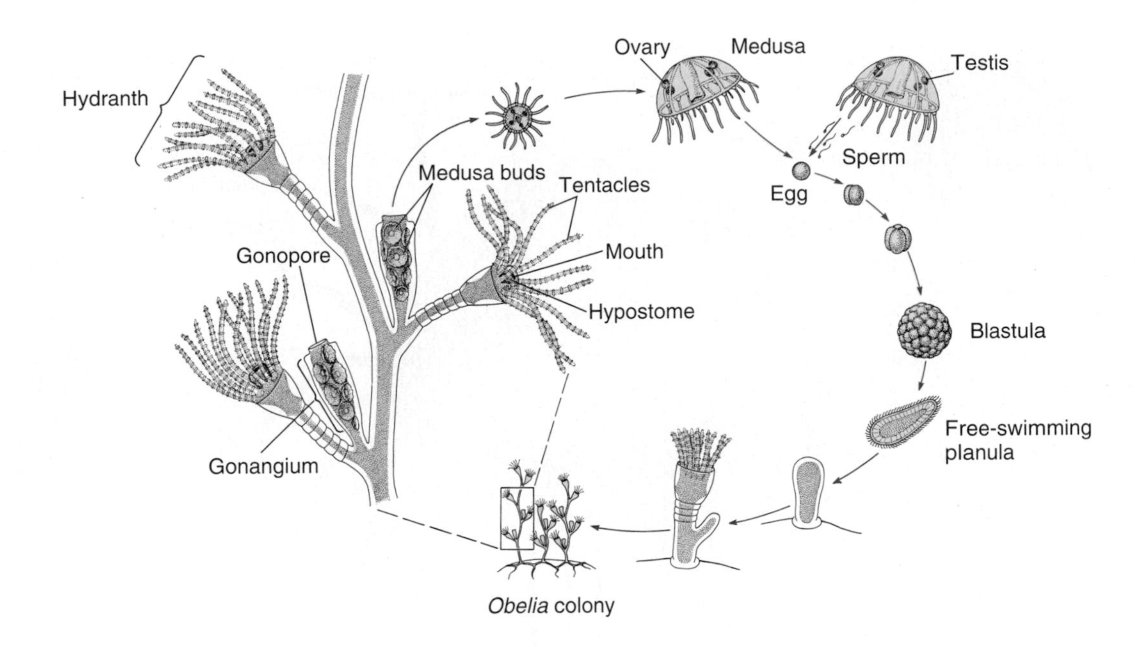

figure 18-9

Life cycle of *Obelia,* showing alternation of polyp (asexual) and medusa (sexual) stages. *Obelia* is a calyptoblastic hydroid; that is, its polyps as well as its stems are protected by continuations of the nonliving covering.

In contrast to hydras, new individuals that bud do not detach from the parent; thus the size of the colony increases. The new polyps may be hydranths or reproductive polyps known as **gonangia.** Medusae are produced by budding within the gonangia. The young medusae leave the colony as free-swimming individuals that mature and produce gametes (eggs and sperm) (Figure 18-9). In some species the medusae remain attached to the colony and shed their gametes there. In other species the medusae never develop, the gametes being shed by male and female gonophores. Development of the zygote results in a ciliated planula larva that swims about for a time. Then it settles down to a substratum to develop into a minute polyp that gives rise, by asexual budding, to the hydroid colony, thus completing the life cycle.

Hydroid medusae are usually smaller than their scyphozoan counterparts, ranging from 2 or 3 mm to several centimeters in diameter (Figure 18-10). The margin of the bell projects inward as a shelflike **velum,** which partly closes the open side of the bell and is used in swimming (Figure 18-11). Muscular pulsations that alternately fill and empty the bell propel the animal forward, aboral side first, with a sort of "jet propulsion." The tentacles attached to the bell margin are richly supplied with nematocysts.

The mouth opening at the end of a suspended **manubrium** leads to a stomach and four radial canals that connect with a ring canal around the margin. This in turn connects with the hollow tentacles. Thus the coelenteron is continuous from mouth to tentacles, and the entire system is lined with gastrodermis. Nutrition is similar to that of the hydranths.

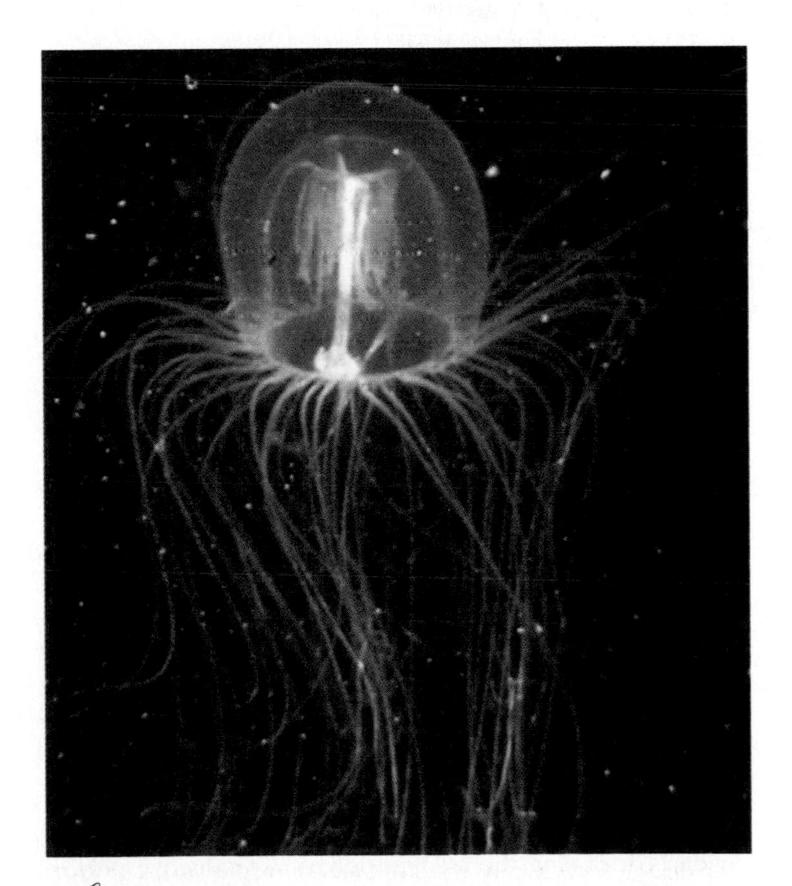

figure 18-10

Bell medusa, *Polyorchis penicillatus,* medusa stage of an unknown attached polyp.

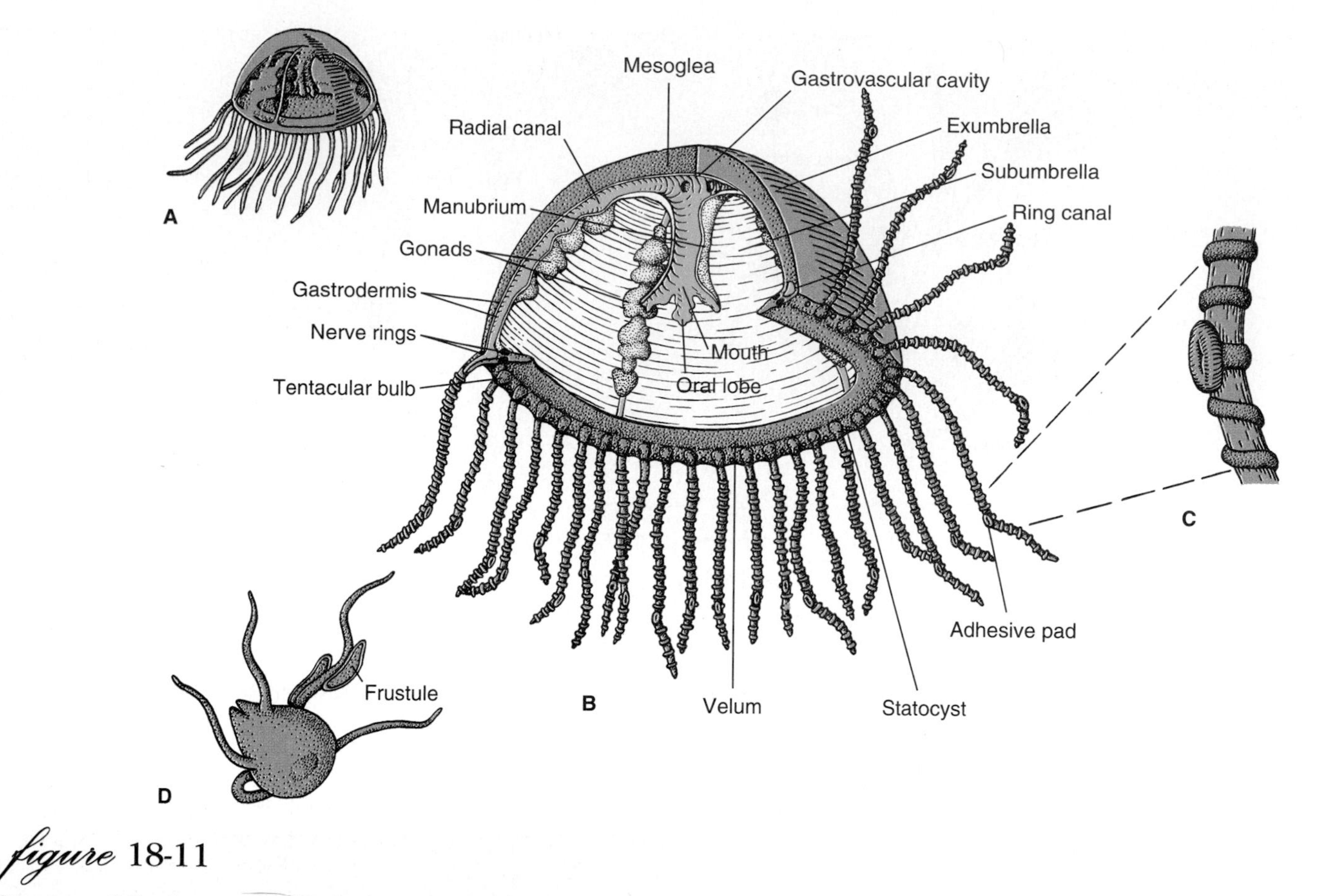

figure 18-11

Structure of *Gonionemus.* **A,** Medusa with typical tetramerous arrangement. **B,** Cutaway view showing morphology. **C,** Portion of a tentacle with its adhesive pad and ridges of nematocysts. **D,** Tiny polyp, or hydroid stage, that develops from the planula larva. It can produce more polyps by budding (frustules) or produce medusa buds.

The nerve net is usually concentrated into two nerve rings at the base of the velum. The bell margin is liberally supplied with sensory cells. It usually also bears two kinds of specialized sense organs: **statocysts,** which are small organs of equilibrium (Figure 18-11), and **ocelli,** which are light-sensitive organs.

Other Hydrozoans

Some hydrozoans form floating colonies, such as *Physalia* (Gr. *physallis,* bladder), the Portuguese man-of-war (Figure 18-12). These colonies include several types of modified medusae and polyps. *Physalia* has a rainbow-hued float, probably a modified polyp, which carries it along at the mercy of the winds and currents. It contains an air sac filled with secreted gas and acts as a carrier for the generations of individuals that bud from it and hang suspended in the water. There are several types of individuals, including the feeding polyps, reproductive polyps, long stinging tentacles, and the so-called jelly polyps. Many swimmers have experienced the uncomfortable sting that these colonial floaters can inflict. The pain, along with the panic of the swimmer, can increase the danger of drowning.

Other hydrozoans secrete massive calcareous skeletons that resemble true corals (Figure 18-13). They are sometimes called **hydrocorals.**

Class Scyphozoa

Class Scyphozoa (si-fo-zo′a) (Gr. *skyphos,* cup) includes most of the larger jellyfishes, or "cup animals." A few, such as *Cyanea* (Gr. *kyanos,* dark-blue substance), may attain a bell diameter exceeding 2 m and tentacles 60 to 70 m long (Figure 18-14). Most scyphozoans, however, range from 2 to 40 cm in diameter. Most are found floating in the open sea, some even at depths of 3000 m, but one unusual order is sessile and attaches by a stalk to seaweeds and other objects on the sea bottom (Figure 18-15). Their coloring may range from colorless to striking orange and pink hues.

Scyphomedusae, unlike hydromedusae, have no velum. The bells of different species vary in depth from a shallow saucer shape to a deep helmet or goblet shape, and in many the margin is scalloped, each notch bearing a sense organ called a **rhopalium** and a pair of lobelike projections called lappets. *Aurelia* (L. *aurum,* gold) has eight such notches (Figure 18-16);

others may have four or sixteen. Each rhopalium bears a statocyst for balance, two sensory pits containing concentrations of sensory cells, and sometimes an ocellus (simple eye) for photoreception. The mesoglea is thick and contains cells as well as fibers. The stomach is usually divided into pouches containing small tentacles with nematocysts.

The mouth is centered on the subumbrellar side. The manubrium is usually drawn out into four frilly **oral lobes** that are used in food capture and ingestion. The marginal tentacles may be many or few and may be short, as in *Aurelia,* or long, as in *Cyanea.* The tentacles, manubrium, and often the entire body surface of scyphozoans are well supplied with nematocysts. Scyphozoans feed on all sorts of small organisms, from protozoa to fishes. Capture of prey involves stinging and manipulation with tentacles and oral arms, but the methods vary. *Aurelia* feeds on small planktonic animals. These are caught in the mucus of the umbrella surface, carried to "food pockets" on the umbrella margin by cilia, and picked up from the pockets by the oral lobes whose cilia carry the food to

 18-12

A Portuguese man-of-war colony, *Physalia physalis* (order Siphonophora, class Hydrozoa). Colonies often drift onto southern ocean beaches, where they are a hazard to bathers. Each colony of medusa and polyp types is integrated to act as one individual. As many as a thousand zooids may be found in one colony. The nematocysts secrete a powerful neurotoxin.

A

B

figure 18-13

These hydrozoans form calcareous skeletons that resemble true coral. **A,** *Stylaster roseus* (order Stylasterina) occurs commonly in caves and crevices in coral reefs. These fragile colonies branch in only a single plane and may be white, pink, purple, red, or red with white tips. **B,** Species of *Millepora* (order Milleporina) form branching or platelike colonies and often grow over the horny skeleton of gorgonians (see Figure 18-26). They have a generous supply of powerful nematocysts that produce a burning sensation on human skin, justly earning the common name fire coral.

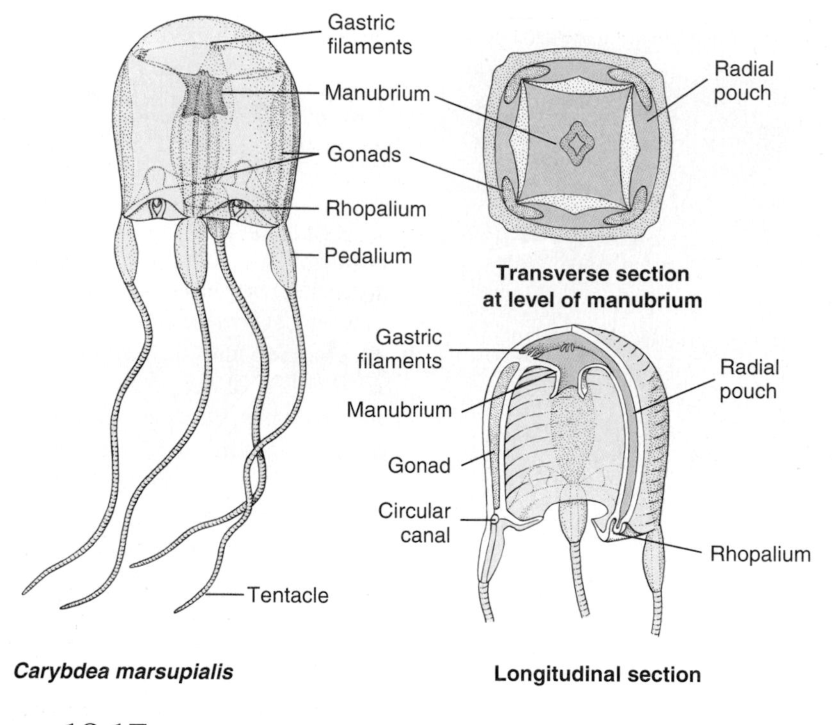

figure 18-17

Carybdea, a cubozoan medusa.

Sea anemones are cylindrical in form with a crown of tentacles arranged in one or more circles around the mouth on the flat **oral disc** (Figure 18-20). The slit-shaped mouth leads into a **pharynx.** At one or both ends of the mouth is a ciliated groove called a **siphonoglyph,** which extends into the pharynx. The siphonoglyphs create water currents directed into the pharynx. The cilia elsewhere on the pharynx direct water outward. The currents thus created carry in oxygen and remove wastes. They also help maintain an internal fluid pressure or a hydrostatic skeleton that serves as a support for opposing muscles.

The pharynx leads into a large **gastrovascular cavity** that is divided into radial chambers by means of pairs of septa that extend vertically from the body wall toward the pharynx (Figure 18-20). These chambers communicate with each other and are open below the pharynx. In many anemones the lower ends of the septal edges are prolonged into **acontia threads,** also provided with nematocysts and gland cells, that can be protruded through the mouth or through pores in the body wall to help overcome prey or provide defense. The pores also aid in the rapid discharge of water from the body when the animal is endangered and contracts to a small size.

figure 18-18

Sea anemones are the familiar and colorful "flower animals" of tide pools, rocks, and pilings of the intertidal zone. Most, however, are subtidal, their beauty seldom revealed to human eyes. These are rose anemones, *Tealia piscivora.*

Anemones form some interesting mutualistic relationships with other organisms. Many anemones house unicellular algae in their tissues (as do reef-building corals), from which they undoubtedly derive some nutrients. Some hermit crabs place anemones on the snail shells in which the crabs live, gaining some protection by the presence of the anemone, while the anemone dines on particles of food dropped by the crab. The anemonefishes (Figure 18-21) of the tropical Indo-Pacific form associations with large anemones. An unknown property of the skin mucus of the fish causes the anemone's nematocysts not to discharge, but if some other fish is so unfortunate as to brush the anemone's tentacles, it is likely to become a meal.

of them are quite colorful. Anemones are found in coastal areas all over the world, especially in the warmer waters, and they attach by means of their pedal discs to shells, rocks, timber, or whatever submerged substrata they can find. Some burrow in the bottom mud or sand.

Sea anemones are carnivorous, feeding on fish or almost any live animals of suitable size. Some species live on minute forms caught by ciliary currents.

The sexes are separate in some sea anemones, and some are hermaphroditic. The gonads are arranged on the

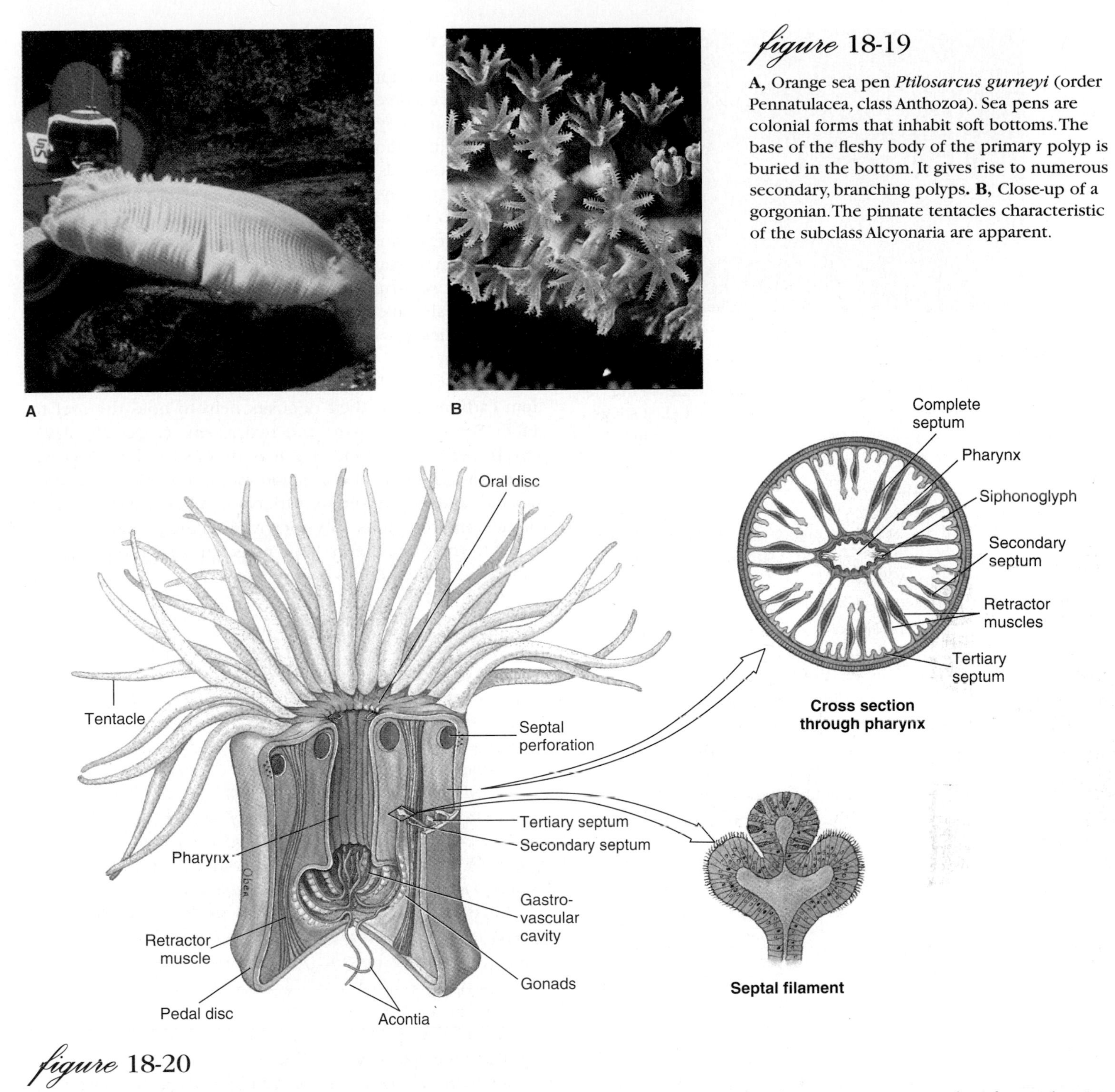

figure 18-19

A, Orange sea pen *Ptilosarcus gurneyi* (order Pennatulacea, class Anthozoa). Sea pens are colonial forms that inhabit soft bottoms. The base of the fleshy body of the primary polyp is buried in the bottom. It gives rise to numerous secondary, branching polyps. **B,** Close-up of a gorgonian. The pinnate tentacles characteristic of the subclass Alcyonaria are apparent.

figure 18-20

Structure of the sea anemone. The free edges of the septa and the acontia threads are equipped with nematocysts to complete the paralyzation of prey begun by the tentacles.

margins of the septa. Fertilization is external in some species, whereas in others the sperm enter the gastrovascular cavity to fertilize the eggs. The zygote develops into a ciliated larva. Asexual reproduction commonly occurs by **pedal laceration.** Small pieces of the pedal disc break off as the animal moves, and each of these regenerates a small anemone.

Zoantharian Corals

The zoantharian corals belong to the order Scleractinia, sometimes known as the true or stony corals. The stony corals might be described as miniature sea anemones that live in calcareous cups they themselves have secreted (Figures 18-23 and 18-24). Like that of the anemones, the coral polyp's gastrovascular

figure 18-21

Orangefin anemonefish *(Amphiprion chrysopterus)* nestles in the tentacles of its sea anemone host. Anemonefishes do not elicit stings from their hosts but may lure unsuspecting other fish to become meals for the anemone.

cavity is subdivided by septa arranged in multiples of six (hexamerous) and its hollow tentacles surround the mouth, but there is no siphonoglyph.

Instead of a pedal disc, the epidermis at the base of the column secretes the limy skeletal cup, including the sclerosepta, which project up into the polyp between the true septa (Figure 18-24). The living polyp can retract into the safety of the cup when not feeding. Since the skeleton is secreted below the living tissue rather than within it, the calcareous material is an exoskeleton. In many colonial corals, the skeleton may become massive, building up over many years, with the living coral forming a sheet of tissue over the surface. The gastrovascular cavities of the polyps are all connected through this sheet of tissue.

Alcyonarian Corals

Alcyonarians are often referred to as octocorals because of their strict octomerous symmetry, with eight pinnate tentacles and eight unpaired, complete septa (Figure 18-25). They are all colonial, and the gastrovascular cavities of the polyps communicate through a system of gastrodermal tubes called **solenia.** The tubes run through an extensive mesoglea in most alcyonarians, and the surface of the colony is covered by epidermis. The skeleton is secreted within the mesoglea and consists of limy spicules, fused spicules, or a horny protein, often in combination. Thus the skeletal support of most alcyonarians is an endoskeleton. The variation in pattern among the species of alcyonarians lends great variety to the form of the colonies.

The graceful beauty of the alcyonarians—in hues of yellow, red, orange, and purple—helps create the "submarine gardens" of the coral reefs (Figure 18-26).

Coral Reefs

Coral reefs are among the most productive of all ecosystems, and they have a diversity of life forms rivaled only by the tropical rain forest. They are large formations of calcium carbonate (limestone) in shallow tropical seas laid down by living organisms over thousands of years; living plants and animals are confined to the top layer of reefs where they add more calcium carbonate to that deposited by their predecessors. The most important organisms that take dissolved calcium and carbonate ions from seawater and precipitate it as limestone to form reefs are the **reef-building corals** and **coralline algae.** Reef-building corals have mutualistic algae **(zooxanthellae)** living in their tissues. Coralline algae are several types of red algae, and they may be encrusting or form upright, branching growths. Not only do they contribute to the total mass of calcium carbonate, but their deposits help to hold the reef together. Some alcyonarians and hydrozoans (especially *Millepora* [L. *mille,* a thousand, + *porus,* pore] spp., the "fire coral," Figure 18-13) contribute in some measure to the calcareous material, and an enormous variety of other organisms contributes small amounts. However, reef-building corals seem essential to the formation of large reefs, since such reefs do not occur where these corals cannot live.

Because zooxanthellae are vital to reef-building corals, and water absorbs light, reef-building corals rarely live below a depth of 30 m (100 feet). Interestingly, some deposits of coral reef limestone, particularly around Pacific islands and atolls, reach great thickness—even thousands of feet. Clearly the corals and other organisms could not have grown from the bottom in the abyssal blackness of the deep sea and reached shallow water where light could penetrate. Charles Darwin was the first to realize that such reefs began their growth in shallow *water around volcanic islands; then, as the islands slowly sank beneath the sea, the growth of the reefs kept up with the rate of sinking, thus accounting for the depth of the deposits.*

Despite their great intrinsic and economic value, coral reefs in many areas are threatened today by a variety of factors, mostly of human origin. These include nutrients from sewage and agricultural fertilizer that wash into the water from land. Agricultural pesticides, as well as sediment from tilled fields, also contribute to reef degradation. Corals in the Persian Gulf have withstood a surprising amount of pollution, high salinity, and temperature swings—much more than reefs in other parts of the world have been able to endure. They apparently have survived the greatest oil slick ever created by humans (in the Gulf War of 1991).

Classification of Phylum Cnidaria

Class Hydrozoa (hi-dro-zo′a) (Gr. *hydra,* water serpent, + *zōon,* animal). Solitary or colonial; asexual polyps and sexual medusae, although one type may be suppressed; hydranths with no mesenteries; medusae (when present) with a velum; both fresh water and marine. Examples: *Hydra, Obelia, Physalia, Tubularia.*

Class Scyphozoa (si-fo-zo′a) (Gr. *skyphos,* cup, + *zōon,* animal). Solitary; polyp stage reduced or absent; bell-shaped medusae without velum; gelatinous mesoglea much enlarged; margin of bell or umbrella typically with eight notches that are provided with sense organs; all marine. Examples: *Aurelia, Cassiopeia, Rhizostoma.*

Class Cubozoa (ku′bo-zo′a) (Gr. *kybos,* a cube, + *zōon,* animal). Solitary; polyp stage reduced; bell-shaped medusae square in cross section, with tentacle or group of tentacles hanging from a bladelike pedalium at each corner of the umbrella; margin of umbrella entire, without velum but with velarium; all marine. Examples: *Tripedalia, Carybdea, Chironex, Chiropsalmus.*

Class Anthozoa (an-tho-zo′a) (Gr. *anthos,* flower, + *zōon,* animal). All polyps; no medusae; solitary or colonial; enteron subdivided by mesenteries or septa bearing nematocysts; gonads endodermal; all marine.

Subclass Zoantharia (zo′an-tha′re-a) (N.L. from Gr. *zōon,* animal, + *anthos,* flower, + L. *aria,* like or connected with) **(Hexacorallia).** With simple unbranched tentacles; mesenteries in pairs, in multiples of six; sea anemones, hard corals, and others. Examples: *Metridium, Anthopleura, Tealia, Astrangia, Acropora.*

Subclass Ceriantipatharia (se′re-ant-ip′a-tha′re-a) (N.L. combination of Ceriantharia and Antipatharia). With simple unbranched tentacles; mesenteries unpaired, initially six; tube anemones and black or thorny corals. Examples: *Cerianthus, Antipathes, Stichopathes.*

Subclass Alcyonaria (al′ce-o-na′re-a) (Gr. *alkonion,* kind of sponge resembling nest of kingfisher [*alkyon,* kingfisher], + L. *aria,* like or connected with) **(Octocorallia).** With eight pinnate tentacles; eight complete, unpaired mesenteries; soft and horny corals. Examples: *Tubipora, Alcyonium, Gorgonia, Plexaura, Renilla.*

figure 18-22

A sea anemone that swims. When attacked by a predatory sea star *Dermasterias,* the anemone *Stomphia didemon* detaches from the bottom and rolls or swims spasmodically to a safer location.

The distribution of coral reefs in the world is limited to locations that offer optimal conditions for their zooxanthellae. They require warmth, light, and the salinity of undiluted seawater, thus limiting coral reefs to shallow waters between 30° N and 30° S latitude and excluding them from areas with upwelling of cold water or areas near major river outflows with attendant low salinity and high turbidity. Photosynthesis and fixation of carbon dioxide by the zooxanthellae furnish food molecules for their hosts, they recycle phosphorus and nitrogenous waste compounds that otherwise would be lost, and they enhance the ability of the coral to deposit calcium carbonate.

figure 18-23

A, Cup coral *Tubastrea* sp. The polyps form clumps resembling groups of sea anemones. Although often found on coral reefs, *Tubastrea* is not a reef-building coral and has no symbiotic zooxanthellae in its tissues. **B,** The polyps of *Montastrea cavernosa* are tightly withdrawn in the daytime but open to feed at night, as in **C.**

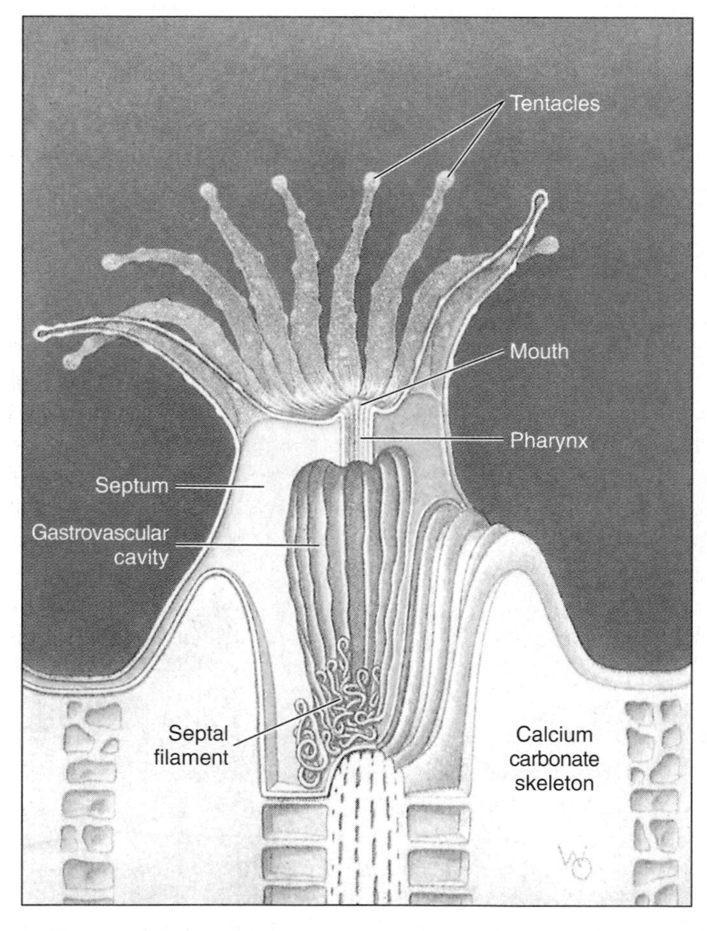

figure 18-24

Polyp of a zoantharian coral (order Scleractinia) showing calcareous cup (exoskeleton), gastrovascular cavity, sclerosepta, septa, and septal filaments.

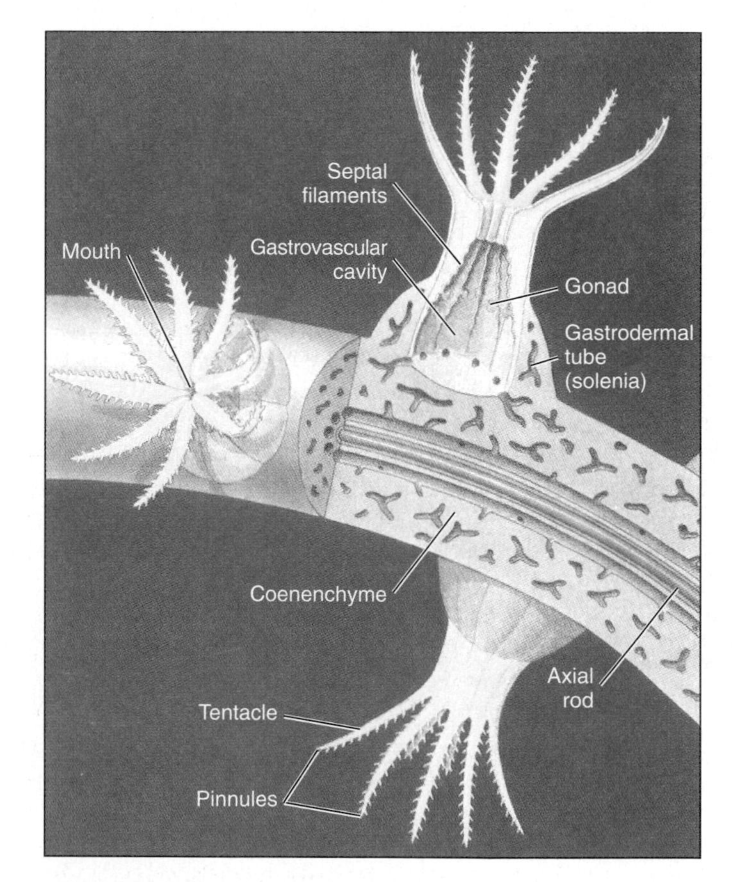

figure 18-25

Polyps of an alcyonarian coral (octocoral). Note the eight pinnate tentacles, coenenchyme, and solenia. They have an endoskeleton of limy spicules often with a horny protein, which may be in the form of an axial rod.

figure 18-26

Sea fan, *Subergorgia mollis* (order Gorgonacea, subclass Alcyonaria), on a Pacific coral reef. The colonial gorgonian, or horny, corals are conspicuous components of reef faunas.

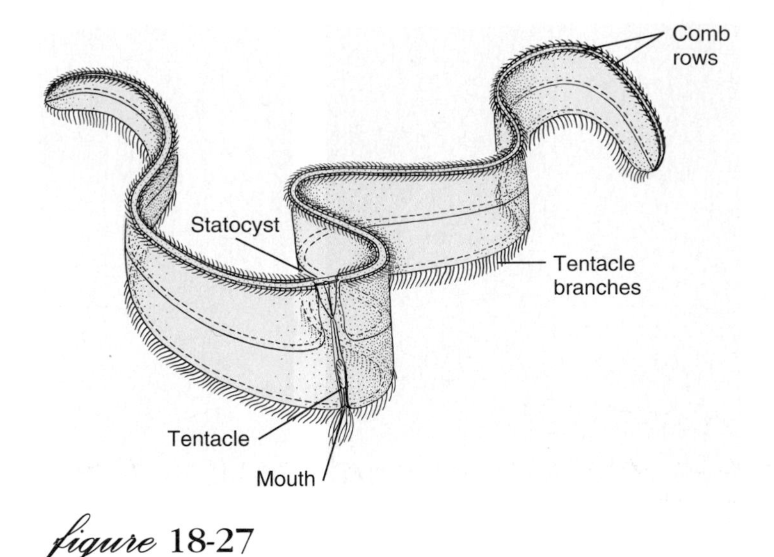

figure 18-27

Venus' girdle (*Cestum* sp.), a highly modified ctenophore. It may reach a length of 5 feet but is usually much smaller.

Phylum Ctenophora

Ctenophora (te-nof′o-ra) (Gr. *kteis, ktenos,* comb, + *phora,* pl. of bearing) is composed of fewer than 100 species. All are marine forms occurring in all seas but especially in warm waters. They take their name from the eight rows of comblike plates they bear for locomotion. Common names for ctenophores are "sea walnuts" and "comb jellies." Ctenophores, along with cnidarians, represent the only two phyla having primary radial symmetry, in contrast to other metazoans, which have primary bilateral symmetry.

Ctenophores do not have nematocysts, except in one species (*Haeckelia rubra,* after Ernst Haeckel, nineteenth-century German zoologist) that carries nematocysts on certain regions of its tentacles but lacks colloblasts. These nematocysts are apparently appropriated from cnidarians on which it feeds.

In common with cnidarians, ctenophores have not advanced beyond the tissue grade of organization. There are no definite organ systems in the strict meaning of the term.

Except for a few creeping and sessile forms, ctenophores are free-swimming. Although they are feeble swimmers and are more common in surface waters, ctenophores are sometimes found at considerable depths. Highly modified forms such as *Cestum* (L. *cestus,* girdle) use sinuous body movements as well as their comb plates in locomotion (Figure 18-27).

The fragile, transparent bodies of ctenophores are easily seen at night when they emit light (luminesce).

Form and Function

Pleurobrachia (Gr. *pleuron,* side, + L. *brachia,* arms), a pretty little sea walnut, is often used as a representative example of the ctenophores (Figure 18-28). Its surface bears eight longitudinal rows of transverse plates bearing long fused cilia and called **comb plates.** The beating of the cilia in each row starts at the aboral end and proceeds along the rows to the oral end, thus propelling the animal forward. All rows beat in unison. A reversal of the wave direction drives the animal backward. Ctenophores may be the largest animals that swim exclusively by cilia.

Two long tentacles are carried in a pair of tentacle sheaths (Figure 18-29) from which they can stretch to a length of perhaps 15 cm. The surface of the tentacles bears specialized glue cells called **colloblasts,** which secrete a sticky substance that facilitates the catching of small prey. When covered with food, the tentacles contract and the food is wiped off on the mouth. The gastrovascular cavity consists of a pharynx, stomach, and a system of gastrovascular canals. Rapid digestion occurs in the pharynx, then partly digested food circulates through the rest of the system where digestion is completed intracellularly. Residues are regurgitated or expelled through small pores in the aboral end.

A nerve net system similar to that of the cnidarians includes a subepidermal plexus concentrated under each comb plate.

The sense organ at the aboral pole is a **statocyst,** or organ of equilibrium, and is also concerned with the beating of the comb rows but does not trigger their beat. Other sensory cells are abundant in the epidermis.

Acoelomate Animals:
Flatworms, Ribbon Worms, and Jaw Worms

chapter | nineteen

Getting Ahead

For animals that spend their lives sitting and waiting, as do most members of the two radiate phyla we considered in the preceding chapter, radial symmetry is ideal. One side of the animal is just as important as any other for snaring prey coming from any direction. But if an animal is active in seeking food, shelter, home sites, and reproductive mates, it requires a different set of strategies and a new body organization. Active, directed movement requires an elongated body form with head (anterior) and tail (posterior) ends. In addition, one side of the body faces up (dorsal) and the other side, specialized for locomotion, faces down (ventral). These conditions resulted in a bilaterally symmetrical animal in which the body could be divided along only one plane of symmetry to yield two halves which were mirror images of each other. Furthermore, because it is better to determine where one is going than where one has been, sense organs and centers for nervous control came to be located at the anterior end. This process is called cephalization. Thus cephalization and primary bilateral symmetry evolved together.

The three acoelomate phyla considered in this chapter are not much more complex in organization than the Radiata except for symmetry. The evolutionary consequence of that difference alone was enormous, however, for bilaterality is the type of symmetry assumed by all more complex animals.

The term "worm" has been applied loosely to elongated, bilateral invertebrate animals without appendages. At one time zoologists considered worms (Vermes) a group in their own right. Such a group included a highly diverse assortment of forms. This unnatural assemblage has been reclassified into various phyla. By tradition, however, zoologists still refer to the various groups of these animals as flatworms, ribbon worms, roundworms, segmented worms, and the like. In this chapter we will consider the Platyhelminthes (Gr. *platys,* flat, + *helmins,* worm), or flatworms, the Nemertea (Gr. *Nemertes,* one of the nereids, unerring one), or ribbon worms, and the Gnathostomulida (Gr. *gnathos,* jaw, + *stoma,* mouth, + L. *ulus,* diminutive), or jaw worms. The Platyhelminthes is by far the most widely prevalent and abundant of the three phyla, and some flatworms are very important pathogens of humans and domestic animals.

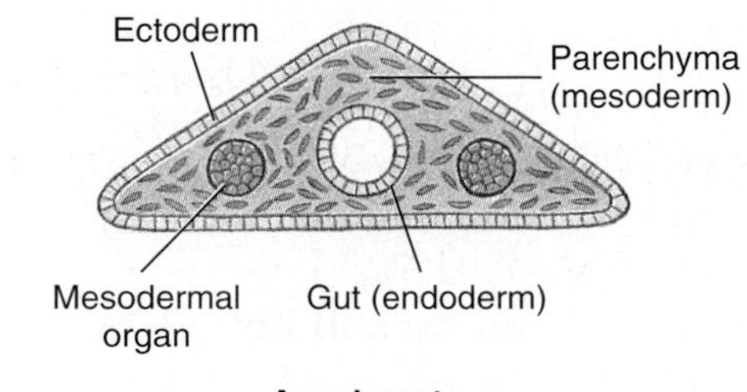

figure 19-1

Acoelomate body plan.

Phylum Platyhelminthes

The Platyhelminthes were derived from an ancestor that probably had many cnidarian-like characteristics, including a gelatinous mesoglea. Nonetheless, replacement of the gelatinous mesoglea with a cellular, mesodermal **parenchyma** laid the basis for a more complex organization. Parenchyma is a form of tissue containing more cells and fibers than the mesoglea of the cnidarians.

Flatworms range in size from a millimeter or less to some of the tapeworms that are many meters in length. Their typically flattened bodies may be slender, broadly leaflike, or long and ribbonlike.

Ecological Relationships

The flatworms include both free-living and parasitic forms, but the free-living members are found exclusively in the class Turbellaria. A few turbellarians are symbiotic (commensals or parasites), but the majority are adapted as bottom dwellers in marine or fresh water or live in moist places on land. Many,

Position in Animal Kingdom

1. The Platyhelminthes, or flatworms, the Nemertea, or ribbon worms, and the Gnathostomulida, or jaw worms, are the simplest animals to have **primary bilateral symmetry.**
2. These phyla have only one internal space, the digestive cavity, with the region between the ectoderm and endoderm filled with mesoderm in the form of muscle fibers and mesenchyme (parenchyma). Since they lack a coelom or a pseudocoelom, they are termed **acoelomate animals** (Figure 19-1), and because they have three well-defined germ layers, they are termed triploblastic.
3. Acoelomates show more specialization and division of labor among their organs than do the radiate animals because having mesoderm makes more elaborate organs possible. Thus the acoelomates are said to have reached the **organ-system level of organization.**
4. They belong to the protostome division of the Bilateria and have spiral cleavage, and at least the platyhelminths and nemerteans have mosaic (determinate) cleavage.

Biological Contributions

1. The acoelomates developed the basic **bilateral** plan of organization that has been widely exploited in the animal kingdom.
2. The **mesoderm** developed into a well-defined embryonic germ layer **(triploblastic),** making available a great source of tissues, organs, and systems.
3. Along with bilateral symmetry, **cephalization** was established. There is some centralization of the nervous system evident in the **ladder type of system** found in flatworms.
4. Along with the subepidermal musculature, there is also a mesenchymal system of muscle fibers.
5. They are the simplest animals with an **excretory system.**
6. The nemerteans are the simplest animals to have a **circulatory system** with blood and a **one-way alimentary canal.** Although not stressed by zoologists, the rhynchocoel cavity in ribbon worms is technically a true coelom, but because it is merely a part of the proboscis mechanism, it probably is not homologous to the coelom of eucoelomate phyla.
7. Unique and specialized structures occur in all three phyla. The parasitic habit of many flatworms has led to many specialized adaptations, such as organs of adhesion.

Characteristics of Phylum Platyhelminthes

1. Three germ layers **(triploblastic)**
2. **Bilateral symmetry;** definite polarity of anterior and posterior ends
3. **Body flattened dorsoventrally** in most; oral and genital apertures mostly on ventral surface
4. Body with multiple reproductive units in one class (Cestoda)
5. Epidermis may be cellular or syncytial (ciliated in some); **rhabdites** in epidermis of most Turbellaria; epidermis a syncytial **tegument** in Monogenea, Trematoda, and Cestoda
6. Muscular system of mesodermal origin, in the form of a sheath of circular, longitudinal, and oblique layers beneath the epidermis or tegument
7. No internal body space (acoelomate) other than digestive tube; spaces between organs filled with parenchyma
8. Digestive system incomplete (gastrovascular type); absent in some
9. Nervous system consisting of a **pair of anterior ganglia** with **longitudinal nerve cords** connected by transverse nerves and located in the parenchyma in most forms; in forms with more primitive characters the nervous system is similar to that of cnidarians
10. Simple sense organs; eyespots in some
11. Excretory system of two lateral canals with branches bearing **flame cells (protonephridia);** lacking in some forms
12. Respiratory, circulatory, and skeletal systems lacking; lymph channels with free cells in some trematodes
13. Most forms monoecious; reproductive system complex, usually with well-developed gonads, ducts, and accessory organs; internal fertilization; life cycle simple in free-swimming forms and those with single hosts; complicated life cycle often involving several hosts in many internal parasites.
14. Class Turbellaria mostly free-living; classes Monogenea, Trematoda, and Cestoda entirely parasitic.

especially of the larger species, are found on the underside of stones and other hard objects in freshwater streams or in the littoral zones of the ocean.

Relatively few turbellarians live in fresh water. Planarians (Figure 19-2) and some others frequent streams and spring pools; others prefer flowing water of mountain streams. Some species occur in moderately hot springs. Terrestrial turbellarians are found in fairly moist places under stones and logs (Figure 19-3).

All members of the classes Monogenea and Trematoda (the flukes) and the class Cestoda (the tapeworms) are parasitic. Most of the Monogenea are ectoparasites, but all the trematodes and cestodes are endoparasitic. Many species have indirect life cycles with more than one host; the first host is often an invertebrate, and the final host is usually a vertebrate. Humans serve as hosts for a number of species. Certain larval stages may be free-living.

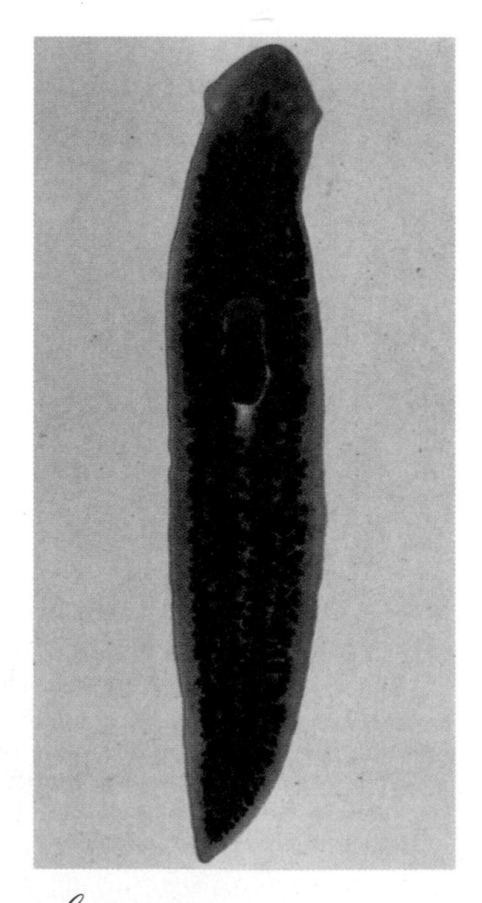

figure 19-2
Stained planarian.

Form and Function

Body Form

The body of turbellarians is covered by a ciliated epidermis resting on a basement membrane. It contains rod-shaped **rhabdites** (Figure 19-4) that, when discharged into water, swell and form a protective mucous sheath around the body. Single-cell mucous glands open on the surface of the epidermis. The body covering in all the other classes is a **tegument** (Figure 19-5), which does not bear cilia in the adult. The cell bodies are sunk beneath the outer layer and superficial muscle layers and communicate with the outer layer **(distal cytoplasm)** by processes extending between the muscles. Because the distal cytoplasm is continuous, with no intervening cell membranes, the tegument is **syncytial.** This peculiar epidermal arrangement probably is related to adaptations for parasitism in ways that are still unclear.

In the body wall beneath the basement membrane are layers of **muscle fibers** that run circularly, longitudinally, and diagonally. Other muscle fibers may cross through the body from one side of the outer muscle sheath to the other (Figure 19-4). A meshwork of **parenchyma** cells, developed from mesoderm, fills the spaces between the muscles

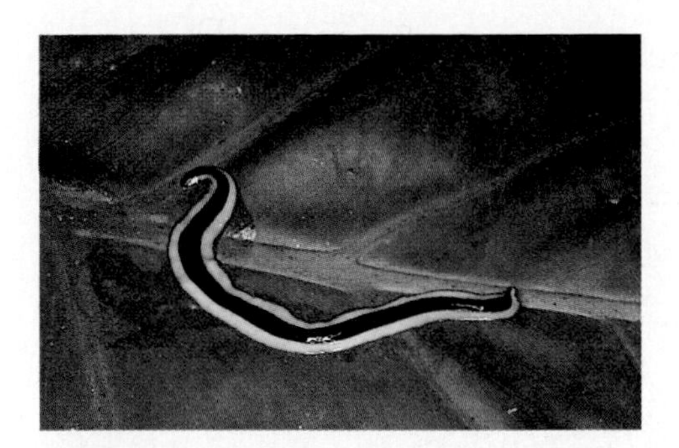

figure 19-3

Terrestrial turbellarian (order Tricladida) from the Amazon Basin, Peru.

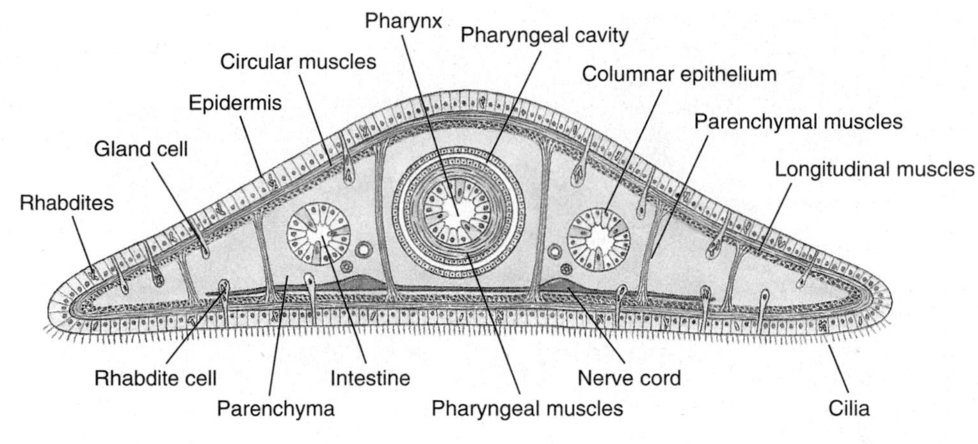

figure 19-4

Cross section of planarian through pharyngeal region, showing relationships of body structures.

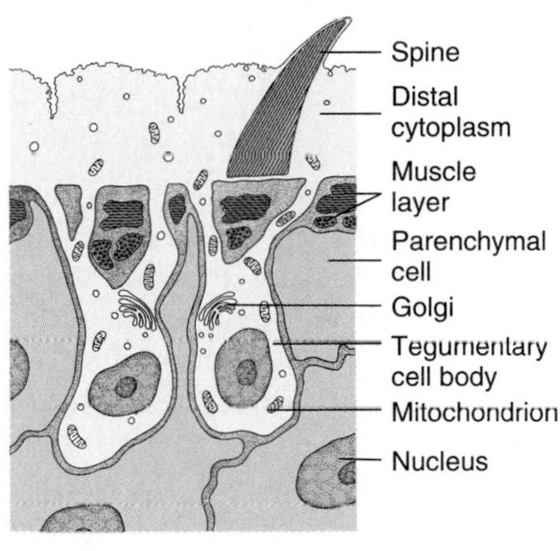

figure 19-5

Diagram of the structure of the tegument of a trematode *Fasciola hepatica.*

and visceral organs. Parenchyma cells in some, perhaps all, flatworms are not a separate cell type, but are the noncontractile portions (cell bodies) of muscle cells.

Earlier investigators had called the distal cytoplasm of parasitic flatworms a cuticle *(L.* cuticula, *dim. of* cutis, *pertaining to the skin). In invertebrate zoology, the term cuticle means an outer, protective layer secreted by an epidermis, which was what the tegment looked like at the light microscope level. When these worms were studied at the electron microscope level, however, biologists found abundant vesicles and mitochondria, that is, it was a living tissue, not a dead, secreted layer. They adopted the word* tegment *(L.* tegmen, *a cover) as being more noncommital. Zoologists often use the word* integument *(also from L.* tegmen*) to refer generally to the body coverings of animals.*

Nutrition and Digestion

The digestive system includes a mouth, pharynx, and intestine. In the turbellarians the muscular **pharynx** opens posteriorly just inside the mouth, through which it can extend (Figure 19-6). The mouth is usually at the anterior end in flukes, and the pharynx is not protrusible. The intestine may be simple or branched.

Intestinal secretions contain proteolytic enzymes for some **extracellular digestion.** Food is sucked into the intestine, where cells of the gastrodermis often phagocytize it and complete the digestion **(intracellular).** Undigested food is egested through the pharynx. The entire digestive system is lacking in the tapeworms. They must absorb all of their nutrients as small molecules (predigested by the host) directly through their tegument.

Excretion and Osmoregulation

Except in some turbellarians, the osmoregulatory system consists of canals with tubules that end in **flame cells (protonephridia)** (Figure 19-6A). The flame cell surrounds a small space into which a tuft of flagella projects. In some turbellarians and in the other classes of flatworms, the protonephridia form a **weir** (Old English *wer,* a fence placed in a stream to catch fish). In a weir the rim of the cup formed by the flame cell bears fingerlike projections that interdigitate with similar projections of a tubule cell. The beat of the flagella (resembling a flickering flame) provides a negative pressure to draw fluid through the weir into the space (lumen) enclosed by the tubule cell. The lumen continues into collecting ducts that finally open to the outside by pores. The wall of the duct beyond the flame cell commonly bears folds or microvilli that probably function in reabsorption of certain ions or molecules. It is likely that this system is osmoregulatory in most forms; it is reduced or absent in marine turbellarians, which do not have to expel excess water.

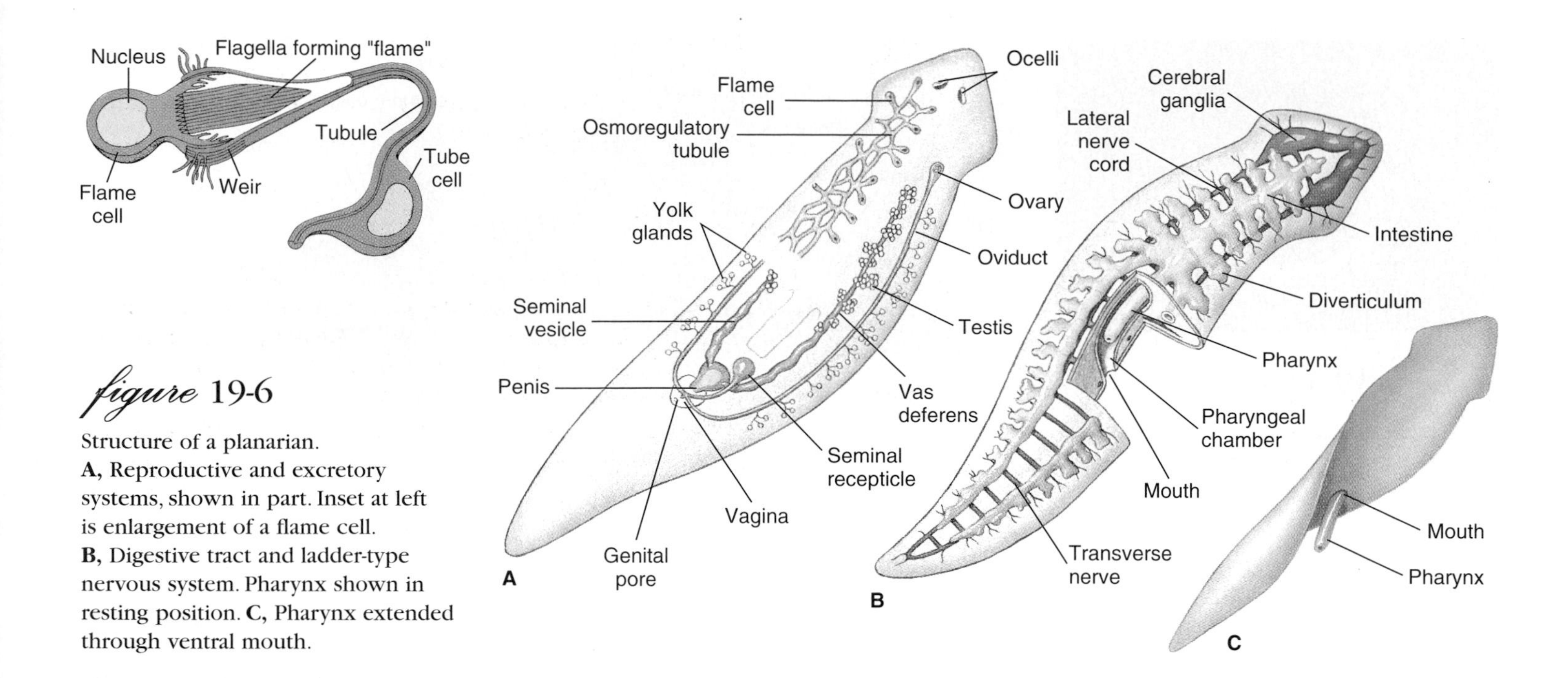

figure 19-6

Structure of a planarian. **A,** Reproductive and excretory systems, shown in part. Inset at left is enlargement of a flame cell. **B,** Digestive tract and ladder-type nervous system. Pharynx shown in resting position. **C,** Pharynx extended through ventral mouth.

Metabolic wastes are largely removed by diffusion through the body wall.

Nervous System

The most primitive flatworm nervous system, found in some turbellarians, is a **subepidermal nerve plexus** resembling the nerve net of the cnidarians. Other flatworms have, in addition to a nerve plexus, one to five pairs of **longitudinal nerve cords** lying under the muscle layer (Figure 19-6B). Connecting nerves form a "ladder-type" pattern. The brain is a mass of ganglion cells arising anteriorly from the nerve cords. Except in simpler turbellarians, which have a diffuse system, the neurons are organized into sensory, motor, and association types—an important advance in the evolution of the nervous system.

Sense Organs

Active locomotion in flatworms has favored not only cephalization in the nervous system but also advancements in the development of sense organs. **Ocelli,** or light-sensitive eyespots, are found in the turbellarians (Figure 19-6A) and some flukes.

Tactile cells and chemoreceptive cells are abundant over the body, and in planarians they form definitive organs on the **auricles** (the earlike lobes on the sides of the head). Some also have **statocysts** for equilibrium and **rheoreceptors** for sensing water current direction.

Reproduction

Many flatworms reproduce both asexually and sexually. Many freshwater turbellarians can reproduce by fission, merely constricting behind the pharynx and separating into two animals, each of which regenerates the missing parts. In some forms such as *Stenostomum* and *Microstomum,* the individuals do not separate at once but remain attached, forming chains of zooids (Figure 19-7B and C). Flukes reproduce asexually in the snail intermediate host (described further below), and some tapeworms, such as *Echinococcus,* can bud off thousands of juveniles in the intermediate host.

Most flatworms are monoecious (hermaphroditic) but practice cross-fertilization. In some turbellarians the yolk for nutrition of the developing embryo is contained within the egg cell itself **(endolecithal),** just as it is normally in other phyla of animals. The endolecithal egg is considered ancestral for flatworms. The other turbellarians plus all trematodes, monogeneans, and cestodes share a derived condition in which the egg cell contains little or no yolk, and the yolk is contributed by cells released from separate organs **(yolk glands).** Usually a number of yolk cells surrounds the zygote within the eggshell **(ectolecithal).** The ectolecithal turbellarians therefore appear to form a clade with the Trematoda, Monogenea, and Cestoda to the exclusion of endolecithal turbellarians.

In some freshwater planarians the capsules are attached by little stalks to the underside of stones or plants, and embryos emerge as juveniles that resemble miniature adults. In other flatworms, the embryo becomes a larva, which may be ciliated or not, according to the group.

Class Turbellaria

Turbellarians are mostly free-living worms that range in length from 5 mm or less to 50 cm. Usually covered with ciliated epi-

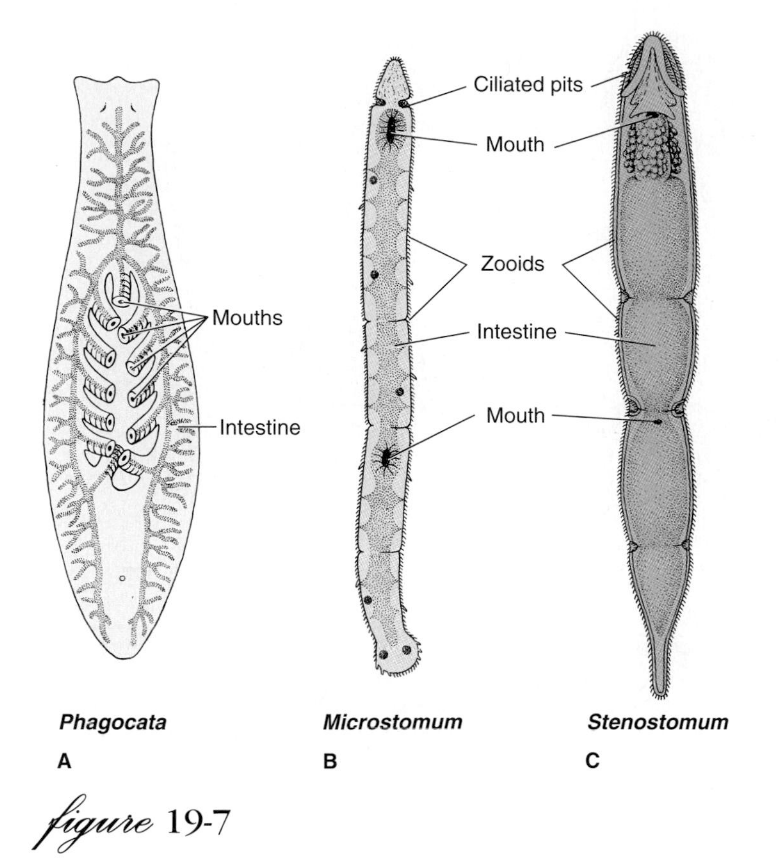

figure 19-7

Some small freshwater turbellarians. **A,** *Phagocata* has numerous mouths and pharynges. **B** and **C,** Incomplete fission results for a time in a series of attached zooids.

dermis, they are typically creeping worms that combine muscular with ciliary movements to achieve locomotion. The mouth is on the ventral side. Unlike the flukes (trematodes) and tapeworms (cestodes), they have simple life cycles.

Members of the order Acoela (Gr. *a,* without, + *koilos,* hollow) are often regarded as having changed least from the ancestral form. Its members are small and have a mouth but no gastrovascular cavity or excretory system. Food is merely passed through the mouth or pharynx into temporary spaces that are surrounded by a syncytial mesenchyme where gastrodermal phagocytic cells digest the food intracellularly. The order has a syncytial epidermis and a diffuse nervous system.

The freshwater planarians, such as *Dugesia* (formerly called *Euplanaria* but changed by priority to *Dugesia* after A.L. Dugès, who first described the form in 1830), belong to one of the more complex orders and are used extensively in introductory laboratories.

Planarians move by gliding, head slightly raised, over a slime track secreted by the marginal adhesive glands. The beating of the epidermal cilia in the slime track drives the animal along. Rhythmical muscular waves can be seen passing backward from the head as it glides.

Planarians are mainly carnivorous, feeding largely on small crustaceans, nematodes, rotifers, and insects. They can detect food from some distance by means of chemoreceptors. They entangle their prey in mucous secretions from the mucous glands and rhabdites. Planarians then grip their prey, encircle it with their bodies, and suck nutrients from it with the proboscis. They also feed on carrion.

Class Trematoda

Trematodes are all parasitic flukes, and as adults they are almost all found as endoparasites of vertebrates. They are chiefly leaflike in form and are structurally similar in many respects to the more complex Turbellaria. A major difference is found in the tegument (described earlier, Figure 19-5).

Other structural adaptations for parasitism are apparent: various penetration glands or glands to produce cyst material; organs for adhesion such as suckers and hooks; and increased reproductive capacity. Otherwise, trematodes share several characteristics with turbellarians, such as a well-developed alimentary canal (but with the mouth at the anterior, or cephalic, end) and similar reproductive, excretory, and nervous systems, as well as a musculature and parenchyma that differ only slightly from those of the Turbellaria. Sense organs are poorly developed.

Of the three subclasses of Trematoda, two are small and poorly known groups, but Digenea (Gr. *dis,* double, + *genos,* descent) is a large group with many species of medical and economic importance.

Digenea

With rare exceptions, digenetic trematodes have a complex life cycle, the first **(intermediate)** host being a mollusc and the final **(definitive)** host being a vertebrate. In some species a second, and sometimes even a third, intermediate host intervenes. The group has many species, and they can inhabit diverse sites in their hosts: all parts of the digestive tract, respiratory tract, circulatory system, urinary tract, and reproductive tract.

One of the world's most amazing biological phenomena is the digenean life cycle. Although the cycles of different species vary widely in detail, a typical example would include the adult, shelled zygote, miracidium, sporocyst, redia, cercaria, and metacercaria stage (Figure 19-8). The shelled embryo or larva usually passes from the definitive host in the excreta and must reach water to develop further. There, it hatches to a free-swimming, ciliated larva, the **miracidium.** The miracidium penetrates the tissues of a snail, where it transforms into a **sporocyst.** The sporocyst reproduces asexually to yield either more sporocysts or a number of **rediae.** The rediae, in turn, reproduce asexually to produce more rediae or to produce **cercariae.** In this way a single zygote can give rise to an enormous number of progeny. The cercariae emerge from the snail and penetrate a second intermediate host or encyst on vegetation or other objects to become **metacercariae,** which are juvenile flukes. The adult grows from the metacercaria when that stage is eaten by the definitive host.

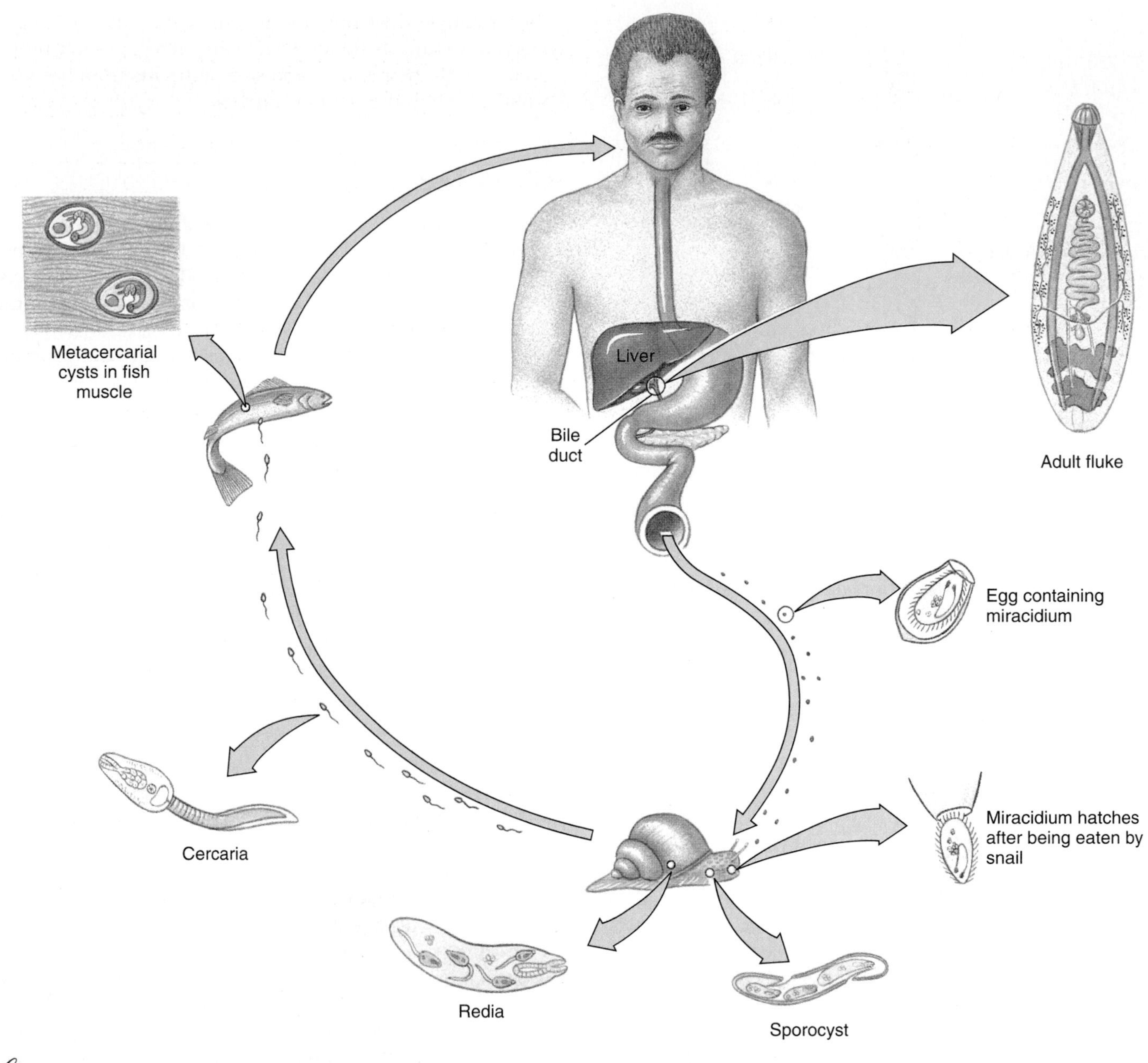

figure 19-8

Life cycle of human liver fluke, *Clonorchis sinensis.*

Some of the most serious parasites of humans and domestic animals belong to the Digenea (Table 19-1).

Clonorchis (Gr. *clon,* branch, + *orchis,* testis) (Figure 19-8) is the most important liver fluke of humans and is common in many regions of the Orient, especially in China, southern Asia, and Japan. Cats, dogs, and pigs are also often infected. The adult lives in the bile passages, and shelled miracidia are shed in the feces. If ingested by certain freshwater snails, the sporocyst and redia stages develop, and free-swimming cercariae emerge. Cercariae that manage to find a suitable fish encyst in the skin or muscles as metacercariae. When the fish is eaten raw, the juveniles migrate up the bile duct to mature and may survive there for 15 to 30 years. The effect of the flukes on humans depends mainly on the extent of the infection. A heavy infection may cause a pronounced cirrhosis of the liver and death.

Schistosomiasis, an infection with blood flukes of the genus *Schistosoma* (Gr. *schistos,* divided, + *soma,* body) (Figure 19-9), ranks as one of the major infectious diseases in the world, with 200 million people infected. The disease is widely prevalent over much of Africa and parts of South America, the West Indies, the Middle East, and the Far East. It is spread when shelled miracidia shed in human feces and urine get into water containing host snails (Figure 19-9B). Cercariae that contact human skin penetrate through the skin to enter blood vessels,

table 19-1
Examples of Flukes Infecting Humans

Common and scientific names	Means of infection; distribution and prevalence in humans
Blood flukes (*Schistosoma* spp.); three widely prevalent species, others reported	Cercariae in water penetrate skin; 200 million people infected with one or more species
S. mansoni	Africa, South and Central America
S. haematobium	Africa
S. japonicum	Eastern Asia
Chinese liver flukes (*Clonorchis sinensis*)	Eating metacercariae in raw fish; about 30 million cases in Eastern Asia
Lung flukes (*Paragonimus* spp.), seven species, most prevalent is *P. westermani*	Eating metacercariae in raw freshwater crabs, crayfish; Asia and Oceania, sub-Saharan Africa, South and Central America; several million cases in Asia
Intestinal fluke (*Fasciolopsis buski*)	Eating metacercariae on aquatic vegetation; 10 million cases in Eastern Asia
Sheep liver fluke (*Fasciola hepatica*)	Eating metacercariae on aquatic vegetation; widely prevalent in sheep and cattle, occasional in humans

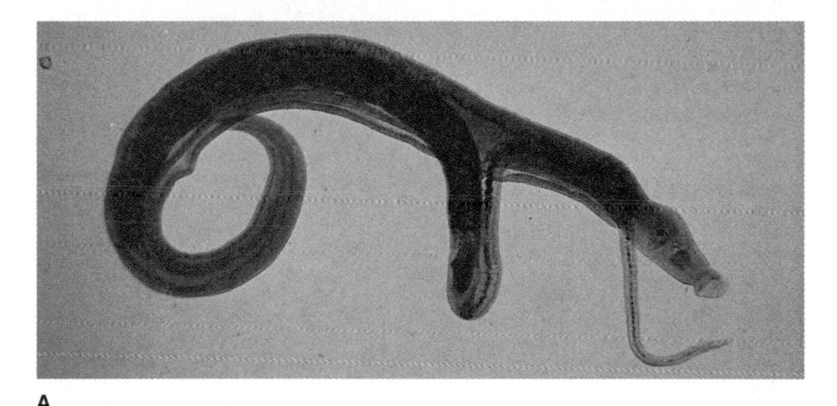
A

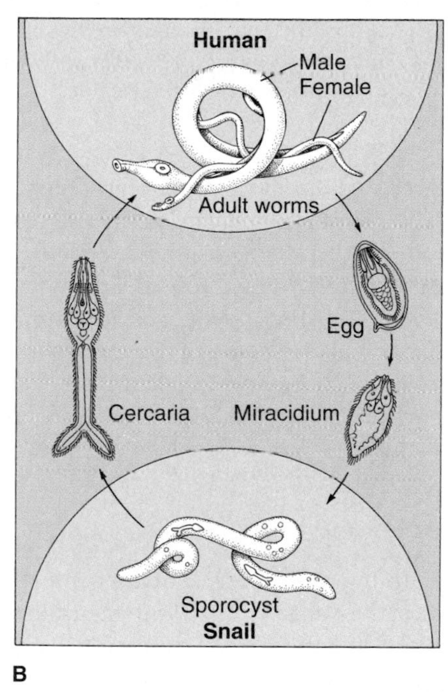

B

figure 19-9

A, Adult male and female *Schistosoma mansoni* in copulation. The blood flukes differ from most other flukes in being dioecious. The male is broader and heavier and has a large, ventral gynecophoric canal, posterior to the ventral sucker. The gynecophoric canal embraces the long, slender female (darkly stained individual) during insemination and oviposition. **B,** Life cycle of *Schistosoma mansoni.*

which they follow to certain favorite regions depending on the type of fluke. *Schistosoma mansoni* lives in the venules draining the large intestine; *S. japonicum* localizes more in the venules draining the small intestine; and *S. haematobium* lives in the venules draining the urinary bladder. In each case many of the eggs released by female worms do not find their way out of the body but lodge in the liver or other organs. There they are sources of chronic inflammation. The inch-long adults may live for years in the human host, and their eggs cause such disturbances as severe dysentery, anemia, liver enlargement, bladder inflammation, and brain damage.

Cercariae of several genera whose normal hosts are birds often enter the skin of human bathers in their search for a suitable bird host, causing a skin irritation known as "swimmer's itch" (Figure 19-10). In this case the human is a dead end in the fluke's life cycle because the fluke cannot develop further in the human.

Class Monogenea

The monogenetic flukes traditionally have been placed as an order of the Trematoda, but they are sufficiently different to deserve a separate class. Cladistic analysis places them closer to the Cestoda. Monogeneans are mostly external parasites that clamp onto the gills and external surfaces of fish using a hooked attachment organ called an **opisthaptor** (Figure 19-11). A few are found in the urinary bladders of frogs and turtles, and one has been reported from the eye of a hippopotamus. Although

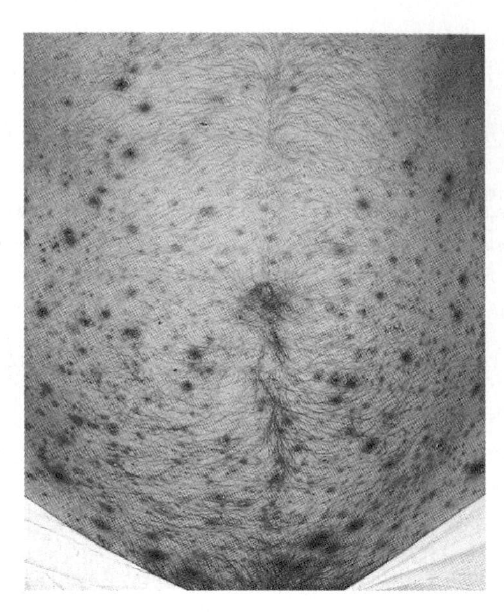

figure 19-10

Human abdomen, showing schistosome dermatitis caused by penetration of schistosome cercariae that are unable to complete development in humans. Sensitization to allergenic substances released by cercariae results in rash and itching.

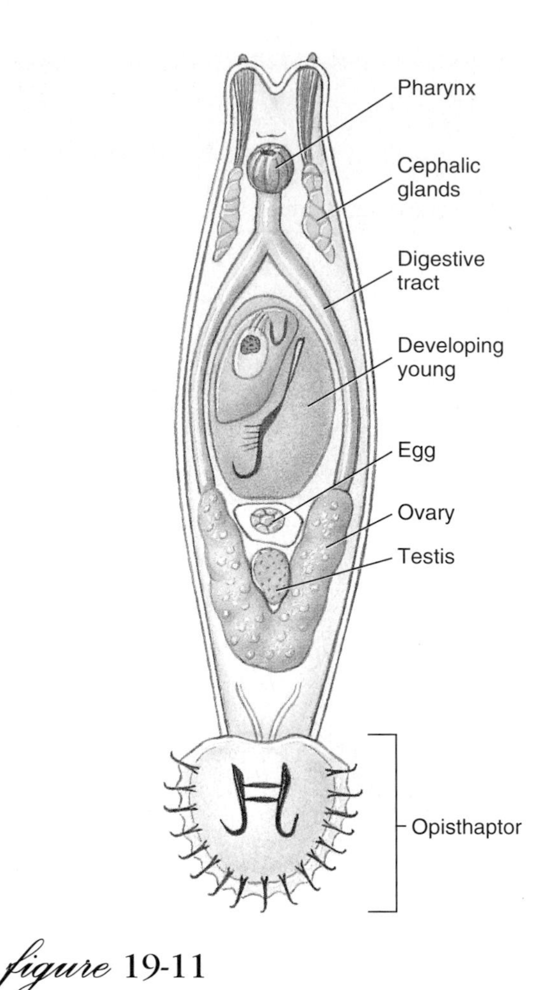

figure 19-11

Gyrodactylus sp. (class Monogenea), ventral view.

widespread and common, monogeneans seem to cause little damage to their hosts under natural conditions. However, like numerous other fish pathogens, they become a serious threat when their hosts are crowded together, as in fish farming.

The life cycles of monogeneans are simple, with a single host, as suggested by the name of the group, which means "single descent." The egg hatches a ciliated larva that attaches to the host or swims around awhile before attachment.

Class Cestoda

Cestoda, or tapeworms, differ in many respects from the preceding classes: they usually have long flat bodies composed of many reproductive units, or **proglottids** (Figure 19-12), and they completely lack a digestive system. As in Monogenea and Trematoda, there are no external, motile cilia in the adult, and the tegument is of a distal cytoplasm with sunken cell bodies beneath the superficial layer of muscle (Figure 19-5). In contrast to the monogeneans and trematodes, however, their entire surface is covered with minute projections that are similar in certain respects to the microvilli of the vertebrate small intestine (Figure 19-13). The microvilli greatly amplify the surface area of the tegument, which is a vital adaptation of the tapeworm, since it must absorb all its nutrients across the tegument.

Tapeworms are nearly all monoecious. They have well-developed muscles, and their excretory system and nervous system are somewhat similar to those of other flatworms. They have no special sense organs but do have sensory endings in the tegument that are modified cilia (Figure 19-13). One of their most specialized structures is the **scolex,** or holdfast, which is the organ of attachment. It is usually provided with suckers or suckerlike organs and often with hooks or spiny tentacles (Figure 19-14).

With rare exceptions, all cestodes require at least two hosts, and the adult is a parasite in the digestive tract of vertebrates. Often one of the intermediate hosts is an invertebrate.

The main body of the worms, the chain of proglottids, is called the **strobila** (Figure 19-14). Typically, there is a **germinative zone** just behind the scolex where new proglottids form. As new proglottids differentiate in front of it, each individual unit moves posteriorly in the strobila, and its gonads mature. The proglottid usually is fertilized by another proglottid in the same or a different strobila. The shelled embryos form in the uterus of the proglottid, and they are either expelled through a uterine pore, or the entire proglottid detaches from the worm as it reaches the posterior end of the strobila.

About 4000 species of tapeworms are known to parasitologists. Almost all vertebrate species are infected. Normally, adult tapeworms do little harm to their hosts. Table 19-2 lists the most common tapeworms in humans.

In the beef tapeworm, *Taeniarhynchus saginatus* (Gr. *tainia,* band, ribbon, + *rhynchos,* snout, beak), shelled larvae

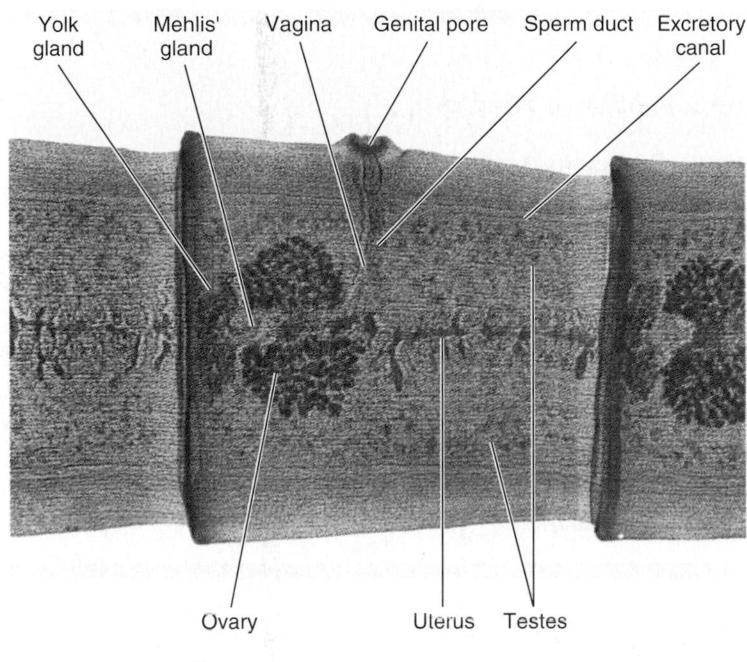

figure 19-12

Mature proglottid of *Taenia pisiformis,* a dog tapeworm. Portions of two other proglottids also shown.

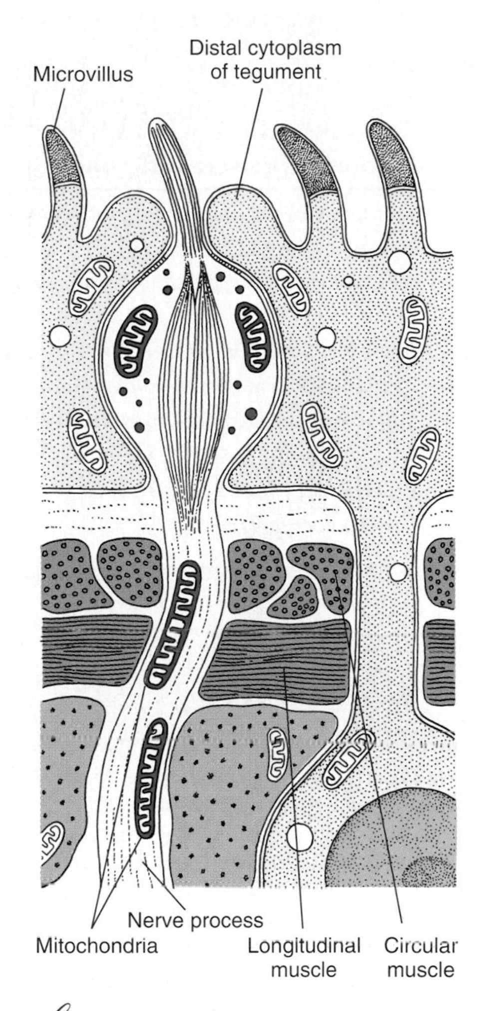

figure 19-13

Schematic drawing of a longitudinal section through a sensory ending in the tegument of *Echinococcus granulosus.*

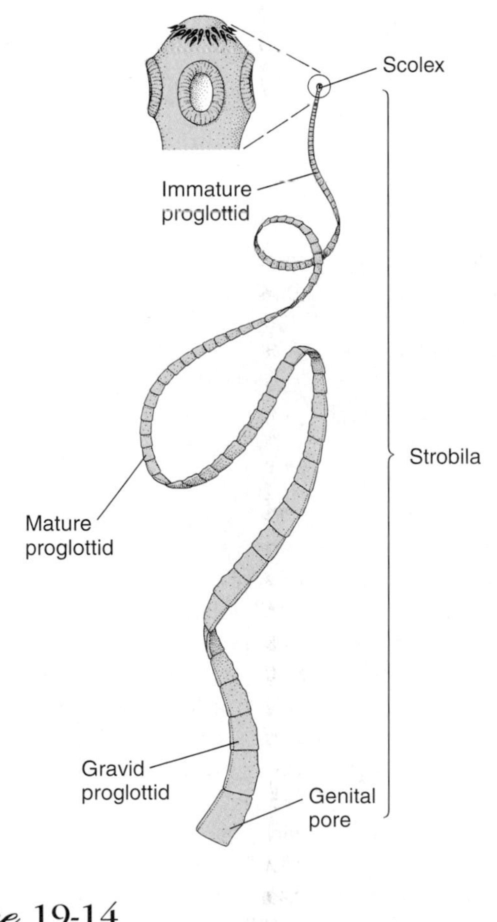

figure 19-14

A tapeworm, showing strobila and scolex. The scolex is the organ of attachment.

shed from the human host are ingested by cattle (Figure 19-15). The six-hooked larvae (oncospheres) hatch, burrow into blood or lymph vessels, and migrate to skeletal muscle where they encyst to become "bladder worms" (cysticerci). Each of these juveniles develops an invaginated scolex and remains quiescent until the uncooked muscle is eaten by humans. In the new host the scolex evaginates, attaches to the intestine, and matures in 2 or 3 weeks; then ripe proglottids may be expelled daily for many years. Humans become infected by eating raw or rare infested ("measly") beef. The adult worm may attain a length of 7 m or more, folded back and forth in the host intestine.

The pork tapeworm, *Taenia solium,* uses humans as definitive hosts and pigs as intermediate hosts. Humans can also serve as intermediate hosts by ingesting the shelled larvae from contaminated hands or food or, in persons harboring an adult worm, by regurgitating segments into the stomach. The juveniles may encyst in the central nervous system, where great damage may result (Figure 19-16).

table 19-2

Common Cestodes of Humans

Common and scientific names	Means of infection; prevalence in humans
Beef tapeworm *(Taeniarhynchus saginatus)*	Eating rare beef; most common of large tapeworms in humans
Pork tapeworm *(Taenia solium)*	Eating rare pork; less common than *T. saginatus*
Fish tapeworm *(Diphyllobothrium latum)*	Eating rare or poorly cooked fish; fairly common in Great Lakes region of United States, and other areas of world where raw fish is eaten
Dog tapeworm *(Dipylidium)*	Unhygienic habits of children (juveniles in flea and louse); moderate frequency
Dwarf tapeworm *(Vampirolepis nana)*	Juveniles in flour beetles; common
Unilocular hydatid *(Echinococcus granulosus)*	Cysts of juveniles in humans; infection by contact with dogs; common wherever humans are in close relationship with dogs and ruminants
Multilocular hydatid *(Echinococcus multilocularis)*	Cysts of juveniles in humans; infection by contact with foxes; less common than unilocular hydatid

 19-15

Life cycle of the beef tapeworm, *Taeniarhynchus.* Ripe proglottids break off in the human intestine, pass out in the feces, crawl out of the feces onto grass, and are ingested by cattle. The larvae hatch in the cow's intestine, freeing oncospheres, which penetrate into muscles and encyst, developing into "bladder worms." A human eats infected rare beef, and the cysticercus is freed in the intestine, where it attaches to the intestine wall, forms a strobila, and matures.

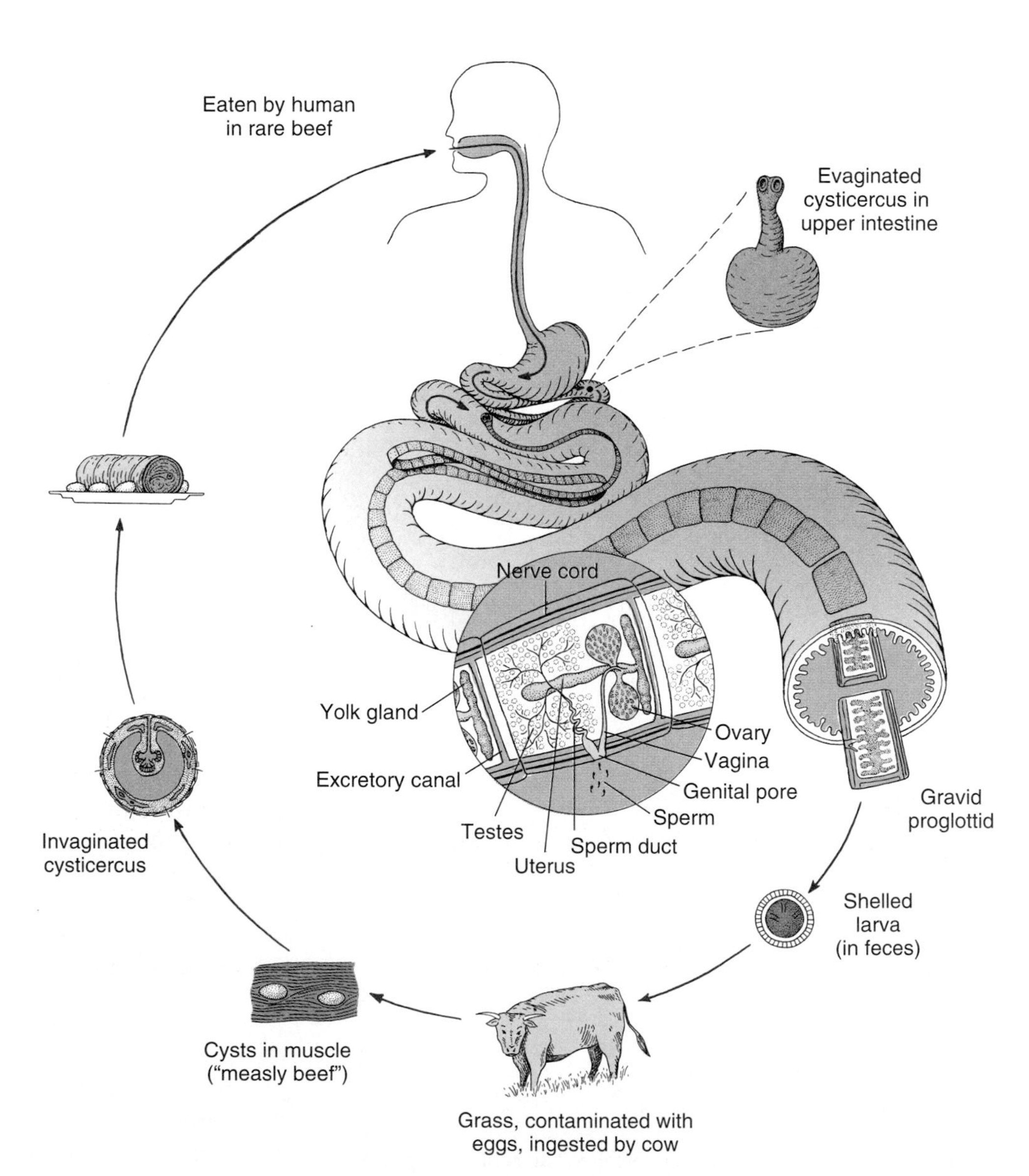

Classification of Phylum Platyhelminthes

Class Turbellaria (tur′bel-lar′e-a) (L. *turbellae* [pl.], stir, bustle, + *-aria,* like or connected with). The turbellarians. Usually free-living forms with soft flattened bodies; covered with ciliated epidermis containing secreting cells and rodlike bodies (rhabdites); mouth usually on ventral surface sometimes near center of body; no body cavity except intercellular spaces in parenchyma; mostly hermaphroditic; some have asexual fission. Examples: *Dugesia* (planaria), *Microstomum, Planocera.*

Class Trematoda (trem′a-to′da) (Gr. *trēma,* a hole, + *eidos,* form). The digenetic flukes. Body covered with a syncytial tegument without cilia; leaflike or cylindrical in shape; usually with oral and ventral suckers, no hooks; alimentary canal usually with two main branches; mostly monoecious; life cycle complex, with first host a mollusc, final host usually a vertebrate; parasitic in all classes of vertebrates. Examples: *Fasciola, Clonorchis, Schistosoma.*

Class Monogenea (mon′o-gen′e-a) (Gr. *mono,* single, + *gonos,* descent). The monogenetic flukes. Body covered with a syncytial tegument without cilia; body usually leaflike to cylindrical in shape; posterior attachment organ with hooks, suckers, or clamps, usually in combination; monoecious; life cycle simple, with single host and usually with free-swimming, ciliated larva; all parasitic, mostly on skin or gills of fish. Examples: *Dactylogyrus, Polystoma, Gyrodactylus.*

Class Cestoda (ses-to′da) (Gr. *kestos,* girdle, + *eidos,* form). The tapeworms. Body covered with nonciliated, syncytial tegument; general form of body tapelike; scolex with suckers or hooks, sometimes both, for attachment; body usually divided into series of proglottids; no digestive organs; usually monoecious; parasitic in digestive tract of all classes of vertebrates; life cycle complex, with two or more hosts; first host may be vertebrate or invertebrate. Examples: *Diphyllobothrium, Hymenolepis, Taeniarhynchus, Taenia.*

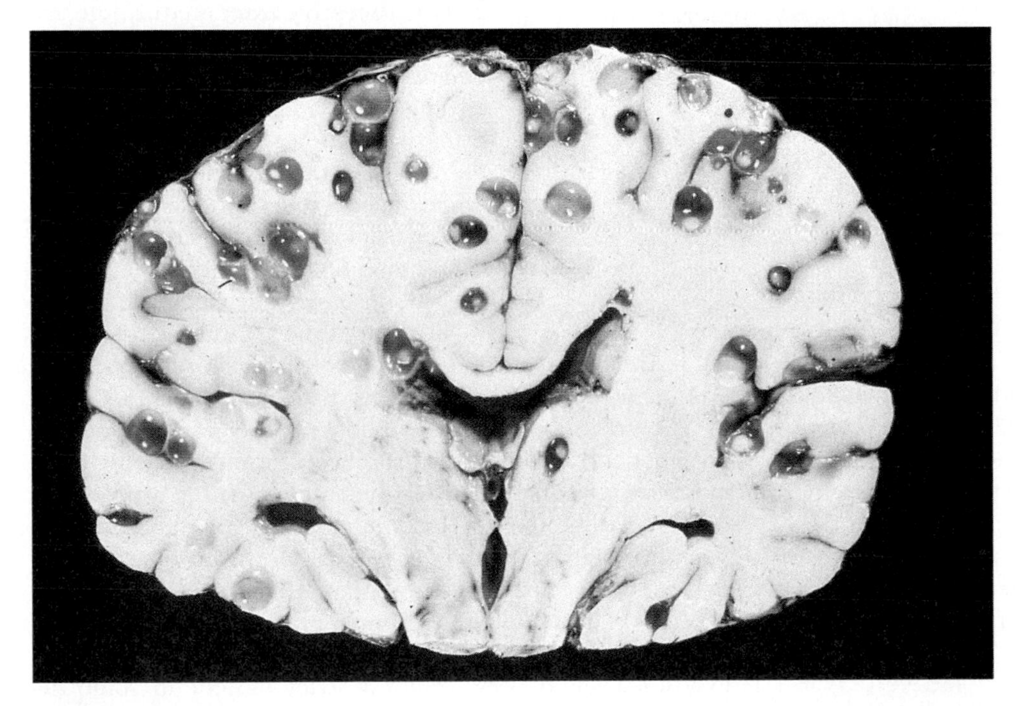

figure 19-16

Section through the brain of a person who died of cerebral cysticercosis, an infection with the cysticerci of *Taenia solium.*

Phylum Nemertea (Rhynchocoela)

Nemerteans are often called the ribbon worms. Their name (Gr. *Nemertes,* one of the nereids, unerring one) refers to the unerring aim of the proboscis, a long muscular tube (Figures 19-17 and 19-18) that can be thrust out swiftly to grasp the prey. The phylum is also called Rhynchocoela (Gr. *rhynchos,* beak, + *koilos,* hollow), which also refers to the proboscis. They are thread-shaped or ribbon-shaped worms. A few of the nemerteans are found in moist soil and fresh water, but by far the larger number are marine. At low tide they are often coiled up under stones. It seems probable that they are active at high tide and quiescent at low tide. Some live in secreted gelatinous tubes. There are about 650 species in the group.

Nemertean worms are usually less than 20 cm long, although a few are several meters in length (Figure 19-19). Some are brightly colored, although most are dull or pallid.

With few exceptions, the general body plan of the nemerteans is similar to that of Turbellaria. Like the latter,

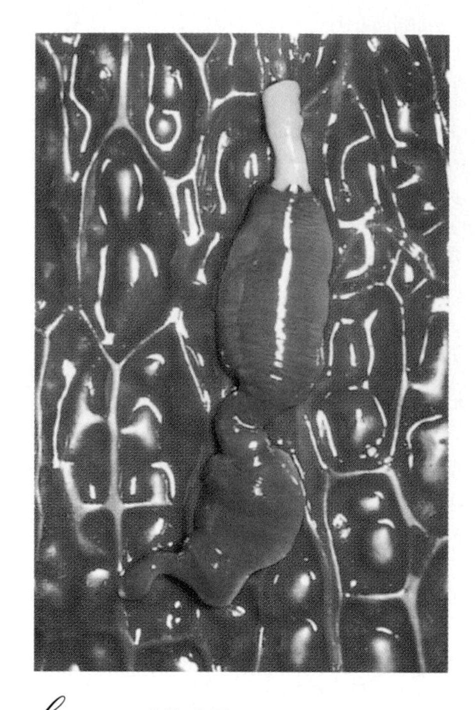

figure 19-17

Ribbon worm *Amphiporus bimaculatus* (phylum Nemertea) is 6 to 10 cm long, but other species range up to several meters. The proboscis of this specimen is partially extended at the top; the head is marked by two brown spots.

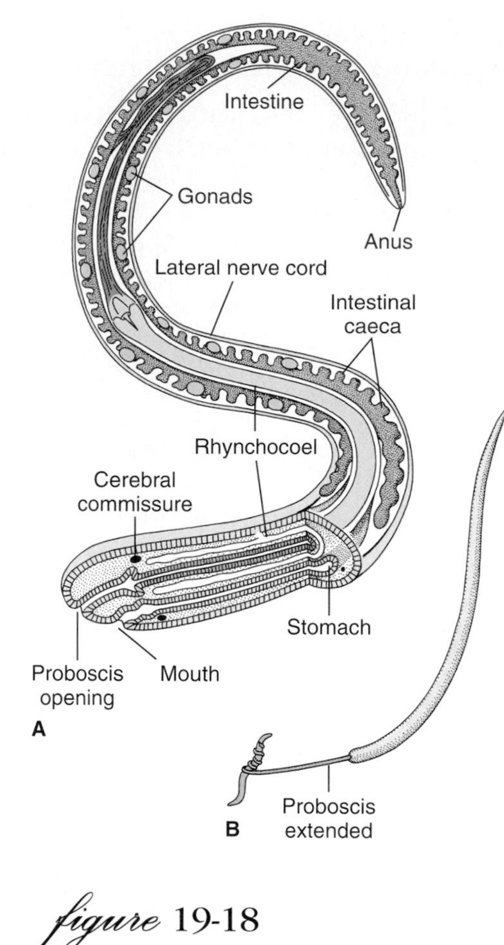

figure 19-18

A, Internal structure of female ribbon worm (diagrammatic). Dorsal view to show proboscis. **B,** *Amphiporus* with proboscis extended to catch prey.

figure 19-19

Baseodiscus is a genus of nemerteans whose members typically measure several meters in length. This *B. mexicanus* from the Galápagos Islands was over 10 m long.

their epidermis is ciliated and has many gland cells. Another striking similarity is the presence of flame cells in the excretory system. Rhabdites occur in several nemerteans. In the marine forms there is a ciliated larva that has some resemblance to the trochophore larva found in annelids and molluscs. Other flatworm characteristics are the presence of bilateral symmetry and a mesoderm and the lack of coelom. All in all, the present evidence seems to indicate that the nemerteans share a close common ancestor with the Platyhelminthes. Nevertheless, they differ from flatworms in several important aspects.

Form and Function

Many nemerteans are difficult to examine because they are so long and fragile. *Amphiporus* (Gr. *amphi,* on both sides, + *poros,* pore), one of the smaller genera that ranges from 2 to 10 cm in length, is fairly typical of the nemertean structure (Figure 19-18). Its body wall consists of ciliated epidermis and layers of circular and longitudinal muscles. Locomotion consists largely of gliding over a slime track, although larger species move by muscular contractions.

The mouth is anterior and ventral, and the **digestive tract** is **complete,** extending the full length of the body and ending at the anus. The development of an anus marks a significant advancement over the gastrovascular systems of the flatworms and radiates. Regurgitation of wastes is no longer necessary; ingestion and egestion can occur simultaneously. Cilia move food through the intestine. Digestion is largely extracellular.

Nemerteans are carnivorous, feeding primarily on annelids and other small invertebrates. They seize their prey with a **proboscis** that lies in an interior cavity of its own, the **rhynchocoel,** above the digestive tract (but not connected with it). The proboscis itself is a long, blind muscular tube that opens at the anterior end at a proboscis pore above the mouth. (In a few nemerteans the esophagus opens through the proboscis pore rather than through a separate mouth.) Muscular pressure on the fluid in the rhynchocoel causes the long tubular proboscis to be everted rapidly through the proboscis pore. Eversion of the proboscis exposes a sharp barb, called a stylet (absent in some nemerteans). The sticky, slime-covered proboscis coils around the prey and stabs it repeatedly with the stylet, while pouring a toxic secretion on the prey. Then retracting the proboscis, the nemertean draws the prey near the mouth, where it is engulfed by the esophagus which is thrust out to meet it.

Unlike other acoelomates, the nemerteans have a true circulatory system, and the irregular flow is maintained by the contractile walls of the vessels. Many flame-bulb protonephridia are closely associated with the circulatory system, so that their function appears to be truly excretory (for disposal of metabolic wastes), in contrast to their apparently osmoregulatory role in Platyhelminthes.

Nemerteans have a pair of nerve ganglia, and one or more pairs of longitudinal nerve cords are connected by transverse nerves.

Some species reproduce asexually by fragmentation and regeneration. In contrast to flatworms, most nemerteans are dioecious.

Phylum Gnathostomulida

The first species of the Gnathostomulida (nath′o-sto-myu′lid-a) (Gr. *gnatho*, jaw, + *stoma*, mouth, + L. *-ulus*, diminutive) was observed in 1928 in the Baltic, but its description was not published until 1956. Since then these animals have been found in many parts of the world, including the Atlantic coast of the United States, and over 80 species in 18 genera have been described.

Gnathostomulids are delicate wormlike animals 0.5 to 1 mm long (Figure 19-20). They live in the interstitial spaces of very fine sandy coastal sediments and silt and can endure conditions of very low oxygen.

Lacking a pseudocoel, a circulatory system, and an anus, the gnathostomulids show some similarities to the turbellarians and were at first included in that group. However, their parenchyma is poorly developed, and their pharynx is reminiscent of the rotifer mastax (p. 434). The pharynx is armed with a pair of lateral jaws used to scrape fungi and bacteria off the substratum. Although the epidermis is ciliated, each epidermal cell has but one cilium, which is a condition not found in other bilateral animals except in some gastrotrichs (p. 435).

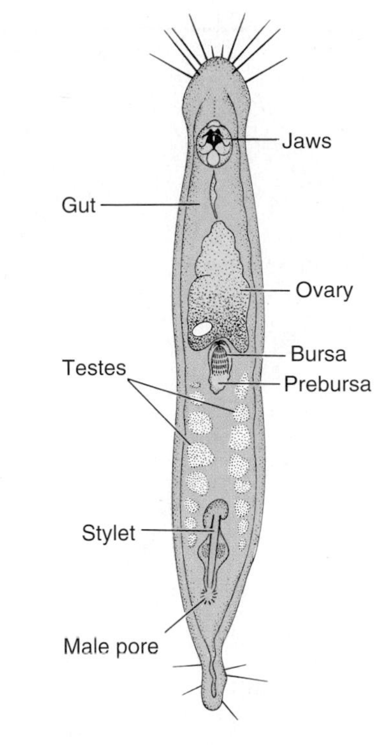

figure 19-20

Gnathostomula jenneri (phylum Gnathostomulida) is a tiny member of the interstitial fauna between grains of sand or mud. Species in this family are among the most commonly encountered jaw worms, found in shallow water and down to depths of several hundred meters.

Phylogeny and Adaptive Radiation

Phylogeny

There can be little doubt that the bilaterally symmetrical flatworms were derived from a radial ancestor, perhaps one very similar to the planula larva of the cnidarians. Some investigators believe that this **planuloid ancestor** may have given rise to one branch of descendants that were sessile or free floating and radial, which became the Cnidaria, and another that acquired a creeping habit and bilateral symmetry. Bilateral symmetry is a selective advantage for creeping or swimming animals because sensory structures are concentrated anteriorly (cephalization), which is the end that first encounters environmental stimuli.

According to Ax,[1] an early branch from the bilateral line would have given rise to the Platyhelminthes and Gnathostomulida, and these two would form a sister group to all the other bilateral animals (Eubilateria). This scheme would separate the Platyhelminthes and Nemertea, traditionally considered closely related, because the Nemertea have a flow-through gut. If this view is correct, the Nemertea must have branched off the Eubilateria line soon after the one-way intestine with anus was established.

Among the Platyhelminthes, it seems clear that the Turbellaria is paraphyletic,[2] but we are retaining the taxon for the present because presentation based on thorough cladistic analysis would require introduction of many more taxa and characteristics beyond the scope of this book. For example, if the possession of separate yolk glands is a derived characteristic, the

[1] Ax, P. 1985. The position of the Gnathostomulida and Platyhelminthes in the phylogenetic system of the Bilateria. In Conway Morris, S., J. D. George, R. Gibson, and H. M. Platt (eds.). The origins and relationships of lower invertebrates. Oxford, Clarendon Press.

[2] Ehlers, U. 1985. Phylogenetic relationships within the Platyhelminthes. In Conway Morris, S., J. D. George, R. Gibson, and H. M. Platt (eds.). The origins and relationships of lower invertebrates. Oxford, Clarendon Press.

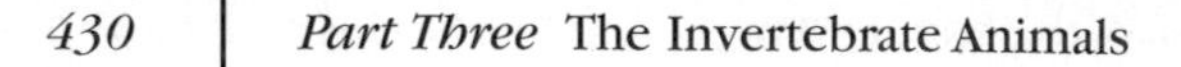

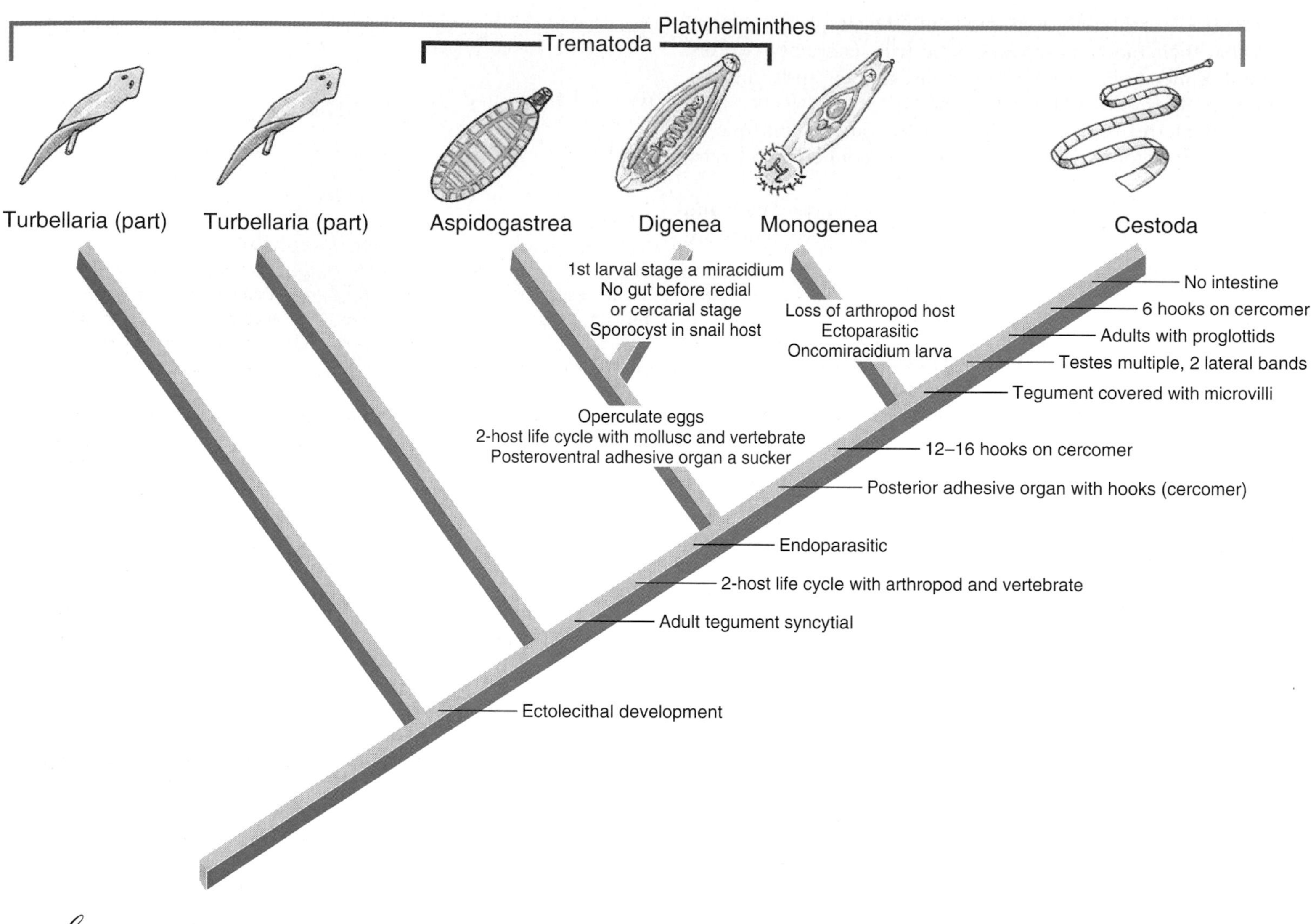

figure 19-21

Phylogenetic relationships among parasitic Platyhelminthes. The traditionally accepted class Turbellaria is paraphyletic. Some turbellarians have ectolecithal development and, together with the Trematoda, Monogenea, and Cestoda, form a monophyletic clade and a sister group of the endolecithal turbellarians. For the sake of simplicity, the synapomorphies of those turbellarians and of the Aspidogastrea, as well as many others given by Brooks (1989) are omitted. All of these organisms comprise a clade (called Cercomeria) with a posterior adhesive organ.

(Source: *After Brooks, D. R. 1989. The phylogeny of the Cercomeria [Platyhelminthes: Rhabdocoela] and general evolutionary principles,* J. Parasitol. *75:606–616.)*

ectolecithal Turbellaria should be placed with the Trematoda and Cestoda, to compose a sister group with the endolecithal Turbellaria. The unique architecture of the tegument in cestodes, monogeneans, and trematodes indicates with high probability that these groups share a common ancestor (Figure 19-21).

Adaptive Radiation

The flatworm body plan, with its creeping adaptation, placed a selective advantage on bilateral symmetry and further development of cephalization, ventral and dorsal regions, and caudal differentiation. Because of their body shape and metabolic requirements, early flatworms must have been well preadapted for parasitism and gave rise to symbiotic lines on numerous occasions. These lines produced descendants that radiated abundantly as parasites, and many flatworms became highly specialized for that mode of existence.

The ribbon worms have stressed the proboscis apparatus in their evolutionary diversity. Its use in capturing prey may have been secondarily evolved from its original function as a highly sensitive organ for exploring the environment. Although the ribbon worms have evolved beyond the flatworms in their complexity of organization, they have been dramatically less abundant as a group.

Likewise, the jaw worms have neither radiated nor become nearly as numerous as the flatworms. However, they have exploited the marine interstitial environment, particularly zones of very low oxygen concentration.

Summary

The Platyhelminthes, the Nemertea, and the Gnathostomulida are the simplest phyla that are bilaterally symmetrical, a condition of adaptive value for actively crawling or swimming animals. They have neither a coelom nor a pseudocoel and are thus acoelomate. They are triploblastic and at the organ-system level of organization.

The body surface of turbellarians is a cellular epithelium, at least in part ciliated, containing mucous cells and rod-shaped rhabdites that function together in locomotion. Members of all other classes of flatworms are covered by a nonciliated, syncytial tegument with a vesicular distal cytoplasm and cell bodies beneath superficial muscle layers. Digestion is extracellular and intracellular in most; cestodes must absorb predigested nutrients across their tegument because they have no digestive tract. Osmoregulation is by flame-cell protonephridia, and removal of metabolic wastes and respiration occur across the body wall. Except for some turbellarians, flatworms have a ladder-type nervous system with motor, sensory, and association neurons. Most flatworms are hermaphroditic, and asexual reproduction occurs in some groups.

The class Turbellaria is a paraphyletic group with mostly free living and carnivorous members. The digenetic trematodes have mollusc intermediate hosts and almost always a vertebrate definitive host. The great amount of asexual reproduction that occurs in the intermediate host helps to increase the chance that some of the offspring will reach a definitive host. Aside from the tegument, digeneans share many basic structural characteristics with the Turbellaria. The Digenea includes a number of important parasites of humans and domestic animals. These contrast with the Monogenea, which are important ectoparasites of fishes and have a direct life cycle (without intermediate hosts).

Cestodes (tapeworms) generally have a scolex at their anterior end, followed by a long chain of proglottids, each of which contains a complete set of reproductive organs of both sexes. Cestodes live as adults in the digestive tract of vertebrates. Their tegument bears microvilli, which increase its surface area for absorption. The shelled larvae are passed in the feces of their host, and the juveniles develop in a vertebrate or invertebrate intermediate host.

Members of the Nemertea have a complete digestive system with an anus and a true circulatory system. They are free-living, mostly marine, and they ensnare prey with their long, eversible proboscis.

The Gnathostomulida are a curious phylum of little wormlike marine animals living among sand grains and silt. They have no anus, and they share certain characteristics with such widely diverse groups as turbellarians and rotifers.

The flatworms and the cnidarians both probably evolved from a common ancestor (planuloid), some of whose descendants became sessile or free floating and radial (cnidarians), while others became creeping and bilateral (flatworms).

Review Questions

1. Why is bilateral symmetry of adaptive value for actively motile animals?
2. Match the terms in the right column with the classes in the left column:

   ___ Turbellaria   a. Endoparasitic
   ___ Monogenea   b. Free-living and commensal
   ___ Trematoda   c. Ectoparasitic
   ___ Cestoda

3. Give several major characteristics that distinguish the Platyhelminthes.
4. Distinguish two mechanisms by which flatworms supply yolk for their embryos. Which system is evolutionarily ancestral for flatworms and which one is derived?
5. How do flatworms digest their food?
6. Briefly describe the osmoregulatory system and the nervous system and sense organs of Platyhelminthes.
7. Contrast asexual reproduction in Turbellaria, Trematoda, and Cestoda.
8. Contrast the typical life cycle of a monogenean with that of a digenetic trematode.
9. Describe and contrast the tegument of turbellarians and the other classes of platyhelminths. Could this be evidence that the trematodes, monogeneans, and cestodes form a clade within the Platyhelminthes? Why?
10. Answer the following questions with respect to both *Clonorchis* and *Schistosoma:* (a) How do humans become infected? (b) What is the general geographical distribution? (c) What are the main disease conditions produced?
11. Define each of the following with reference to cestodes: scolex, microvilli, proglottids, strobila.
12. Why is *Taenia solium* a more dangerous infection than *Taeniarhynchus saginatus?*
13. Give three differences between nemerteans and platyhelminths.
14. Where do gnathostomulids live?
15. Explain how a planuloid ancestor could have given rise to both the Cnidaria and the Bilateria.
16. What is an important character of the Nemertea that might suggest that the phylum is closer to other bilateral protostomes than to the Platyhelminthes? What are some characters of nemerteans suggesting that they share an ancestor with platyhelminths?

Selected References

See also general references for Part Three, p. 562.

Arme, C., and P. W. Pappas (eds.). 1983. Biology of the Eucestoda, 2 vols. New York, Academic Press, Inc. *The most up-to-date reference available on the biology of tapeworms. Advanced; not an identification manual.*

Brooks, D. R. 1989. The phylogeny of the Cercomeria (Platyhelminthes: Rhabdocoela) and general evolutionary principles. J. Parasitol. **75:**606–616. *Cladistic analysis of parasitic flatworms.*

Desowitz, R. S. 1981. New Guinea tapeworms and Jewish grandmothers. New York, W. W. Norton & Company. *Accounts of parasites and parasitic diseases of humans. Entertaining and instructive. Recommended for all students.*

Schell, S. C. 1985. Handbook of trematodes of North America north of Mexico. Moscow, Idaho, University Press of Idaho. *Good for trematode identification.*

Schmidt, G. D. 1985. Handbook of tapeworm identification. Boca Raton, Florida, CRC Press. *Good for cestode identification.*

Strickland, G. T. 1991. Hunter's tropical medicine, ed. 7. Philadelphia, W. B. Saunders Company. *A valuable source of information on parasites of medical importance.*

Pseudocoelomate Animals

chapter | twenty

A World of Nematodes

Without any doubt, nematodes are the most important pseudocoelomate animals, in terms of both numbers and their impact on humans. Nematodes are abundant over most of the world, yet most people are only occasionally aware of them as parasites of humans or of their pets. We are not aware of the billions of these worms in the soil, in ocean and freshwater habitats, in plants, and in all kinds of other animals. Their dramatic abundance moved N.A. Cobb to write in 1914:

> If all the matter in the universe except the nematodes were swept away, our world would still be dimly recognizable, and if, as disembodied spirits, we could then investigate it, we should find its mountains, hills, vales, rivers, lakes and oceans represented by a thin film of nematodes. The location of towns would be decipherable, since for every massing of human beings there would be a corresponding massing of certain nematodes. Trees would still stand in ghostly rows representing our streets and highways. The location of the various plants and animals would still be decipherable, and, had we sufficient knowledge, in many cases even their species could be determined by an examination of their erstwhile nematode parasites.
>
> **From N.A. Cobb. 1914. Yearbook of the United States Department of Agriculture, p. 472.**

Nine distinct phyla of animals belong to the pseudocoelomate category. These are Rotifera, Gastrotricha, Kinorhyncha, Loricifera, Priapulida, Nematoda, Nematomorpha, Acanthocephala, and Entoprocta. They are a heterogeneous assemblage of animals. Most of them are small; some are microscopic; some are fairly large. Some, such as the nematodes, are found in freshwater, marine, terrestrial, and parasitic habitats; others, such as the Acanthocephala, are strictly parasitic. Some have unique characteristics such as the lacunar system of the acanthocephalans and the ciliary corona of the rotifers.

Even in such a diversified grouping, however, a few characteristics are shared. All have a body wall of epidermis (often syncytial) and muscles surrounding the pseudocoel. The digestive tract is complete (except in Acanthocephala), and it, along with the gonads and excretory organs, is within the pseudocoel and bathed in perivisceral fluid. The epidermis in many secretes a nonliving cuticle with some specializations such as bristles and spines.

A constant number of cells or nuclei in the individuals of a species, a condition known as **eutely,** is present in several of the groups, and in most of them there is an emphasis on the longitudinal muscle layer.

Position in Animal Kingdom

In the nine phyla covered in this chapter, the original blastocoel of the embryo persists as a space, or body cavity, between the enteron and body wall. Because this cavity lacks the peritoneal lining found in the true coelomates, it is called a **pseudocoel,** and the animals possessing it are called **pseudocoelomates** (Figure 20-1). Pseudocoelomates belong to the Protostomia division of the bilateral animals, but they may be polyphyletic (not derived from a common ancestor).

Biological Contributions

1. The pseudocoel is a distinct gradation in body plan compared with the solid body structure of the acoelomates. The pseudocoel may be filled with fluid or may contain a gelatinous substance with some mesenchyme cells. In common with a true coelom, it presents certain adaptive potentials, although these are by no means realized in all members: (1) greater freedom of movement; (2) space for the development and differentiation of digestive, excretory, and reproductive systems; (3) a simple means of circulation or distribution of materials throughout the body; (4) a storage place for waste products to be discharged to the outside by excretory ducts; and (5) a hydrostatic organ. Since most pseudocoelomates are quite small, the most important functions of the pseudocoel are probably in circulation and as a means to maintain a high internal hydrostatic pressure.
2. A complete, mouth-to-anus digestive tract is found in these phyla and in all more complex phyla.

Phylum Rotifera

Rotifera (ro-tif′e-ra) (L. *rota,* wheel, + *fero,* to bear) derive their name from their characteristic ciliated crown, or **corona,** which, when beating, often gives the impression of rotating wheels. Rotifers range in size from 40 μm to 3 mm in length, but most are between 100 and 500 μm long. Some have beautiful colors, although most are transparent, and some have odd and bizarre shapes. Their shapes are often correlated with their mode of life. The floaters are usually globular and saclike; the creepers and swimmers are somewhat elongated and wormlike; and the sessile types are commonly vaselike, with a cuticular envelope. Some are colonial.

Rotifers are a cosmopolitan group of about 1800 recognized species, some of which are found throughout the world. Most of the species are freshwater inhabitants, a few are marine, some are terrestrial, and some are epizoic (live on the surface of other animals) or parasitic. Most often they are benthic, occurring on the bottom or in vegetation of ponds or along the shores of large freshwater lakes where they swim or creep about on the vegetation.

The body is usually made up of a **head,** a **trunk,** and a **foot.** The head region bears the corona. The corona may form a ciliated funnel with its upper edges folded into lobes bearing bristles, or the corona may be made up of a pair of ciliated discs (Figure 20-2). The cilia create currents of water toward the mouth that draw in small planktonic forms for food. The corona may be retractile. The mouth, surrounded by some part of the corona, opens into a modified muscular pharynx called a **mastax,** which is a characteristic unique to rotifers. The mastax has a set of intricate jaws used for grasping and chewing. The trunk contains the visceral organs, and the terminal foot, when present, is segmented and, in some, is ringed with joints that can telescope to shorten. The one to four toes secrete a sticky substance from the **pedal glands** for attachment.

Rotifers have a pair of protonephridial tubules with flame bulbs, which eventually drain into a **cloacal bladder** that collects excretory and digestive wastes. They have a bilobed "brain" and sense organs that include eyespots, sensory pits, and papillae.

Female rotifers have one or two syncytial ovaries **(germovitellaria)** that produce yolk as well as oocytes. Although rotifers are dioecious, males are unknown in many species; in these reproduction is entirely parthenogenetic. In parthenogenetic species females produce only diploid eggs that have not undergone reduction division, cannot be fertilized, and de-

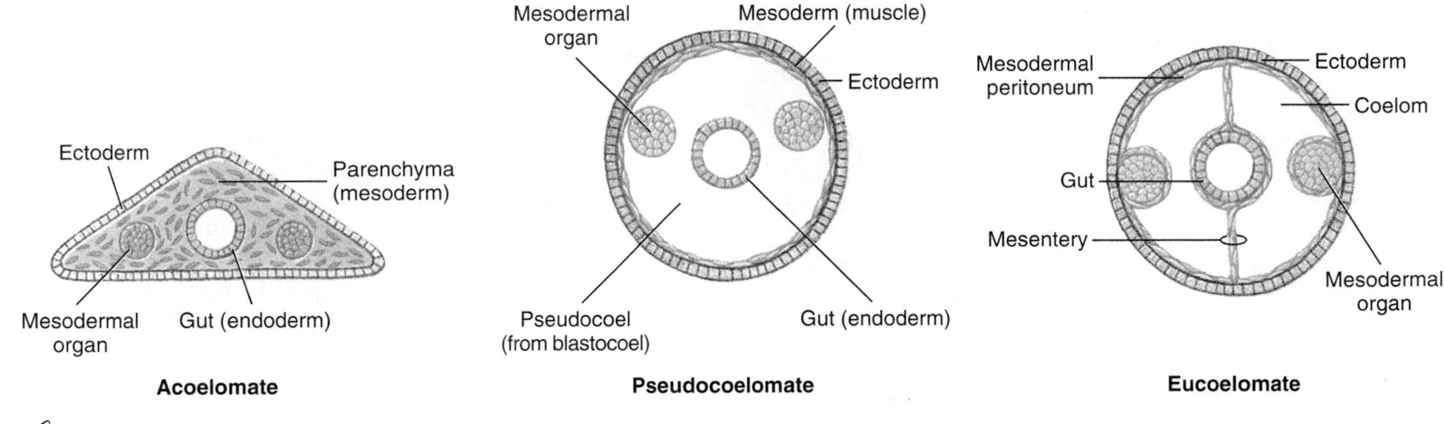

figure 20-1

Acoelomate, pseudocoelomate, and eucoelomate body plans.

velop only into females. Such eggs are called **amictic eggs.** Other species can produce two kinds of eggs—amictic eggs, which develop parthenogenetically into females, and **mictic eggs,** which have undergone meiosis and are haploid. Mictic eggs, if unfertilized, develop quickly and parthenogenetically into males; if fertilized, they secrete a thick shell and become dormant for several months before developing into females. Such dormant eggs can withstand desiccation and other adverse conditions and permit rotifers to live in temporary ponds that dry up during certain seasons.

Mictic (Gr., miktos, *mixed, blended) refers to the capacity of the haploid eggs to be fertilized (that is, "mixed") with the male's sperm nucleus to form a diploid embryo. Amictic ("without mixing") eggs are already diploid and can develop only parthenogenetically.*

Phylum Gastrotricha

The phylum Gastrotricha (gas-trot′re-ka) (Gr. *gaster,* belly, + *thrix,* hair) is a small group (about 460 species) of microscopic animals, approximately 65 to 500 µm long (Figure 20-3). They are usually bristly or scaly in appearance, are flattened on the ventral side, and many move by gliding on ventral cilia. Others move in a leechlike fashion by briefly attaching the posterior end by means of adhesive glands. There are both marine and freshwater species, and they are common in lakes, ponds, and seashore sands. They feed on bacteria, diatoms, and small protozoa. Gastrotrichs are hermaphroditic. However, the male system of many freshwater species is nonfunctional, and the female system produces offspring parthenogenetically.

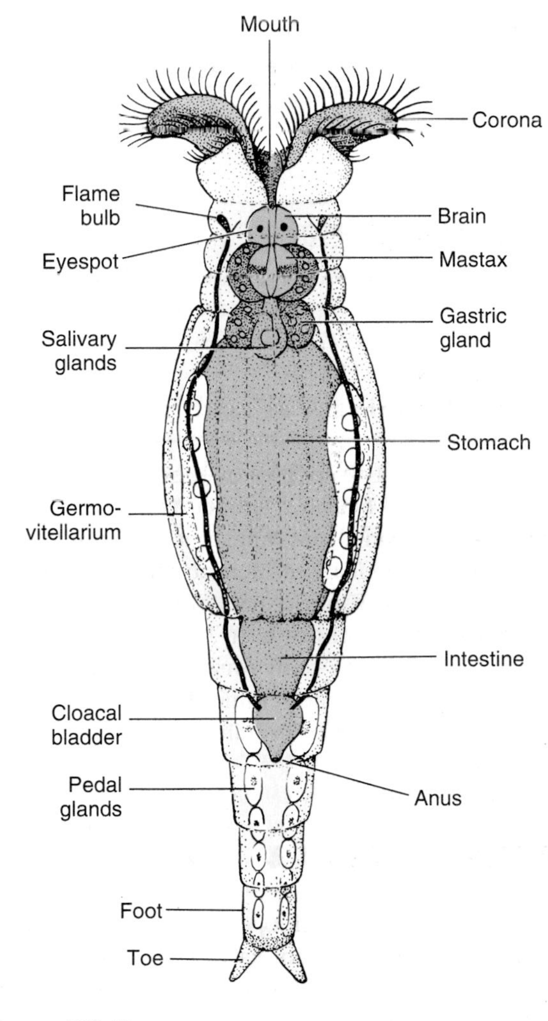

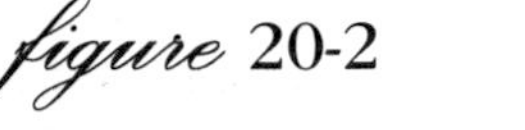 20-2

Structure of *Philodina,* a common rotifer.

table 20-1

Common Parasitic Nematodes of Humans in North America

Common and scientific names	Mode of infection; prevalence
Hookworm (*Ancylostoma duodenale* and *Necator americanus*)	Contact with juveniles in soil that burrow into skin; common in southern states
Pinworm (*Enterobius vermicularis*)	Inhalation of dust with ova and by contamination with fingers; most common worm parasite in United States
Intestinal roundworm (*Ascaris lumbricoides*)	Ingestion of embryonated ova in contaminated food; common in rural areas of Appalachia and southeastern states
Trichina worm (*Trichinella spiralis*)	Ingestion of infected muscle; occasional in humans throughout North America
Whipworm (*Trichuris trichiura*)	Ingestion of contaminated food or by unhygienic habits; usually common wherever *Ascaris* is found

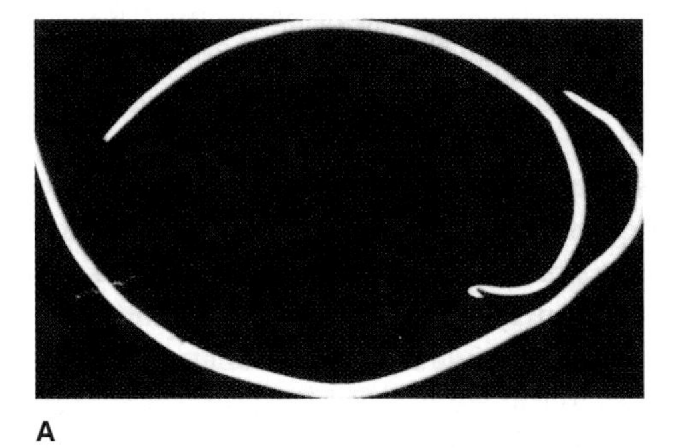

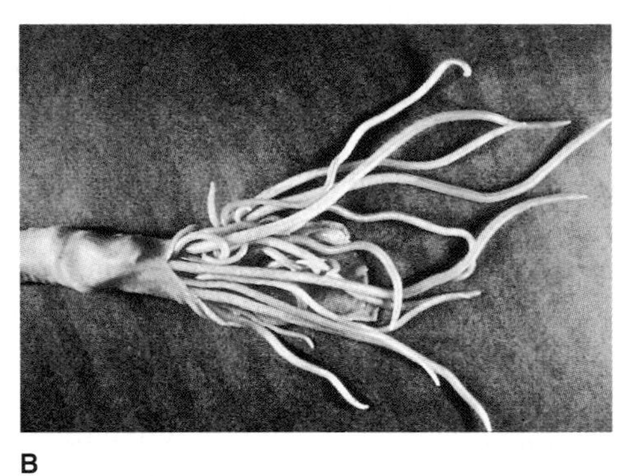

figure 20-9

A, Intestinal roundworm *Ascaris lumbricoides,* male and female. Male *(top)* is smaller and has characteristic sharp kink in the end of the tail. The females of this large nematode may be over 30 cm long. **B,** Intestine of a pig, nearly completely blocked by *Ascaris suum.* Such heavy infections are also fairly common with *A. lumbricoides* in humans.

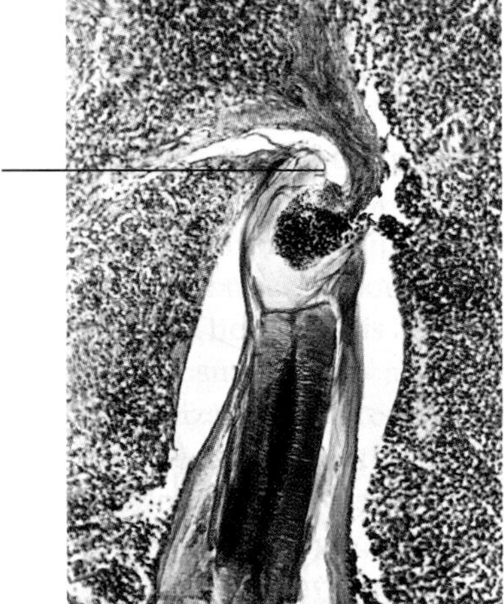

figure 20-10

Section through anterior end of hookworm attached to dog intestine. Note cutting plates of mouth pinching off bit of mucosa from which the thick muscular pharynx sucks blood. Esophageal glands secrete an anticoagulant to prevent blood clotting.

Hookworms

Hookworms are so named because the anterior end curves dorsally, suggesting a hook. The most common species is *Necator americanus* (L. *necator,* killer), whose females are up to 11 mm long. The males can reach 9 mm in length. Large plates in their mouths (Figure 20-10) cut into the intestinal mucosa of the host where they suck blood and pump it through their intestines, partially digesting it and absorbing the nutrients. They suck much more blood than they need for food, and heavy infections cause anemia in the patient. Hookworm disease in children may result in retarded mental and physical growth and general loss of energy.

Shelled embryos are passed in the feces, and the juveniles hatch in the soil, where they live on bacteria. When human skin comes in contact with infested soil, the juveniles burrow through the skin to the blood, and reach the lungs and finally the intestine in a manner similar to that described for *Ascaris.*

Trichina Worm

Trichinella spiralis (Gr. *trichinos,* of hair, + *-ella,* diminutive) is one of the species of tiny nematodes responsible for the potentially lethal disease trichinosis. Adult worms burrow in the mucosa of the small intestine where the female produces living young. The juveniles penetrate into blood vessels and are carried throughout the body, where they may be found in almost any tissue or body space. Eventually, they penetrate skeletal muscle cells, becoming one of the largest known intracellular parasites. The juvenile causes astonishing redirection of gene expression in its host cell, which loses its striations and becomes a **nurse cell** that nourishes the worm (Figure 20-11). When meat containing live juveniles is swallowed, the worms are liberated into the intestine where they mature.

Trichinella spp. can infect a wide variety of mammals in addition to humans, including hogs, rats, cats, and dogs. Hogs become infected by eating garbage containing pork scraps with juveniles or by eating infected rats. In addition to *T. spiralis,* we now know there are four other sibling species in the genus. They differ in geographic distribution, infectivity to different host species, and freezing resistance. Heavy infections may cause death, but lighter infections are much more common—about 2.4% of the population of the United States is infected.

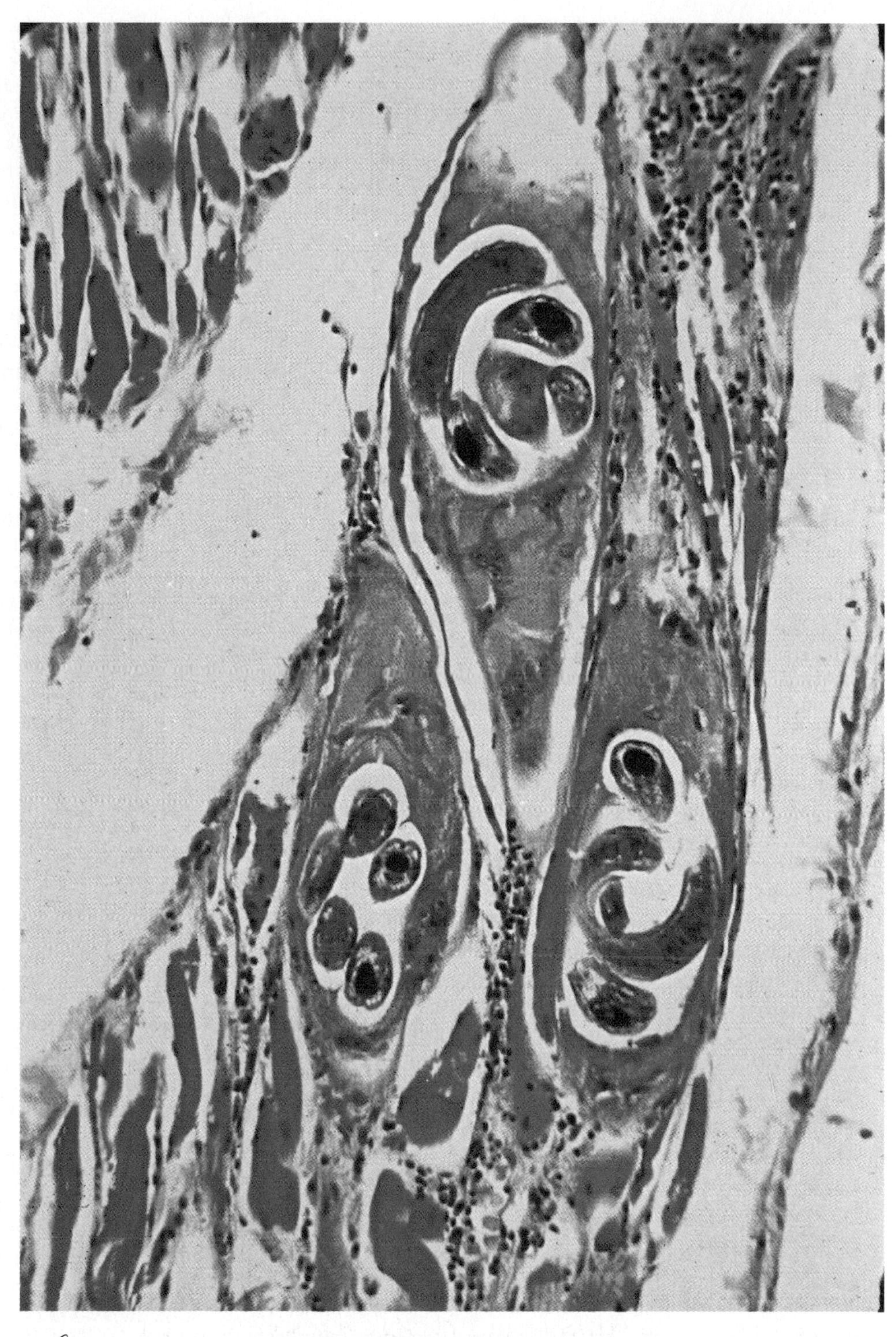

figure 20-11

Section of human muscle infected with trichina worm *Trichinella spiralis.* The juveniles lie within muscle cells that the worms have induced to transform into nurse cells (commonly called cysts). An inflammatory reaction is evident around the nurse cells. The juveniles may live 10 to 20 years, and the nurse cells may eventually calcify.

Pinworms

The pinworm, *Enterobius vermicularis* (Gr. *enteron,* intestine, + *bios,* life), causes relatively little disease, but it is the most common helminth parasite in the United States, estimated at 30% in children and 16% in adults. The adult parasites (Figure 20-12) live in the large intestine and cecum. The females, up to about 12 mm in length, migrate to the anal region at night to lay their eggs (Figure 20-12). Scratching the resultant itch effectively contaminates the hands and bedclothes. Eggs develop rapidly and become infective within 6 hours at body temperature. After they are swallowed, they hatch in the duodenum, and the worms mature in the large intestine.

Members of this order of nematodes have **haplodiploidy,** a characteristic shared with a few other animal groups, notably many hymenopteran insects (p. 312). Males are haploid and are produced parthenogenetically; females are diploid and arise from fertilized eggs.

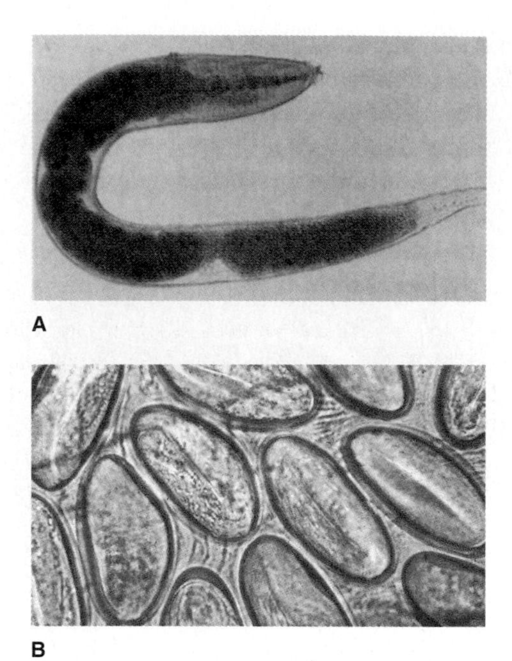

figure 20-12

Pinworms, *Enterobius vermicularis.* **A,** Female worm from human large intestine (slightly flattened in preparation), magnified about 20 times. **B,** Group of shelled juveniles of pinworms, which are usually discharged at night around the anus of the host, who, by scratching during sleep, may get fingernails and clothing contaminated. This may be the most common and widespread of all human helminth parasites.

Diagnosis of most intestinal roundworms is usually by examination of a small bit of feces under the microscope and finding characteristic shelled embryos or juveniles. However, pinworm eggs are not often found in the feces because the female deposits them on the skin around the anus. The "Scotch tape method" is more effective. The sticky side of cellulose tape is applied around the anus to collect the shelled embryos, then the tape is placed on a glass slide and examined under the microscope. Several drugs are effective against this parasite, but all members of a family should be treated at the same time, since the worm easily spreads through a household.

Filarial Worms

At least eight species of filarial nematodes infect humans, and some of these are major causes of diseases. Some 250 million people in tropical countries are infected with *Wuchereria bancrofti* (named for Otto Wucherer) or *Brugia malayi* (named for S.L. Brug), which places these species among the scourges of humanity. The worms live in the lymphatic system, and the females are as long as 100 mm. The disease symptoms are associated with inflammation and obstruction of the lymphatic system. The females release live young, the tiny microfilariae, into the blood and lymph. The microfilariae are picked up by mosquitos as the insects feed, and they develop in the mosquitos to the infective stage. They escape from the mosquito when it is feeding again on a human and penetrate the wound made by the mosquito bite.

The dramatic manifestations of elephantiasis are occasionally produced after long and repeated exposure to the worms. The condition is marked by an excessive growth of connective tissue and enormous swelling of affected parts, such as the scrotum, legs, arms, and more rarely, the vulva and breasts (Figure 20-13).

Another filarial worm causes river blindness (onchocerciasis) and is carried by blackflies. It infects more than 30 million people in parts of Africa, Arabia, Central America, and South America.

The most common filarial worm in the United States is probably the dog heartworm, *Dirofilaria immitis* (Figure 20-14). Carried by mosquitos, it also can infect other canids, cats, ferrets, sea lions, and occasionally humans. Along the Atlantic and Gulf Coast states and northward along the Mississippi River throughout the midwestern states, prevalance in dogs is up to 45%. It occurs in other states at a lower prevalence. This worm causes a very serious disease among dogs, and no responsible owner should fail to provide "heartworm pills" for a dog during mosquito season.

Phylum Nematomorpha

The popular name for the Nematomorpha (nem′a-to-mor′fa) (Gr. *nema, nematos,* thread, + *morphē,* form) is "horsehair worms," based on an old superstition that the worms arise from horsehairs that happen to fall into the water; and indeed they resemble hairs from a horse's tail. They were long included with the nematodes, with which they share the structure of the cuticle, presence of epidermal cords, longitudinal muscles only, and pattern of nervous system. However, since the early larval form of some species has a striking resemblance to the Priapulida, it is impossible to say to what group the nematomorphs are most closely related.

About 250 species of horsehair worms have been named. Worldwide in distribution, horsehair worms are free-living as adults and parasitic in arthropods as juveniles. Adults have a vestigial digestive tract and do not feed, but they can live almost anywhere in wet or moist surroundings if oxygen is adequate. Juveniles do not emerge from the arthropod host unless there is water nearby, and adults are often seen wriggling slowly about in ponds or streams. Juveniles of freshwater forms use various terrestrial insects as hosts, while marine forms use certain crabs.

Phylum Acanthocephala

The members of the phylum Acanthocephala (akan′tho-sef′a-la) (Gr. *akantha,* spine or thorn, + *kephalē,* head) are commonly known as "spiny-headed worms." The phylum derives its name from one of its most distinctive features, a cylindrical

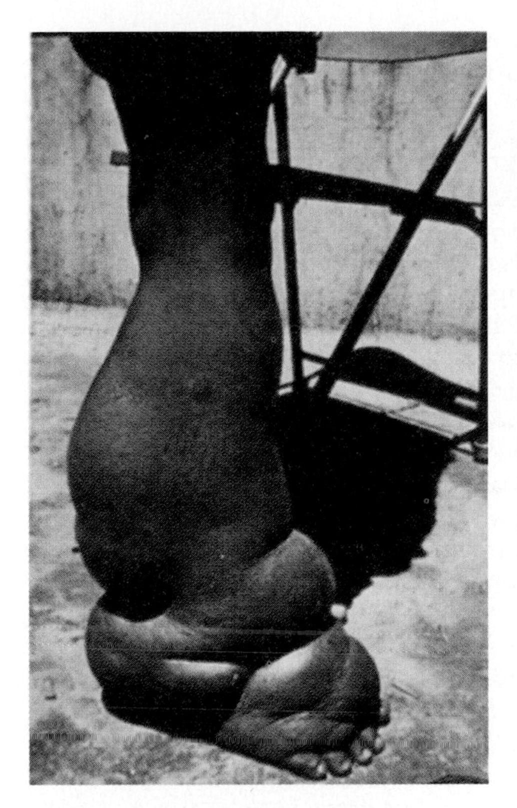

figure 20-13

Elephantiasis of leg caused by adult filarial worms of *Wuchereria bancrofti,* which live in lymph passages and block the flow of lymph. Tiny juveniles, called microfilariae, are picked up in a blood meal of a mosquito, where they develop to the infective stage and are transmitted to a new host.

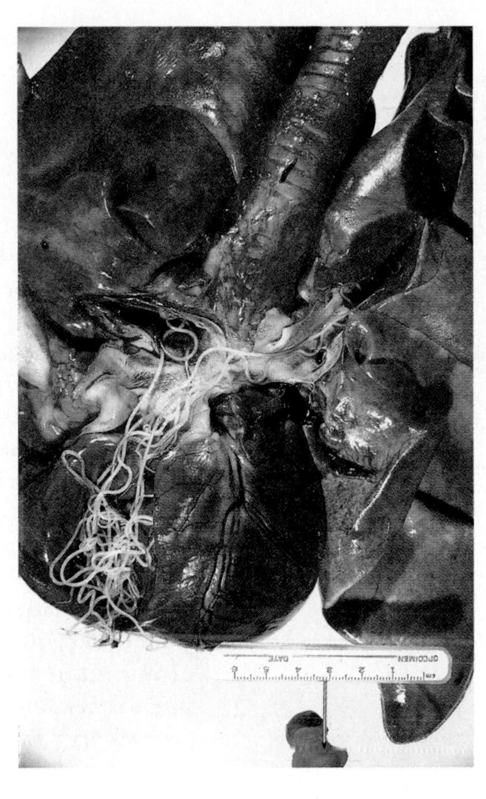

figure 20-14

Dirofilaria immitis in a dog's heart. This nematode is a major menace to the health of dogs in North America. The adults live in the heart, and the juveniles circulate in the blood where they are picked up and transmitted by mosquitos.

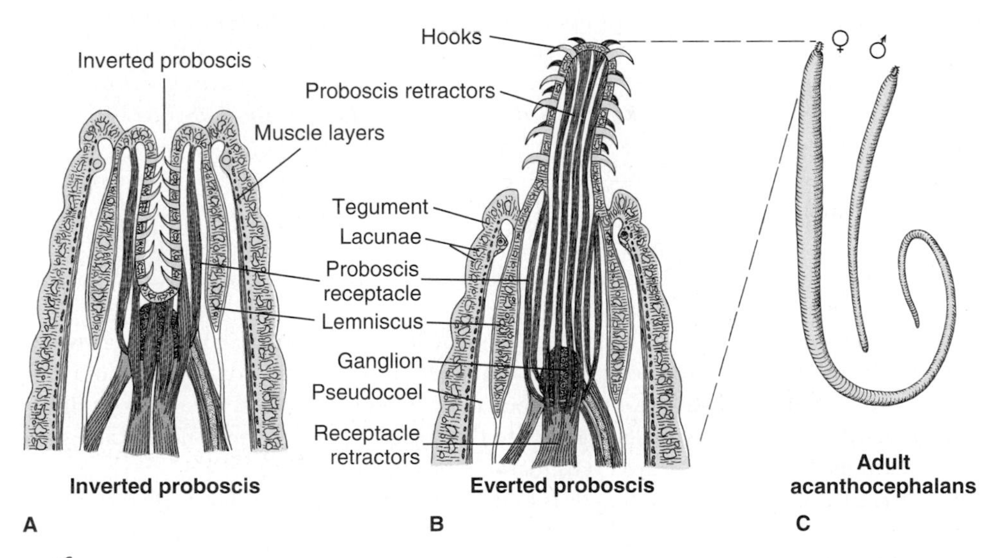

figure 20-15

Structure of a spiny-headed worm (phylum Acanthocephala). **A** and **B,** Eversible spiny proboscis by which the parasite attaches to the intestine of the host, often doing great damage. Because they lack a digestive tract, food is absorbed through the tegument. **C,** Male is typically smaller than female.

invaginable **proboscis** (Figure 20-15) bearing rows of recurved spines, by which the worms attach themselves to the intestine of their hosts. All acanthocephalans are endoparasitic, living as adults in the intestines of vertebrates.

Over 500 species are known, most of which parasitize fishes, birds, and mammals, and the phylum is worldwide in distribution. However, no species is normally a parasite of humans, although rarely humans are infected with species that usually occur in other hosts.

Various species range in size from less than 2 mm to over 1 m in length, with the females of a species usually being larger than the males. In life, the body is usually bilaterally flattened, with numerous transverse wrinkles. The worms are usually cream color but may be yellowish or brown from absorption of pigments from the intestinal contents.

The body wall is syncytial, and its surface is punctuated by minute crypts 4 to 6 μm deep, which greatly increase the surface area of the tegument. About 80% of the thickness of the tegument is the radial fiber zone, which contains a **lacunar system** of ramifying fluid-filled canals (Figure 20-15). The function of the lacunar system is unclear, but it may serve in distribution of nutrients to the peculiar, tubelike muscles in the body wall of these organisms.

Excretion is across the body wall in most species. In one family there is a pair of **protonephridia** with flame cells that unite to form a common tube that opens into the sperm duct or uterus.

Because acanthocephalans have no digestive tract, they must absorb all nutrients through their tegument. They can absorb various molecules by specific membrane transport mechanisms, and their tegument can carry out pinocytosis.

Sexes are separate. The male has a protrusible penis, and at copulation the sperm travel up the genital duct and escape into the pseudocoel of the female. The zygotes develop into shelled acanthor larvae. The shelled larvae escape from the vertebrate host in the feces, and, if eaten by a suitable arthropod, they hatch and work their way into the hemocoel where they grow to juvenile acanthocephalans. Either development ceases until the arthropod is eaten by a

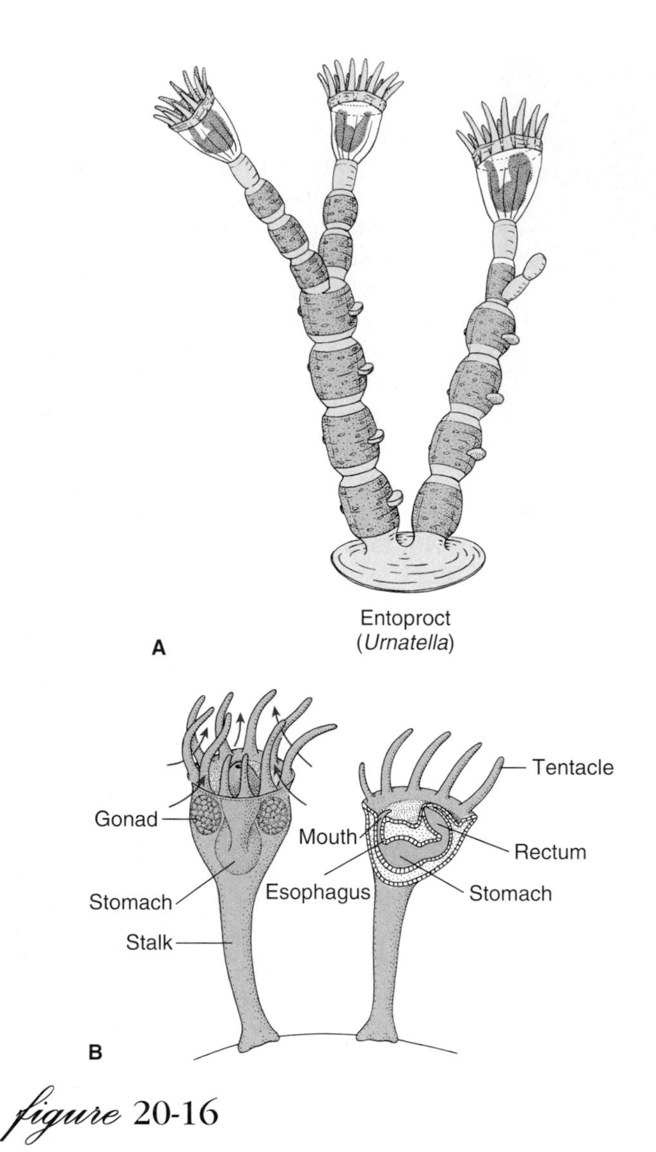

figure 20-16

A, *Urnatella,* a freshwater entoproct, forms small colonies of two or three stalks from a basal plate. **B,** *Loxosomella,* a solitary entoproct. Both solitary and colonial entoprocts can reproduce asexually by budding, as well as sexually.

suitable host, or they may pass through several transport hosts in which they encyst until eaten.

Phylum Entoprocta

Entoprocta (en′to-prok′ta) (Gr. *entos,* within, + *proktos,* anus) is a small phylum of less than 100 species of tiny, sessile animals that, superficially, look much like hydroid cnidarians, but their tentacles are ciliated and tend to roll inward (Figure 20-16). Most entoprocts are microscopic, and none is more than 5 mm long. They are all stalked and sessile forms; some are colonial, and some are solitary. All are ciliary feeders.

With the exception of one genus all entoprocts are marine forms that have a wide distribution from the polar regions to the tropics. Most marine species are restricted to coastal and brackish waters and often grow on shells and algae. Some are commensals on marine annelid worms. Freshwater entoprocts (Figure 20-16) occur on the underside of rocks in running water.

*In December, 1995, P. Funch and R. M. Kristensen reported that they had found some very strange little creatures clinging to the mouthparts of the Norway lobster (*Nephrops norvegicus*), so strange that they did not fit into any known phylum (Nature **378:** 711–714). Funch and Kristensen concluded that the organisms, only 0.35 mm long, represented a new phylum, which was named* **Cycliophora.** *The name refers to a crown of compound cilia, reminiscent of rotifers, with which the organisms feed. They were described as "acoelomate," although whether they might have a pseudocoel is unclear, and they do have a cuticle. Their life cycle seems bizarre. The sessile feeding stages on the lobster's mouthparts undergo internal budding to produce motile stages: (1) larvae containing new feeding stages; (2) dwarf males, which become attached to feeding stages that contain developing females; and (3) females, which also attach to the lobster's mouthparts, then produce dispersive larvae and degenerate.*

Whether the proposed new phylum will withstand the scrutiny of further research is unknown, and its possible relationships to other phyla are quite unclear. Funch and Kristensen think that the organisms are protostomes and see affinities with the Entoprocta and Ectoprocta. Little short of astonishing, however, is their abundance on the mouthparts of a host as well known as the Norway lobster. How could biologists have failed to notice them before? At a time when habitat destruction drives many species to extinction every year, we wonder if there are phyla *suffering the same fate. S. Conway Morris pondered the possibility of further undiscovered phyla (Nature* ***378:****661–662), suggesting you may need a couple of zoology textbooks and a decent microscope when you next dine at your favorite seafood restaurant: "Who knows what might be found lurking under the lettuce?"*

The body of entoprocts is cup shaped, bears a crown, or circle, of ciliated tentacles, and attaches to a substratum by a stalk in solitary species. In colonial species, the stalks of the individuals connect to the stolon of the colony. The tentacles and stalk are continuations of the body wall.

The gut is U shaped and ciliated, and both the mouth and anus open within the circle of tentacles. They capture food particles in the current created by the tentacular cilia and then pass the particles along the tentacles to the mouth. Entoprocts have a pair of protonephridia but no circulatory or respiratory organs.

Some species are monoecious, some are dioecious, and some seem to be protandrous hermaphrodites; that is, the gonad at first produces sperm and then eggs. Cleavage is modi-

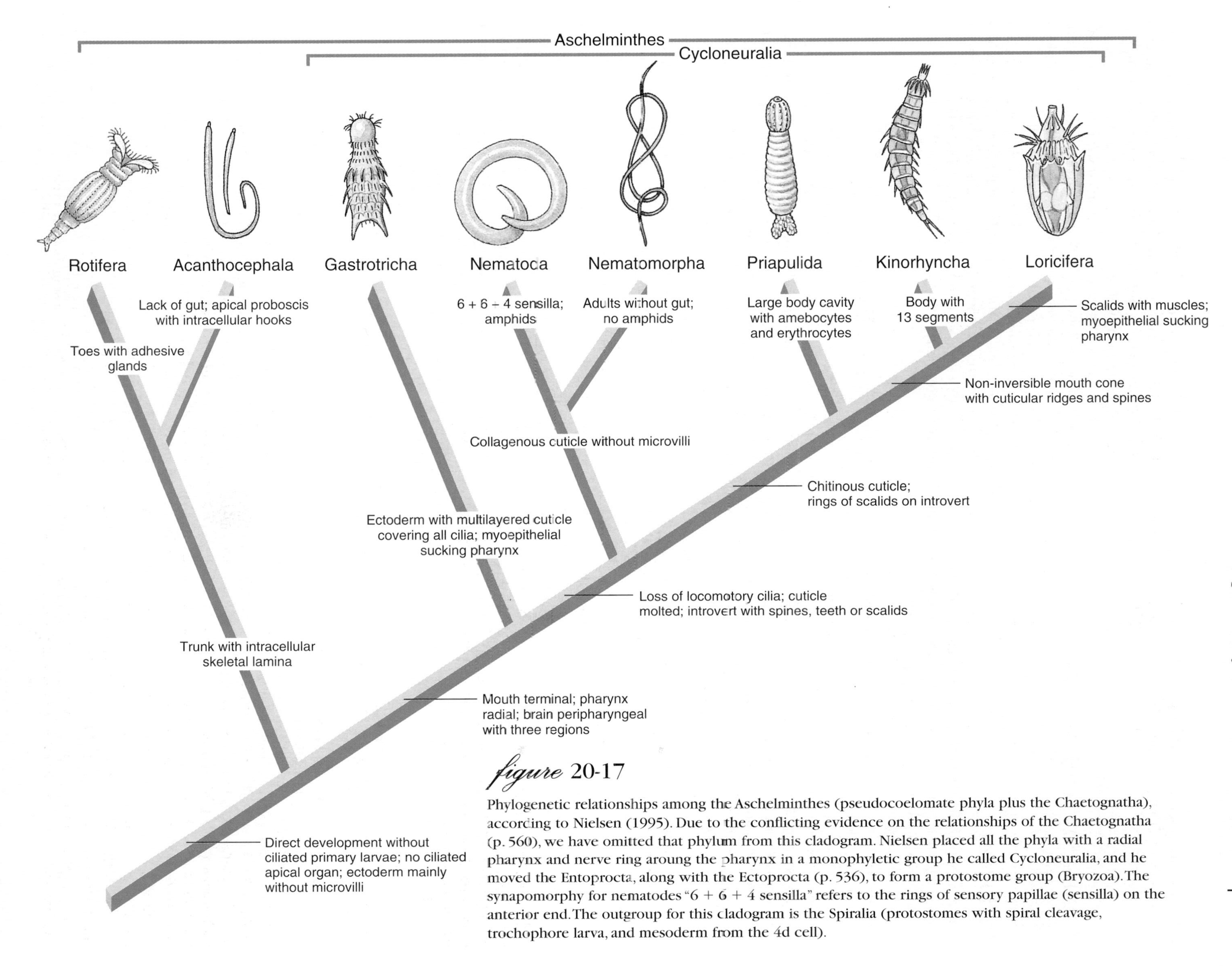

figure 20-17

Phylogenetic relationships among the Aschelminthes (pseudocoelomate phyla plus the Chaetognatha), according to Nielsen (1995). Due to the conflicting evidence on the relationships of the Chaetognatha (p. 560), we have omitted that phylum from this cladogram. Nielsen placed all the phyla with a radial pharynx and nerve ring around the pharynx in a monophyletic group he called Cycloneuralia, and he moved the Entoprocta, along with the Ectoprocta (p. 536), to form a protostome group (Bryozoa). The synapomorphy for nematodes "6 + 6 + 4 sensilla" refers to the rings of sensory papillae (sensilla) on the anterior end. The outgroup for this cladogram is the Spiralia (protostomes with spiral cleavage, trochophore larva, and mesoderm from the 4d cell).

fied spiral and mosaic, and a trochophore-like larva is produced, similar to that of some molluscs and annelids (p. 453).

Phylogeny and Adaptive Radiation

Phylogeny

Hyman[2] grouped the Rotifera, Gastrotricha, Kinorhyncha, Priapulida, Nematoda, and Nematomorpha into a single phylum (Aschelminthes). All of these phyla share a certain combination of characteristics, and Hyman contended that the evidences of relationships were so concrete and specific that they could not be disregarded. Nevertheless, most authors now consider that differences between the groups are sufficient to merit phylum status for each, although some accept the concept of the Aschelminthes as a superphylum. The phyla may well have been derived originally from the protostome line via an acoelomate common ancestor resembling certain flatworms in many features.

Loriciferans bear some similarity to kinorhynchs, larval Priapulida, larval nematomorphs, rotifers, and tardigrades (p. 535). Although the loriciferans are poorly known, cladistic analysis suggests that they form a sister group to the kinorhynchs and that these two phyla together are a sister group of the priapulids (Figure 20-17).

Acanthocephalans are highly specialized parasites with a unique structure and have doubtless been so for millions of years. Any ancestral species or related group that would shed a clue to the phyletic relationships of the Acanthocephala probably went extinct long ago. Like the cestodes, acanthocephalans have no digestive tract and must absorb all nutrients across the tegument, but the tegument of the two groups is quite different in structure.

[2]Hyman, L. H. 1951. The invertebrates, vol. III. Acanthocephala, Aschelminthes, and Entoprocta, the pseudocoelomate Bilateria. New York, McGraw-Hill Book Company.

The entoprocts were once included with the phylum Ectoprocta in a phylum called Bryozoa, but the ectoprocts are true coelomate animals, and many zoologists prefer to place them in a separate group. Some biologists still refer to ectoprocts as bryozoans. The Entoprocta may be distantly related to the Ectoprocta, but there is little evidence of close relationship. The entoprocts may have arisen as an early offshoot of the same line that led to the ectoprocts. These relationships remain controversial.

Adaptive Radiation

Certainly the most impressive adaptive radiation in this group of phyla is shown by the nematodes. They are by far the most numerous in terms of both individuals and species, and they have been able to adapt to almost every habitat available to animal life. Their basic pseudocoelomate body plan, with the cuticle, hydrostatic skeleton, and longitudinal muscles, has proved generalized and plastic enough to adapt to an enormous variety of physical conditions. Free-living lines gave rise to parasitic forms on at least several occasions, and virtually all potential hosts have been exploited. All types of life cycle occur: from the simple and direct to the complex, with intermediate hosts; from normal dioecious reproduction to parthenogenesis, hermaphroditism, and alternation of free-living and parasitic generations. A major factor contributing to the evolutionary opportunism of the nematodes has been their extraordinary capacity to survive suboptimal conditions, for example, the developmental arrests in many free-living and animal parasitic species and the ability to undergo cryptobiosis (survival in harsh conditions by assuming a very low metabolic rate) in many free-living and plant parasitic species.

Summary

The phyla covered in this chapter possess a body cavity called a pseudocoel, which is derived from the embryonic blastocoel, rather than a secondary cavity in the mesoderm (coelom). Several of the groups exhibit eutely, a constant number of cells or nuclei in adult individuals of a given species.

The phylum Rotifera is composed of small, mostly freshwater organisms with a ciliated corona, which creates currents of water to draw planktonic food toward the mouth. The mouth opens into a muscular pharynx, or mastax, that is equipped with jaws.

The Gastrotricha, Kinorhyncha, and Loricifera are small phyla of tiny, aquatic pseudocoelomates. Gastrotrichs move by cilia or adhesive glands, and kinorhynchs anchor and then pull themselves by the spines on their head. Loriciferans can withdraw their bodies into the lorica. Priapulids are marine burrowing forms.

By far the largest and most important of this group of phyla are the nematodes, of which there may be as many as 500,000 species in the world. They are more or less cylindrical, tapering at the ends, and covered with a tough, secreted cuticle. The body-wall muscles are longitudinal only, and to function well in locomotion, such an arrangement must enclose a volume of fluid in the pseudocoel at high hydrostatic pressure. This fact of nematode life has a profound effect on most of their other physiological functions, for example, ingestion of food, egestion of feces, excretion, and copulation. Most nematodes are dioecious, and there are four juvenile stages, each separated by a molt of the cuticle. Almost all invertebrate and vertebrate animals and many plants have nematode parasites, and many other nematodes are free-living in soil and aquatic habitats. Some parasitic nematodes have part of their life cycle free-living, some undergo a tissue migration in their host, and some have an intermediate host in their life cycle.

The Nematomorpha or horsehair worms are related to the nematodes and have parasitic juvenile stages in arthropods, followed by a free-living, aquatic, nonfeeding adult stage.

Acanthocephalans are all parasitic in the intestines of vertebrates as adults, and their juvenile stages develop in arthropods. They have an anterior, invaginable proboscis armed with spines, which they embed in the intestinal wall of their host. They do not have a digestive tract and so must absorb all nutrients across their tegument.

The Entoprocta are small, sessile, aquatic animals with a crown of ciliated tentacles encircling both the mouth and anus.

The Rotifera, Gastrotricha, Kinorhyncha, Pripulida, Nematoda, and Nematomorpha have been included by some workers in one phylum, but most biologists believe that the groups are not sufficiently related to be encompassed by a single phylum. It is possible that they are derived from a common ancestor in the protostome line. Phylogenetic relationships of the Acanthocephala and Entoprocta are even more obscure. Of all these phyla, the Nematoda have achieved enormous evolutionary success and undergone great adaptive radiation.

Review Questions

1. Explain the difference between a true coelom and a pseudocoel.
2. What is the normal size of a rotifer; where is it found; and what are its major body features?
3. Explain the difference between mictic and amictic eggs of rotifers, and tell the adaptive value of each.
4. What is eutely?
5. What are the approximate lengths of loriciferans, priapulids, gastrotrichs, and kinorhynchs? Where are they found?
6. A skeleton is a supportive structure. Explain how a hydrostatic skeleton supports an animal.
7. What feature of body-wall muscles in nematodes requires a high hydrostatic pressure in the pseudocoelomic fluid for efficient function?
8. Explain how the high pseudocoelomic pressure affects feeding and defecation in nematodes. How could ameboid sperm be an adaptation to the high hydrostatic pressure in the pseudocoel?
9. Explain the interaction of the cuticle, body-wall muscles, and pseudocoelomic fluid in the locomotion of nematodes.
10. Outline the life cycle of each of the following: *Ascaris lumbricoides*, hookworm, *Enterobius vermicularis*, *Trichinella spiralis*, *Wuchereria bancrofti*.
11. Where in the human body are the adults of each species in question 10 found?
12. Where are juveniles and adults of nematomorphs found?
13. The evolutionary ancestry of acanthocephalans is particularly obscure. Describe some characters of acanthocephalans that support that statement.
14. How do acanthocephalans get food?
15. What characteristics distinguish entoprocts among the pseudocoelomates?

Selected References

See also general references for Part Three, p. 562.

Bird, A. F., and J. Bird. 1991. The structure of nematodes, ed. 2. New York, Academic Press. *The most authoritative reference available on nematode morphology. Highly recommended.*

Bundy, D. A. P., and E.S. Cooper. 1989. *Trichuris* and trichuriasis in humans. In J. R. Baker and R. Muller (eds.), Advances in parasitology, vol. 28. London, Academic Press. *A recent estimate of people in the world infected with whipworm* (Trichuris) *is 687 million. This number is exceeded among nematodes only by hookworms (932 million) and* Ascaris *(1.26 billion).*

Despommier, D. D. 1990. *Trichinella spiralis:* the worm that would be virus. Parasitol. Today **6:**193–196. *The first-stage juveniles of* Trichinella *are among the largest of all intracellular parasites.*

Duke, B. O. L. 1990. Onchocerciasis (river blindness)—can it be eradicated? Parasitol. Today **6:**82–84. *Despite the introduction of a very effective drug, the author predicts that this parasite will not be eradicated in the foreseeable future.*

Ogilvie, B. M., M. E. Selkirk, and R. M. Maizels. 1990. The molecular revolution and nematode parasitology: yesterday, today, and tomorrow. J. Parasitol. **76:**607–618. *Modern molecular biology has wrought enormous changes in investigations on nematodes.*

Poinar, G. O., Jr. 1983. The natural history of nematodes. Englewood Cliffs, New Jersey, Prentice-Hall, Inc. *Contains a great deal of information about these fascinating creatures, including free-living and plant and animal parasites.*

Roberts, L. 1990. The worm project. Science **248:**1310–1313. *A free-living nematode,* Caenorhabditis elegans, *has been of great value in studies of development and genetics.*

Molluscs

chapter | twenty-one

A Significant Space

Long ago in the Precambrian era, the most complex animals populating the seas were acoelomate. They must have been inefficient burrowers, and they were unable to exploit the rich subsurface ooze. Any that developed fluid-filled spaces within the body would have had a substantial advantage because these spaces could serve has a hydrostatic skeleton and improve burrowing efficiency.

The simplest, and probably the first, mode of achieving a fluid-filled space within the body was retention of the embryonic blastocoel, as in the pseudocoelomates. This evolutionary solution was not ideal because the organs lay loose in the body cavity. Improved efficiency of the pseudocoel as a hydrostatic skeleton depended on increasingly high hydrostatic pressure, a condition that severely limited the potential for adaptive radiation.

Some descendants of the Precambrian acoelomate organisms evolved a more elegant arrangement: a fluid-filled space within the mesoderm, the **coelom.** This space was lined with mesoderm, and the organs were suspended by mesodermal membranes, the **mesenteries.** Not only could the coelom serve as an efficient hydrostatic skeleton, with circular and longitudinal body-wall muscles acting as antagonists, but a more stable arrangement of organs resulted in less crowding. The mesenteries provided an ideal location for networks of blood vessels, and the alimentary canal could become more muscular, more highly specialized, and more diversified without interfering with other organs.

Development of the coelom was a major step in the evolution of larger and more complex forms. All major groups in the chapters to follow are coelomates.

Next to Arthropoda the phylum Mollusca (mol-lus′ka) (L. *mollusca,* soft) has the most named species in the animal kingdom—probably about 50,000 living species, not to mention some 35,000 fossil species discovered to date. The name Mollusca indicates one of their distinctive characteristics, a soft body.

This very diverse group includes organisms as different as chitons, snails, clams, and octopuses (Figure 21-1). The group ranges from fairly simple organisms to some of the most complex of invertebrates, and in size from almost microscopic to the giant squid *Architeuthis harveyi* (Gr. *archi,* primitive, + *teuthis,* squid). The body of this huge species may grow up to 18 m long if its tentacles are extended. It may weigh up to 454 kg (1000 pounds). The shells of some of the giant clams *Tridacna gigas* (Gr. *tridaknos,* eaten at three bites) (see Figure 21-21), which inhabit the Indo-Pacific coral reefs, reach 1.5 m in length and weigh over 225 kg. These are extremes, however, since probably 80% of all molluscs are less than 5 cm in maximum shell size.

The enormous variety, great beauty, and availability of the shells of molluscs have made shell collecting a popular pastime. However, many amateur shell collectors, although able to name hundreds of the shells that grace our beaches, know very little about the living animals that created those shells and once lived in them. The largest classes of molluscs are the Gastropoda (snails and their relatives), Bivalvia (clams, oysters, and others), Polyplacophora (chitons), and Cephalopoda (squids, octopuses, nautiluses). The Monoplacophora, Scaphopoda (tusk shells), Caudofoveata, and Solenogastres are much smaller classes.

Ecological Relationships

Molluscs are found in a great range of habitats, from the tropics to polar seas, at altitudes exceeding 7000 m, in ponds, lakes, and streams, on mud flats, in pounding surf, and in open ocean from the surface to the abyssal depths. Most of them live in the sea, and they represent a variety of life-styles, including bottom feeders, burrowers, borers, and pelagic forms. The phylum includes some of the most sluggish and some of the swiftest and most active of the invertebrates. It includes herbivorous grazers, predaceous carnivores, and ciliary filter feeders.

According to the fossil evidence, the molluscs originated in the sea, and most of them have remained there. Much of their evolution occurred along shores, where food was abundant and habitats were varied. Only the bivalves and gastropods moved on to brackish and freshwater habitats. As filter feeders, the bivalves were unable to leave aquatic surroundings; however, the snails (gastropods) actually invaded the land and may have been the first animals to do so. Terrestrial snails are limited in range by their need for humidity, shelter, and the presence of calcium in the soil.

Economic Importance

A group as large as the molluscs would naturally affect humans in some way. A wide variety of molluscs are used as food. Pearls, both natural and cultured, are produced in the shells of clams and oysters, most of them in a marine oyster, found around eastern Asia (Figure 21-4B).

Position in Animal Kingdom

1. The molluscs are one of the major groups of true **coelomate** animals.
2. They belong to the **protostome** branch, or schizocoelous coelomates, and have spiral cleavage and mosaic development.
3. Many molluscs have a **trochophore larva** similar to the trochophore larva of marine annelids and other marine protostomes. Developmental evidence thus indicates that molluscs and annelids share a common ancestor.
4. Because molluscs are not metameric, they must have diverged from their common ancestor with the annelids before the advent of metamerism.
5. All the **organ systems** found in more derived invertebrates are present and well developed.

Biological Contributions

1. In molluscs gaseous exchange occurs not only through the body surface as in phyla discussed previously, but also in specialized **respiratory organs** in the form of **gills** or a **lung.**
2. Most classes have an **open circulatory system** with pumping **heart,** vessels, and blood sinuses. In most cephalopods the circulatory system is closed.
3. The efficiency of the respiratory and circulatory systems in the cephalopods has made greater body size possible. Invertebrates reach their largest size in some of the cephalopods.
4. They have a fleshy **mantle** that in most cases secretes a shell and is variously modified for a number of functions. Other features unique to the phylum are the **radula** and the muscular **foot.**
5. The highly developed direct **eye** of cephalopods is similar to the indirect eye of the vertebrates but arises as a skin derivative in contrast to the brain eye of vertebrates.

Characteristics of Phylum Mollusca

1. Body bilaterally symmetrical (bilateral asymmetry in some); unsegmented; usually with definite head
2. Ventral body wall specialized as a **muscular foot,** variously modified but used chiefly for locomotion
3. Dorsal body wall forms the **mantle,** which encloses the **mantle cavity,** is modified into **gills** or a **lung,** and secretes the **shell** (shell absent in some)
4. Surface epithelium usually ciliated and bearing mucous glands and sensory nerve endings
5. Coelom mainly limited to area around heart
6. Complex digestive system; rasping organ **(radula)** usually present (Figure 21-3); anus usually emptying into mantle cavity
7. **Open circulatory system** (mostly closed in cephalopods) of heart (usually three chambered, two in most gastropods), blood vessels, and sinuses; respiratory pigments in blood
8. Gaseous exchange by gills, lung, mantle, or body surface
9. Usually one or two kidneys **(metanephridia)** opening into the pericardial cavity and usually emptying into the mantle cavity
10. Nervous system of paired cerebral, pleural, pedal, and visceral ganglia, with nerve cords and subepidermal plexus; ganglia centralized in nerve ring in polyplacophorans, gastropods, and cephalopods
11. Sensory organs of touch, smell, taste, equilibrium, and vision (in some); eyes highly developed in cephalopods

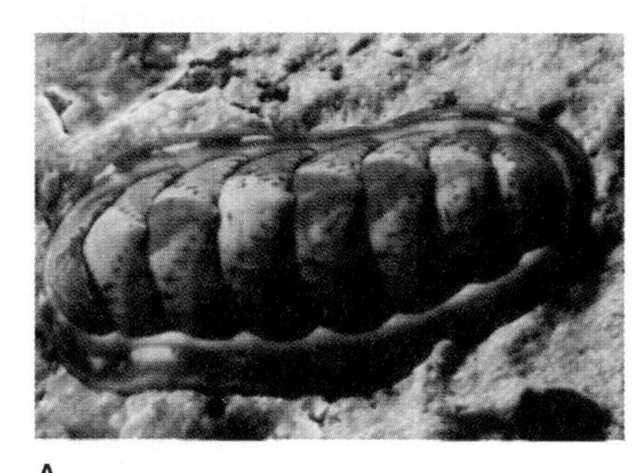

A

B

C

D

E

figure 21-1

Molluscs: a diversity of life forms. The basic body plan of this ancient group has become variously adapted for different habitats. **A,** A chiton *(Tonicella lineata)*, class Polyplacophora. **B,** A marine snail *(Calliostoma annulata)*, class Gastropoda. **C,** A nudibranch *(Chromodoris kuniei),* class Gastropoda. **D,** Pacific giant clam *(Panope abrupta),* with siphons to the left, class Bivalvia. **E,** An octopus *(Octopus briareus),* class Cephalopoda, forages at night on a Caribbean coral reef.

Some molluscs are destructive. The burrowing shipworms (Figure 21-24), a kind of clam, do great damage to wooden ships and wharves. To prevent the ravages of shipworms, wharves must be either creosoted or built of concrete. Snails and slugs often damage garden and other vegetation. In addition, many snails serve as intermediate hosts for serious parasites. A certain boring snail, the oyster drill, is second only to the sea star in destroying oysters.

Pearl production is the by-product of a protective device used by the animal when a foreign object, such as a grain of sand or a parasite, becomes lodged between the shell and mantle. The mantle secretes many layers of nacre around the irritating object (Figure 21-4). Pearls are cultured by inserting small spheres, usually made from pieces of the shells of freshwater clams, in the mantle of a certain species of oyster and by maintaining the oysters in enclosures. The oyster deposits its own nacre around the "seed" in a much shorter time than would be required to form a pearl normally.

Form and Function

Body plan

Reduced to its simplest dimensions, the mollusc body may be said to consist of a **head-foot** portion and a **visceral mass** portion (Figure 21-2).The head-foot is the more active area, containing the feeding, cephalic sensory, and locomotor organs. It depends primarily on muscular action for its function. The visceral mass is the portion containing digestive, circulatory, respiratory, and reproductive organs, and it depends primarily on ciliary tracts for its functioning. Two folds of skin, outgrowths of the dorsal body wall, make up a protective **mantle,** which encloses a space between the mantle and body wall called the **mantle cavity.** The mantle cavity houses the **gills** or lung, and in some molluscs the mantle secretes a protective **shell** over the visceral mass and head-foot. Modifications of the structures that make up the head-foot and the visceral mass produce the great profusion of different patterns making up this major group of animals.

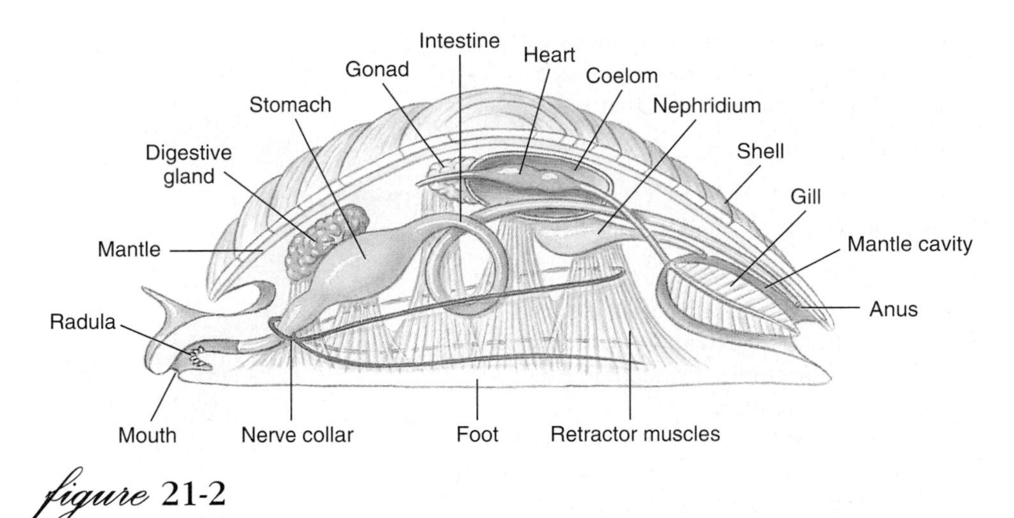

figure 21-2

Generalized mollusc. Although this construct is often presented as a "hypothetical ancestral mollusc" (HAM), most experts now agree that it never actually existed. It is an abstraction used to facilitate description of the general body plan of molluscs.

Head-Foot

Most molluscs have a well-developed head, which bears the mouth and some specialized sensory organs. Photosensory receptors range from fairly simple to the complex eyes of the cephalopods. Tentacles are often present. Within the mouth is a structure unique to molluscs, the radula, and usually posterior to the mouth is the chief locomotor organ, or foot.

Radula The radula is a rasping, protrusible, tonguelike organ found in all molluscs except the bivalves and some nudibranchs and their relatives. It is a ribbonlike membrane on which are mounted rows of tiny teeth that point backward (Figure 21-3). Complex muscles move the radula and its supporting cartilages **(odontophore)** in and out while the membrane is partly rotated over the tips of the cartilages. There may be a few or as many as 250,000 teeth, which, when protruded, can scrape, pierce, tear, or cut particles of food material, and the radula may serve as a rasping file for carrying the particles in a continuous stream toward the digestive tract.

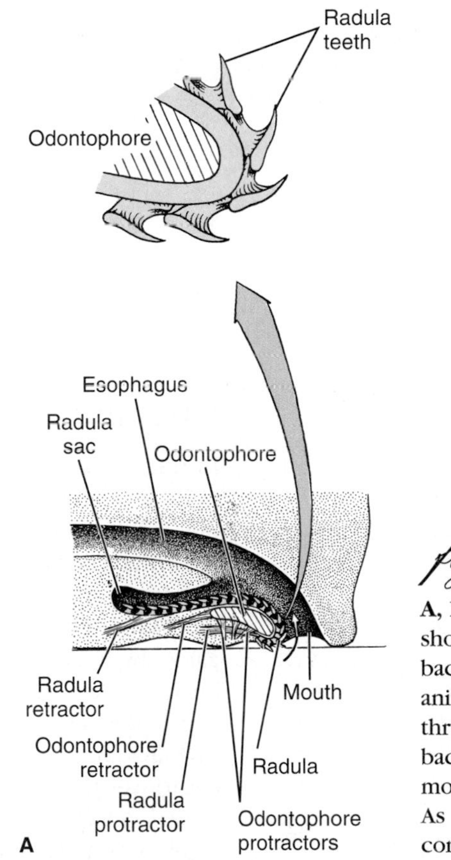

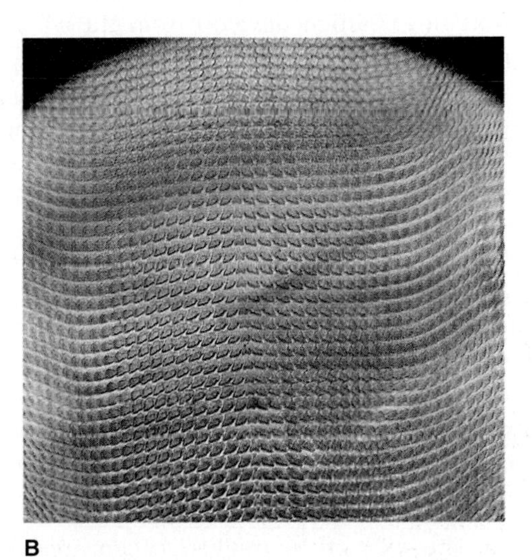

figure 21-3

A, Diagrammatic longitudinal section of gastropod head showing the radula and radula sac. The radula moves back and forth over the odontophore cartilage. As the animal grazes, the mouth opens, the odontophore is thrust forward, the radula gives a strong scrape backward bringing food into the pharynx, and the mouth closes. The sequence is repeated rhythmically. As the radula ribbon wears out anteriorly, it is continually replaced posteriorly. **B,** Radula of a snail prepared for microscopic examination.

Foot The molluscan foot may be variously adapted for locomotion, for attachment to a substratum, or for a combination of functions. It is usually a ventral, solelike structure in which waves of muscular contraction effect a creeping locomotion. However, there are many modifications, such as the attachment disc of the limpets, the laterally compressed "hatchet foot" of the bivalves, or the siphon for jet propulsion in the squids and octopuses. Secreted mucus is often used as an aid to adhesion or as a slime track by small molluscs that glide on cilia.

Visceral Mass

Mantle and Mantle Cavity The mantle is a sheath of skin extending from the visceral hump that hangs down on each side of the body, protecting the soft parts and creating between itself and the visceral mass the space called the mantle cavity. The surface of the mantle secretes the shell externally or internally.

The mantle cavity plays an enormous role in the life of the mollusc. It usually houses the respiratory organs (gills or lung), which develop from the mantle, and the mantle's own exposed surface serves also for gaseous exchange. The products from the digestive, excretory, and reproductive systems empty into the mantle cavity. In aquatic molluscs a continuous current of water, kept moving by surface cilia or by muscular pumping, brings in oxygen, and in some forms, food; flushes out wastes; and carries reproductive products out to the environment. In aquatic forms the mantle usually has sensory receptors for sampling the environmental water. In cephalopods (squids and octopuses) the muscular mantle and its cavity create the jet propulsion used in locomotion.

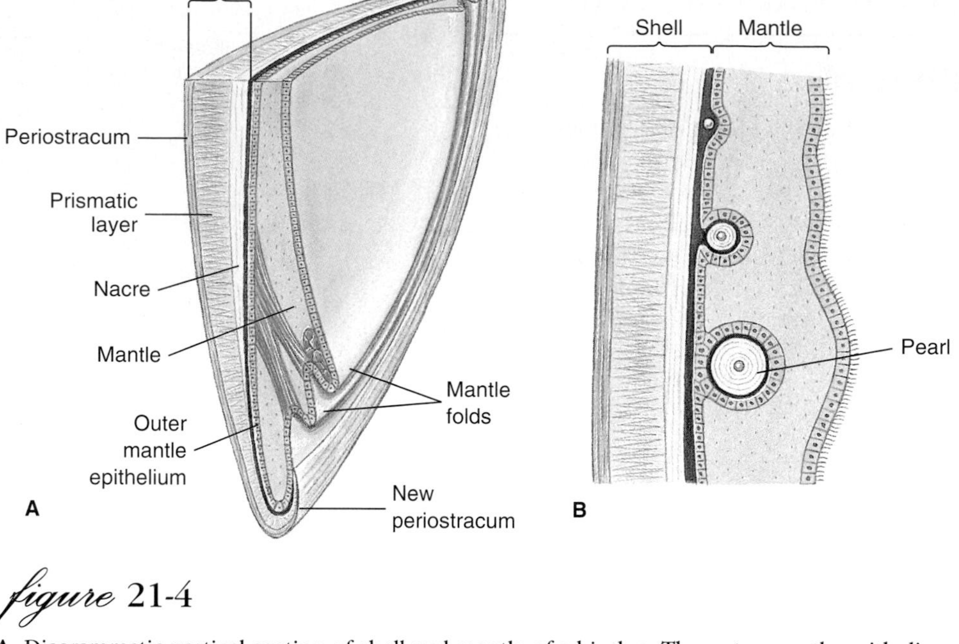

figure 21-4

A, Diagrammatic vertical section of shell and mantle of a bivalve. The outer mantle epithelium secretes the shell; the inner epithelium is usually ciliated. **B,** Formation of pearl between mantle and shell as a parasite or bit of sand under the mantle becomes covered with nacre.

Shell The shell of the mollusc, when present, is secreted by the mantle and is lined by it. Typically there are three layers (Figure 21-4). The **periostracum** is the outer horny layer, composed of an organic substance called conchiolin, which is a resistant protein. It helps protect the underlying calcareous layers from erosion by boring organisms. It is secreted by a fold of the mantle edge, and growth occurs only at the margin of the shell. On the older parts of the shell the periostracum often becomes worn away. The middle **prismatic layer** is composed of densely packed prisms of calcium carbonate laid down in a protein matrix. It is secreted by the glandular margin of the mantle, and increase in shell size occurs at the shell margin as the animal grows. The inner **nacreous layer** of the shell is composed of calcium carbonate sheets laid down over a thin protein matrix. This layer is secreted continuously by the mantle surface, so that it becomes thicker during the life of the animal.

Freshwater molluscs usually have a thick periostracum that gives some protection against the acids produced in the water by the decay of leaf litter. In some marine molluscs the periostracum is thick, but in some it is relatively thin or absent. There is a great range in variation in shell structure. Calcium for the shell comes from the environmental water or soil or from food. In most molluscs the first shell appears during the larval period and grows continuously throughout life.

Internal Structure and Function

In the molluscs, oxygen-carbon dioxide exchange occurs not only through the body surface, particularly that of the mantle, but in specialized respiratory organs such as gills or lungs, which are derivatives of the mantle. There is an **open circulatory system** with a pumping **heart,** blood vessels, and blood sinuses (rather than capillaries), which permeate the organs. Most cephalopods have a closed blood system with heart, vessels, and capillaries. The digestive tract is complex and highly specialized according to the feeding habits of the various molluscs. Most molluscs have a pair of **kidneys (metanephridia),** which connect with the coelom; the ducts of the kidneys in many forms serve also for the discharge of eggs and sperm. The **nervous system** consists of several pairs of ganglia with connecting nerve cords. There are various types of highly specialized sense organs.

Most molluscs are dioecious, although some of the gastropods are hermaphroditic. Many aquatic molluscs pass through free-swimming **trochophore** (Figure 21-5) and **veliger** (Figure 21-6) larval stages. The veliger is the free-swimming larva of most marine snails, tusk shells, and bivalves. It develops from the trochophore and has the beginning of a foot, shell, and mantle.

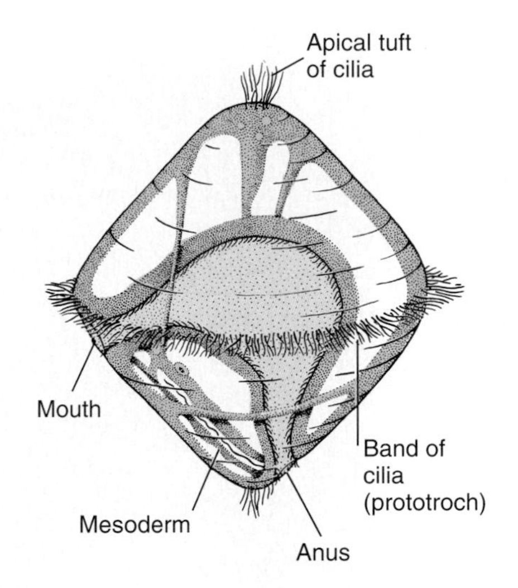

figure 21-5

A generalized trochophore larva. Molluscs and annelids with primitive embryonic development have trochophore larvae, as do several other phyla.

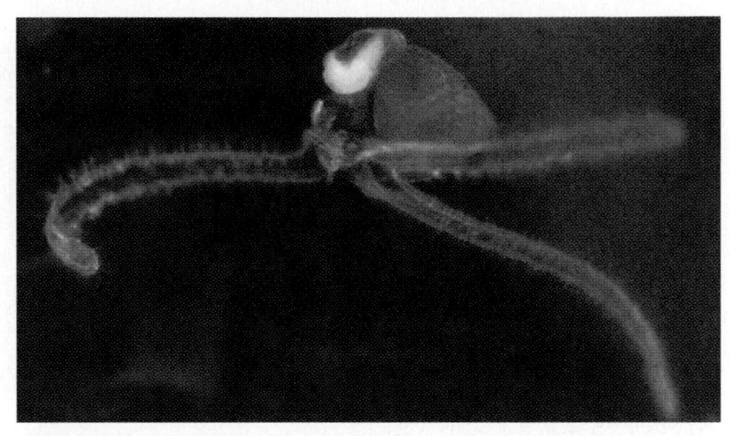

figure 21-6

Veliger of a snail, *Pedicularia,* swimming. The adults are parasitic on corals. The ciliated process (velum) develops from the prototroch of the trochophore (Figure 21-5).

The trochophore larva (Figure 21-5) is minute, translucent, more or less top shaped, and has a prominent circlet of cilia (prototroch) and sometimes one or two accessory circlets. It is found in molluscs and annelids with primitive embryonic development and is considered one of the evidences for common phylogenetic origin of the two phyla. Some form of trochophore-like larva also occurs in marine turbellarians, nemertines, brachiopods, phoronids, sipunculids, and echiurids, and it probably reflects some phylogenetic relationship among all these phyla.

Classes Caudofoveata and Solenogastres

The caudofoveates and the solenogasters (see Figure 21-35, p. 470) are often united in the class Aplacophora, and they are both wormlike and shell-less, with calcareous scales or spicules in their integument, with reduced head, and without nephridia. In contrast to the caudofoveates, the solenogasters usually have no true gills, and they are hermaphroditic. The caudofoveates are burrowing marine animals, feeding on microorganisms and detritus, whereas the solenogasters live freely on the bottom and often feed on cnidarians. The caudofoveates may have more features closer to those of the ancestral mollusc than do any other living groups.

Class Monoplacophora

Until 1952 the Monoplacophora (mon-o-pla-kof′o-ra) were known only from Paleozoic shells. However, in that year living specimens of *Neopilina* (Gr. *neo,* new, + *pilos,* felt cap) were dredged up from the ocean bottom near the west coast of Costa Rica. These molluscs are small and have a low, rounded shell and a creeping foot (Figure 21-7). They have a superficial resemblance to the limpets, but unlike most other molluscs, a number of organs are serially repeated. Serial repetition occurs to a more limited extent in the chitons. Some authors have considered the monoplacophorans truly metameric (p. 358), indicating that molluscs descended from a metameric, annelid-like ancestor and that metamerism was lost secondarily in other molluscs. Others believe that *Neopilina* shows only pseudometamerism and that molluscs did not have a metameric ancestor. However, the phylogenetic affinity of annelids is strongly supported by embryological evidence.

Class Polyplacophora: Chitons

The chitons are somewhat flattened and have a convex dorsal surface that bears eight articulating limy **plates,** or **valves,** which give them their name (Figures 21-8 and 21-9). The term Polyplacophora means "bearing many plates," in contrast to the Monoplacophora, which bear one shell (*mono,* single). The plates overlap posteriorly and are usually dull colored to match the rocks to which the chitons cling.

Most chitons are small (2 to 5 cm); the largest rarely exceeds 30 cm. They commonly occur on rocky surfaces in intertidal regions, although some live at great depths. Chitons are stay-at-home organisms, straying only very short distances for feeding. In feeding, a sensory subradular organ protrudes from the mouth to explore for algae or colonial organisms. When

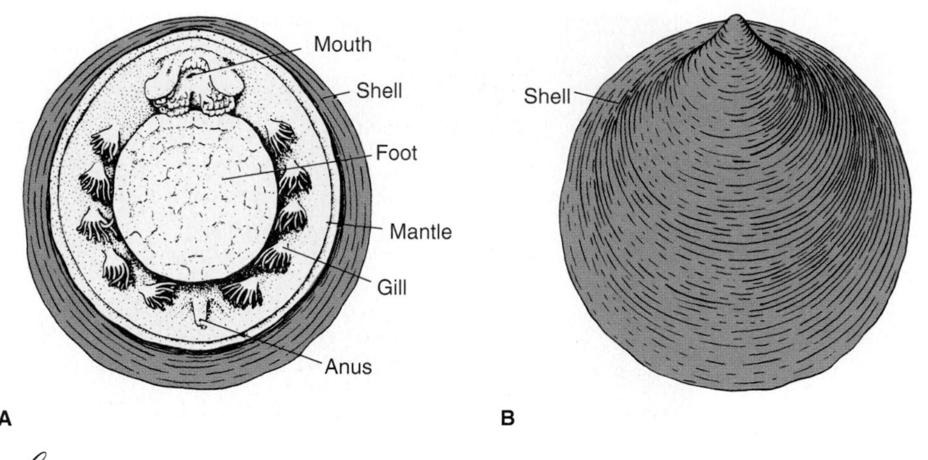

figure 21-7

Neopilina, class Monoplacophora. Living specimens range from 3 mm to about 3 cm in length. **A,** Ventral view. **B,** Dorsal view.

figure 21-8

Mossy chiton, *Mopalia muscosa.* The upper surface of the mantle, or "girdle," is covered with hairs and bristles, an adaptation for defense.

some are found, the radula projects to scrape algae off the rocks. A chiton clings tenaciously to its rock with the broad flat foot. If detached, it can roll up like an armadillo for protection.

The mantle forms a **girdle** around the margin of the plates, and in some species mantle folds cover part or all of the plates. On each side of the broad ventral foot and lying between the foot and the mantle is a row of gills suspended from the roof of the mantle cavity. With the foot and the mantle margin adhering tightly to the substrate, these grooves become closed chambers, open only at the ends. Water enters the grooves anteriorly, flows across the gills, and leaves posteriorly, thus bringing a continuous supply of oxygen to the gills.

Blood pumped by the three-chambered heart reaches the gills by way of an aorta and sinuses. Two kidneys carry waste from the pericardial cavity to the exterior. Two pairs of longitudinal nerve cords are connected in the buccal region. Sense organs include shell eyes on the surface of the shell (in some) and a pair of **osphradia** (chemosensory organs for sampling water).

Sexes are separate in chitons. Sperm shed by males in the excurrent water enter the gill grooves of the females by incurrent openings. Eggs are shed into the sea singly or in strings or masses of jelly. The trochophore larva metamorphoses directly into a juvenile, without an intervening veliger stage.

Class Scaphopoda

The Scaphopoda (ska-fop′o-da), commonly called the tusk shells or tooth shells, are sedentary marine molluscs that have a slender body covered with a mantle and a tubular shell open at both ends. Here the molluscan body plan has taken a new direction, with the mantle wrapped around the viscera and fused to form a tube. Most scaphopods are 2.5 to 5 cm long, although they range from 4 mm to 25 cm long.

The foot, which protrudes through the larger end of the shell, functions in burrowing into mud or sand, always leaving the small end of the shell exposed to the water above (Figure 21-10). Respiratory water circulates through the mantle cavity both by movements of the foot and by ciliary action. Gaseous exchange occurs in the mantle. Most of the food is detritus and protozoa from the substrate. Cilia of the foot or on the mucus-covered, ciliated knobs of long tentacles catch the food.

Class Gastropoda

Among the molluscs the class Gastropoda (gas-trop′o-da) (Gr. *gastēr,* stomach, + *pous, podos,* foot) is by far the largest and most diverse, containing about 40,000 living and 15,000 fossil species. Its members differ so widely that there is no single general term in our language that can apply to them as a group. They include snails, limpets, slugs, whelks, conchs, periwinkles, sea slugs, sea hares, sea butterflies, and others. They range from some marine molluscs with many primitive characters to highly evolved, air-breathing snails and slugs.

Gastropods are often sluggish, sedentary animals because most of them have heavy shells and slow locomotor organs. When present, the shell is almost always of one piece (univalve) and may be coiled or uncoiled. Some snails have an

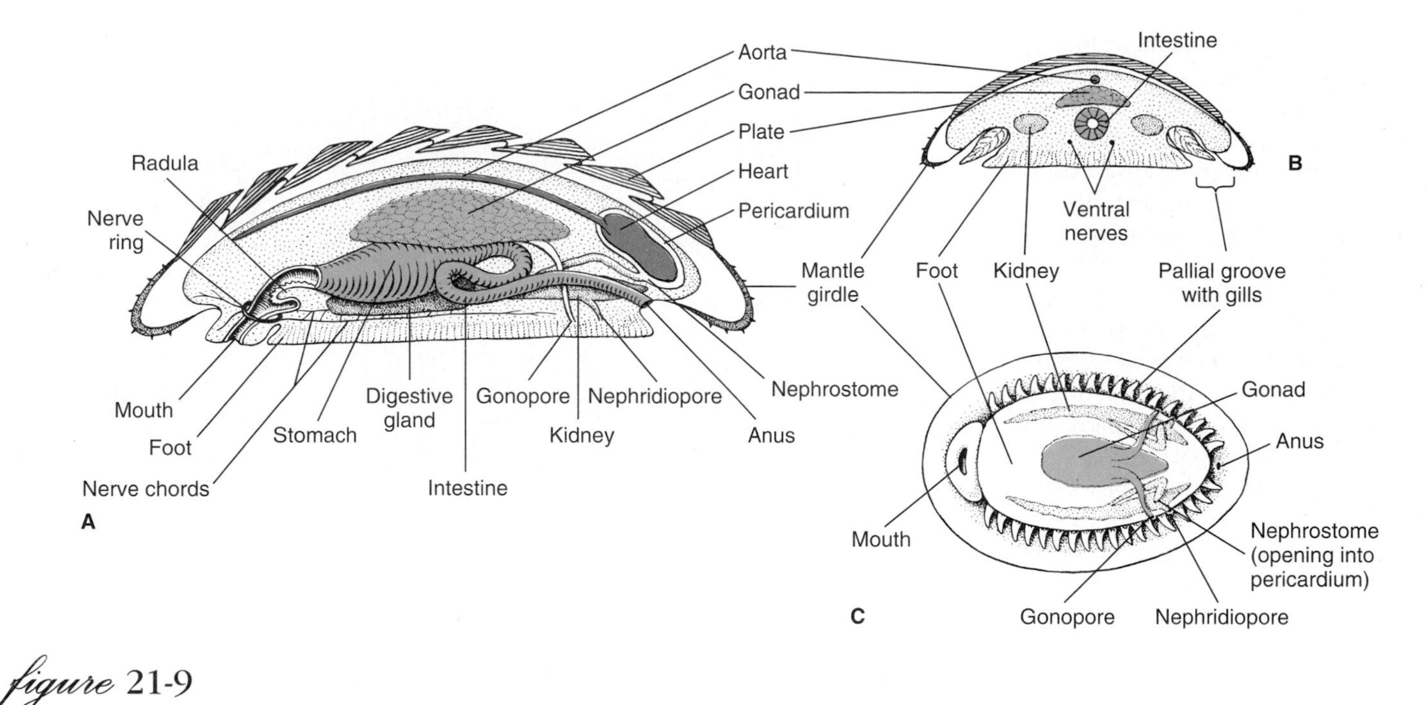

figure 21-9

Anatomy of a chiton (class Polyplacophora). **A,** Longitudinal section. **B,** Transverse section. **C,** External ventral view.

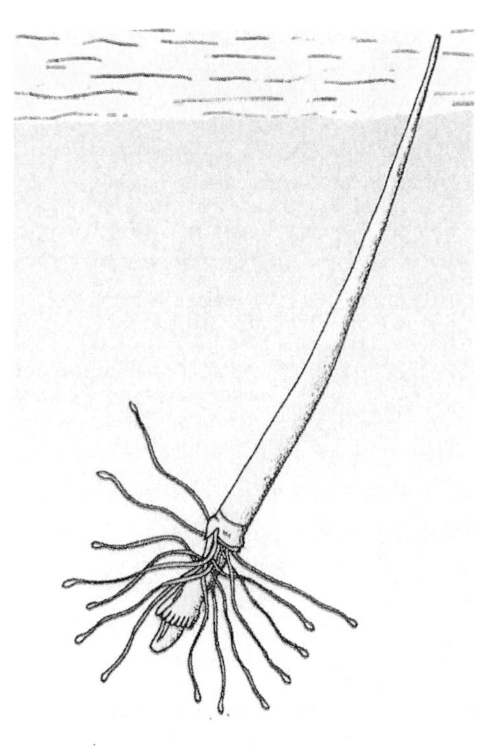

figure 21-10

The tusk shell, *Dentalium,* a scaphopod. It burrows into soft mud or sand and feeds by means of its prehensile tentacles. Respiratory currents of water are drawn in by ciliary action through the small open end of the shell, then expelled through the same opening by muscular action.

operculum, a horny plate that covers the shell aperture when the body withdraws into the shell. It protects the body and prevents water loss. These animals are basically bilaterally symmetrical, but because of **torsion,** a twisting process that occurs in the veliger stage, the visceral mass has become asymmetrical.

Form and Function

Torsion

Of all the molluscs, only gastropods undergo torsion. Torsion is a peculiar phenomenon that moves the mantle cavity to the front of the body, thus twisting the visceral organs as well through a 90- to 180-degree rotation. Torsion occurs during the veliger stage, and in some species the first part may take only a few seconds. The second 90 degrees typically takes a longer period. Before torsion occurs, the embryo is bilaterally symmetrical with an anterior mouth and a posterior anus and mantle cavity (Figure 21-11). The change comes about by an uneven growth of the right and left muscles that attach the shell to the head-foot.

After torsion, the anus and mantle cavity become anterior and open above the mouth and head. The left gill, kidney, and heart auricle are now on the right side, whereas the original right gill, kidney, and heart auricle (lost in most modern gastropods) are now on the left, and the nerve cords have been twisted into a figure eight. Because of the space available in the mantle cavity, the animal's sensitive head end can now be withdrawn into the protection of the shell, with the tougher foot forming a barrier to the outside.

The curious arrangement that results from torsion poses a serious sanitation problem by creating the possibility of wastes being washed back over the gills **(fouling)** and causes us to wonder what evolutionary factors favored such a strange realignment of the body. Several explanations have been proposed, none entirely satisfying. For example, sense organs of the mantle cavity (osphradia) would better sample water when turned in the direction of travel, and as mentioned already, the head could be withdrawn into the shell. Certainly the consequences of torsion and the resulting need to avoid fouling have been very important in the subsequent evolution of gastropods. We cannot explore these consequences, however, until we describe another unusual feature of gastropods—coiling.

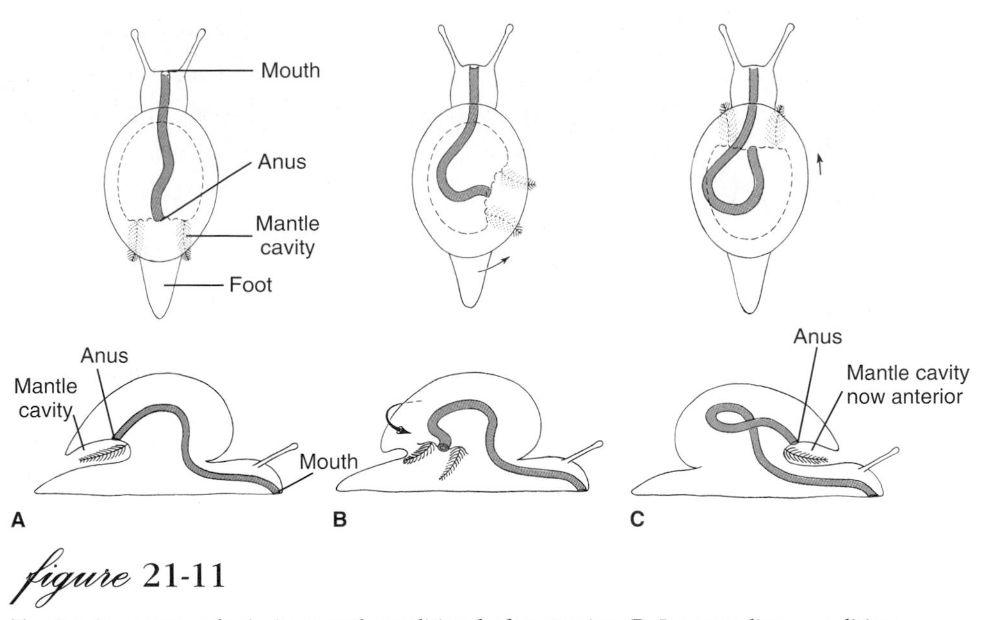

figure 21-11

Torsion in gastropods. **A,** Ancestral condition before torsion. **B,** Intermediate condition. **C,** Early gastropod, torsion complete; direction of crawling now tends to carry waste products back into the mantle cavity, resulting in fouling.

Coiling

The coiling, or spiral winding, of the shell and visceral hump is not the same as torsion. Coiling may occur in the larval stage at the same time as torsion, but the fossil record shows that coiling was a separate evolutionary event and originated in gastropods earlier than torsion did. Nevertheless, all living gastropods have descended from coiled, torted ancestors, whether or not they now show these characteristics.

Early gastropods had a bilaterally symmetrical shell with all the whorls lying in a single plane (Figure 21-12A). Such a shell was not very compact, since each whorl had to lie completely outside the preceding one. Curiously, a few modern species have secondarily returned to that form. The compactness problem of the planospiral shell was solved by a shape in which each succeeding whorl was at the side of the preceding one (Figure 21-12B). However, this shape clearly was unbalanced, hanging as it did with much weight over to one side. They achieved better weight distribution by shifting the shell upward and posteriorly, with the shell axis oblique to the longitudinal axis of the foot (Figure 21-12C and D). The weight and bulk of the main body whorl, the largest whorl of the shell, pressed on the right side of the mantle cavity, however, and apparently interfered with the organs on that side. Accordingly, the gill, auricle, and kidney of the right side have been lost in all except a few living gastropods, leading to a condition of **bilateral asymmetry.**

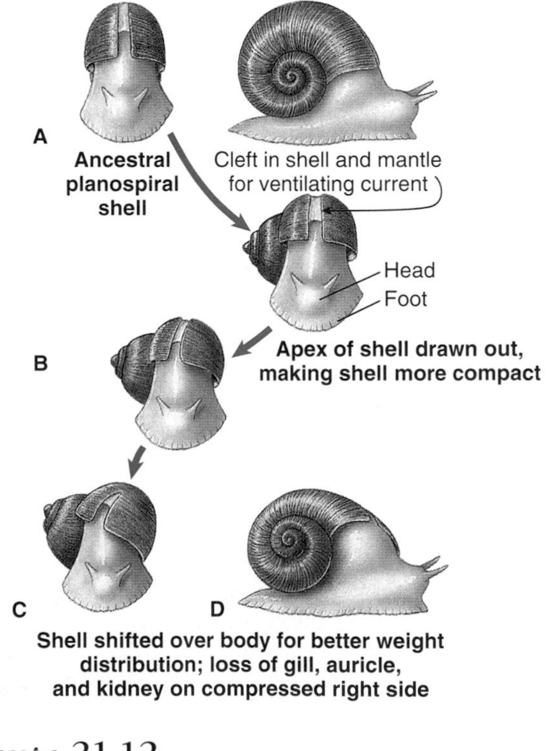

figure 21-12

Evolution of shell in gastropods. **A,** Earliest coiled shells were planospiral, each whorl lying completely outside the preceding whorl. Interestingly, the shell has become planospiral secondarily in some living forms. **B,** Better compactness was achieved by snails in which each whorl lay partially to the side of the preceding whorl. **C** and **D,** Better weight distribution resulted when shell was moved upward and posteriorly.

Adaptations to Avoid Fouling

Although the loss of the right gill was probably an adaptation to the mechanics of carrying the coiled shell, that condition made possible a way to avoid fouling, which is displayed in most modern gastropods. Water is brought into the left side of

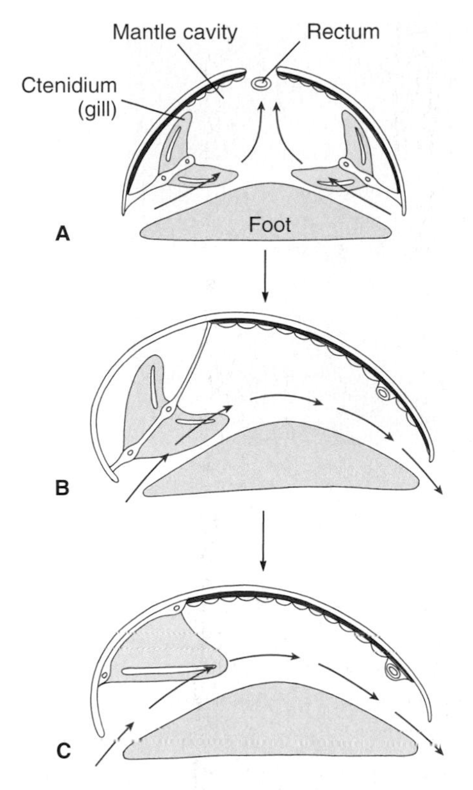

figure 21-13

Evolution of the gills in gastropods. **A,** Primitive condition in prosobranchs with two gills and excurrent water leaving the mantle cavity by a dorsal slit or hole. **B,** Condition after one gill had been lost. **C,** Advanced condition in prosobranchs, in which filaments on one side of remaining gill are lost, and axis is attached to mantle wall.

the mantle cavity and out the right side, carrying with it the wastes from the anus and nephridiopore, which lie near the right side (Figure 21-13). Some gastropods with primitive characteristics (those with two gills, such as abalone) (Figure 21-14A) avoid fouling by venting the excurrent water through a dorsal slit or hole in the shell above the anus (Figure 21-12). The opisthobranchs (nudibranchs and others) have evolved an even more curious "twist"; after undergoing torsion as larvae, they develop various degrees of *detorsion* as adults. The pulmonates (most freshwater and terrestrial snails) have lost the gill altogether, and the vascularized mantle wall has become a lung. The anus and nephridiopore open near the opening of the lung to the outside (pneumostome), and waste is expelled forcibly with air or water from the lung.

Feeding Habits

Feeding habits of gastropods are as varied as their shapes and habitats, but all include the use of some adaptation of the radula. Many gastropods are herbivorous, rasping off particles of algae. Some herbivores are grazers, some are browsers, some are planktonic feeders. The abalone (Figure 21-14) holds seaweed with the foot and breaks off pieces with the radula. Some snails are scavengers, living on dead and decayed flesh; others are carnivorous, tearing their prey apart with their radular teeth. Some, such as the oyster borer and the moon snail (Figure 21-14B), have an extensible proboscis for drilling holes in the shells of the bivalves whose soft parts they find delectable. Some even have a spine for opening the shells. Most of the pulmonates (air-breathing snails) (Figure 21-20, p. 461) are herbivorous, but some live on earthworms and other snails.

Among the most interesting predators are the poisonous cone shells (Figure 21-15), which feed on vertebrates or other invertebrates, depending on the species. When Conus *senses the presence of its prey, a single radular tooth slides into position at the tip of the probscis. When the proboscis strikes the prey, it expels the tooth like a harpoon, and the poison tranquilizes or kills the prey at once. Some species can deliver very painful stings, and the sting of several species is lethal to humans. The venom consists of a series of toxic peptides, and each* Conus *species carries peptides* **(conotoxins)** *that are specific for the neuroreceptors of its preferred prey.*

Some of the sessile gastropods, such as the slipper shells, are ciliary feeders that use the gill cilia to draw in particulate matter, which they roll into a mucous ball and carry to their mouth. Some sea butterflies secrete a mucous net to catch small planktonic forms and then draw the web into the mouth.

After maceration by the radula or by some grinding device, such as the so-called gizzard in sea hares (Figure 21-16) and in others, digestion is usually extracellular in the lumen of the stomach or digestive glands. In ciliary feeders the stomachs are sorting regions and most of the digestion is intracellular in the digestive gland.

Internal Form and Function

Respiration in most gastropods is carried out by a gill (two gills in a few), although some aquatic forms lack gills and depend on the skin. The pulmonates have a lung. Freshwater pulmonates must surface to expel a bubble of gas from the lung and curl the edge of the mantle around the **pneumostome** (pulmonary opening in the mantle cavity) (Figure 21-20B) to form a siphon for taking in air.

Most gastropods have a single nephridium (kidney). The circulatory and nervous systems are well developed (Figure 21-17). The nervous system includes three pairs of ganglia connected by nerves. Sense organs include eyes, statocysts, tactile organs, and chemoreceptors.

There are both dioecious and hermaphroditic gastropods. During copulation in hermaphroditic species there is sometimes an exchange of **spermatophores** (bundles of

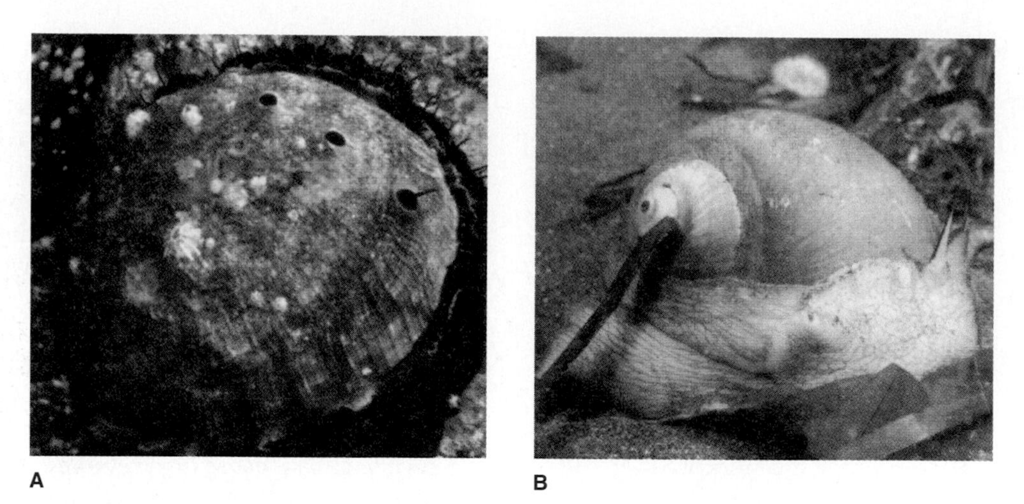

figure 21-14

A, Red abalone, *Haliotus rufescens.* This huge, limpetlike snail is prized as food and extensively marketed. Abalones are strict vegetarians, feeding especially on sea lettuce and kelp. **B,** Moon snail, *Polinices lewisii.* A common inhabitant of West Coast sand flats, the moon snail is a predator of clams and mussels. It uses its radula to drill neat holes through its victim's shell, through which the proboscis is then extended to eat the bivalve's fleshy body.

figure 21-15

Conus extends its long, wormlike proboscis. When the fish attempts to consume this tasty morsel, the *Conus* stings it in the mouth and kills it. The snail engulfs the fish with its distensible stomach, then regurgitates the scales and bones some hours later.

sperm), so that self-fertilization is avoided. Many forms perform courtship ceremonies. Most land snails lay their eggs in holes in the ground or under logs. Some aquatic gastropods lay their eggs in gelatinous masses; others enclose them in gelatinous capsules or in parchment egg cases. Most marine gastropods go through a free-swimming veliger larval stage during which torsion and coiling occur. Others develop directly into a juvenile within the egg capsule.

Major Groups of Gastropods

Traditional classification of the class Gastropoda recognized three subclasses: Prosobranchia, much the largest subclass, almost all of which are marine; Opisthobranchia, an assemblage including sea slugs, sea hares, nudibranchs, and canoe shells—all marine; and Pulmonata, containing most freshwater and terrestrial species. Currently, gastropod taxonomy is in a state of flux, and some workers regard any attempt to present a classification of the class as premature.[1] Nevertheless, present evidence suggests that the Prosobranchia is paraphyletic. The Opisthobranchia may or may not be paraphyletic, but the Opisthobranchia and Pulmonata together apparently form a monophyletic grouping.

[1] Bieler, R. 1992. Ann. Rev. Ecol. Syst. **23:**311–338.

figure 21-16

A, The sea hare, *Aplysia dactylomela,* crawls and swims across a coral reef, assisted by large, winglike parapodia, here curled above the body. **B,** When attacked, the sea hare squirts a copious protective secretion from its "purple gland" in the mantle cavity.

Familiar examples of marine gastropods are periwinkles, limpets (Figure 21-18A), whelks, conchs, abalones (Figure 21-14A), slipper shells, oyster borers, rock shells, and cowries.

At present 8 to 12 groups of opisthobranchs are recognized. Some have a gill and a shell, although the latter may be vestigial, and some have no shell or true gill. The large sea hare *Aplysia* (Figure 21-16) has large earlike anterior tentacles and a vestigial shell. Nudibranchs have no shell as adults and rank among the most beautiful and colorful of molluscs (Figure 21-19). Having lost the gill, the body surface of some nudibranchs is often increased for gaseous exchange by small projections **(cerata),** or a ruffling of the mantle edge.

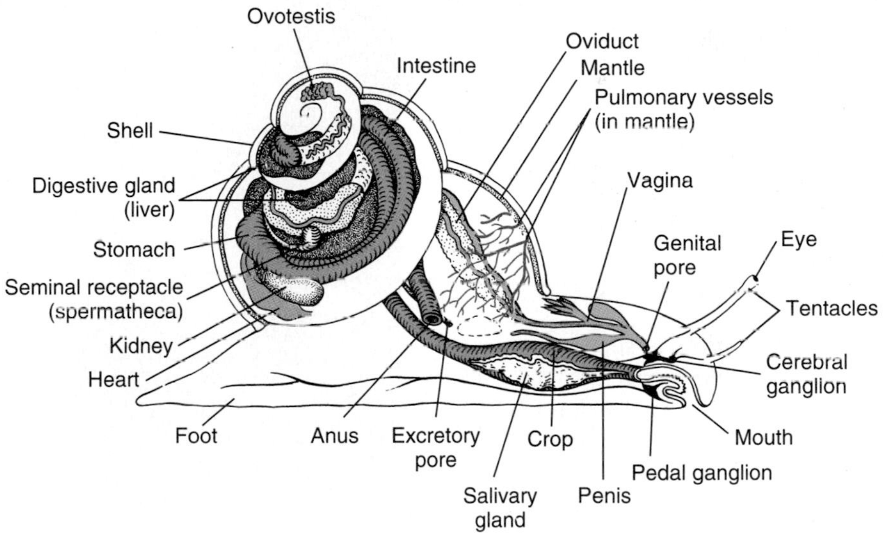

The third major group (Pulmonata) contains most land and freshwater snails and slugs. Usually lacking gills, their mantle cavity has become a lung, which fills with air by contraction of the mantle floor. Aquatic and a few terrestrial species have one pair of nonretractile tentacles, at the base of which are the eyes; land forms usually have two pairs of tentacles, with the posterior pair bearing the eyes (Figures 21-17 and 21-20). The few nonpulmonate species of gastropods that live in fresh water usually can be distinguished from pulmonates because they have an operculum, which is lacking in pulmonates.

figure 21-17

Anatomy of a pulmonate snail.

Class Bivalvia (Pelecypoda)

The Bivalvia (bi-val′ve-a) are also known as Pelecypoda (pel-e-sip′o-da) (Gr. *pelekus,* hatchet, + *pous, podus,* foot). They are bivalved (two-shelled) molluscs that include the mussels, clams, scallops, oysters, and shipworms and range in size from tiny seed shells 1 to 2 mm in length to the giant, South Pacific clams *Tridacna,* mentioned previously (Figure 21-21). Most ivalves are sedentary **suspension feeders** that depend on ciliary currents produced by the gills to bring in food materials. Unlike the gastropods, they have no head, no radula, and very little cephalization (Figure 21-22).

Most bivalves are marine, but many live in brackish water and in streams, ponds, and lakes.

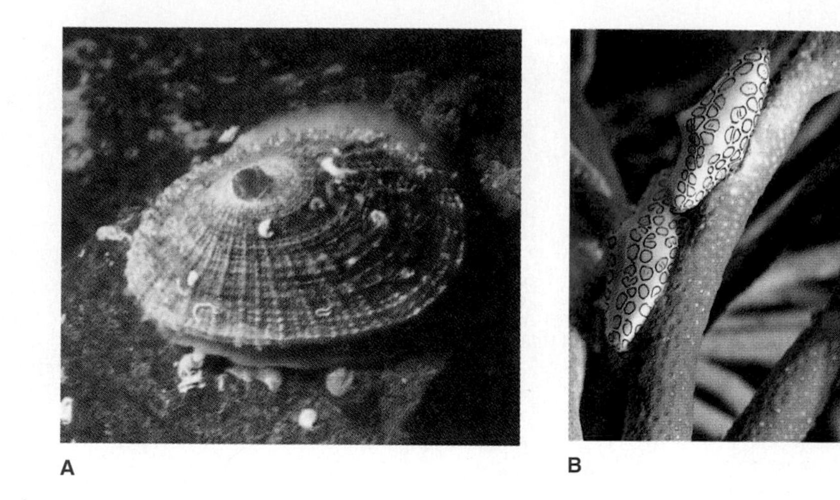

figure 21-18

A, Keyhole limpet, *Diodora aspera,* a prosobranch gastropod with a hole in the apex through which the water leaves the shell. **B,** The flamingo tongue, *Cyphoma gibbosum,* is a showy inhabitant of Caribbean coral reefs, where it is associated with gorgonians. This snail has a smooth creamy orange to pink shell that is normally covered by the brightly marked mantle.

Form and Function

Shell

Bivalves are laterally compressed, and their two shells **(valves)** are held together dorsally by a **hinge ligament** that causes the valves to gape ventrally. **Adductor muscles** work in opposition to the hinge ligament and draw the valves together (Figure 21-23C and D). Projecting above the hinge ligament on each valve is the **umbo,** which is the oldest part of the shell. The valves function largely for protection, but those of shipworms (Figure 21-24) have microscopic teeth for rasping wood, and rock borers use spiny valves for boring into rock. A few bivalves such as scallops (Figure 21-25) use their shells for locomotion by clapping the valves together so that they move in spurts.

figure 21-19

An aeolid nudibranch, *Flabellina iodinea.* Its long, dorsal cerata contain nematocysts, which the animal obtains from its cnidarian diet.

Body and Mantle

The **visceral mass** is suspended from the dorsal midline, and the muscular foot is attached to the visceral mass anteroventrally. The gills hang down on each side, each covered by a fold of the mantle. The posterior edges of the mantle folds form dorsal excurrent and ventral incurrent openings (Figures 21-23A, and 21-26). In some marine bivalves part of the mantle is drawn out into long muscular siphons to allow the clam to burrow into the mud or sand and extend the siphons to the water above. Cilia on the gills and inner surface of the mantle direct the flow of water over the gills.

Locomotion

Most bivalves move by extending the slender muscular foot between the valves (Figure 21-23A). They pump blood into the foot, causing it to swell and to act as an anchor in the mud or sand, then longitudinal muscles contract to shorten the foot and pull the animal forward. In most bivalves the foot is used

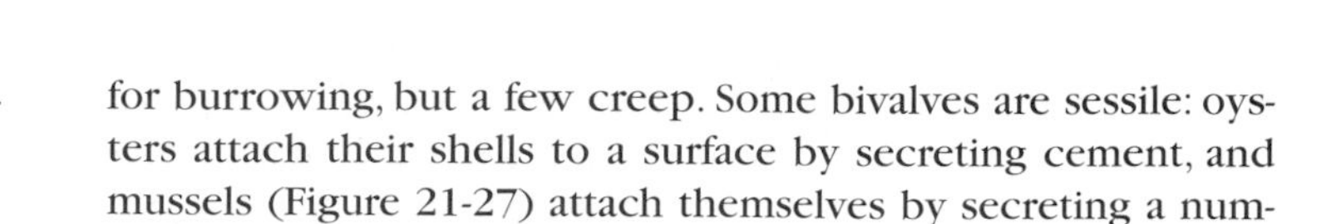

for burrowing, but a few creep. Some bivalves are sessile: oysters attach their shells to a surface by secreting cement, and mussels (Figure 21-27) attach themselves by secreting a number of slender **byssal threads.**

Feeding and Digestion

Most bivalves are suspension feeders. The respiratory currents bring both oxygen and organic materials to the gills where ciliary tracts direct them to the tiny pores of the gills. Gland cells

A B

figure 21-20

A, Pulmonate land snail. Note two pairs of tentacles; the second larger pair bears the eyes. **B,** Banana slug, *Ariolimax columbianus.*

on the gills and labial palps secrete copious amounts of mucus, which entangles particles suspended in the water going through gill pores. Ciliary tracts move the particle-laden mucus to the mouth (Figure 21-22).

In the stomach the mucus and food particles are kept whirling by a rotating gelatinous rod, called a **crystalline style.** Solution of layers of the rotating style frees digestive enzymes for extracellular digestion. Ciliated ridges of the stomach sort food particles and direct suitable particles to the **digestive gland** for intracellular digestion.

Shipworms (Figure 21-24) feed on the particles they excavate as they burrow in wood. Symbiotic bacteria live in a special organ in these bivalves and produce cellulase to digest the wood. Other bivalves such as giant clams gain much of their nutrition from the photosynthetic products of symbiotic algae living in their mantle tissue (Figure 21-21).

figure 21-21

Giant clam, *Tridacna gigas,* lies buried in coral rock with only the richly colored fluted siphonal edge visible. These bivalves bear enormous numbers of symbiotic single-cell algae (zooxanthellae) that provide much of the clam's nutrition.

Internal Features and Reproduction

Bivalves have a three-chambered heart that pumps blood through the gills and mantle for oxygenation and to the kidneys for waste elimination (Figure 21-28). They have three pairs of widely separated ganglia and poorly developed sense organs. A few bivalves have ocelli. The steely blue eyes of some scallops (Figure 21-25), located around the mantle edge, are remarkably complex, equipped with cornea, lens, and retina.

Freshwater clams were once abundant and diverse in streams throughout the eastern United States, but they are now easily the most jeopardized group of animals in the country. Of the more than 300 species once present, 12 are extinct, 42 are listed as threatened or endangered, and as many as 88 more may be listed soon. A combination of causes is responsible, of which a decline in water quality is among the most important. Pollution and sedimentation from mining, industry, and agriculture are among the culprits. Habitat destruction by altering natural water courses and damming is an important factor. Poaching to supply the Japanese cultured pearl industry is partially to blame (see note on p. 450). And in addition to everything else, the prolific zebra mussels (see next note) attach in great numbers to native clams, exhausting food supplies (phytoplankton) in the surrounding water.

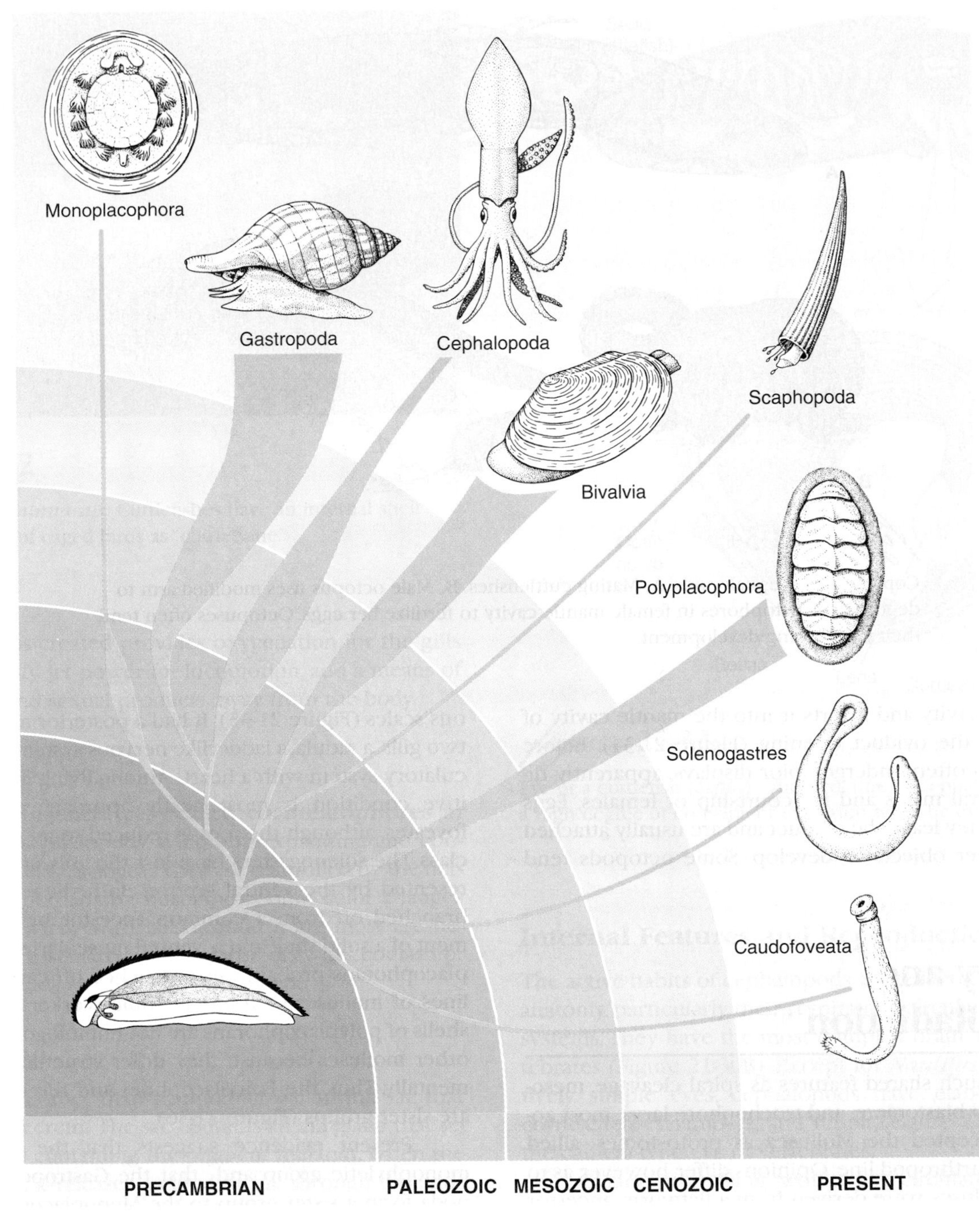

figure 21-35

The classes of Mollusca, showing their derivations and relative abundance.

Most of the diversity among the molluscs is related to their adaptation to different habitats and modes of life and to a wide variety of feeding methods, ranging from sedentary filter feeding to active predation. There are many adaptations for food gathering within the phylum and an enormous variety in radular structure and function, particularly among the gastropods.

The versatile glandular mantle has probably shown more plastic adaptative capacity than any other molluscan structure. Besides secreting the shell and forming the mantle cavity, it is variously modified into gills, lungs, siphons, and apertures, and it sometimes functions in locomotion, in the feeding processes, or in a sensory capacity. The shell, too, has undergone a variety of evolutionary adaptations.

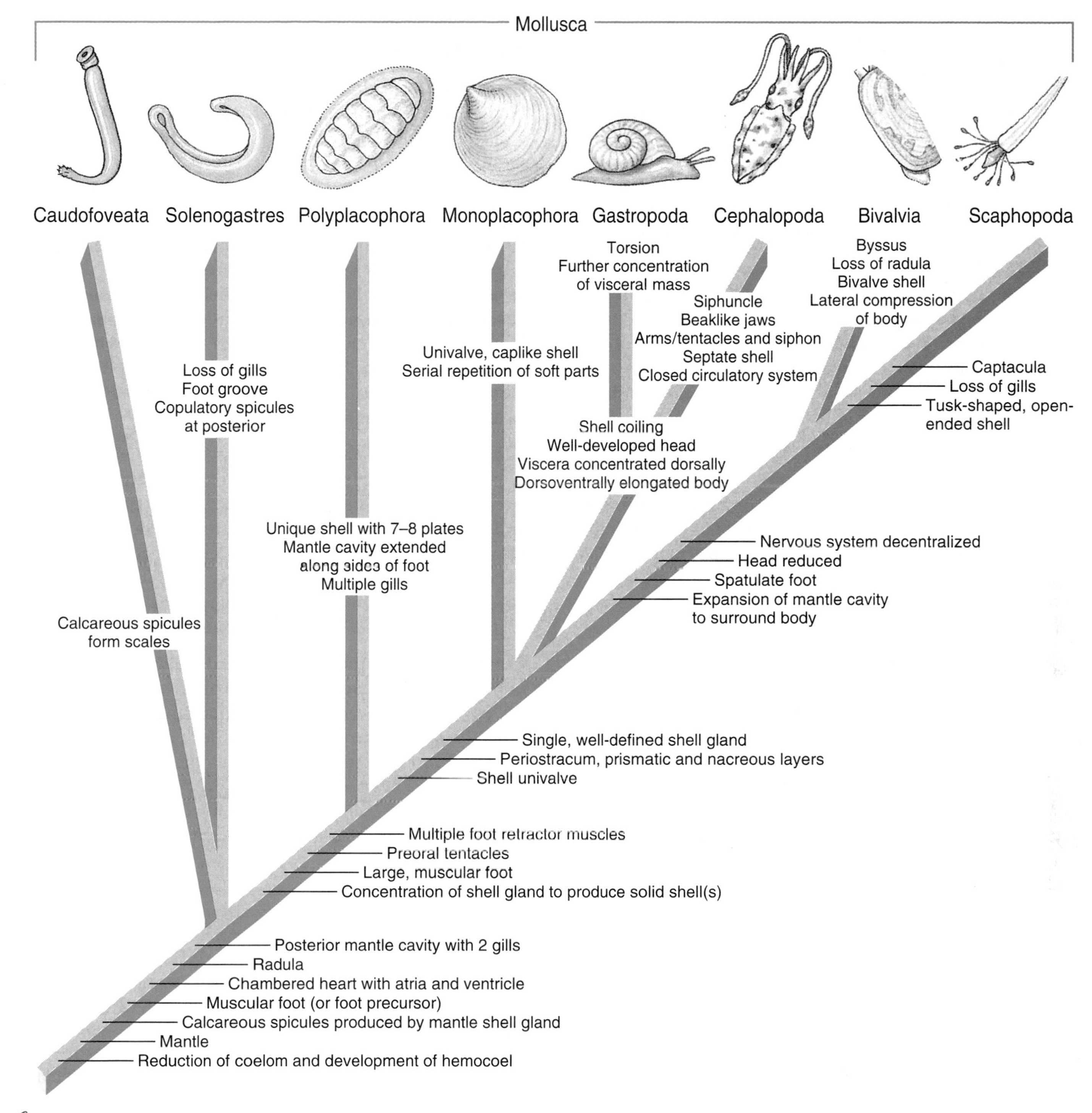

figure 21-36

Cladogram showing hypothetical relationships among the classes of Mollusca. A number of the synapomorphies that identify the various clades have been modified or lost in some descendants. For example, the univalve shell (as well as shell coiling) has been reduced or lost in many gastropods and cephalopods, and many gastropods have undergone detorsion. The bivalve shell of the Bivalvia was derived from the ancestral univalve shell. The byssus is not present in most adult bivalves but functions in larval attachment in many; therefore the byssus is considered a synapomorphy of Bivalvia.

(Source: *After Brusca, R. C., and G. J. Brusca. 1990.* Invertebrates. *Sunderland, MA, Sinauer Associates.*)

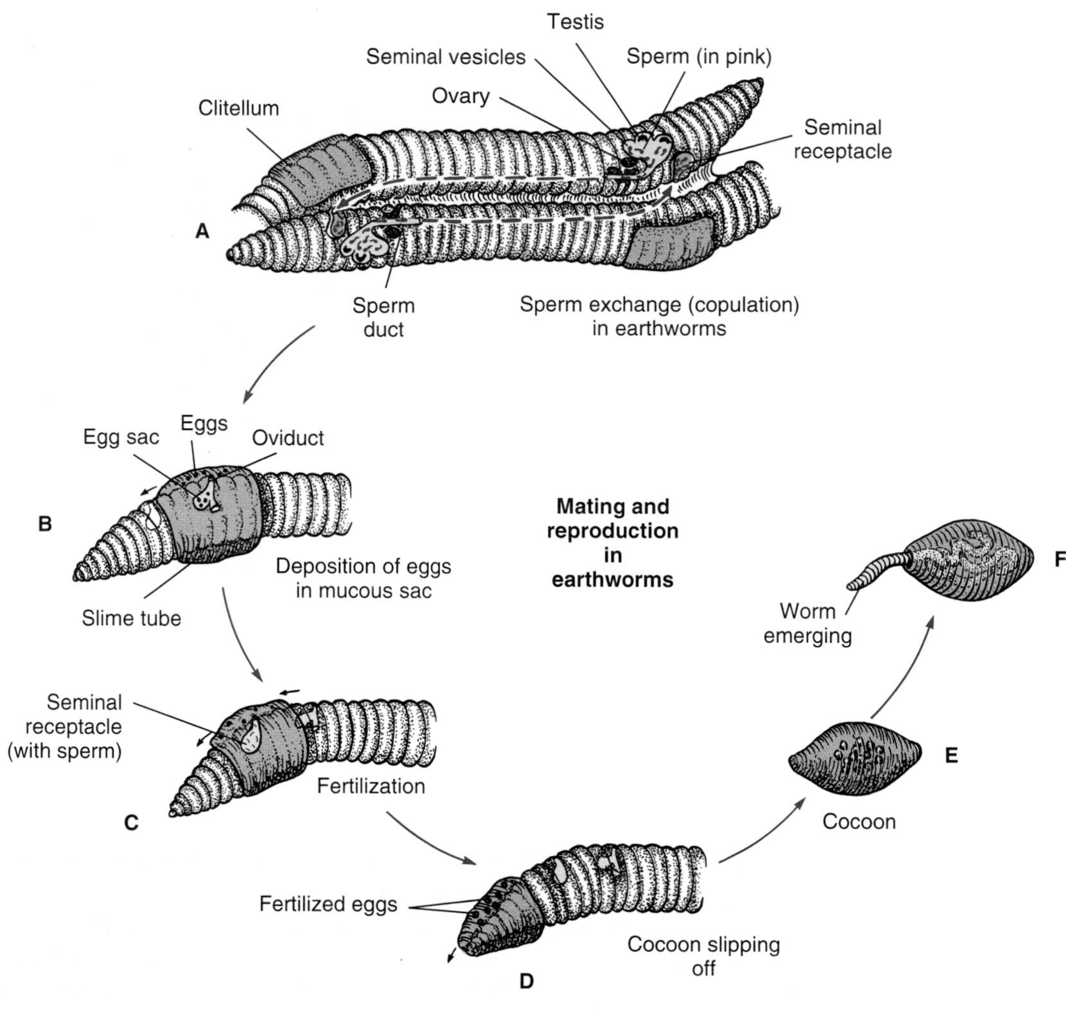

figure 22-15

Earthworm copulation and formation of egg cocoons. **A,** Mutual insemination occurs during copulation; sperm from the genital pore (somite 15) pass along seminal grooves to seminal receptacles (somites 9 and 10) of each mate. **B** and **C,** After worms separate, a slime tube formed over the clitellum passes forward to receive eggs from oviducts and sperm from seminal receptacles. **D,** As the cocoon slips off over the anterior end, its ends close and seal. **E,** The cocoon is deposited near the burrow entrance. **F,** Young worms emerge in two to three weeks.

from the skin glands, and sperm from the mate (stored in the seminal receptacles) are poured into it. Fertilization of the eggs now takes place within the cocoon. When the cocoon leaves the worm, its ends close, producing a lemon-shaped body. Embryonic development occurs within the cocoon, and the form that hatches from the egg is a young worm similar to the adult. It does not develop a clitellum until it is sexually mature.

Freshwater Oligochaetes

Freshwater oligochaetes usually are smaller and have more conspicuous setae than do the earthworms. They are more mobile than earthworms and tend to have better developed sense organs. They are generally benthic forms that creep about on the bottom or burrow into the soft mud. Aquatic oligochaetes provide an important food source for fishes. A few are ectoparasitic.

Some aquatic forms have **gills.** In some the gills are long, slender projections from the body surface. Others have ciliated posterior gills (Figure 22-16D), which they extend from their tubes and use to keep the water moving. Most forms respire through the skin as do the earthworms.

The chief foods are algae and detritus, which worms may pick up by extending a mucus-coated pharynx. Burrowers swallow mud and digest the organic material. Some, such as *Aeolosoma,* are ciliary feeders that use currents produced by cilia at the anterior end of the body to sweep food particles into the mouth (Figure 22-16B).

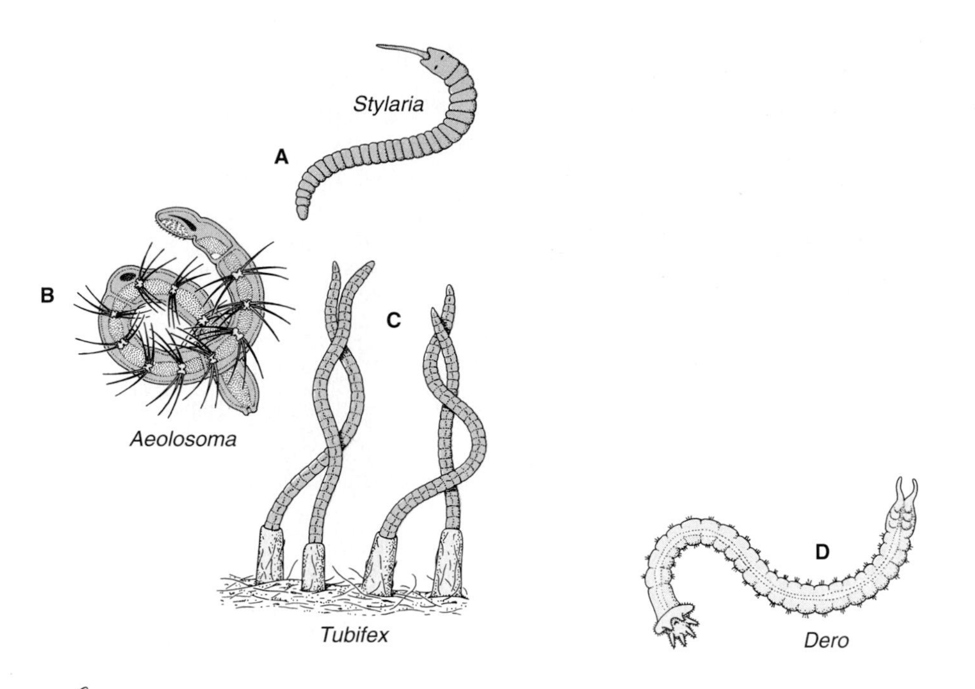

figure 22-16

Some freshwater oligochaetes. **A,** *Stylaria* has the prostomium drawn out into a long snout. **B,** *Aeolosoma* uses cilia around the mouth to sweep in food particles, and it buds off new individuals asexually. **C,** *Tubifex* lives head down in long tubes. **D,** *Dero* has ciliated anal gills.

figure 22-17

The world's largest leech, *Haementeria ghilianii,* on the arm of Dr. Roy K. Sawyer, who found it in French Guiana, South America.

Class Hirudinea

Leeches, numbering over 500 species, are found predominantly in freshwater habitats, but a few are marine, and some have even adapted to terrestrial life in moist, warm areas. Most leeches are between 2 and 6 cm in length, but some are smaller, and some reach 20 cm or more (Figure 22-17). They are found in a variety of patterns and colors—black, brown, red, or olive green. They are usually flattened dorsoventrally (Figure 22-18).

Form and Function

Leeches have a fixed number of segments, usually 34, and typically have both an anterior and a posterior sucker. They have no parapodia and, except in one genus *(Acanthobdella),* they have no setae. Another primitive character of *Acanthobdella* is its five anterior coelomic compartments separated by septa; septa have disappeared in all other leeches. The coelom has become filled with connective tissue and muscle, substantially reducing its effectiveness as a hydrostatic skeleton.

Many leeches live as carnivores on small invertebrates; some are temporary parasites, sucking blood from vertebrates; and some are permanent parasites, never leaving their host. Most leeches have a muscular, protrusible proboscis or a muscular pharynx with three jaws armed with teeth. They feed on the body juices of their prey, penetrating its surface with their proboscis or jaws and sucking the fluids with their powerful, muscular pharynx. Blood-sucking leeches (Figure 22-19) secrete an anticoagulant in their saliva. Predatory leeches feed frequently, but those that feed on blood of vertebrates consume large meals (up to several times their body weight) and digest the food slowly. The slow digestion of their meals results from the lack of secretion of amylases, lipase, or endopeptidases by their gut. In fact, they apparently depend mostly on bacteria in their gut for digestion of the blood meal.

For centuries the "medicinal leech" (Hirudo medicinalis) *was used for blood letting because of the mistaken idea that bodily disorders and fevers were caused by an excess of blood. The use of leeches has an advantage over the lancing of veins; the operation is painless because of anesthetic components in the leech's saliva. A 10 to 12 cm long leech can extend to a much greater length when distended with blood, and the amount of blood it can suck is considerable. Leech collecting and leech culture in ponds were practiced in Europe on a commercial scale during the nineteenth century. Wordsworth's poem "The Leech-Gatherer" was based on this use of the leech.*

Leeches are once again being used medicinally. When fingers or toes are severed, microsurgeons often can reconnect arteries but not the more delicate veins. Leeches are used to relieve congestion until the veins can grow back into the healing digit.

Leeches are hermaphroditic but practice cross-fertilization during copulation. Sperm are transferred by a penis or by hypodermic impregnation. Leeches have a clitellum, but it is evident only during the breeding season. After copulation, the clitellum secretes a cocoon that receives the eggs and sperm. Cocoons are buried in bottom mud, attached to submerged objects or, in terrestrial species, placed in damp soil. Development is similar to that of oligochaetes.

Leeches are highly sensitive to stimuli associated with the presence of a prey or host. They are attracted by and will attempt to attach to an object smeared with appropriate host substances, such as fish scales, oil secretions, or sweat. Those that feed on the blood of mammals are attracted by warmth, and the terrestrial haemadipsid leeches of the tropics will converge on a person standing in one place.

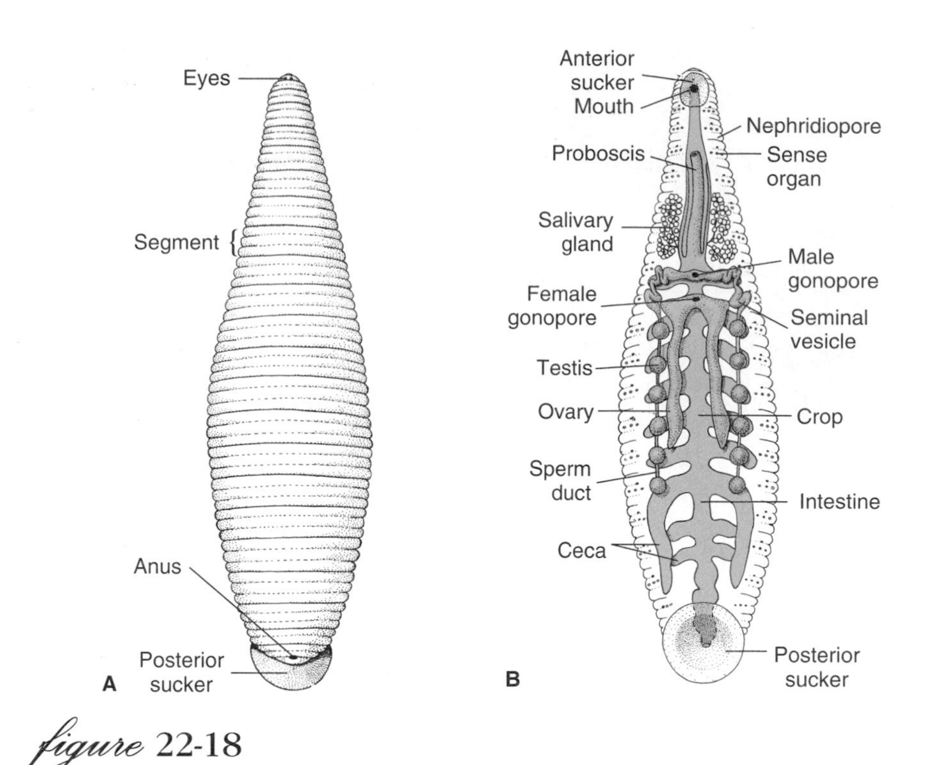

figure 22-18

Structure of a leech, *Placobdella.* **A,** External appearance, dorsal view. **B,** Internal structure, ventral view.

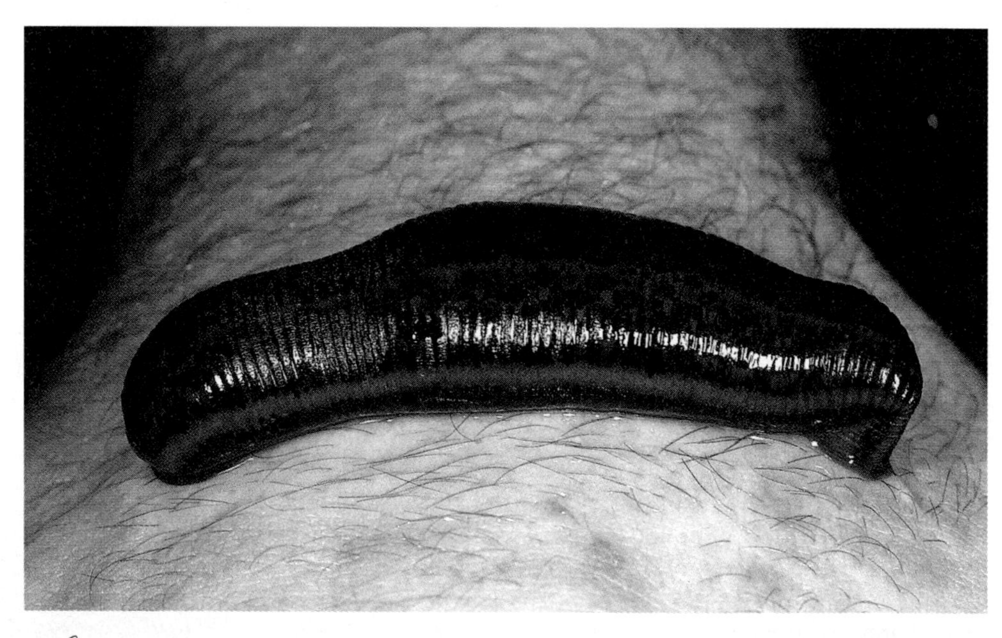

figure 22-19

Hirudo medicinalis feeding on blood from a human arm.

Phylogeny and Adaptive Radiation

Phylogeny

There are so many similarities in the early development of the molluscs, annelids, and primitive arthropods that there seems little doubt about their close relationship. These three phyla apparently form a sister group to the flatworms. Many marine annelids and molluscs have an early embryogenesis typical of protostomes, shared with some marine flatworms, suggesting it is an ancestral trait. Annelids share with the arthropods an outer secreted cuticle and a similar nervous system, and there is a similarity between the lateral appendages (parapodia) of many marine annelids and the appendages of certain arthropods with other primitive characters. The most important resemblance, however, probably lies in the segmented plan of the annelid and the arthropod body structure.

What can we infer about the common ancestor of the annelids? This subject has been a long and continuing debate. Most hypotheses of annelid origin have assumed that metamerism arose in connection with the development of lateral appendages (parapodia) resembling those of the polychaetes. However, the oligochaete body is adapted to vagrant burrowing in the substratum with a peristaltic movement that is highly benefited by a metameric coelom. On the other hand, polychaetes

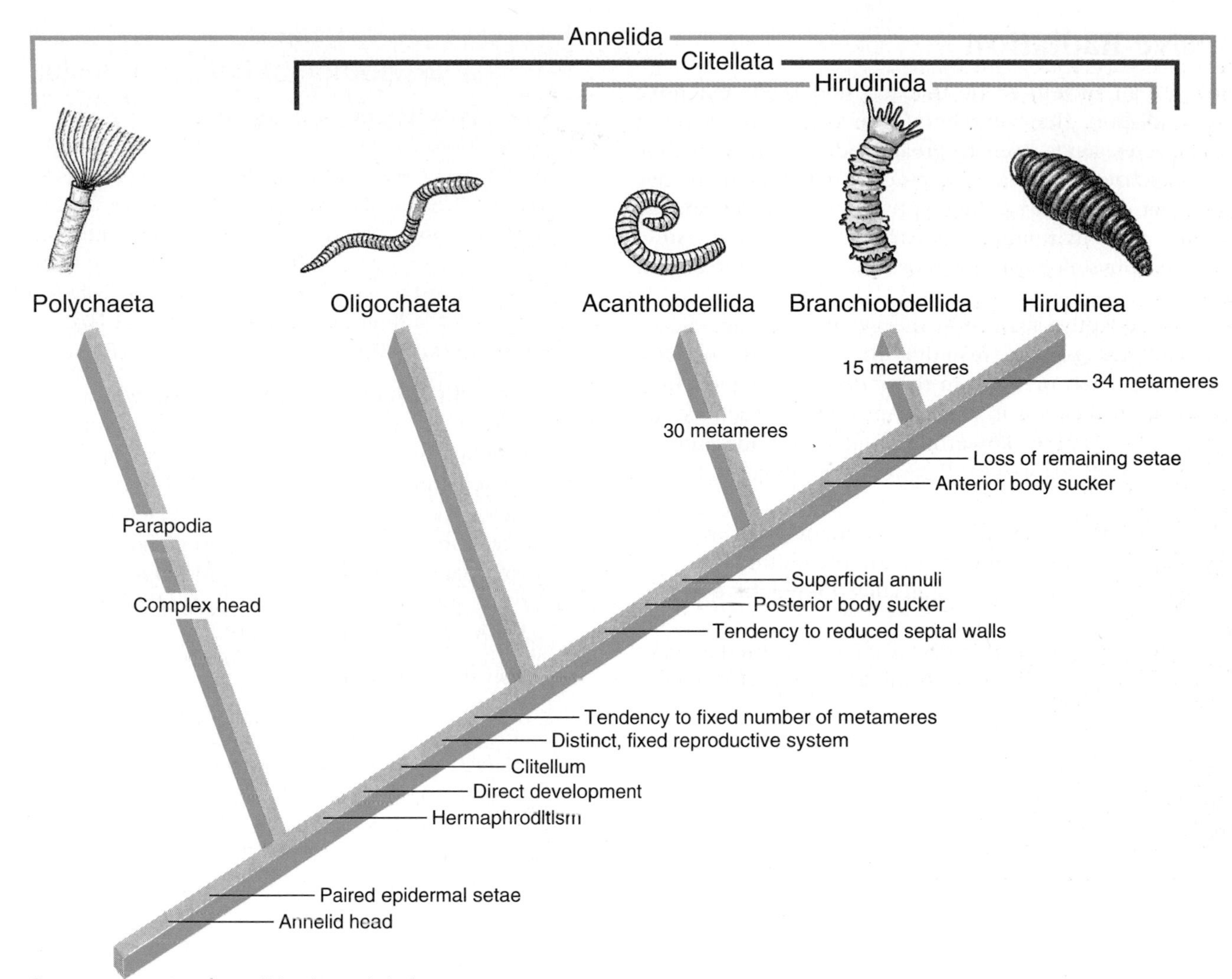

figure 22-20

Cladogram of the annelids, showing the appearance of shared derived characters that specify the five monophyletic groups. The Acanthobdellida and the Branchiobdellida are two small groups discussed briefly in the note on this page. Brusca and Brusca place both groups, together with the Hirudinea ("true" leeches), within a single taxon, the Hirudinida. This clade has several synapomorphies: tendency toward reduction of septal walls, the appearance of a posterior sucker, and the subdivision of body segments by superficial annuli. Note also that, according to this scheme, the Oligochaeta have no defining synapomorphies; that is, they are defined solely by primitive characters that they retain, and thus might be paraphyletic.

(Source: *After Brusca, R. C., and G. J. Brusca. 1990.* Invertebrates. *Sunderland, MA, Sinauer Associates.*)

with well-developed parapodia are generally adapted to swimming and crawling in a medium too fluid for effective peristaltic locomotion. Although the parapodia do not prevent such locomotion, they do little to further it, and it seems likely that they evolved as an adaptation for swimming. Although the polychaetes are more primitive in some characters, such as in their reproductive system, some authorities have argued that the ancestral annelids were more similar to the oligochaetes in overall body plan and that those of the polychaetes and leeches are more evolutionarily derived. The leeches are closely related to the oligochaetes but have diverged from them in connection with a swimming existence and the abandonment of a burrowing mode of life. This relationship is indicated by the cladogram in Figure 22-20.

The Branchiobdellida, a group of small annelids that are parasitic or commensal on crayfish and show similarities to both oligochaetes and leeches, are here placed with the oligochaetes, but they are considered a separate class by some authorities. They have 14 or 15 segments and bear a head sucker.

One genus of leech, Acanthobdella, *has some characteristics of leeches and some of oligochaetes; it is sometimes separated from the other leeches into a special class, Acanthobdellida, that characteristically has 27 somites, setae on the first five segments, and no anterior sucker.*

Adaptive Radiation

Annelids are an ancient group that has undergone extensive adaptive radiation. The basic body structure, particularly of the polychaetes, lends itself to great modification. As marine worms, polychaetes have a wide range of habitats in an environment that is not physically or physiologically demanding. In contrast, the environment of earthworms imposes strict physical and physiological demands, which has limited their radiation.

A basic adaptive feature in the evolution of annelids is their septal arrangement, resulting in fluid-filled coelomic compartments. Fluid pressure in these compartments is used as a hydrostatic skeleton in precise movements such as burrowing and swimming. Powerful circular and longitudinal muscles have been adapted for flexing, shortening, and lengthening the body.

There is wide variation in feeding adaptations, from the sucking pharynx of the oligochaetes and the chitinous jaws of carnivorous polychaetes to the specialized tentacles and radioles of particle feeders.

In polychaetes the parapodia have been adapted in many ways and for a variety of functions, chiefly locomotion and respiration.

In leeches many adaptations, such as suckers, cutting jaws, pumping pharynx, distensible gut, and the secretion of anticoagulants, are related to their predatory and bloodsucking habits.

Classification of Phylum Annelida

Class Polychaeta (pol′e-ke′ta) (Gr. *polys*, many, + *chaitē*, long hair). Mostly marine; head distinct and bearing eyes and tentacles; most segments with parapodia (lateral appendages) bearing tufts of many setae; clitellum absent; sexes usually separate; gonads transitory; asexual budding in some; trochophore larva usually; mostly marine. Examples: *Nereis* (Figure 22-2), *Aphrodita, Glycera, Arenicola* (Figure 22-10), *Chaetopterus* (Figure 22-9), *Amphitrite* (Figure 22-4).

Class Oligochaeta (ol′i-go-ke′ta) (Gr. *oligos*, few, + *chaitē*, long hair). Body with conspicuous segmentation; number of segments variable; setae few per metamere; no parapodia; head absent; coelom spacious and usually divided by intersegmental septa; hermaphroditic; development direct, no larva; chiefly terrestrial and freshwater. Examples: *Lumbricus* (Figure 22-11), *Stylaria* (Figure 22-16A), *Aeolosoma* (Figure 22-16B), *Tubifex* (Figure 22-16C).

Class Hirudinea (hir′u-din′e-a) (L. *hirudo*, leech, + *-ea*, characterized by): **leeches**. Body with fixed number of segments (usually 34) with many annuli; body usually with anterior and posterior suckers; clitellum present; no parapodia; setae absent (except *Acanthobdella*); coelom closely packed with connective tissue and muscle; development direct; hermaphroditic; terrestrial, fresh water, and marine. Examples: *Hirudo, Placobdella* (Figure 22-18), *Macrobdella*.

Summary

The phylum Annelida is a large, cosmopolitan group containing the marine polychaetes, the earthworms and freshwater oligochaetes, and the leeches. Certainly the most important structural advancement of this group is metamerism, the division of the body into a series of similar segments, each of which contains a repeated arrangement of many organs and systems. The coelom is also highly developed in the annelids and has, together with the septal arrangement of fluid-filled compartments and a well-developed body-wall musculature, provided the annelids with an effective hydrostatic skeleton for precise burrowing and swimming. The primitive metamerism of the annelids lays the groundwork for the much more specialized metamerism of the arthropods to be considered in the next chapter.

Polychaetes are the largest class of annelids and are mostly marine. They have on each somite many setae, which are borne on paired parapodia. Parapodia show a wide variety of adaptations among polychaetes, including specialization for swimming, respiration, crawling, maintaining position in a burrow, pumping water through a burrow, and as accessory feeding organs. Some polychaetes are mostly predaceous and have an eversible pharynx with jaws. Other polychaetes rarely leave the burrows or tubes in which they live. Several styles of deposit and suspension feeding are shown among the members of this group. Polychaetes are dioecious with a primitive reproductive system and no clitellum. They practice external fertilization, and their larva is a trochophore.

The class Oligochaeta contains the earthworms and many freshwater forms; they have a small number of setae per segment (compared to the Polychaeta) and no parapodia. The circulatory system is closed, and the dorsal blood vessel is the main pumping organ. There is a pair of nephridia in most somites. Earthworms contain the typical annelid nervous system: dorsal cerebral ganglia connected to a double, ventral nerve cord with segmental ganglia, running the length of the worm. Oligochaetes are hermaphroditic and practice cross-fertilization. The clitellum plays an important role in reproduction, including secretion of mucus to surround the worms during copulation and secretion of a cocoon to receive the eggs and sperm and in which embryonic development occurs. A small, juvenile worm hatches from the cocoon.

The leeches (class Hirudinea) are mostly fresh water, although a few are marine and a few are terrestrial. They feed mostly on fluids; many are predators, some are temporary parasites, and a few are permanent parasites. The hermaphroditic leeches reproduce in a fashion similar to oligochaetes, with cross-fertilization and cocoon formation by the clitellum.

Embryological evidence supports a phylogenetic relationship of the annelids with the molluscs and arthropods.

Review Questions

1. Name the major characteristics that distinguish the phylum Annelida.
2. Distinguish among the classes of the phylum Annelida.
3. Describe the annelid body plan, including the body wall, segments, coelom and its compartments, and coelomic lining.
4. Explain how the hydrostatic skeleton of the annelids helps them to burrow. How is the efficiency for burrowing increased by metamerism?
5. Describe at least three ways that various polychaetes obtain food.
6. Define each of the following: prostomium, peristomium, radioles, parapodium, neuropodium, notopodium.
7. Explain the function of each of the following in earthworms: pharynx, calciferous glands, crop, gizzard, typhlosole, chloragogen tissue.
8. Describe the main features of each of the following in earthworms: circulatory system, nervous system, excretory system.
9. Describe the function of the clitellum and the cocoon.
10. How are freshwater oligochaetes generally different from earthworms?
11. Describe the ways in which leeches obtain food.
12. What are the main differences in reproduction and development among the three classes of annelids?
13. What was the evolutionary significance of metamerism and the coelom to its earliest possessors?
14. What are the phylogenetic relationships between the molluscs, annelids, and arthropods? What is the evidence for these relationships? What do you think is the most important synapomorphy shared by the annelids and arthropods and not the molluscs?

Selected References

See also general references to Part Three, p. 562.

Clark, R. B. 1964. Dynamics in metazoan evolution. The origin of the coelom and segments. Oxford, Clarendon Press. *An important treatise giving the author's hypotheses on the subject.*

Dales, R. P. 1967. Annelids, ed. 2. London, The Hutchinson Publishing Group, Ltd. *A concise account of the annelids.*

Kingman, J., and P. Kingman. 1993. The dance of the luminescent threadworms. Underwater Naturalist **22**(2):36. *Describes the spectacular swarming of the epitokes of* Odontosyllis enopla *off Belize during July and August, the third and fourth nights after a full moon.*

Lent, C. M., and M. H. Dickinson. 1988. The neurobiology of feeding in leeches. Sci. Am. **258:**98–103 (June). *Feeding behavior in leeches is controlled by a single neurotransmitter (serotonin).*

Nicholls, J. C., and D. Van Essen. 1974. The nervous system of the leech. Sci. Am. **230:**38–48 (Jan.). *Because the leech has large nerve cells and only a few neurons perform a given function, its nervous system is particularly appropriate for experimental studies.*

Arthropods

chapter | twenty-three

A Winning Combination

Tunis, Algeria—Treating it as an invading army, Tunisia, Algeria, and Morocco have mobilized to fight the most serious infestation of locusts in over 30 years. Billions of the insects have already caused extensive damage to crops and are threatening to inflict great harm to the delicate economies of North Africa.

From the New York Times, 20 April 1988

The staggering losses occasionally inflicted by the billions of locusts in Africa serve as only one reminder of our ceaseless struggle with the dominant group of animals on earth today: the insects in the phylum Arthropoda. With nearly 1 million species recorded, and probably as many yet remaining to be classified, arthropods far outnumber all the other species of animals in the world combined. Numbers of individuals are equally enormous. Some scientists have estimated that there are 200 million insects for every single human alive today! Insects have an unmatched ability to adapt to all terrestrial environments and to virtually all climates, and crustaceans have radiated likewise in aquatic environments. Having originally evolved as land animals, insects developed wings and invaded the air 150 million years before flying reptiles, birds, or mammals. Many insects have exploited freshwater habitats, where they are now widely prevalent. In the sea the numbers of insects are more limited, but there are vast numbers of crustaceans in marine habitats.

How can we account for the enormous success of these creatures? Arthropods have a combination of valuable structural and physiological adaptations, including a versatile exoskeleton, metamerism, an efficient respiratory system, and highly developed sensory organs. In addition, many have a waterproofed cuticle and have extraordinary abilities to survive adverse environmental conditions. We will describe these adaptations and others in this chapter.

The phylum Arthropoda (ar-throp′o-da) (Gr. *arthron,* joint, + *pous, podos,* foot) embraces the largest assemblage of living animals on earth. It includes the spiders, scorpions, ticks, mites, crustaceans, millipedes, centipedes, insects, and some smaller groups. In addition there is a rich fossil record extending back to the mid-Cambrian period (Figure 23-1).

Arthropods are eucoelomate protostomes with well-developed organ systems, and their cuticular exoskeleton containing chitin is a prominent characteristic. Like the annelids, they are conspicuously segmented; their primitive body pattern is a linear series of similar somites, each with a pair of jointed appendages. However, unlike the annelids, the arthropods have embellished the segmentation theme: variation occurs in the pattern of somites and appendages in the phylum. Often the somites are combined or fused into functional groups, called **tagmata,** for specialized purposes. Appendages, too, are frequently differentiated and specialized for walking, swimming, flying, or eating.

A

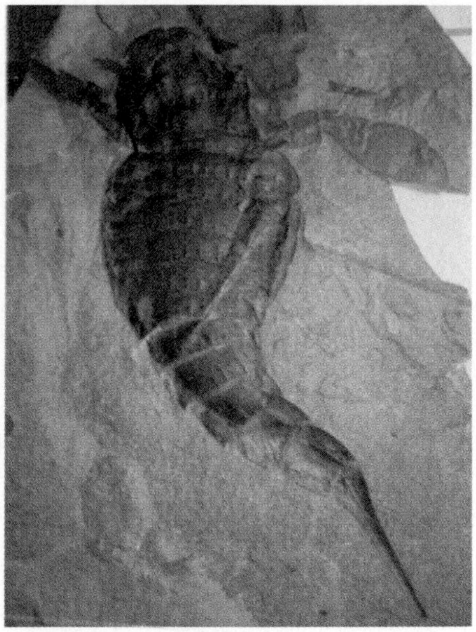

B

figure 23-1

Fossils of early arthropods. **A,** Trilobite fossiles, dorsal view. These animals were abundant in the mid-Cambrian period. **B,** Eurypterid fossil; eurypterids flourished in Europe and North America from Ordovician to Permian periods.

Few arthropods exceed 60 cm in length, and most are far below this size. The largest is a Japanese crab (*Macrocheira kaempferi*), which has approximately a 3.7 m span; the smallest is a parasitic mite, which is less than 0.1 mm long.

Arthropods are usually active, energetic animals. However we judge them, whether by their great diversity or their wide ecological distribution or their vast numbers of species, the answer is the same: they are the most abundant and diverse of all animals.

Although arthropods compete with us for food supplies and spread serious diseases, they are essential in pollination of many food plants, and they also serve as food, yield drugs and dyes, and create such products as silk, honey, and beeswax.

Ecological Relationships

Arthropods are found in all types of environment from low ocean depths to very high altitudes and from the tropics far into both north and south polar regions. Some species are adapted for life in the air; others for life on land or in fresh, brackish, and marine waters; others live in or on the bodies of plants and other animals. Some live in places where no other animal could survive.

Although all types—carnivorous, omnivorous, and herbivorous—occur in this vast group, the majority are herbivorous. Most aquatic arthropods depend on algae for their nourishment, and the majority of land forms live chiefly on plants. There are many parasites. In diversity of ecological distribution the arthropods have no rivals.

Why Have Arthropods Achieved Such Great Diversity and Abundance?

Arthropods have achieved a great diversity, number of species, wide distribution, variety of habitats and feeding habits, and power of adaptation to changing conditions. These are some of the structural and physiological patterns that have been helpful to them:

1. **A versatile exoskeleton.** The arthropods possess an exoskeleton that is highly protective without sacrificing mobility. The skeleton is the **cuticle,** an outer covering secreted by the underlying epidermis.

The cuticle consists of an inner and thicker **procuticle** and an outer, relatively thin **epicuticle.** The procuticle is divided into the **exocuticle,** which is secreted before a molt, and **endocuticle,** which is secreted after molting. Both layers of the procuticle contain **chitin** bound with protein. Chitin is a tough, resistant, nitrogenous polysaccharide that is insoluble in water, alkalis, and weak acids. Thus the procuticle not only is flexible and lightweight but also affords protection, particularly against dehydration. In most crustaceans, the procuticle in some areas is also impregnated with **calcium salts,** which reduce its flexibility. In the hard shells of lobsters and crabs, for instance, this calcification is extreme. The outer epicuticle is

Characteristics of Phylum Arthropoda

1. Bilateral symmetry; metameric body, **tagmata** of head and trunk; head, thorax, and abdomen; or cephalothorax and abdomen
2. **Appendages jointed;** primitively, one pair to each somite (metamere), but number often reduced; appendages often modified for specialized functions
3. **Exoskeleton of cuticle** containing protein, lipid, chitin, and often calcium carbonate secreted by underlying epidermis and shed (molted) at intervals
4. Muscular system complex, with exoskeleton for attachment; striated muscles for rapid action; smooth muscles for visceral organs; **no cilia**
5. Coelom reduced; most of body cavity consisting of **hemocoel** (sinuses, or spaces, in the tissues) filled with blood
6. Complete digestive system; mouthparts modified from appendages and adapted for different methods of feeding
7. **Circulatory system open,** with dorsal contractile heart, arteries, and hemocoel
8. Respiration by body surface, gills, tracheae (air tubes), or book lungs
9. Paired excretory glands called coxal, antennal, or maxillary glands present in some, homologous to metameric nephridial system of annelids; some with other excretory organs, called malpighian tubules
10. Nervous system of annelid plan, with dorsal brain connected by a ring around the gullet to a double nerve chain of ventral ganglia; fusion of ganglia in some species; well-developed sensory organs
11. Sexes usually separate, with paired reproductive organs and ducts; usually internal fertilization; oviparous or ovoviviparous; often with metamorphosis; parthenogenesis in a few forms; **growth with ecdysis**

composed of protein and often lipids. The protein is stabilized and hardened by a chemical process called tanning, adding further protection. Both the procuticle and the epicuticle are laminated, that is, composed of several layers each (Figure 6-9A, p. 148).

The cuticle may be soft and permeable or may form a veritable coat of armor. Between body segments and between the segments of appendages it is thin and flexible, permitting free movement of the joints. In crustaceans and insects the cuticle forms ingrowths for muscle attachment. It may also line the foregut and hindgut, line and support the trachea, and be adapted for a variety of purposes.

The nonexpansible cuticular exoskeleton does, however, impose important conditions on growth. To grow, an arthropod must shed its outer covering at intervals and grow a larger one—a process called **ecdysis,** or **molting.** Arthropods molt from four to seven times before reaching adulthood, and some continue to molt after that. Much of an arthropod's physiology centers on molting, particularly in young animals—preparation, molting itself, and then all the processes that must be completed in the postmolt period.

An exoskeleton is also relatively heavy and becomes proportionately heavier with increasing size. Weight of the exoskeleton tends to limit the ultimate body size.

2. **Segmentation and appendages for more efficient locomotion.** Typically each somite has a pair of jointed appendages, but this arrangement is often modified, with both segments and appendages specialized for adaptive functions. The limb segments are essentially hollow levers that are moved by muscles, most of which are striated for rapid action. The jointed appendages are equipped with sensory hairs and are variously modified for sensory functions, food handling, and swift and efficient walking or swimming.

3. **Air piped directly to cells.** Most land arthropods have a highly efficient tracheal system of air tubes, which delivers oxygen directly to tissues and cells and makes a high metabolic rate possible. Aquatic arthropods breathe mainly by some form of gill.

4. **Highly developed sensory organs.** Sensory organs are found in great variety, from the compound (mosaic) eye to the senses of touch, smell, hearing, balancing, and chemical reception. Arthropods are keenly alert to what goes on in their environment.

5. **Complex behavior patterns.** Arthropods exceed most other invertebrates in complexity and organization of their activities. Innate (unlearned) behavior unquestionably controls much of what they do, but learning also plays an important part in the lives of many arthropods.

6. **Reduced competition through metamorphosis.** Many arthropods pass through metamorphic changes, including a larval form quite different from the adult in structure. The larval form is often adapted for eating a different kind of food from that of the adult and occupies a different space, resulting in less competition within a species.

Subphylum Trilobita

The trilobites (Figure 23-1A) probably had their beginnings a million or more years before the Cambrian period in which they flourished. They have been extinct some 200 million years, but were abundant during the Cambrian and Ordovician periods. Their name refers to the trilobed shape of the body, caused by a pair of longitudinal grooves. They were bottom dwellers, probably scavengers. Most of them could roll up like pill bugs.

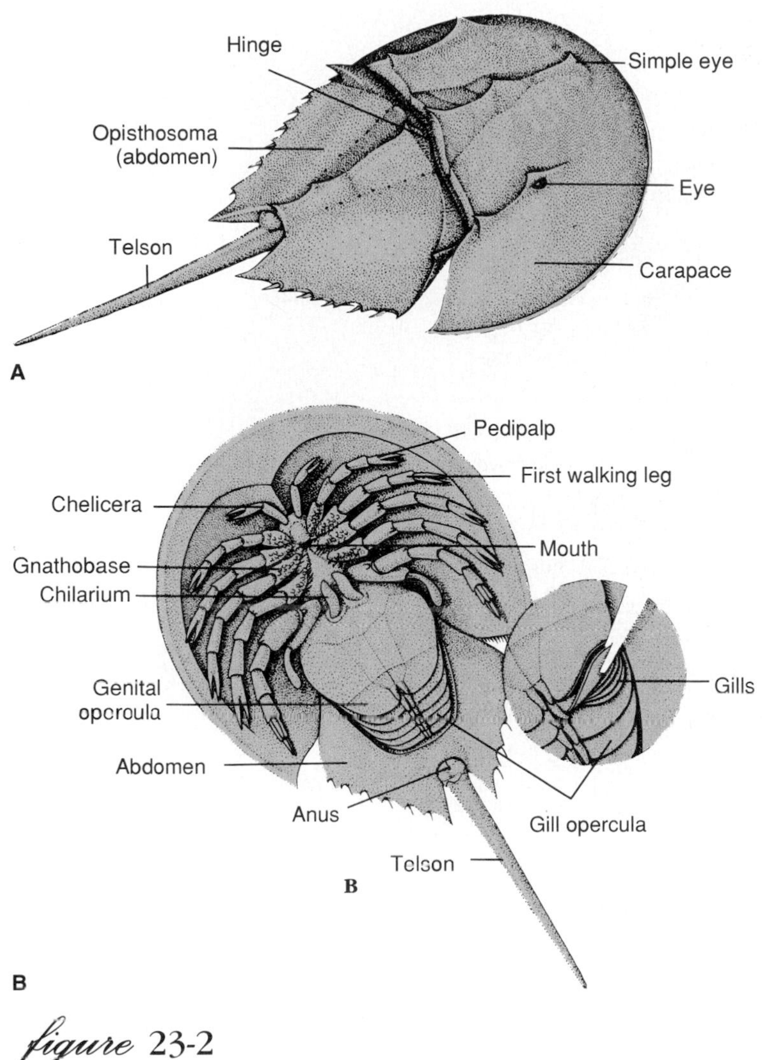

figure 23-2

A, Dorsal view of horseshoe crab *Limulus* (class Merostomata). They grow to 0.5 m in length. **B,** Ventral view.

Subphylum Chelicerata

Chelicerate arthropods are a very ancient group that includes the eurypterids (extinct), horseshoe crabs, spiders, ticks, and mites, scorpions, sea spiders, and others. They are characterized by having six pairs of appendages that include a pair of **chelicerae,** a pair of **pedipalps,** and **four pairs of walking legs** (a pair of chelicerae and five pairs of walking legs in horseshoe crabs). They have **no mandibles** and **no antennae.** Most chelicerates suck liquid food from their prey.

Class Merostomata

Subclass Eurypterida

The eurypterids, or giant water scorpions (Figure 23-1B), lived 200 to 500 million years ago and some were perhaps the largest of all arthropods, reaching a length of 3 m. They

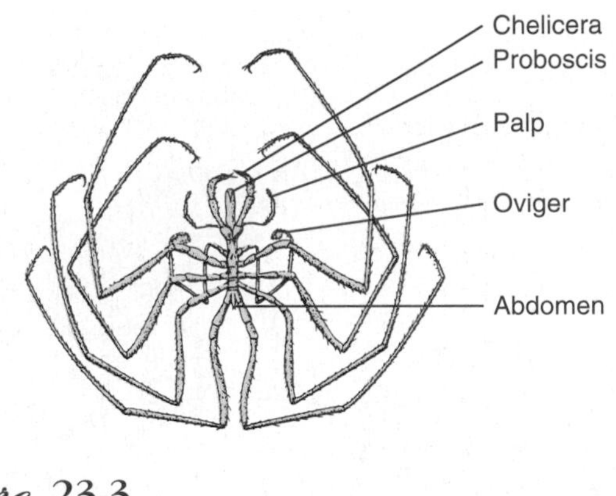

figure 23-3

Pycnogonid, *Nymphon* sp. In this genus all the anterior appendages (chelicerae, palps, and ovigers) are present in both sexes, although ovigers are often not present in females of other genera.

had some resemblances to the marine horseshoe crabs (Figure 23-2) and to the scorpions, their terrestrial counterparts.

Subclass Xiphosurida: Horseshoe Crabs

The xiphosurids are an ancient marine group that dates from the Cambrian period. There are only three genera (five species) living today. *Limulus* (L. *limus,* sidelong, askew) (Figure 23-2), which lives in shallow water along the North American Atlantic coast, goes back practically unchanged to the Triassic period. Horseshoe crabs have an unsegmented, horseshoe-shaped **carapace** (hard dorsal shield) and a broad abdomen, which has a long spinelike **telson,** or tailpiece. On some of the abdominal appendages **book gills** (flat leaflike gills) are exposed. Horseshoe crabs can swim awkwardly by means of the abdominal plates and can walk on their walking legs. They feed at night on worms and small molluscs and are harmless to humans.

Class Pycnogonida: Sea Spiders

Pycnogonids are curious little marine animals that are much more common than most of us realize. They stalk about on their four pairs of long, thin walking legs, sucking juices from hydroids and soft-bodied animals with their large suctorial proboscis (Figure 23-3). They often have a pair of ovigerous legs **(ovigers)** with which males carry the egg masses. Their odd appearance is enhanced by the much reduced abdomen attached to the elongated cephalothorax. Most are only a few millimeters long, although some are much larger. They are common in all oceans.

Class Arachnida

Arachnids (Gr. *arachnē,* spider) are a numerous and diverse group, with over 50,000 species described so far.

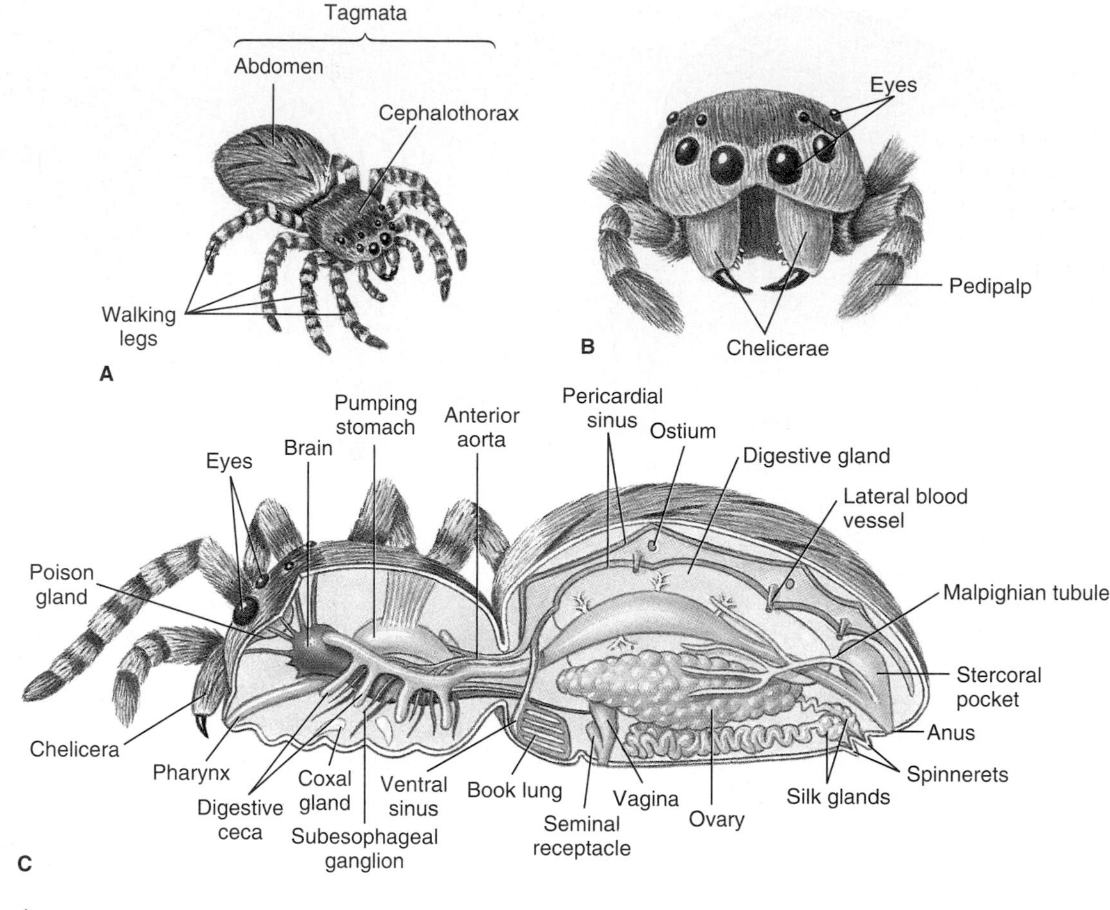

figure 23-4

A, External anatomy of a jumping spider. **B,** Anterior view of head. **C,** Internal anatomy of a spider.

They include the spiders, scorpions, pseudoscorpions, whip scorpions, ticks, mites, harvestmen (daddy longlegs), and others. The arachnid tagmata are a cephalothorax and an abdomen.

Order Araneae: Spiders

The spiders are a large group of 35,000 recognized species, distributed all over the world. The cephalothorax and abdomen show no external segmentation, and the tagmata are joined by a narrow, waistlike **pedicel** (Figure 23-4).

All spiders are predaceous and feed largely on insects (Figure 23-5). Their chelicerae function as fangs and bear ducts from their poison glands, with which they effectively dispatch their prey. Some spiders chase their prey, others ambush them, and many trap them in a net of silk. After the spider seizes its prey with its chelicerae and injects venom, it liquefies the tissues with a digestive fluid and sucks up the resulting broth into the stomach. Spiders with teeth at the bases of the chelicerae crush or chew up the prey, aiding digestion by enzymes from the mouth. Many spiders provision their young with previously captured prey.

Spiders breathe by means of **book lungs** or **tracheae** or both. Book lungs, which are unique in spiders, consist of many parallel air pockets extending into a blood-filled chamber (Figure 23-4). Air enters the chamber by a slit in the body wall. The tracheae are a system of air tubes that carry air directly to the tissues from openings called **spiracles.** The tracheae are similar to those in insects (p. 513), but are much less extensive.

Spiders and insects have a unique excretory system of **malpighian tubules** (Figure 23-4), which work in conjunction with specialized rectal glands. Potassium and other solutes and waste materials are secreted into the tubules, which drain the fluid, or "urine," into the intestine. The rectal glands reabsorb most of the potassium and water, leaving behind such wastes as uric acid. By this cycling of water and potassium, species living in dry environments conserve body fluids, producing a nearly dry mixture of urine and feces. Many spiders also have **coxal glands,** which are modified nephridia that open at the coxa, or base, of the first and third walking legs

Spiders usually have eight **simple eyes,** each provided with a lens, optic rods, and a retina (Figure 23-4B). Chiefly they perceive moving objects, but some, such as those of the hunting and jumping spiders, may form images. Because vision is usually poor, a spider's awareness of its environment depends especially on its hairlike **sensory setae.** Every seta on its surface is useful in communicating some information about the surroundings, air currents, or changing tensions in the spider's web. By sensing the vibrations of its web, the spider can judge the size and activity of its entangled prey or can receive the message tapped out on a silk thread by a prospective mate.

figure 23-5

A, A camouflaged crab spider, *Misumenoides* sp., awaits its insect prey. Its coloration matches the petals among which it lies, thus deceiving insects that visit the flowers in search of pollen or nectar. **B,** A jumping spider, *Eris aurantius.* This species has excellent vision and stalks an insect until it is close enough to leap with unerring precision, fixing its chelicerae into its prey.

Web-Spinning Habits The ability to spin silk is an important factor in the lives of spiders, as it is in some other arachnids. Two or three pairs of spinnerets containing hundreds of microscopic tubes connect to special abdominal **silk glands** (Figure 23-4C). A protein secretion emitted as a liquid hardens on contact with air to form the silk thread. Spiders' silk threads are stronger than steel threads of the same diameter and are said to be second in tensional strength only to fused quartz fibers. The threads will stretch one-fifth of their length before breaking.

The spider web used for trapping insects is familiar to most people. The webs of some species consist merely of a few strands of silk radiating out from a spider's burrow or place of retreat. Other species spin beautiful, geometric orb webs. However, spiders use silk threads for many purposes besides web making. They use silk threads to line their nests; form sperm webs or egg sacs; build draglines; make bridge lines, warning threads, molting threads, attachment discs, or nursery webs; or to wrap up prey securely (Figure 23-6). Not all spiders spin webs for traps. Some, such as the wolf spiders, jumping spiders (Figure 23-5B), and fisher spiders (Figure 23-7), simply chase and catch their prey.

figure 23-6

A grasshopper, snared and helpless in the web of a golden garden spider *(Argiope aurantia),* is wrapped in silk while still alive. If the spider is not hungry, the prize is saved for a later meal.

Reproduction Before mating, the male spins a small web, deposits a drop of sperm on it, and then picks the sperm up and stores it in special cavities of his pedipalps. When he mates, he inserts the pedipalps into the female genital opening to store the sperm in his mate's seminal receptacles. A courtship ritual usually precedes mating. The female lays her fertilized eggs in a silken cocoon, which she may carry about or may attach to a web or plant. A cocoon may contain hundreds of eggs, which hatch in approximately two weeks. The young usually remain in the egg sac for a few weeks and molt once before leaving it. Several molts occur before adulthood.

Are Spiders Really Dangerous? It is truly amazing that such small and helpless creatures as spiders have generated so much unreasoned fear in the human heart. Spiders are timid creatures, which, rather than being dangerous enemies to humans, are actually allies in our continuing conflict with insects. The venom produced to kill prey is usually harmless to humans. Even the most poisonous spiders bite only when threatened or when defending their eggs or young. American

figure 23-7

A fisher spider, *Dolomedes triton,* feeds on a minnow. This handsome spider feeds mostly on aquatic and terrestrial insects but occasionally captures small fishes and tadpoles. It pulls its paralyzed victim from the water, pumps in digestive enzymes, then sucks out the predigested contents.

figure 23-8

A tarantula, *Rhechostica hentzi.*

tarantulas (Figure 23-8), despite their fearsome appearance, are *not* dangerous. They rarely bite, and their bite is not considered serious.

Two genera in the United States can give severe or even fatal bites: *Latrodectus* (L. *latro,* robber, + *dektes,* biter), and *Loxosceles* (Gr. *loxos,* crooked, + *skelos,* leg). The most important species are *Latrodectus mactans,* the **black widow,** and *Loxosceles reclusa,* the **brown recluse.** The black widow is moderate to small in size and shiny black, with a bright orange or red "hourglass" on the underside of the abdomen (Figure 23-9A). The venom is neurotoxic; that is, it acts on the nervous system. About four or five out of each 1000 bites reported were fatal.

The brown recluse, which is smaller than the black widow, is brown, and bears a violin-shaped dorsal stripe on its cephalothorax (Figure 23-9B). Its venom is hemolytic rather than neurotoxic, destroying tissues and skin surrounding the bite. Its bite can be mild to serious and occasionally fatal.

Some spiders in other parts of the world are dangerous, for example, the funnel-web spider *Atrax robustus* in Australia. Most dangerous of all are certain ctenid spiders in South America, for example, *Phoneutria fera.* In contrast to most spiders, these are quite aggressive.

Order Scorpionida: Scorpions

Although scorpions are more common in tropical and subtropical regions, some occur in temperate zones. Scorpions are generally secretive, hiding in burrows or under objects by day and feeding at night. They feed largely on insects and spiders, which they seize with clawlike pedipalps and tear up with jawlike chelicerae.

The scorpion's body consists of a rather short cephalothorax, which bears the appendages and from one to six pairs of eyes and a clearly segmented abdomen. The abdomen is divided into a broader **preabdomen** and tail-like **postabdomen,** which ends in a stinging apparatus used to inject venom (Figure 23-10A). The venom of most species is not harmful to humans, although that of certain species of *Androctonus* in Africa and *Centruroides* in Mexico, Arizona, and New Mexico can be fatal unless antivenom is available.

Scorpions bear living young, which their mother carries on her back until after the first molt.

Order Opiliones: Harvestmen

Harvestmen, often known as "daddy longlegs," are common in the United States and other parts of the world (Figure 23-10B). These curious creatures are easily distinguished from spiders by the fact that their abdomen and cephalothorax are broadly joined, without constriction of the pedicel, and by the presence of external segmentation of the abdomen. They have four pairs of long, spindly legs, and without apparent ill effect, they can cast off one or more legs if they are grasped by a predator (or human hand). The ends of their chelicerae are pincerlike, and they feed much more as scavengers than do most arachnids.

Order Acari: Ticks and Mites

Acarines differ from all other arachnids in having their cephalothorax and abdomen completely fused, with no sign of external division or segmentation (Figure 23-11). Their mouthparts are carried on a little anterior projection, the **capitulum.**

They are found almost everywhere—in both fresh and salt water, on vegetation, on the ground, and parasitic in vertebrates and invertebrates. Over 25,000 species have been described, many of which are of importance to humans, but this is probably only a fraction of the species that exist.

Many species of mites are entirely free living. *Dermatophagoides farinae* (Gr. *dermatos,* skin, + *phago,* to eat, + *eidos,* likeness of form) (Figure 23-12) and related species are denizens of house dust all over the world, sometimes causing allergies and dermatoses. Some mites are marine, but most aquatic species are found in fresh water. They have long, hairlike setae on their legs for swimming, and their larvae may be parasitic on aquatic invertebrates. Such abundant organisms must be important ecologically, but many acarines have more direct effects on our food supply and health. The spider mites (family Tetranychidae) are serious agricultural pests on fruit trees, cotton, clover, and many other plants. Larvae of the genus *Trombicula* are called chiggers or redbugs. They feed on the dermal tissues of terrestrial vertebrates, including humans, and cause an irritating dermatitis; some species of chiggers transmit a disease called Asiatic scrub typhus. The hair-follicle mite, *Demodex* (Figure 23-13), is apparently nonpathogenic in humans; it infects most of us although we are unaware of it. Other species of *Demodex* and other genera of mites cause mange in domestic animals.

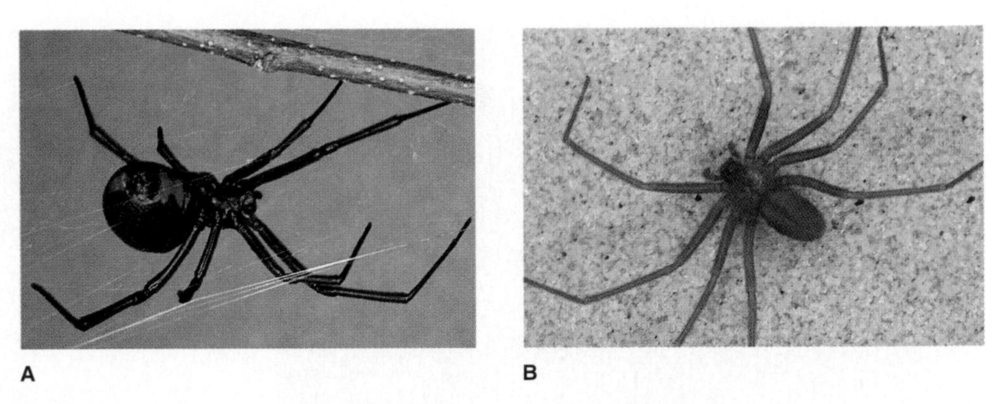

figure 23-9

A, A black widow spider, *Latrodectus mactans,* suspended on her web. Note the orange "hourglass" on the ventral side of her abdomen. **B,** The brown recluse spider, *Loxosceles reclusa,* is a small venomous spider. Note the small violin-shaped marking on its cephalothorax. The venom is hemolytic and dangerous.

figure 23-10

A, Scorpion (order Scorpionida) with young, which stay with the mother until the first molt. **B,** A harvestman (order Opiliones). Harvestmen run rapidly on their stiltlike legs. They are especially noticeable during the harvesting season, hence the common name.

The inflamed welt and intense itching that follows a chigger bite is not the result of the chigger burrowing into the skin, as is popularly believed. Rather the chigger bites through the skin with its chelicerae and injects a salivary secretion containing powerful enzymes that liquefy skin cells. Human skin responds defensively by forming a hardened tube that the larva uses as a sort of drinking straw and through which it gorges itself with host cells and fluid. Scratching usually removes the chigger but leaves the tube, which is a source of irritation for several days.

figure 23-11

A wood tick, *Dermacentor variabilis* (order Acarina).

Ticks are usually larger than mites. They pierce the skin of vertebrates and suck blood until enormously distended;

figure 23-12

Scanning electron micrograph of house dust mite, *Dermatophagoides farinae.*

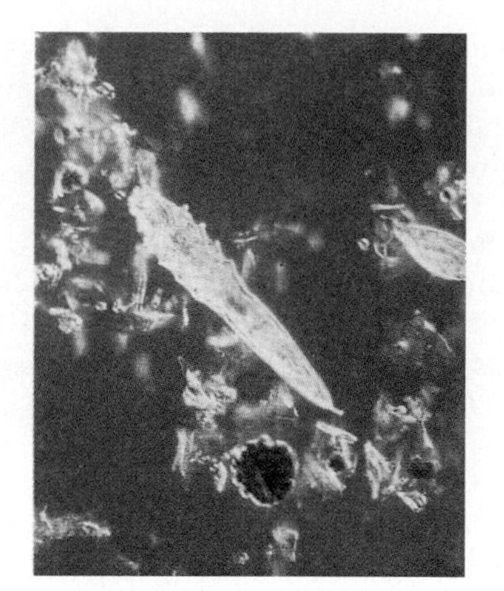

figure 23-13

Demodex follicuorum, the human follicle mite. This tiny mite (100 to 400 μm) lives in follicles, particularly around the nose and eyes. Its prevalence ranges from about 20% in persons 20 years of age or younger to nearly 100% in the aged.

then they drop off and digest their meal. After molting, they are ready for another meal. In addition to disease conditions that they themselves cause, ticks are among the world's premier disease vectors, ranking second only to mosquitos. They carry a greater variety of infectious agents than any other arthropods; such agents include protozoan, rickettsial, viral, bacterial, and fungal organisms. Species of *Ixodes* carry the most common arthropod-borne infection in the United States, Lyme disease (see note below). Species of *Dermacentor* (Figure 23-11) and other ticks transmit Rocky Mountain spotted fever, a poorly named disease because most cases occur in the eastern United States. *Dermacentor* also transmits tularemia and the agents of several other diseases. Texas cattle fever, also called red-water fever, is caused by a protozoan parasite transmitted by the cattle tick *Boophilus annulatus.* Many more examples could be cited.

In the 1970s people in the town of Lyme, Connecticut, experienced an epidemic of arthritis. Subsequently known as Lyme disease, it is caused by a bacterium and carried by ticks of the genus Ixodes. *Now thousands of cases are reported each year in Europe and North America, and other cases have been reported from Japan, Australia, and South Africa. Many people bitten by infected ticks recover spontaneously or do not suffer any ill effects. Others, if not treated at an early stage, develop a chronic, disabling disease. Lyme disease is now the leading arthropod-borne disease in the United States.*

Subphylum Crustacea

Crustaceans traditionally have been included as a class in the subphylum Mandibulata, along with insects and myriapods. Members of all of these groups have, at least, a pair of antennae, mandibles, and maxillae on the head. Whether the Mandibulata constitutes a monophyletic grouping has been debated, and we discuss this question further on p. 526.

The 30,000 or more species of Crustacea (L. *crusta,* shell) include lobsters, crayfish, shrimp, crabs, water fleas, copepods, and barnacles. It is the only arthropod class that is primarily aquatic; they are mainly marine, but many freshwater and a few terrestrial species are known. The majority are free living, but many are sessile, commensal, or parasitic. Crustaceans are often very important components of aquatic ecosystems, and several have considerable economic importance.

Crustaceans are the only arthropods with **two pairs of antennae** (Figure 23-14). In addition to the antennae and **mandibles,** they have **two pairs of maxillae** on the head, followed by a pair of appendages on each body segment (although appendages on some somites are absent in some groups). All appendages, except perhaps the first antennae (antennules), are primitively **biramous** (two main branches), and at least some of the appendages of all present-day adults show that condition. Organs specialized for respiration, if present, are in the form of gills. Crustaceans lack malpighian tubules.

Crustaceans primitively have 60 segments or more, but most tend to have between 16 and 20 somites and increased tagmatization. The major tagmata are the **head, thorax,** and **abdomen,** but these are not homologous throughout the sub-

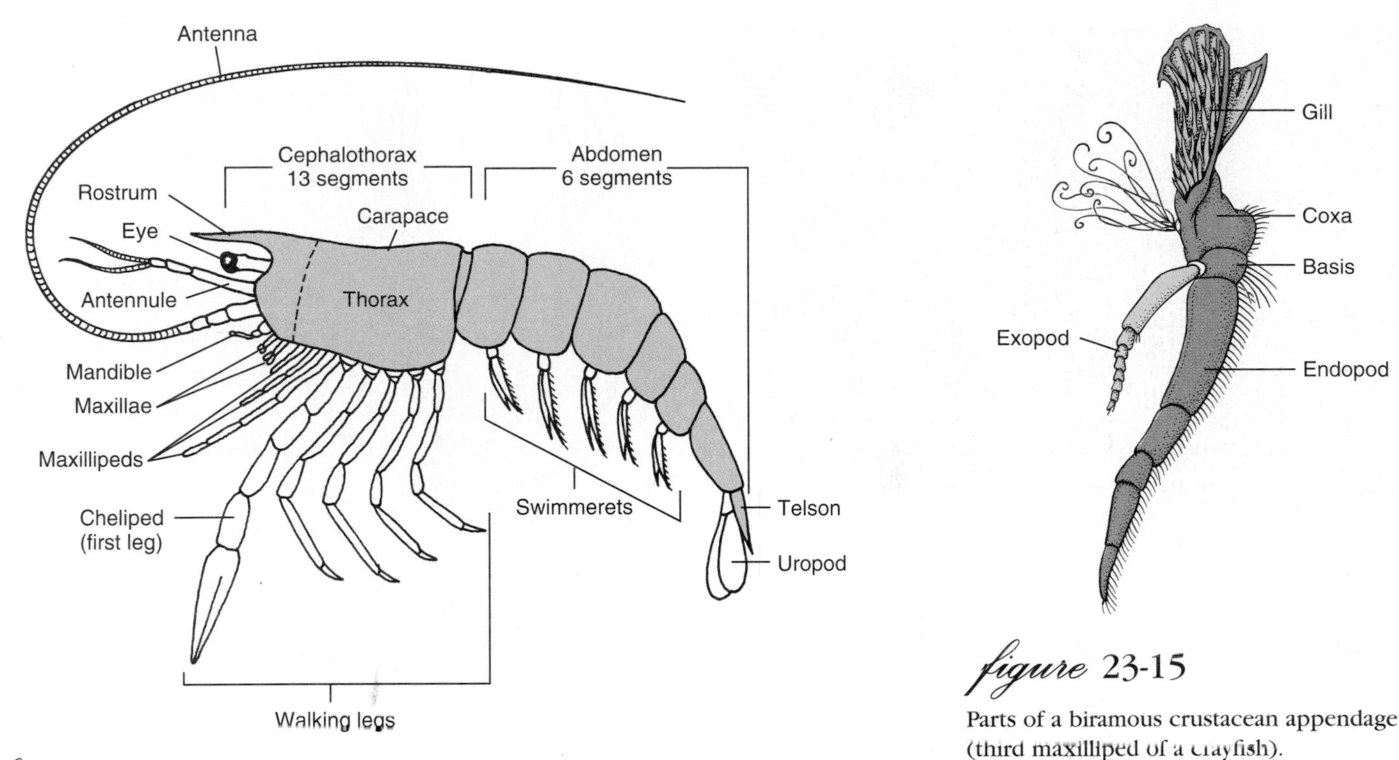

figure 23-14

Archetypical plan of the Malacostraca. Note that the maxillae and maxillipeds have been separated diagrammatically to illustrate general plan. Typically in the living animal only the third maxilliped is visible externally. In the order Decapoda the carapace covers the cephalothorax, as shown here.

figure 23-15

Parts of a biramous crustacean appendage (third maxilliped of a crayfish).

phylum (or even within some classes) because of varying degrees of fusion of somites, for example, as in the cephalothorax.

In many crustaceans, the dorsal cuticle of the head extends posteriorly and around the sides of the animal to cover or fuse with some or all of the thoracic and abdominal somites. This covering is the **carapace.** In some groups the carapace forms clamshell-like valves that cover most or all of the body. In the decapods (including lobsters, shrimp, crabs, and others) the carapace covers the entire cephalothorax but not the abdomen.

Form and Function

Appendages

Some of the modifications of crustacean appendages may be illustrated by those of crayfishes and lobsters (class Malacostraca, order Decapoda, p. 506). The **swimmerets,** or abdominal appendages, retain the primitive biramous condition. Such an appendage consists of inner and outer branches, called the **endopod** and **exopod,** which are attached to one or more basal segments collectively called the **protopod** (Figure 23-15).

There are many modifications of this plan. In the primitive character state for crustaceans, all of the trunk appendages are rather similar in structure and adapted for swimming. The evolutionary trend, shown in the crayfishes, has been toward reduction in number of appendages and toward a variety of modifications that fit them for many functions. Some are foliaceous (flat and leaflike), as are the maxillae; some are biramous, as are the swimmerets, maxillipeds, uropods, and antennae; some have lost one branch and are **uniramous,** as are the walking legs.

The terminology applied by various workers to crustacean appendages has not been blessed with uniformity. At least two systems are in wide use. Alternative terms to those we use, for example, are protopodite, exopodite, endopodite, basipodite, coxopodite, and epipodite. The first and second pairs of antennae may be called antennules and antennae, and the first and second maxillae are often called maxillules and maxillae. A rose by any other name. . . .

In crayfishes we find the first three pairs of thoracic appendages, called **maxillipeds,** serving along with the two pairs of maxillae as food handlers; the other five pairs of appendages are lengthened and strengthened for walking and defense (Figure 23-16). The first pair of walking legs, called

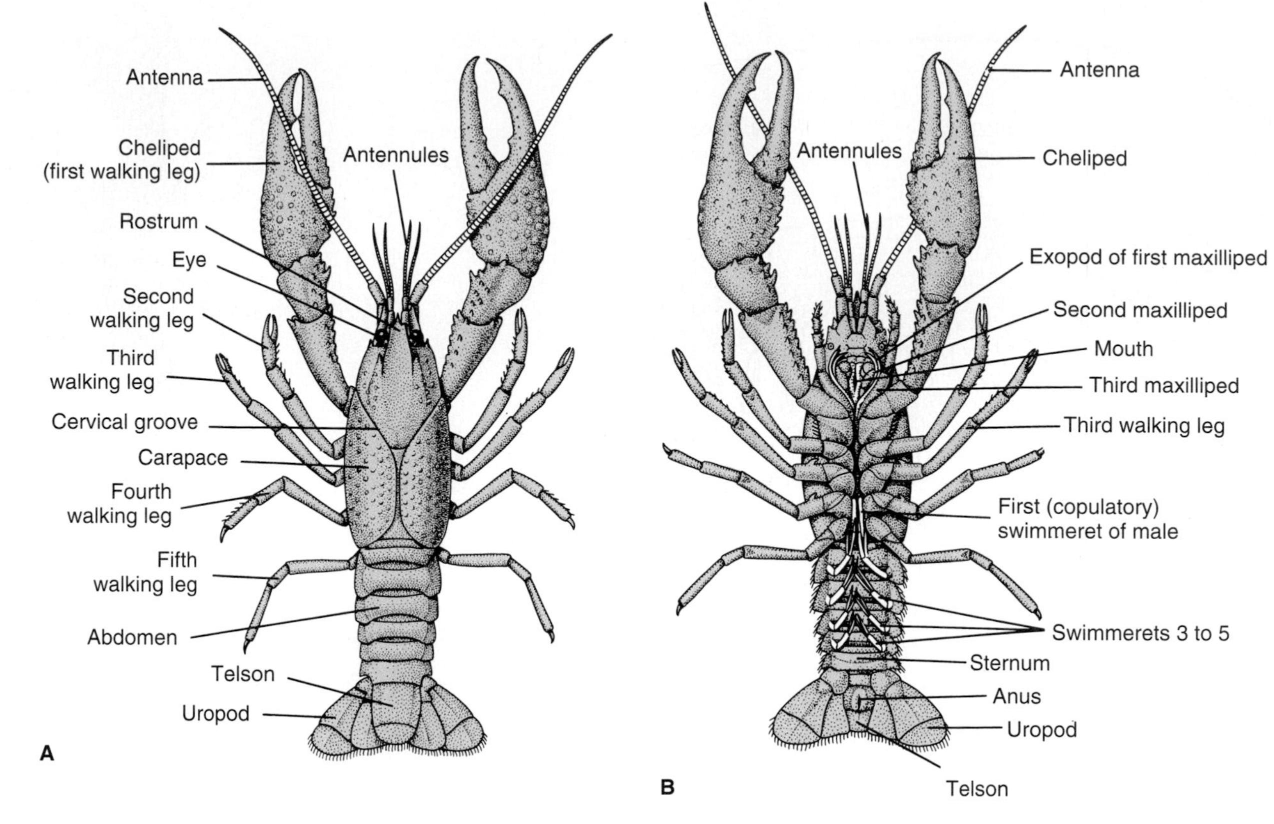

figure 23-16

External structure of crayfishes. **A,** Dorsal view. **B,** Ventral view.

chelipeds, are enlarged with a strong claw, or chela, for defense. The abdominal swimmerets serve not only for locomotion, but in the male the first pair is modified for copulation, and in the female they all serve as a nursery for attached eggs and young. The last pair of appendages, the **uropods,** are wide and serve as paddles for swift backward movements, and, with the telson, they form a protective device for eggs or young on the swimmerets.

Ecdysis

The problem of growth despite a restrictive exoskeleton is solved in crustaceans, as in other arthropods, by ecdysis (Gr. *ekdysis,* to strip off), the periodic shedding of the old cuticle and the formation of a larger new one. Molting occurs most frequently during larval stages and less often as the animal reaches adulthood. Although the actual shedding of the cuticle is periodic, the molting process and the preparations for it, involving the storage of reserves and changes in the integument, are a continuous process going on during most of the animal's life.

During each **premolt** period the old cuticle becomes thinner as inorganic salts are withdrawn from it and stored in the tissues. Other reserves, both organic and inorganic, also accumulate and are stored. The underlying epidermis begins to grow by cell division; it secretes first a new inner layer of epicuticle and then enzymes that digest away the inner layers of old endocuticle (Figure 23-17). Gradually a new cuticle forms inside the degenerating old one. Finally the actual ecdysis occurs as the old cuticle ruptures, usually along the middorsal line, and the animal backs out (Figure 23-18). By taking in air or water the animal swells to stretch the new larger cuticle to its full size. During the **postmolt** period the cuticle thickens, the outer layer hardens by tanning, and the inner layer is strengthened as salvaged inorganic salts and other constituents are redeposited. Usually the animal is very secretive during the postmolt period when its defenseless condition makes it particularly vulnerable to predation.

That ecdysis is under hormonal control has been demonstrated in both crustaceans and insects, but the process is often initiated by a stimulus perceived by the central nervous system. The action of the stimulus in decapods is to decrease the production of a **molt-inhibiting hormone** from neurosecretory cells in the **X-organ** of the eyestalk. The sinus gland, also in the eyestalk, releases the hormone. When the level of molt-inhibiting hormone drops, the **Y-organs** near the mandibles produce **molting hormone.** This hormone initiates the processes leading to premolt. The Y-organs are homologous to the prothoracic glands of insects, which produce ecdysone (p. 269).

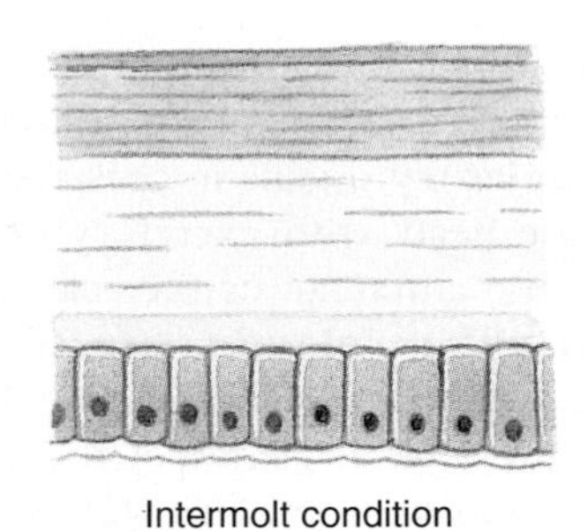

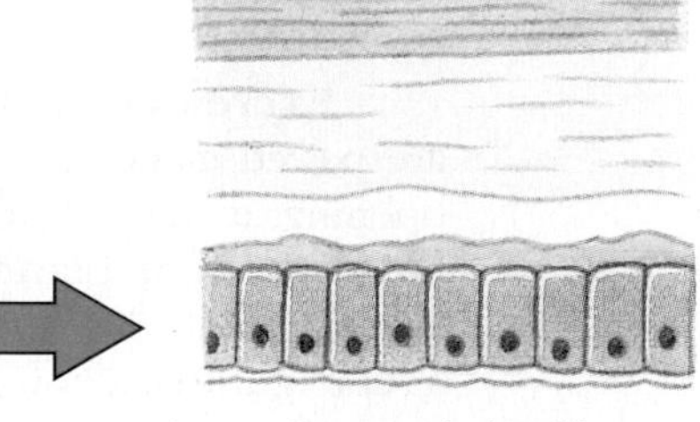

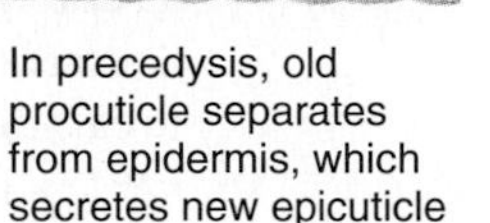

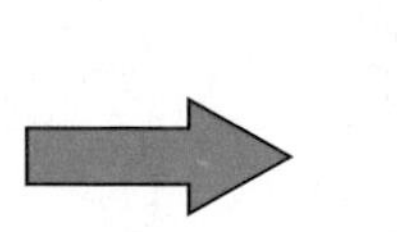

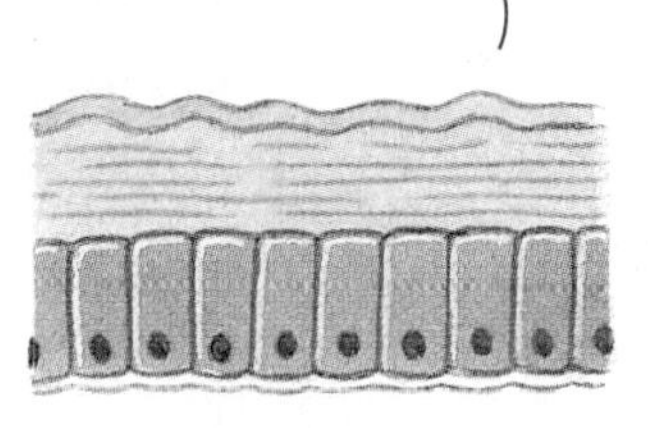

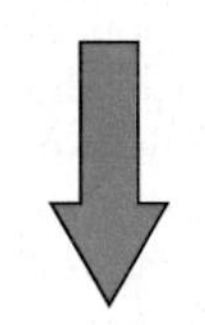

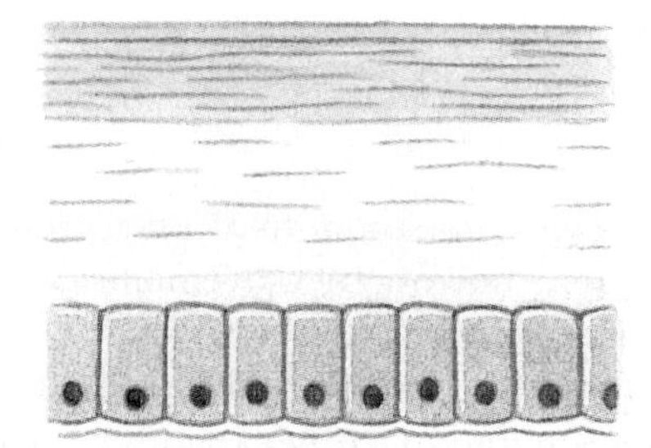

figure 23-17
Cuticle secretion and reabsorption in ecdysis.

> *Neurosecretory cells are modified nerve cells that secrete hormones. They are widespread in invertebrates and also occur in vertebrates. Cells in the vertebrate hypothalamus and in the posterior pituitary are good examples (see p. 272).*

Other Endocrine Functions

Body color of crustaceans is largely a result of pigments in special branched cells **(chromatophores)** in the epidermis. The chromatophores change color by concentrating the pigment granules in the center of the cells, which causes a lightening effect, or by dispersing pigment throughout the cells, which causes darkening. Neurosecretory cells in the eye stalk control pigment behavior. Neurosecretory hormones also control pigment in the eyes for light and dark adaptation, and other neurosecretory hormones control the rate and amplitude of the heartbeat.

Androgenic glands, which are not neurosecretory, occur in male malacostracans, and their secretion stimulates expression of male sexual characteristics. If androgenic glands are artificially implanted in a female, her ovaries transform to testes and begin to produce sperm, and her appendages begin to take on male characteristics at the next molt.

Feeding Habits

Feeding habits and adaptations for feeding vary greatly among crustaceans. Many forms can shift from one type of feeding to another depending on environment and food availability, but fundamentally the same set of mouthparts is used by all. The mandibles and maxillae are involved in the actual ingestion; maxillipeds hold and crush food. In predators the walking legs, particularly the chelipeds, serve in food capture.

Many crustaceans, both large and small, are predatory, and some have interesting adaptations for killing their prey. One shrimplike form, *Lygiosquilla,* has on one of its walking legs a specialized digit that can be drawn into a groove and released suddenly to pierce a passing prey. The pistol shrimp *Alpheus* has one enormously enlarged chela that can be cocked like the hammer of a gun and snapped with a force that stuns its prey.

The food of crustaceans ranges from plankton, detritus, and bacteria, used by **suspension feeders,** to larvae, worms, crustaceans, snails, and fishes, used by **predators,** and dead animal and plant matter, used by **scavengers.** Suspension feeders, such as fairy shrimps, water fleas, and barnacles, use their legs, which bear a thick fringe of setae, to create water currents that sweep food particles through the setae. The mud shrimp *Upogebia* uses long setae on its first two pairs of thoracic

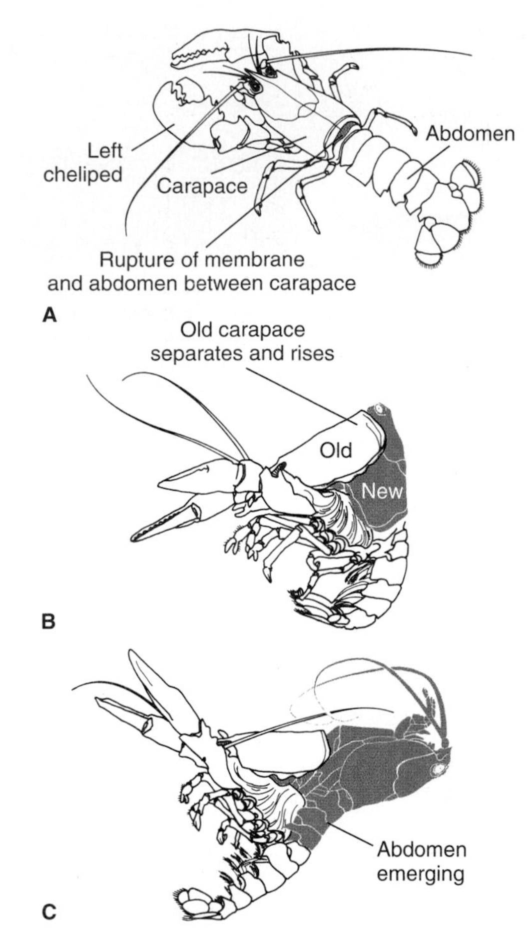

figure 23-18

Molting sequence in the northern lobster, *Homarus americanus*. **A,** Membrane between the carapace and abdomen ruptures, and the carapace begins a slow elevation. This step may take up to two hours. **B** and **C,** Head, thorax, and finally abdomen are withdrawn. This process usually takes no more than 15 minutes. Immediately after ecdysis, the chelipeds are desiccated and the body is very soft. The lobster now begins rapid absorption of water so that within 12 hours the body increases about 20% in length and 50% in weight. Water will be replaced by living tissue in succeeding weeks.

appendages to strain food material from water circulated through its burrow by movements of its swimmerets.

Crayfishes have a two-part stomach. The first contains a **gastric mill** in which food, already torn up by the mandibles, can be ground up further by three calcareous teeth into particles fine enough to pass through a filter of setae in the second part of the stomach; the food particles then pass into the intestine for chemical digestion.

Respiration, Excretion, and Circulation

The **gills** of crustaceans vary in shape—treelike, leaflike, or filamentous—all provided with blood vessels or sinuses. They usually are attached to appendages and kept ventilated by the movement of appendages in the water. The overlapping carapace usually protects the **branchial chambers.** Some smaller crustaceans breathe through the general body surface.

Excretory and **osmoregulatory** organs in crustaceans are paired glands located in the head, with excretory pores opening at the base of either the antennae or the maxillae, thus **antennal glands** or **maxillary glands,** respectively (Figure 23-19). The antennal glands of decapods are also called **green glands.** They resemble the coxal glands of the chelicerates. The waste product is mostly ammonia with some urea and uric acid. Some wastes diffuse through the gills as well as through the excretory glands.

Circulation, as in other arthropods, is an **open system** consisting of a heart, either compact or tubular, and arteries, which transport blood to different areas of the hemocoel. Some smaller crustaceans lack a heart. An open circulatory system depends less on heartbeats for circulation because the movement of organs and limbs circulates the blood more effectively in open sinuses than in capillaries. The blood may contain as respiratory pigments either hemocyanin or hemoglobin (hemocyanin in decapods), and it has the property of clotting to prevent loss of blood in minor injuries.

Nervous and Sensory Systems

A cerebral ganglion above the esophagus sends nerves to the anterior sense organs and connects to a subesophageal ganglion by a pair of connectives around the esophagus. A double ventral nerve cord has a ganglion in each segment that sends nerves to the viscera, appendages, and muscles (Figure 23-19). Giant fiber systems are common among the crustaceans.

Sensory organs are well developed. There are two types of eyes—the median, or nauplius, eye and compound eyes. The **median eye** consists usually of a group of three pigment cups containing retinal cells; the eye may or may not have a lens. Median eyes are found in nauplius larvae and in some adult forms, and they may be the only adult eye, as in copepods.

Most crustaceans have **compound eyes** similar to insect eyes. In crabs and crayfishes they also are on the ends of movable eyestalks (Figure 23-19). Compound eyes are precise instruments, different from vertebrate eyes, yet especially adept at detecting motion; they are able to analyze polarized light. The convex corneal surface gives a wide visual field, particularly in the stalked eyes where the surface may cover an arc of 200 degrees or more.

Compound eyes are composed of many tapering units called **ommatidia** set close together (Figure 23-20). The facets, or corneal surfaces, of the ommatidia give the surface of the eye the appearance of a fine mosaic. Most crustacean eyes are adapted either to bright or to dim light, depending on their diurnal or nocturnal habits, but some are able, by means of screening pigments, to adapt, to some extent at least, to both bright and dim light. The number of ommatidia varies from a dozen or two in some small crustaceans to 15,000 or more in a large lobster. Some insects have approximately 30,000.

Other sensory organs include statocysts, tactile setae on the cuticle of most of the body, and chemosensitive setae, especially on the antennae, antennules, and mouthparts.

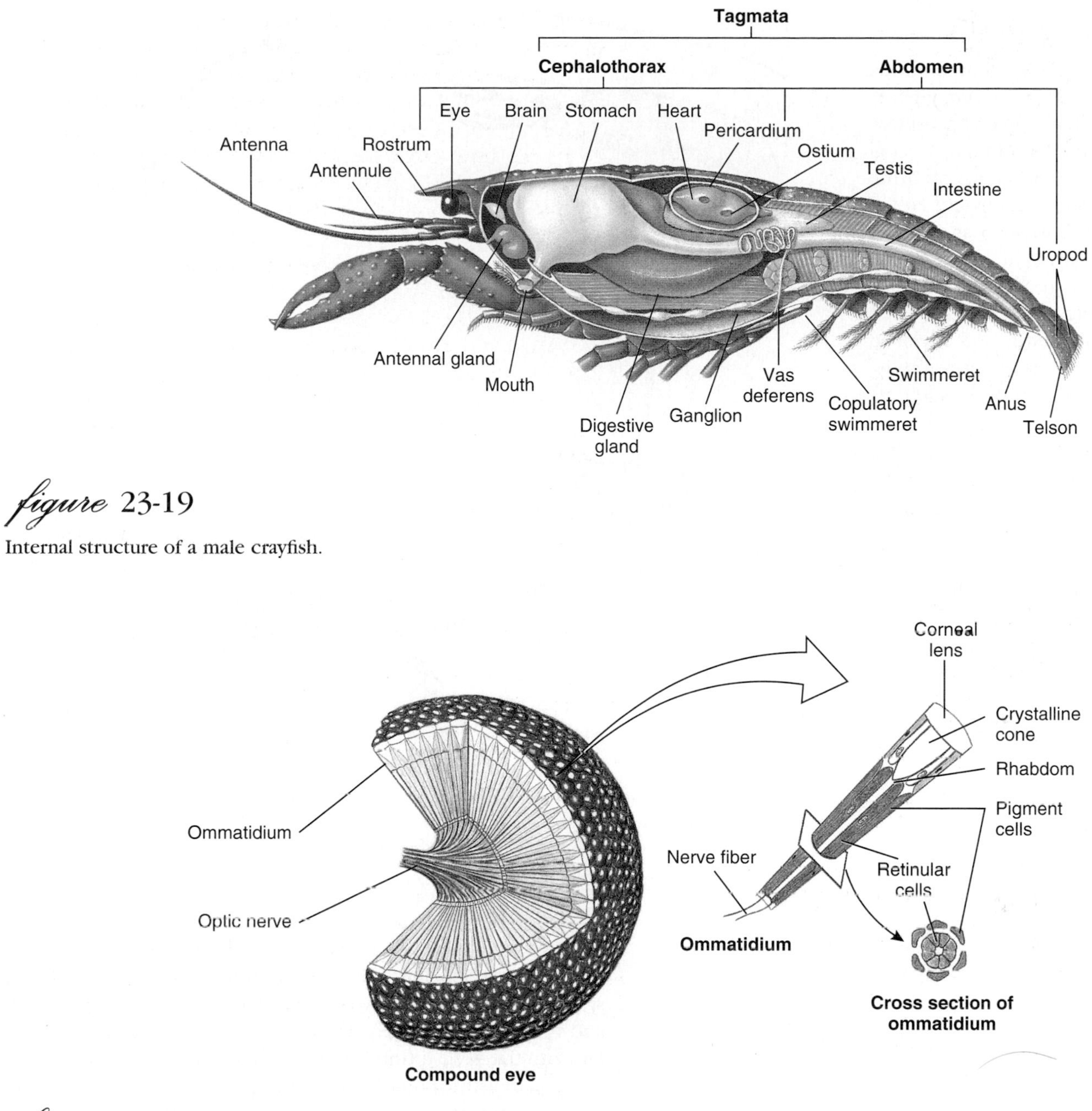

figure 23-19

Internal structure of a male crayfish.

figure 23-20

Compound eye of an insect. A single ommatidium is shown enlarged to the right.

Reproduction and Life Cycles

Most crustaceans have separate sexes, and numerous specializations for copulation occur among the different groups. Barnacles are monoecious but generally practice cross-fertilization. In some ostracods males are scarce, and reproduction is usually parthenogenetic. Most crustaceans brood their eggs in some manner—branchiopods and barnacles have special brood chambers, copepods have egg sacs attached to the sides of the abdomen (see Figure 23-24), and malacostracans usually carry eggs and young attached to their appendages.

The organism that hatches from an egg of a crayfish is a tiny juvenile with the same form as the adult and a complete set of appendages and somites. However, most crustaceans produce larvae that must go through a series of changes, either gradual or abrupt over the course of the series of molts, to assume the adult form (metamorphosis). The primitive larva of the crustaceans is the **nauplius** (Figure 23-21). It has an unsegmented body, a frontal eye, and three pairs of appendages, representing the two pairs of antennae and the mandibles. The form of the developmental stages and postlarvae of different groups of Crustacea are varied and have special names.

Class Branchiopoda

Members of the class Branchiopoda (bran′kee-op′o-da) (Gr. *branchia,* gills, + *pous, podos,* foot) have several primitive characteristics. Four orders are recognized: **Anostraca** (fairy shrimp and brine shrimp), which lack a carapace; **Notostraca** (tadpole shrimp such as *Triops*), whose carapace forms a large dorsal shield covering most of the trunk somites; **Conchostraca** (clam shrimp such as *Lynceus*), whose carapace is bivalved and usually encloses the entire body; and **Cladocera** (water fleas such as *Daphnia,* Figure 23-22), with a carapace typically covering the entire body but not the head. Branchiopods have reduced first antennae and second maxillae. Their legs are flattened and leaflike (**phyllopodia**) and are the chief respiratory organs (hence, the name branchiopods). The legs also are used in suspension feeding in most branchiopods, and in groups other than the cladocerans, they are used for locomotion as well. The most important and diverse order is the Cladocera, which often forms a large segment of the freshwater zooplankton.

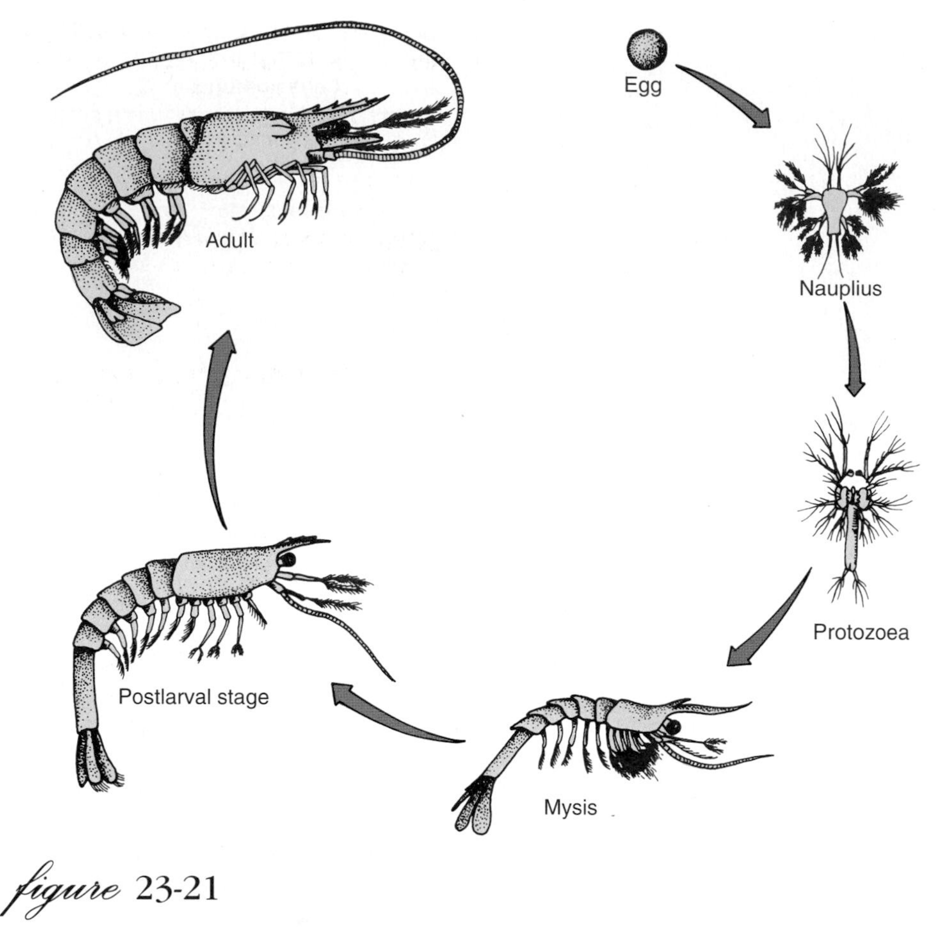

figure 23-21

Life cycle of the Gulf shrimp *Penaeus.* Penaeids spawn at depths of 40 to 90 m. The young larval forms make up part of the plankton fauna and work their way inshore to water of lower salinity to develop as juveniles. Older shrimp return to deeper water offshore.

Class Maxillopoda

The class Maxillopoda includes a number of crustacean groups traditionally considered classes themselves. Specialists have recognized evidence that these groups descended from a common ancestor and thus form a clade within the Crustacea. They basically have five cephalic, six thoracic, and usually four abdominal somites plus a telson, but reductions are common. No typical appendages occur on the abdomen. The eye of the nauplius (when present) has a unique structure and is referred to as a **maxillopodan eye.**

Members of the subclass **Ostracoda** (os-trak′o-da) (Gr. *ostrakodes,* testaceous, that is, having a shell) are, like conchostracans, enclosed in a bivalved carapace and resemble tiny clams, 0.25 to 8 mm long (Figure 23-23). Ostracods show considerable fusion of trunk somites, and numbers of thoracic appendages are reduced to two or none.

The subclass **Copepoda** (ko-pep′o-da) (Gr. *kōpē,* oar, + *pous, podos,* foot) is an important group of Crustacea, second only to Malacostraca in number of species. Copepods are small (usually a few millimeters or less in length), rather elongate, tapering toward the posterior end, lacking a carapace, and retaining the simple, median, nauplius eye in the adult (Figure 23-24). They have four pairs of rather flattened, biramous, thoracic swimming appendages, and a fifth, reduced pair. The abdomen bears no legs. Many symbiotic as well as free-living species are known. Many of the parasites are highly modified, and the adults may be so highly modified (and may depart so far from the description just given) that they can hardly be recognized as arthropods. Ecologically, free-living copepods are of extreme importance, often dominating the primary consumer level (herbivore, p. 123) in aquatic communities.

The subclass **Branchiura** (bran-kee-u′ra) (Gr. *branchia,* gills, + *ura,* tail) is a small group of primarily fish parasites, which, despite its name, has no gills (Figure 23-25). Members of this group are usually between 5 and 10 mm long and may be found on marine or freshwater fish. They typically have a broad, shieldlike carapace, compound eyes, four biramous thoracic appendages for swimming, and a short, unsegmented abdomen. The second maxillae have become modified as suction cups.

The subclass **Cirripedia** (sir-i-ped′i-a) (L. *cirrus,* curl of hair, + *pes, pedis,* foot) includes barnacles, which are usually enclosed in a shell of calcareous plates, as well as three smaller orders of burrowing or parasitic forms. Barnacles are sessile as adults and may be attached to the substrate by a stalk (gooseneck barnacles) (Figure 23-26B) or directly (acorn barnacles)

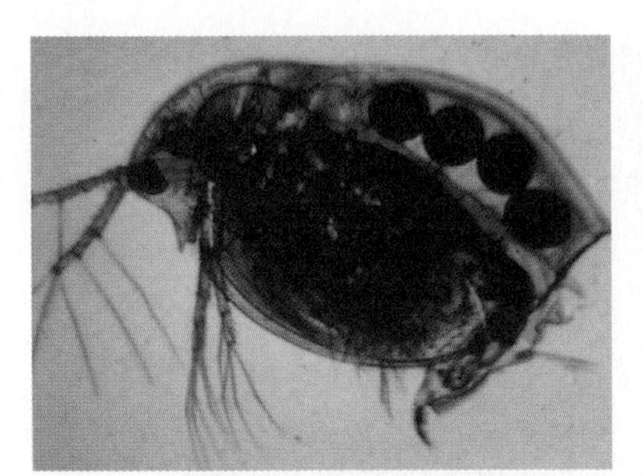

figure 23-22

A water flea, *Daphnia* (order Cladocera), photographed with polarized light. These microscopic forms occur in great numbers in northern lakes and are an important component of the food chain leading to fishes.

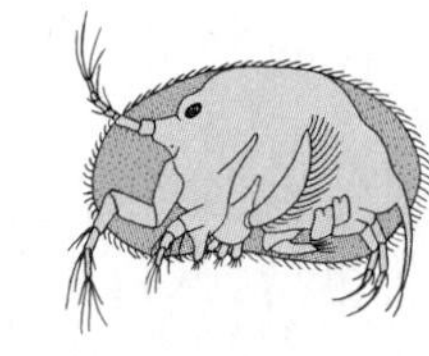

figure 23-23

An ostracod (subclass Ostracoda, class Maxillopoda).

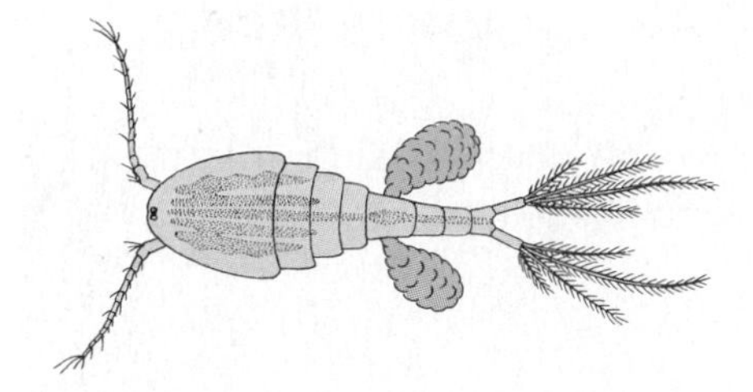

figure 23-24

A copepod with attached ovisacs (subclass Copepoda, class Maxillopoda).

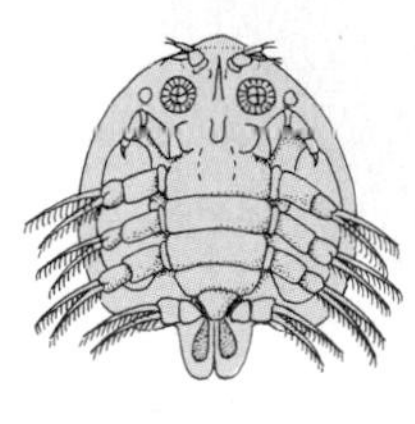

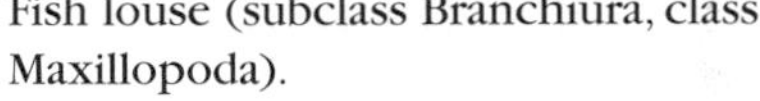

figure 23-25

Fish louse (subclass Branchiura, class Maxillopoda).

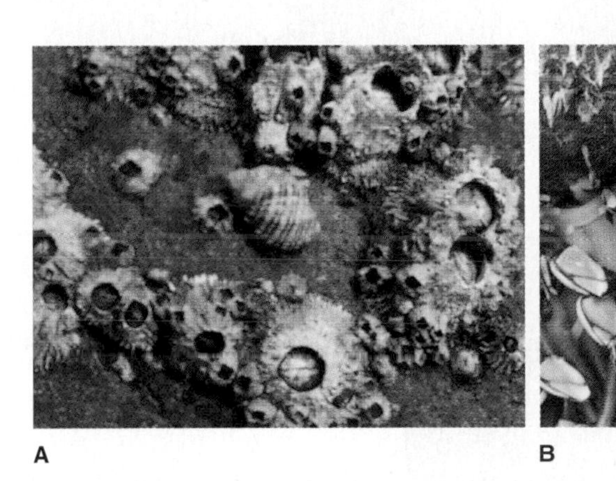

figure 23-26

A, Acorn barnacles, *Semibalanus cariosus* (subclass Cirripedia) are found on rocks along the Pacific Coast of North America. **B,** Common gooseneck barnacles, *Lepas anatifera.* Note the feeding legs, or cirri, on *Lepas.* Barnacles attach themselves to a variety of firm substrates, including rocks, pilings, and boat bottoms.

(Figure 23-26A). Typically, the carapace (mantle) surrounds the body and secretes a shell of calcareous plates. The head is reduced, the abdomen absent, and the thoracic legs are long, many-jointed cirri with hairlike setae. The cirri are extended through an opening between the calcareous plates to filter from the water the small particles on which the animal feeds (Figure 23-26B).

> *Barnacles frequently foul ship bottoms by settling and growing there. So great may be their number that the speed of the ship may be reduced 30% to 40%, necessitating expensive drydocking of the ship to clean them off.*

Class Malacostraca

The class Malacostraca (mal′a-kos′tra-ka) (Gr. *malakos,* soft, + *ostrakon,* shell) is the largest class of Crustacea and shows great diversity. We will mention only 4 of its 12 to 13 orders. The trunk of malacostracans usually has eight thoracic and six abdominal somites, each with a pair of appendages. There are many marine and freshwater species.

The **Isopoda** (i-sop′o-da) (Gr. *isos,* equal, + *pous, podos,* foot) are commonly dorsoventrally flattened, lack a carapace, and have sessile compound eyes. Their abdominal appendages bear gills. Common land forms are sow bugs or pill bugs (*Porcellio* and *Armadillidium,* Figure 23-27A), which live under stones and in damp places. *Asellus* is common in fresh water, and *Ligia* is abundant on sea beaches and rocky shores. Some isopods are parasites of other crustaceans or of fish (Figure 23-28).

The **Amphipoda** (am-fip′o-da) (Gr. *amphis,* on both sides, + *pous, podos,* foot) resemble isopods in that the members have no carapace and have sessile compound eyes. However, they are usually compressed laterally, and their gills are in the thoracic position, as in other malacostracans. There are many marine amphipods (Figure 23-29), such as the beach flea, *Orchestia,* and numerous freshwater species.

A

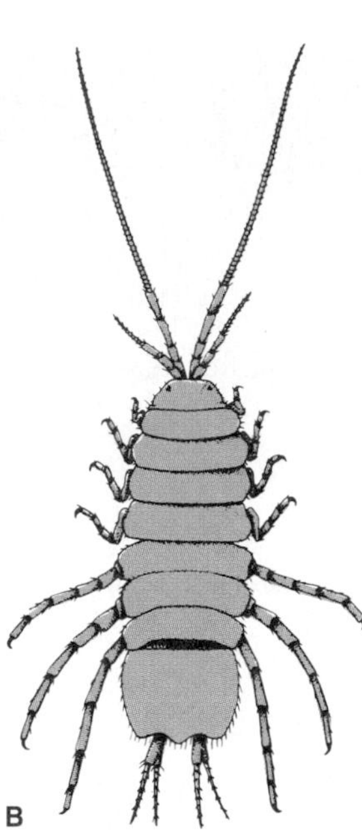

B

figure 23-27

A, Four pill bugs, *Armadillidium vulgare* (order Isopoda), common terrestrial forms. **B,** Freshwater sow bug, *Caecidotea* sp., an aquatic isopod.

A

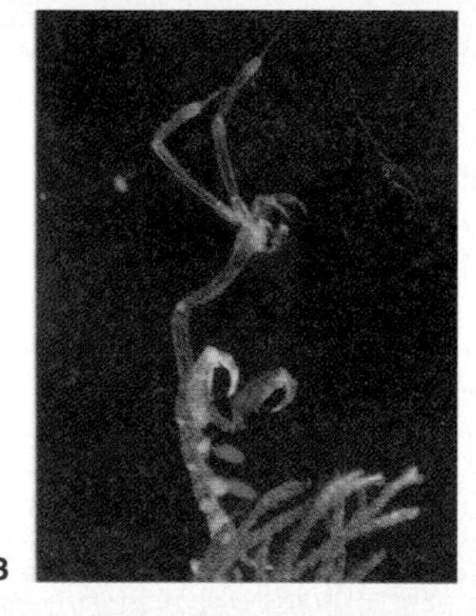

B

figure 23-28

An isopod parasite (*Anilocra* sp.) on a coney (*Cephalopholis fulvus*) inhabiting a Caribbean coral reef.

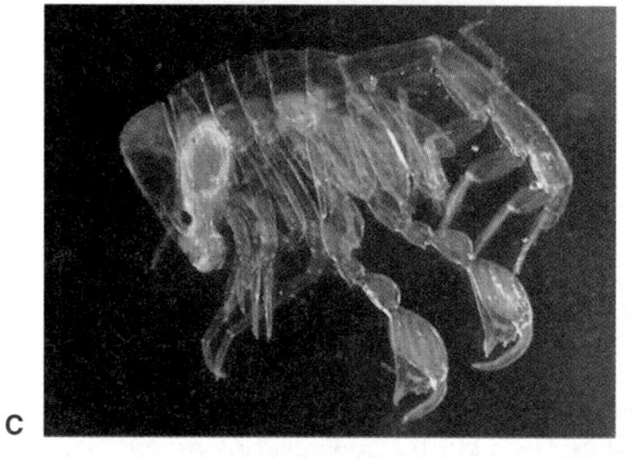

C

figure 23-29

Marine amphipods. **A,** Free-swimming amphipod, *Anisogammarus* sp. **B,** Skeleton shrimp, *Caprella* sp., shown on a bryozoan colony, resemble praying mantids. **C,** *Phronima,* a marine pelagic amphipod, takes over the tunic of a salp (subphylum Urochordata, Chapter 26). Swimming by means of its abdominal swimmerets, which protude from the opening of the barrel-shaped tunic, the amphipod maneuvers to catch its prey.

The **Euphausiacea** (yu-faws′i-a′se-a) (Gr. *eu,* well, + *phausi,* shining bright, + *acea,* L. suffix, pertaining to) is a group of only about 90 species, but they are important as the oceanic plankton known as "krill." They are about 3 to 6 cm long (Figure 23-30) and commonly occur in great oceanic swarms, where they are eaten by baleen whales and many fishes.

The **Decapoda** (de-cap′o-da) (Gr. *deka,* ten, + *pous, podos,* foot) have five pairs of walking legs of which the first is often modified to form pincers (**chelae**) (Figures 23-14 and 23-16). These are the lobsters, crayfishes (see Figure 23-14), shrimps (Figure 23-21), and crabs, the largest of the crustaceans (Figure 23-31). True crabs differ from the others in having a broader carapace and a much reduced abdomen (Figure 23-31A and C). Familiar examples are the fiddler crabs *Uca,* which burrow in sand just below the high-tide level (Figure 23-31C), decorator crabs, which cover their carapaces with sponges and sea anemones for camouflage, and spider crabs, such as *Libinia.* Hermit crabs (Figure 23-31B) have become adapted to live in snail shells; their abdomens, which lack a hard exoskeleton, are protected by the snail shell.

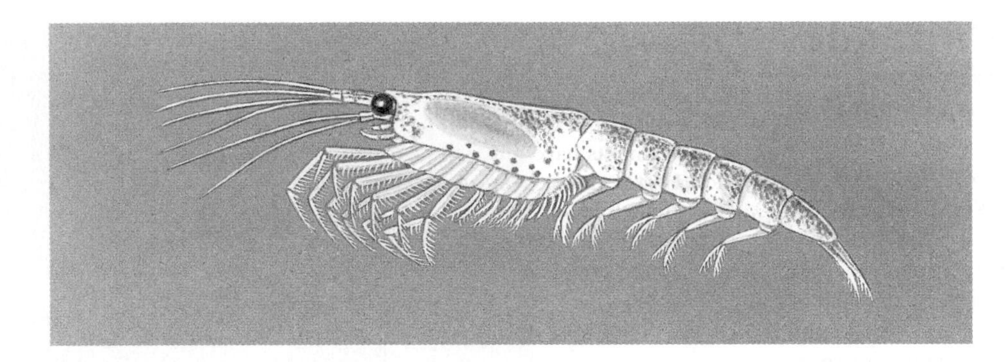

figure 23-30

Meganyctiphanes, order Euphausiacea, the northern krill.

figure 23-31

Decapod crustaceans. **A,** The bright orange tropical rock crab, *Grapsus grapsus,* is a conspicuous exception to the rule that most crabs bear cryptic coloration. **B,** The hermit crab, *Elassochirus gilli,* which has a soft abdominal exoskeleton, lives in a snail shell that it carries about and into which it can withdraw for protection. **C,** The male fiddler crab, *Uca* sp., uses its enlarged cheliped to wave territorial displays and in threat and combat. **D,** The red night shrimp, *Rhynchocinetes rigens,* prowls caves and overhangs of coral reefs, but only at night. **E,** Spiny lobster *Panulirus argus.*

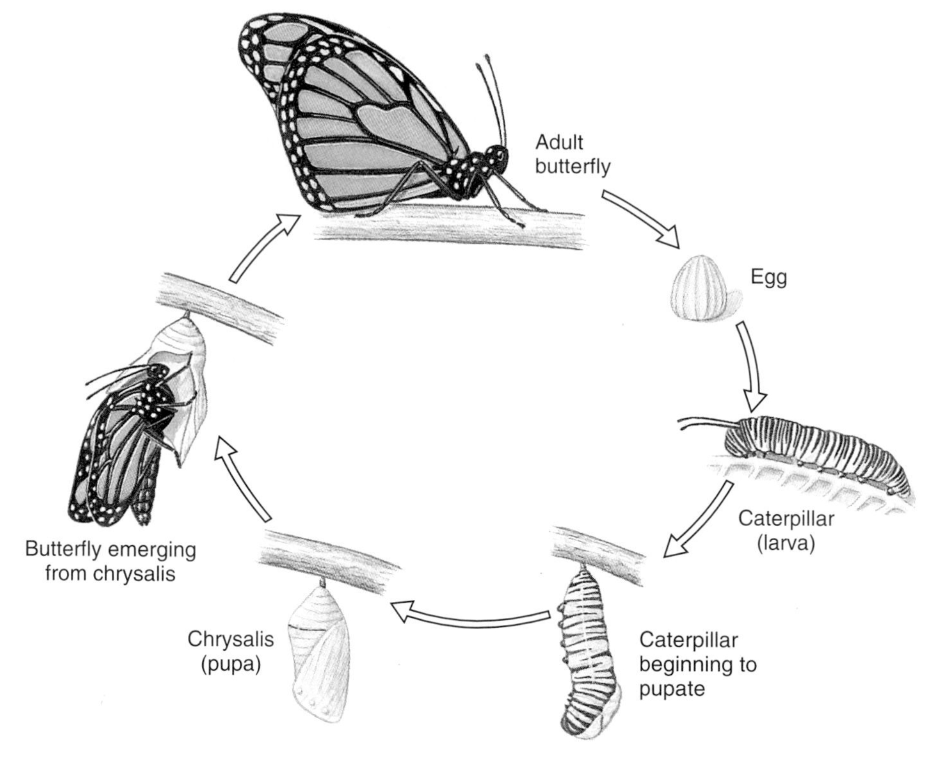

figure 23-42

Holometabolous (complete) metamorphosis in a butterfly, *Danaus plexippus* (order Lepidoptera). Eggs hatch to produce first of several larval instars. The last larval instar molts to become a pupa. The adult emerges at the pupal molt.

A

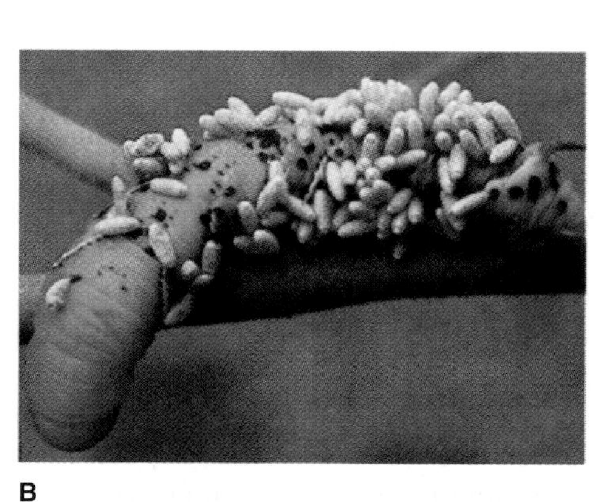

B

figure 23-43

A, Hornworm, larval stage of a sphinx moth (order Lepidoptera). The more than 100 species of North American sphinx moths are strong fliers and mostly nocturnal feeders. Their larvae, called hornworms because of the large fleshy posterior spine, are often pests of tomatoes, tobacco, and other plants. **B,** Hornworm parasitized by a tiny wasp, *Apanteles,* which laid its eggs inside the caterpillar. The wasp larvae have emerged, and their pupae are on the caterpillar's skin. Young wasps emerge in 5 to 10 days, and the caterpillar usually dies.

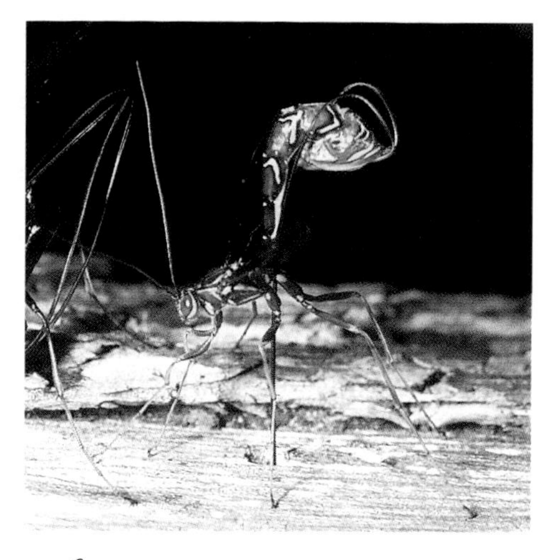

figure 23-44

An ichneumon wasp with the end of the abdomen raised to thrust her long ovipositor into wood to find a tunnel made by the larva of a wood wasp or wood-boring beetle. She can bore 13 mm or more into the wood to lay her eggs in the larva of the wood-boring beetle, which will become host for the ichneumon larvae. Other ichneumon species attack spiders, moths, flies, crickets, caterpillars, and other insects.

The biological meaning of the word "bug" is a great deal more restrictive than in common English usage. People often refer to all insects as "bugs," even extending the word to include such nonanimals as bacteria, viruses, and glitches in computer programs. Strictly speaking, however, a bug is a member of the order Hemiptera and nothing else.

A few insects, such as silverfish (see Figure 23-57) and springtails, undergo direct development. The young, or juveniles, are similar to the adults except in size and sexual maturation. The stages are egg, juvenile, and adult. Such insects include the primitively wingless insects.

Hormones control and regulate metamorphosis in insects. Three major endocrine organs are involved in development through the juvenile instars and eventually the emergence of the adult.

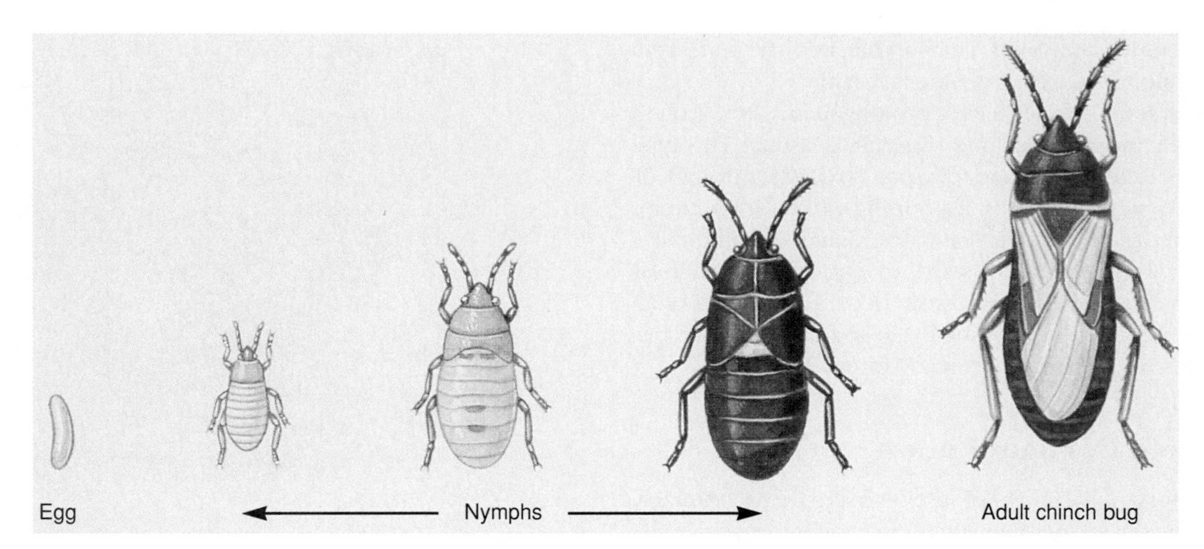

figure 23-45

Life history of a hemimetabolous insect.

figure 23-46

A, A stonefly, *Perla* sp. (order Plecoptera). **B,** A ten-spot dragonfly, *Libellula pulchella* (order Odonata). **C,** Nymph (larva) of a dragonfly. Both stoneflies and dragonflies have aquatic larvae that undergo gradual metamorphosis.

These organs are the **brain, ecdysial (prothoracic) glands,** and **corpora allata.** We described the action of the hormones produced on p. 269. Hormonal control of molting and metamorphosis is the same in holometabolous and hemimetabolous insects.

Diapause Many animals can enter a state of dormancy during adverse conditions, and there are periods in the life cycle of many insects when a particular stage can remain dormant for a long time because external climatic conditions are too harsh for normal activity. Most insects enter such a stage facultatively when some factor of the environment,

figure 23-47

A, Ecdysis in a cicada, *Tibicen davisi* (order Homoptera). The old cuticle splits along a dorsal midline as a result of increased blood pressure and of air forced into the thorax by muscle contraction. The emerging insect is pale, and its new cuticle is soft. The wings will be expanded by blood pumped into veins, and the insect enlarges by taking in air. **B,** An adult *Tibicen davisi.*

such as temperature, becomes unfavorable, and the state continues until conditions again become favorable.

However, some species have a prolonged arrest of growth that is internally programmed and is usually seasonal. This type of dormancy is called **diapause** (di′a-poz) (Gr. *dia,* through, dividing into two parts, + *pausis,* a stopping), and it is an important adaptation to survive adverse environmental conditions. Diapause usually is triggered by some external signal, such as shortening day length. Diapause always occurs at the end of an active growth stage of the molting cycle so that, when the diapause period is over, the insect is ready for another molt.

Behavior and Communication

The keen sensory perceptions of insects make them extremely responsive to many stimuli. The stimuli may be internal (physiological) or external (environmental), and the responses are governed by both the physiological state of the animal and the pattern of nerve pathways involved. Many of the responses are simple, such as orientation toward or away from the stimulus, for example, attraction of a moth to light, avoidance of light by a cockroach, or attraction of carrion flies to the odor of dead flesh.

Much of the behavior of insects, however, is not a simple matter of orientation but involves a complex series of responses. A pair of tumble bugs, or dung beetles, chew off a bit of dung, roll it into a ball, and roll the ball laboriously to where they intend to bury it, after laying their eggs in it (Figure 23-48). A female cicada slits the bark of a twig and then lays an egg in each of the slits. A female potter wasp *Eumenes* scoops up clay into pellets, carries them one by one to her building site, and fashions them into dainty little narrow-necked clay pots, into each of which she lays an egg. Then she hunts and paralyzes a number of caterpillars, pokes them into the opening of a pot, and closes up the opening with clay. Each egg, in its own protective pot, hatches to find a well-stocked larder of food awaiting it.

Some insects can memorize and perform in sequence tasks involving multiple signals in various sensory areas. Worker honeybees have been trained to walk through mazes that involved five turns in sequence, using such clues as the color of a marker, the distance between two spots, or the angle of a turn. The same is true of ants. Workers of one species of Formica *learned a six-point maze at a rate only two or three times slower than that of laboratory rats. The foraging trips of ants and bees often wind and loop about in a circuitous route, but once the forager has found food, the return trip is relatively direct. One investigator suggested that the continuous series of calculations necessary to figure the angles, directions, distance, and speed of the trip and to convert it into a direct return could involve a stopwatch, a compass, and integral vector calculus. How the insect does it is unknown.*

figure 23-48

Tumble bugs, or dung beetles, *Canthon pilularis* (order Coleoptera), chew off a bit of dung, roll it into a ball, and then roll it to where they will bury it in soil. One beetle pushes while the other pulls. Eggs are laid in the ball, and the larvae feed on the dung. Tumble bugs are black, an inch or less in length, and common in pasture fields.

Much of such behavior is "innate," that is, entire sequences of actions apparently have been genetically programmed. However, a great deal more learning is involved than we once believed. A potter wasp, for example, must learn where she has left her pots if she is to return to fill them with caterpillars one at a time. Social insects, which have been studied extensively, are capable of most of the basic forms of learning used by mammals. The exception is insight learning. Apparently insects, when faced with a new problem, cannot reorganize their memories to construct a new response.

Insects communicate with other members of their species by means of chemical, visual, auditory, and tactile signals. **Chemical signals** take the form of **pheromones,** which are substances secreted by one individual that affect the behavior or physiological processes of another individual. Examples of pheromones include sex attractants, releasers of certain behavior patterns, trail markers, alarm signals, and territorial markers. Like hormones, pheromones are effective in minute quantities. Social insects, such as bees, ants, wasps, and termites, can recognize a nestmate—or an alien in the nest—by means of identification pheromones. Pheromones determine caste in termites, and to some extent in ants and bees. In fact, pheromones are probably a primary integrating force in populations of social insects. Many insect pheromones have been extracted and chemically identified.

Sound production and **reception** (phonoproduction and phonoreception) in insects have been studied extensively, and although a sense of hearing is not present in all insects, this means of communication is meaningful to insects that use it. Sounds serve as warning devices, advertisement of territorial claims, or courtship songs. The sounds of crickets and grasshoppers seem to be concerned with courtship and aggression. Male crickets scrape the rough edges of the forewings together to produce their characteristic chirping. Male cicadas produce the long, drawn-out sound of their recruitment call by vibrating membranes in a pair of organs located on the ventral side of the basal abdominal segment.

There are many forms of **tactile communication,** such as tapping, stroking, grasping, and antennae touching, which evoke responses varying from recognition to recruitment and alarm. Certain kinds of flies, springtails, and beetles manufacture their own **visual signals** in the form of **bioluminescence.** The best known of the luminescent beetles are fireflies, or lightning bugs (which are neither flies nor bugs, but beetles), in which the flash of light is a means of locating a prospective mate. Each species has its own characteristic flashing rhythm produced on the ventral side of the last abdominal segments. Females flash an answer to the species-specific pattern to attract the males. This interesting "love call" has been adopted by species of *Photuris,* which prey on the male fireflies of other species they attract (Figure 23-49).

figure 23-49

Firefly femme fatale, *Photuris versicolor,* eating a male *Photinus tanytoxus,* which she has attracted with false mating signals.

figure 23-50

An ant (order Hymenoptera) tending a group of aphids (order Homoptera). The aphids feed copiously on plant juices and excrete the excess as a clear liquid rich in carbohydrates ("honey-dew"), which is cherished as a food by ants.

Social Behavior Insects rank very high in the animal kingdom in their organization of social groups, and cooperation within the more complex groups depends heavily on chemical and tactile communication. Social communities are not all complex, however. Some community groups are temporary and uncoordinated, as are the hibernating associations of carpenter bees or the feeding gatherings of aphids (Figure 23-50). Some are coordinated for only brief periods, such as the tent caterpillars *Malacosoma,* that join in building a home web and a feeding net. However, all these are still open communities with social behavior.

In the true societies of some orders, such as Hymenoptera (honeybees and ants) and Isoptera (termites), a complex social life is necessary for the perpetuation of the species. Such societies are closed. In them all stages of the life cycle are involved, the communities are usually permanent, all activities are collective, and there is reciprocal communication. There is a high degree of efficiency in the division of labor. Such a society is essentially a family group in which the mother or perhaps both parents remain with the young, sharing the duties of the group in a cooperative manner. The society usually demonstrates polymorphism, or **caste** differentiation.

Honeybees have one of the most complex social organizations in the insect world. Instead of lasting one season, their organization continues for a more or less indefinite period. As many as 60,000 to 70,000 honeybees may live in a single hive. Of these, there are three castes—a single sexually mature female, or **queen,** a few hundred **drones,** which are sexually mature males, and thousands of **workers,** which are sexually inactive genetic females (Figure 23-51).

Workers take care of the young, secrete wax with which they build the six-sided cells of the honeycomb, gather the nectar from flowers, manufacture honey, collect pollen, and ventilate and guard the hive. One drone, sometimes more, fertilizes the queen during the mating flight, at which time enough sperm are stored in her seminal receptacle to last her lifetime.

Castes are determined partly by fertilization and partly by what is fed to the larvae. Drones develop from unfertilized eggs (and consequently are haploid); queens and workers develop from fertilized eggs (and thus are diploid; see haplodiploidy, p. 312). Female larvae that will become queens are fed **royal jelly,** a secretion from the salivary glands of the nurse workers. Royal jelly differs from the "worker jelly" fed to ordinary larvae, but the components in it that are essential for queen determination have not yet been identified. Honey and pollen are added to the worker diet about the third day of larval life. Pheromones in the "queen substance," which is produced by the queen's mandibular glands, prevent the female workers from maturing sexually. Workers produce royal jelly only when the level of "queen substance" pheromone in the colony drops. This drop occurs when the queen becomes too old, dies, or is removed. Then the workers start enlarging a larval cell and feeding the larva the royal jelly that produces a new queen.

Honeybees have evolved an efficient system of communication by which, through certain body movements, their scouts inform the workers of the location and quantity of food sources (p. 302).

Termite colonies contain several castes, consisting of fertile individuals, both males and females, and sterile individuals (Figure 23-52). Some of the fertile individuals may have wings and may leave the colony, mate, lose their wings, and as **king**

figure 23-51

Queen bee surrounded by her court. The queen is the only egg layer in the colony. The attendants, attracted by her pheromones, constantly lick her body. As food is transferred from these bees to others, the queen's presence is communicated throughout the colony.

figure 23-52

A, Termite workers, *Reticulitermes flavipes* (order Isoptera), eating yellow pine. Workers are wingless sterile adults that tend the nest and care for the young. **B,** The termite queen becomes a distended egg-laying machine. The queen and several workers and soldiers are shown here.

figure 23-53

A weaver ant nest in Australia.

and **queen** start a new colony. Wingless fertile individuals may under certain conditions substitute for the king or queen. Sterile members are wingless and become **workers** and **soldiers.** Soldiers have large heads and mandibles and serve for the defense of the colony. As in bees and ants, extrinsic factors cause caste differentiation. Reproductive individuals and soldiers secrete inhibiting pheromones that pass throughout the colony to the nymphs through a mutual feeding process, called **trophallaxis,** so that they become sterile workers. Workers also produce pheromones, and if the level of "worker substance" or "soldier substance" falls, as might happen after an attack by marauding predators, for example, the next generation produces compensating proportions of the appropriate caste.

Ants also have highly organized societies. Superficially, they resemble termites, but they are quite different (belong to a different order) and can be distinguished easily. In contrast to termites, ants are usually dark in color, are hard bodied, and have a constriction between the thorax and abdomen.

In ant colonies males die soon after mating and the queen either starts her own new colony or joins some established colony and does the egg laying. The sterile females are wingless workers and soldiers that do the work of the colony—gather food, care for the young, and protect the colony. In many larger colonies there may be two or three types of individuals within each caste.

Ants have evolved some striking patterns of "economic" behavior, such as making slaves, farming fungi, herding "ant cows" (aphids or other homopterans, Figure 23-50), sewing their nests together with silk (Figure 23-53), and using tools.

Insects and Human Welfare

Beneficial Insects Although most of us think of insects primarily as pests, humanity would have great difficulty in surviving if all insects were suddenly to disappear. Insects are necessary for the cross-fertilization of many crops. Bees pollinate almost $10 billion worth of food crops per year in the United States alone, and this value does not include pollination of forage crops for livestock or pollination by other insects. In addition, some insects produce useful materials: honey and beeswax from bees, silk from silkworms, and shellac from a wax secreted by the lac insects.

Very early in their evolution insects and flowering plants formed a relationship of mutual adaptations that have been to each other's advantage. Insects exploit flowers for food, and flowers exploit insects for pollination. Each floral development of petal and sepal arrangement is correlated with the sensory adjustment of certain pollinating insects. Among these mutual adaptations are amazing devices of allurements, traps, specialized structure, and precise timing.

Many predaceous insects, such as tiger beetles, aphid lions, ant lions, praying mantids, and ladybird beetles, destroy harmful insects (Figure 23-54A, B). Some insects control harmful ones by parasitizing them or by laying their eggs where their young, when hatched, may devour the host (Figure 23-54C). Dead animals are quickly consumed by maggots hatched from eggs laid on carcasses.

Insects and their larvae serve as an important source of food for many birds, fish, and other animals.

Harmful Insects Harmful insects include those which eat and destroy plants and fruits, such as grasshoppers, chinch bugs, corn borers, boll weevils, grain weevils, San Jose scale, and scores of others (Figure 23-55). Practically every cultivated crop has some insect pest. Lice, bloodsucking flies, warble flies, botflies, and many others attack humans or domestic animals or both. Malaria, carried by the *Anopheles* mosquito (Figure 23-56), is still one of the world's worst killers; mosquitos also transmit yellow fever and filariasis. Fleas carry plague, which at many times in history has almost wiped out whole human populations. Houseflies are the vector for typhoid and lice for typhus fever; tsetse flies carry African sleeping sickness; and certain bloodsucking bugs are carriers of Chagas' disease. In addition there is a tremendous destruction of food, clothing, and property by weevils, cockroaches, ants, clothes moths, termites, and carpet beetles. Not the least of the insect pests are bedbugs, *Cimex,* bloodsucking hemipterous insects that humans contracted, probably early in their evolution, from bats that shared their caves.

A

B

C

figure 23-54

Some beneficial insects. **A,** A predaceous stink bug (order Hemiptera) feeds on a caterpillar. Note the sucking proboscis of the bug. **B,** A ladybird beetle ("ladybug," order Coleoptera). Adults (and larvae of most species) feed voraciously on plant pests such as mites, aphids, scale insects, and thrips. **C,** A parasitic wasp (*Larra bicolor*) attacking a mole cricket. The wasp drives the cricket from its burrow, then stings and paralyzes it. After the wasp deposits her eggs, the mole cricket recovers and resumes an active life—until it is killed by the developing wasp larvae.

A

B

C

figure 23-55

Insect pests. **A,** Japanese beetles, *Popillia japonica* (order Coleoptera) are serious pests of fruit trees and ornamental shrubs. They were introduced into the United States from Japan in 1917. **B,** Longtailed mealybug, *Pseudococcus longispinus* (order Homoptera). Many mealybugs are pests of commercially valuable plants. **C,** Corn ear worms, *Heliothis zea* (order Lepidoptera). An even more serious pest of corn is the infamous corn borer, an import from Europe in 1908 or 1909.

Control of Insects Because all insects are an integral part of the ecological communities to which they belong, their total destruction would probably do more harm than good. Food chains would be disturbed, some of our most loved birds would disappear, the biological cycles by which dead animal and plant matter disintegrates and returns to enrich the soil would be seriously impeded, and we would lose many flowering plants, including many food-crop plants. We often have overlooked the beneficial role of insects in our environment, and in our zeal to control the pests we have indiscriminately

Classification of Class Insecta

Insects are divided into orders chiefly on the basis of morphology and developmental features. Entomologists do not all agree on the names of the orders or on the limits of each order. Some tend to combine and others to divide the groups. However, the following synopsis of the major orders is one that is rather widely accepted.

Order Protura (pro-tu′ra) (Gr. *protos,* first, + *oura,* tail). Minute (1 to 1.5 mm); no eyes or antennae; appendages on abdomen as well as thorax; live in soil and dark, humid places; slight, gradual metamorphosis.

Order Diplura (dip-lu′ra) (Gr. *diploos,* double, + *oura,* tail): **japygids.** Usually less than 10 mm; pale, eyeless; a pair of long terminal filaments or pair of caudal forceps; live in damp humus or rotting logs; development direct.

Order Collembola (col-lem′bo-la) (Gr. *kolla,* glue, + *embolon,* peg, wedge): **springtails** and **snow fleas.** Small (5 mm or less); no eyes; respiration by trachea or body surface; a springing organ folded under the abdomen for leaping; abundant in soil; sometimes swarm on pond surface film or on snowbanks in spring; development direct.

Order Thysanura (thy-sa-nu′ra) (Gr. *thysanos,* tassel, + *oura,* tail): **silverfish** (Figure 23-57) and **bristletails.** Small to medium size; large eyes; long antennae; three long terminal cerci; live under stones and leaves and around human habitations; development direct.

Order Ephemeroptera (e-fem-er-op′ter-a) (Gr. *ephēmeros,* lasting but a day, + *pteron,* wing): **mayflies** (Figure 23-58). Wings membranous; forewings larger than hindwings; adult mouthparts vestigial; nymphs aquatic, with lateral tracheal gills, hemimetabolous development.

Order Odonata (o-do-na′ta) (Gr. *odontos,* tooth, + *ata,* characterized by): **dragonflies** (Figure 23-46B), **damselflies.** Large; membranous wings are long, narrow, net veined, and similar in size; long and slender body; aquatic nymphs with aquatic gills and prehensile labium for capture of prey; hemimetabolous development.

Order Orthoptera (or-thop′ter-a) (Gr. *orthos,* straight, + *pteron,* wing): **grasshoppers, locusts, crickets, cockroaches, walkingsticks, praying mantids** (Figures 23-35 and 23-36). Wings when present, with forewings thickened and hindwings folded like a fan under forewings; chewing mouthparts; hemimetabolous development.

Order Isoptera (i-sop′ter-a) (Gr. *isos,* equal, + *pteron,* wing): **termites** (Figure 23-52). Small; membranous, narrow wings similar in size with few veins; wings shed at maturity; erroneously called "white ants"; distinguishable from true ants by broad union of thorax and abdomen; complex social organization; hemimetabolous development.

Order Mallophaga (mal-lof′a-ga) (Gr. *mallos,* wool, + *phagein,* to eat): **biting lice** (Figure 23-59). As large as 6 mm; wingless; chewing mouthparts; legs adapted for clinging to host; live on birds and mammals; hemimetabolous development.

Order Anoplura (an-o-plu′ra) (Gr. *anoplos,* unarmed, + *oura,* tail): **sucking lice** (Figure 23-60). Depressed body; as large as 6 mm; wingless; mouthparts for piercing and sucking; adapted for clinging to warm-blooded host; includes the head louse, body louse, crab louse, others; hemimetabolous development.

Order Hemiptera (he-mip′ter-a) (Gr. *hemi,* half + *pteron,* wing) **(Heteroptera): true bugs** (Figure 23-54A). Size 2 to 100 mm; wings present or absent; forewings with basal portion leathery, apical portion membranous; hindwings membranous; at rest, wings held flat over abdomen; piercing-sucking mouthparts; many with odorous scent

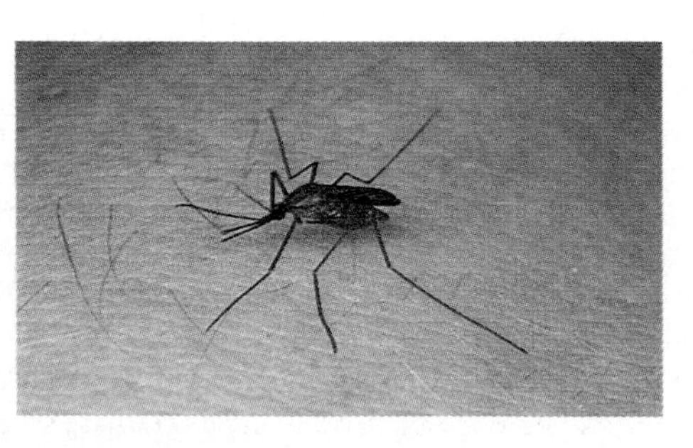

figure 23-56

A mosquito, *Anopheles quadrimaculatus* (order Diptera). *Anopheles* spp. are vectors of malaria.

figure 23-57

Silverfish *Lepisma* (order Thysanura) is often found in homes.

glands; include water scorpions, water striders, bedbugs, squash bugs, assassin bugs, chinch bugs, stinkbugs, plant bugs, lace bugs, others; hemimetabolous.

Order Homoptera (ho-mop′ter-a) (Gr. *homos,* same, + *pteron,* wing): **cicadas** (Figure 23-47), **aphids** (Figure 23-50), **scale insects, mealybugs** (Figure 23-55B), **leafhoppers, treehoppers** (Figure 23-61). (Often included as suborder under Hemiptera.) If winged, either membranous or thickened forewings and membranous hindwings; wings held rooflike over body; piercing-sucking mouthparts; all plant eaters; some destructive; a few serving as source of shellac, dyes, etc.; some with complex life histories; hemimetabolous.

Order Neuroptera (neu-rop′ter-a) (Gr. *neuron,* nerve, + *pteron,* wing): **dobsonflies, ant lions** (Figure 23-62), **lacewings.** Medium to large size; similar, membranous wings with many cross veins; chewing mouthparts; dobsonflies with greatly enlarged mandibles in males, and with aquatic larvae; ant lion larvae (doodlebugs) make craters in sand to trap ants; holometabolous development.

Order Coleoptera (ko-le-op′ter-a) (Gr. *koleos,* sheath, + *pteron,* wing): **beetles** (Figures 23-54B; 23-55A), **fireflies** (Figure 23-49), **weevils.** The largest order of animals in the world; forewings (elytra) thick, hard, opaque; membranous hindwings folded under forewings at rest; mouthparts for biting and chewing; includes ground beetles, carrion beetles, whirligig beetles, darkling beetles, stag beetles, dung beetles (Figure 23-48), diving beetles, boll weevils, fireflies, ladybird beetles (ladybugs), others; holometabolous.

Order Lepidoptera (lep-i-dop′ter-a) (Gr. *lepidos,* scale, + *pteron,* wing): **butterflies** and **moths** (Figures 23-42 and 23-55C). Membranous wings covered with overlapping scales, wings coupled at base; mouthparts a sucking tube, coiled when not in use; larvae (caterpillars) with chewing mandibles for plant eating, stubby prolegs on the abdomen, and silk glands for spinning cocoons; antennae knobbed in butterflies and usually plumed in moths; holometabolous.

Order Diptera (dip′ter-a) (Gr. *dis,* two, + *pteron,* wing): **true flies.** Single pair of wings, membranous and narrow; hindwings reduced to inconspicuous balancers (halteres); sucking mouthparts or adapted for sponging or lapping or piercing; legless larvae called maggots or, when aquatic, wigglers; include crane flies, mosquitos (Figure 23-56), moth flies, midges, fruit flies, flesh flies, houseflies, horseflies (Figure 23-37), botflies, blowflies, gnats, and many others; holometabolous.

Order Trichoptera (tri-kop′ter-a) (Gr. *trichos,* hair, + *pteron,* wing): **caddis flies.** Small, soft bodied; wings well-veined and hairy, folded rooflike over hairy body; chewing mouthparts; aquatic larvae construct cases of leaves, sand, gravel, bits of shell, or plant matter, bound together with secreted silk or cement; some make silk feeding nets attached to rocks in streams; holometabolous.

Order Siphonaptera (si-fon-ap′ter-a) (Gr. *siphon,* a siphon, + *apteros,* wingless): **fleas** (Figure 23-63). Small; wingless; bodies laterally compressed; legs adapted for leaping; no eyes; ectoparasitic on birds and mammals; larvae legless and scavenger; holometabolous.

Order Hymenoptera (hi-men-op′ter-a) (Gr. *hymen,* membrane, + *pteron,* wing): **ants, bees** (Figure 23-51), **wasps.** Very small to large; membranous, narrow wings coupled distally; subordinate hindwings; mouthparts for biting and lapping up liquids; ovipositor sometimes modified into stinger, piercer, or saw; both social and solitary species; most larvae legless, blind, and maggotlike; holometabolous.

figure 23-58

Mayfly (order Ephemeroptera). **A,** Nymph. **B,** Adult.

sprayed the landscape with extremely effective "broad-spectrum" insecticides that eradicate beneficial, as well as harmful, insects. We have also found, to our chagrin, that many of the chemicals we have used persist in the environment and accumulate as residues in the bodies of animals higher up in food chains. Also, many strains of insects have developed resistance to insecticides in common use.

In recent years, methods of control other than chemical insecticides have been under intense investigation and experimentation. Economics, concern for the environment, and consumer demand are causing thousands of farmers across the United States to use alternatives to strict dependence on chemicals.

their cuticular exoskeleton and their jointed appendages, thus resulting in a wide variety of locomotor and feeding adaptations. Whether it be in the area of habitat, feeding adaptations, means of locomotion, reproduction, or general mode of living, the adaptive achievements of the arthropods are truly remarkable.

Summary

The Arthropoda is the largest, most abundant and diverse phylum in the world. Arthropods are metameric, coelomate protostomes with well-developed organ systems. Most show marked tagmatization. They are extremely diverse and occur in all habitats capable of supporting life. Perhaps more than any other single factor, the success of the arthropods is explained by the adaptations made possible by their cuticular exoskeleton. Other important elements in their success are jointed appendages, tracheal respiration, efficient sensory organs, complex behavior, metamorphosis, and the ability to fly.

Members of the subphylum Chelicerata have no antennae, and their main feeding appendages are chelicerae. In addition, they have a pair of pedipalps (which may be similar to walking legs) and four pairs of walking legs. The great majority of living chelicerates are in the class Arachnida: spiders (order Araneae), scorpions (order Scorpionida), harvestmen (order Opiliones), and ticks and mites (order Acarina).

The tagmata of spiders (cephalothorax and abdomen) show no external segmentation and join by a waistlike pedicel. Chelicerae of spiders have poison glands for paralyzing or killing their prey. Spiders can spin silk, which they use for a variety of purposes.

The cephalothorax and abdomen of ticks and mites are completely fused, and the anterior capitulum bears the mouthparts. Ticks and mites are the most numerous of any arachnids; some are important disease carriers, and others are serious plant pests.

The Crustacea is a large, primarily aquatic subphylum of arthropods. Crustaceans bear two pairs of antennae, mandibles, and two pairs of maxillae on the head. Their appendages are primitively biramous, and the major tagmata are the head, thorax, and abdomen. Many have a carapace and respire by means of gills.

All arthropods must periodically cast off their cuticle (ecdysis) and grow larger before the newly secreted cuticle hardens. Premolt and postmolt periods are hormonally controlled, as are several other structures and functions.

There are many predators, scavengers, filter feeders, and parasites among the Crustacea. Respiration is through the body surface or by gills, and excretory organs take the form of maxillary or antennal glands. Circulation, as in other arthropods, is through an open system of sinuses (hemocoel), and a dorsal, tubular heart is the chief pumping organ. Most crustaceans have compound eyes composed of units called ommatidia.

Members of the class Maxillopoda, subclass Copepoda, lack a carapace and abdominal appendages. They are abundant and are among the most important of the primary consumers in many freshwater and marine ecosystems. The Malacostraca are the largest crustacean class, and the most important orders are the Isopoda, Amphipoda, Euphausiacea, and Decapoda. All have both abdominal and thoracic appendages. Decapods include crabs, shrimp, lobster, crayfish, and others; they have five pairs of walking legs (including the chelipeds) on their thorax.

Members of the subphylum Uniramia have uniramous appendages and bear one pair of antennae, a pair of mandibles, and two pairs of maxillae (one pair of maxillae in millipedes) on the head. The tagmata are the head and trunk in the myriapods and head, thorax, and abdomen in the insects.

The Insecta is the largest class of the world's largest phylum. Insects are easily recognized by the combination of their tagmata and the possession of three pairs of thoracic legs.

The radiation and abundance of insects are largely explained by several features allowing them to exploit terrestrial habitats, such as waterproof cuticle and other mechanisms to minimize water loss and the ability to become dormant during adverse conditions.

Feeding habits vary greatly among insects, and there is an enormous variety of specialization of mouthparts reflecting the particular feeding habits of a given insect. Insects breathe by means of a tracheal system, which is a system of tubes that opens by spiracles on the thorax and abdomen. Excretory organs are malpighian tubules.

Sexes are separate in insects, and fertilization is usually internal. Almost all insects undergo metamorphosis during development. In hemimetabolous (gradual) metamorphosis, the larval instars are called nymphs, and the adult emerges at the last nymphal molt. In holometabolous (complete) metamorphosis, the last larval molt gives rise to a nonfeeding stage (pupa). A winged adult emerges at the final, pupal, molt. Both types of metamorphosis are hormonally controlled.

Insects are important to human welfare, particularly because they pollinate food crop plants, control populations of other, harmful insects by predation and parasitism, and serve as food for other animals. Many insects are harmful to human interests because they feed on crop plants, and many are carriers of important diseases affecting humans and domestic animals.

Arthropods share a common ancestor with annelids, and many zoologists believe that the Arthropoda is monophyletic. Present evidence suggests that the crustaceans and uniramians form a mandibulate clade. The endognathous insects have a number of primitive characters and show similarities with the myriapods. Wings, hemimetabolous metamorphosis, and holometabolous metamorphosis evolved among the ectognathous insects.

Adaptive radiation of the arthropods has been enormous, and they are extremely abundant.

Review Questions

1. Give the characteristics of arthropods that most clearly distinguish them from the Annelida.
2. Name the subphyla of arthropods, and give a few examples of each.
3. Much of the success of the arthropods has been attributed to their cuticle. Why do you think this is so? Describe some other factors that probably contributed to their success.
4. What is a trilobite?
5. What appendages are characteristic of chelicerates?
6. Briefly describe the appearance of each of the following: eurypterids, horseshoe crabs, pycnogonids.
7. Tell the mechanism of each of the following with respect to spiders: feeding, excretion, sensory reception, webspinning, reproduction.
8. Distinguish each of the following orders from each other: Araneae, Scorpionida, Opiliones, Acarina.
9. People fear spiders and scorpions, but ticks and mites are far more important medically and economically. Why? Give examples.
10. What are the tagmata and the appendages on the head of crustaceans? What are some other important characteristics of Crustacea?
11. Of the classes of Crustacea, the Branchiopoda, Maxillopoda, and Malacostraca are the most important. Distinguish them from each other.
12. Distinguish among the subclasses Ostracoda, Copepoda, Branchiura, and Cirripedia of the crustacean class Maxillopoda.
13. Copepods sometimes have been called "insects of the sea" because marine planktonic copepods probably are the most abundant animals in the world. What is their ecological importance?
14. Define each of the following: swimmeret, endopod, exopod, maxilliped, cheliped, uropod, nauplius.
15. Describe the molting process in Crustacea, including the action of the hormones.
16. Explain the mechanism of each of the following with respect to crustaceans: feeding, respiration, excretion, circulation, sensory reception, reproduction.
17. Distinguish the following from each other: Diplopoda, Chilopoda, Insecta.
18. Define each of the following with respect to insects: sclerite, notum, tergum, sternum, pleura, labrum, labium, hypopharynx, haltere, instar, diapause.
19. Explain why wings powered by indirect flight muscles can beat much more rapidly than those powered by direct flight muscles.
20. What different modes of feeding are found in insects, and how are these reflected in their mouthparts?
21. Describe each of the following with respect to insects: respiration, excretion and water balance, sensory reception, reproduction.
22. Explain the difference between holometabolous and hemimetabolous metamorphosis in insects, including the stages in each.
23. Describe and give an example of each of the four ways insects communicate with each other.
24. What are the castes found in honeybees and in termites, and what is the function of each? What is trophallaxis?
25. Name several ways in which insects are beneficial to humans and several ways they are detrimental.
26. For the past 50 or more years, people have relied on toxic insecticides for control of harmful insects. What problems have arisen resulting from such reliance on insecticides? What are the alternatives? What is integrated pest management?
27. Some biologists suggest that the Arthropoda is polyphyletic. Explain why this could be so despite the characteristics shared by all arthropods.
28. We believe that the earliest insects were wingless, that is, the lack of wings is the primitive condition, and this was observed in the traditional subclass Apterygota. We now consider the Apterygota paraphyletic. Why?

Selected References

See also general references for Part Three, p. 562.

Berenbaum, M. R. 1995. Bugs in the system. Reading, Massachusetts, Addison-Wesley Publishing Company. *How insects impact human affairs. Well written for a wide audience, highly recommended.*

Blum, M. S., ed. 1985. Fundamentals of insect physiology. New York, John Wiley & Sons. *Good, multiauthored text on insect physiology. Recommended.*

Borror, D. J., D. M. Delong, and C. A. Triplehorn. 1989. An introduction to the study of insects, ed. 6. Philadelphia, Saunders College Publishing. *A good entomology text.*

Cronin, T. W., N. J. Marshall, and M. F. Land. 1994. The unique visual system of the mantis shrimp. Am. Sci. **82:**356–365. *The ancestors of mantis shrimps diverged from other crustaceans about 400 million years ago. Accuracy in the raptorial strike of these aggressive predators requires a highly refined visual system.*

Foelix, R. F. 1982. Biology of spiders. Cambridge, Massachusetts, Harvard University Press. *Attractive, comprehensive book with extensive references; of interest to both amateurs and professionals.*

Hadley, N. F. 1986. The arthropod cuticle. Sci. Am. **234:**100–107 (March). *Modern studies on the chemistry and structure of arthropod cuticle help to explain its remarkable properties.*

Heinrich, B., and H. Esch. 1994. Thermoregulation in bees. Am. Sci. **82:**164–170. *Fascinating behavioral and physiological adaptations for increasing and decreasing body temperature allow bees to function in a surprisingly wide range of environmental temperatures.*

Hölldobler, B. K., and E. O. Wilson. 1990. The ants. Cambridge, Massachusetts, Harvard University Press. *The fascinating story of social organization in ants.*

Kaston, B. J. 1978. How to know the spiders, ed. 3. Dubuque, Iowa, William C. Brown Company, Publishers. *Spiral-bound identification manual.*

Lane, R. P., and R. W. Crosskey, eds. 1993. Medical insects and arachnids. London, Chapman and Hall. *This is the best book currently available on medical entomology.*

McDaniel, B. 1979. How to know the ticks and mites. Dubuque, Iowa, William C. Brown Company, Publishers. *Useful, well-illustrated keys to genera and higher categories of ticks and mites in the United States.*

McMasters, J. H. 1989. The flight of the bumblebee and related myths of entomological engineering. Am. Sci. **77:**164-169. *There is a popular myth about an aerodynamicist who "proved" that a bumblebee cannot fly—but his assumptions were wildly wrong.*

Moffat, A. S. 1991. Research on biological pest control moves ahead. Science **252:**211–212. *A report on the current status of biological pest control, including the contributions of genetic engineering.*

Polis, G. A., ed. 1990. The biology of scorpions. Stanford, California, Stanford University Press. *The editor brings together a readable summary of what is known about scorpions.*

Schram, F. R. 1986. Crustacea. New York, Oxford University Press. *The most recent comprehensive account.*

Shear, W. A. 1994. Untangling the evolution of the web. Am. Sci. **82:**256–266. *Fossil spider webs are nonexistent. Evolution of the web must be studied by comparing modern spider webs to each other and correlating studies of spider anatomy.*

Topoff, H. 1990. Slave-making ants. Am. Sci. **78:**520–528. *An amazing type of social parasitism in which certain species of ants raid the colonies of related species, abduct their pupae, then exploit them to do all the work in the host colony.*

Vollrath, F. 1992. Spider webs and silks. Sci. Am. **266:**70–76 (Mar.). *Spider web design and silk must obey the same constraints as materials used in human structural engineering; we can learn useful lessons from spiders.*

Wooley, T. A. 1988. Acarology. Mites and human welfare. New York, John Wiley.

Wootton, R. J. 1990. The mechanical design of insect wings. Sci. Am. **263:**114–120 (Nov.). *The ingenious architecture of insect wings and how they are adapted to flight.*

Lesser Protostomes and Lophophorates

chapter | twenty-four

Some Evolutionary Experiments

The early Cambrian period, about 570 million years ago, was the most fertile time in evolutionary history. For 3 billion years before this period, evolution had forged little more than bacteria and blue-green algae. Then, within the space of a few million years, all of the major phyla, and probably all of the smaller phyla, became established. This was the Cambrian explosion, the greatest evolutionary "bang" the world has known. In fact, the fossil record suggests that more phyla existed in the Paleozoic era than exist now, but some disappeared during major extinction events that punctuated the evolution of life on earth. The greatest of these disruptions was the Permian extinction about 230 million years ago. Thus evolution has led to many "experimental models." Some of these models failed because they were unable to survive in changing conditions. Others gave rise to abundant and dominant species and individuals that inhabit the world today. Still others radiated but little, with small numbers of species persisting, whereas others were formerly more abundant but are now in decline.

The great evolutionary flow that began with the appearance of the coelom and led to the three huge phyla of molluscs, annelids, and arthropods produced other lines as well. Those that have survived are small and lack great economic and ecological importance; they are usually grouped together as "lesser protostomes." They probably diverged at different times from different ancestors, but in all likelihood each shares a common ancestor with annelids or arthropods or both.

Three phyla—Phoronida, Ectoprocta, and Brachiopoda—are included in this chapter mainly for convenience. They are apparently related to each other because they all possess a crown of cilated tentacles, called a lophophore, used in food capture and respiration. The brachiopods were abundant in the Paleozoic but began to decline thereafter. The exception to the common theme of this chapter is the phylum Ectoprocta. It arose in the Cambrian, became widespread in the Paleozoic, and remains a prevalent group today.

This chapter includes brief discussions of nine coelomate phyla whose position in the phylogenetic lines of the animal kingdom is somewhat problematic. Other than having derived from the main protostome line, we can say little of the relationships of six of the phyla. The Sipuncula, Echiura, and Pogonophora have some annelid-like characters. The Pentastomida, Onychophora, and Tardigrada show some arthropod-like characteristics; they often have been grouped together and called Pararthropoda because they have unjointed limbs with claws (at some stage) and a cuticle that undergoes molting. The relationships of the lophophorate phyla are also obscure, but cladistic analysis places them with the deuterostomes.

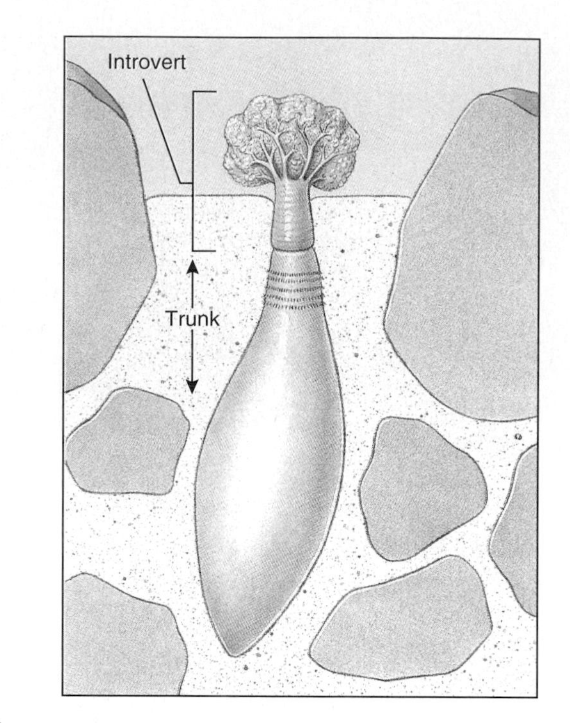

figure 24-1

Themiste, a sipunculan.

The Lesser Protostomes

Three of the lesser protostome phyla are benthic (bottom-dwelling) marine worms. The Pentastomida is an entirely parasitic group, and the onychophorans are terrestrial (but limited to damp areas). Tardigrades are found in marine, freshwater, and terrestrial habitats.

Phylum Sipuncula

The phylum Sipuncula (sy-pun′kyu-la) (L. *sipunculus,* little siphon) consists of about 330 species of benthic marine worms, predominantly littoral or sublittoral. Sometimes called "peanut worms," they live sedentary lives in burrows in mud or sand (Figure 24-1), occupy borrowed snail shells, or live in coral crevices or among vegetation. Some species construct their own rock burrows by chemical and perhaps mechanical means. More than half the species are restricted to tropical zones. Some are tiny, slender worms, but the majority range from 15 to 30 cm in length.

Sipunculans are not metameric, nor do they possess setae. Their head is in the form of an **introvert,** which is crowned by ciliated tentacles surrounding the mouth (Figure 24-1). They are largely deposit feeders, extending the introvert and tentacles from their burrow to explore and feed. They have a cerebral ganglion, nerve cord, and pair of nephridia; the coelomic fluid contains red blood cells bearing a respiratory pigment known as hemerythrin.

Sipunculan larvae are trochophores (Figure 21-5, p. 453), and their early embryological development indicates affinities to the Annelida and Echiura. They appear to have diverged from a common ancestor of the three phyla before metamerism evolved.

Phylum Echiura

The phylum Echiura (ek-ee-yur′a) (Gr. *echis,* viper, + *oura,* tail) consists of marine worms that burrow into mud or sand or live in empty snail shells, sand dollar tests, or rocky crevices. They are found in all oceans, most commonly in littoral zones of warm waters. They vary in length from a few millimeters to 40 to 50 cm.

Although there are only about 140 species, echiurans are more diverse than sipunculans. Their bodies are cylindrical. Anterior to the mouth is a flattened, extensible proboscis, which, unlike the introvert of sipunculans, cannot be retracted into the trunk. Echiurans are often called "spoonworms" because of the shape of the contracted proboscis in some of them. The proboscis has a ciliated groove leading to the mouth. While the animal lies buried, the proboscis can extend out over the mud for exploration and deposit feeding (Figure 24-2). *Urechis* (Gr. *oura,* tail, + *echis,* viper), however, secretes a mucous net in a U-shaped burrow through which it pumps water and strains out food particles. *Urechis* is sometimes called the "fat innkeeper" because it has characteristic species of commensals living with it in its burrow, including a crab, fish, clam, and polychaete annelid.

Echiurans, with the exception of *Urechis,* have a **closed circulatory system** with a contractile vessel; most have one to three pairs of nephridia (some have many pairs), and all have a nerve ring and ventral nerve cord. A pair of anal sacs arises from the rectum and opens into the coelom; they are probably respiratory in function and possibly accessory nephridial organs.

Early cleavage and trochophore stages are very similar to those of annelids and sipunculans.

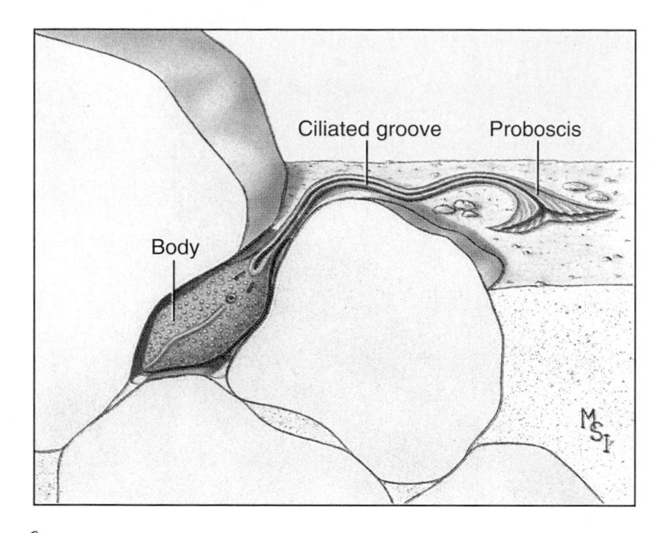

figure 24-2

Bonellia (phylum Echiura) is a detritus feeder. Lying in its burrow, it explores the surface with its long proboscis, which picks up organic particles and carries them along a ciliated groove to the mouth.

In some species of echiurans sexual dimorphism is pronounced, with the female being much the larger of the two. Bonellia *has an extreme sexual dimorphism, and sex is determined in a very interesting way. At first the free-swimming larvae are sexually undifferentiated. Those that come into contact with the proboscis of a female become tiny males (1 to 3 mm long) that migrate to the female uterus. About 20 males are usually found in a single female. Larvae that do not contact a female proboscis metamorphose into females. The stimulus for development into males is apparently a pheromone produced by the female proboscis.*

Phylum Pogonophora

The phylum Pogonophora (po-go-nof′e-ra) (Gr. *pōgōn,* beard, + *pherō,* to bear), or beardworms, was entirely unknown before the twentieth century. The first specimens to be described were collected from deep-sea dredgings in 1900 off the coast of Indonesia. They have since been discovered in several seas, including the western Atlantic off the eastern coast of the United States. Some 80 species have been described so far.

Most pogonophores live in the bottom ooze on the ocean floor, usually at depths of more than 200 m. Their usual length varies from 5 to 85 cm, with a diameter usually of less than a millimeter. They are sessile and secrete very long chitinous tubes in which they live, probably extending the anterior end only for feeding.

The body has a short forepart, a long, very slender trunk, and a small, metameric opisthosoma (Figure 24-3). They are covered with a cuticle similar in structure to that of annelids and sipunculans, and they bear setae on the trunk and opisthosoma similar to those of annelids. A series of coelomic compartments divides the body. The forepart bears from one to many tentacles.

Pogonophores are remarkable in having no mouth or digestive tract, making their mode of nutrition a puzzling matter. They absorb some nutrients dissolved in seawater, such as glucose, amino acids, and fatty acids, through the pinnules and microvilli of their tentacles. They apparently derive most of their energy, however, from a mutualistic association with chemoautotrophic bacteria. These bacteria oxidize hydrogen sulfide to provide the energy to produce organic compounds from carbon dioxide. An organ called the **trophosome** bears the bacteria. The trophosome develops embryonically from the midgut (all traces of the foregut and hindgut are absent in the adult).

There is a well-developed, closed, blood vascular system. Sexes are separate.

Pogonophores have photoreceptor cells very similar to those of annelids (oligochaetes and leeches), and the structure of the cuticle, the makeup of the setae, and the segmentation of the opisthosoma all point strongly toward a common ancestor with the annelids.

Among the most amazing animals found in the deep-water, Pacific rift communities (Chapter 5, p. 125) are the giant pogonophorans, Riftia pachyptila. *Much larger than any pogonophores reported before, they measure up to 1.5 m in length and 4 cm in diameter. Some authors consider them a separate phylum, the* ***Vestimentifera.*** *The trophosome of other pogonophores is confined to the posterior part of the trunk, which is buried in sulfide-rich sediments, but the trophosome of* Riftia *occupies most of its large trunk. It has a much larger supply of hydrogen sulfide, enough to nourish its large body, in the effluent of the hydrothermal vents.*

Phylum Pentastomida

The wormlike Pentastomida (pen-ta-stom′i-da) (Gr. *pente,* five, + *stoma,* mouth) are parasites, 2.5 to 12 cm long, that are found in the lungs and nasal passages of carnivorous vertebrates—most commonly in reptiles (Figure 24-4). Some human infections have been found in Africa and Europe. The intermediate host is usually a vertebrate that is eaten by the final host.

They have four clawlike appendages at their anterior end, and their body is covered by a chitinous cuticle, which they periodically molt during juvenile stages. Pentastomids

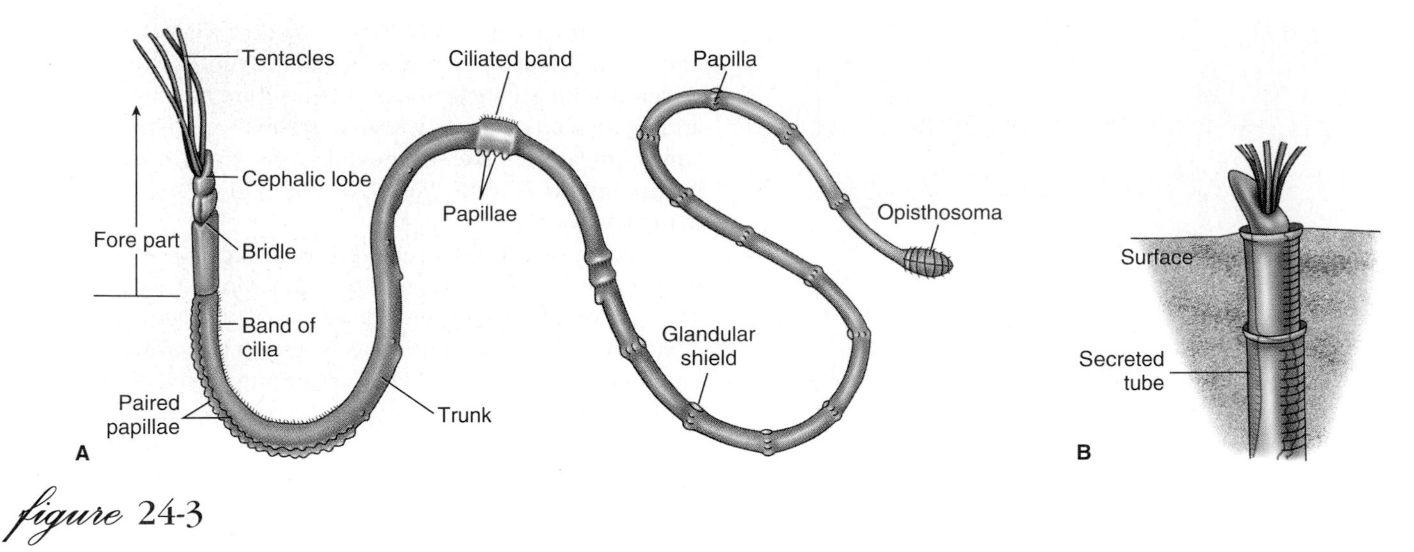

figure 24-3

Diagram of a typical pogonophoran. **A,** External features. The body, in life, is much more elongated than shown in this diagram. **B,** Position in its tube.

show arthropod affinities, but there is little agreement as to where they fit in that phylum. On the basis of the structure of their spermatozoa, pentastomids seem to be closest to the crustacean subclass Branchiura. Some authorities consider Pentastomida a subphylum of Arthropoda. Their extensive modifications for parasitic life make their ancestry difficult to determine.

Phylum Onychophora

Members of the phylum Onychophora (on-i-kof′o-ra) (Gr. *onyx,* claw, + *pherō,* to bear) are called "velvet worms" or "walking worms." They are about 70 species of caterpillar-like animals, 1.4 to 15 cm long, that live in rain forests and other tropical and semitropical leafy habitats.

The fossil record of the onychophorans shows that they have changed little in their 500-million-year history. They were originally marine animals and were probably far more common than they are now. Today they are all terrestrial and are extremely retiring, coming out only at night or when the air is nearly saturated with moisture.

Onychophorans are covered by a soft cuticle, which contains chitin and protein. Their wormlike bodies are carried on 14 to 43 pairs of stumpy, unjointed legs, each ending with a flexible pad and two claws (Figure 24-5). The head bears a pair of flexible antennae with annelid-like eyes at the base.

They are air breathers, using a **tracheal system** that connects with pores scattered over the body. The tracheal system, although similar to that of arthropods, probably evolved independently. Other arthropod-like characteristics include an open circulatory system with a tubular heart, a hemocoel for a body cavity, and a large brain. Annelid-like characteristics include segmentally arranged nephridia, a muscular body wall, and pigment-cup ocelli.

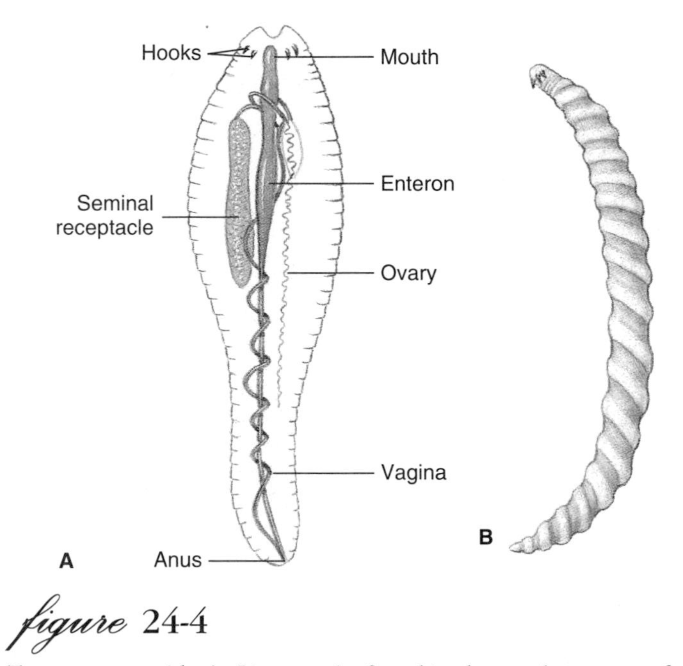

figure 24-4

Two pentastomids. **A,** *Linguatula,* found in the nasal passages of carnivorous mammals. The female is shown with some internal structures. **B,** Female *Armillifer,* a pentastomid with pronounced body rings. In parts of Africa and Asia, humans are parasitized by immature stages; adults (10 cm long or more) live in the lungs of snakes. Human infection may occur from eating snakes or from contaminated food and water.

Onychophorans are dioecious. In some species there is a placental attachment between mother and young, and the young are born as juveniles (viviparous); in others the young are also not encased in a shell when released from the mother, but they develop in the uterus without attachment (ovoviviparous). Two Australian genera are oviparous and lay shell-covered eggs in moist places.

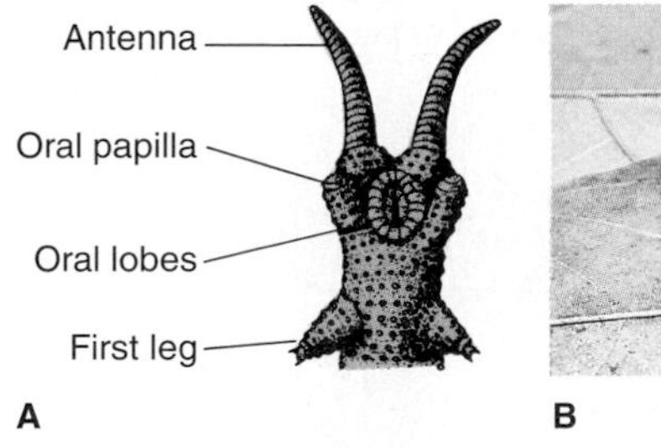

figure 24-5

Peripatus, a caterpillar-like onychophoran that has characteristics in common with both annelids and arthropods. **A,** Ventral view of head. **B,** In natural habitat.

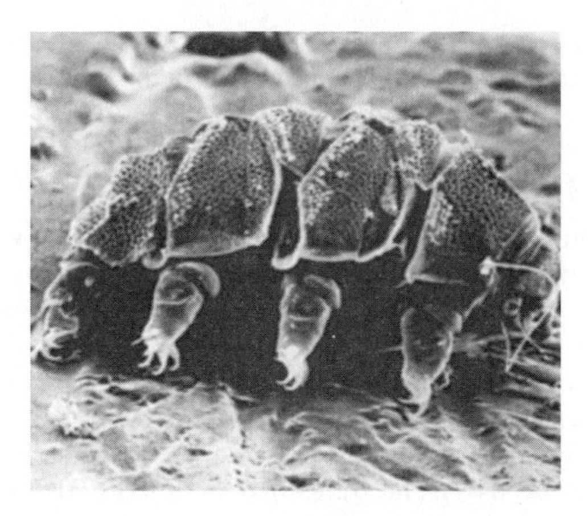

figure 24-6

Scanning electron micrograph of *Echiniscus maucci,* phylum Tardigrada. This species is 300 to 500 μm long. Unable to swim, it clings to moss or water plants with its claws, and if the environment dries up, it goes into a state of suspended animation and "sleeps away" the drought.

Phylum Tardigrada

Tardigrada (tar-di-gray′da) (L. *tardus,* slow, + *gradus,* step), or "water bears," are minute forms usually less than a millimeter in length. Most of the 300 to 400 species are terrestrial forms that live in the water film that surrounds mosses and lichens. Some live in fresh water, and a few are marine.

The body bears eight short, **unjointed legs,** each with claws (Figure 24-6). Unable to swim, they creep about awkwardly, clinging to the substrate with their claws. A pair of sharp stylets and a sucking pharynx adapt them for piercing and sucking plant cells or small prey such as nematodes and rotifers.

There is a body covering of nonchitinous **cuticle** that is molted several times during the life cycle. As in arthropods, muscle fibers are attached to the cuticular exoskeleton, and the body cavity is a hemocoel.

The annelid-type nervous system is surprisingly complex, and in some species there is a pair of eyespots. Circulatory and respiratory organs are lacking.

Females may deposit their eggs in the old cuticle as they molt or attach them to a substrate. Embryonic formation of the coelom is enterocoelous, a deuterostome characteristic. Nevertheless, their numerous arthropod-like characteristics strongly suggest a common ancestry with the Arthropoda (see Figure 24-13).

One of the most intriguing features of terrestrial tardigrades is their capacity to enter a state of suspended animation, called **cryptobiosis,** during which metabolism is virtually imperceptible. Under gradual drying conditions they reduce the water content of their body from 85% to only 3%, movement ceases, and the body becomes barrel shaped. In a cryptobiotic state tardigrades can withstand harsh environmental conditions: temperature extremes, ionizing radiation, oxygen deficiency, etc., and many survive for years. Activity resumes when moisture is again available.

The Lophophorates

The three lophophorate phyla may appear to have little in common. The **phoronids** (phylum Phoronida) are wormlike marine forms that live in secreted tubes in sand or mud or attached to rocks or shells. The **ectoprocts** (phylum Ectoprocta) are minute forms, mostly colonial, whose protective cases often form encrusting masses on rocks, shells, or plants. The **brachiopods** (phylum Brachiopoda) are bottom-dwelling marine forms that superficially resemble molluscs because of their bivalved shells. All have a free-swimming larval stage but are sessile as adults.

One may wonder why these three apparently widely different types of animals are considered together. They are all coelomate; all have some deuterostome and some protostome characteristics; all are sessile; and none has a distinct head. But these characteristics are also shared by other phyla. What really sets them apart from other phyla is the common possession of a ciliary feeding device called a **lophophore** (Gr. *lophos,* crest or tuft, + *phorein,* to bear).

A lophophore is a unique arrangement of ciliated tentacles borne on a ridge (a fold of the body wall), which surrounds the mouth but not the anus. The lophophore with its crown of tentacles contains within it an extension of the coelom, and the thin, ciliated walls of the tentacles not only constitute an efficient feeding device but also serve as a respiratory surface for exchange of gases between environmental water and coelomic fluid. The lophophore can usually be extended for feeding or withdrawn for protection.

In addition, all three phyla have a U-shaped alimentary canal, with the anus placed near the mouth but outside the lophophore. The coelom is divided primitively into three compartments, the **protocoel,** the **mesocoel** and the **metacoel,** and the mesocoel extends into the hollow tentacles of the lophophore. The protocoel, where present, forms a cavity in a flap over the mouth, the **epistome.** The portion of the body that contains the mesocoel is known as the **mesosome,** and that containing the metacoel is the **metasome.**

Phylum Phoronida

The phylum Phoronida (fo-ron′i-da) (Gr. *phoros,* bearing, + L. *nidus,* nest) comprises approximately 10 species of small wormlike animals that live on the bottom of shallow coastal waters, especially in temperate seas. The phylum name refers to the tentacled lophophore. Phoronids range from a few millimeters to 30 cm in length. Each worm secretes a leathery or chitinous tube in which it lies free, but which it never leaves (Figure 24-7). The tubes may be anchored singly or in a tangled mass on rocks, shells, or pilings or buried in the sand. The tentacles on the lophophore are thrust out for feeding, but if the animal is disturbed it can withdraw completely into its tube.

The lophophore has two parallel ridges curved in a horseshoe shape, the bend located ventrally and the mouth lying between the two ridges. The cilia on the tentacles direct a water current toward a groove between the two ridges, which leads toward the mouth. Plankton and detritus caught in this current become entangled in mucus and are carried by the cilia to the mouth.

Mesenteric partitions divide the coelomic cavity into proto-, meso-, and metacoel, similar to the compartments of deuterostomes. Phoronids have a closed system of contractile blood vessels but no heart; the red blood contains hemoglobin. There is a pair of metanephridia. A nerve ring sends nerves to the tentacles and body wall.

There are both monoecious (the majority) and dioecious species of Phoronida, and at least one species reproduces asexually. Cleavage seems to be related to both spiral and radial types.

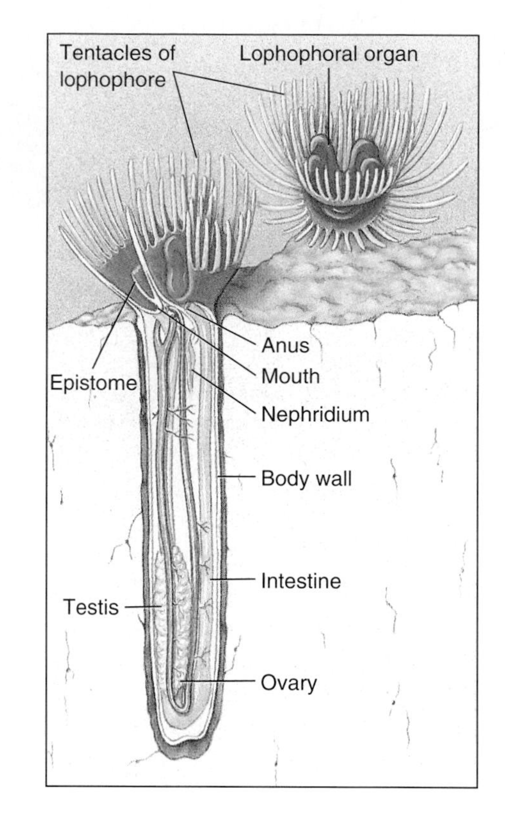

figure 24-7

Internal structure of *Phoronis* (phylum Phoronida), in diagrammatic vertical section.

Phylum Ectoprocta

The Ectoprocta (ek′to-prok′ta) (Gr. *ektos,* outside, + *proktos,* anus) have long been called bryozoans (Gr. *bryon,* moss, + *zoōn,* animal), or moss animals, a term that originally included the Entoprocta also.

Of the 4000 or so species of ectoprocts, few are more than 0.5 mm long; all are aquatic, both fresh water and marine, but they largely occur in shallow waters; and most, with very few exceptions, are colony builders. Ectoprocts, unlike the other phyla considered in this chapter, were abundant and widespread in the past and remain so today. They left a rich fossil record since the Ordovician era. Modern marine forms exploit all kinds of firm surfaces, such as shells, rock, large brown algae, mangrove roots, and ship bottoms. They are one of the most important groups of fouling organisms on boat hulls; they decrease efficiency of the hull passing through the water and make periodic scraping of the hull necessary.

Each member of a colony lives in a tiny chamber, called a **zoecium,** which is secreted by its epidermis (Figure 24-8A). Each individual, or **zooid,** consists of a feeding **polypide** and a case-forming **cystid.** The polypide includes the lophophore, digestive tract, muscles, and nerve centers. The cystid is the body wall of the animal, together with its secreted exoskeleton. The exoskeleton, or zoecium, may, according to the species, be gelatinous, chitinous, or stiffened with calcium and possibly also impregnated with sand. The shape may be boxlike, vaselike, oval, or tubular.

Some colonies form limy encrustations on seaweed, shells, and rocks (Figure 24-9); others form fuzzy or shrubby growths or erect, branching colonies that look like seaweed. Some ectoprocts might easily be mistaken for hydroids but can be distinguished under a microscope by the presence of ciliated tentacles (Figure 24-10). In some freshwater forms individuals are borne on finely branching stolons that form delicate tracings on the underside of rocks or plants. Other freshwater ectoprocts are embedded in large masses of gelatinous material. Although the zooids are minute, the colonies may be several centimeters in diameter; some encrusting colonies may be a meter or more in width, and erect forms may reach 30 cm or more in height.

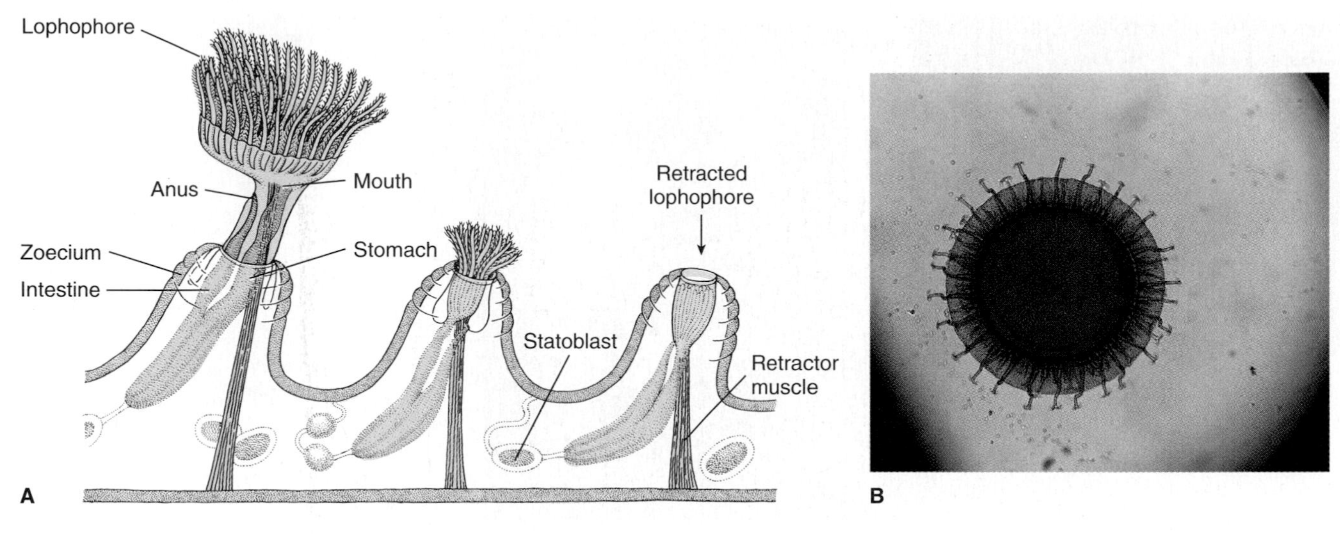

figure 24-8

A, Small portion of freshwater colony of *Plumatella* (phylum Ectoprocta), which grows on the underside of rocks. These tiny individuals disappear into their chitinous zoecia when disturbed. **B,** Statoblast of a freshwater ectoproct, *Cristatella.* Statoblasts are a kind of bud that survives over winter when the colony dies in the autumn. This one is about 1 mm in diameter and bears hooked spines.

The polypide lives in a type of jack-in-the-box existence, popping up to feed and then quickly withdrawing into its little chamber, which often has a tiny trapdoor (operculum) that shuts to conceal its inhabitant. To extend the tentacular crown, certain muscles contract, which increases the hydrostatic pressure within the body cavity and pushes out the lophophore. Other muscles can contract to withdraw the crown to safety with great speed.

The lophophore ridge tends to be circular in marine ectoprocts and U-shaped in freshwater species. When feeding, the animal extends the lophophore, and the tentacles spread out to a funnel shape (Figure 24-10). Cilia on the tentacles draw water into the funnel and out between the tentacles. Food particles trapped in mucus in the funnel are drawn into the mouth, both by the pumping action of the muscular pharynx and by the action of cilia in the pharynx.

Respiratory, vascular, and excretory organs are absent. Gaseous exchange occurs through the body surface, and, since the ectoprocts are small, the coelomic fluid is adequate for internal transport. Coelomocytes engulf and store waste materials. There is a ganglionic mass and a nerve ring around the pharynx, but no sense organs are present. A septum divides the coelom into an anterior mesocoel in the lophophore and a larger posterior metacoel. The protocoel and epistome are present only in freshwater ectoprocts. Pores in the walls between adjoining zooids permit exchange of materials by way of the coelomic fluid.

Feeding individuals make up most colonies, but polymorphism also occurs. One type of modified zooid resembles a bird beak that snaps at small invading organisms that might foul a colony. Another type has a long bristle that sweeps away foreign particles.

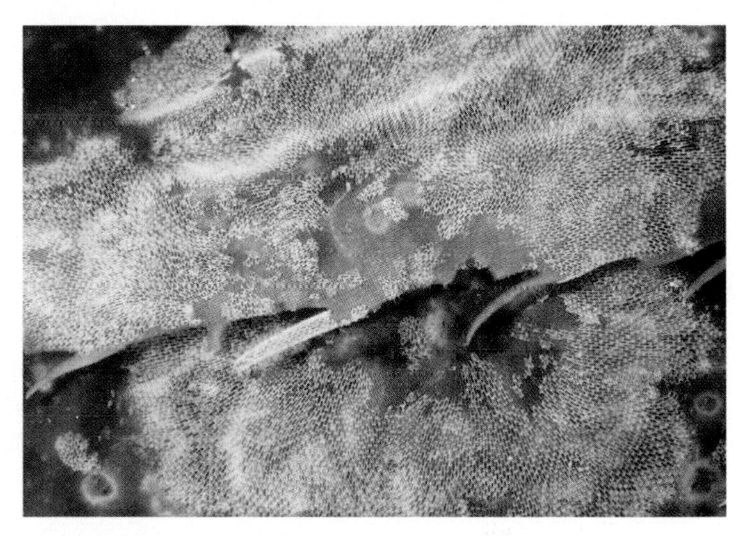

figure 24-9

Skeletal remains of a colony of *Membranipora,* a marine encrusting form of Ectoprocta. Each little oblong zoecium is the calcareous former home of a tiny ectoproct.

Most ectoprocts are hermaphroditic. Some species shed eggs into the seawater, but most brood their eggs, some within the coelom and some externally in a special zoecium in which the embryo develops. Cleavage is radial but apparently determinate.

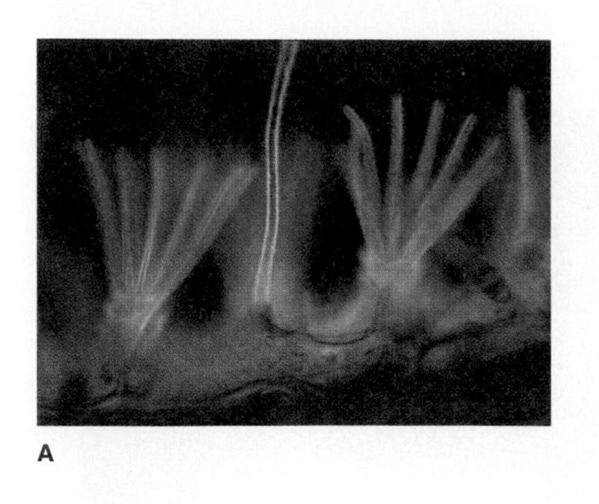

A

B

figure 24-10

A, Ciliated lophophore of *Electra pilosa,* a marine ectoproct. **B,** *Plumatella repens,* a freshwater ectoproct. It grows on the underside of rocks and vegetation in lakes, ponds, and streams.

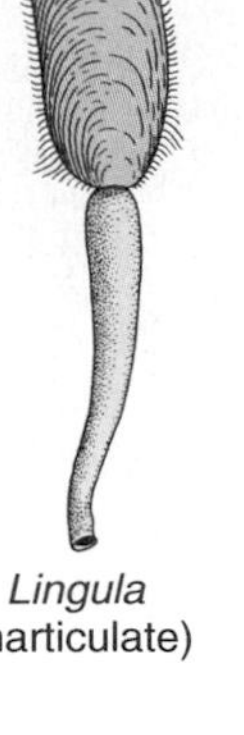

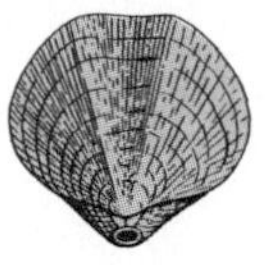

figure 24-11

Brachipods. **A,** *Lingula,* an inarticulate brachiopod that normally occupies a burrow. The contractile pedicel can withdraw the body into the burrow. **B,** An articulate brachiopod, *Terebratella.* The valves have a tooth-and-socket articulation and a short pedicel projects through one valve to attach to the substratum (pedicel shown in Figure 24-12).

Brooding is often accompanied by degeneration of the lophophore and gut of the adults, the remains of which contract into minute dark balls, or **brown bodies.** Later, new internal organs regenerate in the old chambers. The brown bodies may remain passive or may be taken up and eliminated by the new digestive tract—an unusual kind of storage excretion.

Freshwater species reproduce both sexually and asexually. Asexual reproduction is by budding or by means of **statoblasts,** which are hard, resistant capsules containing a mass of germinative cells that form during the summer and fall (Figure 24-8B). When the colony dies in late autumn, the statoblasts are released, and in spring they can give rise to new polypides and eventually to new colonies.

Phylum Brachiopoda

The Brachiopoda (brak′i-op′o-da) (Gr. *brachiōn,* arm, + *pous, podos,* foot), or lamp shells, is an ancient group. Compared with the fewer than 300 species now living, some 30,000 fossil species, which flourished in the Paleozoic and Mesozoic seas, have been described. The brachiopods were once very abundant, but they are now apparently in decline. Some modern forms have changed little from the early ones. The genus *Lingula* (L., little tongue) (Figure 24-11A) has existed virtually unchanged for over 400 million years. Most modern brachiopod shells range between 5 and 80 mm, but some fossil forms reached 30 cm in length.

Brachiopods are all attached, bottom-dwelling, marine forms that mostly prefer shallow water. Externally brachiopods resemble the bivalved molluscs in having two calcareous shell valves secreted by the mantle. They were, in fact, classed with the molluscs until the middle of the nineteenth century, and their name refers to the arms of the **lophophore,** which were thought homologous to the mollusc foot. Brachiopods, however, have **dorsal** and **ventral valves** instead of right and left lateral valves as do the bivalve molluscs and, unlike the bivalves, most of them are attached to a substrate either directly or by means of a fleshy stalk called a **pedicel** (or pedicle).

In most brachiopods the ventral (pedicel) valve is slightly larger than the dorsal (brachial) valve, and one end projects in the form of a short pointed beak that is perforated where the fleshy stalk passes through (Figure 24-11B). In many the shape of the pedicel valve is that of the classical oil lamp of Greek and Roman times, so that the brachiopods came to be known as the "lamp shells."

There are two classes of brachiopods based on shell structure. The shell valves of Articulata are connected by a hinge with an interlocking tooth-and-socket arrangement (articular process); those of Inarticulata lack the hinge and are held together by muscles only.

The body occupies only the posterior part of the space between the valves (Figure 24-12A), and extensions of the body wall form mantle lobes that line and secrete the shell. The large horseshoe-shaped lophophore in the anterior mantle cavity bears long ciliated tentacles used in respiration and feeding. Ciliary water currents carry food particles between

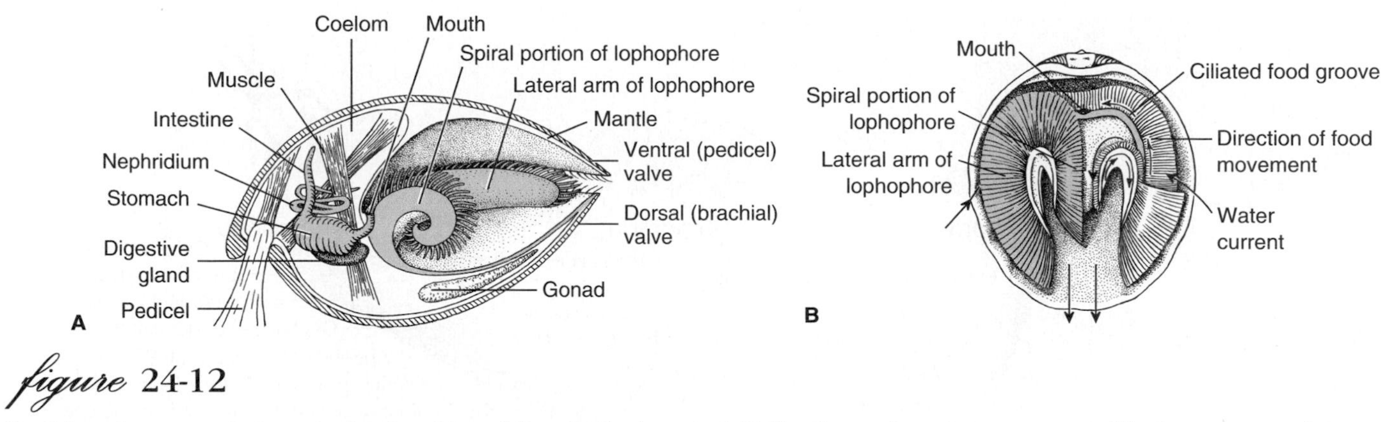

figure 24-12

Brachiopod anatomy. **A,** An articulate brachiopod (longitudinal section). **B,** Feeding and respiratory currents. The large arrows show water flow over the lophophore; the small arrows indicate food movement toward the mouth in the ciliated food groove.

the gaping valves and over the lophophore. Food is caught in mucus on the tentacles and carried in a ciliated food groove along the arm of the lophophore to the mouth (Figure 24-12).

There is no cavity in the epistome of articulates, but in inarticulates there is a protocoel in the epistome that opens into the mesocoel. As in the other lophophorates, the posterior metacoel bears the viscera. One or two pairs of nephridia open into the coelom and empty into the mantle cavity. There is an open circulatory system with a contractile heart. There is a nerve ring with a small dorsal and a larger ventral ganglion.

Sexes are separate and paired gonads discharge gametes through the nephridia. The development of brachiopods is similar in some ways to that of the deuterostomes, with radial, mostly equal, holoblastic cleavage and the coelom forming enterocoelically in the articulates. The free-swimming larva of the articulates resembles a trochophore.

Phylogeny

The early embryological development of sipunculans, echiurans, and annelids is almost identical, showing a very close relationship among the three phyla. Sipunculans and echiurans are not metameric and thus are more primitive in that characteristic than annelids. They probably represent collateral evolutionary lines that branched from protoannelid stock before the origin of metamerism.

Several characters suggest relationship of the Pogonophora to the Annelida, as we noted previously.

The phylogenetic affinities of the Pentastomida are uncertain. Most modern taxonomists align them with the arthropods, however, and evidence is accumulating that they are most closely related to the crustacean subclass Branchiura (p. 504). This evidence includes similarities in morphology of their sperm and in base sequences of ribosomal RNA. If the pentastomids really are close to the branchiurans, then their status as a phylum should be revoked, and they should be classified as crustacean arthropods.

Onychophorans share a number of characteristics with the annelids: metamerically arranged nephridia, muscular body wall, pigment cup ocelli, and ciliated reproductive ducts. Characteristics shared with arthropods include the cuticle, tubular heart and hemocoel with open circulatory system, presence of tracheae (probably not homologous), and large size of the brain. Unique characteristics include oral papillae, slime glands, body tubercles, and suppression of external segmentation. Some authors believe the onychophorans should be included with the arthropods, but that would require redefining the phylum Arthropoda. Most authors believe that the differences seem to warrant keeping them in a separate phylum (Figure 24-13).

The affinities of tardigrades are among the most puzzling of all animal groups. They have some similarities to rotifers, particularly in their reproduction and their cryptobiotic tendencies, and some authors have called them pseudocoelomates. Their embryogenesis, however, would seem to put them among the coelomates. The enterocoelic origin of the mesoderm is a deuterostome characteristic. Other authors identify several important synapomorphies that suggest an ancestor in common with the arthropods (Figure 24-13).

The phylogenetic position of the lophophorates has been the subject of much controversy and debate. Sometimes they have been considered protostomes with some deuterostome characters, and at other times deuterostomes with some protostome characters. Brusca and Brusca[1] contended that there is overwhelming evidence that they are a monophyletic clade and are deuterostomes. Their common possession of a lophophore is a unique synapomorphy. Other features, such as the U-shaped digestive tract, metanephridia (except in ectoprocts), and tendency to secrete outer casings may be homologous within the clade, but they are convergent with many other taxa.

[1]Brusca, R. C., and G. J. Brusca. 1990. Invertebrates. Sunderland, Massachusetts, Sinauer Associates.

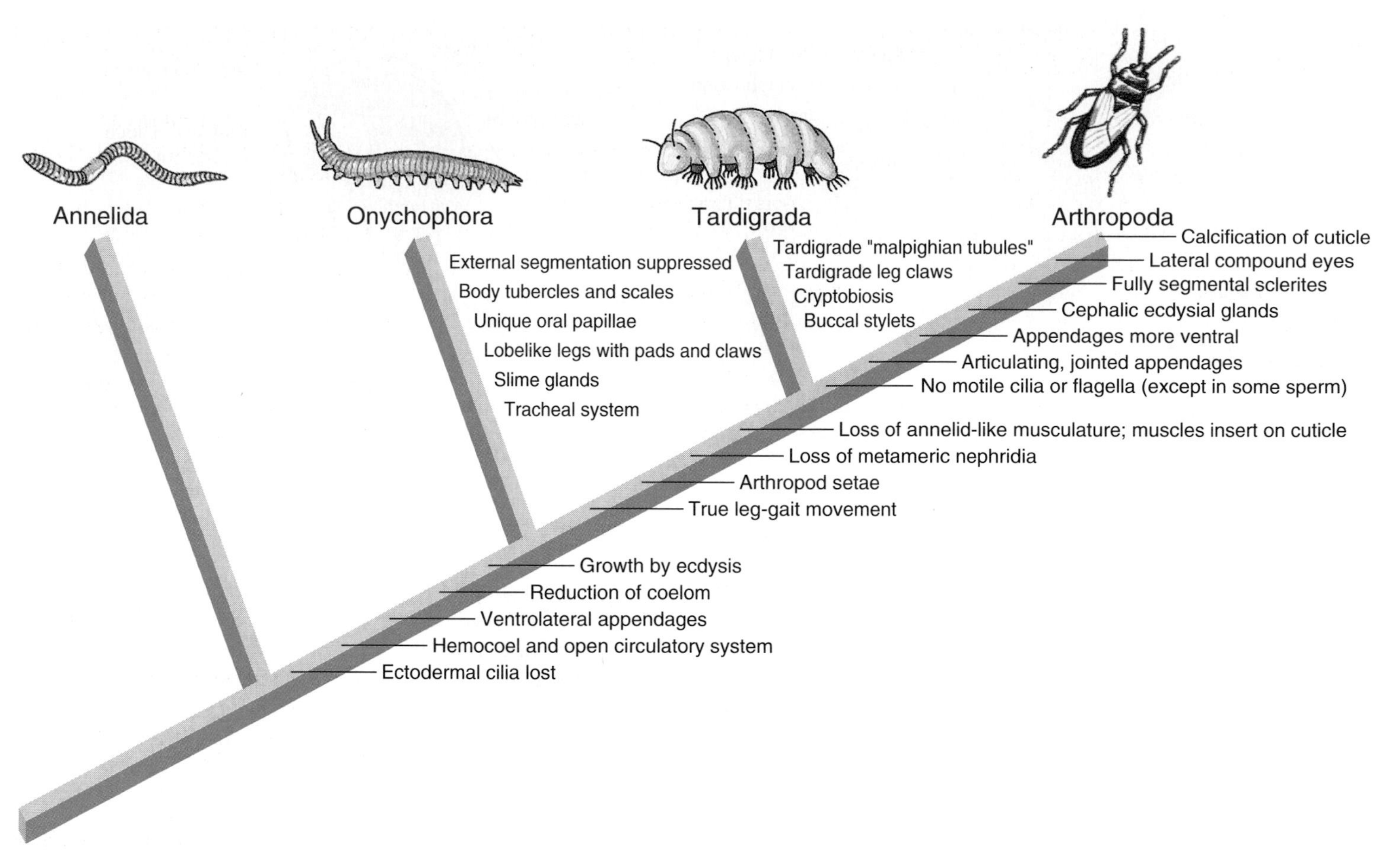

figure 24-13

Cladogram depicting hypothetical relationships of Onychophora and Tardigrada to annelids and arthropods. Onychophorans diverged from the arthropod line after the development of such synapomorphies as hemocoel and growth by ecdysis, but they share several primitive characters with the annelids, such as the metameric arrangement of the nephridia. Note that the tracheal system of onychophorans is not homologous to that of arthropods but represents a convergence. The phylogenetic relationships of other phyla covered in this chapter are too difficult to evaluate to permit construction of a cladogram that includes them.

Summary

The nine small, coelomate phyla considered in this chapter are grouped together here for convenience. The Sipuncula, Echiura, and Pogonophora probably share an ancestor with annelids; the Pentastomida, Onychophora, and Tardigrada apparently have a common ancestor with the arthropods. The lophophorates (Phoronida, Ectoprocta, and Brachiopoda) have some characteristics of both protostomes and deuterostomes.

Sipunculans and echiurans are burrowing marine worms; neither shows metamerism. Pogonophorans live in tubes on the deep ocean floor and have no digestive tract. They are metameric.

Although the Pentastomida have certain arthropod-like characteristics, their specialized modifications complicate determination of their phylogenetic relationships. They may be related to the crustacean subclass Branchiura. Onychophora are caterpillar-like animals that are metameric and show some annelid and arthropod characteristics. Tardigrades are minute, mostly terrestrial animals that have a hemocoel, as do arthropods.

The Phoronida, Ectoprocta, and Brachiopoda all bear a lophophore, which is a crown of ciliated tentacles surrounding the mouth but not the anus and containing an extension of the mesocoel. They are also sessile as adults, have a U-shaped digestive tract, and have free-swimming larvae. The lophophore functions both as a respiratory and a feeding structure.

Phoronida are the least abundant of the lophophorates, living in tubes mostly in shallow coastal waters.

Ectoprocts are abundant in marine habitats, living on a variety of submerged substrata, and a number of species are common in fresh water. Ectoprocts are colonial, and although each individual is quite small, the colonies are commonly several centimeters or more in width or height.

Brachiopods were a widely prevalent phylum in the Paleozoic era but have been declining since the early Mesozoic era. They have a dorsal and a ventral shell and usually attach to the substrate directly or by means of a pedicel.

Because of their possession of a unique synapomorphy, the lophophore, the lophophorates appear to form a monophyletic clade.

Review Questions

1. Give the main distinguishing characteristics of the following, and tell where each lives: Sipuncula, Echiura, Pogonophora, Pentastomida, Onychophora, Tardigrada.
2. What do the members of each of the aforementioned groups eat?
3. What is the evidence that the Sipuncula and Echiura share an ancestor with the annelids?
4. What is the largest pogonophoran known? Where is it found, and how is it nourished?
5. Some investigators regard the Onychophora as a "missing link" between the Annelida and the Arthopoda. Give evidence for and against this hypothesis.
6. What is the survival value of cryptobiosis in tardigrades?
7. Some investigators consider the lophophorates a monophyletic clade within the deuterostomes. What are some synapomorphies that distinguish this clade, and what are some synapomorphies that diagnose each phylum of lophophorates?
8. Define each of the following: lophophore, zoecium, zooid, polypide, cystid, brown bodies, statoblasts.
9. What are the coelomic compartments found in the lophophorates?
10. Brachiopods superficially resemble bivalve molluscs. How would you explain the difference to a layperson?
11. The lophophorates are sometimes placed in a phylogenetic position between the protostomes and the deuterostomes. How would you justify such placement?

Selected References

See also general references for Part Three, p. 562.

American Society of Zoologists. 1977. Biology of lophophorates. Am. Zool. **17**(1):3–150. *A collection of 13 papers.*

Childress, J. J., H. Felbeck, and G. N. Somero. 1987. Symbiosis in the deep sea. Sci. Am. **256**:114–120 (May). *The amazing story of how the animals around deep-sea vents, including* Riftia pachyptila, *manage to absorb hydrogen sulfide and transport it to their mutualistic bacteria. For most animals, hydrogen sulfide is highly toxic.*

Crowe, J. H., and A. F. Cooper, Jr. 1971. Cryptobiosis. Sci. Am. **225**:30–36 (Dec.). *Cryptobiotic nematodes, rotifers, and tardigrades can withstand adverse conditions of astonishing rigor, yet perceptible metabolism continues in their state of suspended animation.*

Gould, S. J. 1995. Of tongue worms, velvet worms, and water bears. Natural History **104**(1):6–15. *Intriguing essay on affinities of Pentastomida, Onychophora, and Tardigrada and how they, along with larger phyla, were products of the Cambrian explosion.*

Haugerud, R. E. 1989. Evolution in the pentastomids. Parasitol. Today **5**:126–132. *Much remains to be learned of this puzzling group, but there is strong evidence of its crustacean affinities.*

Rice, M. E., and M. Todorovic, eds. 1975. Proceedings of the International Symposium on the biology of the Sipuncula and Echiura, 2 vols. Washington, D.C., National Museum of Natural History. *A series of technical articles, but much of interest for further reading on these two phyla.*

Richardson, J. R. 1986. Brachiopods. Sci. Am. **255**:100–106 (Sept.). *Reviews brachiopod biology and adaptations and contends that in the next few million years there may be an increase in the number of species, rather than a further decline.*

Echinoderms, Hemichordates, and Chaetognaths

chapter | twenty-five

A Design To Puzzle the Zoologist

The distinguished American zoologist Libbie Hyman once described the echinoderms as a "noble group especially designed to puzzle the zoologist." With a combination of characteristics that should delight the most avid reader of science fiction, the echinoderms would seem to confirm Lord Byron's observation that

> Tis strange—but true;
> for truth is always strange;
> Stranger than fiction.

Despite the adaptive value of bilaterality for free-moving animals, and the merits of radial symmetry for sessile animals, echinoderms confounded the rules by becoming free moving but radial. That they evolved from a bilateral ancestor there can be no doubt, for their larvae are bilateral. They undergo a bizarre metamorphosis to a radial adult in which there is a 90° reorientation in body axis, with a new mouth arising on the left side, and a new anus appearing on the right side.

A compartment of the coelom has been transformed in echinoderms into a unique water-vascular system that uses hydraulic pressure to power a multitude of tiny tube feet used in food gathering and locomotion. An endoskeleton of dermal ossicles may fuse together to invest the echinoderm in armor, or it may be reduced in some to microscopic bodies. Many echinoderms have miniature jawlike pincers (pedicellariae) scattered on their body surface, often stalked and some equipped with poison glands, that keep their surface clean by snapping at animals that would settle there.

This constellation of characteristics is unique in the animal kingdom. It has both defined and limited the evolutionary potential of the echinoderms. Despite the vast amount of research that has been devoted to them, we are still far from understanding many aspects of echinoderm biology.

The Echinodermata, along with the chordates, the lophophorates, and the Hemichordata (acorn worms and pterobranchs) are deuterostomes. Typical deuterostome embryogenesis is shown only by some chordates such as amphioxus (p. 572), but these shared characters support monophyly of the Deuterostomia. Nonetheless, their evolutionary history has taken the echinoderms to the point where they are very much unlike any other animal group. The phylum Chaetognatha (arrowworms) traditionally has been included among the deuterostomes, but this arrangement is not supported by recent molecular evidence.[1] Until the issue is clarified, we will cover the chaetognaths in this chapter for convenience.

Phylum Echinodermata

The Echinodermata are marine forms and include the classes Asteroidea (sea stars [or starfishes]), Ophiuroidea (brittle stars), Echinoidea (sea urchins), Holothuroidea (sea cucumbers), and Crinoidea (sea lilies). Echinoderms have a combination of characteristics that are found in no other phylum: (1) an endoskeleton of plates or ossicles, usually spiny, (2) the water-vascular system, (3) the pedicellariae, (4) the dermal branchiae, and (5) secondary radial or biradial symmetry. The water-vascular system and the dermal ossicles have been particularly important in determining the evolutionary potential and limitations of this phylum. Their larvae are bilateral and undergo a metamorphosis to a radial adult.

[1]Telford, M. J., and P. W. H. Holland. 1993. Mol. Biol. Evol. **10:**660–676; Wada, H., and N. Satoh. 1994. Proc. Natl. Acad. Sci. **91:**1801–1804.

Ecological Relationships

Echinoderms are all marine; they have no ability to osmoregulate and so are rarely found in waters that are brackish. Virtually all benthic as adults, they are found in all oceans of the world and at all depths, from the intertidal to the abyssal regions.

Some sea stars (Figure 25-1) are particle feeders, but many are predators, feeding particularly on sedentary or sessile prey. Brittle stars (see Figure 25-7) are the most active echinoderms, moving by their arms; and they may be scavengers, browsers, or deposit or filter feeders. Some brittle stars are commensals with sponges. Compared to other echinoderms, sea cucumbers (see Figure 25-14) are greatly extended in the oral-aboral axis and are oriented with that axis more or less parallel to the substrate and lying on one side. Most are suspension or deposit feeders. "Regular" sea urchins (see Figure 25-10), which are radially symmetrical, prefer hard bottoms and feed chiefly on algae or detritus. "Irregular" urchins (sand dollars and heart urchins) (see Figure 25-11), which have become secondarily bilateral, are usually found on sand and feed on small particles. Sea lilies and feather stars (see Figures 25-17 and 25-18) stretch their arms out and up like a flower's petals and feed on plankton and suspended particles.

Class Asteroidea: Sea Stars

Sea stars, often called starfishes, demonstrate the basic features of echinoderm structure and function very well, and they are easily obtainable. We shall consider them first, then comment on the major differences shown by the other groups.

Position in Animal Kingdom

1. Phylum Echinodermata (e-ki′no-der′ma-ta) (Gr. *echinos,* sea urchin, hedgehog, + *derma,* skin, + *ata,* characterized by) belongs to the **Deuterostomia** branch of the animal kingdom, the members of which are enterocoelous coelomates. The other phyla traditionally assigned to this group are Chaetognatha, Hemichordata, and Chordata, but recent evidence questions placement of chaetognaths in the deuterostomes. We are also placing the lophophorate phyla (Phoronida, Ectoprocta, and Brachiopoda) in the Deuterostomia.
2. Primitively, deuterostomes have the following embryological features in common: anus developing from or near the blastopore, and mouth developing elsewhere; coelom budded off from the archenteron (enterocoel); radial and regulative (indeterminate) cleavage; and endomesoderm (mesoderm derived from or with the endoderm) from enterocoelic pouches.

Biological Contributions

1. There is one word that best describes the echinoderms: strange. They have a unique constellation of characteristics found in no other phylum. Among the more striking of the features shown by the echinoderms are as follows:
   a. The system of channels composing the **water-vascular system,** derived from a coelomic compartment.
   b. The **dermal endoskeleton** composed of calcareous ossicles.
   c. The **hemal system,** whose function remains mysterious, also enclosed in a coelomic compartment.
   d. Their **metamorphosis,** which changes a bilateral larva to a radial adult.

Characteristics of Phylum Echinodermata

1. Body not metameric, adult with **radial, pentamerous symmetry** characterized by five or more radiating areas
2. No head or brain; few specialized sensory organs
3. Nervous system with circumoral ring and radial nerves
4. **Endoskeleton** of **dermal calcareous ossicles** with **stereom** structure; covered by an epidermis (ciliated in most); pedicellariae (in some)
5. A **water-vascular system** of coelomic origin that extends from the body surface as a series of tentacle-like projections (podia or tube feet)
6. **Locomotion by tube feet,** which project from the **ambulacral areas,** or by movement of spines, or by movement of arms, which project from central disc of the body
7. Digestive system usually complete; axial or coiled; anus absent in ophiuroids
8. Coelom extensive, forming the perivisceral cavity and the cavity of the water-vascular system; coelom of enterocoelous type
9. So-called **hemal system** present, of uncertain function but playing little, if any, role in circulation of body fluids, and surrounded by extensions of the coelom (perihemal sinuses)
10. Respiration by dermal branchiae, tube feet, respiratory tree (holothuroids), and bursae (ophiuroids)
11. Excretory organs absent
12. Sexes separate (except a few hermaphroditic); fertilization usually external
13. Development through free-swimming, bilateral, larval stages (some with direct development); metamorphosis to radial adult or subadult form

A

B

C

D

figure 25-1

Some sea stars (class Asteroidea) from the Pacific. **A,** Cushion star *Pteraster tesselatus* can secrete incredible quantities of mucus as a defense. **B,** *Choriaster granulatus* scavenges dead animals on shallow Pacific reefs. **C,** *Tosia queenslandensis* from the Great Barrier Reef browses encrusting organisms. **D,** *Crossaster papposus,* one of the sun stars, feeds on other sea stars.

Sea stars are familiar along shorelines, where sometimes large numbers may aggregate on the rocks. They also live on muddy or sandy bottoms and among coral reefs. They are often brightly colored and range in size from a centimeter in greatest diameter to about a meter across from arm tip to opposite arm tip.

Form and Function

External Features Reflecting their radial pentamerous symmetry, sea stars typically have five arms (rays), but there may be more (Figure 25-1D). The arms merge gradually with the central disc (Figure 25-2A). Beneath the epidermis of sea stars is a mesodermal endoskeleton of small calcareous plates, or **ossicles,** bound together with connective tissue. From these ossicles project the spines and tubercles that are responsible for the spiny surface. The ossicles are penetrated by a meshwork of spaces, usually filled with fibers and dermal cells. This internal meshwork structure is described as **stereom** (see Figure 25-15) and is unique to the echinoderms.

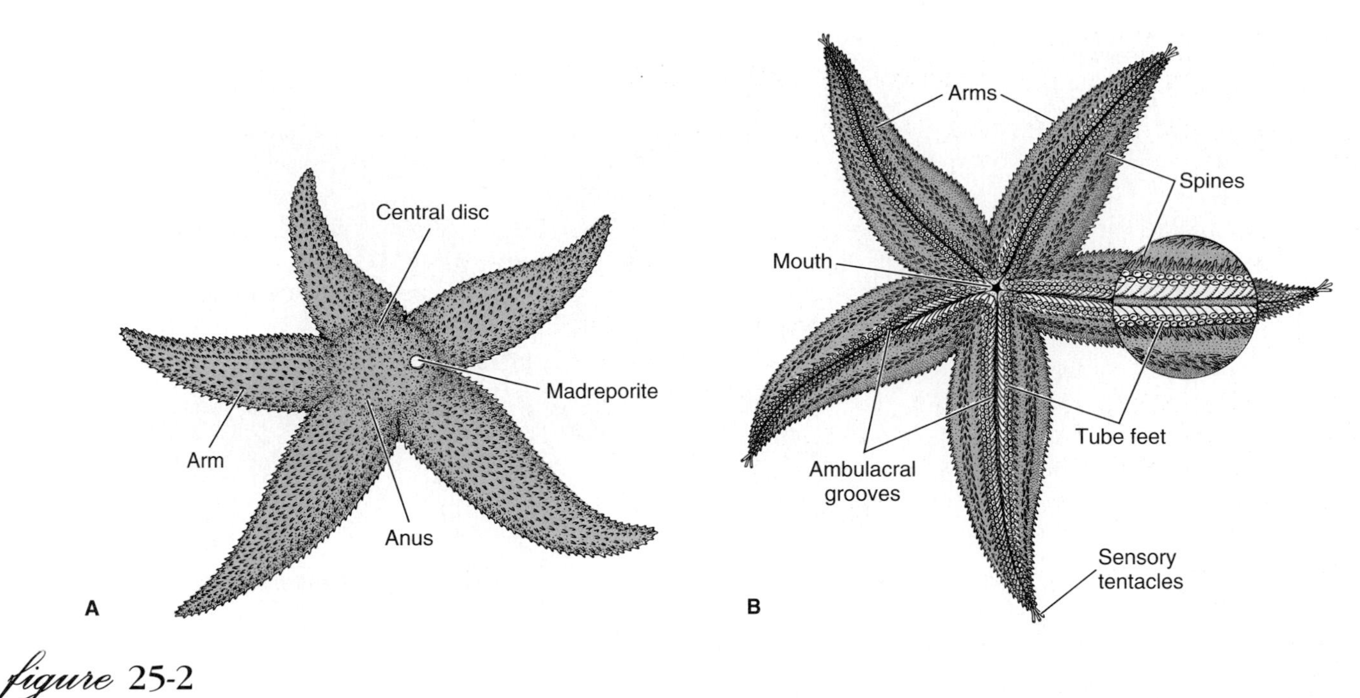

figure 25-2
External anatomy of asteroid. **A,** Aboral view. **B,** Oral view.

Ambulacral (am-bu-la′kral) **grooves** (Figure 25-2B) radiate out along the arms from the centrally located mouth on the under, or oral, side of the animal. The **tube feet (podia)** project from the grooves, which are bordered by movable spines. Viewed from the oral side, the large **radial nerve** can be seen in the center of each ambulacral groove (Figure 25-3C), between the rows of tube feet. The nerve is very superficially located, covered only by thin epidermis. Under the nerve is an extension of the coelom and the radial canal of the water-vascular system (Figure 25-3C). In all other classes of living echinoderms except crinoids, ossicles or other dermal tissue cover over these structures; thus the ambulacral grooves in asteroids and crinoids are **open,** and those of the other groups are **closed.**

The aboral surface is usually rough and spiny, although the spines of many species are flattened, so that the surface appears smooth. Around the bases of the spines in many sea stars are groups of minute pincerlike **pedicellariae** (ped-e-cell-ar′e-ee), bearing tiny jaws manipulated by muscles (Figure 25-4). These help keep the body surface free of debris, protect the papulae, and sometimes aid in food capture. The **papulae** (pap′u-lee) **(dermal branchiae** or **skin gills)** are soft, delicate projections of the coelomic cavity, covered only with epidermis and lined internally with peritoneum; they extend out through spaces between the ossicles (Figure 25-3C) and are concerned with respiration. Also on the aboral side are the inconspicuous **anus** and the circular **madreporite** (Figure 25-3A), a calcareous sieve leading to the water-vascular system.

The function of the madreporite is still obscure. One suggestion is that it allows rapid adjustment of hydrostatic pressure within the water-vascular system in response to changes in external hydrostatic pressure resulting from depth changes, as in tidal fluctuations.

The coelomic compartments of larval echinoderms give rise to several structures in adults, one of which is a spacious body **coelom** filled with fluid. The coelomic fluid circulates around the body cavity and into the papulae, propelled by cilia on the peritoneal lining. Exchange of respiratory gases and excretion of nitrogenous waste, principally ammonia, take place by diffusion through the thin walls of the papulae and tube feet.

Water-Vascular System The water-vascular system is another coelomic compartment and is unique to the echinoderms. It is a system of canals and specialized tube feet that shows exploitation of hydraulic mechanisms to a greater degree than in any other animal group. In sea stars the primary functions of the water-vascular system are locomotion and food gathering, as well as those of respiration and excretion.

Structurally, the water-vascular system opens to the outside through small pores in the madreporite. The madreporite of asteroids is on the aboral surface (Figure 25-2A), and leads into the **stone canal,** which descends toward the **ring canal** around the mouth (Figure 25-3B).

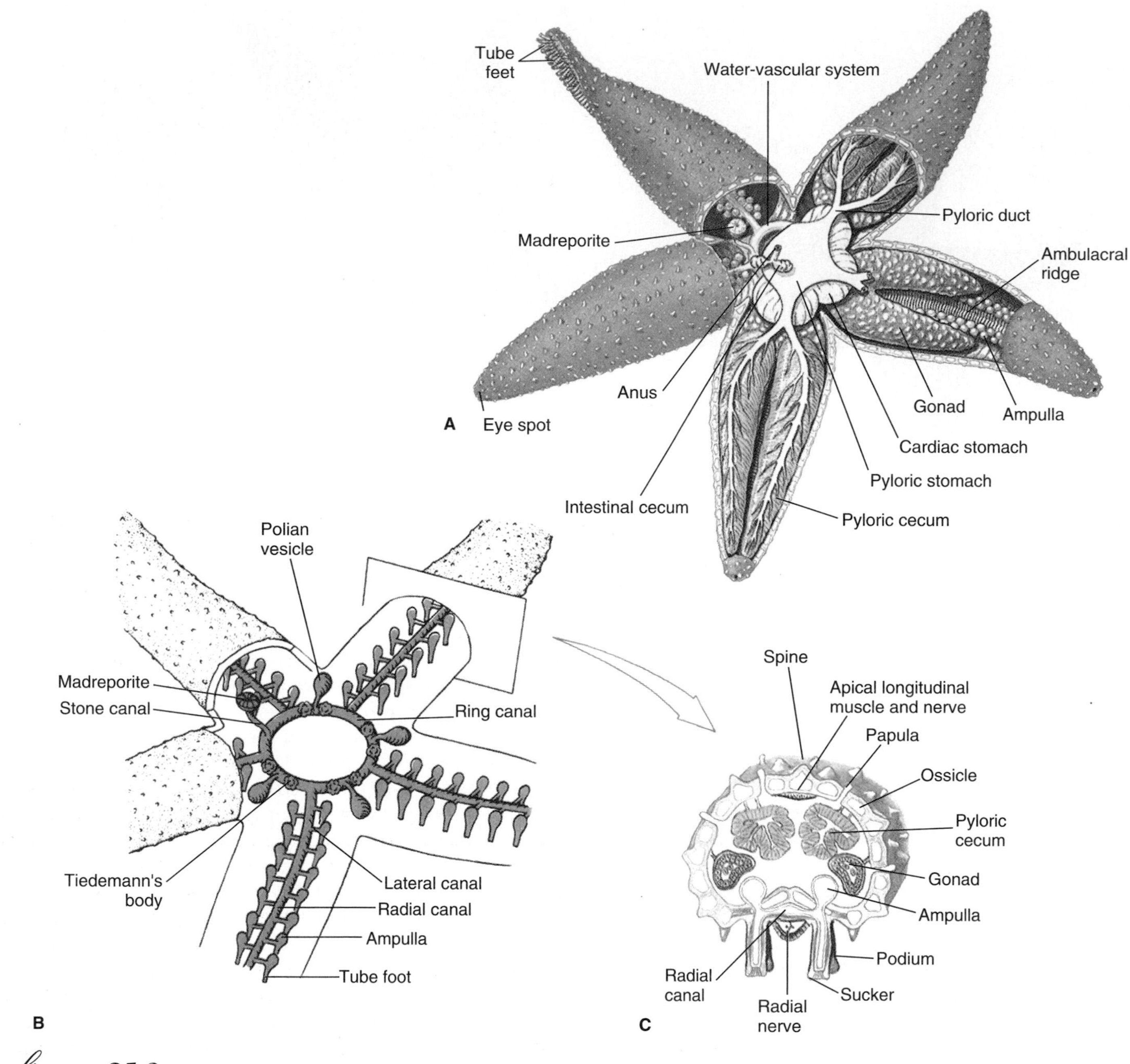

figure 25-3

A, Internal anatomy of a sea star. **B,** Water-vascular system. **C,** Cross section of arm at level of gonads, illustrating open ambulacral groove. Podia (tube feet) penetrate between ossicles. (Polian vesicles are not present in *Asterias.*)

Radial canals diverge from the ring canal, one into the ambulacral groove of each ray. **Polian vesicles** are also attached to the ring canal of most asteroids (but not *Asterias*) and apparently serve as fluid reservoirs for the water-vascular system.

A series of small **lateral canals,** each with a one-way valve, connects the radial canal to the cylindrical **podia** or **tube feet,** along the sides of the ambulacral groove in each ray. Each podium is a hollow, muscular tube, the inner end of which is a muscular sac, the **ampulla,** that lies within the body coelom (Figure 25-3), and the outer end of which usually bears a sucker. Some species lack the suckers. The podia pass to the outside between the ossicles in the ambulacral groove.

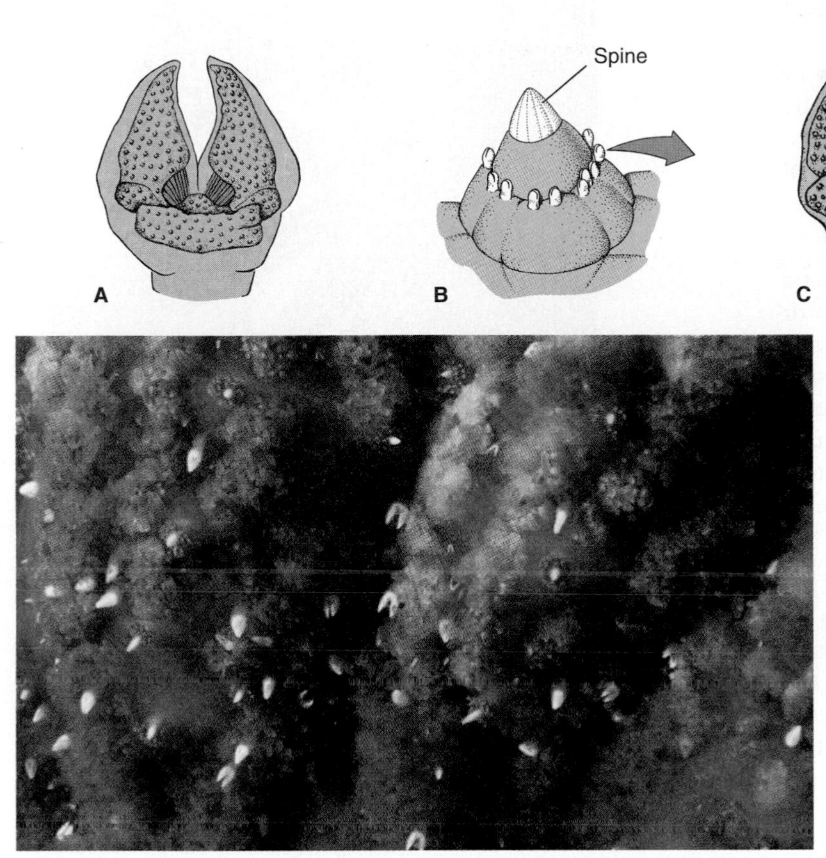

figure 25-4

Pedicellariae of sea stars and sea urchins. **A,** Forceps-type pedicellaria of *Asterias.* **B** and **C,** Scissors-type pedicellariae of *Asterias;* size relative to spine is shown in **B. D,** Tridactyl pedicellaria of *Strongylocentrotus,* cutaway showing muscle. **E,** Globiferous pedicellaria of *Strongylocentrotus.* **F,** Close-up view of the aboral surface of the sea star *Pycnopodia helianthoides.* Note the large pedicellariae, as well as the groups of small pedicellariae around the spines. Many thin-walled papulae can be seen.

Locomotion by means of tube feet illustrates the interesting exploitation of hydraulic mechanisms by echinoderms. The valves in the lateral canals prevent backflow of fluid into the radial canals. The tube foot has in its walls connective tissue that maintains the cylinder at a relatively constant diameter. On contraction of muscles in the ampulla, fluid is forced into the tube foot, extending it. Conversely, contraction of the longitudinal muscles in the tube foot retracts the podium, forcing fluid back into the ampulla. Contraction of muscles in one side of the tube foot bends the organ toward that side. Small muscles at the end of the tube foot can raise the middle of the disclike end, thus creating a suction-cup effect when the end is applied to the substrate. We can estimate that by combining mucous adhesion with suction, a single tube foot can exert a pull equal to 25 to 30 g. Coordinated action of all or many of the tube feet is sufficient to draw the animal up a vertical surface or over rocks. On a soft surface, such as muck or sand, the suckers are ineffective (and numerous sand-dwelling species have no suckers), so the tube feet are employed as legs.

Feeding and Digestive System The mouth on the oral side leads into a two-part stomach located in the central disc (Figure 25-3). In some species, the large, lower **cardiac stomach** can be everted. The smaller upper **pyloric stomach** connects with **digestive ceca** located in the arms. Digestion is largely extracellular, occurring in the digestive ceca. A short **intestine** leads from the stomach to the inconspicuous **anus** on the aboral side. Some species lack an intestine and anus.

Many sea stars are carnivorous and feed on molluscs, crustaceans, polychaetes, echinoderms, other invertebrates, and sometimes small fish, but many show particular preferences. Some feed on brittle stars, sea urchins, or sand dollars, swallowing them whole and later regurgitating undigestible ossicles and spines. Some attack other sea stars, and if the predator is small compared to its prey, it may attack and begin eating at the end of one of the prey's arms.

Many asteroids feed heavily on molluscs, and *Asterias* is a significant predator on commercially important clams and oysters. When feeding on a bivalve, a sea star will hump over its prey, attaching its podia to the valves, and then exert a steady pull, using its feet in relays. A force of some 1300 g can be exerted. In half an hour or so the adductor muscles of the bivalve fatigue and relax. With a very small gap available, the star inserts its soft everted stomach into the space between the valves, wraps it around the soft parts of the bivalve, and secretes digestive juices to start digesting them. After feeding, the sea star draws in its stomach by contraction of the stomach muscles and relaxation of body-wall muscles.

Some sea stars feed on small particles, either entirely or in addition to carnivorous feeding. Plankton or other organic particles coming in contact with the animal's oral or aboral surface are carried by the epidermal cilia to the ambulacral grooves and then to the mouth.

Hemal System Although the hemal system is characteristic of echinoderms, its function remains unclear. It has little or nothing to do with circulation of body fluids, despite its name, which means "blood." It is a system of tissue strands enclosing unlined channels and is itself enclosed in another coelomic compartment, the perihemal channels or sinuses. The main channel of the hemal system connects aboral, gastric, and oral rings that give rise to branches to the gonads, stomach ceca, and arms, respectively.

Nervous and Sensory System The nervous system in echinoderms comprises three subsystems, each made up of a nerve ring and radial nerves placed at different levels in the disc and arms. An epidermal nerve plexus, or nerve net, connects the systems. Sense organs include ocelli at the arm tips and sensory cells scattered all over the epidermis.

Reproductive System and Regeneration and Autotomy Most sea stars have separate sexes. A pair of gonads lies in each interradial space (Figure 25-3), and fertilization is external.

Echinoderms can regenerate lost parts. Sea star arms can regenerate readily, even if all are lost. Stars also have the power of **autotomy** (the ability to break off part of their own bodies) and can cast off an injured arm near the base. An arm may take months to regenerate.

If an arm is broken off or removed, and it contains a part of the central disc (about one-fifth), the arm can regenerate a complete new sea star! In former times fishermen used to dispatch sea stars they collected from their oyster beds by chopping them in half with a hatchet—a worse than futile activity. Some sea stars reproduce asexually under normal conditions by cleaving the central disc, each part regenerating the rest of the disc and missing arms (Figure 25-5).

figure 25-5

Pacific sea star *Echinaster luzonicus* can reproduce itself by splitting across the disc, then regenerating missing arms. The one shown here has evidently regenerated six arms from the longer one at top left.

Development Some species brood their eggs, either under the oral side of the animal or in specialized aboral structures, and development is direct; but in most species the embryonating eggs are free in the water and hatch to free-swimming larvae.

Early embryogenesis shows the typical primitive deuterostome pattern. Gastrulation is by invagination, and the anterior end of the archenteron pinches off to become the coelomic cavity, which expands in a U shape to fill the blastocoel. Each of the legs of the U, at the posterior end, constricts to become a separate vesicle, and these eventually give rise to the main coelomic compartments of the body (metacoels). The anterior portions of the U (protocoels and mesocoels) give rise to the water-vascular system and the perihemal channels. The free-swimming larva has cilia arranged in bands, and these tracts extend onto the larval arms as development continues. The larva grows three adhesive arms and a sucker at its anterior end and attaches to the substratum. While it is thus attached by this temporary stalk, it undergoes metamorphosis.

In echinoderms the metacoel, mesocoel, and protocoel are called the somatocoel, hydrocoel, and axocoel, respectively. During metamorphosis of sea stars, the paired somatocoels become the oral and aboral coelomic cavities, the right axocoel and hydrocoel are lost, and the left axocoel and hydrocoel become the water-vascular system and perihemal channels.

Metamorphosis involves a dramatic reorganization of a bilateral larva into a radial juvenile. The anteroposterior axis of the larva is lost. *What was the left side becomes the oral surface, and the larval right side becomes the aboral surface* (Figure 25-6). Correspondingly, the larval mouth and anus disappear, and a new mouth and anus form on what were originally the left and right sides, respectively. As internal reorganization proceeds, short, stubby arms and the first podia appear. The animal then detaches from its stalk and begins life as a young sea star.

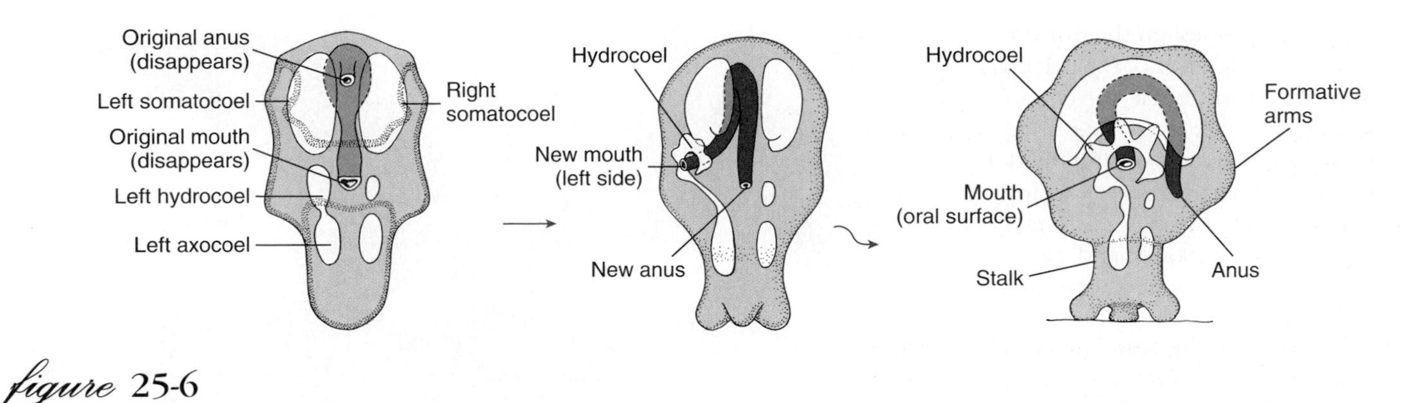

figure 25-6

Asteroid metamorphosis. The left somatocoel becomes the oral coelom, and the right somatocoel becomes the aboral coelom. The left hydrocoel becomes the water-vascular system and the left axocoel the stone canal and perihemal channels. The right axocoel and hydrocoel are lost.

Class Ophiuroidea: Brittle Stars

The brittle stars are the largest of the major groups of echinoderms in numbers of species, and they are probably the most abundant also. They abound in all types of benthic marine habitats, even carpeting the abyssal sea floor in many areas.

Apart from the typical possession of five arms, brittle stars are surprisingly different from asteroids. The arms of brittle stars are slender and sharply set off from the central disc (Figure 25-7). They have no pedicellariae or papulae, and their ambulacral grooves are closed, covered with arm ossicles. The tube feet are without suckers; they aid in feeding but are of limited use in locomotion. In contrast to that in the asteroids, the madreporite of the ophiuroids is located on the oral surface, on one of the oral-shield ossicles (Figure 25-8).

figure 25-7

A, Brittle star *Ophiura lutkeni* (class Ophiuroidea). Brittle stars do not use their tube feet for locomotion but can move rapidly (for an echinoderm) by means of their arms. **B,** Basket star *Astrophyton muricatum* (class Ophiuroidea). Basket stars extend their many-branched arms to filter feed, usually at night.

Each of the jointed arms consists of a column of articulated ossicles connected by muscles and covered by plates. Locomotion is by arm movement.

Five movable plates surround the mouth, serving as **jaws** (Figure 25-8). There is no anus. The skin is leathery, with dermal plates and spines arranged in characteristic patterns. Surface cilia are mostly lacking.

The visceral organs are all in the central disc, since the arms are too slender to contain them. The **stomach** is saclike and there is no intestine. Indigestible material is cast out of the mouth.

Five pairs of **bursae** (peculiar to ophiuroids) open toward the oral surface by **genital slits** at the bases of the arms. Water circulates in and out of these sacs for exchange of gases. On the coelomic wall of each bursa are small **gonads** that discharge into the bursa their ripe sex cells, which pass through the genital slits into the water for fertilization. Sexes are usually separate; a few ophiuroids are hermaphroditic. The ciliated bands of the larva extend onto delicate, beautiful larval arms, like those of larval echinoids (Figure 25-9C). During metamorphosis to the juvenile, there is no temporarily attached phase, as in asteroids.

Water-vascular, nervous, and hemal systems are similar to those of sea stars.

Brittle stars tend to be secretive, living on hard bottoms where no light penetrates. They are generally negatively phototropic and work themselves into small crevices between rocks, becoming more active at night. They are commonly fully exposed on the bottom in the permanent darkness of the deep sea. Ophiuroids feed on a variety of small particles, either browsing food from the bottom or suspension feeding. Podia are important in transferring food to the mouth. Some brittle stars extend arms into the water and catch suspended particles in mucous strands between the arm spines.

Regeneration and autotomy are even more pronounced in brittle stars than in sea stars. Many seem very fragile, releasing an arm or even part of the disc at the slightest provocation. Some can reproduce asexually by cleaving the disc; each progeny then regenerates the missing parts.

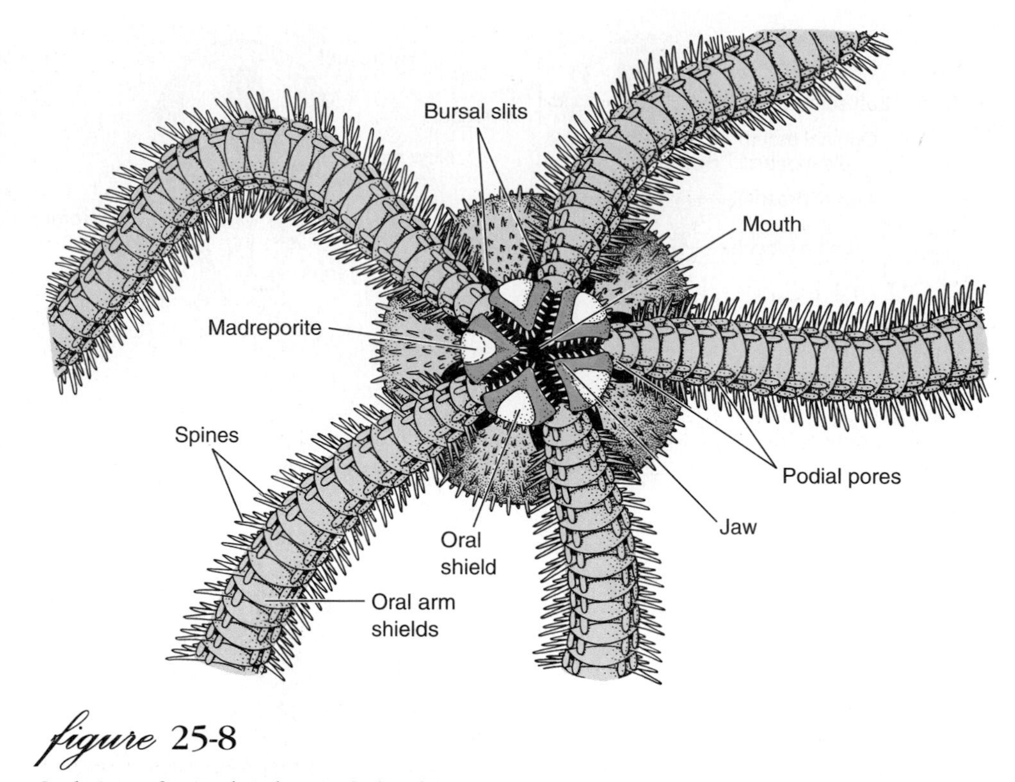

figure 25-8

Oral view of spiny brittle star *Ophiothrix.*

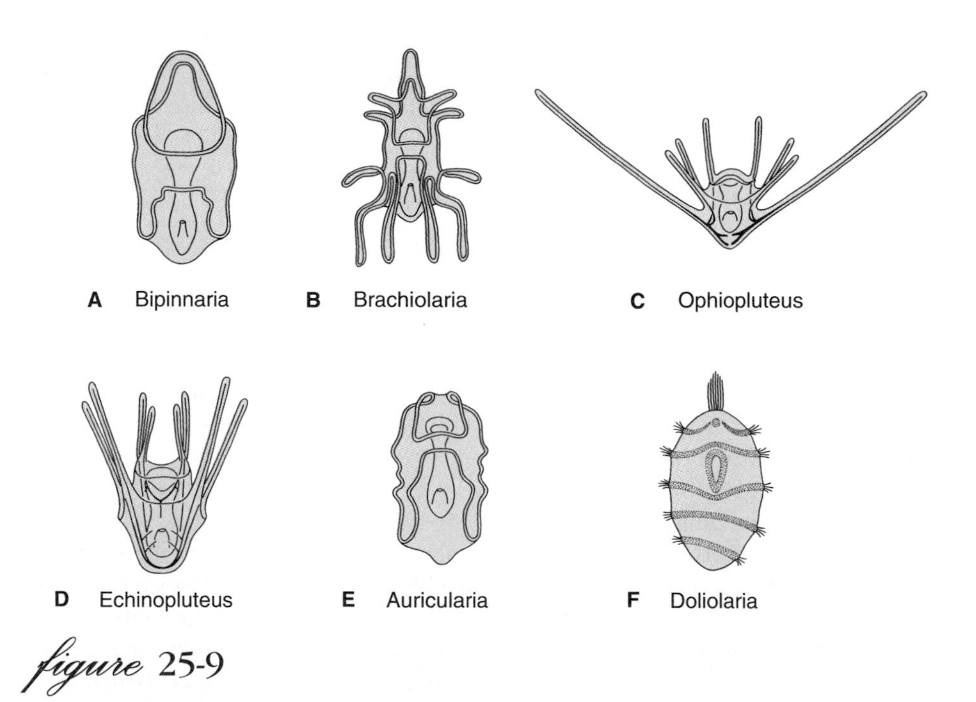

figure 25-9

Larvae of echinoderms. **A,** Bipinnaria of asteroids. **B,** Brachiolaria of asteroids. **C,** Ophiopluteus of ophiuroids. **D,** Echinopluteus of echinoids. **E,** Auricularia of holothuroids. **F,** Doliolaria of crinoids.

Class Echinoidea: Sea Urchins, Sand Dollars, and Heart Urchins

Echinoids have a compact body enclosed in an endoskeletal **test,** or shell. The dermal ossicles, which have become closely fitting plates, make up the test. Echinoids lack arms, but their tests reflect the typical five-part plan of echinoderms in their five ambulacral areas. Rather than extending from the oral surface to the tips of the arms, as in asteroids, the ambulacral areas follow the contours of the test from the mouth around to the aboral side, ending at the area around the anus **(periproct).** The majority of living species of sea urchins are termed "regular"; they are hemispherical in shape, radially symmetrical, and have medium to long spines (Figure 25-10). Sand dollars and heart urchins (Figure 25-11) are "irregular" because the orders to which they belong have become secondarily bilateral; their spines are usually very short. Regular urchins move by means of their tube feet, with some assistance from their spines, and irregular urchins move chiefly by their spines. Some echinoids are quite colorful.

Echinoids have wide distribution in all seas, from the intertidal regions to the deep oceans. Regular urchins often prefer rocky or hard bottoms, whereas sand dollars and heart urchins like to burrow into a sandy substrate.

The echinoid test is a compact skeleton of 10 double rows of plates that bear movable, stiff spines (Figure 25-12). The five pairs of ambulacral rows have pores (Figure 25-12) through which the long tube feet extend. The spines are moved by small muscles around the bases.

There are several kinds of **pedicellariae,** the most common of which have three jaws and are mounted on long stalks (Figure 25-4D, E).

Five converging teeth surround the mouth of regular urchins and sand dollars. In some sea urchins branched **gills** (modified podia) encircle the peristome, although these are of little importance in respiratory gas exchange. The **anus, genital openings,** and **madreporite** are aboral in the periproct region (Figure 25-12). The mouth of sand dollars is located at about the center of the oral side, but the anus has shifted to the margin or even the oral side of the disc, so that an anteroposterior axis and bilateral symmetry can be recognized. Bilateral symmetry is even more accentuated in the heart urchins, with the anus near the posterior end on the oral side and the mouth moved away from the oral pole toward the anterior end (Figure 25-11).

Inside the test (Figure 25-12) is a coiled digestive system and a complex chewing mechanism (in regular urchins and in sand dollars), called **Aristotle's lantern,** to which the teeth are attached (Figure 25-13). A ciliated siphon connects the esophagus to the intestine and enables the water to bypass the stomach to concentrate the food for digestion in the intestine. Sea urchins eat algae and other organic material, and sand dollars collect fine particles on ciliated tracts.

The hemal and nervous systems are basically similar to those of the asteroids. The ambulacral grooves are closed, and the radial canals of the water-vascular system run just beneath the test, one in each of the ambulacral radii (Figure 25-12).

Sexes are separate, and both eggs and sperm are shed into the sea for external fertilization. The larvae may live a planktonic existence for several months and then metamorphose quickly into young urchins. Sea urchins have been used extensively as models in studies of development.

figure 25-10

Purple sea urchin *Strongylocentrotus purpuratus* is common along the Pacific Coast of North America where there is heavy wave action.

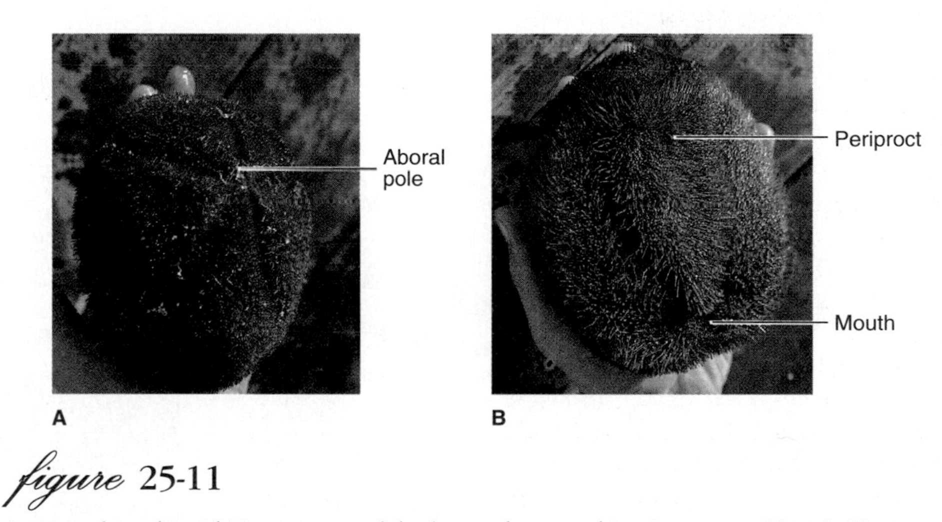

figure 25-11

An irregular echinoid *Meoma,* one of the largest heart urchins (test up to 18 cm). *Meoma* occurs in the West Indies and from the Gulf of California to the Galápagos Islands. **A,** Aboral view. **B,** Oral view. Note curved mouth at anterior end and periproct at posterior end.

Class Holothuroidea: Sea Cucumbers

In a phylum characterized by odd animals, class Holothuroidea contains members that both structurally and physiologically are among the strangest. These animals have a remarkable resemblance to the vegetable after which they are named (Figure 25-14). Compared to the other echinoderms, holothurians are greatly elongated in the oral-aboral axis, and the ossicles (Figure 25-15) are much reduced in most, so that the animals are soft bodied. Some species characteristically crawl on the surface of the sea bottom; others are found beneath rocks, and some are burrowers.

The body wall is usually leathery, with the tiny ossicles embedded in it, although a few species have large ossicles forming a dermal armor. Because of the elongate body form of the sea cucumbers, they characteristically lie on one side. In most species the tube feet are well developed only in the ambulacra normally applied to the substratum. Thus

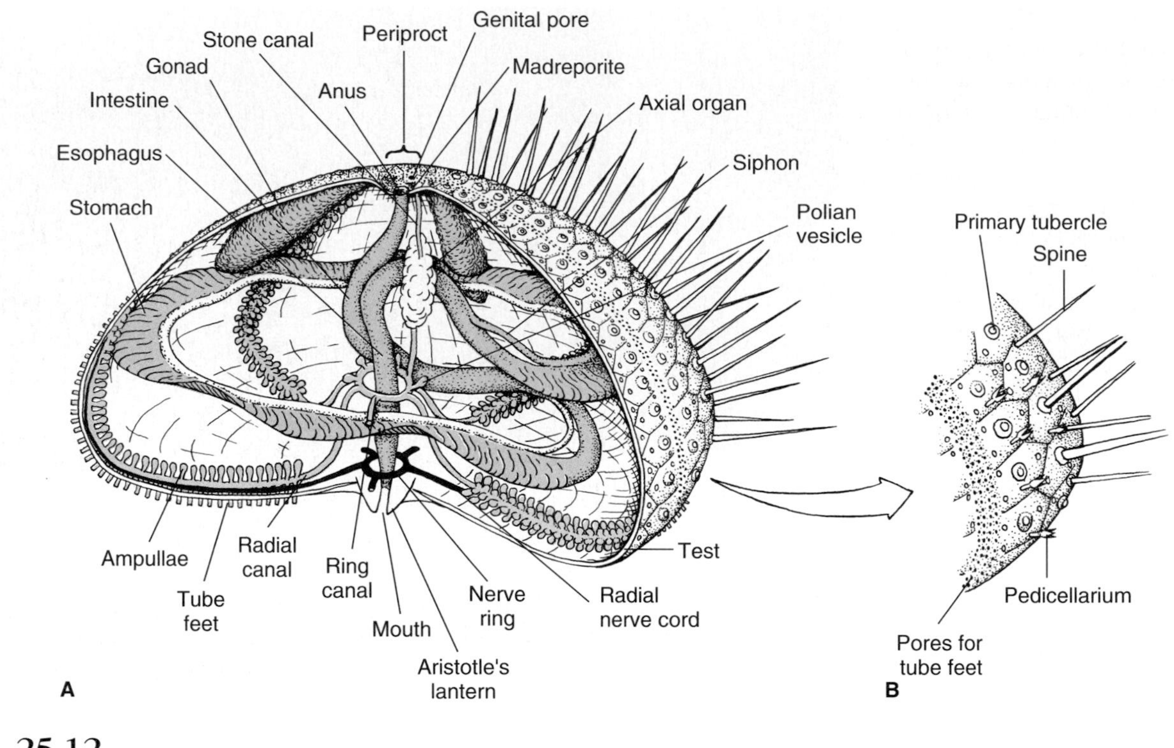

figure 25-12

A, Internal structure of the sea urchin; water-vascular system in orange. **B,** Detail of portion of test.

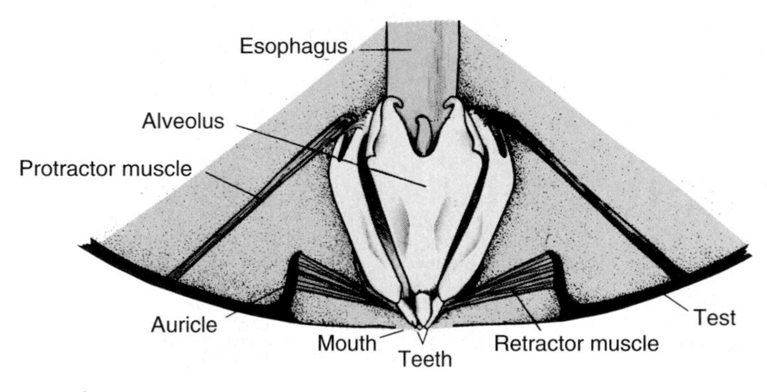

figure 25-13

Aristotle's lantern, the complex mechanism used by the sea urchin for masticating its food. Five pairs of retractor muscles draw the lantern and teeth up into the test; five pairs of protractors push the lantern down and expose the teeth. Other muscles produce a variety of movements. Only major skeletal parts and muscles are shown in this diagram.

figure 25-14

Sea cucumber (class Holothuroidea). Common along the Pacific Coast of North America, *Parastichopus californicus* grows up to 50 cm in length. Its tube feet on the dorsal side are reduced to papillae and warts.

a secondary bilaterality is present, albeit of quite different origin from that of the irregular urchins.

The **oral tentacles** are 10 to 30 retractile, modified tube feet around the mouth. The body wall contains circular and longitudinal muscles along the ambulacra.

The **coelomic cavity** is spacious and filled with fluid. The digestive system empties posteriorly into a muscular **cloaca** (Figure 25-16). A **respiratory tree** composed of two long, many-branched tubes also empties into the cloaca, which pumps seawater into it. The respiratory tree serves both for respiration and excretion and is not found in any other group of living echinoderms. Gas exchange also occurs through the skin and tube feet.

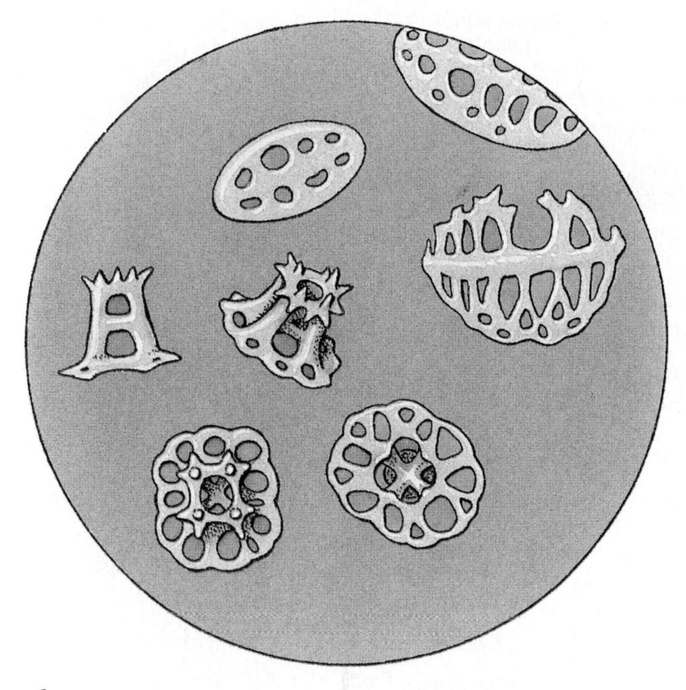

figure 25-15

Ossicles of sea cucumbers are usually microscopic bodies buried in the leathery dermis. They can be extracted from the tissue with commercial bleach and are important taxonomic characteristics. The ossicles shown here, called tables, buttons, and plates, are from the sea cucumber *Holothuria difficilis.* They illustrate the meshwork (stereom) structure observed in ossicles of all echinoderms at some stage in their development. (×250)

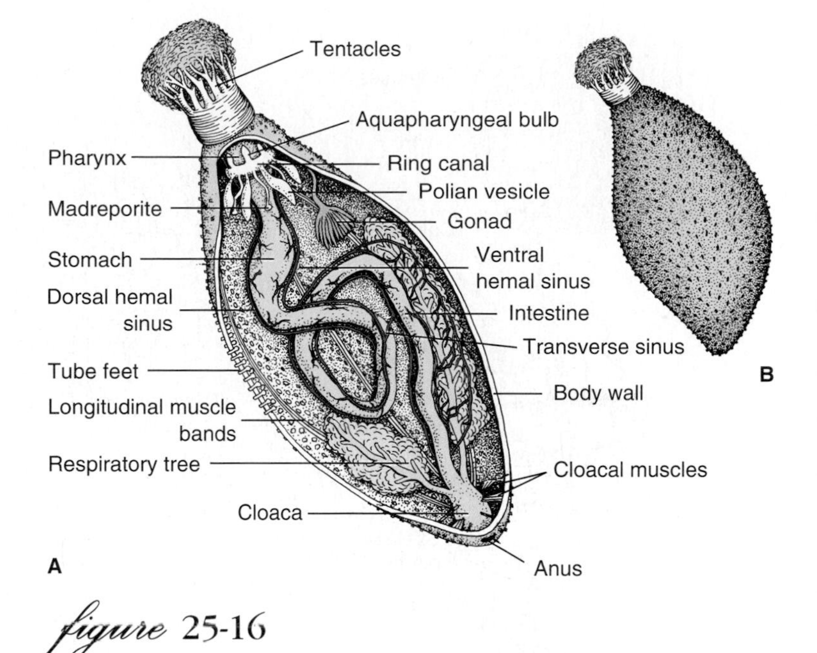

figure 25-16

Anatomy of the sea cucumber *Sclerodactyla.* **A,** Internal view; hemal system in red. **B,** External view.

figure 25-17

Comantheria briareus are crinoids found on Pacific coral reefs. They extend their arms into the water to catch food particles both during the day and at night.

The hemal system is better developed in holothurians than in other echinoderms. The water-vascular system is peculiar in that the madreporite lies free in the coelom.

The sexes are separate, but some holothurians are hermaphroditic. Among the echinoderms, only sea cucumbers have a single gonad, which is considered a primitive character. Fertilization is external.

Sea cucumbers are sluggish, moving partly by means of their ventral tube feet and partly by waves of contraction in the muscular body wall. The more sedentary species trap suspended food particles in the mucus of their outstretched oral tentacles or pick up particles from the surrounding bottom. They then stuff the tentacles into their pharynx, one by one, sucking off the food material. Others crawl along, grazing the bottom with their tentacles.

Class Crinoidea: Sea Lilies and Feather Stars

Crinoids have several primitive characters. As fossil records reveal, crinoids were once far more numerous than now. They differ from other echinoderms by being attached during a substantial part of their lives. Many crinoids are deep-water forms, but feather stars may inhabit shallow waters, especially in the Indo-Pacific and the West Indian–Caribbean regions, where the largest numbers of species are found.

The body disc has a leathery skin containing calcareous plates. The epidermis is poorly developed. Five flexible arms branch to form many more arms, each with many lateral **pinnules** arranged like barbs on a feather (Figure 25-17). Sessile forms have a long, jointed **stalk** attached to the aboral side of the body **(calyx)** (Figure 25-18). This stalk is made up of plates, appears jointed, and may bear **cirri.** Madreporite, spines, and pedicellariae are absent.

The upper (oral) surface bears the mouth and the anus. With the aid of tube feet and mucous nets, crinoids feed on small organisms that they catch in the ambulacral grooves. The **ambulacral grooves** are open and ciliated and serve to carry food to the mouth. Tube feet in the form of tentacles are also

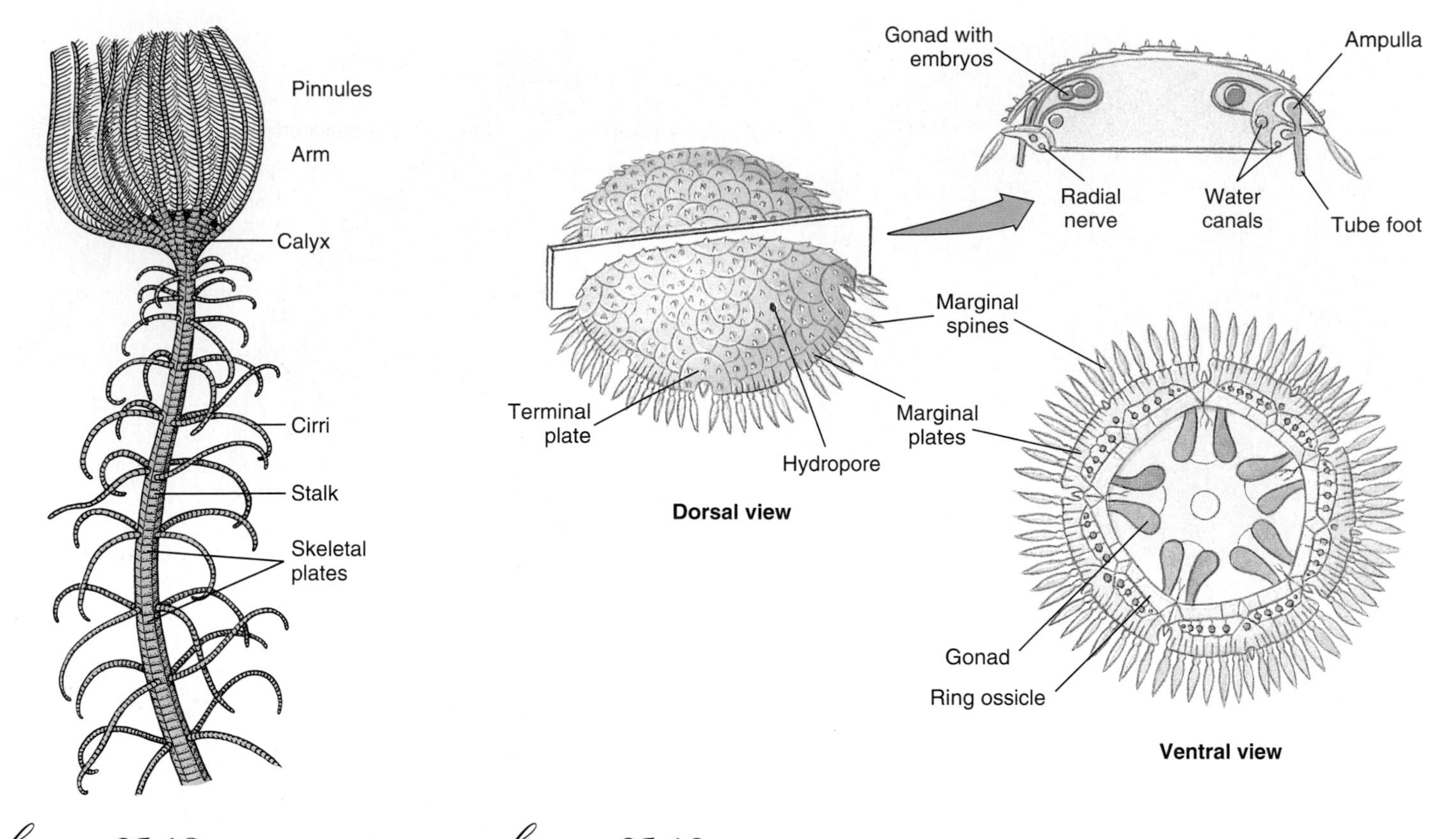

figure 25-18

A stalked crinoid with portion of stalk. Modern crinoid stalks rarely exceed 60 cm, but fossil forms were as much as 20 m long.

figure 25-19

Xyloplax spp. (class Concentricycloidea) are peculiar little disc-shaped echinoderms. With their podia around the margin, they are the only echinoderms not having podia distributed along ambulacral areas.

found in the grooves. The **water-vascular system** has the echinoderm plan. Sense organs are scanty and primitive.

The flower-shaped body of the sea lily is attached to the substratum by a stalk. During metamorphosis feather stars also become sessile and attached, but after several months they detach and become free moving. Although they may remain attached in the same location for long periods, they are capable of crawling and swimming short distances. They swim by alternate sweeping of their long, feathery arms.

The sexes are separate, and the gonads are primitive. The larvae swim freely for a time before they become attached and metamorphose. Most living crinoids are from 15 to 30 cm long, but some fossil species had stalks 25 m in length.

Class Concentricycloidea: Sea Daisies

Strange little (less than 1 cm diameter), disc-shaped animals (Figure 25-19) were discovered in water over 1000 m deep off New Zealand. They are the most recently described (1986) class of echinoderms, and only two species are known so far. They have no arms, and their tube feet are around the periphery of the disc, rather than along ambulacral areas. Their water-vascular system includes two concentric ring canals; the outer ring may represent the radial canals because the podia arise from it. A hydropore, homologous to the madreporite, connects the inner ring canal to the aboral surface.

Phylogeny and Adaptive Radiation

Phylogeny

Despite the existence of an extensive fossil record, there have been contesting hypotheses on echinoderm phylogeny. Based on the embryological evidence of the bilateral larvae of echinoderms, there can be little doubt that their ancestors were bilateral and that their coelom had three pairs of spaces (trimeric). Some investigators have held that radial symmetry arose in a free-moving echinoderm ancestor and that sessile groups were derived several times independently from free-moving ancestors. However, this view does not account for the adaptive significance of radial symmetry as an adaptation for

Classification of Phylum Echinodermata

There are about 6000 living and 20,000 extinct or fossil species of Echinodermata. The traditional classification placed all the free-moving forms that were oriented with oral side down in the subphylum Eleutherozoa, containing most of the living species. The other subphylum, Pelmatozoa, contained mostly forms with stems and oral side up; most of the extinct classes and the living Crinoidea belong to this group. Although alternative schemes have strong supporters, cladistic analysis provides evidence that the two traditional subphyla are monophyletic groups.[2] The following includes only groups with living members.

Subphylum Pelmatozoa (pel-ma′to-zo′a) (Gr. *pelmatos,* a stalk, + *zōon,* animal). Body in form of a cup or calyx, borne on aboral stalk during part or all of life; oral surface directed upward; open ambulacral grooves; madreporite absent; both mouth and anus on oral surface; several fossil classes plus living Crinoidea.

Class Crinoidea (krin-oy′de-a) (Gr. *krinon,* lily, + *eidos,* form, + *-ea,* characterized by): **sea lilies** and **feather stars.** Five arms branching at base and bearing pinnules; ciliated ambulacral grooves on oral surface with tentacle-like tube feet for food gathering; spines, madreporite, and pedicellariae absent. Examples: *Antedon, Nemaster, Comantherla* (Figure 25-17).

Subphylum Eleutherozoa (e-lu′ther-o-zo′a) (Gr. *eleutheros,* free, not bound, + *zōon,* animal). Body form star-shaped, globular, discoidal, or cucumber shaped; oral surface directed toward substratum or oral-aboral axis parallel to substratum; body with or without arms; ambulacral grooves open or closed.

Class Concentricycloidea (kon-sen′tri-sy-kloy′de-a) (L. *cum,* together, + *centrum,* center [having a common center], + Gr., *kyklos,* circle, + *eidos,* form, + *-ea,* characterized by): **sea daisies.** Disc-shaped body, with marginal spines but no arms; concentrically arranged skeletal plates; ring of suckerless podia near body margin; hydropore present; gut present or absent, no anus. Example: *Xyloplax* (Figure 25-19).

Class Asteroidea (as′ter-oy′de-a) (Gr. *aster,* star, + *eidos,* form, + *-ea,* characterized by): **sea stars.** Star shaped, with arms not sharply demarcated from the central disc; ambulacral grooves open, with tube feet on oral side; tube feet often with suckers; anus and madreporite aboral; pedicellariae present. Examples: *Asterias, Pisaster.*

Class Ophiuroidea (o′fe-u-roy′de-a) (Gr. *ophis,* snake, + *oura,* tail, + *eidos,* form, + *-ea,* characterized by): **brittle stars** and **basket stars.** Star shaped, with arms sharply demarcated from central disc; ambulacral grooves closed, covered by ossicles; tube feet without suckers and not used for locomotion; pedicellariae absent. Examples: *Ophiura* (Figure 25-7), *Astrophyton* (Figure 25-7).

Class Echinoidea (ek′i-noy′de-a) (Gr. *echinos,* sea urchin, hedgehog, + *eidos,* form, + *-ea* characterized by): **sea urchins, sea biscuits,** and **sand dollars.** More or less globular or disc-shaped, with no arms; compact skeleton or test with closely fitting plates; movable spines; ambulacral grooves closed; tube feet often with suckers; pedicellariae present. Examples: *Arbacia, Strongylocentrotus* (Figure 25-10), *Lytechinus, Meoma* (Figure 25-11).

Class Holothuroidea (hol′o-thu-roy′de-a) (Gr. *holothourion,* sea cucumber, + *eidos,* form, + *−ea,* characterized by): **sea cucumbers.** Cucumber-shaped, with no arms; spines absent; microscopic ossicles embedded in muscular body wall; anus present; ambulacral grooves closed; tube feet with suckers; circumoral tentacles (modified tube feet); pedicellariae absent; madreporite plate internal. Examples: *Sclerodactyla, Parastichopus* (Figure 25-14), *Cucumaria.*

[2]Brusca, R. C., and G. J. Brusca. 1990. Invertebrates. Sunderland Massachusetts, Sinauer Associates. Meglitsch, P. A., and F. R. Schram. 1991. Invertebrate zoology, ed. 3. New York, Oxford University Press. Paul, C.R.C., and A.B. Smith. 1984. The early radiation and phylogeny of echinoderms. Biol. Rev. **59:**443–481.

sessile existence. The more traditional view is that the first echinoderms were sessile, became radial as an adaptation to that existence, and then gave rise to the free-moving groups. Figure 25-20 is consistent with this hypothesis. It views the evolution of endoskeletal plates with stereom structure and of external ciliary grooves for feeding as early echinoderm (or pre-echinoderm) developments. The extinct carpoids (Homalozoa, Figure 25-20) had stereom ossicles but were not radially symmetrical, and the status of their water-vascular system, if any, is uncertain. Some investigators regard carpoids as a separate subphylum of echinoderms (Homalozoa), and others believe they represent a group of pre-echinoderms that shows affinities to the chordates (Calcichordata, p. 570). The fossil helicoplacoids (Figure 25-20) show evidence of three, true ambulacral grooves, and their mouth was on the side of the body.

Attachment to the substratum by the aboral surface would have led to radial symmetry and the origin of the Pelmatozoa. An ancestor that became free moving and applied its oral surface to the substratum would have given rise to the

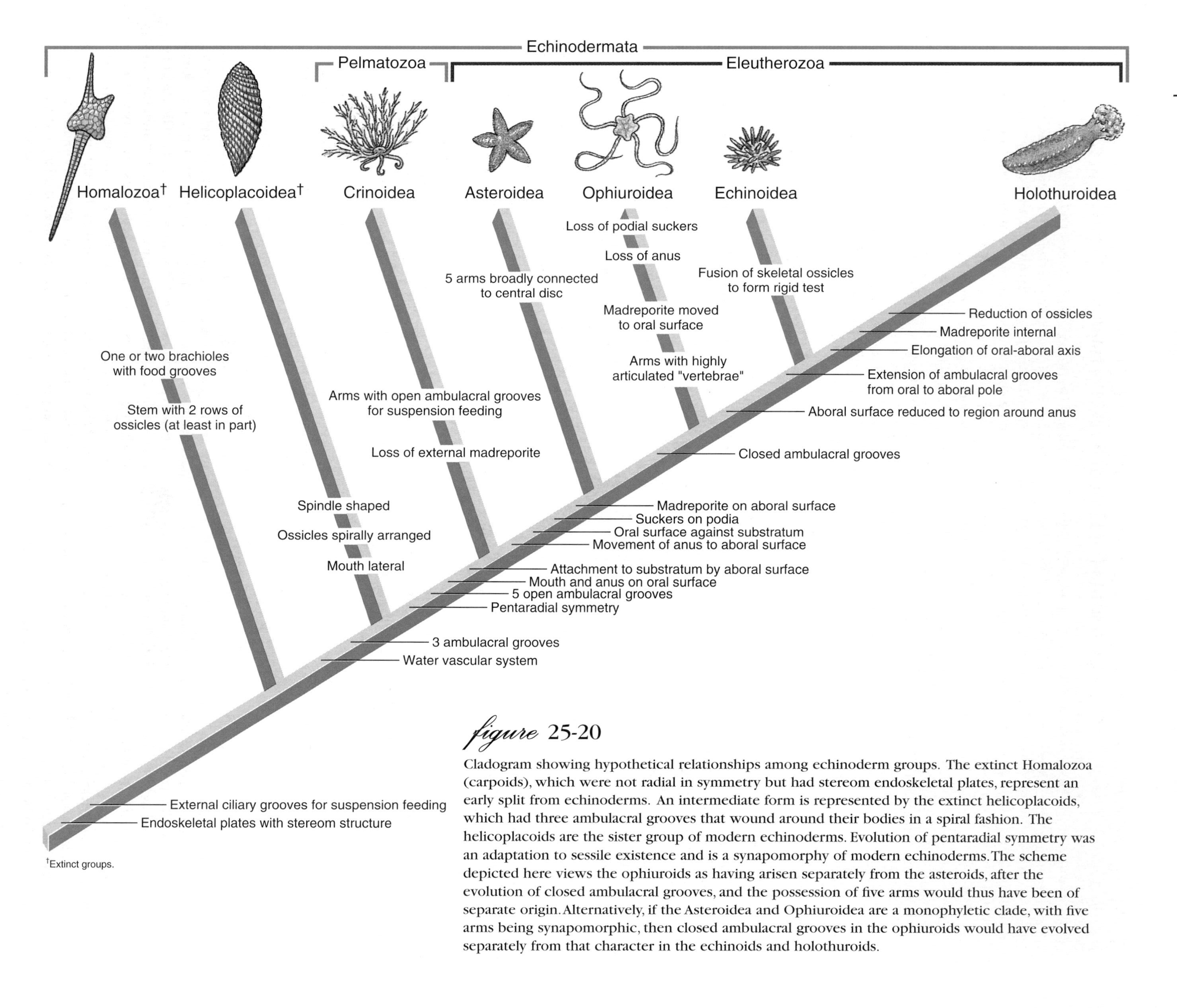

figure 25-20

Cladogram showing hypothetical relationships among echinoderm groups. The extinct Homalozoa (carpoids), which were not radial in symmetry but had stereom endoskeletal plates, represent an early split from echinoderms. An intermediate form is represented by the extinct helicoplacoids, which had three ambulacral grooves that wound around their bodies in a spiral fashion. The helicoplacoids are the sister group of modern echinoderms. Evolution of pentaradial symmetry was an adaptation to sessile existence and is a synapomorphy of modern echinoderms. The scheme depicted here views the ophiuroids as having arisen separately from the asteroids, after the evolution of closed ambulacral grooves, and the possession of five arms would thus have been of separate origin. Alternatively, if the Asteroidea and Ophiuroidea are a monophyletic clade, with five arms being synapomorphic, then closed ambulacral grooves in the ophiuroids would have evolved separately from that character in the echinoids and holothuroids.

Eleutherozoa. Phylogeny within the Eleutherozoa is controversial. Most investigators agree that the echinoids and holothuroids are related and form a single clade, but opinions diverge on the relationship of the ophiuroids and asteroids. Figure 25-20 illustrates the view that the ophiuroids arose after the closure of ambulacral grooves, but this scheme treats the evolution of five ambulacral rays (arms) in the ophiuroids and asteroids as independently evolved. Alternatively, if the ophiuroids and asteroids are a single clade, then closed ambulacral grooves must have evolved separately in ophiuroids and in the common ancestor of echinoids and holothuroids.

Data on the Concentricycloidea are insufficient to place this group on a cladogram.

Adaptive Radiation

The radiation of the echinoderms has been determined by the limitations and potentials of their most important characteristics: radial symmetry, the water-vascular system, and their dermal endoskeleton. If their ancestors had a brain and specialized sense organs, these were lost in the adoption of radial symmetry. Thus, it is not surprising that there are large numbers of creeping, benthic forms with filter-feeding, deposit-feeding, scavenging, and herbivorous habits, comparatively few predators, and very few pelagic species. In this light the relative success of the asteroids as predators is impressive and probably attributable to the extent to which they have exploited the hydraulic mechanism of the tube feet.

Phylum Hemichordata: Acorn Worms

The Hemichordata (hem′i-kor-da′ta) (Gr. *hemi,* half, + *chorda,* string, cord) are marine animals that were formerly considered a subphylum of the chordates, based on their possession of gill slits, a rudimentary notochord, and a dorsal nerve cord. However, zoologists now agree that the so-called hemichordate "notochord" is really an evagination of the mouth cavity and not homologous with the chordate notochord, so the hemichordates are given the rank of a separate phylum.

Hemichordates are wormlike bottom dwellers, living usually in shallow waters. Some are colonial and live in secreted tubes. Many are sedentary or sessile. They are widely distributed, but their secretive habits and fragile bodies make collecting them difficult.

Members of class Enteropneusta (acorn worms) range from 20 mm to 2.5 m in length and 3 to 200 mm in breadth. Members of class Pterobranchia (pterobranchs) are smaller, usually from 5 to 14 mm, not including the stalk. About 70 species of enteropneusts and three small genera of pterobranchs have been described.

Hemichordates have the typical tricoelomate structure of deuterostomes.

Position in Animal Kingdom

1. Hemichordates belong to the deuterostome branch of the animal kingdom and are enterocoelous coelomates with radial cleavage.
2. Hemichordates show some of both echinoderm and chordate characteristics.
3. A chordate plan of structure is suggested by gill slits and a restricted dorsal tubular nerve cord.
4. Similarity to the echinoderms is shown in larval characteristics.

Biological Contributions

1. A tubular dorsal nerve cord in the collar zone may represent an early stage of the condition in chordates; a diffused net of nerve cells is similar to the uncentralized, subepithelial plexus of echinoderms.
2. The gill slits in the pharynx, also characteristic of chordates, serve primarily for filter feeding and only secondarily for breathing and are thus comparable to gill slits in the protochordates.

Class Enteropneusta

The enteropneusts, or acorn worms (Figure 25-21), are sluggish wormlike animals that live in burrows or under stones, usually in mud or sand flats of intertidal zones.

The mucus-covered body is divided into a tonguelike **proboscis,** a short **collar,** and a long **trunk** (protosome, mesosome, and metasome.)

In the posterior end of the proboscis is a small coelomic sac (protocoel) into which extends the buccal diverticulum, a slender, blindly ending pouch of the gut that reaches forward into the buccal region and was formerly believed to be a notochord. A row of gill pores extends dorsolaterally on each side of the trunk just behind the collar (Figure 25-21). These gill pores open from a series of gill chambers that in turn connect with a series of gill slits in the sides of the pharynx. The primary function of these structures is not respiration, but food gathering (Figure 25-22). Food particles caught in mucus and brought to the mouth by ciliary action on the proboscis and collar are strained out of the branchial water that leaves through the gill slits. The particles are then directed along the ventral part of the pharynx and esophagus to the intestine.

A mid-dorsal vessel carries blood forward to a heart above the buccal diverticulum. The blood then flows into a network of sinuses that may have an excretory function, then posteriorly through a ventral blood vessel and through a network of sinuses in the gut and body wall. Acorn worms have dorsal and ventral nerve cords, and the dorsal cord is hollow in some. Sexes are separate. In some enteropneusts there is a free-swimming larva that closely resembles the larva of sea stars.

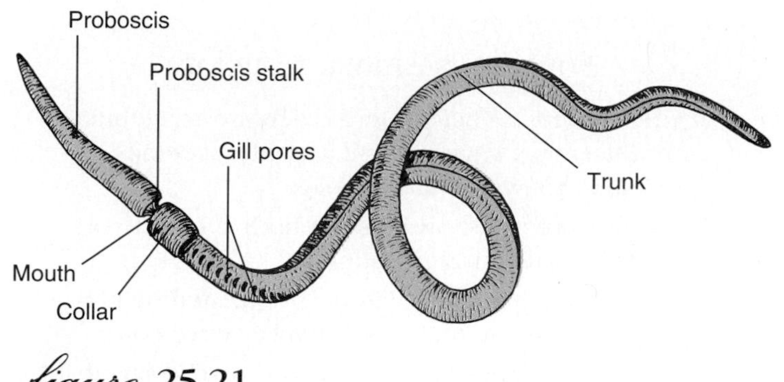

figure 25-21

External lateral view of the acorn worm, *Saccoglossus* (phylum Hemichorlata).

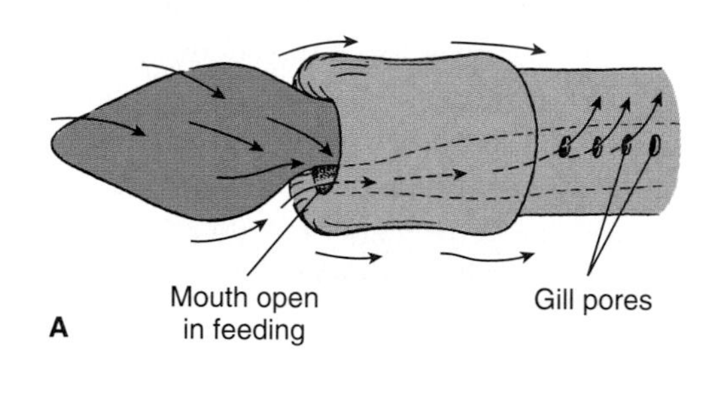

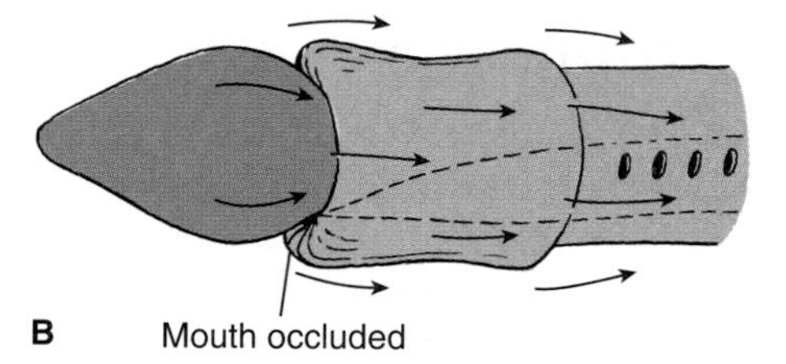

figure 25-22

Food currents of enteropneust hemichordate. **A,** Side view of acorn worm with mouth open, showing direction of currents created by cilia on proboscis and collar. Food particles are directed toward mouth and digestive tract. Rejected particles move toward outside of collar. Water leaves through gill pores. **B,** When mouth is occluded, all particles are rejected and passed onto the collar. Nonburrowing and some burrowing hemichordates use this feeding method.

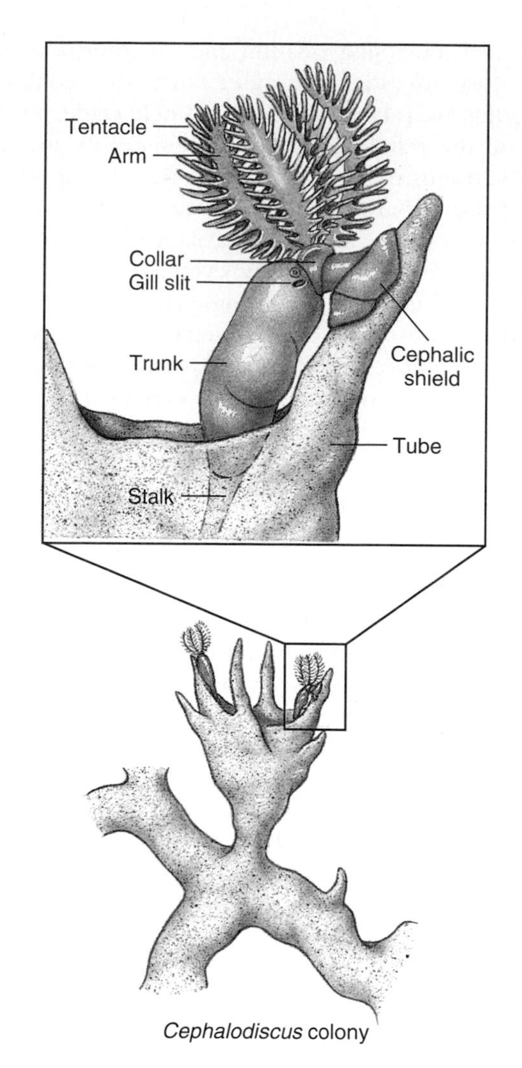

figure 25-23

Cephalodiscus, a pterobranch hemichordate. These tiny (5 to 7 mm) forms live in tubes in which they can move freely. Ciliated tentacles and arms direct currents of food and water toward mouth.

Class Pterobranchia

The basic plan of the class Pterobranchia is similar to that of the Enteropneusta, but certain differences are correlated with the sedentary mode of life of pterobranchs. Only two genera are known in any detail. In both genera there are arms with tentacles containing an extension of the coelomic compartment of the mesosome, as in a lophophore (Figure 25-23). One genus has a single pair of gill slits, and the other has none. Both live in tubes from which they project their proboscis and tentacles to feed by mucociliary mechanisms. Some species are dioecious, others are monoecious, and asexual reproduction occurs by budding.

Phylogeny

Hemichordates share characteristics with both the echinoderms and the chordates. With chordates they share the gill slits, which serve primarily for filter feeding and secondarily for breathing, as they do in some of the protochordates. In addition, a short dorsal, somewhat hollow nerve cord in the collar zone may be homologous to the nerve cord of the chordates. Their early embryogenesis is remarkably like that of echinoderms, and the early larva is almost identical with the bipinnaria larva of asteroids (Figure 25-24). However, the hypothetical relationships shown in Figure 25-25 place the lophophorates as sister groups of the hemichordates and chordates, required by the proposed synapomorphy for all these groups of a crown of ciliated tentacles containing extensions of the mesocoel.

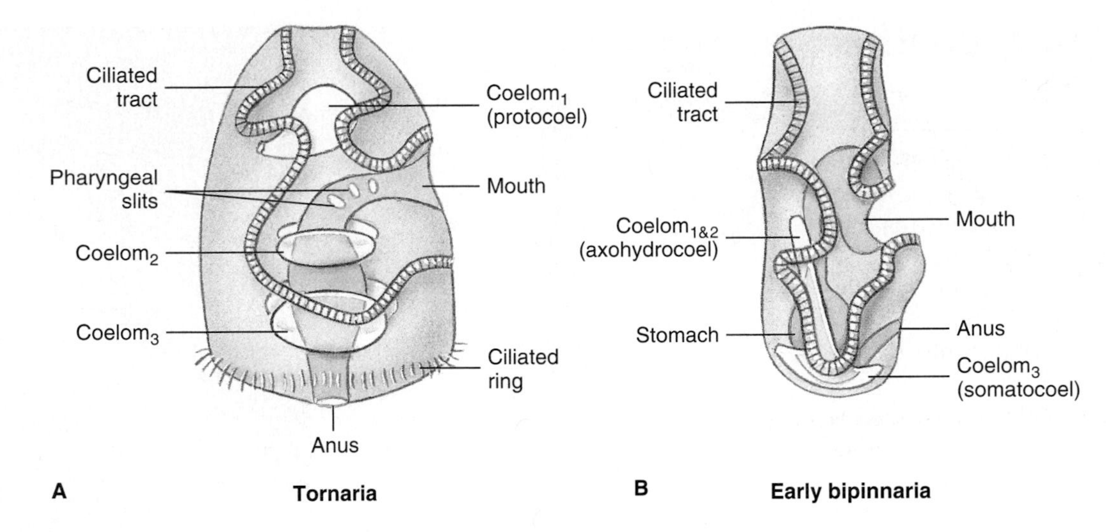

figure 25-24

Comparison of a hemichordate tornaria **(A)**, to an echinoderm bipinnaria **(B)**.

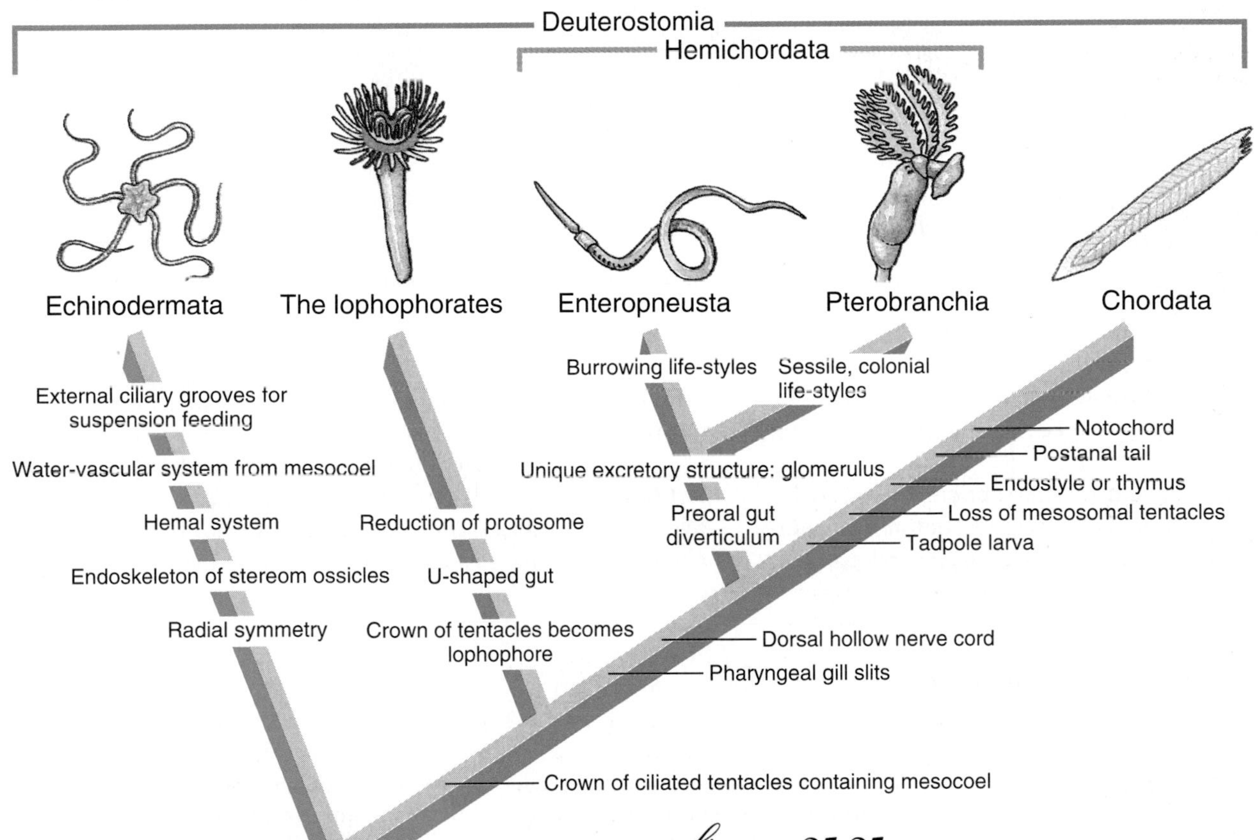

figure 25-25

Cladogram showing hypothetical relationships among deuterostome phyla. The crown of ciliated tentacles (containing extensions of the mesocoel) is here considered a character borne by the ancestors of the lophophorates, hemichordates, and chordates. The crown would have been lost in the line leading to the chordates, and among the hemichordates, the enteropneusts. The pterobranchs retain the primitive character, whereas in the lophophorate phyla it is modified into a lophophore. Uncertainties regarding the relationships of the Chaetognatha preclude placement in this cladogram. The Protostomia serves as outgroup.

(Source: *After Brusca, R.C., and G.J. Brusca. 1990.* Invertebrates. *Sunderland, MA, Sinauer Associates.*

Phylum Chaetognatha: Arrowworms

The Chaetognatha (ke-tog′na-tha) (Gr. *chaitē*, long flowing hair, + *gnathos*, jaw) is a small group (65 species) of marine animals highly specialized for a planktonic existence. Their relationship to other groups is obscure, but their embryology suggests a relationship to the deuterostomes.

Their small, straight bodies resemble miniature torpedoes, or darts, ranging from 2.5 to 10 cm in length.

Arrowworms usually swim to the surface at night and descend during the day. Much of the time they drift passively, but they can dart forward in swift spurts, using the caudal fin and longitudinal muscles—a fact that no doubt contributes to their success as planktonic predators. Horizontal fins bordering the trunk function in flotation rather than in active swimming.

The body of arrowworms is unsegmented and is made up of the head, trunk, and postanal tail (Figure 25-26). They have teeth and chitinous spines on the head around the mouth. When the animal captures prey, the teeth and raptorial spines spread apart and then snap shut with startling speed. Arrowworms are voracious feeders, living on other planktonic forms, especially copepods, and even small fish (Figure 25-26B).

The body is covered with a thin cuticle. They have a complete digestive system, well-developed coelom, and nervous system with a nerve ring containing several ganglia. Vascular, respiratory, and excretory systems, however, are entirely lacking.

Arrowworms are hermaphroditic with either cross-fertilization or self-fertilization. Juveniles develop directly without metamorphosis. There is no true peritoneum lining the coelom. Cleavage is radial, complete, and equal.

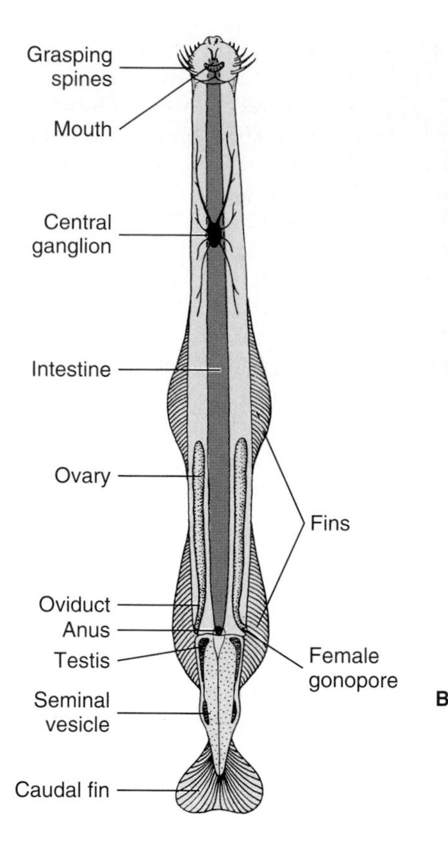

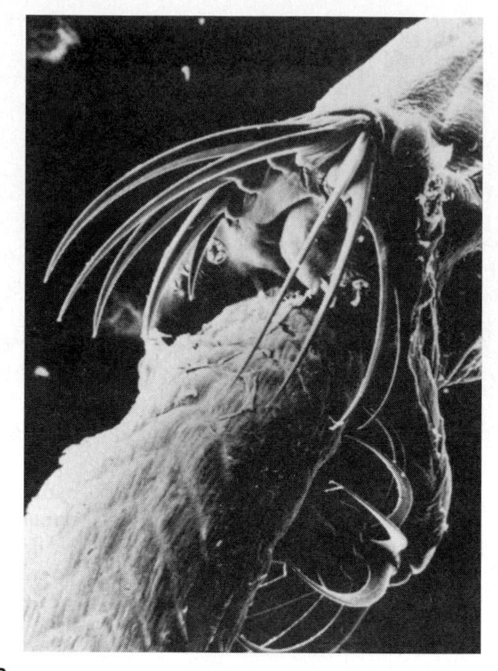

figure 25-26

Arrowworms. **A,** Internal structure of *Sagitta*. **B,** Scanning electron micrograph of a juvenile arrowworm, *Flaccisagitta hexaptera* (35 mm length) eating a larval fish.

Summary

The phyla Echinodermata, Chordata, and Hemichordata show the characteristics of the Deuterostomia division of the animal kingdom. Molecular evidence does not support placement of the Chaetognatha among the deuterostomes, but this issue is yet to be clarified.

The echinoderms are an important marine group sharply distinguished from other phyla of animals. They have a pentaradial symmetry but were derived from bilateral ancestors. They fill a variety of benthic niches, including particle feeders, browsers, scavengers, and predators.

The sea stars (class Asteroidea) usually have five arms, and the arms merge gradually with a central disc. They have no head and few specialized sensory organs. The mouth is central on the under (oral) side of the body. They have an endoskeleton of dermal ossicles, open ambulacral areas, pedicellariae, and papulae. The water-vascular system is an elaborate hydraulic system derived from one of the coelomic cavities. The madreporite, opening to the outside, connects to the ring canal around the esophagus by way of the stone canal, and radial canals extend from the ring canal along each ambulacral area. Branches from the radial canals lead to the many tube feet, structures that are important in locomotion, food gathering, respiration, and excretion. Many sea stars are predators, whereas others feed on particulate organic matter. Sexes are separate, and the reproductive systems are very simple. The bilateral, free-swimming larva becomes attached, transforms to a radial juvenile, then detaches and becomes a motile sea star.

Brittle stars (class Ophiuroidea) have slender arms that are sharply set off from the central disc; they have no pedicellariae or ampullae, and their ambulacral grooves are closed. Their madreporite is on the oral side. They crawl by means of their arms, and they can move around more rapidly than other echinoderms.

In sea urchins (class Echinoidea), the dermal ossicles have become closely fitting plates, and there are no arms. Some urchins (sand dollars and heart urchins) have evolved a return to bilateral symmetry.

Sea cucumbers (class Holothuroidea) have very small dermal ossicles and a soft body wall. They are greatly elongated in the oral-aboral axis and lie on their side. Because three of their ambulacral areas characteristically lie against the substratum, sea cucumbers also have undergone some return to bilateral symmetry.

Sea lilies and feather stars (class Crinoidea) are the only group of living echinoderms, other than asteroids, with open ambulacral areas. They are mucociliary particle feeders and lie with their oral side upward. They are attached to the substratum by a stem for a substantial part of their lives.

Members of the phylum Hemichordata are marine worms that were formerly considered chordates because their buccal diverticulum was considered a notochord. In common with the chordates some of them do have gill slits and a hollow, dorsal nerve cord. The hemichordates are important phylogenetically because they show affinities with the chordates, echinoderms, and lophophorates, and they are the likely sister group of the chordates.

The arrowworms (phylum Chaetognatha) are a small group but are important as a component of marine plankton.

Review Questions

1. What is the constellation of characteristics possessed by echinoderms that is found in no other phylum?
2. How do we know that echinoderms were derived from an ancestor with bilateral symmetry?
3. Distinguish the following groups of echinoderms from each other: Crinoidea, Asteroidea, Ophiuroidea, Echinoidea, Holothuroidea, Concentricycloidea.
4. What is an ambulacral groove, and what is the difference between open and closed ambulacral grooves? Open ambulacra is considered the primitive condition, and closed ambulacra is considered derived. Can you suggest a reason why this is probably correct?
5. Trace or make a rough copy of Figure 25-3B, without the labels, then from memory label the parts of the water-vascular system of the sea star.
6. Name the structures involved in the following functions in sea stars, and briefly describe the action of each: respiration, feeding and digestion, excretion, reproduction.
7. Compare the structures and functions in question 6 as they are found in brittle stars, sea urchins, sea cucumbers, and crinoids.
8. Match the groups in the left column with *all* correct answers in the right column.

___ Crinoidea	a. Closed ambulacral grooves
___ Asteroidea	b. Oral surface generally upward
___ Ophiuroidea	c. With arms
___ Echinoidea	d. Without arms
___ Holothuroidea	e. Approximately globular or disc shaped
	f. Elongated in oral-aboral axis
	g. With pedicellariae
	h. Madroporite internal
	i. Madreporite on oral plate

9. Define the following: pedicellariae, madreporite, respiratory tree, Aristotle's lantern.
10. That the ancestor of the eleutherozoan groups was a radial, sessile organism is a widely held hypothesis concerning echinoderm evolution. What is the reasoning for this hypothesis?
11. Give three examples of how echinoderms are important to humans.
12. Distinguish the Enteropneusta from the Pterobranchia.
13. The Hemichordata were once considered chordates. Why?
14. Although hemichordates are not now considered chordates, these two phyla appear to form a monophyletic group. What is evidence for this relationship?
15. What are four morphological characteristics of Chaetognatha? Why is their status as deuterostomes now in doubt?
16. What is the ecological importance of arrowworms?

Selected References

See also general references for Part Three, p. 562.

Baker, A. N., F. W. E. Row, and H. E. S. Clark. 1986. A new class of Echinodermata from New Zealand. Nature **321:**862–864. *The strange class Concentricycloidea is described.*

Bieri, R., and E. V. Thuesen. 1990. The strange worm *Bathybelos*. Am. Sci. **78:**542–549. Bathybelos *is a peculiar chaetognath with a dorsal nervous system, a characteristic shared in the animal kingdom only with the Hemichordata and Chordata. The authors contend that the character in chaetognaths is convergent with that in the hemichordates and chordates.*

Birkeland, C. 1989. The Faustian traits of the crown-of-thorns starfish. Am. Sci. **77:**154–163. *The fast growth in the early years of the life of an* Acanthaster planci *results in loss of body integrity in later life.*

Davidson, E. H., B. R. Hough-Evans, and R. J. Britten. 1982. Molecular biology of the sea urchin embryo. Science **217:**17–26. *Many fundamental insights into the process of embryogenesis have been revealed through studies of sea urchins.*

Gilbert, S. F. 1994. Developmental biology, ed. 4. Sunderland, Massachusetts, Sinauer Associates. *Any modern text in developmental biology, such as this one, provides a multitude of examples in which studies on echinoderms have contributed (and continue to contribute) to our knowledge of development.*

Hughes, T. P. 1994. Catastrophes, phase shifts and large-scale degradation of a Caribbean coral reef. Science **265:**1547–1551. *Describes the sequence of events, including the die-off of sea urchins, leading to the destruction of the coral reefs around Jamaica.*

Lawrence, J. 1987. A functional biology of echinoderms. Baltimore, The Johns Hopkins University Press. *Treats many aspects of the fascinating biology of these organisms.*

Moran, P. J. 1990. *Acanthaster planci* (L.): biographical data. Coral Reefs 9:95–96. *Presents a summary of essential biological data on* A. planci. *This entire issue of Coral Reefs is devoted to* A. planci.

Thuesen, E. V., and K. Kogure. 1989. Bacterial production of tetrodotoxin in four species of Chaetognatha. Biol. Bull. **176:**191–194. *Chaetognaths use venom to enhance prey capture, and the venom (tetrodotoxin) is produced by bacteria* (Vibrio alginolyticus).

References to Part III

The references below pertain to groups covered in more than one chapter of Part III. They include a number of very valuable field manuals that aid in identification, as well as general texts.

Barrington, E. J. W. 1979. Invertebrate structure and function, ed. 2. New York, John Wiley & Sons, Inc. *Excellent account of function in major invertebrate groups.*

Brusca, R. C., and G. J. Brusca. 1990. Invertebrates. Sunderland, Massachusetts, Sinauer Associates, Inc. *Invertebrate text organized around the bauplan ("body plan") concept—structural range, architectural limits, and functional aspects of a design—for each phylum. Includes cladistic analysis of phylogeny for most groups.*

Conway Morris, S., J. D. George, K. Gibson, and H. M. Platt (eds.). 1985. The origins and relationships of lower invertebrates. Oxford, Clarendon Press. *Technical discussions of phylogenetic relationships among lower invertebrates; essential for serious students of this topic.*

Fotheringham, N. 1980. Beachcomber's guide to Gulf Coast marine life. Houston, Texas, Gulf Publishing Company. Coverage arranged by habitats. *No keys, but common forms that occur near shore can be identified.*

Gosner, K. L. 1979. A field guide to the Atlantic seashore: invertebrates and seaweeds of the Atlantic coast from the Bay of Fundy to Cape Hatteras. The Peterson Field Guide Series. Boston, Houghton Mifflin Company. *A helpful aid for students of invertebrates found along the northeastern coast of the United States.*

Humann, P. 1992. Reef creature identification. Florida, Caribbean, Bahamas. Jacksonville, Florida, New World Publications. *Excellent field guide to aid identification of Atlantic reef invertebrates except corals.*

Hyman, L. H. 1940–1967. The invertebrates, 6 vols. New York, McGraw-Hill Book Company. *Informative discussions on the phylogenies of most invertebrates are treated in this outstanding series of monographs. Volume I contains a discussion of the colonial hypothesis of the origin of metazoa, and volume 2 contains a discussion of the origin of bilateral animals, body cavities, and metamerism.*

Kaplan, E. H. 1988. A field guide to southeastern and Caribbean seashores: Cape Hatteras to the Gulf Coast, Florida, and the Caribbean. A Peterson Field Guide Series. Boston, Houghton Mifflin Company. *More than just a field guide, this comprehensive book is filled with information on the biology of seashore animals; complements Gosner's field guide which covers animals norlh of Cape Hatteras.*

Kozloff, E. N. 1987. Marine invertebrates of the Pacific Northwest. Seattle, University of Washington Press. *Contains keys for many marine groups.*

Kozloff, E. N. 1990. Invertebrates. Philadelphia, Saunders College Publishing. *A good invertebrate text, less exhaustive than Ruppert and Barnes (1994) or Brusca and Brusca (1990). Cladistic analysis not emphasized.*

Lane, R. P., and H. W. Crosskey. 1992. Medically important insects and arachnids. London, Chapman and Hall. *The most up-to-date medical entomology text available.*

Meglitsch, P. A., and F. R. Schram. 1991. Invertebrate zoology, ed. 3. New York, Oxford University Press. *Schram's thorough revision of the older Meglitsch text. Includes cladistic treatments.*

Morris, R. H., D. P. Abbott, and E. C. Haderlie. 1980. Intertidal invertebrates of California. Stanford, Stanford University Press. *An essential reference on the most important invertebrates of the intertidal zone in California. Contains 900 color photographs.*

Nielson, C. 1995. Animal evolution: interrelationships of the living phyla. New York, Oxford University Press. *Cladistic analysis of morphology is used to develop sister-group relationships of the living Metazoa. An advanced but essential reference.*

Pennak, R. W. 1989. Freshwater invertebrates of the United States, ed. 3. New York, John Wiley & Sons, Inc. *Contains keys for identification of freshwater invertebrates with brief accounts of each group. Indispensable for freshwater biologists.*

Ricketts, E. F., J. Calvin, and J. W. Hedgpeth (revised by D. W. Phillips). 1985. Between Pacific tides, ed. 5. Stanford, Stanford University Press. *A revision of a classic work in marine biology. It stresses the habits and habitats of the Pacific coast invertebrates, and the illustrations are revealing. It includes an excellent, annotated systematic index and bibliography.*

Roberts, L. S., and J. Janovy, Jr. 1996. Foundations of parasitology, ed. 5. Dubuque, Wm. C. Brown Publishers. *Highly readable and up-to-date account of parasitic protistans, worms and arthropods.*

Ruppert, E. E., and R. D. Barnes. 1994. Invertebrate zoology, ed. 6. Philadelphia, Saunders College Publishing. *Authoritative, detailed coverage of invertebrate phyla.*

Smith, D. L. 1977. A guide to marine coastal plankton and marine invertebrate larvae. Dubuque, Iowa, Kendall/Hunt Publishing Company. *Valuable manual for identification of marine plankton, which is usually not covered in most field guides.*

Willmer, P. 1990. Invertebrate relationships. Patterns in animal evolution. Cambridge, Cambridge University Press. *Articulate statement of invertebrate phylogeny from a non-cladist. Good account of the polyphyletic hypothesis for the origin of arthropods.*

four

The Vertebrate Animals

26 Vertebrate Beginnings: The Chordates
27 Fishes
28 The Early Tetrapods and Modern Amphibians
29 Reptiles
30 Birds
31 Mammals

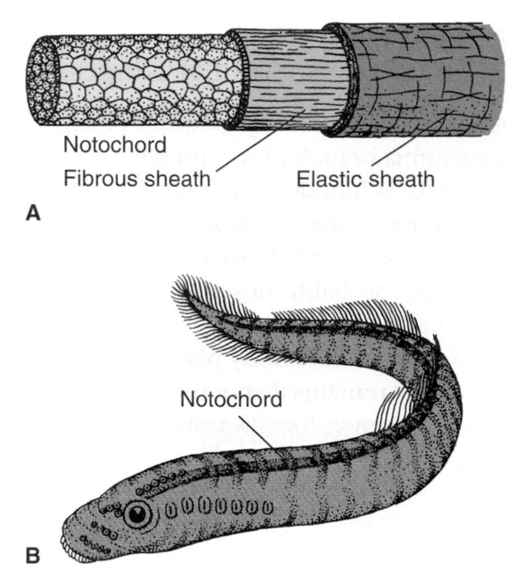

figure 26-1

A, Structure of the notochord and its surrounding sheaths. Cells of the notochord proper are thick walled, pressed together closely, and filled with semifluid. Stiffness is caused mainly by turgidity of fluid-filled cells and surrounding connective tissue sheaths. This primitive type of endoskeleton is characteristic of all chordates at some stage of the life cycle. The notochord provides longitudinal stiffening of the main body axis, a base for trunk muscles, and an axis around which the vertebral column develops. **B,** In hagfishes and lampreys it persists throughout life, but in other vertebrates it is largely replaced by the vertebrae. In mammals slight remnants are found in nuclei pulposi of intervertebral discs. The method of notochord formation is different in the various groups of vertebrates. In amphioxus it originates from the endoderm; in birds and mammals it arises as an anterior outgrowth of the embryonic primitive streak.

embryological, behavioral, chromosomal, or molecular in nature. Although the cladogram shows the *relative* time of origin of the novel properties of taxonomic groups and their specific positions in the hierarchical system of evolutionary common descent, it contains no timescale or information on ancestral lineages. By contrast, the branches of a phylogenetic tree are intended to represent real lineages that occurred in the evolutionary past. Geological information regarding ages of lineages is added to the information from the cladogram to generate a phylogenetic tree for the same taxa.

In our treatment of the chordates, we have retained the traditional Linnaean classification (p. 579) because of its conceptual usefulness and because the alternative—thorough revision following cladistic principles—would require extensive change and the virtual abandonment of familiar rankings. However, we have tried to use monophyletic taxa as much as possible, because such usage is consistent with both evolutionary and cladistic taxonomy (see p. 346).

Several of the traditional divisions of the phylum Chordata used in Linnaean classifications are shown in Table 26-1. A fundamental separation is the Protochordata from the Vertebrata. Since the former lack a well-developed head, they are also called Acraniata. All vertebrates have a well-developed skull case enclosing the brain and are called Craniata. The vertebrates (craniates) may be variously subdivided into groups based on shared possession of characteristics. Two such subdivisions shown in Table 26-1 are: (1) Agnatha, vertebrates lacking jaws (hagfishes and lampreys), and Gnathostomata, vertebrates having jaws (all other vertebrates) and (2) Amniota, vertebrates whose embryos develop within a fluid-filled sac, the amnion (reptiles, birds, and mammals), and Anamniota, vertebrates lacking this adaptation (fishes and amphibians). The Gnathostomata in turn can be subdivided into Pisces, jawed vertebrates with limbs (if any) in the shape of fins; and the Tetrapoda (Gr. *tetras,* four, + *podos,* foot), jawed vertebrates with two pairs of limbs. Note that several of these groupings are paraphyletic (Protochordata, Acraniata, Agnatha, Anamniota, Pisces) and consequently are not accepted in cladistic classifications. Accepted monophyletic taxa are shown at the top of the cladogram in Figure 26-3 as a nested hierarchy of increasingly more inclusive groupings.

Characteristics of Phylum Chordata

1. Bilateral symmetry; segmented body; three germ layers; well-developed coelom
2. **Notochord** (a skeletal rod) present at some stage in life cycle
3. **Single, dorsal, tubular nerve cord;** anterior end of cord usually enlarged to form a brain
4. **Pharyngeal pouches** present at some stage in life cycle; in aquatic chordates these develop into gill slits
5. **Postanal tail,** usually projecting beyond the anus at some stage but may or may not persist
6. **Segmented muscles** in an unsegmented trunk
7. **Ventral heart,** with dorsal and ventral blood vessels; closed blood system
8. Complete digestive system
9. A cartilaginous or bony **endoskeleton** present in the majority of members (vertebrates)

Four Chordate Hallmarks

The four distinctive characteristics that, taken together, set chordates apart from all other phyla are the **notochord; single, dorsal, tubular nerve cord; pharyngeal pouches;** and **postanal tail.** These characteristics are always found at some embryonic stage, although they may be altered or may disappear altogether in later stages of the life cycle.

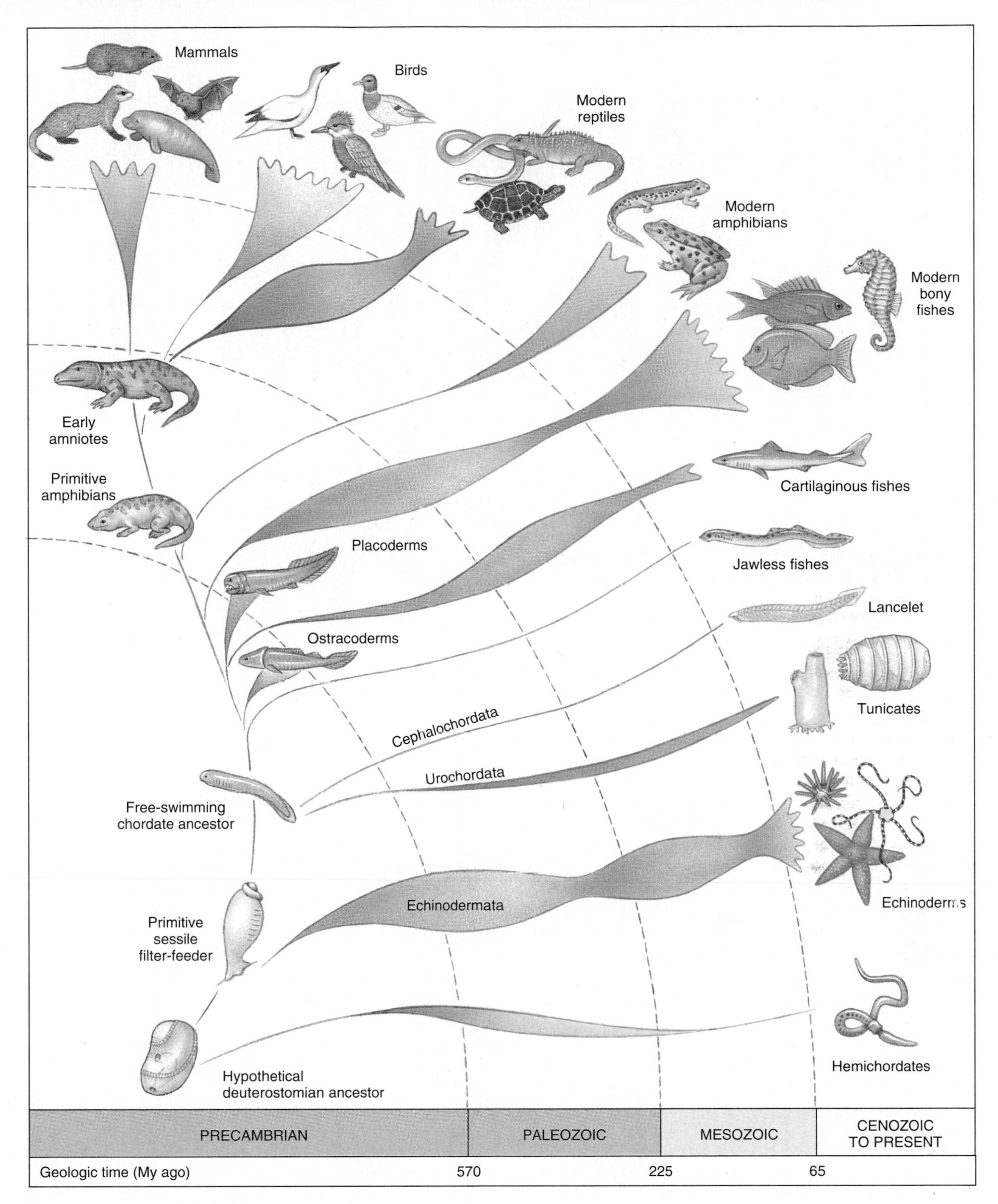

figure 26-2

Phylogenetic tree of the chordates, suggesting probable origin and relationships. Other schemes have been suggested and are possible. The relative abundance in numbers of species of each group through geological time, as indicated by the fossil record, is suggested by the bulging and thinning of that group's line of descent.

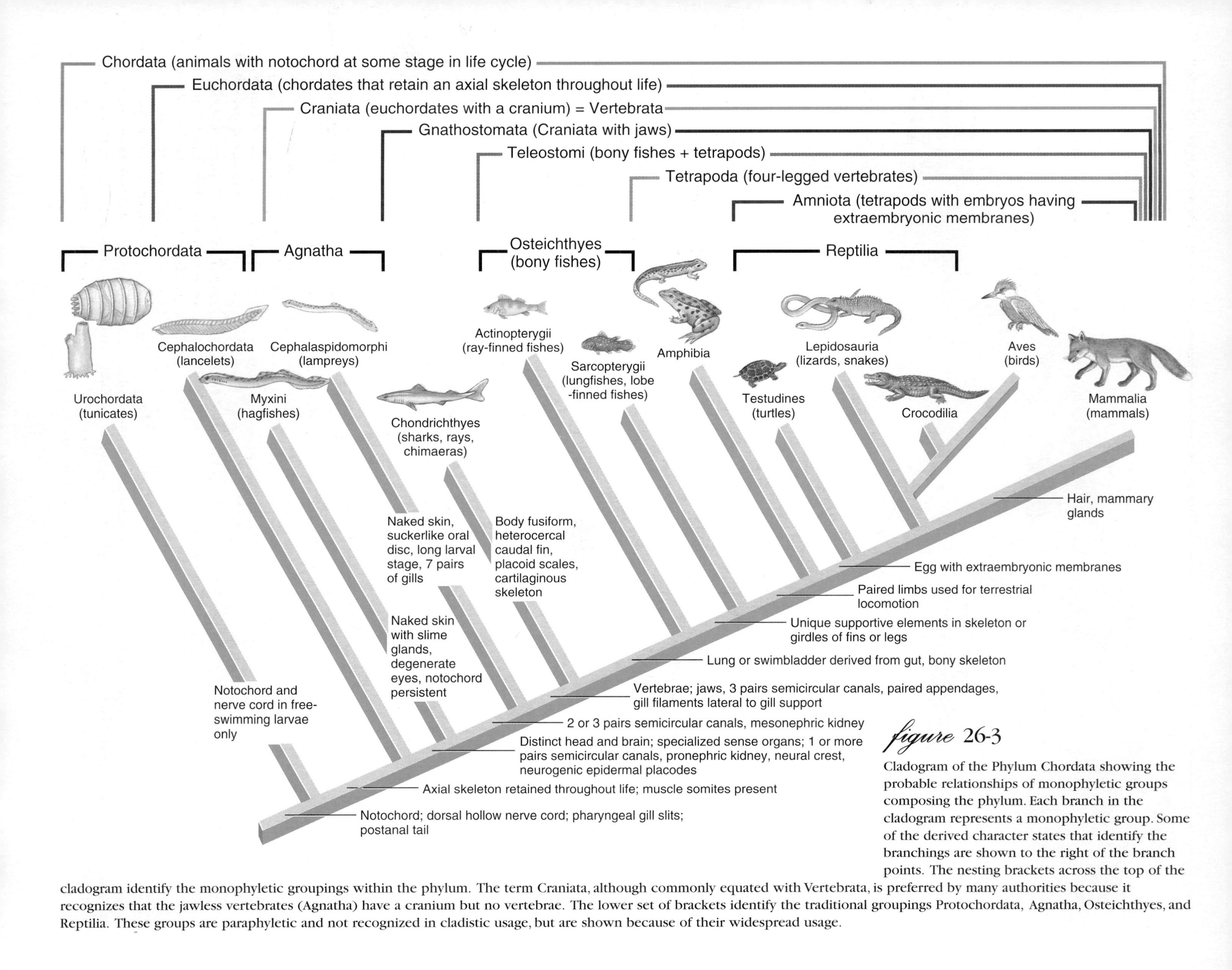

figure 26-3

Cladogram of the Phylum Chordata showing the probable relationships of monophyletic groups composing the phylum. Each branch in the cladogram represents a monophyletic group. Some of the derived character states that identify the branchings are shown to the right of the branch points. The nesting brackets across the top of the cladogram identify the monophyletic groupings within the phylum. The term Craniata, although commonly equated with Vertebrata, is preferred by many authorities because it recognizes that the jawless vertebrates (Agnatha) have a cranium but no vertebrae. The lower set of brackets identify the traditional groupings Protochordata, Agnatha, Osteichthyes, and Reptilia. These groups are paraphyletic and not recognized in cladistic usage, but are shown because of their widespread usage.

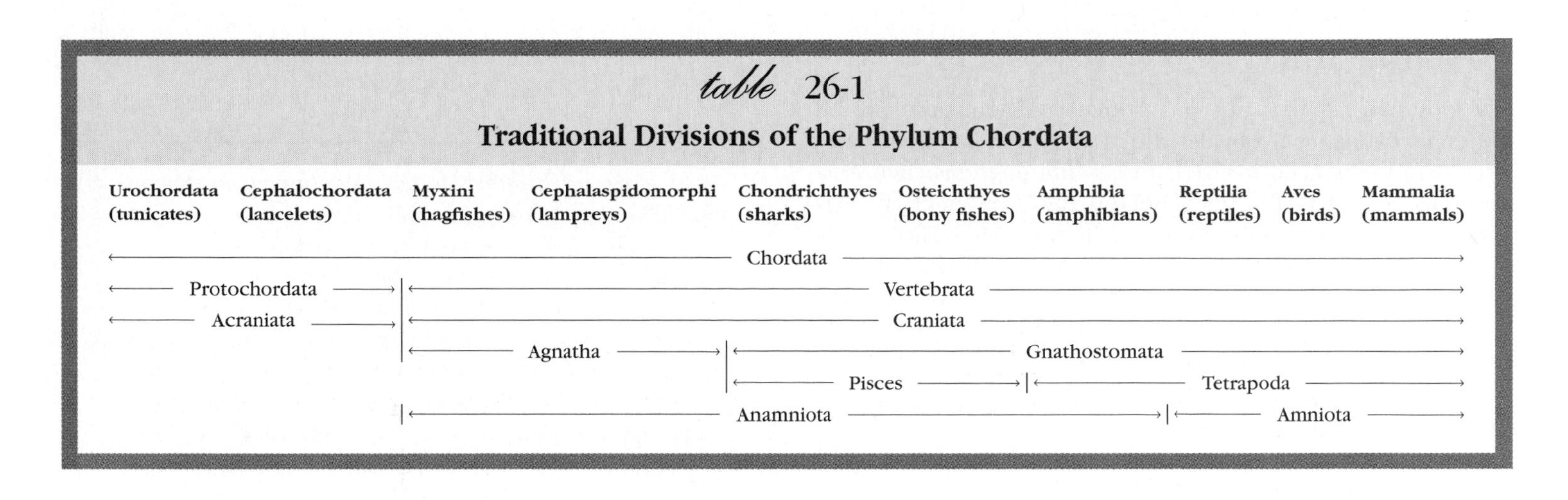

table 26-1

Traditional Divisions of the Phylum Chordata

Urochordata (tunicates)	Cephalochordata (lancelets)	Myxini (hagfishes)	Cephalaspidomorphi (lampreys)	Chondrichthyes (sharks)	Osteichthyes (bony fishes)	Amphibia (amphibians)	Reptilia (reptiles)	Aves (birds)	Mammalia (mammals)
Chordata									
Protochordata		Vertebrata							
Acraniata		Craniata							
		Agnatha		Gnathostomata					
				Pisces		Tetrapoda			
		Anamniota					Amniota		

Notochord

The notochord is a flexible, rodlike structure, extending the length of the body; it is the first part of the endoskeleton to appear in the embryo. The notochord is an axis for muscle attachment, and because it can bend without shortening, it permits undulatory movements of the body. In most protochordates and in jawless vertebrates, the notochord persists throughout life (Figure 26-1). In all jawed vertebrates a series of cartilaginous or bony vertebrae is formed from the connective tissue sheath around the notochord and replaces the notochord as the chief mechanical axis of the body.

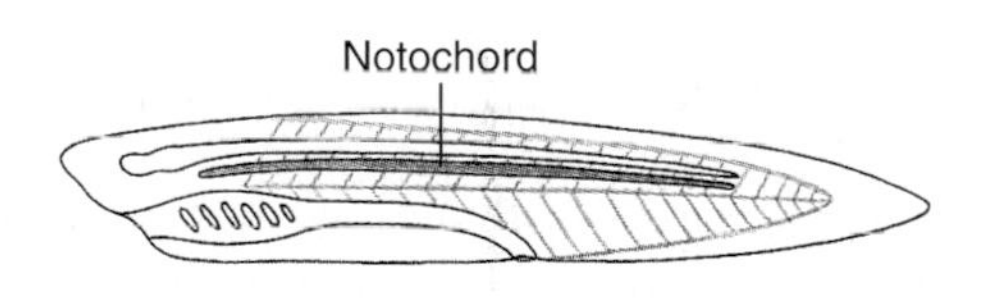

Dorsal, Tubular Nerve Cord

In most invertebrate phyla that have a nerve cord, it is ventral to the alimentary canal and is solid, but in the chordates the single cord is dorsal to the alimentary canal and notochord and is a tube (although the hollow center may be nearly obliterated during growth). In vertebrates the anterior end becomes enlarged to form the brain. The hollow cord is produced in the embryo by the infolding of the ectodermal cells on the dorsal side of the body above the notochord (see Figure 14-20, p. 327). The nerve cord passes through the protective neural arches of the vertebrae, and the anterior brain is surrounded by a bony or cartilaginous cranium.

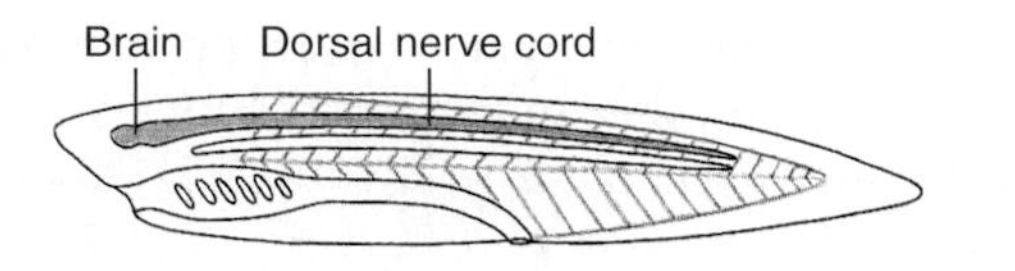

Pharyngeal Pouches and Slits

Pharyngeal slits are perforated slitlike openings that lead from the pharyngeal cavity to the outside. They are formed by the inpocketing of the outside ectoderm (pharyngeal grooves) and the evagination, or outpocketing, of the endodermal lining of the pharynx (pharyngeal pouches). In aquatic chordates, the two pockets break through the pharyngeal cavity where they meet to form the pharyngeal slit. In amniotes these pockets may not break through the pharyngeal cavity and only grooves are formed instead of slits. In tetrapod vertebrates the pharyngeal pouches give rise to several different structures, including Eustachian tube, middle ear cavity, tonsils, and parathyroid glands (see p. 328).

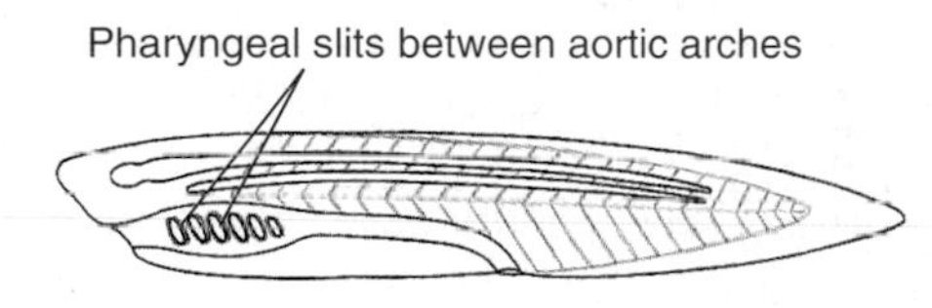

The perforated pharynx evolved as a filter-feeding apparatus and is used as such in the protochordates. Water with suspended food particles is drawn by ciliary action through the mouth and flows out through the pharyngeal slits, where food is trapped in mucus. In the vertebrates, ciliary action is replaced by a muscular pump that drives water through the pharynx by expanding and contracting the pharyngeal cavity. Also modified are the blood vessels that carry blood through the pharyngeal bars. In protochordates these are simple vessels surrounded by connective tissue. In the early fishes a capillary network was added with only thin, gas-permeable walls separating water outside from blood inside. This improved efficiency of gas transfer. These adaptations led to the development of **internal gills,** completing the conversion of the pharynx from a filter-feeding apparatus in protochordates to a respiratory organ in aquatic vertebrates.

Postanal Tail

The postanal tail, together with somatic musculature and the stiffening notochord, provides the motility that larval tunicates and amphioxus need for their free-swimming existence. As a structure added to the body behind the anus, it clearly has evolved specifically for propulsion in water. Its efficiency is later increased in fishes with the addition of fins. The tail is evident in humans only as a vestige (the coccyx, a series of small vertebrae at the end of the spinal column) but most other mammals have a waggable tail as adults.

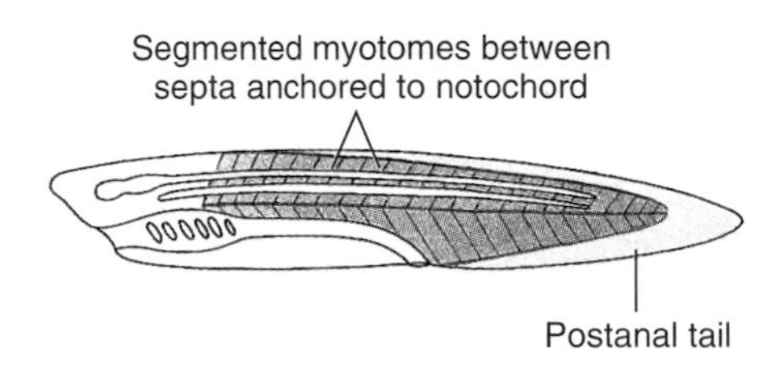

Ancestry and Evolution of the Chordates

Since the middle of the nineteenth century when the theory of organic evolution became the focal point for ferreting out relationships among groups of living organisms, zoologists have debated the question of chordate origins. It has been difficult to reconstruct lines of descent because the earliest protochordates were in all probability soft-bodied creatures that stood little chance of being preserved as fossils even under the most ideal conditions. Consequently, such reconstructions largely come from the study of living organisms, especially from an analysis of early developmental stages, which tend to be more insulated from evolutionary change than the differentiated adult forms that they become.

Zoologists at first speculated that the chordates evolved within the protostome lineage (annelids and arthropods) but discarded such ideas when they realized that supposed morphological similarities had no developmental basis. Early in this century when further theorizing became rooted in developmental patterns of animals, it became apparent that the chordates must have originated within the deuterostome branch of the animal kingdom. As explained earlier (p. 323 and Figure 14-17), the Deuterostomia, a grouping that includes the echinoderms, hemichordates, lophophorates, and chordates, has several important embryological features that clearly separate it from the Protostomia and establish its monophyly. Accordingly the deuterostomes are almost certainly a natural grouping of interrelated animals that have their common origin in ancient Precambrian seas. There are several lines of anatomical, developmental, and molecular evidence suggesting that somewhat later, at the base of the Cambrian period some 570 million years ago, the first distinctive chordates arose from a lineage related to echinoderms and hemichordates (Figure 26-2; see also Figure 25-25, p. 559).

Most of the early efforts to pin together invertebrate and chordate kinship are now recognized as based on similarities due to analogy rather than homology. Analogous structures are those that perform similar functions but have altogether different origins (such as wings of birds and butterflies). Homologous structures, on the other hand, share a common origin but may look quite different (at least superficially) and perform quite different functions. For example, all vertebrate forelimbs are homologous because they are derived from a pentadactyl limb of the same ancestor, even though they may be modified as differently as the human arm and a bird's wing. Homologous structures share a genetic heritage; analogous structures do not. Obviously, only homologous similarities have any bearing in genealogical connections.

While modern echinoderms look nothing at all like modern chordates, evolutionary affinity between chordates and echinoderms gains support from fossil evidence. One curious group of fossil echinoderms, the "Calcichordata," have pharyngeal slits and possibly other chordate attributes (Figure 26-4; see also p. 555). These small, nonsymmetrical forms have a head resembling a long-toed medieval boot, a series of branchial slits covered with flaps much like the gill openings of sharks, a postanal tail, and structures that are doubtfully interpreted as notochord and muscle blocks. These creatures apparently used their pharyngeal slits for filter feeding, as do the protochordates today. Although the calcichordates seem to have some of the right chordate characters based on soft anatomy, there is no convincing similarity between the hard skeleton of calcichordates (which was calcium carbonate) and that of vertebrates (which is composed of a complex of calcium and phosphate). Thus, while such fossils bring us closer to an understanding of chordate origins, we are not yet in a position to know the precise characteristics of the long-sought chordate ancestor. We do, however, know a great deal about two living protochordate groups that also descended from it. These we will now consider.

Subphylum Urochordata (Tunicata)

The urochordates ("tail-chordates"), more commonly called tunicates, number some 3000 species. They are found in all seas from near the shoreline to great depths. Most of them are sessile as adults, although some are free living. The name "tunicate" is suggested by the usually tough, nonliving **tunic,** or test, that surrounds the animal (Figure 26-5). As adults, tunicates are highly specialized chordates, for in most species only the larval form, which resembles a microscopic tadpole, bears all the chordate hallmarks. During adult metamorphosis, the notochord (which, in the larva, is restricted to the tail, hence the group name Urochordata) and the tail disappear altogether, while the dorsal nerve cord becomes reduced to a single ganglion.

Urochordata is divided into three classes—**Ascidiacea** (Gr. *askiolion,* little bag, + *acea,* suffix), **Larvacea** (L. *larva,* ghost, + *acea,* suffix), and **Thaliacea** (Gr. *thalia,* luxuriance, + *acea,* suffix). Of these, the members of **Ascidiacea,** commonly known as the ascidians, or sea squirts, are by far the most common and are the best known. Ascidians may be solitary, colonial, or compound. Each of the solitary and colonial forms has its own test, but among the compound forms many individuals may share the same test. In some of these compound ascidians each member has its own incurrent siphon, but the excurrent opening is common to the group.

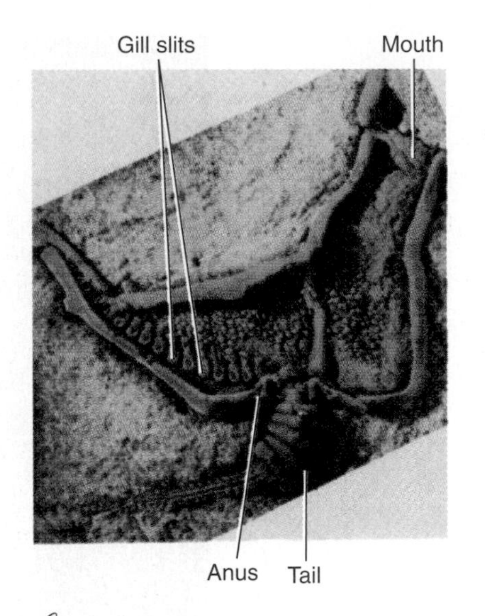

figure 26-4

Fossil of an early echinoderm, a calcichordate, that lived during the Ordovician period (450 million years BP). It shows affinities with both echinoderms and chordates and may belong to a lineage that was ancestral to the chordates.

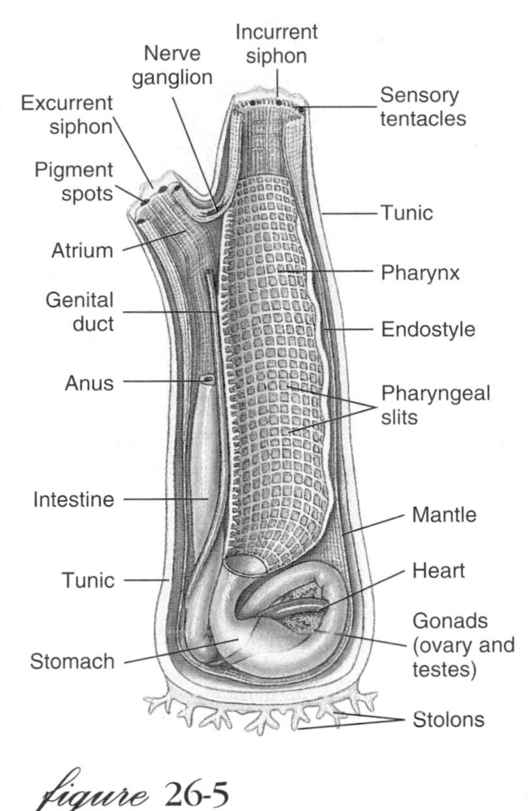

figure 26-5

Structure of a common tunicate, *Ciona* sp.

The typical solitary ascidian (Figure 26-5) is a spherical or cylindrical form that is attached by its base to hard substrates such as rocks, pilings, or the bottoms of ships. Lining the tunic is an inner membrane, the **mantle.** On the outside are two projections: the **incurrent** and **excurrent siphons** (Figure 26-5). Water enters the incurrent siphon and passes into the branchial sac (pharynx) through the mouth. On the midventral side of the branchial sac is a groove, the **endostyle,** which is ciliated and secretes mucus. As the mucous sheet is carried by cilia across the inner surface of the pharynx to the dorsal side, it sieves small food particles from the water passing through the slits in the wall of the branchial sac. Then the mucus with its entrapped food is collected and passed posteriorly to the esophagus. The water, now largely cleared of food particles, is driven by cilia into the atrial cavity and finally out the excurrent siphon. The intestine leads to the anus near the excurrent siphon.

The circulatory system consists of a ventral **heart** near the stomach and two large vessels, one on either side of the heart. An odd feature found in no other chordate is that the heart drives the blood first in one direction for a few beats, then pauses, reverses, and drives the blood in the opposite direction. The excretory system is a type of nephridium near the intestine. The nervous system is restricted to a **nerve ganglion** and a few nerves that lie on the dorsal side of the pharynx. A notochord is lacking. The animals are hermaphroditic. The germ cells are carried out the excurrent siphon into the surrounding water, where cross-fertilization occurs.

Of the four chief characteristics of chordates, adult sea squirts have only one, the pharyngeal gill slits. However, the larval form gives away the secret of their true relationship. The tiny tadpole larva (Figure 26-6) is an elongate, transparent form with a head and all four chordate characteristics: a notochord, hollow dorsal nerve cord, propulsive postanal tail, and a large pharynx with endostyle and gill slits. The larva does not feed but swims for several hours before fastening itself vertically by its adhesive papillae to some solid object. It then metamorphoses to become the sessile adult.

The remaining two classes of the Urochordata—**Larvacea** and **Thaliacea**—are mostly small, transparent animals of the open sea (Figure 26-7). Some are small, tadpole-like forms resembling the larval stage of ascidians. Others are spindle shaped or cylindrical forms surrounded by delicate muscle bands. They are mostly carried along by the ocean currents and as such form a part of the plankton. Many are provided with luminous organs and emit a beautiful light at night.

Subphylum Cephalochordata

The cephalochordates are the marine lancelets: slender, laterally compressed, translucent animals about 5 to 7 cm in length (Figure 26-8) that inhabit the sandy bottoms of coastal waters

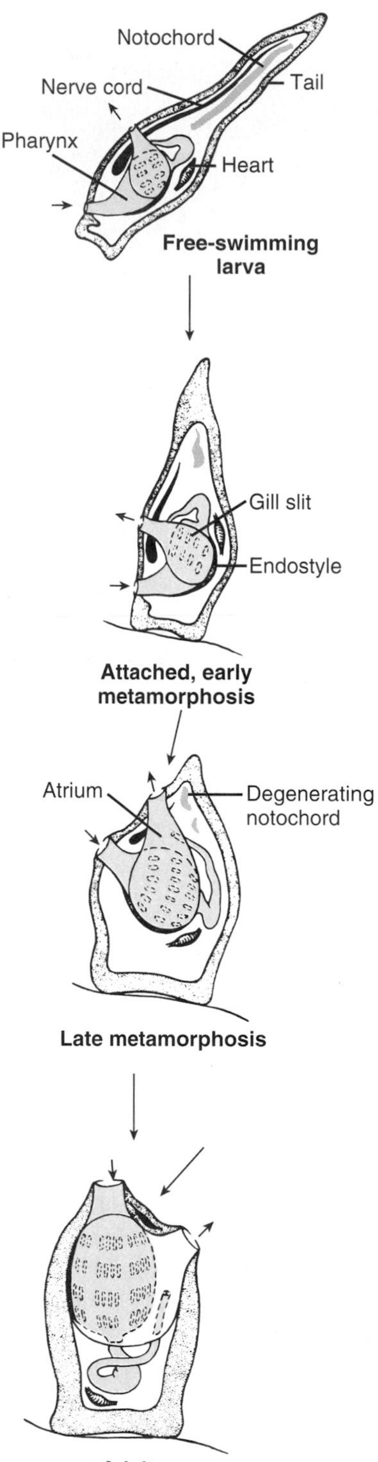

figure 26-6

Metamorphosis of a solitary ascidian from a free-swimming larval stage.

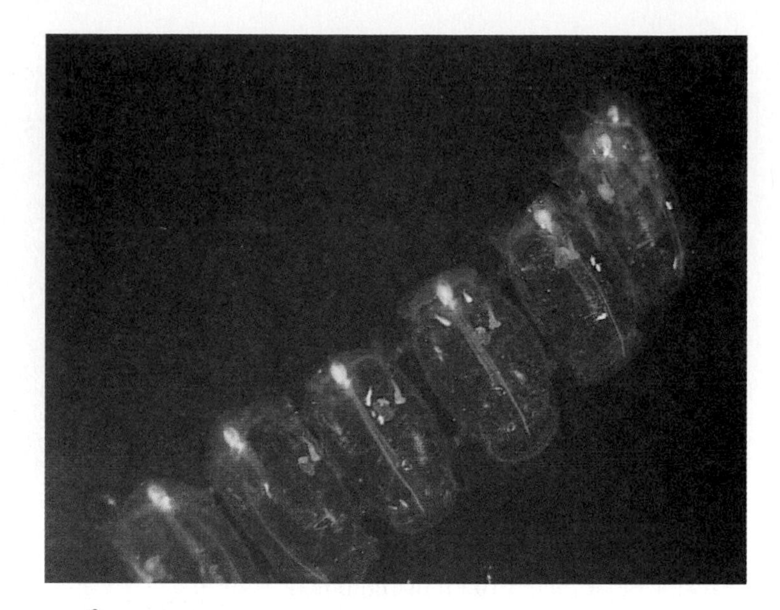

figure 26-7

Colonial thaliacean. The transparent individuals of this delicate, planktonic tunicate are grouped in a chain. Visible within each individual is an orange gonad, an opaque gut, and a long serrated gill bar.

around the world. Lancelets originally bore the generic name *Amphioxus* (Gr. *amphi,* both ends, + *oxys,* sharp), later surrendered by priority to *Branchiostoma* (Gr. *branchia,* gills, + *stoma,* mouth). Amphioxus is still used, however, as a convenient common name for all of the approximately 26 species in this diminutive subphylum. Four species of amphioxus (lancelets) are found in North American coastal waters.

Amphioxus is especially interesting because it has the four distinctive characteristics of chordates in simple form. Water enters the mouth, driven by cilia in the buccal cavity, and then passes through numerous pharyngeal slits in the pharynx, where food is trapped in mucus, which is then moved by cilia into the intestine. Here the smallest food particles are separated from the mucus and passed into the midgut caecum (diverticulum), where they are phagocytized and digested intracellularly. As in tunicates, the filtered water passes first into an atrium, and then leaves the body by an atriopore (equivalent to the excurrent siphon of tunicates).

The closed circulatory system is complex for so simple a chordate. The flow pattern is remarkably similar to that of fishes, although there is no heart. Blood is pumped forward in the ventral aorta by peristaltic-like contractions of the vessel wall, and then passes upward through the branchial arteries (aortic arches) in the gill bars to the dorsal aorta. From here the blood is distributed to the body tissues by capillary plexi and then is collected in veins, which return it to the ventral aorta. The blood is colorless, lacking both erythrocytes and hemoglobin.

The nervous system is centered around a hollow nerve cord lying above the notochord. Sense organs are simple, unpaired bipolar receptors located in various parts of the body. The "brain" is a simple vesicle at the anterior end of the nerve cord.

Sexes are separate in amphioxus. The sex cells are set free in the atrium, and then pass out the atriopore to the outside, where fertilization occurs. The larvae hatch soon after the eggs are fertilized and gradually assume the shape and size of adults.

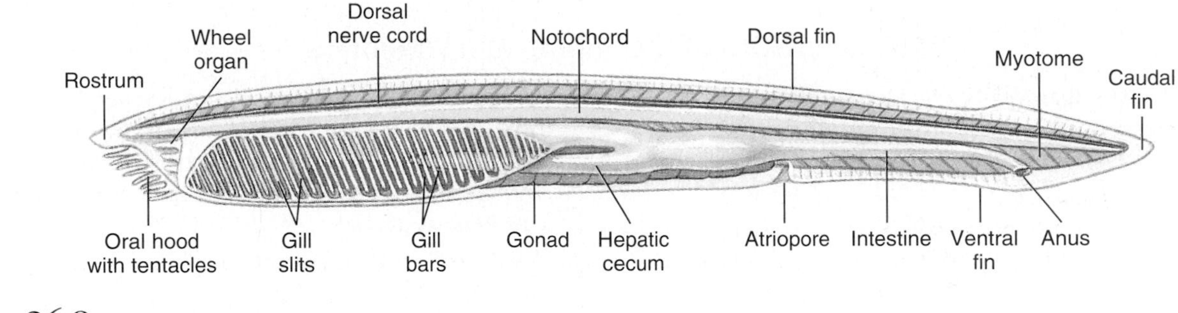

figure 26-8

Amphioxus. This interesting bottom-dwelling cephalochordate illustrates the four distinctive chordate characteristics (notochord, dorsal nerve cord, pharyngeal gill slits, and postanal tail). The vertebrate ancestor probably had a similar body plan.

No other chordate shows the basic diagnostic characteristics of the chordates so well. In addition to the four chordate anatomical hallmarks, amphioxus possesses several structural features that foreshadow the vertebrate plan. Among these are the midgut diverticulum, which secretes digestive enzymes, segmented trunk musculature, and the basic circulatory pattern of more advanced chordates.

Subphylum Vertebrata

The third subphylum of the chordates is the large and diverse Vertebrata, the subject of the next five chapters of this book. This monophyletic group shares the basic chordate characteristics with the other two subphyla, but in addition it reveals a number of novel homologies that the others do not share. The alternative name of the subphylum, Craniata, more accurately describes the group since all have a cranium (bony or cartilaginous braincase) whereas the jawless fishes lack vertebrae.

Adaptations That Have Guided Vertebrate Evolution

From the earliest fishes to the mammals, the evolution of the vertebrates has been guided by the basic adaptations of the living endoskeleton, pharynx and efficient respiration, advanced nervous system, and paired limbs.

Living Endoskeleton

The endoskeleton of vertebrates, as in the echinoderms, is an internal supportive structure and framework for the body. This condition is a departure in animal architecture, since invertebrate skeletons are more commonly exoskeletons. Exoskeletons and endoskeletons have their own particular set of advantages and limitations relating to size (see note on p. 151). For vertebrates the living endoskeleton possesses an overriding advantage over the secreted, nonliving exoskeleton of arthropods: growing with the body as it does, the endoskeleton permits almost unlimited body size with much greater economy of building materials. Some vertebrates have become the most massive animals on earth. The endoskeleton forms an excellent jointed scaffolding for muscles, and the muscles in turn protect the skeleton and cushion it from potentially damaging impact.

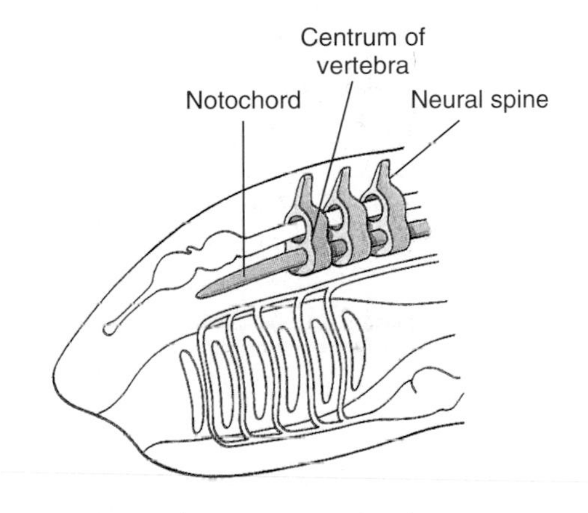

We should note that the vertebrates have not wholly lost the protective function of a firm external covering. The skull and the thoracic rib cage enclose and protect vulnerable organs. Most vertebrates are further protected with a tough integument, often bearing nonliving structures such as scales, hair, and feathers that may provide insulation as well as physical security.

Pharynx and Efficient Respiration

The perforated pharynx (gill slits), present as pharyngeal pouches in all chordates at some stage in their life cycle, evolved as an apparatus for suspension feeding. In primitive chordates, water with suspended food particles is drawn through the mouth by ciliary action and flows out through the gill slits, where food is trapped in mucus. This condition is observed today in amphioxus. As the protovertebrates shifted from suspension feeding to a predatory life habit, the pharynx became modified into a muscular feeding apparatus through

Characteristics of the Subphylum Vertebrata

1. Chief diagnostic features of chordates—**notochord, dorsal nerve cord, pharyngeal pouches,** and **postanal tail**—all present at some stage of the life cycle
2. **Integument** basically of two divisions, an outer **epidermis** of stratified epithelium from the ectoderm and an inner **dermis** of connective tissue derived from the mesoderm; many modifications of skin among the various classes, such as glands, scales, feathers, claws, horns, and hair
3. Distinctive **endoskeleton** consisting of vertebral column (notochord persistent in jawless fishes which lack vertebrae), limb girdles, and two pairs of jointed appendages derived from somatic mesoderm, and a head skeleton (cranium and pharyngeal skeleton) derived largely from neural crest cells.
4. Muscular, perforated pharynx; in fishes pharyngeal slits possess gills and muscular aortic arches; in tetrapods the much reduced pharynx is an embryonic source of glandular tissue
5. **Many muscles** attached to the skeleton to provide for movement
6. Complete digestive system ventral to the spinal column and provided with large digestive glands, liver, and pancreas
7. Circulatory system consisting of a **ventral heart** of two to four chambers; a closed blood vessel system of arteries, veins, and capillaries; blood fluid containing red corpuscles with hemoglobin and white corpuscles; paired aortic arches connecting the ventral and dorsal aortas and branching to the gills in the gill-breathing vertebrates; in the terrestrial types modification of the aortic arch into pulmonary and systemic systems
8. Well-developed **coelom** largely filled with the visceral systems
9. Excretory system consisting of **paired kidneys** (mesonephric or metanephric types in adults) provided with ducts to drain the waste to cloaca or anal region
10. Highly differentiated **brain;** 10 or 12 pairs of **cranial nerves** usually with both motor and sensory functions; a pair of spinal nerves for each primitive myotome; an **autonomic nervous system** in control of involuntary functions of internal organs; **paired special sense organs**
11. **Endocrine system** of ductless glands scattered through the body
12. Nearly always separate sexes; each sex containing paired gonads with ducts that discharge their products either into the cloaca or into special openings near the anus
13. **Body plan** consisting typically of **head, trunk,** and **postanal tail; neck** present in some, especially terrestrial forms; usually two pairs of appendages, although entirely absent in some; coelom divided into a pericardial space and a general body cavity; mammals with a thoracic cavity

which water could be pumped by expanding and contracting the pharyngeal cavity. Circulation to the internal gills was improved by the addition of capillary beds (lacking in protochordates) and the development of a ventral heart and muscular aortic arches. All of these changes supported an increased metabolic rate that would have to accompany the switch to an active life of selective predation.

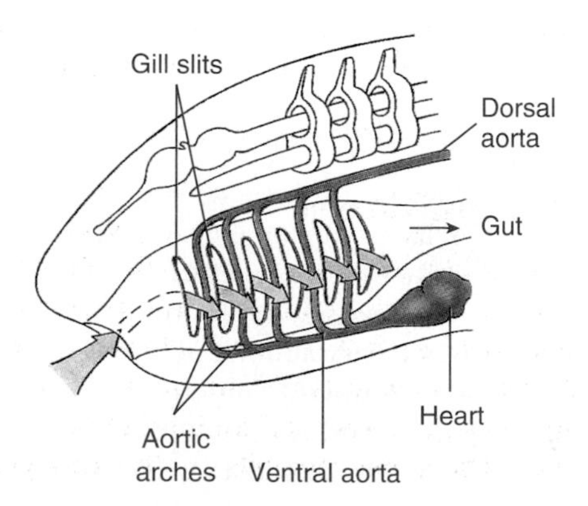

Advanced Nervous System

No single system in the body is more strongly associated with functional and structural advancement than is the nervous system. The prevertebrate nervous system consisted of a brainless nerve cord and rudimentary sense organs, which were mostly chemosensory in function. When the protovertebrates switched to a predatory life-style, new sensory, motor, and in-

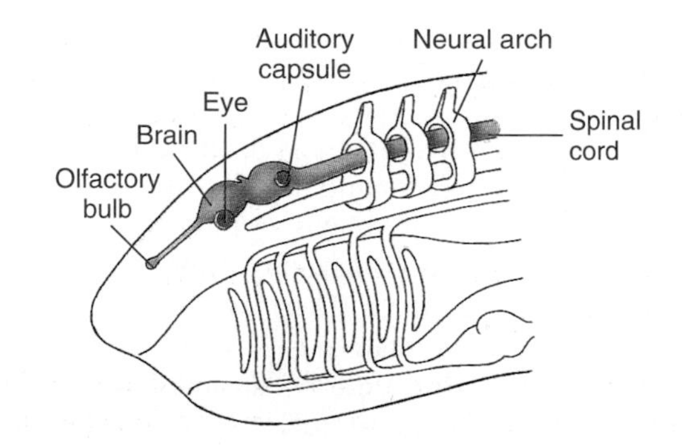

tegrative controls were now essential for the location and capture of larger prey items. In short, the protovertebrates developed a new head, complete with a brain and external paired sense organs especially designed for distance reception. These included paired eyes with lenses and inverted retinas; pressure receptors, such as paired ears designed for equilibrium and later redesigned to include sound reception; and chemical receptors, including taste receptors and exquisitely sensitive olfactory organs.

Paired Limbs

Pectoral and pelvic appendages are present in most vertebrates in the form of paired fins or legs. They originated as swimming stabilizers and later became prominently developed into legs for travel on land. Jointed limbs are especially suited for life on land because they permit finely graded movement against a substrate.

The Search for the Vertebrate Ancestral Stock

Early Chordate Fossils

The earliest Paleozoic vertebrate fossils, the jawless ostracoderm fishes to be considered later in this chapter, share many novel features of organ-system development with living vertebrates. These organ systems therefore must have originated either in an early vertebrate or invertebrate chordate lineage. With one exception, hardly any invertebrate chordates are known as fossils. The exception is *Pikaia gracilens,* a ribbon-shaped, somewhat fishlike creature about 5 cm in length discovered in the famous Burgess Shale of British Columbia (Figure 26-9; see also Figure 4-11, p. 93). *Pikaia* is a mid-Cambrian form that precedes the earliest vertebrate fossils by many millions of years. Although this fossil has not yet been described in detail, we know that it possessed both a notochord and the characteristic chordate >-shaped muscle bands (myotomes). Without question *Pikaia* is a chordate. It shows a remarkable resemblance to living amphioxus, at least in overall body organization, and may in fact be an early cephalochordate. *Pikaia* is a provocative fossil but, until other Cambrian chordate fossils are discovered, its relationship to the earliest vertebrates remains uncertain. In the absence of additional fossil evidence, most speculations on vertebrate ancestry have focused on the living cephalochordates and tunicates, since it is widely believed that the vertebrates must have emerged from a lineage resembling one of these protochordate groups.

Garstang's Hypothesis of Chordate Larval Evolution

At first glance, tunicates seem unlikely candidates for the sister group of vertebrates. The adult tunicate, which spends its life anchored to some marine surface, lacks a notochord, tubular nerve cord, postanal tail, sense organs, and segmented musculature. Their larvae, however, bear all the right qualifications for chordate membership. Called "tadpole" larvae because of their superficial resemblance to larval frogs, these tiny, site-seeking forms have a notochord, hollow dorsal nerve cord, pharyngeal gill slits, and postanal tail, as well as a brain and sense organs.

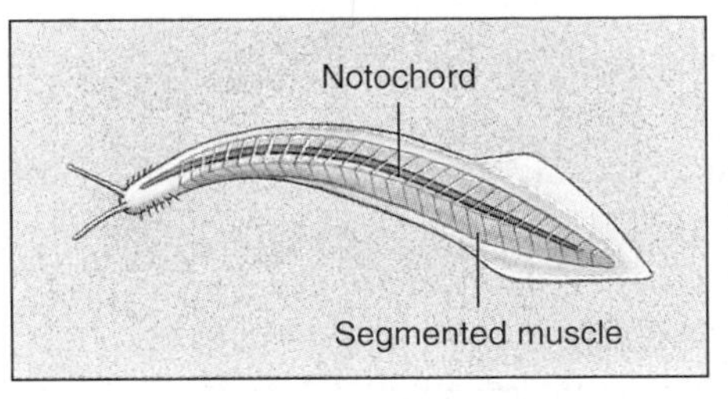

figure 26-9

Pikaia, the earliest known chordate, from the Burgess Shale of British Columbia, Canada.

At the time of its discovery in 1869, the tadpole larva was considered the descendant of an ancient free-swimming chordate ancestry of the tunicates. The adults then were regarded as degenerate, sessile descendants of the free-swimming form. In 1928, Walter Garstang in England introduced fresh thinking into the vertebrate ancestor debate by turning this sequence around: rather than the ancestral tadpole larva giving rise to a degenerative tunicate sessile adult, he suggested that the sessile adults *were* the ancestral stock. The tadpole larvae then evolved as an adaptation for spreading to new habitats. Next, Garstang suggested that at some point the tadpole failed to metamorphose into an adult but developed gonads and reproduced in the larval stage. With continued larval evolution, a new group of free-swimming animals appeared (Figure 26-10).

Garstang called this process **paedomorphosis** (Gr. *pais,* child, + *morphē,* form), a term that describes the evolutionary retention of juvenile or larval traits in the adult body. Garstang departed from previous thinking by suggesting that evolution may occur in the larval stages of animals—and in this case, lead to the vertebrate lineage. Paedomorphosis is a well-known phenomenon in several different animal groups (paedomorphosis in amphibians is described on p. 613). Furthermore, Garstang's hypothesis agrees with the embryological evidence. Nevertheless, it remains untested and thus speculative.

Position of Amphioxus

For many years zoologists believed that the cephalochordate amphioxus is the closest living relative of vertebrates. No other protochordate shows the basic diagnostic characteristics of the chordates so well. However, amphioxus lacks a brain and all of the specialized sensory equipment that characterize the vertebrates. There are no gills in the pharynx and no mouth or pharyngeal musculature for pumping water through the gill slits; movement of water is entirely by the action of cilia. Despite these specializations and others peculiar to modern cephalochordates, many zoologists believe that amphioxus has retained the primitive pattern of the immediate prevertebrate condition. Thus the cephalochordates are the probable sister group of the vertebrates (Figure 26-3).

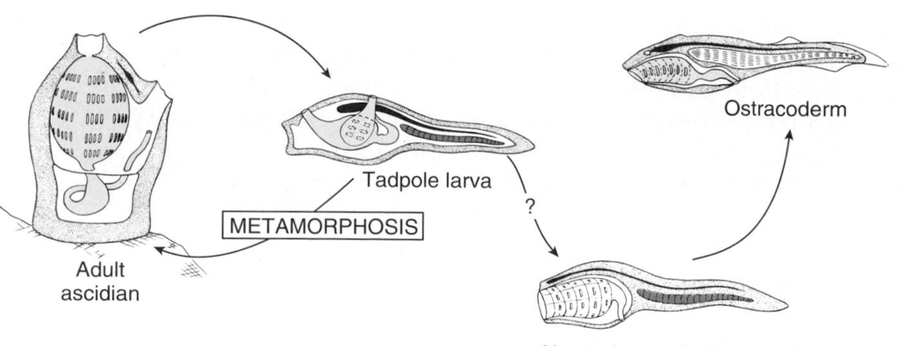

figure 26-10

Garstang's hypothesis of larval evolution. Adult tunicates live on the sea floor but reproduce through a free-swimming tadpole larva. More than 500 million years ago, some larvae began to reproduce in the swimming stage. These gave rise to the ostracoderms, the first known vertebrates.

Paedomorphosis, the displacement of ancestral larval or juvenile features into a descendant adult, can be produced by three different evolutionary-developmental processes: neoteny, progenesis, and post-displacement. In neoteny, the growth rate of body form is slowed so that the animal does not attain the ancestral adult form at the time it reaches reproductive maturity. Progenesis is the precocious maturation of gonads in a larval (or juvenile) body that then stops growing and never attains the adult body form. In post-displacement, the onset of a developmental process is delayed relative to reproductive maturation, so that the ancestral adult form is not attained at the time of reproductive maturation. Neoteny, progenesis and post-displacement thus describe different ways in which paedomorphosis can happen. Zoologists use the inclusive term paedomorphosis to describe the results of these evolutionary-developmental processes.

The Earliest Vertebrates: Jawless Ostracoderms

The earliest vertebrate fossils are late Cambrian articulated skeletons from the United States, Bolivia, and Australia. They were small, jawless creatures collectively called ostracoderms (os-trak′o-derm) (Gr. *ostrakon,* shell, + *derma,* skin), which belong to the Agnatha division of the vertebrates. The earliest ostracoderms, called **heterostracans,** lacked paired fins, which later fishes found so important for stability (Figure 26-11). Their swimming movements must have been clumsy, although sufficient to propel them along the ocean bottom where they searched for food. With fixed circular or slitlike mouth openings, they probably filtered small food particles from the water or ocean bottom. However, unlike the ciliary filter-feeding protochordates, the ostracoderms sucked water into the pharynx by muscular pumping, an important innovation that suggests to some authorities that the ostracoderms may have been mobile predators that fed on soft-bodied animals.

The term "ostracoderm" does not describe a natural evolutionary assemblage but rather is a term of convenience for describing several groups of heavily armored, extinct jawless fishes.

During the Devonian period, the ostracoderms underwent a major radiation, resulting in the appearance of several peculiar-looking forms varying in shape and length of the snout, dorsal spines, and dermal plates. One group, the **osteostracans** (Figure 26-11), improved the efficiency of their benthic life by evolving paired pectoral fins. These fins, located just behind the head, provided control over pitch and yaw, which ensured well-directed forward movement. Another group of ostracoderms, the **anaspids,** (Figure 26-11) were more streamlined and more closely resembled modern-day jawless fishes (the lamprey, for example) than any other ostracoderm.

As a group, the ostracoderms were basically fitted for a simple, bottom-feeding life. Yet, despite their anatomical limitations, they enjoyed a respectable radiation in the Silurian and Devonian periods.

For decades, geologists have used strange microscopic, toothlike fossils called **conodonts** (Gr. *kōnos,* cone, + *odontos,* tooth) to date Paleozoic marine sediments without having any idea what kind of creature originally possessed these elements. The discovery in the early 1980s of the fossils of complete conodont animals showed that conodont elements belonged to a small early marine vertebrate (Figure 26-12). It is widely believed that as more is learned about conodont animals they will play an important role in understanding the origin of vertebrates. At present, however, their position in the vertebrate phylogeny is a matter of debate.

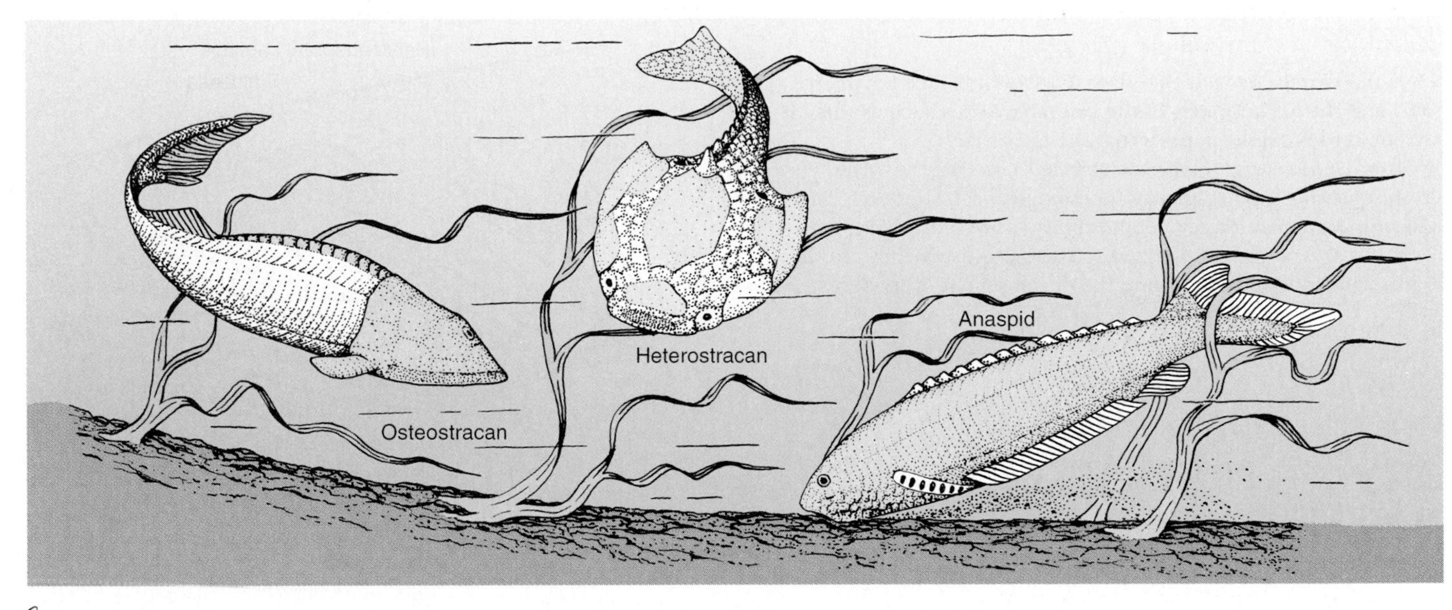

figure 26-11

Three ostracoderms, jawless fishes of Silurian and Devonian times. They are shown as they might have appeared while searching for food on the floor of a Devonian sea. All were probably suspension-feeders, but employed a strong pharyngeal pump to circulate water rather than the much more limiting mode of ciliary feeding used by their protovertebrate ancestors and by amphioxus today. Modern lampreys are believed to be derived from the anaspid group.

Early Jawed Vertebrates

All jawed vertebrates, whether extinct or living, are collectively called **gnathostomes** ("jaw mouth") in contrast to the jawless vertebrates, the **agnathans** ("without jaw"). The living agnathans, the naked hagfishes and lampreys, also are often called cyclostomes ("circle mouth"). The gnathostomes are almost certainly a monophyletic group since the presence of jaws is a derived character state shared by all jawed fishes and tetrapods. The agnathans, however, are defined principally by the absence of a feature—jaws—that characterize the gnathostomes, and the superclass Agnatha therefore may be paraphyletic.

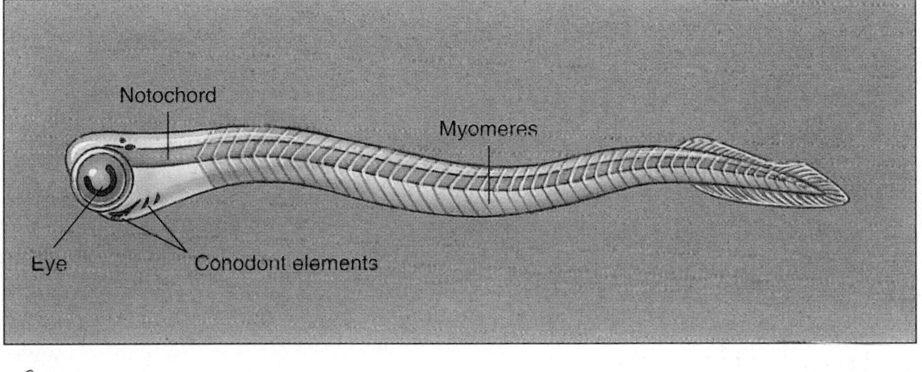

figure 26-12

Restoration of a living conodont animal. The conodont superficially resembled amphioxus, but it possessed a much greater degree of encephalization (large paired eyes, possible auditory capsules) and bone-like mineralized elements—all indicating that the conodont animal was a vertebrate. The conodont elements were probably gill-supporting structures or part of a suspension-feeding apparatus.

The origin of jaws was one of the most important events in vertebrate evolution. The utility of jaws is obvious: they allow predation on large and active forms of food not available to the jawless vertebrates. There is ample evidence that jaws arose through modifications of the first two of the serially repeated cartilaginous gill arches. The beginnings of this trend can be seen in some of the jawless ostracoderms, where the mouth becomes bordered by strong dermal plates that could be manipulated somewhat like jaws with the gill-arch musculature. Later, the anterior gill arches became hinged and bent forward into the characteristic position of vertebrate jaws (Figure 26-13). Nearly as remarkable as this drastic morphological remodeling is the subsequent evolutionary fate of the many jawbone elements—their transformation into the ear ossicles of the mammalian middle ear (see note, p. 258).

Among the first jawed vertebrates were the heavily armored **placoderms** (plak′o-derm) (Gr. *plax,* plate, + *derma,* skin); these forms first appear in the fossil record in the early *figure* 26-13

How the vertebrates got their jaw. The resemblance between jaws and the gill supports of the primitive fishes such as this carboniferous shark suggests that the upper jaw (palatoquadrate) and lower jaw (Meckel's cartilage) evolved from structures that originally functioned as gill supports. The gill supports immediately behind the jaws are hinged like the jaws and served to link the jaws to the braincase. Relics of this transformation are seen during the development of modern sharks.

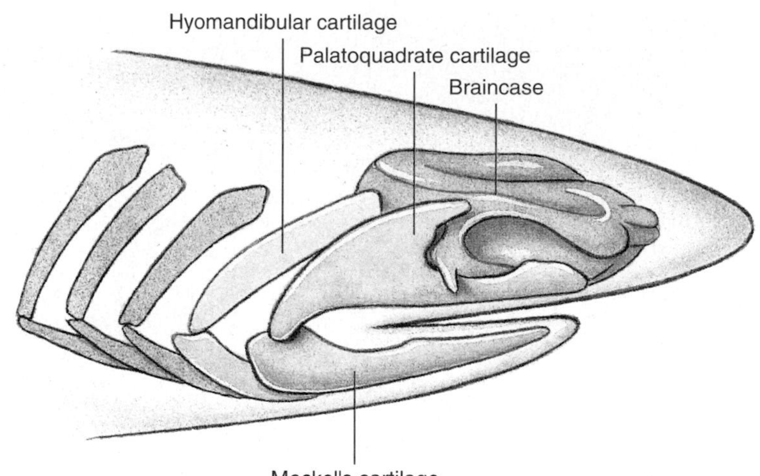

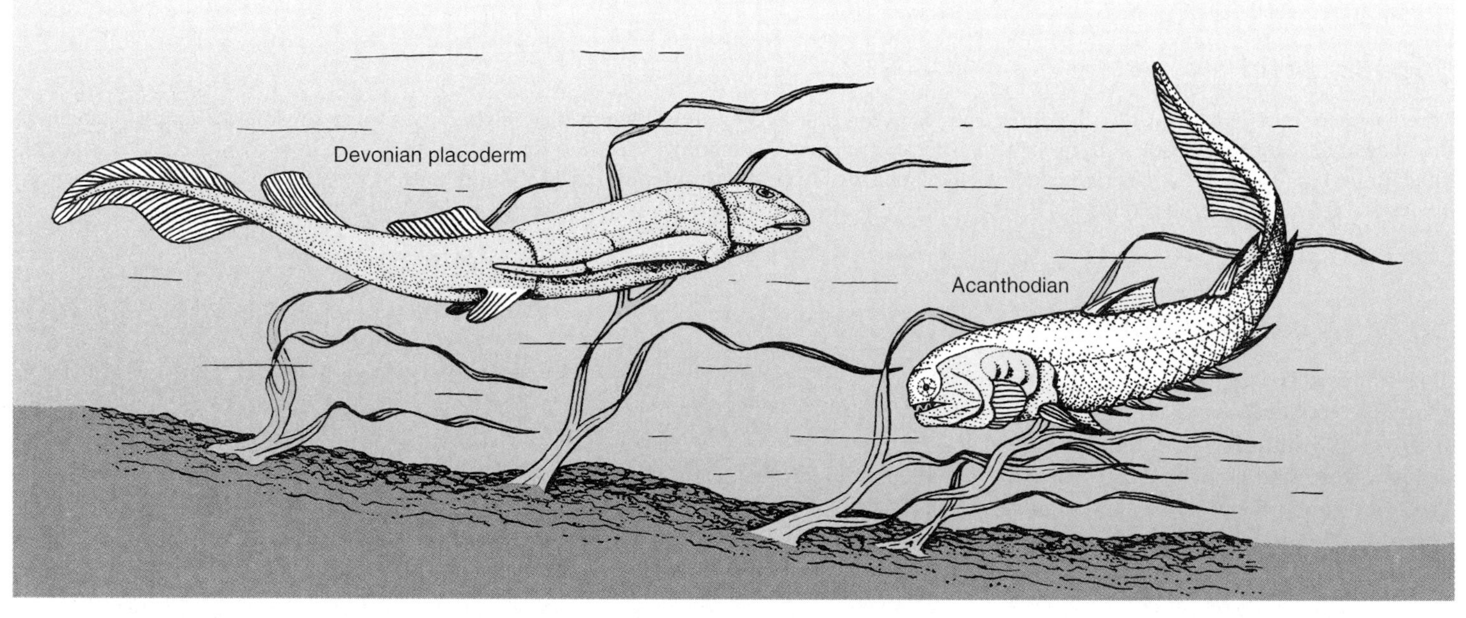

figure 26-14

Early jawed fishes of the Devonian period, 400 million years ago. Shown are a placoderm *(left)* and a related acanthodian *(right)*. The jaws and the gill supports from which the jaws evolved develop from neural crest cells, a diagnostic character of vertebrates. Most placoderms were bottom dwellers that fed on detritus although some were active predators. The acanthodians, the earliest-known true jawed fishes, carried less armor than the placoderms. Most were marine but several species entered fresh water.

Devonian period (Figure 26-14). Placoderms evolved a great variety of forms, some very large (one was 10 m in length!) and grotesque in appearance. They were armored fish covered with diamond-shaped scales or with large plates of bone. All became extinct by the end of the Paleozoic era and appear to have left no descendants. However, the **acanthodians** (Figure 26-14), a group of early jawed fishes that were contemporary with the placoderms, may have given rise to the great radiation of bony fishes that dominate the waters of the world today.

Evolution of Modern Fishes and Tetrapods

Reconstruction of the origins of the vast and varied assemblage of modern living vertebrates is based, as we have seen, largely on fossil evidence. Unfortunately the fossil evidence for the earliest vertebrates is often incomplete and tells us much less than we would like to know about subsequent trends in evolution. Affinities become much easier to establish as the

Classification of Phylum Chordata

Phylum Chordata

Group Protochordata (Acrania)

Subphylum Urochordata (u′ro-kor-da′ta) (Gr. *oura,* tail, + L. *chorda,* cord, + *ata,* characterized by) **(Tunicata): tunicates.** Notochord and nerve cord in free-swimming larva only; ascidian adults sessile, encased in tunic.

Subphylum Cephalochordata (sef′a-lo-kor-da′ta) (Gr. *kephalē,* head, + L. *chorda,* cord): **lancelets (amphioxus).** Notochord and nerve cord found along entire length of body and persist throughout life; fishlike in form.

Group Craniata

Subphylum Vertebrata (ver′te-bra′ta) (L. *vertebratus,* backboned). Bony or cartilaginous vertebrae surrounding spinal cord (vertebrae absent in agnathans); notochord only in embryonic stages, persisting in some fishes; also may be divided into two groups (superclasses) according to presence of jaws.

Superclass Agnatha (ag′na-tha) (Gr. *a,* without, + *gnathos,* jaw) **(Cyclostomata): hagfishes, lampreys.** Without true jaws or paired appendages. Probably a paraphyletic group.

Class Myxini (mik-sin′y) (Gr. *myxa,* slime): **hagfishes.** Terminal mouth with four pairs of tentacles; buccal funnel absent; nasal sac with duct to pharynx; 5 to 15 pairs of gill pouches; partially hermaphroditic.

Class Cephalaspidomorphi (sef-a-lass′pe-do-morf′e) (Gr. *kephalē,* head, + *aspidos,* shield, *morphē,* form) **(Petromyzones): lampreys.** Suctorial mouth with horny teeth; nasal sac not connected to mouth; seven pairs of gill pouches.

Superclass Gnathostomata (na′tho-sto′ma-ta) (Gr. *gnathos,* jaw, + *stoma,* mouth): **jawed fishes, all tetrapods.** With jaws and (usually) paired appendages.

Class Chondrichthyes (kon-drik′thee-eez) (Gr. *chondros,* cartilage, + *ichthys,* a fish): **sharks, skates, rays, chimaeras.** Streamlined body with heterocercal tail; cartilaginous skeleton; five to seven gills with separate openings, no operculum, no swim bladder.

Class Osteichthyes (ost′e-ik′thee-eez) (Gr. *osteon,* bone, + *ichthys,* a fish): **bony fishes.** Primitively fusiform body but variously modified; mostly ossified skeleton; single gill opening on each side covered with operculum; usually swim bladder or lung. A paraphyletic group.

Class Amphibia (am-fib′e-a) (Gr. *amphi,* both or double, + *bios,* life): **amphibians.** Ectothermic tetrapods; respiration by lungs, gills, or skin; development through larval stage; skin moist, containing mucous glands, and lacking scales.

Class Reptilia (rep-til′e-a) (L. *repere,* to creep): **reptiles.** Ectothermic tetrapods possessing lungs; embryo develops within shelled egg; no larval stage; skin dry, lacking mucous glands, and covered by epidermal scales. A paraphyletic group.

Class Aves (ay′veez) (L. pl. of *avis,* bird): **birds.** Endothermic vertebrates with front limbs modified for flight; body covered with feathers; scales on feet.

Class Mammalia (ma-may′lee-a) (L. *mamma,* breast): **mammals.** Endothermic vertebrates possessing mammary glands; body more or less covered with hair; well-developed neocerebrum.

fossil record improves. For instance, the descent of birds and mammals from early tetrapod ancestors has been worked out in a highly convincing manner from the relatively abundant fossil record available. By contrast, the ancestry of modern fishes is shrouded in uncertainty.

Despite the difficulty of clarifying early lines of descent for the vertebrates, they are clearly a natural, monophyletic group, distinguished by a large number of shared derived characteristics. They almost certainly have descended from a common ancestor, although we still do not know from which invertebrate group the vertebrate lineage originated. Early in their evolution, the vertebrates divided into the agnathans and the gnathostomes. These two groups differ from each other in many fundamental ways, in addition to the absence of jaws in the former group and their presence in the latter. The appearance of both jaws and paired fins were major innovations in vertebrate evolution, perhaps the most important reasons for the subsequent major radiations of the vertebrates that produced the modern fishes and all of the tetrapods—including you, the reader of this book.

Summary

The phylum Chordata is named for the rodlike notochord that forms a stiffening body axis at some stage in the life cycle of every chordate. All chordates share four distinctive hallmarks that set them apart from all other phyla: notochord, dorsal tubular nerve cord, pharyngeal pouches, and postanal tail. Two of the three chordate subphyla are invertebrates and lack a well-developed head. They are the Urochordata (tunicates), most of which are sessile as adults, but all of which have a free-swimming larval stage; and the Cephalochordata (lancelets), fishlike forms that include the famous amphioxus.

The chordates have evolutionary affinities to echinoderms, but the exact evolutionary origin of the chordates is not yet, and may never be, known with certainty. Taken as a whole, the chordates have a greater fundamental unity of organ systems and body plan than have many of the invertebrate phyla.

The subphylum Vertebrata includes the backboned members of the animal kingdom (the living jawless vertebrates, the hagfishes and lampreys, actually lack vertebrae but are included with the Vertebrata by tradition because they share numerous homologies with the vertebrates). As a group the vertebrates are characterized by having a well-developed head and by their comparatively large size, high degree of motility, and distinctive body plan, which embodies several distinguishing features that have permitted exceptional adaptive radiation. Most important of these are the living endoskeleton, which allows continuous growth and provides a sturdy framework for efficient muscle attachment and action; a pharynx perforated with gill slits (lost or greatly modified in the reptiles, birds, and mammals) with vastly increased respiratory efficiency; a complex nervous system with clear separation of the brain and spinal cord; and paired limbs.

Review Questions

1. What characteristics are shared by the six deuterostome phyla that indicate a natural grouping of interrelated animals?
2. Explain how the use of a cladistic classification for the vertebrates results in important regroupings of the traditional vertebrate taxa (refer to Figure 26-3). Why are certain traditional groupings such as Reptilia and Agnatha not recognized in cladistic usage?
3. Name four hallmarks shared by all chordates, and explain the function of each.
4. In debating the question of chordate origins, zoologists eventually agreed that the chordates must have evolved within the deuterostome assemblage rather than from a protostome group as earlier argued. What embryological evidences support this view? What characteristics does the fossil echinoderm group Calcichordata possess that suggest it might closely resemble the ancestor of the chordates?
5. Offer a description of an adult tunicate that would identify it as a chordate, yet distinguish it from any other chordate group.
6. Amphioxus long has been of interest to zoologists searching for a vertebrate ancestor. Explain why amphioxus captured such interest and why it no longer is considered to resemble closely the direct ancestor of the vertebrates.
7. Both sea squirts (urochordates) and lancelets (cephalochordates) are suspension-feeding organisms. Describe the suspension-feeding apparatus of a sea squirt and explain in what ways its mode of feeding is similar to, and different from, that of amphioxus.
8. Explain why it is necessary to know the life history of a tunicate to understand why tunicates are chordates.
9. List four adaptations that guided vertebrate evolution, and explain how each has contributed to the success of vertebrates.
10. In 1928 Walter Garstang hypothesized that tunicates resemble the ancestral stock of the vertebrates. Explain this hypothesis.
11. Distinguish between ostracoderms and placoderms. What important evolutionary advances did each contribute to vertebrate evolution? What are conodonts?
12. Explain how we think the vertebrate jaw evolved.

Selected References

Bone, Q. 1979. The origin of chordates. Oxford Biology Readers, No. 18. New York, Oxford University Press. *Synthesis of hypotheses and disagreements bearing on an unsolved riddle.*

Carroll, R. L. 1988. Vertebrate paleontology and evolution. New York, W.H. Freeman & Company. *Authoritative treatment of the vertebrate fossil record. The first two chapters contain discussions of cladistic classification of the vertebrates, the vertebrate body plan, and the origin of vertebrate characters.*

Gans, C. 1989. Stages in the origin of vertebrates: analysis by means of scenarios. Biol. Rev. **64**:221–268. *Reviews the diagnostic characters of protochordates and ancestral vertebrates and presents a scenario for the protochordate-vertebrate transition.*

Gould, S. J., ed. 1993. The book of life. New York, W.W. Norton & Company. *A sweeping, handsomely illustrated view of (almost entirely) vertebrate life.*

Jeffries, R. P. S. 1986. The ancestry of the vertebrates. Cambridge, Cambridge University Press. *Jeffries argues that the Calcichordata are the direct ancestors of the vertebrates, a view that most zoologists are not willing to accept. Still, this book is an excellent summary of the deuterostome groups and of the various competing hypotheses of vertebrate ancestry.*

Long, J. A. 1995. The rise of fishes: 500 million years of evolution. Baltimore, The Johns Hopkins University Press. *An authoritative, liberally illustrated evolutionary history of fishes.*

Norman, D. 1994. Prehistoric life: the rise of the vertebrates. New York, Macmillan USA. *Although this beautifully illustrated volume focuses on the evolution of land vertebrates, the early chapters deal with the origins of chordates and vertebrates.*

Fishes

chapter | twenty-seven

What Is a Fish?

In common (and especially older) usage, the term fish has often been used to describe a mixed assortment of water-dwelling animals. We speak of jellyfish, cuttlefish, starfish, crayfish, and shellfish, knowing full well that when we use the word "fish" in such combinations, we are not referring to a true fish. In earlier times, even biologists did not make such a distinction. Sixteenth-century natural historians classified seals, whales, amphibians, crocodiles, even hippopotamuses, as well as a host of aquatic invertebrates, as fish. Later biologists were more discriminating, eliminating first the invertebrates and then the amphibians, reptiles, and mammals from the narrowing concept of a fish. Today we recognize a fish as a gill-breathing, ectothermic, aquatic vertebrate that possesses fins, and skin that is usually covered with scales. Even this modern concept of the term "fish" is controversial, at least as a taxonomic unit, because fishes do not compose a monophyletic group. The common ancestor of the fishes is also an ancestor to the land vertebrates, which we exclude from the term "fish," unless we use the term in an exceedingly nontraditional way. Because fishes live in a habitat that is basically alien to humans, people have rarely appreciated the remarkable diversity of these vertebrates. Nevertheless, whether appreciated by humans or not, the world's fishes have enjoyed an effusive proliferation that has produced an estimated 24,600 living species—more than all other species of vertebrates combined—with adaptations that have fitted them to almost every conceivable aquatic environment. No other animal group threatens their domination of the seas.

The life of a fish is bound to its body form. Mastery of river, lake, and ocean is revealed in the many ways that fishes have harmonized their design to the physical properties of their aquatic surroundings. Suspended in a medium that is 800 times more dense than air, a trout or pike can remain motionless, varying its neutral buoyancy by adding or removing air from the swim bladder. Or it may dart forward or at angles, using its fins as brakes and tilting rudders. With excellent organs for salt and water exchange, bony fishes can steady and finely tune their body fluid composition in their chosen freshwater or seawater environment. Their gills are the most effective respiratory devices in the animal kingdom for extracting oxygen from a medium that contains less than $\frac{1}{20}$ as much oxygen as air. Fishes have excellent olfactory and visual senses and a unique lateral line system, which with its exquisite sensitivity to water currents and vibrations provides a "distance touch" in water. Thus in mastering the physical problems of their element, early fishes evolved a basic body plan and set of physiological strategies that both shaped and constrained the evolution of their descendants.

The use of fishes *as the plural form of* fish *may sound odd to most people accustomed to using* fish *in both the singular and the plural. Both plural forms are correct but zoologists use* fishes *to mean more than one kind of fish.*

Ancestry and Relationships of Major Groups of Fishes

The fishes are of ancient ancestry, having descended from an unknown free-swimming protochordate ancestor (hypotheses of chordate and vertebrate origins are discussed in Chapter 26). Whatever their origin, during the Cambrian period, or perhaps even in the Precambrian, the earliest fishlike vertebrates branched into the jawless **agnathans** and the jawed **gnathostomes** (Figure 27-1). All vertebrates have descended from one of these two ancestral groups.

The jawless agnathans, the least derived of the two groups, include the extinct ostracoderms and the living **hagfishes** and **lampreys,** fishes adapted as scavengers or parasites. The agnathans have no vertebrae but are nevertheless included within the subphylum Vertebrata because they have a cranium and many other vertebrate homologies. The ancestry of hagfishes and lampreys is uncertain; they bear little resemblance to the extinct ostracoderms. Although the hagfishes and the more derived lampreys superficially look much alike, they are in fact so different from each other that they have been assigned to separate classes by ichthyologists.

All remaining fishes have paired appendages and jaws and are included, along with the tetrapods (land vertebrates), in the monophyletic lineage of gnathostomes. They appear in the fossil record in the Silurian period with fully formed jaws, and no forms intermediate between agnathans and gnathostomes are known. By the Devonian period, the Age of Fishes, several distinct groups of jawed fishes were well represented. One of these, the placoderms (Figure 26-14, p. 578), became extinct, leaving no descendants, in the Carboniferous period, which followed the Devonian. A second group, the **cartilaginous fishes** of the class Chondrichthyes (sharks, rays, and chimaeras), lost the heavy dermal armor of the early jawed fishes and adopted cartilage rather than bone for the skeleton. Most are active predators with a sharklike body form that has undergone only minor changes over the ages.

Of all the gnathostomes, the **bony fishes** of the class Osteichthyes radiated most extensively and are the dominant fishes today (Figure 27-1). We can recognize two distinct lineages of bony fishes. Of these two, by far the most diverse are the **ray-finned fishes** (subclass Actinopterygii), which radiated to form the modern bony fishes. The other lineage, the **fleshy-finned fishes** (subclass Sarcopterygii), although a relic group today, carry the distinction of including the closest living relatives of the tetrapods. The fleshy-finned fishes are represented today by the **lungfishes** and the **coelacanth**—meager remnants of important stocks that flourished in the Devonian period (Figure 27-1). A classification of the major fish taxa is on p. 603.

Superclass Agnatha: Jawless Fishes

The living jawless fishes are represented by approximately 84 species almost equally divided between two classes: Myxini (hagfishes) and Cephalaspidomorphi (lampreys) (Figures 27-2 and 27-3). Members of both groups lack jaws, internal ossification, scales, and paired fins, and both groups share porelike gill openings and an eel-like body form. In other respects, however, the two groups are morphologically very different. Lampreys bear many derived morphological characters that place them phylogenetically much closer to the jawed bony fishes than to the hagfishes. Because of these differences, hagfishes and lampreys have been assigned to separate vertebrate classes, leaving the grouping "Agnatha" as a paraphyletic assemblage of jawless fishes.

Hagfishes: Class Myxini

The hagfishes are an entirely marine group that feed on dead or dying fishes, annelids, molluscs, and crustaceans. They are neither parasitic like lampreys nor predaceous, but are scavengers. There are only 43 species of hagfishes, of which the best known in North America are the Atlantic hagfish *Myxine glutinosa* (Gr. *myxa,* slime) (Figure 27-3) and the Pacific hagfish *Eptatretus stouti* (NL, *ept*<Gr. *hepta*, seven, + *tretos,* perforated). Although almost completely blind, the hagfish is

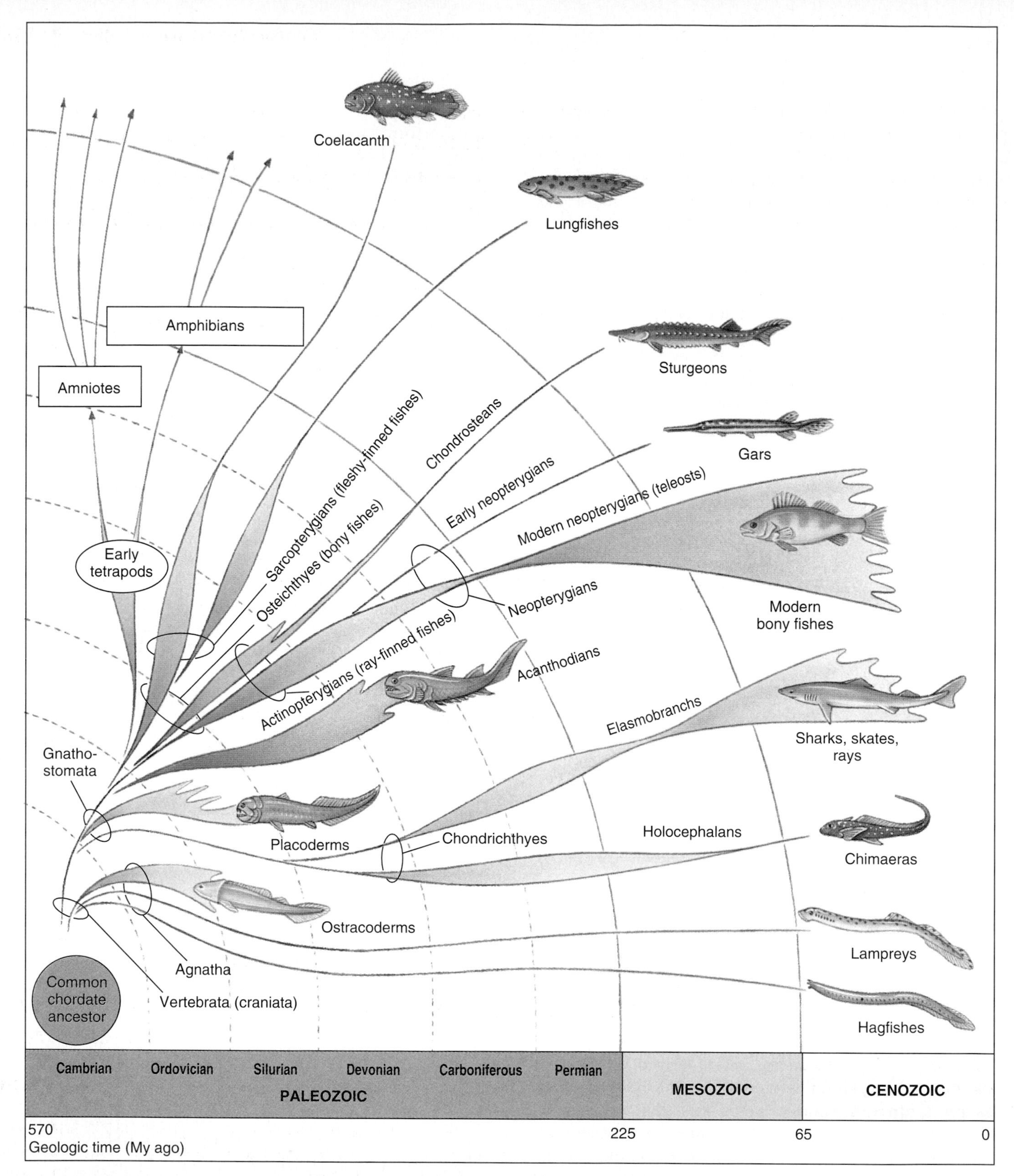

figure 27-1

Graphic representation of the family tree of fishes, showing the evolution of major groups through geological time. Numerous lineages of extinct fishes are not shown. Widened areas in the lines of descent indicate periods of adaptive radiation and the relative number of species in each group. The fleshy-finned fishes (sarcopterygians), for example, flourished in the Devonian period, but declined and are today represented by only four surviving genera (lungfishes and coelacanth). Homologies shared by the sarcopterygians and tetrapods suggest that they are sister groups. The sharks and rays radiated during the Carboniferous period. They came dangerously close to extinction during the Permian period but staged a recovery in the Mesozoic era and are a secure group today. Johnny-come-latelies in fish evolution are the spectacularly diverse modern fishes, or teleosts, which include most living fishes. Note that the class Osteichthyes is a paraphyletic group because it does not include their descendants, the tetrapods; in cladistic usage the Osteichthyes includes the tetrapods (see Figure 27-2).

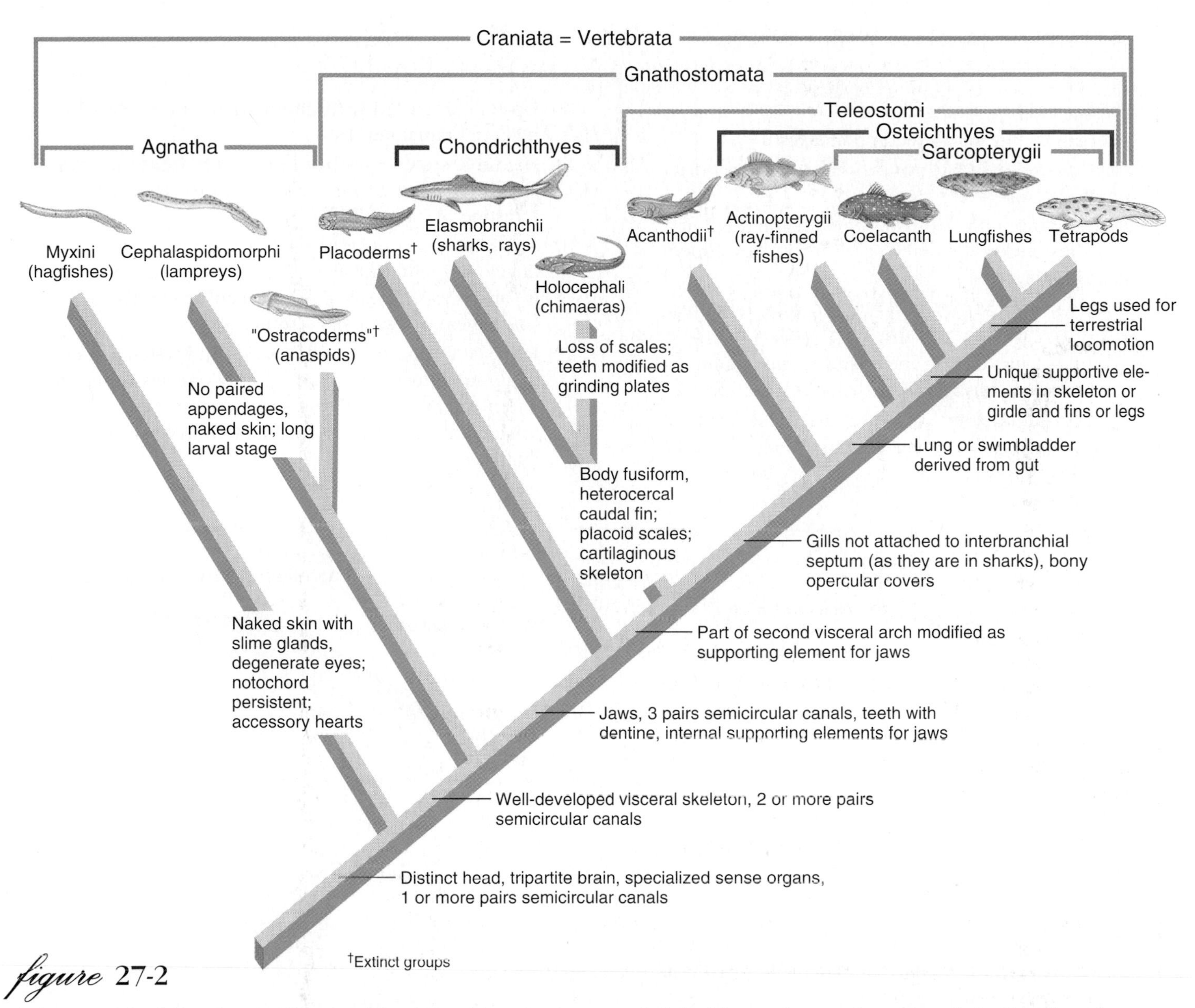

figure 27-2

Cladogram of the fishes, showing the probable relationships of major monophyletic fish taxa. Several alternative relationships have been proposed. Extinct groups are designated by a dagger (†). Some of the shared derived characters that mark the branchings are shown to the right of the branch points. Agnatha is a paraphyletic structural grade not recognized in cladistic classification.

quickly attracted to food, especially dead or dying fish, by its keenly developed senses of smell and touch. After attaching itself to its prey by means of toothed plates, the hagfish thrusts the tongue forward to rasp off pieces of tissue. For extra leverage, the hagfish often ties a knot in its tail, then passes it forward along the body until it is pressed securely against the side of its prey (Figure 27-4).

Unlike any other vertebrate, the body fluids of hagfishes are in osmotic equilibrium with seawater, as in most marine invertebrates. Hagfishes have several other anatomical and physiological peculiarities, including a low-pressure circulatory system served by three accessory hearts in addition to the main heart positioned behind the gills. Hagfishes are also renowned for their ability to generate enormous quantities of slime.

While the unique anatomical and physiological features of the strange hagfishes are of interest to biologists, hagfishes have not endeared themselves to either sports or commercial fishermen. In earlier days of commercial fishing mainly by gill nets and set lines, hagfish often bit into the bodies of captured fish and ate out the contents, leaving behind a useless sack of skin and bones. But as large and efficient otter trawls came into use, hagfishes ceased to be an important pest.

The reproductive biology of hagfishes remains largely a mystery. Both male and female gonads are found in each

Characteristics of the Jawless Fishes

1. Slender, **eel-like** body
2. Median fins but **no** paired appendages
3. **Fibrous** and **cartilaginous skeleton;** notochord persistent; no vertebrae
4. Biting mouth with two rows of eversible teeth in hagfishes; suckerlike oral disc with well-developed teeth in lampreys
5. Heart with one atrium and one ventricle; hagfishes with three accessory hearts; aortic arches in gill region
6. Five to 16 pairs of gills and a single pair of gill apertures in hagfishes; 7 pairs of gills in lampreys
7. **Pronephric kidney** anteriorly and **mesonephric kidney** posteriorly in hagfishes; mesonephric kidney only in lampreys
8. Dorsal nerve cord with differentiated brain; 8 to 10 pairs of cranial nerves
9. Digestive systems **without stomach;** intestine with spiral valve and cilia in lampreys; both lacking in intestine of hagfishes
10. Sense organs of taste, smell, hearing; eyes poorly developed in hagfishes but moderately well developed in lampreys; one pair of **semicircular canals** (hagfishes) or two pairs (lampreys)
11. External fertilization; both ovaries and testes present in an individual but gonads of only one sex functional and no larval stage in hagfishes; separate sexes and long larval stage with radical metamorphosis in lampreys

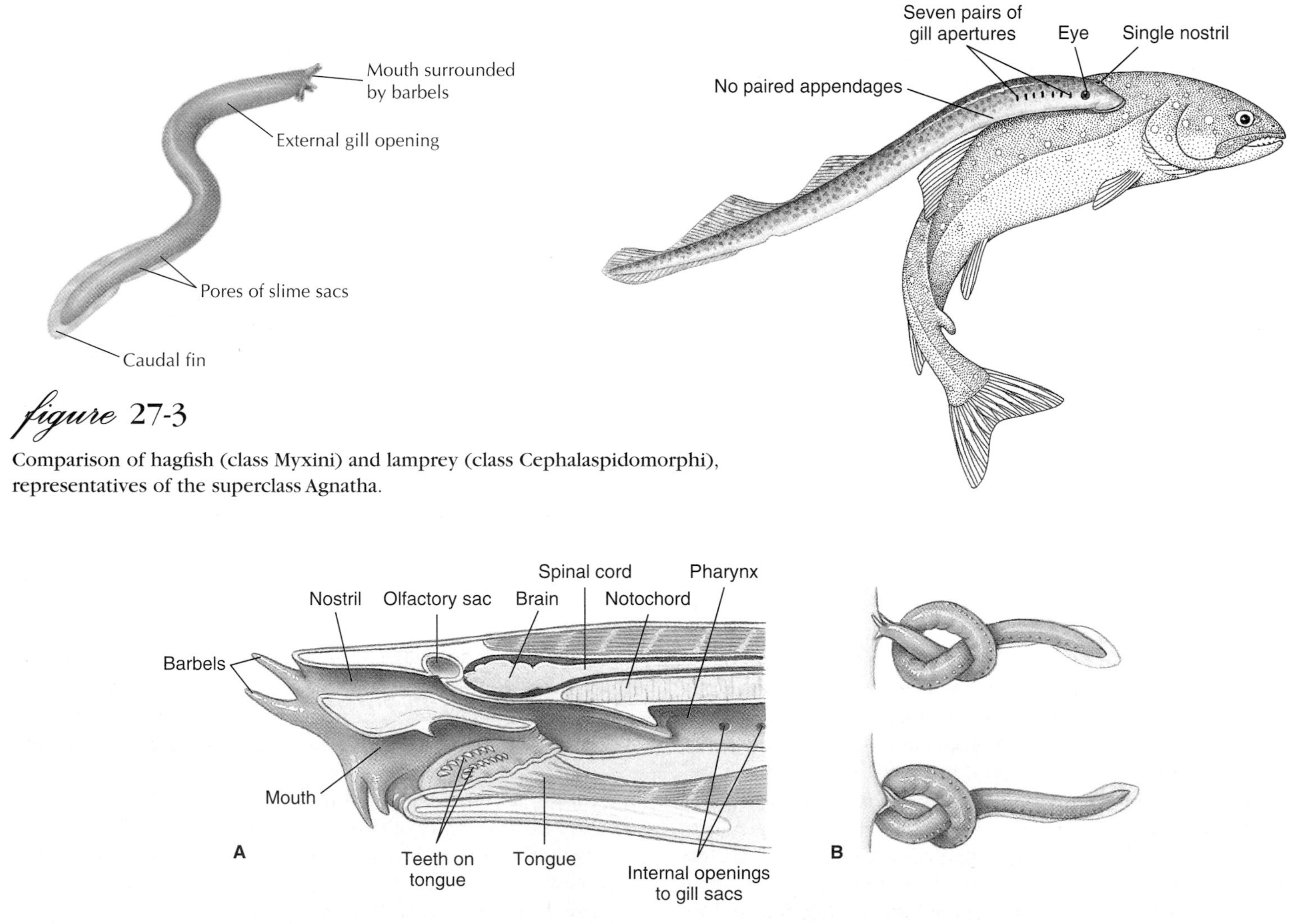

figure 27-3

Comparison of hagfish (class Myxini) and lamprey (class Cephalaspidomorphi), representatives of the superclass Agnatha.

figure 27-4

The Atlantic hagfish *Myxine glutinosa* (Class Myxini). **A,** Sagittal section of the head region showing the rasping tongue (retracted) and internal openings to gill sacs; **B,** Hagfish knotting, showing how it obtains leverage to tear flesh from prey.

animal, but only one gonad becomes functional. The females produce small numbers of surprisingly large, yolky eggs up to 3 cm in diameter. There is no larval stage and growth is direct.

Lampreys: Class Cephalaspidomorphi

Of the 41 described species of lampreys distributed around the world, by far the best known to North Americans is the destructive marine lamprey, *Petromyzon marinus,* of the Great Lakes (Figure 27-3). The name *Petromyzon* (Gr. *petros,* stone, + *myzon,* sucking) refers to the lamprey's habit of grasping a stone with its mouth to hold position in a current. There are 17 species of lampreys in North America of which about half are parasitic; the rest are species that never feed after metamorphosis and die soon after spawning.

In North America all lampreys, marine as well as freshwater forms, spawn in the winter or spring in shallow gravel and sand in freshwater streams. The males begin nest building and are joined later by females. Using their oral discs to lift stones and pebbles and using vigorous body vibrations to sweep away light debris, they form an oval depression. At spawning, with the female attached to a rock to maintain position over the nest, the male attaches to the dorsal side of her head. As the eggs are shed into the nest, they are fertilized by the male. The sticky eggs adhere to pebbles in the nest and soon become covered with sand. The adults die soon after spawning.

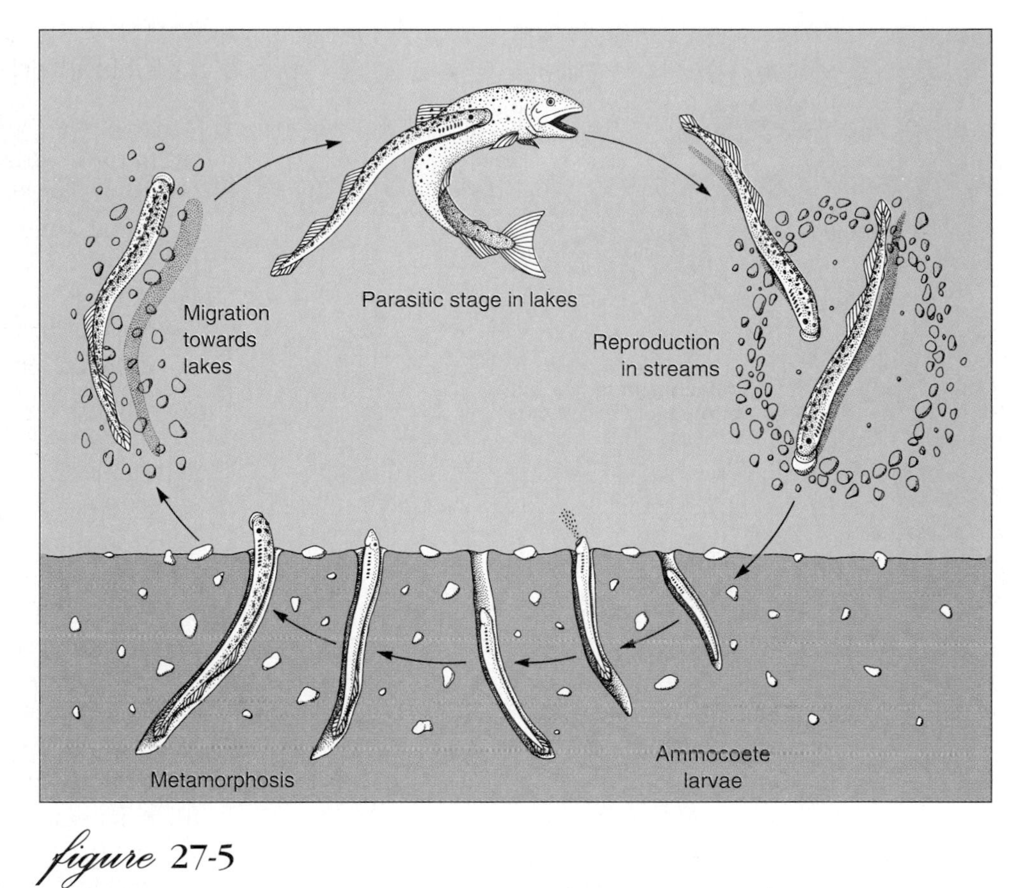

figure 27-5

Life cycle of the "landlocked" form of the sea lamprey *Petromyzon marinus.*

The eggs hatch in approximately 2 weeks, releasing small larvae **(ammocoetes)** (Figure 27-5), which stay in the nest until they are approximately 1 cm long; they then burrow into mud or sand and emerge at night to feed on small invertebrates, detritus, and other particulate matter in the water. The larval period lasts from 3 to 17 or more years before the larva rapidly metamorphoses into an adult.

Parasitic lampreys either migrate to the sea, if marine, or remain in fresh water, where they attach themselves by their suckerlike mouth to fish and with their sharp horny teeth rasp through flesh and suck the body fluids (Figure 27-6). To promote the flow of blood, the lamprey injects an anticoagulant into the wound. When gorged, the lamprey releases its hold but leaves the fish with a wound that may prove fatal. The parasitic freshwater adults live a year or more before spawning and then die; the marine forms may live longer.

Nonparasitic lampreys do not feed after emerging as adults, since their alimentary canal degenerates to a nonfunctional strand of tissue. Within a few months and after spawning, they die.

The invasion of the Great Lakes above Lake Ontario by the landlocked sea lamprey, *Petromyzon marinus,* in this century had a devastating effect on the fisheries. Lampreys first entered the Great Lakes after the Welland Canal around Niagara Falls, a barrier to further western migration, was deepened between 1913 and 1918. Moving first through Lake Erie to Lakes Huron, Michigan, and Superior, sea lampreys, accompanied by overfishing, caused the total collapse of a multimillion dollar lake trout fishery in the early 1950s. Other less valuable fish species were attacked and destroyed in turn. After reaching a peak abundance in 1951 in Lakes Huron and Michigan and in 1961 in Lake Superior, the sea lampreys began to decline, due in part to depletion of their food and in part to the effectiveness of control measures (mainly chemical larvicides placed in selected spawning streams). Lake trout, aided by a restocking program, are now recovering, but wounding rates are still high in some lakes. Fishery organizations are now experimenting with the release into spawning streams of sterilized male lampreys; when fertile females mate with sterilized males, the female's eggs fail to develop.

Cartilaginous Fishes: Class Chondrichthyes

There are more than 850 living species in the class Chondrichthyes, an ancient, compact, and highly developed group. Although a much smaller and less diverse assemblage than the

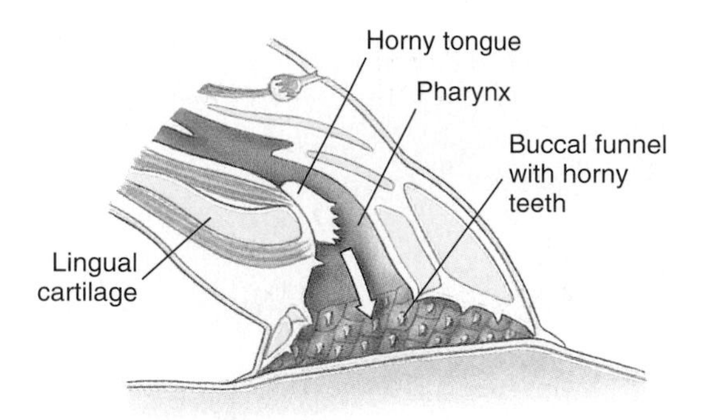

Attachment to fish with horny teeth and suction

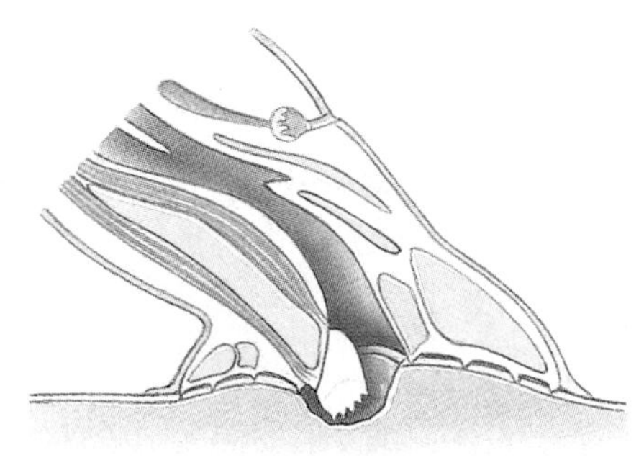

Tongue protruded for rasping flesh

figure 27-6

How the lamprey uses its horny tongue to feed. After firmly attaching to a fish by its sucker, the protrusable tongue rapidly rasps an opening through the fish's integument. Body fluid, abraded skin and muscle are eaten.

Characteristics of the Sharks and Rays (Elasmobranchii)

1. **Body fusiform** (except rays) with a **heterocercal** caudal fin (Figure 27-14)
2. **Mouth ventral** (Figure 27-7); two olfactory sacs that do not connect to the mouth cavity; jaws present
3. Skin with **placoid scales** (Figure 27-16) and mucous glands; teeth of modified placoid scales
4. **Endoskeleton entirely cartilaginous**
5. Digestive system with a J-shaped stomach and **intestine with spiral valve** (Figure 27-8)
6. Circulatory system of several pairs of aortic arches; two-chambered heart
7. Respiration by means of five to seven pairs of gills with **separate and exposed gill slits,** no operculum
8. No swim bladder or lung
9. Mesonephric kidney and rectal gland (Figure 27-8); blood isosmotic or slightly hyperosmotic to seawater; **high concentrations of urea and trimethylamine oxide in blood**
10. Brain of two olfactory lobes, two cerebral hemispheres, two optic lobes, a cerebellum, and a medulla oblongata; 10 pairs of cranial nerves; **three pairs of semicircular canals;** senses of smell, vibration reception (lateral line system), and electroreception well developed
11. Separate sexes; oviparous, ovoviviparous, or viviparous; direct development; **internal fertilization**

bony fishes, their impressive combination of well-developed sense organs, powerful jaws and swimming musculature, and predaceous habits ensures them a secure and lasting niche in the aquatic community. One of their distinctive features is their cartilaginous skeleton. Although there is some limited calcification, bone is entirely absent throughout the class—a curious feature, since the Chondrichthyes are derived from ancestors having well-developed bone.

Sharks and Rays: Subclass Elasmobranchii

Sharks, which make about 45% of the approximately 815 species in the subclass Elasmobranchii, are typically predaceous fishes with five to seven gill slits and gills on each side and (usually) a spiracle behind each eye. Sharks track their prey using their lateral line system and large olfactory organs, since their vision is not well developed. The larger sharks, such as the massive (but harmless) plankton-feeding whale shark, may reach 15 m in length, the largest of all fishes. The dogfish sharks so widely used in zoological laboratories rarely exceed 1 m. More than half of all elasmobranchs are rays, specialized for a bottom-feeding life-style. Unlike sharks, which swim with thrusts of the tail, rays propel themselves by wave-like motions of the "wings," or pectoral fins.

Although to most people sharks have a sinister appearance and a fearsome reputation, they are at the same time among the most gracefully streamlined of all fishes (Figure 27-7). Sharks are heavier than water and will sink if not swimming forward. The asymmetrical **heterocercal tail,** in which the vertebral column turns upward and extends into the dorsal lobe of the tail (see Figure 27-14), provides lift and thrust as it sweeps to and fro in the water, and the broad head and flat pectoral fins act as planes to provide head lift.

Sharks are well equipped for their predatory life. The tough leathery skin is covered with numerous dermal **pla-**

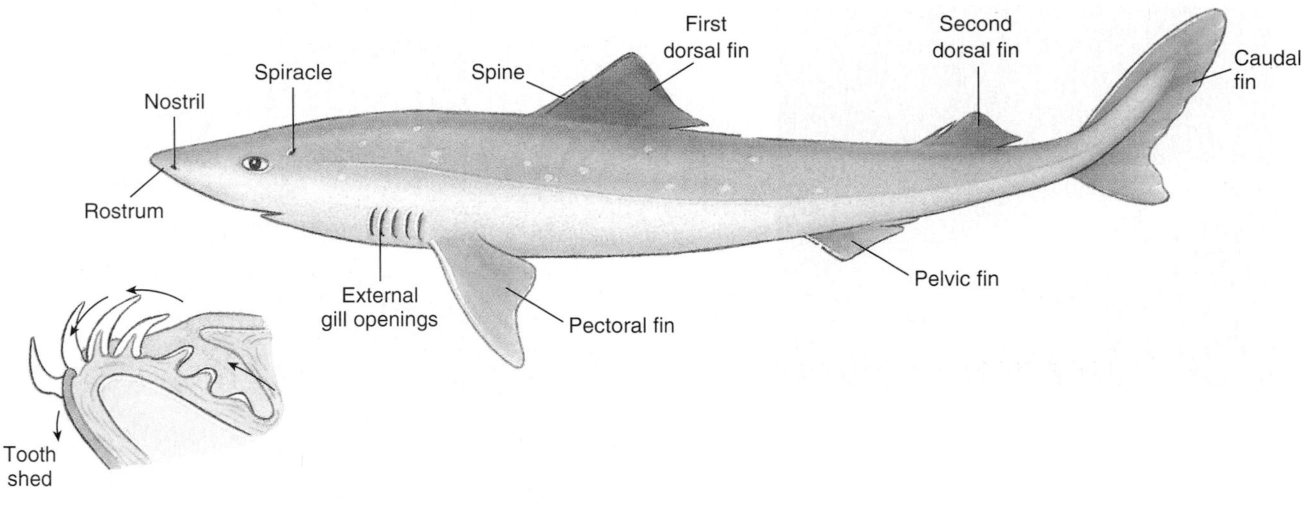

figure 27-7

Dogfish shark, *Squalus acanthias.* Inset: Section of lower jaw shows the formation of new teeth developing inside the jaw. These move forward to replace lost teeth. Rate of replacement varies in different species.

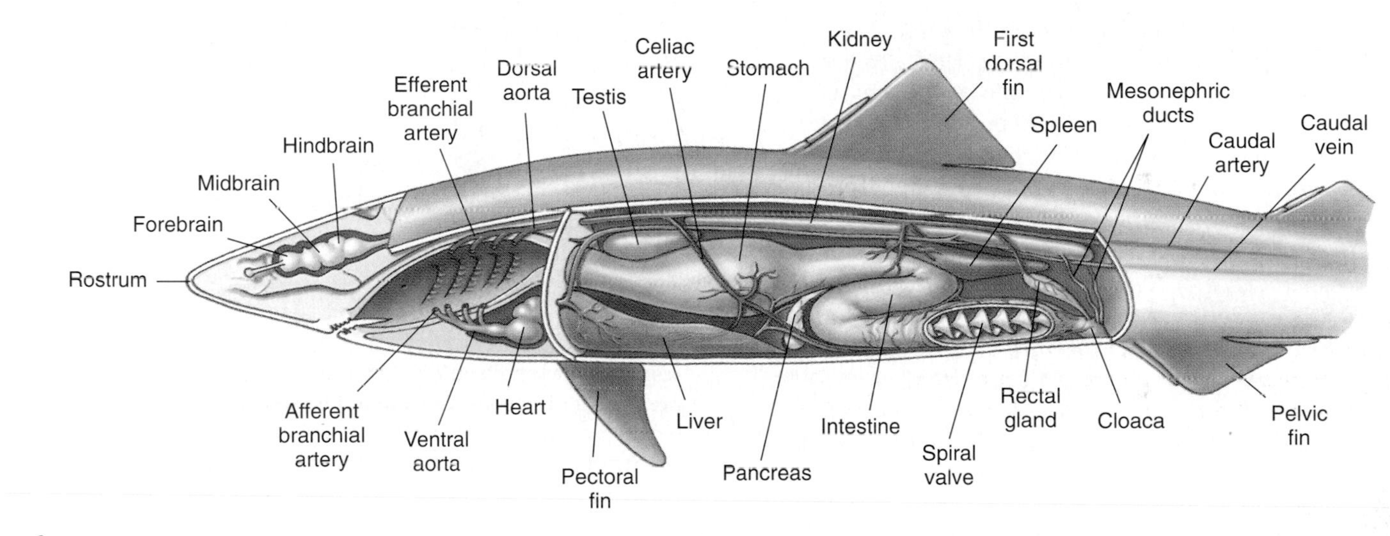

figure 27-8

Internal anatomy of dogfish shark *Squalus acanthias.*

coid scales (see Figure 27-16) that are modified anteriorly to form replaceable rows of teeth in both jaws (Figure 27-9). Placoid scales in fact consist of dentine enclosed by an enamel-like substance, and they very much resemble the teeth of other vertebrates. Sharks have a keen sense of smell used to guide them to food. Vision is less acute than in most bony fishes, but a well-developed **lateral line system** is used for detecting and locating objects and moving animals (predators, prey, and social partners). It is composed of a canal system extending along the side of the body and over the head (Figure 27-10). Inside are special receptor organs **(neuromasts)** that are extremely sensitive to vibrations and currents in the water. Sharks can also detect and aim attacks at prey buried in the sand by sensing the bioelectric fields that surround all animals. The receptors, the **ampullary organs of Lorenzini,** are located on the shark's head.

Rays belong to a separate order from the sharks. Rays are distinguished by their dorsoventrally flattened bodies and the much-enlarged pectoral fins that behave as wings in swimming. The gill openings are on the underside of the head, and the **spiracles** (on top of the head) are unusually large. Respiratory water enters through these spiracles to prevent clogging the gills, because the mouth is often buried in sand. The teeth are adapted for crushing the prey—mainly molluscs, crustaceans, and an occasional small fish.

figure 27-9

Head of sand tiger shark *Carcharias* sp. Note the series of successional teeth. Also visible in a row below the eye are the ampullae of Lorenzini *(arrow).*

The worldwide shark fishery is experiencing unprecedented pressure, driven by the high price of shark fins for shark-fin soup, an oriental delicacy (which commonly sells for $50.00 per bowl). Coastal shark populations in general have declined so rapidly that "finning" is to be outlawed in the United States; other countries, too, are setting quotas to protect threatened shark populations. Even in the Marine Resources Reserve of the Galápagos Islands, one of the world's exceptional wild places, tens of thousands of sharks have been killed illegally for the Asian shark-fin market. That illegal fishery continues at this writing. Contributing to the threatened collapse of shark fisheries worldwide is the long time required by most sharks to reach sexual maturity; some species take as long as 15 years.

In the order containing the rays (Rajiformes), we commonly refer to members of one family (Rajidae) as skates. Alone among members of the Rajiformes, skates do not bear living young but lay large, yolky eggs enclosed within a horny covering (the "mermaid's purse") that often washes up on beaches. Although the tail is slender, skates have a somewhat more muscular tail than most rays, and they usually have two dorsal fins and sometimes a caudal fin.

In the stingrays, the caudal and dorsal fins have disappeared, and the tail is slender and whiplike. The stingray tail is armed with one or more saw-toothed spines that can inflict dangerous wounds. Electric rays have certain dorsal muscles modified into powerful electric organs, which can give severe shocks to stun their prey.

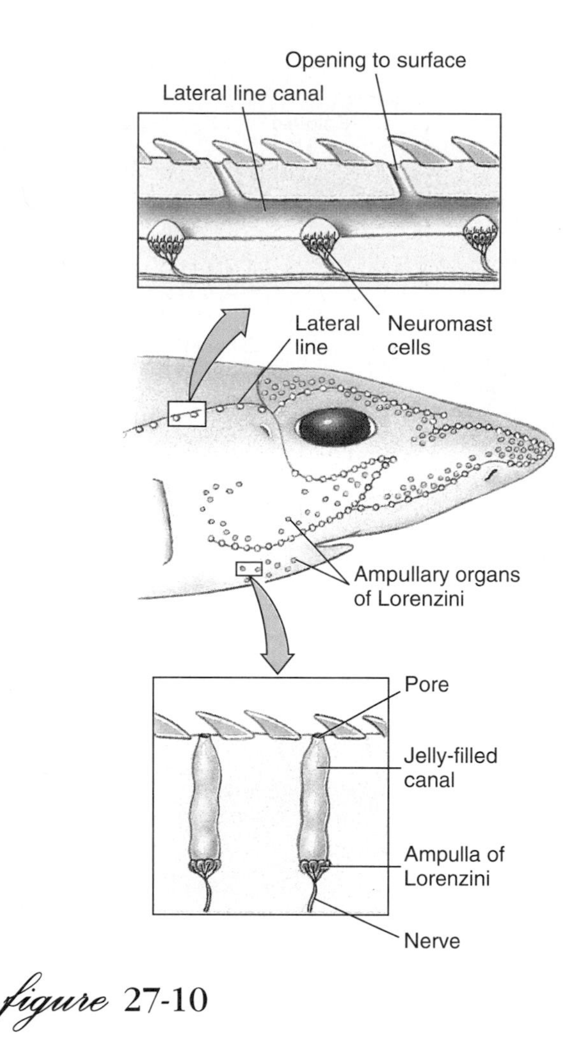

figure 27-10

Sensory canals and receptors in a shark. The ampullae of Lorenzini respond to weak electric fields, and possibly to temperature, water pressure, and salinity. The lateral line sensors, called neuromasts, are sensitive to disturbances in the water, enabling the shark to detect nearby objects by reflected waves in the water.

Chimaeras: Subclass Holocephali

The approximately 30 species of chimaeras (ky-meer′uz; L. monster), distinguished by such suggestive names as ratfish (Figure 27-12), rabbitfish, spookfish, and ghostfish, are remnants of an aberrant line that diverged from the placoderms at least 350 million years ago (Devonian period of the Paleozoic era). Fossil chimaeras first appeared in the Jurassic period, reached their zenith in the Cretaceous and early Tertiary periods (120 million to 50 million years ago), and have declined ever since. Anatomically they present an odd mixture of sharklike and bony fishlike features. Their food is a mixed diet of seaweed, molluscs, echinoderms, crustaceans, and fishes. Chimaeras are not commercial species

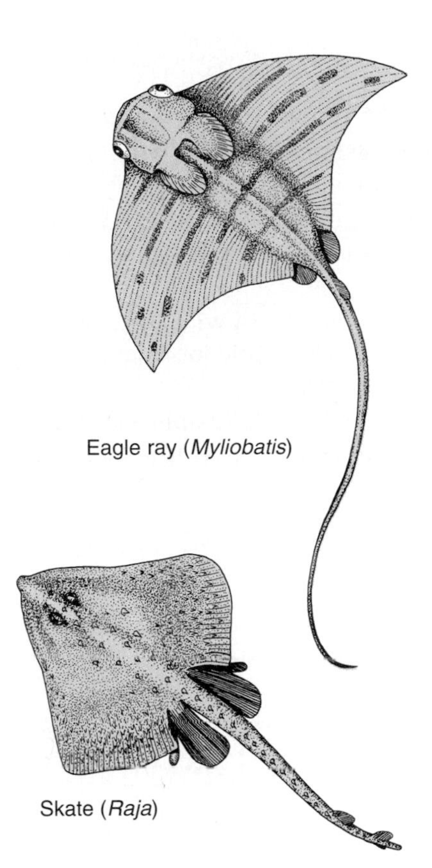

figure 27-11

Rays are specialized for life on the sea floor. They are flattened dorsoventrally and move by undulations of greatly expanded winglike pectoral fins. One group of rays, the skates, are distinguishd by laying their eggs in a horny capsule, the "mermaid's purse."

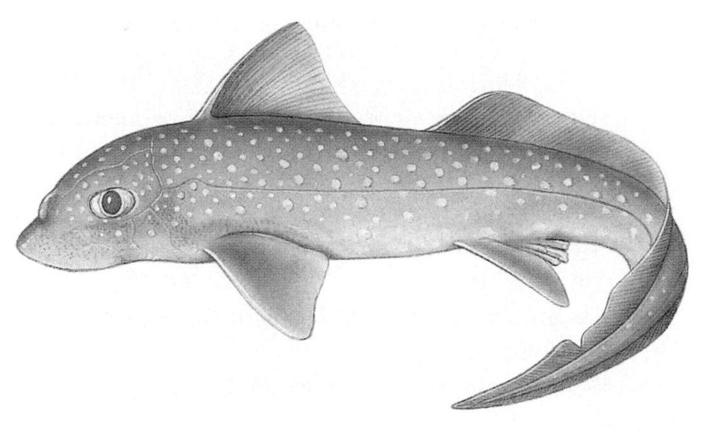

figure 27-12

Chimaera, or ratfish, of North American west coast. This species is one of the most handsome of chimaeras, which tend toward bizarre appearances.

and are seldom caught. Despite their grotesque shape, they are beautifully colored with a pearly iridescence.

Bony Fishes: Class Osteichthyes

Origin, Evolution, and Diversity

The bony fishes, the largest and most diverse taxon of all vertebrates, originated in the late Silurian period, approximately 410 million years ago. Details of head structure of earliest complete bony fishes fossils indicate that they probably descended from an ancestor shared with the acanthodians (p. 578). By the middle of the Devonian the bony fishes had developed several key adaptations that contributed to an extensive adaptive radiation. An **operculum** over the gill slits, composed of bony plates attached to the first gill arch, served to increase the efficiency of drawing water across the gill surfaces. These earliest bony fishes also had a pair of lungs, which served as accessory breathing structures. The fin pattern established at that time persists in bony fishes today: **pectoral and pelvic fins** supported by bony girdles embedded in the body musculature, and median dorsal and anal fins (Figure 27-13; see also Figure 6-14, p. 153). Progressive specialization of jaw structure and feeding mechanisms is another key feature in bony fish evolution. Bony fishes have high levels of activity, supported by efficient gill design for gas exchange, rapid metabolic oxidation of food, and an effective form of undulatory locomotion that persisted in many tetrapods (for example, salamanders, snakes, and many lizards).

By the middle Devonian, the Osteichthyes had divided into two distinct lineages. One lineage, the **ray-finned fishes** (Actinopterygii), includes the modern bony fishes, the largest of all vertebrate radiations. The other lineage is the **fleshy-finned fishes** (Sarcopterygii), a remnant group represented today by the **lungfishes** and the **coelacanth.** Their evolutionary history is of great interest because their descendents include all the land vertebrates (tetrapods) (Figure 27-2).

Ray-Finned Fishes: Subclass Actinopterygii

Ray-finned fishes are an enormous assemblage containing all of our familiar bony fishes—more than 24,600 species. The group had its beginnings in the Devonian freshwater lakes and streams. The ancestral forms were small, bony fishes, heavily armored with ganoid scales (Figure 27-16), and had functional lungs as well as gills.

From these earliest ray-finned fishes, two major groups emerged. Those bearing the most primitive characteristics are the **chondrosteans** (Gr. *chondros,* cartilage, + *osteon,* bone), represented today by the sturgeons, paddlefishes, and bichir *Polypterus* (Gr. *poly,* many, + *pteros,* winged) of African rivers (Figure 27-17). *Polypterus* is an interesting relic with a

Characteristics of Bony Fishes (Osteichthyes)

1. **Skeleton more or less bony,** vertebrae numerous; **tail usually homocercal** (Figure 27-14)
2. Skin with mucous glands and embedded **dermal scales** (Figure 27-15) of three types: **ganoid, cycloid,** or **ctenoid;** some without scales, no placoid scales (Figure 27-16)
3. Fins both median and paired with **fin rays of cartilage or bone**
4. **Mouth terminal** with many teeth (some toothless); jaws present; olfactory sacs paired and may or may not open into mouth
5. Respiration by gills supported by bony gill arches and covered by a **common operculum**
6. **Swim bladder** often present with or without duct connected to pharynx
7. Circulation consisting of a two-chambered heart, arterial and venous systems, and four pairs of aortic arches
8. Nervous system of brain with small olfactory lobes and cerebrum; large optic lobes and cerebellum; 10 pairs of cranial nerves
9. Sexes separate (some hermaphroditic), gonads paired; fertilization usually external; larval forms may differ greatly from adults

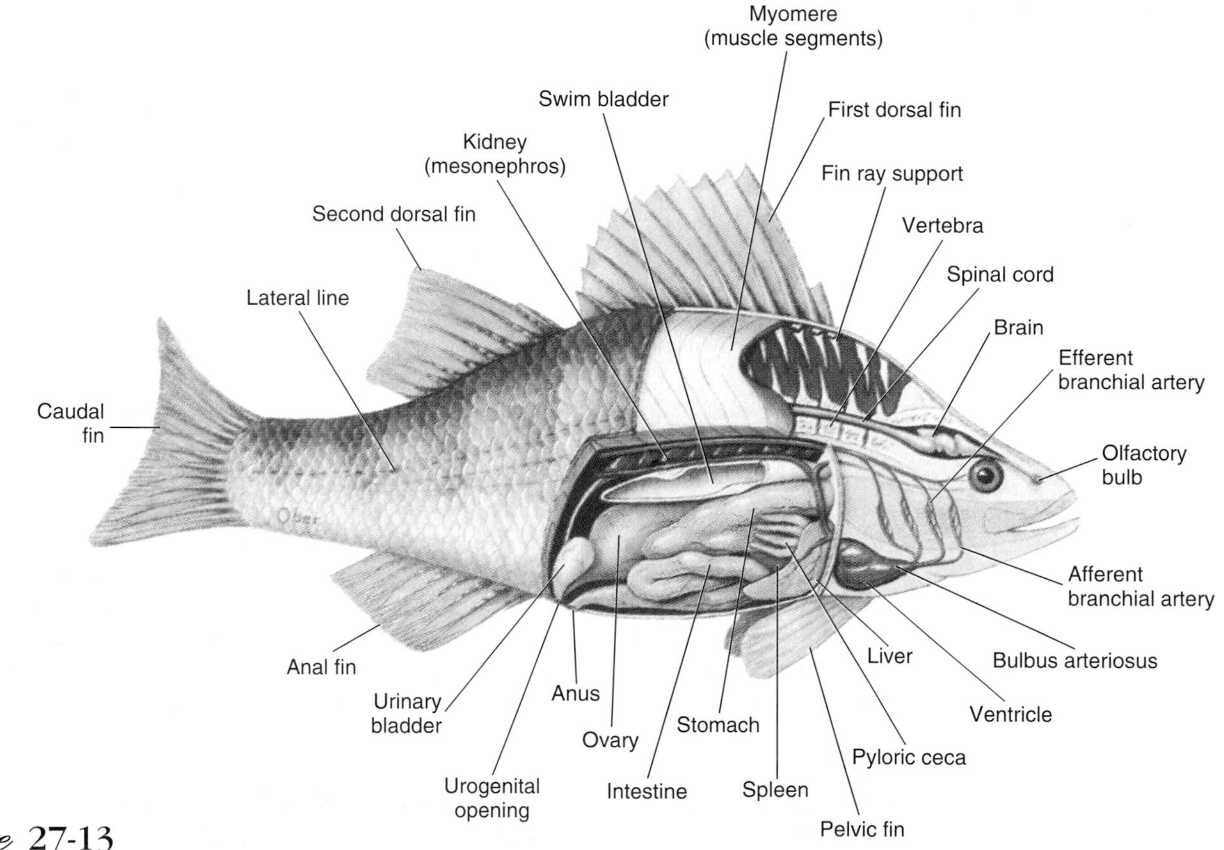

figure 27-13

Internal anatomy of the yellow perch *Perca flavescens,* a freshwater teleost fish.

lunglike swim bladder and many other primitive characteristics; it resembles an ancestral ray-finned fish more than it does any other living descendant. There is no satisfactory explanation for the survival to the present of this fish and the coelacanth *Latimeria* when all of their kin perished millions of years ago.

The second major group to emerge from the early ray-finned stock were **neopterygians** (Gr. *neos,* new, + *pteryx,* fin). The neopterygians appeared in the late Permian and radiated extensively during the Mesozoic era. During the Mesozoic one lineage gave rise to a secondary radiation that led to the modern bony fishes, the teleosts. The two surviving genera

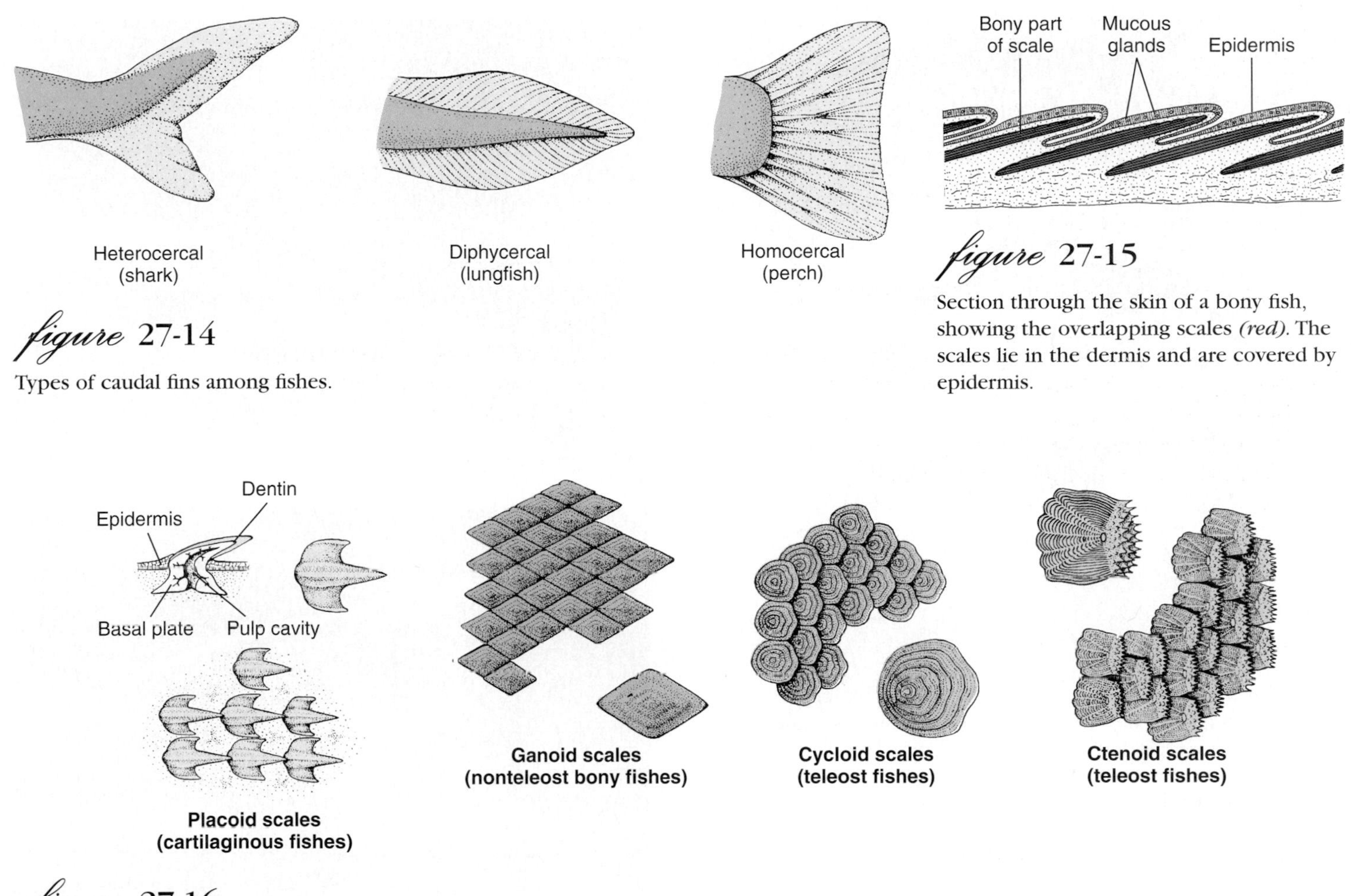

figure 27-14

Types of caudal fins among fishes.

figure 27-15

Section through the skin of a bony fish, showing the overlapping scales *(red)*. The scales lie in the dermis and are covered by epidermis.

figure 27-16

Types of fish scales. Placoid scales are small, conical toothlike structures characteristic of Chondrichthyes. Diamond-shaped ganoid scales, present in primitive bony fishes such as the gar, are composed of layers of silvery enamel (ganoin) on the upper surface and bone on the lower. Advanced bony fishes have either cycloid or ctenoid scales. These are thin and flexible and are arranged in overlapping rows.

of nonteleost neopterygians are the bowfin *Amia* (Gr. tunalike fish) of shallow, weedy waters of the Great Lakes and Mississippi Valley, and the gars *Lepisosteus* (Gr. *lepidos,* scale, + *osteon,* bone) of eastern and southern North America (Figure 27-18). Gars are large predators that belie their lethargic appearance by suddenly dashing forward to grasp their prey with needle-sharp teeth.

The major lineage of neopterygians are the teleosts (Gr. *teleos,* perfect, + *osteon,* bone), the modern bony fishes (Figure 27-13). Diversity appeared early in teleost evolution, foreshadowing the truly incredible variety of body forms among teleosts today. The heavy armorlike scales of the early ray-finned fishes have been replaced in teleosts by light, thin, and flexible **cycloid** and **ctenoid** scales. These look much alike (Figure 27-16) except that ctenoid scales have comblike ridges on the exposed edge that may be an adaptation for reducing frictional drag. Some teleosts, such as some catfishes and sculpins, lack scales altogether. Nearly all teleosts have a **homocercal tail,** with the upper and lower lobes of about equal size (Figure 27-14). The lungs of early forms were transformed in the teleosts to a swim bladder with a buoyancy function. Teleosts have highly maneuverable fins for control of body movement. In small teleosts the fins are often provided with stout, sharp spines, thus making themselves prickly mouthfuls for would-be predators. With these adaptations (and many others), teleosts have become the most diverse of fishes.

The Fleshy-Finned Fishes: Subclass Sarcopterygii

The fleshy-finned fishes are today represented by only seven species: six species (three genera) of lungfishes and a single lobe-finned fish, the coelacanth (seal′a-canth)—survivors of a group once abundant during the Devonian period of the Paleozoic. All of the early sarcopterygians had lungs as well as gills and strong, fleshy, paired lobed fins (pectoral and pelvic) that may have been used like four legs to scuttle along the bottom. They had powerful jaws, a skin covered with heavy, enameled scales, and a **diphycercal** tail (Figure 27-14).

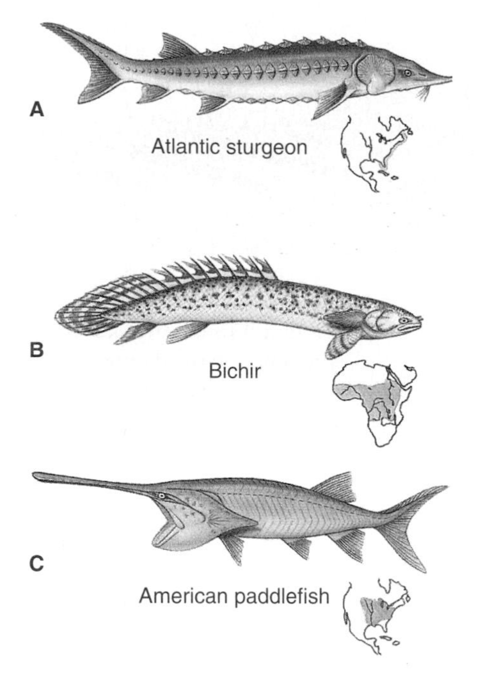

figure 27-17

Chondrostean ray-finned fishes of the subclass Actinopterygii. **A,** Atlantic sturgeon, *Acipenser oxyrhynchus* (now uncommon) of Atlantic coast rivers. **B,** Bichir *Polypterus bichir* of the African Congo. It is a nocturnal predator. **C,** Paddlefish *Polyodon spathula* of the Mississippi River reaches a length of 2 m and a weight of 90 kg.

figure 27-18

Nonteleost neopterygian fishes. **A,** Bowfin *Amia calva.* **B,** Longnose gar *Lepisosteus osseus.* The bowfin lives in the Great Lakes region and Mississippi basin. Gars are common fishes of eastern and southern North America. They frequent slow-moving streams where they may hang motionless in the water, ready to

Of the surviving lungfishes, the least specialized is *Neoceratodus* (Gr. *neos,* new, + *keratos,* horn, + *odes,* form), the living Australian lungfish, which may attain a length of 1.5 m (Figure 27-19). This lungfish is able to survive in stagnant, oxygen-poor water by coming to the surface and gulping air into its single lung, but it cannot live out of water. The South American lungfish, *Lepidosiren* (L. *lepidus,* pretty, + *siren,* Siren, mythical mermaid), and the African lungfish, *Protopterus* (Gr. *protos,* first, + *pteron,* wing), can live out of water for long periods of time. *Protopterus* lives in African streams and rivers that run completely dry during the dry season, with their mud beds baked hard by the hot tropical sun. The fish burrows down at the approach of the dry season and secretes a copious slime that mixes with mud to form a hard cocoon in which it remains dormant until the rains return.

The lobe-finned fishes consist of two groups: the **rhipidistians** which flourished in the late Paleozoic era then became extinct; and the **coelacanths,** a group that also radiated in the Paleozoic and later disappeared except for one remarkable species, the famous coelacanth *Latimeria chalumnae* (Figure 27-20). Since the last coelacanths were believed to have become extinct 70 million years ago, the astonishment of the scientific world can be imagined when the remains of a coelacanth were found on a trawler off the coast of South Africa in 1938. An intensive search was begun in the Comoro Islands area near Madagascar, where, it was learned, native Comoran fishermen occasionally caught them with hand lines at great depths. Numerous specimens have now been caught, many in excellent condition, although none has been kept alive beyond a few hours after capture. The "modern" marine coelacanth is a descendant of the Devonian freshwater stock, which reached its evolutionary peak in the Mesozoic era and then disappeared—or so it was believed until 1938.

The fleshy-finned fishes occupy an important position in vertebrate evolution because they include the closest living relatives of the tetrapods. Of the living fleshy-finned fishes, the lungfishes are the sister group of the tetrapods. Because the fleshy-finned fishes as traditionally recognized form a paraphyletic group, cladists include tetrapods as well as fleshy-finned fishes in the Sarcopterygii (Figure 27-2).

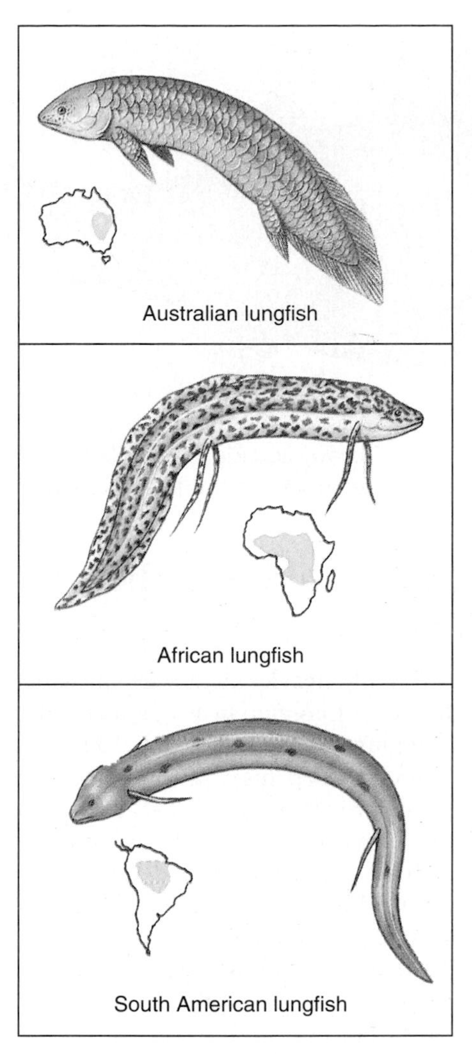

figure 27-19

Lungfishes are fleshy-finned fishes of the subclass Sarcopterygii. The Australian lungfish *Neoceratodus forsteri* is the least specialized of three lungfish genera. The African lungfish *Protopterus* is best adapted of the three for remaining dormant in mucous-lined cocoons breathing air during prolonged periods of drought.

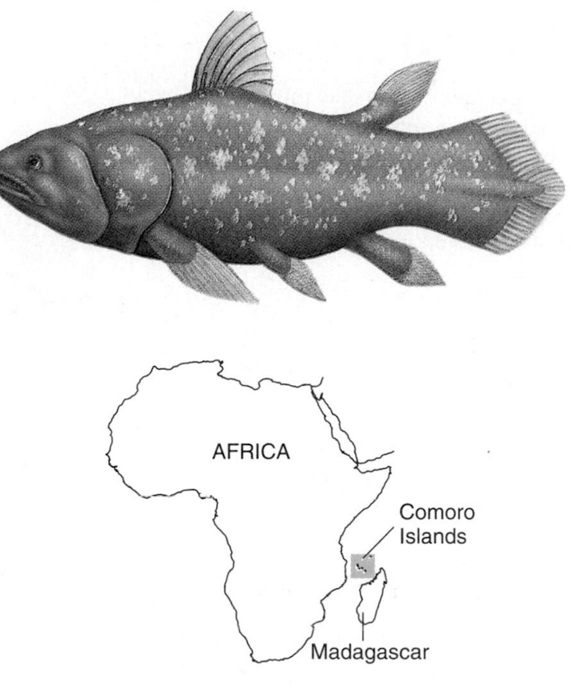

figure 27-20

The coelacanth *Latimeria chalumnae* is a surviving marine relic of a group of lobe-finned fishes that flourished some 350 million years ago.

Structural and Functional Adaptations of Fishes

Locomotion in Water

To the human eye, some fishes appear capable of swimming at extremely high speeds. But our judgment is unconsciously tempered by our own experience that water is a highly resistant medium through which to move. Most fishes, such as a trout or a minnow, can swim maximally about 10 body lengths per second, obviously an impressive performance by human standards. Yet when these speeds are translated into kilometers per hour it means that a 30 cm (1 foot) trout can swim only about 10.4 km (6.5 miles) per hour. As a general rule, the larger the fish the faster it can swim.

Measuring fish cruising speeds accurately is best done in a "fish wheel," a large ring-shaped channel filled with water that is turned at a speed equal and opposite to that of the fish. Much more difficult to measure are the sudden bursts of speed that most fish can make to capture prey or to avoid being captured. A hooked bluefin tuna was once "clocked" at 66 km per hour (41 mph); swordfish and marlin may be capable of incredible bursts of speed approaching, or even exceeding, 110 km per hour (68 mph). They can sustain such high speeds for no more than 1 to 5 seconds.

The propulsive mechanism of a fish is its trunk and tail musculature. The axial, locomotory musculature is composed of zigzag bands, called **myomeres.** The muscle fibers in each myomere are relatively short and connect the tough connective tissue partitions that separate each myomere from the next. On the surface the myomeres take the shape of a W lying on its side (Figure 27-21) but internally the bands are complexly folded and nested so that the pull of each myomere extends over several vertebrae. This arrangement produces more

figure 27-21

Trunk musculature of a teleost fish, partly dissected to show internal arrangement of the muscle bands (myomeres). The myomeres are folded into a complex, nested grouping, an arrangement that favors stronger and more controlled swimming.

power and finer control of movement since many myomeres are involved in bending a given segment of the body.

Understanding how fishes swim can be approached by studying the motion of a very flexible fish such as an eel (Figure 27-22). The movement is serpentine, not unlike that of a snake, with waves of contraction moving backward along the body by alternate contraction of the myomeres on either side. The anterior end of the body bends less than the posterior end, so that each undulation increases in amplitude as it travels along the body. While undulations move backward, the bending of the body pushes laterally against the water, producing a **reactive force** that is directed forward, but at an angle. It can be analyzed as having two components: **thrust,** which is used to overcome drag and propels the fish forward, and **lateral force,** which tends to make the fish's head "yaw," or deviate from the course in the same direction as the tail. This side-to-side head movement is very obvious in a swimming eel or shark, but many fishes have a large, rigid head with enough surface resistance to minimize yaw.

The movement of an eel is reasonably efficient at low speed, but its body shape generates too much frictional drag for rapid swimming. Fishes that swim rapidly, such as trout, are less flexible and limit the body undulations mostly to the caudal region (Figure 27-22). Muscle force generated in the large anterior muscle mass is transferred through tendons to the relatively nonmuscular caudal peduncle and tail where thrust is generated. This form of swimming reaches its highest development in the tunas, whose bodies do not flex at all. Virtually all the thrust is derived from powerful beats of the tail fin (Figure 27-23). Many fast oceanic fishes such as marlin, swordfish, amberjacks, and wahoo have swept-back tail fins shaped much like a sickle. Such fins are the aquatic counterpart of the high-aspect ratio wings of the swiftest birds (p. 653).

Swimming is the most economical form of animal locomotion, largely because aquatic animals are almost perfectly supported by their medium and need expend little energy to overcome the force of gravity. If we compare the energy cost per kilogram of body weight of traveling 1 km by different forms of locomotion, we find swimming costs only 0.39 kcal (salmon) as compared with 1.45 kcal for flying (gull) and 5.43 for walking (ground squirrel). However, part of the unfinished business of biology is understanding how fish and aquatic mammals are able to move through the water while creating almost no turbulence. The secret lies in the way aquatic animals bend their bodies and fins (or flukes) to swim and in the friction-reducing properties of the body surface.

Neutral Buoyancy and the Swim Bladder

All fishes are slightly heavier than water because their skeletons and other tissues contain heavy elements that are present only in trace amounts in natural waters. To keep from sinking, sharks must always keep moving forward in the water. The asymmetrical (heterocercal) tail of a shark provides the necessary tail lift as it sweeps to and fro in the water, and the broad head and flat pectoral fins (Figure 27-8) act as angled planes to provide head lift. Sharks are also aided in buoyancy by having very large livers containing a special fatty hydrocarbon called **squalene** that has a density of only 0.86. The liver thus acts like a large sack of buoyant oil that helps to compensate for the shark's heavy body.

By far the most efficient flotation device is a gas-filled space. Swim bladders are present in most pelagic bony fishes but are absent in tunas, most abyssal fishes, and most bottom dwellers, such as flounders and sculpins. By adjusting the volume of gas in the swim bladder, a fish can achieve neutral buoyancy and remain suspended indefinitely at any depth with no muscular effort. There are severe technical problems, however. If the fish descends to a greater depth, the swim bladder gas is compressed so that the fish becomes heavier and tends to sink. Gas must be added to the bladder to establish a new equilibrium buoyancy. If the fish swims upward, the gas in the bladder expands, making the fish lighter. Unless gas is removed, the fish will rise with ever-increasing speed while the bladder continues to expand.

Fishes adjust gas volume in the swim bladder in two ways. The less specialized fishes (trout, for example) have a **pneumatic duct** that connects the swim bladder to the esophagus; these forms must come to the surface and gulp air to charge the bladder and obviously are restricted to relatively shallow depths. More specialized teleosts have lost the pneumatic duct. In these fishes, the gas must originate in the blood and be secreted into the swim bladder. Gas exchange depends on two highly specialized areas: a

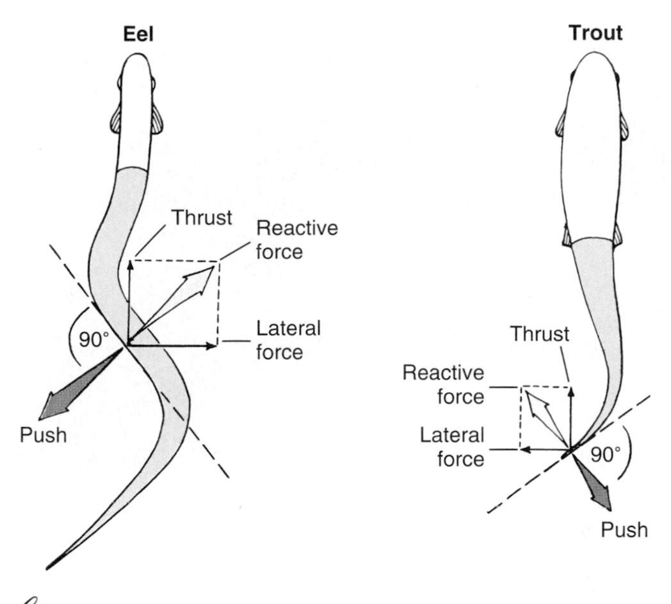

figure 27-22

Movements of swimming fishes, showing the forces developed by an eel-shaped and spindle-shaped fish. (From *Vertebrate life,* 4/e by Pough et al., 1996. Reprinted by permission of Prentice-Hall, Inc., Upper Saddle River, NJ.)

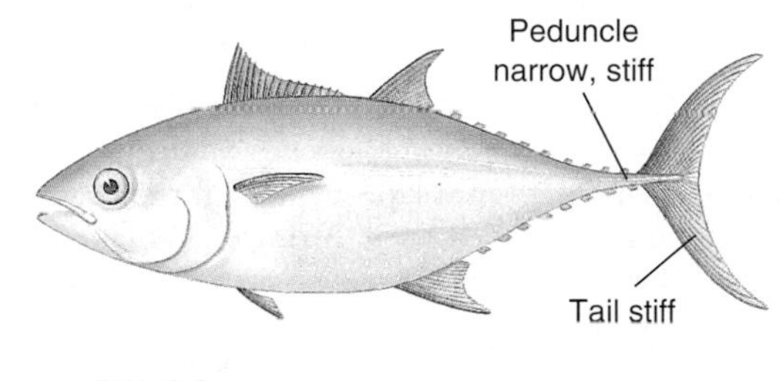

figure 27-23

Bluefin tuna, showing adaptations for fast swimming. Powerful trunk muscles pull on the slender tail stalk. Since the body does not bend, all of the thrust comes from beats of the stiff, sickle-shaped tail.

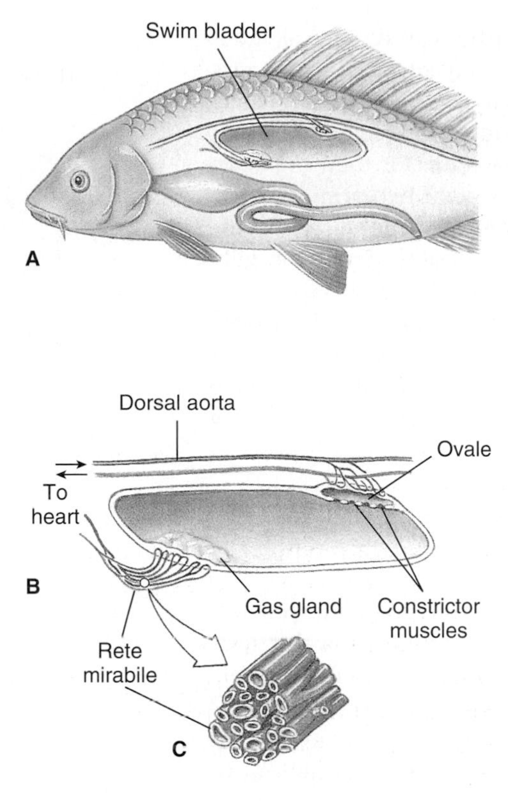

figure 27-24

Swim bladder of a teleost fish. The swim bladder **(A)** lies in the coelom just beneath the vertebral column. Gas is secreted into the swim bladder by the gas gland **(B).** Gas from the blood is moved into the gas gland by the rete mirabile, a complex array of tightly-packed capillaries that act as a countercurrent multiplier to build up the oxygen concentration. The arrangement of venous and arterial capillaries in the rete is shown in **C.** To release gas during ascent, a muscular valve opens, allowing gas to enter the ovale from which the gas is removed by the circulation.

gas gland that secretes gas into the bladder and a **resorptive area,** or "ovale," that can remove gas from the bladder. The gas gland is supplied by a remarkable network of blood capillaries, called the **rete mirabile** ("marvelous net") that functions as a countercurrent exchange system to trap gases, especially oxygen, and prevent their loss to the circulation (Figure 27-24).

The amazing effectiveness of this device is exemplified by a fish living at a depth of 2400 m (8000 feet). To keep the bladder inflated at that depth, the gas inside (mostly oxygen, but also variable amounts of nitrogen, carbon dioxide, argon, and even some carbon monoxide) must have a pressure exceeding 240 atmospheres, which is much greater than the pressure in a fully charged steel gas cylinder. Yet the oxygen pressure in the fish's blood cannot exceed 0.2 atmosphere—equal to the oxygen pressure at the sea surface.

Respiration

Fish gills are composed of thin filaments, each covered with a thin epidermal membrane that is folded repeatedly into plate-like **lamellae** (Figure 27-25). These are richly supplied with blood vessels. The gills are located inside the pharyngeal cavity and are covered with a movable flap, the **operculum.** This arrangement provides excellent protection to the delicate gill filaments, streamlines the body, and makes possible a pumping system for moving water through the mouth, across the gills, and out the operculum. Instead of opercular flaps as in bony fishes, the elasmobranchs have a series of **gill slits** out of which the water flows. In both elasmobranchs and bony fishes the branchial mechanism is arranged to pump water continuously and smoothly over the gills, although to an observer it appears that fish breathing is pulsatile.

The flow of water is opposite to the direction of blood flow (countercurrent flow), the best arrangement for extracting the greatest possible amount of oxygen from the water.

Some bony fishes can remove as much as 85% of the oxygen from water passing over their gills. Very active fishes, such as herring and mackerel, can obtain sufficient water for their high oxygen demands only by swimming forward continuously to force water into the open mouth and across the gills. This process is called ram ventilation. Such fish will be asphyxiated if placed in an aquarium that restricts free swimming movements, even if the water is saturated with oxygen.

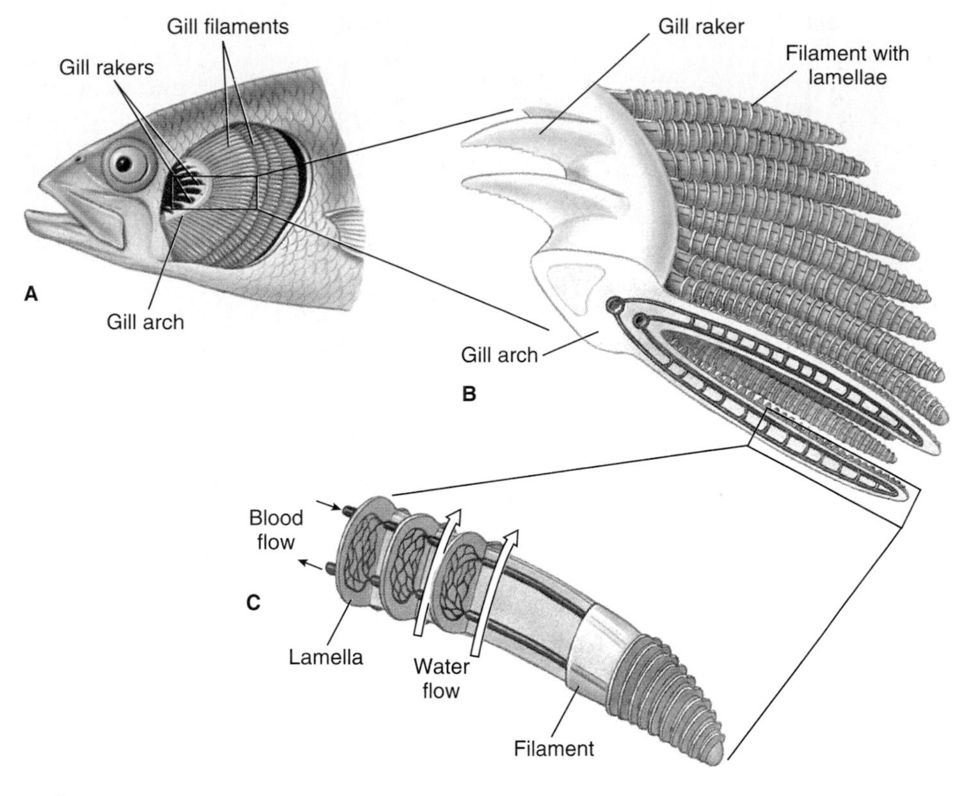

figure 27-25

Gills of fish. Bony, protective flap covering the gills (operculum) has been removed, **A,** to reveal branchial chamber containing the gills. There are four gill arches on each side, each bearing numerous filaments. A portion of gill arch, **B,** shows gill rakers that project forward to strain food and debris, and gill filaments that project to the rear. A single gill filament, **C,** is dissected to show the blood capillaries within the platelike lamellae. Direction of water flow *(large arrows)* is opposite the direction of blood flow.

Migration

Eel

For centuries naturalists had been puzzled about the life history of the freshwater eel *Anguilla* (an-gwil′a) (L. eel), a common and commercially important species of coastal streams of the North Atlantic. Eels are **catadromous** (Gr. *kata,* down, + *dromos,* running), meaning that they spend most of their lives in fresh water but migrate to the sea to spawn. Each fall, people saw large numbers of eels swimming down the rivers toward the sea, but no adults ever returned. Each spring countless numbers of young eels, called "elvers" (Figure 27-26), each about the size of a wooden matchstick, appeared in the coastal rivers and began swimming upstream. Beyond the assumption that eels must spawn somewhere at sea, the location of their breeding grounds was completely unknown.

The first clue was provided by two Italian scientists, Grassi and Calandruccio, who in 1896 reported that elvers were not larval eels but rather were relatively advanced juveniles. The true larval eels, they discovered, were tiny, leaf-shaped, completely transparent creatures that bore absolutely no resemblance to an eel. They had been called **leptocephali** (Gr. *leptos,* slender, + *kephalē,* head) by early naturalists, who never suspected their true identity. In 1905 Johann Schmidt, supported by the Danish government, began a systematic study of eel biology that he continued until his death in 1933. With the cooperation of captains of commercial vessels plying the Atlantic, thousands of the leptocephali were caught in different areas of the Atlantic with the plankton nets Schmidt supplied. By noting where larvae in different stages of development were captured, Schmidt and his colleagues eventually reconstructed the spawning migrations.

When the adult eels leave the coastal rivers of Europe and North America, they swim steadily and apparently at great depth for one to two months until they reach the Sargasso Sea, a vast area of warm oceanic water southeast of Bermuda (Figure 27-26). Here, at depths of 300 m or more, the eels spawn and die. The minute larvae then begin an incredible journey back to the coastal rivers of Europe. Drifting with the Gulf Stream, those not eaten by numerous predators reach the middle of the Atlantic after two years. By the end of the third year they reach the coastal waters of Europe where the leptocephali metamorphose into elvers, with an unmistakable eel-like body form (Figure 27-26). Here the males and females part company; the males remain in the brackish waters of coastal rivers and estuaries while the females continue up the rivers, often traveling hundreds of miles upstream. After 8 to 15 years of growth, the females, now 1 m or more long, return to the sea to join the smaller males; both return to the ancestral breeding grounds thousands of miles away to complete the life cycle. Since the Sargasso Sea is much closer to the American coastline than it is to Europe, American eel larvae require only about eight months to make the journey.

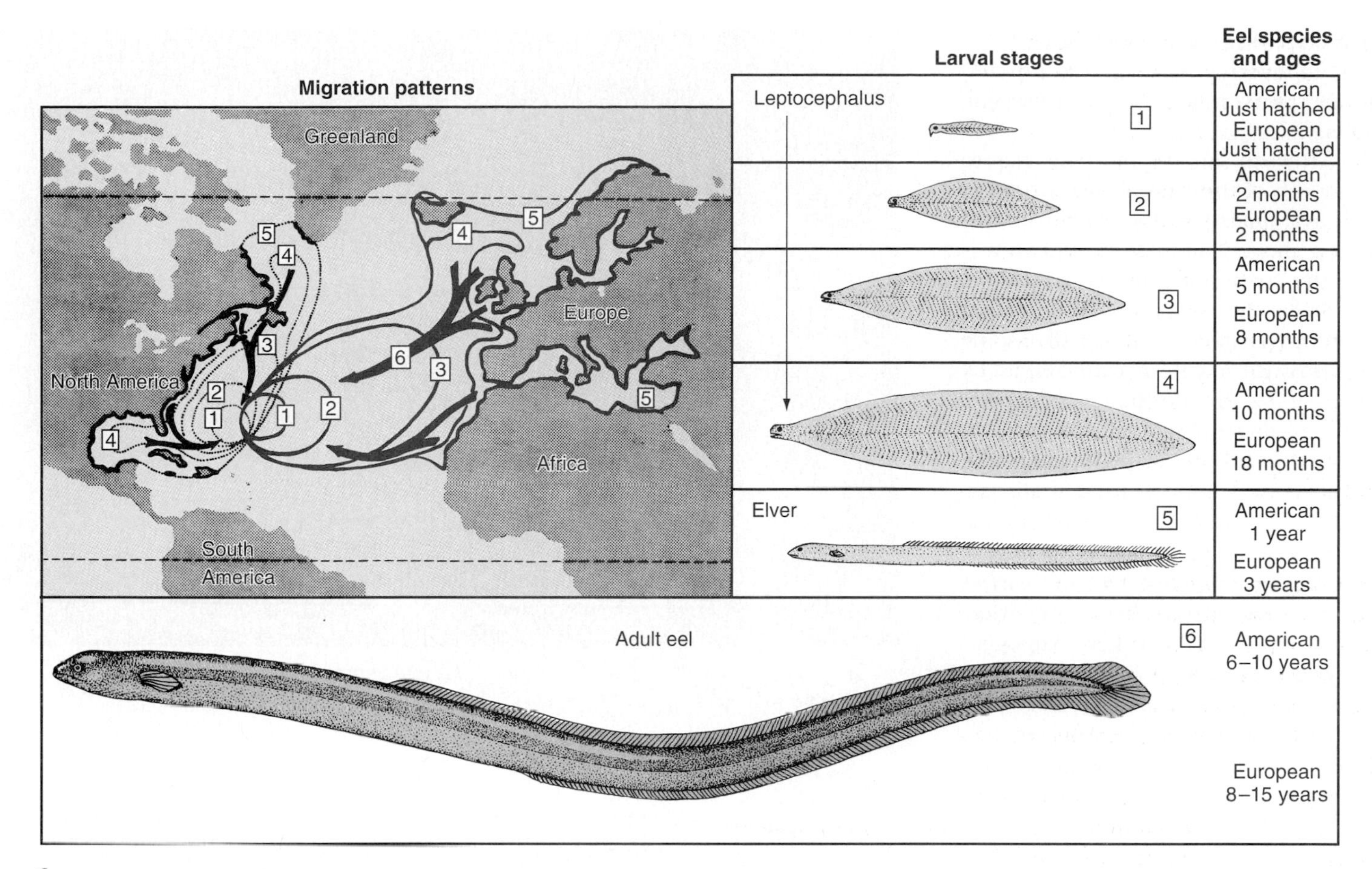

figure 27-26

Life histories of the European eel, *Anguilla anguilla,* and American eel, *Anguilla rostrata. Red,* Migration patterns of European species. *Black,* Migration patterns of American species. Boxed numbers refer to stages of development. Note that the American eel completes its larval metamorphosis and sea journey in one year. It requires nearly three years for the European eel to complete its much longer journey.

Recent enzyme electrophoresis analysis of eel larvae confirmed not only the existence of separate European and American species but also Schmidt's belief that the European and American eels spawn in partially overlapping areas of the Sargasso Sea.

Homing Salmon

The life history of salmon is nearly as remarkable as that of the freshwater eel and certainly has received far more popular attention. Salmon are **anadromous;** that is, they spend their adult lives at sea but return to fresh water to spawn. The Atlantic salmon *(Salmo salar)* and the Pacific salmon (six species of the genus *Oncorhynchus* [on-ko-rink′us]) have this practice, but there are important differences among the seven species. The Atlantic salmon (as well as the closely related steelhead trout) make upstream spawning runs year after year. The six Pacific salmon species (king, sockeye, silver, humpback, chum, and Japanese masu) each make a single spawning run (Figures 27-27 and 27-28), after which they die.

The virtually infallible homing instinct of the Pacific species is legendary. After migrating downstream as a smolt (a juvenile stage, Figure 27-28), a sockeye salmon ranges many hundreds of miles over the Pacific for nearly four years, grows to 2 to 5 kg in weight, and then returns almost unerringly to spawn in the headwaters of its parent stream. Some straying does occur and is an important means of increasing gene flow and populating new streams.

Experiments by A. D. Hasler and others have shown that homing salmon are guided upstream by the characteristic odor of their parent stream. When the salmon finally reach the spawning beds of their parents (where they themselves were hatched), they spawn and die. The following spring, the newly hatched fry transform into smolts before and during the downstream migration. At this time they are imprinted (p. 296) with the distinctive odor of the stream, which is apparently a mosaic of compounds released by the characteristic vegetation and soil in the watershed of the parent stream. They also seem to imprint on the odors of other streams they pass while migrating downriver and use these odors in reverse sequence as a map during the upriver migration as returning adults.

How do salmon find their way to the mouth of the coastal river from the trackless miles of the open ocean? Salmon move hundreds of miles away from the coast, much too far to be able to detect the odor of their parent stream. Experiments suggest that some migrating fish, like birds, can navigate by orienting to the position of the sun. However, migrant salmon can navigate on cloudy days and at night, indicating that solar navigation, if used at all, cannot be the salmon's only navigational cue. Fish also (again, like birds, see p. 656) appear able to detect the earth's magnetic field and to navigate by orientating to it. Finally, fishery biologists concede that salmon may not require precise navigational abilities at all, but instead may use ocean currents, temperature gradients, and food availability to reach the general coastal area where "their" river is located. From this point, they would navigate by their imprinted odor map, making correct turns at each stream junction until they reach their natal stream.

figure 27-27
Migrating Pacific sockeye salmon.

Salmon runs in the Pacific Northwest have been devasted by a lethal combination of spawning stream degradation by logging, pollution and, especially, by more than 50 hydroelectric dams which obstruct upstream migration of adult salmon and kill downstream migrants as they pass through the dams' power-generating turbines. In addition, the chain of reservoirs behind the dams, which has converted the Columbia and Snake Rivers into a series of lakes, increases mortality of young salmon migrating downstream by slowing their passage to the sea. The result is that the annual run of wild salmon is today only about 3% of the 10 to 16 million fish that ascended the rivers 150 years ago. While recovery plans have been delayed by the power industry, environmental groups argue that in the long run losing the salmon will be more expensive to the regional economy than making the changes now that will allow salmon stocks to recover.

Reproduction and Growth

In a group as diverse as the fishes, it is no surprise to find extraordinary variations on the basic theme of sexual reproduction. Most fishes favor a simple theme: they are **dioecious,** with **external fertilization** and **external development** of the eggs and embryos. This mode of reproduction is called **oviparous** (meaning "egg-producing"). However, as tropical fish enthusiasts are well aware, the ever-popular guppies and mollies of home aquaria bear their young alive after development in the ovarian cavity of the mother (Figure 27-29). These fish are said to be **ovoviviparous,** meaning "live egg-producing." Some sharks develop a kind of placental attachment through which the young are nourished during gestation. These forms, like placental mammals, are **viviparous** ("alive-producing") (the different forms of maternal support of the embryo are described in more detail on p. 333).

Let us return to the much more common oviparous mode of reproduction. Many marine fishes are extraordinarily profligate egg producers. Males and females aggregate in great schools and, without elaborate courtship behavior, release vast numbers of germ cells into the water to drift with the current. Large female cod may release 4 to 6 million eggs at a single spawning. Less than one in a million will survive the numerous perils of the ocean to reach reproductive maturity.

Unlike the minute, buoyant, transparent eggs of pelagic marine teleosts, those of many near-shore and bottom-dwelling (benthic) species are larger, typically yolky, nonbuoyant, and adhesive. Some bury their eggs, many attach them to vegetation, some deposit them in nests, and some even incubate them in their mouths (Figure 27-30). Many benthic spawners guard their eggs. Intruders expecting an easy meal of eggs may

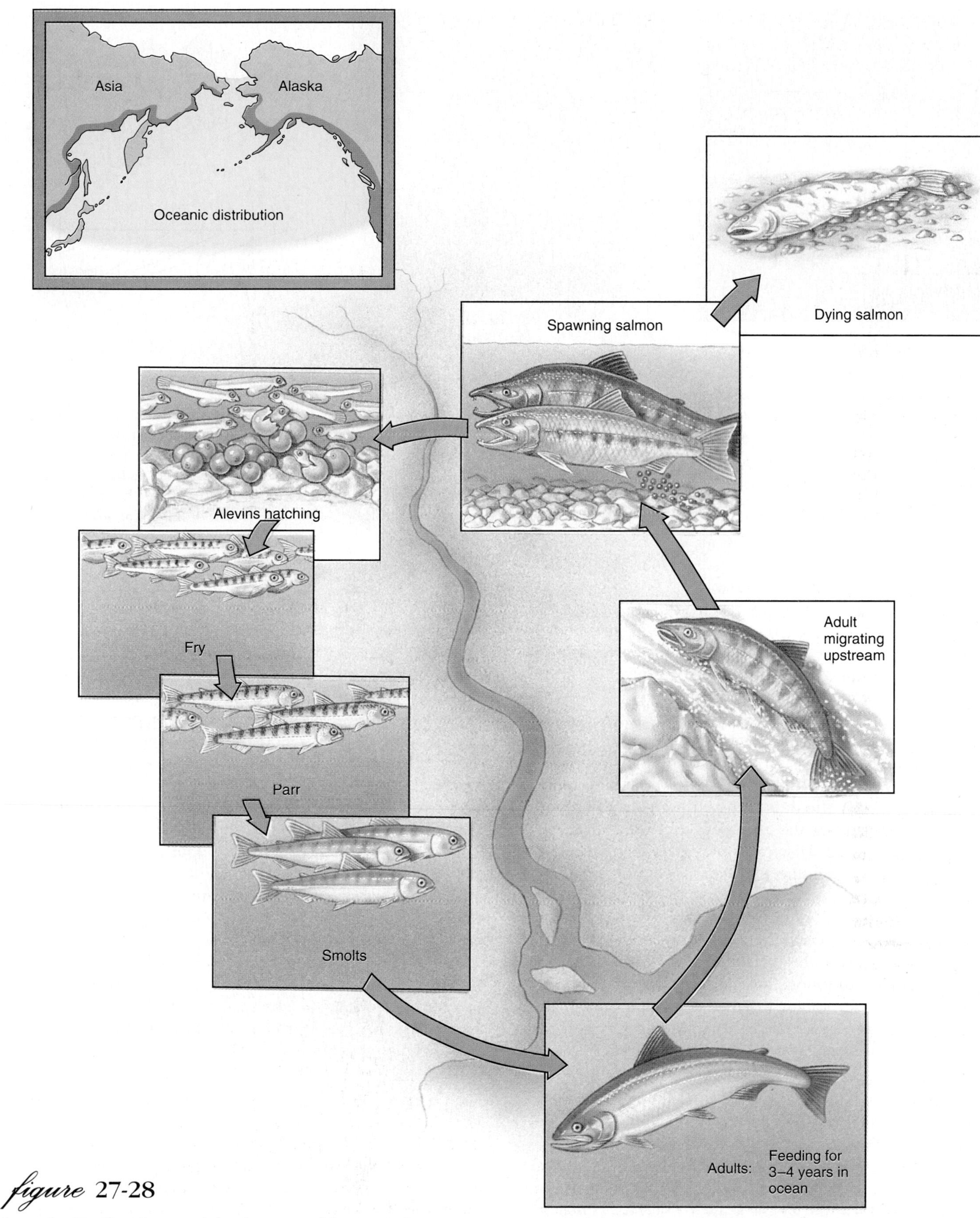

figure 27-28

Spawning Pacific salmon and development of the egg and young.

figure 27-29

Rainbow surfperch *Hypsurus caryi* giving birth. All of the West Coast surfperches (family Embiotocidae) are ovoviviparous.

figure 27-30

Male banded jawfish *Opistognathus macrognathus* orally brooding its eggs. The male retrieves the female's spawn and incubates the eggs until they hatch. During brief periods when the jawfish is feeding, the eggs are left in the burrow.

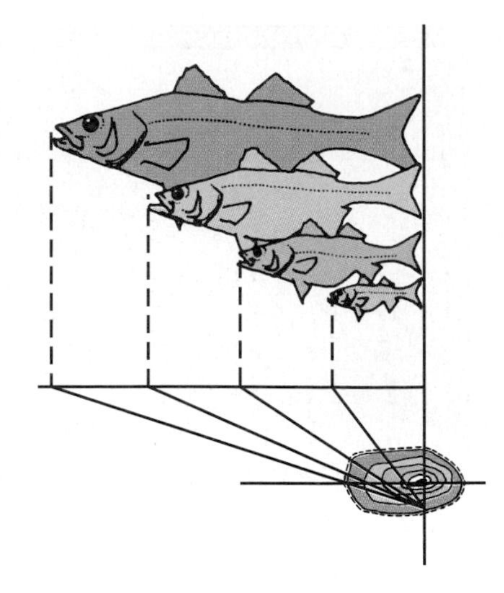

figure 27-31

Scale growth. Fish scales disclose seasonal changes in growth rate. Growth is interrupted during winter, producing year marks (annuli). Each year's increment in scale growth is a ratio to the annual increase in body length. Otoliths (ear stones) and certain bones can also be used in some species to determine age and growth rate.

be met with a vivid and often belligerent display by the guard, which is almost always the male.

Freshwater fishes almost invariably produce nonbuoyant eggs. Those, such as perch, that provide no parental care simply scatter their myriads of eggs among weeds or along the bottom. Freshwater fishes that do provide some form of egg care produce fewer, larger eggs that enjoy a better chance for survival.

Elaborate preliminaries to mating are the rule for freshwater fishes. The female Pacific salmon, for example, performs a ritualized mating "dance" with her breeding partner after arriving at the spawning bed in a fast-flowing, gravel-bottomed stream (Figure 27-28). She then turns on her side and scoops out a nest with her tail. As the eggs are laid by the female, they are fertilized by the male (Figure 27-28). After the female covers the eggs with gravel, the exhausted fish dies and drifts downstream.

Soon after the egg of an oviparous species is laid and fertilized, it takes up water and the outer layer hardens. Cleavage follows, and the blastoderm is formed, sitting astride a relatively enormous yolk mass. Soon the yolk mass is enclosed by the developing blastoderm, which then begins to assume a fishlike shape. The fish hatches as a larva carrying a semitransparent sac of yolk which provides its food supply until the mouth and digestive tract have developed. The larva then begins searching for its own food. After a period of growth the larva undergoes a metamorphosis, especially dramatic in many marine species such as the freshwater eel described previously (Figure 27-26). Body shape is refashioned, fin and color patterns change, and the animal becomes a juvenile bearing the unmistakable definitive body form of its species.

Growth is temperature dependent. Consequently, fish living in temperate regions grow rapidly in summer when temperatures are high and food is abundant but nearly stop growing in winter. Annual rings in the scales reflect this seasonal growth (Figure 27-31), a distinctive record of convenience to fishery biologists who wish to determine a fish's age. Unlike birds and mammals, which stop growing after reaching adult size, most fishes after attaining reproductive maturity continue to grow for as long as they live. This may be a selective advantage for the species, since the larger the fish, the more germ cells it produces and the greater its contribution to future generations.

Classification of Living Fishes

The following Linnaean classification of major fish taxa mostly follows that of Nelson (1994). The probable relationships of these traditional groupings together with the major extinct groups of fishes are shown in a cladogram in Figure 27-2. Other schemes of classification have been proposed. Because of the difficulty of determining relationships among the numerous living and fossil species, we can appreciate why fish classification has undergone, and will continue to undergo, continuous revision.

Phylum Chordata

Subphylum Vertebrata

Superclass Agnatha (ag′na-tha) (Gr. *a,* not, + *gnathos,* jaw) **(Cyclostomata).** No jaws; cartilaginous skeleton; paired fins absent; one or two semicircular canals; notochord persistent. A paraphyletic group that is retained because of traditional usage.

Class Myxini (mik-sy′ny) (Gr. *myxa,* slime): **hagfishes.** Mouth terminal with four pairs of tentacles; nasal sac with duct to pharynx; gill pouches, 5 to 15 pairs; partially hermaphroditic.

Class Cephalaspidomorphi (sef-a-lass′pe-do-morf′e) (Gr. *kephalē,* head, + *aspidos,* shield, + morphē, form) **(Petromyzontes): lampreys.** Mouth suctorial with horny teeth; nasal sac not connected to mouth; gill pouches, seven pairs.

Superclass Gnathostomata (na′tho-sto′ma-ta) (Gr. *gnathos,* jaw, + *stoma,* mouth). Jaws present; usually paired limbs; three pairs of semicircular canals; notochord persistent or replaced by vertebral centra. A paraphyletic group.

Class Chondrichthyes (kon-drik′thee-eez) (Gr. *chondros,* cartilage, + *ichthys,* fish): **cartilaginous fishes.** Cartilaginous skeleton; teeth not fused to jaws; no swim bladder; intestine with spiral valve.

Subclass Elasmobranchii (e-laz′mo-bran′kee-i) (Gr. *elasmos,* plated, + *branchia,* gills): **sharks and rays.** Placoid scales or no scales; five to seven gill arches and gills in separate clefts along pharynx. Examples: *Squalus, Raja.*

Subclass Holocephali (hol′o-sef′a-li) (Gr. *holos,* entire, + *kephalē,* head): **chimaeras,** or **ghostfish.** Gill slits covered with operculum; jaws with tooth plates; single nasal opening; without scales; accessory clasping organs in male; lateral line an open groove.

Class Osteichthyes (os′te-ik′thee-eez) (Gr. *osteon,* bone, + *ichthys,* a fish): **bony fishes.** Body primitively fusiform but variously modified; skeleton mostly ossified; single gill opening on each side covered with operculum; usually swim bladder or lung. A paraphyletic group as traditionally defined; in cladistic usage the Osteichthyes includes the tetrapods to be covered in later chapters.

Subclass Actinopterygii (ak′ti-nop-te-rij′ee-i) (Gr. *aktis,* ray, + *pteryx,* fin, wing): **ray-finned fishes.** Paired fins supported by dermal rays and without basal lobed portions; nasal sacs open only to outside.

Superorder Chondrostei (kon-dros′tee-i) (Gr. *chondros,* cartilage, + *osteon,* bone): **chondrostean ray-finned fishes.** Skeleton mostly cartilaginous; tail heterocercal; notochord persists in adults; intestine with spiral valve. Two living orders containing the bichir *(Polypterus),* sturgeons, and paddlefish.

Superorder Neopterygii (nee-op-te-rij′ee-i) (Gr. *neos,* new, + *pteryx,* fin, wing): **modern bony fishes.** Skeleton mostly bone; body covered with thin scales without bony layer (cycloid or ctenoid) or scaleless; caudal fin mostly homocercal; mouth terminal; notochord a mere vestige; swim bladder mainly a hydrostatic organ and usually not opened to the esophagus. Living neopterygeans are broadly divided into the nonteleosts, two orders represented by the gars *(Lepisosteus)* and the bowfin *(Amia)* ("holosteans" in older classifications); and the teleosts, represented by 38 living orders. There are approximately 23,640 living named species of neopterygians (96% of all living fishes).

Subclass Sarcopterygii (sar-cop-te-rij′ee-i) (Gr. *sarkos,* flesh, + *pteryx,* fin, wing): **fleshy-finned fishes.** Heavy bodied; paired fins with sturdy internal skeleton of basic tetrapod type and musculature; muscular lobes at bases of anal and second dorsal fins; diphycercal tail; intestine with spiral valve. Ten extinct orders and three living orders containing the coelacanth, *Latimeria chalumnae,* and three genera of lungfishes: *Neoceratodus, Lepidosiren,* and *Protopterus.*

Summary

Fishes are poikilothermic, gill-breathing aquatic vertebrates with fins. They include the oldest vertebrate groups, having originated from an unknown chordate ancestor in the Cambrian period or possibly earlier. Four classes of fishes are recognized. The jawless hagfishes (class Myxini) and lampreys (class Cephalaspidomorphi), are ancient groups having an eel-like body form without paired fins; a cartilaginous skeleton (although their ancestors, the ostracoderms, had bony skeletons); a notochord that persists throughout life; and a disclike mouth adapted for sucking or biting. All other vertebrates have jaws, a major development in vertebrate evolution.

Members of the class Chondrichthyes (sharks, rays, and chimaeras) are a compact group having a cartilaginous skeleton (a derived feature), paired fins, excellent sensory equipment, and an active, characteristically predaceous habit. To the fourth class of fishes belong the bony fishes (class Osteichthyes), which may be subdivided into two stems of descent. One stem is a relic group, the fleshy-finned fishes of the subclass Sarcopterygii, represented today by the lungfishes and the coelacanth. The terrestrial vertebrates arose from within one lineage of this group. The second stem is the ray-finned fishes (subclass Actinopterygii), a huge and diverse modern assemblage containing nearly all of the familiar freshwater and marine fishes.

The modern bony fishes (teleost fishes) have radiated into approximately 24,600 species that reveal an enormous diversity of adaptations, body form, behavior, and habitat preference. Most fishes swim by undulatory contractions of the body muscles, which generate thrust (propulsive force) and lateral force. Flexible fishes oscillate the whole body, but in more rapid swimmers the undulations are limited to the caudal region or tail fin alone.

Most pelagic bony fishes achieve neutral buoyancy in water using a gas-filled swim bladder, the most effective gas-secreting device known in the animal kingdom. The gills of fishes, having efficient countercurrent flow between water and blood, facilitate high rates of oxygen exchange.

Many fishes are migratory to some extent, and some, such as freshwater eels and anadromous salmon, make remarkable migrations of great length and precision. Fishes reveal an extraordinary range of sexual reproductive strategies. Most fishes are oviparous, but ovoviviparous and viviparous fishes are not uncommon. The reproductive investment may be in large numbers of germ cells with low survival (many marine fishes) or in fewer germ cells with greater parental care for better survival (freshwater fishes).

Review Questions

1. Provide a brief description of the fishes citing characteristics that would distinguish them from all other animals.
2. What characteristics distinguish the hagfishes and lampreys (superclass Agnatha) from all other fishes?
3. Describe feeding behavior in hagfishes and lampreys. How do they differ?
4. Describe the life cycle of the sea lamprey, *Petromyzon marinus,* and the history of its invasion of the Great Lakes.
5. In what ways are sharks well equipped for the predatory life habit?
6. The lateral line system has been described as a "distant touch" system for sharks. What function does the lateral line system serve? Where are the receptors located?
7. Explain how bony fishes differ from sharks and rays in the following systems or features: skeleton, tail shape, scales, buoyancy, and position of mouth.
8. Match the ray-finned fishes in the right column with the group to which each belongs in the left column:

____ Chondrosteans	a. Perch
____ Nonteleost neopterygians	b. Sturgeon
	c. Gar
____ Teleosts	d. Salmon
	e. Paddlefish
	f. Bowfin

9. Although the chondrosteans are today a relic group, they were one of two major lineages that emerged from early ray-finned fishes of the Devonian period. Give examples of living chondrosteans. What does the term Actinopterygii, the subclass to which the chondrosteans belong, literally mean (refer to the Classification of living fishes on p. 603)?
10. What is the other major lineage of actinopterygians? What are some distinguishing characteristics of modern bony fishes?
11. Only seven species of fleshy-finned fishes are alive today, remnants of a group that flourished in the Devonian period of the Paleozoic. What morphological characteristics distinguish the fleshy-finned fishes? What is the literal meaning of Sarcopterygii, the subclass to which the fleshy-finned fishes belong?
12. Give the geographical locations of the three surviving genera of lungfishes and explain how they differ in their ability to survive out of water. Which of the three is the least specialized?
13. Describe the discovery of the living coelacanth. What is the evolutionary significance of the group to which it belongs?
14. Compare the swimming movements of the eel with that of the trout, and explain why the latter is more efficient for rapid locomotion.

15. Sharks and bony fishes approach or achieve neutral buoyancy in different ways. Describe the methods evolved in each group. Why must a teleost fish adjust the gas volume in its swim bladder when it swims upward or downward? How is gas volume adjusted?
16. What is meant by "countercurrent flow" as it applies to fish gills?
17. Describe the life cycle of the European eel. How does the life cycle of the American eel differ from that of the European?
18. How do adult Pacific salmon find their way back to their parent stream to spawn?
19. What mode of reproduction in fishes is described by each of the following terms: oviparous, ovoviviparous, viviparous?
20. Reproduction in marine pelagic fishes and in freshwater fishes is distinctively different. How and why do they differ?

Selected References

Bone, Q., and N. B. Marshall. 1982. Biology of fishes. New York, Chapman & Hall. *Concise, well-written, and well-illustrated primer on the functional processes of fishes.*

Conniff, R. 1991. The most disgusting fish in the sea. Audubon **93**(2):100–108 (March). *Recent discoveries shed light on the life history of the enigmatic hagfish that fishermen loathe.*

Horn, M. H., and R. N. Gibson. 1988. Intertidal fishes. Sci. Am. **258:**64–70 (Jan.). *Describes the special adaptations of fishes living in a demanding environment.*

Long, J. A. 1995. The rise of fishes: 500 million years of evolution. Baltimore, The Johns Hopkins University Press. *A lavishly illustrated evolutionary history of fishes.*

Moyle, P. B. 1993. Fish: an enthusiast's guide. Berkeley, University of California Press. *Textbook written in a lively style and stressing function and ecology rather than morphology; abbreviated treatment of the fish groups.*

Nelson, J. S. 1994. Fishes of the world, ed. 3. New York, John Wiley & Sons, Inc. *Authoritative classification of all major groups of fishes.*

Stevens, J. D., ed. 1987. Sharks. New York, Facts on File Publications. *Evolution, biology, and behavior of sharks, handsomely illustrated.*

Thomson, K. S. 1991. Living fossil. The story of the coelacanth. New York, W. W. Norton.

Webb, P. W. 1984. Form and function in fish swimming. Sci. Am. **251:**72–82 (July). *Specializations of fish for swimming and analysis of thrust generation.*

The Early Tetrapods and Modern Amphibians

chapter | twenty-eight

Vertebrate Landfall

The chorus of frogs beside a pond on a spring evening heralds one of nature's dramatic events. Masses of frog eggs soon hatch into limbless, gill-breathing, fishlike tadpole larvae. Warmed by the late spring sun, they feed and grow. Then, almost imperceptibly, a remarkable transformation takes place. Hindlegs appear and gradually lengthen. The tail shortens. The larval teeth are lost, and the gills are replaced by lungs. Eyelids develop. The forelegs emerge. In a matter of weeks the aquatic tadpole has completed its metamorphosis to an adult frog.

The evolutionary transition from water to land occurred not in weeks but over millions of years. A lengthy series of alterations cumulatively fitted the vertebrate body plan for life on land. The origin of land vertebrates is no less a remarkable feat for this fact—a feat that incidentally would have a poor chance of succeeding today because well-established competitors make it impossible for a poorly adapted transitional form to gain a foothold.

Amphibians are the only living vertebrates that have a transition from water to land in both their ontogeny and phylogeny. Even after some 350 million years of evolution, few amphibians are completely land adapted; most are quasiterrestrial, hovering between aquatic and land environments. This double life is expressed in their name. Even the amphibians that are best adapted for a terrestrial existence cannot stray far from moist conditions. Many, however, have developed ways to keep their eggs out of open water where the larvae would be exposed to enemies.

Adaptation for life on land is a major theme of the remaining vertebrate groups treated in this and the following chapters. These animals form a monophyletic unit known as the **tetrapods.** The amphibians and the amniotes (including reptiles, birds, and mammals) represent the two major extant branches of tetrapod phylogeny. In this chapter, we review what is known about the origins of terrestrial vertebrates and discuss the amphibian lineage in detail. We discuss the major amniote groups in Chapters 29 through 31.

Movement onto Land

The movement from water to land is perhaps the most dramatic event in animal evolution, because it involves the invasion of a habitat that in many respects is more hazardous for life. Life originated in water. Animals are mostly water in composition, and all cellular activities occur in water. Nevertheless, organisms eventually invaded land, carrying their watery composition with them. Vascular plants, pulmonate snails, and tracheate arthropods made the transition much earlier than vertebrates, and winged insects were diversifying at approximately the same time that the earliest terrestrial vertebrates evolved. Although the invasion of land required modification of almost every system in the vertebrate body, aquatic and terrestrial vertebrates retain many basic structural and functional similarities. We see the transition between the aquatic and terrestrial vertebrates most clearly today in the many living amphibians that make this transition during their own life histories.

Beyond the obvious difference in water content, there are several important physical differences that animals must accommodate when moving from water to land. These include (1) oxygen content, (2) density, (3) temperature regulation, and (4) habitat diversity. Oxygen is at least 20 times more abundant in air and it diffuses much more rapidly through air than through water. Consequently, terrestrial animals can obtain oxygen far more easily than aquatic ones once they possess the appropriate adaptations, such as lungs. Air, however, has approximately 1000 times less buoyant density than water and is approximately 50 times less viscous. It therefore provides relatively little support against gravity, requiring the terrestrial animal to develop strong limbs and to remodel the skeleton to achieve adequate structural support. Air fluctuates in temperature more readily than water does, and terrestrial environments therefore experience harsh and unpredictable cycles of freezing, thawing, drying, and flooding. Terrestrial animals require behavioral and physiological strategies to protect themselves from thermal extremes; one such important strategy is the homeothermy (regulated constant body temperature) of birds and mammals.

Despite its hazards, the terrestrial environment offers a great variety of new habitats including coniferous, temperate, and tropical forests, grasslands, deserts, mountains, oceanic islands, and polar regions (Chapter 5, pp. 128–131). The provision of safe shelter for the protection of vulnerable eggs and young may be accomplished much more readily in many of these terrestrial habitats than in aquatic ones.

Early Evolution of Terrestrial Vertebrates

Devonian Origin of the Tetrapods

The Devonian period, beginning some 400 million years ago, was a time of mild temperatures and alternating droughts and floods. During this period some primarily aquatic vertebrates evolved two features that would be important for permitting the subsequent evolution for life on land: lungs and limbs.

The Devonian freshwater environment was unstable. During dry periods, many pools and streams evaporated, water became foul, and the dissolved oxygen disappeared. Only those fishes able to acquire atmospheric oxygen survived such conditions. Gills were unsuitable because in air the filaments collapsed, dried, and quickly lost their function. Virtually all freshwater fishes surviving this period, including the lobe-finned (rhipidistian) fishes and the lungfishes (p. 594), had a kind of lung that developed as an outgrowth of the pharynx. It was relatively simple to enhance the efficiency of the air-filled cavity by improving its vascularity with a rich capillary network, and by supplying it with arterial blood from the last (sixth) pair of aortic arches. Oxygenated blood returned directly to the heart by a pulmonary vein to form a complete pulmonary circuit. Thus the **double circulation** characteristic of all tetrapods originated: a systemic circulation serving the body and a pulmonary circulation supplying the lungs.

Vertebrate limbs also arose during the Devonian period. Although fish fins at first appear very different from the jointed limbs of tetrapods, an examination of the bony elements of the paired fins of the lobe-finned fishes shows that they broadly resemble the equivalent limbs of amphibians. In *Eusthenopteron,* a Devonian lobe-fin, we can recognize an upper arm bone (humerus) and two forearm bones (radius and ulna) as well as other elements that we can homologize with the wrist bones of tetrapods (Figure 28-1). *Eusthenopteron* could walk—more accurately flop—along the bottom mud of pools with its fins, since backward and forward movement of the fins was limited to about 20–25 degrees. *Acanthostega,* one of the earliest known Devonian tetrapods, had well-formed tetrapod legs with clearly formed digits on both fore- and hindlimbs, but the limbs were too weakly constructed to enable the animal to hoist its body off the surface for proper walking on land. *Ichthyostega,* however, with its fully developed shoulder girdle, bulky limb bones, well-developed muscles, and other adaptations for terrestrial life, must have been able to pull itself onto land, although it probably did not walk very well. Thus, the tetrapods evolved their legs underwater and only then, for reasons unknown, began to pull themselves onto land.

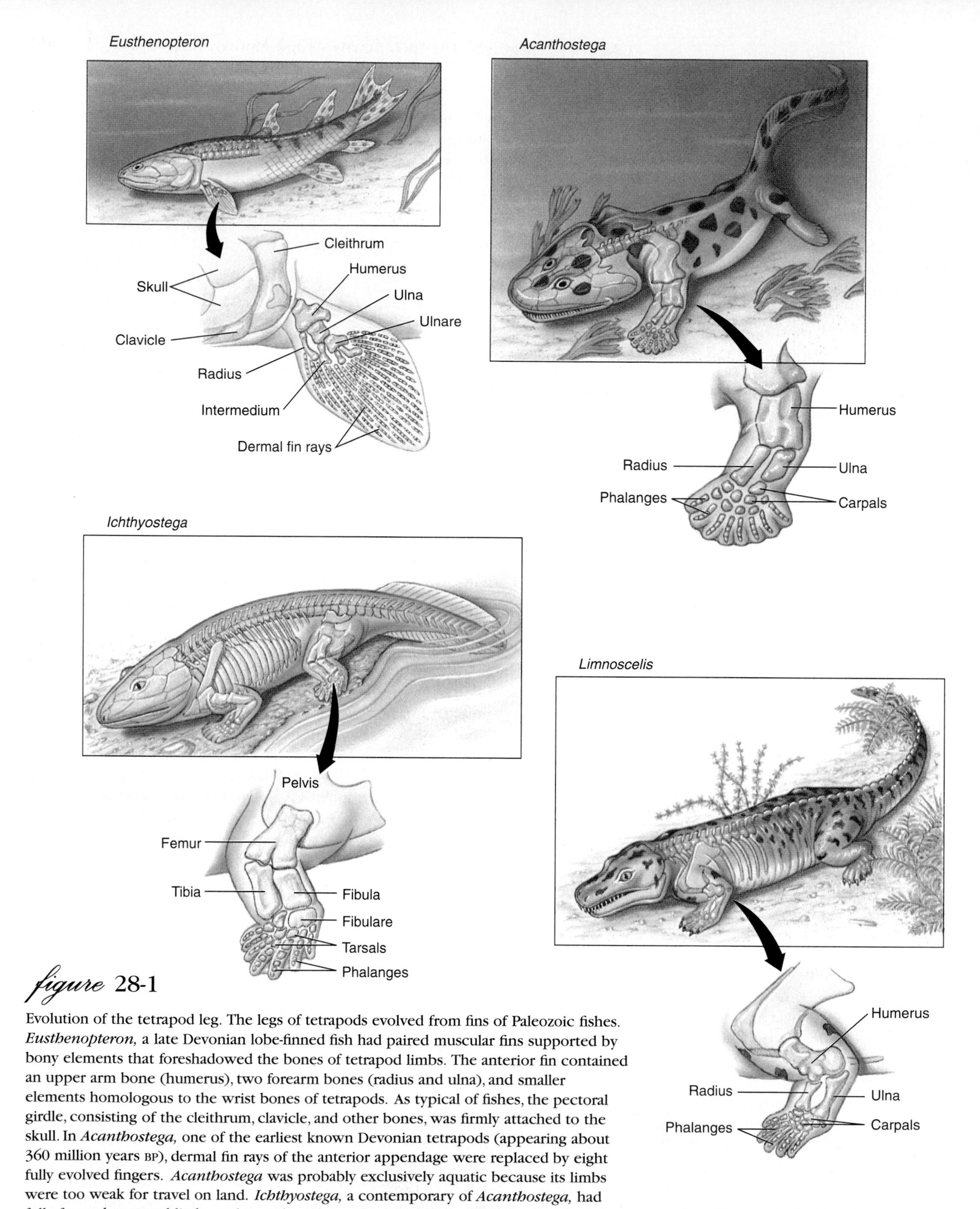

figure 28-1

Evolution of the tetrapod leg. The legs of tetrapods evolved from fins of Paleozoic fishes. *Eusthenopteron,* a late Devonian lobe-finned fish had paired muscular fins supported by bony elements that foreshadowed the bones of tetrapod limbs. The anterior fin contained an upper arm bone (humerus), two forearm bones (radius and ulna), and smaller elements homologous to the wrist bones of tetrapods. As typical of fishes, the pectoral girdle, consisting of the cleithrum, clavicle, and other bones, was firmly attached to the skull. In *Acanthostega,* one of the earliest known Devonian tetrapods (appearing about 360 million years BP), dermal fin rays of the anterior appendage were replaced by eight fully evolved fingers. *Acanthostega* was probably exclusively aquatic because its limbs were too weak for travel on land. *Ichthyostega,* a contemporary of *Acanthostega,* had fully formed tetrapod limbs and must have been able to walk on land. The hindlimb bore seven toes (the number of forelimb digits is unknown). *Limnoscelis,* an anthracosaur amphibian of the Carboniferous (about 300 million years BP) had five digits on both fore- and hindlimbs, the basic pentadactyl model which became the tetrapod standard.

Sources: Carroll, R. L. 1988. Vertebrate paleontology and evolution, New York, W.H. Freeman and Company; Coates, M. I., and J. A. Clack. 1990. ***347:****66–69; Edwards, J. L. 1989. Amer. Zool.* ***29:****235–254; Jarvik, E. 1955. Scient. Monthly, (Mar.)* ***1955:****141–154; Zimmer, C. N. 1995. Discover* ***16****(6):118–127.*

As noted above, evidence points to the lobe-finned fishes as the closest relatives of the tetrapods; in cladistic terms they are the sister group of tetrapods (Figures 28-2 and 28-3). Both the lobe-finned fishes and early tetrapods such as *Acanthostega* and *Ichthyostega* shared several characteristics of skull, teeth, and pectoral girdle. *Ichthyostega* (Gr. *ichthys,* fish, + *stegē*, roof, or covering, in reference to the roof of the skull which was shaped like that of a fish) represents an early offshoot of tetrapod phylogeny that possessed several adaptations, in addition to jointed limbs, that equipped it for life on land (Figure 28-1). These include a stronger backbone and associated muscles to support the body in air, new muscles to elevate the head, strengthened shoulder and hip girdles, a protective rib cage, a more advanced ear structure for detecting airborne sounds, a foreshortening of the skull, and a lengthening of the snout that improved olfactory powers for detecting dilute airborne odors. Yet *Ichthyostega* still resembled aquatic forms in retaining a tail complete with fin rays and in having opercular (gill) bones.

The bones of Ichthyostega, *the most thoroughly studied of all early tetrapods, were first discovered on an east Greenland mountainside in 1897 by Swedish scientists looking for three explorers lost two years earlier during an ill-fated attempt to reach the North Pole by hot-air balloon. Later expeditions by Gunnar Säve-Söderberg uncovered skulls of* Ichthyostega *but Säve-Söderberg died, at age 38, before he was able to make a thorough study of the skulls. After Swedish paleontologists returned to the Greenland site where they found the remainder of* Ichthyostega*'s skeleton, Erik Jarvik, one of Säve-Söderberg's assistants, assumed the task of examining the skeleton in detail. This study became his life's work, resulting in the detailed description of* Ichthyostega *available to us today.*

Carboniferous Radiation of the Tetrapods

The capricious Devonian period was followed by the Carboniferous period, characterized by a warm, wet climate during which mosses and large ferns grew in profusion on a swampy landscape. Tetrapods radiated quickly in this environment to produce a great variety of forms, feeding on the abundance of insects, insect larvae, and aquatic invertebrates available. The evolutionary relationships of the early tetrapod groups are still very controversial. We present a tentative cladogram (Figure 28-2) that almost certainly will undergo future revision as new data are collected. Several extinct lineages plus the **Lissamphibia,** which contains the modern amphibians, are placed in a group termed the **temnospondyls** (see Figures 28-2 and 28-3). This group is distinguished by having generally only four digits on the forelimb rather than the five characteristic of most tetrapods.

The lissamphibians diversified during the Carboniferous to produce the ancestors of the three major groups of amphibians alive today, **frogs** (Anura or Salientia), **salamanders** (Caudata or Urodela), and **caecilians** (Apoda or Gymnophiona). The early amphibians improved their adaptations for living in water during this period. Their bodies became flatter for moving about in shallow water. Early salamanders developed weak limbs and the tail became better developed as a swimming organ. Even the anurans (frogs and toads), which are now largely terrestrial as adults, developed specialized hindlimbs with webbed feet better suited for swimming than for movement on land. All amphibians use their porous skin as a primary or accessory breathing organ. This specialization was encouraged by the swampy surroundings of the Carboniferous period but presented serious desiccation problems for life on land.

The Modern Amphibians

The three living amphibian orders comprise more than 3900 species. Most share general adaptations for life on land, including skeletal strengthening and a shifting of special sense priorities from the ancestral lateral line system to the senses of smell and hearing. For this, both the olfactory epithelium and the ear are redesigned to improve sensitivities to airborne odors and sounds.

Nonetheless, most amphibians meet the problems of independent life on land only halfway. In the ancestral life history of amphibians, eggs are aquatic and hatch to produce an aquatic larval form that uses gills for breathing. A metamorphosis follows in which gills are lost and lungs, which are present throughout larval life, are then activated for respiration. Many amphibians retain this general pattern but there are some important exceptions. Some salamanders lack a complete metamorphosis and retain a permanently aquatic, larval morphology throughout life. Others live entirely on land and lack the aquatic larval phase completely. Both of these are evolutionarily derived conditions. Some frogs also have acquired a strictly terrestrial existence by eliminating the aquatic larval stage.

Even the most terrestrial amphibians remain dependent on very moist if not aquatic environments. Their skin is thin, and it requires moisture for protection against desiccation in air. An intact frog loses water nearly as rapidly as a skinless frog. Amphibians also require moderately cool environments. Being ectothermic, their body temperature is determined by and varies with the environment, greatly restricting where they can live. This restriction is especially important for reproduction. Eggs are not well protected from desiccation, and they must be shed directly into the water or onto moist terrestrial surfaces. Completely terrestrial amphibians may lay eggs under logs or rocks, in the moist forest floor, in flooded

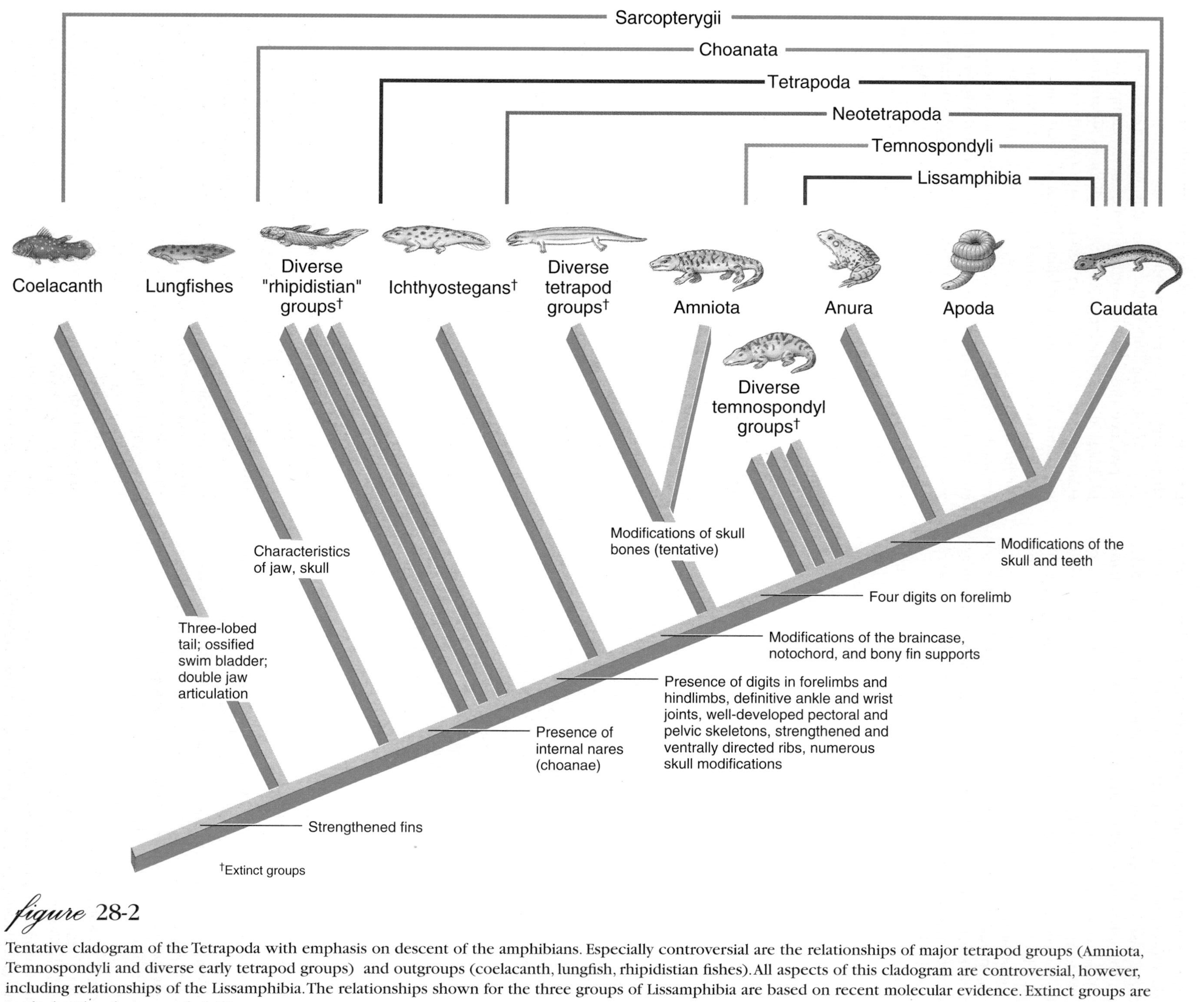

figure 28-2

Tentative cladogram of the Tetrapoda with emphasis on descent of the amphibians. Especially controversial are the relationships of major tetrapod groups (Amniota, Temnospondyli and diverse early tetrapod groups) and outgroups (coelacanth, lungfish, rhipidistian fishes). All aspects of this cladogram are controversial, however, including relationships of the Lissamphibia. The relationships shown for the three groups of Lissamphibia are based on recent molecular evidence. Extinct groups are marked with a dagger symbol (†).

Source: Modified from E.W. Gaffney in the Bulletin of the Carnegie Museum of Natural History ***13****:92–105 (1979).*

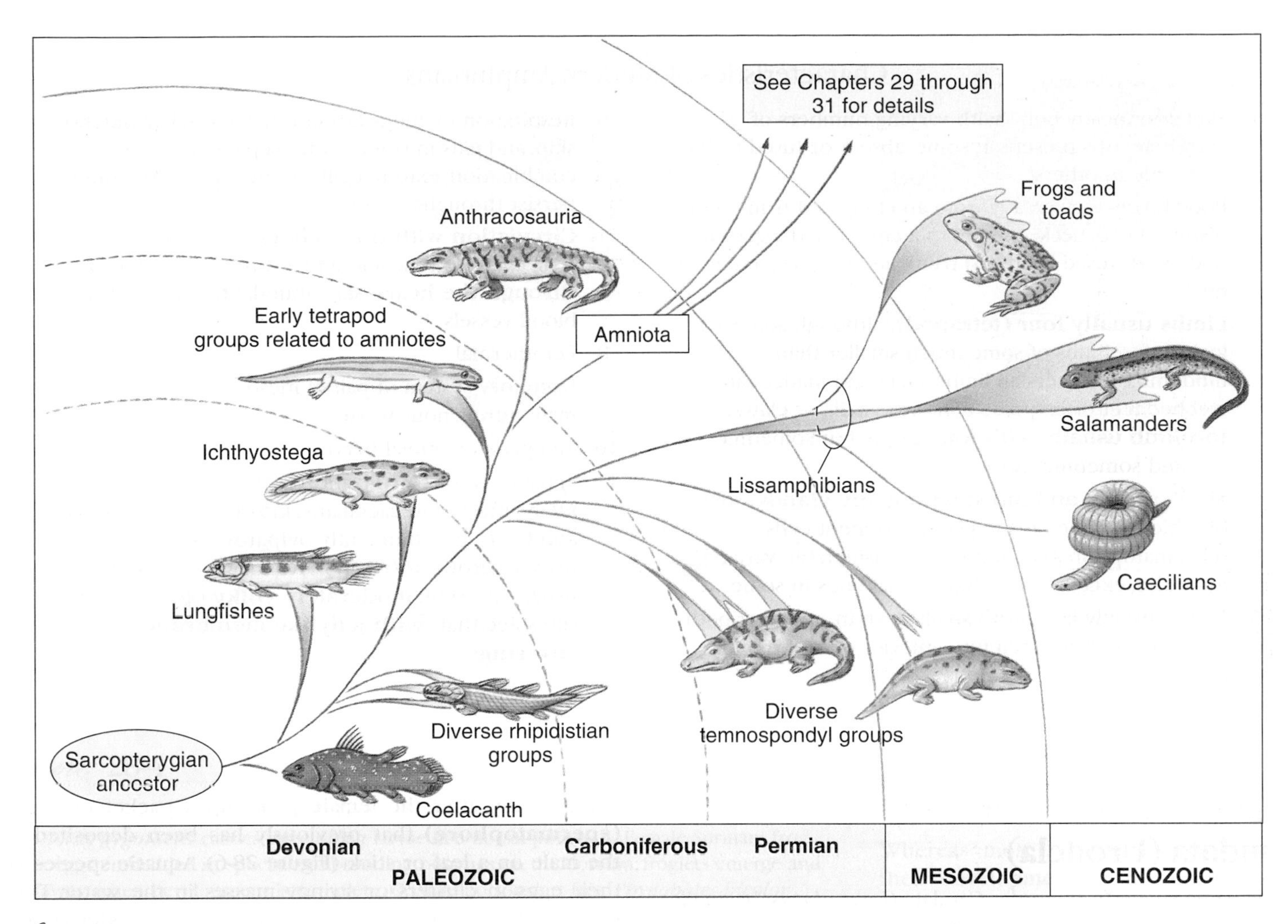

figure 28-3

Early tetrapod evolution and the descent of amphibians. The tetrapods share most recent common ancestry with the extinct Devonian rhipidistian fishes (of *living* groups, the tetrapods are most closely related to the lungfishes). The amphibians share most recent common ancestry with the diverse temnospondyls of the Carboniferous and Permian periods of the Paleozoic, and Triassic period of the Mesozoic.

tree holes, in pockets on the mother's back (Figure 28-4), or in folds of the body wall. One species of Australian frog even broods its young in its vocal pouch (see Figure 28-4).

We now highlight the special characteristics of the three major groups of amphibians. We will expand the coverage of general amphibian features when discussing the groups in which particular features have been studied most extensively. For most features, this group will be the frogs.

Caecilians: Order Gymnophiona (Apoda)

The order Gymnophiona (jim′no-fy′o-na) (Gr. *gymnos,* naked, + *ophineos,* of a snake) contains approximately 160 species of elongate, limbless, burrowing creatures commonly called **caecilians** (Figure 28-5). They occur in tropical forests of South America (their principal home), Africa, and Southeast Asia. They possess a long, slender body, small scales in the skin of some, many vertebrae, long ribs, no limbs, and a terminal anus. The eyes are small, and most species are totally blind as adults. Their food consists mostly of worms and small invertebrates, which they find underground. Fertilization is internal, and the male is provided with a protrusible copulatory organ. The eggs are usually deposited in moist ground near the water. In some species the eggs are carefully guarded in folds of the body during their development. Viviparity (p. 333) also is common in some caecilians, with the embryos obtaining nourishment by eating the wall of the oviduct.

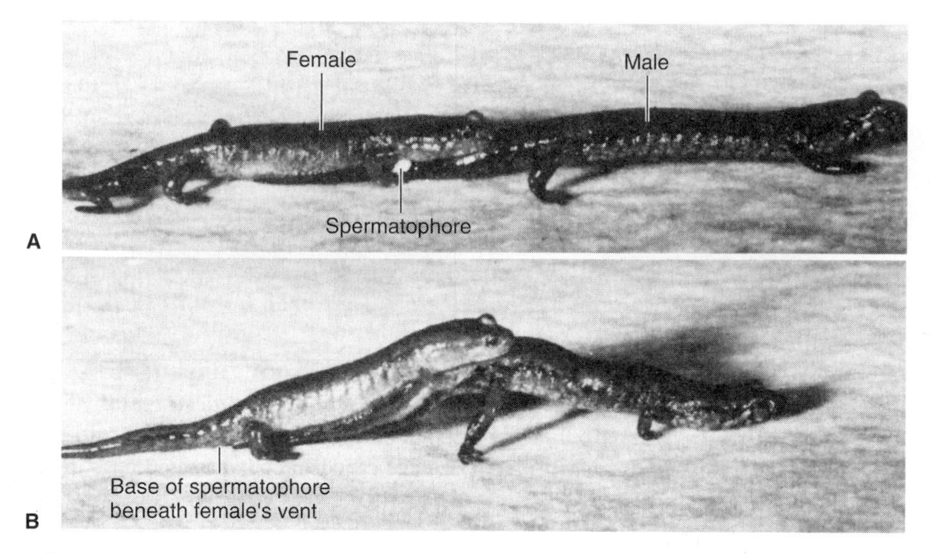

figure 28-6

Courtship and sperm transfer in the pygmy salamander, *Desmognathus wrighti.* After judging the female's receptivity by the presence of her chin on his tail base, the male deposits a spermatophore on the ground, then moves forward a few paces. **A,** The white mass of the sperm atop a gelatinous base is visible at the level of the female's forelimb. The male moves ahead, the female following until the spermatophore is at the level of her vent. **B,** The female has recovered the sperm mass in her vent, while the male arches his tail, tilting the female upward and presumably facilitating recovery of the sperm mass.

figure 28-7

Female dusky salamander (*Desmognathus* sp.) attending eggs. Many salamanders exercise parental care of the eggs, which includes rotating the eggs and protecting them from fungal infections and predation by various arthropods and other salamanders.

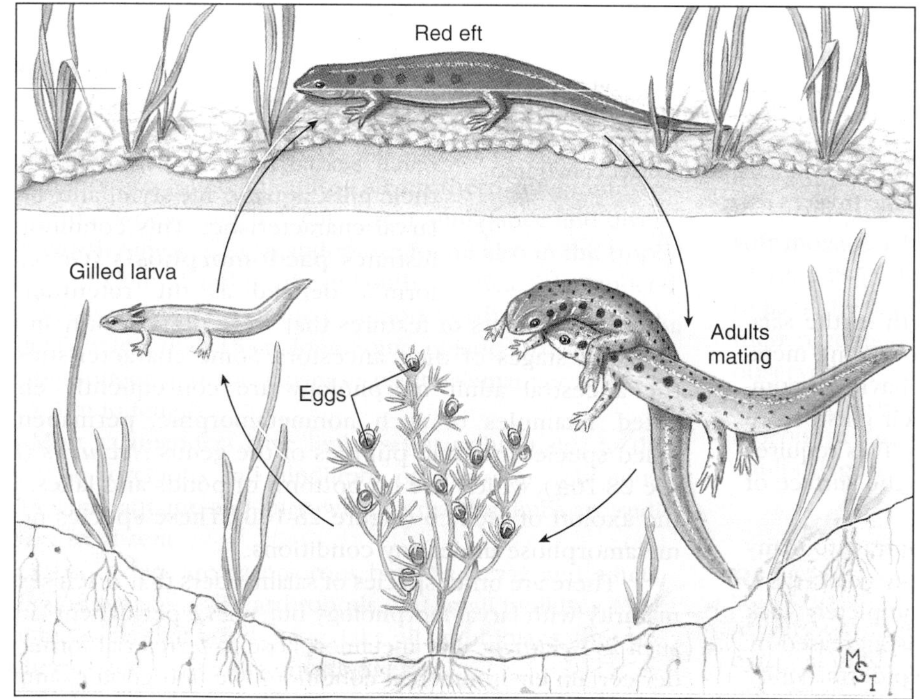

figure 28-8

Life history of the red-spotted newt, *Notophthalmus viridescens* of the family Salamandridae. In many habitats the aquatic larva metamorphoses into a brightly colored "red eft" stage, which remains on land from one to three years before transforming into a secondarily aquatic adult.

figure 28-9

Longtail salamander *Eurycea longicauda,* a common plethodontid salamander.

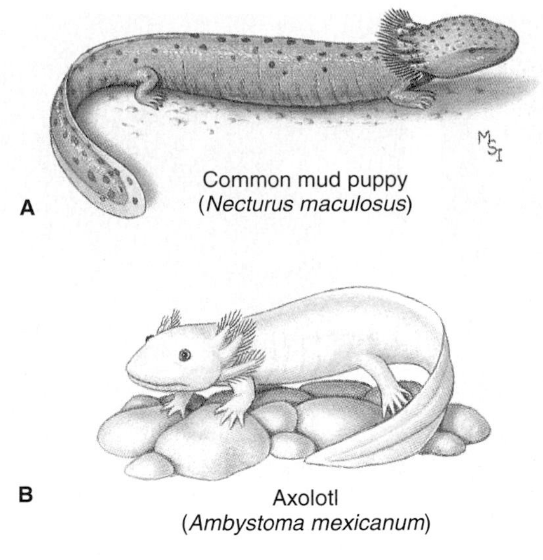

figure 28-10

Paedomorphosis in salamanders. **A,** The mud puppy *Necturus* sp. and **B,** the axolotl *(Ambystoma mexicanum)* are permanently gilled aquatic forms.

In addition to their importance in biomedical research and education, frogs have long served the epicurean frog-leg market. Mainstay of this market are bullfrogs, which are in such heavy demand in Europe (especially France) and the United States—the worldwide harvest is an estimated 200 million bullfrogs (about 10,000 metric tons) annually—that its populations have fallen drastically as the result of excessive exploitation and the draining and pollution of wetlands. Most are Asian bullfrogs imported from India and Bangladesh, some 80 million collected each year from rice fields in Bangladesh alone. With so many insect-eating frogs removed from the ecosystem, rice production is threatened from uncontrolled, flourishing insect populations. In the United States, attempts to raise bullfrogs in farms have not been successful, mainly because bullfrogs are voracious eating machines that normally will accept only living prey, such as insects, crayfish, and other frogs.

Frogs and Toads: Order Anura (Salientia)

The more than 3450 species of frogs and toads that comprise the order Anura (Gr. *an,* without, + *oura,* tail) are for most people the most familiar amphibians. The Anura are an old group, known from the Jurassic period, 150 million years ago. Frogs and toads occupy a great variety of habitats, despite their aquatic mode of reproduction and water-permeable skin, which prevent them from wandering too far from sources of water, and their ectothermy, which bars them from polar and subarctic habitats. The name of the order, Anura, refers to an obvious group characteristic, the absence of tails in adults (although all pass through a tailed larval stage during development). Frogs and toads are specialized for jumping, as suggested by the alternative order name, Salientia, which means leaping.

We see in the appearance and life-style of their larvae further distinctions between the Anura and Caudata. The eggs of most frogs hatch into a tadpole ("polliwog") stage, having a long, finned tail, both internal and external gills, no legs, specialized mouthparts for herbivorous feeding (salamander larvae and some tadpoles are carnivorous), and a highly specialized internal anatomy. They look and act altogether differently from adult frogs. The metamorphosis of a frog tadpole to an adult frog is thus a striking transformation. The permanently gilled larval condition never occurs in frogs and toads as it does in salamanders.

Frogs and toads are divided into 21 families. The best-known frog families in North America are the Ranidae, which contains most of our familiar frogs (Figure 28-11A), and the Hylidae, the tree frogs (Figure 28-11B). True toads, belonging to the family Bufonidae, have short legs, stout bodies, and thick skins usually with prominent warts (Figure 28-12). However, the term "toad" is used rather loosely to refer also to more or less terrestrial members of several other families.

The largest anuran is the West African *Conraua goliath,* which is more than 30 cm long from tip of nose to anus (Figure 28-13). This giant eats animals as big as rats and ducks. The smallest frog recorded is *Phyllobates limbatus,* which is only approximately 1 cm long. This tiny frog, which can be more than covered by a dime, is found in Cuba. The largest American frog is the bullfrog, *Rana catesbeiana* (Figure 28-11A), which reaches a head and body length of 20 cm.

Habitats and Distribution

Probably the most abundant frogs are the approximately 260 species of the genus *Rana* (Gr. frog), found over all the temperate and tropical regions of the world except in New Zealand, the oceanic islands, and southern South America. They are usually found near water, although some, such as the wood frog *R. sylvatica,* spend most of their time on damp forest floors. The larger bullfrogs, *R. catesbeiana,* and green frogs, *R. clamitans,* are nearly always found in or near permanent water or swampy regions. The leopard frogs, *Rana pipiens* and related species, are found in nearly every state and Canadian province and are the most widespread of all North American frogs. The northern leopard frog, *R. pipiens,*

A

B

figure 28-11

Two common North American frogs. **A,** Bullfrog, *Rana catesbeiana,* largest American frog and mainstay of the frog-leg epicurean market (family Ranidae). **B,** Green tree frog *Hyla cinerea,* a common inhabitant of swamps of the southeastern United States (family Hylidae). Note adhesive pads on the feet.

figure 28-12

American toad *Bufo americanus* (family Bufonidae). This principally nocturnal yet familiar amphibian feeds on large numbers of insect pests and on snails and earthworms. The warty skin contains numerous glands that produce a surprisingly poisonous milky fluid, providing the toad excellent protection from a variety of potential predators.

is the species most commonly used in biology laboratories and for classical electrophysiological research.

Within the range of any species, frogs are often restricted to certain localities (for instance, to specific streams or pools) and may be absent or scarce in similar habitats elsewhere. The pickerel frog *(R. palustris)* is especially noteworthy in this respect because it is known to be abundant only in certain localized regions. Recent studies have shown that many populations of frogs worldwide may be suffering declines in numbers and becoming even more patchy than usual in their distributions. In most declining populations the causes are unknown.[1]

Most of the larger frogs are solitary in their habits except during the breeding season. During the breeding period most of them, especially the males, are very noisy. Each male usually takes possession of a particular perch near water, where he may remain for hours or even days, trying to attract a female to that spot. At times frogs are mainly silent, and their presence is not detected until they are disturbed. When they enter the water, they dart about swiftly and reach the bottom of the pool, where they kick up a cloud of muddy water. In swimming, they hold the forelimbs near the body and kick backward with their webbed hindlimbs, which propel them forward. When they come to the surface to breathe, only the head and foreparts are exposed and, since they usually take advantage of any protective vegetation, they are difficult to see.

What is responsible for the widely reported decline in amphibian, especially frog, populations around the world? Puzzling is the evidence that whereas amphibian populations are falling rapidly in many parts of the world, in other areas they are doing well. No single explanation fits all instances of declines. In some instances, changes in sizes of populations are simply random fluctuations caused by periodic droughts and other naturally occurring phenomena. However, several other environmental and human-related factors have been implicated in amphibian declines: habitat destruction and modification; rises in environmental pollutants such as acid rain, fungicides, herbicides, and industrial chemicals; diseases; introduction of nonnative predators and competitors; and depletion of the ozone shield in the stratosphere resulting in severe losses in developing frog embryos from increased ultraviolet radiation. These explanations are radically different from each other. One or more explanations do seem to explain certain population declines; in other instances the reasons for the declines are unknown.

During the winter months most frogs hibernate in the soft mud of the bottoms of pools and streams. Their life processes are at a very low ebb during their hibernation period, and such energy as they need is derived from the glycogen and

[1]Sarkar, S. 1996. Ecological theory and anuran declines. *BioScience* **46**(3):199–207.

figure 28-13

Conraua (Gigantorana) goliath (family Ranidae) of West Africa, the world's largest frog. This specimen weighed 3.3 kg (approximately 7½ pounds).

fat stored in their bodies during the spring and summer months. The more terrestrial frogs, such as tree frogs, hibernate in the humus of the forest floor. They are tolerant of low temperatures, and many actually survive prolonged freezing of all the extracellular fluid, representing 35% of the body water. Such frost-tolerant frogs prepare for winter by accumulating glucose and glycerol in body fluids, which protects tissues from the normally damaging effects of ice crystal formation.

Adult frogs have numerous enemies, such as snakes, aquatic birds, turtles, raccoons, and humans; fish prey upon tadpoles and only a few survive to maturity. Although usually defenseless, in the tropics and subtropics many frogs and toads are aggressive, jumping and biting at predators. Some defend themselves by feigning death. Most anurans can inflate their lungs so that they are difficult to swallow. When disturbed along the margin of a pond or brook, a frog often remains quite still. When it thinks it is detected it jumps, not always into the water where enemies may be lurking but into grassy cover on the bank. When held in the hand, a frog may cease its struggles for an instant to put its captor off guard and then leap violently, at the same time voiding its urine. Their best protection is their ability to leap and their use of poison glands. Bullfrogs in captivity do not hesitate to snap at tormenters and are capable of inflicting painful bites.

While native American amphibians continue to disappear as wetlands are drained, an exotic frog introduced into southern California has found the climate quite to its liking. The African clawed frog Xenopus laevis *(Figure 28-14) is a voracious, aggressive, primarily aquatic frog that is rapidly displacing native frogs and fish from several waterways and is spreading rapidly. The species was introduced into North America in the 1940s when it was used extensively in human pregnancy tests. When more efficient tests appeared in the 1960s, some hospitals simply dumped surplus frogs into nearby streams, where the prolific breeders have become almost indestructible pests. As is so often the case with alien wildlife introductions, benign intentions frequently lead to serious problems.*

Reproduction

Frogs and toads living in temperate regions of the world breed, feed, and grow only during the warmer seasons of the year, usually in a predictable annual cycle. With warming spring temperatures, in combination with sufficient rainfall, males croak and call vociferously to attract females. After a brief courtship, females enter the water and are clasped by the males in a process called **amplexus** (Figure 28-15). As the female lays the eggs, the male discharges seminal fluid containing sperm over the eggs to fertilize them. After fertilization, the jelly layers absorb water and swell (Figure 28-16). The eggs are laid in large masses, often anchored to vegetation, then abandoned by the parents. The eggs begin development immediately (early development of the leopard frog embryo is illustrated in Figure 14-15B, p. 322). Within a few days the embryos have developed into tadpoles (Figure 28-16) and hatch, often to face a precarious existence if the eggs were laid in a temporary pond or puddle. In such instances, it becomes a race against time to complete development before the habitat dries up.

At the time of hatching, the tadpole has a distinct head and body with a compressed tail. The mouth is located on the ventral side of the head and is provided with horny jaws for scraping vegetation from objects for food. Behind the mouth is a ventral adhesive disc for clinging to objects. In front of the mouth are two deep pits, which later develop into the nostrils. Swellings are found on each side of the head, and these later become external gills. There are three pairs of external gills, which are transformed into internal gills that become covered with a flap of skin (the operculum) on each side. On the right side the operculum completely fuses with the body wall, but on the left side a small opening, the **spiracle** (L. *spiraculum,* air hole) remains, through which water flows after entering the mouth and passing the internal gills. The hindlegs appear first, whereas the forelimbs are hidden for a time by the folds of the operculum. During metamorphosis the tail is absorbed,

figure 28-14

African clawed frog, *Xenopus laevis.* The claws, an unusual feature in frogs, are on the hind feet. This frog has been introduced into California, where it is considered a serious pest.

figure 28-15

A male green frog, *Hyla cinerea,* clasps a larger female during the breeding season in a South Carolina swamp. Clasping (amplexus) is maintained until the female deposits her eggs. Like most tree frogs, these are capable of rapid and marked color changes; the male here, normally green, has darkened during amplexus.

the intestine becomes much shorter, the mouth undergoes a transformation into the adult condition, lungs develop, and the gills are absorbed. The leopard frog usually completes its metamorphosis within three months; the bullfrog takes much longer to complete the process.

The reproductive mode just described, while typical of most temperate zone anurans, is only one of a great variety of reproductive patterns in tropical anurans. Some of these remarkable strategies are illustrated in Figure 28-4 (p. 613). Some species lay their eggs in foam masses that float on the surface of the water; some deposit their eggs on leaves overhanging ponds and streams into which the emerging tadpoles will drop; some lay their eggs in damp burrows; and others place their eggs in water trapped in tree cavities or in water-filled chambers of some bromeliads (epiphytic plants in the tropical forest canopy). While most frogs abandon their eggs, some, such as the tropical dendrobatids (a family that includes the poison-dart frogs), tend the eggs. When the tadpoles hatch, they squirm up on the parent's back to be carried for varying lengths of time (Figure 28-4C). The marsupial frogs carry the developing eggs in a pouch on the back (Figure 28-4A).

Although most frogs develop through a larval stage (the tadpole), many tropical frogs have evolved direct development. In direct development the tadpole stage is bypassed and the froglet that emerges is a miniture replica of the adult.

Classification of Class Amphibia

Order Gymnophiona (jim′no-fy′o-na) (Gr. *gymnos,* naked, + *ophioneos,* of a snake) **(Apoda): caecilians.** Body elongate; limbs and limb girdle absent; mesodermal scales present in skin of some; tail short or absent; 95 to 285 vertebrae; pantropical, 6 families, 34 genera, approximately 160 species.

Order Caudata (caw-dot′uh) (L. *caudatus,* having a tail) **(Urodela): salamanders.** Body with head, trunk, and tail; no scales; usually two pairs of equal limbs; 10 to 60 vertebrae; predominantly holarctic; 9 living families, 62 genera, approximately 360 species.

Order Anura (uh-nur′uh) (Gr. *an,* without, + *oura,* tail) **(Salientia): frogs, toads.** Head and trunk fused; no tail; no scales; two pairs of limbs; large mouth; lungs; 6 to 10 vertebrae including urostyle (coccyx); cosmopolitan, predominantly tropical; 21 living families; 301 genera; approximately 3450 species.

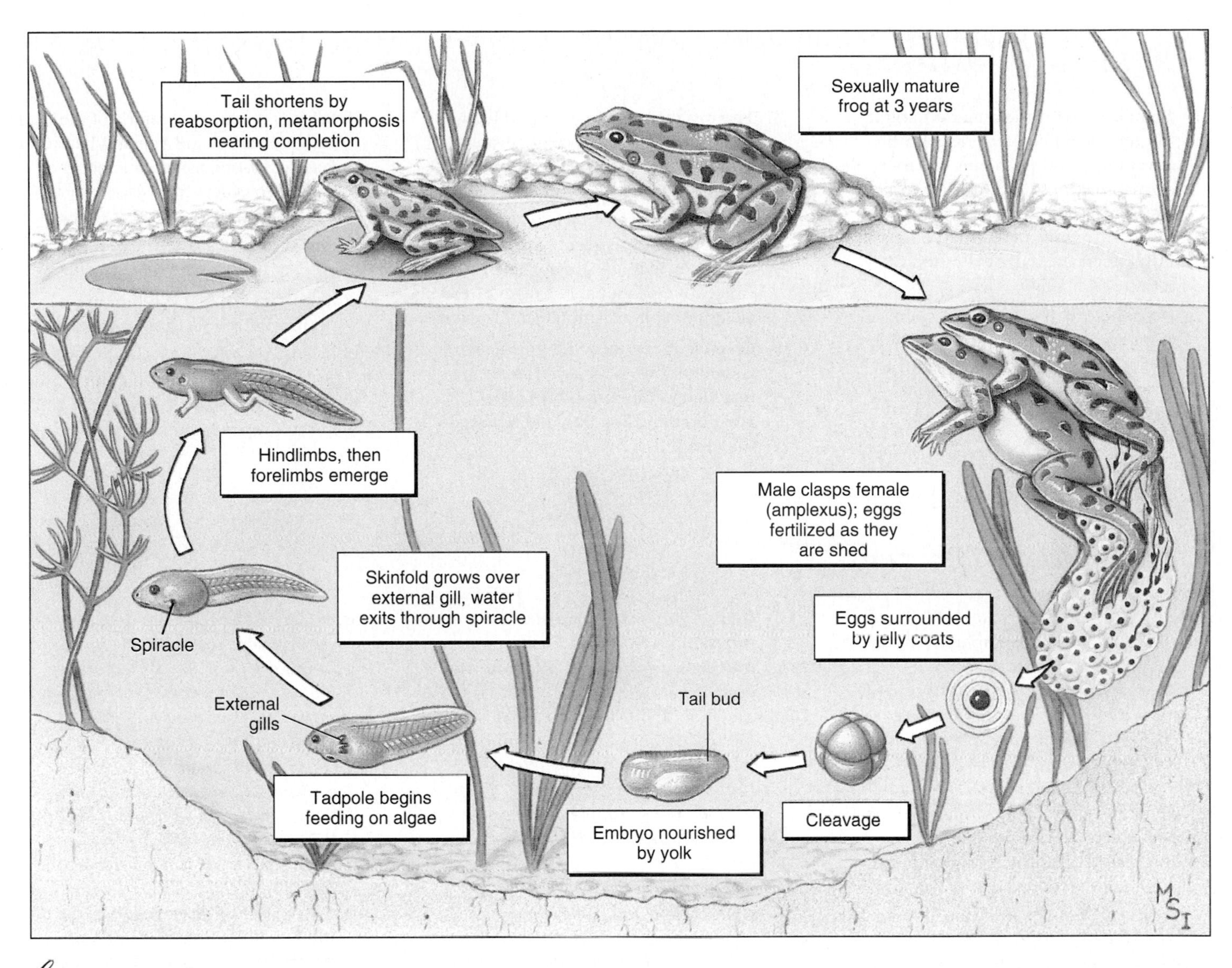

figure 28-16

Life cycle of a leopard frog.

Summary

Amphibians are ectothermic, primitively quadrupedal vertebrates that have glandular skin and that breathe by lungs, gills, or skin. They are the survivors of one of two major branches of tetrapod phylogeny, the other one being represented today by the amniotes. The modern amphibians consist of three major evolutionary groups. The caecilians (order Gymnophiona) are a small tropical group of limbless, elongate forms. The salamanders (order Caudata) are tailed amphibians that have retained the generalized four-legged body plan of their Paleozoic ancestors. The frogs and toads (order Anura) are the largest group of modern amphibians, all of which are specialized for a jumping mode of locomotion.

Most amphibians have a biphasic life cycle that begins with an aquatic larva that later metamorphoses to produce a terrestrial adult that returns to the water to lay eggs. Some frogs, salamanders, and caecilians have evolved direct development that omits the aquatic larval stage and some caecilians have evolved viviparity. Salamanders are unique among amphibians in having evolved several permanently gilled species that retain a larval morphology throughout life, eliminating the terrestrial phase completely. The permanently gilled condition is obligate in some species, but others will metamorphose to a terrestrial form if the pond habitat dries up.

Despite their adaptations for terrestrial life, the adults and eggs of all amphibians require cool, moist environments if not actual pools or streams. The eggs and adult skin have no effective protection against very cold, hot, or dry conditions, greatly restricting the adaptive radiation of amphibians to environments that have moderate temperatures and abundant water.

Review Questions

1. Compared with the aquatic habitat, the terrestrial habitat offers both advantages and problems for an animal making the transition from water to land. Summarize how these differences might have influenced the early evolution of tetrapods.
2. Describe the different modes of respiration used by amphibians. What paradox do the amphiumas and terrestrial plethodontids present regarding the association of lungs with life on land?
3. The evolution of the tetrapod leg was one of the most important advances in vertebrate history. Describe the supposed sequence in its evolution.
4. Compare the general life history patterns of salamanders with those of frogs. Which group shows the greater variety of evolutionary changes of the ancestral biphasic amphibian life cycle?
5. Give the literal meaning of the name Gymnophiona. What animals are included in this amphibian order, what do they look like, and where do they live?
6. What is the literal meaning of the order names Caudata and Anura? What major features distinguish the members of these two orders from each other?
7. Describe the breeding behavior of a typical woodland salamander.
8. How has paedomorphosis been important to the evolution of permanently aquatic salamanders?
9. Briefly describe the reproductive behavior of frogs. In what important ways do frogs and salamanders differ in their reproduction?

Selected References

Blaustein, A. R., and D. B. Wake. 1995. The puzzle of declining amphibian populations. Sci. Am. **272:**52–57 (Apr.). *Amphibian populations are dwindling in many areas of the world. The causes are multiple but all derive from human activities.*

Conant, R., and J. T. Collins. 1991. A field guide to reptiles and amphibians, ed. 3. The Peterson field guide series. Boston, Houghton Mifflin Company. *Updated version of a popular field guide; color illustrations and distribution maps for all species.*

del Pino, E. M. 1989. Marsupial frogs. Sci. Am. **260:**110–118 (May). *Several species of tropical frogs incubate their eggs on the female's back, often in a special pouch, and emerge as advanced tadpoles or fully formed froglets.*

Duellman, W. E. 1992. Reproductive strategies of frogs. Sci. Am. **267:**80–87 (July). *Many frogs have evolved improbable reproductive strategies that have permitted colonization of land.*

Duellman, W. E., and L. R. Trueb. 1994. Biology of amphibians. Baltimore, Johns Hopkins University Press. *Important comprehensive sourcebook of information on amphibians, extensively referenced and illustrated.*

Halliday, T. R., and K. Adler, eds. 1986. The encyclopedia of reptiles and amphibians. New York, Facts on File, Inc. *Excellent authoritative reference work with high-quality illustrations.*

Hanken, J. 1989. Development and evolution in amphibians. Am. Sci. **77:**336–343 (July–Aug.). *Explains how the diversity in amphibian morphology has been achieved by modifications in development.*

Lewis, S. 1989. Cane toads: an unnatural history. New York, Dolphin/Doubleday. *Based on an amusing and informative film of the same title that describes the introduction of cane toads to Queensland, Australia and the unexpected consequences of their population explosion there. "If Monty Python teamed up with National Geographic, the result would be* Cane Toads.*"*

Moffett, M. W. 1995. Poison-dart frogs: lurid and lethal. National Geographic **187**(5):98–111 (May). *Photographic essay of frogs that can be lethal even to the touch.*

Narins, P. M. 1995. Frog communication. Sci. Am. **273:**78–83 (Aug.). *Frogs employ several strategies to hear and be heard amidst the cacophony of chorusing of many frogs.*

Reptiles

chapter | twenty-nine

Enclosing the Pond

Amphibians, with well-developed legs, redesigned sensory and respiratory systems, and modifications of the postcranial skeleton for supporting the body in air, have made a notable conquest of land. But, with shell-less eggs and often gill-breathing larvae, their development remains hazardously tied to water. The lineage containing reptiles, birds, and mammals developed an egg that could be laid on land. This shelled egg, perhaps more than any other adaptation, unshackled the early reptiles from the aquatic environment by freeing the developmental process from dependence upon aquatic or very moist terrestrial environments. In fact, the "pond-dwelling" stages were not eliminated but enclosed within a series of extraembryonic membranes that provided complete support for embryonic development. One membrane, the amnion, encloses a fluid-filled cavity, the "pond," within which the developing embryo floats. Another membranous sac, the allantois, serves both as a respiratory surface and as a chamber for the storage of nitrogenous wastes. Enclosing these membranes is a third membrane, the chorion, through which oxygen and carbon dioxide freely pass. Finally, surrounding and protecting everything is a porous, parchmentlike or leathery shell.

With the last ties to aquatic reproduction severed, conquest of land by the vertebrates was ensured. The Paleozoic tetrapods that developed this reproductive pattern were ancestors of a single, monophyletic assemblage called the Amniota, named after the innermost of the three extraembryonic membranes, the amnion. Before the end of the Paleozoic era the amniotes had diverged into multiple lineages that gave rise to all the reptilian groups, the birds, and the mammals.

Members of the paraphyletic class Reptilia (rep-til′e-a) (L. *repto,* to creep) include the first truly terrestrial vertebrates. With nearly 7000 species (approximately 300 species in the United States and Canada) occupying a great variety of aquatic and terrestrial habitats, they are diverse and abundant. Nevertheless, reptiles are perhaps remembered best for what they once were, rather than for what they are now. The Age of Reptiles, which lasted for more than 165 million years, saw the appearance of a great radiation of reptilian lineages into a bewildering array of terrestrial and aquatic forms. Among these were the herbivorous and carnivorous dinosaurs, many of huge stature and awesome appearance, that dominated animal life on land—the ruling reptiles. Then, during a mass extinction at the end of Mesozoic era, they suddenly declined. Among the few reptilian lineages to emerge from the Mesozoic extinction are today's reptiles. One of these, the tuatara *(Sphenodon)* of New Zealand, is the sole survivor of a group that otherwise disappeared 100 million years ago. But others, especially the lizards and snakes, have radiated since the Mesozoic extinction into diverse and abundant groups (Figure 29-1). Understanding the 300 million-year-old history of reptilian life on earth has been complicated by widespread convergent and parallel evolution among the many lineages and by large gaps in the fossil record.

Origin and Adaptive Radiation of Reptiles

As mentioned in the prologue to this chapter, the amniotes are a monophyletic group that evolved in the late Paleozoic. Most paleontologists agree that the amniotes arose from a group of amphibian-like tetrapods, the anthracosaurs, during the early Carboniferous period of the Paleozoic. By the late Carboniferous (approximately 300 million years ago), the amniotes had separated into three lineages. The first lineage, the **anapsids** (Gr. *an,* without, + *apsis,* arch), was characterized by a skull having no temporal opening behind the orbits, that is, the skull behind the orbits was completely roofed with dermal bone (Figure 29-2). This group is represented today only by the turtles. Their morphology is an odd mix of ancestral and derived characters that has scarcely changed at all since the turtles first appeared in the fossil record in the Triassic some 200 million years ago.

The second lineage, the **diapsids** (Gr. *di,* double, + *apsis,* arch), gave rise to all other reptilian groups and to the birds (Figure 29-1). The diapsid skull was characterized by the presence of two temporal openings: one pair located low on the cheeks, and a second pair positioned above the lower pair and separated from them by a bony arch (Figure 29-2). Three subgroups of diapsids appeared. The **lepidosaurs** include the extinct marine ichthyosaurs and all of the modern reptiles with the exception of the turtles and crocodilians. The more derived **archosaurs** comprised the dinosaurs and their relatives, and the living crocodilians and birds. A third, smaller lineage of diapsids, the **sauropterygians** included several extinct aquatic groups, the most conspicuous of which were the large, long-necked plesiosaurs.

The third lineage was the **synapsids** (Gr. *syn,* together, + *apsis,* arch), the mammal-like reptiles. The synapsid skull had a single pair of temporal openings located low on the cheeks and bordered by a bony arch (Figure 29-2). The synapsids were the first group of amniotes to diversify, giving rise first to the pelycosaurs, later to the therapsids, and finally to mammals (Figure 29-1).

Changes in Traditional Classification of Reptiles

With increasing use of cladistic methodology in zoology, and its insistence on hierarchical arrangement of monophyletic groups (see p. 346 and following), important changes have been made in the traditional classification of reptiles. The class Reptilia is no longer recognized by cladists as a valid taxon because it is not monophyletic. As customarily defined, the class Reptilia excludes the birds which descend from the most recent common ancestor of the reptiles. Consequently, the reptiles are a paraphyletic group because they do not include all descendants of their most recent common ancestor. Reptiles can be identified only as amniotes that are not birds or mammals. This is clearly shown in the phylogenetic tree of the amniotes (Figure 29-1).

An example of this problem is the shared ancestry of birds and crocodilians. Based solely on shared derived characteristics, crocodilians and birds are sister groups; that is they are more recently descended from a common ancestor than either is from any other living reptilian lineage. In other words, birds and crocodilians belong to a monophyletic group apart from other reptiles and, according to the rules of cladism, should be assigned to a clade that separates them from the remaining reptiles. This clade is in fact recognized; it is the Archosauria (Figures 29-1 and 29-2), a grouping that also includes the extinct dinosaurs. Therefore birds should be classified as reptiles. The archosaurs plus their sister group, the lepidosaurs (tuataras, lizards, and snakes) comprise a monophyletic group that some taxonomists call the Reptilia. The term Reptilia is thereby redefined to include birds in contrast to its traditional usage. However, evolutionary taxonomists argue that birds represent a novel adaptive zone and grade of organization whereas crocodilians remain within the traditionally recognized reptilian adaptive zone and grade. In this view, the morphological and ecological novelty of birds has been recognized by maintaining the traditional classification that places crocodilians in the class Reptilia and birds in the class Aves. Such conflicts of opinion between proponents of the two major competing schools of taxonomy (cladistics and evolutionary taxonomy) have had the healthy effect of forcing zoologists to reevaluate their views of amniote genealogy and how vertebrate classifications should represent genealogy and degree of divergence. In our treatment we retain the traditional class Reptilia because this is still standard taxonomic practice, but we emphasize that this taxonomy is likely to be discontinued.

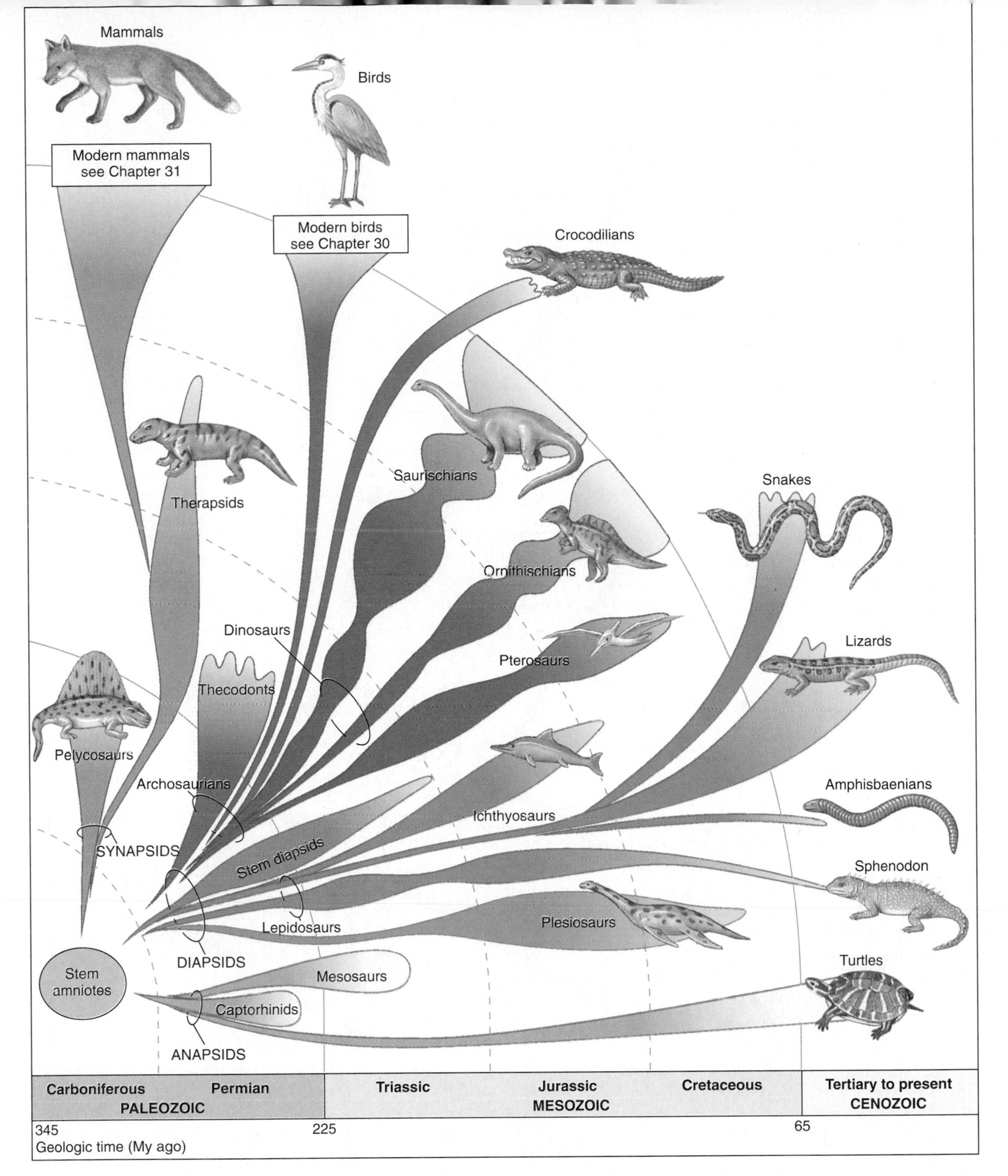

figure 29-1

Evolution of the amniotes. The evolutionary origin of amniotes occurred by the evolution of an amniotic egg that made reproduction on land possible, although this egg may well have developed before the earliest amniotes had ventured far on land. The amniote assemblage, which includes the reptiles, birds, and mammals, evolved from a lineage of small, lizardlike forms that retained the skull pattern of the early tetrapods. First to diverge from the primitive stock were the mammal-like reptiles, characterized by a skull pattern termed the synapsid condition. All other amniotes, including the birds and all living reptiles except the turtles, have a skull pattern known as diapsid. The turtles have a skull pattern known as anapsid. The great Mesozoic radiation of reptiles may have resulted partly by the increased variety of ecological habitats into which the amniotes could move.

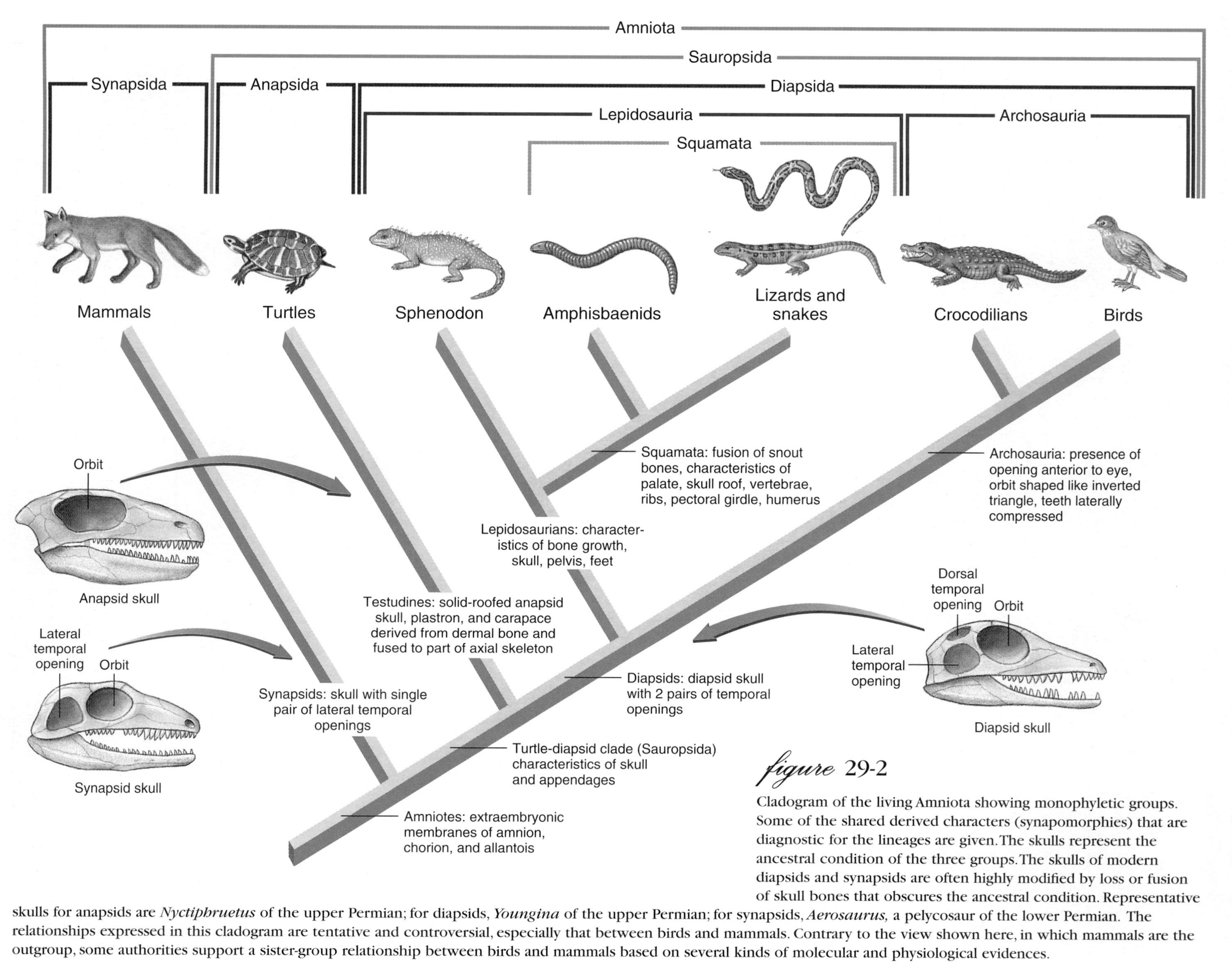

figure 29-2

Cladogram of the living Amniota showing monophyletic groups. Some of the shared derived characters (synapomorphies) that are diagnostic for the lineages are given. The skulls represent the ancestral condition of the three groups. The skulls of modern diapsids and synapsids are often highly modified by loss or fusion of skull bones that obscures the ancestral condition. Representative skulls for anapsids are *Nyctiphruetus* of the upper Permian; for diapsids, *Youngina* of the upper Permian; for synapsids, *Aerosaurus,* a pelycosaur of the lower Permian. The relationships expressed in this cladogram are tentative and controversial, especially that between birds and mammals. Contrary to the view shown here, in which mammals are the outgroup, some authorities support a sister-group relationship between birds and mammals based on several kinds of molecular and physiological evidences.

Source: Data from Pough, F.H., J.B. Heiser, and W.N. McFarland. 1989. Vertebrate life, *ed.3, New York, Macmillan.*

Characteristics of Class Reptilia

1. Body varied in shape, compact in some, elongated in others; **body covered with an exoskeleton of horny epidermal scales** with the addition sometimes of bony dermal plates; **integument with few glands**
2. **Limbs paired, usually with five toes,** and adapted for climbing, running, or paddling; absent in snakes and some lizards
3. Skeleton well ossified; ribs with sternum (sternum absent in snakes) forming a complete thoracic basket; **skull with one occipital condyle**
4. Respiration by lungs; **no gills;** cloaca used for respiration by some; branchial arches in embryonic life
5. Three-chambered heart; **crocodilians with four-chambered heart;** usually one pair of aortic arches; systemic and pulmonary circuits functionally separated
6. Ectothermic; many thermoregulate behaviorally
7. **Metanephric kidney (paired); uric acid main nitrogenous waste**
8. Nervous system with the optic lobes on the dorsal side of brain; **12 pairs of cranial nerves** in addition to nervus terminalis
9. Sexes separate; **fertilization internal**
10. **Eggs covered with calcareous or leathery shells; extraembryonic membranes (amnion, chorion,** and **allantois)** present during embryonic life; **no larval stages**

Characteristics of Reptiles That Distinguish Them from Amphibians

1. **Reptiles have tough, dry, scaly skin offering protection against desiccation and physical injury.** The skin consists of a thin epidermis, shed periodically, and a much thicker, well-developed **dermis** (Figure 29-3). The dermis is provided with **chromatophores,** the color-bearing cells that give many lizards and snakes their colorful hues. It is also the layer that, unfortunately for their bearers, is converted into alligator and snakeskin leather, so esteemed for expensive pocketbooks and shoes. The characteristic **scales** of reptiles are formed largely of keratin. Scales are derived mostly from the epidermis; they are not homologous to fish scales, which are bony, dermal structures. In some reptiles, such as alligators, the scales remain throughout life, growing gradually to replace wear. In others, such as snakes and lizards, new scales grow beneath the old, which are shed at intervals. Turtles add new layers of keratin under the old layers of the platelike scutes, which are modified scales. In snakes the old skin (epidermis and scales) is turned inside out when discarded; lizards split out of the old skin leaving it mostly intact and right side out, or it may slough off in pieces.
2. **The shelled (amniotic) egg of reptiles contains food and protective membranes for supporting embryonic development on land.** Reptiles lay their eggs in sheltered locations on land. The young hatch as lung-breathing juveniles rather than as aquatic larvae. The appearance of the shelled egg (Figure 29-4)

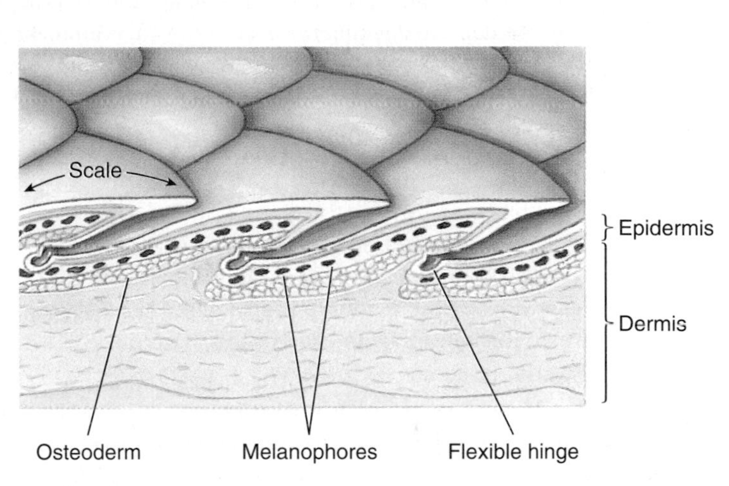

figure 29-3

Section of the skin of a reptile showing the overlapping epidermal scales.

widened the division between the evolving amphibians and reptiles and, probably more than any other adaptation, contributed to the evolutionary establishment of reptiles.

3. **Reptilian jaws are efficiently designed for applying crushing or gripping force to prey.** The jaws of fish and amphibians are designed for quick jaw closure, but once the prey is seized, little static force can be applied. In reptiles jaw muscles became larger, longer, and arranged for much better mechanical advantage.
4. **Reptiles have some form of copulatory organ, permitting internal fertilization.** Internal fertilization is obviously a requirement for a shelled egg, because the sperm must reach the egg before the

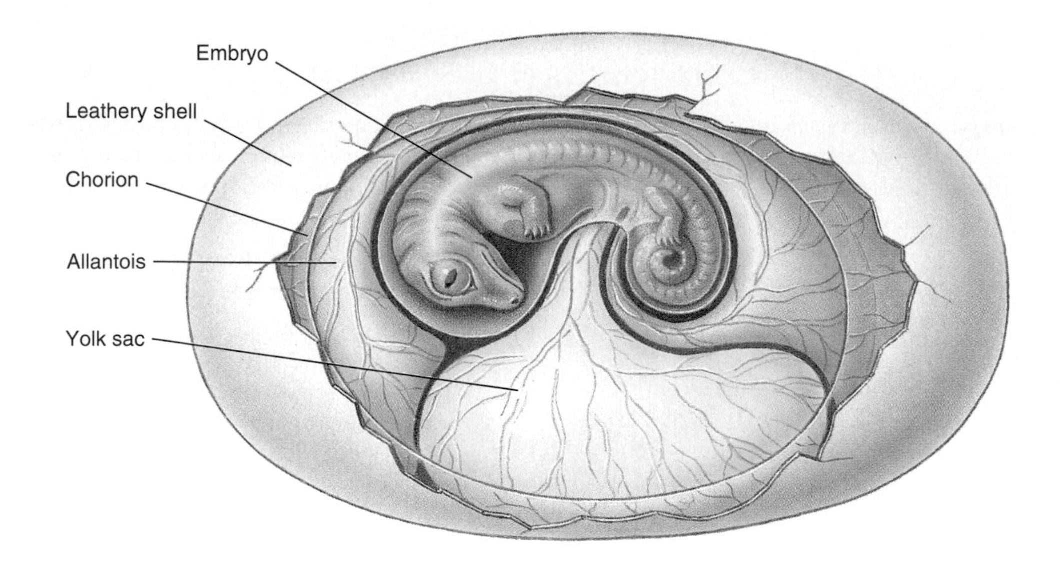

figure 29-4

Amniotic egg. The embryo develops within the amnion and is cushioned by amniotic fluid. Food is provided by yolk from the yolk sac and metabolic wastes are deposited within the allantois. As development proceeds, the allantois fuses with the chorion, a membrane lying against the inner surface of the shell; both membranes are supplied with blood vessels that assist in the exchange of oxygen and carbon dioxide across the porous shell. Because this kind of egg is an enclosed, self-contained system, it is often called a "cleidoic" egg (Gr. *kleidoun,* to lock in).

egg is enclosed. Sperm from the paired testes are carried by the vasa deferentia to the copulatory organ, which is an evagination of the cloacal wall. The female system consists of paired ovaries and oviducts. The glandular walls of the oviducts secrete albumin (source of amino acids, minerals, and water for the embryo) and shells for the large eggs.

5. **Reptiles have a more efficient circulatory system and higher blood pressure than amphibians.** In all reptiles the right atrium, which receives unoxygenated blood from the body, is completely partitioned from the left atrium, which receives oxygenated blood from the lungs. Crocodilians have two completely separated ventricles as well (Figure 29-5); in other reptiles the ventricle is incompletely separated. Even in reptiles with incomplete separation of the ventricles, flow patterns within the heart prevent admixture of pulmonary (oxygenated) and systemic (unoxygenated) blood; all reptiles therefore have two functionally separate circulations.
6. **Reptilian lungs are better developed than those of amphibians.** Reptiles depend almost exclusively on lungs for gas exchange, supplemented by respiration through the pharyngeal membranes in some aquatic turtles. Unlike amphibians, which *force* air into the lungs with mouth muscles, reptiles *suck* air into the lungs by enlarging the pleural cavity, either by expanding the rib cage (snakes and lizards) or by movement of internal organs (turtles and crocodilians). Reptiles have no muscular diaphragm, a structure found only in mammals. Cutaneous respiration (gas exchange across the skin), so important to amphibians, has been completely abandoned by reptiles.
7. **Reptiles have evolved efficient strategies for water conservation.** All amniotes have a metanephric kidney which is drained by its own passageway, the ureter. However, the nephrons of the reptilian metanephros lack the specialized intermediate section of the tubule, the loop of Henle (p. 177–179), that enables the kidney to concentrate solutes in the urine. To remove salts from the blood, many reptiles have salt glands located near the nose or eyes (in the tongue of saltwater crocodiles) which secrete a salty fluid that is strongly hyperosmotic to the body fluids. Nitrogenous wastes are excreted by the kidney as uric acid, rather than urea or ammonia. Uric acid has a low solubility and precipitates out of solution readily, allowing water to be conserved; the urine of many reptiles is a semisolid suspension.
8. **All reptiles, except the limbless members, have better body support than the amphibians and more efficiently designed limbs for travel on land.** Nevertheless, most modern reptiles walk with their legs splayed outward and their belly close to the ground. Most dinosaurs, however, (and some modern lizards) walked on upright legs held beneath the body, the best arrangement for rapid movement and for the support of body weight. Many dinosaurs walked on powerful hindlimbs alone.
9. **The reptilian nervous system is considerably more complex than the amphibian system.** Although the reptile's brain is small, the cerebrum is

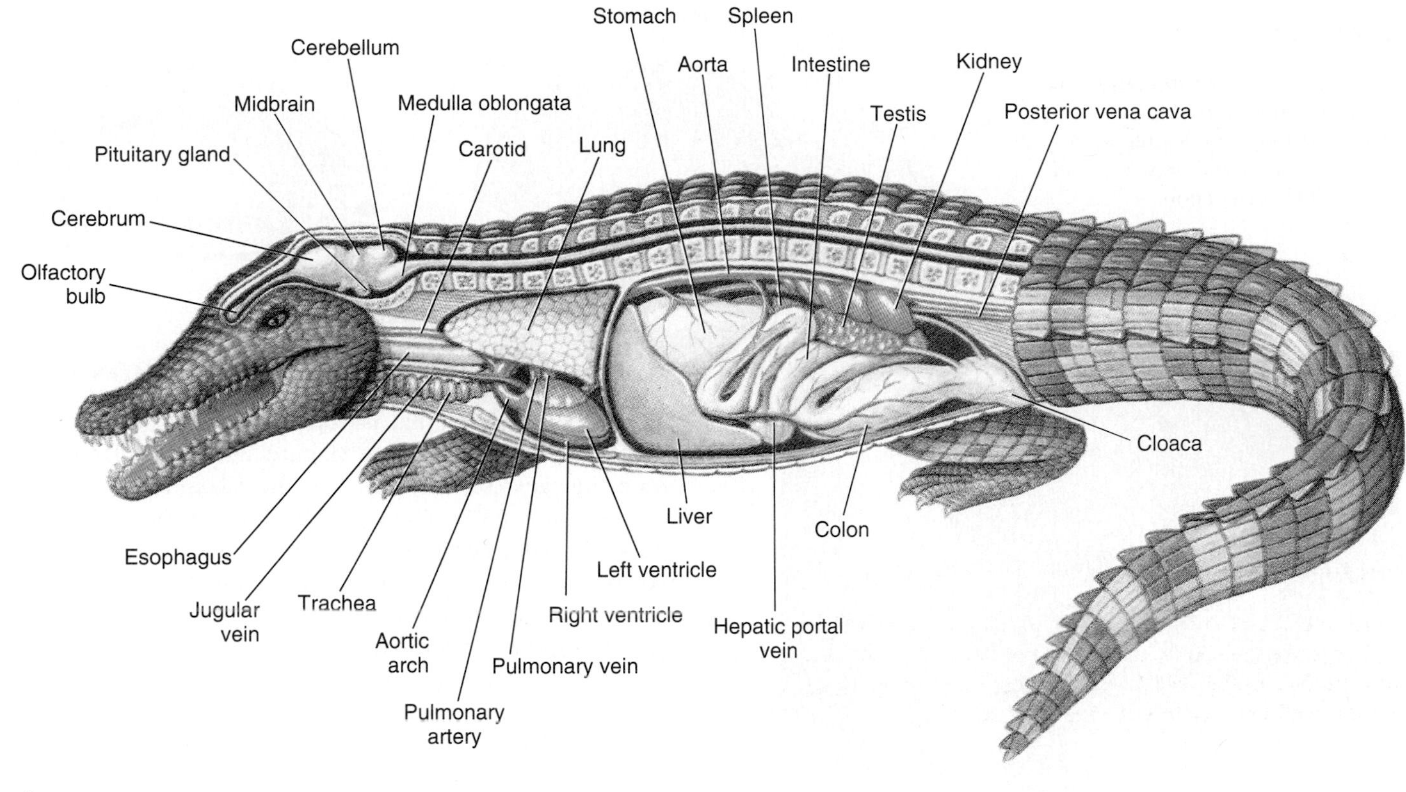

figure 29-5

Internal structure of a male crocodile.

larger relative to the rest of the brain. Connections to the central nervous system are more advanced, permitting complex kinds of behavior unknown in amphibians. With the exception of hearing, sense organs in general are well developed. Jacobson's organ, a specialized olfactory chamber present in many tetrapods, is highly developed in lizards and snakes. Odors are carried to Jacobson's organ by the tongue.

Characteristics and Natural History of Reptilian Orders

Anapsid Reptiles: Subclass Anapsida

Turtles: Order Testudines

Turtles descended from one of the earliest anapsid lineages, probably a group known as the procolophonids of the late Permian, but turtles themselves do not appear in the fossil record until the Upper Triassic, some 200 million years ago. From the Triassic, turtles plodded on to the present with very little change in their early morphology. They are enclosed in shells consisting of a dorsal **carapace** (Fr., from Sp. *carapacho,* covering) and ventral **plastron** (Fr., breastplate). Clumsy and unlikely as they appear to be within their protective shells, they are nonetheless a varied and ecologically diverse group that seems able to adjust to human presence. The shell is so much a part of the animal that it is fused to thoracic vertebrae and ribs (Figure 29-6). Like a medieval coat of armor, the shell offers protection for the head and appendages, which, in most turtles, can be retracted into it. But because the ribs are fused to the shell, the turtle cannot expand its chest to breathe. Instead, turtles employ certain abdominal and pectoral muscles as a "diaphragm." Air is drawn inward by contracting limb flank muscles to make the body cavity larger. Exhalation is also active: the shoulder girdle is drawn back into the shell, thus compressing the viscera and forcing air out of the lungs.

The terms "turtle," "tortoise," and "terrapin" are applied variously to different members of the turtle order. In North American usage, they are all correctly called turtles. The term "tortoise" is frequently given to land turtles, especially the large forms. British usage of the terms is different: "tortoise" is the inclusive term, whereas "turtle" is applied only to the aquatic members.

Lacking teeth, the turtle jaw is provided with tough, horny plates for gripping food (Figure 29-7). Sound perception

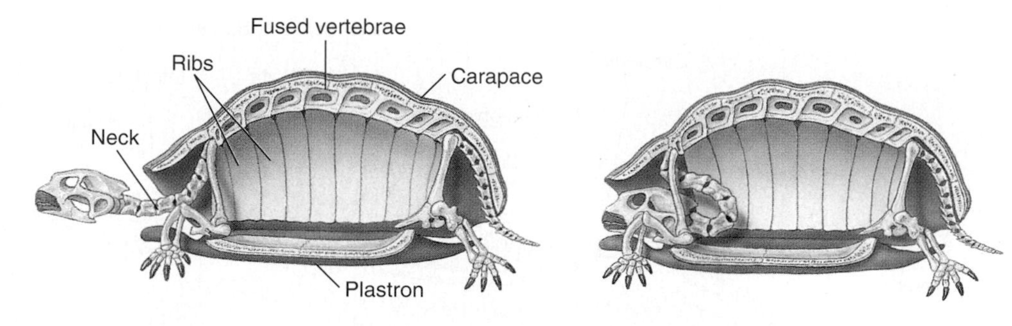

figure 29-6

Skeleton and shell of a turtle, showing fusion of vertebrae and ribs with the carapace. The long and flexible neck allows the turtle to withdraw its head into its shell for protection.

figure 29-7

Snapping turtle, *Chelydra serpentina,* showing the absence of teeth. Instead, the jaw edges are covered with a horny plate.

is poor in turtles, and most turtles are mute (the biblical "voice of the turtle" refers to the turtledove, a bird). Compensating for poor hearing is a good sense of smell and color vision. Turtles are oviparous, and fertilization is internal. All turtles, even the marine forms, bury their shelled, amniotic eggs in the ground. An odd feature of turtle reproduction is that in some turtle families, as in all crocodilians and some lizards, the nest temperature determines the sex of the hatchlings. In turtles, low temperatures during incubation produce males and high temperatures produce females.

The great marine turtles, buoyed by their aquatic environment, may reach 2 m in length and 725 kg in weight. One is the leatherback. The green turtle, so named because of its greenish body fat, may exceed 360 kg, although most individuals of this economically valuable and heavily exploited species seldom live long enough to reach anything approaching this size. Some land tortoises may weigh several hundred kilograms, such as the giant tortoises of the Galápagos Islands (Figure 29-8) that so intrigued Darwin during his visit there in 1835. Most tortoises are rather slow moving; one hour of determined trudging carries a large Galápagos tortoise approximately 300 m. A low metabolism probably explains in part the longevity of turtles, for some are believed to live more than 150 years.

Diapsid Reptiles: Subclass Diapsida

The diapsid reptiles, that is, reptiles having a skull with two pairs of temporal openings (Figure 29-2), are classified into three lineages (superorders; see the Classification of Living Reptiles on p. 637). The two with living representatives are the superorder Lepidosauria, containing the lizards, snakes, worm lizards (a small group not considered in our treatment), and *Sphenodon;* and the superorder Archosauria, containing the crocodilians.

Lizards, Snakes, and Worm Lizards: Order Squamata

The squamates are the most recent and diverse products of diapsid evolution, making up approximately 95% of all known living reptiles. Lizards appeared in the fossil record as early as the Permian, but they did not begin their radiation until the Cretaceous period of the Mesozoic when the dinosaurs were at the climax of their radiation. Snakes appeared during the late Cretaceous period, probably from a group of lizards whose descendants include the Gila monster and monitor lizards. Two specializations in particular characterize snakes: extreme elongation of the body and accompanying displacement and rearrangement of internal organs; and specializations for eating large prey.

The diapsid skulls of the squamates are modified from the ancestral diapsid condition by the loss of dermal bone ventral and posterior to the lower temporal opening. This modification has allowed the evolution in most lizards of a **kinetic skull** having movable joints (Figure 29-9). The quadrate, which in other reptiles is fused to the skull, has a joint at its dorsal end, as well as its usual articulation with the lower jaw. In addition, there are joints in the palate and across the roof of the skull that allow the snout to be tilted upward. The specialized mobility of the skull enables lizards to seize and manipulate their prey. It also increases the effective closing force of the jaw musculature. The skull of snakes is even more kinetic than that of lizards. Such exceptional skull mobility is considered a major factor in the diversification of lizards and snakes.

Lizards: Suborder Sauria Lizards are an extremely diverse group, including terrestrial, burrowing, aquatic, arboreal and

figure 29-8

Mating Galápagos tortoises. The male has a concave plastron that fits over the highly convex carapace of the female, helping to provide stability during mating. Males utter a roaring sound during mating, the only time they are known to emit vocalizations.

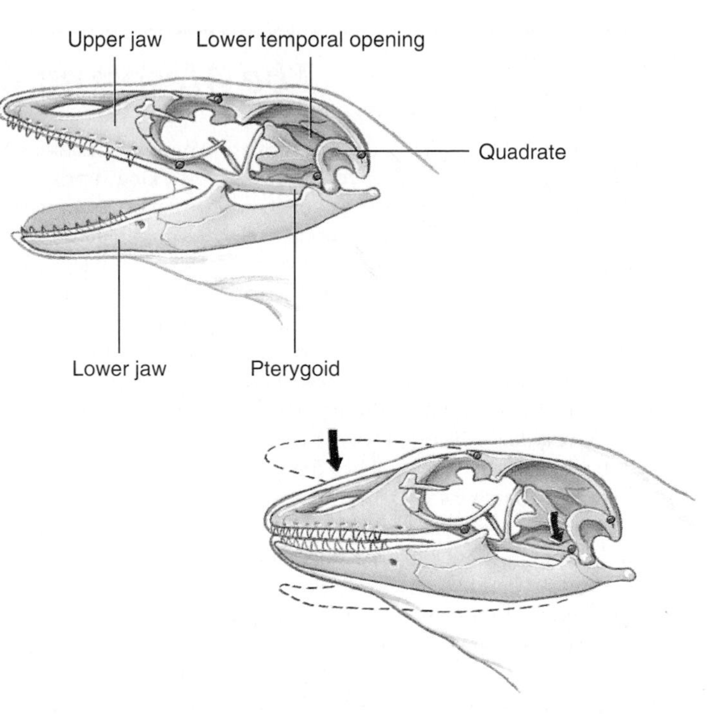

figure 29-9

Kinetic diapsid skull of a modern lizard (monitor lizard, *Varanus* sp.) showing the joints (indicated by dots) that allow the snout and upper jaw to move on the rest of the skull. The quadrate can move at its dorsal end and ventrally at both the lower jaw and the pterygoid. The front part of the braincase is also flexible, allowing the snout to be raised. Note that the lower temporal opening is very large with no lower border; this modification of the diapsid condition, common in modern lizards, provides space for expansion of large jaw muscles. The upper temporal opening lies dorsal and medial to the postorbital-squamosal arch and is not visible in this drawing.

aerial members. Among the more familiar groups in this varied suborder are the **geckos** (Figure 29-10), small, agile, mostly nocturnal forms with adhesive toe pads that enable them to walk upside down and on vertical surfaces; the **iguanas,** often brightly colored New World lizards with ornamental crests, frills, and throat fans, and a group that includes the remarkable marine iguana of the Galápagos Islands (Figure 29-11); **skinks,** with elongate bodies and reduced limbs; and **chameleons,** a group of arboreal lizards, mostly of Africa and Madagascar. The chameleons are entertaining creatures that catch insects with a sticky-tipped tongue that can be flicked accurately and rapidly to a distance greater than the length of their body (Figure 29-12). The great majority of lizards have four limbs and relatively short bodies, but in many the limbs are degenerate, and a few such as the glass lizards (Figure 29-13) are completely limbless.

Unlike turtles, snakes, and crocodilians, which have distinctive body forms and ways of life, lizards have radiated extensively into a variety of habitats and reveal an array of functional and behavioral specializations. Most lizards have movable eyelids, whereas a snake's eyes are permanently covered with a transparent cap. Lizards have keen vision for daylight (retinas rich in both cones and rods; see p. 263 for discussion of color vision), although one group, the nocturnal geckos, has pure rod retinas for night vision. Most lizards have an external ear that snakes lack. However, as with other reptiles, hearing does not play an important role in the lives of most lizards. Geckos are exceptions because the males are strongly vocal (to announce territory and discourage the approach of other males), and they must, of course, hear their own vocalizations.

Many lizards have successfully invaded the world's hot and arid regions, aided by characteristics that make desert life possible. Because their skin lacks glands, water loss by this avenue is much reduced. They produce a semisolid urine with a high content of crystalline uric acid, a feature well suited for conserving water also found in other groups that live successfully in arid habitats (birds, insects, and pulmonate snails). Some, such as the Gila monster of the southwestern United States deserts, store fat in their tails, which they draw on during drought to provide both energy and metabolic water (Figure 29-14). Many lizards keep their body temperature relatively constant by behavioral thermoregulation (described in Chapter 7).

Worm Lizards: Suborder Amphisbaenia The somewhat inappropriate common name "worm lizards" describes a group of highly specialized, burrowing forms that are neither worms nor true lizards but certainly are related to the latter. They

The Mesozoic World of Dinosaurs

When, in 1841, the English anatomist Richard Owen coined the term *dinosaur* ("terrible lizard") to describe fossil Mesozoic reptiles of gigantic size, only three poorly known dinosaur genera were distinguished. But with new and marvelous fossil discoveries quickly following, by 1887 zoologists were able to distinguish two groups of dinosaurs based on differences in the structure of the pelvic girdles. The Saurischia ("lizard-hipped") had a simple, three-pronged pelvis with the hip bones arranged much as they are in other reptiles. The large bladelike ilium is attached to the backbone by stout ribs. The pubis and ischium extend ventrally and posteriorly, respectively, and all three bones meet at the hip socket, a deep opening on the side of the pelvis. The Ornithischia ("bird-hipped") had a somewhat more complex pelvis. The ilium and ischium were arranged similarly in ornithischians and saurischians, but the ornithischian pubis was a narrow, rod-shaped bone with anteriorly and posteriorly directed processes lying alongside the ischium. Oddly, while the ornithischian pelvis, as the name suggests, was similar to that of birds, birds are of the saurischian lineage.

Dinosaurs and their living relatives, the birds, are archosaurs ("ruling lizards"), a group that includes thecodonts (early archosaurs restricted to the Triassic), crocodiles, and pterosaurs (refer to the classification of the reptiles on p. 637). As traditionally recognized, the dinosaurs are a paraphyletic group because they do not include birds which are descended from the most recent common ancestor of dinosaurs.

From among the various archosaurian radiations of the Triassic there emerged a thecodont lineage with limbs drawn under the body to provide an upright posture. This lineage gave rise to the earliest dinosaurs of the Late Triassic. In *Herrerasaurus,* a bipedal dinosaur from Argentina, we see one of the most distinctive characteristics of dinosaurs: walking upright on pillar-like legs, rather than on legs splayed outward as with modern amphibians and reptiles. This arrangement allowed the legs to support the great weight of the body while providing an efficient and rapid stride.

Although their ancestry is unclear, two groups of saurischian dinosaurs have been proposed based on differences in feeding habits and locomotion: the carnivorous and bipedal theropods, and the herbivorous and quadrupedal sauropods (sauropodomorphs). *Coelophysis* was an early theropod with a body form typical of all theropods: powerful hindlegs with three-toed feet; long, heavy counterbalancing tail; slender, grasping forelimbs; flexible neck; and a large head with jaws armed with dagger-like teeth. Large predators such as *Allosaurus,* common during the Jurassic, were replaced by even more massively built carnivores of the Cretaceous, such as *Tyrannosaurus,* which reached a length of 14.5 m (47 ft), stood nearly 6 m high, and weighed more than 7200 kg (8 tons). Not all predatory saurischians were massive; several were swift and nimble, such as *Velociraptor* ("speedy predator") of the Upper Cretaceous.

Herbivorous saurischians, the quadrupedal sauropods, appeared in the Late Triassic. Although early sauropods were small- and medium-sized dinosaurs, those of the Jurassic and Cretaceous attained gigantic proportions, the largest terrestrial vertebrates ever to have lived. *Brachiosaurus* reached 25 m (82 ft) in length and may have weighed in excess of 30,000 kg (33 tons). Even larger sauropods have been discovered; *Supersaurus* was 43 m (140 ft) long. With long necks and long front legs, the sauropods were the first vertebrates adapted to feed on trees. They reached their greatest diversity in the Jurassic and began to decline in overall abundance and diversity during the Cretaceous.

The second group of dinosaurs, the Ornithischia, were all herbivorous. Although more varied, even grotesque, in appearance than saurischians, the ornithischians are united by several derived skeletal features that indicate common ancestry. The huge back-plated *Stegosaurus* of the Jurassic is a well known example of armored ornithischians which comprised two of the five major groups of ornithischians. Even more shielded with bony plates than the stegosaurs were the heavily built ankylosaurs, "armored tanks" of the dinosaur world. As the Jurassic gave way to the Cretaceous, several groups of unarmored ornithischians appeared, although many bore impressive horns. The steady increase in ornithiscian diversity in the Cretaceous paralleled a concurrent gradual decline in giant sauropods which had flourished in the Jurassic. *Triceratops* is representative of horned dinosaurs that were common in the Upper Cretaceous. Even more prominent in the Upper Cretaceous were the duck-billed dinosaurs (hadrosaurs) which are believed to have lived in large herds. Many hadrosaurs had skulls elaborated with crests that probably functioned as vocal resonators to produce species-specific calls.

Sixty-five million years ago, the last of the Mesozoic dinosaurs became extinct, leaving birds as the only surviving lineage of archosaurs. There is increasingly convincing evidence that the demise of dinosaurs coincided with the impact on earth of a large asteroid that produced devastating worldwide environmental upheaval. We continue to be fascinated by the awe-inspiring, often staggeringly large creatures that dominated the Mesozoic era for 165 million years—an incomprehensibly long period of time. Today, inspired by clues from fossils and footprints from a lost world, scientists continue to piece together the puzzle of how the various dinosaur groups arose, behaved, and diversified.

SAURISCHIANS
ORNITHISCHIANS
65 My ago
CRETACEOUS
136 My ago
JURASSIC
190 My ago
TRIASSIC
275 My ago
Titanosaurus 12m (40ft)
Velociraptor 1.8m (6ft)
Hadrosaur (duck-billed dinosaur) 10m (33ft)
Triceratops 9m (30ft)
Brachiosaurus 25m (82ft)
Allosaurus 11m (35ft)
Stegosaurus 9m (30ft)
Ilium
Ischium
Pubis
Coelophysis 3m (10ft)
Pubis
Ischium
Ilium
Herrerasaurus 4m (13ft)
One of the oldest known dinosaurs. Has characteristics of both saurischians and ornithischians.

figure 29-10

Tokay, *Gekko gecko,* of Southeast Asia has a true voice and is named after the strident repeated *to-kay, to-kay* call.

figure 29-11

A large male marine iguana, *Amblyrhynchus cristatus,* of the Galápagos Islands, feeding underwater on algae. This is the only marine lizard in the world. It has special salt-removing glands in the eye orbits and long claws that enable it to cling to the bottom while feeding on small red and green algae, its principal diet. It may dive to depths exceeding 10 m (33 feet) and remain submerged more than 30 minutes.

figure 29-12

A chameleon snares a dragonfly. After cautiously edging close to its target, the chameleon suddenly lunges forward, anchoring its tail and feet to the branch. A split second later, it launches its sticky-tipped, foot-long tongue to trap the prey. The eyes of this common European chameleon *(Chamaeleo chamaeleon)* are swiveled forward to provide binocular vision and excellent depth perception.

have elongate, cylindrical bodies of nearly uniform diameter, and most lack any trace of external limbs (Figure 29-15). With soft skin divided into numerous rings, and eyes and ears hidden under skin, the amphisbaenians superficially resemble earthworms—a kind of structural convergence that often occurs when two very distantly related groups come to occupy similar habitats. The amphisbaenians have an extensive distribution in South America and tropical Africa.

Snakes: Suborder Serpentes Snakes are entirely limbless and lack both pectoral and pelvic girdles (the latter persists as a vestige in pythons and boas). The numerous vertebrae of snakes, shorter and wider than those of legged vertebrates, permit quick lateral undulations through grass and over rough terrain. The ribs increase rigidity of the vertebral column, providing more resistance to lateral stresses. The elevation of the neural spine gives the numerous muscles more leverage.

figure 29-13

A glass lizard, *Ophisaurus* sp., of the southeastern United States. This legless lizard feels stiff and brittle to the touch and has an extremely long, fragile tail that readily fractures when the animal is struck or seized. Most specimens, such as this one, have only a partly regenerated tip to replace a much longer tail previously lost. Glass lizards can be readily distinguished from snakes by the deep, flexible groove running along each side of the body. They feed on worms, insects, spiders, birds' eggs, and small reptiles.

figure 29-14

Gila monster, *Heloderma suspectum,* of southwestern United States desert regions and the related Mexican bearded lizard are the only venomous lizards known. These brightly colored, clumsy-looking lizards feed principally on birds' eggs, nesting birds, mammals, and insects. Unlike venomous snakes, the Gila monster secretes venom from glands in its lower jaw. The chewing bite is painful to humans but seldom fatal.

In addition to the highly kinetic skull that enables snakes to swallow prey several times their own diameter (Figure 29-16), snakes differ from lizards in having no movable eyelids (snakes' eyes are permanently covered with upper and lower transparent eyelids fused together) and no external ears. Most snakes have relatively poor vision, the tree-living snakes of the tropical forest being a conspicuous exception (Figure 29-17). In fact, some arboreal snakes possess excellent binocular vision, which they use to track prey through the branches where scent trails would be difficult to follow. Snakes are totally deaf, although they are sensitive to low-frequency vibrations conducted through the ground.

Nevertheless, most snakes employ the chemical senses rather than vision or vibration detection to hunt their prey. In addition to the usual olfactory areas in the nose, which are not well developed, snakes have a pair of pit-like **Jacobson's organs** in the roof of the mouth. These organs are lined with an olfactory epithelium and are richly innervated. The **forked tongue,** flicked through the air, picks up scent particles (Figure 29-18); the tongue is then withdrawn and sampled molecules are delivered to Jacobson's organs. Information is transmitted to the brain, where scents are identified.

Snakes of the subfamily Crotalinae within the family Viperidae are called **pit vipers** because they possess special heat-sensitive **pit organs** on their heads, located between the nostrils and the eyes (Figures 29-18, 29-19, and 29-20). All of the best-known North American venomous snakes are pit vipers, such as the several species of rattlesnakes, water moccasins, and copperheads. The pits are supplied with a

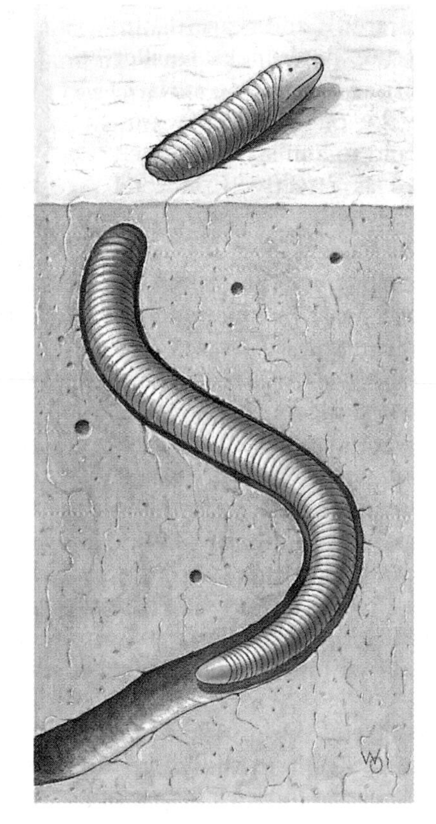

figure 29-15

A worm lizard of the suborder Amphisbaenia. Worm lizards are burrowing forms with a solidly constructed skull used as a digging tool. The species pictured, *Amphisbaena alba,* is widely distributed in South America.

figure 30-1

Archaeopteryx, a 147-million-year-old relative of modern birds. **A,** Cast of the second and most nearly perfect fossil of *Archaeopteryx,* which was discovered in a Bavarian stone quarry. Six specimens of *Archaeopteryx* have been discovered, the most recent one in 1987. **B,** Reconstruction of *Archaeopteryx.*

was named, was an especially fortunate discovery because it proved beyond reasonable doubt the phylogenetic relatedness of birds and reptiles.

The controversy over the dinosaur origin of birds was refueled in late 1996 with announcement of the discovery in China of fossil birds of the Late Jurassic and Early Cretaceous that its discoverers believe were too highly derived to have descended from dinosaurs. The authors assert that Archaeopteryx, *rather than being the ancestor of modern birds, represented a dead-end lineage that became extinct. Modern birds, argue the authors, descended separately from archosaurian ancestors that predated dinosaurs. However, this scenario is strongly rejected by critics who note the uncertain dating of the Chinese discoveries and who point to another fossil discovery from China that, paradoxically, appears to strengthen the dinosaur origin of birds. For the moment,* Archaeopteryx *remains on its perch as a descendant of dinosaurs that gave rise to birds.*

Zoologists had long recognized the similarity of birds and reptiles because of their many shared morphological, developmental, and physiological homologies. The distinguished English zoologist Thomas Henry Huxley was so impressed with these affinities that he called birds "glorified reptiles" and classified them with a group of dinosaurs called theropods that displayed several bird-like characteristics (Figures 30-2 and 30-3). Theropod dinosaurs share many derived characters with birds, the most obvious of which is the elongate, mobile, S-shaped neck. As shown in the cladogram (Figure 30-3), theropods belong to a lineage of diapsid reptiles, the archosaurians, that includes crocodilians and pterosaurs, as well as the dinosaurs. There is now overwhelming evidence that Huxley was correct: birds' closest phylogenetic affinity is to the theropod dinosaurs. The only anatomical feature required to link bird ancestry with the theropod dinosaurs was feathers, and this was provided by the discovery of *Archaeopteryx.* However, recent fossil discoveries have complicated the picture of bird origins and renewed the debate over which amniote lineage was ancestral to birds.

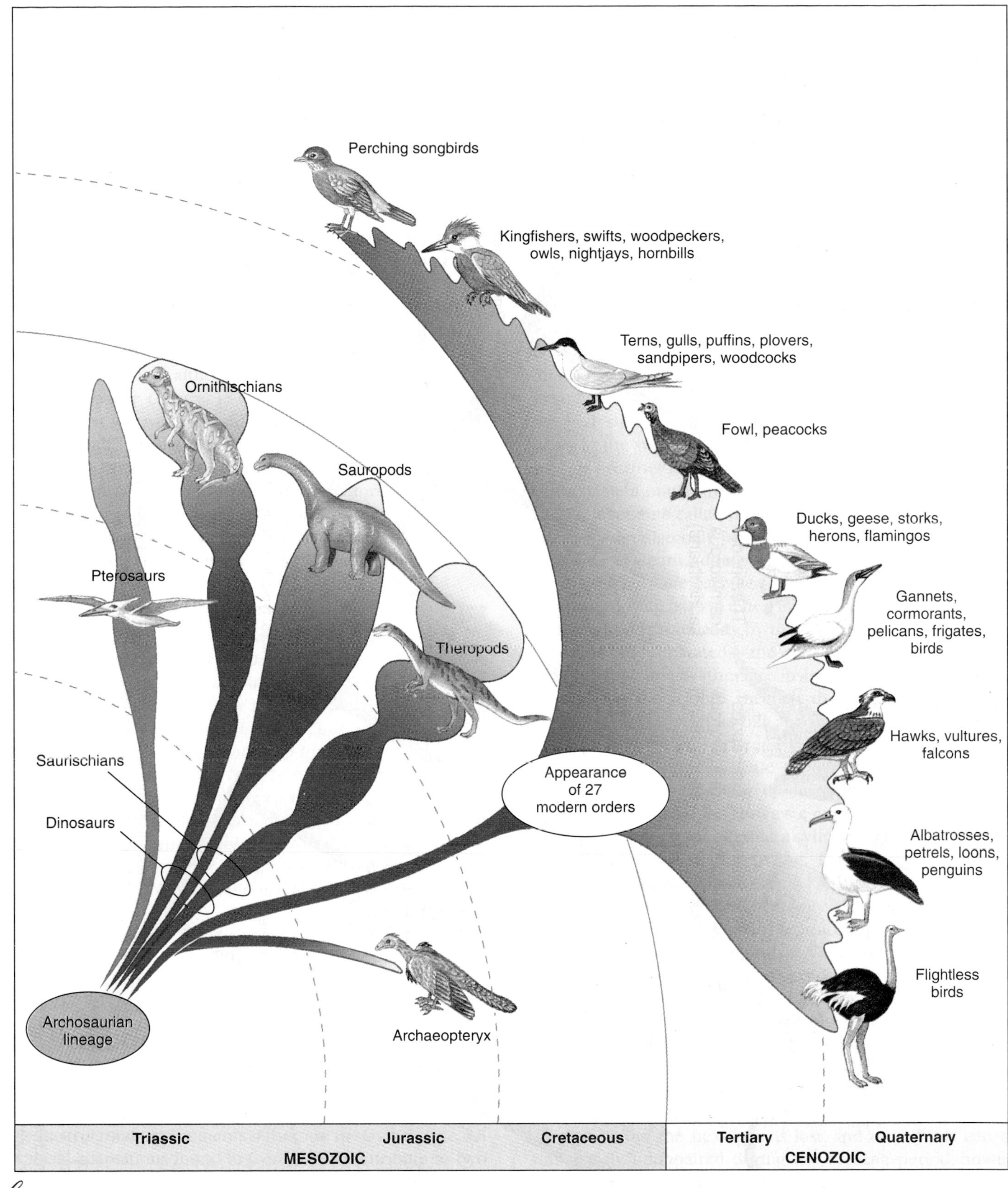

figure 30-2

Evolution of modern birds. Of 27 living bird orders, 9 of the largest are shown. The earliest known bird, *Archaeopteryx,* lived in the Upper Jurassic, about 147 million years ago. *Archaeopteryx* uniquely shares many specialized aspects of its skeleton with the smaller theropod dinosaurs and is considered to have evolved within the theropod lineage. Evolution of modern bird orders occurred rapidly during the Cretaceous and early Tertiary periods.

figure 30-4

Contour feather. Inset enlargement of the vane shows the minute hooks on the barbules that cross-link loosely to form a continuous surface of vane.

Skeleton

One of the major structural requirements for flight is a light, yet sturdy, skeleton (Figure 30-5A). As compared with the earliest known bird, *Archaeopteryx* (Figure 30-5B), the bones of modern birds are phenomenally light, delicate, and laced with air cavities. Such **pneumatized** bones (Figure 30-6) are nevertheless strong. The skeleton of a frigate bird with a 2.1 m (7-foot) wingspan weighs only 114 grams (4 ounces), less than the weight of all its feathers.

As archosaurs, birds evolved from ancestors with diapsid skulls (p. 622). However, the skulls of modern birds are so specialized that it is difficult to see any trace of the original diapsid condition. The bird skull is built lightly and mostly fused into one piece. A pigeon skull weighs only 0.21% of its body weight; by comparison the skull of a rat weighs 1.25% of its body weight. The braincase and orbits are large to accommodate a bulging brain and large eyes needed for quick motor coordination and superior vision.

In *Archaeopteryx,* both jaws contained teeth set in sockets, an archosaurian characteristic. Modern birds are completely toothless, having instead a horny (keratinous) beak molded around the bony jaws. The mandible is a complex of several bones hinged to provide a double-jointed action which permits the mouth to gape widely. Most birds have kinetic skulls (kinetic skulls of lizards are described on p. 628) with a flexible attachment between upper jaw and skull. This attachment allows the upper jaw to move slightly, thus increasing the gape.

The most distinctive feature of the vertebral column is its rigidity. Most of the vertebrae except the **cervicals** (neck vertebrae) are fused together and with the pelvic girdle to form a stiff but light framework to support the legs and provide rigidity for flight. To assist in this rigidity, the ribs are mostly fused with the vertebrae, pectoral girdle, and sternum. Except in flightless birds, the sternum bears a large, thin keel that provides for attachment of the powerful flight muscles. Because *Archaeopteryx* had no sternum (Figure 30-5B), there was no anchorage for the flight muscles equivalent to that of modern birds. This is one of the principal reasons why *Archaeopteryx* could not have done any strenuous wing-beating. *Archaeopteryx* did, however, have a furcula (wishbone) on which enough pectoral muscle could have attached to permit weak flight.

The bones of the forelimbs are highly modified for flight. They are reduced in number, and several are fused together. Despite these alterations, the bird wing is clearly a rearrangement of the basic vertebrate tetrapod limb from which it arose (Figure 28-1, p. 608), and all the elements—upper arm, forearm, wrist, and fingers—are represented in modified form (Figure 30-5A). The birds' legs have undergone less pronounced modification than the wings, since they are still designed principally for walking, as well as for perching, scratching, food gathering, and occasionally for swimming, as were those of their archosaurian ancestors.

Muscular System

The locomotor muscles of the wings are relatively massive to meet the demands of flight. The largest of these is the **pectoralis,** which depresses the wings in flight. Its antagonist is the **supracoracoideus** muscle, which raises the wing (Figure 30-7). Surprisingly, perhaps, this latter muscle is not located on the backbone (anyone who has been served the back of a chicken knows that it offers little meat) but is positioned under the pectoralis on the breast. It is attached by a tendon to the upper side of the humerus of the wing so that it pulls from below by an ingenious "rope-and-pulley" arrangement. Both pectoralis and supracoracoideus are anchored to the keel. With the main muscle mass low in the body, aerodynamic stability is improved.

The main leg muscle mass is located in the thigh, surrounding the femur, and a smaller mass lies over the tibiotarsus (shank or "drumstick"). Thin but strong tendons extend

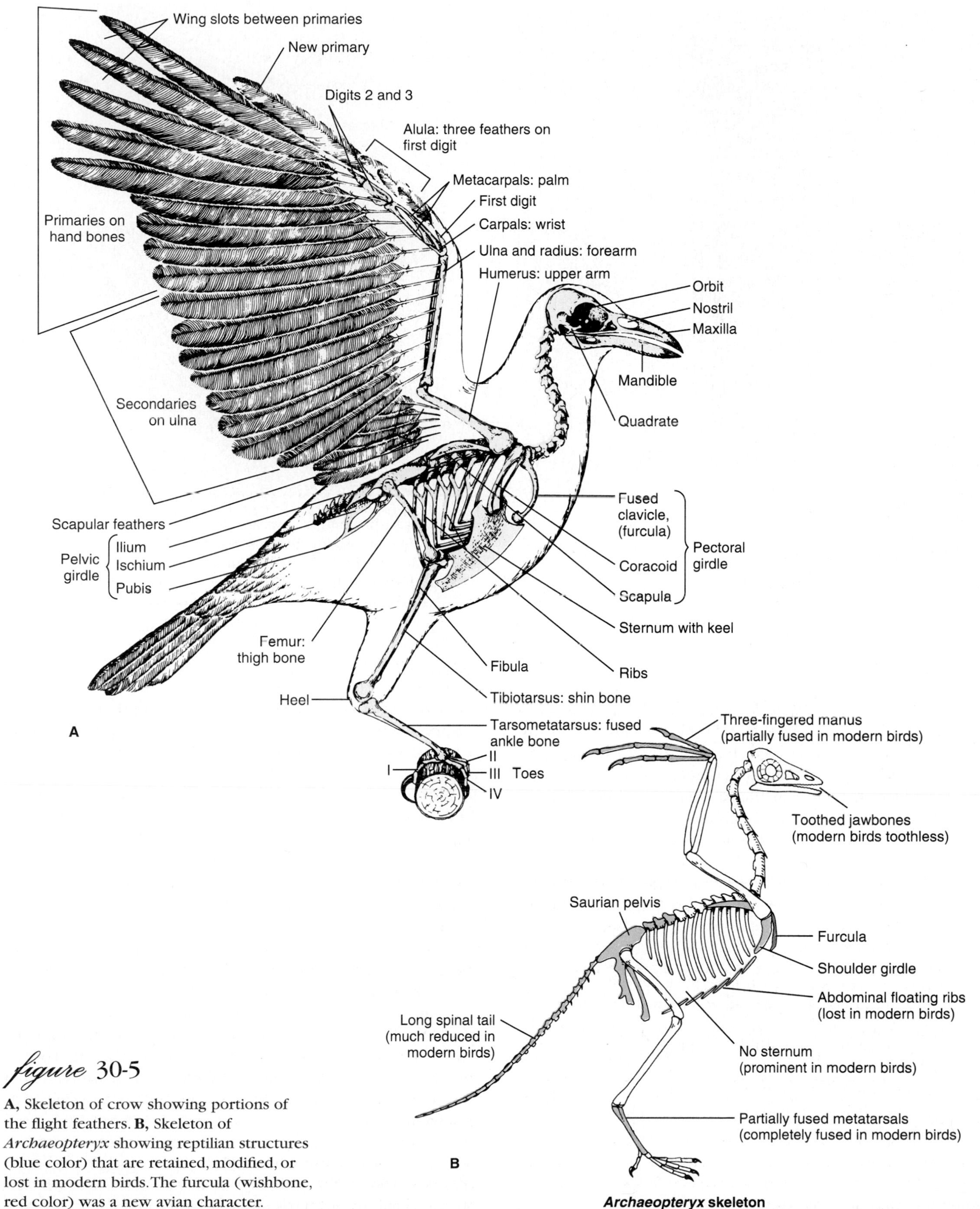

figure 30-5

A, Skeleton of crow showing portions of the flight feathers. **B,** Skeleton of *Archaeopteryx* showing reptilian structures (blue color) that are retained, modified, or lost in modern birds. The furcula (wishbone, red color) was a new avian character.

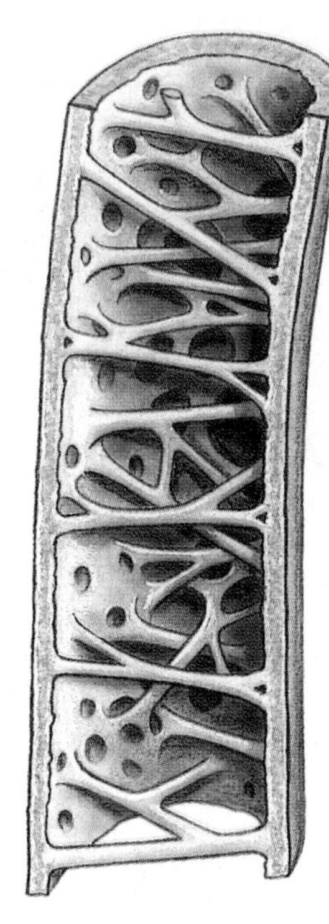

figure 30-6

Hollow wing bone of a songbird showing the stiffening struts and air spaces that replace bone marrow. Such "pneumatized" bones are remarkably light and strong.

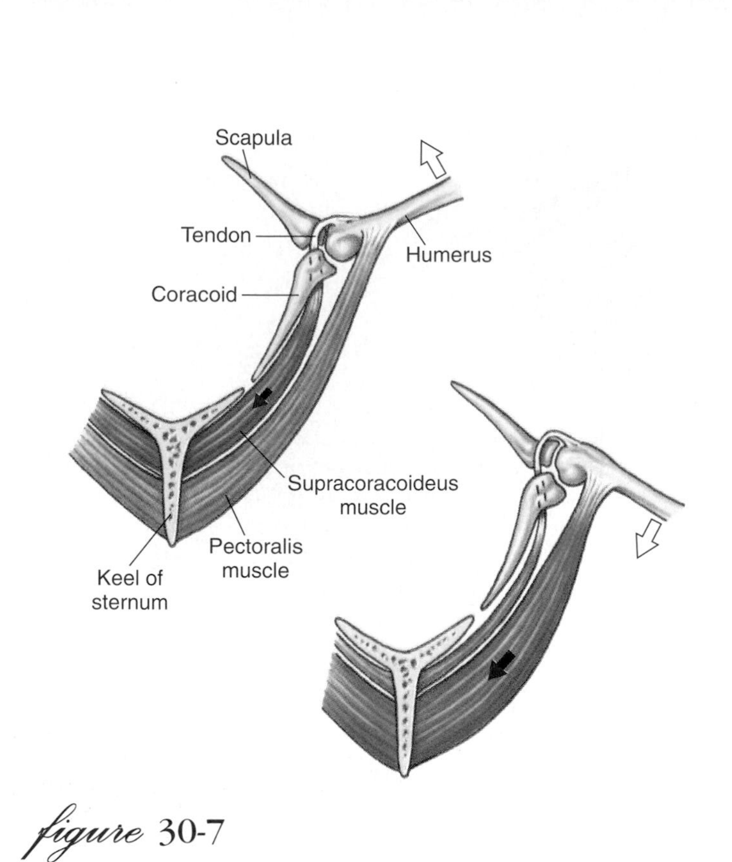

figure 30-7

Flight muscles of a bird are arranged to keep the center of gravity low in the body. Both major flight muscles are anchored on the sternum keel. Contraction of the pectoralis muscle pulls the wing downward. Then, as the pectoralis relaxes, the supracoracoideus muscle contracts and, acting as a pulley system, pulls the wing upward.

downward through sleevelike sheaths to the toes. Consequently the feet are nearly devoid of muscles, explaining the thin, delicate appearance of the bird leg. This arrangement places the main muscle mass near the bird's center of gravity and at the same time allows great agility to the slender, lightweight feet. Because the feet are composed mostly of bone, tendon, and tough, scaly skin, they are highly resistant to damage from freezing. When a bird perches on a branch, an ingenious toe-locking mechanism (Figure 30-8) is activated, which prevents the bird from falling off its perch when asleep. The same mechanism causes the talons of a hawk or owl automatically to sink deeply into its prey as the legs bend under the impact of the strike. The powerful grip of a bird of prey was described by L. Brown.[1]

> When an eagle grips in earnest, one's hand becomes numb, and it is quite impossible to tear it free, or to loosen the grip of the eagle's toes with the other hand. One just has to wait until the bird relents, and while waiting one has ample time to realize that an animal such as a rabbit would be quickly paralyzed, unable to draw breath, and perhaps pierced through and through by the talons in such a clutch.

[1]Brown, L. 1970, Eagles, New York, Arco Publishing.

Digestive System

Birds process an energy-rich diet rapidly and thoroughly with efficient digestive equipment. A shrike can digest a mouse in 3 hours, and berries pass completely through the digestive tract of a thrush in just 30 minutes. Although many animal foods find their way into the diet of birds, insects comprise by far the largest component. Because birds lack teeth, foods that require grinding are reduced in the gizzard (see below). The poorly developed salivary glands mainly secrete mucus for lubricating the food and the slender, horn-covered **tongue.** There are few taste buds, although all birds can taste to some extent. From the short **pharynx** a relatively long, muscular, elastic **esophagus** extends to the **stomach.** In many birds there is an enlargement **(crop)** at the lower end of the esophagus, which serves as a storage chamber.

In pigeons, doves, and some parrots the crop not only stores food but, during the nesting season, produces "milk" by the breakdown of epithelial cells of the crop lining. For the first few days after hatching, the helpless young are fed regurgitated crop milk by both parents. Crop milk is especially rich in fat and protein.

The stomach proper consists of a **proventriculus,** which secretes gastric juice, and the muscular **gizzard,** a region specialized for grinding food. To assist the grinding process, grain-eating birds swallow gritty objects or pebbles,

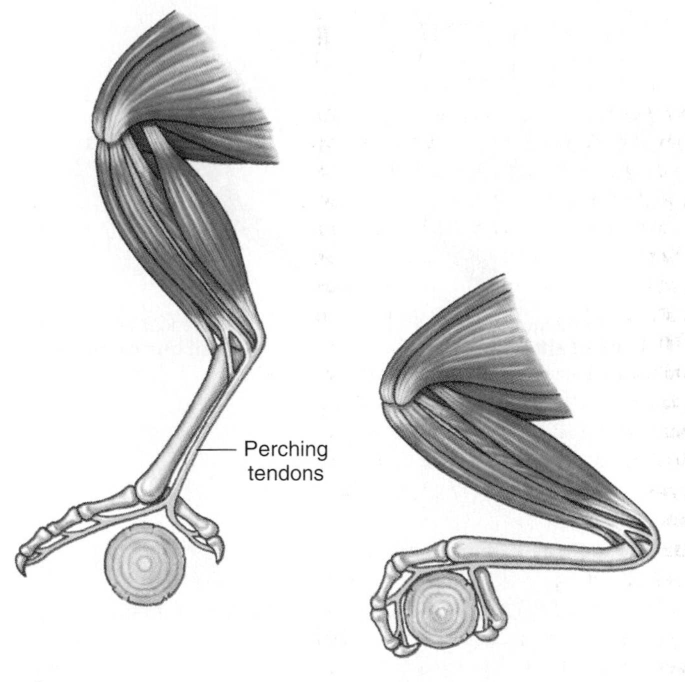

figure 30-8

Perching mechanism of a bird. When a bird settles on a branch, tendons automatically tighten, closing the toes around the perch.

which lodge in the gizzard. Certain birds of prey, such as owls, form pellets of indigestible materials, mainly bones and fur, in the proventriculus and eject them through the mouth. At the junction of the intestine with the rectum there are paired **ceca;** these are well developed in herbivorous birds in which they serve as fermentation chambers. The terminal part of the digestive system is the **cloaca,** which also receives the genital ducts and ureters.

The beaks of birds are strongly adapted to specialized food habits—from generalized types, such as the strong, pointed beaks of crows and ravens, to grotesque, highly specialized ones in flamingos, pelicans, and avocets (Figure 30-9). The beak of a woodpecker is a straight, hard chisel-like device. Anchored to a tree trunk with its tail serving as a brace, the woodpecker delivers powerful, rapid blows to excavate nest cavities or expose the burrows of wood-boring insects. It then uses its long, flexible, barbed tongue to seek insects in their galleries. The woodpecker's skull is especially thick to absorb shock.

Circulatory System

The general plan of circulation in birds is not greatly different from that of mammals. The four-chambered heart is large with strong ventricular walls; thus birds share with mammals a complete separation of the respiratory and systemic circulations. The heartbeat is extremely fast, and as in mammals there is an inverse relationship between heart rate and body weight. For example, a turkey has a heart rate at rest of approximately 93 beats per minute, a chicken has a rate of 250 beats per minute, and a black-capped chickadee has a heart rate of 500 beats per minute when asleep, which may increase to a phenomenal 1000 beats per minute during exercise. Blood pressure in birds is roughly equivalent to that in mammals of similar size. Birds' blood contains **nucleated, biconvex erythrocytes.** (Mammals, the only other endothermic vertebrates, have biconcave erythrocytes without nuclei that are somewhat smaller than those of birds.) The **phagocytes,** or mobile ameboid cells of the blood, are particularly efficient in birds in the repair of wounds and in destroying microbes.

Respiratory System

The respiratory system of birds differs radically from the lungs of reptiles and is marvelously adapted for meeting the high metabolic demands of flight. In birds the finest branches of the bronchi, rather than ending in saclike alveoli as in mammals, are tubelike **parabronchi** through which air flows continuously. Also unique is the extensive system of nine interconnecting **air sacs** that are located in pairs in the thorax and abdomen and even extend by tiny tubes into the centers of the long bones (Figure 30-10A). The air sacs are connected to the lungs in such a way that perhaps 75% of the inspired air bypasses the lungs and flows directly into the air sacs, which serve as reservoirs for fresh air. On expiration, some of this fully oxygenated air is shunted through the lung, while the used air passes directly out (Figure 30-10B). The advantage of such a system is obvious: the lungs receive fresh air during both inspiration and expiration. An almost continuous stream of oxygenated air passes through a system of richly vascularized parabronchi. Although many details of a bird's respiratory system are not fully understood, it is clearly the most efficient respiratory system of any vertebrate.

The remarkable efficiency of the bird respiratory system is emphasized by bar-headed geese that routinely migrate over the Himalayan mountains and have been sighted flying over Mt. Everest (8848 meters or 29,141 feet) under conditions that are severly hypoxic to humans. They reach altitudes of 9000 meters in less than a day, without the acclimatization that is absolutely essential for humans even to approach the upper reaches of Mt. Everest.

Excretory System

The relatively large paired metanephric kidneys are composed of many thousands of **nephrons,** each consisting of a renal corpuscle and a nephric tubule. As in other vertebrates, urine

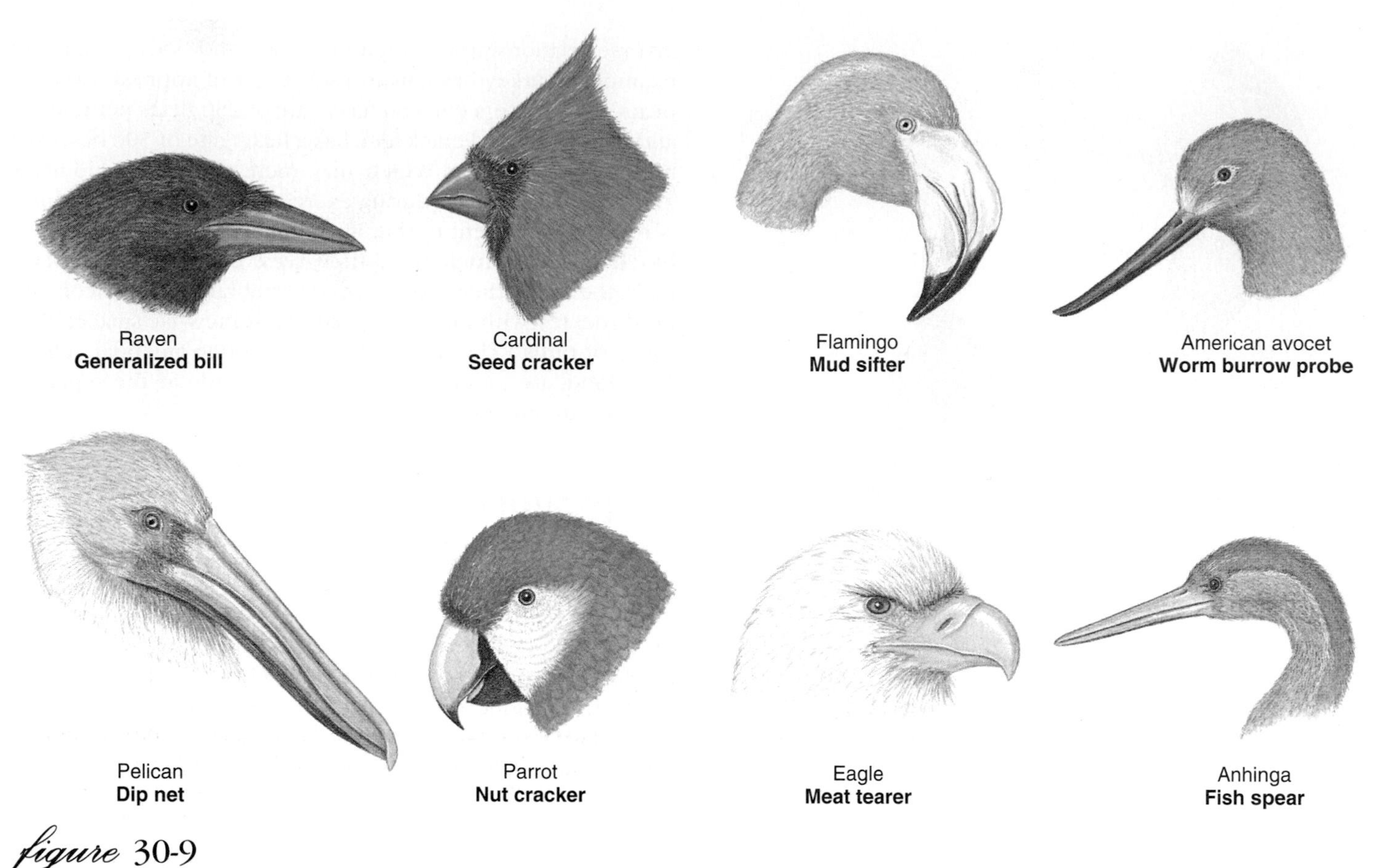

figure 30-9

Some bills of birds showing variety of adaptations.

is formed by glomerular filtration followed by the selective modification of the filtrate in the tubule (the details of this sequence are found on pp. 175–178).

Birds, like reptiles, excrete their nitrogenous wastes as uric acid rather than urea, an adaptation that originated with the evolution of the shelled (amniotic) egg. In shelled eggs, all excretory products must remain within the eggshell with the growing embryo. If urea were produced, it would quickly accumulate in solution to toxic levels. Uric acid, however, crystallizes out of solution and can be stored harmlessly within the egg shell. Thus, from an embryonic necessity was born an adult virtue. Because of uric acid's low solubility, a bird can excrete 1 g of uric acid in only 1.5 to 3 ml of water, whereas a mammal may require 60 ml of water to excrete 1 g of urea. Uric acid is combined with fecal material in the cloaca. Excess water is reabsorbed in the cloaca, resulting in the formation of a white paste. Thus, despite having kidneys that are less effective in true concentrative ability than mammalian kidneys, birds can form urine containing uric acid nearly 3000 times more concentrated than that in the blood. Even the most effective mammalian kidneys, those of certain desert rodents, can excrete urea only about 25 times the plasma concentration.

Marine birds (also marine turtles) have evolved a unique method for excreting the large loads of salt eaten with their food and in the seawater they drink. Seawater contains approximately 3% salt and is three times saltier than a bird's body fluids. Because the bird kidney cannot concentrate salt in urine above approximately 0.3%, excess salt is removed from the blood by special **salt glands,** one located above each eye (Figure 30-11) .These glands are capable of excreting a highly concentrated solution of sodium chloride—up to twice the concentration of seawater. The salt solution runs out the internal or external nostrils, giving gulls, petrels, and other sea birds a perpetual runny nose.

Nervous and Sensory System

The design of a bird's nervous and sensory system reflects the complex problems of flight and a highly visible existence, in which it must gather food, mate, defend territory, incubate and rear young, and correctly distinguish friend from foe. The brain of a bird has well-developed **cerebral hemispheres, cerebellum,** and **midbrain tectum** (optic lobes). The **cerebral cortex**—chief coordinating center of the mammalian

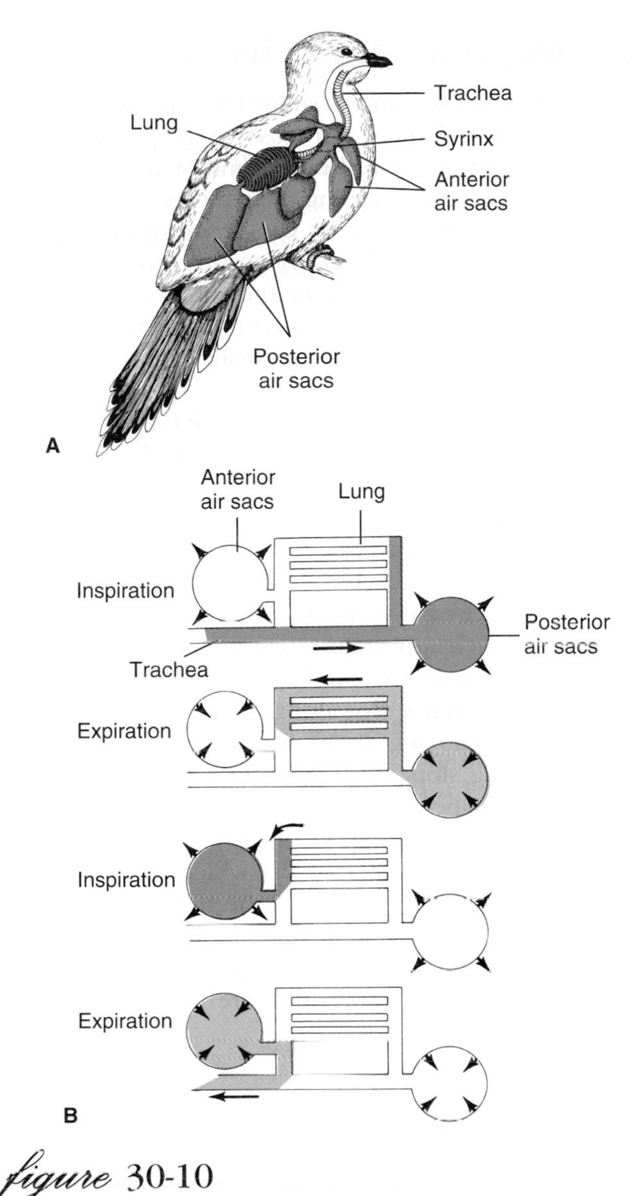

figure 30-10

Respiratory system of a bird. **A,** Lungs and air sacs. One side of the bilateral air sac system is shown. **B,** Movement of a single volume of air through the bird's respiratory system. Two full respiratory cycles are required to move the air through the system.

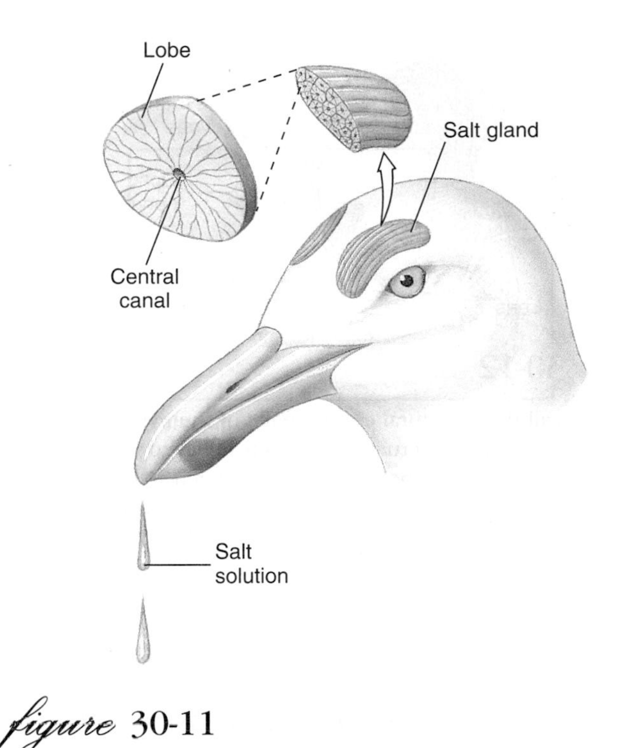

figure 30-11

Salt glands of a marine bird (gull). One salt gland is located above each eye. Each gland consists of several lobes arranged in parallel. One lobe is shown in cross section, much enlarged. Salt is secreted into many radially arranged tubules, then flows into a central canal that leads into the nose.

brain—is thin, unfissured, and poorly developed in birds. But the core of the cerebrum, the **corpus striatum,** has enlarged into the principal integrative center of the brain, controlling such activities as eating, singing, flying, and all complex instinctive reproductive activities. Relatively intelligent birds, such as crows and parrots, have larger cerebral hemispheres than do less intelligent birds, such as chickens and pigeons. The **cerebellum** is a crucial coordinating center where muscle-position sense, equilibrium sense, and visual cues are assembled and used to coordinate movement and balance. The **optic lobes,** laterally bulging structures of the midbrain, form a visual apparatus comparable to the visual cortex of mammals.

Except in flightless birds, ducks, and vultures, the senses of smell and taste are poorly developed in birds. This deficiency, however, is more than compensated by good hearing and superb vision, the keenest in the animal kingdom. The organ of hearing, the **cochlea,** is much shorter than the coiled mammalian cochlea, yet birds can hear roughly the same range of sound frequencies as humans. Actually the bird ear far surpasses our capacity to distinguish differences in intensities and to respond to rapid fluctuations in pitch.

The bird eye resembles that of other vertebrates in gross structure but is relatively larger, less spherical, and almost immobile; instead of turning their eyes, birds turn their heads with their long flexible necks to scan the visual field. The light-sensitive **retina** (Figure 30-12) is generously equipped with rods (for dim light vision) and cones (for color vision). Cones predominate in day birds, and rods are more numerous in nocturnal birds. A distinctive feature of the bird eye is the **pecten,** a highly vascularized organ attached to the retina and jutting into the vitreous humor (Figure 30-12). The pecten is thought to provide nutrients to the eye. It may do more, but its function remains largely a mystery.

figure 30-19

Cooperative feeding behavior by the white pelican, *Pelecanus onocrotalus.* **A,** Pelicans form a horseshoe to drive fish together. **B,** Then they plunge simultaneously to scoop fish in their huge bills. These photographs were taken 2 seconds apart.

figure 30-20

Copulation in birds. In most bird species the male lacks a penis. The male copulates by standing on the back of the female, pressing his cloaca against that of the female, and passing sperm to the female.

Nesting and Care of Young

To produce offspring, all birds lay eggs that must be incubated by one or both parents. The eggs of most songbirds (order Passeriformes) require approximately 14 days for hatching; those of ducks and geese require at least twice that long. Most of the duties of incubation fall on the female, although in many instances both parents share the task, and occasionally only the male incubates the eggs.

Most birds build some form of nest in which to rear their young. Some birds simply lay their eggs on the bare ground or rocks, making no pretense of nest building. Others build elaborate nests such as the pendant nests constructed by orioles, the delicate lichen-covered mud nests of hummingbirds (Figure 30-23) and flycatchers, the chimney-shaped mud nests of cliff swallows, the floating nests of rednecked grebes, and the huge brush-pile nests of Australian brush turkeys. Most birds take considerable pains to conceal their nests from enemies. Woodpeckers, chickadees, bluebirds, and many others place their nests in tree hollows or other cavities; kingfishers excavate tunnels in the banks of streams for their nests; and birds of prey build high in lofty trees or on inaccessible cliffs. Nest parasites such as the brown-headed cowbird and the European cuckoo build no nests at all but simply lay their eggs in the nests of birds smaller than themselves. When the eggs hatch, the foster parents care for the cowbird young which outcompete the host's own hatchlings.

Newly hatched birds are of two types: **precocial** and **altricial.** Precocial young, such as quail, fowl, ducks, and most water birds, are covered with down when hatched and can run or swim as soon as their plumage is dry (Figure 30-24). The altricial ones, on the other hand, are naked and helpless at birth and remain in the nest for a week or more. The young of both types require care from parents for some time after hatching. They must be fed, guarded, and protected against rain and sun. The parents of altricial species must carry food to their young almost constantly, for most young birds will eat more than their weight each day. This enormous food consumption explains the rapid growth of the young and their quick exit from the nest. The food of the young, depending on the species, includes worms, insects, seeds, and fruit.

Nesting success is very low with many birds, especially in altricial species. One investigation several years ago of 170 altricial bird nests reported that only 21% produced at least one young. The annual censusing of birds shows that nesting success is even lower today. Of the many causes of nesting failures, predation by raccoons, skunks, opossums, blue jays, crows, and others, especially in suburban and rural woodlots,

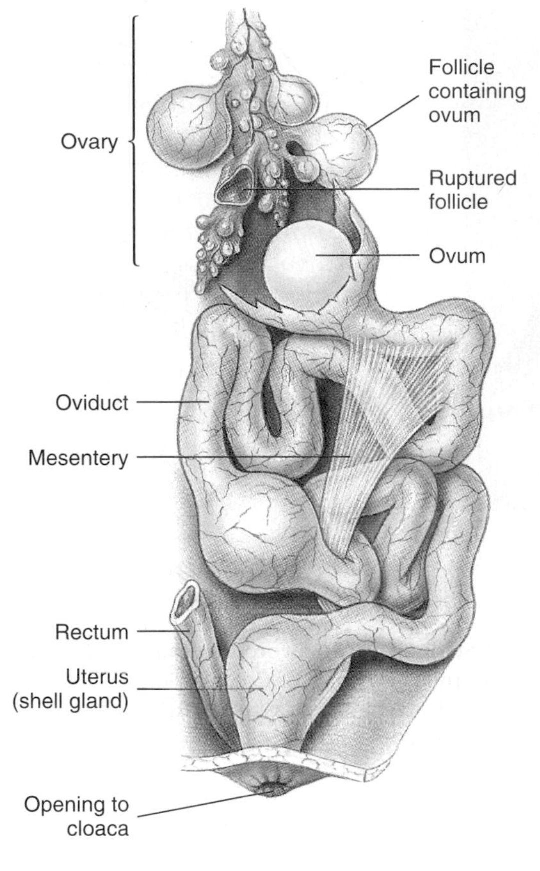

figure 30-21

Reproductive system of a female bird.

figure 30-22

Dominant male sage grouse, *Centrocercus urophasianus,* surrounded by several hens that have been attracted by his "booming" display.

figure 30-23

Anna's hummingbird, *Calypte anna,* feeding its young in its nest of plant down and spider webs and decorated on the outside with lichens. The female builds the nest, incubates the two pea-sized eggs, and rears the young with no assistance from the male. Anna's hummingbird is a common resident of California. It is the only hummer to overwinter in the United States.

and nest parasitism by the brown-headed cowbird are the most important factors.

Bird Populations

Bird populations, like those of other animal groups, vary in size from year to year. Snowy owls, for example, are subject to population cycles that closely follow cycles in their food supply, mainly rodents. Voles, mice, and lemmings in the north have a fairly regular 4-year cycle of abundance (p. 680); at population peaks, predator populations of foxes, weasels, and buzzards, as well as snowy owls, increase because there is abundant food for rearing their young. After a crash in the rodent population, snowy owls move south, seeking alternative food supplies. They occasionally appear in large numbers in southern Canada and the northern United States, where their total absence of fear of humans makes them easy targets for thoughtless hunters.

Occasionally the activities of people bring about spectacular changes in bird distribution. Both starlings (Figure 30-25) and house sparrows have been accidentally or deliberately introduced into numerous countries, to become the two most abundant bird species on earth, with the exception of domestic fowl.

Humans also are responsible for the extinction of many bird species. More than 80 species of birds have, since 1695, followed the last dodo to extinction. Many were victims of changes in their habitat or competition with better-adapted species. But several have been hunted to extinction, among them the passenger pigeon, which only a century ago darkened the skies over North America in incredible numbers estimated in the billions (Figure 30-26).

Summary

The more than 9600 species of living birds are egg-laying, endothermic vertebrates covered with feathers and having the forelimbs modified as wings. Birds are closest phylogenetically to the theropods, a group of Mesozoic dinosaurs with several birdlike characteristics. The oldest known fossil bird, *Archaeopteryx* from the Jurassic period of the Mesozoic era, had numerous reptilian characteristics and was almost identical to certain theropod dinosaurs except that it had feathers. It is probably not in the direct lineage leading to modern birds but can be considered a sister group to modern birds.

The adaptations of birds for flight are of two basic kinds: those reducing body weight and those promoting more power for flight. Feathers, the hallmark of birds, are complex derivatives of reptilian scales and combine lightness with strength, water repellency, and high insulative value. Body weight is further reduced by elimination of some bones, fusion of others (to provide rigidity for flight), and the presence in many bones of hollow, air-filled spaces. The light, horny bill, replacing the heavy jaws and teeth of reptiles, serves as both hand and mouth for all birds and is variously adapted for different feeding habits.

Adaptations that provide power for flight include a high metabolic rate and body temperature coupled with an energy-rich diet; a highly efficient respiratory system consisting of a system of air sacs arranged to pass air through the lungs during both inspiration and expiration; powerful flight and leg muscles arranged to place muscle weight near the bird's center of gravity; and an efficient, high-pressure circulation.

Birds have keen eyesight, good hearing, poorly developed sense of smell, and superb coordination for flight. The metanephric kidneys produce uric acid as the principal nitrogenous waste.

Birds fly by applying the same aerodynamic principles as an airplane and using similar equipment: wings for lift and support, a tail for steering and landing control, and wing slots for control at low flight speed. Flightlessness in birds is unusual but has evolved independently in several bird orders, usually on islands where terrestrial predators are absent; all are derived from flying ancestors.

Bird migration refers to regular movements between summer nesting places and wintering regions. Spring migration to the north, where more food is available for nestlings, enhances reproductive success. Many cues are used for finding direction during migration, including innate sense of direction and ability to navigate by the sun, the stars, or the earth's magnetic field.

The highly developed social behavior of birds is manifested in vivid courtship displays, mate selection, territorial behavior, and incubation of eggs and care of the young.

Review Questions

1. Explain the significance of the discovery of *Archaeopteryx*. Why did this fossil prove beyond reasonable doubt that birds share an ancestor with some reptilian groups?
2. Birds are broadly divided into two groups: ratite and carinate. Explain what these terms mean and briefly discuss the appearance of flightlessness in birds.
3. The special adaptations of birds all contribute to two essentials for flight: more power and less weight. Explain how each of the following contributes to one or both of these two essentials: feathers, skeleton, muscle distribution, digestive system, circulatory system, respiratory system, excretory system, reproductive system.
4. How do marine birds rid themselves of excess salt?
5. In what ways are the bird's ears and eyes specialized for the demands of flight?
6. Explain how the bird wing is designed to provide lift. What design features help to prevent stalling at low flight speeds?
7. Describe the four basic forms of bird wings. How does wing shape correlate with bird size and nature of flight (whether powered or soaring)?
8. What are the advantages of seasonal migration for birds?
9. Describe the different navigational resources birds may use in long-distance migration.
10. What are some of the advantages of social aggregation among birds?
11. More than 90% of all bird species are monogamous. Explain why monogamy is so much more common among birds than among mammals.
12. Briefly describe an example of polygyny among birds.
13. Define the terms precocial and altricial as they relate to birds.
14. Offer some examples of how human activities have affected bird populations.

Selected References

Brooke, M., and T. Birkhead, eds. 1991. The Cambridge encyclopedia of ornithology. New York, Cambridge University Press. *Comprehensive, richly illustrated treatment that includes a survey of all modern bird orders.*

Burton, R. 1985. Bird behavior. New York, Alfred A. Knopf, Inc. *Well-written and well-illustrated summary of bird behavior.*

Elphick, J. ed. 1995. The atlas of bird migration: tracing the great journeys of the world's birds. New York, Random House. *Lavishly illustrated collection of maps of birds' breeding and wintering areas, migration routes, and many facts about each bird's migration journey.*

Emlen, S. T. 1975. The stellar-orientation system of a migratory bird. Sci. Am. **233:**102–111 (Aug.). *Describes fascinating research with indigo buntings, revealing their ability to navigate by the center of celestial rotation at night.*

Feduccia, A. 1996. The origin and evolution of birds. New Haven, Yale University Press. *An updated successor to the author's* The Age of Birds *(1980) but more comprehensive; rich source of information on the evolutionary relationships of birds.*

Norbert, U. M. 1990. Vertebrate flight. New York, Springer-Verlag. *Detailed review of the mechanics, physiology, morphology, ecology, and evolution of flight. Covers bats as well as birds.*

Proctor, N. S., and P. J. Lynch. 1993. Manual of ornithology: avian structure and function. New Haven, Connecticut, Yale University Press.

Sibley, C. G., and J. E. Ahlquist. 1990. Phylogeny and classification of birds: a study in molecular evolution. New Haven, Yale University Press. *A comprehensive application of DNA annealing experiments to the problem of resolving avian phylogeny.*

Terborgh, J. 1992. Why American songbirds are vanishing. Sci. Am. **266:**98–104 (May). *The number of songbirds in the United States has been dropping sharply. The author suggests the reasons why.*

Waldvogel, J. A. 1990. The bird's eye view. Am. Sci. **78:**342–353 (July–Aug.). *Birds possess visual abilities unmatched by humans. So how can we know what they really see?*

Wellnhofer, P. 1990. *Archaeopteryx.* Sci. Am. **262:**70–77 (May). *Description of perhaps the most important fossil ever discovered.*

Welty, J. C., and L. Baptista. 1988. The life of birds, ed. 4. Philadelphia, Saunders College Publishing. *Among the best of the ornithology texts; lucid style and well illustrated.*

Mammals

chapter | thirty-one

The Tell-Tale Hair

If Fuzzy Wuzzy, the bear that had no hair (according to the children's rhyme), was truly hairless, he could not have been a mammal or a bear. For hair is as much an unmistakable characteristic of mammals as feathers are of birds. If an animal has hair it is a mammal; if it lacks hair it must be something else. It is true that many aquatic mammals are nearly hairless (whales, for example) but hair can usually be found (with a bit of searching) at least in vestigial form somewhere on the body of the adult. Unlike feathers, which evolved from converted reptilian scales, mammalian hair is a completely new epidermal structure. Mammals use their hair for protection from the elements, for protective coloration and concealment, for waterproofing and buoyancy, and for behavioral signaling; they have turned hairs into sensitive vibrissae on their snouts and into prickly quills. Perhaps most important of all, mammals use their hair for thermal insulation, which allows them to enjoy the great advantages of homeothermy. Warm-blooded animals in most climates and at sunless times benefit from this natural and controllable protective insulation.

Hair, of course, is only one of several features that together characterize a mammal and help us to understand the mammalian evolutionary achievement. Among these are a highly developed placenta for feeding the embryo; mammary glands for nourishing the newborn; and a surpassingly advanced nervous system that far exceeds in performance that of any other animal group. It is doubtful, however, that even with this winning combination of adaptations, the mammals could have triumphed as they have without their hair.

Mammals, with their highly developed nervous system and numerous ingenious adaptations, occupy almost every environment on earth that supports life. Although not a large group (about 4450 species as compared with more than 9000 species of birds, approximately 24,600 species of fishes, and 800,000 species of insects), the class Mammalia (mam-may′lee-a) (L. *mamma,* breast) is overall the most biologically differentiated group in the animal kingdom. Many potentialities that dwell more or less latently in other vertebrates are highly developed in mammals. Mammals are exceedingly diverse in size, shape, form, and function. They range in size from the recently discovered Kitti's hognosed bat, weighing only 1.5 g, to the whales, some of which exceed 100 tons.

Yet, despite their adaptability and in some instances because of it, mammals have been influenced by the presence of humans more than any other group of animals. We have domesticated numerous mammals for food and clothing, as beasts of burden, and as pets. We use millions of mammals each year in biomedical research. We have introduced alien mammals into new habitats, occasionally with benign results but more frequently with unexpected disaster. Although history provides us with numerous warnings, we continue to overcrop valuable wild stocks of mammals. The whale industry has threatened itself with total collapse by exterminating its own resource—a classic example of self-destruction in the modern world, in which competing segments of an industry are intent only on reaping all they can today as though tomorrow's supply were of no concern whatever. In some cases destruction of a valuable mammalian resource has been deliberate, such as the officially sanctioned (and tragically successful) policy during the Indian wars of exterminating the bison to drive the Plains Indians into starvation. Although commercial hunting has declined, the ever-increasing human population with the accompanying destruction of wild habitats has harassed and disfigured the mammalian fauna.

We are becoming increasingly aware that our presence on this planet as the most powerful product of organic evolution makes us responsible for the character of our natural environment. Since our welfare has been and continues to be closely related to that of the other mammals, it is clearly in our interest to preserve the natural environment of which all mammals, ourselves included, are a part. We need to remember that nature can do without us but we cannot exist without nature.

Origin and Evolution of Mammals

The evolutionary descent of mammals from their earliest amniote ancestors is perhaps the most fully documented transition in vertebrate history. From the fossil record, we can trace the derivation over 150 million years of endothermic, furry mammals from their small, ectothermic, hairless ancestors. The structure of the skull roof permits us to identify three major groups of amniotes that diverged in the Carboniferous period of the Paleozoic era, the **synapsids, anapsids,** and **diapsids** (p. 622). The synapsid group, which includes the mammals and their ancestors, has a pair of openings in the skull roof for the attachment of jaw muscles (Figure 31-2). Synapsids were the first amniote group to radiate widely into terrestrial habitats. The anapsid group is characterized by solid skulls and includes the turtles and their ancestors (see p. 627). The diapsids have two pairs of openings in the skull roof (Figure 29-2, p. 624) and this group contains the dinosaurs, lizards, snakes, crocodilians, birds, and their ancestors.

The earliest synapsids radiated extensively into diverse herbivorous and carnivorous forms that are often collectively called **pelycosaurs** (Figures 31-1 and 31-2). The pelycosaurs share a general outward resemblance to lizards, but this resemblance is misleading. The pelycosaurs are not closely related to lizards, which are diapsids, nor are they a monophyletic group. From one group of early carnivorous synapsids arose the **therapsids** (Figure 31-2), the only synapsid group to survive beyond the Paleozoic. In the therapsids we see for the first time an efficient erect gait with upright limbs positioned beneath the body. Since stability was reduced by raising the animal from the ground, the muscular coordination center of the brain, the cerebellum, took on an expanded role. The therapsids radiated into numerous herbivorous and carnivorous forms but most disappeared during the great extinction at the end of the Permian.

Only the last therapsid subgroup to evolve, the **cynodonts,** survived to enter the Mesozoic. The cynodonts evolved several novel features including a high metabolic rate, which supported a more active life; increased jaw musculature, permitting a stronger bite; several skeletal changes, supporting greater agility; and a secondary bony palate (Figure 31-3), enabling the animal to breathe while holding prey or chewing food. The secondary palate would be important to subsequent mammalian evolution by permitting the young to breathe while suckling. Toward the end of the Triassic certain cynodont groups arose that closely resembled mammals, sharing with them several derived features of the skull and teeth.

> *Fishes, amphibians, most reptiles, and birds have a* ***primary palate,*** *which is the roof of the mouth cavity formed by the ventral skull bones. In these vertebrates, there is no separation of nasal passages from the mouth cavity. In mammals and crocodilians the nasal passages are completely separated from the mouth by the development of a secondary bony roof, the* ***secondary palate.*** *Mammals extend the separation of oral and nasal cavities even farther backward by adding to this "hard palate" a fleshy soft palate; these structures are shown in Figure 31-3.*

The earliest mammals of the late Triassic were small mouse- or shrew-sized animals with enlarged cranium, jaws redesigned for a shearing action, and a new type of dentition in

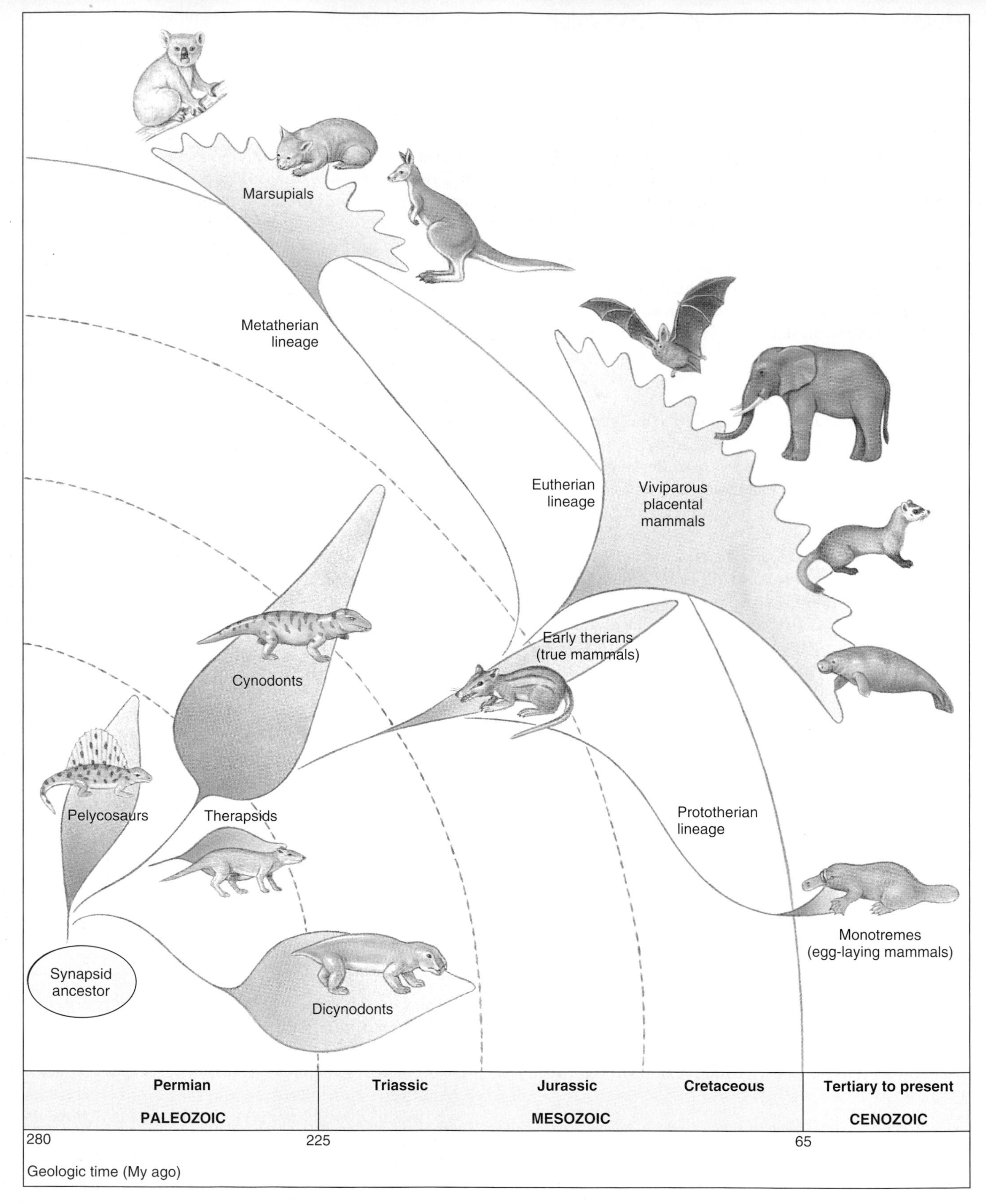

figure 31-1

Evolution of the major groups of synapsids. The synapsid lineage, characterized by lateral temporal openings in the skull, began with the pelycosaurs, early mammal-like amniotes of the Permian. The pelycosaurs radiated extensively and evolved changes in the jaws, teeth, and body form that presaged several mammalian characteristics. These trends continued in their successors, the therapsids, especially in the cynodonts. One lineage of cynodonts gave rise in the Triassic to the therians, the true mammals. Fossil evidence, as currently interpreted, indicates that all three groups of living mammals—monotremes, marsupials, and placentals—are derived from the same lineage. The great radiation of modern placental orders occurred during the Cretaceous and Tertiary periods.

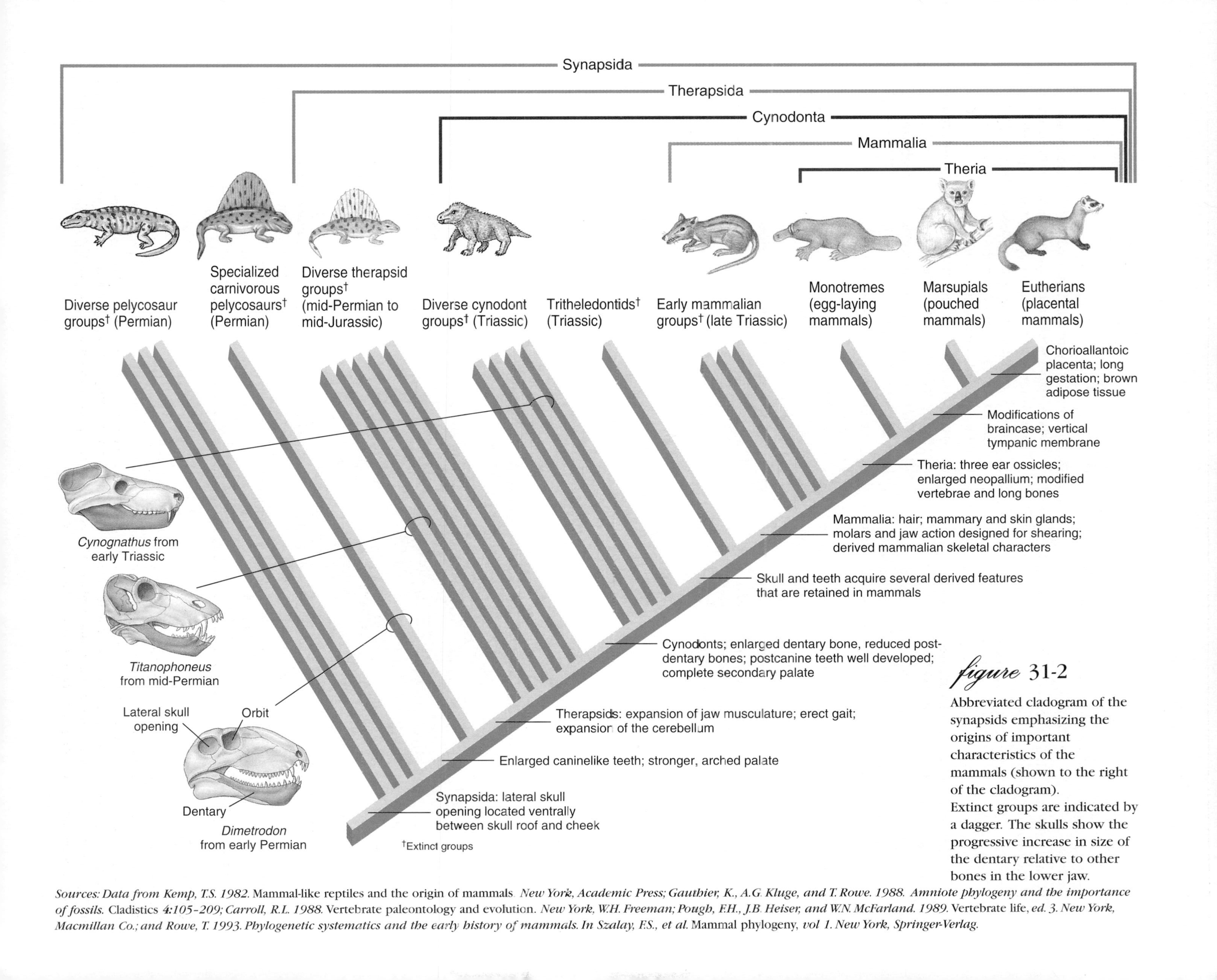

figure 31-2

Abbreviated cladogram of the synapsids emphasizing the origins of important characteristics of the mammals (shown to the right of the cladogram).
Extinct groups are indicated by a dagger. The skulls show the progressive increase in size of the dentary relative to other bones in the lower jaw.

Sources: Data from Kemp, T.S. 1982. Mammal-like reptiles and the origin of mammals. *New York, Academic Press; Gauthier, K., A.G. Kluge, and T. Rowe. 1988. Amniote phylogeny and the importance of fossils.* Cladistics *4:105–209; Carroll, R.L. 1988.* Vertebrate paleontology and evolution. *New York, W.H. Freeman; Pough, F.H., J.B. Heiser, and W.N. McFarland. 1989.* Vertebrate life, *ed. 3. New York, Macmillan Co.; and Rowe, T. 1993. Phylogenetic systematics and the early history of mammals. In Szalay, F.S., et al.* Mammal phylogeny, *vol 1. New York, Springer-Verlag.*

Characteristics of Mammals

1. **Body covered with hair,** but reduced in some
2. **Integument** with **sweat, scent, sebaceous,** and **mammary glands**
3. Skull with **two occipital condyles** and **secondary bony palate;** middle ear with **three ossicles** (malleus, incus, stapes); **seven cervical vertebrae** (except some xenarthrans [edentates] and the manatee); **pelvic bones fused**
4. Mouth with **diphyodont teeth** (milk, or deciduous, teeth replaced by a permanent set of teeth); teeth heterodont in most (varying in structure and function); lower jaw a **single enlarged bone (dentary)**
5. Movable eyelids and **fleshy external ears (pinnae)**
6. Four limbs (reduced or absent in some) adapted for many forms of locomotion
7. Circulatory system of a four-chambered heart, **persistent left aorta,** and **nonnucleated, biconcave red blood corpuscles**
8. Respiratory system of lungs with alveoli, and voice box (larynx); **secondary palate** (anterior bony palate and posterior continuation of soft tissue, the soft palate) separates air and food passages (Figure 31-3); **muscular diaphragm** for air exchange separates thoracic and abdominal cavities
9. Excretory system of metanephros kidneys and ureters that usually open into a bladder
10. Brain highly developed, especially **neocerebrum;** 12 pairs of cranial nerves
11. Endothermic and homeothermic
12. Separate sexes
13. Internal fertilization; **embryos develop in a uterus** with **placental attachment** (placenta rudimentary in marsupials and absent in monotremes); **fetal membranes (amnion, chorion, allantois)**
14. Young nourished by **milk from mammary glands**

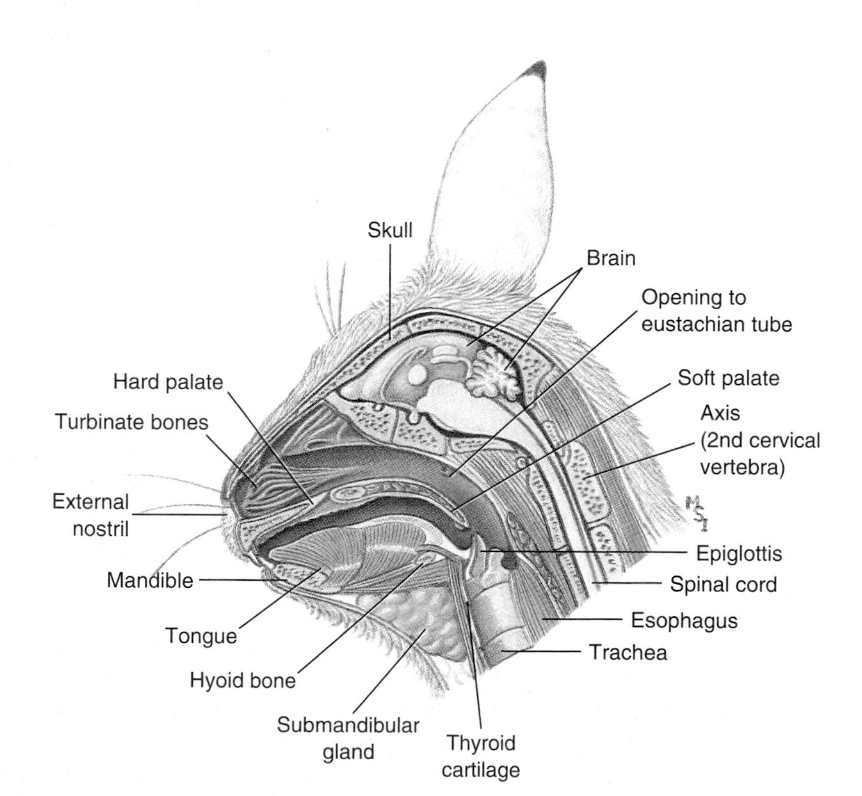

figure 31-3

Sagittal section through the head of a rabbit. The hard and soft palates together form the secondary palate, a roof that separates mouth and nasal cavities, and a characteristic of all mammals and some reptiles.

which the teeth were replaced only once (deciduous and permanent teeth). This event contrasts with the primitive amniote pattern of continual tooth replacement throughout life. The earliest mammals were almost certainly endothermic, although their body temperature would have been rather lower than modern placental mammals. Hair was essential for insulation, and the presence of hair implies that sebaceous and sweat glands must also have evolved at this time to lubricate the hair and promote heat loss. The fossil record is silent on the appearance of mammary glands, but they must have evolved before the end of the Triassic.

Oddly, the early mammals of the mid-Triassic, having developed nearly all of the novel attributes of modern mammals, had to wait for another 150 million years before they could achieve their great diversity. In the meantime the dinosaurs became diverse and abundant, while all nonmammalian synapsid groups became extinct. But mammals survived, first as shrewlike, probably nocturnal, creatures. Then, in the Tertiary, especially during the Eocene epoch that began about 54 million years ago, the modern mammals began to expand rapidly. The great Cenozoic radiation of mammals is partly attributed to the numerous environments vacated by the many amniote groups that became extinct at the end of the Cretaceous. The mammalian radiation was almost certainly promoted by the fact that mammals were agile, endothermic, intelligent, adaptable, and gave birth to living young, which they protected and nourished from their own milk supply, thus dispensing with vulnerable eggs laid in nests.

The class Mammalia includes 21 orders: one order containing the monotremes, one order containing the marsupials, and 19 orders of placentals. A complete classification is on pp. 685–686.

Structural and Functional Adaptations of Mammals

Integument and Its Derivatives

The mammalian skin and its modifications especially distinguish mammals as a group. As the interface between the animal and its environment, the skin is strongly molded by the animal's way of life. In general the skin is thicker in mammals than in other classes of vertebrates, although as in all vertebrates it is made up of **epidermis** and **dermis** (see Figure 6-9, p. 148). Among the mammals the dermis becomes much thicker than the epidermis. The epidermis is relatively thin where it is well protected by hair, but in places that are subject to much contact and use, such as the palms or soles, its outer layers become thick and cornified with keratin.

Hair

Hair is especially characteristic of mammals, although humans are not very hairy creatures, and in whales hair is reduced to only a few sensory bristles on the snout. A hair grows out of a hair follicle that, although an epidermal structure, is sunk into the dermis of the skin (Figure 31-4). The hair grows continuously by rapid proliferation of cells in the follicle. As the hair shaft is pushed upward, new cells are carried away from their source of nourishment and die, turning into the same dense type of fibrous protein, called **keratin,** that constitutes nails, claws, hooves, and feathers.

Mammals characteristically have two kinds of hair forming the **pelage** (fur coat): (1) dense and soft **underhair** for insulation and (2) coarse and longer **guard hair** for protection against wear and to provide coloration. The underhair traps a layer of insulating air. In aquatic animals, such as the fur seal, otter, and beaver, it is so dense that it is almost impossible to wet. In water the guard hairs become wet and mat down, forming a protective blanket over the underhair (Figure 31-5).

When a hair reaches a certain length, it stops growing. Normally it remains in the follicle until a new growth starts, whereupon it falls out. In most mammals, there are periodic molts of the entire coat. In humans, hair is shed and replaced throughout life (although balding males confirm that replacement is not assured!).

A hair is more than a strand of keratin. It consists of three layers: the medulla or pith in the center of the hair, the cortex with pigment granules next to the medulla, and the outer cuticle composed of imbricated scales. The hair of different mammals shows a considerable range of structure. It may be deficient in cortex, such as the brittle hair of deer, or it may be deficient in medulla, such as the hollow, air-filled hairs of the wolverine. The hairs of rabbits and some others are scaled to interlock when pressed together. Curly hair, such as that of sheep, grows from curved follicles.

In the simplest cases, such as foxes and seals, the coat is shed once each year during the summer months. Most mammals have two annual molts, one in the spring and one in the fall. The summer coat is always much thinner than the winter coat and in some mammals it may be a different color. Several northern mustelid carnivores, such as weasels, have white winter coats and brown-colored summer coats. It was once believed that the white inner pelage of arctic animals conserves body heat by reducing radiation loss; in fact, dark and white pelages radiate heat equally well. The winter white of arctic animals is simply camouflage in a land of snow. The varying hare of North America has three annual molts: the white winter coat is replaced by a brownish gray summer coat, and this is replaced in autumn by a grayer coat, which is soon shed to reveal the winter white coat beneath (Figure 31-6).

Outside the Arctic, most mammals wear somber colors that are protective. Often the species is marked with "salt-and-pepper" coloration or a disruptive pattern that helps make it inconspicuous in its natural surroundings. Examples are the spots of leopards and fawns and the stripes of tigers. Other mammals, such as skunks, advertise their presence with conspicuous warning coloration.

The hair of mammals has become modified to serve many purposes. The bristles of hogs, the spines of porcupines and their kin, and the vibrissae on the snouts of most mammals are examples. **Vibrissae,** commonly called "whiskers," are really sensory hairs that provide a tactile sense to many mammals. The slightest movement of a vibrissa generates impulses in sensory nerve endings that travel to special sensory areas in the brain. Vibrissae are especially long in nocturnal and burrowing animals.

Porcupines, hedgehogs, echidnas, and a few other mammals have developed an effective and dangerous spiny armor. When cornered, the common North American porcupine turns its back toward the attacker and lashes out with the barbed tail. The lightly-attached quills break off at their bases

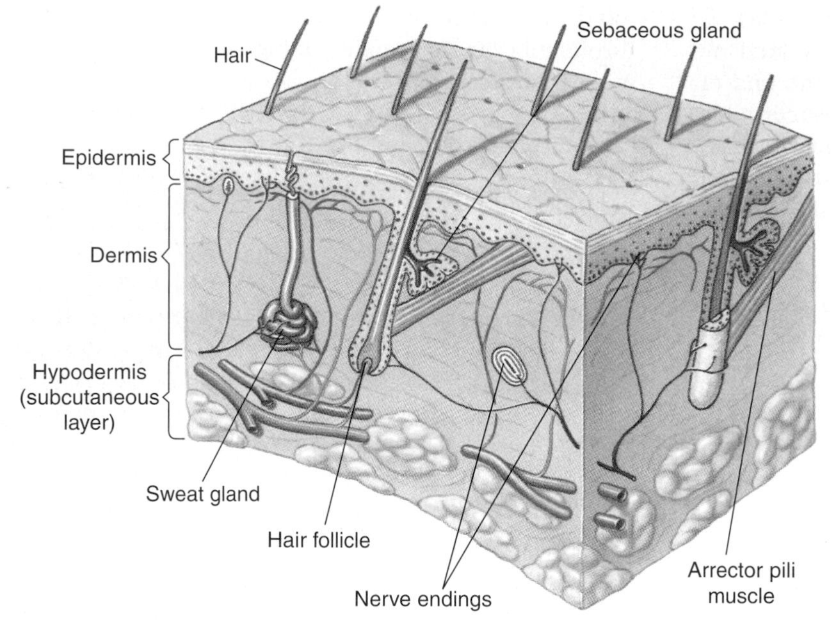

figure 31-4

Structure of human skin (epidermis and dermis) and hypodermis, showing hair and glands.

figure 31-5

American beaver, *Castor canadensis,* about to cut down an aspen tree. This second largest rodent (the South American capybara is larger) has a heavy waterproof pelage consisting of long, tough guard hairs overlying the thick, silky underhair so valued in the fur trade. Order Rodentia, family Castoridae.

when they enter the skin and, aided by backward-pointing hooks on the tips, work deeply into tissues. Dogs are frequent victims (Figure 31-7) but fishers, wolverines, and bobcats are able to flip the porcupine onto its back to expose vulnerable underparts.

Horns and Antlers

Three kinds of horns or hornlike substances are found in mammals. **True horns,** found in **ruminants** (for example, cud-chewers such as sheep and cattle), are sheaths of keratinized epidermis that embrace a hollow core of bone arising from the skull. Horns are not normally shed, usually are not branched (although they may be greatly curved), and are found in both sexes (except pronghorn antelope in which they occur only in the male).

Antlers of the deer family are solid bone when mature. During their annual growth, antlers develop beneath a covering of highly vascular soft skin called **"velvet"** (Figure 31-8). When growth of the antlers is complete just before the breeding season, blood vessels constrict and the stag tears off the velvet by rubbing the antlers against trees. Antlers are dropped after the breeding season. New buds appear a few months later to herald the next set of antlers. For several years each new pair of antlers is larger and more elaborate than the previous set. The annual growth of antlers places a strain on the mineral metabolism, since during the growing season a large moose or elk must accumulate 50 or more pounds of calcium salts from its vegetable diet.

The **rhinoceros horn** is the third kind of horn. Hairlike keratinized filaments that arise from dermal papillae are cemented together to form a single horn.

The escalating trade in rhinoceros products—especially rhinoceros horn—during the last three decades, is pushing Asian and African rhinos to the brink of extinction. Rhinoceros horn is valued in China as an agent for reducing fever, and for treating heart, liver, and skin diseases; and in North India as an aphrodisiac. Such supposed medicinal values are totally without pharmacological basis. The principal use of rhinoceros horns, however, is to fashion handles for daggers in the Middle East. Because of their phallic shape, rhinoceros horn daggers are traditional gifts at puberty rites. Between 1969 and 1977, horns from 8000 slaughtered rhinos were imported into North Yemen alone.

Glands

Of all vertebrates, mammals have the greatest variety of integumentary glands. Most fall into one of four classes: sweat, scent, sebaceous, and mammary. All are derivatives of the epidermis.

figure 31-6

Snowshoe, or varying, hare, *Lepus americanus* in **A,** brown summer coat and, **B,** white winter coat. In winter, extra hair growth on the hind feet broadens the animal's support in snow. Snowshoe hares are common residents of the taiga (northern coniferous forests) and are an important food for lynxes, foxes, and other carnivores. Population fluctuations of hares and their predators are closely related. Order Lagomorpha.

figure 31-7

Dogs are frequent victims of the porcupine's impressive armor. Unless removed (usually by a veterinarian) the quills will continue to work their way deeper in the flesh causing great distress and may lead to the victim's death.

Sweat glands are simple, tubular, highly coiled glands that occur over much of the body in most mammals. They are not present in other vertebrates. Two kinds of sweat glands may be distinguished: eccrine and apocrine (Figure 31-4). **Eccrine glands** secrete a watery sweat that, when evaporated on the skin's surface, draws heat from the skin and cools it. Eccrine glands occur in hairless regions, especially the foot pads, in most mammals, although in horses and most primates they are scattered all over the body. Eccrine glands are much reduced or absent in rodents, rabbits, whales, and others. **Apocrine glands,** the second type of sweat glands, are larger than eccrine glands and have longer and more winding ducts. Their secretory coil is in the dermis and extends deep into the hypodermis. They always open into the follicle of a hair or where a hair has been. Apocrine glands develop approximately at sexual puberty and are restricted (in the human species) to the axillae (armpits), mons pubis, breasts, external auditory canals, prepuce, scrotum, and a few other places. Their secretion is not watery, like ordinary sweat of eccrine glands, but is a milky, whitish or yellow secretion that dries on the skin to form a plasticlike film. Apocrine glands are not involved in heat regulation, but their activity is correlated with certain aspects of the sex cycle, among other possible functions.

Scent glands are present in nearly all mammals. Their location and functions vary greatly. They are used in communication with members of the same species, to mark territorial boundaries, for warning, or for defense. Scent-producing glands are located in orbital, metatarsal, and interdigital regions (deer); behind the eyes and on the cheek (pica and woodchuck); penis (muskrats, beavers, and many canines); base of the tail (wolves and foxes); back of the head (dromedary); and anal region (skunks, minks, and weasels). The latter, the most odoriferous of all glands, open by ducts into the anus; their secretions can be discharged forcefully for several feet. During the mating season many mammals give off strong scents for attracting the opposite sex. Humans also are endowed with scent glands. But civilization has taught us to dislike our own scent, a concern that has stimulated a lucrative deodorant industry to produce an endless output of soaps and odor-masking compounds.

Sebaceous glands are intimately associated with hair follicles, although some are free and open directly onto the surface. The cellular lining of the gland itself is discharged in the secretory process and must be renewed for further secretion. These gland cells become distended with a fatty accumulation, then die, and are expelled as a greasy mixture called **sebum** into the hair follicle. Called a "polite fat" because it does not turn rancid, it serves as a dressing to keep the skin and hair pliable and glossy. Most mammals have sebaceous glands all over the body; in humans they are most numerous in the scalp and on the face.

Mammary glands, which provide the name for mammals, are probably modified apocrine glands. Whatever their evolutionary origin, they occur on all female mammals and in a rudimentary form on all male mammals. They develop by the thickening of the epidermis to form a milk line along each side of the abdomen in the embryo. On certain parts of these lines the mammae appear while the intervening parts of the ridge disappear. In the human female the mammary glands begin to increase in size at puberty because of fat accumulation and reach their maximum development in approximately the twentieth year. The breasts (or mammae) undergo additional

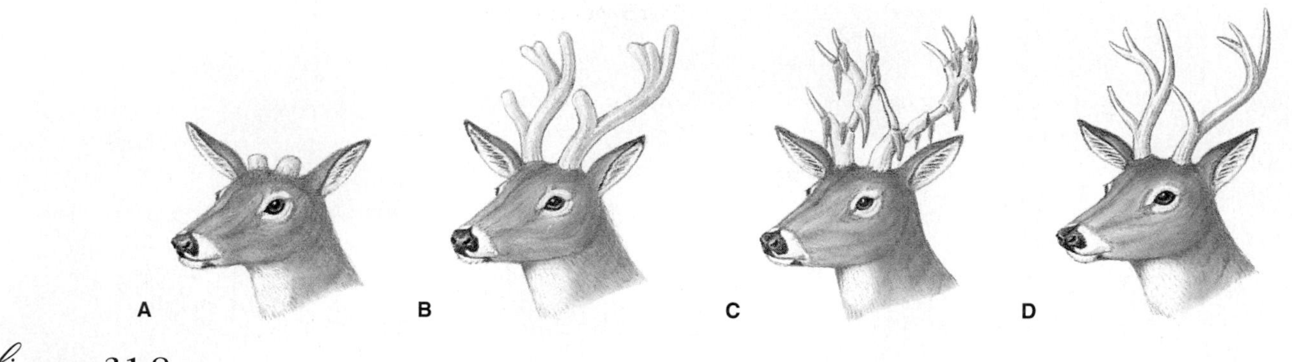

figure 31-8

Annual growth of buck deer antlers. **A,** Antlers begin growth in late spring, stimulated by pituitary gonadotropins. **B,** The bone grows very rapidly until halted by a rapid rise in testosterone production by the testes. **C,** The skin (velvet) dies and sloughs off. **D,** Testosterone levels peak during the fall breeding season. The antlers are shed in January as testosterone levels subside.

development during pregnancy. In other mammals the mammae are swollen only periodically when they are distended with milk during pregnancy and subsequent nursing of the young.

Food and Feeding

Mammals have exploited an enormous variety of food sources; some mammals require highly specialized diets, whereas others are opportunistic feeders that thrive on diversified diets. For all mammals, food habits and physical structure are inextricably linked. A mammal's adaptations for attack and defense and its specializations for finding, capturing, chewing, swallowing, and digesting food all determine a mammal's shape and habits.

Teeth, perhaps more than any other single physical characteristic, reveal the life habit of a mammal (Figure 31-9). All mammals have teeth, except monotremes, anteaters, and certain whales, and their modifications are correlated with what the mammal eats.

As the mammals evolved during the Mesozoic, major changes occurred in the teeth and jaws. Unlike the uniform **homodont** dentition of the reptiles, mammalian teeth became differentiated to perform specialized functions such as cutting, seizing, gnawing, tearing, grinding, or chewing. Teeth differentiated in this manner are called **heterodont.** Typically, the mammalian dentition is differentiated into four types: **incisors,** with simple crowns and sharp edges, used mainly for snipping or biting; **canines,** with long conical crowns, specialized for piercing; **premolars,** with compressed crowns and one or two cusps, suited for shearing and slicing; and **molars,** with large bodies and variable cusp arrangement, used for crushing and grinding. The primitive tooth formula, which expresses the number of each tooth type in one-half of the upper and lower jaw, was I 3/3, C 1/1, PM 4/4, M 3/3. Members of the order Insectivora (e.g., shrews), some omnivores, and carnivores come closest to this primitive pattern (Figure 31-9).

Unlike reptiles, mammals do not continuously replace their teeth throughout their lives. Most mammals grow just two sets of teeth: a temporary set, called **deciduous,** or **milk,** teeth, which is replaced by a permanent set when the skull has grown large enough to accommodate a full set. Only the incisors, canines, and premolars are deciduous; the molars are never replaced and the single permanent set must last a lifetime.

Feeding Specializations

The feeding apparatus of a mammal—the teeth and jaws, tongue, and alimentary canal—are adapted to its particular feeding habits. Mammals are customarily divided among four basic categories—insectivores, carnivores, omnivores, and herbivores—but many other feeding specializations have evolved in mammals, as in other living organisms, and the feeding habits of many mammals defy exact classification. The principal feeding specializations of mammals are shown in Figure 31-9.

Insectivores are small mammals, usually opportunistic feeders, that feed on a variety of small invertebrates, such as worms and grubs, as well as insects. Examples are shrews, moles, anteaters, and most bats. The insectivorous category is not a sharply distinguished one because carnivores and omnivores often include insects in their diets.

Herbivorous mammals that feed on grasses and other vegetation form two main groups: the **browsers** and **grazers,** such as the ungulates (hooved mammals including horses, deer, antelope, cattle, sheep, and goats), and the **gnawers,** such as the rodents, and rabbits and hares. In herbivores, the canines are reduced in size or absent, whereas the molars, which are adapted for grinding, are broad and usually high-crowned. Rodents have chisel-sharp incisors that grow throughout life and must be worn away to keep pace with their continual growth (Figure 31-9).

Herbivorous mammals have a number of interesting adaptations for dealing with their fibrous diet of plant food.

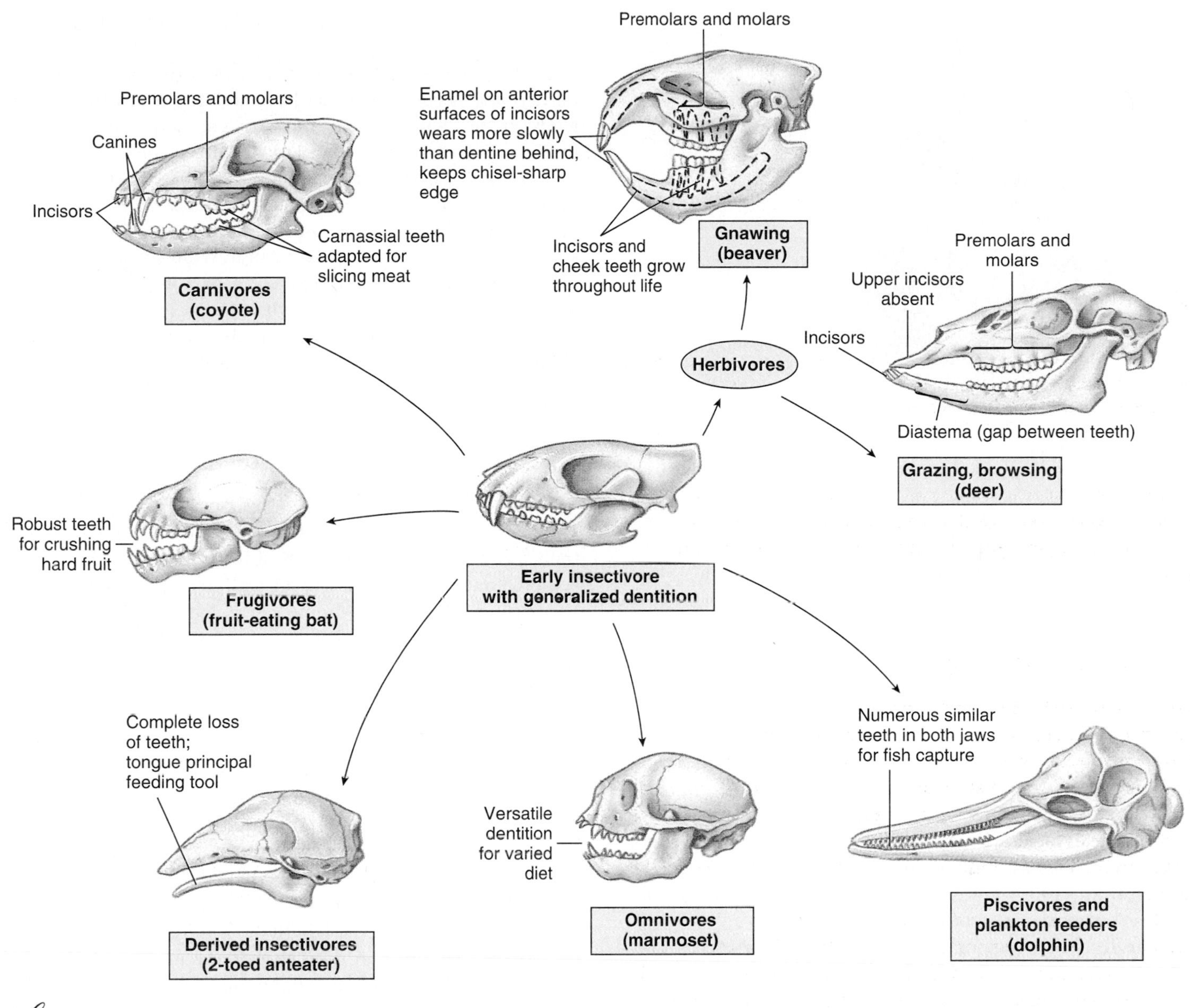

figure 31-9

Feeding specializations of major trophic groups of eutherian mammals. The early eutherians were insectivores; all other types are descended from them.

Cellulose, the structural carbohydrate of plants, is composed of long chains of glucose units linked by a type of chemical bond that few enzymes can attack. No vertebrates synthesize cellulose-splitting enzymes. Instead, herbivorous vertebrates harbor anaerobic bacteria and protozoa in large fermentation chambers in the gut. These microorganisms break down and metabolize the cellulose, releasing a variety of fatty acids, sugars, and starches that the host animal can absorb and use.

In some herbivores, such as horses and zebras, rabbits and hares, elephants, and many rodents, the gut has a spacious sidepocket, or diverticulum, called a **cecum,** which serves as a fermentation chamber and absorptive area. Hares, rabbits, and some rodents often eat their fecal pellets **(coprophagy),** giving the food a second pass through the fermenting action of the intestinal bacteria.

The **ruminants** (cattle, bison, buffalo, goats, antelopes, sheep, deer, giraffes, and okapis) have a huge **four-chambered stomach** (Figure 31-10). When a ruminant feeds, grass passes down the esophagus to the **rumen,** where it is broken down by bacteria and protozoa and then formed into small balls of cud. At its leisure the ruminant returns the cud to its mouth where the cud is deliberately chewed at length to crush the fiber. Swallowed again, the food returns to the rumen where the cellulolytic bacteria and protozoa continue fermentation. The pulp

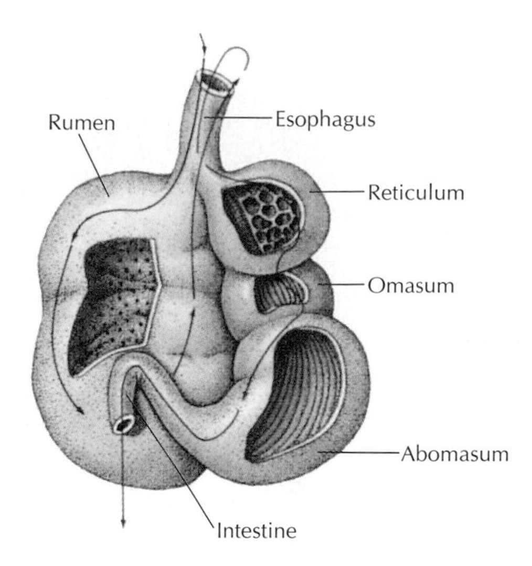

figure 31-10

Ruminant's stomach. Food passes first to the rumen (sometimes through the reticulum) and then is returned to the mouth for chewing (chewing the "cud," or rumination) (*black arrow*). After reswallowing, food returns to the rumen or passes directly to reticulum, omasum, and abomasum for final digestion (*red arrow*).

passes to the **reticulum,** then to the **omasum,** where water, soluble food, and microbial products are absorbed. The remainder proceeds to the **abomasum** ("true" acid stomach), where proteolytic enzymes are secreted and normal digestion takes place.

Herbivores in general have large, long digestive tracts and must eat a considerable amount of plant food to survive. A large African elephant weighing 6 tons must consume 135 to 150 kg (300 to 400 pounds) of rough fodder each day to obtain sufficient nourishment for life.

Carnivorous mammals feed mainly on herbivores. This group includes foxes, dogs, weasels, wolverines, fishers, cats, lions, and tigers. Carnivores are well-equipped with biting and piercing teeth and powerful clawed limbs for killing their prey. Since their protein diet is much more easily digested than is the woody food of herbivores, their digestive tract is shorter and the cecum small or absent. Carnivores organize their feeding into discrete meals rather than feeding continuously (as do most herbivores) and therefore have much more leisure time for play and exploration (Figure 31-11).

Note that the terms "insectivores" and "carnivores" have two different uses in mammals: to describe diet and to denote specific taxonomic orders of mammals. For example, not all carnivores belong to the order Carnivora (many marsupials, pinnipeds, cetaceans, and all insectivores are carnivorous) and not all members of the order Carnivora are carnivorous. Many are opportunistic feeders and some, such as the panda, are strict vegetarians.

Omnivorous mammals live on both plants and animals for food. Examples are pigs, raccoons, rats, bears, and most primates (including humans). Many carnivorous forms also eat fruits, berries, and grasses when hard pressed. The fox, which usually feeds on mice, small rodents, and birds, eats frozen apples, beechnuts, and corn when its normal food sources are scarce.

For most mammals, searching for food and eating occupy most of their active life. Seasonal changes in food supplies are considerable in temperate zones. Living may be easy in the summer when food is abundant, but in winter many carnivores must range far and wide to eke out a narrow existence. Some mammals migrate to regions where food is more abundant, while others hibernate and sleep the winter months away. Many mammals, such as squirrels, chipmunks, gophers, and certain mice, build up stores of food during periods of plenty for use during the winter (Figure 31-12).

Migration

Migration is a much more difficult undertaking for mammals than for birds. Not surprisingly, few mammals make regular seasonal migrations, preferring instead to center their activities in a defined and limited home range. Nevertheless, there are some striking examples of mammalian migrations. More migrators are found in North America than on any other continent.

An example is the barren-ground caribou of Canada and Alaska, which undertakes direct and purposeful mass migrations spanning 160 to 1100 km (100 to 700 miles) twice annually (Figure 31-13). From winter ranges in the boreal forests (taiga), they migrate rapidly in late winter and spring to calving ranges on the barren grounds (tundra). The calves are born in mid-June. As summer progresses, they are increasingly harassed by warble and nostril flies that bore into their flesh, by mosquitos that drink their blood (estimated at a liter per caribou each week during the height of the mosquito season), and by wolves that prey on the calves. They move southward in July and August, feeding little along the way. In September they reach the forest and feed there almost continuously on low ground vegetation. Mating (rut) occurs in October.

Caribou have suffered a drastic decline in numbers since the nineteenth century when there were several million of them. By 1958 less than 200,000 remained in Canada. The decline has been attributed to several factors, including habitat alteration from exploration and development in the North, but especially to excessive hunting. For example the Western Arctic herd in Alaska exceeded 250,000 caribou in 1970. Following five years of heavy unregulated hunting, a 1976 census revealed only about 65,000 animals left. After restricting hunting, the herd had increased to 140,000 by 1980 and was expected to reach its original population of 250,000 in the 1990s. However, the proposed scheme to open the Arctic National Wildlife Refuge to petroleum development threatens this recovery.

figure 31-11

Lionesses, *Panthera leo,* eating a wildebeest. Lions stalk prey and then charge suddenly to surprise the victim. They lack stamina for a long chase. Lions gorge themselves with the kill, then sleep and rest for periods as long as one week before eating again. Order Carnivora, family Felidae.

figure 31-12

Eastern chipmunk, *Tamias striatus,* with cheek pouches stuffed with seeds to be carried to a hidden cache. It will try to store at least a half-bushel of food for the winter. It hibernates but awakens periodically to eat some of its cached food. Order Rodentia, family Sciuridae.

The plains bison, before its deliberate near extinction by humans, made huge circular migrations to separate summer and winter ranges.

The longest mammalian migrations are made by the oceanic seals and whales. One of the most remarkable migrations is that of fur seals, which breed on the Pribilof Islands approximately 300 km (185 miles) off the coast of Alaska and north of the Aleutian Islands. From wintering grounds off southern California the females journey as much as 2800 km (1740 miles) across open ocean, arriving in the spring at the Pribilofs where they congregate in enormous numbers (Figure 31-14). The young are born within a few hours or days after arrival of the cows. Then the bulls, having already arrived and established territories, collect harems of cows, which they guard with vigilance. After the calves have been nursed for approximately three months, cows and juveniles leave for their long migration southward. The bulls do not follow but remain in the Gulf of Alaska during the winter.

Although we might expect bats, the only winged mammals, to use their gift of flight to migrate, few of them do. Most spend the winter in hibernation. The four species of American bats that do migrate spend their summers in the northern or western states and their winters in the southern United States or Mexico.

Flight and Echolocation

Mammals have not exploited the skies to the same extent that they have the terrestrial and aquatic environments. However, many mammals scamper about in trees with amazing agility; some can glide from tree to tree, and one group, the bats, is capable of full flight. Gliding and flying evolved independently in several groups of mammals, including marsupials, rodents, flying lemurs, and bats. Anyone who has watched a gibbon perform in a zoo realizes there is something akin to flight in this primate, too. Among the arboreal squirrels, all of which are nimble acrobats, by far the most efficient is the flying squirrel (Figure 31-15). These forms actually glide rather than fly, using the gliding skin that extends from the sides of the body.

Bats, the only group of flying mammals, are nocturnal and thus hold a niche unoccupied by most birds. Their achievement is attributed to two things: flight and the capacity to navigate by echolocation. Together these adaptations enable bats to fly and avoid obstacles in absolute darkness, to locate and catch insects with precision, and to find their way deep into caves (a habitat largely ignored by both mammals and birds) where they sleep away the daytime hours.

When in flight, bats emit short pulses 5 to 10 milliseconds in duration in a narrow directed beam from the mouth or nose (Figure 31-16). Each pulse is frequency modulated; that is, it is highest at the beginning, up to 100,000 hertz (Hz, cycles per second), and sweeps down to perhaps 30,000 Hz at the end. Sounds of this frequency are ultrasonic to the human ear, which has an upper limit of about 20,000 Hz. When a bat is searching for prey, it produces about 10 pulses per second. If a prey is detected, the rate increases rapidly up to 200 pulses per second in the final phase of approach and capture. The pulses are spaced so that the echo of each is received before

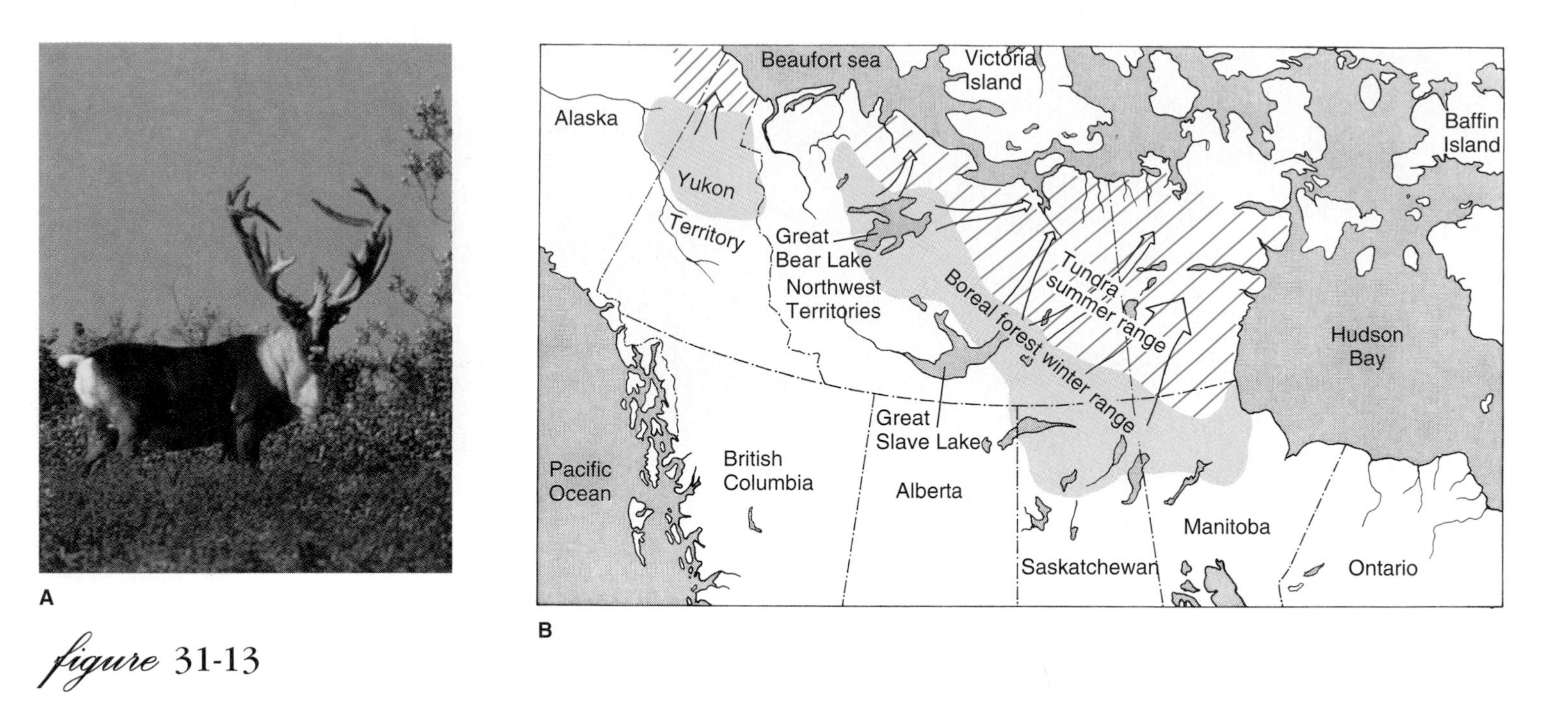

figure 31-13

Barren-ground caribou, *Rangifer tarandus,* of Canada and Alaska. **A,** Adult male caribou in autumn pelage and antlers in velvet. **B,** Summer and winter ranges of some major caribou herds in Canada and Alaska (other herds not shown occur on Baffin Island and in western and central Alaska). The principal spring migration routes are indicated by arrows; routes vary considerably from year to year. The same species is known as reindeer in Europe. Order Artiodactyla, family Cervidae.

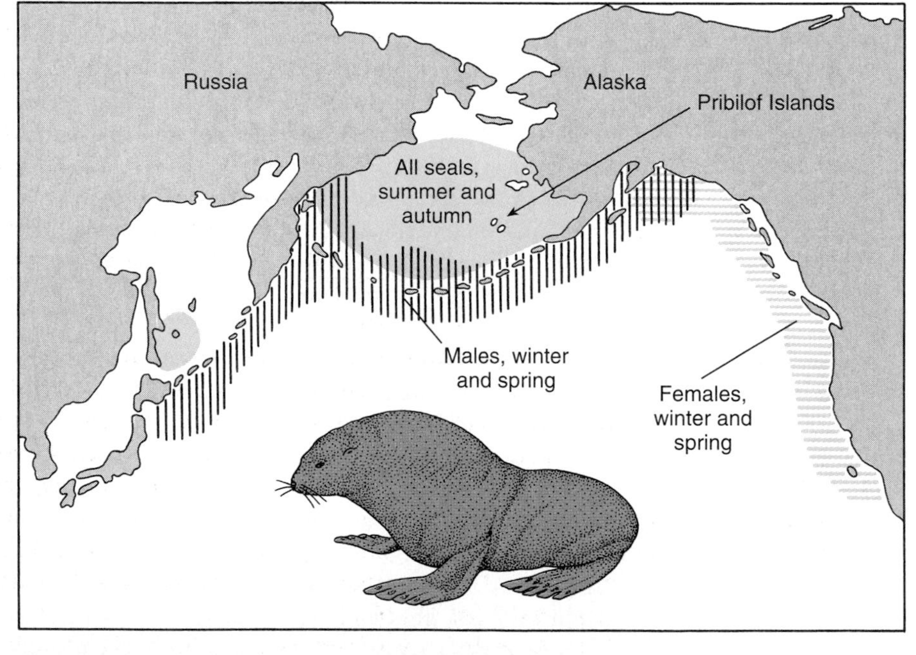

figure 31-14

Annual migrations of the fur seal, showing the separate wintering grounds of males and females. Both males and females of the large Pribilof population migrate in early summer to the Pribilof Islands, where the females give birth to their pups and then mate with the males. Order Pinnipedia, family Otariidae.

the next pulse is emitted, an adaptation that prevents jamming. Since the transmission-to-reception time decreases as the bat approaches an object, it can increase pulse frequency to obtain more information about the object. Pulse length is also shortened as the bat nears the object. It is interesting that some prey of bats, certain nocturnal moths for example, have evolved ultrasonic detectors used to detect and avoid approaching bats (p. 256).

The external ears of bats are large, like hearing trumpets, and shaped variously in different species. Less is known about the inner ear of bats, but it obviously is capable of receiving the ultrasonic sounds emitted. Biologists believe that bat navigation is so refined that a bat builds up a mental image of its surroundings from echo scanning that approaches the resolution of a visual image from eyes of diurnal animals.

For reasons not fully understood, all bats are nocturnal, even the fruit-eating bats that use vision and olfaction instead of echolocation to find their food. The tropics and subtropics have many kinds of bats, including the famed vampire bat.

figure 31-15

Flying squirrel, *Glaucomys sabrinus,* coming in for a landing. Area of undersurface is nearly trebled when gliding skin is spread. Glides of 40 to 50 m are possible. Good maneuverability during flight is achieved by adjusting the position of the gliding skin with special muscles. Flying squirrels are nocturnal and have superb night vision. Order Rodentia, family Sciuridae.

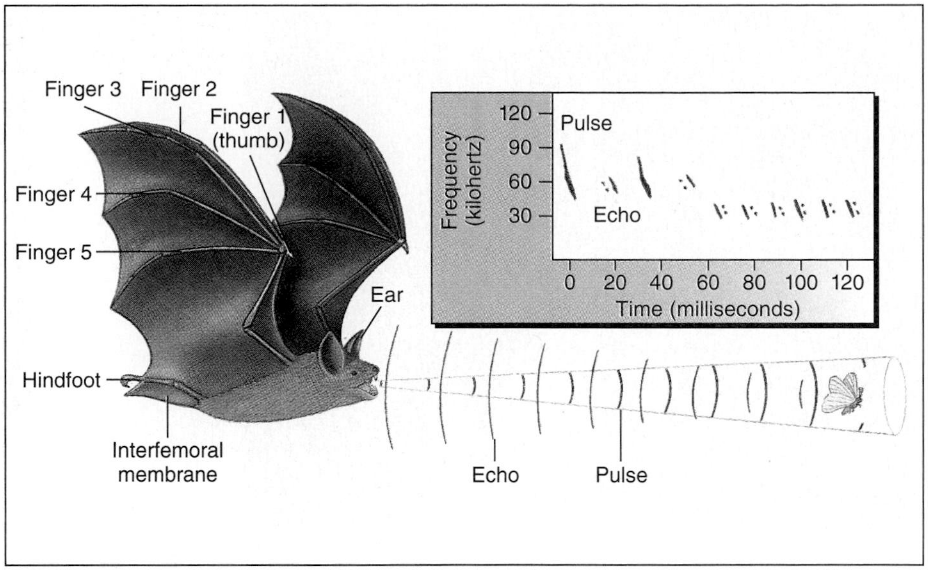

figure 31-16

Echolocation of an insect by the little brown bat *Myotis lucifugus.* Frequency modulated pulses are directed in a narrow beam from the bat's mouth. As the bat nears its prey, it emits shorter, lower signals at a faster rate. Order Chiroptera.

Vampire bats are provided with razor-sharp incisors used to shave away the epidermis of their prey, exposing underlying capillaries. After infusing an anticoagulant to keep the blood flowing, the bat laps up its meal and stores it in a specially modified stomach.

Reproduction

Most mammals have definite mating seasons, usually in the winter or spring and timed to coincide with the most favorable time of the year for rearing the young after birth. Many male mammals are capable of fertile copulation at any time, but the female mating function is restricted to a time during a periodic cycle, known as the **estrous cycle.** The female receives the male only during a relatively brief period known as **estrus,** or heat (Figure 31-17).

There are three different patterns of reproduction in mammals. One pattern is represented by the egg-laying (oviparous) mammals, the **monotremes.** The duck-billed platypus has one breeding season each year. The ovulated eggs, usually two, are fertilized in the oviduct. As they continue down the oviduct, various glands add albumin and then a thin, leathery shell to each egg. When laid, the eggs are about the size of a robin's egg. The platypus lays its eggs in a burrow nest where they are incubated for about 12 days. After hatching, the young suck milk from the openings of the mother's mammary glands for a prolonged period. Thus in monotremes there is no gestation (period of pregnancy) and the developing embryo draws on nutrients stored in the egg, much as do the embryos of reptiles and birds. But in common with all other mammals, monotremes rear their young on milk.

The **marsupials** are pouched, viviparous mammals that exhibit a second pattern of reproduction. Although only the eutherians (p. 685) are called "placental mammals," the marsupials do have a primitive type of yolk sac placenta. The embryo (blastocyst) of a marsupial is at first encapsulated by shell membranes and floats free for several days in the uterine fluid. After "hatching" from the shell membranes, the embryo does not implant, or "take root" in the uterus as in eutherians, but it does erode a shallow depression in the uterine wall in which it lies and absorbs nutrient secretions from the mucosa by way of the vascularized yolk sac. Gestation (the intrauterine period of development) is brief in marsupials, and all marsupials give birth to tiny young that are effectively still embryos, both anatomically and physiologically (Figure 31-18). However, early birth is followed by a prolonged interval of lactation and parental care (Figure 31-19).

The third pattern of reproduction is that of the viviparous **placental mammals,** the eutherians. In placentals, the reproductive investment is in prolonged gestation, unlike marsupials in which the reproductive investment is in prolonged lactation (Figure 31-19). The embryo remains in the mother's uterus, nourished by food supplied through a chorioallantoic type of placenta (described on p. 336), an intimate connection between mother and young. The length of gestation is longer in placentals than marsupials, and in large mammals it is much

figure 31-17

African lions *Panthera leo* mating. Lions breed at any season, although predominantly in spring and summer. During the short period a female is receptive, she may mate repeatedly. Three or four cubs are born after gestation of 100 days. Once the mother introduces the cubs into the pride, they are treated with affection by both adult males and females. Cubs go through an 18- to 24-month apprenticeship learning how to hunt and then are frequently driven from the pride to manage themselves. Order Carnivora, family Felidae.

figure 31-18

Opossums, *Didelphis marsupialis,* 15 days old, fastened to teats in mother's pouch. When born after a gestation period of only 12 days, they are the size of honeybees. They remain attached to the nipples for 50 to 60 days. Order Marsupialia, family Didelphidae.

longer. For example, mice have a gestation period of 21 days; rabbits and hares, 30 to 36 days; cats and dogs, 60 days; cattle, 280 days; and elephants, 22 months. But there are important exceptions (nature seldom offers perfect correlations). Baleen whales, the largest mammals, carry their young for only 12 months, while bats, no larger than mice, have gestation periods of 4 to 5 months. The condition of the young at birth also varies. An antelope bears its young well furred, eyes open, and able to run about. Newborn mice, however, are blind, naked, and helpless. We all know how long it takes a human baby to gain its footing. Human growth is in fact slower than that of any other mammal, and this is one of the distinctive attributes that sets us apart from other mammals.

A curious phenomenon that lengthens the gestation period of many mammals is delayed implantation. The blastocyst remains dormant while its implantation in the uterine wall is postponed for periods of a few weeks to several months. For many mammals (for example, bears, seals, weasels, badgers, bats, and many deer) delayed implantation is a device for extending gestation so that the young are born at the time of year that is best for their survival.

Mammalian Populations

A population of animals includes all the members of a species that share a particular space and potentially interbreed (see Chapter 5). All mammals (like other organisms) live in ecological communities, each composed of numerous populations of different animal and plant species. Each species is affected by the activities of other species and by other changes, especially climatic, that occur. Thus populations are always changing in size. Populations of small mammals are lowest before the breeding season and greatest just after the addition of the new members. Beyond these expected changes in population size, mammalian populations may fluctuate from other causes.

Irregular fluctuations are commonly produced by variations in climate, such as unusually cold, hot, or dry weather, or by natural catastrophes, such as fires, hailstorms, and hurricanes. These are **density-independent** causes because they affect a population whether it is crowded or dispersed. However, the most spectacular fluctuations are **density dependent;** that is, they correlate with population crowding (density-dependent and density-independent causes of growth limitation are discussed on p. 118).

Cycles of abundance are common among many rodent species. One of the best known examples is the mass migrations of the Scandinavian and arctic North American lemmings following population peaks. Lemmings (Figure 31-20) breed all year, although more in the summer than in the winter. The gestation period is only 21 days; young born at the beginning of the summer are weaned in 14 days and are capable of reproducing by the end of the summer. At the peak of their population density, having devastated the vegetation by tunneling and grazing, lemmings begin long, mass migrations to find new undamaged

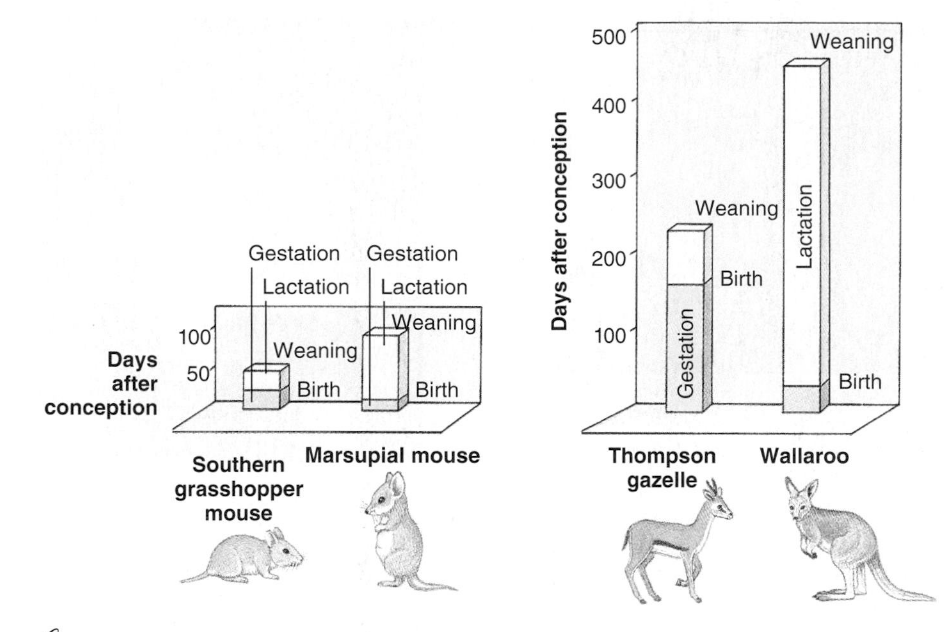

figure 31-19

Comparison of gestation and lactation periods between matched pairs of ecologically similar species of marsupial and placental mammals. The graph shows that marsupials have shorter intervals of gestation and much longer intervals of lactation than in similar species of placentals.

figure 31-20

Collared lemming, *Dicrostonyx* sp., a small rodent of the far north. Populations of lemmings fluctuate widely. Order Rodentia, family Muridae.

habitats for food and space. They swim across streams and small lakes as they go but cannot distinguish these from large lakes, rivers, and the sea, in which they drown. Since lemmings are the main diet of many carnivorous mammals and birds, any change in lemming population density affects all their predators as well.

The renowned fecundity of meadow mice, and the effect of removing the natural predators from rodent populations, is felicitously expressed in this excerpt from Thornton Burgess's "Portrait of a Meadow Mouse."

He's fecund to the nth degree
In fact this really seems to be
His one and only honest claim
To anything approaching fame.
In just twelve months, should all survive,
A million mice would be alive—
His progeny. And this, 'tis clear,
Is quite a record for a year.
Quite unsuspected, night and day
They eat the grass that would be hay.
On any meadow, in a year,
The loss is several tons, I fear.
Yet man, with prejudice for guide,
The checks that nature doth provide
Destroys. The meadow mouse survives
And on stupidity he thrives.

In his book The Arctic *(1974. Montreal, Infacor, Ltd.), Canadian naturalist Fred Bruemmer describes the growth of lemming populations in arctic Canada: "After a population crash one sees few signs of lemmings; there may be only one to every 10 acres. The next year, they are evidently numerous; their runways snake beneath the tundra vegetation, and frequent piles of rice-sized droppings indicate the lemmings fare well. The third year one sees them everywhere. The fourth year, usually the peak year of their cycle, the populations explode. Now more than 150 lemmings may inhabit each acre of land and they honeycomb it with as many as 4000 burrows. Males meet frequently and fight instantly. Males pursue females and mate after a brief but ardent courtship. Everywhere one hears the squeak and chitter of the excited, irritable, crowded animals. At such times they may spill over the land in manic migrations."*

Varying hares (snowshoe rabbit) of North America show 10-year cycles in abundance. The well-known fecundity of rabbits enables them to produce litters of three or four young as many as five times per year. The density may increase to 4000 hares competing for food in each square mile of northern forest. Predators (owls, minks, foxes, and especially lynxes) also increase (Figure 31-21). Then the population crashes precipitously

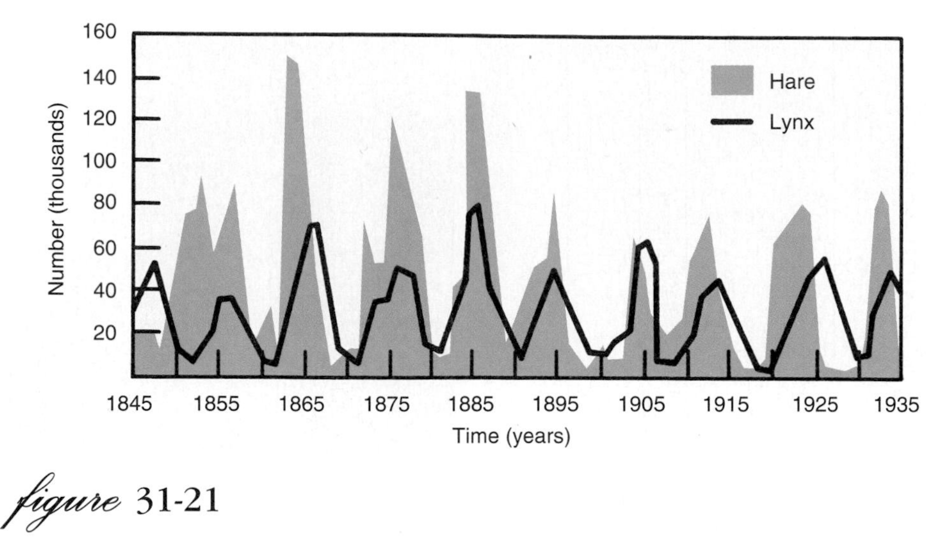

figure 31-21

Changes in population size of varying hare and lynx in Canada as indicated by pelts received by the Hudson's Bay Company. The abundance of lynx (predator) follows that of the hare (prey).

figure 31-22

A prosimian, the Mindanao tarsier, *Tarsius syrichta carbonarius* of Mindanao Island in the Philippines.

for reasons that have long been a puzzle to scientists. Rabbits die in great numbers, not from lack of food or from an epidemic disease (as was once believed) but evidently from some density-dependent psychogenic cause. As crowding increases, hares become more aggressive, show signs of fear and defense, and stop breeding. The entire population reveals symptoms of pituitary-adrenal gland exhaustion, an endocrine imbalance called "shock disease," which results in death. These dramatic crashes are not well understood. Whatever the causes, population crashes that follow superabundance, although harsh, permit the vegetation to recover, thus providing the survivors with a much better chance for successful breeding.

Human Evolution

Darwin devoted an entire book, *The Descent of Man and Selection in Relation to Sex,* largely to human evolution. The idea that humans shared common descent with apes and other animals was repugnant to the Victorian world, which responded with predictable outrage (Figure 4-16, p. 97). When Darwin's views were first debated, few human fossils had been unearthed, but the current accumulation of fossil evidence has strongly vindicated Darwin's belief that humans descended from primate ancestors. All primates share certain significant characteristics: grasping fingers on all four limbs, flat fingernails instead of claws, and forward-pointing eyes with binocular vision and excellent depth perception. The following synopsis highlights the current hypotheses of the evolutionary descent of the primates.

The earliest primate was probably a small, nocturnal animal similar in appearance to tree shrews. This ancestral primate stock split into two major lineages, one of which gave rise to the **prosimians,** which include lemurs, tarsiers (Figure 31-22), and lorises, and the other to the **simians,** which include the monkeys (Figure 31-23) and apes (Figure 31-24). Prosimians and many simians are arboreal (tree-dwellers), which is probably the ancestral life-style for both groups. Arboreality probably selected for increased intelligence. Flexible limbs are essential for active animals moving through trees. Grasping hands and feet, in contrast to the clawed feet of squirrels and other rodents, enable the primates to grip limbs, hang from branches, seize food and manipulate it, and, most significantly, use tools. Highly developed sense organs, especially good vision, and proper coordination of limb and finger muscles are essential for an active arboreal life. Of course, sense organs are no better than the brain processing the sensory information. Precise timing, judgment of distance, and alertness require a large cerebral cortex.

The earliest simian fossils appeared in Africa some 40 million years ago. Many of these primates became day-active rather than nocturnal, making vision the dominant special sense, now enhanced by color vision. We recognize three major simian groups whose precise phylogenetic relationships are unknown. These are (1) the New World monkeys of South America (ceboids), including the howler monkey (Figure 31-23A), spider monkey, and the tamarins; (2) the Old World monkeys (cercopithecoids), including baboons (Figure 31-23B), mandrills, and colobus monkeys; and (3) the anthropoid apes (Figure 31-24). In addition to their geographic separation, Old World monkeys differ from New World monkeys in lacking a grasping tail while having close-set nostrils, better opposable, grasping thumbs, and more advanced teeth. Apes first appear in 25-million-year-old fossils. At this time the woodland savannas were arising in Africa, Europe, and North America. Perhaps motivated by the greater abundance of food on the ground, these apes left the trees and became largely terrestrial. Because of the benefits of standing upright (better view of predators, freeing of hands for using tools and gathering food), emerging hominids gradually evolved upright posture.

figure 31-23

Monkeys. **A,** Red-howler monkeys, an example of the New World monkeys. **B,** The olive baboon, an example of the Old World monkeys.

figure 31-24

The gorilla, an example of the anthropoid apes.

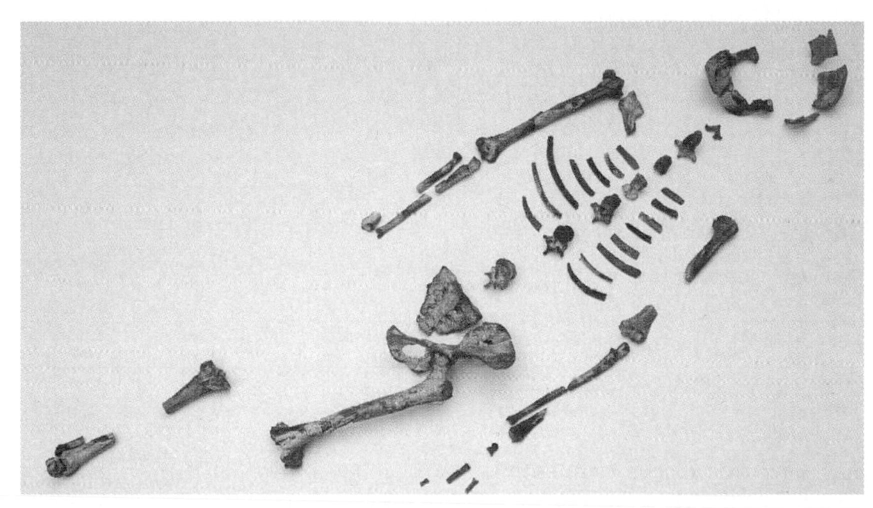

figure 31-25

Lucy *(Australopithecus afarensis),* the most nearly complete skeleton of an early hominid ever found. Lucy is dated at 2.9 million years old. A nearly complete skull of *A. afarensis* was discovered in 1994.

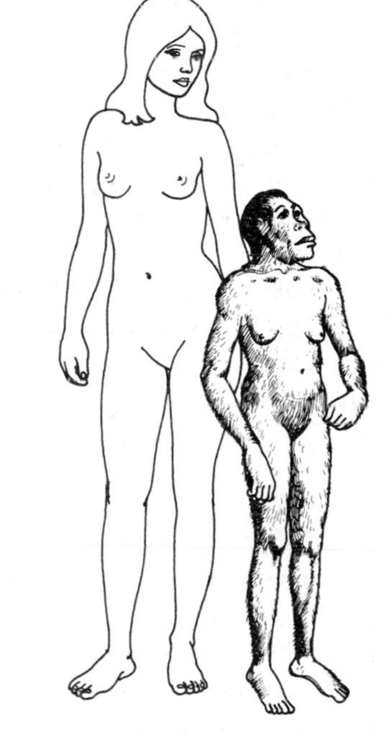

figure 31-26

A reconstruction of the appearance of Lucy *(right)* compared with a modern human *(left).*

Evidence of the earliest hominids of this period is sparse. Not until about 4.4 million years ago, after a lengthy fossil gap, do the first "near humans" appear in the fossil record. The best documented of early hominid fossils is *Australopithecus afarensis,* a short, bipedal hominid with a face and brain size resembling those of a chimpanzee. Numerous fossils of this species have now been unearthed, the most celebrated of which was the 40% complete skeleton of a female discovered in 1974 by Donald Johanson and named "Lucy" (Figures 31-25 and 31-26). Many paleoanthropologists believe that *Australopithecus afarensis* represents the ancestral stock of all human and humanlike forms that followed (Figure 31-27).

Until very recently, *A. afarensis,* dated to nearly 4 million years ago, was the oldest known hominid. However, in 1994 a 4.4 million-year-old hominid fossil named *Ardipithecus ramidus* was uncovered in Africa. *Ardipithecus ramidus,* a mosaic of primitive ape-like traits and derived hominid traits, appears to be ancestral to the australopithecine species. In 1995 a new species called *Australopithecus anamensis* was discovered, dated to 4.2 million years ago. Some paleoanthropologists

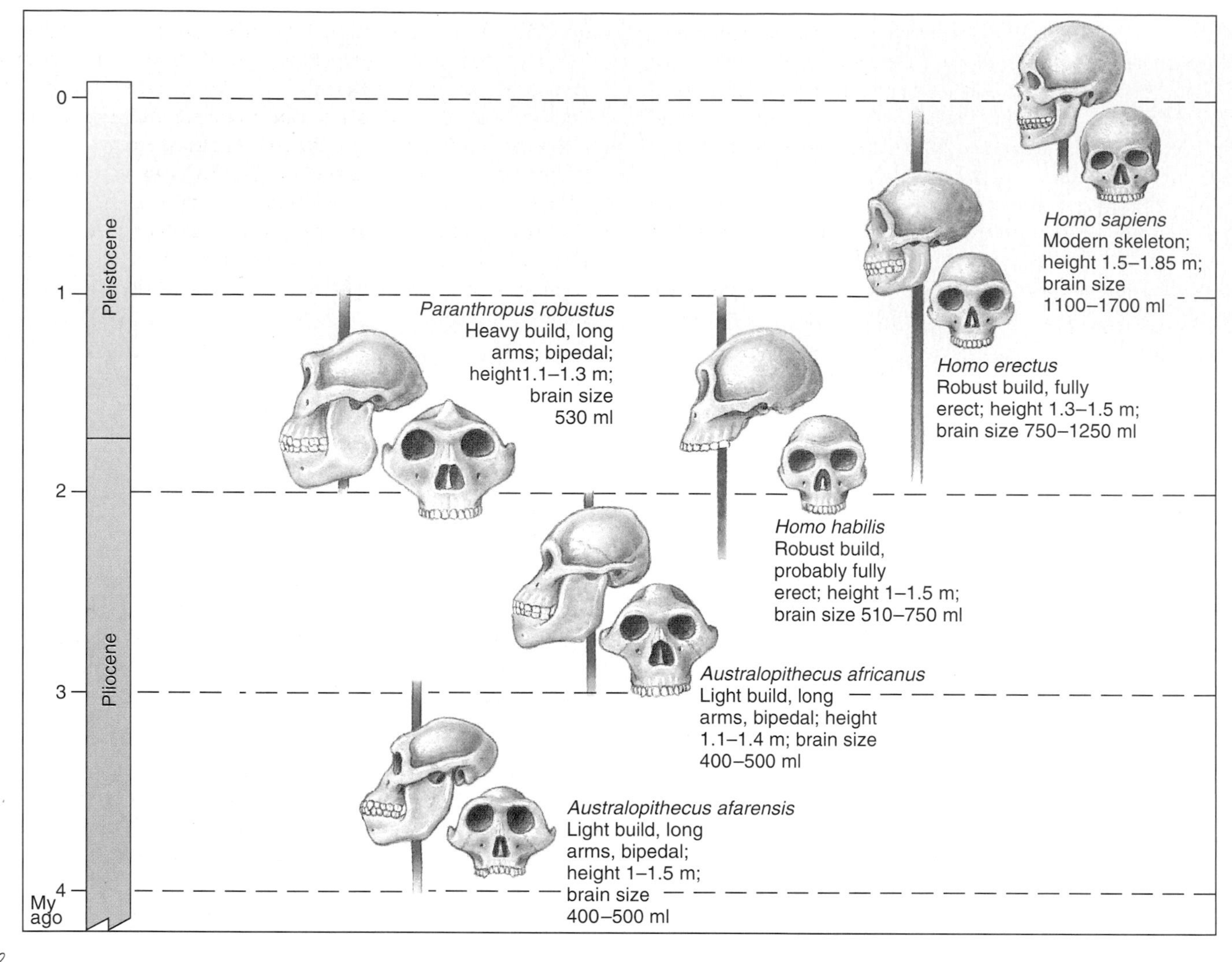

figure 31-27

Hominid skulls, showing several of the best-known hominid lines preceding modern humans *(Homo sapiens).* The time span of existence for each species, as indicated by the fossil record, is suggested by the vertical red lines.

believe that this fossil may be intermediate in a lineage leading from *A. ramidus* to *A. afarensis;* others argue for caution and await new fossil discoveries.

Between 3 and 4 million years ago two quite separate hominid lines emerged that coexisted for at least 2 million years. One was the bipedal *Australopithecus africanus,* the "southern African ape," with a brain size only about one-third as large as that of modern humans. A different line of australopithecines was large and robust (*Paranthropus robustus,* Figure 31-27) and probably approached the size of a gorilla. Until they became extinct between 1.75 and 1 million years ago, the australopithecines shared the countryside with an advanced, fully erect hominid, *Homo habilis,* the first true human (Figure 31-27). *Homo habilis,* meaning "able man," was more lightly built but larger brained than the australopithecines and unquestionably used stone and bone tools. This species appeared about 2 million years ago and survived for perhaps half a million years.

About 1.5 million years ago *Homo erectus* appeared, probably as a descendant of *Homo habilis. Homo erectus* was a large hominid standing 150 to 170 cm (5 to 5½ feet) tall, with a low but distinct forehead, strong brow ridges, and a brain capacity of around 1000 cc (about intermediate between the brain capacity of *Homo habilis* and modern humans) (Figure 31-27). *Homo erectus* was a social species living in tribes of 20 to 50, had a successful and complex culture, and became widespread throughout the tropical and temperate Old World.

After the disappearance of *Homo erectus* about 300,000 years ago, subsequent human evolution and the establishment of *Homo sapiens* ("wise man") threaded a complex course. From among the many early subcultures of *Homo sapiens,* the Neanderthals emerged about 130,000 years ago. With a brain capacity well within the range of modern humans, the Neanderthals were proficient hunters and tool users. They dominated the Old World in the late Pleistocene epoch. About 30,000 years ago the Neanderthals were replaced and quite possibly exterminated by modern humans, tall people with a culture very different from that of the Neanderthals. Implement crafting developed rapidly, and human culture became enriched with aesthetics, artistry, and sophisticated language.

Biologically, *Homo sapiens* is a product of the same processes that have directed the evolution of every organism from the time of life's origin. Mutation, isolation, genetic drift, and natural selection have operated for us as they have for other animals. Yet we have what no other animal has, a nongenetic cultural evolution that provides a constant feedback between past and future experience. Our symbolic languages, capacities for conceptual thought, knowledge of our history, and abilities to manipulate our environment emerge from this nongenetic cultural endowment. Finally, we owe much of our cultural and intellectual achievements to our arboreal ancestry which bequeathed us with binocular vision, superb visuotactile discrimination, and manipulative skills in the use of our hands. If the horse (with one toe instead of five fingers) had human mental capacity, could it have accomplished what humans have?

Classification of Living Mammalian Orders[1]

All modern mammals are placed in two subclasses, the Prototheria, containing the monotremes, and the Theria, containing the marsupials and the placentals. Of the 19 recognized placental orders, seven of the small orders are omitted from the following classification.

Class Mammalia

Subclass Prototheria (pro′to-thir′ee-a) (Gr. *prōtos,* first, + *thēr,* wild animal). Cretaceous and early Cenozoic mammals. Extinct except for the egg-laying monotremes.

Infraclass Ornithodelphia (or′ni-tho-del′fee-a) (Gr. *ornis,* bird, + *delphys,* womb). Monotreme mammals.

Order Monotremata (mon′o-tre′ma-tah) (Gr. *monos,* single, + *trēma,* hole): **egg-laying (oviparous) mammals: duck-billed platypus, spiny anteater.** Three species in this order from Australia, Tasmania, and New Guinea; most noted member of order is the duck-billed platypus *(Ornithorhynchus anatinus);* spiny anteater, or echidna *(Tachyglossus),* has a long, narrow snout adapted for feeding on ants, its chief food.

Subclass Theria (thir′ee-a) (Gr. *thēr,* wild animal). Marsupial and placental mammals.

Infraclass Metatheria (met′a-thir′e-a) (Gr. *meta,* after, + *thēr,* wild animal). Marsupial mammals.

Order Marsupialia (mar-su′pe-ay′le-a) (Gr. *marsypion,* little pouch): **viviparous pouched mammals: opossums, kangaroos, koalas, Tasmanian wolves, wombats, bandicoots, numbats, and others.** Mammals characterized by an abdominal pouch, the **marsupium,** in which they rear their young; young nourished via a yolk-sac placenta; mostly Australian with representatives in the Americas; 260 species.

Infraclass Eutheria (yu-thir′e-a) (Gr. *eu,* true, + *thēr,* wild animal). The viviparous placental mammals.

Order Insectivora (in-sec-tiv′o-ra) (L. *insectum,* an insect, + *vorare,* to devour): **insect-eating mammals: shrews** (Figure 31-28), **hedgehogs, tenrecs, moles.** Small, sharp-snouted animals with primitive characters that feed principally on insects; 390 species.

Order Chiroptera (ky-rop′ter-a) (Gr. *cheir,* hand, + *pteron,* wing): **bats.** Flying mammals with forelimbs modified into wings; use of echolocation by most bats; most nocturnal; second largest mammalian order, exceeded in species numbers only by the order Rodentia; 986 species.

Order Primates (pry-may′teez) (L. *prima,* first). **prosimians, monkeys, apes, humans.** First in the animal kingdom in brain development with especially large cerebral hemispheres; mostly arboreal, apparently derived from insectivores with retention of many primitive characteristics; five digits (usually provided with flat nails) on both forelimbs and hindlimbs; group singularly lacking in claws, scales, horns, and hooves; two suborders[2]; 233 species.

Suborder Prosimii (pro-sim′ee-i) (Gr. *pro,* before, + *simia,* ape): **lemurs, bush babies, tarsiers, lorises, pottos.** Arboreal, mostly nocturnal, primates restricted to the tropics of the Old World.

Suborder Anthropoidea (an′thro-poy′de-a) (Gr. *anthropos,* man): **monkeys, gibbons, apes, humans.**

Order Xenarthra (ze-nar′thra) (Gr. *xenos,* intrusive, + *arthron,* joint) (formerly Edentata

Continued on next page

[L. *edentatus,* toothless]): **anteaters, armadillos, sloths.** Either toothless (anteaters) or with simple peglike teeth (sloths and armadillos); restricted to South and Central America with the nine-banded armadillo in the southern United States; 30 species.

Order Lagomorpha (lag′o-mor′fa) (Gr. *lagos,* hare; + *morphē,* form): **rabbits, hares, pikas** (Figure 31-29). Dentition resembling that of rodents but with four upper incisors rather than two as in rodents; 69 species.

Order Rodentia (ro-den′che-a) (L. *rodere,* to gnaw): **gnawing mammals: squirrels** (Figure 31-30), **rats, woodchucks.** Most numerous of all mammals both in numbers and species; dentition with two upper and two lower chisel-like incisors that grow continually and are adapted for gnawing; 1814 species.

Order Cetacea (see-tay′she-a) (L. *cetus,* whale): **whales** (Figure 31-31), **dolphins, porpoises.** Anterior limbs of cetaceans modified into broad flippers; posterior limbs absent; nostrils represented by a single or double blowhole on top of the head; teeth, when present, all alike and lacking enamel; hair limited to a few hairs on muzzle, no skin glands except the mammary and those of eye; no external ear; 79 species.

Order Carnivora (car-niv′o-ra) (L. *caro,* flesh, + *vorare,* to devour): **flesh-eating mammals: dogs, wolves, cats, bears** (Figure 31-32), **weasels.** All with predatory habits; teeth especially adapted for tearing flesh; in most, canines used for killing prey; worldwide except in the Australian and Antarctic regions; 240 species.

Order Pinnipedia (pi-ni-peed′e-a) (L. *pinna,* feather, + *ped,* foot): **sea lions, seals, and walruses.** Aquatic carnivores with limbs modified as flippers for swimming; nearly all marine; food consists mostly of fish; 34 species.

Order Proboscidea (pro′ba-sid′e-a) (Gr. *proboskis,* elephant's trunk, from *pro,* before, + *boskein,* to feed): **proboscis mammals: elephants.** Living land animals, have two upper incisors elongated as tusks, and the molar teeth are well developed; two extant species: the Indian elephant, with relatively small ears, and the African elephant, with large ears.

Order Perissodactyla (pe-ris′so-dak′ti-la) (Gr. *perissos,* odd, + *dactylos,* toe): **odd-toed hoofed mammals: horses, asses, zebras, tapirs, rhinoceroses.** Mammals with an odd number (one or three) of toes and with well-developed hooves (Figure 31-33); all herbivorous; both Perissodactyla and Artiodactyla often referred to as ungulates, or hoofed mammals, with teeth adapted for chewing; 17 species.

Order Artiodactyla (ar′te-o-dak′ti-la) (Gr. *artios,* even, + *daktylos,* toe): **even-toed hoofed mammals: swine, camels, deer and their allies, hippopotamuses, antelopes, cattle, sheep, goats.** Each toe sheathed in a cornified hoof; most have two toes, although the hippopotamus and some other have four (Figure 31-33); many, such as the cow, deer, and sheep, with horns or antlers; many are ruminants, that is, herbivores with partitioned stomachs; 211 species.

---

[1]Based on Nowak, R.M. 1991. Walker's Mammals of the world, ed. 5. Baltimore, The Johns Hopkins University Press.

[2]G.G. Simpson's (1945) division of the primates into prosimian and anthropoid suborders is followed here, but it is no longer recognized by many mammalogists who hold conflicting views on classification, especially between the order and family levels.

figure 31-28

The shorttail shrew, *Blarina brevicauda,* eating a grasshopper. This tiny but fierce mammal, with a prodigious appetite for insects, mice, snails, and worms, spends most of its time underground and so is seldom seen by humans. Shrews are believed to resemble the insectivorous ancestors of placental mammals. Order Insectivora, family Soricidae.

figure 31-29

A pika, *Ochotona princeps,* atop a rockslide in Alaska. This little rat-sized mammal does not hibernate but prepares for winter by storing dried grasses beneath boulders. Order Lagomorpha.

figure 31-30

Eastern gray squirrel, *Sciurus carolinensis.* This common resident of Eastern towns and hardwood forests serves as an important reforestation agent by planting numerous nuts that sprout into trees. Order Rodentia, family Sciuridae.

figure 31-31

Humpback whale, *Megaptera novaeangliae,* breaching. Among the most acrobatic of whales, humpbacks appear to breach to stun fish schools or to communicate information to other herd members. Order Cetacea, family Balaenopteridae.

figure 31-32

Grizzly bear, *Ursus horribilis,* of Alaska. Grizzlies, once common in the lower 48 states, are now confined largely to northern wilderness areas. Order Carnivora, family Ursidae.

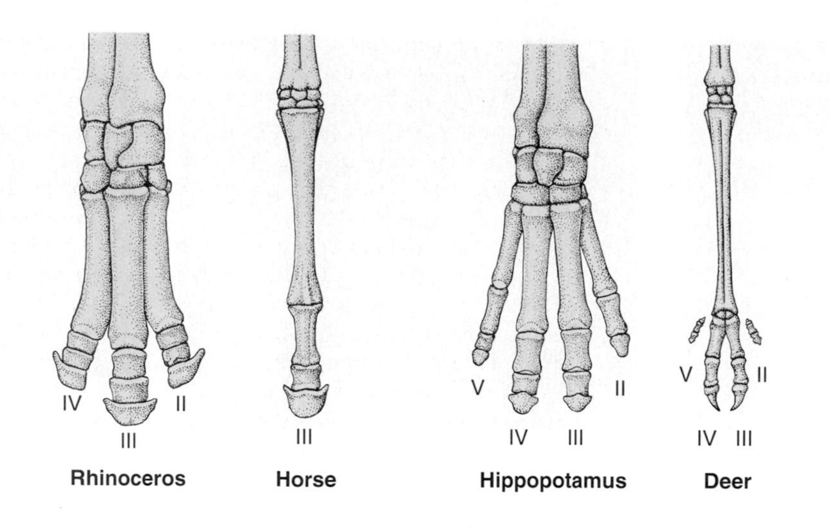

figure 31-33

Odd-toed and even-toed ungulates. The rhinoceros and horse (order Perissodactyla) are odd-toed; the hippopotamus and deer (order Artiodactyla) are even-toed. The lighter, faster mammals run on only one or two toes.

Summary

Mammals are endothermic and homeothermic vertebrates whose bodies are insulated by hair and who nurse their young with milk. The approximately 4450 species of mammals are descended from the synapsid lineage of amniotes that arose in the Carboniferous period of the Paleozoic era. Their evolution can be traced from the pelycosaurs of the Permian period to the therapsids of the late Permian and Triassic periods of the Mesozoic era. One group of the therapsids, the cynodonts, gave rise during the Triassic to the therians, the true mammals. Mammalian evolution was accompanied by the appearance of many important derived characteristics, among these the enlarged brain with greater sensory integration, high metabolic rate, endothermy, and many changes in the skeleton that supported a more active life. Mammals diversified rapidly during the Tertiary period of the Cenozoic era.

Mammals are named for the glandular milk-secreting organs of the female (rudimentary in the male), a unique adaptation which, combined with prolonged parental care, buffers the infants from the demands of foraging for themselves and eases the transition to adulthood. Hair, the integumentary outgrowth that covers most mammals, serves variously for mechanical protection, thermal insulation, protective coloration, and waterproofing. Mammalian skin is rich in glands: sweat glands that function in evaporative cooling, scent glands used in social interactions, and sebaceous glands that secrete lubricating skin oil. All placental mammals have deciduous teeth that are replaced by permanent teeth (diphyodont dentition). The four groups of teeth—incisors, canines, premolars, and molars—may be highly modified in different mammals for specialized feeding tasks, or they may be absent.

The food habits of mammals strongly influence their body form and physiology. Insectivores feed mainly on insects and other small invertebrates. Herbivorous mammals have special adaptations for harboring the intestinal bacteria that break down cellulose of plant materials, and they have developed adaptations for detecting and escaping predators. Carnivorous mammals feed mainly on herbivores, have a simple digestive tract, and have developed adaptations for a predatory life. Omnivores feed on both plant and animal foods.

Some marine, terrestrial, and aerial mammals migrate; some migrations, such as those of fur seals and caribou, are extensive. Migrations are usually made toward favorable climatic and optimal food and calving conditions, or to bring the sexes together for mating.

Mammals with true flight, the bats, are nocturnal and thus avoid direct competition with birds. Most employ ultrasonic echolocation to navigate and feed in darkness.

The living mammals with the most primitive characters are the egg-laying monotremes of the Australian region. After hatching, the young are nourished with the mother's milk. All other mammals are viviparous. Embryos of marsupials have brief gestation periods, are born underdeveloped, and complete their early growth in the mother's pouch, nourished by milk. The remaining 19 of the 21 orders of mammals are eutherians, mammals that develop an advanced placental attachment between mother and embryos through which the embryos are nourished for a prolonged period.

Mammal populations fluctuate from both density-dependent and density-independent causes and some mammals, particularly rodents, may experience extreme cycles of abundance in population density. The unqualified success of mammals as a group cannot be attributed to greater organ system perfection, but rather to their impressive overall adaptability—the capacity to fit more perfectly in total organization to environmental conditions and thus exploit virtually every habitat on earth.

Darwinian evolutionary principles give us great insight into our own origins. Humans are primates, a mammalian group that descended from a shrewlike ancestor. The common ancestor of all modern primates was arboreal and had grasping fingers and forward-facing eyes capable of binocular vision. Primates radiated over the last 80 million years to form two major lines of descent: the prosimians (lemurs, lorises, and tarsiers) and the simians (monkeys, apes, and hominids). The earliest hominids appeared about 4.4 million years ago. The bipedal australopithecines that appeared about 4 million years ago gave rise to, and coexisted with, the species *Homo habilis,* the first user of stone tools. *Homo erectus* appeared about 1.5 million years ago and was eventually replaced by *Homo sapiens* some 300,000 years ago.

Review Questions

1. Describe the evolution of mammals, tracing the synapsid lineage from early amniote ancestors to true mammals. How would you distinguish the skull of a synapsid from that of diapsid?
2. Describe some of the structural and functional adaptations that appeared in the early amniotes that foreshadowed the mammalian body plan. Which mammalian attributes do you think were especially important to the successful radiation of mammals?
3. Hair is believed to have evolved in the therapsids as an adaptation for insulation, but modern mammals have adapted hair for several other purposes. Describe these.
4. What is distinctive about each of the following: ruminant horns, deer antlers, and the rhinoceros horn? Briefly describe the growth cycle of antlers.
5. Describe the location and principal function(s) of each of the following skin glands: sweat glands (of two kinds, eccrine and apocrine), scent glands, sebaceous glands, and mammary glands.
6. Define the terms "diphyodont" and "heterodont" and explain how both terms apply to mammalian dentition.
7. Describe the food habits of each of the following groups: insectivores, herbivores, carnivores, and omnivores. Can you give the common names of some mammals belonging to each group?
8. Most herbivorous mammals depend on cellulose as their main energy source, yet no mammal synthesizes cellulose-splitting enzymes. How are the digestive tracts of mammals specialized for digestion of cellulose?
9. Describe the annual migrations of barren-ground caribou and fur seals.
10. Explain what is distinctive about the life habit and mode of navigation in bats.
11. Describe and distinguish the patterns of reproduction in monotremes, marsupials, and placental mammals. What aspects of mammalian reproduction are present in *all* mammals but in no other class of vertebrates?
12. What is the difference between density-dependent and density-independent causes of fluctuations in the size of mammalian populations?
13. Describe the hare-lynx population cycle, considered a classic example of a prey-predator relationship (Figure 31-21). From your examination of the cycle, can you formulate a hypothesis to explain the oscillations?
14. What anatomical characteristics set the primates apart from other mammals?
15. What role does the fossil named "Lucy" play in the reconstruction of human evolutionary history?
16. In what ways do the genera *Australopithecus* and *Homo,* which coexisted for at least 2 million years, differ?
17. When approximately did the different species of *Homo* appear and how did they differ socially?
18. What major attributes make the human position in animal evolution unique?

Selected References

Eisenberg, J. F. 1981. The mammalian radiations: an analysis of trends in evolution, adaptation, and behavior. Chicago, University of Chicago Press. *Wide-ranging, authoritative synthesis of mammalian evolution and behavior.*

Grzimek's encyclopedia of mammals. 1990. vol. 1-5. New York, McGraw-Hill Publishing Company. *Valuable source of information on all mammalian orders.*

Jones, S., R. D. Martin, and D. Pilbeam. 1992. Cambridge encyclopedia of human evolution. Cambridge, England, Cambridge University Press. *Comprehensive and informative encyclopedia written for the nonspecialist. Highly readable and highly recommended.*

Macdonald, D., ed. 1984. The encyclopedia of mammals. New York, Facts on File Publications. *Coverage of all mammalian orders and families, enhanced with fine photographs and color artwork.*

Nowak, R. M. 1991. Walker's mammals of the world, ed. 5. Baltimore, The Johns Hopkins University Press. *The definitive illustrated reference work on mammals, with descriptions of all extant and recently extinct species.*

Preston-Mafham, R., and K. Preston-Mafham. 1992. Primates of the world. New York, Facts on File Publications. *A small "primer" with high quality photographs and serviceable descriptions.*

Rice, J. A., ed. 1994. The marvelous mammalian parade. Natural History **103**(4):39-91. *A special multi-authored section on mammalian evolution.*

Rismiller, P. D., and R. S. Seymour. 1991. The echidna. Sci. Am. **294:**96-103 (Feb.). *Recent studies of this fascinating monotreme have revealed many secrets of its natural history and reproduction.*

Savage, R. J. G., and M. R. Long. 1986. Mammal evolution: an illustrated guide. New York, Facts on File Publications. *Profusely illustrated survey of fossil mammals.*

Stringer, C. B. 1990. The emergence of modern humans. Sci. Am. **263:**98-104 (Dec.). *A review of the geographical origins of modern humans.*

Suga, N. 1990. Biosonar and neural computation in bats. Sci. Am. **262:**60-68 (June). *How the bat nervous system processes echolocation signals.*

Basic Structure of Matter

For the convenience of students who have not had a course in basic chemistry, or for those who wish to review the material, we are providing a brief introduction here.

table A-1

Some of the Most Important Elements in Living Organisms

Elements	Symbol	Atomic Number	Approximate Atomic Weight
Carbon	C	6	12
Oxygen	O	8	16
Hydrogen	H	1	1
Nitrogen	N	7	14
Phosphorus	P	15	31
Sodium	Na	11	23
Sulfur	S	16	32
Chlorine	Cl	17	35
Potassium	K	19	39
Calcium	Ca	20	40
Iron	Fe	26	56
Iodine	I	53	127

Elements and Atoms

All matter is composed of **elements,** which are substances that cannot be subdivided further by ordinary chemical reactions. Only 92 elements occur naturally, but the elements may be combined by chemical bonds into a vast number of different compounds. The elements are designated by one or two letters derived from their Latin or English names (Table A-1). The elements are composed of discrete units called **atoms,** which are the smallest components into which an element can be subdivided by normal chemical means. Combination of the atoms of an element with each other or with those of other elements by chemical bonds creates **molecules.** When molecules are composed of the atoms of two or more different kinds of elements, they are a **compound.**

In a chemical formula, the symbol for an element stands for one atom of the element, with additional atoms indicated by appropriately placed numbers. Thus atmospheric nitrogen is N_2 (each molecule is composed of two atoms of nitrogen), and water is H_2O (two atoms of hydrogen and one of oxygen in each molecule), and so on.

Subatomic Particles

Each atom is composed of subatomic particles, of which there are three with which we need concern ourselves: protons, neutrons, and electrons. Every atom consists of a positively charged nucleus surrounded by a negatively charged system of electrons (Figure A-1). The nucleus, containing most of the atom's mass, is made up of protons and neutrons clustered together in a very small volume. These two particles have about the same mass, each being about 2000 times heavier than an

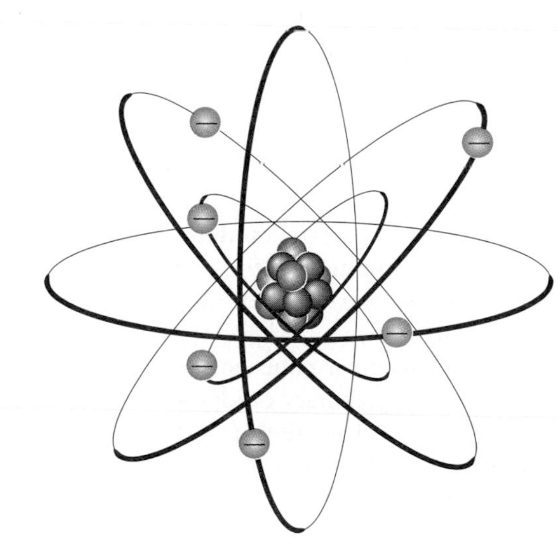

figure A-1

Structure of an atom. Planetary system of negatively charged electrons around a dense nucleus of positively charged protons and uncharged neutrons.

electron. The protons bear positive charges, and the neutrons are uncharged (neutral). Although the number of protons in the nucleus is the same as the number of electrons around the nucleus, the number of neutrons may vary. For every positively charged proton in the nucleus, there is a negatively charged electron; the total charge of the atom is thus neutral.

The **atomic number** of an element is equal to the number of protons in the nucleus, whereas the **atomic mass** is nearly equal to the number of protons plus the number of neutrons (explanation of why atomic mass is not exactly equal to protons plus neutrons can be found in any introductory chemistry text). The mass of the electrons may be neglected because it is only $\frac{1}{1836}$ that of a proton or neutron.

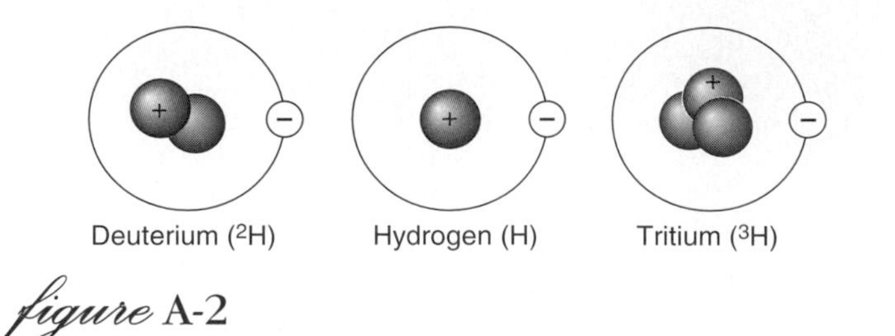

figure A-2

Three isotopes of hydrogen. Of the three isotopes, hydrogen 1 makes up about 99.98% of all hydrogen, and deuterium (heavy hydrogen) makes up about 0.02%. Tritium is radioactive and is found only in traces in water. Numbers indicate approximate atomic weights. Most elements are mixtures of isotopes. Some elements (for example, tin) have as many as 10 isotopes.

Isotopes

It is possible for two atoms of the same element to have the same number of protons in their nuclei but have a different number of neutrons. Such different forms, having the same number of protons but different atomic masses, are called **isotopes.** For example, the predominant form of hydrogen in nature has 1 proton and no neutron (^{1}H) (Figure A-2). Another form (deuterium [^{2}H]) has 1 proton and 1 neutron. Tritium (^{3}H) has 1 proton and 2 neutrons. Some isotopes are unstable and undergo a spontaneous disintegration with the emission of one or more of three types of particles, or rays: gamma rays (a form of electromagnetic radiation), beta rays (electrons), and alpha rays (positively charged helium nuclei stripped of their electrons). These unstable isotopes are said to be **radioactive.** Using radioisotopes, biologists are able to trace movements of elements and tagged compounds through organisms. Our present understanding of metabolic pathways in animals and plants is in large part the result of this powerful analytical tool. Among the commonly used radioisotopes are carbon 14 (^{14}C), tritium, and phosphorus 32 (^{32}P).

Electron "Shells" of Atoms

According to Niels Bohr's planetary model of the atom, the electrons revolve around the nucleus of an atom in circular orbits of precise energy and size. All of the orbits of any one energy and size compose an electron shell. This simplified picture of the atom has been greatly modified by more recent experimental evidence; definite electron pathways are no longer hypothesized, and an electron shell is more vaguely understood as a thick region of space around the nucleus rather than a narrow shell of a particular radius surrounding the nucleus.

However, the old planetary model with the idea of electronic shells is still useful in interpreting chemical phenomena. The number of concentric shells required to contain an element's electrons varies with the element. Each shell can hold a maximum number of electrons. The first shell next to the atomic nucleus can hold a maximum of 2 electrons, and the second shell can hold 8; other shells also have a maximum number, but no atom can have more than 8 electrons in its outermost shell. Inner shells are filled first, and if there are not enough electrons to fill all the shells, the outer shell is left incomplete. Hydrogen has 1 proton in its nucleus and 1 electron in its single orbit but no neutron. Since its shell can hold 2 electrons, it has an incomplete shell. Helium has 2 electrons in its single shell, and its nucleus is made up of 2 protons and 2 neutrons. Since the 2-electron arrangement in helium's shell is the maximum number for this shell, the shell is closed and precludes all chemical activity. There is no known compound of helium. Neon is another inert (chemically inactive) gas because its outer shell contains 8 electrons, the maximum number (Figure A-3). However, stable compounds of xenon (an inert gas) with fluorine and oxygen are formed under special conditions. Oxygen has an atomic number of 8. Its 8 electrons are arranged with 2 in the first shell and 6 in the second shell (Figure A-3). It is active chemically, forming compounds with almost all elements except inert gases.

Chemical Bonds

As we noted above, atoms joined to each other by chemical bonds form molecules, and atoms of each element form molecules with each other or with atoms of other elements in particular ways, depending on the number of electrons in their outer orbits.

Ionic Bonds

Elements react in such a way as to gain a stable configuration of electrons in their outer shells. The number of electrons in the outer shell varies from 0 to 8. With either 0 or 8 in this shell, the element is chemically inactive. When there are fewer than 8 electrons in the outer shell, the atom will tend to lose or gain electrons to have an outer shell of 8. This will give the atom a net electrical charge, and the atom is now an **ion.** Atoms with 1 to 3 electrons in the outer shell tend to lose them to other atoms and to become positively charged ions because of the excess protons in the nucleus. Atoms with 5 to 7 electrons in the outer orbit tend to gain electrons from other atoms and to become negatively charged ions because of the

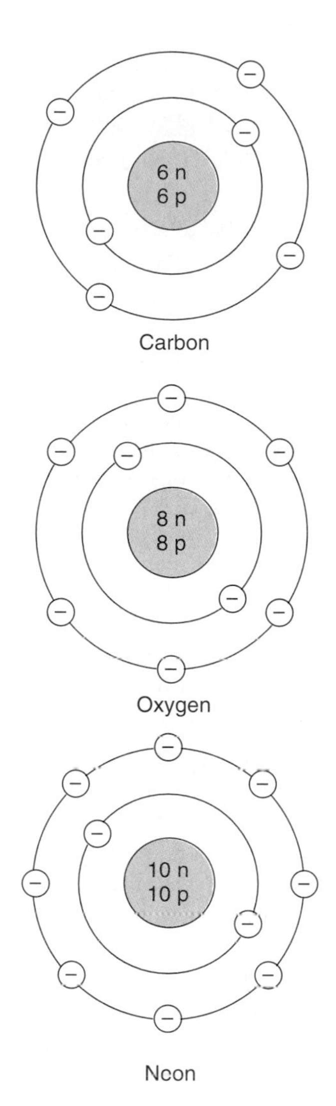

figure A-3

Electron shells of three common atoms. Since no atom can have more than 8 electrons in its outermost shell and 2 electrons in its innermost shell, neon is chemically inactive. However, the second shells of carbon and oxygen, with 4 and 6 electrons, respectively, are open so that these elements are electronically unstable and react chemically whenever appropriate atoms come into contact. Chemical properties of atoms are determined by their outermost electron shells.

greater number of electrons than protons. Positive and negative ions tend to unite.

Every atom has a tendency to complete its outer shell to increase its stability in the presence of other atoms. Let us examine how two atoms with incomplete outer shells, sodium and chlorine, can interact to fill their outer shells. Sodium, with 11 electrons, has 2 electrons in its first shell, 8 in its second shell, and only 1 in the third shell. The third shell is highly incomplete; if this third-shell electron were lost, the second shell would be the outermost shell and would produce a stable atom. Chlorine, with 17 electrons, has 2 in the first shell, 8 in the second, and 7 in the incomplete third shell. Chlorine must gain an electron to fill the outer shell and become a stable atom. Clearly, the transfer of the third-shell sodium electron to the incomplete chlorine third shell would yield simultaneous stability to both atoms.

Sodium, now with 11 protons but only 10 electrons, becomes electropositive (Na^+). In gaining an electron from sodium, chlorine contains 18 electrons but only 17 protons and thus becomes an electronegative chloride ion (Cl^-). Since unlike charges attract, a strong electrostatic force, called an **ionic bond** (Figure A-4), is formed. The ionic compound formed, sodium chloride, can be represented in electron dot notation ("fly-speck formulas") as:

$$Na\bullet + \bullet\ddot{\underset{\cdot\cdot}{Cl}}: \longrightarrow Na^+ + (:\ddot{\underset{\cdot\cdot}{Cl}}:)^-$$

The number of dots shows the number of electrons present in the outer shell of the atom: 7 in the case of the neutral chlorine atom and 8 for the chloride ion; 1 in the case of the neutral sodium atom and none for the sodium ion.

If an element with 2 electrons in its outer shell, such as calcium, reacts with chlorine, it must give them both up, one to each of two chlorine atoms, and calcium becomes doubly positive:

$$Ca: + 2\bullet\ddot{\underset{\cdot\cdot}{Cl}}: \longrightarrow Ca^{2+} + 2(:\ddot{\underset{\cdot\cdot}{Cl}}:)^-$$

Processes that involve the **loss of electrons** are called **oxidation** reactions; those that involve the **gain of electrons** are **reduction** reactions. Since oxidation and reduction always occur simultaneously, each of these processes is really a "half-reaction." The entire reaction is called an **oxidation-reduction** reaction, or simply a **redox** reaction. The terminology is confusing because oxidation-reduction reactions involve electron transfers, rather than (necessarily) any reaction with oxygen. However, it is easier to learn the system than to try to change accepted usage.

Covalent Bonds

Stability can also be achieved when two atoms share electrons. Let us again consider the chlorine atom, which, as we have seen, has an incomplete 7-electron outer shell. Stability is attained by gaining an electron. One way this can be done is for two chlorine atoms to *share* one pair of electrons (Figure A-5). To do this, the two chlorine atoms must *overlap* their third shells so that the electrons in these shells can now spread themselves over both atoms, thereby completing the filling of both shells. Many other elements can form covalent (or electron-pair) bonds. Examples are hydrogen (H_2)

$$H\bullet + H\bullet \longrightarrow H:H$$

and oxygen (O_2)

$$\ddot{\underset{\cdot\cdot}{O}}: + :\ddot{\underset{\cdot\cdot}{O}} \longrightarrow \ddot{\underset{\cdot\cdot}{O}}::\ddot{\underset{\cdot\cdot}{O}}$$

In this case oxygen must share two pairs of electrons to achieve stability. Each atom now has 8 electrons available to its outer shell, the stable number.

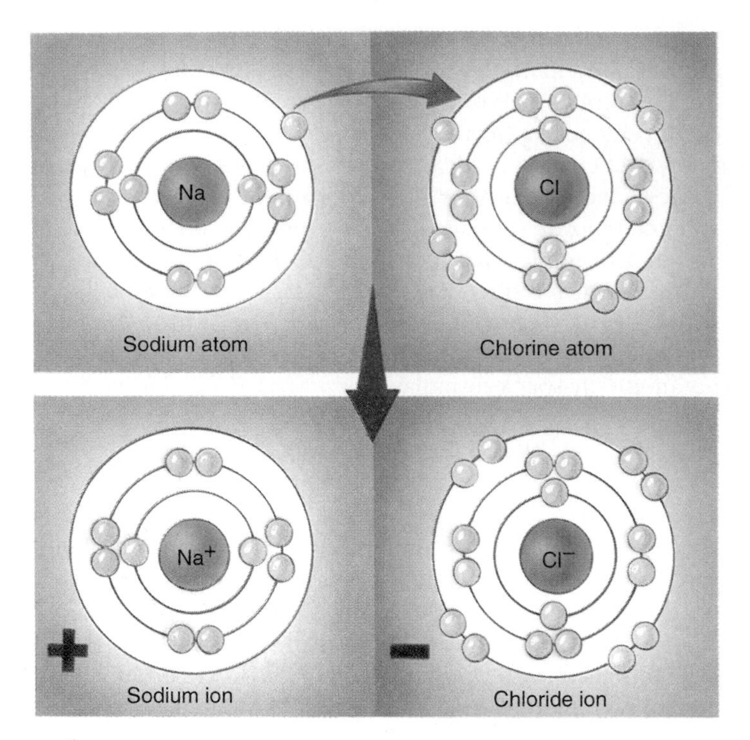

figure A-4

Ionic bond. When one atom of sodium and one of chlorine react to form a molecule, a single electron in the outer shell of sodium is transferred to the outer shell of chlorine. This causes the outer or second shell (third shell is now empty) of sodium to have 8 electrons and also chlorine to have 8 electrons in its outer or third shell. The compound thus formed is sodium chloride (NaCl). By losing 1 electron, sodium becomes a positive ion, and by gaining 1 electron, chlorine becomes a negative ion (chloride). This ionic bond is the strong electrostatic force acting between positively and negatively charged ions.

Covalent bonds are critically important to living systems, since the major elements of living matter (carbon, oxygen, nitrogen, hydrogen) almost always share electrons in strong covalent bonds. The stability of these bonds is essential to the integrity of DNA and other macromolecules, which, if easily dissociated, would result in biological disorder.

The outer shell of carbon contains 4 electrons. This element is endowed with great potential for forming a variety of atomic configurations with itself and other molecules. It can, for example, share its electrons with hydrogen to form methane (Figure A-6). Carbon now achieves stability with 8 electrons, and each hydrogen atom becomes stable with 2 electrons. Carbon can also bond with itself (and hydrogen) to form, for example, ethane:

```
   H  H              H  H
   ..  ..            |  |
 H:C:C:H    or    H—C—C—H
   ..  ..            |  |
   H  H              H  H
```

Carbon also forms covalent bonds with oxygen:

```
•C• + 2 O: → O::C::O
```

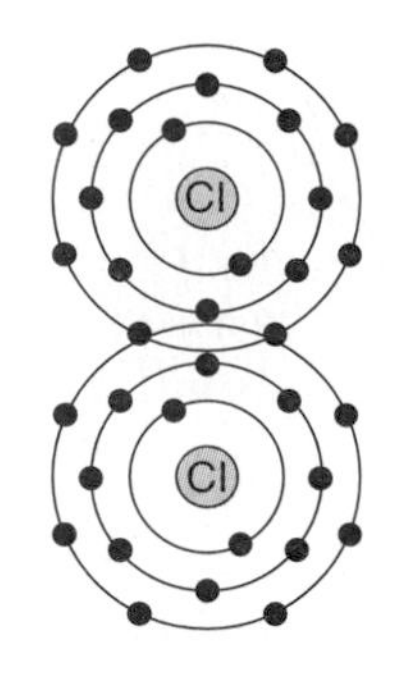

figure A-5

Covalent bond. Each chlorine atom has 7 electrons in its outer shell, and by sharing one pair of electrons, each atom acquires a complete outer shell of 8 electrons, thus forming a molecule of chlorine (Cl_2).

This is a "double-bond" configuration usually written as O=C=O. Carbon can even form triple bonds as, for example, in acetylene:

```
H:C:::C:H or H—C≡C—H
```

The significant aspect of each of these molecules is that each carbon gains a share in 4 electrons from atoms nearby, thus attaining the stability of 8 electrons. The sharing may occur between carbon and other elements or other carbon atoms, and in many instances the 8-electron stability is achieved by means of multiple bonds.

These examples only begin to illustrate the amazing versatility of carbon. It is a part of virtually all compounds comprising living substance, and without carbon, life as we know it would not exist.

Hydrogen Bonds

Hydrogen bonds are described as "weak" bonds because they require little energy to break. They do not form by transfer or sharing of electrons, but result from unequal charge distribution on a molecule, so that the molecule is polar. For example, the two hydrogen atoms that share electrons with an oxygen atom to form water (H_2O) are not 180 degrees away from each other around the oxygen, but form an angle of about 105 degrees (Figure A-7). Thus the side of the molecule away from the hydrogen atoms is more negative, and the hydrogen side is more positive (contrast the methane molecule [Figure A-6], in which the equidistant placement of hydrogen atoms cancels out charge displacements). The electrostatic attraction between the electropositive part of one molecule forms a hydrogen bond with the electronegative part of an adjacent molecule. The ability of water molecules to form hydrogen bonds with each other (Figure A-8) accounts for many unusual properties of this unique substance. Hydrogen bonds are important in the formation and function of other biologically active substances, such as proteins and nucleic acids (pp. 10-11).

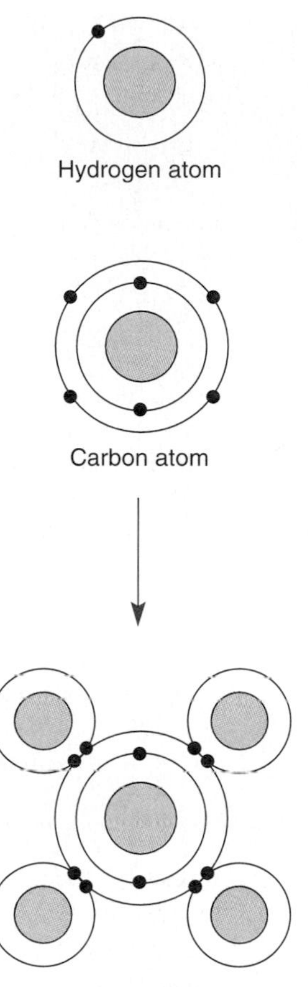

figure A-6

In methane the four hydrogen atoms each share an electron with a carbon atom. They are arranged symmetrically around the carbon atom and form a pyramid-shaped tetrahedron in which each of the hydrogen atoms is equally distant from the others.

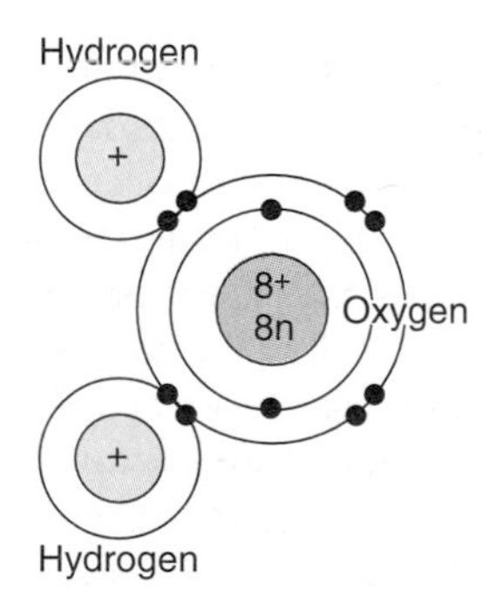

figure A-7

Molecular structure of water. The two hydrogen atoms bonded covalently to an oxygen atom are arranged at an angle of about 105 degrees to each other. Since the electrical charge is not symmetrical, the molecule is polar with positively and negatively charged ends.

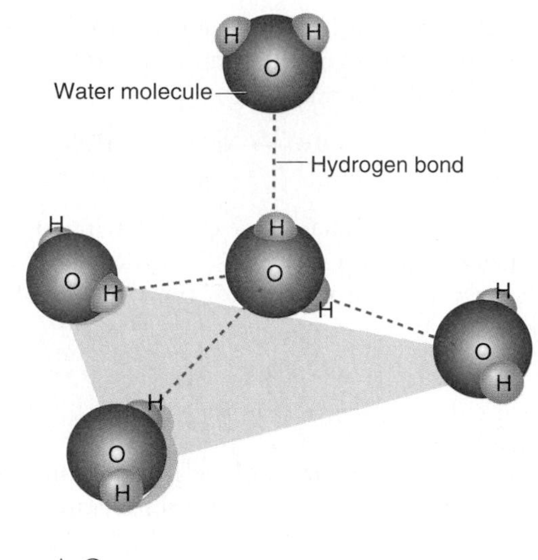

figure A-8

Geometry of water molecules. Each water molecule is linked by hydrogen bonds *(dashed lines)* to four other water molecules. If imaginary lines are used to connect the divergent oxygen atoms, a tetrahedron is obtained. In ice, the individual tetrahedrons associate to form an open lattice structure.

Acids, Bases, and Salts

The hydrogen ion (H^+) is one of the most important ions in living organisms. The hydrogen atom contains a single electron. When this electron is completely transferred to another atom (not just shared with another atom as in the covalent bonds with carbon), only the hydrogen nucleus with its positive proton remains. Any molecule that dissociates in solution and gives rise to a hydrogen ion is an **acid.** An acid may be strong or weak, depending on the extent to which the acid molecule dissociates in solution. Examples of strong acids that dissociate completely in water are hydrochloric acid ($HCl \rightarrow H^+ + Cl^-$) and nitric acid ($HNO_3 \rightarrow H^+ + NO_3^-$). Weak acids, such as carbonic acid ($H_2CO_3 \rightarrow H^+ + HCO_3^-$), dissociate only slightly. A solution of carbonic acid is mostly undissociated carbonic acid molecules with only a small number of bicarbonate (HCO_3^-) and hydrogen ions (H^+) present.

A **base** contains negative ions called hydroxide ions and may be defined as a molecule or ion that will accept a proton (hydrogen ion). Bases are produced when compounds containing them are dissolved in water. Sodium hydroxide (NaOH) is a strong base because it will dissociate completely in water into sodium (Na^+) and hydroxide (OH^-) ions. Among the characteristics of bases is their ability to combine with hydrogen ions, thus decreasing the concentration of the hydrogen ions. Like acids, bases vary in the extent to which they dissociate in aqueous solutions into hydroxide ions.

A **salt** is a compound resulting from the chemical interaction of an acid and a base. Common salt, sodium chloride (NaCl), is formed by the interaction of hydrochloric acid (HCl) and sodium hydroxide (NaOH). In water the HCl is dissociated into H^+ and Cl^- ions. The hydrogen and hydroxide ions combine to form water (H_2O), and the sodium and chloride ions remain as dissolved salt (Na^+Cl^-):

$$\underset{\text{Acid}}{H^+Cl^-} + \underset{\text{Base}}{Na^+OH^-} \longrightarrow \underset{\text{Salt}}{Na^+Cl^-} + H_2O$$

Organic acids are usually characterized by having in their molecule the carboxyl group (—COOH). They are weak acids because a relatively small proportion of the H^+ reversibly dissociates from the carboxyl:

$$\begin{matrix} R\text{—}C\text{=}O \\ | \\ O\text{—}H \end{matrix} \rightleftharpoons \begin{matrix} R\text{—}C\text{=}O \\ | \\ O\text{—} \end{matrix} + H^+$$

R refers to an atomic grouping unique to the molecule. Some common organic acids are acetic, citric, formic, lactic, and oxalic.

Hydrogen Ion Concentration (pH)

Solutions are acidic, basic, or neutral according to the proportion of hydrogen (H^+) and hydroxide (OH^-) ions they possess. In acid solutions there is an excess of hydrogen ions; in alkaline, or basic, solutions the hydroxide ion is more common; and in neutral solutions both hydrogen and hydroxide ions are present in equal numbers.

To express the acidity or alkalinity of a substance, we use a logarithmic scale, a type of mathematical shorthand, from 1 to 14. This is the pH, defined as (where $[H^+]$ is the hydrogen ion concentration):

$$pH = \log_{10}\frac{1}{[H^+]}$$

or

$$pH = -\log_{10}[H^+]$$

Thus pH is the negative logarithm of the hydrogen ion concentration in moles per liter. In other words, when the hydrogen ion concentration is expressed exponentially, pH is the exponent, but with the *opposite* sign; if $[H^+] = 10^{-2}$, then $pH = -(-2) = +2$. Unfortunately, pH can be a confusing concept because, as the $[H^+]$ decreases, the pH increases. Numbers below 7 indicate an acid range, and numbers above 7 indicate alkalinity (Figure A-9). The number 7 indicates neutrality, that is, there are equal numbers of H^+ and OH^- ions present. Because this is a logarithmic scale, a pH of 3 is 10 times more acid than one of 4; a pH of 9 is 10 times more alkaline than one of 8.

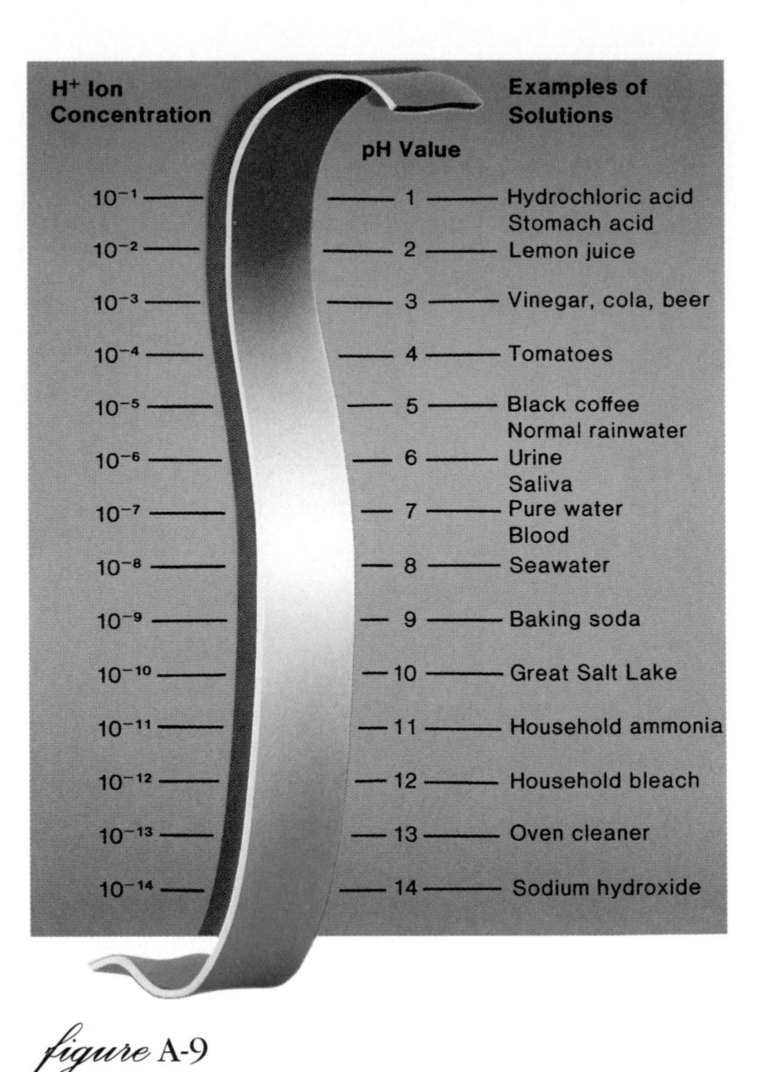

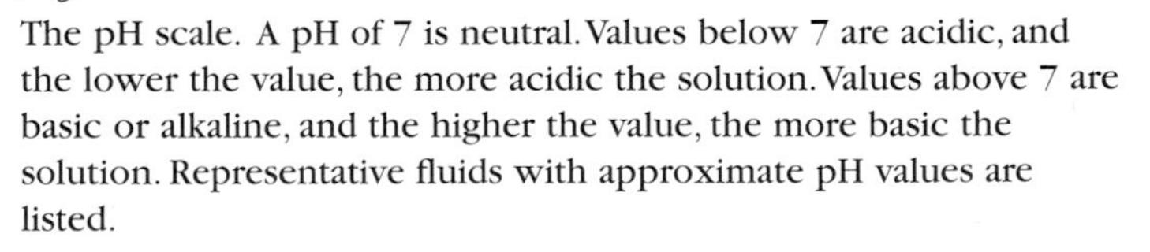

figure A-9

The pH scale. A pH of 7 is neutral. Values below 7 are acidic, and the lower the value, the more acidic the solution. Values above 7 are basic or alkaline, and the higher the value, the more basic the solution. Representative fluids with approximate pH values are listed.

Buffer Action

The hydrogen ion concentration in extracellular fluids must be regulated so that metabolic reactions within the cell will not be adversely affected by a constantly changing hydrogen ion concentration, to which they are extremely sensitive. A change in pH of only 0.2 from the normal mammalian blood pH of about 7.35 can cause serious metabolic disturbances. To maintain pH within physiological limits, there are certain substances in cells and organisms that tend to compensate for any change in the pH when acids or alkalies are produced in metabolic reactions or are added to body fluids. These are called **buffers.** A buffer is a mixture of slightly ionized weak acid and its completely ionized salt. In such a system, added H^+ combines with the anion of the salt to form undissociated acid, and added OH^- combines with H^+ from the weak acid molecule to form water. The most important buffers in blood and other extracellular fluids are the bicarbonates and phosphates, and organic molecules such as amino acids and proteins are important buffers within cells. The bicarbonate buffer system consists of carbonic acid (H_2CO_3, a weak acid) and its salt, sodium bicarbonate ($NaHCO_3$). Sodium bicarbonate is strongly ionized into sodium ions (Na^+) and bicarbonate ions

(HCO_3^-). When a strong acid (for example, HCl) is added to the fluid, the H^+ ions of the dissociated acid will react with the bicarbonate ion (HCO_3^-) to form a very weak acid, carbonic acid, which dissociates only slightly. Thus the H^+ ions from the HCl are removed and the pH is little altered. When a strong base (for example, NaOH) is added to the fluid, the OH^- ions of the strong base will react with the carbonic acid by removing H^+ ions from the H_2CO_3, to form water and bicarbonate ions. Again the H^+ ion concentration in solution is little altered and the pH remains nearly unchanged.

glossary

This glossary lists definitions, pronunciations, and derivations of the most important recurrent technical terms, units, and names (excluding taxa) used in the text.

A

abiotic (ā′bī-äd′ik) (Gr. *a*, without, + *biōtos,* life, livable). Characterized by the absence of life.

abomasum (ab′ō-mā′səm) (L. *ab,* from, + *omasum,* paunch). Fourth and last chamber of the stomach of ruminant mammals.

aboral (ab-ō′rəl) (L. *ab,* from, + *os,* mouth). A region of an animal opposite the mouth.

abscess (ab′ses) (L. *abscessus,* a going away). Dead cells and tissue fluid confined in a localized area, causing swelling.

acanthodians (a′kan-thō′dē-əns) (Gr. *akantha,* prickly, thorny). A group of the earliest known true jawed fishes from Lower Silurian to Lower Permian.

acanthor (ə-kan′thor) (Gr. *akantha,* spine or thorn, + *or*). First larval form of acanthocephalans in the intermediate host.

acclimatization (ə-klī′mə-də-zā-shən) (L. *ad,* to, + Gr. *klima,* climate). Gradual physiological adaptation in response to relatively long-lasting environmental changes.

acetabulum (as′ə-tab′ū-ləm) (L. a little saucer for vinegar). True sucker, especially in flukes and leeches; the socket in the hip bone that receives the thigh bone.

acicula (ə-sik′ū-lə) (L. *acicula,* a small needle). Needlelike supporting bristle in parapodia of some polychaetes.

acid A molecule that dissociates in solution to produce a hydrogen ion (H^+).

acinus (as′ə-nəs), pl. **acini** (as′ə-nī) (L. grape). A small lobe of a compound gland or a saclike cavity at the termination of a passage.

acoelomate (ā-sēl′ə-māt′) (Gr. *a,* not, + *koilōma,* cavity). Without a coelom, as in flatworms and proboscis worms.

acontium (ə-kän′chē-əm), pl. **acontia** (Gr. *akontion,* dart). Threadlike structure bearing nematocysts located on mesentery of sea anemone.

acrocentric (ak′rō-sen′trək) (Gr. *akros,* tip, + *kentron,* center). Chromosome with centromere near the end.

acron (a′crän) (Gr. *akron,* mountaintop, fr. *akros,* tip). Preoral region of an insect.

actin (Gr. *aktis,* ray). A protein in the contractile tissue that forms the thin myofilaments of striated muscle.

active transport Mediated transport in which a permease transports a molecule across a cell membrane against a concentration gradient; requires expenditure of energy; contrast with **facilitated diffusion.**

adaptation (L. *adaptatus,* fitted). An anatomical structure, physiological process, or behavioral trait that evolved by natural selection and improves an organism's ability to survive and leave descendants.

adaptive radiation Evolutionary diversification that produces numerous ecologically disparate lineages from a single ancestral one, especially when this diversification occurs within a short interval of geological time.

adaptive value Degree to which a characteristic helps an organism to survive and reproduce or lends greater fitness in its environment; selective advantage.

adaptive zone A characteristic reaction and mutual relationship between environment and organism ("way of life") demonstrated by a group of evolutionarily related organisms.

adductor (ə-duk′tər) (L. *ad,* to, + *ducere,* to lead). A muscle that draws a part toward a median axis, or a muscle that draws the two valves of a mollusc shell together.

adenine (ad′nēn, ad′ə-nēn) (Gr. *adēn,* gland, + *ine,* suffix). A purine base; component of nucleotides and nucleic acids.

adenosine (ə-den′ə-sen) **(di-, tri) phosphate** (ADP and ATP). A nucleotide composed of adenine, ribose sugar, and two (ADP) or three (ATP) phosphate units; ATP is an energy-rich compound that, with ADP, serves as a phosphate bond–energy transfer system in cells.

adipose (ad′ə-pōs) (L. *adeps,* fat). Fatty tissue; fatty.

adrenaline (ə-dren′ə-lən) (L. *ad,* to, + *renalis,* pertaining to kidneys). A hormone produced by the adrenal, or suprarenal, gland; epinephrine.

adsorption (ad-sorp′shən) (L. *ad,* to, + *sorbeo,* to absorb). The adhesion of molecules to solid bodies.

aerobic (a-rō′bik) (Gr. *aēr,* air, + *bios,* life). Oxygen-dependent form of respiration.

afferent (af′ə-rənt) (L. *ad,* to, + *ferre,* to bear). Adjective meaning leading or bearing toward some organ, for example, nerves conducting impulses toward the brain or blood vessels carrying blood toward an organ; opposed to **efferent.**

aggression (ə-gres′hən) (L. *aggressus,* attack). An offensive action or procedure.

agonistic behavior (Gr. *agōnistēs,* combatant). An offensive action or threat directed toward another organism.

alate (ā′lāt) (L. *alatus,* wing). Winged.

albumin (al-bū′mən) (L. *albumen,* white of egg). Any of a large class of simple proteins that are important constituents of vertebrate blood plasma and tissue fluids and also present in milk, whites of eggs, and other animal substances.

bat/āpe/ärmadillo/herring/fēmale/finch/līce/crocodile/crōw/duck/ūnicorn/tüna/ə indicates unaccented vowel sound "uh" as in mammal, fishes, cardinal, heron, vulture/stress as in bi-ol′o-gy, bi′o-log′i-cal

alimentary (al′ə-mən′tə-rē) (L. *alimentum,* food, nourishment). Having to do with nutrition or nourishment.

allantois (ə-lan′tois) (Gr. *allas,* sausage, + *eidos,* form). One of the extraembryonic membranes of the amniotes that functions in respiration and excretion in birds and reptiles and plays an important role in the development of the placenta in most mammals.

allele (ə-lēl′) (Gr. *allēlōn,* of one another). Alternative forms of genes coding for the same trait; situated at the same locus in homologous chromosomes.

allograft (a′lō-graft) (Gr. *allos,* other, + graft). A piece of tissue or an organ transferred from one individual to another individual of the same species, not identical twins; homograft.

allometry (ə-lom′ə-trē) (Gr. *allos,* other, + *metry,* measure). Relative growth of a part in relation to the whole organism.

allopatric (Gr. *allos,* other, + *patra,* native land). In separate and mutually exclusive geographical regions.

alpha-helix (Gr. *alpha,* first, + L. *helix,* spiral). Literally the first spiral arrangement of the genetic DNA molecule; regular coiled arrangement of polypeptide chain in proteins; secondary structure of proteins.

altricial (al-tri′shəl) (L. *altrices,* nourishers). Referring to young animals (especially birds) having the young hatched in an immature, dependent condition.

alula (al′yə-lə) (L. dim. of *ala,* wing). The first digit or thumb of a bird's wing, much reduced in size.

alveolus (al-vē′ə-ləs) (L. dim. of *alveus,* cavity, hollow). A small cavity or pit, such as a microscopic air sac of the lungs, terminal part of an alveolar gland, or bony socket of a tooth.

ambulacra (am′byə-lak′rə) (L. *ambulare,* to walk). In echinoderms, radiating grooves where podia of water-vascular system characteristically project to outside.

amebocyte (ə-mē′bə-sīt) (Gr. *amoibē,* change, + *kytos,* hollow vessel). Cell in metazoan invertebrate, often functioning in defense against invading particles.

bat/āpe/ärmadillo/herring/fēmale/finch/līce/crocodile/crōw/duck/ūnicorn/tüna/ə indicates unaccented vowel sound "uh" as in mammal, fishes, cardinal, heron, vulture/stress as in bi-ol′o-gy, bi′o-log′i-cal

ameboid (ə-mē′boid) (Gr. *amoibē,* change, + *oid,* like). Ameba-like in putting forth pseudopodia.

amictic (ə-mik′tic) (Gr. *a,* without, + *miktos,* mixed or blended). Pertaining to female rotifers, which produce only diploid eggs that cannot be fertilized, or to the eggs produced by such females. Compare with **mictic.**

amino acid (ə-mē′nō) (amine, an organic compound). An organic acid with an amino group (—NH_2). Makes up the structure of proteins.

amitosis (ā′mī-tō′səs) (Gr. *a,* not, + *mitos,* thread). A form of cell division in which mitotic nuclear changes do not occur; cleavage without separation of daughter chromosomes.

amniocentesis (am′nē-ō-sin-tē′səs) (Gr. *amnion,* membrane around the fetus, + *centes,* puncture). Procedure for withdrawing a sample of fluid around the developing embryo for examination of chromosomes in the embryonic cells and other tests.

amnion (am′nē-än) (Gr. *amnion,* membrane around the fetus). The innermost of the extraembryonic membranes forming a fluid-filled sac around the embryo in amniotes.

amniote (am′nē-ōt). Having an amnion; as a noun, an animal that develops an amnion in embryonic life, that is, reptiles, birds, and mammals.

amphiblastula (am′fə-blas′chə-lə) (Gr. *amphi,* on both sides, + *blastos,* germ, + L. *ula,* small). Free-swimming larval stage of certain marine sponges; blastula-like but with only the cells of the animal pole flagellated; those of the vegetal pole unflagellated.

amplexus (am-plek′səs) (L. embrace). The copulatory embrace of frogs or toads.

ampulla (am-pūl′ə) (L. flask). Membranous vesicle; dilation at one end of each semicircular canal containing sensory epithelium; muscular vesicle above tube foot in water-vascular system of echinoderms.

amylase (am′ə-lās′) (L. *amylum,* starch, + *ase,* suffix meaning enzyme). An enzyme that breaks down starch into smaller units.

anadromous (an-ad′rə-məs) (Gr. *anadromos,* running upward). Refers to fishes that migrate up streams from the sea to spawn.

anaerobic (an′ə-rō′bik) (Gr. *an,* not, + *aēr,* air, + *bios,* life). Not dependent on oxygen for respiration.

analogy (L. *analogus,* ratio). Similarity of function but not of origin.

anapsid (ə-nap′səd) (Gr. *an,* without, + *apsis,* arch). Amniotes in which the skull lacks temporal openings, with turtles the only living representatives.

androgen (an′drə-jən) (Gr. *anēr, andros,* man, + *genēs,* born). Any of a group of vertebrate male sex hormones.

androgenic gland (an′drō-jen′ək) (Gr. *anēr,* male, + *gennaein,* to produce). Gland in Crustacea that causes development of male characteristics.

aneuploidy (an′ū-ploid′ē) (Gr. *an,* without, not, + *eu,* good, well, + *ploid,* multiple of). Loss or gain of a chromosome, cells of the organism have one fewer than normal chromosome number, or one extra chromosome, for example, trisomy 21 (Down syndrome).

angiotensin (an′jē-o-ten′sən) (Gr. *angeion,* vessel, + L. *tensio,* to stretch). Blood protein formed from the interaction of renin and a liver protein, causing increased blood pressure and stimulating release of aldosterone and ADH.

Angstrom (after Ångström, Swedish physicist). A unit of one ten-millionth of a millimeter (one ten-thousandth of a micrometer); it is represented by the symbol Å.

anhydrase (an′hī′drās (Gr. *an,* not, + *hydōr,* water, + *ase,* enzyme suffix). An enzyme involved in the removal of water from a compound. Carbonic anhydrase promotes the conversion of carbonic acid into water and carbon dioxide.

annulus (an′yəl-əs) (L. ring). Any ringlike structure, such as superficial rings on leeches.

antenna (L. sail yard). A sensory appendage on the head of arthropods, or the second pair of the two such pairs of structures in crustaceans.

antennal gland Excretory gland of Crustacea located in the antennal metamere.

anterior (L. comparative of *ante,* before). The head end of an organism, or (as an adjective) toward that end.

anthracosaurs (an-thrak′ə-sors) (Gr. *anthrax,* coal, carbon, + *sauros,* lizard). A group of Paleozoic labyrinthodont amphibians.

antibodies (an′tē-bod′ēz). Proteins (immunoglobulins) in cell surfaces and dissolved in blood, capable of combining with the antigens that stimulated their production.

anticodon (an′tī-kō′don). A sequence of three nucleotides in transfer RNA that is complementary to a codon in messenger RNA.

antigen (an′ti-jən). Any substance capable of stimulating an immune response, most often a protein.

aperture (ap′ər-chər) (L. *apertura* from *aperire,* to uncover). An opening; the opening into the first whorl of a gastropod shell.

apex (ā′peks) (L. summit). Highest or uppermost point; the lower pointed end of the heart.

apical (ā′pə-kəl) (L. *apex,* tip). Pertaining to the tip or apex.

apical complex A certain combination of organelles found in the protozoan phylum Apicomplexa.

apocrine (ap′ə-krən) (Gr. *apo,* away, + *krinein,* to separate). Applies to a type of mammalian sweat gland that produces a viscous secretion by breaking off a part of the cytoplasm of secreting cells.

apoptosis (a′pə-tō′səs) (Gr. *apo-,* prefix meaning away from, + *ptōsis,* a falling). Genetically determined cell death, "programmed" cell death.

apopyle (ap′ə-pīl) (Gr. *apo,* away from, + *pylē,* gate). In sponges, opening of the radial canal into the spongocoel.

appendicular (L. *ad,* to, + *pendere,* to hang). Pertaining to appendages; pertaining to vermiform appendix.

arboreal (är-bōr′ē-əl) (L. *arbor,* tree). Living in trees.

archaeocytes (ärk′ē-ō-sīts) (Gr. *archaios,* beginning, + *kytos,* hollow vessel). Ameboid cells of varied function in sponges.

archenteron (ärk-en′tə-rän) (Gr. *archē,* beginning, + *enteron,* gut). The main cavity of an embryo in the gastrula stage; it is lined with endoderm and represents the future digestive cavity.

archinephros (ärk′ē-nəf′rōs) (Gr. *archaois,* ancient, + *nephros,* kidney). Ancestral vertebrate kidney, existing today only in the embryo of hagfishes.

archosaur (är′kə-sor) (Gr. *archōn,* ruling, + *sauros,* lizard). Advanced diapsid vertebrates, a group that includes the living crocodiles and the extinct pterosaurs and dinosaurs.

areolar (a-rē′ə-ler) (L. *areola,* small space). A small area, such as spaces between fibers of connective tissue.

arginine phosphate Phosphate storage compound (phosphagen) found in many invertebrates and used to regenerate stores of ATP.

Aristotle's lantern Masticating apparatus of some sea urchins.

arteriole (är-tir′ē-ōl) (L. *arteria,* artery). A small arterial branch that delivers blood to a capillary network.

artery (ärt′ə-rē) (L. *arteria,* artery). A blood vessel that carries blood away from the heart and toward a peripheral capillary.

artiodactyl (är′tē-o-dak′təl) (Gr. *artios,* even, + *daktylos,* toe). One of an order of mammals with two or four digits on each foot.

asconoid (Gr. *askos,* bladder). Simplest form of sponges, with canals leading directly from the outside to the interior.

asexual Without distinct sexual organs; not involving formation of gametes.

assimilation (L. *assimilatio,* bringing into conformity). Absorption and building up of digested nutriments into complex organic protoplasmic materials.

atherosclerosis (a′thə-rō-sklə-rō′səs) (Gr. *atherōma,* tumor full of gruel-like material, + *sklēros,* hard). Disease characterized by fatty plaques forming in the inner lining of arteries.

atoke (ā′tōk) (Gr. *a,* without, + *tokos,* offspring). Anterior, nonreproductive part of a marine polychaete, as distinct from the posterior, reproductive part (epitoke) during the breeding season.

atoll (ə-tol′) (Maldivian, *atolu*). A coral reef or island surrounding a lagoon.

atom The smallest unit of an element, composed of a dense nucleus of protons and (usually) neutrons surrounded by a system of electrons.

ATP Adenosine triphosphate. In biochemistry, an ester of adenosine and triphosphoric acid.

atrium (ā′trē-əm) (L. *atrium,* vestibule). One of the chambers of the heart; also, the tympanic cavity of the ear; also, the large cavity containing the pharynx in tunicates and cephalochordates.

auricle (aw′ri-kəl) (L. *auricula,* dim. of *auris,* ear). One of the less muscular chambers of the heart; atrium; the external ear, or pinna; any earlike lobe or process.

auricularia (ə-rik′u-lar′ē-ə) (L. *auricula,* a small ear). A type of larva found in Holothuroidea.

autogamy (aw-täg′ə-me) (Gr. *autos,* self, + *gamos,* marriage). Condition in which the gametic nuclei produced by meiosis fuse within the same organism that produced them to restore the diploid number.

autosome (aw′tə-sōm) (Gr. *autos,* self, + *sōma,* body). Any chromosome that is not a sex chromosome.

autotomy (aw-täd′ə-mē) (Gr. *autos,* self, + *tomos,* a cutting). The breaking off of a part of the body by the organism itself.

autotroph (aw′tə-trōf) (Gr. *autos,* self, + *trophos,* feeder). An organism that makes its organic nutrients from inorganic raw materials.

autotrophic nutrition (Gr. *autos,* self, + *trophia,* denoting nutrition). Nutrition characterized by the ability to use simple inorganic substances for the synthesis of more complex organic compounds, as in green plants and some bacteria.

avicularium (L. *avicula,* small bird, + *aria,* like or connected with). Modified zooid that is attached to the surface of the major zooid in Ectoprocta and resembles a bird's beak.

axial (L. *axis,* axle). Relating to the axis, or stem; on or along the axis.

axocoel (ak′sə-cēl) (Gr. *axon,* an axle, + *koilos,* hollow). The most anterior of three coelomic spaces that appear during larval echinoderm development.

axolotl (ak′sə-lot′əl) (Nahuatl, *atl,* water, + *xolotl,* doll, servant, spirit). Larval stage of any of several species of the genus *Ambystoma* (such as *Ambystoma tigrinum*) exhibiting neotenic reproduction.

axon (ak′sän) (Gr. *axōn*). Elongate extension of a neuron that conducts impulses away from the cell body and toward the synaptic terminals.

axoneme (aks′ə-nēm) (L. *axis,* axle, + Gr. *nēma,* thread). The microtubules in a cilium or flagellum, usually arranged as a circlet of nine pairs enclosing one central pair; also, the microtubules of an axopodium.

axopodium (ak′sə-pō′di-um) (Gr. *axon,* an axis, + *podion,* small foot). Long, slender, more or less permanent pseudopodium found in certain sarcodine protozoa. (Also **axopod.**)

B

B cell A type of lymphocyte that is most important in the humoral immune response.

barrier reef A coral reef that runs approximately parallel to the shore and is separated from the shore by a lagoon.

basal body Also known as kinetosome and blepharoplast, a cylinder of nine triplets of microtubules found basal to a flagellum or cilium; same structure as a centriole.

base A molecule that dissociates in solution to produce a hydroxide ion.

basis, basipodite (bā′səs, bā-si′pə-dīt) (Gr. *basis,* base, + *pous, podos,* foot). The distal or second joint of the protopod of a crustacean appendage.

bathypelagic (bath′ə-pe-laj′ik) (Gr. *bathys,* deep, + *pelagos,* open sea). Relating to or inhabiting the deep sea.

benthos (ben′thäs) (Gr. depth of the sea). Organisms that live along the bottom of the seas and lakes; adj., **benthic.** Also, the bottom itself.

bilirubin (bil′ə-rü-bən) (L. *bilis,* bile, + *rubeo,* to be red). A breakdown product of the heme group of hemoglobin, excreted in the bile.

binary fission A mode of asexual reproduction in which the animal splits into two approximately equal offspring.

biogenesis (bī′ō-jen′ə-səs) (Gr. *bios,* life, + *genesis,* birth). The doctrine that life originates only from preexisting life.

biological species concept A reproductive community of populations (reproductively isolated from others) that occupies a specific niche in nature.

bioluminescence Method of light production by living organisms in which usually certain proteins (luciferins), in the presence of oxygen and an enzyme (luciferase), are converted to oxyluciferins with the liberation of light.

biomass (Gr. *bios,* life, + *maza,* lump or mass). The weight of total living organisms or of a species population per unit of area.

biome (bī′ōm) (Gr. *bios,* life, + *ōma,* abstract group suffix). Complex of plant and animal communities characterized by climatic and soil conditions; the largest ecological unit.

biosphere (Gr. *bios,* life, + *sphaira,* globe). That part of earth containing living organisms.

biotic (bī-äd′ik) (Gr. *biōtos,* life, livable). Of or relating to life.

bipinnaria (L. *bi,* double, + *pinna,* wing, + *aria,* like or connected with). Free-swimming, ciliated, bilateral larva of the asteroid echinoderms; develops into the brachiolaria larva.

biramous (bī-rām′əs) (L. *bi,* double, + *ramus,* a branch). Adjective describing appendages with two distinct branches, contrasted with uniramous, unbranched.

bivalent (bī-vāl′ənt) (L. *bi,* double, + *valen,* strength, worth). The pairs of homologous chromosomes at synapsis in the first meiotic division, a tetrad.

blastocoel (blas′tə-sēl) (Gr. *blastos,* germ, + *koilos,* hollow). Cavity of the blastula.

blastocyst (blast′ō-sist) (Gr. *blastos,* germ, + *kystis,* bladder). Mammalian embryo in the blastula stage.

blastomere (Gr. *blastos,* germ, + *meros,* part). An early cleavage cell.

blastopore (Gr. *blastos,* germ, + *poros,* passage, pore). External opening of the archenteron in the gastrula.

blastula (Gr. *blastos,* germ, + L. *ula,* dim.). Early embryological stage of many animals; consists of a hollow mass of cells.

blending See **polygenic inheritance.**

blepharoplast (blə-fā′rə-plast) (Gr. *blepharon,* eyelid, + *plastos,* formed). See **basal body.**

blood plasma The liquid, noncellular fraction of blood, including dissolved substances.

blood type Characteristic of human blood given by the particular antigens on the membranes of the erythrocytes, genetically determined, causing agglutination when incompatible groups are mixed; the blood types are designated A, B, O, AB, Rh negative, Rh positive, and others.

Bohr effect A characteristic of hemoglobin that causes it to dissociate from oxygen in greater degree at higher concentrations of carbon dioxide.

boreal (bōr′ē-əl) (L. *boreas,* north wind). Relating to a northern biotic area characterized by a predominance of coniferous forests and tundra.

B.P. Before the present.

brachial (brak′ē-əl) (L. *brachium,* forearm). Referring to the arm.

brain hormone See **ecdysiotropin.**

branchial (brank′ē-əl) (Gr. *branchia,* gills). Referring to gills.

bronchiole (brän′kē-ōl) (Gr. *bronchion,* dim. of *bronchos,* windpipe). Small, thin-walled branch of the bronchus.

bronchus (brän′kəs) pl. **bronchi** (Gr. *bronchos,* windpipe). Either of two primary divisions of the trachea that lead to the right and left lung.

brown fat Mitochondria-rich, heat-generating adipose tissue of endothermic vertebrates.

buccal (buk′əl) (L. *bucca,* cheek). Referring to the mouth cavity.

budding Reproduction in which the offspring arises as an outgrowth from the parent and is initially smaller than the parent. Failure of the offspring to separate from the parent leads to colony formation.

buffer Any substance or chemical compound that tends to keep pH levels constant when acids or bases are added.

bursa pl. **bursae** (M. L. *bursa,* pouch, purse made of skin). A saclike cavity. In ophiuroid echinoderms, pouches opening at bases of arms and functioning in respiration and reproduction (genitorespiratory bursae).

C

calciferous glands (kal-si′fə-rəs). Glands in an earthworm that secrete calcium ions into the gut.

calorie (kal′ə-rē) (L. *calere,* to be warm). Unit of heat defined as the amount of heat required to heat 1 g of water from 14.5 to 15.5° C; 1 cal 4.184 joules in the International System of Units.

calyx (kā-liks) (L. bud cup of a flower). Any of various cup-shaped zoological structures.

cancellous (kan′səl-əs) (L. *cancelli,* latticework, + *osus,* full of). Having a spongy or porous structure.

capitulum (ka-pi′tū-ləm) (L. small head). Term applied to small, headlike structures of various organisms, including projection from body of ticks and mites carrying mouthparts.

carapace (kar′ə-pās) (F. from Sp. *carapacho,* shell). Shieldlike plate covering the cephalothorax of certain crustaceans; dorsal part of the shell of a turtle.

carbohydrate (L. *carbo,* charcoal, + Gr. *hydōr,* water). Compounds of carbon, hydrogen, and oxygen having the generalized formula $(CH_2O)_n$; aldehyde or ketone derivatives of polyhydric alcohols, with hydrogen and oxygen atoms attached in a 2:1 ratio.

carboxyl (kär-bäk′səl) (carbon + oxygen + yl, chemical radical suffix). The acid group of organic molecules —COOH.

cardiac (kär′dē-ak) (Gr. *kardia,* heart). Belonging or relating to the heart.

carinate (kar′ə-nāt) (L. *carina,* keel). Having a keel, in particular the flying birds with a keeled sternum for the insertion of flight muscles.

bat/āpe/ärmadillo/herring/fēmale/finch/līce/crocodile/crōw/duck/ūnicorn/tüna/ə indicates unaccented vowel sound "uh" as in mammal, fishes, cardinal, heron, vulture/stress as in bi-ol′o-gy, bi′o-log′i-cal

carnivore (kar′nə-vōr′) (L. *carnivorous,* flesh eating). One of the flesh-eating mammals of the order Carnivora. Also, any organism that eats animals. Adj., **carnivorous.**

carotene (kär′ə-tēn) (L. *carota,* carrot, + *ene,* unsaturated, straight-chain hydrocarbons). A red, orange, or yellow pigment belonging to the group of carotenoids; precursor of vitamin A.

carrying capacity The maximum number of individuals that can persist under specified environmental conditions.

cartilage (L. *cartilago;* akin to L. *cratis,* wickerwork). A translucent elastic tissue that makes up most of the skeleton of embryos, very young vertebrates, and adult cartilaginous fishes, such as sharks and rays; in other gnathostomes much of it is converted into bone.

caste (kast) (L. *castus,* pure, separated). One of the polymorphic forms within an insect society, each caste having its specific duties, as queen, worker, soldier, and so on.

catabolism (Gr. *kata,* downward, + *bol,* to throw, + *ism,* suffix meaning state of condition). Destructive metabolism; process in which complex molecules are reduced to simpler ones.

catadromous (kə-tad′rə-məs) (Gr. *kata,* down, + *dromos,* a running). Refers to fishes that migrate from fresh water to the ocean to spawn.

catalyst (kad′ə-ləst) (Gr. *kata,* down, + *lysis,* a loosening). A substance that accelerates a chemical reaction but does not become a part of the end product.

caudal (käd′əl) (L. *cauda,* tail). Constituting, belonging to, or relating to a tail.

caveolae (ka-vē′ə-lē) (L. *cavea,* a cave, + dim. suffix). The invaginated vesicles and pits in potocytosis.

cDNA See **complementary DNA.**

cecum, caecum (sē′kəm) (L. *caecus,* blind). A blind pouch at the beginning of the large intestine; any similar pouch.

cell-mediated immune response Immune response involving cell surfaces only, not antibody production, specifically the T_H1 arm of the immune response. Contrast **humoral immune response.**

cellulose (sel′ū-lōs) (L. *cella,* small room). Chief polysaccharide constituent of the cell wall of green plants and some fungi; an insoluble carbohydrate $(C_6H_{10}O_5)_n$ that is converted into glucose by hydrolysis.

centriole (sen′trē-ol) (Gr. *kentron,* center of a circle, + L. *ola,* small). A minute cytoplasmic organelle usually found in the cetrosome and considered to be the active division center of the animal cell; organizes spindle fibers during mitosis and meiosis. Same structure as basal body or kinetosome.

centromere (sen′trə-mir) (Gr. *kentron,* center, + *meros,* part). A localized constriction in a characteristic position on a given chromosome, bearing the kinetochore.

centrosome (sen′trə-sōm) (Gr. *kentron,* center, + *sōma,* body). Microtubule organizing center in nuclear division in most eukaryotic cells; in animals and many protists it surrounds the centrioles.

cephalization (sef′ə-li-zā-shən) (Gr. *kephalē,* head). The process by which specialization, particularly of the sensory organs and appendages, become localized in the head end of animals.

cephalothorax (sef′ə-lə-thō-raks) (Gr. *kephalē,* head, + thorax). A body division found in many Arachnida and higher Crustacea, in which the head is fused with some or all of the thoracic segments.

cercaria (ser′kar′ē-ə) (Gr. *kerkos,* tail, + L. *aria,* like or connected with). Tadpolelike larva of trematodes (flukes).

cervical (sər′və-kəl) (L. *cervix,* neck). Relating to a neck.

character (kar′ik-tər). A component of phenotype (including specific molecular, morphological, behavioral or other features) used by systematists to diagnose species or higher taxa, or to evaluate phylogenetic relationships among different species or higher taxa, or relationships among populations within a species.

charging In protein synthesis, a reaction catalyzed by tRNA synthetase, in which an amino acid is attached to its particular tRNA molecule.

chelicera (kə-lis′ə-rə) pl. **chelicerae** (Gr. *chēlē,* claw, + *keras,* horn). One of a pair of the most anterior head appendages on the members of the subphylum Chelicerata.

chelipeds (kēl′ə-peds) (Gr. *chēlē,* claw, + L. *pes,* foot). Pincerlike first pair of legs in most decapod crustaceans; specialized for seizing and crushing.

chemoautotroph (ke′mə-aw′tə-trōf) (Gr. *chemeia,* transmutation, + *autos,* self, + *trophos,* feeder). An organism utilizing inorganic compounds as a source of energy.

chemotaxis (kē′mə-tak′səs) (Gr. *chēmeia,* an infusion, + *taxō* > *tassō,* to put in order). Orientation movement of cells or organisms in response to a chemical stimulus.

chemotroph (kem′ə-trōf) (Gr. *chēmeia,* an infusion, + *tropē,* to turn). An organism that derives nourishment from inorganic substances without using chlorophyll.

chiasma (kī-az′mə), pl. **chiasmata** (Gr. cross). An intersection or crossing, as of nerves; a connection point between homologous chromatids where crossing over has occurred at synapsis.

chitin (kī′tən) (Fr. *chitine,* from Gr. *chitōn,* tunic). A horny substance that forms part of the cuticle of arthropods and is found sparingly in certain other invertebrates; a nitrogenous polysaccharide insoluble in water, alcohol, dilute acids, and digestive juices of most animals.

chlorocruorin (klō′rō-krü′ə-rən) (Gr. *chlōros,* light green, + L. *cruor,* blood). A greenish iron-containing respiratory pigment dissolved in the blood plasma of certain marine polychaetes.

chlorogogen cells (klōr′ə-gog-ən) (Gr. *chlōros,* light green, + *agōgos,* a leading, a guide). Modified peritoneal cells, greenish or brownish, clustered around the digestive tract of certain annelids; apparently they aid in elimination of nitrogenous wastes and in food transport.

chlorophyll (klō′rə-fil) (Gr. *chlōros,* light green, + *phyllōn,* leaf). Green pigment found in plants and in some animals; necessary for photosynthesis.

chloroplast (klō′rə-plast) (Gr. *chlōros,* light green, + *plastos,* molded). A plastid containing chlorophyll and usually other pigments, found in cytoplasm of plant cells.

choanocyte (kō-an′ə-sīt) (Gr. *choanē,* funnel, + *kytos,* hollow vessel). One of the flagellate collar cells that line cavities and canals of sponges.

chorion (kō′rē-on) (Gr. *chorion,* skin). The outer of the double membrane that surrounds the embryo of reptiles, birds, and mammals; in mammals it contributes to the placenta.

choroid (kōr′oid) (Gr. *chorion,* skin, + *eidos,* form). Delicate, highly vascular membrane; in vertebrate eye the layer between the retina and sclera.

chromatid (krō′mə-tid) (Gr. *chromato,* from *chrōma,* color, + L. *id,* feminine stem for particle of specified kind). A replicated chromosome joined to its sister chromatid by the centromere; separates and becomes daughter chromosome at anaphase of mitosis or anaphase of the second meiotic division.

chromatin (krō′mə-tin) (Gr. *chrōma,* color). The nucleoprotein material of a chromosome; the hereditary material containing DNA.

chromatophore (krō-mat′ə-fōr) (Gr. *chrōma,* color, + *pherein,* to bear). Pigment cell, usually in the dermis, in which usually the pigment can be dispersed or concentrated.

chromomere (krō′mə-mir) (Gr. *chrōma,* color, + *meros,* part). One of the chromatin granules of chracteristic size on the chromosome; may be identical with a gene or a cluster of genes.

chromonema (krō-mə-nē′mə) (Gr. *chrōma,* color, + *nēma,* thread). A convoluted thread in prophase of mitosis or the central thread in a chromosome.

chromoplast (krō′mə-plast) (Gr. *chrōma,* color, + *plastos,* molded). A plastid containing pigment.

chromosome (krō′mə-sōm) (Gr. *chrōma,* color, + *sōma,* body). A complex body, spherical or rod shaped, that arises from the nuclear network during mitosis, splits longitudinally, and carries a part of the organism's genetic information as genes composed of DNA.

chrysalis (kris′ə-lis) (L. from Gr. *chrysos,* gold). The pupal stage of a butterfly.

chyme (kīm) (Gr. *chymos,* juice). Semifluid mass of partly digested food in stomach and small intestine as digestion proceeds.

cilium (sil′ē-əm), pl. **cilia** (L. hair). A hairlike, vibratile organelle process found on many animal cells. Cilia may be used in moving particles along the cell surface or, in ciliate protozoans, for locomotion.

cinclides (sing′klid-əs), sing. **cinclis** (sing′kləs) (Gr. *kinklis,* latticed gate or partition). Small pores in the external body wall of sea anemones for extrusion of acontia.

circadian (sər-kād′ē-ən) (L. *circa,* around, + *dies,* day). Occuring at a period of approximately 24 hours.

cirrus (sir′əs) (L. curl). A hairlike tuft on an insect appendage; locomotor organelle of fused cilia; male copulatory organ of some invertebrates.

cisternae (sis-ter′nē) (L. *cista,* box). Space between membranes of the endoplasmic reticulum within cells.

cistron (sis′trən) (L. *cista,* box). A series of codons in DNA that code for an entire polypeptide chain.

clade (klād) (Gr. *klados,* branch). A taxon or other group consisting of an ancestral species and all of its descendants, forming a distinct branch on a phylogenetic tree.

cladistics (klad-is′-təks) (Gr. *klados,* branch, sprout). A system of arranging taxa by analysis of evolutionarily derived characteristics so that the arrangement will reflect phylogenetic relationships.

cladogram (klād′ə-gram) (Gr. *klados,* branch, + *gramma,* letter). A branching diagram showing the pattern of sharing of evolutionarily derived characters among species or higher taxa.

clathrin (kla′thrən) (L. *clathri,* latticework). A protein forming a lattice structure lining the invaginated pits during receptor-mediated endocytosis.

cleavage (O.E. *cleofan,* to cut). Process of nuclear and cell division in animal zygote.

climax (klī′maks) (Gr. *klimax,* ladder). Stage of relative stability attained by a community of organisms, often the culminating development of a natural succession. Also, orgasm.

climax community (Gr. *klimax,* ladder, staircase, climax). A self-perpetuating, more-or-less stable community of organisms that continues as long as environmental conditions under which it developed prevail.

clitellum (klī-tel′əm) (L. *clitellae,* packsaddle). Thickened saddlelike portion of certain midbody segments of many oligochaetes and leeches.

cloaca (klō-ā′kə) (L. sewer). Posterior chamber of digestive tract in many vertebrates, receiving feces and urogenital products. In certain invertebrates, a terminal portion of digestive tract that serves also as respiratory, excretory, or reproductive tract.

clone (klōn) (Gr. *klōn,* twig). All descendants derived by asexual reproduction from a single individual.

cnidoblast (nī′də-blast) (Gr. *knidē,* nettle, + *blastos,* germ). See **cnidocyte.**

cnidocil (nī′də-sil) (Gr. *knidē,* nettle, + L. *cilium,* hair). Modified cilium on nematocyst-bearing cnidocytes in cnidarians; triggers nematocyst.

cnidocyte (nī′də-sīt) (Gr. *knidē,* nettle, + *kytos,* hollow vessel). Modified interstitial cell that holds the nematocyst; during development of the nematocyst, the cnidocyte is a cnidoblast.

coacervate (kō′ə-sər′vət) (L. *coacervatus,* to heap up). An aggregate of colloidal droplets held together by electrostatic forces.

coagulation (kō-ag′ū-lā-shən). Process in which a series of enzymes are activated, resulting in clotting of blood.

cochlea (kōk′lēə) (L. snail, from Gr. *kochlos,* a shellfish). A tubular cavity of the inner ear containing the essential organs of hearing; occurs in crocodiles, birds, and mammals; spirally coiled in mammals.

cocoon (kə-kün′) (Fr. *cocon,* shell). Protective covering of a resting or developmental stage, sometimes used to refer to both the covering and its contents; for example, the cocoon of a moth or the protective covering for the developing embryos in some annelids.

codominance See **intermediate inheritance.**

codon (kō′dän) (L. code, + on). In messenger RNA a sequence of three adjacent nucleotides that codes for one amino acid.

coelenteron (sē-len′tər-on) (Gr. *koilos,* hollow, + *enteron,* intestine). Internal cavity of a cnidarian; gastrovascular cavity; archenteron.

coelom (sē′lōm) (Gr. *koilōma,* cavity). The body cavity in triploblastic animals, lined with mesodermal peritoneum.

coelomocyte (sē′lō′mə-sīt) (Gr. *koilōma,* cavity, + *kytos,* hollow vessel). Another name for amebocyte; undifferentiated cell of the coelom and the water-vascular system.

coelomoduct (sē-lō′mə-dukt) (Gr. *koilos,* hollow, + L. *ductus,* a leading). A duct that carries gametes or excretory products (or both) from the coelom to the exterior.

coenenchyme (sēn′ən-kīm) (Gr. *koinos,* shared in common, + *enchyma,* something poured in). Extensive mesogleal tissue between the polyps of an alcyonarian (phylum Cnidaria) colony.

coenzyme (kō-en′zīm) (L. prefix, *co,* with, + Gr. *enzymos,* leavened, from *en,* in, + *zymē,* leaven). A required substance in the activation of an enzyme; a prosthetic or nonprotein constituent of an enzyme.

collagen (käl′ə-jən) (Gr. *kolla,* glue, + *genos,* descent). A structural protein, the most abundant protein in the animal kingdom, characterized by high content of the amino acids glycine, alanine, proline, and hydroxyproline.

collencyte (käl′lən-sīt) (Gr. *kolla,* glue, + *en,* in, + *kytos,* hollow vessel). A type of cell in sponges that is star shaped and apparently contractile.

colloblast (käl′ə-blast) (Gr. *kolla,* glue, + *blastos,* germ). A glue-secreting cell on the tentacles of ctenophores.

colloid (kä′loid) (Gr. *kolla,* glue, + *eidos,* form). A two-phase system in which particles of one phase are suspended in the second phase.

columella (kä′lə-mel′ə) (L. small column). Central pillar in gastropod shells.

comb plate One of the plates of fused cilia that are arranged in rows for ctenophore locomotion.

commensalism (kə-men′səl-iz′əm) (L. *cum,* together with, + *mensa,* table). A relationship in which one individual lives close to or on another and benefits, and the host is unaffected; often symbiotic.

community (L. *communitas,* community, fellowship). An assemblage of organisms that are associated in a common environment and interact with each other in a self-sustaining and self-regulating relation.

competition Some degree of overlap in ecological niches of two populations in the same community, such that both depend on the same food source, shelter, or other resources, and negatively affect each other's survival.

complement Collective name for a series of enzymes and activators in the blood, some of which may bind to antibody, and may lead to rupture of a foreign cell.

complementary DNA (cDNA) DNA prepared by transcribing the base sequence from mRNA into DNA by reverse transcriptase; also called **copy DNA.**

compound A substance whose molecules are composed of atoms of two or more elements.

condensation reaction A chemical reaction in which reactant molecules are combined by the removal of a water molecule (a hydrogen from one and a hydroxyl from the other reactant).

condyle (kon′dīl) (Gr. *kondylos,* bump). A process on a bone used for articulation.

conjugation (kon′ju-ga′shən) (L. *conjugare,* to yoke together). Temporary union of two ciliate protozoa while they are exchanging chromatin material and undergoing nuclear phenomena resulting in binary fission. Also, formation of cytoplasmic bridges between bacteria for transfer of plasmids.

conspecific (L. *com,* together, + *species*). A member of the same species.

contractile vacuole A clear fluid-filled cell vacuole in protozoa and a few lower metazoa; takes up water and releases it to the outside in a cyclical manner, for osmoregulation and some excretion.

control That part of a scientific experiment to which the experimental variable is not applied but which is similar to the experimental group in all other respects.

coprophagy (kə-prä′fə-jē) (Gr. *kopros,* dung, + *phagein,* to eat). Feeding on dung or excrement as a normal behavior among animals; reinjestion of feces.

copulation (Fr. from L. *copulare,* to couple). Sexual union to facilitate the reception of sperm by the female.

copy DNA See **complementary DNA.**

coralline algae Algae that precipitate calcium carbonate in their tissues; important contributors to coral reef mass.

cornea (kor′nē-ə) (L. *corneus,* horny). The outer transparent coat of the eye.

cornified (kor′nə-fīd) (L. *corneus,* horny). Adjective for conversion of epithelial cells into nonliving, keratinized cells.

corona (kə-rō′nə) (L. crown). Head or upper portion of a structure; ciliated disc on anterior end of rotifers.

corpora allata (kor′pə-rə əl-la′tə) (L. *corpus,* body, + *allatum,* aided). Endocrine glands in insects that produce juvenile hormone.

cortex (kor′teks) (L. bark). The outer layer of a structure.

covalent bond A chemical bond in which electrons are shared between atoms.

coxa, coxopodite (kox′ə, kəx-ä′pə-dīt) (L. *coxa,* hip, + Gr. *pous, podos,* foot). The proximal joint of an insect or arachnid leg; in crustaceans, the proximal joint of the protopod.

creatine phosphate High-energy phosphate compound found in the muscle of vertebrates and some invertebrates, used to regenerate stores of ATP.

cretin (krēt′n) (Fr. *crétin,* [dialect], fr. L. *christianus,* Christian, to indicate idiots so afflicted were also human). A human with severe mental, somatic, and sexual retardation resulting from hypothyroidism during early stages of development.

crista (kris′ta), pl. **cristae** (L. *crista,* crest). A crest or ridge on a body organ or organelle; a platelike projection formed by the inner membrane of mitochondrion.

crossing over Exchange of parts of nonsister chromatids at synapsis in the first meiotic division.

cryptobiotic (Gr. *kryptos,* hidden, + *biōticus,* pertaining to life). Living in concealment; refers to insects and other animals that live in secluded situations, such as underground or in wood; also tardigrades and some nematodes, rotifers, and others that survive harsh environmental conditions by assuming for a time a state of very low metabolism.

ctenidia (te-ni′dē-ə) (Gr. *kteis,* comb). Comblike structures, especially gills of molluscs; also applied to comb plates of Ctenophora.

ctenoid scales (ten′oyd) (Gr. *kteis, ktenos,* comb). Thin, overlapping dermal scales of the more advanced fishes; exposed posterior margins have fine, toothlike spines.

cupula (kū′pū-lə) (L. little tub). Small inverted cuplike structure housing another structure; gelatinous matrix covering hair cells in lateral line and equilibrium organs.

cuticle (kū′ti-kəl) (L. *cutis,* skin). A protective, noncellular, organic layer secreted by the external epithelium (hypodermis) of many invertebrates. In higher animals the term refers to the epidermis or outer skin.

cyanobacteria (sī-an-ō-bak-ter′ē-ə) (Gr. *kyanos,* a dark-blue substance, + *bakterion,* dim, of *baktron,* a staff). Photosynthetic prokaryotes, also called blue-green algae, cyanophytes.

cyanophyte (sī-an′ō-fīt) (Gr. *kyanos,* a dark-blue substance, + *phyton,* plant). A cyanobacterium, blue-green alga.

cyclin A protein important in the control of the cell division cycle and mitosis.

cycloid scales (sī′-kloid) (Gr. *kyklos,* circle). Thin, overlapping dermal scales of the more primitive fishes; posterior margins are smooth.

cynodonts (sin′ə-dänts) (Gr. *kynodōn,* canine tooth). A group of mammal-like carnivorous synapsids of the Upper Permian and Triassic.

cystacanth (sis′tə-kanth) (Gr. *kystis,* bladder, pouch, + *akantha,* thorn). Juvenile stage of an acanthocephalan that is infective to the definitive host.

cysticercoid (sis′tə-ser′koyd) (Gr. *kystis,* bladder, + *kerkos,* tail, + *eidos,* form). A type of juvenile tapeworm composed of a solid-bodied cyst containing an invaginated scolex; contrast with **cysticercus.**

cysticercus (sis′tə-ser′kəs) (Gr. *kystis,* bladder, + *kerkos,* tail). A type of juvenile tapeworm in which an invaginated and introverted scolex is contained in a fluid-filled bladder; contrast with **cysticercoid.**

cystid (sis′tid) (Gr. *kystis,* bladder). In an ectoproct, the dead secreted outer parts plus the adherent underlying living layers.

cytochrome (sī′tə-krōm) (Gr. *kytos,* hollow vessel, + *chrōma,* color). Several iron-containing pigments that serve as electron carriers in aerobic respiration.

cytokine (sī′tə-kīn) (Gr. *kytos,* hollow vessel, + *kinein,* to move). A molecule secreted by an activated or stimulated cell, for example, macrophages, that causes physiological changes in certain other cells.

cytokinesis (sī′tə-kin-ē′sis) (Gr. *kytos,* hollow vessel, + *kinesis,* movement). Division of the cytoplasm of a cell.

cytopharynx (Gr. *kytos,* hollow vessel, + *pharynx,* throat). Short tubular gullet in ciliate protozoa.

cytoplasm (sī′tə-plazm) (Gr. *kytos,* hollow vessel, + *plasma,* mold). The living matter of the cell, excluding the nucleus.

cytoproct (sī′tə-prokt) (Gr. *kytos,* hollow vessel, + *prōktos,* anus). Site on a protozoan where undigestible matter is expelled.

cytopyge (sī′tə-pīj) (Gr. *kytos,* hollow vessel, + *pyge,* rump or buttocks). In some protozoa, localized site for expulsion of wastes.

cytosol (sī′tə-sol) (Gr. *kytos,* hollow vessel, + L. *sol,* from *solutus,* to loosen). Unstructured portion of the cytoplasm in which the organelles are bathed.

cytosome (sī′tə-sōm) (Gr. *kytos,* hollow vessel, + *sōma,* body). The cell body inside the plasma membrane.

cytostome (sī′tə-stōm) (Gr. *kytos,* hollow vessel, + *stoma,* mouth). The cell mouth in many protozoa.

cytotoxic T cells (Gr. *kytos,* hollow vessel, + toxin). A special T cell activated during cell-mediated immune responses that recognizes and destroys virus-infected cells.

bat/āpe/ärmadillo/herring/fēmale/finch/līce/crocodile/crōw/duck/ūnicorn/tüna/ə indicates unaccented vowel sound "uh" as in mammal, fishes, cardinal, heron, vulture/stress as in bi-ol′o-gy, bi′o-log′i-cal

D

dactylozooid (dak-til′ə-zō′id) (Gr. *dakos,* bite, sting, + *tylos,* knob, + *zōon,* animal). A polyp of a colonial hydroid specialized for defense or killing food.

Darwinism Theory of evolution emphasizing common descent of all living organisms, gradual change, multiplication of species and natural selection.

data sing. **datum** (Gr. *dateomai,* to divide, cut in pieces). The results in a scientific experiment, or descriptive observations, upon which a conclusion is based.

deciduous (də-sij′ə-wəs) (L. *decidere,* to fall off). Shed or falling off at end of a growing period.

deduction (L. *deductus,* led apart, split, separated). Reasoning from the general to the particular, that is, from given premises to their necessary conclusion.

definitive host The host in which sexual reproduction of a symbiont takes place; if no sexual reproduction, then the host in which the symbiont becomes mature and reproduces; contrast **intermediate host.**

delayed type hypersensitivity Inflammatory reaction based primarily on cell-mediated immunity.

deme (dēm) (Gr. populace). A local population of closely related animals.

demography (də-mog′rə-fē) (Gr. *demos,* people, + *graphy*). The properties of the rate of growth and the age structure of populations.

dendrite (den′drīt) (Gr. *dendron,* tree). Any of nerve cell processes that conduct impulses toward the cell body.

deoxyribonucleic acid (DNA) The genetic material of all organisms, characteristically organized into linear sequences of genes.

deoxyribose (dē-ok′sē-rī′bōs) (L. *deoxy,* loss of oxygen, + *ribose,* a pentose sugar). A 5-carbon sugar having 1 oxygen atom less than ribose; a component of deoxyribonucleic acid (DNA).

dermal (Gr. *derma,* skin). Pertaining to the skin; cutaneous.

dermis The inner, sensitive mesodermal layer of skin.

desmosome (dez′mə-sōm) (Gr. *desmos,* bond, + *sōma,* body). Buttonlike plaque serving as an intercellular connection.

determinate cleavage Mosaic cleavage, usually spiral, in which the fate of the blastomeres is determined very early in development.

detritus (də-trī′tus) (L. that which is rubbed or worn away). Any fine particulate debris of organic or inorganic origin.

Deuterostomia (dü′də-rō-stō′mē-ə) (Gr. *deuteros,* second, secondary, + *stoma,* mouth). A group of higher phyla in which cleavage is indeterminate (regulative) and primitively radial. The endomesoderm is enterocoelous, and the mouth is derived away from the blastopore. Includes Echinodermata, Chordata, and a number of minor phyla. Compare with Prostostomia.

dextral (dex′trəl) (L. *dexter,* right-handed). Pertaining to the right; in gastropods, shell is dextral if opening is to right of columella when held with spire up and facing observer.

diapause (dī′ə-pawz) (Gr. *diapausis,* pause). A period of arrested development in the life cycle of insects and certain other animals in which physiological activity is very low and the animal is highly resistant to unfavorable external conditions.

diapsids (dī-ap′sədz) (Gr. *di,* two, + *apsis,* arch). Amniotes in which the skull bears two pairs of temporal openings; includes reptiles (except turtles) and birds.

diastole (dī-as′tə-lē) (Gr. *diastolē,* dilation). Passive relaxation and expansion of the heart during which the chambers are filled with blood.

diffusion (L. *diffusus,* dispersion). The movement of particles or molecules from area of high concentration of the particles or molecules to area of lower concentration.

digitigrade (dij′ə-də-grād) (L. *digitus,* finger, toe, + *gradus,* step, degree). Walking on the digits with the posterior part of the foot raised; compare plantigrade.

dihybrid (dī′hī′brəd) (Gr. *dis,* twice, + L. *hibrida,* mixed offspring). A hybrid whose parents differ in two distinct characters; an offspring having two different alleles at two different loci, for example, *A/a B/b.*

dimorphism (dī-mor′fizm) (Gr. *di,* two, + *morphē,* form). Existence within a species of two distinct forms according to color, sex, size, organ structure, and so on. Occurrence of two kinds of zooids in a colonial organism.

dioecious (dī-ē′shəs) (Gr. *di,* two, + *oikos,* house). Having male and female organs in separate individuals.

diphycercal (dif′i-ser′kəl) (Gr. *diphyēs,* twofold, + *kerkos,* tail). A tail that tapers to a point, as in lungfishes; vertebral column extends to tip without upturning.

diphyodont (di′fi-ə-dänt) (Gr. *diphyēs,* twofold, + *odous,* tooth). Having deciduous and permanent sets of teeth successively.

diploblastic (di′plə-blak′tək) (Gr. *diploos,* double, + *blastos,* bud). Organism with two germ layers, endoderm and ectoderm.

diploid (dip′loid) (Gr. *diploos,* double, + *eidos,* form). Having the somatic (double, or 2N) number of chromosomes or twice the number characteristic of a gamete of a given species.

disaccharides (dī-sak′ə-rīds) (Gr. *dis,* twice, + L. *saccharum,* sugar). A class of sugars (such as lactose, maltose, and sucrose) that yield two monosaccharides on hydrolysis.

distal (dis′təl). Farther from the center of the body than a reference point.

DNA See **deoxyribonucleic acid.**

dominance hierarchy A social ranking, formed through agonistic behavior, in which individuals are associated with each other so that some have greater access to resources than do others.

dominant An allele that is expressed regardless of the nature of the corresponding allele on the homologous chromosome.

dorsal (dor′səl) (L. *dorsum,* back). Toward the back, or upper surface, of an animal.

Down syndrome A congenital syndrome including mental retardation, caused by the cells in a person's body having an extra chromosome 21; also called trisomy 21.

dual-gland adhesive organ Organs in the epidermis of most turbellarians, with three cell types; viscid and releasing gland cells and anchor cells.

duodenum (dü-ə-dēn′əm) (L. *duodeni,* twelve each, fr. its length, about 12 fingers' width). The first and shortest portion of the small intestine lying between the pyloric end of the stomach and the jejunum.

dyad (dī′əd) (Gr. *dyas,* two). One of the groups of two chromosomes formed by the division of a tetrad during the first meiotic division.

E

eccrine (ek′rən) (Gr. *ek,* out of, + *krinein,* to separate). Applies to a type of mammalian sweat gland that produces a watery secretion.

ecdysiotropin (ek-dē′zē-ō-trō′pən) (Gr. *ekdysis,* to strip off, escape, + *tropos,* a turn, change). Hormone secreted in brain of insects that stimulates prothoracic gland to secrete molting hormone. Prothoracicotropic hormone; brain hormone.

ecdysis (ek′də-sis) (Gr. *ekdysis,* to strip off, escape). Shedding of outer cuticular layer; molting, as in insects or crustaceans.

ecdysone (ek-dī′sōn) (Gr. *ekdysis,* to strip off). Molting hormone of arthropods, stimulates growth and ecdysis, produced by prothoracic glands in insects and Y organs in crustaceans.

ecocline (ek′ō-klīn) (Gr. *oikos,* home, + *klino,* to slope, recline). The gradient between adjacent biomes; a gradient of environmental conditions.

ecology (Gr. *oikos,* house, + *logos,* discourse). Part of biology that deals with the relationship between organisms and their environment.

ecosystem (ek′ō-sis-təm) (eco[logy] from Gr. *oikos,* house, + system). An ecological unit consisting of both the biotic communities and the nonliving (abiotic) environment, which interact to produce a stable system.

ectoderm (ek′tō-derm) (Gr. *ektos,* outside, + *derma,* skin). Outer layer of cells of an early embryo (gastrula stage); one of the germ layers, also sometimes used to include tissues derived from ectoderm.

ectognathous (ek′tə-na′thəs) (Gr. *ektos,* outside, without, + *gnathos,* jaw). Derived character of most insects; mandibles and maxillae not in pouches.

ectolecithal (ek′tō-les′ə-thəl) (Gr. *ektos,* ouside, + *lekithos,* yolk). Yolk for nutrition of the embryo contributed by cells that are separate from the egg cell and are combined with the zygote by envelopment within the eggshell.

ectoneural (ek′tə-nü′rəl) (Gr. *ektos,* outside, without, + *neuron,* nerve). Oral (chief) nervous system in echinoderms.

ectoplasm (ek′tō-plazm) (Gr. *ektos,* outside, + *plasma,* form). The cortex of a cell or that part of cytoplasm just under the cell surface; contrasts with **endoplasm.**

ectothermic (ek′tō-therm′ic) (Gr. *ektos,* outside, + *thermē,* heat). Having a variable body temperature derived from heat acquired from the environment; contrasts with **endothermic.**

edema (ē-dē′mə) (Gr. *oidēma,* swelling). Escape of fluid from blood into interstitial space, causing swelling.

effector (L. *efficere,* bring to pass). An organ, tissue, or cell that becomes active in response to stimulation.

efferent (ef′ə-rənt) (L. *ex,* out, + *ferre,* to bear). Leading or conveying away from some organ, for example, nerve impulses conducted away from the brain, or blood conveyed away from an organ; contrasts with **afferent.**

egestion (ē-jes′chən) (L. *egestus,* to discharge). Act of casting out indigestible or waste matter from the body by any normal route.

electron A subatomic particle with a negative charge and a mass of 9.1066 × 10^{-28} gram.

eleocyte (el′ē-ə-sīt) (Gr. *elaion,* oil, + *kytos,* hollow vessel). Fat-containing cells in annelids that originate from the chlorogogen tissue.

elephantiasis (el-ə-fən-tī′ə-səs). Disfiguring condition caused by chronic infection with certain filarial worms.

embryogenesis (em′brē-ō-jen′ə-səs) (Gr. *embryon,* embryo, + *genesis,* origin). The origin and development of the embryo; embryogeny.

emergence (L. *e,* out, + *mergere,* to plunge). The appearance of properties in a biological system (at the molecular, cellular, organismal, or species levels) that cannot be deduced from knowledge of the component parts taken separately or in partial combinations; such properties are termed **emergent properties.**

emigrate (L. *emigrare,* to move out). To move *from* one area to another to take up residence.

emulsion (ə-mul′shən) (L. *emulsus,* milked out). A colloidal system in which both phases are liquids.

endemic (en-dem′ik) (Gr. *en,* in, + *demos,* populace). Peculiar to a certain region or country; native to a restricted area; not introduced.

endergonic (en-dər-gän′ik) (Gr. *endon,* within, + *ergon,* work). Used in reference to a chemical reaction that requires energy; energy absorbing.

endite (en′dīt) (Gr. *endon,* within). Medial process on an arthropod limb.

endochondral (en′dō-kän′drəl) (Gr. *endon,* within, + *chondros,* cartilage). Occurring with the substance of cartilage, especially bone formation.

endocrine (en′də-krən) (Gr. *endon,* within, + *krinein,* to separate). Refers to a gland that is without a duct and that releases its product directly into the blood or lymph.

endocytosis (en′dō-sī-tō-səs) (Gr. *endon,* within, + *kytos,* hollow vessel). The engulfment of matter by phagocytosis, potocytosis, receptor-mediated endocytosis, and by bulk-phase (nonspecific) endocytosis.

endoderm (en′də-dərm) (Gr. *endon,* within, + *derma,* skin). Innermost germ layer of an embryo, forming the primitive gut; also may refer to tissues derived from endoderm.

endognathous (en′də-nā-thəs) (Gr. *endon,* within, + *gnathous,* jaw). Ancestral character in insects, found in orders Diplura, Collembola, and Protura, in which the mandibles and maxillae are located in pouches.

endolecithal (en′də-les′ə-thəl) (Gr. *endon,* within, + *lekithos,* yolk). Yolk for nutrition of the embryo incorporated into the egg cell itself.

endolymph (en′də-limf) (Gr. *endon,* within, + *lympha,* water). Fluid that fills most of the membranous labyrinth of the vertebrate ear.

endometrium (en′də-mē′trē-əm) (Gr. *endon,* within, + *mētra,* womb). The mucous membrane lining the uterus.

endoplasm (en′də-pla-zm) (Gr. *endon,* within, + *plasma,* mold or form). The portion of cytoplasm that immediately surrounds the nucleus.

endoplasmic reticulum A complex of membranes within a cell; may bear ribosomes (rough) or not (smooth).

endopod, endopodite (en′də-päd, en-dop′ə-dīt) (Gr. *endon,* within, + *pous, podos,* foot). Medial branch of a biramous crustacean appendage.

endopterygote (en′dəp-ter′i-gōt) (Gr. *endon,* within, + *pteron,* feather, wing). Insect in which the wing buds develop internally; has holometabolous metamorphosis.

endorphin (en-dor′fin) (Contraction of endogenous morphine). Group of opiate-like brain neuropeptides that modulate pain perception and are implicated in many other functions.

endoskeleton (Gr. *endon,* within, + *skeletos,* hard). A skeleton or supporting framework within the living tissues of an organism; contrasts with **exoskeleton.**

endosome (en′də-sōm) (Gr. *endon,* within, + *sōma,* body). Nucleolus in nucleus of some protozoa that retains its identity through mitosis.

endostyle (en′də-stīl) (Gr. *endon,* within, + *stylos,* a pillar). Ciliated groove(s) in the floor of the pharynx of tunicates, cephalochordates, and larval jawless fishes useful for accumulating and moving food particles to the stomach.

endothelium (en-də-thē′lē-əm) (Gr. *endon,* within, + *thēlē,* nipple). Squamous epithelium lining internal body cavities such as heart and blood vessels. Adj., **endothelial.**

endothermic (en′də-therm′ic) (Gr. *endon,* within, + *thermē,* heat). Having a body temperature determined by heat derived from the animal's own oxidative metabolism; contrasts with **ectothermic.**

enkephalin (en-kef′ə-lin) (Gr. *endon,* within, + *kephalē,* head). Group of small brain neuropeptides with opiate-like qualities.

enterocoel (en′tər-ō-sēl′) (Gr. *enteron,* gut, + *koilos,* hollow). A type of coelom formed by the outpouching of a mesodermal sac from the endoderm of the primitive gut.

enterocoelic mesoderm formation Embryonic formation of mesoderm by a pouchlike outfolding from the archenteron, which then expands and obliterates the blastocoel, thus forming a large cavity, the coelom, lined with mesoderm.

enterocoelomate (en′ter-ō-sēl′ō-māte) (Gr. *enteron,* gut, + *koilōma,* cavity, + Eng. *ate,* state of). An animal having an enterocoel, such as an echinoderm or a vertebrate.

enteron (en′tə-rän) (Gr. intestine). The digestive cavity.

entomology (en′tə-mol′ə-jē) (Gr. *entoma,* an insect, + *logos,* discourse). Study of insects.

entozoic (en-tə-zō′ic) (Gr. *entos,* within, + *zōon,* animal). Living within another animal; internally parasitic (chiefly parasitic worms).

entropy (en′trə-pē). (Gr. *en,* in, on, + *tropos,* turn, change in manner). A quantity that is the measure of energy in a system not available for doing work.

enzyme (en′zīm) (Gr. *enzymos,* leavened, from *en,* in, + *zyme,* leaven). A substance, produced by living cells, that is capable of speeding up specific chemical transformations, such as hydrolysis, oxidation, or reduction, but is unaltered itself in the process; a biological catalyst.

eocytes (ē′ə-sīts) (Gr. *ēos,* the dawn, + *kytos,* hollow vessel). A group of prokaryotes currently classified among the Archaebacteria but possibly a sister group of eukaryotes.

ephyra (ef′ə-rə) (Gr. *Ephyra,* Greek city). Refers to castlelike appearance. Medusa bud from a scyphozoan polyp.

epidermis (ep′ə-dər′məs) (Gr. *epi,* on, upon, + *derma,* skin). The outer, nonvascular layer of skin of ectodermal origin; in invertebrates, a single layer of ectodermal epithelium.

epididymis (ep′ə-did′ə-məs) (Gr. *epi,* on, upon, + *didymos,* testicle). Part of the sperm duct that is coiled and lying near the testis.

epigenesis (ep′ə-jen′ə-sis) (Gr. *epi,* on, upon, + *genesis,* birth). The embryological (and generally accepted) view that an embryo is a new creation that develops and differentiates step by step from an initial stage; the progressive production of new parts that were nonexistent as such in the original zygote.

epigenetics (ep′ə-je-net′iks) (Gr. *epi,* on, upon, + *genesis,* birth). Study of the relationship between genotype and phenotype as mediated by developmental processes.

epipod, epipodite (ep′ē-päd, e-pip′ə-dīt) (Gr. *epi,* on, upon, + *pous, podos,* foot). A lateral process on the protopod of a crustacean appendage, often modified as a gill.

epistome (ep′i-stōm) (Gr. *epi,* on, upon, + *stoma,* mouth). Flap over the mouth in some lophophorates bearing the protocoel.

epithelium (ep′ə-thē′lē-əm) (Gr. *epi,* on, upon, + *thēlē,* nipple). A cellular tissue covering a free surface or lining a tube or cavity.

epitoke (ep′ə-tōk) (Gr. *epitokos,* fruitful). Posterior part of a marine polychaete when swollen with developing gonads during the breeding season; contrast with **atoke.**

erythroblastosis fetalis (ə-rith′rə-blas-tō′səs fə-tal′əs) (Gr. *erythros,* red, + *blastos,* germ, + *osis,* a disease; L. *fetalis,* relating to a fetus). A disease of newborn infants caused when Rh-negative mothers develop antibodies against the Rh-positive blood of the fetus. See **blood type.**

erythrocyte (ə-rith′rə-sīt) (Gr. *erythros,* red, + *kytos,* hollow vessel). Red blood cell; has hemoglobin to carry oxygen from lungs or gills to tissues; during formation in mammals, erythrocytes lose their nuclei, those of other vertebrates retain the nuclei.

estrus (es′trəs) (L. *oestrus,* gadfly, frenzy). The period of heat, or rut, especially of the female during ovulation of the egg. Associated with maximum sexual receptivity.

estuary (es′chə-we′rē) (L. *aestuarium,* estuary). An arm of the sea where the tide meets the current of a freshwater drainage.

ethology (e-thäl′-ə-jē) (Gr. *ethos,* character, + *logos,* discourse). The study of animal behavior in natural environments.

euchromatin (ū′krō-mə-tən) (Gr. *eu,* good, well, + *chrōma,* color). Part of the chromatin that takes up stain less than heterochromatin, contains active genes.

eukaryotic, eucaryotic (ū′ka-rē-ot′ik) (Gr. *eu,* good, true, + *karyon,* nut, kernel). Organisms whose cells characteristically contain a membrane-bound nucleus or nuclei; contrasts with **prokaryotic.**

euploidy (ū′ployd′ē) (Gr. *eu,* good, well, + *ploid,* multiple of). Change in chromosome number from one generation to the next in which there is an addition or deletion of a complete set of chromosomes in the progeny; the most common type is polyploidy.

euryhaline (u′-rə-ha′līn) (Gr. *eurys,* broad, + *hals,* salt). Able to tolerate wide ranges of saltwater concentrations.

eutely (ū′te-lē) (Gr. *euteia,* thrift). Condition of a body composed of a constant number of cells or nuclei in all adult members of a species, as in rotifers, acanthocephalans, and nematodes.

evagination (ē-vaj′ə-nā′shən) (L. *e,* out, + *vagina,* sheath). An outpocketing from a hollow structure.

evolution (L. *evolvere,* to unfold). Organic evolution encompasses all changes in the characteristics and diversity of life on earth throughout its history.

evolutionary duration The length of time that a species or higher taxon exists in geological time.

evolutionary species concept A single lineage of ancestral-descendant populations that maintains its identity from other such lineages and has its own evolutionary tendencies and historical fate; differs from the biological species concept by explicitly including a time dimension and including asexual lineages.

evolutionary taxonomy A system of classification, formalized by George Gaylord Simpson, that groups species into Linnean higher taxa representing a hierarchy of distinct adaptive zones; such taxa may be monophyletic or paraphyletic but not polyphyletic.

excision repair Means by which cells are able to repair certain kinds of damage (dimerized pyrimidines) in their DNA.

exergonic (ek′sər-gān′ik) (Gr. *exō,* outside of, + *ergon,* work). An energy-yielding reaction.

exite (ex′īt) (Gr. *exō,* outside). Process from lateral side of an arthropod limb.

exocrine (ek′sə-krən) Gr. *exō,* outside, + *krinein,* to separate). A type of gland that releases its secretion through a duct; contrasts with **endocrine.**

exocytosis (eks′ə-sī-tō′səs) (Gr. *exo,* outside, + *kytos,* hollow vessel). Transport of a substance from inside a cell to the outside.

exon (ex′ən) (Gr. *exō,* outside). Part of the mRNA as transcribed from the DNA that contains a portion of the information necessary for final gene product.

exopod, exopodite (ex′ə-päd, ex-äp′ə-dīt) (Gr. *exō,* outside, + *pous, podos,* foot). Lateral branch of a biramous crustacean appendage.

exopterygote (ek′səp-ter′i-gōt) (Gr. *exō,* without, + *pteron,* feather, wing). Insect in which the wing buds develop externally during nymphal instars; has hemimetabolous metamorphosis.

exoskeleton (ek′sō-skel′ə-tən) (Gr. *exō,* outside, + *skeletos,* hard). A supporting structure secreted by ectoderm or epidermis; external, not enveloped by living tissue, as opposed to **endoskeleton.**

experiment (L. *experiri,* to try). A trial made to support or disprove a hypothesis.

exteroceptor (ek′stər-ō-sep′tər) (L. *exter,* outward, + *capere,* to take). A sense organ excited by stimuli from the external world.

F

facilitated diffusion Mediated transport in which a permease makes possible diffusion of a molecule across a cell membrane in the direction of a concentration gradient; contrast with **active transport.**

FAD Abbreviation for flavine adenine dinucleotide, an electron acceptor in the respiratory chain.

fascicle (fas′ə-kəl) (L. *fasciculus,* small bundle). A small bundle, usually referring to a collection of muscle fibers or nerve axons.

fatty acid Any of a series of saturated organic acids having the general formula $C_nH_{2n}O_2$, occurs in natural fats of animals and plants.

fermentation (L. *fermentum,* ferment). Enzymatic transformation, without oxygen, of organic substrates, especially carbohydrates, yielding products such as alcohols, acids, and carbon dioxide.

fiber (L. *fibra,* thread). A fiberlike cell or strand of protoplasmic material produced or secreted by a cell and lying outside the cell.

fibril (L. *fibra,* thread). A strand of protoplasm produced by a cell and lying within the cell.

fibrillar (fī′brə-lər) (L. *fibrilla,* small fiber). Composed of or pertaining to fibrils or fibers.

fibrin Protein that forms a meshwork, trapping erythrocytes, to become blood clot. Precursor is fibrinogen.

fibrosis (fī-brō′səs). Deposition of fibrous connective tissue in a localized site, during process of tissue repair or to wall off a source of antigen.

filipodium (fi′li-pō′dē-əm) (L. *filum,* thread, + Gr. *pous, podos,* a foot). A type of pseudopodium that is very slender and may branch but does not rejoin to form a mesh.

filter feeding Any feeding process by which particulate food is filtered from water in which it is suspended.

fission (L. *fissio,* a splitting). Asexual reproduction by a division of the body into two or more parts.

fitness Degree of adjustment and suitability for a particular environment. Genetic fitness is relative contribution of one genetically distinct organism to the next generation; organisms with high genetic fitness are naturally selected and become prevalent in a population.

flagellum (flə-jel′əm) pl. **flagella** (L. a whip). Whiplike organelle of locomotion.

flame cell Specialized hollow excretory or osmoregulatory structure of one or several small cells containing a tuft of flagella (the "flame") and situated at the end of a minute tubule; connected tubules ultimately open to the outside. See **solenocyte, protonephridium.**

fluke (O.E. *flōc,* flatfish). A member of class Trematoda or class Monogenea. Also, certain of the flatfishes (order Pleuronectiformes).

FMN Abbreviation for flavin mononucleotide, the prosthetic group of a protein (flavoprotein) and a carrier in the electron transport chain in respiration.

food vacuole A digestive organelle in the cell.

foraminiferan (for′əm-i-nif′-ər-ən) (L. *foramin,* hole, perforation, + *fero,* to bear). A member of the class Granuloreticulosea (phylum Sarcomastigophora) bearing a test with many openings.

fossil (fos′əl). Any remains or impression of an organism from a past geological age that has been preserved by natural processes, usually by mineralization in the earth's crust.

fossorial (fo-sōr′ē-əl) (L. *fossor,* digger). Characterized by digging or burrowing.

fouling Contamination of feeding or respiratory areas of an organism by excrement, sediment, or other matter. Also, accumulation of sessile marine organisms on the hull of a boat or ship so as to impede its progress through the water.

founder event Establishment of a new population by a small number of individuals (sometimes a single female carrying fertile eggs) that disperse from their parental population to a new location geographically isolated from the parental population.

fovea (fō′vē-ə) (L. small pit). A small pit or depression; especially the fovea centralis, a small rodless pit in the retina of some vertebrates, a point of acute vision.

free energy The energy available for doing work in a chemical system.

frontal plane. A plane parallel to the main axis of the body and at right angles to the sagittal plane.

fusiform (fū′zə-form) (L. *fustus,* spindle, + *forma,* shape). Spindle shaped; tapering toward each end.

G

gamete (ga′mēt, gə-mēt′) (Gr. *gamos,* marriage). A mature haploid sex cell; usually, male and female gametes can be distinguished. An egg or a sperm.

gametic meiosis Meiosis that occurs during formation of the gametes, as in humans and other metazoa.

gametocyte (gə-mēt′ə-sīt) (Gr. *gametēs,* spouse, + *kytos,* hollow vessel). The mother cell of a gamete, that is, immature gamete.

ganglion (gang′lē-ən) pl. **ganglia** (Gr. little tumor). An aggregation of nerve tissue containing nerve cells.

ganoid scales (ga′noid) (Gr. *ganos,* brightness). Thick, bony, rhombic scales of some primitive bony fishes; not overlapping.

gap junction An area of tiny canals communicating the cytoplasm between two cells.

gastrodermis (gas′trə-dər′mis) (Gr. *gastēr,* stomach, + *derma,* skin). Lining of the digestive cavity of cnidarians.

gastrolith (gas′trə-lith) (Gr. *gastēr,* stomach, + *lithos,* stone). Calcareous body in the wall of the cardiac stomach of crayfish and other Malacostraca, preceding the molt.

gastrovascular cavity (Gr. *gastēr,* stomach, + L. *vasculum,* small vessel). Body cavity in certain lower invertebrates that functions in both digestion and circulation and has a single opening serving as both mouth and anus.

gastrozooid (gas′trə-zō′id) (Gr. *gastēr,* stomach, + *zōon,* animal). The feeding polyp of a hydroid, a hydranth.

gastrula (gas′trə-lə) (Gr. *gastēr,* stomach, + L. *ula,* dim.). Embryonic stage, usually cap or sac shaped, with walls of two layers of cells surrounding a cavity (archenteron) with one opening (blastopore).

gastrulation (gas′trə-lā′shən) (Gr. *gastēr,* stomach). Process by which an early metazoan embryo becomes a gastrula, acquiring first two and then three layers of cells.

gel (jel) (from gelatin, from L. *gelare,* to freeze). That state of a colloidal system in which the solid particles form the continuous phase and the fluid medium the discontinuous phase.

gemmule (je′mūl) (L. *gemma,* bud, + *ula,* dim.). Asexual, cystlike reproductive unit in freshwater sponges; formed in summer or autumn and capable of overwintering.

gene (Gr. *genos,* descent). A nucleic acid sequence (usually DNA) that encodes a functional polypeptide or RNA sequence.

gene pool A collection of all of the alleles of all of the genes in a population.

genetic drift Random change in allelic frequencies in a population occurring by chance. In small populations, genetic variation at a locus may be lost by chance fixation of a single allelic variant.

genome (jē′nōm) (Gr. *genos,* offspring, + *ōma,* abstract group). All the DNA in a haploid set of chromosomes (nuclear genome), organelle (mitochondrial genome, chloroplast genome) or virus (viral genome, which in some viruses consists of RNA rather than DNA).

genotype (jēn′ə-tīp) (Gr. *genos,* offspring, + *typos,* form). The genetic constitution, expressed and latent, of an organism; the total set of genes present in the cells of an organism; contrasts with **phenotype.**

genus (jē-nəs), pl. **genera** (L. race). A group of related species with taxonomic rank between family and species.

germ layers In the animal embryo, one of three basic layers (ectoderm, endoderm, mesoderm) from which the various organs and tissues arise in the multicellular animal.

germovitellarium (jer′mə-vit-əl-ar′ē-əm) (L. *germen,* a bud, offshoot, + *vitellus,* yolk). Closely associated ovary (germarium) and yolk-producing structure (vitellarium) in rotifers.

germ plasm Cell lineages giving rise to the germ cells of a multicellular organism, as distinct from the somatoplasm.

gestation (jes-tā′shən) (L. *gestare,* to bear). The period in which offspring are carried in the uterus.

globulins (glo′bū-lənz) (L. *globus,* a globe, ball, + *-ulus,* ending denoting tendency). A large group of compact proteins with high molecular weight; includes immunoglobulins (antibodies).

glochidium (glō-kid′ē-əm) (Gr. *glochis,* point, + *idion,* dim.). Bivalved larval stage of freshwater mussels.

glomerulus (glo-mer′ə-ləs) (L. *glomus,* ball). A tuft of capillaries projecting into a renal corpuscle in a kidney. Also, a small spongy mass of tissue in the proboscis of hemichordates, presumed to have an excretory function. Also, a concentration of nerve fibers situated in the olfactory bulb.

gluconeogenesis (glü-cō-nē-ō-gen′ə-səs) (Gr. *glykys,* sweet, + *neos,* new, *genesis,* origin). Synthesis of glucose from protein or lipid precursors.

glycogen (glī′kə-jən) (Gr. *glykys,* sweet, *genēs,* produced). A polysaccharide constituting the principal form in which carbohydrate is stored in animals; animal starch.

glycolysis (glī-kol′ə-səs) (Gr. *glykys,* sweet, + *lysis,* a loosening). Enzymatic breakdown of glucose (especially) or glycogen into phosphate derivatives with release of energy.

gnathobase (nath′ə-bās′) (Gr. *gnathos,* jaw, + base). A median basic process on certain appendages in some arthropods, usually for biting or crushing food.

gnathostomes (nath′ə-stōmz) (Gr. *gnathos,* jaw, + *stoma,* mouth). Vertebrates with jaws.

Golgi complex (gōl′jē) (after Golgi, Italian histologist). An organelle in cells that serves as a collecting and packaging center for secretory products.

gonad (gō′nad) (N.L. *gonas,* primary sex organ). An organ that produces gametes (ovary in the female and testis in the male).

gonangium (gō-nan′jē-əm) (N.L. *gonas,* primary sex organ, + *angeion,* dim. of vessel). Reproductive zooid of hydroid colony (Cnidaria).

gonoduct (Gr. *gonos,* seed, progeny, + duct). Duct leading from a gonad to the exterior.

gonopore (gon′ə-pōr) (Gr. *gonos,* seed, progeny, + *poros,* an opening). A genital pore found in many invertebrates.

grade (L. *gradus,* step). A level of organismal complexity or adaptive zone characteristic of a group of evolutionarily related organisms.

gradualism (graj′ə-wal-iz′əm). A component of Darwin's evolutionary theory postulating that evolution occurs by the temporal accumulation of small, incremental changes, usually across very long periods of geological time; it opposes claims that evolution can occur by large, discontinuous or macromutational changes.

granulocytes (gran′ū-lə-sīts) (L. *granulus,* small grain, + Gr. *kytos,* hollow vessel). White blood cells (neutrophils, eosinophils, and basophils) bearing "granules" (vacuoles) in their cytoplasm that stain deeply.

green gland Excretory gland of certain Crustacea; the antennal gland.

gregarious (L. *grex,* herd). Living in groups or flocks.

guanine (gwä′nēn) (Sp. from Quechua, *huanu,* dung). A white crystalline purine base, $C_5H_5N_5O$, occurring in various animal tissues and in guano and other animal excrements.

guild (gild) (M.E. *gilde,* payment, tribute). In ecology, a group of species that exploit the same class of environment in a similar way.

gynecophoric canal (gī′nə-kə-fōr′ik) (Gr. *gynē,* woman, + *pherein,* to carry). Groove in male schistosomes (certain trematodes) that carries the female.

H

habitat (L. *habitare,* to dwell). The place where an organism normally lives or where individuals of a population live.

habituation A kind of learning in which continued exposure to the same stimulus produces diminishing responses.

halter (hal′tər), pl. **halteres** (hal-ti′rēz) (Gr. leap). In Diptera, small club-shaped structure on each side of the metathorax representing the hindwings; believed to be sense organs for balancing; also called balancer.

haplodiploidy (Gr. *haploos,* single, + *diploos,* double, + *eidos,* form). Reproduction in which haploid males are produced parthenogenetically, and diploid females are from fertilized eggs.

haploid (Gr. *haploos,* single). The reduced, or N, number of chromosomes, typical of gametes, as opposed to the diploid, or 2N, number found in somatic cells. In certain groups, mature organisms may have a haploid number of chromosomes.

Hardy-Weinberg equilibrium Mathematical demonstration that the Mendelian hereditary process does not change the populational frequencies of alleles or genotypes across generations, and that change in allelic or genotypic frequencies requires factors such as natural selection, genetic drift in finite populations, recurring mutation, migration of individuals among populations, and nonrandom mating.

hectocotylus (hek-tə-kät′ə-ləs) (Gr. *hekaton,* hundred, + *kotylē,* cup). Specialized, and sometimes autonomous, arm that serves as a male copulatory organ in cephalopods.

hemal system (hē′məl) (Gr. *haima,* blood). System of small vessels in echinoderms; function unknown.

hemerythrin (hē′mə-rith′rin) (Gr. *haima,* blood, + *erythros,* red). A red, iron-containing respiratory pigment found in the blood of some polychaetes, sipunculids, priapulids, and brachiopods.

hemimetabolous (he′mi-mə-ta′bə-ləs) (Gr. *hēmi,* half, + *metabolē,* change). Refers to gradual metamorphosis during development of insects, without a pupal stage.

hemocoel (hēm′ə-sēl) (Gr. *haima,* blood, + *koiloma,* cavity). Major body space in arthropods replacing the coelom, contains the blood (hemolymph).

hemoglobin (Gr. *haima,* blood, + L. *globulus,* globule). An iron-containing respiratory pigment occurring in vertebrate red blood cells and in blood plasma of many invertebrates; a compound of an iron porphyrin heme and globin proteins.

hemolymph (hē′mə-limf) (Gr. *haima,* blood, + L. *lympha,* water). Fluid in the coelom or hemocoel of some invertebrates that represents the blood and lymph of vertebrates.

hepatic (hə-pat′ik) (Gr. *hēpatikos,* of the liver). Pertaining to the liver.

herbivore ([h]erb′ə-vōr′) (L. *herba,* green crop, + *vorare,* to devour). Any organism subsisting on plants. Adj., **herbivorous.**

heredity (L. *heres,* heir). The faithful transmission of biological traits from parents to their offspring.

hermaphrodite (hər-maf′rə-dīt) (Gr. *hermaphroditos,* containing both sexes; from Greek mythology. Hermaphroditos, son of Hermes and Aphrodite). An organism with both male and female functional reproductive organs. **Hermaphroditism** may refer to an aberration in dioecious animals; **monoecy** implies that this is the normal condition for the species.

hermatypic (hər-mə-ti′pik) (Gr. *herma,* reef, + *typos,* pattern). Relating to reef forming corals.

heterocercal (het′ər-ō-ser′kəl) (Gr. *heteros,* different, + *kerkos,* tail). In some fishes, a tail with the upper lobe larger than the lower, and the end of the vertebral column somewhat upturned in the upper lobe, as in sharks.

heterochromatin (he′tə-rō-krōm′ə-tən) (Gr. *heteros,* different, + *chrōma,* color). Chromatin that stains intensely and appears to represent inactive genetic areas.

heterochrony (hed′ə-rō-krōn-y) (Gr. *heteros,* different, + *chronos,* time). Evolutionary change in the relative time of appearance or rate of development of characteristics from ancestor to descendant.

heterodont (hed′ə-ro-dänt) (Gr. *heteros,* different, + *odous,* tooth). Having teeth differentiated into incisors, canines, and molars for different purposes.

heterotroph (hət′ə-rō-trōf) (Gr. *heteros,* different, + *trophos,* feeder). An organism that obtains both organic and inorganic raw materials from the environment in order to live; includes most animals and those plants that do not carry on photosynthesis.

heterozygote (het′ə-rō-zī′gōt) (Gr. *heteros,* different, + *zygōtos,* yoked). An organism in which homologous chromosomes contain different allelic forms (often dominant and recessive) of a locus; derived from a zygote formed by union of gametes of dissimilar allelic constitution.

hexamerous (hek-sam′ər-əs) (Gr. *hex,* six, + *meros,* part). Six parts, specifically, symmetry based on six or multiples thereof.

hibernation (L. *hibernus,* wintry). Condition, especially of mammals, of passing the winter in a torpid state in which the body temperature drops nearly to freezing and the metabolism drops close to zero.

hierarchical system A scheme arranging organisms into a series of taxa of increasing inclusiveness, as illustrated by Linnean classification.

histogenesis (his-tō-jen′ə-sis) (Gr. *histos,* tissue, + *genesis,* descent). Formation and development of tissue.

histone (hi′stōn) (Gr. *histos,* tissue). Any of several simple proteins found in cell nuclei and complexed at one time or another with DNA. Histones yield a high proportion of basic amino acids on hydrolysis; characteristic of eukaryotes.

holoblastic cleavage (Gr. *holo,* whole, + *blastos,* germ). Complete and approximately equal division of cells in early embryo. Found in mammals, amphioxus, and many aquatic invertebrates that have eggs with a small amount of yolk.

holometabolous (hō′lō-mə-ta′bə-ləs) (Gr. *holo,* complete, + *metabolē,* change). Complete metamorphosis during development of insects, including larval instars, pupa, and adult.

holophytic nutrition (hōl′ō-fit′ik) (Gr. *holo,* whole, + *phyt,* plant). Occurs in green plants and certain protistans and involves synthesis of carbohydrates from carbon dioxide and water in the presence of light, chlorophyll, and certain enzymes.

holozoic nutrition (hōl′ō-zō′ik) (Gr. *holo,* whole, + *zoikos,* of animals). Type of nutrition involving ingestion of liquid or solid organic food particles.

homeobox (hō′mē-ō-box) (Gr. *homoios,* like, resembling, + L. *buxus,* boxtree [used in the sense of enclosed, contained]). A highly conserved 180-base pair sequence found in regulatory sequences of protein-coding genes that regulate development.

homeostasis (hō′-mē-ō-stā′sis) (Gr. *homeo,* alike, + *stasis,* state or standing). Maintenance of an internal steady state by means of self-regulation.

homeothermic (hō′-mē-ō-thər′mik) (Gr. *homeo,* alike, + *thermē,* heat). Having a nearly uniform body temperature, regulated independent of the environmental temperature; "warm blooded."

homeotic genes (hō-mē-ät′ik) (Gr. *homoios,* like, resembling). Genes, identified through mutations, that give developmental identity to specific body segments.

home range The area over which an animal ranges in its activities. Unlike territories, home ranges are not defended.

hominid (häm′ə-nid) (L. *homo, hominis,* man). A member of the family Hominidae, now represented by one living species, *Homo sapiens.*

hominoid (häm′ə-noyd). Relating to the Hominoidea, a superfamily of primates to which the great apes and humans are assigned.

homocercal (hō′mə-ser′kəl) (Gr. *homos,* same, common, + *kerkos,* tail). A tail with the upper and lower lobes symmetrical and the vertebral column ending near the middle of the base, as in most teleost fishes.

homodont (hō′mō-dänt) (Gr. *homos,* same, + *odous,* tooth). Havng all teeth similar in form.

homograft See **allograft.**

homology (hō-mäl′ə-jē) (Gr. *homologos,* agreeing). Similarity of parts or organs of different organisms caused by evolutionary derivation from a corresponding part or organ in a remote ancestor, and usually having a similar embryonic origin. May also refer to a matching pair of chromosomes. Serial homology is the correspondence in the same individual of repeated structures having the same origin and development, such as the appendages of arthropods. Adj., **homologous.**

homoplasy (hō′mə-plā′sē). Phenotypic similarity among characteristics of different species or populations (including molecular, morphological, behavioral, or other features) that does not accurately represent patterns of common evolutionary descent (= nonhomologous similarity); it is produced by evolutionary parallelism, convergence and/or reversal, and is revealed by incongruence among different characters on a cladogram or phylogenetic tree.

homozygote (hō-mə-zī′gōt) (Gr. *homos,* same, + *zygotos,* yoked). An organism in which the pair of alleles for a trait is composed of the same genes (either dominant or recessive but not both). Adj., **homozygous.**

humoral (hū′mər-əl) (L. *humor,* a fluid). Pertaining to an endocrine secretion.

humoral immune response Immune response involving production of antibodies, specifically the T_H2 arm of the immune response. Contrast **cell-mediated immune response.**

hyaline (hī′ə-lən) (Gr. *hyalos,* glass). Adj., glassy, translucent. Noun, a clear, glassy, structureless material occurring, for example, in cartilage, vitreous body, mucin, and glycogen.

hybridoma (hī-brid-ō′mah) (contraction of hybrid + myeloma). Fused product of a normal and a myeloma (cancer) cell, which has some of the characteristics of the normal cell.

hydatid cyst (hī-da′təd) (Gr. *hydatis,* watery vesicle). A type of cyst formed by juveniles of certain tapeworms *(Echinococcus)* in their vertebrate hosts.

hydranth (hī′dranth) (Gr. *hydōr,* water, + *anthos,* flower). Nutritive zooid of hydroid colony.

hydrocoel (hī-drə-sēl) (Gr. *hydōr,* water, + *koilos,* hollow). Second or middle coelomic compartment in echinoderms; left hydrocoel gives rise to water vascular system.

hydrogen bond A relatively weak chemical bond resulting from unequal charge distribution within molecules, in which a hydrogen atom covalently bonded to another atom is attracted to the electronegative portion of another molecule.

hydroid The polyp form of a cnidarian as distinguished from the medusa form. Any cnidarian of the class Hydrozoa, order Hydroida.

hydrolysis (Gr. *hydōr,* water, + *lysis,* a loosening). The decomposition of a chemical compound by the addition of water; the splitting of a molecule into its groupings so that the split products acquire hydrogen and hydroxyl groups.

hydrorhiza (hī′drə-rī′zə) (Gr. *hydōr,* water, + *rhiza,* a root). Rootlike stolon that attaches a hydroid to its substrate.

hydrosphere (Gr. *hydōr,* water, + *sphaira,* ball, sphere). Aqueous envelope of the earth.

hydrostatic pressure The pressure exerted by a fluid (gas or liquid), defined as force per unit area. For example, the hydrostatic pressure of one atmosphere (1 atm) is 14.7 lb/in^2.

hydrostatic skeleton A mass of fluid or plastic parenchyma enclosed within a muscular wall to provide the support necessary for antagonistic muscle action; for example, parenchyma in acoelomates and perivisceral fluids in pseudocoelomates serve as hydrostatic skeletons.

hydrothermal vent A submarine hot spring; seawater seeping through the sea bottom is heated by magma and expelled back into the sea through the hydrothermal vent.

hydroxyl (hydrogen + oxygen, + yl). Containing an OH^- group, a negatively charged ion formed by alkalies in water.

hyomandibular (hī-ō-mən-dib′yə-lər) (Gr. *hyoeides* [shaped like the Gr. letter upsilon Υ, + *eidos,* form], + L. *mandere,* to chew). Bone derived from the hyoid gill arch, forming part of articulation of the lower jaw of fishes, and forming the stapes of the ear of amniotic vertebrates.

hyperosmotic (Gr. *hyper,* over, + *ōsmos,* impulse). Refers to a solution whose osmotic pressure is greater than that of another solution to which it is compared; contains a greater concentration of dissolved particles and gains water through a selectively permeable membrane from a solution containing fewer particles; contrasts with **hypoosmotic.**

hyperparasitism (hī′pər-par′ə-sid-iz-əm) (Gr. *hyper,* over, + *para,* beside, + *sitos,* food). Parasitism of a parasite by another parasite.

hypertrophy (hī-per′trə-fē) (Gr. *hyper,* over, + *trophē,* nourishment). Abnormal increase in size of a part or organ.

hypodermis (hī′pə-dər′mis) (Gr. *hypo,* under, + L. *dermis,* skin). The cellular layer lying beneath and secreting the cuticle of annelids, arthropods, and certain other invertebrates.

hypoosmotic (Gr. *hypo,* under, + *ōsmos,* impulse). Refers to a solution whose osmotic pressure is less than that of another solution with which it is compared or taken as a standard; contains a lesser concentration of dissolved particles and loses water during osmosis; contrasts with **hyperosmotic.**

hypophysis (hī-pof′ə-sis) (Gr. *hypo,* under, + *physis,* growth). Pituitary body.

hypostome (hī′pə-stōm) (Gr. *hypo,* under, + *stoma,* mouth). Name applied to structure in various invertebrates (such as mites and ticks), located at posterior or ventral area of mouth.

hypothalamus (hī-pə-thal′ə-məs) (Gr. *hypo,* under, + *thalamos,* inner chamber). A ventral part of the forebrain beneath the thalamus; one of the centers of the autonomic nervous system.

hypothesis (Gr. *hypothesis,* foundation, supposition). A statement or proposition that can be tested by experiment.

hypothetico-deductive (Gr. *hypotithenai,* to suppose, + L. *deducere,* to lead). Scientific process of making a conjecture and then seeking empirical tests that potentially lead to its rejection.

I

immediate hypersensitivity Inflammatory reaction based on humoral immunity.

immunoglobulin (im′yə-nə-glä′byə-lən) (L. *immunis,* free, + *globus,* globe). Any of a group of plasma proteins, produced by plasma cells, that participates in the immune response by combining with the antigen that stimulated its production. Antibody.

imprinting (im′print-ing) (L. *imprimere,* to impress, imprint). Rapid and usually stable learning pattern appearing early in the life of a member of a social species and involving recognition of its own species; may involve attraction to the first moving object seen.

inbreeding The tendency among members of a population to mate preferentially with close relatives.

incomplete dominance See **intermediate inheritance.**

incus (in′kəs) (L. *incus,* anvil). The middle of a chain of three bones of the mammalian middle ear.

indeterminate cleavage A type of embryonic development in which the fate of the blastomeres is not determined very early as to tissues or organs, for example, in echinoderms and vertebrates; regulative cleavage.

induction (L. *inducere, inductum,* to lead). Reasoning from the particular to the general, that is, deriving a general statement (hypothesis) based on individual observations. In embryology, the alteration of cell fates as the result of interaction with neighboring cells.

inductor (in-duk′tər) (L. *inducere,* to introduce, lead in). In embryology, a tissue or organ that causes the differentiation of another tissue or organ.

inflammation (in′fləm-mā′shən) (L. *inflammare,* from *flamma,* flame). The complicated physiological process in mobilization of body defenses against foreign substances and infectious agents and repair of damage from such agents.

infraciliature (in-frə-sil′ē-ə-chər) (L. *infra,* below, + *cilia,* eyelashes). The organelles just below the cilia in ciliate protozoa.

infundibulum (in′fun-dib′u-ləm) (L. funnel). Stalk of the neurohypophysis linking the pituitary to the diencephalon.

innate (i-nāt′) (L. *innatus,* inborn). A characteristic based partly or wholly on genetic or epigenetic constitution.

instar (inz′tär) (L. form). Stage in the life of an insect or other arthropod between molts.

instinct (L. *instinctus,* impelled). Stereotyped, predictable, genetically programmed behavior. Learning may or may not be involved.

integument (ən-teg′ū-mənt) (L. *integumentum,* covering). An external covering or enveloping layer.

intercellular (in-tər-sel′yə-lər) (L. *inter,* among, + *cellula,* chamber). Occurring between body cells.

interferons Several cytokines encoded by different genes, important in mediation of natural immunity and inflammation.

interleukins A series of cytokines produced primarily by various leukocytes, such as macrophages and T cells, whose target cells are various leukocytes and other cells. Given the name "interleukins" when it was believed that they were produced only by leukocytes and their target cells were limited to leukocytes.

intermediary meiosis Meiosis that occurs neither during gamete formation nor immediately after zygote formation, resulting in both haploid and diploid generations, such as in foraminiferan protozoa.

intermediate host A host in which some development of a symbiont occurs, but in which maturation and sexual reproduction do not take place.

intermediate inheritance Neither of alternate alleles of a gene are completely dominant, and heterozygote shows a condition intermediate between or different from homozygotes for each allele.

interstitial (in-tər-sti′shəl) (L. *inter,* among, + *sistere,* to stand). Situated in the interstices or spaces between structures such as cells, organs, or grains of sand.

intracellular (in-trə-sel′yə-lər) (L. *intra,* inside, + *cellula,* chamber). Occurring within a body cell or within body cells.

intrinsic growth rate Exponential growth rate of a population, that is, the difference between the density-independent components of the birth and death rates of a natural population with stable age distribution.

intron (in′trän) (L. *intra,* within). Portion of mRNA as transcribed from DNA that will not form part of mature mRNA, and therefore does not encode an amino-acid sequence in the protein product.

introvert (L. *intro,* inward, + *vertere,* to turn). The anterior narrow portion that can be withdrawn (introverted) into the trunk of a sipunculid worm.

invagination (in-vaj′ə-nā′shən) (L. *in,* in, + *vagina,* sheath). An infolding of a layer of tissue to form a saclike structure.

inversion (L. *invertere,* to turn upside down). A turning inward or inside out, as in embryogenesis of sponges; also, reversal in order of genes or reversal of a chromosome segment.

ion An atom or group of atoms with a net positive or negative electrical charge because of the loss or gain of electrons.

ionic bond A chemical bond formed by transfer of one or more electrons from one atom to another; characteristic of salts.

iridophore (ī-rid′ə-fōr) (Gr. *iris,* rainbow, or iris of eye). Iridescent or silvery chromatophores containing crystals or plates of guanine or other purine.

bat/āpe/ärmadillo/herring/fēmale/finch/līce/crocodile/crōw/duck/ūnicorn/tüna/ə indicates unaccented vowel sound "uh" as in mammal, fishes, cardinal, heron, vulture/stress as in bi-ol′o-gy, bi′o-log′i-cal

irritability (L. *irritare,* to provoke). A general property of all organisms involving the ability to respond to stimuli or changes in the environment.

isogametes (ī′so-gam′ēts) (Gr. *isos,* equal, + *gametēs,* spouse). Gametes of a species in which gametes of both sexes are alike in size and appearance.

isosmotic A liquid having the same osmotic pressure as another, reference liquid.

isotonic (Gr. *isos,* equal, + *tonikos,* tension). Pertaining to solutions having the same or equal osmotic pressure; isosmotic.

isotope (Gr. *isos,* equal, + *topos,* place). One of several different forms (species) of a chemical element, differing from each other in atomic mass but not in atomic number.

J

juvenile hormone Hormone produced by the corpora allata of insects; among its effects are maintenance of larval or nymphal characteristics during development.

juxtaglomerular apparatus (jək′stə-glä-mer′yə-lər) (L. *juxta,* close to, + *glomus,* ball). Complex of sensory cells located in the afferent arteriole adjacent to the glomerulus and a loop of the distal tubule, which produces the enzyme renin.

K

keratin (ker′ə-tən) (Gr. *kera,* horn, + *in,* suffix of proteins). A scleroprotein found in epidermal tissues and modified into hard structures such as horns, hair, and nails.

keystone species A species (typically a predator) whose removal leads to reduced species diversity within the community.

kinesis (kə-nē′səs) (Gr. *kinēsis,* movement). Movements by an organism in random directions in response to stimulus.

kinetochore (kī-nēt′ə-kōr) (Gr. *kinein,* to move, + *choris,* asunder, apart). A disc of proteins located on the centromere, specialized to interact with the spindle fibers during mitosis.

kinetodesma (kə-nē′tə-dez′mə). pl. **kinetodesmata** (Gr. *kinein,* to move, + *desma,* bond). Fibril arising from the kinetosome of a cilium in a cilate protozoan, and passing along the kinetosomes of cilia in that same row.

kinetosome (kən-ēt′ə-sōm) (Gr. *kinētos,* moving, + *sōma,* body). The self-duplicating granule at the base of the flagellum or cilium; similar to centriole, also called basal body or blepharoplast.

kinety (kə-nē′tē) (Gr. *kinein,* to move). All the kinetosomes and kinetodesmata of a row of cilia.

kinin (kī′nin) (Gr. *kinein,* to move, + *in,* suffix of hormones). A type of local hormone that is released near its site of origin; also called parahormone or tissue hormone.

K-selection (from the K term in the logistic equation). Natural selection under conditions that favor survival when populations are controlled primarily by density-dependent factors.

kwashiorkor (kwash-ē-or′kər) (from Ghana). Malnutrition caused by diet high in carbohydrate and extremely low in protein.

L

labium (lā′bē-əm) (L. a lip). The lower lip of the insect formed by fusion of the second pair of maxillae.

labrum (lā′brəm) (L. a lip). The upper lip of insects and crustaceans situated above or in front of the mandibles; also refers to the outer lip of a gastropod shell.

labyrinth (L. *labyrinthus,* labyrinth). Vertebrate internal ear, composed of a series of fluid-filled sacs and tubules (membranous labyrinth) suspended within bone cavities (osseous labyrinth).

labyrinthodont (lab′ə-rin′thə-dänt) (Gr. *labyrinthos,* labyrinth, + *odous, odontos,* tooth). A group of Paleozoic amphibians containing the temnospondyls and the anthracosaurs.

lacteal (lak′tē-əl) (L. *lacteus,* of milk). Noun, one of the lymph vessels in the villus of the intestine. Adj., relating to milk.

lacuna (lə-kū′nə), pl. **lacunae** (L. pit, cavity). A sinus; a space between cells; a cavity in cartilage or bone.

lagena (lə-jē′nə) (L. large flask). Portion of the primitive ear in which sound is translated into nerve impulses; evolutionary beginning of cochlea.

Lamarckism Hypothesis, as expounded by Jean Baptiste de Lamarck, of evolution by the acquisition during an organism's lifetime of characteristics that are transmitted directly to offspring.

lamella (lə-mel′ə) (L. dim. of *lamina,* plate). One of the two plates forming a gill in a bivalve mollusc. One of the thin layers of bone laid concentrically around an osteon (Haversian canal). Any thin, platelike structure.

lappets Lobes around the margin of scyphozoan medusae (phylum Cnidaria).

larva (lar′və), pl. **larvae** (L. a ghost). An immature stage that is quite different in body form from the adult.

larynx (lar′inks) (Gr. the larynx, gullet). Modified upper portion of respiratory tract of air-breathing vertebrates, bounded by the glottis above and the trachea below; voice box; adj., **laryngeal** (lə-rin′jē-əl), relating to the larynx.

lateral (L. *latus,* the side, flank). Of or pertaining to the side of an animal; a *bilateral* animal has two sides.

laterite (lad′ə-rīt) (L. *later,* brick). Group of hard, red soils from topical areas that show intense weathering and leaching of bases and silica, leaving aluminum hydroxides and iron oxides; adj. **lateritic.**

lek (lek) (Sw. play, game). An area where animals assemble for communal courtship display and mating.

lemniscus (lem-nis′kəs) (L. ribbon). One of a pair of internal projections of the epidermis from the neck region of Acanthocephala, which functions in fluid control in the protrusion and invagination of the proboscis.

lentic (len′tik) (L. *lentus,* slow). Of or relating to standing water such as swamp, pond, or lake.

lepidosaurs (lep′ə-dō-sors) (L. *lepidos,* scale, + *sauros,* lizard). A lineage of diapsid reptiles that appeared in the Permian and that includes the modern snakes, lizards, amphisbaenids, and tuataras, and the extinct ichthyosaurs.

leptocephalus (lep′tə-sef′ə-ləs) pl. **leptocephali** (Gr. *leptos,* thin, + *kephalē,* head). Transparent, ribbonlike migratory larva of the European or American eel.

leukemism (lü′kə-mi-zəm) (Gr. *leukos,* white, + *ismos,* condition of). Presence of white pelage or plumage in animals with normally pigmented eyes and skin.

leukocyte (lü′kə-sīt) (Gr. *leukos,* white, + *kytos,* hollow vessel). Any of several kinds of white blood cells (for example, granulocytes, lymphocytes, monocytes), so-called because they bear no hemoglobin, as do red blood cells.

library In molecular biology, a set of clones containing recombinant DNA. Obtained from and representing the genome of the organism.

ligament (lig′ə-mənt) (L. *ligamentum,* bandage). A tough, dense band of connective tissue connecting one bone to another.

ligand (lī′gənd) (L. *ligo,* to bind). A molecule that specifically binds to a receptor; for example, a hormone (ligand) binds specifically to its receptor on the cell surface.

lipase (lī′pās) (Gr. *lipos,* fat, + *ase,* enzyme suffix). An enzyme that accelerates the hydrolysis or synthesis of fats.

lipid, lipoid (li′pid, lə-poyd′) (Gr. *lipos,* fat). Certain fatlike substances, often containing other groups such as phosphoric acid; lipids combine with proteins and carbohydrates to form principal structural components of cells.

lithosphere (lith′ə-sfir) (Gr. *lithos,* rock, + *sphaira,* ball). The rocky component of the earth's surface layers.

littoral (lit′ə-rəl) (L. *litoralis,* seashore). Adj., pertaining to the shore. Noun, that portion of the sea floor between the extent of high and low tides, intertidal; in lakes, the shallow part from the shore to the lakeward limit of aquatic plants.

lobopodium (lō′bə-pō′dē-əm) (Gr. *lobos,* lobe, + *pous, podos,* foot). Blunt, lobelike pseudopodium.

locus (lō′kəs), pl. **loci** (lō′sī) (L. place). Position of a gene in a chromosome.

logistic equation A mathematical expression describing an idealized sigmoid curve of population growth.

lophophore (lōf′ə-fōr) (Gr. *lophos,* crest, + *phoros,* bearing). Tentacle-bearing ridge or arm within which is an extension of the coelomic cavity in lophophorate animals (ectoprocts, brachiopods, and phoronids).

lorica (lo′rə-kə) (L. corselet). Protective external case found in some protozoa, rotifers, and others.

lotic (lō′tik) (L. *lotus,* action of washing or bathing). Of or pertaining to running water, such as a brook or river.

lumbar (lum′bär) (L. *lumbus,* loin). Relating to or near the loins or lower back.

lumen (lü′mən) (L. light). The cavity of a tube or organ.

lymph (limf) (L. *lympha,* water). The interstitial (intercellular) fluid in the body, also the fluid in the lymphatic system.

lymphocyte (lim′fō-sīt) (L. *lympha,* water, goddess of water, + Gr. *kytos,* hollow vessel). Cell in blood and lymph that has central role in immune responses. See **T cell** and **B cell.**

lymphokine (limf′ə-kīn) (L. *lympha,* water, + Gr. *kinein,* to move). A molecule secreted by an activated or stimulated lymphocyte that causes physiological changes in certain other cells.

lysosome (lī′sə-sōm) (Gr. *lysis,* loosing, + *sōma,* body). Intracellular organelle consisting of a membrane enclosing several digestive enzymes that are released when the lysosome ruptures.

M

macroevolution (L. *makros,* long, large, + *evolvere,* to unfold). Evolutionary change on a grand scale, encompassing the origin of novel designs, evolutionary trends, adaptive radiation, and mass extinction.

macrogamete (mak′rə-gam′ēt) (Gr. *makros,* long, large, + *gamos,* marriage). The larger of the two gamete types in a heterogametic organism, considered the female gamete.

macromolecule A very large molecule, such as a protein, polysaccharide, or nucleic acid.

macronucleus (ma′krō-nü′klē-əs) (Gr. *makros,* long, large, + *nucleus,* kernel). The larger of the two kinds of nuclei in ciliate protozoa; controls all cell functions except reproduction.

macrophage (mak′rə-fāj) (Gr. *makros,* long, large, + *phagō,* to eat). A phagocytic cell type in vertebrates that performs crucial functions in the immune response and inflammation, such as presenting antigenic epitopes to T cells and producing several cytokines.

madreporite (ma′drə-pōr′īt) (Fr. *madrépore,* reef-building coral, + *ite,* suffix for some body parts). Sievelike structure that is the intake for the water-vascular system of echinoderms.

major histocompatibility complex (MHC) Complex of genes coding for proteins inserted in the cell membrane; the proteins are the basis of self-nonself recognition by the immune system.

malacostracan (mal′ə-käs′trə-kən) (Gr. *malako,* soft, + *ostracon,* shell). Any member of the crustacean subclass Malacostraca, which includes both aquatic and terrestrial forms of crabs, lobsters, shrimps, pillbugs, sand fleas, and others.

malleus (mal′ē-əs) (L. hammer). The ossicle attached to the tympanum in middle ears of mammals.

malpighian tubules (mal-pig′ē-ən) (Marcello Malpighi, Italian anatomist, 1628-1694). Blind tubules opening into the hindgut of nearly all insects and some myriapods and arachnids, and functioning primarily as excretory organs.

mantle Soft extension of the body wall in certain invertebrates, for example, brachiopods and molluscs, which usually secretes a shell; thin body wall of tunicates.

manubrium (man-ü′bri-əm) (L. handle). The portion projecting from the oral side of a jellyfish medusa, bearing the mouth; oral cone; presternum or anterior part of sternum; handle-like part of malleus of ear.

marasmus (mə-raz′məs) (Gr. *marasmos,* to waste away). Malnutrition, especially of infants, caused by a diet deficient in both calories and protein.

marsupial (mär-sü′pē-əl) (Gr. *marsypion,* little pouch). One of the pouched mammals of the subclass Metatheria.

mastax (mas′təx) (Gr. jaws). Pharyngeal mill of rotifers.

matrix (mā′triks) (L. *mater,* mother). The intercellular substance of a tissue, or that part of a tissue into which an organ or process is set.

maturation (L. *maturus,* ripe). The process of ripening; the final stages in the preparation of gametes for fertilization.

maxilla (mak-sil′ə) (L. dim. of *mala,* jaw). One of the upper jawbones in vertebrates; one of the head appendages in arthropods.

maxilliped (mak-sil′ə-ped) (L. *maxilla,* jaw, + *pes,* foot). One of the pairs of head appendages located just posterior to the maxilla in crustaceans, a thoracic appendage that has become incorporated into the feeding mouthparts.

medial (mē′dē-əl). Situated, or occurring, in the middle.

bat/āpe/ärmadillo/herring/fēmale/finch/līce/crocodile/crōw/duck/ūnicorn/tüna/ə indicates unaccented vowel sound "uh" as in mammal, fishes, cardinal, heron, vulture/stress as in bi-ol′o-gy, bi′o-log′i-cal

mediated transport Transport of a substance across a cell membrane mediated by a carrier molecule in the membrane.

medulla (mə-dul′ə) (L. marrow). The inner portion of an organ in contrast to the cortex or outer portion. Also, hindbrain.

medusa (mə-dü-sə) (Gr. mythology, female monster with snake-entwined hair). A jellyfish, or the free-swimming stage in the life cycle of cnidarians.

meiosis (mī-ō′səs) (Gr. from *mieoun,* to make small). The nuclear changes by means of which the chromosomes are reduced from the diploid to the haploid number; in animals, usually occurs in the last two divisions in the formation of the mature egg or sperm.

melanin (mel′ə-nin) (Gr. *melas,* black). Black or dark-brown pigment found in plant or animal structures.

melanophore (mel′ə-nə-fōr, mə-lan′ə-fōr) (Gr. *melania,* blackness, + *pherein,* to bear). Black or brown chromatophore containing melanin.

memory cells Population of long-lived B lymphocytes remaining after initial immune response that provides for the secondary response.

meninges (mə-nin′jez), sing. **meninx** (Gr. *mēninx,* membrane). Any of three membranes (arachnoid, dura mater, pia mater) that envelop the vertebrate brain and spinal cord. Also, solid connective tissue sheath enclosing the central nervous system of some vertebrates.

merozoite (me′rə-zō′īt) (Gr. *meros,* part, + *zōon,* animal). A very small trophozoite at the stage just after cytokinesis has been completed in multiple fission of a protozoan.

mesenchyme (me′zən-kīm) (Gr. *mesos,* middle, + *enchyma,* infusion). Embryonic connective tissue; irregular or amebocytic cells often embedded in gelatinous matrix.

mesentery (mes′ən-ter′ē) (L. *mesenterium,* mesentery). Peritoneal fold serving to hold the viscera in position.

mesocoel (mez′ō-sēl) (Gr. *mesos,* middle, *koilos,* hollow). Middle body coelomic compartment in some deuterostomes, anterior in lophophorates, corresponds to hydrocoel in echinoderms.

mesoderm (me′zə-dərm) (Gr. *mesos,* middle, + *derma,* skin). The third germ layer, formed in the gastrula between the ectoderm and endoderm; gives rise to connective tissues, muscle, urogenital and vascular systems, and the peritoneum.

mesoglea (mez′ō-glē′ə) (Gr. *mesos,* middle, + *glia,* glue). The layer of jellylike or cement material between the epidermis and gastrodermis in cnidarians and ctenophores; also may refer to jellylike matrix between epithelial layers in sponges.

mesohyl (me′sə-hil) (Gr. *mesos,* middle, + *hylē,* a wood). Gelatinous matrix surrounding sponge cells; mesoglea, mesenchyme.

mesonephros (me-zō-nef′rōs) (Gr. *mesos,* middle, + *nephros,* kidney). The middle of three pairs of embryonic renal organs in vertebrates. Functional kidney of fishes and amphibians; its collecting duct is a Wolffian duct. Adj., **mesonephric.**

mesosome (mez′ə-sōm) (Gr. *mesos,* middle, + *sōma,* body). The portion of the body in lophophorates and some deuterostomes that contains the mesocoel.

messenger RNA (mRNA) A form of ribonucleic acid that carries genetic information from the gene to the ribosome, where it determines the order of amino acids as a polypeptide is formed.

metabolism (Gr. *metabolē,* change). A group of processes that includes digestion, production of energy (respiration), and synthesis of molecules and structures by organisms; the sum of the constructive (anabolic) and destructive (catabolic) processes.

metacercaria (me′tə-sər-ka′rē-ə) (Gr. *meta,* between, among, after, + *kerkos,* tail, + L. *aria,* connected with). Fluke juvenile (cercaria) that has lost its tail and has become encysted.

metacoel (met′ə-sēl) (Gr. *meta,* between, among, after, + *koilos,* hollow). Posterior coelomic compartment in some deuterostomes and lophophorates; corresponds to somatocoel in echinoderms.

metamere (met′ə-mēr) (Gr. *meta,* after, + *meros,* part). A repeated body unit along the longitudinal axis of an animal; a somite, or segment.

metamerism (mə-ta′mə-ri′zəm) (Gr. *meta,* between, among, after, + *meros,* part). Condition of being made up of serially repeated parts (metameres); serial segmentation.

metamorphosis (Gr. *meta,* between, among, after, + *morphē,* form, + *osis,* state of). Sharp change in form during postembryonic development, for example, tadpole to frog or larval insect to adult.

metanephridium (me′tə-nə-fri′di-əm) (Gr. *meta,* between, among, after, + *nephros,* kidney). A type of tubular nephridium with the inner open end draining the coelom and the outer open end discharging to the exterior.

metanephros (me′tə-ne′fräs) (Gr. *meta,* between, among, after, + *nephros,* kidney). Embryonic renal organs of vertebrates arising behind the mesonephros; the functional kidney of reptiles, birds, and mammals. It is drained from a ureter.

metasome (met′ə-som) (Gr. *meta,* after, behind, + *sōma,* body). The portion of the body in lophophorates and some deuterostomes that contains the metacoel.

metazoa (met-ə-zō′ə) (Gr. *meta,* after, + *zōon,* animal). Multicellular animals.

MHC See **major histocompatibility complex.**

microevolution (mī-krō-ev-ə-lü′shən). (L. *mikros,* small, + *evolvere,* to unfold). A change in the gene pool of a population across generations.

microfilament (mī′krō-fil′ə-mənt) (Gr. *mikros,* small, + L. *filum,* a thread). A thin, linear structure in cells; of actin in muscle cells and others.

microfilariae (mīk′rə-fil-ar′ē-ē) (Gr. *mikros,* small, + L. *filum,* a thread). Partially developed juveniles borne alive by filarial worms (phylum Nematoda).

microgamete (mīk′rə-gam′et) (Gr. *mikros,* small, + *gamos,* marriage). The smaller of the two gamete types in a heterogametic organism, considered the male gamete.

micron (μ) (mī′krän) (Gr. neuter of *mikros,* small). One one-thousandth of a millimeter; about 1/25,000 of an inch. Now largely replaced by micrometer (μm).

microneme (mī′krə-nēm) (Gr. *mikros,* small, + *nēma,* thread). One of the types of structures composing the apical complex in the phylum Apicomplexa, slender and elongate, leading to the anterior and thought to function in host cell penetration.

micronucleus A small nucleus found in ciliate protozoa; controls the reproductive functions of these organisms.

micropyle (mīk′rə-pīl) (Gr. *mikros,* small, + *pileos,* a cap). The small opening through which the cells emerge from a gemmule (phylum Porifera).

microthrix See **microvillus.**

microtubule (Gr. *mikros,* small, + L. *tubule,* pipe). A long, tubular cytoskeletal element with an outside diameter of 20 to 27 μm. Microtubules influence cell shape and play important roles during cell division.

microvillus (Gr. *mikros,* small, + L. *villus,* shaggy hair). Narrow, cylindrical cytoplasmic projection from epithelial cells; microvilli from the brush border of several types of epithelial cells. Also, microvilli with unusual structure cover the surface of cestode tegument (also called **microthrix** [pl. **microtriches**]).

mictic (mik′tik) (Gr. *miktos,* mixed or blended). Pertaining to haploid egg of rotifers or the females that lay such eggs.

mineralocorticoids (min′ə-rəl-ō-kord′ə-koids) (M.E. *minerale,* ore, + L. *cortex,* bark, + *oid,* suffix denoting likeness of form). Hormones of the adrenal cortex, especially aldosterone, that regulate salt balance.

miracidium (mir′ə-sid′ē-əm) (Gr. *meirakidion,* youthful person). A minute ciliated larval stage in the life of flukes.

mitochondrion (mīd′ə-kän′drē-ən) (Gr. *mitos,* a thread, + *chondrion,* dim. of *chondros,* corn, grain). An organelle in the cell in which aerobic metabolism takes place.

mitosis (mī-tō′səs) (Gr. *mitos,* thread, + *osis,* state of). Nuclear division in which there is an equal qualitative and quantitative division of the chromosomal material between the two resulting nuclei; ordinary cell division.

molecule A configuration of atomic nuclei and electrons bound together by chemical bonds.

monocyte (mon′ə-sīt) (Gr. *monos,* single, + *kytos,* hollow vessel). A type of leukocyte that becomes a phagocytic cell (macrophage) after moving into tissues.

monoecious (mə-nē′shəs) (Gr. *monos,* single, + *oikos,* house). Having both male and female gonads in the same organism; hermaphroditic.

monogamy (mə-näg′ə-mē) adj. **monogamous** (Gr. *monos,* single, + *gamos,* marriage). The condition of having a single mate at any one time.

monohybrid (Gr. *monos,* single, + L. *hybrida,* mongrel). A hybrid offspring of parents different in one specified character.

monomer (mä′nə-mər) (Gr. *monos,* single, + *meros,* part). A molecule of simple structure, but capable of linking with others to form polymers.

monophyly (män′ə-fī-lē) (Gr. *monos,* single, + *phyle,* tribe). The condition that a taxon or other group of organisms contains the most recent common ancestor of the group and all of its descendants; contrasts with **polyphyly** and **paraphyly.**

monosaccharide (män′nə-sa′kə-rīd) (Gr. *monos,* one, + *sakcharon,* sugar from Sanskrit *sarkarā,* gravel, sugar). A simple sugar that cannot be decomposed into smaller sugar molecules; the most common are pentoses (such as ribose) and hexoses (such as glucose).

morphogenesis (mor′fə-je′nə-səs) (Gr. *morphē,* form, + *genesis,* origin). Development of the architectural features of organisms; formation and differentiation of tissues and organs.

morphology (Gr. *morphē,* form, + L. *logia,* study, from Gr. *logos,* work). The science of structure. Includes cytology, the study of cell structure; histology, the study of tissue structure; and anatomy, the study of gross structure.

morula (mär′u-lə) (L. *morum,* mulberry, + *ula,* dim.). Solid ball of cells in early stage of embryonic development.

mosaic cleavage Embryonic development characterized by independent differentiation of each part of the embryo; determinate cleavage.

mucin (mū′sən) (L. *mucus,* nasal mucus). Any of a group of glycoproteins secreted by certain cells, especially those of salivary glands.

mucus (mū′kəs) (L. *mucus,* nasal mucus). Viscid, slippery secretion rich in mucins produced by secretory cells such as those in mucous membranes. Adj., **mucous.**

multiple fission A mode of asexual reproduction in some protistans in which the nuclei divide more than once before cytokinesis occurs.

mutation (mū-tā′shən) (L. *mutare,* to change). A stable and abrupt change of a gene; the heritable modification of a characteristic.

mutualism (mū′chə-wə-li′zəm) (L. *mutuus,* lent, borrowed, reciprocal). A type of interaction in which two different species derive benefit from their association and in which the association is necessary to both; often symbiotic.

myelin (mī′ə-lən) (Gr. *myelos,* marrow). A fatty material forming the medullary sheath of nerve fibers.

myocyte (mī′ə-sīt) (Gr. *mys,* muscle, + *kytos,* hollow vessel). Contractile cell (pinacocyte) in sponges.

myofibril (Gr. *mys,* muscle, + L. dim. of *fibra,* fiber). A contractile filament within muscle or muscle fiber.

myogenic (mī′o-jen′ik) (Gr. *mys,* muscle, + N.L., *genic,* giving rise to). Originating in muscle, such as heartbeat arising in vertebrate cardiac muscle because of inherent rhythmical properties of muscle rather than because of neural stimuli.

myomere (mī′ə-mer) (Gr. *mys,* muscle, + *meros,* part). A muscle segment of successive segmental trunk musculature.

myosin (mī′ə-sin) (Gr. *mys,* muscle, + *in,* suffix, belonging to). A large protein of contractile tissue that forms the thick myofilaments of striated muscle. During contraction it combines with actin to form actomyosin.

myotome (mī′ə-tōm) (Gr. *mys,* muscle, + *tomos,* cutting). That part of a somite destined to form muscles; the muscle group innervated by a single spinal nerve.

N

nacre (nā′kər) (F. mother-of-pearl). Innermost lustrous layer of mollusc shell, secreted by mantle epithelium. Adj., **nacreous.**

NAD Abbreviation of nicotinamide adenine dinucleotide, an electron acceptor or donor in many metabolic reactions.

nares (na′rēz), sing. **naris** (L. nostrils). Openings into the nasal cavity, both internally and externally, in the head of a vertebrate.

natural killer cells Lymphocyte-like cells that can kill virus-infected cells and tumor cells in the absence of antibody.

natural selection A nonrandom reproduction of varying organisms in a population that results in the survival of those best adapted to their environment and elimination of those less well adapted; leads to evolutionary change if the variation is heritable.

nauplius (naw′plē-əs) (L. a kind of shellfish). A free-swimming microscopic larval stage of certain crustaceans, with three pairs of appendages (antennules, antennae, and mandibles) and median eye. Characteristic of ostracods, copepods, barnacles, and some others.

nekton (nek′tən) (Gr. neuter of *nēktos,* swimming). Term for actively swimming organisms, essentially independent of wave and current action. Compare with **plankton.**

nematocyst (ne-mad′ə-sist′) (Gr. *nēma,* thread, + *kystis,* bladder). Stinging organelle of cnidarians.

neo-Darwinism (nē′ō′där′wə-niz′əm). A modified version of Darwin's evolutionary theory that eliminates elements of the Lamarckian inheritance of acquired characteristics and pangenesis that were present in Darwin's formulation; this theory originated with August Weissmann in the late nineteenth century and, after incorporating Mendelian genetic principles, has become the currently favored version of Darwinian evolutionary theory.

neopterygian (nē-äp′tə-rij′ē-ən) (Gr. *neos,* new, + *pteryx,* fin). Any of a large group of bony fishes that includes most modern species.

neoteny (nē′ə-tē′nē, nē-ot′ə-nē) (Gr. *neos,* new, + *teinein,* to extend). An evolutionary process by which organismal development is retarded relative to sexual maturation; produces a descendant that reaches sexual maturity while retaining a morphology characteristic of the preadult or larval stage of an ancestor.

nephridiopore (nə-frid′ē-ə-pōr) (Gr. *nephros,* kidneys, + *porus,* pore). An external excretory opening in invertebrates.

nephridium (nə-frid′ē-əm) (Gr. *nephridios,* of the kidney). One of the segmentally arranged, paired excretory tubules of many invertebrates, notably the annelids. In a broad sense, any tubule specialized for excretion and/or osmoregulation; with an external opening and with or without an internal opening.

nephron (ne′frän) (Gr. *nephros,* kidney). Functional unit of kidney structure of vertebrates, consisting of a Bowman's capsule, an enclosed glomerulus, and the attached uriniferous tubule.

nephrostome (nef′rə-stōm) (Gr. *nephros,* kidney, + *stoma,* mouth). Ciliated, funnel-shaped opening of a nephridium.

neritic (nə-rid′ik) (Gr. *nērites,* a mussel). Portion of the sea overlying the continental shelf, specifically from the subtidal zone to a depth of 200 m.

nested hierarchy A pattern in which species are ordered into a series of increasingly more inclusive clades according to the taxonomic distribution of synapomorphies.

neurogenic (nü-rä-jen′ik) (Gr. *neuron,* nerve, + N.L. *genic,* give rise to). Originating in nervous tissue, as does the rhythmical beat of some arthropod hearts.

neuroglia (nü-räg′le-ə) (Gr. *neuron,* nerve, + *glia,* glue). Tissue supporting and filling the spaces between the nerve cells of the central nervous system.

neurolemma (nü-rə-lem′ə) (Gr. *neuron,* nerve, + *lemma,* skin). Delicate nucleated outer sheath of a nerve cell; sheath of Schwann.

neuromast (Gr. *neuron,* sinew, nerve, + *mastos,* knoll). Cluster of sense cells on or near the surface of a fish or amphibian that is sensitive to vibratory stimuli and water.

neuron (Gr. nerve). A nerve cell.

neuropodium (nü′rə-pō′de-əm) (Gr. *neuron,* nerve, + *pous, podos,* foot). Lobe of parapodium nearer the ventral side in polychaete annelids.

neurosecretory cell (nü′rō-sə-krēd′ə-rē). Any cell (neuron) of the nervous system that produces a hormone.

neutron A subatomic particle lacking an electrical charge and having a mass 1839 times that of an electron and found in the nucleus of atoms.

niche The role of an organism in an ecological community; its unique way of life and its relationship to other biotic and abiotic factors.

nitrogen fixation (Gr. *nitron,* soda, + *gen,* producing). Reduction of molecular nitrogen to ammonia by some bacteria and cyanobacteria, often followed by **nitrification,** the oxidation of ammonia to nitrites and nitrates by other bacteria.

nondisjunction Failure of a pair of homologous chromosomes to separate during meiosis, leading to one gamete with n + 1 chromosomes (see **trisomy**) and another gamete with n−1 chromosomes.

notochord (nōd′ə-kord′) (Gr. *nōtos,* back, + *chorda,* cord). An elongated cellular cord, enclosed in a sheath, which forms the primitive axial skeleton of chordate embryos and adult cephalochordates.

notopodium (nō′tə-pō′de-əm) (Gr. *nōtos,* back, + *pous, podos,* foot). Lobe of parapodium nearer the dorsal side in polychaete annelids.

nucleic acid (nü-klē′ik; nu-klā′ik) (L. *nucleus,* kernel). One of a class of molecules composed of joined nucleotides; chief types are deoxyribonucleic acid (DNA), found in cell nuclei (chromosomes) and mitochondria, and ribonucleic acid (RNA), found both in cell nuclei (chromosomes and nucleoli) and in cytoplasmic ribosomes.

nucleoid (nü'klē-oid) (L. *nucleus,* kernel, + *oid,* like). The region in a prokaryotic cell where the chromosome is found.

nucleolus (nü-klē'ə-ləs) (dim. of L. *nucleus,* kernel). A deeply staining body within the nucleus of a cell and containing RNA; nucleoli are specialized portions of certain chromosomes that carry multiple copies of the information to synthesize ribosomal RNA.

nucleoplasm (nü'klē-ə-plazm') (L. *nucleus,* kernel, + Gr. *plasma,* mold). Protoplasm of nucleus, as distinguished from cytoplasm.

nucleoprotein A molecule composed of nucleic acid and protein; occurs in the nucleus and cytoplasm of all cells.

nucleosome (nü'klē-ə-som) (L. *nucleus,* kernel, + *sōma,* body). A repeating subunit of chromatin in which one and three-quarter turns of the double-helical DNA are wound around eight molecules of histones.

nucleotide (nü'klē-ə-tīd). A molecule consisting of phosphate, 5-carbon sugar (ribose or deoxyribose), and a purine or a pyrimidine; the purines are adenine and guanine, and the pyrimidines are cytosine, thymine, and uracil.

nucleus (nü'klē-əs) (L. *nucleus,* a little nut, the kernel). The organelle in eukaryotes that contains the chromatin and which is bounded by a double membrane (nuclear envelope).

nuptial flight (nup'shəl). The mating flight of insects, especially that of the queen with male or males.

nurse cells Single cells or layers of cells surrounding or adjacent to other cells or structures for which the nurse cells provide nutrient or other molecules (for example, for insect oocytes or *Trichinella* spp. juveniles).

nymph (L. *nympha,* nymph, bride). An immature stage (following hatching) of a hemimetabolous insect that lacks a pupal stage.

O

ocellus (ō-sel'əs) (L. dim. of *oculus,* eye). A simple eye or eyespot in many types of invertebrates.

octomerous (ok-tom'ər-əs) (Gr. *oct,* eight, + *meros,* part). Eight parts, specifically, symmetry based on eight.

odontophore (ō-don'tə-fōr') (Gr. *odous,* tooth, + *pherein,* to carry). Tooth-bearing organ in molluscs, including the radula, radular sac, muscles, and cartilages.

olfactory (äl-fakt'ə-rē) (L. *olor,* smell, + *factus,* to bring about). Pertaining to the sense of smell.

omasum (ō-mā'səm) (L. paunch). The third compartment of the stomach of a ruminant mammal.

ommatidium (ä'mə-tid'ē-əm) (Gr. *omma,* eye, + *idium,* small). One of the optical units of the compound eye of arthropods.

omnivore (äm'nə-vōr) (L. *omnis,* all, + *vorare,* to devour). An animal that uses a variety of animal and plant material in its diet.

oncogene (än'kə-jen) (Gr. *onkos,* protuberance, tumor, + *genos,* descent). Any of a number of genes that are associated with neoplastic growth (cancer). The gene in its benign state, either inactivated or carrying on its normal role, is a **proto-oncogene.**

oncosphere (än'kō-sfir) (Gr. *onkinos,* a hook, + *sphaira,* ball). Rounded larva common to all cestodes, bears hooks.

ontogeny (än-tä'jə-nē) (Gr. *ontos,* being, + *geneia,* act of being born, from *genēs,* born). The course of development of an individual from egg to senescence.

oocyst (ō'ə-sist) (Gr. *ōion,* egg, + *kystis,* bladder). Cyst formed around zygote of malaria and related organisms.

oocyte (o'ə-sīt) (Gr. *ōion,* egg, + *kytos,* hollow). Stage in formation of ovum, just preceding first meiotic division (primary oocyte) or just following first meiotic division (secondary oocyte).

oogenesis (ō-ə-jen'ə-səs) (Gr. *ōion,* egg, + *genesis,* descent). Formation, development, and maturation of a female gamete or ovum.

oogonium (ō'ə-gōn'ē-əm) (Gr. *ōion,* egg, + *gonos,* offspring). A cell that, by continued division, gives rise to oocytes; an ovum in a primary follicle immediately before the beginning of maturation.

ootid (ō-ə-tid') (Gr. *ōion,* egg, + *idion,* dim.). Stage of formation of ovum after second meiotic division following expulsion of second polar body.

operculum (ō-per'kū-ləm) (L. cover). The gill cover in bony fishes; horny plate in some snails.

ophthalmic (äf-thal'mik) (Gr. *ophthalamos,* an eye). Pertaining to the eye.

opisthaptor (ō'pəs-thap'tər) (Gr. *opisthen,* behind, + *haptein,* to fasten). Posterior attachment organ of a monogenetic trematode.

opisthosoma (ō-pis'thə-sō'mə) (Gr. *opisthe,* behind, + *sōma,* body). Posterior body region in arachnids and pogonophorans.

opsonization (op'sən-i-zā'shən) (Gr. *opsonein,* to buy victuals, to cater). The facilitation of phagocytosis of foreign particles by phagocytes in the blood or tissues, mediated by antibody bound to the particles.

organelle (Gr. *organon,* tool, organ, + L. *ella,* dim.) Specialized part of a cell; literally, a small organ that performs functions analogous to organs of multicellular animals.

organizer (or'gan-ī-zer) (Gr. *organos,* fashioning). Area of an embryo that directs subsequent development of other parts.

orthogenesis (ōr'thō-jen'ə-səs). An undirectional trend in the evolutionary history of a lineage as revealed by the fossil record; also, a now discredited, anti-Darwinian evolutionary theory, popular around 1900, postulating that genetic momentum forced lineages to evolve in a predestined linear direction that was independent of external factors and often led to decline and extinction.

osculum (os'kū-ləm) (L. *osculum,* a little mouth). Excurrent opening in a sponge.

osmole Molecular weight of a solute, in grams, divided by the number of ions or particles into which it dissociates in solution. Adj., **osmolar.**

osmoregulation Maintenance of proper internal salt and water concentrations in a cell or in the body of a living organism; active regulation of internal osmotic pressure.

osmosis (oz-mō'sis) (Gr. *ōsmos,* act of pushing, impulse). The flow of solvent (usually water) through a semipermeable membrane.

osmotic potential Osmotic pressure.

osmotroph (oz'mə-trōf) (Gr. *ōsmos,* a thrusting, impulse, + *trophē,* to eat). A heterotrophic organism that absorbs dissolved nutrients.

osphradium (äs-frā'dē-əm) (Gr. *osphradion,* small bouquet, dim. of *osphra,* smell). A sense organ in aquatic snails and bivalves that tests incoming water.

ossicles (L. *ossiculum,* small bone). Small separate pieces of echinoderm endoskeleton. Also, tiny bones of the middle ear of vertebrates.

osteoblast (os′tē-ō-blast) (Gr. *osteon,* bone, + *blastos,* bud). A bone-forming cell.

osteoclast (os′tē-ō-clast) (Gr. *osteon,* bone, + *klan,* to break). A large, multinucleate cell that functions in bone dissolution.

osteocyte (os′tē-ə-sīt) (Gr. *osteon,* bone, + *kytos,* hollow). A bone cell that is characteristic of adult bone, has developed from an osteoblast, and is isolated in a lacuna of the bone substance.

osteon (os′tē-on) (Gr. bone). Unit of bone structure; Haversian system.

ostium (L. door). Opening.

otolith (ōd′əl-ith′) (Gr. *ous, otos,* ear, + *lithos,* stone). Calcareous concretions in the membranous labyrinth of the inner ear of lower vertebrates, or in the auditory organ of certain invertebrates.

outgroup In phylogenetic systematic studies, a species or group of species closely related to but not included within a taxon whose phylogeny is being studied, and used to polarize variation of characters and to root the phylogenetic tree.

oviger (ō′vi-jər) (L. *ovum,* egg, + *gerere,* to bear). Leg that carries eggs in pycnogonids.

oviparity (ō′və-pa′rəd-ē) (L. *ovum,* egg, + *parere,* to bring forth). Reproduction in which eggs are released by the female; development of offspring occurs outside the maternal body. Adj., **oviparous** (ō′vip′ə-rəs).

ovipositor (ō′və-päz′əd-ər) (L. *ovum,* egg, + *positor,* builder, placer, + *or,* suffix denoting agent or doer). In many female insects a structure at the posterior end of the abdomen for laying eggs.

ovoviviparity (ō′vo-vī-və-par′ə-dē) (L. *ovum,* egg, + *vivere,* to live, + *parere,* to bring forth). Reproduction in which eggs develop within the maternal body without additional nourishment from the parent and hatch within the parent, or immediately after laying. Adj., **ovoviviparous** (ō′vo-vī-vip′ə-rəs).

ovum (L. *ovum,* egg). Mature female germ cell (egg).

bat/āpe/ärmadillo/herring/fēmale/finch/līce/crocodile/crōw/duck/ūnicorn/tüna/ə indicates unaccented vowel sound "uh" as in mammal, fishes, cardinal, heron, vulture/stress as in bi-ol′o-gy, bi′o-log′i-cal

oxidation (äk′sə-dā-shən) (Fr. *oxider,* to oxidize, from Gr. *oxys,* sharp, + *ation*). The loss of an electron by an atom or molecule; sometimes addition of oxygen chemically to a substance. Opposite of reduction, in which an electron is accepted by an atom or molecule.

oxidative phosphorylation (äk′sə-dād′iv fäs′fər-i-lā′shən). The conversion of inorganic phosphate to energy-rich phosphate of ATP, involving electron transport through a respiratory chain to molecular oxygen.

P

p53 protein A tumor suppressor protein with critical functions in normal cells. A mutation in the gene that encodes it, *p53,* can result in loss of control over cell division and thus cancer.

paedomorphosis (pē-dō-mor′fə-səs) (Gr. *pais,* child, + *morphē,* form). Displacement of ancestral juvenile features to later stages of the ontogeny of descendants.

pair bond An affiliation between an adult male and an adult female for reproduction. Characteristic of monogamous species.

pallium (pal′e-əm) (L. mantle). Mantle of a mollusc or brachiopod.

papilla (pə-pil′ə) (L. nipple). A small nipplelike projection. A vascular process that nourishes the root of a hair, feather, or developing tooth.

papula (pa′pū-lə) (L. pimple). Respiratory processes on skin of sea stars; also, pustules on skin.

paraphyly (par′ə-fī-lē) (Gr. *para,* beside, + *phyle,* tribe). The condition that a taxon or other group of organisms contains the most recent common ancestor of all members of the group but excludes some descendants of that ancestor; contrasts with **monophyly** and **polyphyly.**

parapodium (pa′rə-pō′dē-əm) (Gr. *para,* beside, + *pous, podos,* foot). One of the paired lateral processes on each side of most segments in polychaete annelids; variously modified for locomotion, respiration, or feeding.

parasitism (par′ə-sīd′iz-əm) (Gr. *parasitos,* from *para,* beside, + *sitos,* food). The condition of an organism living in or on another organism (host) at whose expense the parasite is maintained; destructive symbiosis.

parasympathetic (par′ə-sim-pə-thed′ik) (Gr. *para,* beside, + *sympathes,* sympathetic, from *syn,* with, + *pathos,* feeling). One of the subdivisions of the autonomic nervous system, whose fibers originate in the brain and in anterior and posterior parts of the spinal cord.

parenchyma (pə-ren′kə-mə) (Gr. anything poured in beside). In lower animals, a spongy mass of vacuolated mesenchyme cells filling spaces between viscera, muscles, or epithelia; in some, cell bodies of muscle cells. Also, the specialized tissue of an organ as distinguished from the supporting connective tissue.

parenchymula (pa′rən-kīm′yə-lə) (Gr. *para,* beside, + *enchyma,* infusion). Flagellated, solid-bodied larva of some sponges.

parietal (pä′rī-ə-təl) (L. *paries,* wall). Something next to, or forming part of, a wall of a structure.

parthenogenesis (pär′thə-nō-gen′ə-sis). (Gr. *parthenos,* virgin, + L. from Gr. *genesis,* origin). Unisexual reproduction involving the production of young by females not fertilized by males; common in rotifers, cladocerans, aphids, bees, ants, and wasps. A parthenogenetic embryo may be diploid or haploid.

pathogenic (path′ə-jen′ik) (Gr. *pathos,* disease, + N.L. *genic,* giving rise to). Producing or capable of producing disease.

PCR See **polymerase chain reaction.**

peck order A hierarchy of social privilege in a flock of birds.

pecten (L. comb). Any of several types of comblike structures on various organisms, for example, a pigmented, vascular, and comblike process that projects into the vitreous humor from the retina at a point of entrance of the optic nerve in the eyes of all birds and many reptiles.

pectoral (pek′tə-rəl) (L. *pectoralis,* from *pectus,* the breast). Of or pertaining to the breast or chest; to the pectoral girdle; or to a pair of horny shields of the plastron of certain turtles.

pedalium (pə-dal′ē-əm) (L. *pedalis,* of or belonging to the foot). Flattened blade at the base of the tentacles in cubozoan medusae (Cnidaria).

pedal laceration Asexual reproduction found in sea anemones, a form of fission.

pedicel (ped′ə-kəl) (L. *pediculus,* little foot). A small or short stalk or stem. In insects, the second segment of an antenna or the waist of an ant.

pedicellaria (ped′ə-sə-lar′ē-ə) (L. *pediculus,* little foot, + *aria,* like or connected with). One of many minute pincerlike organs on the surface of certain echinoderms.

pedipalps (ped′ə-palps′) (L. *pes, pedis,* foot, + *palpus,* stroking, caress). Second pair of appendages of arachnids.

pelage (pel′ij) (Fr. fur). Hairy covering of mammals.

pelagic (pə-laj′ik) (Gr. *pelagos,* the open sea). Pertaining to the open ocean.

pellicle (pel′ə-kəl) (L. *pellicula,* dim. of *pellis,* skin). Thin, translucent, secreted envelope covering many protozoa.

pelvic (pel′vik) (L. *pelvis,* a basin). Situated at or near the pelvis, as applied to girdle, cavity, fins, and limbs.

pelycosaur (pel′ə-kō-sor) (Gr. *pelyx,* basin, + *sauros,* lizard). Any of a group of carnivorous Permian synapsids distinguished by powerful jaws, stabbing teeth, and a large skin-covered sail on the back.

pentadactyl (pen-tə-dak′təl) (Gr. *pente,* five, + *daktylos,* finger). With five digits, or five fingerlike parts, to the hand or foot.

pentamerous symmetry (pen-tam′ər-əs) (Gr. *pente,* five, + *meros,* part). A radial symmetry based on five or multiples thereof.

peptidase (pep′tə-dās) (Gr. *peptein,* to digest, + *ase,* enzyme suffix). An enzyme that breaks down simple peptides, releasing amino acids.

peptide bond A bond that binds amino acids together into a polypeptide chain, formed by removing an OH from the carboxyl group of one amino acid and an H from the amino group of another to form an amide group —CO—NH—.

pericardium (per-ə-kär′dē-əm) (Gr. *peri,* around, + *kardia,* heart). Area around heart; membrane around heart.

periostracum (pe-rē-äs′trə-kəm) (Gr. *peri,* around, + *ostrakon,* shell). Outer horny layer of a mollusc shell.

peripheral (pə-ri′fər-əl) (Gr. *peripherein,* to move around). Structure or location distant from center, near outer boundaries.

periproct (per′ə-präkt) (Gr. *peri,* around, + *prōktos,* anus). Region of aboral plates around the anus of echinoids.

perissodactyl (pə-ris′ə-dak′təl) (Gr. *perissos,* odd, + *daktylos,* finger, toe). Pertaining to an order of ungulate mammals with an odd number of digits.

peristalsis (per′ə-stal′səs) (Gr. *peristaltikos,* compressing around). The series of alternate relaxations and contractions that serve to force food through the alimentary canal.

peristomium (per′ə-stō′mē-əm) (Gr. *peri,* around, + *stoma,* mouth). Foremost true segment of an annelid; it bears the mouth.

peritoneum (per′ə-tə-nē′əm) (Gr. *peritonaios,* stretched around). The membrane that lines the coelom and covers the coelomic viscera.

permease A transporter molecule; a molecule in the cell membrane that makes it possible for another molecule (to which the membrane is not otherwise permeable) to be transported across the membrane, that is, mediated transport.

petaloids (pe′tə-loids) (Gr. *petalon,* leaf, + *eidos,* form). Describes flowerlike arrangement of respiratory podia in irregular sea urchins.

pH (*p*otential of *h*ydrogen). A symbol referring to the relative concentration of hydrogen ions in a solution; pH values are from 0 to 14, and the lower the value, the more acid or hydrogen ions in the solution. Equal to the negative logarithm of the hydrogen ion concentration.

phagocyte (fag′ə-sīt) (Gr. *phagein,* to eat, + *kytos,* hollow vessel). Any cell that engulfs and devours microorganisms or other particles.

phagocytosis (fag′ə-sī-tō-səs) (Gr. *phagein,* to eat, + *kytos,* hollow vessel). The engulfment of a particle by a phagocyte or a protozoan.

phagosome (fa′gə-sōm) (Gr. *phagein,* to eat, + *sōma,* body). Membrane-bound vesicle in cytoplasm containing food material engulfed by phagocytosis.

phagotroph (fag′ə-trōf) (Gr. *phagein,* to eat, + *trophē,* food). A heterotrophic organism that ingests solid particles for food.

pharynx (far′inks), pl. **pharynges** (Gr. *pharynx,* gullet). The part of the digestive tract between the mouth cavity and the esophagus that, in vertebrates, is common to both digestive and respiratory tracts. In cephalochordates the gill slits open from it.

phasmid (faz′mid) (Gr. *phasma,* apparition, phantom, + *id*). One of a pair of glands or sensory structures found in the posterior end of certain nematodes.

phenetic (fə-ne′tik) (Gr. *phaneros,* visible, evident). Refers to the use of a criterion of overall similarity to classify organisms into taxa; contrasts with classifications based explicitly on a reconstruction of phylogeny.

phenotype (fē′nə-tīp) (Gr. *phainein,* to show). The visible or expressed characteristics of an organism, controlled by the genotype, but not all genes in the genotype are expressed.

phenotypic gradualism The hypothesis that new traits, even those that are strikingly different from ancestral ones, evolve by a long series of small, incremental steps.

pheromone (fer′ə-mōn) (Gr. *pherein,* to carry, + *hormōn,* exciting, stirring up). Chemical substance released by one organism that influences the behavior or physiological processes of another organism.

phosphagen (fas′fə-jən) (phosphate + gen). A term for creatine phosphate and arginine phosphate, which store and may be sources of high-energy phosphate bonds.

phosphatide (fäs′fə-tīd′) (phosphate + ide). A lipid with phosphorus, such as lecithin. A complex phosphoric ester lipid, such as lecithin, found in all cells. Phospholipid.

phosphorylation (fäs′fə-rə-lā′shən). The addition of a phosphate group, that is, $—PO_3$, to a compound.

photoautotroph (fōt-ō-aw′tō-trōf) (Gr. *phōtos,* light, + *autos,* self, + *trophos,* feeder). An organism requiring light as a source of energy for making organic nutrients from inorganic raw materials.

photosynthesis (fōt-ō-sin′thə-sis) (Gr. *phōs,* light, + *synthesis,* action or putting together). The synthesis of carbohydrates from carbon dioxide and water in chlorophyll-containing cells exposed to light.

phototaxis (fōt′ō-tak′sis) (Gr. *phōs,* light, + *taxis,* arranging, order). A taxis in which light is the orienting stimulus. An involuntary tendency for an organism to turn toward (positive) or away from (negative) light.

phototrophs (fōt′ō-trōfs) (Gr. *phōs, phōtos,* light, + *trophē,* nourishment). Organisms capable of using CO_2 in the presence of light as a source of metabolic energy.

phyletic gradualism A model of evolution in which morphological evolutionary change is continuous and incremental and occurs mainly within unbranched species or lineages over long periods of geological time; contrasts with **punctuated equilibrium.**

phylogenetic species concept An irreducible (basal) cluster of organisms, diagnosably distinct from other such clusters, and within which there is a parental pattern of ancestry and descent.

phylogenetic systematics See **cladistics.**

phylogeny (fī-loj′ə-nē) (Gr. *phylon,* tribe, race, + *geneia,* origin). The origin and diversification of any taxon, or the evolutionary history of its origin and diversification, usually presented in the form of dendrogram.

phylum (fī′ləm), pl. **phyla** (N.L. from Gr. *phylon,* race, tribe). A chief category, between kingdom and class, of taxonomic classifications into which are grouped organisms of common descent that share a fundamental pattern of organization.

physiology (L. *physiologia,* natural science). A branch of biology dealing with the organic processes and phenomena of an organism or any of its parts or of a particular bodily process.

phytoflagellates (fī-tə-flájə-lāts). Members of the class Phytomastigophorea, plantlike flagellates.

phytophagous (fī-täf′ə-gəs) (Gr. *phyton,* plant, + *phagein,* to eat). Organisms that feed on plants.

pinacocyte (pin′ə-kō-sīt′) (Gr. *pinax,* tablet, + *kytos,* hollow vessel). Flattened cells composing dermal epithelium in sponges.

pinna (pin′ə) (L. feather, sharp point). The external ear. Also a feather, wing, or fin or similar part.

pinocytosis (pin′o-sī-tō′sis) (Gr. *pinein,* to drink, + *kytos,* hollow vessel, + *osis,* condition). Taking up of fluid by endocytosis; cell drinking.

placenta (plə-sen′tə) (L. flat cake). The vascular structure, embryonic and maternal, through which the embryo and fetus are nourished while in the uterus.

placode (pla′kōd) (Gr. *plakos,* flat round plate). Localized, platelike thickening of vertebrate head ectoderm from which a specialized structure develops; such structures include eye lens, special sense organs, and certain neurons.

placoderms (plak′ə-dərmz) (Gr. *plax,* plate, + *derma,* skin). A group of heavily armored jawed fishes of the Lower Devonian to Lower Carboniferous.

placoid scale (pla′koid) (Gr. *plax, plakos,* tablet, plate). Type of scale found in cartilaginous fishes, with basal plate of dentin embedded in the skin and a backward-pointing spine tipped with enamel.

plankton (plank′tən) (Gr. neuter of *planktos,* wandering). The passively floating animal and plant life of a body of water; compares with **nekton.**

plantigrade (plan′tə-grād′) (L. *planta,* sole, + *gradus,* step, degree). Pertaining to animals that walk on the whole surface of the foot (for example, humans and bears); compares with **digitigrade.**

planula (plan′yə-lə) (N.L. dim. from L. *planus,* flat). Free-swimming, ciliated larval type of cnidarians; usually flattened and ovoid, with an outer layer of ectodermal cells and an inner mass of endodermal cells.

planuloid ancestor (plan′yə-loid) (L. *planus,* flat, + Gr. *eidos,* form). Hypothetical form representing ancestor of Cnidaria and Platyhelminthes.

plasma cell (plaz′mə) (Gr. *plasma,* a form, mold). A descendant cell of a B cell, functions to secrete antibodies.

plasmalemma (plaz′mə-lem-ə) (Gr. *plasma,* a form, mold, + *lemma,* rind, sheath). The cell membrane.

plasma membrane (plaz′mə) (Gr. *plasma,* a form, mold). A living, external, limiting, protoplasmic structure that functions to regulate exchange of nutrients across the cell surface.

plasmid (plaz′məd) (Gr. *plasma,* a form, mold). A small circle of DNA that may be carried by a bacterium in addition to its genomic DNA.

plasmodium (plaz-mō′dē-əm) (Gr. *plasma,* a form, mold, + *eidos,* form). Multinucleate ameboid mass, syncytial.

plastid (plas′təd) (Gr. *plast,* formed, molded, + L. *id,* feminine stem for particle of specified kind). A membranous organelle in plant cells functioning in photosynthesis and/or nutrient storage, for example, chloroplast.

plastron (plast′trən) (Fr. *plastron,* breast plate). Ventral body shield of turtles; structure in corresponding position in certain arthropods; thin film of gas retained by epicuticle hairs of aquatic insects.

platelet (plāt′lət) (Gr. dim. of *plattus,* flat). A tiny, incomplete cell in the blood that releases substances initiating blood clotting.

pleiotropic (plī-ə-trō′pic) (Gr. *pleiōn,* more, + *tropos,* to turn). Pertaining to a gene producing more than one effect; affecting multiple phenotypic characteristics.

pleopod (plē′ə-päd) (Gr. *plein,* to sail, + *pous, podos,* foot). One of the swimming appendages on the abdomen of a crustacean.

plesiomorphic (plē′sē-ə-mōr′fik). An ancestral condition of a variable character.

pleura (plü′rə) (Gr. side, rib). The membrane that lines each half of the thorax and covers the lungs.

plexus (plek′səs) (L. network, braid). A network, especially of nerves or blood vessels.

pluteus (plü′dē-əs), pl. **plutei** (L. *pluteus,* movable shed, reading desk). Echinoid or ophiuroid larva with elongated processes like the supports of a desk; originally called "painter's easel larva."

pneumostome (nü′mə-stōm) (Gr. *pneuma,* breathing, + *stoma,* mouth). The opening of the mantle cavity (lung) of pulmonate gastropods to the outside.

podium (pō′de-əm) (Gr. *pous, podos,* foot). A footlike structure, for example, the tube foot of echinoderms.

poikilothermic (poi-ki′lə-thər′mik) (Gr. *poikilos,* variable, + thermal). Pertaining to animals whose body temperature is variable and fluctuates with that of the environment; cold blooded; compares with **ectothermic.**

polarity (Gr. *polos,* axis). In systematics, the ordering of alternative states of a taxonomic character from ancestral to successively derived conditions in an evolutionary transformation series. In developmental biology, the tendency for the axis of an ovum to orient corresponding to the axis of the mother. Also, condition of having opposite poles; differential distribution of gradation along an axis.

polarization (L. *polaris,* polar, Gr. *iz,* make). The arrangement of positive electrical charges on one side of a surface membrane and negative electrical charges on the other side (in nerves and muscles).

Polian vesicles (pōl′le-ən) (from G. S. Poli, Italian naturalist). Vesicles opening into ring canal in most asteroids and holothuroids.

polyandry (pol′ē-an′drē) (Gr. *polys,* many, + *anēr,* man). Condition of having more than one male mate at one time.

polygamy (pə-lig′ə-mē) (Gr. *polys,* many, + *gamos,* marriage). Condition of having more than one mate at a time.

polygenic inheritance Inheritance of traits influenced by multiple alleles; traits show continuous variation between extremes; offspring are usually intermediate between the two parents; also known as **blending** and **quantitative inheritance.**

polygyny (pə-lij′ə-nē) (Gr. *polys,* many, + *gynē,* woman). Condition of having more than one female mate at one time.

polymer (pä′lə-mər) (Gr. *polys,* many, + *meros,* part). A chemical compound composed of repeated structural units called monomers.

polymerase chain reaction (PCR) A technique for preparing large quantities of DNA from tiny samples, making it easy to clone a specific gene as long as part of the sequence of the gene is known.

polymerization (pə lim′ər-ə-zā′shən). The process of forming a polymer or polymeric compound.

polymorphism (pä′lē-mor′fī-zəm) (Gr. *polys,* many, + *morphē,* form). The presence in a species of more than one structural type of individual.

polynucleotide (Gr. *polys,* many + nucleotide). A nucleotide of many mononucleotides combined.

polyp (pä′lip) (Gr. *polypous,* many-footed). Individual of the phylum Cnidaria, generally adapted for attachment to the substratum at the aboral end, often form colonies.

polypeptide (pä-lē-pep′tīd) (Gr. *polys,* many, + *peptein,* to digest). A molecule consisting of many joined amino acids, not as complex as a protein.

polyphyly (päl′ē-fī′lē) (Gr. *polys,* many, + *phylon,* tribe). The condition that a taxon or other group of organisms does not contain the most recent common ancestor of all members of the group, implying that it has multiple evolutionary origins; such groups are not valid as formal taxa and are recognized as such only through error. Contrasts with **monophyly** and **paraphyly.**

polyphyodont (pä′lē-fī′ə-dänt) (Gr. *polyphyes,* manifold, + *odous,* tooth). Having several sets of teeth in succession.

polypide (pä′li-pīd) (L. *polypus,* polyp). An individual or zooid in a colony, specifically in ectoprocts, which has a lophohore, digestive tract, muscles, and nerve centers.

polyploid (pä′lə-ploid′) (Gr. *polys,* many, + *ploidy,* number of chromosomes). An organism possessing more than two full homologous sets of chromosomes.

polysaccharide (pä′lē-sak′ə-rid, -rīd). (Gr. *polys,* many, + *sakcharon,* sugar, from Sanskrit *sarkarā,* gravel, sugar). A carbohydrate composed of many monosaccharide units, for example, glycogen, starch, and cellulose.

polysome (polyribosome) (Gr. *polys,* many, + *sōma,* body). Two or more ribosomes connected by a molecule of messenger RNA.

polyzoic (pä′lē-zō′ik) (Gr. *polys,* many, + *zōon,* animal). A tapeworm forming a strobila of several to many proglottids; also, a colony of many zooids.

pongid (pän′jəd) (L. *Pongo,* type genus of orangutan). Of or relating to the primate family Pongidae, comprising the anthropoid apes (gorillas, chimpanzees, gibbons, orangutans).

population (L. *populus,* people). A group of organisms of the same species inhabiting a specific geographical locality.

populational gradualism The observation that new genetic variants become established in a population by increasing their frequencies across generations incrementally, initially from one or a few individuals and eventually characterizing a majority of the population.

porocyte (pō′rə-sīt) (Gr. *porus,* passage, pore, + *kytos,* hollow vessel). Type of cell found in asconoid sponges through which water enters the spongocoel.

portal system (L. *porta,* gate). System of large veins beginning and ending with a bed of capillaries; for example, hepatic portal and renal portal system in vertebrates.

posterior (L. latter). Situated at or toward the rear of the body; situated toward the back; in human anatomy the upright posture makes posterior and dorsal identical.

potocytosis (pä′tə-sī-tō′səs) (Gr. *potos,* a drinking, + *kytos,* hollow vessel). Endocytosis of certain small molecules and ions bound to specific receptors limited to small areas on the cell surface. The areas of the receptors are invaginated and pinch off to form tiny vesicles. See **caveolae.**

preadaptation The possession of a trait that coincidentally predisposes an organism for survival in an environment different from those encountered in its evolutionary history.

prebiotic synthesis The chemical synthesis that occurred before the emergence of life.

precocial (prē-kō′shəl) (L. *praecoquere,* to ripen beforehand). Referring (especially) to birds whose young are covered with down and are able to run about when newly hatched.

predaceous, predacious (prē-dā′shəs) (L. *praedator,* a plunderer, *pradea,* prey). Living by killing and consuming other animals; predatory.

predator (pred′ə-tər) (L. *praedator,* a plunderer, *praeda,* prey). An organism that preys on other organisms for its food.

prehensile (prē-hen′səl) (L. *prehendere,* to seize). Adapted for grasping.

primary bilateral symmetry Usually applied to a radially symmetrical organism descended from a bilateral ancestor and developing from a bilaterally symmetrical larva.

primary radial symmetry Usually applied to a radially symmetrical organism that did not have a bilateral ancestor or larva, in contrast to a secondarily radial organism.

primate (prī-māt) (L. *primus,* first). Any mammal of the order Primates, which includes the tarsiers, lemurs, marmosets, monkeys, apes, and humans.

primitive (L. *primus,* first). Primordial; ancient; little evolved; said of characteristics closely approximating those possessed by early ancestral types.

proboscis (prō-bäs′əs) (Gr. *pro,* before, + *boskein,* feed). A snout or trunk. Also, tubular sucking or feeding organ with the mouth at the end as in planarians, leeches, and insects. Also, the sensory and defensive organ at the anterior end of certain invertebrates.

producers (L. *producere,* to bring forth). Organisms, such as plants, able to produce their own food from inorganic substances.

production In ecology, the energy accumulated by an organism that becomes incorporated into new biomass.

progesterone (prō-jes′tə-rōn′) (L. *pro,* before, + *gestare,* to carry). Hormone secreted by the corpus luteum and the placenta; prepares the uterus for the fertilized egg and maintains the capacity of the uterus to hold the embryo and fetus.

proglottid (prō-gläd′əd) (Gr. *proglōttis,* tongue tip, from *pro,* before, + *glōtta,* tongue, + *id,* suffix). Portion of a tapeworm containing a set of reproductive organs; usually corresponds to a segment.

prohormone (prō'hor-mōn) (Gr. *pro,* before, + *hormaein,* to excite). A precursor of a hormone, especially a peptide hormone.

prokaryotic, procaryotic (pro-kar'ē-ät'ik) (Gr. *pro,* before, + *karyon,* kernel, nut). Not having a membrane-bound nucleus or nuclei. Prokaryotic cells characterize the bacteria and cyanobacteria.

promoter A region of DNA to which the RNA polymerase must have access for transcription of a structural gene to begin.

pronephros (prō-nef'rəs) (Gr. *pro,* before, + *nephros,* kidney). Most anterior of three pairs of embryonic renal organs of vertebrates, functional only in adult hagfishes and larval fishes and amphibians, and vestigial in mammalian embryos. Adj., **pronephric.**

proprioceptor (prō'prē-ə-sep'tər) (L. *proprius,* own, particular, + receptor). Sensory receptor located deep within the tissues, especially muscles, tendons, and joints, that is responsive to changes in muscle stretch, body position, and movement.

prosimian (prō-sim'ē-ən) (Gr. *pro,* before, + L. *simia,* ape). Any member of a group of arboreal primates including lemurs, tarsiers, and lorises, but excluding monkeys, apes, and humans.

prosopyle (präs'ə-pīl) (Gr. *prosō,* forward, + *pyle,* gate). Connections between the incurrent and radial canals in some sponges.

prostaglandins (präs'tə-glan'dəns). A family of fatty-acid hormones, originally discovered in semen, known to have powerful effects on smooth muscle, nerves, circulation, and reproductive organs.

prostomium (prō-stōm'ē-əm) (Gr. *protos,* first, + *stoma,* mouth, + *-idion,* dim. ending). Anterior closure of a metameric animal, anterior to the mouth.

protandrous (prō-tan'drəs) (Gr. *prōtos,* first, + *anēr,* male). Condition of hermaphroditic animals and plants in which male organs and their products appear before the corresponding female organs and products, thus preventing self-fertilization.

bat/āpe/ärmadillo/herring/fēmale/finch/līce/crocodile/crōw/duck/ūnicorn/tüna/ə indicates unaccented vowel sound "uh" as in mammal, fishes, cardinal, heron, vulture/stress as in bi-ol'o-gy, bi'o-log'i-cal

protease (prō'tē-ās) (Gr. *protein,* + *ase,* enzyme). An enzyme that digests proteins; includes proteinases and peptidases.

protein (prō'tēn, prō'tē-ən) (Gr. *protein,* from *proteios,* primary). A macromolecule of carbon, hydrogen, oxygen, and nitrogen and sometimes sulfur and phosphorus; composed of chains of amino acids joined by peptide bonds; present in all cells.

prothoracic glands Glands in the prothorax of insects that secrete the hormone ecdysone.

prothoracicotropic hormone See **ecdysiotropin.**

prothrombin (prō-thräm'bən) (Gr. *pro,* before, + *thrombos,* clot). A constituent of blood plasma that is changed to thrombin by a catalytic sequence that includes thromboplastin, calcium, and plasma globulins; involved in blood clotting.

protist (prō'tist) (Gr. *protos,* first). A member of the kingdom Protista, generally considered to include the protozoa and eukaryotic algae.

protocoel (prō'tə-sēl) (Gr. *protos,* first, + *koilos,* hollow). The anterior coelomic compartment in some deuterostomes, corresponds to the axocoel in echinoderms.

protocooperation A mutually beneficial interaction between organisms in which the interaction is not physiologically necessary to the survival of either.

proton A subatomic particle with a positive electrical charge and having a mass of 1836 times that of an electron; found in the nucleus of atoms.

protonephridium (prō'tə-nə-frid'ē-əm) (Gr. *protos,* first, + *nephros* kidney). Primitive osmoregulatory or excretory organ consisting of a tubule terminating internally with flame bulb or solenocyte; the unit of a flame bulb system.

proto-oncogene See **oncogene.**

protoplasm (prō'tə-plazm) (Gr. *protos,* first, + *plasma,* form). Organized living substance; cytoplasm and nucleoplasm of the cell.

protopod, protopodite (prō'tə-päd, prō-top'ə-dīt) (Gr. *protos,* first, + *pous, podos,* foot). Basal portion of crustacean appendage, containing coxa and basis.

Protostomia (prō'tə-stō'mē-ə) (Gr. *protos,* first, + *stoma,* mouth). A group of phyla in which cleavage is determinate, the coelom (in coelomate forms) is formed by proliferation of mesodermal bands (schizocoelic formation), the mesoderm is formed from a particular blastomere (called 4d), and the mouth is derived from or near the blastopore. Includes the Annelida, Arthropoda, Mollusca, and a number of minor phyla. Compares with **Deuterostomia.**

proventriculus (prō'ven-trik'ū-ləs) (L. *pro,* before, + *ventriculum,* ventricle). In birds the glandular stomach between the crop and gizzard. In insects, a muscular dilation of foregut armed internally with chitinous teeth.

proximal (L. *proximus,* nearest). Situated toward or near the point of attachment; opposite of distal, distant.

proximate cause (L. *proximus,* nearest, + *causa*). The factors that underlie the functioning of a biological system at a particular place and time, including those responsible for metabolic, physiological, and behavioral functions at the molecular, cellular, organismal, and population levels.

pseudocoel (sü'də-sēl) (Gr. *pseudēs,* false, + *koilōma,* cavity). A body cavity not lined with peritoneum and not a part of the blood or digestive systems, embryonically derived from the blastocoel.

pseudopodium (sü'də-pō'dē-əm) (Gr. *pseudēs,* false, + *podion,* small foot, + *eidos,* form). A temporary cytoplasmic protrusion extended out from a protozoan or ameboid cell, and serving for locomotion or for taking up food.

puff Strands of DNA spread apart at certain locations on giant chromosomes of some flies where the DNA is being transcribed.

pulmonary (pul'mən-ner-ē) (L. *pulmo,* lung, + *aria,* suffix denoting connected to). Relating to or associated with lungs.

punctuated equilibrium A model of evolution in which morphological evolutionary change is discontinuous, being associated primarily with discrete, geologically instantaneous events of speciation leading to phylogenetic branching; morphological evolutionary stasis characterizes species between episodes of speciation; contrasts with **phyletic gradualism.**

pupa (pū'pə) (L. girl, doll, puppet). Inactive quiescent stage of the holometabolous insects. It follows the larval stages and precedes the adult stage.

purine (pū'rēn) (L. *purus,* pure, + *urina,* urine). Organic base with carbon and nitrogen atoms in two interlocking rings.

The parent substance of adenine, guanine, and other naturally occurring bases.

pyrimidine (pi-rim′ə-dēn) (alter. of pyridine, from Gr. *pyr,* fire, + *id,* adj. suffix, *ine*). An organic base composed of a single ring of carbon and nitrogen atoms; parent substance of several bases found in nucleic acids.

Q

quantitative inheritance See **polygenic inheritance.**

queen In entomology, the single fully developed female in a colony of social insects such as bees, ants, and termites, distinguished from workers, nonreproductive females, and soldiers.

R

radial canals Canals along the ambulacra radiating from the ring canal of echinoderms; also choanocyte-lined canals in syconoid sponges.

radial cleavage Embryonic development in which early cleavage planes are symmetrical to the polar axis, each blastomere of one tier lying directly above the corresponding blastomere of the next layer; indeterminate cleavage.

radial symmetry A morphological condition in which the parts of an animal are arranged concentrically around an oral-aboral axis, and more than one imaginary plane through this axis yields halves that are mirror images of each other.

radiolarian (rā′dē-ə-la′rē-ən) (L. *radius,* ray, spoke of a wheel, + *Lar,* tutelary god of house and field). Members of the classes Acantharea, Phaeodarea, and Polycystinea (phylum Sarcomastigophora) with actinopodia and beautiful tests.

radioles (rā′dē-ōlz) (L. *radius,* ray, spoke of a wheel). Featherlike processes from the head of many tubicolous polychaete worms (phylum Annelida), used primarily for feeding.

radula (ra′jə-lə) (L. scraper). Rasping tongue found in most molluscs.

Ras protein A protein that initiates a cascade of reactions leading to cell division when a growth factor is bound to the cell surface. The gene encoding Ras becomes an oncogene when a mutation produces a form of Ras protein that initiates the cascade even in the absence of the growth factor.

ratite (ra′tīt) (L. *ratis,* raft). Referring to birds having an unkeeled sternum; compares with **carinate.**

recapitulation Summing up or repeating; hypothesis that an individual repeats its phylogenetic history in its development.

receptor-mediated endocytosis Endocytosis of large molecules, which are bound to surface receptors in clathrin-coated pits.

recessive An allele that must be homozygous for the allele to be expressed.

recombinant DNA DNA from two different species, such as a virus and a mammal, combined into a single molecule.

redia (rē′dē-ə), pl. **rediae** (rē′dē-ē) (from Redi, Italian biologist). A larval stage in the life cycle of flukes; it is produced by a sporocyst larva, and in turn gives rise to many cercariae.

reduction In chemistry, the gain of an electron by an atom or molecule of a substance; also the addition of hydrogen to, or the removal of oxygen from, a substance.

regulative development Progressive determination and restriction of initially totipotent embryonic material.

releaser (L. *relaxare,* to unloose). Simple stimulus that elicits an innate behavior pattern.

renin (rē′nən) (L. *ren,* kidney). An enzyme produced by the kidney juxtaglomerular apparatus that initiates changes leading to increased blood pressure and increased sodium reabsorption.

rennin (re′nən) (M.E. *renne,* to run). A milk-clotting endopeptidase secreted by the stomach of some young mammals, including bovine calves and human infants.

replication (L. *replicatio,* a folding back). In genetics, the duplication of one or more DNA molecules from the preexisting molecule.

reproductive barrier (L. *re* + *producere,* to lead forward; M.F. *barriere,* bar). The factors that prevent one sexually propagating population from interbreeding and exchanging genes with another population.

repugnatorial glands (L. *repugnare,* to resist). Glands secreting a noxious substance for defense or offense, for example, as in the millipedes.

respiration (L. *respiratio,* breathing). Gaseous interchange between an organism and its surrounding medium. In the cell, the release of energy by the oxidation of food molecules.

restriction endonuclease An enzyme that cleaves a DNA molecule at a particular base sequence.

rete mirabile (rē′tē mə-rab′ə-lē) (L. wonderful net). A network of small blood vessels so arranged that the incoming blood runs countercurrent to the outgoing blood and thus makes possible efficient exchange between the two bloodstreams. Such a mechanism serves to maintain the high concentration of gases in the fish swim bladder.

reticular (rə-tik′ū-lər) (L. *reticulum,* small net). Resembling a net in appearance or structure.

reticuloendothelial system (rə-tic′ū-lō-en-dō-thēl′i-əl) (L. *reticulum,* dim. of net, + Gr. *endon,* within, + *thele,* nipple). The fixed phagocytic cells in the tissues, especially the liver, lymph nodes, spleen, and others; also called RE system.

reticulopodia (rə-tik′ū-lə-pō′dē-ə) (L. *reticulum,* dim. of *rete,* net, + *podos, pous,* foot). Pseudopodia that branch and rejoin extensively.

retina (ret′nə, ret′ən-ə) (L. *rete,* net). The posterior sensory membrane of the eye that receives images.

rhabdite (rab′dīt) (Gr. *rhabdos,* rod). Rodlike structures in the cells of the epidermis or underlying parenchyma in certain turbellarians. They are discharged in mucous secretions.

rheoreceptor (rē′ə-rē-cep′tər) (Gr. *rheos,* a flowing, receptor). A sensory organ of aquatic animals that responds to water current.

rhinophore (rī′nə-fōr) (Gr. *rhis,* nose, + *pherein,* to carry). Chemoreceptive tentacles in some molluscs (opisthobranch gastropods).

rhopalium (rō-pā′lē-əm) (N.L. from Gr. *rhopalon,* a club). One of the marginal, club-shaped sense organs of certain jellyfishes; tentaculocyst.

rhoptries (rōp′trēz) (Gr. *rhopalon,* club, + *tryō,* to rub, wear out). Club-shaped bodies in Apicomplexa composing one of the structures of the apical complex; open at anterior and apparently functioning in penetration of host cell.

rhynchocoel (ring′kō-sēl) (Gr. *rhynchos,* snout, + *koilos,* hollow). In nemerteans, the dorsal tubular cavity that contains the inverted proboscis. It has no opening to the outside.

ribosome (rī′bə-sōm). Subcellular structure composed of protein and ribonucleic acid. May be free in the cytoplasm or attached to the membranes of the endoplasmic reticulum; functions in protein synthesis.

ritualization In ethology, the evolutionary modification, usually intensification, of a behavior pattern to serve communication.

RNA Ribonucleic acid, of which there are several different kinds, such as messenger RNA, ribosomal RNA, and transfer RNA (mRNA, rRNA, tRNA).

RNA world Hypothetical stage in the evolution of life on earth in which both catalysis and replication were performed by RNA, not protein enzymes and DNA.

rostellum (räs-tel′ləm) (L. small beak). Projecting structure on scolex of tapeworm, often with hooks.

rostrum (räs′trəm) (L. ship's beak). A snoutlike projection on the head.

rumen (rü′mən) (L. cud). The large first compartment of the stomach of ruminant mammals.

ruminant (rüm′ə-nənt) (L. *ruminare,* to chew the cud). Cud-chewing artiodactyl mammals with a complex four-chambered stomach.

S

saccule (sa′kūl) (L. *sacculus,* small bag). Small chamber of the membranous labyrinth of the inner ear.

sacrum Adj. **sacral** (sā′krəm, sā′krəl) (L. *sacer,* sacred). Bone formed by fused vertebrae to which pelvic girdle is attached; pertaining to the sacrum.

sagittal (saj′ə-dəl) (L. *sagitta,* arrow). Pertaining to the median anteroposterior plane that divides a bilaterally symmetrical organism into right and left halves.

salt (L. *sal,* salt). The reaction product of an acid and a base; dissociates in water solution to negative and positive ions, but not H^+ or OH^-.

saprophagous (sə-präf′ə-gəs) (Gr. *sapros,* rotten, + *phagos,* from *phagein,* to eat). Feeding on decaying matter; saprobic; saprozoic.

bat/āpe/ärmadillo/herring/fēmale/finch/līce/crocodile/crōw/duck/ūnicorn/tüna/ə indicates unaccented vowel sound "uh" as in mammal, fishes, cardinal, heron, vulture/stress as in bi-ol′o-gy, bi′o-log′i-cal

saprophyte (sap′rə-fīt) (Gr. *sapros,* rotten, + *phyton,* plant). A plant living on dead or decaying organic matter.

saprozoic nutrition (sap-rə-zō′ik) (Gr. *sapros,* rotten, + *zōon,* animal). Animal nutrition by absorption of dissolved salts and simple organic nutrients from surrounding medium; also refers to feeding on decaying matter.

sarcolemma (sär′kə-lem′ə) (Gr. *sarx,* flesh, + *lemma,* rind). The thin, noncellular sheath that encloses a striated muscle fiber.

sarcomere (sär′kə-mir) (Gr. *sarx,* flesh, + *meros,* part). Transverse segment of striated muscle believed to be the fundamental contractile unit.

sarcoplasm (sär′kə-plaz′əm) (Gr. *sarx,* flesh, + *plasma,* mold). The clear, semifluid cytoplasm between the fibrils of muscle fibers.

sauropterygians (so-räp′tə-rij′ē-əns) (Gr. *sauros,* lizard, + *pteryginos,* winged). Mesozoic marine reptiles.

schizocoel (skiz′ō-sēl) (Gr. *schizo,* from *schizein,* to split, + *koilōma,* cavity). A coelom formed by the splitting of embryonic mesoderm. Noun, **schizocoelomate,** an animal with a schizocoel, such as an arthropod or mollusc. Adj., **schizocoelous.**

schizocoelous mesoderm formation (skiz′ō-sēl-ləs). Embryonic formation of the mesoderm as cords of cells between ectoderm and endoderm; splitting of these cords results in the coelomic space.

schizogony (skə-zä′gə-nē) (Gr. *schizein,* to split, + *gonos,* seed). Multiple asexual fission.

sclerite (skler′īt) (Gr. *sklēros,* hard). A hard chitinous or calcareous plate or spicule; one of the plates making up the exoskeleton of arthropods, especially insects.

scleroblast (skler′ə-blast) (Gr. *sklēros,* hard, + *blastos,* germ). An amebocyte specialized to secrete a spicule, found in sponges.

sclerocyte (skler′ə-sīt) (Gr. *sklēros,* hard, + *kytos,* hollow vessel). An amebocyte in sponges that secretes spicules.

sclerotic (skler-äd′ik) (Gr. *sklēros,* hard). Pertaining to the tough outer coat of the eyeball.

sclerotization (sklər′ə-tə-zā′shən). Process of hardening of the cuticle of arthropods by the formation of stabilizing cross linkages between peptide chains of adjacent protein molecules.

scolex (skō′leks) (Gr. *skōlex,* worm, grub). The holdfast, or so-called head, of a tapeworm; bears suckers and, in some, hooks, and posterior to it new proglottids are differentiated.

scrotum (skrō′təm) (L. bag). The pouch that contains the testes in most mammals.

scyphistoma (sī-fis′tə-mə) (Gr. *skyphos,* cup, + *stoma,* mouth). A stage in the development of scyphozoan jellyfish just after the larva becomes attached, the polyp form of a scyphozoan.

sebaceous (sə-bāsh′əs) (L. *sebaceus,* made of tallow). A type of mammalian epidermal gland that produces a fatty substance.

sedentary (sed′ən-ter-ē). Stationary, sitting, inactive; staying in one place.

selectively permeable Permeable to small particles, such as water and certain inorganic ions, but not to larger molecules.

seminiferous (sem-ə-nif′rəs) (L. *semen,* semen, + *ferre,* to bear). Pertains to the tubules that produce or carry semen in the testes.

semipermeable (L. *semi,* half, + *permeabilis,* capable of being passed through). Permeable to small particles, such as water and certain inorganic ions, but not to larger molecules.

sensillum, pl. **sensilla** (sin-si′ləm) (L. *sensus,* sense). A small sense organ, especially in the arthropods.

septum, pl. **septa** (L. fence). A wall between two cavities.

serial homology See **homology.**

serosa (sə-rō′sə) (N.L. from L. *serum,* serum). The outer embryonic membrane of birds and reptiles; chorion. Also, the peritoneal lining of the body cavity.

serotonin (sir′ə-tōn′ən) (L. *serum,* serum). A phenolic amine, found in the serum of clotted blood and in many other tissues, that possesses several poorly understood metabolic, vascular, and neural functions; 5-hydroxytryptamine.

serous (sir′əs) (L. *serum,* serum). Watery, resembling serum; applied to glands, tissue, cells, fluid.

serum (sir′əm) (L. whey, serum). The liquid that separates from the blood after coagulation; blood plasma from which fibrinogen has been removed. Also, the clear portion of a biological fluid separated from its particulate elements.

sessile (ses′əl) (L. *sessilis,* low, dwarf). Attached at the base; fixed to one spot, not able to move about.

seta (sēd′ə), pl. **setae** (sē′tē) (L. bristle). A needlelike chitinous structure of the integument of annelids, arthropods, and others.

sex chromosomes Chromosomes that determine gender of an animal. They may bear a few or many other genes.

sibling species Reproductively isolated species that are so similar morphologically that they are difficult or impossible to distinguish using morphological characters.

sickle cell anemia A condition that causes the red blood cells to collapse (sickle) under oxygen stress. The condition becomes manifest when an individual is homozygous for the gene for hemoglobin-S (HbS).

siliceous (sə-li′shəs) (L. *silex,* flint). Containing silica.

simian (sim′ē-ən) (L. *simia,* ape). Pertaining to monkeys or apes.

sinistral (si′nə-strəl, sə-ni′stral) (L. *sinister,* left). Pertaining to the left; in gastropods, shell is sinistral if opening is to left of columella when held with spire up and facing observer.

sinus (sī′nəs) (L. curve). A cavity or space in tissues or in bone.

siphonoglyph (sī′fän′ə-glif′) (Gr. *siphōn,* reed, tube, siphon, + *glyphē,* carving). Ciliated furrow in the gullet of sea anemones.

siphuncle (sī′fun-kəl) (L. *siphunculus,* small tube). Cord of tissue running through the shell of a nautiloid, connecting all chambers with body of animal.

sister group The relationship between a pair of species or higher taxa that are each other's closest phylogenetic relatives.

sociobiology Ethological study of social behavior in humans or other animals.

soma (sō′mə) (Gr. body). The whole of an organism except the germ cells (germ plasm).

somatic (sō-mat′ik) (Gr. *sōma,* body). Refers to the body, for example, somatic cells in contrast to germ cells.

somatocoel (sə-mat′ə-sēl) (Gr. *sōma,* the body, + *koilos,* hollow). Posterior coelomic compartment of echinoderms; left somatocoel gives rise to oral coelom, and right somatocoel becomes aboral coelom.

somatoplasm (sō′mə-də-pla′zəm) (Gr. *sōma,* body, + *plasma,* anything formed). The living matter that makes up the mass of the body as distinguished from germ plasm, which makes up the reproductive cells. The protoplasm of body cells.

somite (sō′mīt) (Gr. *soma,* body). One of the blocklike masses of mesoderm arranged segmentally (metamerically) in a longitudinal series beside the neural tube of the embryo; metamere.

sorting Differential survival and reproduction among varying individuals; often confused with natural selection which is one possible cause of sorting.

speciation (spē′sē-ā′shən) (L. *species,* kind). The evolutionary process or event by which new species arise.

species (spē′shez, spē′sēz) sing. and pl. (L. particular kind). A group of interbreeding individuals of common ancestry that are reproductively isolated from all other such groups; a taxonomic unit ranking below a genus and designated by a binomen consisting of its genus and the species name.

spermatheca (spər′mə-thē′kə) (Gr. *sperma,* seed, + *thēkē,* case). A sac in the female reproductive organs for the reception and storage of sperm.

spermatid (spər′mə-təd) (Gr. *sperma,* seed, + *eidos,* form). A growth stage of a male reproductive cell arising by division of a secondary spermatocyte; gives rise to a spermatozoon.

spermatocyte (spər-mad′ə-sīt) (Gr. *sperma,* seed, + *kytos,* hollow vessel). A growth stage of a male reproductive cell; gives rise to a spermatid.

spermatogenesis (spər-mad′ə-jen′-ə-səs) (Gr. *sperma,* seed, + *genesis,* origin). Formation and maturation of spermatozoa.

spermatogonium (spər′mad-ə-gō′nē-əm) (Gr. *sperma,* seed, + *gonē,* offspring). Precursor of mature male reproductive cell; gives rise directly to a spermatocyte.

spermatophore (spər-mad′ə-fōr′) (Gr. *sperma, spermatos,* seed, + *pherein,* to bear). Capsule or packet enclosing sperm, produced by males of several invertebrate groups and a few vertebrates.

sphincter (sfingk′tər) (Gr. *sphinkter,* band, sphincter, from *sphingein,* to bind tight). A ring-shaped muscle capable of closing a tubular opening by constriction.

spicule (spi′kūl) (L. dim. *spica,* point). One of the minute calcareous or siliceous skeletal bodies found in sponges, radiolarians, soft corals, and sea cucumbers.

spiracle (spi′rə-kəl) (L. *spiraculum,* from *spirare,* to breathe). External opening of a trachea in arthropods. One of a pair of openings on the head of elasmobranchs for passage of water. Exhalent aperture of tadpole gill chamber.

spiral cleavage A type of embryonic cleavage in which cleavage planes are diagonal to the polar axis and unequal cells are produced by the alternate clockwise and counterclockwise cleavage around the axis of polarity; determinate cleavage.

spongin (spun′jin) (L. *spongia,* sponge). Fibrous, collagenous material making up the skeletal network of horny sponges.

spongioblast (spun′jē-o-blast) (Gr. *spongos,* sponge, + *blastos,* bud). Cell in a sponge that secretes spongin, a protein.

spongocoel (spun′jō-sēl) (Gr. *spongos,* sponge, + *koilos,* hollow). Central cavity in sponges.

spongocyte (spun′jō-sīt) (Gr. *spongos,* sponge, + *kytos,* hollow vessel). A cell in sponges that secretes spongin.

sporocyst (spō′rə-sist) (Gr. *sporos,* seed, + *kystis,* pouch). A larval stage in the life cycle of flukes; it originates from a miracidium.

sporogony (spor-äg′ə-nē) (Gr. *sporos,* seed, + *gonos,* birth). Multiple fission to produce sporozoites after zygote formation.

sporozoite (spō′rə-zō′īt) (Gr. *sporos,* seed, + *zōon,* animal, + *ite,* suffix for body part). A stage in the life history of many sporozoan protozoa; released from oocysts.

squalene (skwā′lēn) (L. *squalus,* a kind of fish). A liquid acyclic triterpene hydrocarbon found especially in the liver oil of sharks.

squamous epithelium (skwā′məs) (L. *squama,* scale, + *osus,* full of). Simple epithelium of flat, nucleated cells.

stapes (stā′pēz) (L. stirrup). Stirrup-shaped innermost bone of the middle ear.

statoblast (stad′ə-blast) (Gr. *statos,* standing, fixed, + *blastos,* germ). Biconvex capsule containing germinative cells and produced by most freshwater ectoprocts by asexual budding. Under favorable conditions it germinates to give rise to new zooid.

statocyst (Gr. *statos,* standing, + *kystis,* bladder). Sense organ of equilibrium; a fluid-filled cellular cyst containing one or more granules (statoliths) used to sense direction of gravity.

statolith (Gr. *statos,* standing, + *lithos,* stone). Small calcareous body resting on tufts of cilia in the statocyst.

stenohaline (sten-ə-hā'līn, -lən) (Gr. *stenos,* narrow, + *hals,* salt). Pertaining to aquatic organisms that have restricted tolerance to changes in environmental saltwater concentration.
stereom (ster'ē-ōm) (Gr. *steros,* solid, hard, firm). Meshwork structure of endoskeletal ossicles of echinoderms.
stereotyped behavior A pattern of behavior repeated with little variation in performance.
sternum (ster'nəm) (L. breastbone). Ventral plate of an arthropod body segment; breastbone of vertebrates.
sterol (ste'rōl), **steroid** (ste'royd) (Gr. *stereos,* solid, + L. *ol,* from *oleum,* oil). One of a class of organic compounds containing a molecular skeleton of four fused carbon rings; it includes cholesterol, sex hormones, adrenocortical hormones, and vitamin D.
stigma (Gr. *stigma,* mark, tatoo mark). Eyespot in certain protozoa. Spiracle of certain terrestrial arthropods.
stolon (stō'lən) (L. *stolō, stolonis,* a shoot, or sucker of plant). A rootlike extension of the body wall giving rise to buds that may develop into new zooids, thus forming a compound animal in which the zooids remain united by the stolon. Found in some colonial anthozoans, hydrozoans, ectoprocts, and ascidians.
stoma (stō'mə) (Gr. mouth). A mouthlike opening.
stomochord (stō'mə-kord) (Gr. *stoma,* mouth, + *chordē,* cord). Anterior evagination of the dorsal wall of the buccal cavity into the proboscis of hemichordates; the buccal diverticulum.
strobila (strō'bə-lə (Gr. *strobilē,* lint plug like a pine cone *[strobilos]*). A stage in the development of the scyphozoan jellyfish. Also, the chain of proglottids of a tapeworm.
strobilation (strō'bə-lā'shən) (Gr. *strobilos,* a pine cone). Repeated, linear budding of individuals, as in scyphozoans (phylum Cnidaria), or sets of reproductive organs, as in tapeworms (phylum Platyhelminthes).
stroma (strō'mə) (Gr. *strōma,* bedding). Supporting connective tissue framework of an animal organ; filmy framework of red blood corpuscles and certain cells.

bat/āpe/ärmadillo/herring/fēmale/finch/līce/crocodile/crōw/duck/ūnicorn/tūna/ə indicates unaccented vowel sound "uh" as in mammal, fishes, cardinal, heron, vulture/stress as in bi-ol'o-gy, bi'o-log'i-cal

structural gene A gene carrying the information to construct a protein.
subnivean (səb-ni'vē-ən) (L. *sub,* under, below, + *nivis,* snow). Applied to environments beneath snow, in which snow insulates against a colder atmospheric temperature.
substrate The substance upon which an enzyme acts; also, a base or foundation (substratum); and the substance or base on which an organism grows.
sycon (sī'kon) (Gr. *sykon,* fig). A type of canal system in certain sponges. Sometimes called syconoid.
symbiosis (sim'bī-ōs'əs, sim'bē-ōs'əs) (Gr. *syn,* with, + *bios,* life). The living together of two different species in an intimate relationship. Symbiont always benefits; host may benefit, may be unaffected, or may be harmed (mutualism, commensalism, and parasitism).
sympatric (sim'pa'-trik) (Gr. *syn,* with, + *patra,* native land). Having the same or overlapping regions of geographical distribution. Noun, **sympatry.**
symplesiomorphy (sim-plē'sē-ə-mōr'fē). Sharing among species of ancestral characteristics, not indicative that the species comprise a monophyletic group.
synapomorphy (sin-ap'ə-mor'fē) (Gr. *syn,* together with, + *apo,* of, + *morphē,* form). Shared, evolutionarily derived character states that are used to recover patterns of common descent among two or more species.
synapse (si'naps, si-naps') (Gr. *synapsis,* contact, union). The place at which a nerve impulse passes between neuron processes, typically from an axon of one nerve cell to a dendrite of another nerve cell.
synapsids (si-nap'sədz) (Gr. *synapsis,* contact, union). An amniote lineage comprising the mammals and the ancestral mammal-like reptiles, having a skull with a single pair of temporal openings.
synapsis (si-nap'səs) (Gr. *synapsis,* contact, union). The time when the pairs of homologous chromosomes lie alongside each other in the first meiotic division.
synaptonemal complex (sin-ap'tə-nē'məl) (Gr. *synapsis,* a joining together, + *nēma,* thread). The structure that holds homologous chromosomes together during synapsis in prophase of meiosis I.
syncytium (sən-sish'ē-əm) adj. **syncytial** (Gr. *syn,* with, + *kytos,* hollow). A multinucleated cell.

syndrome sin'drōm) (Gr. *syn,* with, + *dramein,* to run). A group of symptoms characteristic of a particular disease or abnormality.
syngamy (sin'gə-mē) (Gr. *syn,* with, + *gamos,* marriage). Fertilization of one gamete with another individual gamete to form a zygote, found in most animals with sexual reproduction.
syrinx (sir'inks) (Gr. shepherd's pipe). The vocal organ of birds located at the base of the trachea.
systematics (sis-tə-mat'iks). Science of classification and reconstruction of phylogeny.
systole (sis'tə-lē) (Gr. *systolē,* drawing together). Contraction of heart.

T

T cell A type of lymphocyte important in cellular immune response and in regulation of most immune responses.
tactile (tak'til) (L. *tactilis,* able to be touched, from *tangere,* to touch). Pertaining to touch.
tagma, pl. **tagmata** (Gr. *tagma,* arrangement, order, row). A compound body section of an arthropod resulting from embryonic fusion of two or more segments; for example, head, thorax, abdomen.
tagmatization, tagmosis Organization of the arthropod body into tagmata.
taiga (tī'gä) (Russ.). Habitat zone characterized by large tracts of coniferous forests, long, cold winters, and short summers; most typical in Canada and Siberia.
tantulus (tan'tə-ləs) (Gr. *tantulus,* so small). Larva of a tantulocaridan (subphylum Crustacea).
taxis (tak'sis), pl. **taxes** (Gr. *taxis,* arrangement). An orientation movement by a (usually) simple organism in response to an environmental stimulus.
taxon (tak'son), pl. **taxa** (Gr. *taxis,* arrangement). Any taxonomic group or entity.
taxonomy (tak-sän'ə-mi) (Gr. *taxis,* arrangement, + *nomos,* law). Study of the principles of scientific classification; systematic ordering and naming of organisms.
tectum (tek'təm) (L. roof). A rooflike structure, for example, dorsal part of capitulum in ticks and mites.

tegmen (teg′mən) (L. *tegmen,* a cover). External epithelium of crinoids (phylum Echinodermata).

tegument (teg′ū-ment) (L. *tegumentum,* from *tegere,* to cover). An integument: specifically external covering in cestodes and trematodes, formerly believed to be a cuticle.

telencephalon (tel′en-sef′ə-lon) (Gr. *telos,* end, + *encephalon,* brain). The most anterior vesicle of the brain; the anteriormost subdivision of the prosencephalon that becomes the cerebrum and associated structures.

teleology (tel′ē-äl′ə-jē) (Gr. *telos,* end, + L. *logia,* study of, from Gr. *logos,* word). The philosophical view that natural events are goal directed and are preordained, as opposed to the scientific view of mechanical determinism.

telocentric (tē′lō-sen′trək) (Gr. *telos,* end, + *kentron,* center). Chromosome with centromere at the end.

telolecithal (te-lō-les′ə-thəl) (Gr. *telos,* end, + *lekithos,* yolk, + *al*). Having the yolk concentrated at one end of an egg.

telson (tel′sən) (Gr. *telson,* extremity). Posterior projection of the last body segment in many crustaceans.

temnospondyls (tem-nō-spän′dəls) (Gr. *temnō,* to cut, + *spondylos,* vertebra). A large lineage of amphibians that extended from the Carboniferous to the Triassic.

template (temp′plət). A pattern or mold guiding the formation of a duplicate; often used with reference to gene duplication.

tendon (ten′dən) (L. *tendo,* tendon). Fibrous band connecting muscle to bone or other movable structure.

tergum (ter′gəm) (L. back). Dorsal part of an arthropod body segment.

territory (L. *territorium,* from *terra,* earth). A restricted area preempted by an animal or pair of animals, usually for breeding purposes, and guarded from other individuals of the same species.

test (L. *testa,* shell). A shell or hardened outer covering.

tetrad (te′trad) (Gr. *tetras,* four). Group of two pairs of chromatids at synapsis and resulting from the replication of paired homologous chromosomes; the bivalent.

tetrapods (te′trə-päds) (Gr. *tetras,* four, + *pous, podos,* foot). Four-footed vertebrates; the group includes amphibians, reptiles, birds, and mammals.

thecodonts (thēk′ə-dänts) (Gr. *thēkē,* box, + *odontos,* tooth). A large assemblage of Triassic archosaurian diapsids of the order Thecodontia and characterized by having teeth set in sockets.

therapsids (thə-rap′sidz) (Gr. *theraps,* an attendant). Extinct Mesozoic mammal-like reptiles from which true mammals evolved.

thermocline (thər′mō-klīn) (Gr. *thermē,* heat, + *klinein,* to swerve). Layer of water separating upper warmer and lighter water from lower colder and heavier water in a lake or sea; a stratum of abrupt change in water temperature.

thoracic (thō-ra′sək) (L. *thōrax,* chest). Pertaining to the thorax or chest.

thrombin Enzyme catalyzing fibrinogen transformation into fibrin. Percursor is **prothrombin.**

Tiedemann's bodies (tēd′ə-mənz) (from F. Tiedemann, German anatomist). Four or five pairs of pouchlike bodies attached to the ring canal of sea stars, apparently functioning in production of coelomocytes.

tight junction Region of actual fusion of cell membranes between two adjacent cells.

tissue (ti′shü) (M.E. *tissu,* tissue). An aggregation of cells, usually of the same kind, organized to perform a common function.

titer (ī′tər) (Fr. *titrer,* to titrate). Concentration of a substance in a solution as determined by titration.

torsion (L. *torquere,* to twist). A twisting phenomenon in gastropod development that alters the position of the visceral and pallial organs by 180 degrees.

toxicyst (toks′i-sist) (Gr. *toxikon,* poison, + *kystis,* bladder). Structures possessed by predatory ciliate protozoa, which on stimulation expel a poison to subdue the prey.

trabecular net (trə-bek′ū-lər) (L. *trabecula,* a small beam). Network of living tissue formed by pseudopodia of amebocytes in Hexactinellida (phylum Porifera).

trachea (trā′kē-ə) (M.L. windpipe). The windpipe. Also, any of the air tubes of insects.

transcription Formation of messenger RNA from the coded DNA.

transduction Condition in which bacterial DNA (and the genetic characteristics it bears) is transferred from one bacterium to another by the agent of viral infection.

transfer RNA (tRNA) A form of RNA of about 70 or 80 nucleotides, which are adapter molecules in the synthesis of proteins. A specific amino acid molecule is carried by transfer RNA to a ribosome-messenger RNA complex for incorporation into a polypeptide.

transformation Condition in which DNA in the environment of bacteria somehow penetrates them and is incorporated into their genetic complement, so that their progeny inherit the genetic characters so acquired.

translation (L. a transferring). The process in which the genetic information present in messenger RNA is used to direct the order of specific amino acids during protein synthesis.

transporter See **permease.**

transverse plane (L. *transversus,* across). A plane or section that lies or passes across a body or structure.

trichinosis (trik-ən-o′səs). Disease caused by infection with the nematode *Trichinella spiralis.*

trichocyst (trik′ə-sist) (Gr. *thrix,* hair, + *kystis,* bladder). Sac-like protrusible organelle in the ectoplasm of ciliates, which discharges as a threadlike weapon of defense.

triglyceride (trī-glis′ə-rīd) (Gr. *tria,* three, + *glykys,* sweet, + *ide,* suffix denoting compound). A triester of glycerol with one, two, or three acids.

triploblastic (trip′lō-blas′tik) (Gr. *triploos,* triple, + *blastos,* germ). Pertaining to metazoa in which the embryo has three primary germ layers—ectoderm, mesoderm, and endoderm.

trisomy 21 See **Down syndrome.**

trochophore (trōk′ə-fōr) (Gr. *trochos,* wheel, + *pherein,* to bear). A free-swimming ciliated marine larva characteristic of most molluscs and certain ectoprocts, brachiopods, and marine worms; an ovoid or pyriform body with preoral circlet of cilia and sometimes a secondary circlet behind the mouth.

trophallaxis (trōf′ə-lak′səs) (Gr. *trophē,* food, + *allaxis,* barter, exchange). Exchange of food between young and adults, especially certain social insects.

trophic (trō′fək) (Gr. *trophē,* food). Pertaining to feeding and nutrition.

trophoblast (trōf′ə-blast) (Gr. *trephein,* to nourish, + *blastos,* germ). Outer ectodermal nutritive layer of blastodermic vesicle; in mammals it is part of the chorion and attaches to the uterine wall.

trophosome (trof′ə-sōm) (Gr. *trophē,* food, + *sōma,* body). Organ in poganophorans bearing mutualistic bacteria, derived from midgut.

trophozoite (trōf′ə-zō′īt) (Gr. *trophē,* food, + *zōon,* animal). Adult stage in the life cycle of a protozoan in which it is actively absorbing nourishment.

tropic (trä′pik) (Gr. *tropē,* to turn toward). Related to the tropics (tropical); in endocrinology, a hormone that influences the action of another hormone or endocrine gland (usually pronounced trō′pik).

tropomyosin (trōp′ə-mī′ə-sən) (Gr. *tropos,* turn, + *mys,* muscle). Low-molecular weight protein surrounding the actin filaments of striated muscle.

troponin (trə-pōn′in). Complex of globular proteins positioned at intervals along the actin filament of skeletal muscle; thought to serve as a calcium-dependent switch in muscle contraction.

tube feet (podia) Numerous small, muscular, fluid-filled tube projections from body of echinoderms; part of water-vascular system; used in locomotion, clinging, food handling, and respiration.

tubercle (tü′bər-kəl) (L. *tuberculum,* small hump). Small protuberance, knob, or swelling.

tubulin (tü′bū-lən) (L. *tubulus,* small tube, + *in,* belonging to). Globular protein forming the hollow cylinder of microtubules.

tumor necrosis factor A cytokine, the most important source of which is macrophages, that is a major mediator of inflammation.

tundra (tun′drə) (Russ. from Lapp, *tundar,* hill). Terrestrial habitat zone, located between taiga and polar regions; characterized by absence of trees, short growing season, and mostly frozen soil during much of the year.

tunic (L. *tunica,* tunic, coat). In tunicates, a cuticular, cellulose-containing covering of the body secreted by the underlying body wall.

tympanic (tim-pan′ik) (Gr. *tympanon,* drum). Relating to the tympanum that separates the outer and middle ear (eardrum).

type specimen A specimen deposited in a museum that formally defines the name of the species that it represents.

typhlosole (tif′lə-sōl′) (Gr. *typhlos,* blind, + *sōlēn,* channel, pipe). A longitudinal fold projecting into the intestine in certain invertebrates such as the earthworm.

typology (tī-päl′ə-jē) (L. *typus,* image). A classification of organisms in which members of a taxon are perceived to share intrinsic, essential properties, and variation among organisms is regarded as uninteresting and unimportant.

U

ulcer (ul-sər) (L. *ulcus,* ulcer). An abscess that opens through the skin or a mucous surface.

ultimate cause (L. *ultimatus,* last, + *causa*). The evolutionary factors responsible for the origin, state of being, or purpose of a biological system.

umbilical (L. *umbilicus,* navel). Refers to the navel, or umbilical cord.

umbo (um′bō), pl. **umbones** (əm-bō′nēz) (L. boss of a shield). One of the prominences on either side of the hinge region in a bivalve mollusc shell. Also, the "beak" of a brachiopod shell.

ungulate (un′gū-lət) (L. *ungula,* hoof). Hooved. Noun, any hooved mammal.

uniformitarianism (ū′nə-fōr′mə-ter′ē-ə-niz′əm). Methodological assumptions that the laws of chemistry and physics have remained constant throughout the history of the earth, and that past geological events occurred by processes that can be observed today.

ureter (ūr′ə-tər) (Gr. *ourētēr,* ureter). Duct carrying urine from kidney to bladder.

urethra (ū-rē′thrə) (Gr. *ourethra,* urethra). The tube from the urinary bladder to the exterior in both sexes.

uropod (ū′rə-pod) (Gr. *oura,* tail, + *pous, podos,* foot). Posteriormost appendage of many crustaceans.

utricle (ū′trə-kəl) (L. *utriculus,* little bag). That part of the inner ear containing the receptors for dynamic body balance; the semicircular canals lead from and to the utricle.

V

vacuole (vak′yə-wōl) (L. *vacuus,* empty, + Fr. *ole,* dim.). A membrane-bounded, fluid-filled space in a cell.

valence (vā′ləns) (L. *valere,* to have power). Degree of combining power of an element as expressed by the number of atoms of hydrogen (or its equivalent) that the element can hold (if negative) or displace in a reaction (if positive). The oxidation state of an element in a compound. The number of electrons gained, shared, or lost by an atom when forming a bond with one or more other atoms.

valve (L. *valva,* leaf of a double door). One of the two shells of a typical bivalve mollusc or brachiopod.

variation (L. *varius,* various). Differences among individuals of a group or species that cannot be ascribed to age, sex, or position in the life cycle.

vector (L. a bearer, carrier, from *vehere, vectum,* to carry). Any agent that carries and transmits pathogenic microorganisms from one host to another host. Also, in molecular biology, an agent such as bacteriophage or plasmid that carries recombinant DNA.

veins (vānz) (L. *vena,* a vein). Blood vessels that carry blood toward the heart; in insects, fine extensions of the tracheal system that support the wings.

velarium (və-la′rē-əm) (L. *velum,* veil, covering). Shelf-like extension of the subumbrella edge in cubozoans (phylum Cnidaria).

veliger (vēl′ə-jər, vel-) (L. *velum,* veil, covering). Larval form of certain molluscs; develops from the trochophore and has the beginning of a foot, mantle, shell, and so on.

velum (vē′ləm) (L. veil, covering). A membrane on the subumbrella surface of jellyfish of class Hydrozoa. Also, a ciliated swimming organ of the veliger larva.

ventral (ven′trəl) (L. *venter,* belly). Situated on the lower or abdominal surface.

venule (ven′ūl) (L. *venula,* dim. of *vena,* vein). Small vessel conducting blood from capillaries to vein; small vein of insect wing.

vermiform (ver′mə-form) (L. *vermis,* worm, + *forma,* shape). Adjective to describe any wormlike organism; an adult (nematogen) rhombozoan (phylum Mesozoa).

vestige (ves′tij) (L, *vestigium,* footprint). A rudimentary organ that may have been well developed in some ancestor or in the embryo.

vibrissa (vī′bris′ə), pl. **vibrissae** (L. nostril-hair). Stiff hairs that grow from the nostrils or other parts of the face of

bat/āpe/ärmadillo/herring/fēmale/finch/līce/crocodile/crōw/duck/ūnicorn/tüna/ə indicates unaccented vowel sound "uh" as in mammal, fishes, cardinal, heron, vulture/stress as in bi-ol′o-gy, bi′o-log′i-cal

many mammals and that serve as tactile organs; "whiskers."

vicariance (vī-kar′ē-ənts) (L. *vicarius,* a substitute). Geographical separation of populations, especially as imposed by discontinuities in the physical environment that fragmented populations that were formerly geographically continuous.

villus (vil′əs), pl. **villi** (L. tuft of hair). A small fingerlike, vascular process on the wall of the small intestine. Also one of the branching, vascular processes on the embryonic portion of the placenta.

virus (vī′rəs) (L. slimy liquid, poison). A submicroscopic noncellular particle composed of a nucleoprotein core and a protein shell; parasitic; will grow and reproduce in a host cell.

viscera (vis′ər-ə) (L. pl. of *viscus,* internal organ). Internal organs in the body cavity.

visceral (vis′ər-əl). Pertaining to viscera.

vitalism (L. *vita,* life). The view that natural processes are controlled by supernatural forces and cannot be explained through the laws of physics and chemistry alone, as opposed to mechanism.

vitamin (L. *vita,* life, + *amine,* from former supposed chemical origin). An organic substance required in small amounts for normal metabolic function; must be supplied in the diet or by intestinal flora because the organism cannot synthesize it.

vitellaria (vi′təl-lar′ē-ə) (L. *vitellus,* yolk of an egg). Structures in many flatworms that produce vitelline cells, that is, cells that provide eggshell material and nutrient for the embryo.

vitelline gland See **vitellaria.**

vitelline membrane (və-tel′ən, vī′təl-ən) (L. *vitellus,* yolk of an egg). The noncellular membrane that encloses the egg cell.

viviparity (vī′və-par′ə-dē) (L. *vivus,* alive, + *parere,* to bring forth). Reproduction in which eggs develop within the female body, with nutritional aid of maternal parent as in therian mammals, many reptiles, and some fishes; offspring are born as juveniles. Adj., **viviparous** (vī-vip′ə-rəs).

W

water-vascular system System of fluid-filled closed tubes and ducts peculiar to echinoderms; used to move tentacles and tube feet that serve variously for clinging, food handling, locomotion, and respiration.

weir (wēr) (Old English *wer,* a fence placed in a stream to catch fish). Interlocking extensions of a flame cell and a collecting tubule cell in some protonephridia.

X

xanthophore (zan′thə-fōr) (Gr. *xanthos,* yellow, + *pherein,* to bear). A chromatophore containing yellow pigment.

X-organ Neurosecretory organ in eyestalk of crustaceans that secretes molt-inhibiting hormone.

Y

Y-organ Gland in the antennal or maxillary segment of some crustaceans that secretes molting hormone.

Z

zoecium, zooecium (zō-ē′shē-əm) (Gr. *zōon,* animal, + *oikos,* house). Cuticular sheath or shell of Ectoprocta.

zoochlorella (zō′ə-klōr-el′ə) (Gr. *zōon,* life, + *Chlorella*). Any of various minute green algae (usually *Chlorella*) that live symbiotically within the cytoplasm of some protozoa and other invertebrates.

zooflagellates (zō′ə-fla′jə-lāts). Members of the Zoomastigophora, the animal-like flagellates (phylum Sarcomastigophora).

zooid (zō-id) (Gr. *zōon,* life). An individual member of a colony of animals, such as colonial cnidarians and ectoprocts.

zooxanthella (zo′ə-zan-thəl′ə) (Gr. *zōon,* animal, + *xanthos,* yellow). A minute dinoflagellate alga living in the tissues of many types of marine invertebrates.

zygote (Gr. *zygōtos,* yoked). The fertilized egg.

zygotic meiosis Meiosis that takes place within the first few divisions after zygote formation; thus all stages in the life cycle other than the zygote are haploid.

credits

Photos

Chapter 1

Opener: NOAA; 1.1A: © William C. Ober; 1.1B: Cleveland P. Hickman, Jr.; 1.1C: Duke University Marine Laboratory; 1.1D,E: Cleveland P. Hickman, Jr.; 1.2: © John D. Cunningham/Visuals Unlimited; 1.3A: © A.C. Barrington Brown/Photo Researchers, Inc.; 1.4A: © M. Abbey/Visuals Unlimited; 1.4B: © S. Dalton/National Audubon Society Collection/Photo Researchers, Inc.; p. 14: Courtesy Foundations For Biomedical Research; 1.15: The Bettman Archive; 1.17B: Courtesy Kevin Walsh, U.S.C.D.; 1.18: Courtesy R.M. Syren and S.W. Fox, Institute of Molecular Evolution/Univ. of Miami, Coral Gables, Florida; 1.19: Cleveland P. Hickman, Jr.; 1.20: From Schopf, J.W., 1971 J. Paleontology 45:925–60

Chapter 2

Opener: Photo Disc/Vol. 6; 2.1A: © John D. Cunningham/Visuals Unlimited; 2.1B: From Morgan, C.R. and R.A. Jersild, Jr., 1970 Anat. Rec. 166:575–586; 2.6: Courtesy G.E. Palade, University of California School of Medicine; 2.7B: Courtesy Richard Rodewald; 2.8B: Courtesy Charles Flickinger; 2.10B: Courtesy of Charles Flickinger; 2.11: © K.G. Murti/Visuals Unlimited; 2.12B: Courtesy Kent McDonald

Chapter 3

Opener: Larry S. Roberts; 3.1B: Courtesy Mendel Museum, Brno, Czechoslovakia; 3.8A: © Peter J. Bryant/Biological Photo Service

Chapter 4

4.1A: Courtesy American Museum of Natural History, Neg. #326662; 4.1B: Courtesy The Natural History Museum, London; 4.2: Courtesy The Natural History Museum, London; 4.3: Courtesy The Natural History Museum, London; 4.5A: © Bridgeman/Art Resource; 4.5B: © Stock Montage; 4.6: Cleveland P. Hickman, Jr.; 4.7: Cleveland P. Hickman, Jr.; 4.8: Courtesy Harvard University Press, Mike Kelley; p. 89: © The Natural History Museum, London; 4.10A: © Ken Lucas/Biological Photo Service; 4.10B: © A.J. Copley/Visuals Unlimited; 4.10C: © G.O. Poinar, Oregon State University, Corwallis; 4.10D: © Roberta Hess Poinar; 4.11A: © Boehm Photography; 4.12: Cleveland P. Hickman, Jr.; 4.16: Courtesy Library of Congress; 4.21B: Cleveland P. Hickman, Jr.; 4.22: Courtesy of Storrs Agricultural Experiment Station, University of Connecticut at Storrs; 4.25A,B: © Michael Tweedie/Photo Researchers, Inc.; 4.27: © Timothy W. Ransom/ Biological Photo Service; 4.28: © S. Krasemann/ Photo Researchers, Inc.; 4.30: Courtesy Natural Resources Canada

Chapter 5

Opener: Photo Disc/Vol. 6; 5.6A,B: Cleveland P. Hickman, Jr.; 5.10A, B, C: © James L. Castner; 5.19: Cleveland P. Hickman, Jr.; 5.21: © William J. Weber/Visuals Unlimited, p. 125: © D. Foster/ WHOI/Visuals Unlimited

Chapter 6

Opener: © Stephen Dalton/Photo Researchers, Inc.; 6.4A: © Ed Reschke; 6.5A: © Ed Reschke; 6.6A, B: © Ed Reschke; 6.6C: Cleveland P. Hickman, Jr.; 6.6D: © Ed Reschke; 6.7A, B, C: © Ed Reschke

Chapter 7

Opener: Cleveland P. Hickman, Jr.; 7.1: From J.F. Fulton & J.G. Wilson: *Selected Readings in the History of Physiology,* 1966. Courtesy Charles C. Thomas, Publisher, Springfield, IL.; 7.11: © R.G. Kessel and R.H. Kardon, *Tissues and Organs: A Text-Atlas of Scanning Electron Microscopy,* 1979 W.H. Freeman and Co.; 7.21: © Leonard Lee Rue, III

Chapter 8

Opener: Cleveland P. Hickman, Jr.; 8.2: From J.F. Fulton and L.G. Wilson, *Selected Readings in the History of Physiology,* 1966. Courtesy of Charles C. Thomas, Publisher, Springfield, IL.; 8.4: © David M. Phillips/Visuals Unlimited

Chapter 9

Opener: © David M. Phillips/Photo Researchers, Inc.

Chapter 10

Opener: © Beth Davidow/Visuals Unlimited; 10.3A,B: Courtesy Carl Gans; 10.6: Cleveland P. Hickman, Jr.; 10.11: Courtesy Wyeth-Ayerst; 10.12: © R.G. Kessel and R.H. Kardon, *Tissues and Organs: A Text-Atlas of Scanning Electron Microscopy,* 1979 W.H. Freeman and Co.; 10.13D: Courtesy Jerry D. Berlin; 10.16: Courtesy Medical Tribune, from Hosp. Tribune 8:1, Oct. 14, 1974

Chapter 11

Opener: © Virginia P. Weinland/Photo Researchers, Inc.; 11.28A: © James L. Castner

Chapter 12

Opener: © Ed Reschke; 12.1A, B: From J.F. Fulton and L.G. Wilson, *Selected Readings in the History of Physiology,* 1966. Courtesy of Charles C. Thomas, Publisher, Springfield, IL.; 12.10: Courtesy Helen Prior, from J.A. Prior, et al., *Physical Diagnosis,* 1981 Mosby-Year Book, Inc.; 12.16: From J.F. Fulton and L.G. Wilson, *Selected Readings in the History of Physiology,* 1966. Courtesy of Charles C. Thomas, Publisher, Springfield, IL.

Chapter 13

Opener: Cleveland P. Hickman, Jr.; 13.1A: © Thomas McAvoy/Life, Inc. Time 1955; 13.1C: Courtesy Lary Shaffer; 13.7: Cleveland P. Hickman, Jr.; 13.9: © Gary W. Carter/Visuals Unlimited; 13.11: Cleveland P. Hickman, Jr.; 13.12: © Daniel J. Cox/ Tony Stone Images; 13.14: © Renee Lynn/ Tony Stone Images; 13.15: © Tom & Pat Leeson/ Photo Researchers, Inc.; 13.17: Cleveland P. Hickman, Jr.; 13.18: © James Castner

Chapter 14

Opener: Photo Disc/Vol. 6; 14.5: From R.G. Kessel and R.H. Kardon, *Tissues and Organs: A Text-Atlas of Scanning Electron Microscopy,* 1979, W.H. Freeman and Co.; 14.14: © Dr. Gerald Schatten/SPL/Photo Researchers, Inc.; 14.26: © F.R. Turner/Biological Photo Service

Chapter 15

Opener: Cleveland P. Hickman, Jr.; 15.1: Courtesy Library of Congress; 15.5: Courtesy American Museum of Natural History, Neg. #334101; 15.6A: © M. Cole/Animals, Animals/Earth Scenes; 15.6B: © D. Allen/Animals Animals/Earth Scenes; 15.8: Courtesy of Dr. George W. Byers, University of Kansas

Chapter 16

Opener: © M. Abbey/Visuals Unlimited; 16.2B: Courtesy Dr. Ian R. Gibbons; 16.4A,B,C: © M. Abbey/Visuals Unlimited; 16.5B: Courtesy L. Evans Roth; 16.15A: Courtesy Gustaaf M. Hallegraeff; 16.15B: © A.M. Siegelman/Visuals Unlimited; 16.16: Courtesy J. and M. Cachon. From Lee, J.J., S.H. Hutner, and E.C. Bovee (editors). 1985. *An Illustrated Guide to the Protozoa.* Society of Protozoologists, Allen Press, Lawrence, KS.

Chapter 17

Opener: Photogear; 17.2: Larry S. Roberts; 17.4: Larry S. Roberts; 17.11A,B,C: Larry S. Roberts

Chapter 18

Opener: Larry S. Roberts; 18.1 A: © William Ober; 18.6: © Rick Harbo; 18.7: © Carolina Biological Supply/Phototake; 18.8: © Cabisco/Visuals Unlimited; 18.10: © Daniel W. Gotshall; 18.12: © Peter Parks/OSF/Animals, Animals/Earth Scenes; 18.13A,B: Larry S. Roberts; 18.14: © Rick Harbo; 18.15: © Rick Harbo; 18.18: © Daniel W. Gotshall; 18.19A: © Jeff L. Rotman; 18.19B: Larry S. Roberts; 18.21: Larry S. Roberts; 18.22A,B: © Rick Harbo; 18.23: Cleveland P. Hickman, Jr.; 18.23 B,C: Larry S. Roberts; 18.28A: © Jeff Rotman; 18.28B: © Kjell Sandved/Butterfly Alphabet

Chapter 19

Opener: © Alex Kerstitch/Visuals Unlimited; 19.2: © Cabisco/Visuals Unlimited; 19.3: © James Castner; 19.9A: Larry S. Roberts; 19.10: R.E. Kuntz, From H. Zaiman *A Pictorial Presentation of Parasites;* 19.12: © Cabisco/Visuals Unlimited; 19.16: Larry S. Roberts; 19.17: © Stan Elems/Visuals Unlimited; 19.19: Cleveland P. Hickman, Jr.

Chapter 20

Opener: Photo #1206 by D. Despommier, from *A Pictorial Presentation of Parasites,* edited by H. Zaiman, M.D.; 20.9A: Frances M. Hickman; 20.9B: G.W. Kelley, Jr./From H. Zaiman *A Pictorial Presentation of Parasites;* 20.10: E. Pike/From H. Zaiman *A Pictorial Presentation of Parasites;* 20.11: H. Zaiman/From *A Pictorial Presentation of Parasites;* 20.12A: © R. Calentine/Visuals Unlimited; 20.12B: Courtesy H. Zaiman/From *A Pictorial Presentation of Parasites;* 20.13: Contributed by E.L. Schiller, AFIP; 20.14: Larry S. Roberts

Chapter 21

Opener: Photogear; 21.1A,B,C: © Rick Harbo; 21.1D: © Daniel W. Gothshall; 21.1E: Larry S. Roberts; 21.3B: Larry S. Roberts; 21.6: © Kjell Sandved/Butterfly Alphabet; 21.8: © Rick Harbo; 21.14A,B: © Daniel W. Gotshall; 21.15A,B: © Alex Kerstitch/Sea of Cortez Enterprises; 21.16A,B: Cleveland P. Hickman, Jr.; 21.18A: © Rick Harbo; 21.18 B: Larry S. Roberts; 21.19: © Tom Phillipp Underwater Photography; 21.20A: Larry S. Roberts; 21.20B: Cleveland P. Hickman, Jr.; 21.21: Larry S. Roberts; 21.24A,B: Larry S. Roberts; 21.25: Larry S. Roberts; 21.26: © Rick Harbo; 21.27: © Rick Harbo; 21.29B: Richard J. Neves; 21.30A: © Dave Fleetham/ Tom Stack & Associates; 21.31A: Courtesy M. Butschler, Vancouver Public Aquarium; 21.32: Courtesy Captain Louis Usie

Chapter 22

Opener: Photogear; 22.4A,B: Larry S. Roberts; 22.5: © S. Elems/Visuals Unlimited; 22.6: Larry S. Roberts; 22.9: © W.C. Jorgensen/Visuals Unlimited; 22.14: © G.L. Twiest/Visuals Unlimited; 22.17: © Timothy Branning; 22.19: Cleveland P. Hickman, Jr.

Chapter 23

Opener: Photogear; 23.1A,B: © A.J. Copely/Visuals Unlimited; 23.5 A: © James Castner; 23.5 B: © John H. Gerard/Nature Press; 23.6: © John H. Gerard/Nature Press; 23.7: © John H. Gerard/Nature Press; 23.8: © J.H. Gerard/Nature Press; 23.9: © James Castner; 23.10A: © J.A. Alcock/Visuals Unlimited; 23.11A: © John H. Gerard/Nature Press; 23.12: Larry S. Roberts; 23.13: © D.S. Snyder/Visuals Unlimited; 23.22: Carolina Biological Supply/Phototake; 23.26A,B: © Rick Harbo; 23.27A: Cleveland P. Hickman, Jr.; 23.28: Larry S. Roberts; 23.29A: © Rick Harbo; 23.29B,C: © Kjell Sandved/Butterfly Alphabet; 23.31A: Cleveland P. Hickman, Jr.; 23.31B: © Rick Harbo; 23.31C: Cleveland P. Hickman, Jr.; 23.31D,E: Larry S. Roberts; 23.32A: © James Castner; 23.33A: © James Castner; 23.35B: © James Castner; 23.36A,B: © Ron West/Nature Photography; 23.37: © James Castner; 23.41C: © James Castner 23.43A: Cleveland P. Hickman, Jr.; 23.43B: © John H. Gerard/Nature Press; 23.44: Cleveland P. Hickman, Jr.; 23.46A: Cleveland P. Hickman, Jr.; 23.46B: © John H. Gerard/Nature Press; 23.46C: © Carolina Biological Supply/Phototake; 23.47A: © James Castner; 23.48: © John H. Gerard/Nature Press; 23.49: Courtesy James E. Lloyd; 23.50: © James L. Castner; 23.51: K. Lorenzen © 1979, Educational Images, Andromeda Productions; 23.52A: © John H. Gerard/Nature Press; 23.52B: © James Castner; 23.53: Larry S. Roberts; 23.54A,B,C: © James Castner; 23.55A: © Leonard Lee Rue, III; 23.55B: © James Castner; 23.55C: © John H. Gerard/Nature Press; 23.56: © James Castner; 23.57: © James Castner; 23.58B: © James Castner; 23.59: Courtesy Jay Georgi; 23.60: © James Castner; 23.61: © James Castner; 23.62: © James Castner; 23.63: © John D. Cunningham/ Visuals Unlimited

Chapter 24

Opener: © William Ober; 24.5B: © James Castner; 24.6: Courtesy Diane R. Nelson; 24.8B: Larry S. Roberts; 24.9: © Ken Lucas/Biological Photo Service; 24.10A,B: © Robert Brons/Biological Photo Service

Chapter 25

Opener: © Paul Gier/Visuals Unlimited; 25.1A: © Rick Harbo; 25.1B,C: Larry S. Roberts; 25.1D: © Rick Harbo; 25.4F: © Rick Harbo; 25.5: Larry S. Roberts; 25.7A: © Rick Harbo; 25.7B: © Daniel W. Gotshall; 25.10: © Rick Harbo; 25.11A,B: Larry S. Roberts; 25.14: © Rick Harbo; 25.17: Larry S. Roberts; 25.26B: Thuesen, E.V., and R. Bieri, 1987. Canad. J. Zool. 65:181–187

Chapter 26

Opener: © Heather Angel/Biofoto; 26.4: Courtesy of R.P.S. Jeffries, The Natural History Museum, London; 26.7: © Dave B. Fleetham/Visuals Unlimited

Chapter 27

Opener: Larry S. Roberts; 27.9: © Jeff L. Rotman; 27.18A,B: © John G. Shedd Aquarium/Patrice Ceisel; 27.27: © Will Troyer/Visuals Unlimited; 27.29: © Daniel W. Gotshall; 27.30: © Frederick McConnaughey

Chapter 28

Opener: © Gary Meszaros/Visuals Unlimited; 28.6A,B: Courtesy L. Houck; 28.9: Cleveland P. Hickman, Jr.; 28.11A: © Ken Lucas/Biological Photo Service; 28.11B: Cleveland P. Hickman, Jr.; 28.12: Cleveland P. Hickman, Jr.; 28.13: Courtesy American Museum of Natural History, Neg. #125617; 28.14: Cleveland P. Hickman, Jr.; 28.15: Cleveland P. Hickman, Jr.

Chapter 29

Opener: Courtesy National Zoological Park, Smithsonian Institution, Jessie Cohen, photographer; 29.7: Cleveland P. Hickman, Jr.; 29.8: Cleveland P. Hickman, Jr.; 29.10: © John Mitchell/Photo Researchers, Inc.; 29.11: Cleveland P. Hickman, Jr.; 29.12: © Javier Andrada; 29.13: © Leonard Lee Rue, III; 29.14: © Leonard Lee Rue, III; 29.16: © Leonard Lee Rue, III; 29.17: Cleveland P. Hickman, Jr.; 29.18: © Leonard Lee Rue, III; 29.21: Cleveland P. Hickman, Jr.; 29.24A,B: Cleveland P. Hickman, Jr.

Chapter 30

Opener: Photo Disc/Vol. 6; 30.1A: Courtesy American Museum of Natural History, Neg. #125065; 30.15: © J.L. McAlonay/Visuals Unlimited; 30.19A: © Leonard Lee Rue, III; 30.22: © John Gerland/Visuals Unlimited; 30.23: © Richard R Hansen/Photo Researchers, Inc.; 30.25A: © Leonard Lee Rue, III; 30.26: Courtesy Culver Pictures; 30.27: Cleveland P. Hickman, Jr.; 30.28: Cleveland P. Hickman, Jr.; 30.29: Cleveland P. Hickman, Jr.; 30.30: Cleveland P. Hickman, Jr.

Chapter 31

Opener: © Jeff Lepore/Photo Researchers, Inc.; 31.5: © Leonard Lee Rue, III; 31.6A,B: © Leonard Lee Rue, III; 31.7: Robert E. Treat; 31.11: © Leonard Lee Rue, III; 31.12: © J. Gerlach/Visuals Unlimited; 31.13A: Cleveland P. Hickman, Jr.; 31.15: © S. Maslowski/Visuals Unlimited; 31.17: © Kjell Sandved/ Visuals Unlimited; 31.18: © Leonard Lee Rue, III; 30.19 B: © Leonard Lee Rue, III; 31.20: © G. Herben/Visuals Unlimited; 31.22: Courtesy San Diego Zoo; 31.23A: Courtesy San Diego Zoo; 31.23B: Cleveland P. Hickman, Jr.; 31.24: © J. McDonald/Visuals Unlimited; 31.25: © John Reader; 31.28: © J. Gerlach/Visuals Unlimited; 31.29: Cleveland P. Hickman, Jr.; 31.30: Cleveland P. Hickman, Jr.; 31.31: © William Ober; 31.32: Cleveland P. Hickman, Jr.

Line Art

Chapter 1

Figure 1.14: From Raven, P.H., and G.B. Johnson. 1995. *Understanding biology,* ed. 3. New York, McGraw-Hill. Reprinted by permission of The McGraw-Hill Companies.
Figure 1.21: From Raven, P.H., and G.B. Johnson. 1996. *Biology,* ed. 4. New York, McGraw-Hill. Reprinted by permission of The McGraw-Hill Companies.

Chapter 2

Figures 2.4, 2.7a, 2.8a, 2.10a, 2.12a, 2.14, 2.15a: From Raven, P.H., and G.B. Johnson. 1995. *Understanding biology,* ed. 3. New York, McGraw-Hill. Reprinted by permission of The McGraw-Hill Companies.
Figure 2.16: From Raven, P.H., and G.B. Johnson. 1992. *Biology,* ed. 3, New York, McGraw-Hill. Reprinted by permission of The McGraw-Hill Companies.
Figure 2.18: Source: After Darnell, J., H. Lodish, and D. Baltimore. 1986. *Molecular cell biology.* New York, Scientific American Books.
Figure 2.21: Source: After Murray, A.W., and M.W. Kirschner. 1991. *Sci. Am.* 264:56–63.
Figure, 2.24, 2.28: From Raven, P.H., and G.B. Johnson. 1996. *Biology,* ed. 4. New York, McGraw-Hill. Reprinted by permission of The McGraw-Hill Companies.

Chapter 3

Figure 3.16: Etkin, W. 1973. A representation of the structure of DNA. *Figure. BioScience* 23:653. © 1973 American Institute of Biological Sciences.
Figure 3.18: Source: After Chambon, P. 1981. *Sci. Am.* 244:60–71.
Figure 3.20: From Raven, P.H., and G.B. Johnson. 1995. *Understanding biology,* ed. 3. New York, McGraw-Hill. Reprinted by permission of The McGraw-Hill Companies.

Chapter 4

Figure 4.4: Source: After Moorehead, A. 1969. *Darwin and the Beagle.* New York, Harper & Row.
Page 90: Source: Mayr, E. 1991. *One long argument.* Cambridge, Harvard University Press.
Figure 4.14: Source: Sepkoski, J.J., Jr. 1981. *Paleobiology* 7:36–53.
Figure 4.15: From Raven, P.H., and G.B. Johnson. 1996. *Biology,* ed. 4. New York, McGraw-Hill. Reprinted by permission of The McGraw-Hill Companies.
Figure 4.17: Source: After Cracraft, J. 1974. *Ibis* 116:294–521.
Figure 4.20: Source: After Grant, P.R. 1981. Speciation and adaptive radiation of Darwin's finches. *Amer. Sci.* 69:653–663.
Figure 4.21a: From Raven, P.H., and G.B. Johnson. 1996. *Biology,* ed. 4. New York, McGraw-Hill. Reprinted by permission of The McGraw-Hill Companies.
Figure 4.25c: Source: After Brakefield, P.M. 1987. Industrial melanism: Do we have the answers? *Tr. Ecol. Evo.* 2:117–122.
Figure 4.26: Source: After Mourant, A.E. 1954. *The distribution of human blood.* Toronto, Ryerson Press.
Figure 4.29: Source: After Raup, D.M., and J.J. Sepkoski Jr. 1982. Mass extinctions in the marine fossil record. *Science* 215:1502–1504.

Chapter 5

Figure 5.3: Source: Data from Bos, E., et al. 1994. "World population projections 1994–95." Baltimore, Johns Hopkins University Press for the World Bank.
Figure 5.7: Source: Data from Lack, D. 1947. *Darwin's finches.* Cambridge University Press.
Figure 5.9: Source: Data from Gause, G.F. 1934. *The struggle for existence.* New York, Williams and Wilkins.
Figures 5.11, 5.17: From Castro, P., and M.E. Huber. 1997. *Marine biology,* ed. 2. New York, McGraw-Hill. Reprinted by permission of The McGraw-Hill Companies.
Figure 5.18: From *Natural History,* March 1990, Copyright the American Museum of Natural History.
Figure 5.22: From Castro, P., and M.E. Huber. 1997. *Marine biology,* ed. 2. New York, McGraw-Hill. Reprinted by permission of The McGraw-Hill Companies.

Chapter 6

Figure 6.1: Source: After Bonner, J.T. 1988. *The evolution of complexity.* Princeton University Press.
Figure 6.2: Source: After Taylor, C.R., K. Schmidt-Nielsen, and J.L. Raab. 1970. Scaling of energetic cost of running to body size in animals, *Amer. J. Physiol.* 219(4):1106.
Figures 6.8, 6.13, 6.15: From Van De Graaff, K.M., and S.I. Fox. 1995. *Concepts of human anatomy & physiology,* ed. 4. New York, McGraw-Hill. Reprinted by permission of The McGraw-Hill Companies.
Figure 6.16: Source: After Biewener, A.A. 1989. Mammalian terrestrial locomotion and size. *BioScience* 39(11):776–783; and Alexander, R.M. 1991. How dinosaurs ran. *Sci. Am.* 264:130–136.
Figure 6.18: Source: After Sleigh, M.A. 1962. *The biology of cilia and flagella.* Oxford, Pergamon Press.
Figure 6.21: From Raven, P.H., and G.B. Johnson. 1995. *Understanding biology,* ed. 3. New York, McGraw-Hill. Reprinted by permission of The McGraw-Hill Companies.
Figure 6.24: Source: After Eckert, R., D. Randall, and G. Augustine. 1988. *Animal physiology,* ed 3. New York, W. H. Freeman.

Chapter 7

Figures 7.3, 7.4: Source: After Webster, D., and M. Webster. 1974. *Comparative vertebrate morphology.* New York, Academic Press.
Figure 7.13: Source: After Pitts, R.F. 1974. *Physiology of the kidney and body fluids,* ed 3. St. Louis, Mosby-Year Book.
Figure 7.15a: Source: After Wirz, H., B. Hargitay, and W. Kuhn. 1951. *Helv. Physiol. Acta* 9:196–207.
Figure 7.15b: Source: After Ullrich, K.J., and K.H. Jarausch. 1956. *Pflugers Archiv.* 262:537–550.
Figure 7.20: Source: After Lasiewski, R.C. 1963. *Physiol. Zool.* 36:122–140.

Chapter 8

Figure 8.22: Source: After Gordon, M.S., et al. 1968. *Animal function: Principles and adaptations.* New York, Macmillan.

Chapter 9

Figure 9.3: From Raven, P.H., and G.B. Johnson. 1995. *Understanding biology,* ed. 3. New York, McGraw-Hill. Reprinted by permission of The McGraw-Hill Companies.

Chapter 10

Opener text: Source: Berrill, N.J. 1958. *You and the universe.* New York, Dodd, Mead & Co.
Figure 10.10: From Raven, P.H., and G.B. Johnson. 1996. *Biology,* ed. 4. New York, McGraw-Hill. Reprinted by permission of The McGraw-Hill Companies.
Figure 10.15: Source: After Winick, M. 1976. *Malnutrition and brain development.* New York, Oxford University Press.

Chapter 11

Figure 11.9: From Raven, P.H., and G.B. Johnson. 1992. *Biology,* ed. 3. New York, McGraw-Hill. Reprinted by permission of The McGraw-Hill Companies.
Figure 11.12: From Van De Graaff, K.M., and S.I. Fox. 1995. *Concepts of human anatomy & physiology,* ed. 4. New York, McGraw-Hill. Reprinted by permission of The McGraw-Hill Companies.
Figure 11.13: From Raven, P.H., and G.B. Johnson. 1995. *Understanding biology,* ed. 3. New York, McGraw-Hill. Reprinted by permission of The McGraw-Hill Companies.
Figure 11.15: From Raven, P.H., and G.B. Johnson. 1996. *Biology,* ed. 4. New York, McGraw-Hill. Reprinted by permission of The McGraw-Hill Companies.
Figure 11.16: Source: After Wilson, E.O., and W.H. Bossert. 1963. *Res. Prog. Horm. Res.* 19:673–716.
Figure 11.27: Source: After Lenci, F., and G. Colombetti. 1980. *Photoreception and sensory transduction in aneural organisms.* New York, Plenum Press.
Figure 11.29: From Raven, P.H., and G.B. Johnson. 1996. *Biology,* ed. 4. New York, McGraw-Hill. Reprinted by permission of The McGraw-Hill Companies.
Figure 11.30: From Raven, P.H., and G.B. Johnson. 1995. *Understanding biology,* ed. 3. New York, McGraw-Hill. Reprinted by permission of The McGraw-Hill Companies.

Chapter 12

Figure 12.9: Source: After Bentley, P.J. 1982. *Comparative vertebrate endocrinology,* ed 2. Cambridge University Press.
Figure 12.11: Source: After Copp, D.H. 1969. *J. Endocrinol.* 43:137–161.
Figure 12.20: Source: After Ulmann, A., G. Teutsch, and D. Philbert. 1990. RU 486. *Sci. Am.* 262:42–48.

Chapter 13

Figure 13.2: Source: After Lorenz, K., and N. Tinbergen. 1938. *Zeit. Tierpsychol.* 2:1–29.
Figure 13.4: From N. Tinbergen, *The study of instinct,* Oxford University Press, Oxford, England, 1951; modified from D. Lack, *The life of the robin,* H. F. & G. Witherby Ltd., London, England, 1943. Reprinted by permission.
Figure 13.5: Source: After Rothenbuhler, N. 1964. Behavior genetics of nest cleaning honey bees. IV. Responses of F1 and backcross generations to disease-killed brood. *Am. Zool.* 4:111–123.
Figure 13.6: Source: After Dilger, W.C. 1962. The behavior of lovebirds. *Sci. Am.* 206:89–98.
Figure 13.8: Source: After Kandel, E.R. 1979. Small systems of neurons. *Sci. Am.* 241:66–76.
Figure 13.10: Source: After J. Alcock, *Animal behavior: An evolutionary approach,* 3d ed., Sinauer Associates, Sunderland, MA, 1984, from a photograph by Masakasu Konishi.
Figure 13.16: Source: From Darwin, C. 1872. *Expression of the emotions in man and animals.* New York, Appleton and Co.
Text ch 13: Source: Nelson, B. 1968. *Galapagos: Islands of birds.* London, Longmans, Green & Company.
Text ch 13: Source: DeVore, I. 1972. *The marvels of animal behavior.* Washington, DC, National Geographic Society.

Chapter 14

Figure 14.4: Source: After Cole, C.J. 1984. Unisexual lizards. *Sci. Am.* 250:94–100.
Figure 14.12: Source: After Epel, D. 1977. The program of fertilization. *Sci. Am.* 237:128–138.
Figure 14.17: Source: After Gilbert, S.F. 1994. *Developmental biology,* ed 4. Sunderland, MA, Sinauer Associates; and other sources.
Figure 14.18: Source: After Browder, L.W., C.A. Erickson, and W.R. Jeffrey. 1991. *Developmental biology.* Philadelphia, Saunders College Publishing.
Figure 14.24: Source: After Gilbert, S.F. 1991. *Developmental biology,* ed. 3. Sunderland, MA, Sinauer Associates; and King, T.J. 1966. Nuclear transplantation in amphibia. *Meth. Cell Physiol.* 2:1–36.
Figure 14.27: William McGinnis; adapted by the studio of Wood Ronsaville Harlin, Inc., for Howard Hughes Medical Institute as published in *From Egg to Adult* © 1992.
Figure 14.28: Source: After De Robertis, E.M., O. Guillermo, and C. V. E. Wright. 1990. Homeobox genes and the vertebrate body plan. *Sci. Am.* 263:46–52.
Figure 14.30: Source: After Patten, B.M. 1951. The first heart beats and the beginning of embryonic circulation. *Am. Sci.* 39:225–243.
Figure 14.31: Source: Hickman, C.P., Jr. 1959. The larval development of the sand sole, Paralichthys melanostictus. Washington State Fisheries Research Papers 2:38–47.

Chapter 15

Figure 15.3: Source: After Wiley, E.O. 1981. *Phylogenetics.* New York, John Wiley & Sons.

Chapter 16

Figure 16.1: Source: After Lasman, M. 1977. *J. Protozool.* 24:244–248.
Figure 16.2a: From Raven, P.H., and G.B. Johnson. 1996. *Biology,* ed. 4. New York, McGraw-Hill. Reprinted by permission of The McGraw-Hill Companies.

Chapter 19

Figure 19.5: Source: Based on a drawing by L.T. Threadgold, from Roberts, L.S., and J. Janovy Jr. 1996. *Foundations of parasitology,* ed. 5. New York, McGraw-Hill.
Figure 19.11: Source: Mueller, J.F., and H.J. Van Cleave. 1932. *Roosevelt wildlife annals.*
Figure 19.13: Source: After Morseth, D.J. 1967. *J. Parasitol.* 53:492–500.
Figure 19.20: From W. E. Sterrer, Systematics and evolution within the Gnathostomulida, *System. Zool.* 21:151, 1972.

Chapter 20

Opener text: Source: Cobb, N.A. 1914. *Yearbook of the United States Department of Agriculture,* p. 472.
Figure 20.5: Source: After Kristensen, R.M. 1983. Loricifera, a new phylum with Aschelminthes characters from the meiobenthos. *Zeitsch. Zool. Syst. Evol.* 21:163.
Figure 20.16: Source: After Con, C. 1936. Kamptozoa. In H.G. Bronn, ed. *Klassen und Ordnungen des Tier-Reichs,* vol. 4, part 2. Leipzig, Akademische Verlagsgesselschaft.

Chapter 22

Figure 22.7: Source: From Fauvel, P. 1959. Annelides polychetes. Reproduction. In P.P. Grasse, ed. *Traite de Zoologie,* vol. 5, part 1. Paris, Masson et Cie. Modified from W.M. Woodworth, 1907.

Chapter 23

Opener text: Source: *New York Times,* 20 April 1988.
Figure 23.3: From *Synopsis and classification of living organisms,* edited by S. P. Parker. Copyright (c) 1982 McGraw-Hill, Inc. Reprinted by permission.
Figure 23.40: From Raven, P.H., and G.B. Johnson. 1996. *Biology,* ed. 4. New York, McGraw-Hill. Reprinted by permission of The McGraw-Hill Companies.

Chapter 24

Figure 24.12b: Source: After Russell-Hunter, W.D. 1969. *A biology of higher invertebrates.* New York, Macmillan.
Figure 25.4: Source: Drawings by Tim Doyle.
Figure 25.19: Source: After Baker, A.N., F.W.E. Row, and H.E.S. Clark. 1986. A new class of Echinodermata from New Zealand. *Nature* 321:862–864.
Figure 25.22: Source: After Russell-Hunter, W.D. 1969. *A biology of higher invertebrates.* New York, Macmillan.

Chapter 26

Opener text: Reprinted by permission of Alpha Music Inc.
Figure 26.9: Source: After Gould, S.J. 1989. *Wonderful life.* New York, W.W. Norton.
Figure 26.12: Source: After Aldridge, R.J., D.E.G. Briggs, M.P. Smith, E.N.K. Clarkson, and N.D.L. Clark. 1993. The anatomy of conodonts. *Phil. Trans. Roy. Soc. London B* 340:405–421.
Figure 26.13: Source: After Zangerl, R., and M.E. Williams. 1975. *Paleontology* 18:333–341.

Chapter 27

Figure 27.4a: Source: After Conniff, R. 1991. *Audubon,* March.
Figure 27.4b: Source: After Pough, F.H., et al. 1966. *Vertebrate life.* New York, Macmillan; and Jensen, D. 1966. *Sci. Am.* 214(2):82–90.
Figure 27.20: From Castro, P., and M.E. Huber. 1997. *Marine biology,* ed. 2. New York, McGraw-Hill. Reprinted by permission of The McGraw-Hill Companies.

Chapter 28

Figure 28.1: Sources: Carroll, R.L. 1988. *Vertebrate paleontology and evolution.* New York, W.H. Freeman; Coates, M.I., and J.A. Clack. 1990. 347:66–69; Edwards, J.L. 1989. *Amer. Zool.* 29:235–254; Jarvik, E. 1955. *Scient. Monthly,* March, 141–154; Zimmer, C. 1995. *Discover* 16(6):118–127.
Figure 28.2: Source: After Gaffney, E.W. 1979. Bulletin of the Carnegie Museum of Natural History 13:92–105.
Figure 28.5: Source: After Duellman, W.E., and L. Trueb. 1986. *Biology of amphibians.* New York, McGraw-Hill.

Chapter 29

Figure 29.9: Source: After Alexander, R.M. 1975. *The chordates.* England, Cambridge University Press.

Chapter 30

Figure 30.5b: Source: After Wellenhofer, P. 1990. Archaeopteryx. *Sci. Am.* 262:70–77.
Text ch 30: Source: Brown, L. 1970. *Eagles.* New York, Arco Publishing.
Figure 30.19b: Source: After Schmidt-Nielsen, K. 1990. *Animal physiology,* ed 4. Cambridge University Press.
Figure 30.25b: Source: After Johnson, S.R., and I.T. McCowan. 1974. Thermal adaptation as a factor affecting colonizing success of introduced Sturnidae (Aves) in North America. *Can. J. Zool.* 52:1559–1576.

Chapter 31

Figure 31.1: Source: From Carroll, R.L. 1988. *Vertebrate paleontology and evolution.* New York, W.H. Freeman.
Figure 31.3: Source: After Young, J.Z. 1975. *The life of mammals.* Oxford University Press.
Figure 31.16: Source: After Suga, N. 1990. Biosonar and neural computation in bats. *Sci. Am.* 262:60–68.
Figure 31.19: Source: After Lillegraven, J.A., et al. 1987. The origin of eutherian mammals. *Biol. J. Linn. Soc.* 32:281–336.
Figures A.4, A.9: From Raven, P.H., and G.B. Johnson. 1996. *Biology,* ed. 4. New York, McGraw-Hill. Reprinted by permission of The McGraw-Hill Companies.

index

Page numbers in *italic* indicate figures; those followed by *t* indicate tables.

A

Abalone, 459
  red, *458*
Abdomen
  of crustaceans, 498, *499*
  of insects, 510, *510*
ABO blood types, 219, 219*t*
Abomasum, of ruminant stomach, 676, *676*
Absorption
  alimentary system for, *229,* 231–234
  of carbohydrates, 234
  of fats, 234
  of nutrients, 223, 234
  of protein, 234
  water, *229,* 234
Abyssopelagic layer, of oceans, 133, *133*
Acanthamoeba palestinensis, structure of, *363*
Acanthobdella, 485, 487, 488
Acanthobdellida, 487, *487*
Acanthocephala, 353, 434, 442–444, *443, 445*
Acanthodians (Acanthodii), 578, *578, 585*
Acanthostega, 607, *608,* 609
Acari, 496–498, *497, 498*
Accelerator nerves, in cardiac control, 195
Accessory molecules, of T lymphocytes, 215, *216*
Acetospora, 378
Acetyl coenzyme A
  in cellular respiration, 45, *45*
  in fat metabolism, 50–51
  formation of, 45, *45*
  in Krebs cycle, 44, *44,* 46, *46, 48*
Acetylcholine
  in myoneural junction, 160, *161*
  neurotransmitter function of, 244, *245,* 252
Acetylcholinesterase, 244
Achilles tendon, energy storage in, 163, *163*
Acid(s), 695
  organic, 696
Acid rain, 616
Acineta, 375
Acipenser oxyrhynchus, 594
Acoela, 421
Acoelomate animals, 352, 357, *357,* 416–432, *435*
  adaptive radiation of, 430
  body plan of, *355,* 357, *357,* 416, 417, *417*
  phylogeny, 429–430, *430*
  position in animal kingdom, 417
Acontia threads (acontium), of sea anemones, 406, *407*
Acorn worms, 557, *558*
Acquired immune deficiency syndrome, 218
  apoptosis in, 38
  toxoplasmosis in, 375
Acraniata, 566, 569*t,* 579
Acropora, 409
ACTH. *See* Adrenocorticotropic hormone
Actin, 29–30, 156
  in excitation-contraction coupling, 160–161, *162*
  myofilaments, *158,* 159, *159*
Actin-binding protein, 365
Actinophrys, 372, 378
Actinopoda, 365, 378
Actinopterygians (Actinopterygii), 583, *584, 585,* 591–593, *593, 594,* 603
Actinosphaerium, 378
  axopodium, *366*
Action potential(s)
  in excitation-contraction coupling, 160–161, *162*
  of nerve impulse, 243
    conduction of, 243, *243*
Activation energy, 40
Active transport, 33–34, *34*
  in osmotic regulation, 168
  in renal tubular reabsorption, 176
Actomyosin system, 156
Adaptation, 89–90
  of touch receptors, 255
Adaptive radiation, 100, *100, 101*
  of acoelomate animals, 430
  of annelids, 488
  of arthropods, 526–528
  of cnidarians, 413
  of ctenophores, 413
  of echinoderms, 557
  of molluscs, 469–470
  of protozoa, 377
  of pseudocoelomate animals, 446
  of reptiles, 622, *623*
  of sponges, 389
Adaptive zone, 347, 348, *348*
ADCC. *See* Antibody-dependent cell-mediated cytotoxicity
Adders, 634
Adductor muscles
  of bivalves, 460, *462, 463*
  of molluscs, 157–158
Adenine, synthesis in prebiotic conditions, 14
Adenohypophysis, 270
Adenosine diphosphate
  ATP formation from, 42, *42*
  in excitation-contraction coupling, 161, *162*
Adenosine triphosphate
  and contractile proteins, 156
  in coupled reactions, 42–43, *43*
  as energy-coupling agent, 43, *43*
  in excitation-contraction coupling, 160–161, *162*
  formation of, 45, *45*
    from ADP, 42, *42*
    in anaerobic glycolysis, *49,* 49–50
    in fat metabolism, 50–51
    in oxidative phosphorylation, 46–47, *47, 48,* 49*t*
  production, in cell, 29
  structure of, 42, *42*
Adenylate cyclase, 267–268
ADH. *See* Antidiuretic hormone
Adipose tissue, 50, 234–235
ADP. *See* Adenosine diphosphate
Adrenal cortex, 279, *279*
  hormones of, 279–280, *280*
Adrenal glands, anatomy of, 279, *279*
Adrenal medulla, 279, *279*
  embryology of, *326,* 327
  hormones of, 280–281
Adrenaline. *See* Epinephrine
Adrenocortical hormones, 10
Adrenocorticotropic hormone (corticotropin), 271, 273*t*
  and adrenal cortical hormone production, 279–280
  secretion of, negative feedback system for, 268, *269*
Adult, insect, 515, *516*
Aeolosoma, 484, *485,* 488
Aerobes, 43–44
Aerobic metabolism, in muscle contraction, 161
Afferent arterioles, 175, *175*
African boomslang, 635
African clawed frog, 617, *618*
African house snake, *636*
African sleeping sickness, 369, 521
African twig snake, 635
Agapornis, nest building, effects of crossbreeding on, 295, *295*
Age of Reptiles, 622
Age structure, of populations, 114–115, *116*
Agglutination, 219
Aggression, 299–300
Agnathans (Agnatha), 565, 566, *568,* 569*t,* 577, 579, 583–587, *584, 585, 586,* 603
Agonistic behavior, 299–300
AIDS. *See* Acquired immune deficiency syndrome
Air
  gas concentrations in, 204, 204*t*
  inspired, 204, *205*
  partial pressures in, 204, 204*t*
Air capillaries, 202
Air sacs, *201,* 202
  avian, 649, *651*
Airfoil
  angle of attack, 652, *653*
  lift-to-drag ratio of, 652, *653*
Albatross, 661
  wings of, 653, *654*
Albinism, genetics of, 105–106
Albumin, 657
  plasma, 189
Alcoholic fermentation, 49
Alcyonaria, 404, 409
Alcyonium, 409
Aldosterone, 280, *280*
  and sodium reabsorption, 177
Algae
  in cnidarians, 398
  coralline, 408
Alimentary system, 227
  conduction region, 229, *229*
  development of, 328, *328*
  motility in, 228, *228*
  of nematodes, 438, *439*

  organization of, 228–234
  receiving region, 228–229, *229, 230*
  region of grinding and early digestion, *229,* 229–231
  region of terminal digestion and absorption, *229,* 231–234
  region of water absorption and concentration of solids, *229,* 234
  regional function of, 228–234
  storage region, 229, *229*
Alkaline phosphatase, 233
Allantoic mesoderm, mammalian, *336*
Allantois, 334, *335,* 621
  mammalian, 336, *336*
Allele(s), 58
  multiple, 65–66
Allelic frequency, 104, *104*
Alligator(s), 636–638, *637*
  brain of, *248*
  sex determination in, 60
Alligator mississipiensis, 637, 638
Allograft, 212
Allopatric speciation, 99–100
All-or-none phenomenon, 242
Allosaurus, 630, *631*
Alpha cells, pancreatic, 281, *281*
Alpha rays, 692
Alpha-helix, 11
Alpheus, 501
Altricial young, avian, 658
Alula, *647,* 652, *655*
Alvarez, Walter, 108
Alveolar ducts, 202, *202*
Alveolus (pl., alveoli), pulmonary, 201, 202, *202*
Amblyrhynchus cristatus, 632
Ambulacral grooves
  of crinoids, 545, 553
  of sea stars, 545, *545, 546*
Ambystoma mexicanum, 615
Ambystoma tigrinum, 613
Ameba. *See also Amoeba*
  feeding, 7
  locomotion, 365, *365*
Amebic dysentery, 373
Ameiotic parthenogenesis, 312
Amensalism, 119
Amia, 593, *594,* 603
Amictic egg(s), 434–435, *435*
Amino acid(s), *10,* 10–11
  DNA coding for, 72, *74t*
  essential, *235t,* 236
  metabolism of, 51–52
  synthesis of
    Miller experiment on, 13–15, *14*
    in prebiotic conditions, 13–15, *14*
Amino acid pool, *51,* 51–52
Aminopeptidase, 233
Ammocoetes, of lampreys, 587, *587*
Ammonia
  excretion of, 170
  formation of, in amino acid metabolism, 51
Ammonoids, 465
Amniocentesis, 69
Amnion, *322,* 334, *335,* 565, 621
  mammalian, 336, *336*
Amniotes (Amniota), 334–335, 565, 566, *568,* 569*t, 610,* 621
  evolution of, 622, *623*
Amniotic cavity, *322*
  mammalian, *336*
Amniotic egg, 334–335, *335,* 625, *626*
Amoeba, 372, 378. *See also* Ameba
  feeding method of, *367*
  life cycle of, 368
Amoeba proteus
  contractile vacuoles of, 171, *367*
  excretion of water, *367*
Amphibian(s) (Amphibia), *567,* 579
  circulatory system of, *193,* 193–194, *194*
  classification of, 618
  evolution of, 606
  homeotic genes, 333, *333*
  integument of (skin), *148,* 609
  larva, *325*
  lungs of, 200, 201, *201*
  metamorphosis, 609
  modern, 606–620
    characteristics of, 612
  ontogeny of, 606
  population decline, 616
  reproduction, 609–611, *613*
  and reptiles, comparison of, 625–627
  salt balance in, 168, *169*
  ventilation in, 202, *203*
Amphioxus, 572–573, *573,* 579
  and vertebrate ancestry, 564, 575–576
Amphipoda, 505, *506*
Amphiporus, 428, *428*
Amphiporus bimaculatus, 428
Amphisbaena alba, 633
Amphisbaenia, 629–632, *633,* 637
Amphitrite, 478, *479,* 488
  deposit feeding by, *225*
Amphiumas, respiration in, 613, *614*
Amplexus, *309,* 617, *618, 619*
Ampulla, 259, *260*
  of sea stars, 546, *546*
Ampullary organs of Lorenzini, 589, *590*
Amylase
  pancreatic, 233
  salivary, 228–229
Anabolic steroids, 280
Anadromous organisms, 599
Anaerobes, 43–44
Anaerobic glycolysis, in muscle contraction, 161, 163
Analogous structures, 570
Anamnestic response, 217, *218*
Anamniota, 566, 569*t*
Anaphase, *36,* 37
Anapsids (Anapsida), 576, *577,* 622, *623, 624,* 667
  characteristics of, 627–628, 637
Ancestral character state, 345
Ancon sheep, 101, *102*
Ancylostoma duodenale, 440*t*
Androctonus, 496
Androgen(s), adrenal, 280
Androgenic glands, in crustaceans, 501
Anemonefishes, 406, *408*
Aneuploidy, 68
Angstrom, 24
Anguilla anguilla, life cycle of, 598, *599*
Anguilla rostrata, life cycle of, 598, *599*
Anhingus, 661
Anilocra, 506
Animal(s)
  architectural patterns of, evolution of, 353, *355*
  body plan of, 353–358, *355*
    tube-within-a-tube, *355,* 357, *357*
  classification of, 342–360
  complexity of
    and body size, 140, *141*
    hierarchical organization of, 139*t,* 139–140
  diversity of, 342
    evolution of, 83–111, *96*
  organization of, levels of, 139*t,* 139–140
  phylogeny of, 342–360
  symmetry of, 353–354, *356*
Animal behavior, 290–307. *See also* Social behavior
  aggressive, 299–300
  in arthropods, 492
  control of, 293–297
  dominance, 299–300, *300*
  genetics of, 294–295
  innate, 293–294
  in insects, 518–520
  learning and, 295–297
  science of, 291–292
  stereotyped, 292–293
  territorial, 301–302
Animal communication, 302–305
  chemical, 302
  by displays, 303–304
  human–animal, 304–305
  in insects, 518–520
  by symbolic language, 303, 304
Animal kingdom (Animalia), 351, 352
  subdivisions of, 352–353
Animal research, 14–15
Animal rights, 14–15
Anisogammarus, 506
Anna's hummingbird, *659*
Annelids (Annelida), 353, *354,* 474–489
  adaptive radiation, 488
  biological contributions of, 475
  body plan of, *355, 476,* 476–477
  characteristics of, 476
  cladogram of, *487*
  classification of, 488
  ecological relationships of, 475
  economic importance of, 475–476
  immunity in, 212, 213*t*
  metamerism, 358, *358*
  nervous system of, 246, *246*
  number of species of, 475
  phylogeny, 486–487, *487,* 539, *540*
  position in animal kingdom, 475
Anopheles mosquito, and malaria, 374, *374,* 521, *522*
Anoplura, 522, *524,* 526
Anostraca, 504
Ant(s), 523
  and aphids, *519,* 520
  learning and memory in, 518
  pheromone-producing glands, 253, *253*
  social behavior of, *519,* 520, *520*
Ant lions, 523, *524*
Anteater, 686
  feeding specializations of, *675*
  spiny, 685
    reproduction in, 335
Antedon, 555
Antelope(s), 686
Antennae
  of crustaceans, 498, *499*
  of myriapods and insects, 508, *508*
  of polychaetes, *477,* 478
Antennal glands, 172, *173*
  of crustaceans, 502, *503*
Anterior, definition of, 354, *356*
Anthopleura, 409
Anthozoa, 394, 409
  characteristics of, 404–409, *406, 407*
  phylogeny of, *413*
Anthracosaurs, 622, *623*
Antibody(ies), 214
  heavy chains, 214, *214,* 215
  in host defense, 217–218
  light chains, 214, *214,* 215
  molecular structure of, 214, *214*
  production of, 73, 76, 215
  secondary (anamnestic) response, 217, *218*
  titer, 217
Antibody-dependent cell-mediated cytotoxicity, 217
Anticodon, 75, *76*
Antidiuretic hormone (vasopressin), 272, 273*t,* 274, *274,* 275
  actions of, 178–179
Antigen(s), definition of, 213
Antigen-presenting cell(s), 216, *216, 217*
Antipathes, 409
Antithesis, principle of, 300, *301*
Antlers, 672, *674*
  velvet, 672, *674*
Anura, 609, 615–618
  reproduction, *613*
Anus, *354*
  of insects, 512, *512*
  of nematodes, 438, *439*
  of sea stars, 545
  of sea urchins, 551, *552*

Apatosaurus, posture of, and locomotion, 155, *156*
APCs. *See* Antigen-presenting cell(s)
Ape(s), 685
  anthropoid, 682, *683*
  language studies using, 304, 305
Aphid(s), *521,* 523
Aphrodita, 488
Apical complex, of apicomplexans, 373, *373*
Apicomplexa, 373–375, 377, 378
Aplacophora, 453, 472
Aplysia, 295, 472
  memory in, 250
  neurophysiological and behavioral studies using, 295–296
Aplysia dactylomela, 459, *459*
Apocrine glands, 673
Apoda, 609, 611, *613,* 618
Apoptosis, 38
Apopyle, of sponges, *385, 386*
Appendages
  of arthropods, 492
  of crustaceans, 498, *499,* 499–500
  of vertebrates, 575
Appendicular skeleton, 153
Apterygota, 526
Aqueous humor, 261, *261*
Arachnids (Arachnida), 493–498, 525
Arbacia, 555
Arboreal life-style, 682
Arcella, 372, 378
  binary fission in, *368*
Archaea, 352, *352*
Archaebacteria, 18
Archaeocytes, of sponges, 384–385, *386*
Archaeopteryx lithographica, 641–642, *642, 643,* 646, *647*
Archaeornithes, 661
Archenteron, *322,* 324, 328, *357*
Archinephric duct, 173
Archinephros, 173, *174*
Architeuthis, 465
Architeuthis harveyi, 449
Archosaurs (Archosauria), 622, *623, 624,* 628, 630, 637
  cladogram of, 642, *644*
Arctic tern, migration of, 654
Ardipithecus ramidus, 683, 684
Arenicola, 150, 475, 478, 479, *485,* 488
Arginine phosphate, in muscle contraction, 161
Argiope aurantia, 495, 525
Argopecten irradians, 464
Argulus, 525
Ariolimax columbianus, 461
Aristotle, 343
Aristotle's lantern, 551, *552*
Arkansas, Balanced Treatment for Creation-Science and Evolution-Science Act (1981), 3
Armadillidium, 505, *506,* 525
Armadillo, 686
Armillifer, 534
Arrector pili muscle, *672*
Arrowworms, 542, 560, *560*
Arteriole(s), 193, 196
Arteriosclerosis, 196
Artery(ies), 193, 195–197, *197*
  layers of, 195, *197*
Arthropod(s) (Arthropoda), 353, *354,* 490–530
  adaptive radiation of, 526–528
  behavior of, 492
  body plan of, *355*
  characteristics of, 492
  circulatory system in, 192, *193*
  classification of, 525
  ecological relationships of, 491
  exoskeleton, 150–151
  fossils of, 491, *491*
  integument of, 147–149, *148*
  kidneys of, 172, *173*
  metamerism in, 358, *358*
  metamorphosis in, 492
  nervous system of, 246, *246*
  phylogeny, 526, *527,* 539, *540*
  sensory organs of, 492
  success of, 490–492
Artiodactyla, 686, *688*
Ascaris, female, *439*
Ascaris lumbricoides, 439, *440,* 440*t*
Ascaris suum, 439, *440*
Aschelminthes
  cladogram of, *445*
  phylogeny, *445,* 446
Ascidians (Ascidiacea), 571, *571, 572*
  metamorphosis, 571, *572*
Asconoids, 383, *385*
Asellus, 505
Aspidogastrea, *430*
Asses, 686
Association areas, 250
Aster(s), 30, *36,* 37
Asterias, 555
  feeding, 547
  pedicellariae of, *547*
Asteroidea, 543–548, 555, *556. See also* Sea stars
Asteroids, mass extinctions and, 108–109
Astrangia, 409
Astrocytes, 242, 244
Astrophyton muricatum, 549, 555
Astrosclera, 390
Asymmetric competition, 119
Atherosclerosis, 196, 236
Atlas, 155
Atmosphere, 127
Atmospheric pressure, 204
Atokes, 479, *480*
Atom(s), 691
  electron shells of, 692, *693*
  structure of, 691, *691*
Atomic mass, 691*t,* 692
Atomic number, 691*t,* 692
ATP. *See* Adenosine triphosphate
Atrax robustus, 496
Atrioventricular bundle, 194, *196*
Atrioventricular node, *196*
Atrium (pl., atria), 193, *193, 194*
  left, 194, *194, 195, 196*
  right, 194, *194, 195, 196*
Atrophy, muscle, 160
Attach-pull-release cycle, 160–161, *162*
Auditory canal, *257,* 258
Auk(s), 662
Aurelia, 402, 403, 409
  life cycle of, *405*
Auricle(s)
  of bivalves, *465*
  of flatworms, 420, *420*
Auricularia, *550*
Australian tiger snake, 635
Australopithecus afarensis, 683, *683,* 684, *684*
Australopithecus africanus, 684, *684*
Australopithecus anamensis, 683
Authority, citation in species nomenclature, 344
Autogamy, 368
Autoimmune disease, 214
Autonomic nervous system, *251,* 251–252, *252*
Autosomes, 60
Autotomy
  of brittle stars, 549
  of sea stars, 548
Autotrophs, 17, *17,* 20, 39, 223
  protozoan, 365–366
Aves, 579, 622, 641, 641*t. See also* Bird(s)
  classification of, 661–662
Avocet(s), 662
Axial skeleton, 153
Axis, 155
Axocoel, of echinoderms, 548, *549*
Axolotl, 613, *615*
Axon(s), *147,* 241, *241,* 244
  formation of, 327, *328*
  growth cone of, 327–328, *328*
Axoneme, 363–364
Axopodium (pl., axopodia), 365, *366, 372*
Axosome, *364*

B

B lymphocytes (B cells), 215
Babesia, 378
Bacillus thuringiensis, 524
Bacteria, 18, 351, 352, *352. See also* Prokaryotes
  chemoautotrophic, 125
  colonic, 234
  fossil, 17, *18*
  genetic recombination in, 20
Bacteriophage, 77
Baculum, 317
Balance organs, 259
Balanced Treatment for Creation-Science and Evolution-Science Act (1981), 3
Balantidium, 379
Balanus, 525
Baleen, 223, *224*
Banana slug, *461*
Bandicoots, 685
Bankia, 472
Banting, Frederick, 282, *282*
Barb(s), of contour feather, 645
Barbules, of contour feather, 645
Barnacles, 504–505, *505,* 525
  competition among, 119
  feeding, *224*
  reproduction of, *309*
Basal body, 31, 364
Basal granule, 364
Base(s), 695
  nitrogenous, 11
    of nucleic acids, 70, 70*t, 71*
Base pairing, 70, *71, 72*
Baseodiscus, 428
Baseodiscus mexicanus, 428
Basilar membrane, of inner ear, *257, 258,* 258–259, *259*
Basket star, *549,* 555
Basophils, 190, *190,* 212
Bat(s), 685
  echolocation, 677–678, *679*
  flight of, 677
  food and feeding of, 678–679
  fruit-eating, feeding specializations of, *675*
  migration of, 677
Bathypelagic layer, of oceans, 133, *133*
Bayliss, W. H., 266, *267*
Beach flea, 505
Beagle, voyage of, 85–88, *86*
Bear(s), 686
Beardworms, 533
Beaumont, William, 230–231, *231*
Beaver, feeding specializations of, *675*
Bedbugs, 521
Bee(s), *520,* 523. *See also* Honeybee(s)
Beef tapeworm, 424–425, *426,* 426*t*
Beetle(s), 518, *518, 519, 521,* 523
Behavior. *See* Animal behavior
Benthic organisms, 132
  independent of solar energy, 125, 133
Benthos, 132, 133
Bernard, Claude, 166, 188, *189*
Berrill, N. J., 223
Berson, Solomon, 275
Best, Charles H., 282, *282*
Beta cells, pancreatic, 281, *281*
Beta rays, 692
Bicarbonate, in body fluids, 188, *189*
Bicarbonate buffer system, 696–697
Bichir, 591, *594,* 603
Bicuspid valve, 194, *195*
Bighorn sheep, social dominance in, 300, *300*
Bilateral symmetry, 353, *355, 356,* 429
  of ctenophores, 411
  in gastropods, 456

- primary, 416
- of sea urchin, 551, *551*

Bilateria, 352, 354
Bile, 232, 233, *233*
Bile duct, 233
Bile pigments, 233
Bile salts, 233
Bilirubin, 190
Binary fission, *309,* 310, 367, *368, 369*
- of ciliates, 376, *377*

Binomial nomenclature, 343
Biogenesis, 12
Biogenetic law, 98
Biogeochemical cycles, 126–127, *127*
Biological concept of species, 350–351
Biological hierarchy, 5, *5*
Biology, 3
Bioluminescence, in insects, 519
Biomass, 124
- ecological pyramids of, 125, 126, *126*

Biome(s), *128,* 128–131
Biosphere, 112, 113, 127
- subdivisions of, 127

Bipinnaria, *550,* 558, *559*
Biradial symmetry, 393
- in echinoderms, 542, 543

Biramous appendages, of crustaceans, 498, *499*
Bird(s), *567,* 579, 640–665
- adaptive hypothermia in, 183–184, *184*
- air sacs, 649, *651*
- altricial young, 658
- anatomy of, 641
- ancestry of, 622
- aquatic, taxonomy of, 347, *348*
- beaks, 649, *650*
- care of young, 658–659, *660*
- carinate, 645
  - evolution of, *644*
- characteristics of, 641, 641*t*
- circulatory system of, 649
- classification of, 661–662
- copulation, 657, *658*
- digestive system of, 648–649
- evolution of, 622, *623, 624,* 641–645, *643, 644*
- excretory system of, 649–650
- extinction of, 659, *660*
- flight, 652–654
  - adaptations for, 645–652
  - flapping, 653–654, *655*
- flightless, 96–97, 645, 661, *663*
- fossil record of, 641–642, *642*
- gastrulation in, *322,* 325
- guilds of, *120,* 120–121
- lead poisoning of, 660
- lungs of, *201*
- mating systems of, 657, *659*
- migration of, 640
  - direction finding in, 655–656, *657*
  - routes, 654, *656*
  - stimulus for, 654–655
- molting of, 645
- muscular system of, 646–648, *648*
- nervous system of, 650–652
- nesting, 658, *659*
  - effects of crossbreeding on, 295, *295*
- nesting colonies of, 301, *302*
- nesting failure, 658–659
- origin of, 641–645
- perching mechanism of, 648, *649*
- populations of, 659–663
- precocial young, 658, *660*
- ratite, 645, 661
  - evolution of, *644*
  - phylogeny of, 96–97, *97*
- relationships of, 641–645
- reproductive system of, 656–657, *659*
- and reptiles, relationship of, 622, *623, 624*
- respiratory system of, 649, *651*
- salt glands, 650, *651*
- sensory system of, 650–652
- skeleton of, 646, *647, 648*
- skull of, 646
- social behavior of, 656–659
- structure of, 641
- territoriality in, 301–302
- tinamous, 661
- ultraviolet light sensitivity in, 652
- wings of, 646, *647*
  - bones of, 155
  - elliptical, 652–653, *654*
  - forms of, 652–653
  - high-lift, 653, *654, 655*
  - high-speed, 653, *654*
  - as lift device, 652, *653, 655*
  - soaring, 653, *654*
  - tip vortex, 653, *653*

Birth, human, hormones of, *285,* 285–286
Bispira brunnea, 478
Biston betularia, industrial melanism in, 102, *103*
Bittern(s), 661
Bivalent, 59, *59*
Bivalves (Bivalvia), 449, *450,* 459–464, *461, 462,* 469, *470,* 472
- adductor muscles of, 460, *462, 463*
- digestion, 460–461
- feeding habits of, 460–461, *462, 463*
- form and function of, 460–464
- internal features of, 461–464, *464, 465*
- locomotion, 460, *463, 464*
- reproduction, 461–464
- shell of, 460, *463, 464*

Black widow spider, 496, *497*
Bladder worms, of beef tapeworm, 425, *426*
Blarina brevicauda, 687
Blastocoel, *322,* 323, *357,* 448
- persistent, 192, *192*

Blastocyst, human, 336
Blastomere(s), 320, 322, *354*
Blastopore(s), *322,* 324, 325, *354, 357,* 358
Blastula, *322*
- formation of, 323–324

Blepharisma, 379
Blepharoplast, 364
Blood, 143
- circulation of, 187, 188, 192
- coagulation, 190–191, *191*
  - abnormalities of, 191
- composition of, 188–190, *189*
- formed elements in, 189–190, *190*
- gas transport in, 204–206
- volume, 192

Blood flukes, 422–423, *423, 423t*
Blood group antigens, 219–220
Blood plasma, 140–141, 188, *189*
- composition of, 188
- proteins in, 188, 189, *189*

Blood pressure
- arterial, 195–196
- and body size, 197
- capillary, 197–198
- measurement of, 197
- venous, 198

Bluefin tuna, adaptations for swimming, *597*
Blue-green algae. *See* Cyanobacteria
Boaedon fuluginosus, 636
Bobolink, migration of, *656*
Body cavities, 354–358, *355*
Body cells, formation of, 35
Body fluid(s), 140–141, 143, 188
- composition of, 188
- extracellular, 188, *189*
- intracellular, 188, *189*

Body, of bivalves, 460, *463, 464*
Body size, and bone stress, 155, *156*
Body temperature, regulation of, 179–184
Body tissues, gas exchange in, 204
Body wall, of cnidarians, 397, *397*
Bohr effect, 205, *206*
Bombykol, 302
Bombyx mori, chemical sex attraction in, 302, *302*
Bond(s)
- chemical, 692–694
- covalent, 693–694, *694, 695*
- high-energy, 42, *42*
- hydrogen, 694, *695*
  - in DNA, 70, *71, 72*
- ionic, 692–693, *694*
- in protein structure, 11

Bone(s), 144, *145,* 151–152
- avian, 646, *648*
- compact, 152, *152*
- endochondral (replacement), 152
- growth of, 153
- loss, postmenopausal, 153
- membranous, 152
- microscopic structure of, 151–152, *152,* 152–153
- mineral metabolism in, 277–278
- numbers of, in various species, 153
- pneumatized, 646, *648*
- spongy (cancellous), 152
- stress on, body size and, 155, *156*

Bonellia, 533, *533*
Boobies, 661
- blue-footed, pair-bonding behavior of, 304, *305*

Book gills, of horseshoe crabs, 493
Book lungs, of spiders, 494, *494*
Boophilus annulatus, 498
Boreal forest, 130
Bowfin, 593, *594,* 603
Bowman's capsule, 174, *175, 176*
Brachiolaria, *550*
Brachiopods (Brachiopoda), 353, *354,* 453, 531, 535, *538,* 538–539, *539*
Brachiosaurus, 630, *631*
Brain, 248–250
- avian, 650–652
- cephalopod, *467,* 468
- control (cardiac) center in, 194–195
- embryology of, *326,* 327
- human, 248
- neuropeptides of, 275
- of oligochaetes, *482,* 482–483
- of rotifers, 434, *435*
- vertebrate, 246–247
  - variation in, 248, *248*

Brain hormone, 269, *270*
Branchial chambers, of crustaceans, 502
Branchial tufts, 200
Branchiobdellida, 487, *487*
Branchiopoda (branchiopods), characteristics of, 504, *505,* 525
Branchiostoma, 572
Branchiura (branchiurans), 504, *505,* 525
Branta canadensis, imprinting in, *297*
Breathing, regulation of, 203–204
Brenner, Sydney, 438
Briggs, R., 330
Bristle worms, 475
Bristletails, 522
Brittle stars, *389,* 543, *549,* 555
- autotomy of, 549
- characteristics of, *549,* 549–550, *550*
- regeneration of, 549

Bronchi, 201, 202, *202*
Bronchiole(s), 201, 202, *202*
Brown fat, 234–235
Brown recluse spider, 496, *497*
Browsers, 674, *675*
Bruemmer, Fred, 681
Brugia malayi, 442
Bruneria borealis, parthenogenesis in, 115
Bryozoa (bryozoans), 446, 536
Buccal cavity, 228
Budding, *309,* 310, 367
- of hydras, 400, *400*

Buffer(s)

action of, 696–697
definition of, 696
Buffon, Georges Louis, 84
Bufo americanus, 616
Bufonidae, 615, *616*
Bug(s)
definition of, 516
true, *521,* 522
Bulbourethral glands, *317,* 318
Bullfrogs, 615, *616*
epicurean market for, 615, *616*
populations of, 615
Burgess, Thornton, "Portrait of a Meadow Mouse," 681
Burrowing, of annelids, 275, 476–477
Bursae, of brittle stars, 549
Bush babies, 685
Busycon, 472
Butterfly(ies), 523
Batesian mimicry in, *122*
metamorphosis, holometabolous (complete), 515, *516,* 526
mouthparts of, 513, *513*
Müllerian mimicry in, *122*
reproduction, 515
Buzzard(s), 661
Byssal threads, of bivalves, 460

C

Caddis flies, 523
Caecidotea, 506
Caecilians, 611, *613,* 618
evolution of, 609, *610*
Caenorhabditis elegans, 438
apoptosis in, 38
Calamus, of contour feather, 645
Calcarea, 382, 385, *385,* 387, 390
characteristics of, *387,* 387–388
Calcichordates (Calcichordata), 570, *571*
Calcifibrospongia, 390
Calcification, 149
of procuticle, in arthropods, 491
Calcispongiae, 390
Calcitonin
and bone growth, 153
and calcium metabolism, *278,* 279
Calcium
in body fluids, *189*
and bone mass, 153
in excitation-contraction coupling, 160–161, *162*
metabolism, hormonal regulation of, 277–279
Calliostoma annulata, 450
Callyspongia, 387
Calypte anna, 659
Calyx, of crinoids, 553, *554*
Cambered surface, of avian wing, 652
Cambrian explosion, 17, 531
Camel(s), 686
Canaliculi, of bone, *145,* 152, *152*
Cancer
growth of, 37
molecular genetics of, 79–80
Canine teeth, 225, *226,* 674
Cannibalism, 226
Cannon, Walter B., 166, *167,* 281
Canoe shells, 458
Canthon pilularis, 518, *518*
Capillary(ies), 193, *197,* 197–198
Capillary exchange, 197–198, *198*
Capitulum, of ticks and mites, 496
Caprella, 506
Carapace
of crustaceans, 499, *499*
of horseshoe crabs, 493
of turtles, 627, *628*
Carbohydrate(s), 8–9
absorption, 234
digestion of, 228
as fuels, 8–9
requirements for, 235, 236
Carbon, 7–8
covalent bond formation, 694, *695*
Carbon dioxide
atmospheric, 129, *129*
in cutaneous respiration, 200
diffusion, 204, *205*
and hemoglobin satruation curve, 205, *206*
and respiratory rate, 203–204
in tracheal system, 200
transport, in blood, 206, *207*
Carbonic acid, 206, *207*
Carbonic anhydrase, 206, *207*
Carboniferous period, radiation of tetrapods in, 609, *610*
Carboxypeptidase, 233
Carcharias, 590
Carchesium, 379
Cardiac muscle, 144, *146,* 194
Cardiac output, 194
Cardiac sphincter, 230
Cardiac stomach, of sea stars, *546,* 547
Caribou
migration of, 676, *678*
population of, 676
Carnivora, 676, 686
Carnivores, 123, 223, 676
mammalian, *675,* 676, *677,* 686
origin of, 20
Carrier(s), 66
Carrier-mediated transport, 32
Carrying capacity (K)
of environment, *116,* 116–117, *118*
for human species, 117–118, *118*
Cartilage, 143–144, *145,* 151
of invertebrates, 151
Carybdea, 409
Carybdea marsupialis, 406
Cassiopeia, 409
Cassowary, 645, 661
Castes, of social insects, 519
Cat(s), 686
Catadromous organisms, 598
Catalysts, 40
Catastrophic species selection, 109
Caterpillar, insect, 515, *516*
Cattle, 686
Cattle tick, 498
Caudata, 609, 612, 618
Caudofoveata (caudofoveates), 449, 453, 469, *470,* 472
Causes
proximate (immediate), 4, 291
ultimate, 5, 291
Caveolae, 34, *35*
CCK. *See* Cholecystokinin
CD4 lymphocyte(s), 215, 216, *216*
CD8 lymphocyte(s), 215, 216, *216*
cdk's. *See* Cyclin-dependent kinase(s)
Cecum (pl., ceca), 675
avian, 649
digestive, of sea stars, *546,* 547
gastric, of insects, 512, *512*
Cell(s), 139*t,* 140. *See also* specific cell types
ameboid, 156, 364–365
as basic unit of life, 5, *5,* 23–55
components of, 24
in connective tissue, 142, 143
determined, 319
discovery of, 24
division, 35–38
energy transfer in, 42–43
eukaryotic, 26*t,* 26–35
components of, *27,* 27–30
metabolism, 39–52
multinucleate, 36
organelles of, 18, 19, *19,* 24, 27, *27*
density gradient separation of, 24, *25*
organization of, 26–35
primitive, 16
prokaryotic, 26, 26*t*
respiration in, 43–50, *44*
specializations of, 30–31
study of, 24–26
surface of, 30–31
Cell aggregates, *355*
Cell cycle, 37–38
G_0 period, *38*
G_1 period, *38, 38*
S period, 38, *38*
Cell membrane, *19,* 24, 27, *27*
fluid-mosaic model, 27, *27*
functions of, 31–35
Cell theory, 24
Cell-mediated immunity, 218
Cellular respiration, 199
Cellulase, 229–230
Cellulose, 8
digestion of, 229–230, 675
Centipedes, 508, *508,* 525
Central nervous system, 246
in reflex arcs, 247, *247*
Centriole(s), 27, *27,* 30, *30,* 364
Centrocercus urophasianus, 659
Centromere, 35, *36,* 37
Centrosome(s), 30, *36,* 37
Centruroides, 496, 525
Cephalaspidomorphi, 579, 583, *585, 586,* 587, *587, 588,* 603
Cephalization, 358, 416, 429
Cephalochordates (Cephalochordata), 571–573, *573*
Cephalodiscus, 558
Cephalopholis fulvus, 506
Cephalopoda, 449, *450,* 464–469, *470, 471,* 472
color changes in, 468
external features of, 466–468, *467*
eyes of, 468, *468*
feeding habits of, 468
form and function of, 465–469
ink production, 468
internal features of, *467, 468,* 468–469
locomotion, 466, *467, 468*
nutrition, 468
reproduction, 468–469, *469*
shell of, 465, 466, *467*
Cerata, of nudibranchs, 459
Ceratium, 371, 378
Cercariae, of flukes, 421, *422*
Cercomeria, *430*
Cerebellum, 249, *249, 250*
avian, 650, 651
Cerebral cortex, avian, 650–651
Cerebral ganglia, of oligochaetes, *482,* 482–483
Cerebral hemispheres, avian, 650, 651
Cerebrum, *249,* 250
Cerianthus, 409
Ceriantipatharia, 404, 405, 409
Cermatia, 525
Cervix, uterine, 318, *318*
Cestoda, 418, 424–425, *425, 430*
characteristics of, 424–425, *425, 426,* 427, *427*
Cestum, 411, *411*
Cetacea, 686
Chaetoderma, 472
Chaetognatha, 353, *354, 445,* 543, 560, *560*
Chaetonotus, 436
Chaetopleura, 472
Chaetopterus, 479, *481,* 488
Chagas' disease, 369, 521
Challenge, antigenic, 217
Challenger expedition, *373*
Chameleon(s), 629, *632*
feeding, *7, 632*
Character(s), taxonomic, 344–346
Character displacement, 119–120, *120*
Charging, 75
Cheetah, genetic variability, erosion of, 106, *106*
Chelae, of decapods, *499, 500,* 506
Chelicerae, 493
of spiders, 494, *494*
Chelicerata (chelicerates), 493–498, 525
aquatic, 525
Chelipeds, of crustaceans, 499, *499, 500*
Chelonia, 637

Chelydra serpentina, 628
Chemical bonds, 692–694
Chemical mutagens, 80
Chemiosmotic coupling, 47
Chemoautotrophic bacteria, 125
Chemoreception, 253–254
Chemotaxis, 253
Chemotrophs, 223
Chick, embryonic cleavage in, *322*
Chickens
intermediate inheritance in, 62–63, *63*
social dominance (peck-order) in, 300
Chief cells, 230
Chiggers, 497
Chilomonas, 371, 378
Chilopoda, 508, *508,* 525, *527*
Chimaeras, 579, *585,* 590–591, *591,* 603
Chimpanzee(s), taxonomy of, 347–348, 349, *349*
Chinese liver flukes, *423,* 423*t*
Chipmunk
food and feeding of, *677*
posture of, and locomotion, 155, *156*
Chironex, 409
Chironex fleckeri, 404
Chiropsalmus, 409
Chiroptera, 685
Chitin, 8, 147
of arthropods, 491
Chiton(s), 449, *450,* 453–454, *454, 455,* 472
Chlamydomonas, 371
Chlamydophrys, 372
Chloragogen tissue, of earthworms, 481, *482*
Chloride
in body fluids, 188, *189*
in nerve impulse conduction, 242–243, *243*
Chlorocruorin, 205
Chloroplast(s), 20, 77, 369, *371*
origin of, 361
Choanocytes, of sponges, 383, 384, *385, 386*
Cholecystokinin, 275, 282, *283*
Cholesterol, 10, *10*
in cell membrane, 27, *27*
in steroid hormone synthesis, 279, *280*
Chondrichthyes, 579, 583, *584, 585,* 587–591, 603
Chondrocytes, *145,* 151
Chondrosteans (Chondrostei), 591, 603
Chordates (Chordata), 353, *354, 559,* 564–581
ancestry of, *567,* 570
biological contributions of, 565
characteristics of, 566–570
cladogram of, *559,* 565, *568*
classification of, 579
cladistic, 565–566
traditional, 565–566, 569*t*
evolution of, 564, 565, 570, *571*
fossil, 575, *575*
Garstang's hypothesis of larval evolution, 575, *576*
metamerism, 358, *358*
phylogenetic tree of, 565, *567*
position in animal kingdom, 565
structure of, 565
Choriaster granulatus, 544
Chorioallantoic membrane, 334–335
Chorion, 334, *335,* 621
mammalian, 336, *336*
Chorionic villi, 336, *336*
Choroid coat, 261, *261*
Christmas tree worm, *384, 478*
Chromatids, *36,* 37
Chromatin, 28, 35
Chromatophores
of cephalopods, 468
in crustaceans, 501
of reptiles, 625, *625*
Chromodoris kuniei, 450
Chromosomal aberrations, 68–69
Chromosomal theory of inheritance, 103
Chromosome(s)
definition of, 35
and inheritance, 57–60
structure of, 35–36
Chromosome number, 35, 58, 68
Chyme, 232
Chymotrypsin, 232–233, *233*
Cicada, *517,* 523
Cilia, 31, 156–157, *157*
in alimentary canal, 228
of ciliates, 376, *376*
of *Paramecium, 377*
protozoan, 363–364
structure of, 31
of tapeworms, 424, *425*
of trochophore larva, 453, *453*
Ciliary muscles, 261, *261*
Ciliata, 352
Ciliates, *375,* 375–376, *376,* 377, *377*
life cycles of, 376, *377*
reproduction, 376, *377*
Ciliophora, *375,* 375–376, *376, 377,* 379
binary fission in, *369*
Ciona, 571
Circadian rhythms, 274
Circular muscle
of annelids, *476,* 476–477, *477*
gastrointestinal, *232*
Circulation, 191–199
coronary, 195
double, *193,* 193–194, 607
pulmonary circuit, *193,* 193–194
systemic circuit, *193,* 193–194
Circulatory system(s), 187, 188, 191–199
of amphioxus, 572, *573*
avian, 649
of bivalves, 461, *465*
closed, *192,* 192–193
of crustaceans, 502
of earthworms, 481–482
echiuran, 532
embryology of, 329
of molluscs, 452
of nemerteans, *428,* 428–429
open, *192,* 192–193
reptilian, 626, *627*
of vertebrates, *193,* 193–194, *194*
Cirri
of ciliates, *375,* 376
of crinoids, 553, *554*
Cirripedia, 504–505, *505*
Cisternae, in endoplasmic reticulum, 28, *29*
Citric acid cycle. *See* Krebs cycle
Clades, 345
nested hierarchy of, 345–346
Cladistics, 346, 348–349, 622
Cladocera, 504
Cladogram, *345,* 345–346
of annelids, *487*
of archosaurs, 642, *644*
of Aschelminthes, *445*
of chordates (Chordata), *559,* 565, *568*
of cnidarians, *413*
construction of, 346, *346*
of deuterostomes, *559*
of echinoderms, *556*
of fishes, *585*
of molluscs, *471*
of synapsids, *669*
of tetrapods, *610*
Clam(s), 449, *450,* 459, *463, 464*
freshwater
decline in, 461
feeding mechanism of, *462*
glochidium of, 464, *466*
Clam worms, 475
Clathrin, 34
Clathrina, 383, 388, 390
Clathrina canariensis, 385
Clathrin-coated pits, 34, *35*
Clathrulina, 372
Cleavage furrow, 37
Cleavage, of zygote, 320–323
patterns of, 321–323, *322*
Climate, global variation in, 129, *129*
Climex, 521
Cliona, 390
Clitellum, of earthworms, *482,* 483, 484, *484*
Clitoris, 318, *318*
Cloaca
avian, 649, 650
of sea cucumbers, 552, *553*
vertebrate, 316
Cloacal bladder, of rotifers, 434, *435*
Clone(s), 78, 310
Cloning, asexual, reproduction by, 115
Clonorchis, 422, 427
Clonorchis sinensis, 422, 423*t*
Clotting. *See* Blood, coagulation
Clotting factors, 190–191
Cnidaria (cnidarians), 353, 392, 393–409
adaptive radiation of, 413
biological contributions of, 393
body structure of, 397–398
characteristics of, 393–394
classification of, 409
digestion, 398–399
dimorphism, 394–395
ecological relationships of, 394, *395*
feeding by, 398–399
immunity in, 212, 213*t*
locomotion, 398
medusa form, 395, *396*
nematocysts of, 392, 393, 395–396, *397*
nerve net of, 396–397
neuromuscular system of, 397
phylogeny of, 412–413, *413,* 429
polymorphism, 394–395
polyps, 394–395, *396*
position in animal kingdom, 393
reproduction, 399, *401*
Cnidoblast, 396
Cnidocil, 396, *397*
Cnidocyte(s), 393, 396, *397,* 398
Coagulation. *See* Blood, coagulation
Cobb, N. A., 433
Cobra(s), 634, 635, *636*
Coccidians, 375
Coccyx, 155
Cochlea, 256, *257,* 258
avian, 651
Cochlear duct, *257,* 258
Cockroaches, 522
Cocoon, of earthworms, 483–484, *484*
Codfish, brain of, *248*
Codominance, 62–63
Codon, 72
Codosiga, feeding method of, *367*
Coelacanth, 583, *584, 585,* 591, 592, 593–594, *595,* 603, *610*
Coelenterata, 393
Coelom, 357, 448
of annelids, 476
development of, 448
extraembryonic, mammalian, *336*
formation of, *324,* 326, *354*
of sea stars, 545
Coelomates, 448, 449
Coelomic cavity, of sea cucumbers, 552, *553*
Coelomic compartments, 324
Coelomic fluid, of earthworms, 481–482
Coelomic vesicles, *322,* 324
Coelophysis, 630, *631*
Coenenchyme, of alcyonarian corals, *410*
Coenzyme(s), 40
Coenzyme A, 40
in cellular respiration, 45, *45*
Coevolution
of parasite and host, 122–123
of predator and prey, 121
Cofactors, 40
Cohort, 115
Coleoptera, 523
Collagen
in connective tissue, 142–143, *145*
in nematodes, 438
Collar, of acorn worms, 557, *558*

Collecting duct(s)
- renal, 174, *175*
- and urine concentrating mechanism, 178, *178*

Collembola, 522, 526
Collenchyme, of ctenophores, *412*
Collencytes, of sponges, 385, *386*
Colloblasts, of ctenophores, 411
Colloid osmotic pressure, 198
Colonies
- of ectoprocts, 536, 537, *537*
- of modular animals, 115

Color blindness, inheritance of, 66, *67*
Color vision, 263
Colpoda, 379
- cysts, 368

Colubridae, 634–635
Comantheria, 555
Comantheria briareus, 553
Comb jelly, 411, *412*
Comb plates, of ctenophores, 411, *411, 412*
Comets, mass extinctions and, 108–109
Commensalism, 119
Common descent, Darwin's theory of, 88–89
- evidence for, 94–98

Communication. *See* Animal communication
Community(ies), definition of, 113, 118
Community ecology, 118–123
Comparative biochemistry, 346
Comparative biology, 343
Comparative cytology, 346
Comparative method, 5
Comparative morphology, 346
Comparative psychology, 291
Competition, 113, 119
- asymmetric, 119

Competitive exclusion, 119
Complement, 217
Compound, definition of, 691
Concentration gradient, 32
Concentricycloidea, 554, *554,* 555
Conchostraca, 504
Conchs, 454, 459
Condensation reaction(s), 8, *9*
- in polymer formation, 16

Condor(s), 661
Cone opsin, 262
Cone shells, 457, *458*
Cones, 261, 262, *262*
- avian, 651, *652*

Coney, *506*
Conjugation, 20
- of ciliates, 376, *377*
- of *Paramecium, 377*
- protozoan, 368

Connective tissue, 142–144, *145*
- dense, 142, *145*
- loose, 142, *145*

Connell, Joseph, 119
Conodonts, 576, *577*
Conotoxins, 457
Conraua goliath, 615, *617*
Conscious mind, 250
Constant region, of antibody, 214
Consumers, 123
Contact chemical receptors, 253
Contraceptives, 285
Contractile proteins, 156
Contractile vacuoles, 29, 171
- of ciliates, 376, *377*
- protozoan, *365,* 366, *367, 371, 377*

Control(s), 4
Control (cardiac) center, in brain, 194–195
Conus, 457
Coots, 662
Copepoda (copepods), 504, *505,* 525
Coprolites, 91
Coprophagy, 675
Coral(s), *384,* 395
- alcyonarian, 408, *410, 411*
- cup, *410*
- hard, 404
- horny, 404, 405
- reef-building, 394, 408–409
- reefs of, 408–409
- soft, 404
- stony, 407–408, *410*
- thorny, 404
- true, 407–408, *410*
- zoantharian, 407–408, *410*

Coral snakes, 634
Coralline algae, 408
Coreceptors, of T lymphocytes, 215, *216*
Cormorant(s), 661
Corn borer, 521, *521*
Corn ear worms, *521*
Cornea, 261, *261*
Cornified cells, 149
Corona, of rotifers, 434, *435*
Coronary artery disease, 195, 196
Coronary circulation, 195
Corpora allata, 517
Corpus luteum, 284, *284*
- hormones released from, 285, *286*

Corpus striatum, avian, 651
Cortical reaction, in fertilization, 320, *321*
Corticosterone, 279
Corticotropin-releasing hormone, 279
- and adrenal cortical hormone production, 279–280

Cortisol, 279, *280*
Countercurrent flow, 200
- over fish gills, 597–598, *598*

Countercurrent multiplication, 178
Coupled septa, of Anthozoa, 404
Covalent bonds, 693–694, *694, 695*
Cowries, 459
Coxal glands, of spiders, 494, *494*
Coyote, feeding specializations of, *675*
Crab(s), 506, *507,* 525
- hyperosmotic regulation in, *167,* 167–168

Crab spider, *495*
Cranes, 662
Craniata, 566, *568,* 569*t,* 573, 579
Crayfish, 506, 525
- antennal glands, *173*
- external structure of, 499–500, *500*
- internal structure of, 502, *503*
- reproduction, 503
- stomach of, 502, *503*

Creatine phosphate, in muscle contraction, 161
Creationism, versus evolution, 4
Creation-science, 3
Cretaceous extinction, 108, *108,* 109, *109*
Cretin, 276
Crick, Francis, *6,* 56, 71
Cricket(s), 522
- reproductive system of, *317*
- song of, 302

Crinoids (Crinoidea), 543, *553,* 553–554, *554,* 555, *556*
- larvae of, *550*

Cristae, of mitochondria, 29, *30*
Cristatella, 537
Crocodiles, 630, 636–638, *637*
Crocodilians (Crocodilia), 628, 630, 636–638, *637*
- ancestry of, 622, *624*
- heart of, 626, *627*
- internal structure of, 626, *627*

Crocodylus acutus, 638
Crocodylus niloticus, 636–638, *637*
Crocodylus porosus, 636
Crop, 229, 648
- of insects, 512, *512*

Crop milk, 648
Crossaster papposus, 544
Crossing over, 66, *68*
Crotalus horridus, 634
Crow, skeleton of, *647*
Crustaceans (Crustacea), 498–506, 525, *527*
- appendages of, 498, *499,* 499–500
- circulation in, 502
- ecdysis in, 500, *501, 502*
- endocrine functions in, 500, 501
- evolution of, mandibulate hypothesis, 526, *527*
- excretion, 502
- feeding habits, 501–502
- immunity in, 212, 213*t*
- life cycles of, 503, *504*
- nervous system of, 502, *503*
- reproduction, 503, *505*
- respiration in, 502
- sensory system of, 502, *503*

Cryptic defenses, 121
Cryptobiosis, 535
Cryptosporidium parvum, 375
Crystalline style, of bivalves, 461
Ctenoid scales, 593, *593*
Ctenophora, 353, 392, 411–412
- adaptive radiation of, 413
- biological contributions of, 393
- function in, 411–412
- luminescence, 411
- phylogeny of, 412–413
- position in animal kingdom, 393
- structure of, 411–412

CTLs. *See* T lymphocytes, cytotoxic
Cubozoa, 394, 409
- characteristics of, 404, *406*
- phylogeny of, *413*

Cuckoos, 662
Cucumaria, 555
Cud chewing, 675–676, *676*
Cupula, 256, *256,* 259, *260*
Cushion star, *544*
Cutaneous respiration, 200
Cuticle, 419
- of annelids, 476, *476*
- of arthropods, 491
- invertebrate, 147, *148*
- of nematodes, 438
- of tardigrades, 535

Cuttlefish, 464, 465, *467*
- locomotion, 466, *468*

Cyanea, 402, 403
Cyanea capillata, 404
Cyanobacteria, 18, 20, 351
Cyclic AMP, as second messenger, 267–268
Cyclin(s), 38, *38*
Cyclin-dependent kinase(s), 38, *38*
Cycliophora, 444
Cycloid scales, 593, *593*
Cycloneuralia, *445*
Cyclops, 525
Cyclospora cayetanensis, 375
Cyclosporine, 218
Cyclostomata, 579, 603
Cynodonts, 667, *668, 669*
Cypris, 525
Cyst(s), protozoan, *363,* 368
Cysteine, *10,* 11
Cysticerci, of beef tapeworm, 425, *426*
Cysticercosis, 425, 426*t, 427*
Cystid, of ectoprocts, 536
Cytogenetics, 57
Cytokines, 210, 215–216, *216,* 275–276
Cytokinesis, 36, 37
Cytopharynx, of ciliates, 376, *377*
Cytoplasm, distal, of flatworms, 418, 419
Cytoplasmic division, 37
Cytoplasmic localization, 319, 329
Cytoproct
- of *Paramecium, 377*
- protozoan, 366, *377*

Cytopyge, protozoan, 366, *377*
Cytoskeleton, 29, *30*
Cytostome
- of *Paramecium, 377*
- protozoan, 366, *367, 373, 377*

D

Dactylogyrus, 427
Dactylozooids, 400
Daily torpor, 183, *184*
Damselflies, 522
Danaus plexippus

metamorphosis, *516*
migration of, 512
Daphnia, 504, *505*
Darwin, Charles Robert, 2, 84, *84, 86,* 408
The Descent of Man and Selection in Relation to Sex, 95–96, 682
evolutionary theory of, 88–91
The Expression of the Emotions of Man and Animals, 290, 300
The Formation of Vegetable Mould Through the Action of Worms, 480
influence of, 290
On the Origin of Species, 88, 89, 90, 350
and voyage of *Beagle,* 85–88, *86, 87*
Darwinism, 88–91
evidence for, 91–103
modern, emergence of, 103–104
Darwin's frog, *613*
Dash (-), 65
Dasypeltis scaber, 225
Dead space, pulmonary, 201
Deamination, 51
Decapoda, *499, 500, 504,* 506, *507*
Deciduous (milk) teeth, 674
Decomposers, 123, 126–127
Decorator crabs, 506
Deer, 686, *688*
antlers of, *674*
feeding specializations of, *675*
Defecation, 234
Definitive host, of trematodes, 421
Dehydration, in polymer formation, 16
Delayed implantation, 680
Deletion, chromosomal, 69
Demes, 114
Demodex, 497, *498*
Demography, 114–115
Demospongiae, 382, *384,* 385, 387, *387, 388,* 390
characteristics of, 388, *389*
Dendrites, *147,* 241, *241*
Dentalium, 455, 472
Deoxyribonucleic acid, 6, 11, 56
chemical components of, 70, 70*t*
coding of, by base sequence, 71–72
coding strand, 73
double helix, *6,* 70, *72*
eukaryotic, 19, *19,* 35
5′ end, 70, *71*
methylation of, 76, *77*
mitochondrial, 20, 77
modification of, gene regulation by, 76, *77*
packaging, in chromosomes, 36
plasmid, 20, 73, 77
plastid, 20
prokaryotic, 18
replication, in interphase, 37–38
replication of, 71, *73*
sequencing, 18
structure of, 70, *71, 72*
template, 71, *73*
3′ end, 70, *71*
Deoxyribose, 70, *70*
Depolarization, 243
Deposit feeding, 223–225, *225*
Derived character state, 345
Dermacentor variabilis, 497, 498
Dermal branchiae, of sea stars, 545
Dermal ostia, of sponges, 383, *385, 386*
Dermal papulae, 200
Dermaptera, 526
Dermatophagoides farinae, 496, *498*
Dermis
reptilian, 625, *625*
structure of, *672*
vertebrate, *148,* 149, *150,* 671
Dero, 485
Desert, *128,* 131
Desmognathus wrighti, 614
Desmosomes, 31, *31*
Detorsion, in gastropods, 457, *471*
Deuterium, 692, *692*
Deuterostomes (Deuterostomia), 323, *324,* 352–353, *354,* 539, 543, 565, 570
cladogram of, *559*
Development, 6, 318–326
definition of, 318
of deuterostomes, 323, *324,* 325
direct
in frogs, 618
of salamanders, 612, *614*
embryonic, 320–326
events in, *319*
gene expression during, 331–333
mechanisms of, 329–333
of protostomes, 323, *324,* 325
of sponges, 387, *388*
of systems and organs, *326,* 326–329
vertebrate, 333–336
Devonian period, tetrapod evolution in, 607–609, *608*
DeVore, Ivan, 305
Diabetes mellitus, 176, 281–282
Diadema antillarum, as keystone species, 121–122
Diapause, in insects, 509, 517–518
Diaphragm, 203
Diapsids (Diapsida), 622, *623, 624,* 637, 667
characteristics of, 628–638
Diastole, 194, *196*
Diatoms, 223
Dicrostonyx, 681
Didelphis marsupialis, reproduction, *680*
Didinium, 376
feeding, *367*
Diencephalon, 274
Dientamoeba, 378
Difflugia, 372
Diffusion, 32
gas exchange by, 199, 200
Digenea, *430*
characteristics of, 421–423, *422*
life cycle of, 421, *422*
Digestion, 226–228, *227*
cnidarian, 398–399
extracellular, 227, 399
in flatworms, 419
hormones of, 282, *283*
intracellular, 226–227, *227,* 399
in flatworms, 419
in protozoa, 365–366
in vertebrate small intestine, 231–234
Digestive ceca, of sea stars, *546,* 547
Digestive gland, of bivalves, 461, *462*
Digestive system
avian, 648–649
of nemerteans, 428, *428*
of sea stars, *546,* 547–548
Diglyceride, 9
1,25-Dihydroxyvitamin D, and calcium metabolism, *278,* 278–279
Dileptus, 375
Dilger, W. C., 295
Dinobryon, 371
Dinoflagellates, 369, *371*
Dinosaurs, 622
characteristics of, 630
locomotion, 155
Dioecious organisms, 312, 438, 452, 534, 536, 600
Diphyllobothrium, 427
Diphyllobothrium latum, 426*t*
Diploblastic animals, 326
Diploid number, of chromosomes, 35, 58, 311
Diplopoda, 508, *509,* 525, *527*
Diplura, 522, 526
Diptera, 510, 511, *511,* 523
Dipylidium, 426*t*
Dirofilaria immitis, 442, *443*
Disaccharidases, 233
Disaccharides, 8
formation of, 8, *9*
Distal convoluted tubule, 174, *175*
and urine concentrating mechanism, 178, *178*
Distal, definition of, 354
Distance chemical receptors, 253
Disulfide bond, 11
Diverticula, digestive, 229, 231
Diving petrel, *348*
DNA. *See* Deoxyribonucleic acid
DNA ligase, 77
DNA polymerase, 71
Dobsonflies, 523
Dobzhansky, Theodosius, 350
Dog(s), 686
principle of antithesis in, *301*
Dog tapeworm, 426*t*
Dogfish shark
external anatomy of, *589*
internal anatomy of, *589*
Doliolaria, *550*
Dolomedes triton, 496
Dolphin(s), 686
feeding specializations of, *675*
Domestic fowl, 661
Dominance, 299–300, *300*
Dominant allele(s), 61, 62, 65–66, 104
Doodlebugs, 523
Dorsal blood vessel, of earthworms, 481, *482*
Dorsal, definition of, 354, *356*
Double circulation, *193,* 193–194
Dove(s), 662
Down syndrome, 69
Dragonfly, 522
metamorphosis, 515, *517*
Dreissena polymorpha, 464
Drone(s), of social insects, 519, *520*
Drosophila melanogaster, 57
eye color, sex-linked inheritance of, 66, *67, 68*
gene maps for, 66
linkage groups in, 66
Duck(s), 661
flight, *655*
Dugesia, 421, 427
Dung beetles, 518, *518,* 523
Duodenum, 230, 232
Duplication, chromosomal, 69
Dwarf tapeworm, 426*t*
Dwarfism, 271
Dyads, 59
Dynein, 157, *157*

E

Eagle(s), 661
grip of, 648
Eagle ray, *591*
Ear(s), 256, *257*
external, *257*
human, 256–258, *257*
inner, *257,* 258
middle, *257,* 258
Earth
climate of, 129, *129*
environmental conditions of (*see* Environment)
temperature of, 129, *129*
Earthworm(s), 475, 480–484, *482*
anatomy of, 481–482, *482*
cocoon formation, 483–484, *484*
copulation, 483, *483*
dorsal median giant fiber of, 483
form and function, 481–484, *482, 483*
locomotion, 150, *151,* 481
nervous system of, *246*
vascular system of, 192, *192*
Eccrine glands, 673
Ecdysial (prothoracic) glands, 269, 517
Ecdysis
of arthropods, 492
in crustaceans, 500, *501, 502*
in insects, *517*

Ecdysone, 269, 500
Echidnas, armor of, 671–672
Echinaster luzonicus, 548
Echinococcus granulosus, 426*t*
Echinococcus multilocularis, 426*t*
Echinococcus, reproduction, 420
Echinoderes, 436
Echinoderm(s) (Echinodermata), 353, *354, 567,* 570
  adaptive radiation of, 557
  biological contributions, 543
  body plan of, *355,* 542
  characteristics of, 542–557
  cladogram of, *556*
  classification of, 555
  ecological relationships of, 543, *544, 549, 551, 552, 553, 554*
  fossil, *571*
  immunity in, 212, 213*t*
  larvae, 549, *550*
  phylogeny, 554–557, *556, 559*
  position in animal kingdom, 543
Echinoidea, 543, 550–551, *551, 552, 555, 556*
  larvae of, *550*
Echinopluteus, *550*
Echiura, 532, *533,* 539
Echiurids, 353, 453
Echolocation, mammalian, 677–679
Ecoline, 128
Ecological pyramids, 125–126, *126*
Ecology, 112
  definition of, 113
  hierarchy of, 113–127
Ecosystems, 113, 123–127
Ectoderm, *355, 357*
  of acoelomates, 417, *417*
  derivatives of, *326,* 326–328
  development of, *322,* 324, 325
Ectognathous insects, 526
Ectolecithal egg, 420
Ectoparasites, 122, 226
Ectoplasm, 156
  of *Paramecium, 377*
  protozoan, 365, *365*
Ectoprocts (Ectoprocta), 353, *354, 445,* 446, 531, 535, 536–538, *537, 538*
Ectotherms, 113, 180
  body temperature regulation in, 180–181
    by behavioral adjustments, 180, *181*
Ectyoplasia ferox, 389
Edema, 218
Eel(s)
  life cycle of, 598, *599*
  locomotion in water, 596, *597*
  metamorphosis in, *599,* 602
  migration of, 598, *599*
Effectors, 241
  in reflex arcs, 247, *247*
Efferent arterioles, 175, *175*
Egg(s), 35, 310
  activation of, 320
  amictic, of rotifers, 434–435, *435*
  amniotic, 334–335, *335,* 625, *626*
  amphibian, 617–618
  avian, 657, 658
  of earthworms, 483, *484*
  ectolecithal, 420
  endolecithal, 420
  of fish, 600–602, *601, 602*
  of flatworms, 420
  formation of, 313–314
  insect, 515, *516*
  mictic, 435
  morphological determinants in, 319
  preparation for fertilization, 319
  shelled, 621, 625, *626*
  of snakes, 635
  and sperm, contact and recognition between, 319–320, *321*
Eimeria, 378
Eland(s), temperature regulation in, 182, *182*
Elaphe obsoleta obsoleta, 634
Elapidae, 634
Elasmobranchs (Elasmobranchii), *585,* 588–590, 603
  characteristics of, 588
  osmotic regulation in, 170
  urea in, 170
Elassochirus gilli, 506, *507*
Elastin, *145*
Eldredge, Niles, 101
Electra pilosa, 538
Electric rays, 590
Electrolytes
  in body fluids, 188, *189*
  renal tubular reabsorption of, 176–177
  urinary excretion of, 176–177
Electron(s), *691,* 691–692
  gain of, 693
  loss of, 693
  shells, of atoms, 692, *693*
  transport, and oxidation-reduction reactions, 43, *43*
  transport chain, *45,* 46–47, *47, 48*
Electron microscopy, 24, *25*
Elements, 691, 691*t*
  atomic number of, 691, 691*t*
  atomic weight of, 691, 691*t*
  symbols for, 691, 691*t*
Elephant(s), 686
  posture of, and locomotion, 155, *156*
  trunk of, as muscular hydrostat, 150, *151*
  tusks of, 225, *227*
Elephantbird, 645
Elephantiasis, 442, *443*
Eleutherozoa, 555–557, *556*
Elton, Charles, 125, 126
Eltonian pyramid, 125–126, *126*
Elvers, 598, *599*
Embioptera, 526
Embolus, 196
Embryo(s)
  anteroposterior axis of, 324, *325*
  cleavage (*see* Zygote, cleavage)
  cortex, developmental significance of, 329
  developing, protection of, 333
  dorsoventral axis of, 324, *325*
  gill arches of, *98*
  human, early development of, 336, *337*
  left-right axis of, 324, *325*
  polarity of, 324, *325*
Embryonic period, of embryonic development, 336
Emergence, 5
Emergent properties, 5
Emlen, S., 656
Emu(s), 645, 661
Encephalization, 246
Encystment, protozoan, 368
Endergonic reactions, 39–40
Endocrine glands, 267
  of vertebrates, 270–282
Endocrine system, 250, 266–289
Endocrinology, founders of, 266, *267*
Endocuticle, of arthropods, 491
Endocytosis, 32, 34
Endoderm, *355, 357*
  of acoelomates, 417, *417*
  derivatives of, *326,* 328
  development of, *322,* 324, 325
Endognathy, 526
Endolecithal egg, 420
Endolymph, 259, *260*
Endomesoderm, *354*
Endometrium, 318, 336
  in menstrual cycle, 284, *284*
Endoparasites, 122, 226
Endoplasm, 156
  of *Paramecium, 377*
  protozoan, 365, *365*
Endoplasmic reticulum, 18, *19,* 27, *27,* 28, *28*
  cisternae, 28, *29*
  of *Paramecium, 377*
  rough, *25,* 28, *28*
  smooth, 28, *28*
Endopod, of crustaceans, 499, *499*
Endorphin(s), 256, 275
Endoskeleton, 150–151
  of alcyonarian corals, 408, *410*
  vertebrate, 573
Endostyle, of ascidians, 571, *571*
Endothelial cells, of capillary wall, 197
Endotherms, 113, 180
  body temperature regulation, *181,* 181–182
    in cold environments, 182–183, *183*
  in hot environments, 182, *182*
Energy
  chemical-bond, 23, 39
  ecological pyramids of, 125, 126, *126*
  efficiency, and body size, 140, *141*
  flow of, 123–126
  kinetic, 39
  and life, *39,* 39–40
  for muscle contraction, 161–163
  potential, 39
  storage of, in tendons, 163, *163*
Energy budget, 123–124
Energy-coupling agent, adenosine triphosphate as, 43, *43*
Enkephalin(s), 256, 275
Ensatina eschscholtzii, speciation, 99, *99*
Entamoeba, 378
Entamoeba histolytica
  cysts, 368
  parasitism, 373
Enterobius vermicularis, 440*t,* 441, *442*
Enterocoelomate body plan, *355*
Enterocoelous coelom formation, *324,* 326, *354,* 357, *357*
Enteropneusta, 557, *558, 559*
Entodesma saxicola, 464
Entomology, 508
Entoprocta, 353, 434, *444,* 444–446
Entropy, 23
Environment
  aquatic, 132–133
  carrying capacity of, *116,* 116–117, *118*
  versus ecology, 113
  interaction with, 113–114
  subnivean, 183
  terrestrial, 128–131, 607
  tolerance of, 114
Enzyme(s)
  activity of, 41, *41*
    catalytic, 40, *40*
  in cellular metabolism, 40–42
  coupled reactions, 41, 42–43, *43*
  digestive (membrane), 227–228, 233, *233*
  functions of, 11
  hydrolytic, 227–228
  nature of, 40
  one gene–one enzyme hypothesis for, 69–70
  pancreatic, 232–233, *233*
  in protocells, 16, 17
  reactions catalyzed by, 42
  regulation of, 52, *52*
  RNA as, 16
  specificity of, 41–42, *42*
  structure of, 11
Enzyme-substrate complex, 41
Eocytes, 18
Eon(s), 92
Eosinophilia, 212
Eosinophils, 190, *190,* 212
Ephelota, 379
Ephemeroptera, 522, 526
Ephyrae, of scyphozoans, 404
Epicuticle, 147–148
  of arthropods, 491–492
Epidermis
  of cnidarians, *397,* 397–398, *398*
  invertebrate, 147, *148*
  reptile, 625, *625*
  structure of, *672*
  vertebrate, *148,* 149, 671

Epididymis, *315*, 317, *317*
Epidinium, 379
Epinephrine, 280-281, *281*
Epipelagic layer, of oceans, 133, *133*
Epistome
of brachiopods, 539
of lophophorates, 535, *536*
Epistylis, *375*
Epitheliomuscular cells, of cnidarians, *397*, 397-398, *398*
Epithelium, 142, *142*, *143*, *144*
embryology of, *326*, 327
olfactory, 254, *255*
simple, 142, *142*, *143*
stratified, 142, *142*, *144*
transitional, *144*
Epitoke, 479, *480*
Epitope, 216
Epoch(s), 92
Eptatretus stouti, 583
Equilibrium, sense of, 259-262
ER. *See* Endoplasmic reticulum
Era(s), 92
Ergasilus, 525
Eris aurantius, *495*
Erythroblastosis fetalis, 219-220
Erythroblasts, 189
Erythrocyte(s), 189-190, *190*
avian, 649
ES complex. *See* Enzyme-substrate complex
Esophagus, 229, *229*
avian, 648
of insects, 512
Essential nutrients, 236
Estradiol, *283*
Estrogen(s), *283*, 283-284
in birth, 286, *286*
in menstrual cycle, 284, *284*
in pregnancy, *285*
Estrous cycle, 283, 679
Estrus, 283, 679
Estuarine crocodile, 636
Ethology
classical, principles of, 292-293
definition of, 291
founders of, 291, *291*
science of, 291-292
Eubacteria, 18
Eubilateria, 429
Eucarya, 352, *352*
Euchordata, *568*
Eucoelomates, 353, 357-358, *435*
body plan of, *355*, *357*, 357-358
Euglena, 369, *371*, 378
binary fission in, *309*, *368*
reproduction, *309*
Euglypha, 365
binary fission in, *368*
Eukaryotes, 351
cells, *26*, 26-35
components of, 19, *19*
emergence of, 361
evolution of, *19*, 19-20
gene regulation in, 76
microtubules in, 20
origin of, 18, *19*
reproduction, 19
Eumetazoa, 352, *355*, 381, 382
Eunice viridis, *480*
Euphausiacea, 506, *507*
Euplectella, *387*, 388, 390
Euploidy, 68
Euplotes, *375*, 379
Eurycea longicauda, *614*
Euryhaline organisms, 167
Eurypterids (Eurypterida)
characteristics of, 493, 525
fossil, *491*
Eustachian tube, *257*, 258
Eusthenopteron, 607, *608*
Eutely, 434
Eutherians, 679, 685
Evaginations, 199-200
Evaporative cooling, *181*, 182, *182*
Evolution
abiogenic molecular, 13
chemical, and origin of life, 11-16
clock of biological time for, *19*
definition of, 84
of diversity of animals, 83-111, *96*
genetic basis of, 56-82
human, 682-685
timescale of, 83
trends in, 94
Evolutionary concept of species, 351
Evolutionary sciences, 5
Evolutionary taxonomy, 346-348
Evolutionary theory, 4
versus creationism, 4
Darwinian, 88-91
evidence for, 91-103
revisions of, 103-104
historical perspective on, 84-88
of Lamarck, 85, 98, 103
of Lyell, 85, 100
of neo-Darwinism, 103
pre-Darwinian, 84-85
transformational, 85
variational, 85
Excitation-contraction coupling, 160-161, *162*
Excretion
of ammonia, 170
in crustaceans, 502
of electrolytes, 176-177
in insects, 514
of nutrients, 223
in platyhelminthes, 419-420, *420*
in protozoa, 366
of salts, 176-177
of uric acid, 177
by birds, 650
of water, 177-179
by *Amoeba proteus*, *367*
Excretory system
avian, 649-650
of crustaceans, 502
in invertebrates, 171-172
Exercise, effects on skeletal muscle, 159
Exergonic reactions, 39, 40, *40*
Exocrine glands, 267
Exocuticle, of arthropods, 491
Exocytosis, 34-35
Exons, 73
Exopod, of crustaceans, 499, *499*
Exoskeleton, 147, 150-151
of arthropods, 491-492
of zoantharian coral, 408, *410*
Experimental method, 4-5
Expiration, 203, *203*
Exponential growth, 116, *116*, 117
External nares, 202, *203*
External respiration, 199
Exteroceptors, 253
Extinction(s)
avian, 659, *660*
mass, *108*, 108-109, *109*, 531, 622, 630
through geological time, 108
Extracellular space, 140
Extraembryonic membrane(s), 334, *335*, 621
mammalian, 335, 336, *336*
Eye(s), 260
avian, 651, *652*
camera-type, 260
cephalopod, 468, *468*
of chameleons, *632*
compound, 260, *261*, 502, *503*
of insects, 514-515
of crustaceans, 502, *503*
human, structure of, 261, *261*
maxillopodan, 504
median, 502
of polychaetes, *477*, 478
of scallops, 461, *464*
of spiders, *494*, 494-495

F

Fab region, of antibody, 214
Facilitated transport, 33, *33*
FAD. *See* Flavin adenine dinucleotide
Falcons, 661
Fanworms, 478
Fascia, *145*
Fasciola hepatica, *419*, 423*t*, 427
Fasciolopsis buski, 423*t*
Fat(s)
absorption of, 234
dietary, requirements for, 235, 236
digestion of, 228
as fuels, 50
metabolism of, *50*, 50-51
neutral, 9, *9*, 10*t*
Fatty acid(s)
polyunsaturated, requirements for, 235*t*
saturated, 9
in triglycerides, 9, *9*, 50, *50*
unsaturated, 9, *10*
Fc region, of antibody, 214
Feather(s), 641
contour, 645, *646*
development of, 645
primary, slotting between, *647*, 652, *655*
structure of, 645, *646*
Feather stars, 543, *554*
characteristics of, 553-554, 555
Featherduster worms, 477, 478, *480*
Fectonotis pygmaeus, *613*
Feeding
of acorn worms, 557, *558*
of ameba, *7*, *367*
of anteater, *675*
of *Asterias*, 547
of barnacles, *224*
of bats, 678-679
of beaver, *675*
of bivalves, 460-461, *462*, *463*
of cephalopods, 468
of chameleon, *7*, *632*
of chipmunk, *677*
of cnidarians, 398-399
of crustaceans, 501-502
deposit, 223-225, *225*
by *Amphitrite*, *225*
on fluids, 226
on food masses, *225*, 225-226
of gastropods, 457, *458*, *459*, *461*
of lampreys, 587, *588*
of mammals, 225, 674-676, *675*
mechanisms for, 223-226
on particulate matter, 223-225, *224*
of protozoa, 366, *367*
of sea stars, 547-548
of shipworms, 461, *463*
of snakes, 225, *225*
suspension, 223, *224*
in bivalves, 459-461
in crustaceans, 501
of whales, 223, *224*
Fertilization, 309, 311, 319-320
avian, 657
external
in fish, 600
in invertebrates, 315
internal
in invertebrates, 315
in reptiles, 625-626
preparations for, 319
Fertilization cone, 320, *321*
Fertilization membrane, 320, *321*
Fetal period, of embryonic development, 336
Fetus, 336
Fiber(s)
extracellular, in connective tissue, 142, *145*
muscle. *See* Muscle fiber(s)
nerve, *147*
Fibrillar muscle, 158
Fibrin, formation of, 190, *191*
Fibrinogen, 189, 190, *191*
Fiddler crabs, 506, *507*
Fight or flight response, 281
Filarial worms, 442
Filopodia, 365, *372*
of axons, 327-328, *328*
Filosea, 365

- Fin(s)
  - anal, *592*
  - caudal, *592*
    - of arrowworm, 560, *560*
    - diphycercal, 593, *593*
    - heterocercal, 588, *593*
    - homocercal, 593, *593*
  - dorsal, *592*
  - pectoral, *153,* 155, 591
  - pelvic, *153,* 155, 591
- Final electron acceptor, 43, 44
- Fire coral, *403,* 408
- Fireflies, 519, *519,* 523
- Fireworms, 478, *479*
- First messenger, 267–268
- Fish louse, *505*
- Fish tapeworm, 426*t*
- Fisher spider, 495, *496*
- Fishes, 582–605
  - ancestry of, 583, *584,* 607
  - bony, 566, *567, 568,* 579, 583, *584,* 591–594
    - characteristics of, 592, *593,* 603
    - diversity of, 591
    - evolution of, 591
    - marine, osmotic regulation in, 168–170, *169*
    - origin of, 591
    - scales of, 592, *593*
    - skin of, 592, *593*
  - cartilaginous, 566, *567,* 583, *584,* 587–591, 603
  - circulatory system of, *193,* 193–194
  - cladogram of, *585*
  - classification of, 583, 603
  - definition of, 582, 583
  - evolution of, 583, *584,* 607
  - fleshy-finned, 583, *584, 585,* 591, *593,* 593–594, *595,* 603
  - freshwater, osmotic regulation in, 168, *169*
  - functional adaptations of, 595–602
  - growth of, 602, *602*
  - jawed, 579, 583, *584*
  - jawless, 566, *567,* 583–587, *584, 585, 586*
    - characteristics of, 586
    - earliest, 576, *577,* 583
  - larvae, with yolk sac, 334, *335*
  - lateral line system of, 256, *256*
  - locomotion in water, 595–596, *596, 597*
  - metamorphosis in, *599,* 602
  - migration, 598–600
  - modern, evolution of, 578–579
  - neutral buoyancy of, 596–597
  - number of species of, 582
  - ray-finned, 583, *584, 585,* 591–593, *593, 594,* 603
  - reproduction, 600–602, *602*
  - respiration in, 597–598, *598,* 607
  - rhipidistian (lobe-finned), 607, *610, 611*
    - and tetrapods, 607, 609
  - sex determination in, 60
  - skeleton of, *153,* 153–155
  - structural adaptations of, 595–602
  - suspension feeding of, *224*
  - swim bladder of, 596–597, *597*
- Fission, 367. *See also* Binary fission; Multiple fission
  - in flatworms, 420, *421*
  - for reproduction, *309,* 310
- *Flaccisagitta hexaptera, 560*
- Flagella (sing., flagellum), 31, 157, *157*
  - protozoan, 363–364
  - structure of, 31
- Flagellates (Flagellata), 352, 369–371, *371,* 377
  - binary fission in, *368*
- Flame cells, of flatworms, 171–172, *172,* 419–420, *420*
- Flamingo, 661, *663*
- Flatworms, 417–427
  - body plan of, *355*
  - flame cells of, 171–172, *172*
  - nervous system of, 246, *246,* 420, *420*
- Flavin adenine dinucleotide, 40
  - in cellular respiration, 46, *46*
- Flea(s), 523, *524*
  - disease transmission by, 521
- Flies, true, *511, 522,* 523
- Flight
  - avian, 652–654
    - adaptations for, 645–652
    - flapping, 653–654, *655*
  - mammalian, 677–679
- Flight muscles
  - of birds, 646–648, *648*
  - of insects, 510–512, *511*
    - asynchronous, 511–512
    - synchronous, 511–512
- Fluid-mosaic model, of cell membrane, 27, *27*
- Flukes, 418
  - characteristics of, *419,* 421–423, 427
  - human infection with, 422–423, *423,* 423*t*
  - reproduction, 420
- Flycatcher, wings of, *654*
- Flying squirrel, 677, *679*
- Follicle(s), ovarian, 284, *284,* 318
- Follicle-stimulating hormone, 271, 273*t,* 283, 284
  - in menstrual cycle, 284, *284*
- Food(s). *See also* Nutrient(s)
  - intake, regulation of, 234–235
  - worldwide supply of, 237, *237*
- Food chains, 125–126
- Food vacuole, 29
  - of *Paramecium, 377*
  - protozoan, *365,* 366
- Food webs, 123, *124*
- Foot
  - molluscan, *451,* 451–452
  - of rotifers, 434, *435*
- Foraminiferans, *372,* 372–373, 377
- Forebrain, 248, *249,* 250
- Foregut
  - development of, 328, *328*
  - of insects, 512
- Forest
  - boreal, 130
  - lake, 130
  - southern evergreen, *128,* 130
  - temperate coniferous, *128,* 130, *130*
  - temperate deciduous, *128,* 129–130
  - tropical, *128, 130,* 130–131
- Fossil(s), 91, *92*
- Fossil record, 2, 83
  - of arthropods, 491, *491*
  - of birds, 641–642, *642*
  - Cambrian, 94
  - evolutionary trends seen in, 94, *95, 96*
  - information gained from, 346
  - interpretation of, 91–92, *92–94*
  - of onychophorans, 534
  - Precambrian, 17, *18,* 94
  - of protozoa, 377
  - of sponges, 382
- Fossorial mammals, 182
- Fouling
  - adaptations to avoid, *456,* 456–457, *457, 458*
  - definition of, 456
- Fovea, avian, 652, *652*
- Fovea centralis, 261, 262
- Fox, Sidney, 16
- Fragmentation, for reproduction, 310
- Franklin, Rosalind, 71
- Free energy, 39–40
- Freshwater worms, 475, 480–484, 484, *485*
- Frigatebirds, 661
- Frog(s), 615–618, *616,* 618
  - amplexus, *309*
  - brain of, *248*
  - defenses of, 617
  - distribution of, 615–617
  - eggs, nuclear transplantation experiments with, 330, *330*
  - embryo, polarity of, *325*
  - embryonic cleavage in, *322*
  - evolution of, 609, *610*
  - habitats of, 615–617
  - hibernation, 616–617
  - marsupial, *613,* 618
  - metamorphosis in, 276, *277*
  - reproduction, *613,* 617–618, *618, 619*
  - reproduction in, *309*
- Frontal plane, definition of, 354, *356*
- Fructose, structure of, 8, *8*
- Fruit fly, 57
  - *Antennepedia* homeotic gene, *331*
  - eye color, sex-linked inheritance of, 66, *67, 68*
  - gene maps for, 66
  - homeobox genes, 332, *332*
  - linkage groups in, 66
- FSH. *See* Follicle-stimulating hormone
- Fuels
  - carbohydrates as, 8–9, 50
  - fats as, 50
- Fulmars, 661
- Funch, P., 444
- Fungi, 351, 352
- Funnel, of cephalopods, 466, *467*
- Funnel-web spider, 496
- Fur seal(s), migration of, 677, *678*
- Furcula, 646

G

- Galactose, structure of, 8, *8*
- Galápagos Islands, 85–87, *87*
  - Darwin's finches on
    - displacement of beak size among, 119–120, *120*
    - evolution of, 100, *100, 101*
  - giant tortoises of, 628, *629*
    - social behavior of, 299
  - marine iguanas of, 629, *632*
- Galen, 187
- Galileo, 138
- Gallbladder, 233
- Gallinule(s), 662
- Gametes, 35, 57, 58, 309, 310
  - formation of, 313–314
- Gametogenesis, 314
- Gamma rays, 692
- Gamma-aminobutyric acid, neurotransmitter function of, 244
- *Gammarus,* 525
- Ganglia (sing., ganglion), 241
  - cerebral, of oligochaetes, *482,* 482–483
  - nerve, of ascidians, 571, *571*
  - sensory, embryology of, *326,* 327
  - sympathetic, embryology of, *326,* 327
- Ganglion cells, 261, *262*
- Gannet(s), 661
  - automatic release of inappropriate behavior in, 304
  - nesting colonies of, 301, *302*
- Ganoid scales, 592, *593*
- Gap junction(s), 31, *31,* 244
- Gar(s), 593, *594,* 603
- Garden spider, *495*
- Garstang, Walter, hypothesis of larval evolution, 575, *576*
- Gas chambers, of cephalopods, 465, 466, *467*
- Gas exchange, 199
  - in body tissues, 204
  - by direct diffusion, 200
  - in insects, 513–514, *514*
  - in lungs, 204
  - respiratory, 204, *205*
  - through tubes, 200
  - in water, 200
- Gas gland, of fishes, 596–597, *597*

Gastric ceca, of insects, 512, *512*
Gastric juice, 230, *233*
Gastric mill, of crayfish, 502
Gastric pouches, of scyphozoans, 404
Gastrin, 275, 282, *283*
Gastrodermis, of cnidarians, 397, *397,* 398
Gastrointestinal tract, development of, 328, *328*
Gastropods (Gastropoda), 449, *450,* 454-459, 469, *470, 471,* 472
- adaptations to avoid fouling, *456,* 456-457, *457, 458*
- classification of, 458-459
- coiling, 456, *456*
- feeding habits of, 457, *458, 459, 461*
- form and function of, 455-458
- internal form and function of, 457-458, *459, 461*
- torsion in, 455-456, *456*

Gastrotricha, 353, 434, 435, *436, 445,* 446
Gastrovascular cavity
- of cnidarians, 397
- of corals, 408, *410*
- of hydras, 400
- of sea anemones, 406, *407*

Gastrozooids, 400
Gastrula, *322,* 324
Gastrulation, 324-326
Gause, G. E., predator-prey experiment by, *121*
Geckos, 629, *632*
Geese, 661
- brain of, *248*
- young, imprinting in, 296-297, *297*

Gemmulation, 310
Gemmules, of sponges, 387, *388*
Gene(s), 56, 57, 69-70
- allelic, 58
- definition of, 69-70
- expression during development, 331-333
- linked, 66
- mutations (*see* Mutation(s))
- one gene-one enzyme hypothesis for, 69-70
- rearrangement, 76
- regulation of, 75-77
  - in eukaryotes, 76

Gene pool, 104
Genetic code, 6, 11, 56, 72, 74*t*
Genetic drift, 106
- and selection, interactions of, 107

Genetic engineering, 77-78
Genetic equilibrium, 105-106
- disturbance of, 106-107

Genetic information, storage and transfer of, 70-78
Genetics, 57
- Mendelian laws of, 61-69

Genital openings, of sea urchins, 551, *552*
Genital slits, of brittle stars, 549
Genome, 35-36
Genotypes, 61
Genus, 343-344
Geological time, 92-94
Geophilus, 525
Geospiza fulginosa, 663
Germ cell line, 314
Germ cells, 57, 58, 309
- formation of, 35, 313-314
- male, 314, *314, 315*
- primordial, 313-314

Germ layers, *322,* 324, 325
- derivatives of, *326,* 326-329

Germinal period, of embryonic development, 336
Germinal vesicle, 319
Germinative zone, of tapeworms, 424, *425*
Germovitellaria, of rotifers, 434, *435*
Gestation, in mammals, 679-680, *680, 681*
GH. *See* Growth hormone
Ghostfish. *See* Chimaeras
Giant axons, of earthworms, 483
Giant clam(s), 449, 459, *461*
Giant squid, 449, 465
Giardia, 371
Gibbon(s), *349,* 685
Gila monster, 629, *633*
Gill arches
- development of, 328, *334*
- embryology of, *326,* 327

Gill pores, of acorn worms, 557, *558*
Gill slits, of fish, 597
Gills, 199
- of bivalves, 461, *465*
- of cephalopods, 466, *467*
- of crustaceans, *499,* 502
- external, 200
- of fish, 597-598, *598*
- of freshwater worms, 484, *485*
- functions of, 200, *201*
- in gastropods, evolution of, 457, *457*
- of insects, 514
- internal, 200, 569, 574
- of molluscs, 451, *451*
- of polychaetes, 478, *479*
- of salamanders, 612-613
- of sea urchins, 551

Giraffe(s), social dominance in, 300, *300*
Girdle, of chitons, 454, *454, 455*
Gizzard, 229, 648-649
- of insects, 512

Gland(s), mammalian, 672-674
Gland cells, of cnidarians, *397,* 398
Glass lizards, 629, *633*
Glaucomys sabrinus, 679
Glial cells, 241-242
Gliding, 677, *679*
Gliricola porcelli, 524
Globin, 205
Globogerina, 372
Globulins, 189
Glochidium, of bivalves, 464, *466*
Glomerular filtration, 175-176
Glomerulus, 174, *175, 176*
Glottis, 202
Glucagon
- actions of, 282
- secretion of, 281, *281*

Glucocorticoid(s), 279
Gluconeogenesis, 279
Glucose, 8
- influx into cells, 33
- oxidation of, 47-49, *48*
- renal tubular reabsorption of, 176, *177*
- structure of, 8, *8*
- transport maximum for, 176

Glutamic acid, *10*
Glutathione, and feeding of hydras, 400
Glycera, 488
Glycine, *10*
Glycogen, 8
- in muscle contraction, 161

Glycogen granules, *25*
Glycolysis, 44-45, *45, 48*
- anaerobic, *49,* 49-50
  - in muscle contraction, 161, 163

Glycoprotein(s), *27,* 27-28
Gnathostomes (Gnathosomata), 565, 566, *568,* 569*t,* 577, 579, 583, *584,* 603
Gnathostomula jenneri, 429
Gnathostomulida, 352, 417, 429, *429*
Gnawers, 674, *675,* 686
GnRH. *See* Gonadotropin-releasing hormone
Goats, 686
Goatsuckers, 662
Goiter, 277, *278*
Golden plover, migration of, *656*
Golgi apparatus (Golgi complex), 18, *19,* 27, *27,* 28, *29*
Gonadotropin(s), 271, 283
Gonadotropin-releasing hormone, 274, 283
Gonads, 311
- of brittle stars, 549
- of sea stars, *546,* 548

Gonangia, 401
Gonionemus, 402
Gonium, 371
Gonopore(s), *399, 401*
Gorgonia, 409
Gorgonian(s), *407, 411*
Gorilla(s), 682, *683*
- taxonomy of, 347-348, 349, *349*

Gould, Stephen Jay, 101
Grade(s), taxonomic, 347, 348
Gradualism
- Darwin's theory of, 89
  - evidence for, 100-102, *102*
- phenotypic, 100-102, *102*
- phyletic, 101-102

Granulocytes, 190, 212
Grapsus, 507, 525
Grasshopper, *510,* 522
- nervous system of, *246*

Grassland, *128,* 131, *131*
Gray crescent, 324
Gray matter, 247, *247*
Grazers, 674, *675*
Great Lakes, lampreys in, 587
Grebe(s), 661
Green frog, 615, *618*
Green glands, of crustaceans, 502
Green tree frog, *616*
Green turtle, 628
Greenhouse effect, 129, *129*
Gregarina, 378
Gregarines, 375, 378
Greylag goose, egg-retrieval behavior of, 292, *292*
Grizzly bear, *687*
Gross productivity, 124
Ground finch, *663*
Ground substance, in connective tissue, 142
Grouse, 661
- mating system of, 657, *659*

Growth hormone, 271, 273*t*
Guard hair, 671, *672*
Guild(s), *120,* 120-121
Gulf shrimp, life cycle of, *504*
Gull(s), 662, *663*
- social behavior of, 298

Gut
- of acoelomates, 417, *417*
- primitive, *326,* 328, *328, 354, 357*

Gymnophiona, 609, 611, *613,* 618
Gypsy moth, 524
Gyrodactylus, 424, 427

H

Habitat, definition of, 114
Habituation
- *Aplysia* experiments, 296, *296*
- definition of, 296

Hadal pelagic layer, of oceans, 133, *133*
Hadrosaur, 631
Haeckel, Ernst, *89,* 98, 113, 351, *373*
Haecklia rubra, 411
Haementeria ghilianii, 485
Hagfishes, 565, 566, *566,* 579, 583-587, *584, 585, 586,* 603
Hair, 670*t,* 671-672, *672*
- functions of, 666
- structure of, 671

Hair cells
- in balance, 259, *260*
- of inner ear, *257,* 258

Haldane, J.B.S., 13
Hales, Stephen, 197
Haliotus rufescens, 458
Halteres, 510
Haplodiploidy, 312, 441
Haploid number, of chromosomes, 35, 58, 311
Hardy-Weinberg equilibrium, 105, 106, 107

Hare(s), 686
snowshoe
population fluctuations, 681–682, *682*
seasonal coat color variation in, 671, *673*
shock disease, 682
Harrison, Ross G., 327
Harvestmen, 496, *497*, 525
Harvey, William, 187, 308
Hasler, A. D., 599
Haversian system, 152, *152*
Hawk(s), 661
eye of, *652*
visual acuity of, 652
wings of, 653, *654*
Head
of crustaceans, 498–499, *499*
of insects, 510, *510*
of rotifers, 434, *435*
Head-foot, of molluscs, *451*, 451–452
Hearing, 256–259
avian, 651
Heart, 187, 193
amphibian, 193, *194*
of ascidians, 571, *571*
auxiliary, in arthropods, 192, *193*
of bivalves, 461, *465*
of cephalopods, *467*, 468
control of, 194–195, *196*
of crocodilians, 626, *627*
embryology of, 329
excitation of, 194–195
mammalian, 194–195, *195*
of molluscs, *451*, 452
myogenic, 195
neurogenic, 195
reptilian, 626
Heart rate, 194
Heart urchins, 543, *551*
body plan of, 551, *551*
characteristics of, 550–551, *551*, *552*
Heart valve(s), 194, *195*
Heartbeat, avian, 649
Heartworm, 442, *443*
Heat
exchange
countercurrent, in leg of arctic wolf, 182, *183*
with environment, by endotherms, *181*, 181–182
peripheral system for, 182
loss, by endotherms, *181*, 181–182
production, by endotherms, 181–182
Hedgehog(s), 685
armor of, 671–672
Helicobacter pylori, 230
Helicoplacoidea, *556*
Heliothis zea, *521*
Helix, 11, *12*, 472
Heloderma suspectum, *633*
Hemal system, of sea stars, 548
Heme, 205
Hemerythrin, 205
Hemichordates (Hemichordata), 353, *354*, 557–558, *567*
biological contributions of, 557
phylogeny, 558, *559*
position in animal kingdom, 557
Hemiptera, 516, 522, 526
Hemocoel, 192, *192*
Hemocyanin, 205
Hemoglobin, 189, 190, 205
in earthworms, 482
oxygen-carrying capacity, 205
sickle cell, 206
structure of, 11
Hemoglobin saturation curves, 205, *206*
Hemolymph, 188, 192, 212
Hemolytic disease of the newborn, 219–220
Hemophilia, 191
Hemostasis, 190–191
Hench, P. S., 279
Hennig, Willi, 348, *350*
Herbivores, 123, 223, 225–226
mammalian, 674–676, *675*
origin of, 20
Herbivory, 119
Heredity, 6, 56, 57
Hermaphroditism, 312
Hermit crab(s), 506, *507*
with cnidarian mutuals, 394, *395*
Hermodice carunculata, *479*
Herons, 661
Herrerasaurus, 630, *631*
Herring, feeding of, *224*
Hesperonoe adventor, *479*
Heterochrony, 98
Heterodont dentition, 674
Heteroptera, *521*, 522
Heterostracans, 576, *577*
Heterotrophs, 16–17, *17*, 20, 39, 223
primary, 17
protozoan, 366
Heterozygote, 61
Hexacorallia, 404, 409
Hexactinellida, 382, *387*, 390
characteristics of, *387*, 388
Hexamerous body plan, 404
Hibernation, 183–184, *184*
of frogs, 616–617
Hierarchical classification, 343
Hierarchy, of developmental decisions, 318
High-energy bond(s), 42, *42*
Hindbrain, 248, 249, *249*
Hindgut, of insects, 512, *512*
Hinge ligament, of bivalves, 460
Hippopotamus, 686, *688*
Hirudinea, 475, *485*, 485–486, *486*, *487*, 488
Hirudo medicinalis, 485, *486*, 488
Histology, 141
Histones, 28
HIV. *See* Human immunodeficiency virus
Holocephali, *585*, 590–591, *591*, 603
Holothuria difficilis, *553*
Holothuroids (Holothuroidea), 543, 551–553, *552*, *553*, 555, *556*
larvae of, *550*
Holozoic feeders, protozoan, *365*, 366, *367*
Homalozoa, *556*
Homarus, *502*, 525
Home range(s), of mammals, 301–302
Homeobox genes, 332, *332*
Homeodomain, 332, *333*
Homeostasis, 166–186. *See also* Osmoregulation; Water balance
origin of concept, 166
Homeothermic, 180
Homeothermy
in cold environments, 182–183, *183*
in hot environments, 182, *182*
Homeotic genes, *331*, 331–333
Hominidae, taxonomy of, 347–348, 349, *349*
Homo. *See also* Human(s)
taxonomy of, 347–348, 349, *349*
Homo erectus, 684, *684*
Homo habilis, 684, *684*
Homo sapiens, 684–685
Homodont dentition, 674
Homologous chromosomes, 58
Homologous structures, 570
Homology, 95–98, *96*
definition of, 345
nested hierarchical patterns of, 97–98
Homoplasy, 345
Homoptera, *517*, *521*, 523, *524*, 526
Homozygote, 61
Honeybee(s)
hygienic behavior in, genetics of, *294*, 294–295
language of, 303, *303*, 304
learning and memory in, 518
mouthparts of, 513
round dance of, 303
social behavior of, 519, *520*
waggle dance of, 303, *303*, 304
Honeyguides, 662
Hooke, Robert, 24
Hookworm, 440, *440*, 440*t*
Hormonal communication, 241
Hormone(s)
actions of, 210
of adrenal cortex, *279*, 279–280, *280*
of adrenal medulla, 280–281
definition of, 267
discovery of, 266
gastrointestinal, 282, *283*
of human birth, *285*, 285–286
of human pregnancy, *285*, 285–286
of invertebrates, 269–270
mechanisms of action, 267–268, *268*
of metabolism, 276–282
neurosecretory, 269
nonendocrine, 275–276
pituitary, 273*t*
secretion of
negative feedback systems for, 268, *269*
regulation of, 268
tropic, 271
of vertebrate reproduction, 282–286
of vertebrates, 270–282
Hormone receptor(s), 267, *268*
membrane-bound, 267–268, *268*
nuclear, 267, 268, *268*
Horn(s), 672
true, 672
Hornbill(s), 662
Hornworms, 515, *516*
Horse(s), 686, *688*
brain of, *248*
evolution of, 94, *95*
limbs of, 155
posture of, and locomotion, 155, *156*
Horsefly, *511*, 513
Horsehair worms, 442
Horseshoe crab, 493, *493*, 525
Housefly
disease transmission by, 521
mouthparts of, *513*
Human(s), 685. *See also Homo*
birth, hormones of, *285*, 285–286
brain of, 248
communication with other animals, 304–305
ears of, 256–258, *257*
embryo, early development of, 336, *337*
evolution of, 682–685
female reproductive system, 318, *318*
kidneys, 174, *175*
male reproductive system, 316–318, *317*
menstrual cycle of, 284, *284*
olfactory epithelium, 254, *255*
placenta of, 336
pregnancy, hormones of, *285*, 285–286
skin, structure of, *672*
swallowing in, *229*, *230*
urinary system, 174, *175*
ventilatory function in, 203, *203*
water balance in, 170, 171*t*
Human chorionic gonadotropin, *286*
in birth, 286, *286*
in pregnancy, 285, *285*
Human immunodeficiency virus, 218
Human placental lactogen, 285
Human population
growth of, 117–118
size of
agriculture and, 117
worldwide, 117, *118*
Hummingbird(s), 662
daily torpor in, 183, *184*
Hutchinson, G. E., *The Ecological Theater and the Evolutionary Play*, 112

Huxley, A. F., 159
Huxley, H. E., 159
Huxley, Thomas Henry, 89, 350, 642
Hyaline cap, 365, *365*
Hyaline cartilage, 151, *152*
Hyalonema, 390
Hyalospongiae, 390
Hybridoma, 214
Hydatid
  multilocular, 426*t*
  unilocular, 426*t*
Hydra(s), 394, *397,* 399–400, *400,* 409
  budding of, *309*
  reproduction of, *309*
Hydractinia milleri, 395
Hydranths, 400
Hydrocephalus, 248
Hydrochloric acid, secretion of, 230
Hydrocoel, of echinoderms, 548, *549*
Hydrocorals, 402, *403*
Hydrogen bonds, 694, *695*
Hydrogen ion concentration, 696
Hydrogen, isotopes of, 692, *692*
Hydroids, 394
  colonies of, 400–402, *401*
  medusae, 401, *401*
Hydrolase(s), 227–228
Hydrolysis, 8, 15, 227–228
Hydrolytic reaction, 8, *9*
Hydrophiidae, 634
Hydrorhiza, *395*
Hydrosphere, 127
Hydrostatic pressure, 33
  and nematocyst discharge, 396
  in nematodes, 437–438
Hydrozoa, 394, 395
  calcareous skeletons of, 402, *403*
  characteristics of, *399,* 399–402
  phylogeny of, *413*
Hyla cinerea, 616, 618
Hylidae, *616*
Hylobatidae, *349*
Hyman, Libbie Henrietta, 437, 542
Hymen, 318
Hymenolepis, 427
Hymenoptera, 511, *520,* 523
Hyperglycemia, 176
Hyperosmotic regulation, 167–168
Hyperparasitism, 512
Hypodermis, 147
  of nematodes, 438, *439*
  structure of, *672*
Hypoosmotic regulation, 169
Hypopharynx, of insects, 510, *510*
Hypophysis, 270–271
Hypostome, of hydras, 399
Hypothalamus, *249,* 250, 271, *271*
  and neurosecretion, 272,
    272, 273
Hypothetico-deductive method, 3–4

I

Ibises, 661
Ichthyostega, 607, *608,* 609, *610*
ICSH. *See* Luteinizing hormone
IFN-γ. *See* Interferon-γ
Iguanas, 629, *632*
Immune response
  cells of, 215
    lineages of, *212*
  cellular, 213
  humoral, 213
    generation of, *216,*
      216–218, *217*
Immunity, 210–221
  acquired, 211
    in vertebrates, 213–219
  cell signaling in, 210
  definition of, 211
  innate, 211–213
    in vertebrates, 212–213
  in invertebrates, 212, *213*
Immunoglobulin(s), 214
  classes of, 214, *214*
  molecular structure of, 214, *214*
Immunosuppression, 218
Implantation, 285, 336
  delayed, 680
Imprinting, 296–297, *297*
Inbreeding, 106–107
Incisors, 225, *226,* 674
Incomplete dominance, 62–63
Incus, *257,* 258
Individual distance, 301
Induction, embryonic, 319,
  330–331, *331*
Industrial melanism, 102, *103*
Inflammation, 218–219
Infraciliature, of ciliates, 376, *376*
Infundibulum, 271, *271*
Inheritances, chromosomal basis of,
  57–60
Inland waters, 132
Innate behavior, 293–294
Inner cell mass, *322,* 323
Insect(s) (Insecta), 508–526
  adaptability of, 509
  behavior, 518–520
  beneficial, 520–521, *521*
  brain hormone of, 269, *270*
  circulatory system in, 192, *193*
  classification of, 522–523
  communication, 518–520
  control of, 521–526
  distribution of, 509
  excretion in, 514
  external structure of, 508–509,
    510, *510, 511*
  flight muscles of, 158,
    510–512, *511*
  flight speeds of, 512
  gas exchange in, 513–514, *514*
  growth of, 515–518, *516*
  guilds of, 121
  harmful, 521, *521, 522*
  homeobox genes, 332, *332*
  and human welfare, 520–526
  immunity in, 212, 213*t*
  internal structure of, *512,*
    512–515, *513, 514*
  learning and memory in, 518
  Malpighian tubules, 172, *173*
  metamorphosis, 269, 515–518,
    516, 517
  molting in, 269, *270*
  mouthparts of, 512–513, *513*
  nervous system of, *246, 512,* 514
  numbers of species of, 508, *509*
  nutrition, *512,* 512–513
  parasitic, 512
  phylogeny, 526, *527*
  predaceous, 512, 521, *521*
  reproduction, *510,* 515, *516*
  reproductive systems of,
    315–316, *317*
  sense organs of, 514–515
  social behavior of, 519–520
  tagmata of, 508
  tracheal system of, 200, *200*
  water balance in, 514
  wings of, 508–509, *510,* 510–512,
    511, 516, 518
Insecticides, 523–524
Insectivora, 685
Insectivores, 676
  mammalian, 674, *675,* 676
Inspiration, 203, *203*
Instar(s), 515–516, *516*
Instinct theory, 294
Insulin, 176
  actions of, 281–282
  first extraction of, 282
  radioimmunoassay for, 275
  recombinant, 281
  secretion of, *281,* 281–282
Integrated pest management, 526
Integument, 145–150
  amphibian, *148*
  definition of, 419
  functions of, 145–147
  human, *148*
  invertebrate, 147–149, *148*
  vertebrate, *148,* 149–150
Intercellular, definition of, 141
Intercostal muscles, external, 203
Interferon-γ, 216, *216*
Interleukin(s)
  IL-1, 215, 216, *216, 217*
  IL-2, 215, *216*
  IL-3, 215, *216*
  IL-4, 215, *216,* 217, *217*
  IL-5, 216, *216,* 217, *217*
  IL-6, 216, *216,* 217, *217*
  IL-8, 216, *216*
  IL-10, 216, *216*
Intermediate filaments, 30
Intermediate host, of
  trematodes, 421
Intermediate inheritance, 62–63, *63*
Intermediate neurons, of retina,
  261, *262*
Internal nares, 202
Interneurons, 241
Interphase, *36,* 36–38
Interstitial cells, 283
  of cnidarians, *397,* 398
Interstitial (intercellular) fluid,
  140–141, 188, *189*
  hydrostatic pressure, 198
  osmotic pressure, 198
Interstitial tissue, 316
Interstitial-cell-stimulating hormone.
  See Luteinizing hormone
Intertidal zone, 132, *133*
Intestinal fluke, 423*t*
Intestine(s), *229,* 231–234
  of flatworms, 419, *419, 420*
  of insects, 512, *512*
  of nematodes, 438, *439*
Intracellular, definition of, 141
Intracellular space, 140
Intrinsic rate of increase (r),
  115–116, *116*
Introns, 69, 73–75
Introvert, of sipunculans, 532, *532*
Invagination(s), 199–200, 324
Inversion, chromosomal, 69
Invertebrates
  cartilage in, 151
  excretory structures of,
    171–172
  hemolymph of, 188
  hormones of, 269–270
  immunity in, 212, *213*
  integument of, 147–149, *148*
  leukocytes in, 212, 213*t*
  muscle, types of, 157–158
  nervous systems of,
    244–246, *246*
    evolution of, 244–246
  reproduction, 313
  reproductive systems of,
    315–316, *317*
Iodine
  deficiency, 277, *278*
  in thyroid, 276
Ionic bonds, 692–693, *694*
Iris, 261, *261*
Irritability, 6, 241
Islets of Langerhans, 281, *281*
Isopoda, 505, *506*
Isoptera, 522, 526
Isotopes, 692, *692*
Ixodes, 498

J

Jacobson's organ, 627, 633
Japanese beetles, *521*
Japanese macaques, acquisition and
  passing-on of tradition in,
  298–299, *299*
Japygids, 522
Jarvik, Erik, 609
Jaw(s)
  of brittle stars, 549, *550*
  of cephalopods, *467,* 468
  reptilian, 625
  of turtles, 627
Jaw worms, 417
Jellyfish, 392, 394, 395
Johnson, Ben, 280
Julus, 525
Jumping spider, *494,* 495, *495*
Juvenile hormone, 269, 270, *270*

K

- K (carrying capacity)
  - of environment, *116,* 116-117, *118*
  - for human species, 117-118, *118*
- Kandel, E. R., *Aplysia* experiments, 295-296
- Kangaroo, 685
  - reproduction in, 335
- Kangaroo rat, water balance in, 170, 171*t*
- Keeton, W.T., 656
- Keratin, 149, 671
- Keratinization, 149
- Keystone species, 121-122, *123*
- Kidney(s)
  - arthropod, 172, *173*
  - avian, 649-650
  - of bivalves, 461, *465*
  - filtrate, 174, 175-176
  - filtration function, 174, 175-176
  - mammalian, osmotic concentration of tissue fluid in, 178, *179*
  - of molluscs, *451,* 452
  - secretory function, 174, 177
  - tubular reabsorption in, 174, 176-177
  - vertebrate, 173-179
    - ancestral, 173
    - embryology of, 173, *174*
    - function of, 173-175
- Kinetochore, 35, *36,* 37
- Kinetosome(s), 31, 364
  - of ciliates, 376, *376*
- King, T. J., 330
- King cobra, 635
- Kingdom(s), 344*t,* 351
- Kingfishers, 662
- Kinorhyncha, 353, 434, 436, *436, 445,* 446
- Kiwi(s), 661
- Knee-jerk reflex, 247, *247*
- Koala, *17,* 685
- Kraits, 634
- Kramer, G., experiments with bird navigation, 656, *657*
- Krebs cycle, 44, 46, *46*
- Krebs, Hans, 44
- Krill, 132, 223, *224,* 506, *507*
- Kristensen, R. M., 444
- Kupffer cells, 212
- Kwashiorkor, 236

L

- Labia majora, 318, *318*
- Labia minora, 318, *318*
- Labium, of insects, 510, *510*
- Labor, in birth, 286
- Labrum, of insects, 510, *510*
- Labyrinth, 256, 259
- Labyrinthomorpha, 378
- Lacertilia, 637
- Lacewings, 523
- Lack, David, 119
- Lactase, 233
  - deficiency of, 233
- Lactation, duration of, 679-680, *681*
- Lacteals, *232,* 234
- Lactic acid, accumulation, in muscle contraction, 163
- Lactose, 8
  - intolerance, 233
- Lacunae, *145*
  - of bone, 152, *152*
- Lacunar system, of spiny-headed worm, 443, *443*
- Ladybird beetle, 521, *521*
- Lagena, 256
- Lagomorpha, 686
- Lake(s), lifespans of, 132
- Lake forest, 130
- Lamarck, Jean Baptiste de, 85, *85,* 98, 103
- Lamellae, of fish gills, 597, *598*
- Lamp shells, 538
- Lampreys, 565, 566, *566,* 579, 583, *584, 585, 586,* 603
  - brain of, *248*
  - feeding habits of, 587, *588*
  - life cycle of, 587, *587*
- *Lampsilis ovata, 466*
- Lancelet(s), *567,* 571-573, *573,* 579
- Lansteiner, Karl, 219
- Lappets, 402
- Lapwings, 662
- Large intestine, functions of, 234
- *Larus atricilla, 663*
- Larva
  - amphibian, *325,* 606, 617-618
  - of amphioxus, 572
  - ascidian, 571, *572*
  - of echinoderms, 549, *550*
  - of fish, *599,* 602
  - insect, 515, *516*
  - of lampreys, 587, *587*
  - of molluscs, 449, 452, *453*
  - of salamanders, 612, *614*
  - trochophore, 449, 452, 453, *453,* 464, 532
  - tunicate, Garstang's hypothesis of evolution for, 575, *576*
- Larvacea, 571
- Larynx, 202
- Lateral canals, of sea stars, 546, *546,* 547
- Lateral, definition of, 354
- Lateral force, in swimming fishes, 596
- Lateral line system, of fishes, 256, *256,* 589, *590*
- Laterite, 131
- *Latimeria,* 592
- *Latimeria chalumnae,* 594, *595,* 603
- *Latrodectus mactans,* 496, *497*
- Laughing gulls, *663*
- Law of independent assortment, Mendel's, 63-65
- Law of segregation, Mendel's, 61-63
- Lead poisoning, of waterfowl, 660
- Leafhoppers, 523
- Learning, and behavior, 295-297
- Lecithin, 10, *10*
- Leeches, 475, 476, *485,* 485-486, *486,* 488
  - medicinal uses for, 476, 485
- Leg(s)
  - of insects, 508-509, 510, *510, 511*
  - of tardigrades, 535, *535*
- *Leidyopsis,* feeding method of, *367*
- *Leishmania,* 378
- Lek, 657
- Lemmings, populations of, fluctuations in, 680-681, *681*
- Lemur(s), 685
- Lens, 261, *261*
- Lentic habitat, 132
- Leopard frog, 615-616, *619*
- *Lepas anatifera, 505*
- Lepidoptera, 523
- Lepidosaurs (Lepidosauria), 622, *623, 624,* 628, 637, *644*
- *Lepidosiren,* 594, 603
- *Lepisma, 522*
- *Lepisosteus,* 593, *594,* 603
- Leptin, 235
- Leptocephali, 598
- *Leptophis ahaetulla, 634*
- *Lepus americanus,* seasonal coat color variation in, *673*
- Leuconoids, 384, *384, 385*
- *Leucosolenia,* 383, *385, 387,* 388, 390
- Leukocyte(s), 189, 190, *190*
  - in invertebrates, 212, 213*t*
- LH. *See* Luteinizing hormone
- *Libellula pulchella, 517*
- Lice
  - biting, 522, *524*
  - disease transmission by, 521
  - sucking, 522, *524*
- Life
  - in Cambrian period, 17
  - chemistry of, 5, 6-11
  - definition of, 5-6
  - distribution of, on Earth, 127-133
  - earliest known, 2
  - first evidence for, age of, 361
  - major divisions of, 351-352
  - origin of, 2, 11-16
    - modern experimentation, 13-15, *14*
    - Oparin-Haldane hypothesis, 13
    - theories, historical perspective on, 12-13
  - in Precambrian period, 17-20
- Life cycles
  - of ciliates, 376, *377*
  - protozoan, *363,* 368
- Ligaments, *145*
- Ligands, in receptor-mediated endocytosis, 34, *35*
- Light microscopy, 24, *25*
- Light, properties of, 262
- *Ligia,* 505
- Limb buds, 329, *329*
- Limbic system, 250
- *Limifossor,* 472
- Limiting resource, 116, 119
- Limpet(s), 454, 459, *460*
- *Limulus,* 493, *493*
- *Linguatula, 534*
- *Lingula,* 538, *538*
- Linkage, genetic, 66, 68
- Linnaeus, Carolus, 343, *343*
  - classification system of, 343-344
- Linoleic acid, 9, *10*
- Lion(s)
  - food and feeding of, *677*
  - reproduction, *680*
- Lipase, pancreatic, 233
- Lipid(s), 9-10
  - dietary, requirements for, 235, 236
  - digestion of, 228
  - metabolism of, *50,* 50-51
- Lissamphibia, 609, *610*
- *Lithobius,* 525
- Lithosphere, 127
- Littoral zone, 132, *133*
- Liver, development of, 328, *328*
- Living systems
  - chemical uniqueness of, 5
  - complexity of, 5
  - development, 6
  - environmental interaction, 6, *7*
  - genetic program of, 6
  - hierarchical organization of, 5, *5*
  - metabolism, 6, *7*
  - origin of, 16-17
  - reproduction of, 5-6
- Lizard(s)
  - body temperature regulation in
    - by behavioral adjustments, 180, *181*
    - by metabolic adjustments, 180-181
  - characteristics of, 628-629, 637
  - environmental interaction, *7*
  - kinetic skull of, 628, *629*
  - unisexual, 312, 313, *313*
- Lobopodia, 365, *365, 372*
- Lobster(s), 506, *507,* 525
  - American, *502*
  - Norway, 444
- Locomotion. *See also* Movement
  - of ameba, 365, *365*
  - of annelids, 476-477
  - of bivalves, 460, *463, 464*
  - of cephalopods, 466, *467, 468*
  - cnidarian, 398
  - of cuttlefish, 466, *468*
  - of dinosaurs, 155
  - of earthworms, 150, *151,* 481
  - of eels, in water, 596, *597*
  - of fishes, 595-596, *596, 597*
  - of mammals, *688*
  - of *Nautilus,* 466
  - of nemerteans, 428
  - of octopus, 466
  - of planarians, 421
  - and posture, 155, *156*
  - reptilian, 626
  - of sea cucumbers, 553
  - of sea stars, 547
  - of squid, 466, *468*
- Locus, 59

Locust(s), 522
  African desert, 490, *510*
Logistic growth, *116*, 116–117, 117
Loligo, 465, 472
Longitudinal muscle(s)
  of annelids, *476*, 476–477, *477*
  gastrointestinal, *232*
  of nematodes, 438, *439*
Loon(s), 661
Loop of Henle, 174, *175*
  and urine concentrating mechanism, *178*, 178–179
Lophophorates, 531, 532, 535–539
  phylogeny, 539, *559*
Lophophore, 531, 535
  of brachiopods, 538–539, *539*
  of ectoprocts, 536, 537, *537*
Lorenz, Konrad, 291, *291*, 292
  studies of imprinting, *291*, 296–297
Lorica, 436, *436*
Loricifera, 353, 434, 436, *436*, *445*, 446
Lorises, 685
Lotic habitat, 132
Loudness, of sound, 259
Lovebirds, nest-building behavior, effects of crossbreeding on, 295, *295*
Loxosceles reclusa, 496, *497*
Loxosomella, *444*
Lucy, 683, *683*
Lugworms, 475, 478, 479
Lumbricus, 488
  dorsal median giant fiber of, 483
Lumbricus terrestris, 480, *482*
Lumen, of alimentary canal, 227
Lung(s), 193, *193*, 199, 200–202, *201*
  amphibian, 200, 201, *201*
  avian, *201*
  development of, 328, *328*
  gas exchange in, 204
  human, 201, *202*
    surface area of, 201
  mammalian, 201, *201*, *202*
  reptile, 201, *201*
  reptilian, 626
Lung flukes, 423*t*
Lungfishes, 583, *584*, *585*, 591, 593–594, 603, *610*
  lung of, 200–201
Luteinizing hormone, 271, 273*t*, 283, 284
  in menstrual cycle, 284, *284*
Lyell, Charles, 85, *85*, 100
Lygiosquilla, 501
Lyme disease, 498
Lymph, 143, 198
Lymph capillaries, 198
Lymph nodes, 198–199, *199*
Lymphatic system, 193, 198–199, *199*
Lymphocyte(s), 190, *190*, 215
  activated, 215
Lymphokine-activated killer cells, 215
Lynx, population fluctuations, with hare population, *682*
Lysosomes, *19*, 29, 211, *211*
Lysozyme, activity of, 41, *41*
Lytechinus, 555

M

Macaque monkeys, acquisition and passing-on of tradition in, 298–299, *299*
MacArthur, Robert, 120–121
Macleod, J.J.R., 282
Macrobdella, 488
Macrocheira kaempferi, 491
Macroevolution, 103–104, 107–109
Macronucleus
  ciliate, 375, *375*
  of *Paramecium*, *377*
Macrophages, 190, 212, *212*
Madreporite
  of sea stars, 545, *545*, *546*
  of sea urchins, 551, *552*
Magnesium, in body fluids, 188, *189*
Magnetite, in birds, 656
Major histocompatibility complex, 213–214
Malacostraca, 505–506, 525
  structure of, *499*
Malaria, 374, 521
Malleus, *257*, 258
Mallophaga, 522, *524*, 526
Malnourishment, 236, 237, *237*
Malpighi, Marcello, 197
Malpighian tubules, 172, *173*
  in insects, *512*, 514
  of spiders, 494, *494*
Maltase, 233
Malthus, T. R., 87–88, 90
Maltose, 228
  formation of, 8, *9*
Mambas, 634
Mammal(s) (Mammalia), *567*, 579, 666–680, 685–686
  adaptability of, 667
  adaptations of, 666, 667, 671–680
  adaptive hypothermia in, 183–184, *184*
  carnivorous, *675*, 676, *677*, 686
  characteristics of, 666, 667, 670*t*
  circulatory system of, *193*, 193–194
  classification of, 671, 685–686
  dentition of, 225, *226*
  echolocation, 677–678, *679*
  estrous cycle, 283
  evolution of, 622, *623*, *624*, 667–671
  feeding, 225, 674–676
  feeding specializations in, 674–676, *675*
  flight, 677–679
  frugivorous, *675*
  gastrulation in, *322*, 325
  geographic distribution of, 667
  glands of, 672–674
  gnawing, 674, *675*, 686
  herbivorous, 674–676, *675*
  home ranges of, 301–302
  hoofed
    even-toed, 686, *688*
    odd-toed, 686, *688*
  horns and antlers of, 672, *674*
  insectivorous, 674, *675*, 676
  integument of, 670*t*, 671–674
  locomotion, *688*
  lungs of, 201, *201*, *202*
  menstrual cycle, 283
  migration of, 676–677, *678*
  omnivorous, *675*, 676, *677*
  origin of, 667
  piscivorous, *675*
  placenta, 335–336
  populations of, 680–682, *681*
  proboscis, 686
  reproduction, 679–680, *680*
  reproductive cycles, hormonal control of, 283
  respiratory system of, 202–207
  teeth of, 670*t*, 674, *675*
  territoriality in, 301
Mammary glands, 673–674
Mandibles
  of crustaceans, 498, *499*
  of insects, 510, *510*
  of myriapods and insects, 508
Mandibulata, 498, 526, *527*
Mandibulate hypothesis, 526, *527*
Mangold, Hilde, 330, 331
Mantle
  of ascidians, 571, *571*
  of bivalves, 460, 461, *462*, *463*, *464*, *465*
  of molluscs, 450, 451, *451*, 452, *452*
Mantle cavity, of molluscs, 451, *451*, 452
Manubrium, of hydroids, 401
Marasmus, 236
Margulis, L., 20, 31
Marine fan worms, feeding of, *224*
Marine worms, 475
  echiuran, 532, *533*
  sipunculan, 532, *532*
Marmoset, feeding specializations of, *675*
Marmota monax, hibernation in, *184*
Marsupials, 671, 685
  evolution of, *668*, *669*
  reproduction, 335, 679, *680*, *681*
Marsupium, 685
Mass extinctions, *108*, 108–109, *109*
Mast cells, 215
Mastax, of rotifers, 434, *435*
Mastication, 225
Mastigophora, 369–371, 377, 378
Mating
  nonrandom, 106–107
  positive assortative, 106
Mating types, 376
Maxillae
  of crustaceans, 498, *499*
  of insects, 510, *510*
  of myriapods and insects, 508, *508*
Maxillary glands, of crustaceans, 502
Maxillipeds, of crustaceans, 499, *499*, *500*
Maxillopoda, characteristics of, 504–505, *505*, 525
Mayflies, 522, *523*
Mayr, Ernst, 88, *88*, 350
McClung, C., 59
Mealybugs, *521*, 523
Mechanoreception, 254–263
Medial, definition of, 354
Mediated transport, 33–34
Medulla, 249, *249*
Medusae, 394, 395, *396*
Megacytiphanes, 525
Meganyctiphanes, *507*
Megaptera novaeangliae, *687*
Meiosis, 35, 58–59, *59*, 310–311, *311*
  crossing over during, 66, *68*
  gametic, 368, *370*
  intermediary, 368, *370*
  in oogenesis, 314, *316*, 319
  in spermatogenesis, 314, *315*
  zygotic, 368, *370*
Meiotic divisions, 35
Meiotic parthenogenesis, 312
Melanin, 150
Melanophore-stimulating hormone, 271–272, 273*t*
Melatonin, 274–275
Membrane-bound receptor(s), 267–268, *268*
Membranelles, of ciliates, 376
Membranipora, *537*
Memory cells, 217, *217*
Memory, neural substrate for, 250
Mendel, Gregor Johann, 57, *58*
  law of independent assortment, 63–65, 68
  law of segregation, 61–63
  laws of genetics, 61–69
Meninges, 247, *247*
Menstrual cycle, 283, 284, *284*
Menstruation, 284
Meoma, *551*, 555
Mermaid's purse, 590, *591*
Merostomata, 493, 525
Merozoites, of apicomplexans, 373
Mesencephalon, 248, *249*
Mesentery(ies), *232*, 448
  of annelids, 476, *476*
Mesocoel, of lophophorates, 535
Mesoderm, *355*, *357*, 358
  of acoelomates, 417, *417*
  allantoic, mammalian, *336*
  derivatives of, *326*, 328–329, *329*
  development of, *322*, 324, 325, *354*
Mesoglea, of cnidarians, 395, 397, *397*, 398
Mesohyl, 384, *386*
Mesonephros, 173, *174*
Mesopelagic layer, of oceans, 133, *133*
Mesosome, of lophophorates, 536

Mesothorax, of insects, 510, *510*
Mesotocin, 273*t*
Mesozoa, 352, *355,* 382
Messenger RNA
  cap, 73, *74*
  poly-A tail, 73, *74*
  in transcription, 72-75
Metabolic water, 170
Metabolism, 6
  aerobic versus anaerobic, 43-44
  cellular, 39-52
  enzyme regulation in, 52, *52*
  hormones of, 276-282
  origin of, 16-17
  oxidative (aerobic), 17
  reducing (anaerobic), 15, 17
Metacercariae, of flukes, 421, *422*
Metacoel, of lophophorates, 535
Metamerism, 358, *358*
  definition of, 474
  evolution of, 474
Metamorphosis
  amphibian, 609
  in arthropods, 492
  ascidian, 571, *572*
  in eels, *599,* 602
  in fishes, *599,* 602
  in frogs, 276, *277*
  of insects, 269, 515-518, *516, 517*
    hemimetabolous (gradual, incomplete), 515, *517,* 526
    holometabolous (complete), 515, *516,* 526
  of salamanders, 612, *614*
  in sea stars, 548, *549*
Metanephridium (pl., metanephridia), 172, *172*
  of molluscs, 452
Metanephros, 173, *174*
Metaphase, *36,* 37
Metaphasic plate, *36,* 37
Metasome, of lophophorates, 536
Metatherians, 685
Metathorax, of insects, 510, *510*
Metazoa, 140, 352, 382
  body of, extracellular components of, 140-141
  reproduction of, 311
Methane, molecular structure of, 694, *695*
Metridium, 409
MHC. *See* Major histocompatibility complex
Michaelis, Leonor, 41
Microevolution, 103-107
Microfilaments, 29
Microglia, 242
Microglial cells, 212
Micrometer, 24
Micronemes, of apicomplexans, 373, *373*
Micronucleus
  ciliate, 375-376
  of *Paramecium, 377*
Microscope(s)
  electron, 24
  historical perspective on, 24
  light, 24
Microspora, 378
Microsporidia, 352
Microstomum, 420, *421,* 427
Microtubule(s), 30, *30, 36,* 37
  of cilia and flagella, 157, *157,* 363, *364*
  in eukaryotes, 20
Microvilli, 31, *31*
  intestinal, 231, *232*
  of tapeworms, 424, *425*
Mictic egg(s), 435
Midbrain, 248, *249,* 249-250
  avian, 650, 651
Midgut, of insects, 512
Migration
  avian, 640
    direction finding in, 655-656, *657*
    routes, 654, *656*
    stimulus for, 654-655
  of eels, 598, *599*
  effects on genetic equilibrium, 107
  of fish, 598-600
  and genetic drift and selection, interactions of, 107
  mammalian, 676-677, *678*
  of salmon, 599-600, *600, 601*
Milieu intérieur, 189
Millepora, 403
Miller, Stanley, experiment on origin of life, 13-15, *14*
Millipedes, 508, *509,* 525
Mimic(s), 121, *122*
Mineral(s), requirements for, 235, 235*t*
Mineral salts, requirements for, 235
Mineralocorticoid(s), synthesis of, 280
Minkowski, O., 282
Miracidium, of flukes, 421, *422*
Miscarriage, 286
Misumenoides, 495
Mites, 496-498, *498,* 525
Mitochondria, 18, 19, *19, 25,* 27, *27,* 29, *30,* 77
  cristae, 29, *30*
  DNA, 20
  functions of, 29
  genome of, 29
  origin of, 361
  structure of, 29, *30*
Mitosis, 19, 35, 311
  in oogenesis, 314, *316*
  in spermatogenesis, 314, *315*
  stages in, *36,* 36-37
Mnemiopsis, 412, *412*
Moas, 645
Model(s), for mimicry, 121, *122*
Modular organisms, 115
Molar teeth, 225, *226,* 674
Molecular sequencing, 18
Molecules, 691
Moles, 685
Molluscs (Mollusca), 353, *354,* 448-473
  adaptive radiation, 469-470
  adductor muscles of, 157-158
  biological contributions of, 449
  bivalve, feeding of, *224*
  body plan of, *451,* 451-452
  characteristics of, 450
  cladogram of, *471*
  classification of, 472
  diversity of, 449, *450*
  ecological relationships of, 449
  economic importance of, 449-450
  immunity in, 212, 213*t*
  internal structure and function of, 452, *453*
  nervous system of, 246, 452
  number of species of, 449
  phylogeny, 469-470, *470, 471*
  position in animal kingdom, 449
Molt(s), mammalian, 671
Molting
  of arthropods, 492
  of feathers, 645
Molting hormone, 269, *270,* 500
Molt-inhibiting hormone, 500
Monanchora unguifera, 389
Monarch butterfly
  metamorphosis, *516*
  migration of, 512
Monera, 18, 351, 352, *353*
Monkey(s), 685
  acquisition and passing-on of tradition in, 298-299, *299*
  New World, 682, *683*
  Old World, 682, *683*
Monoclonal antibodies, 214
Monocystis, 378
Monocytes, 190, *190,* 211, *212*
Monoecious organisms, 312, 387, 420, 424, 536
Monogamy
  in birds, 657
  definition of, 657
Monogenea, 418, 423-424, *424,* 427, *430*
Monoglyceride, 9
Mononuclear phagocyte system, 211-212
Monophyly, 346, 347, *347,* 565
Monoplacophora (monoplacophorans), 449, 453, *454,* 469, *470, 471,* 472
Monosaccharides, 8
Monosomy, 68-69
Monotremes, 671, 685
  evolution of, *668, 669*
  reproduction, 335, 679
Montastrea cavernosa, 410
Moon snail, *458*
Mopalia muscosa, 454, 472
Morgan, Thomas Hunt, 57, 66
Morphogenesis, in hydras, 399
Morphogens, 399
Morphological determinants, 319
Morris, S. Conway, 444
Mosquito
  bloodsucking by, 226
  disease transmission by, 442, 521
  mouthparts of, 513, *513*
Moth(s), 523
  Batesian mimicry in, *122*
  chemical sex attraction in, 302, *302*
  ears of, 256, *257*
  industrial melanism in, 102, *103*
  molting in, 269, *270*
  mouthparts of, 513
  reproduction, 515
Motor cortex, 250, *250, 251*
Motor neurons, 160, 241, *241,* 246
Motor unit, 160
  recruitment, 160
Mouse
  embryonic cleavage in, *322*
  homeobox genes, *332,* 332-333
  reproduction in, 680, 681, *681*
Mousebirds, 662
Mouth, *354*
  of flatworms, 419, *419, 420*
  of hydras, 399
  of insects, 512, *512*
  of nematodes, 438, *439*
  of sea stars, *546,* 547
Movement. *See also* Locomotion
  ameboid, 156, 364-365
  of animals, 156-163
  ciliary, 156-157, 364
  flagellar, 156-157, 364
  muscular, 157-163
mRNA. *See* Messenger RNA
MSH. *See* Melanophore-stimulating hormone
Mucin, 230
Mucosa, gastrointestinal, *232*
  cells of, 232, *232*
Mud puppy, 613, *615*
Multicellular organism(s), *355*
  advent of, 381
  classification of, 382
Multinucleate cell, 36
Multiple fission, *309,* 310, 367
Muscle(s)
  embryology of, 328-329, *329*
  invertebrate, types of, 157-158
Muscle fiber(s), 144, 157
  of flatworms, 418, *419*
Muscular hydrostats, 150, *151*
Muscular system, avian, 646-648, *648*
Muscular tissue, *142,* 144, *146*
Musk-oxen, social behavior of, 297-298
Mussels, 459, 461, *464*
Mutagens, 80
Mutation(s), 56, 66, 79
  chromosomal, 68
  frequency of, 79
  and reproduction, 310
  sporting, 101, *102*
Mutualism, 119, *119*
Mycale laevis, 384
Myelin, 242, *242*
Myenia, 390

Myliobatis, 591
Myocardial infarction, 195, 196
Myocytes, of sponges, 384
Myofibrils, 144, *158,* 159
Myofilaments, *158,* 159, *159*
attach-pull-release cycle in, 160–161, *162*
interaction during muscle contraction, 159, *160*
thick, *158,* 159, *159*
thin, *158,* 159, *159*
Myogenic heart, 195
Myomeres, 595–596, *596*
Myoneural junction, 160, *161*
Myosin, 29–30, 156
in excitation-contraction coupling, 160–161, *162*
myofilaments, *158,* 159, *159*
Myotis lucifugus, echolocation, *679*
Myriapod(s), 508, 525, *527*
Mytilus edulis, 464, 472
Myxine glutinosa, 583, *586*
Myxini, 579, 583–587, *585, 586,* 603
Myxozoa, 378

N

Nacreous layer, of molluscs, 452, *452*
NAD. *See* Nicotinamide adenine dinucleotide
Naja naja, 636
Nanoloricus mysticus, 436
Nanometer, 24
Nasal chamber, 202
Natural killer (NK) cells, 215
Natural selection
Darwin's theory of, 89–91, *90*
effects on genetic equilibrium, 107
evidence for, 102–103, *103*
and genetic drift, interactions of, 107
and primitive life, 16, 17
theory of, 3–4
Nauplius, 503, *504*
Nautaloids, 465
Nautilus (nautiluses), 449, 464, 465, 466, *467,* 472
external features of, 466, *467*
locomotion, 466
Necator americanus, 440, *440,* 440*t*
Necturus, 613, *615*
Negative feedback systems, 268, *269*
Nekton, 132
Nemaster, 555
Nematocysts, 392, 393, 395–396, *397*
of ctenophores, 411
Nematodes (Nematoda), 353, 434, 437–442, *445,* 446
abundance of, 433, 437
body plan of, *355*
characteristics of, 437
form and function of, 437–438
male, copulatory spicules of, 438, *439*
parasitic, 438–439, 440*t*
reproduction, 438, *439*
Nematodinium, eyespot of, 260, *260*
Nematomorpha, 353, 434, 442, *445,* 446
Nemertea, 352, *354,* 417, 427–429
body plan of, *355,* 427–428, *428*
characteristics of, 427–429, *428*
embryonic cleavage in, *322*
form and function of, *428,* 428–429
gastrulation in, *322,* 325
Nemertines, 453
immunity in, 212, 213*t*
Neoceratodus, 594, *595,* 603
Neocortex, 250
Neo-Darwinism, 103
Neomenia, 472
Neopilina, 453, *454,* 469, 472
Neoplastic growth, 79
Neopterygians (Neopterygii), 592–593, 603
Neornithes, *644,* 645, 661
Nephridiopore, of earthworms, 482, *483*
Nephridium (pl., nephridia), 171–172, *172*
of earthworms, 482, *482, 483*
of gastropods, 457
Nephron(s), 174, *175*
avian, 649–650
Nephrops norvegicus, 444
Nephrostome, 172
of earthworms, 482, *483*
Nereis, 478, 488
feeding of, 225
Nereis virens, 477
Neritic zone, 132, *133*
Nerve(s), 241, *242*
growth of, embryonic, *326,* 326–328, *328*
motor, embryology of, *326,* 327
sensory, embryology of, *326,* 327
Nerve cell(s)
of cnidarians, *397,* 398
resting membrane potential of, 242–243, *243*
Nerve cord
of amphioxus, 572, *573*
of nematodes, 438, *439*
single dorsal, 566, 569
Nerve fiber, *147*
Nerve ganglion, of ascidians, 571, *571*
Nerve impulse, 242–244
transmission across synapses, 244, *245*
Nerve net, 139*t,* 245–246, *246*
of cnidarians, 396–397
Nerve plexus, subepidermal, of flatworms, 420
Nervous system, 240–265
of annelids, 246, *246*
of arthropods, 246, *246*
autonomic, *251,* 251–252, *252*
avian, 650–652
of crustaceans, 502, *503*
of earthworms, *246*
embryology of, *326,* 326–328, *327, 328*
evolution of, 244–253
of flatworms, 246, *246,* 420, *420*
of insects, *246, 512,* 514
of invertebrates, 244–246, *246*
evolution of, 244–246
of molluscs, 246, 452
of oligochaetes, 482–483
parasympathetic, *251,* 251–252, *252*
peripheral, 246, 251–252
reptilian, 626–627
of sea stars, 548
somatic, 251
sympathetic, *251,* 251–253, *252*
adrenal medullary hormones and, 280–281
of vertebrates, 246–253, 574–575
Nervous tissue, 145
Nest(s), avian, 658, *659*
Nested hierarchy, 97
of clades, 345–346
Net productivity, 124
Neural communication, 241
Neural crest, *326,* 327, *327*
Neural plate, *326,* 327, *327*
Neural tube, *326,* 327, *327*
Neurogenic heart, 195
Neuroglia, 145
Neuroglial cells, 241
Neuromasts, 256, 590
Neuromuscular system, of cnidarians, 397
Neuron(s), 145, 241–244
afferent (sensory), 241, 246
in reflex arcs, 247, *247*
definition of, 241
efferent (motor), 241, 246
in reflex arcs, 247, *247*
functional anatomy of, *147*
intermediate, of retina, 261, *262*
myelinated, 242, *242*
postsynaptic, 244
presynaptic, 244
processes of, 241, *241*
unmyelinated, 242
Neuropeptides, of brain, 275
Neuropodium, of polychaetes, *477,* 478
Neuroptera, 523, *524*
Neurosecretion, 269
hypothalamus and, 272, *272, 273*
Neurosecretory cells, 269, *270*
hypothalamic, *271,* 272, *272*
in invertebrates, 500, 501
Neurotransmitters, 244
Neutral fats, 9, *9,* 10*t*
Neutrons, *691,* 691–692
Neutrophils, 190, *190,* 212
Niche(s), 112, 114, *114*
in biological concept of species, 350
fundamental, 114
realized, 114
Niche overlap, 119, 120
Nicotinamide adenine dinucleotide, 40
in cellular respiration, 45, *45,* 46, *46*
Nighthawks, 662
Nile crocodile, 636–638, *637*
Nitrogenous bases, 11
Nitrogenous wastes, 51–52
Noctiluca, 371
Nocturnal animals, 182
Nodes of Ranvier, *147*
Nonrandom mating, 106–107
Nonshivering thermogenesis, 183
Noradrenaline. *See* Norepinephrine
Norepinephrine, 280–281, *281*
neurotransmitter function of, 244, 252
Nostrils, 202
Notochord, 151, 565, 566, *566,* 569
of amphioxus, *573*
Notophthalamus viridescens, 614
Notopodium, of polychaetes, *477,* 478
Notostraca, 504
Nuclear envelope, 27, *27*
pores in, 28, *28*
Nuclear equivalence, theory of, 329
Nuclear receptor(s), 267, 268, *268*
Nuclear transplantation, 330, *330*
Nuclease(s), 233
Nucleic acids, 11, 70–72
Nucleoid, 18
Nucleolus (pl., nucleoli), *19,* 28
Nucleotides, 11, 70
Nucleus
atomic, 691, *691*
cell, *19, 25,* 27, *27,* 28, *28*
midbrain, 249, *249*
Nudibranchs, 458, 459, *460*
Numbats, 685
Nurse cells, 441, *441*
Nutrient(s)
absorption, 223, 234
active transport of, 234
assimilation, 223
deficiencies of, 236
egestion, 223
essential, 236
excretion, 223
oxidation, 223
requirements for, 235*t,* 235–237
storage, 223
transport, 223
Nutrient cycles, 126–127, *127*
Nutrition, 222–239
of cephalopods, 468
of earthworms, 481, *482*
in flatworms, 419, *420*
insect, *512,* 512–513
of pogonophores, 533
in protozoa, 365–366
in tapeworms, 419
Nutritive-muscular cells, of cnidarians, *397,* 398
Nymphon, 493, *493*
Nymphs, 515, *517*

O

Oak treehoppers, *524*
Obelia, 399, 400, 409
  life cycle of, *401*
Obesity, 50, 234, 235
Oceans, 132–133
Ocelli
  of flatworms, 420, *420*
  of hydroids, 402
  of insects, 514–515
  of scyphozoans, 403
Ochotona princeps, 687
Octocorallia, 404, 409
Octomerous body plan, 404
Octopus *(Octopus),* 449, *450,* 464, 465, 472
  locomotion, 466
Octopus briareus, 450
Odonata, 522, 526
Odontophore, of molluscs, 451, *451*
Odor receptor genes, 254
Oleic acid, 9, *10*
Olfactory epithelium, 254, *255*
Oligochaetes (Oligochaeta), 475, 480–484, *487,* 488
  freshwater, 484, *485*
Omasum, of ruminant stomach, 676, *676*
Ommatidia, 260, *261,* 502, *503*
  of insects, 514–515
Omnivores, 223
  mammalian, 676, *677*
Onchocerciasis, 442
Oncogenes, 79–80
Oncorhynchus, 599
Oncospheres, of beef tapeworm, 425
One gene–one enzyme hypothesis, 69–70
Ontogeny
  definition of, 98
  and phylogeny, parallels between, 98
Onychophora, 353, 526, 532, 534, *535,* 539, *540*
Oocyst, of apicomplexans, 374
Oocyte(s), 318, 319
  primary, 314, *316*
  secondary, 314, *316*
Oogenesis, 314, *316,* 319
Oogonia, 314, *316*
Ootid, 314, *316*
Opalina, 378
Opalinata, 378
Oparin, Alexander I., 13
Operculum
  of amphibians, 617
  of bony fishes, 591
  of cnidarians, 396, *397*
  of fish, 597
  of gastropods, 454–455
Ophiopluteus, *550*
Ophiothrix, 550
Ophisaurus, 633
Ophiura lutkeni, 549, 555
Ophiuroidea, 543, 549–550, 555, *556*
  larvae of, *550*
Opiliones, 496, *497*
Opiothrix suensoni, 389
Opisthaptor, of monogeneans, 423–424, *424*
Opisthobranchia, 458
Opossum, 685
  reproduction, 335, *680*
Opsin, 262
Opsonins, 212
Opsonization, 217
Optic lobes, *249,* 249–250
  avian, 650, 651
Oral contraceptives, 284
Oral disc, of sea anemones, 406, *407*
Oral lobes
  of cnidarians, 397
  of scyphozoans, 403
Orangutan(s), taxonomy of, 347–348, 349, *349*
Orchestia, 505
Organ(s), 139*t,* 140
Organ of Corti, *257,* 258
Organ system(s), 139*t,* 140
Organ transplantation, 218
Organic compounds, 7–8
Organic molecules, 7–8
Organism(s), in ecological hierarchy, 113
Ornithischia, 630, *631*
Ornithodelphia, 685
Orthoptera, 522, 526
Oscula (sing., osculum), of sponges, 382, *384, 385*
Osmolarity, 168
Osmometer, *32,* 32–33
Osmoregulation, 167–171
  in flatworms, 419–420
  in freshwater organisms, 168, *169*
  in marine bony fishes, 168–170, *169*
  in protozoa, 366
  in terrestrial animals, 170–171
Osmoregulatory organs, of crustaceans, 502
Osmosis, 32
Osmotic conformers, 167
Osmotic potential, 33
Osmotic pressure, 33
Osmotrophs, protozoan, 366
Osphradia, 454
Osprey, wings of, 653, *655*
Ossicles
  of sea cucumbers, 551, *553*
  of sea stars, 544
Osteichthyes, *568,* 579, 583, *584, 585,* 591–594, 603
Osteoblasts, 153, 278
Osteoclasts, 153, 278
Osteocytes, *145,* 152, *152*
Osteon, 152, *152*
Osteoporosis, postmenopausal, 153
Osteostracans, 576, *577*
Ostia, of sponges, 382, *385*
Ostracoda (ostracods), 504, *505,* 525
Ostracoderms, 576, *577,* 583, *584*
Ostrich, 645, 661, *663*
Outgroup comparison, 345
Outgroup, definition of, 345
Oval window, *257,* 258
Ovalbumin, gene transcription and maturation of, *74*
Ovary(ies), 311, *316,* 318, *318*
  avian, 657, *659*
  of hydras, 400, *400*
  of rotifers, 434, *435*
Overton, William R., 3
Oviduct(s), 316, 318, *318*
  avian, 657, *659*
Ovigers, 493, *493*
Oviparous reproduction, 333, 534, 600, 602, *602,* 635, 638
Ovis canadensis, social dominance in, 300, *300*
Ovoviviparous reproduction, 333, 534, 600, 635
Ovulation, 284
Ovum (pl., ova), 310
Owen, Richard, 95, 630
Owl(s), 662
  visual acuity of, 652
Oxidation reactions, 693
Oxidation-reduction reaction, 43, *43,* 44, 693
Oxidative (aerobic) metabolism, 17
Oxidative phosphorylation, 46–47, *47*
  efficiency of, 47–49, *48*
Oxygen
  in air, 199
  consumption, in muscle contraction, 163
  in cutaneous respiration, 200
  diffusion, 204, *205*
  in tracheal system, 200
  in water, 199
Oxygen debt, in muscle contraction, 163
Oxytocin, 272, 273*t, 274,* 275, 285
  in birth, 286
Oyster(s), 449, 459
Oyster borers, 459
Oyster catchers, 662
Oyster drill, 450
Ozone, 17

P

p53 protein, 80
Pacemaker cells, of heart, 194
Pacemaker, of heart, 329
Pacinian corpuscle, 255, *255*
Paddlefishes, 591, *594,* 603
Paedomorphosis, 575, 576, *576*
  in salamanders, 613, *615*
Pain, sensation of, 255, 256
Paired septa, of Anthozoa, 404
Palate
  hard, 667, *670*
  primary, 667
  secondary, 667, *670,* 670*t*
    of crocodilians, 636
  soft, 667, *670*
Paleocortex, 250
Palolo worm, 479, *480*
Palp, of polychaetes, *477,* 478
Pan, taxonomy of, 347–348, 349, *349*
Pancreas
  development of, 328, *328*
  endocrine, 281, *281*
  exocrine, 281, *281*
Pancreatic juice, 232–233, *233*
Pancreatic lipase, 233
Pandion haliaetus, 655
Pannulirus argus, 507
Panope abrupta, 450
Panthera leo
  food and feeding of, *677*
  reproduction, *680*
Panulirus, 525
Papulae, of sea stars, 545
Parabronchi, avian, 649
Paragonimus, 423*t*
Parakeets, 662
Paramecium, 379
  binary fission in, *369*
  conjugation of, *377*
  contractile vacuoles in, 366, *377*
Paranthropus robustus, 684, *684*
Paraphyly, 346, *347,* 347–349, 565
Parapodia, of polychaetes, 477, *477,* 478, 486–487, 488
Pararthropoda, 532
Parasites, fluid feeding by, 226
Parasitism, 113, 118, 122–123
  by flukes, 421
Parastichopus, 555
Parastichopus californicus, 552
Parasympathetic nervous system, *251,* 251–252, *252*
Parathyroid gland(s), 277
  development of, 328, *328*
Parathyroid hormone, 278, *278*
  and bone growth, 153
Parazoa, 352, 382
Parchment worms, 475
Parenchyma
  of acoelomates, 357, *357,* 417, *417*
  of flatworms, 418–419, *419*
Parenchymula, of sponges, 387, *388*
Parental care, social behavior and, 298, *299*
Parietal cells, 230
Parietal pleura, *202,* 202–203
Parrot(s), 662
Parrot snake, *634*
Parthenogenesis, 115, 310, 312, 313
  ameiotic, 312
  meiotic, 312
Partial pressure(s), 204, 204*t*
Passenger pigeon, extinction of, 659, *660*
Passeriformes, 662, *663*
Pasteur, Louis, *13*
  refutation of spontaneous generation, 12–13

swan-neck flask experiment, 12–13, *13*
Pauropods (Pauropoda), 508, 525
Pavlov, Ivan, 266
PCR. *See* Polymerase chain reaction
Peanut worms, 532
Pearl(s), 449, 450, *452*
Peas, Mendel's experiments with, 57, *58,* 61–62, 63, *64,* 68
Pecten, 651, *652*
Pectoral, definition of, 354
Pectoral girdle, 155
Pectoralis muscle, avian, 646–648, *648*
Pedal disc, of hydras, 399
Pedal glands, of rotifers, 434, *435*
Pedal laceration, 399
in sea anemones, 407
Pedalium, of Cubozoa, 404, *406*
Pedicel
of brachiopods, 538, *538, 539*
of spiders, 494, *494*
Pedicellariae, of echinoderms, 542, 545, *547,* 550
Pedicularia, veliger of, *453*
Pediculus humanus, 524
Pedipalps, 493
Pelage, 671
Pelagic organisms, 132
Pelagic realm, 132, *133*
Pelecanus onocrotalus, 658
Pelecypoda, 459, 472
Pelican(s), 661
social behavior of, 656, *658*
Pellicle, of ciliates, 376, *376*
Pelmatozoa, 555, *556*
Pelvic, definition of, 354
Pelvic girdle, 155
Pelycosaurs, 622, *623,* 667, *668, 669*
Pen, of squids, 465, *467*
Penaeus, 504
Penguin(s), 347, *348,* 645, 661
Penis, 317, *317*
Pentadactyl limbs, 155
Pentastomida, 353, 532, 533–534, *534,* 539
Pepsin
activity of, 230
reaction catalyzed by, 42
secretion of, 230
Peptide bonds, 11
Peptide hormones
mechanism of action, 267–268, *268*
receptors, 267–268, *268*
Peranema, 371, 378
Perca flavescens, 592
Perch, *592*
skeleton of, *153*
Pericardium, 194
Period(s), geological, 92
Periostracum, of molluscs, 452, *452*
Peripatus, 535
Peripheral nervous system, 246, 251–252
afferent division, 251
efferent division, 251
Periproct, of echinoids, 550, *551*
Perissodactyla, 686, *688*
Peristalsis, 228, *228*
Peristaltic movement
of annelids, 476
of earthworms, 481
Peristomeum, of polychaetes, *477,* 478
Peritoneum
of annelids, 476, *476*
mesodermal, *357,* 358
Periwinkles, 454, 459
Perkinsea, 378
Perla, 517
Permeability, membrane, 32
selective, 32
Permeases, 33, *33, 34*
Permian extinction, 108, *108*
Perpetual change, 88
evidence for, 91–94
Petrel(s), 661
Petromyzon marinus, 587, *587*
Petromyzones, 579, 603
pH, 696, *696*
Phagocata, 421
Phagocyte(s), 211
avian, 649
Phagocytosis, 34, *35*
ameboid, 365, *365,* 366, 372
of antigens, *211,* 211–212
of food particles, 226–227, *227*
Phagosome, 29
protozoan, *365,* 366
Phagotrophs, protozoan, 366
Phalaropes, 662
Pharyngeal bars, 569
Pharyngeal grooves, 569
Pharyngeal pouches, 566, 569
development of, 328, *334*
Pharyngeal slits, 566, 569, 573
of amphioxus, 572, *573*
Pharynx, 202, 228
avian, 648
of earthworms, 481, *482*
of flatworms, 419, *419, 420*
of nematodes, 438, *439*
of sea anemones, 406, *407*
vertebrate, 573–574
Pheasants, 661
Phenetic taxonomy, 348
Phenotypes, 61
Pheromones, 253, 302
insect, 518, 519, 520, *520*
Philodina, 435
Phoenicopterus ruber, 663
Phoneutria fera, 496
Phonoproduction, in insects, 518–519
Phonoreception, in insects, 518–519
Phoronids (Phoronida), 353, *354,* 453, 531, 535, 536, *536*
Phoronis, 536
Phosphate group, of nucleic acids, 70, 70*t*
Phosphate, in body fluids, 188, *189*
Phosphatidylcholine; *see* Lecithin
Phospholipid bilayer, 27, *27*
Phospholipids, 10
Photic zone, of oceans, 132
Photoperiod, 274
Photoreception, 260–263
Photoreceptors, 260
of pogonophores, 533
Photosynthesis, 8, 20, 23, 39, 123
origin of, 17
reactions of, 17
Phototrophs, 223
Photuris, 519, *519*
Phronima, 506
Phyllobates bicolor, 613
Phyllobates limbatus, 615
Phyllopodia, 504
Phylogenetic concept of species, 351
Phylogenetic systematics, 348–349
Phylogenetic tree, 96, *97,* 345–346, *346*
of chordates (Chordata), 565, *567*
Phylogeny, 84, 89, 95, 344–346
of acoelomate animals, 429–430, *430*
of annelids, 486–487, *487,* 539, *540*
of arthropods, 526, *527,* 539, *540*
of aschelminthes, *445,* 446
of cnidarians, 412–413, *413,* 429
of ctenophores, 412–413
of echinoderms, 554–557, *556, 559*
of hemichordates, 558, *559*
information sources for, 346
of insects, 526, *527*
of lophophorates, 539, *559*
of molluscs, 469–470, *470, 471*
of platyhelminthes, 429–430, *430*
of protozoa, 376–377
of pseudocoelomate animals, *445,* 446
of ratite birds, 96–97, *97*
reconstruction of, using character variation, 345–346
of sponges, 389
Phylotypic stage, of vertebrate development, 333, *334*
Physa, 472
Physalia, 402, *403,* 409
Physiological sciences, 4–5
Phytoflagellates, 369, 377
Phytomastigophorea, 369, *371,* 377, 378
Phytophagous organisms, 512
Pickerel frog, 616
Pigeons, 662
Pikaia, 575, *575*
Pikas, 686, *687*
Pill bugs, 505, *506*
Pinacocytes, of sponges, 384, *385, 386*
Pineal gland, 274–275
Pinnipedia, 686
Pinnules, of crinoids, 553, *554*
Pinworm, 440*t,* 441, 442, *442*
Pisaster, 555
Pisaster ochraceus, as keystone species, 122, *123*
Pisces, 566, 569*t*
Pit vipers, 633–634, *635*
Pitch discrimination, place hypothesis of, 258–259
Pituitary gland, 270–271
anterior, 270–272, *271, 273*
anterior lobe, 271, *271*
hormones of, 271
intermediate lobe, 271, *271*
hormones of, 273*t*
posterior, 271, *271,* 272–274, *274*
Placenta, 285, 318
chorioallantoic, 679
hormones released from, 285, *286*
human, 336
mammalian, 335–336
primitive, 679
Placental lactogen, 285
Placental mammals, 335–336, *336,* 671, 685
evolution of, *668, 669*
reproduction, 679–680, *681*
Placobdella, 486, 488
Placoderms, 577–578, *578,* 583, *584, 585*
Placoid scales, 588–589, *593*
Placozoa, 382
Planarians, 418, *418*
characteristics of, 421, 427
reproduction, 420
structure of, *420*
Plankton, 132, 223, *224*
Planocera, 427
Plantae, 351, 352, 361
Planula, of cnidarians, 399, *401*
Plasma cells, 217, *217*
Plasmid(s), 20, 77
Plasmodium, 378
life cycle of, 368, 374, *374*
Plastids, 18, 27, 77
DNA, 20
Plastron, of turtles, 627, *628*
Platelets, 190, *190*
Plates, of chitons, 453, *455*
Platycotis vittata, 524
Platyhelminthes, 352, *354,* 417–427
body plan of, *355,* 417, *417,* 418–419, *419*
characteristics of, 418, 427
digestion in, 419, *420*
ecological relationships of, 417–418
excretion, 419–420, *420*
flame cells of, 419–420, *420*
form and function of, 418–420
nervous system of, 246, *246,* 420, *420*
nutrition, 419, *420*
osmoregulation in, 419–420
phylogeny, 429–430, *430*
position in animal kingdom, 417
reproduction, 420, *421*
sense organs of, 420, *420*
Platypus, duck-billed, 685
reproduction, 335, 679
Pleated sheet, *12*
Plesiosaurs, 622, *623*
Pleura, of insects, 510, *510*
Pleural cavity, *202,* 202–203

Pleurobrachia, 411, *412*
Plexaura, 409
Plovers, 662
Plumatella, 537
Plumatella repens, 538
Plumed worms, 475
PMNs. *See* Polymorphonuclear leukocytes
Pneumatic duct, 596
Pneumostome, 457, *461*
Podia (sing., podium), of sea stars, 545, *545,* 546, *546,* 547
Podophrya, 379
  feeding method of, *367*
Pogonophora, 353, 532, 533, *534,* 539
  tubeworms, 125, *125*
Poikilothermic, 180
Poison-dart frog, *613*
Polar body(ies), 314, *316*
  first, 314, *316*
Polarity
  of embryo, 321, 324, *325,* 358
  of phylogenetic characters, 345
Polarization, of embryo, 324, *325*
Polian vesicles, of sea stars, 546, *546*
Polinices lewisii, 458, 472
Polychaeta, 475, *477,* 477–479, *478, 479, 480, 481, 487,* 488
Polygamy, definition of, 657
Polygyny
  in birds, 657
  definition of, 657
Polymerase chain reaction, 78, *78*
Polymers, 8
  formation of, 15–16
    thermal condensations in, 16
Polymorphism, 104, 113
Polymorphonuclear leukocytes, 212
Polyodon spathula, 594
Polyorchis penicillatus, 401
Polypeptide(s), 11
  synthesis of, 75, *75, 76*
Polyphyly, 346, *347*
Polypide, of ectoprocts, 536, 537
Polyplacophora, 449, *450,* 453–454, *454, 455, 470,* 472
Polyploidy, 68
Polyps, cnidarian, 394–395, *396*
Polypterus, 591–592, *594,* 603
Polyribosome, 75
Polysaccharides, 8
Polysome, 75
Polyspermy
  definition of, 320
  prevention of, 320, *321*
Polystoma, 427
Pongidae, taxonomy of, 347–348, 349, *349*
Pongo, taxonomy of, 347–348, 349, *349*
Pons, 249, *249*
Pope, Philip, 564
Popillia japonica, 521
Population(s), 114–118
  age structure of, 114–115, *116*
  avian, 659–663
  definition of, 113, 680
  growth, 115–118
    density-dependent limits to, 118, 680
    density-independent limits to, 118, 680
    extrinsic limits to, 117, 118
    intrinsic limits to, 116–117 (*see also* Carrying capacity)
    rate of, 114–115
  interactions among, in communities, 118–119
  mammalian, 680–682, *681*
  sex ratio of, 114–115
Population genetics, 103
Porcellio, 505
Porcupines, quills of, 671–672, *673*
Porifera, 381–391. *See also* Sponge(s)
  characteristics of, 383
  classification of, 390
Pork tapeworm, 425, *426t, 427*
Porocytes, of sponges, 384, *385*
Porpoises, 686
Portuguese man-of-war, 394, 396, 402, *403*
Positive assortative mating, 106
Postabdomen, of scorpions, 496, *497*
Posterior, definition of, 354, *356*
Postmolt, in crustaceans, 500, *501*
Postsynaptic neuron(s), 244
Potassium
  in body fluids, 188, *189*
  in nerve impulse conduction, 242–243, *243*
Potassium-argon dating, 92
Poterion, 387
Potocytosis, 34, *35*
Pottos, 685
Prairie dogs, social behavior of, 298
Praying mantis, *511,* 522
  guilds of, 121
  parthenogenesis in, 115
Preabdomen, of scorpions, 496, *497*
Precambrian era, 94
  life in, 17–20
Precocial young, avian, 658, *660*
Predation, 113, 118, 121–122
  keystone species and, 121–122, *123*
Predators, 225
  crustacean, 501
Pregnancy
  human, hormones of, *285,* 285–286
  unwanted, 285
Premolars, 225, *226,* 674
Premolt, in crustaceans, 500, *501*
Prenatal diagnosis, 69
Presynaptic neuron(s), 244
Priapulida, 353, 434, 436–437, *437, 445,* 446
Priapulus, 437
Primary organizer, 330–331, *331*
Primary producers, 123
Primary structure, of protein, 11, *12*
Primate(s), 685
  anthropoid, taxonomy of, 347–348, 349, *349*
  evolution of, 682–685
Primitive streak, *322,* 325
Prismatic layer, of molluscs, 452, *452*
Probability, 65
Proboscidea, 686
Proboscis
  of acorn worms, 557, *558*
  echiuran, 533, *533*
  of ribbon worms, 427, 428, *428*
  of spiny-headed worm, 442–443, *443*
Procuticle, 147
  of arthropods, 491–492
Product rule
  definition of, 65
  for genotype and phenotype ratios in dihybrid cross, 65, *65t*
Productivity, 123
  gross, 124
  net, 124
  trophic levels of, 123
Progesterone, *283,* 283–284
  in birth, 286, *286*
  in menstrual cycle, 284, *284*
  in pregnancy, *285*
Proglottids, of tapeworms, 424, *425*
Prokaryotes, 18, 351. *See also* Bacteria
  cells, 26, *26*
  components of, *19*
  reproduction, 18, 19, 20
Prolactin, 271, 273*t,* 285
Proline, *10*
Pronephros, 173, *174*
Pronucleus, in fertilization, 320
Prophase, *36,* 37
Proprioceptors, 253
Prosencephalon, 248, *249*
Prosimians, 682, *682,* 685
Prosobranchia, 458
Prosopyle, of sponges, *385, 386*
Prostaglandin(s), 275
  in birth, 286
Prostate gland, *317,* 318
Prostomium
  of earthworms, 481, *482*
  of polychaetes, *477,* 477–478
Protease(s), 230
Protein(s)
  absorption, 234
  animal, consumption of, 236
  in body fluids, 188, *189*
  dietary
    complementary, 236
    deficiency, 236, 237, *237*
    requirements for, 235, 236
  digestion of, 228
  as enzymes, 11
  metabolism of, *51,* 51–52
  primary structure of, 11, *12*
  quaternary structure of, 11, *12*
  secondary structure of, 11, *12*
  secretion, from eukaryotic cell, 28–29, *29*
  structure of, 11, *12*
  synthesis of, 75, *75, 76*
  tertiary structure of, 11, *12*
Proteinoid microspheres, synthesis of, 16, *16*
Prothorax, of insects, 510, *510*
Prothrombin, 190, *191*
Protista, 19, 351, 352, *353*
  animal-like, 352, 361–380
  classification of, 378–379
Protochordata, 566, *568,* 569*t,* 579
Protocoel
  of acorn worms, 557
  of lophophorates, 535
Protoctista, 362
Proton(s), *691,* 691–692
Protonephridia, 443
  of flatworms, 171–172, *172,* 419–420, *420*
Proto-oncogenes, 80
Protoopalina, 378
Protoplasm, 24, 139*t*
Protopod, of crustaceans, 499
Protopterus, 594, *595,* 603
Protostomes (Protostomia), 323, *324,* 352–353, *354,* 449, 565, 570
  lesser, 531–541
Prototroch, 453, *453*
Protozoa, 140, 352, 361, 362. *See also* Protista, animal-like
  adaptive radiation of, 377
  biological contributions of, 362
  characteristics of, 362, 363
  cilia, 363–364
  classification of, 378–379
  colonies of, 367, *370, 371*
  conjugation, 368
  contractile vacuoles of, 171
  digestion in, 365–366
  excretion in, 366
  feeding methods of, 366, *367*
  flagella, 363–364
  life cycles of, *363,* 368
  locomotor organelles of, 363–365
  nutrition in, 365–366
  organelles of, 362, *363*
  osmoregulation in, 366
  phyla of, 362, 378–379
  phylogeny of, 376–377
  position relative to animal kingdom, 362
  pseudopodia, 364–365, *365*
  reproduction, 311, 367–368
    asexual, 367, *368, 369, 370, 374*
    sexual, 367–368, *370*
  structure of, 362, *363,* 363–368
Protura, 522, 526
Proventriculus, of avian stomach, 648–649
Proximal convoluted tubule, 174, *175*
  and urine concentrating mechanism, 178, *178*
Proximal, definition of, 354
Proximate (immediate) causes, 4, 291

Pseudoceratina crassa, 389
Pseudococcus longispinus, 521
Pseudocoel, 434, *435,* 448
as hydrostatic skeleton, in nematodes, 437–438
Pseudocoelomate animals, 353, 357, *357,* 433–447, 448. *See also* Nematodes
adaptive radiation of, 446
biological contributions of, 434
body plan of, *355,* 357, *357*
characteristics of, 434, *435*
diversity of, 434
phylogeny, *445,* 446
position in animal kingdom, 434
Pseudopodia, 156
protozoan, 364–365, *365,* 371–372, *372*
Ptarmigan, 661
Pteraster tesselatus, 544
Pterobranchia, 557, 558, *558, 559*
Pterosaurs, 630
PTH. *See* Parathyroid hormone
Ptilosarcus gurneyi, 407
Ptychodiscus, 371
Puffbirds, 662
Puffins, 662
Pulex irritans, 524
Pulmonata, 458, 459
Punctuated equilibrium, 101–102, *102*
Punnett square, 63–65, *64*
Pupa, insect, 515, *516*
Pupil, 261, *261*
Purines, 70, 70*t, 71*
Purkinje, J., 24
Purkinje fibers, 194, *196*
Pycnogonida, 493, *493,* 525
Pycnopodia helianthoides, pedicellariae of, *547*
Pyloric sphincter, 231–232
Pyloric stomach, of sea stars, *546,* 547
Pyrimidines, 70, 70*t, 71*
Pyruvic acid, in glycolysis, 44–45, *45, 48*

Q

Q_{10}, 180
Quail, 661
Quaternary structure, of protein, 11, *12*
Queen(s), of social insects, 519–520, *520*
Quill, of contour feather, 645

R

r (intrinsic rate of increase), 115–116, 116
Rabbit(s), 686
head, anatomy of, *670*
Rachis, of contour feather, 645
Radial canals
of scyphozoans, 404
of sea stars, 546, *546,* 547
Radial nerve, of sea stars, 545, *546*
Radial symmetry, 353, *355, 356,* 393
secondary, in echinoderms, 542, 543
Radiata, 352, 353
Radiate animals, 392–415
Radiation, ionizing, mutagenic effects of, 80
Radioactive isotopes, 692
Radioimmunoassay, 275
Radiolarians, 373, *373,* 377
Radioles, of polychaetes, 478, *480*
Radiometric dating, 92
Radula
of cephalopods, *467,* 468
of molluscs, 451, *451*
Rail(s), 662
Raja, 591, 603
Ram ventilation, 598
Rana catesbeiana, 615, *616*
Rana clamitans, 615
Rana palustris, 616
Rana pipiens, 615–616
eggs, nuclear transplantation experiments with, 330, *330*
Rana sylvatica, 615
Rangifer tarandus, 678
migration of, *678*
Ranidae, 615, *616*
Ras protein, 80
Rat(s), 686
Ratfish, 590, *591*
Rattlesnake, *634, 635*
Ray(s), 579, *585,* 588–590, 590, *591,* 603
characteristics of, 588
Ray, John, 343
Razor clam, *463*
Reaction(s)
coupled, 41, 42–43, *43*
direction of, 42
enzyme-catalyzed, 42
Reactive force, in swimming fishes, 596
Reactive nitrogen intermediates, 211, *216*
Reactive oxygen intermediates, 211, *216*
Recapitulation, 98
Receptor(s)
chemical, 253
contact, 253
distance, 253
hormone, 267, *268*
light, 253
mechanical, 253
pain, 255
sensory, 241, 247
classification of, 253
thermal, 253
touch, 255
Receptor potential, 255
Receptor-mediated endocytosis, 34, *35*
Recessive allele, 61, 65–66, 105–106
Recombinant DNA, 77
Rectal glands, 234
Rectum
of insects, 512, *512*
of nematodes, 438, *439*
Red blood cell(s). *See* Erythrocyte(s)
Red muscles, 163
Red night shrimp, *507*
Red tide, 369
Redia, of flukes, 421, *422*
Redox reactions. *See* Oxidation-reduction reactions
Red-spotted newt, life cycle of, *614*
Reduction reactions, 693
Reflex act, 247–248
Reflex arcs, *247,* 247–248
Regeneration
of brittle stars, 549
of sea stars, 548, *548*
Relative fitness, 107
Relaxin, 285
Release-inhibiting hormones, 272
Releaser,
of stereotypes behavior, 292
Releasing hormones, 272
Renal artery, 174–175, *175*
Renal corpuscle, 174
and urine concentrating mechanism, 178, *178*
Renal pelvis, 174, *175*
Renal tubule, 174, *175*
Renal vein, 175, *175*
Renin, 230
Rentila, 409
Reproduction, 5–6, 308–340
amphibian, 609–611, *613*
asexual, 19, 20, 309, *309,* 310
advantages of, 310
biparental, *311,* 312
cephalopod, 468–469, *469*
of ciliates, 376, *377*
of cnidarians, 399, *401*
in fish, 600–602, *602*
of flatworms, 420, *421*
of frogs, 617–618, *618, 619*
of insects, *510,* 515, *516*
mammalian, 679–680, *680*
nonsexual, 308
process of, 309–313
prokaryotic, 18, 19, 20
protozoan, 367–368
of salamanders, 612, *614*
sexual, 19, 309, *309,* 310–312
advantages and disadvantages of, 313
and genetic variability, 78–79
Harvey's study of, 308
in sponges, 387
social behavior and, 298, 299
of spiders, 495
of sponges, 387
of toads, 617–618, *618, 619*
vertebrate
cycles, hormonal control of, 282–283
hormones of, 282–286
Reproductive barriers, 99, *99*
Reproductive cells, formation of, 313–314
Reproductive organs
female, 311, *316*
male, 311, *315*
Reproductive system(s)
avian, 656–657, *659*
invertebrate, 315–316, *317*
plan of, 314–318
of sea stars, *546,* 548
vertebrate, 316–318
female, 318, *318*
male, 316–318, *317*
Reptiles (Reptilia), *567, 568,* 579, 621–639
adaptive radiation of, 622, *623*
and amphibians, comparison of, 625–627
characteristics of, 625
circulatory system of, 626, *627*
classification of, 622, *623, 624,* 637
diversity of, 622
evolution of, 622, *623*
gastrulation in, *322,* 325
internal structure of, 626, *627*
locomotion, 626
lungs of, 201, *201*
nervous system of, 626–627
number of species of, 622
origin of, 622, *623*
respiratory system of, 626
sense organs of, 627
skin of, 625, *625*
water balance in, 626
Resorptive area, of fishes, 597
Resource(s)
environmental, 114
limiting, 116, 119
Respiration, 124, 199–208
aerial, 199–200
in amphiumas, 613, *614*
aquatic, 199–200
cutaneous, 200
in fishes, 597–598, *598*
in salamanders, 612–613
vertebrate, 573–574
Respiratory gases
exchange of, in lungs and tissues, 204, *205*
transport, in blood, 204–206
Respiratory organs, 200–202
Respiratory pigments, 205
Respiratory system
avian, 649, *651*
of bivalves, 461, *465*
epithelium, development of, 328
mammalian, 202–207
of sea cucumbers, 552
Restriction endonuclease(s), 77
activity of, 77, *77*
Rete mirabile, of fishes, 597, *597*
Reticuloendothelial system, 211–212
Reticulopodia, 365, *372*
Reticulum, of ruminant stomach, 676, *676*

Retina, 261, *261*
avian, 651, *652*
structure of, 261, *262*
Retinal, 262
Rh factor, 219–220
genetics of, 220
Rhabdites, of flatworms, 418, *419*
Rhea(s), 645, 661
Rhechostica hentzi, 496
Rheoreceptors, of flatworms, 420
Rheumatoid arthritis, steroid therapy for, 279
Rhinencephalon, 250
Rhinoceros, 686, *688*
horn, 672
Rhinoderma darwinii, 613
Rhipidistians, 594
Rhizopoda, 378
Rhizostoma, 409
Rhodopsin, 262
Rhombencephalon, 248, *249*
Rhopalium, 402–403, *406*
Rhoptry(ies), of apicomplexans, 373, *373*
Rhynchocinetes rigens, 507
Rhynchocoel, of nemerteans, 428, *428*
Rhynchocoela, 427–429
Rib(s), *153,* 155
Ribbon worms, 417, 427–429
Ribonucleic acid, 11
chemical components of, 70, 70*t*
enzymatic activity, 16
structure of, 71
Ribose, 70, *70*
structure of, 8, *8*
Ribosomal RNA, 28, 73, 75, *75*
among organisms, comparison of, evolutionary relationships inferred from, 351–352, *352,* 389
Ribosome(s), 16, 28
translation on, 75, *75, 76*
Ribozymes, 16
Rickets, 279
Riftia pachyptila, 533
Ring canal(s)
of scyphozoans, 404
of sea stars, 545–546, *546*
Ritualization
definition of, 303
in pair-bonding behavior of blue-footed boobies, 304, *305*
River blindness, 442
RNA. *See* Ribonucleic acid
RNA polymerase, 73
RNA world, 16
RNIs. *See* Reactive nitrogen intermediates
Roadrunner(s), 662
Robin(s)
English, territorial behavior of, 293, *293*
nomenclature for, 343, 344
Rock crab, *507*
Rock shells, 459
Rocky Mountain spotted fever, 498
Rod(s), 261, 262, *262*
avian, 651, *652*
Rodentia, 686
ROIs. *See* Reactive oxygen intermediates
Root, Richard, 120
Rotational acceleration, 259
Rothenbuhler, W. C., 294–295
Rotifera, 353, 434–435, *435, 445,* 446
Rotifers, body plan of, *355*
Round dance, of honeybees, 303
Round window, *257,* 258
Roundworms. *See* Nematodes (Nematoda)
Royal jelly, of social insects, 519
rRNA. *See* Ribosomal RNA
Rumen, 675
Ruminants, 230, 672, 675–676, *676*
Rumination, 675–676, *676*

S

Sabella, 480
Saccule, 256, 259
Sacrum, 155
Sagitta, 560
Sagittal plane, definition of, 354, *356*
St. Martin, Alexis, 231, *231*
Salamander(s), 612–613, *614, 615,* 618
breeding behavior of, 612, *614*
evolution of, 609, *610*
metamorphosis, 612, *614*
paedomorphosis, 613, *615*
respiration, 612–613
speciation, 99, *99*
Salientia, 609, 615–618, 618
Saliva, *233*
Salivary glands, 228
of insects, 512, *512*
Salmo salar, 599
Salmon
development of young, *601*
homing, 599–600, *600, 601*
spawning, *601,* 602
stocks, in Pacific Northwest, 600
Salt(s), 695–696
urinary excretion of, 176–177
Salt balance
in amphibians, 168, *169*
in freshwater fishes, 168, *169*
in marine birds, 650, *651*
in marine bony fishes, 168–170, *169*
in marine invertebrates, *167,* 167–168
in terrestrial animals, 170–171
Salt gland(s), 171
avian, 650, *651*
reptilian, 626
Sand dollars, 543, *551,* 555
characteristics of, 550–551, *551, 552*
Sandpiper(s), 662
Saprophagous organisms, 223, 512
Saprozoic feeders, protozoan, 366
Sarcodina, 369, 371–373, *372,* 377, 378
binary fission in, *368*
pseudopodia, 364–365
Sarcolemma, *158,* 159, 160, *161, 162*
Sarcomastigophora, 369–373, 378
Sarcomere, *158,* 159
Sarcoplasm, 144
Sarcoplasmic reticulum, *158,* 160, *161, 162*
Sarcopterygians (Sarcopterygii), 583, *584, 585,* 591, *593,* 593–594, *595,* 603
Sargasso Sea, 598, 599, *599*
Sauer, E., 656
Sauria, 628–629, 637
Saurischia, 630, *631*
Sauropods, 630, *631*
Sauropterygians, 622, *623*
Säve-Söderberg, Gunnar, 609
Scale(s), of reptiles, 625, *625*
Scale insects, 523
Scaleworms, 475, 478, *479*
Scalids, 436
Scaling, 138
Scallops, 459, 460
eyes of, 461, *464*
Scanning electron microscope, 24
Scaphopoda, 449, 454, *455,* 469, *470,* 472
Scavengers, crustacean, 501
Scent glands, 673
Schistocerca gregaria, 510
Schistocerca obscura, 510
Schistosoma, 423, 423*t*, 427
Schistosoma haematobium, 423, 423*t*
Schistosoma japonicum, 423, 423*t*
Schistosoma mansoni, 423, *423,* 423*t*
Schistosomiasis, 422–423, *423*
Schizocoel, of annelids, 476
Schizocoelomate body plan, *355*
Schizocoelous coelom formation, *324,* 326, *354,* 357, *357*
Schizogony, 367
Schleiden, Matthias, 24
Schmidt, Johann, 598
Schwann cell(s), *147,* 242, *242*
Schwann, Theodor, 24
Science
nature of, 3
principles of, 3–5
Scientific method, 3–4
Sciurus carolinensis, 687
Sclera, 261, *261*
Scleractinia, 407–408, *410*
Sclerites, of insects, 510, *510*
Sclerocytes, of sponges, 384–385
Sclerodactyla, 553, 555
Sclerosepta, of zoantharian coral, 408, *410*
Sclerospongiae, 382, 385, 390
characteristics of, 388–389
Sclerotin, 149
Sclerotization, 149
Scolex, of tapeworms, 424, *425*
Scolopendra, 508, *508*
Scorpions (Scorpionida), 496, *497,* 525
Screwworm flies, control of, 524
Scutigera, 508
Scutigerella, 525
Scyphistoma, 404, *405*
Scyphozoa, 394, 409
characteristics of, 402–404, *404, 405*
phylogeny of, *413*
Sea anemone(s), 385, 394, 404
characteristics of, 405–407, *406, 407*
mutualistic relationships with other organisms, 406
swimming, *409*
Sea biscuits, 555
Sea butterflies, 454, 457
Sea cucumbers, 543, *552,* 555
characteristics of, 551–553, *552, 553*
Sea daisies, 554, *554,* 555
Sea fans, 404, *411*
Sea hares, 454, 457, 458, *459*
Sea lilies, 543, *553*
characteristics of, 553–554, 555
Sea lions, 686
territoriality in, 301
Sea pansies, 404
Sea pens, 404, *407*
Sea slugs, 454, 458
Sea snakes, 634, 635
Sea spiders, 493, *493,* 525
Sea squirts, 571, *571, 572*
Sea stars, 543–548, *544*
autotomy of, 548
characteristics of, 555
development of, 548, *549*
digestive system of, *546,* 547–548
embryo, germ layers of, *322,* 324
embryonic cleavage in, *322*
external features of, *544,* 544–545, *545, 553*
feeding, 547–548
form and function of, 544–548
gastrulation in, *322,* 324
hemal system of, 548
larvae of, *550*
locomotion, 547
metamorphosis in, 548, *549*
nervous system of, 548
pedicellariae of, *547*
regeneration of, 548, *548*
reproductive system of, *546,* 548
sensory system of, 548
water-vascular system of, *545,* 545–546, *546*
Sea urchins, 543, *551*
characteristics of, 550–551, *551, 552,* 555
eggs, fertilization, 319–320, *320, 321*
irregular, 543, 550, *551*
pedicellariae of, *547,* 550
regular, 550, *551*
Sea walnut(s), 411, *412*
Sea wasp, 404

Seal(s), 686
  migration of, 677, *678*
Seasonal affective disorder, 274–275
Sebaceous gland(s), *672,* 673
Sebum, 673
Second messenger(s), 267–268
Secondary structure, of protein, 11, *12*
Secretin, 266, 282, *283*
Segmentation, 358, *358*
  of arthropods, 492
  of crustaceans, 498, *499*
  of gut, 228, *228*
Segmented worms. *See* Annelids (Annelida)
Self/nonself recognition, 211, 213–215
***Semibalanus cariosus,* 505**
Semicircular canals, 256, *257,* 259, *260*
  response to acceleration, 259, *260*
Semilunar valve(s), 194, *195*
Seminal vesicle(s), *317,* 318
  avian, 657
Seminiferous tubule(s), 314, *314, 315*
  human, 316, *317*
Sense(s), 240
Sense organs, 253–263
  of arthropods, 492
  avian, 650–652
  of crustaceans, 502, *503*
  of flatworms, 420, *420*
  of gastropods, 457
  of insects, 514–515
  of polychaetes, *477,* 478
  reptilian, 627
  of snakes, 633
Sensilla, of insects, 515
Sensitization
  Aplysia experiments, 296, *296*
  definition of, 296
Sensory cells, of cnidarians, *397,* 398
Sensory cortex, 250, *250, 251*
Sensory neurons, 241, 246
Sensory setae, of spiders, 495
Sensory system, of sea stars, 548
***Sepia latimanus,* 468,** 472
***Sepioteuthis lessoniana,* 467**
Septa (sing., septum)
  of annelids, 476, *476*
  of zoantharian coral, 408, *410*
Septal filaments, of corals, *410*
Serosa, gastrointestinal, *232*
Serpentes, 632–635, 637
Sertoli cells, *315*
Serum, 189
Setae
  of annelids, 475
  of crustaceans, 502
  of earthworms, 481, *482, 483*
  of oligochaetes, 480
Sex, 19
Sex cells, 57, 309, 310
Sex chromosomes, 60
Sex determination, 59–60
  in crocodilians, 638
  in turtles, 628
  XX-XO, 59–60, *60*
  XX-XY, 60, *60*
Sex hormones, 10, *283,* 283–284
Sex organs
  accessory, 311–312, 315
  primary, 311–312, 315
Sexes, 310
Sex-linked inheritance, 66, *67,* 68
Sexual selection, 107, *107*
Shaft, of contour feather, 645
Shark(s), 579, *585,* 588–590, 603
  brain of, *248*
  characteristics of, 588
  external anatomy of, 588, *589*
  internal anatomy of, 588, *589*
  osmotic regulation in, 170
Shark fisheries, 590
Shaw, George Bernard, 78
Shearwaters, 661
Sheep, 686
Sheep liver fluke, 423*t*
Shell
  of bivalves, 460, *463, 464*
  of cephalopods, 465, 466, *467*
  of molluscs, 451, *451,* 452, *452*
    evolution of, 456, *456*
  of turtles, 627, *628*
Sherrington, Charles, 248
Shipworms, 450, 459, *463*
  feeding habits of, 461, *463*
Shrews, 685, *687*
Shrimp(s), *504,* 506, *507,* 525
Sickle cell anemia, 79, 206
Sign stimulus, 292–293
Silk glands, of spiders, *494,* 495
Silkworm moths, chemical sex attraction in, 302, *302*
Silverfish, 522, *522*
  development of, 516
Simians, 682, *683*
Simple diffusion, 32, 33
Simpson, George Gaylord, 94, 347, *347, 348*
Sinoatrial node, embryology of, 329
Sinus node, of heart, 194, *196*
Sinus venosus, 193, *194*
Siphon(s)
  excurrent, of ascidians, 571, *571*
  intercurrent, of ascidians, 571, *571*
Siphonaptera, 523
Siphonoglyph, of sea anemones, 406, *407*
Siphuncle, of cephalopods, 465, 466, *467*
Sipuncula, 532, *532,* 539
Sipunculida (sipunculids), 353, 453
  immunity in, 212, 213*t*
Sister taxa, 349
Skates, 579, 590, *591*
Skeletal muscle, 144, *146*
  exercise and, 159
  fascicles, 158, *158*
  fibers, 158, *158,* 159
    fast, 163
    slow, 163
  function of, 158–159
  innervation, 160
  performance, 163
  structure of, *158,* 158–159
Skeleton(s), 150–155
  avian, 646, *647, 648*
  human, *154*
  hydrostatic, 150, *151*
    of annelids, 476
    of cnidarians, 398
    of nematodes, 437–438
  rigid, 150–153
  of sponges, 385, *387*
  vertebrate
    appendicular, 153
    axial, 153
    plan of, *153,* 153–155, *154*
Skeleton shrimp, *506*
Skimmers, 662
Skin. *See also* Integument
  amphibian, 609
  human, structure of, *672*
  reptile, 625, *625*
Skin gills, of sea stars, 545
Skinks, 629
Skuas, 662
Skull
  anapsid, 622, *623, 624*
  avian, 646
  diapsid, 622, *623, 624,* 628, *629*
  embryology of, *326,* 327
  kinetic, 628, *629,* 633, *634,* 646
  synapsid, 622, *623, 624*
Sleeping sickness. *See* Trypanosomiasis
Sliding filament model, of striated muscle contraction, 159, *160*
Sliding microtubule hypothesis, 364
Slipper shells, 458, 459
Sloth, 686
Slug(s), 450, 454, 459, *461*
Small intestine
  digestion in, 231–234
  digestive (membrane) enzymes, 233, *233*
Smell, sense of, 253, 254
  avian, 651
Smooth muscle, 144, *146*
  of invertebrates, 157
Snail(s), 449, 450, *450,* 454–455, 459, *461,* 472
  pulmonate, 459, *459*
  speciation in, 102
Snake(s)
  characteristics of, 628, 632–635, 637
  feeding of, 225, *225*
  forked tongue of, 633
  kinetic skull of, 628, *629,* 633, *634*
  nonvenomous, 634, *636*
  oviparous, 635
  ovoviviparous, 635
  sense organs of, 633
  venomous, 633–635, *635, 636*
  viviparous, 635
Snapping turtle, *628*
Snipe, 662
Snow fleas, 522
Social behavior, 292, 297–302
  advantages of, 297–299
  of birds, 656–659
  definition of, 297
  disadvantages of, 299
  of insects, 519–520
  and reproduction, 298, 299
Sociobiology, 292
Sodium. *See also* Salt(s)
  in body fluids, 188, *189*
  in nerve impulse conduction, 242–243, *243*
  renal tubular reabsorption of, 176–177
Sodium chloride. *See also* Salt(s)
  urinary excretion of, 176–177
Sodium-potassium pump, 34, *34*
  in nerve impulse conduction, 242–243, *243,* 244
Soldiers, of social insects, 520, *520*
Solenia, of alcyonarian corals, 408, *410*
Solenogastres (solenogasters), 449, 453, 469, *470, 471,* 472
Soma, of neuron, *147*
Somatic cells, 313
  formation of, 35
Somatic nervous system, 251
Somatocoel, of echinoderms, 548, *549*
Somatotropin, 271, 273*t*
Somites, 328–329, *329*
Songbirds
  learning in, 297, *298*
  perching, 662, *663*
  population decline of, 663
  territoriality in, 301
Sorting, 91
Sound
  production, in insects, 518–519
  reception, in insects, 518–519
South American pygmy marsupial frog, *613*
Sow bugs, 505, *506*
Spaceship Earth, 112
Spat, of bivalves, 464
Speciation, 99, *99*
  through geological time, 108
Species, 350–351
  common descent of, 350
  concepts of, 350–351
    biological, 350–351
    evolutionary, 351
    phylogenetic, 351
    typological, 350
  criteria for recognition of, 350
  definition of, 98–99
  diversity of, 113, 118
  multiplication of
    Darwin's theory of, 89
    evidence for, 98–100
  reproductive community of, 350
  as smallest distinct grouping, 350
Species epithet, 343–344

Species selection, 108
  catastrophic, 109
  and sexual reproduction, 313
Spectacled cobra, *636*
Spemann, Hans, 330, 331
Spencer, Herbert, 91
Sperm, 35, 310, *314*
  of earthworms, 483-484
  in fertilization, 319-320
  formation of, 313-314
  of nematodes, 438
Spermatids, 314, *315*
Spermatocyte(s)
  primary, 314, *315*
  secondary, 314, *315*
Spermatogenesis, 314, *314, 315*
Spermatogonia, 314, *315*
Spermatophore(s)
  of cephalopods, 468-479, *469*
  of gastropods, 457-458
  of salamanders, 612, *614*
Spermatozoon, 310
Sphenodon, 622, 628, 635, *636,* 637
Sphenodonta, 635, *636,* 637
Spherical symmetry, 353, *356*
Sphinx moths, *516*
Sphygmomanometer, 197
Spicules, of sponges, 382, 385, *385, 386, 387*
Spider(s), *494,* 494-496, *495,*525
  predation by, 494, *495,* 495-496, *496*
  reproduction, 495
  web-spinning by, 495, *495*
Spider crabs, 506
Spinal cord, 246-247, *247*
  embryology of, *326,* 327
Spinal nerve(s), 246-247, *247*
Spindle, 37
Spindle fibers, 37
Spiny lobster, *507*
Spiny-headed worms, 353, 434, 442-444, *443*
Spiracle(s), 200, *200*
  of amphibian tadpole, 617, *619*
  of insects, 513, *514*
  of sharks, 589, *589*
  of spiders, 494, *494*
Spirobolus, 525
Spirobranchus giganteus, 384, 478
Sponge(s), *355,* 381-391
  adaptive radiation of, 389
  asconoid, 383, *385*
  biological contributions of, 382
  body plan of, 382
  calcareoys, 385, *387*
  canal systems of, 383-384
    excurrent, 384, *385, 386*
    flagellated, 383-384, *385, 386*
    intercurrent, 383, 384, *385, 386*
    radial, 383, *385, 386*
  cells
    food trapping by, 384, *386*
    types of, 384-385, *386*
  cellular organization, 382
  development of, 387, *388*
  ecological relationships of, 382, *384*
  with flagellated chambers, 384, *384, 385*
  forms of, 382-387, *383*
  fossil record of, 382
  freshwater, contractile vacuoles of, 171
  function in, 382-387
  growth habits of, 382, *383*
  immunity in, 212, 213*t*
  intracellular digestion in, 386-387
  leuconoid, 384, *384, 385*
  oscula, 382, *384, 385*
  ostia, 382, *385*
  phylogeny of, 389
  physiology of, 386-387
  position in animal kingdom, 382
  reproduction, 387
  skeleton of, 385, *387*
  spicules of, 382, 385, *385, 386, 387*
  syconoid, 383-384, *385, 386*
Spongilla, 390
Spongin, 382, 385, *387*
Spongocoel(s), *386*
  flagellated, 383, *385*
  in syconoids, 383-384
Spontaneous abortion, 286
Spontaneous generation, 12-13
Spoonbill(s), 661
Spore, of apicomplexans, 374
Sporocyst(s), of flukes, 421, *422*
Sporogony, 367
Sporozoea, 373-375, 378
Sporozoites, of apicomplexans, 373
Springtails, 522
  development of, 516
Squalene, 596
Squalus, 589, 603
Squamata, *624,* 628-635, 637
Squid(s), 449, 464, *467,* 472
  locomotion, 466, *468*
Squirrel(s), 686
  Eastern gray, *687*
Stalk, of crinoids, 553, 554, *554*
Stapes, *257,* 258
Starches, 8
Starfishes. *See* Sea stars
Starling, E. H., 266, *267*
Starlings, distribution of, 659, *660*
Statoblast(s), of ectoprocts, *537,* 538
Statocyst(s), 259, *259*
  of crustaceans, 502
  of ctenophores, 411, *411, 412*
  of flatworms, 420
  of hydroids, 402, *402*
  of polychaetes, 478
Statoliths, 259, *259*
Stearic acid, 50
Stearin, formation of, 9, *9*
Stegosaurus, 630, *631*
Stenohaline organisms, 167
Stenostomum, 420, *421*
Stentor, 375, 379
Stereom, of echinoderms, 544, *553*
Stereotyped behavior, 292-293
Sternum
  of insects, 510, *510*
  keeled, of carinate birds, 645, 646, *648*
Steroid hormones, 10
  gonadal, *283,* 283-284
  mechanism of action, 268, *268*
  receptors, 268, *268*
  synthesis of, 279-280
  therapy with, 279
Stichopathes, 409
Stickleback, territorial behavior of, 293, *293*
Stigma, of flagellates, 369, *371*
Stimulus, definition of, 253
Stingray, 590
Stink bug, *521*
Stolon, of sponges, 383
Stomach, 229-230
  avian, 648
  of insects, 512, *512*
  of ruminants, 675-676, *676*
  of sea stars, *546,* 547
Stomphia didemon, 409
Stone canal, of sea stars, 545-546, *546*
Stonefly, 515, *517*
Storks, 661
Stratum corneum, 149
Striated muscle, 144
  contraction
    control of, 160
    energy for, 161-163
    sliding filament model of, 159, *160*
  of invertebrates, 157
  of vertebrates, structure of, *158,* 158-159
Strobila
  of scyphozoans, 404
  of tapeworms, 424, *425*
Stroke volume, 194
Strongylocentrotus, 555
  pedicellariae of, *547*
Strongylocentrotus pupuratus, 551
Struthio camelus, 663
Sturgeons, 591, *594,* 603
Sturnus vulgaris, 660
Stylaria, 485, 488
Stylaster roseus, 403
Subatomic particles, 691-692
Subergorgia mollis, 411
Sublittoral zone, 132, *133*
Submucosa, gastrointestinal, *232*
Subnivean environment, 183
Subspecies, 344
Substrate, 41, *41*
Subtidal zone, 132, *133*
Sucrase, 233
Sucrose, 8
Sugar(s), 8
  pentose, of nucleic acids, 70, *70*
Sula nebouxii, pair-bonding behavior of, 304, *305*
Sun-azimuth orientation, 656
Sunlight, injurious effects on integument, 149-150
Supersaurus, 630
Supracoracoideus muscle, avian, 646-648, *648*
Surinam frog, *613*
Survivorship, 115, *115*
Suspension feeding, 223, *224*
  in bivalves, 459-461
  in crustaceans, 501
Swallow, wings of, 653, *654*
Swallowing, in humans, 229, *230*
Swan(s), 661
Sweat glands, *672,* 673
Swift(s), 662
Swim bladder, of fishes, 596-597, *597*
Swimmerets, of crustaceans, 499, *499*
Swimmer's itch, 423, *424*
Swine, 686
Sycon, 384, *385, 386,* 388, 390
Syconoids, 383-384, *385, 386*
Symbiosis, 20, 31, 361
Sympathetic nervous system, *251,* 251-253, *252*
  adrenal medullary hormones and, 280-281
Symphylans (Symphyla), 508, 525
Synapomorphy, 345
Synapses, *147,* 244
  chemical, 244
  electrical, 244
  excitatory, 244
  inhibitory, 244
  nerve impulse transmission across, 244, *245*
Synapsids (Synapsida), 622, *623, 624,* 637
  cladogram of, *669*
  evolution of, 667, *668, 669*
Synapsis, of chromosomes, 59, *59*
Synaptic cleft, 244, *245*
  at myoneural junction, 160, *161*
Synaptic vesicles, 160, *161*
Syncytial tegument, of flatworms, 418
Syncytium, 36
Syndrome, 69
Syngamy, 368
Systematics, 343
Systole, 194, *196*

T

T cells (T lymphocytes), 215
  cytotoxic, 215
  helper, activation of, 216, *217*
  receptors, 214, 215
  subsets of, 215
Tabanus, 511
Tadpoles, frog, 617-618, *619*
Taenia, 427
Taenia solium, 425, 426*t, 427*
Taeniarhynchus, 427
Taeniarhynchus saginatus, 424-425, *426,* 426*t*
Taenidia, of insects, 513, *514*
Tagelus plebius, 463

Tagmata
of arthropods, 491
of insects, 508, 510, *510*
of spiders, 494, *494*
Taiga, *128,* 130
Tail(s)
diphycercal, 593, *593*
heterocercal, 588, *593*
homocercal, 593, *593*
postanal, 566, 570
of amphioxus, *573*
of arrowworm, 560, *560*
Talons, of predatory birds, 648
Tamias striatus, food and feeding of, *677*
Tapeworms, 418
characteristics of, 424–425, *425, 426,* 427, *427*
in humans, 424, 426*t*
nutrition, 419
reproduction, 420
Tapir(s), 686
Tarantula, 496, *496*
Tardigrada, 353, 526, 532, *535,* 539, *540*
Target cells, 210, 267
Tarsier(s), 682, *682,* 685
Tarsius syrichta carbonarius, 682
Tasmanian wolves, 685
Taste buds, 253, *254*
Taste receptors, 253–254, *254*
Taste sense, 253–254
avian, 651
Taxa (sing., taxon), 343
monophyletic, 346, 347, *347,* 565
paraphyletic, 346, *347,* 347–349, 565
polyphyletic, 346, *347*
sister, 349
Taxonomic ranks, 343
Taxonomy, 343. *See also* Cladistics
of animals, 343, 344*t*
current state of, 349–350
definition of, 343
phenetic, 348
theories of, 346–350
traditional evolutionary, 346–348, 622
Tealia, 409
Tealia piscivora, 406
Tectorial membrane, of inner ear, *257,* 258
Tectum, 249, *249*
avian, 650, 651
Teeth, 225, *226*
deciduous (milk), 674
dentin, embryology of, *326,* 327
mammalian, 670*t,* 674, *675*
of molluscs, 451, *451*
permanent, 674
of sharks, 589, *590*
Tegument
definition of, 419
of flatworms, 418, *419,* 430
Teleost fish, *592,* 593
musculature of, 595–596, *596*
Teleostomi, *568, 585*
Telophase, *36,* 37
Telson, of horseshoe crabs, 493
Temnospondyls, 609, *610*
Temperate coniferous forest, *128,* 130, *130*
Temperate deciduous forest, *128,* 129–130
Temperature, global, 129, *129*
Temperature compensation, 180–181
Tendon(s), *145,* 148
energy storage in, 163, *163*
Tenrecs, 685
Tentacles
of ctenophores, 411, *411, 412*
of ectoprocts, 536, 537, *538*
of sea cucumbers, 552
Terebratella, 538
Tergum, of insects, 510, *510*
Termites, 522
social behavior of, 519–520, *520*
Terns, 662
Terrapin, 627
Territoriality, 301–302
in birds, 301
in mammals, 301
Tertiary structure, of protein, 11, *12*
Test(s)
echinoid, 550, 551, *552*
sarcodine, 372, *372,* 373, *373*
Testcross, 62
Testes (sing., testis), 311, *315*
avian, 656–657
human, 316, *317*
of hydras, 400
Testosterone, 280, *283,* 283–284
production of, 316
Testudines, 627–628, 637
Tetrad, 59, *59*
Tetrahymena, 375, 379
Tetramerous symmetry, 395, *402*
Tetrapods (Tetrapoda), 566, *568,* 569*t,* 579, *585,* 607
Carboniferous radiation of, 609, *610*
cladogram of, *610*
Devonian origin of, 607–609
legs, evolution of, 607, *608*
modern, evolution of, 578–579
TGF-β. *See* Transforming growth factor-β
Thalamus, *249,* 250
Thalassicolla, 378
Thaliacea, 571, *572*
Thaumatoscyphus hexaradiatus, 404
Thecodonts, 630
Themiste, 532
Thenea, 390
Theory, 4
Therapsids, 622, *623,* 667, *668, 669*
Therians, 685
evolution of, *668, 669*
Thermal condensation, in polymer formation, 16
Thermal vents, chemoautotrophic bacterial producers living near, 125
Thermodynamics, second law of, 23, 124–125
Thermogenesis
diet-induced, 235
nonshivering, 234, 235
Thermogenin, 234–235
Theropods, 630, *631*
phylogenetic affinity with birds, 642, *643, 644*
Third eye, of tuatara, 635
Thorax
of crustaceans, 498, *499*
of insects, 510, *510*
Threshold current, 255
Thromboplastin, 190
Thrombus, 196
Thrust, in swimming fishes, 596
Thymus, development of, 328, *328*
Thyroid gland, development of, 328, *328*
Thyroid hormone(s), 276–277
Thyroid-stimulating hormone, 271, 273*t,* 277
Thyrotropic hormone, 277. *See also* Thyroid-stimulating hormone
Thyrotropin, 271
Thyrotropin-releasing hormone, 277
Thyroxine, 276–277
effects on growth and metamorphosis of frog, 276, *277*
Thysanoptera, 526
Thysanura, 522, 526
Tibicen davisi, 517
Ticks, 496–498, *497,* 525
Tight junction(s), 31, *31*
Tilde (~), 42, *42*
Timber rattlesnake, *634*
Timbre, 259
Time, geological, 92–94
Tinbergen, Niko, 291, *291,* 292–293
The Study of Instinct, 292
Tissue(s), 139*t,* 140
epithelial, 141, 142, *142, 143, 144*
types of, 141–145, *142*
Tissue culture, 327
Tissue factors, 267
Tissue fluid, 143
Titanosaurus, 631
Titer, 217
TNF. *See* Tumor necrosis factor
Toad(s), 615–618, *616*
reproduction, 617–618, *618, 619*
Tokay, *632*
Tongue, 229, *230*
avian, 648
forked, of snakes, 633
Tonicella lineata, 450
Topi, *298*
Tornaria, hemichordate, 558, *559*
Torsion
definition of, 455
in gastropods, 455–456, *456,* 469, *471*
Tortoise, 627, 628, *629*
Tosia queenslandensis, 544
Toucan(s), 662
Touch sense, 255
Toxicysts, of ciliates, 376
Toxocara, 439
Toxoplasma, 375, 378
Trachea (pl., tracheae), 200, *200,* 201, 202, *202*
of insects, 513, *514*
of spiders, 494, *494*
Tracheal gills, of insects, 514
Tracheal system(s), 200, *200*
of insects, 513, *514*
of onychophorans, 534
Tracheole(s), 200, *200*
of insects, 513, *514*
Transamination, 51
Transcription, 72–75
regulation of, 76
Transcriptional factors, 76
Transduction, 20
Transfer RNA, 73, 75, *75, 76*
Transformation, 20
Transforming growth factor-β, 216
Translation, 75, *75*
regulation of, 76
Translocation, chromosomal, 69
Transmission electron microscope, 24
Transport maximum, 176
Transporters, 33, *33*
Transverse plane, definition of, 354, *356*
Treehoppers, 523, *524*
Trematodes (Trematoda), 418, *419, 430*
characteristics of, *419,* 421–423, 427
Trembley, Abraham, 399
Tricarboxylic acid cycle. *See* Krebs cycle
Triceratops, 631
Trichina worm, 440*t,* 440–441, *441*
Trichinella spiralis, 440*t,* 440–441, *441*
Trichocyst(s)
of ciliates, 376, *376, 377*
of *Paramecium, 377*
Trichodina, 379
Trichomonas, 371, *371,* 378
Trichonympha, 371, 378
Trichoptera, 523
Trichuris trichiura, 440*t*
Tricladida, *419*
Tricuspid valve, 194, *195*
Tridacna, 459, *461*
Tridacna gigas, 449, *461*
Triglycerides, 9, *9*
hydrolysis of, 50, *50*
Triiodothyronine, 276–277
Trilobites (Trilobita), 525
characteristics of, 492
fossil, *93, 491*
Trimethylamine oxide, in sharks, 170
Tripedalia, 409
Triploblastic animals, 326
Trisomy, 69
Trisomy 21, 69

Tritium, 692, *692*
tRNA. *See* Transfer RNA
tRNA synthetase, 75
Trochophore, of bivalves, 464
Trogons, 662
Trombicula, 497
Trophallaxis, 520
Trophic levels, of productivity, 123
Trophoblast, 285, 336
Trophosome, of pogonophores, 533
Tropical forest, *128, 130,* 130-131
Tropicbirds, 661
Tropomyosin, 159, *159*
Troponin, 159, *159*
  in excitation-contraction coupling, 160-161, *162*
Trunk
  of acorn worms, 557, *558*
  of rotifers, 434, *435*
Trypanosoma, 369, *371,* 378
  binary fission in, *368*
Trypanosoma brucei, 369
Trypanosoma cruzi, 369
Trypanosomiasis
  African, 369, 521
  American, 369
Trypsin
  pancreatic, 232-233, *233*
  specificity of, 41-42, *42*
Tryptophan, *10*
Tsetse flies, disease transmission by, 521
TSH. *See* Thyroid-stimulating hormone
T-system, 160, *161, 162*
Tuatara, 622, 635, *636,* 637
Tubastrea, 410
Tube anemones, 404, 405
Tube feet
  of crinoids, 553-554
  of sea stars, 545, *545,* 546, *546,* 547
Tubifex, 485, 488
Tubipora, 409
Tubularia, 409
Tubularia crocea, 399
Tubulin, 20, 30, 36
Tumble bugs, 518, *518*
Tumor(s), growth of, 37
Tumor necrosis factor, 216, *216*
Tumor suppressor genes, 79-80
Tundra, *128,* 131
Tunic, 570-571, *571*
Tunicates (Tunicata), *567,* 570-571, *571,* 579
  immunity in, 212, 213*t*
Turbelleria (turbellarians), 417-418, *418, 419,* 427, 429, *430,* 453
  body plan of, 418-419, *419*
  characteristics of, 420-421
  reproduction, 420
Turdus migratorius, 343, 344
Turkeys, 661
Turnstones, 662
Turtle(s), 637
  characteristics of, 627-628
  evolution of, *623, 624*
  marine, 628
Tusk shells, 449, *455,* 472. *See also* Scaphopoda
Tympanic canal, 258, *258*
Tympanic membrane, *257,* 258
Tympanic organs, of insects, *510,* 515
Typhlosole, 231
Typological concept of species, 350
Tyrannosaurus, 630

U

Uca, 506, *507*
Ultimate causes, 5, 291
Ultraviolet light sensitivity, in birds, 652
Ultraviolet radiation, mutagenic effects of, 80
Umbilical cord, formation of, 336, *336*
Umbo, of bivalves, 460, *463*
Unconscious mind, 250
Underhair, 671
Undernourishment, 236, 237
Undulating membrane, of ciliates, 376
Undulipodia, 31
Ungulates, 182
Unicellular organism(s), *355*
Uniformitarianism, 85, 100
Uniramia, 508-521, 525, 526, *527*
Uniramous appendages
  of crustaceans, 499
  of myriapods and insects, 508
Unitary organisms, 115
Upogebia, 501-502
Upwelling, 132
Urea
  production of, 51
  in shark blood, 170
Urechis, 532
Ureter(s), 173, 174, *175,* 316
Urethra, male, 317, *317*
Uric acid, 170-171
  excretion, 177
    by birds, 650
  production of, 51-52
Urigenital system, epithelium, development of, 328, *334*
Urinary bladder, 174, *175*
Urine
  concentration, renal mechanism for, 177-179, *178*
  formation, 174, 176
    in birds, 649-650
    in lizards, 629
    in metanephridia, 172, *172*
    in protonephridia, 171-172, *172*
Urnatella, 444
Urochordata, 570-571, *571,* 579
Urodela, 609, 612, 618
Urogenital system, 316
Uropods, of crustaceans, *499,* 500, *500*
Ursus horribilis, 687
Ussher, James, 84
Uterus, 318, *318,* 335
Utricle, 256, 259

V

Vagina, 318
Vagus nerves, in cardiac control, 194-195
Valves
  of bivalves, 460
  of brachiopods, 538-539, *539*
  of chitons, 453
Vampire bats, 678-679
Vampirolepis nana, 426*t*
van Leeuwenhoek, A., 24
Vane, of contour feather, 645
Variability/variation, genetic, 6, *6,* 19
  sources of, 78-79
Variable region, of antibody, 214
Vas deferens, *315,* 316, 317, *317*
Vasa efferentia, 317
Vasopressin. *See* Antidiuretic hormone
Vasotocin, 273*t,* 274
Vectors, 77
Vein(s), 193, 198
  layers of, *197*
Velarium, of Cubozoa, 404
Veliger larva, 452, *453*
  of bivalves, 464
Vellella, 394
Velociraptor, 630, *631*
Velum, of hydroid medusa, 401, *402*
Velvet, of antlers, 672, *674*
Velvet worms, 534
Venom
  of Gila monster, *633*
  snake, 633-635
    hemorrhagin type, 635
    neurotoxic, 635
Ventilation
  in arthropods, 492
  gill, 200, *201*
  in humans, 203, *203*
  lung, 200-202, *201*
  by negative pressure, 202
  by positive pressure, 202, *203*
Ventral, definition of, 354, *356*
Ventricle(s), cardiac, *193,* 193-194, *194, 195, 196*
  left, 194, *194, 195, 196*
  right, 194, *194, 195, 196*
Venule(s), 193, 198
Venus, 472
Venus's flower basket. *See Euplectella*
Venus' girdle, 411, *411*
Vermes, 417
Vertebrae, 153-155
  caudal, 155
  cervical, 153-155
  lumbar, 153-155
  numbers of, in various species, 155
  sacral, 153-155
  thoracic, 153-155
Vertebralima striata, 372
Vertebrates (Vertebrata), 566, 569*t,* 573-579
  acquired immunity in, 213-219
  ancestors of, 575-576
  appendages of, 575
  body plan of, *355*
  characteristics of, 574
  circulatory system of, *193,* 193-194, *194*
  classification of, 579
  development of, 333-336
    common pattern of, 333, *334*
    phylotypic stage of, 333, *334*
  earliest, 576, *577*
  endocrine glands of, 270-282
  endoskeleton, 150-151, 573
  evolution of, 573-575
  hormones of, 270-282
  innate immunity in, 212-213
  jawed, evolution of, 577-578, *578*
  kidneys of, 173-179
  nervous system of, 246-253, 574-575
  phagocytes of, *190,* 211-212, *212*
  pharynx of, 573-574
  reproduction, hormones of, 282-286
  reproductive systems of, 316-318
  respiration in, 573-574
  skeleton, plan of, *153,* 153-155, *154*
  terrestrial
    early evolution of, 607-609
    movement onto land, 607
    origin of, 607
Vestibular canal, 258, *258*
Vestimentifera, 533
Vibrissae, 671
Villi, intestinal, 231, *231, 232*
Vipers (Viperidae), 633, 634
Virchow, Rudolf, 24
Visceral larva migrans, 439
Visceral mass
  of bivalves, 460, *463, 464*
  of molluscs, *451,* 451-452
Visceral pleura, *202,* 202-203
Vision, 260-263
  avian, 651
  chemistry of, 262
  human, absorption spectrum of, 263, *263*
Visual pigments, 262
Vitamin(s)
  definition of, 235
  fat-soluble, requirements for, 235*t,* 236
  requirements for, 235*t,* 235-236
  water-soluble, requirements for, 235*t,* 236

Vitamin D, 10
and calcium metabolism, 278
deficiency, 279
Vitreous humor, 261, *261*
Viviparity (viviparous reproduction), 333, 335
in fish, 600
in onychophorans, 534
in snakes, 635
Vocalizations, of crocodilians, 638
Voluntary muscle, 159
Volvox, 5, 378
life cycle of, *370,* 371
von Baer, K. E., 98
von Békésy, Georg, 258
von Frisch, Karl, 291, *291,* 292, 304
Von Mering, J., 282
Von Uexküll, Jakob, 240
Vorticella, 375, 379
Voy, Robert, 280
Vulture(s), 661
Vulva, 318, *318*

W

Waggle dance, of honeybees, 303, *303,* 304
Walking worms, 534
Walkingsticks, 522
Wallace, Alfred Russel, 84, *84,* 88
Walruses, 686
Washoe, 304-305
Wasp(s), 523
reproduction, 515, *516*
Water
body, 188
excretion of, 177-179
molecular structure of, 694, *695*
molecules, geometry of, 694, *695*
requirements for, 235
Water balance, 167-171
in insects, 514
in lizards, 629
in marine bony fishes, 168-170, *169*
in marine invertebrates, *167,* 167-168
in reptiles, 626
in terrestrial animals, 170-171
Water fleas, 504, *505*
Waterfowl, lead poisoning of, 660
Waters, inland, 132
Water-vascular system
of crinoids, 554
of sea cucumbers, 553
of sea stars, *545,* 545-546, *546*
of sea urchin, 551, *552*
Watson, James, *6,* 56, 71
Weasels, 686
Weevils, 523
Weir, 419, *420*
Weissmann, August, 103
Wells, P. H., 304
Wenner, Adrian, 304
Whale(s), 686
feeding of, 223, *224*
humpback, *687*
migration of, 677
Whelks, 454, 459
Whippoorwill(s), 662
Whipworm, 440*t*
White blood cell(s). *See* Leukocyte(s)
White matter, 247, *247*
White muscles, 163
White-crowned sparrow, learning in, 297, *298*
Whittaker, R. H., 351, 352
Wilkins, Maurice H. F., 71
Williamson, Peter, 102
Wilson, E. O., 292
Wing(s)
avian, 646, *647*
forms of, 652-653, *654*
high aspect ratio, 653
as lift device, 652, *653, 655*
low aspect ratio, 652
of insects, 508-509, *510,* 510-512, *511, 516, 518*
Wing slot(s), *647,* 652, *653*
Woese, Carl, 18
Wolf spider, 495
Wolffian duct, 316
Wolves, 686
Wombats, 685
Wood frog, 615
Wood tick, 498
Woodchuck, 686
hibernation in, 183, *184*
Woodcock, 662
Woodpeckers, 662
Worker(s), of social insects, 519, 520, *520*
Worm(s), 417
Worm lizards, characteristics of, 628, 629-632, *633,* 637
Wuchereria bancrofti, 442, *443*

X

X chromosome, 60, *60*
Xenartha, 685-686
Xenograft, 212
Xenopus laevis, 617, *618*
Xiphosurida, 493, *493,* 525
Xyloplax, 554, 555

Y

Y chromosome, 60, *60*
Yalow, Rosalyn, 275
Yolk glands, of flatworms, 420, *420,* 429-430
Yolk sac, *322,* 334, *335*
mammalian, 336, *336*
Y-organs, 500
Young, Thomas, 263

Z

Z line, *158,* 159
Zebra(s), *298,* 686
Zebra mussels, 461, 464
Zoantharia, 404, 409
Zoecium, of ectoprocts, 536, *537*
Zonotrichia leucophrys, learning in, 297, *298*
Zoochlorellae, in cnidarians, 398
Zooflagellates, 369, 377
Zooid(s)
of ectoprocts, 536
of hydroids, 400
Zoology, 3
research in, *4*
Zoomastigophorea, 369, *371,* 378
Zoothamnium, 375
Zooxanthellae
of cnidarians, 398, 408, 409
of giant clams, 461, *461*
Zygote, 35, 61, 309, 311, *316*
cleavage, 320-323
discoidal, *322,* 323
mosaic, 323, *323, 354*
patterns of, 321-323, *322*
radial, 321, *322,* 353, *354*
regulative, 321-322, *323, 354*
rotational, *322,* 323
spiral, *322,* 322-323, *354*
formation of, 319-320
Zygote nucleus, 320

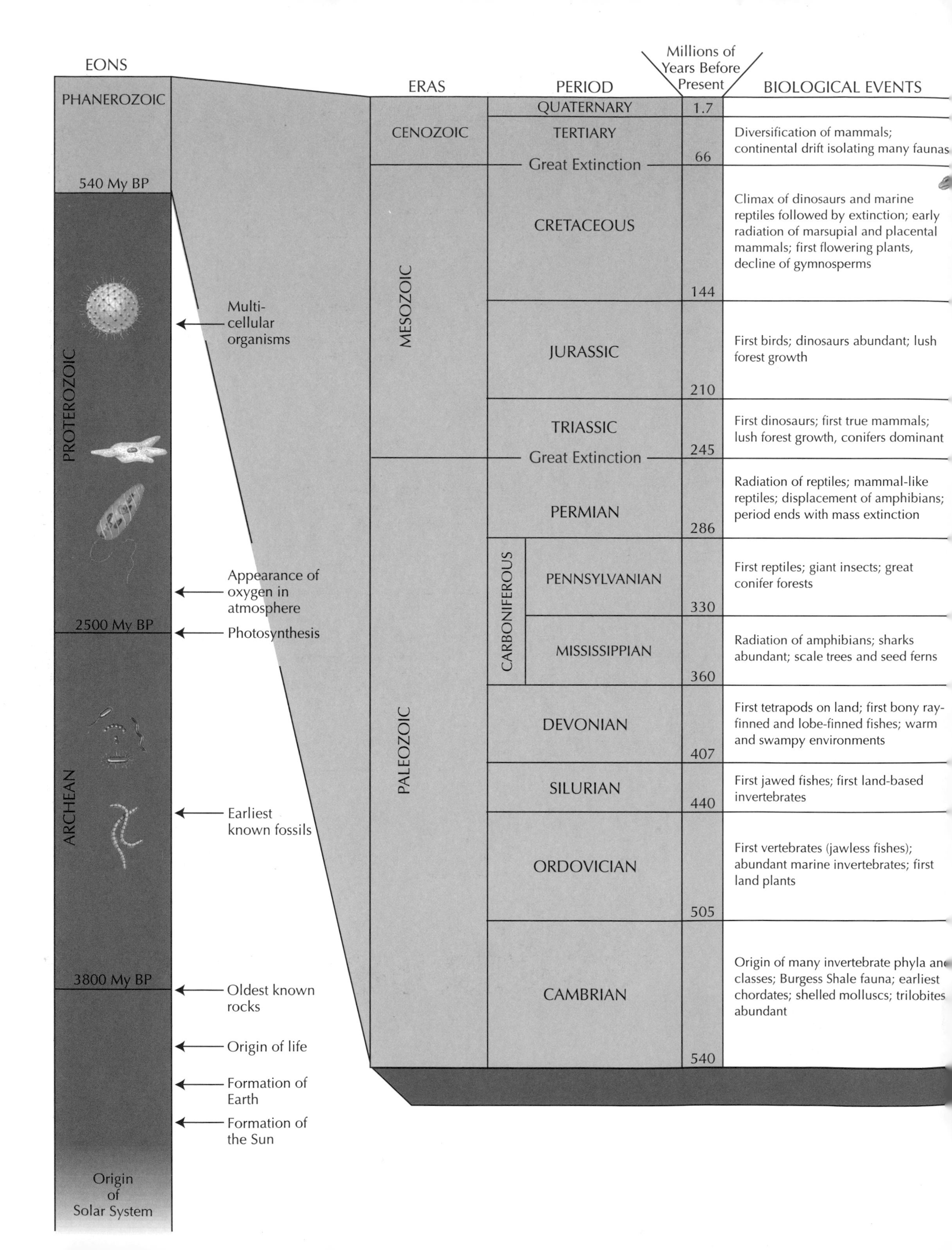

EONS
PHANEROZOIC
540 My BP
PROTEROZOIC
Multi-cellular organisms
Appearance of oxygen in atmosphere
2500 My BP
Photosynthesis
ARCHEAN
Earliest known fossils
3800 My BP
Oldest known rocks
Origin of life
Formation of Earth
Formation of the Sun
Origin of Solar System
ERAS
PERIOD
Millions of Years Before Present
BIOLOGICAL EVENTS
CENOZOIC
QUATERNARY
1.7
TERTIARY
Diversification of mammals; continental drift isolating many faunas
66
Great Extinction
MESOZOIC
CRETACEOUS
Climax of dinosaurs and marine reptiles followed by extinction; early radiation of marsupial and placental mammals; first flowering plants, decline of gymnosperms
144
JURASSIC
First birds; dinosaurs abundant; lush forest growth
210
TRIASSIC
First dinosaurs; first true mammals; lush forest growth, conifers dominant
245
Great Extinction
PALEOZOIC
PERMIAN
Radiation of reptiles; mammal-like reptiles; displacement of amphibians; period ends with mass extinction
286
CARBONIFEROUS
PENNSYLVANIAN
First reptiles; giant insects; great conifer forests
330
MISSISSIPPIAN
Radiation of amphibians; sharks abundant; scale trees and seed ferns
360
DEVONIAN
First tetrapods on land; first bony ray-finned and lobe-finned fishes; warm and swampy environments
407
SILURIAN
First jawed fishes; first land-based invertebrates
440
ORDOVICIAN
First vertebrates (jawless fishes); abundant marine invertebrates; first land plants
505
CAMBRIAN
Origin of many invertebrate phyla and classes; Burgess Shale fauna; earliest chordates; shelled molluscs; trilobites abundant
540

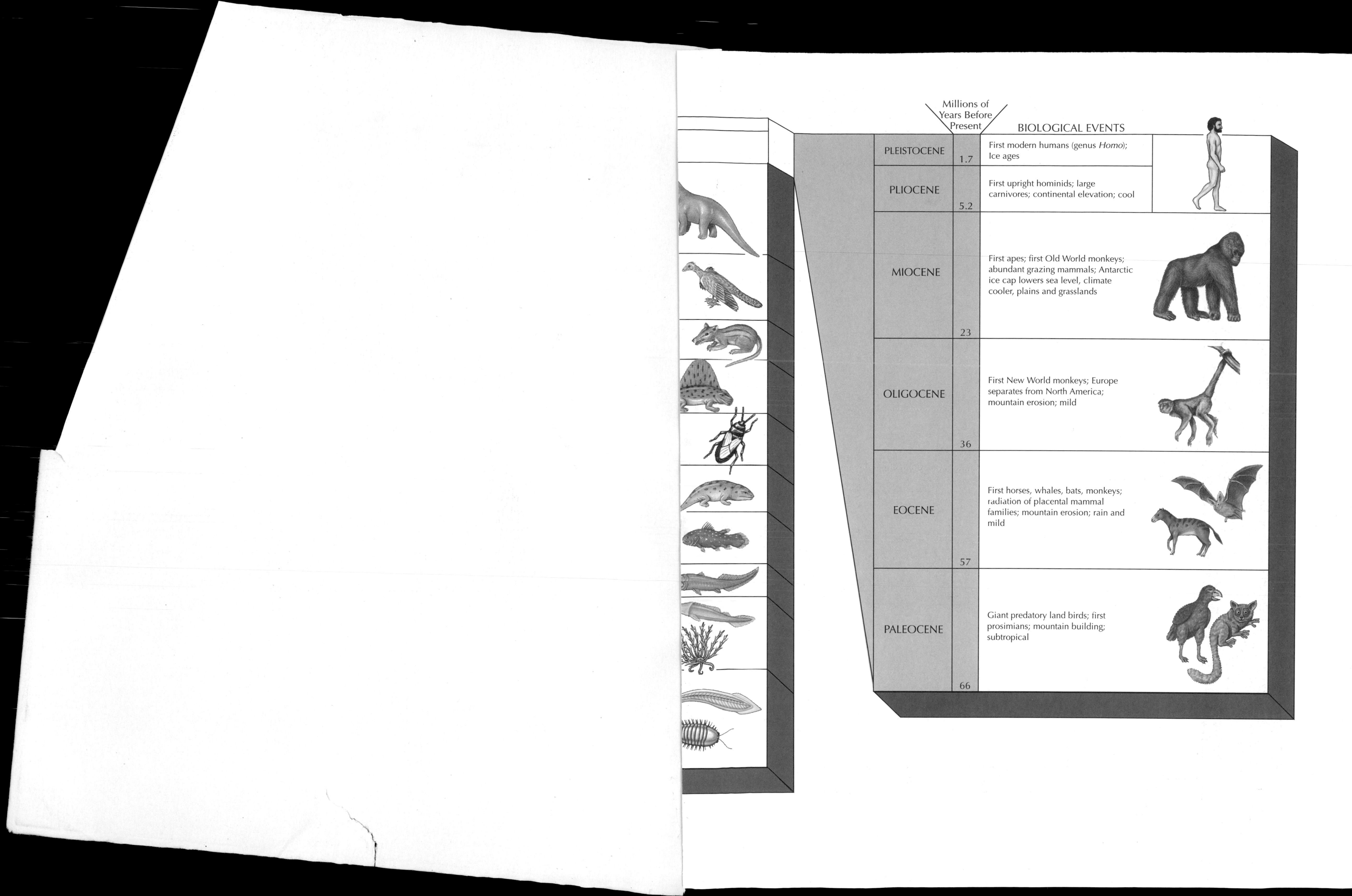
Millions of Years Before Present
BIOLOGICAL EVENTS
PLEISTOCENE
1.7
First modern humans (genus Homo); Ice ages
PLIOCENE
5.2
First upright hominids; large carnivores; continental elevation; cool
MIOCENE
23
First apes; first Old World monkeys; abundant grazing mammals; Antarctic ice cap lowers sea level, climate cooler, plains and grasslands
OLIGOCENE
36
First New World monkeys; Europe separates from North America; mountain erosion; mild
EOCENE
57
First horses, whales, bats, monkeys; radiation of placental mammal families; mountain erosion; rain and mild
PALEOCENE
66
Giant predatory land birds; first prosimians; mountain building; subtropical